Clinical and Laboratory Medicine Textbook

Marcello Ciaccio
Editor

Clinical and Laboratory Medicine Textbook

Springer

Editor
Marcello Ciaccio
School of Medicine and Surgery
University of Palermo
Palermo, Italy

ISBN 978-3-031-24957-0 ISBN 978-3-031-24958-7 (eBook)
https://doi.org/10.1007/978-3-031-24958-7

The translation was done with the help of artificial intelligence (machine translation by the service DeepL.com). A subsequent human revision was done primarily in terms of content.

Translation from the Italian language edition: Trattato di biochimica clinica e medicina di laboratorio. © EDISES Edizioni 2021. All Rights Reserved.

This Springer imprint is published by the registered company Springer Nature Switzerland AG
The registered company address is: Gewerbestrasse 11, 6330 Cham, Switzerland

To Anna Maria,
with great love.

Your father

Foreword

The COVID-19 pandemic, with requirements for social distancing, has accelerated the development of novel teaching methodologies in medical schools. Several medical specialties, including laboratory medicine, have decided to employ novel teaching methodologies such as the use of telemedicine, remote teaching, online curricula, virtual rotations, virtual conferences, simulations, and learning consortia. Traditional teaching methodologies, therefore, are challenged by innovative proposals, but I strongly believe that medical textbooks should still be considered one of the fundamental sources of medical information. This is particularly true for laboratory medicine, one of the youngest medical disciplines, as the number, type, and complexity of laboratory testing have increased markedly during the past 50 years, and healthcare providers increasingly rely on laboratory tests to make diagnoses, assess the efficacy of therapy (or even to predict the outcome of therapy), and follow the natural progression of diseases. In addition, innovative biomarkers and the adoption of "omics" are increasingly used for more personalized medicine, wellness promotion, and disease prevention. This is the first reason for accepting the invitation to write a foreword to the textbook by Marcello Ciaccio *Clinical and Laboratory Medicine Textbook.*

The second reason is that this textbook fully recognizes the development of laboratory medicine and its evolution over the past decades. Today, laboratory medicine is a highly complex set of medical subdisciplines that span the breadth of diagnostic testing needed to support all health care. It includes clinical biochemistry, hematology and hemopathology, coagulation, medical microbiology and virology, molecular pathology and genetics, and transfusion medicine. More recently, the integration between pathology and laboratory medicine, known by the acronym "PALM", has been proposed to ensure a holistic approach to diagnostic testing. PALM services, infact, are cross-cutting, intersectoral, and provide the foundation for safe, effective, and equitable healthcare delivery, population health, and global health security. Until a few years ago, medical education on laboratory testing was found to be inadequate in Italy, as well as in many other countries including the United States. Despite the integral role of laboratory testing in the practice of medicine, formal teaching of laboratory medicine was a relatively neglected component of the medical school curriculum, and for many years, successive editions of traditional textbooks did not reflect the changing paradigm of the discipline. Most textbooks, infact, focused on only the methodological aspects of laboratory tests, omitting fundamental requirements of the extra-analytical phases. In particular, medical students should be trained to appropriately request laboratory tests, as well as to correctly interpret and use laboratory information. Infact, physicians, who use clinical laboratory services, need to be familiar with their role in ensuring quality test results through appropriate test selection, interpretation, and clinical action. This should be recognized as a fundamental issue in teaching laboratory medicine in medical schools and this textbook fully recognizes this strategic aspect of medical education in laboratory medicine. By selecting the appropriate tests and interpreting the results correctly, medical students and physicians using this book should be able to optimize patient outcomes and reduce the cost of achieving a diagnosis. Most chapters begin with a brief description of the disorder/ medical condition followed by a discussion that includes sentences, tables, and figures detailing the laboratory evaluation of specific disorders, and coverage of diagnosis and baseline tests to exclude diagnostic possibilities and clinical indications that warrant further screening and special testing. Obviously, the methodological aspects are included and represent the fundamental information to link clinical needs, analytical performance specifications, and diagnostic accuracy.

The third reason which highlights the value of the textbook is the faculty, as the chapters have been prepared and written by true experts, thus allowing an effective "evidence-based" approach which, in turn, assures an updated, reliable, and accurate handling of the matter. In an interconnected and integrated medicine, laboratory information plays an increasingly dominant role, but medical student knowledge should be improved to allow future physicians to manage the high complexity of the current test menu, to more appropriately require laboratory tests, to interpret better and use laboratory information, and to integrate it into the right clinical context. An in-depth knowledge of laboratory medicine principles is vital to all practicing physicians. Great variation exists in the ways that medical students learn the principles of laboratory medicine: to ensure that important principles are addressed, medical schools should establish goals and objectives specifically related to laboratory medicine and experiment with optimal teaching and assessment methods. Medical students should understand the differences in interpretation that different settings require and that laboratory tests must be interpreted in conjunction with the entire clinical presentation. They should be able to order not only all of the tests necessary in a particular clinical setting but also only the tests that are medically useful for their patients. They should know their limitations of knowledge in this area and how to seek appropriate consultation from clinical pathologists, clinical scientists, midlevel provider laboratory personnel, and other subspecialists to maximize appropriate patient care. They should understand the scientific underpinnings of laboratory medicine, for the purpose of proper use of current testing and also to be able to assess future modalities wisely. They should also understand the therapeutic modalities available from the clinical laboratory, such as the appropriate use of blood products, cell therapy, and other individualized patient treatments, and understand the cost-effective, evidence-based use of the treatments. Structured classroom teaching, even in an era of novel methodologies, still represents an effective approach to transfering knowledge to medical students through the guidance of a mentor who must be an expert, highly skilled in laboratory medicine. However, even the best mentor cannot provide a series of lessons comprehensive of the increasingly complex and huge number of laboratory tests. Medical students, however, should be allowed to know and understand the complexity of current laboratory testing: they need to look up a textbook!

A valuable and updated textbook, such as the textbook by Marcello Ciaccio is, therefore, an invaluable and irreplaceable tool to provide an accurate teaching of laboratory medicine in medical schools. The textbook should not substitute ex cathedra lecturers, nor ward rounds to discuss clinical cases and practical laboratory training in the interpretation of peripheral blood smears and microscopic examination of urine sediment. It should be a source of valuable information for covering the aspects of laboratory testing not treated during the course of study and/or in-depth self-directed learning.

Padua, Italy
Galveston, TX, USA

Mario Plebani

References

Plebani M (2016) Towards a new paradigm in laboratory medicine: the five rights. Clin Chem Lab Med 54:1881–1891

Smith BR, Aguero-Rosenfeld M, Anastasi J, Baron B, Berg A, Bock JL et al (2010) Academy of Clinical Laboratory Physicians and Scientists. Educating medical students in laboratory medicine: a proposed curriculum. Am J Clin Pathol 133:533–542.

Wilson ML, Fleming KA, Kuti MA, Looi LM, Lago N, Ru K (2018) Access to pathology and laboratory medicine services: a crucial gap. Lancet. 391:1927–1938.

Preface

Laboratory Medicine is a highly dynamic discipline which has undergone a significant evolution and revolution over the years. Technological innovation represents the leitmotif of change, characterized not only by the introduction of methods with high analytical performance but also by a mature consideration of the value and significance of Laboratory Medicine. The latter has undergone an essential transition from an essentialist vision, focused exclusively on the analytical aspect, to a holistic vision centered on the patient. The common goal of Laboratory Medicine and Clinical Medicine is to improve patient outcomes. Although Laboratory Medicine is a "faceless" discipline, it has a central role in the Health System as it is an integral and integrated part of clinical-care pathways, from screening to diagnosis to prognosis and therapy monitoring.

In the future, the Clinical Laboratory will acquire an increasingly central role in the panorama of Clinical Medicine, guiding its change from a generalized conception of health and disease to an individualized approach, which involves decisions and interventions tailored to the individual patient, based on the answers.

In the past, medicine was practiced according to the signs and symptoms presented by the patient and was based exclusively on the individual knowledge of the physician. Currently, medicine is based on evidence produced by scientific research, including clinical studies, known as evidence-based medicine. In the near future, medicine will be adapted to the individual patient. Indeed, advances in science and technology, as well as the development of "Omics Sciences", have led to a significant increase in knowledge of the pathophysiology of many diseases. Understanding the differences associated with disease at different levels, including genome, transcriptome, proteome, and metabolome, is the basis for developing Precision Medicine. In this context, Laboratory Medicine plays a leading role by providing a large panel of genetic, epigenetic, and circulating biomarkers, which allow for defining the molecular profile of an individual in order to obtain a "personal" description of the state of health.

The *Clinical and Laboratory Medicine Textbook* is translated from the Italian Text *Trattato di Biochimica Clinica e Medicina di Laboratorio*, which has a huge spread in Italy. Indeed, it represents an innovative text on Clinical and Laboratory Medicine because each topic is presented exhaustively, from the explanation of the pathophysiological mechanisms, with a particular focus on the biochemical-clinical characteristics, to the description of the main clinical pictures, highlighting the importance of the role of Laboratory Medicine and providing the tools for the correct use and interpretation of laboratory data, according to the most recent guidelines and scientific evidence.

In particular, diseases with a high incidence in the population, such as diabetes mellitus, cardiovascular diseases, dyslipidemias, and autoimmune diseases, and rarer diseases, such as hereditary metabolic diseases, are described. In addition, there are original topics, such as clinical biochemistry of the mind, and topics of great relevance, such as biological drugs, biobanks, health technology assessment, and omics sciences. Finally, a chapter is dedicated to the COVID-19 pandemic.

The textbook is aimed at all Medical Science professionals who wish to have a broad, complete, and in-depth knowledge of Clinical Biochemistry and Laboratory Medicine and who wish to deepen specific aspects of the Clinical Laboratory and update themselves on technological and biotechnological advances in the medical field.

Palermo, Italy Marcello Ciaccio

Contents

About the Editor

Marcello Ciaccio, MD, PhD Marcello Ciaccio, specialist in Endocrinology and Metabolic Diseases, is Full Professor of Clinical Biochemistry and Laboratory Medicine and President of the School of Medicine at the University of Palermo. He is Director of the Institute of Clinical Biochemistry, Clinical Molecular Medicine and Clinical Laboratory Medicine, Department of Biomedicine, Neurosciences and Advanced Diagnostics at the University of Palermo. He is also Director of the Department of Laboratory Medicine – A.O.U.P. Palermo "P. Giaccone". He is member of the Commission for National Scientific Qualification (ASN) 2021–2023 for the Scientific Sector 05/E3-Clinical Biochemistry and Clinical Molecular Biology. He is past President (2016–2017) and has been elected for the biennium 2024–2025 President of the Italian Society of Clinical Biochemistry and Clinical Molecular Biology – Laboratory Medicine (SIBioC).

Prof. Ciaccio is engaged in intense research activity, as attested by over 400 scientific publications in journals with national and international relevance. His research mainly aims to identify and clinically validate new biomarkers in the field of cardiovascular, metabolic, infectious, and neurodegenerative diseases. He is also a member of the Editorial Board of numerous international journals and has drafted Guidelines and Consensus Documents on various pathologies and clinical conditions.

The Role of Laboratory Medicine in the Clinical Context

1

Marcello Ciaccio

Introduction

> It is more important to know what kind of person has a disease than to know what kind of disease a person has.
> (Hippocrates)

This is nowadays the essence of laboratory medicine, as well as one of the objectives it sets as a goal to be achieved. It studies biological samples from humans (in vitro analysis of blood, urine, feces, etc.) or in humans themselves (in vivo analysis by nuclear magnetic resonance spectroscopy), those chemical and physical parameters that can provide information on physiological and/or pathological processes occurring at various levels of structural organization and, therefore, of systems, organs, tissues, cells, and even single molecules (DNA, proteins, etc.). Until a little more than a decade ago, laboratory medicine played a supporting role in clinical activities, often in the role of passive service provider.

More recently, it has acquired an active and central role in our health care reality: it provides essential elements for many clinical decisions (often early for the onset of symptoms), allows risk stratification, and represents the basis for Personalized Medicine, a primary objective for a "sustainable" Health Care System.

It has been estimated that laboratory investigations account for less than 2% of the budget of healthcare systems, although they influence more than 70% of clinical decisions. This has happened thanks to a new awareness on the part of professionals working in this sector and a modern approach to patient care, placing the patient at the center of the entire diagnostic pathway. The latter begins with the clinical question formulated by the physician and an initial diagnostic hypothesis, followed by the request for appropriate laboratory tests "tailored" to the patient's characteristics, which the clinical laboratory transforms into a valuable contribution to the resolution of the clinical problem.

George D. Lundberg, in his "Brain-to-brain loop" model, highlights how the request for a laboratory test originates in the brain of a physician in front of a patient and ends only when the laboratory information reaches the brain of the same physician, who can then manage the clinical case in the most appropriate way (Fig. 1.1).

Noteworthy, laboratory medicine is a complex reality: it is enough to consider that a small–medium-sized laboratory performs services in multiple fields (clinical chemistry, hematology, hemostasis and coagulation, immunology, pharmacology, molecular biology, microbiology and virology,

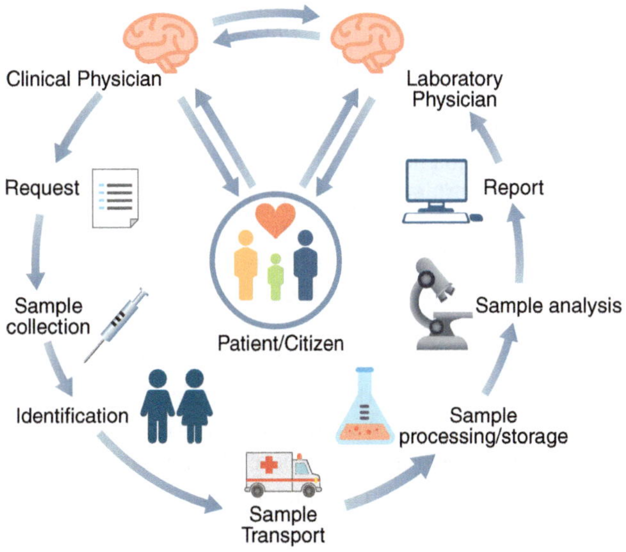

Fig. 1.1 Brain-to-brain loop. (Copyright EDISES 2021. Reproduced with permission)

M. Ciaccio (✉)
Department of Biomedicine, Neurosciences and Advanced Diagnostics, Institute of Clinical Biochemistry, Clinical Molecular Medicine and Clinical Laboratory Medicine, University Hospital "P. Giaccone", Palermo, Italy

Department of Laboratory Medicine, University Hospital "P. Giaccone", Palermo, Italy
e-mail: marcello.ciaccio@unipa.it

M. Ciaccio (ed.), *Clinical and Laboratory Medicine Textbook*, https://doi.org/10.1007/978-3-031-24958-7_1

autoimmunity), which require the support of different professionals (Physician, Biologist, Biotechnologist, Biomedical Laboratory Technician, Chemist) with different but complementary skills, in a socio-political-economic context that is sometimes "schizophrenic."

In addition, clinical laboratories are asked to constantly increase productivity, with fewer and fewer resources, but without compromising quality. Given that health is a priceless asset, the resources used to maintain its integrity should be considered a value and not a cost.

Today, the clinical laboratory is an essential tool for most branches of medicine, a "transversal" discipline that actively participates in the diagnosis, differential diagnosis, prognosis, therapy, monitoring of pathology, and therapy. Thus, laboratory medicine represents a clinical discipline, irreplaceable in hospital and territorial medicine, with a fundamental function in drafting and validating diagnostic-clinical-therapeutic pathways and an active role alongside the patient.

Integration Between Laboratory and Clinic

The perspective of the medicine of the third millennium must necessarily involve the integration among different disciplines, both clinical and diagnostic. Indeed, it would be appropriate to overcome the sectorial approach and to adopt an integrated vision of medicine, with the common goal of improving the patient's outcome.

In medicine, the outcome is the result of a medical intervention in terms of both health and costs; laboratory diagnostics can influence the outcome as it participates in the formulation of faster and more accurate diagnoses, orients toward more appropriate therapeutic choices, contributes to the definition of the prognosis and, therefore, to the start of effective diagnostic and therapeutic pathways; ultimately, therefore, improving the patient's *outcome* implies increasing his/her degree of satisfaction and life quality.

The integration between the laboratory and the clinic can be pursued by different approaches; for example, through the introduction of shared diagnostic algorithms in the laboratory practice or through the inclusion of interpretative comments in the report to ensure the correct use of biochemical data and possibly suggesting the action to be taken. According to a modern vision, the laboratory report should not be considered as a simple report of a numerical data, but the transmission of a biochemical information that arises from precise clinical reasons and has the purpose of improving the accuracy of the diagnosis and therapeutic treatment.

A recent survey of primary care physicians in the United States found that more than 70% of physicians consider it useful to report previous results and interpretive comments in order to use the laboratory result correctly and improve the communication process between the laboratory and the clinic. The same study shows, however, that only 4% of physicians turn to laboratory medicine specialists for scientific or medical advice, although they recognize this practice as extremely useful. This apparent contradiction is probably an expression of the discrepancy between the large number of tests introduced in recent years and the difficulty in transmitting the necessary knowledge for their use, an activity in which laboratory professionals must be protagonists.

From this point of view, the integration between the laboratory and the clinic can be achieved by implementing all the necessary actions to ensure appropriateness in the whole process, from the clinical question to the formulation of a request, the execution of one or more laboratory tests, the production of a report and its interpretation, up to the clinical decision, and the management of the patient.

Appropriateness in Laboratory Medicine

The concept of appropriateness in medicine shows multiple facets because it can be declined into educational, prescriptive, organizational, analytical, diagnostic, hospitalization, and therapeutic appropriateness; it involves, therefore, different aspects and professional figures. It should be emphasized that professional appropriateness is influenced by the available evidence on the efficacy/safety of health care interventions, while organizational appropriateness meets its own "bottleneck" in the available resources. In the last years, several definitions of appropriateness have been formulated, depending on whether it is considered from the point of view of the Patient, Society, Health System, or Health Worker. The first definition, dating back to 1984 by Woodward et al., proposes an "ethical" medicine model in which appropriateness is proper to any act able to contribute to the patient's health in a positive way. Since then, different models of appropriateness have been proposed; today, a balanced view of appropriateness includes any intervention related to the need of the patient or the community, provided in the appropriate manner and time, based on recognized standards, with a positive balance among benefits, risks, and costs.

It is possible to identify almost five variables related to the appropriateness of the professional action: the characteristics of the patient (clinical, distinguishing acute and chronic condition of the disease, cultural, etc.), the characteristics of the service (efficacy, safety, cost, acceptability, continuity of care, etc.), the time required to provide the service in relation to the patient's clinical history, the characteristics of the level of care (high speciality, critical area, ordinary *hospital stay, day hospital*, specialist consultancy, outpatient care), and the characteristics of the professional providing the service (degree of specialist training, specific professional experience, etc.).

In the reality of laboratory medicine, prescriptive appropriateness is of crucial importance: clinical laboratory information comes, in fact, from the appropriateness of the request and becomes valuable if it can influence the clinical decision-making process and the patient's health outcomes.

The issue of appropriateness in laboratory medicine is considered, today, of primary importance: a recent meta-analysis has estimated the rate of underutilization of laboratory tests at 44.8%, while the rate of overutilization at 20.6%, making it necessary to carefully analyze the appropriateness in laboratory medicine. Moreover, it should not be forgotten that a "badly requested" test often generates a series of other inappropriate tests if the result is out of the reference values, thus generating a vicious circle that can have not only health but also economic, legal, and social consequences.

Laboratory medicine professionals have a decisive role in the dissemination of appropriate practices. There are, in fact, tools to full relevance of the laboratory that allow to actively intervene on the appropriateness of prescription, such as the use of *reflex testing*, the introduction of "rules" managed by the computer systems of the laboratory (Laboratory Information Systems, LIS) to "block" inappropriate requests in relation to the time of execution, specific diagnostic suspects or departments of origin, etc..

Evidence-based medicine (EBM) is the basis of the concept of appropriateness. EBM is a discipline of recent development and diffusion based on the systematic review of scientific evidence on a specific clinical situation in relation to its diagnosis, monitoring, and treatment to ensure accurate decisions about patient care, while recognizing the need to integrate evidence with clinical expertise. Several tools derived from EBM have been identified to promote appropriateness, with different areas of application and promoting subjects: guidelines, diagnostic and therapeutic care pathways (DTCP), procedures and operating protocols shared by the Scientific Societies, essential levels of care, notes of the drug Agencies, such as the National Agency for Food and Drug Administration and Control (NAFDAC) or the Agenzia Italiana del Farmaco (AIFA).

Guidelines can be defined as recommendations for clinical behavior developed through a systematic review of the literature and expert opinion, with the aim of helping physicians and patients to choose the most appropriate models of care in specific clinical situations. The primary goals of guidelines are to minimize variability in clinical decisions related to lack of knowledge and subjectivity and facilitate the transfer of research into clinical practice.

DTCPs represent the contextualization of the guidelines related to a pathology or clinical problem in the specific organizational reality of a health care facility, also considering the available resources. They result, therefore, from the integration of two components: the clinical recommendations of the reference guidelines and the elements of the local context that can condition their application. Indeed, one of the main limitations to the application of EBM is that the guidelines refer to a "general" and not to the "particular" patient that the professional must treat in the clinical practice, with its clinical complexity and pathological characteristics.

Another useful tool for knowledge transfer is the dissemination of protocols, i.e., consensus documents that formalize the sequence of actions to be carried out to achieve an objective, thus describing a predefined diagnostic and therapeutic behavioral pattern.

Precision Medicine

Laboratory medicine is going through a moment of radical change, in which the traditional approach to the diagnosis and treatment of pathologies by the evaluation of clinical signs and symptoms has been converted, today, into an individual approach, on the single patient, in which laboratory information has made quantifiable what before was generalized and, sometimes, subjective. This revolution of thought was born in the 19th century when the Canadian physician Sir William Osler highlighted the great variability among individuals. However, the passage of time, the discovery of blood types, the evidence of different responses to certain pharmacological treatments led to the concept of personalization. In this context, the idea of a new, modern, and kaleidoscopic medicine was born, initially called the "4Ps" to indicate a Predictive, Preventive, Personalized, and Participatory medicine. These have become the watchwords for an approach to medicine that considers environmental and genetic factors, lifestyle, individual susceptibility to disease, variability in response to therapy. All these represent the concept of the medicine of the Third Millennium, and the precision medicine plays a key role in the evaluation of clinical, genetic, and environmental information *ad personam*. It involves the integration of multidisciplinary teams to promote the health and the education of the patient and supports disease prevention and treatment. The Public Health System plays an equally important role in the health status of the individual. "*Despite the advances in molecular genetics...as genetic tests are proliferating in the U.S. population, their appropriate usage in the public health setting needs careful scrutiny*," stated Muin J. Khoury, in 1996, emphasizing the ever-increasing advances in molecular diagnostics and their intersection with the Public Health System. Technological progress in health care, the development of new disciplines (e.g., omics, epigenetics), and an individual approach to the patient have generated, in the 20 years following Khoury's statement, a scientific debate that is still topical.

Changes in laboratory medicine have led to changes in public health. Public health refers to a body responsible for solving problems related to the collective health of a population. It aims to increase life expectancy, improve quality of life, and prevent disease.

According to a traditional view, the public health is simply involved in increasing the number of individuals who can benefit from a given intervention (e.g., proper nutrition and/or use of appropriate medications) over those who cannot, without individual personalization. A modern view defines public health by the term "precision public health," characterized by the discovery, validation, and optimization of care strategies for a better population representation and stratification.

This is achieved with the aid and integration of new clinical interventions through the monitoring of the effectiveness of existing health maintenance and prevention programs. The integration between Precision Medicine and Precision Public Health finds application in all branches of medicine, thanks also to the remarkable development of omics sciences. The object of omics is to study genes (genomics), transcripts (transcriptomics), proteins (proteomics), and metabolites (metabolomics) that are expressed by a cell, allowing a global vision that considers its high degree of complexity. They allow the increasingly detailed characterization of biological processes (genetic, cellular, and biochemical) associated with the clinical phenotypes of diseases, thus leading to the identification of inter-individual differences. This increasingly personalized knowledge of the individual allows us to make early diagnoses and manage the course of the disease and its associated complications in the best possible way, leading to the development of personalized, more effective, and safer therapies. This will allow, in the near future, to obtain a large amount of data, making possible a greater stratification of the population and identifying individuals at high, medium, and low risk of a particular disease. Moreover, this background information will make it possible to identify only some individuals to targeted screening (with savings in the request for second/third level examinations) and thus achieve *ad hoc* primary prevention.

The applications of precision medicine are numerous and range from oncology to diabetology to pathologies with a lower incidence, such as epilepsy and liver cirrhosis. In the field of Cancer Precision Medicine (CPM), the application of the most modern diagnostic methods has made it possible to personalize cancer therapies and monitor therapeutic treatments.

Noteworthy, precision medicine does not only evaluate the individual complexity deriving from the DNA of the person, but it also pays attention to the DNA of the "hosts" of the human organism (microbiome). Indeed, it is known that the microbiome influences immunity and metabolism; therefore, its knowledge, through modern sequencing techniques, plays a key role in personalized medicine. Recently, Roy C. Ziegelstein has suggested the term personomics, which considers the individual as the inseparable whole of a genetic imprint and his peculiar social and economic characteristics, both fundamental aspects to know the person as a whole and to concretely apply a personalized medicine *ad personam*. However, in order to make precision medicine a routine practice, accessible to everyone, it will be necessary to enhance the clinical validity of the treatments currently in use, to progress with the discovery of new diagnostic tests and to consolidate the knowledge of health professionals regarding the application of new methods of diagnosis and treatment. It will also be necessary to improve the communication with patients and integrate available information. To achieve all this, a coordinated and integrated approach among the different medical disciplines is required, based on evidence, to assist the patient along the continuum from health to disease and integrate precision medicine with the new and modern vision of precision public health.

Recommended Readings

Arnett AK, Claas SA (2016) Precision medicine, genomics, and public health. Diabetes Care 39:1870–1873

Berti E, Casolari L, Cisbani L et al (2004) Il monitoraggio delle performance cliniche. In: Grilli R, Taroni F (eds) Governo clinico. Governo delle organizzazioni sanitarie e qualità dell'assistenza. Il Pensiero Scientifico Editore, Roma

Brook RH (1994) The RAND/UCLA appropriateness method. In: McCormick KA, Moore SR, Siegel RA (eds) Clinical practice guideline development: methodology perspectives. Agency for Health Care Policy and Research, Rockville, pp 59–70. (AHCPR publication no. 95-0009)

Buetow SA, Sibbald B, Cantrill JA et al (1997) Appropriateness in health care: application to prescribing. Soc Sci Med 45(2):261–271

Goossens N, Nakagawa S, Hoshida Y (2015) Molecular prognostic prediction in liver cirrhosis. World J Gastroenterol 21:10262–10273

Hallworth MJ (2011) The '70% claim': what is the evidence base? Ann Clin Biochem 48:487–488

Hickner J, Thompson PJ, Wilkinson T et al (2014) Primary care physicians' challenges in ordering clinical laboratory tests and interpreting results. J Am Board Fam Med 27(2):268–274

Insel RA, Dunne JL, Atkinson MA et al (2015) Staging presymptomatic type 1 diabetes: a scientific statement of JDRF, the Endocrine Society, and the American Diabetes Association. Diabetes Care 38(10):1964–1974. https://doi.org/10.2337/dc15-1419

Khoury MJ, Genetics Working Group (1996) From genes to public health: the application of genetic technology in disease prevention. Am J Public Health 86:1717–1722

Linee di indirizzo per la riorganizzazione dei servizi di Medicina di Laboratorio nel Servizio Sanitario Nazionale. Agenzia Nazionale per i Servizi Sanitari Regionali (AGENAS). http://www.agenas.it/innovazione-hta-e-dispositivi-medici/linee-di-indirizzo-per-la-riorganizzazione-dei-servizi-di-medicina-di-laboratorio-nel-servizio-sanitario-nazionale

Morosini PL, Perraro F (2001) Enciclopedia della gestione di qualità in sanità, 2ª edn. Centro Scientifico Editore, Torino

Plebani M, Laposata M, Lundberg GD (2011) The brain -to-brain loop concept for laboratory testing 40 years after its introduction. Am J Clin Pathol 136(6):829–833

Report of the Review of NHS Pathology Services in England (2006) An independent review for the Department of Health. Department of Health, London, p 6

Schaffhausen J (in press) What precisely is precision medicine? TIPS 2016; Article

Sharpe VA, Faden AI (1996) Appropriateness in patient care: a new conceptual framework. Milbank Q 74(1):115–138

Vasselli S, Filippetti G, Spizzichino L (2005) Misurare la performance del sistema sanitario. Proposta di una metodologia. Il Pensiero Scientifico Editore, Roma

Woodward RS, Warren-Boulton F (1984) Considering the effects of financial incentives and professional ethics on 'appropriate' medical care. J Health Econ 3(3):223–237

Zhi M, Ding EL, Theisen-Toupal J et al (2013) The landscape of inappropriate laboratory testing: a 15-year meta-analysis. PLoS One 8(11):e78962

Ziegelstein RC (2015) Personomics. JAMA Intern Med 175(6):888–889

The Laboratory Diagnostic Process

2

Mario Plebani

Introduction

The laboratory diagnostic process is a fundamental part of the overall diagnostic process and the evolution of modern medicine, from curative to predictive and preventive, relies more and more on laboratory information for early diagnosis, identification of disease risk, prognosis, and personalization of therapies.

The conceptual model underlying the laboratory diagnostic process remains the brain-to-brain loop, i.e., the coordinated and integrated set of the various steps that make up the overall laboratory testing cycle. In the past decades, the use of analytical quality indicators (basically imprecision and inaccuracy) and tools for their measurement (internal control and external quality assurance) led to a considerable improvement in performance and a reduction in analytical errors. However, to ensure the overall quality of laboratory information, it is necessary to extend quality control to the extra-analytical phases, identifying specific indicators and performance criteria that enable comparison among clinical laboratories and continuous improvement. Therefore, laboratory diagnostic process cannot disregard the pre-pre-analytical quality control, starting from the appropriateness of the request and the quality of the biological sample, and must extend to the post-post-analytical phase, i.e., the timeliness and accuracy of the interpretation and use of the results.

The improvement of the laboratory diagnostic process must be based on two main objectives. The first one is the reduction of "cognitive" errors, which include the errors that occur essentially at the time of request and interpretation of results, and therefore on interventions at the interface between laboratory and clinic. The second objective is the reduction of "system" errors, i.e., all errors in the implementation and monitoring of the various procedures and processes performed in the various phases of the analysis cycle.

M. Plebani (✉)
Department of Laboratory Medicine and Department of Medicine, University of Padova, Padova, Italy

Department of Pathology, University of Texas, Medical Branch, USA
e-mail: mario.plebani@unipd.it

The Diagnostic Process

The most recent conceptualization of the diagnosis process has been produced by the National Academy of Medicine, formerly known as the Institute of Medicine (IOM), which in the document "Improving Diagnosis in Health Care" focused attention on diagnosis for its intrinsic and obvious implications in patient care, research, and social-health policy. There is no doubt that an accurate and timely diagnosis not only ensures the patient the best treatment modalities and increases the probability of positive clinical outcomes, but it also has a positive impact on health care planning processes. Indeed, epidemiological data that collect information on diagnoses performed influence many health policy decisions, including payment and reimbursement policies, resource allocation, and research priorities.

The conceptual model developed by the National Academy of Medicine, and illustrated in Fig. 2.1, is based on overcoming the emphasis on classificatory and categorization schemes, given the evidence that the diagnostic process is "a complex, patient-centered, and collaborative activity involving information gathering and clinical reasoning with the aim of understanding the patient's health problem." According to the National Academy of Medicine committee that prepared the document, the diagnostic process represents an adaptation of the decision-making model based on the cyclical process of collecting, integrating, and interpreting information that leads to the establishment of a working hypothesis. This working hypothesis (working diagnosis) may consist of a list of potential diagnoses (differential diagnosis) or a single potential diagnosis. As the diagnostic process proceeds, the list of possible diagnoses narrows – or the probability that the provisional diagnosis is valid increases, a process masterfully described by Jerome Kassirer et al. as "diagnostic modification and refinement." Traditionally, it

M. Ciaccio (ed.), *Clinical and Laboratory Medicine Textbook*, https://doi.org/10.1007/978-3-031-24958-7_2

was at this stage that laboratory testing intervened, but the evolution of the discipline, technologies, and medicine has changed this traditional view. In emergency medicine, for example, the rapid availability of laboratory information has changed the approach to the critical patient, improving diagnostic accuracy, and overcoming, especially in cases of coma, the impossibility of collecting the anamnesis and the limitations of the semeiological examination. Additionally, the evolution of medicine from curative to the one now known by the acronym "4 P" (predictive, preventive, participatory, and personalized) makes the laboratory test essential to acquire information in the asymptomatic patient and, even more, to identify risk factors in the "healthy" subject and modify the natural history of the disease. The laboratory diagnostic process, therefore, fits into this changed context of the diagnostic process of modern medicine and, in the era of "omics", cannot be reductively confined to the frame of the diagnosis in the symptomatic patient.

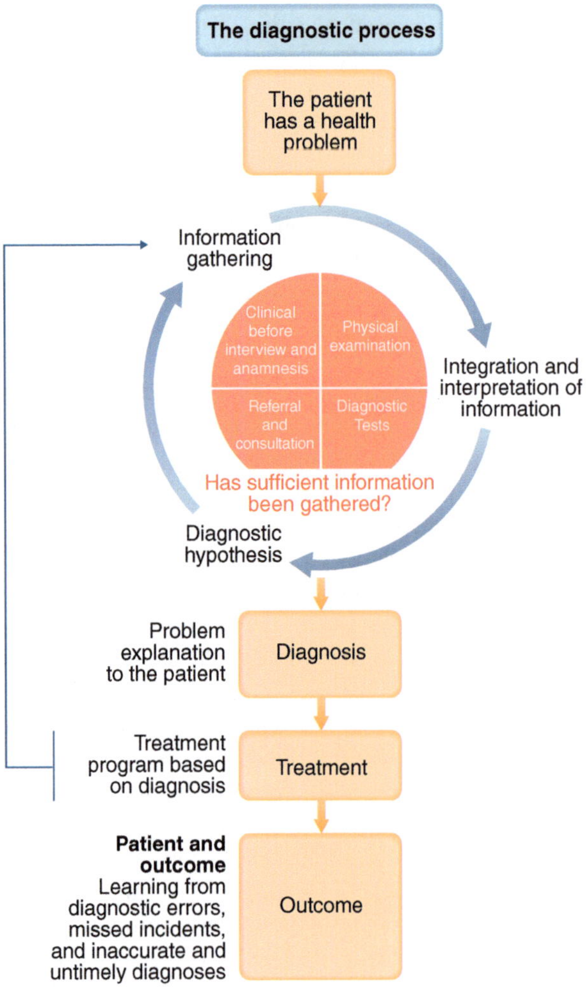

Fig. 2.1 The diagnostic process model according to the National Academy of Medicine. (Copyright EDISES 2021. Reproduced with permission)

The Laboratory Diagnostic Process

Although the document of the National Academy of Medicine is strongly addressed to clarify the traditional diagnostic process and reduce the errors related to it, already in the introduction of the paragraph dedicated to the "diagnostic examination," it clarifies the concept previously illustrated, namely that the diagnostic examination (laboratory, imaging, etc.) can identify clinical conditions "before" they become symptomatic and provide visible signs to clinical observation and physical examination. Even more interestingly, the paper introduces with great force the brain-to-brain loop model, initially described by George Lundberg in 1981, and more recently reviewed by Plebani et al. in 2011. Figure 2.2 shows the brain-to-brain loop in its updated version, which presents, in addition to the physician's brain, the brains of the clinical laboratory and the patient, to recognize the change that occurred in recent years and characterized by the more relevant role assumed by the clinical laboratory and the patient in the overall management of the laboratory test. If the role of the laboratory has progressively expanded from the management of the analytical phase to an ever-greater possibility of governing the initial and final phases of the cycle, the change in the figure of the patient is just as relevant. The patient is increasingly informed and interested in knowing the various aspects of the laboratory test, the usefulness of the test in providing answers to his or her health problems and the problems related to cost (in terms of both price and time required to access the service and obtain the results).

Although the model was designed and developed for laboratory examination, it has absolute utility and validity for

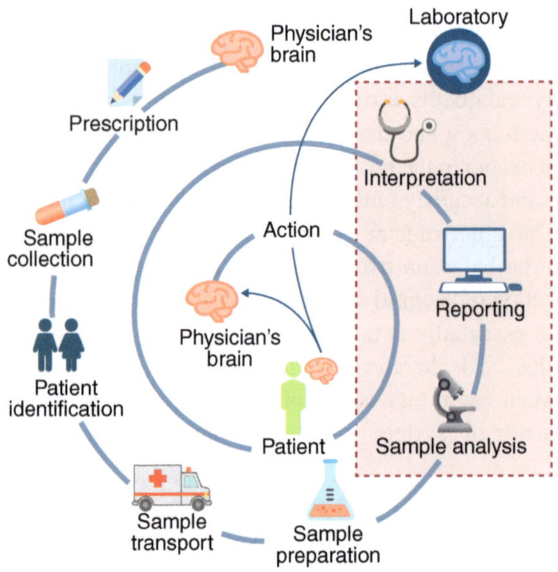

Fig. 2.2 The brain-to-brain loop. (Copyright EDISES 2021. Reproduced with permission)

all types of diagnostic examinations. As is well known, the model comprises nine essential processes, namely: examination selection and request, patient identification, specimen collection, specimen transport, specimen preparation and subsequent analysis, reporting and interpretation of results, and clinical action. These processes have traditionally been grouped into the pre-, intra-, and post-analytical phases. More recently, however, the phases have been redefined and expanded to recognize the existence of an initial phase (pre-pre-analytical), which occurs even before a biological sample arrives at the clinical laboratory, and a pre-analytical phase within the laboratory and is dedicated essentially to the preparation of the sample for analysis (intra-analytical phase). This is followed by the post-analytical phase, which translates into the validation of results, composition and transmission of the report, and the final phase (post-post-analytical), which ranges from the moment in which the clinician (and/or the health professional) receives the laboratory information, to the moment in which it is interpreted and used in the diagnostic-therapeutic process. The last link in the cycle is the evaluation of the clinical and economic outcomes related to the diagnostic process, outcomes that may fall exclusively on the individual patient and/or the population.

The main reasons for extending the phases of the cycle, introducing the pre- and post-post-analytical phase, are basically three.

- Recognition of one of the main domains of quality according to IOM, namely the centrality of the patient. The principle of patient-centeredness brings into play previous visions centered on the organization and the different responsibilities within the cycle. In order to ensure the quality of the entire laboratory examination cycle, it is necessary to remove the danger that lurks in the transition areas of the various phases and procedures. For example, it is evident that in the initial steps many procedures performed outside the clinical laboratory are not strictly under its control. Nevertheless, since the appropriateness of the request, the identification of the patient and that of the biological samples, as well as the quality of the collection and of the initial management of the samples, including transport, have significant repercussions on all the residual activity, on the analytical quality and on the global information, it is clear that the laboratory must take charge and "govern" these processes, in close collaboration with the health professionals and the clinicians who treat the patient. The same reasoning applies to the post-post-analytical phase, which cannot limit the responsibility of the clinical laboratory to the simple communication of results. It is well known the importance of timeliness of reporting and therefore of monitoring turnaround time (TAT), interpretative comments, and communication of critical results. It is now undisputed that there must be a control of the timeliness of the reception of the reports and of the quality of interpretation and use of laboratory information.
- Technological developments, especially thanks to information technology, have made it possible to control the processes outside the laboratory, starting with the identification of the patient and the biological samples, their transport, up to the real-time communication of the reports and the control of their receipt. The automation of sample preparation, thanks to the introduction of pre-analytical stations, has radically reduced the risk of errors due to the creation and identification of aliquots, centrifugation errors, and sample addresses. These improvements, however, have brought out the problems of incorrect patient identification, poor quality of samples, delays in receiving and interpreting laboratory reports.
- Studies on error in laboratory medicine have demonstrated the vulnerability of the early and late phases of the cycle, in the face of a drastic reduction in analytical error. Persisting in focusing only on analytical quality indicators, typically measured by internal control and external quality assessment, risks exposing the patient to even more hidden and serious errors. Therefore, the need to identify ex-analytical quality indicators that can be used both for corrective actions and improvement within the individual laboratory and for comparison with other laboratories (benchmark) has emerged with increasing strength.

Laboratory Information and Diagnostic Reasoning

The laboratory information, which is the result of the articulated and complex set of the various phases of the cycle described above, has the following main clinical objectives:

- Predict susceptibility to diseases
- Prevent diseases by identifying risk factors that can lead to lifestyle changes or other types of clinical interventions
- Diagnose diseases, preferably in the early stages where the chances of cure are best
- Determine the prognosis at the individual patient level
- Monitor diseases, identifying the effectiveness of treatments and evaluating possible changes
- Personalize therapy to achieve better clinical outcomes, reducing the risk of adverse events

In spite of the diversity of the abovementioned clinical objectives, laboratory information has a common denominator: it is the result of the coordinated and integrated whole of the various phases that make up the cycle and is not identified with the "analytical" moment alone. Indeed, the analytical quality, in addition to the intrinsic methodical aspects, is significantly affected by the quality of the biological sample to be analyzed. "Good" biological samples are the necessary premise for reliable results, while "bad" samples determine – even with high analytical quality – clinically misleading and "erroneous" results. However, despite the evidence that 70% of errors in laboratory medicine occur in the pre-analytical phase, the lack of pre-analytical, and in general extra-analytical, quality indicators have not been allowed so far to ensure greater quality and safety to the laboratory information.

In support of this assertion is the evidence showing that few clinical laboratories collect in a systematic and objective way the data on the main factors of pre-analytical non-quality, such as hemolyzed, jaundiced, coagulated samples, with the wrong relation with the anticoagulant, etc. It should be even more emphasized that, even today, there is a high percentage of clinical laboratories that relies on the "visual" recognition of hemolysis (and of jaundiced and lipemic samples), compared to the possibility of using serum indices, which represent objective, automated indicators and do not slow down the analytical productivity of the most widespread instruments in clinical laboratories. Moreover, the criteria for the rejection of biological samples and the consequent operating procedures, also in terms of preventive and corrective actions, are poorly harmonized and often left to the will of individual operators.

Diagnostic reasoning, however, cannot disregard the evaluation and acceptance of the quality of the biological sample, and without this premise, the quality of the laboratory information cannot be used with confidence in the clinical decision-making process.

It is therefore essential to guarantee the quality of the initial phases of the cycle, even if they are performed by health care personnel who are not hierarchically traceable to the responsibilities of the laboratory medicine units, strongly reproposing the concept of process responsibility, independently of the physical locations in which the various phases are performed. Figure 2.3 shows the contribution of the laboratory diagnostic process, in its essential components, to clinical diagnosis, therapeutic action, and again, to patient monitoring. It is clear from this diagram that laboratory information is conditioned not only by analytical accuracy but also by pre-analytical accuracy and by correct interpretation/use in the clinical context of the patient.

The importance of pre-analytical quality has been understood especially by researchers involved in the issue of biobanks and collections of biological samples aimed at validating and developing new biomarkers, while it remains dangerously underestimated in routine activities. This leads to a dangerous dissociation between the quality required in research activities and clinical practice.

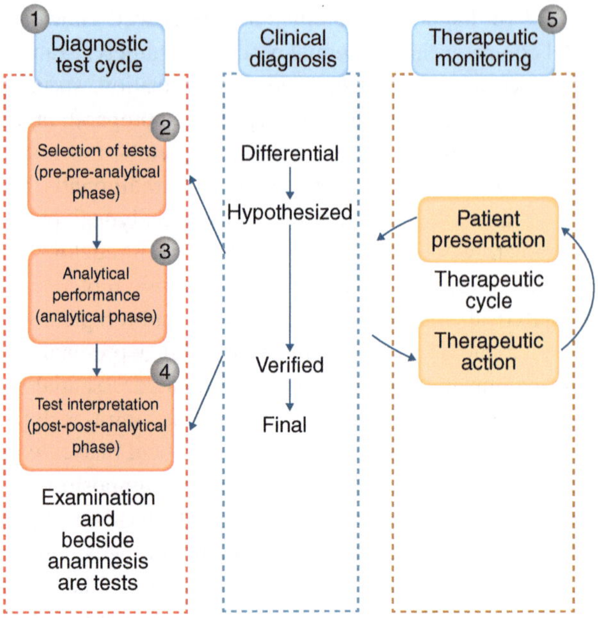

Fig. 2.3 The laboratory diagnostic process in the context of the diagnostic-therapeutic cycle. (Copyright EDISES 2021. Reproduced with permission)

Similarly, the undoubted improvement in analytical quality that has occurred in recent decades must not remain isolated from the post-analytical context, i.e., it must be supplemented with better ways of interpreting and using laboratory information. Data from the literature, and even more from daily practice, underline the increasing difficulties in the interpretation of clinical laboratory reports. The increased complexity of laboratory tests, especially in some diagnostic areas (autoimmunity, coagulation, allergology, molar pathology, etc.), the insufficient training of physicians in undergraduate and graduate courses, and the overload of data that the clinician must manage on a daily basis are at the basis of errors and difficulties in the interpretation and use of laboratory information.

How to Improve the Laboratory Diagnostic Process

A careful analysis of the laboratory test cycle and of the data in the literature allows us to recognize two major categories of errors: cognitive errors and system errors. Cognitive errors include all the problems in clinical reasoning that lead to errors in requesting the appropriate test, to poor interpretation of results, and to the insufficient synthesis of available clinical information. Cognitive diagnostic errors can be summarized in the groups represented in Table 2.1.

Table 2.1 Possible causes of error in diagnostic reasoning

Errors in access to care	Errors due to delays by the patient in recognizing symptoms and clinical signs of the disease and/or barriers to the health system in accessing care
Errors in the information collection process	Errors and/or delays in requesting appropriate diagnostic tests
Interpretation errors	Inaccuracy or inability to interpret clinical information (medical history, laboratory, image, etc.)
Errors in the integration of clinical information	Errors in the generation of diagnostic hypotheses, inability to give due weight and priority to the information collected in the diagnostic process and inability to adequately recognize clinical symptoms and signs
Errors in defining the diagnosis	Inability to weigh the various clinical information, delays in identifying the most plausible diagnosis and in patient follow-up
Communication errors	Inability and delays in communicating the explanation of the patient's clinical problem

This type of error, therefore, has several roots: (1) insufficient preparation and training of those who prescribe and subsequently interpret the results of the clinical laboratory; (2) limits of scientific knowledge; (3) reduction of care time and progressive "industrialization" of customer care activities; (4) concerns typical of defensive medicine and overload of data and diagnostic reports; and (5) difficulty in integrating laboratory data in the complex clinical context of the patient also for the lack of knowledge of probabilistic and Bayesian concepts.

System errors, on the other hand, include all errors that arise from the implementation of the procedures and processes that constitute the laboratory test cycle and that can be minimized through the development and continuous monitoring of the quality system, in particular by following the requirements of the specific international standard for the accreditation of clinical laboratories (ISO 15189).

There are several strategies to reduce cognitive errors in the early phase of the cycle, beginning with more up-to-date physician education in the various domains of laboratory medicine and an increased focus on teaching the principles governing appropriate testing requests. Clinical laboratory professionals must play a role in improving this early phase of the examination cycle and even more so in reclaiming their responsibility at the clinical-laboratory interface.

The tools are various and range from making available "at the point of care" tools such as LabTestOnline and other sources of information on the meaning of the various laboratory test, to the creation of alert systems to restrict the request for expensive and complex tests to particular clinical questions and, again, to computer blocks in case of requests for repetition of tests before the minimum interval of biological plausibility.

Recently, the theme of the appropriateness of the request for diagnostic exams, and in particular of laboratory tests, has assumed ever-increasing importance, even if focused only on the economic cost of test requested in an "inappropriate" way. The topic of the inappropriateness of the request, indeed, deserves an in-depth examination that involves both the real dimension and the motivations and the clinical-management consequences.

Data from the literature show that in the case of laboratory medicine is greater the underuse than the excessive demand for tests, thus contradicting the current thinking too often based on impressions and anecdotal evidence. In reality, the problem is complex because appropriateness, or rather inappropriateness, can only be assessed by knowing the clinical picture of the individual patient, given that any diagnostic test and clinical intervention are considered "appropriate" when the benefit to health exceeds the risk to the patient by an adequate margin. Certainly, however, there is consensus on the definition of "inappropriate request" in terms of requesting examinations outside any consensually accepted form of "guidance." In practice, this means that a request that, even in the absence of a defined guideline, does not take into account the presence of evidence, recommendations, and suggestions from the literature and experts is considered inappropriate. Another problem concerns the reasons that lead to inappropriateness. There are many reasons why laboratory tests are requested in an "inappropriate" way, among which we cannot underestimate the problems deriving from the existence of defensive medicine, the pressure from patients, and the attempt of clinicians to overcome diagnostic uncertainty in many clinical situations. It is important to emphasize that the problem of the cost of inappropriateness should not be seen purely in terms of the cost of the tests themselves, but in terms of the impact, it has on the economy and patient outcomes. Modest variations in the results of laboratory tests, especially if they are requested uncritically and not addressed to a specific clinical question, can lead, by fall, to new requests for the same laboratory tests, but also requests for imaging tests, other diagnostic investigations, specialist visits, and inappropriate hospitalizations. Moreover, they can determine negative psychological fallout for patients due to the anxiety and fear caused by what we could define "asterisk syndrome." Finally, the debate on inappropriateness has shifted to the issue of overdiagnosis and overtreatment that is inherent in modern medicine. Overdiagnosis and overtreatment represent, however, problems of modern medicine in the era of "omics" and the attempt to overcome curative medicine based on the recognition, often late, of clinical symptoms and signs. The improvement of appropriateness is, therefore, a complex but equally important goal for all laboratory medicine professionals and for the relationship with clinicians.

In the final stages of the cycle, the cognitive tools that can improve clinical reasoning – in addition to better preparation and training of physicians – are basically the correct definition of the reference interval and, even more than the critical difference in the case of repeated examinations over time, interpretive comments, and communication of critical results. In addition, the availability of expert systems and computer databases at the point of care presents a further approach to the problem.

Quality and Outcome Indicators

The mantra "you only improve what you measure" has proven its truth in the case of the analytical quality of laboratory testing. There is ample evidence of improved analytical performance in terms of imprecision and inaccuracy since the introduction of analytical quality indicators, i.e., quality specifications, and the tools to measure them, i.e., internal quality control and external quality assessment programs.

Figure 2.4 is emblematic in documenting the improvement in analytical error, expressed as the average coefficient of variation (CV%) since the "discovery" of external quality assessment as a tool for comparison among clinical laboratories and improvement of analytical performance. There is no doubt that analytical quality still represents the core business of the profession and that there is still room for improvement and error reduction. However, the magnitude of analytical error is far less than the risk to the patient of the more serious pre-analytical errors such as misidentification of patient and biological samples or failure to recognize issues that should lead to sample rejection. Similarly, post-analytical errors such as transcription errors and/or errors in units of measure can lead to serious diagnostic errors and real risks to patient health.

For all these reasons, the laboratory diagnostic process cannot be based solely on analytical quality but must be based on ensuring that all stages of the cycle are performed according to strict operating procedures. In addition, there is the need to monitor all phases of the cycle with quality indicators that allow for detecting non-conformities and errors that may determine risks for the final quality of laboratory information. The harmonization of these indicators and the reporting system is an essential tool to make possible not only improvement actions within the individual clinical laboratory but also the comparison and benchmarking among different laboratories. In this way, it will be possible, as already happened for the analytical quality, to identify performance criteria (quality specifications) that can indicate the appropriate corrective actions and continuous improvement. Unlike analytical performance criteria that, being related to measures, can be established according to the hierarchy established in the Stockholm Conference of 1999 and updated in the Milan Conference of 2014, extra-analytical quality specifications do not target a number, be it a percentage, an absolute number or a sigma value, but tend to "zero defects." Indeed, patient identification errors, like that of the transcription of analytical data, cannot be admitted nor justified, also because of the serious repercussions on clinical outcomes and patient safety. However, in order to make improvement goals practical, it is necessary to establish the current "state of the art," i.e., the current level of performance that clinical laboratories should document and continuously monitor. Similar to analytical quality specifications, only the collection of numerically significant data from an adequate number of clinical laboratories can lead to the definition of performance criteria in the extra-analytical phase.

Currently, there is a phenomenon that has been called "the paradox of quality indicators." On the one hand, several quality indicators have been identified for which a harmonization process has been agreed upon. A website has been developed that, in addition to providing all the information on the project of the LEPS (Laboratory Errors and Patient Safety) working group of the International Federation of Clinical Chemistry and Laboratory Medicine (IFCC), allows interested laboratories to enter the data collected on the indicators and a homogeneous reporting system to compare the data of the individual laboratory with the consensus values obtained by all the laboratories participating in the project (benchmark).

On the other hand, there are still very few clinical laboratories that continuously and consistently collect data on quality indicators, especially in the pre-analytical phase.

Overcoming this paradox requires, first, that clinical laboratory professionals understand the need to broaden the scope of action and quality control from the analytical sphere to all phases of the cycle. Secondly, it is necessary that scientific societies, international federations, and clinical laboratory regulatory and accreditation bodies promote the use of quality indicators and the adoption of extra-analytical per-

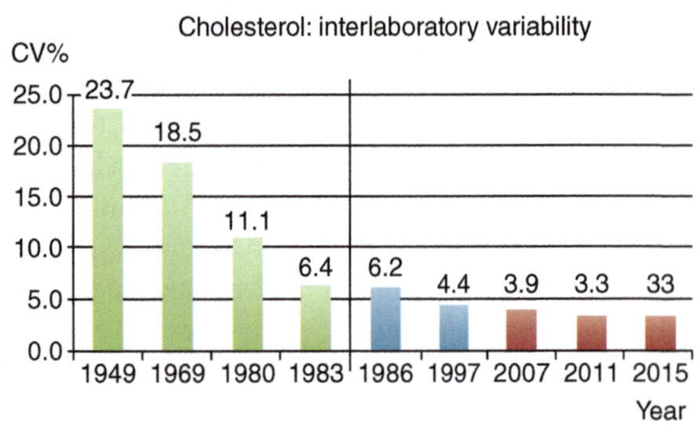

Fig. 2.4 Analytical error in the determination of cholesterol: decrease over time in the average coefficient of variation (CV%), starting from the time of the introduction of the external quality assessment

formance criteria as tools for monitoring and improving service quality.

The next step is the identification of outcome indicators that can be used by clinical laboratories to assess the final quality of their daily operations and to guarantee to users (patients and prescribers) that laboratory information is safe and is really used appropriately in the diagnostic and therapeutic context. In other words, the brain-to-brain concept, 40 years after its formulation, is now living a fundamental moment in which it is recognized the need and the possibility to document and monitor all phases of the cycle and measure the clinical and economic outcomes of the laboratory test.

Recommend Readings

Callen JL, Westbrook JI, Georgiou A, Li J (2012) Failure to follow-up test results for ambulatory patients: a systematic review. J Gen Intern Med 27:1334–1348

Carraro P, Zago T, Plebani M (2012) Exploring the initial steps of the testing process: frequency and nature of pre-preanalytic errors. Clin Chem 58:638–642

Cornes MP, Atherton J, Pourmahram G et al (2016) Monitoring and reporting of preanalytical errors in laboratory medicine: the UK situation. Ann Clin Biochem 53:279–284

Diamandis EP, Li M (2015) The side effects of translational omics: overtesting, overdiagnosis, overtreatment. Clin Chem Lab Med 54:389–396

Ellervik C, Vaught J (2015) Preanalytical variables affecting the integrity of human biospecimens in biobanking. Clin Chem 61:914–934

Fraser CG, Kallner A, Kenny D, Petersen PH (1999) Introduction: strategies to set global quality specifications in laboratory medicine. Scand J Clin Lab Invest 59:477–478

Fryer A, Smellie WSA (2013) Managing demand for laboratory tests: a laboratory toolkit. J Clin Pathol 66:62–72

Gandhi TK, Kachalia A, Thomas EJ et al (2006) Missed and delayed diagnoses in the ambulatory setting: a study of closed malpractice claims. Ann Intern Med 145:488–496

Graber M (2005) Diagnostic errors in medicine: a case of neglect. Jt Comm J Qual Patient Saf 31:106e13

Graber ML, Franklin N, Gordon R (2005) Diagnostic error in internal medicine. Arch Intern Med 165:1493e9

Hood L, Friend SH (2011) Predictive, personalised, preventive, participatory (p4) cancer medicine. Nat Rev Clin Oncol 8:184–187

Institute of Medicine (2001) Crossing the quality chasm: a new health system for the 21st century. National Academies Press, Washington, DC

ISO 15189: 2012 (2012) Medical laboratories – requirements for quality and competence. International Organization for Standardization, Geneva

Kassirer JP, Wong J, Kopelman R (2010) Learning reasoning. Williams and Wilkins, Baltimore

Kellogg MD, Ellervik C, Morrow D et al (2015) Preanalytical considerations in the design of clinical trials and epidemiological studies. Clin Chem 61:797–803

Lippi G, Becan-McBride K, Behúlová D et al (2013) Preanalytical quality improvement: in quality we trust. Clin Chem Lab Med 51:229–241

Lippi G, Banfi G, Church S et al (2015) Preanalytical quality improvement. In pursuit of harmony, on behalf of European Federation for Clinical Chemistry and Laboratory Medicine (EFLM) Working group for Preanalytical Phase (WG-PRE). Clin Chem Lab Med 53:357–370

Lundberg GD (1981) Acting on significant laboratory results [editorial]. JAMA 245:1762–1763

National Academies of Sciences, Engineering, and Medicine (2015) Improving diagnosis in health care. National Academies Press, Washington, DC

Piva E, Pelloso M, Penello L, Plebani M (2014) Laboratory critical values: automated notification supports effective clinical decision making. Clin Biochem 47:1163–1168

Plebani M (2006) Errors in clinical laboratories or errors in laboratory medicine? Clin Chem Lab Med 44:750–759

Plebani M (2007) Errors in laboratory medicine and patient safety: the road ahead. Clin Chem Lab Med 45:700–707

Plebani M (2009) Exploring the iceberg of errors in laboratory medicine. Clin Chim Acta 404:16–23

Plebani M (2010) The detection and prevention of errors • in laboratory medicine. Ann Clin Biochem 47:101–110

Plebani M (2012) Quality indicators to detect pre-analytical errors in laboratory testing. Clin Biochem Rev 33:85–88

Plebani M (2015) Medicina difensiva e appropriatezza: il caso degli esami diagnostici. In: Simonetti G (ed) . Idelson-Gnocchi, Napoli, pp 92–97

Plebani M (2016a) Harmonization in laboratory medicine: requests, samples, measurements and reports. Crit Rev Clin Lab Sci 53:184–196

Plebani M (2016b) The quality indicator paradox. Clin Chem Lab Med 54:1119–1122

Plebani M, Lippi G (2011) Closing the brain-to-brain loop in laboratory testing. Clin Chem Lab Med 49:1131–1133

Plebani M, Panteghini M (2014) Promoting clinical and laboratory interaction by harmonization. Clin Chim Acta 432:15–21

Plebani M, Laposata M, Lundberg GD (2011) The brain-tobrain loop concept for laboratory testing 40 years after its introduction. Am J Clin Pathol 136:829–833

Plebani M, Astion ML, Barth JH et al (2014) Harmonization of quality indicators in laboratory medicine. A preliminary consensus. Clin Chem Lab Med 52:951–958

Plebani M, Sciacovelli L, Aita A et al (2015) Performance criteria and quality indicators for the pre-analytical phase. Clin Chem Lab Med 53:943–948

Sandberg S, Fraser CG, Horvath AR et al (2015) Defining analytical performance specifications: consensus statement from the 1st strategic conference of the European federation of clinical chemistry and laboratory medicine. Clin Chem Lab Med 53:833–835

Schiff GD, Hasan O, Kim S et al (2009) Diagnostic error in medicine: analysis of 583 physician-reported errors. Arch Intern Med 169:1881–1887

Sciacovelli L, Plebani M (2009) The IFCC Working Group on laboratory errors and patient safety. Clin Chim Acta 404:79–85

Sciacovelli L, Zardo L, Secchiero S, Plebani M (2004) Quality specifications in EQA schemes: from theory to practice. Clin Chim Acta 346:87–97

Sciacovelli L, O'Kane M, Skaik YA et al (2011a) Quality indicators in laboratory medicine: from theory to practice. Preliminary data from the IFCC Working Group project £laboratory errors and patient safety. Clin Chem Lab Med 49:835–844

Sciacovelli L, Sonntag O, Padoan A et al (2011b) Monitoring quality indicators in laboratory medicine does not automatically result in quality improvement. Clin Chem Lab Med 50:463–469

Sciacovelli L, Aita A, Padoan A et al (2016) Performance criteria and quality indicators for the post-analytical phase. Clin Chem Lab Med 54:1169–1176

Smellie WS (2012) Demand management and test request rationalization. Ann Clin Biochem 49:323–336

Zhi M, Ding EL, Theisen-Toupal J et al (2013) The landscape of inappropriate laboratory testing: a 15-year meta-analysis. PLoS One 8:e78962

Elements of Biomedical Laboratory Organization

3

Giuseppe Lippi, Camilla Mattiuzzi, and Chiara Bovo

Introduction

The organization of a biomedical laboratory is complex, multifaceted, and represents the result of a series of reforms that over the years have determined its structure and functioning. The aim of this chapter is to provide the basic elements and legislative references to understand the organizational context of a biomedical laboratory, the role of different structures, and various professional profiles in the different articulations and duties.

As rationally defined by the guidelines for the reorganization of laboratory medicine services in the National Health System (NHS), the scope of laboratory medicine, strongly focused until a few years ago on the analytical aspect, has expanded in favor of a more integral and holistic vision, aimed at recognizing the contribution of disciplines from a patient-centered perspective and, more generally, on clinical needs. A biomedical laboratory (historically and conventionally known by the term "biomedical laboratory") has the fundamental task of performing investigations with chemical, physical, or biological methods on materials from the human body and producing analytical results accompanied by diagnostic judgments.

The basic staff of a public biomedical laboratory obviously depends on its size (volume of examinations), its location within the NHS, and its aims. Therefore, the professional figures that work in it include a director, a team of health managers (doctors, biologists, chemists, pharmacists, and other graduates with a specialization that is preparatory to employment in this specialty), assisted by a variable number of biomedical laboratory health technicians), and auxiliary and administrative staff. Nurses may also be present, but they are mainly responsible for collecting biological samples.

Even though the specific description of the techniques and purposes of the analyses are more comprehensively dealt with in other chapters of this volume, it is convenient to formulate some general premises, to better understand the organization of the structure.

Sectors of a Biomedical Laboratory

Strictly dependent on the size, location, and purpose, the structure of a biomedical laboratory includes a series of "diagnostic areas," which are homogeneous and differentiated on the basis of the biological matrix or the characteristics of the analytical techniques and/or instruments.

- Classification (simplified) according to the biological matrix:
  - Serum and/or plasma area (analysis of serum and/or lithium heparin and/or plasma ethylenediaminetetraacetic acid (EDTA) tubes and/or plasma anticoagulated with other anticoagulants)
  - Hematology area (analysis of test tubes containing EDTA as an anticoagulant)
  - Coagulation area (analysis of test tubes containing sodium citrate as an anticoagulant)
  - Urine and other materials area (analysis of test tubes and/or containers containing urine, feces, or other fluids such as effusions, cerebrospinal fluid, pleural fluid, peritoneal fluid, pericardial fluid, bile fluid, etc.)
- Classification (simplified) according to the activities, analytical techniques, and/or instruments used:
  - Sample acceptance area
  - Clinical chemistry area
  - Immunochemistry area

G. Lippi (✉)
Section of Clinical Biochemistry, University Hospital of Verona, Verona, Italy
e-mail: giuseppe.lippi@univr.it

C. Mattiuzzi
Hospital of Rovereto, Provincial Agency for Social and Sanitary Services (APSS), Trento, Italy

C. Bovo
Medical Direction, Hospital of Padova, Padova, Italy

M. Ciaccio (ed.), *Clinical and Laboratory Medicine Textbook*, https://doi.org/10.1007/978-3-031-24958-7_3

- Complete blood count (CBC) area
- Coagulation examination area
- Standard urine test area
- Protein area
- Toxicology area
- Microbiology area
- Immunohematology area
- Molecular biology and/or cytogenetics area
- Other specific areas (e.g., radioimmunology, allergology, autoimmunity, functional testing, etc.)

The connotation of an biomedical laboratory is considerably heterogeneous. There are, indeed, facilities that include all the described areas (or others called "multidisciplinary"), whereas the so-called "specialized" or "small-sized" laboratories may consist of only one area or some of them. There are, for example, laboratories that perform only coagulation tests, as well as small laboratories that perform only a limited panel of routine tests (e.g., clinical chemistry profile, immunochemistry, coagulation, and hematology), sending biological samples to other facilities.

The structural articulation of the different areas is extremely heterogeneous in different laboratories and depends on the degree of integration and automation. In particular, the organization can vary from a model of total laboratory automation (TLA), in which all (or almost all) analyzers in the laboratory are physically connected to each other, to a model of total compartmentalization, in which the analyzers are all stand-alone, i.e., without any physical connection between them. In the middle of these two extremes coexist integration/automation solutions of variable degrees, for homogeneous analytical areas (e.g., integration of all clinical chemistry analyzers), for biological matrix areas (e.g., all instruments using serum or heparinized plasma), or for simple convenience (e.g., integration for lack of personnel). The intermediate solutions are generically defined as "islands" of automation, as they have functional autonomy with respect to the remaining laboratory instrumentation (Fig. 3.1). The physical connection between the different instruments is guaranteed by chains or conveyor belts of various types and articulation, which have the function of conveying the samples from an entry point to the different instruments in the chain up to a terminal module where they are stored for a limited period of time. Modules intended for the management of pre-analytical activities can also be part of the automation component (Fig. 3.2), in particular for:

- Sample loading (input module or bulk module)
- Sample check-in (verification that the sample and test schedule have been captured by the laboratory computer system)
- Centrifugation (separation of serum or plasma from blood cells)
- Uncorking (removing the stopper from the test tube)
- Aliquoting (creation of secondary aliquots from the primary sample, e.g., intended for other analysis on instruments located outside the chain)
- Recorking of the tube
- Transport chain
- Storage (refrigerators of various sizes to keep the samples after analysis for a limited period of time)

A strong element characterizing the activity of a laboratory is represented by the priority with which the tests must be performed and reported. Aware that no definition can ever

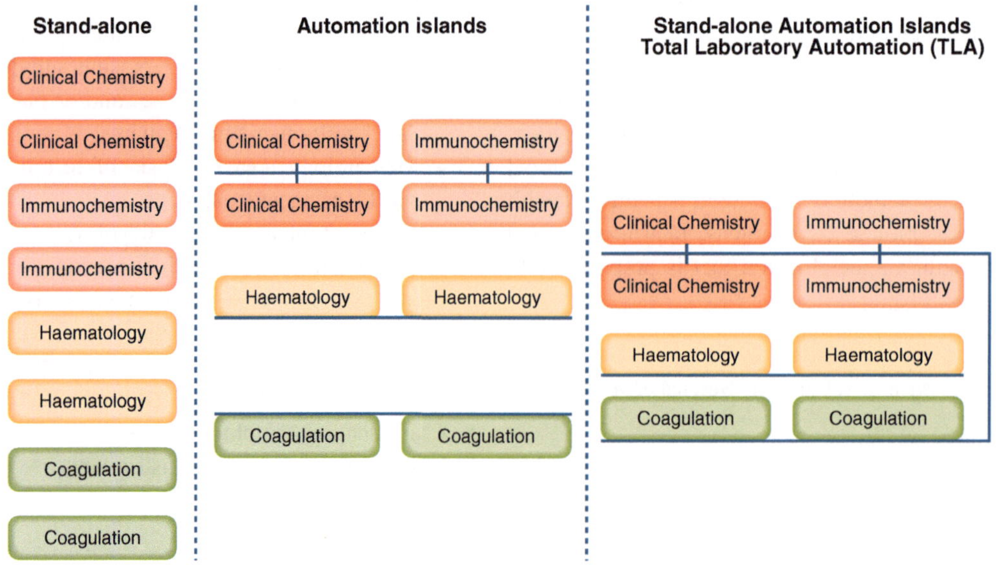

Fig. 3.1 Organizational solutions in laboratory medicine. (Copyright EDISES 2021. Reproduced with permission)

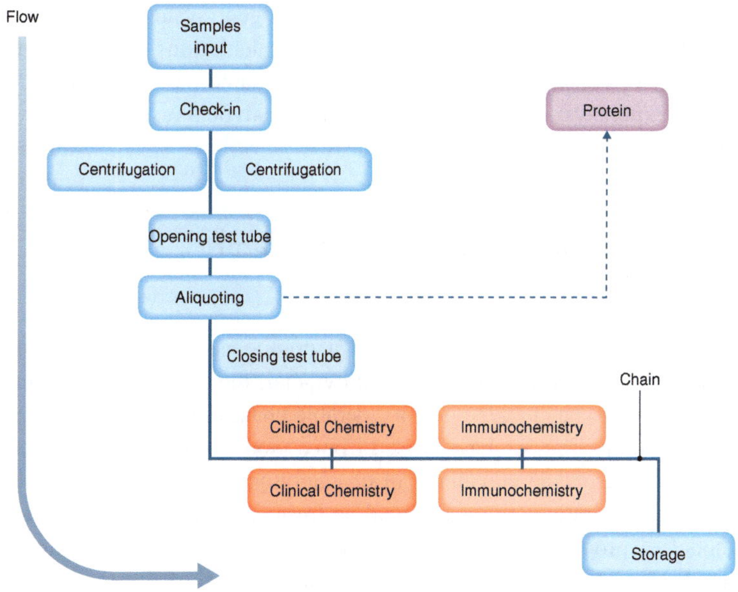

Fig. 3.2 Automation in laboratory medicine. (Copyright EDISES 2021. Reproduced with permission)

be all-encompassing, we can simplify the classification according to the examinations performed in the regime of:

- Routine: These are examinations that usually require a report within a few hours or even days.
- Urgency: These are examinations that require an accelerated report (1–3 hours usually) on the basis of the patient's clinical condition (usually serious) or for organizational reasons (the patient is in a day hospital).
- Emergency: These are typically examinations that have to be reported in the shortest possible time (within 1 hour) due to the (extremely serious) condition of the patient.

The abovementioned priority classification of examinations also reflects the internal organization of a laboratory. Indeed, there are laboratories in which the majority of routine and emergency examinations are performed using the same instrumentation (routine–urgency integration) and those in which routine examinations are performed on instrumentation different from that intended for emergency/urgency examinations. This depends on many variables of a technical and organizational nature, specific to the structure in which the laboratory is located. For example, in large structures (hospitals or others) where the laboratory is located at a long distance from the critical departments (emergency room, emergency medicine, resuscitation), it may be functional to create a satellite laboratory that can meet the needs of the critical departments. Another variable may be represented by the volume of analysis performed by the laboratory, i.e., when the number of analyses (or test tubes) under examination is considerable, leading to the inability to guarantee an effective priority of analysis on the instruments for the processing of examinations required in urgency/emergency, it may be functional to allocate a series of dedicated analyzers for this purpose. This can obviously generate problems of alignment of results performed in routine or in urgency/emergency, especially when the instrumentation is different. It is quite understandable how an increase in the complexity of the system (in reference to the presence of a large number of analyzers) involves problems of an analytical nature for the consistent alignment of the results generated by the different modules.

The Computer Link

In the modern conception of laboratory activity, computerization plays a fundamental role. In almost all modern laboratories, analyzers are connected to a laboratory management system (Laboratory Information System, LIS), from which they receive the programming related to the types of tests to be performed on biological samples and to which they transmit the results, thus allowing a validation process at multiple levels (technical/analytical validation, biological, etc.). This mode of communication is conventionally known as "query host". Indeed, the primary samples have a positive and univocal identification by means of a barcode, which is read by the analyzers and allows the univocal identification of every single sample under examination. There is still the possibility of manual programming of the analyses, which is, however, limited to instruments that are not connected or connectable to the LIS, or in undesirable circumstances in which there are malfunctions in the computer system or its connections. The results of the examinations, after transfer to the LIS, are stored in a database from which they can be extracted for the generation of laboratory reports, whether paper or digital. Modern programs also have a series of additional modules, which allow analyzing productivity, control of financial processes and costs, and "intelligent systems" equipped with logical activities that include calculations and comparisons (e.g., with previous data of the same patient), up to "expert systems" for the validation of analytical results and the management of appropriateness.

Organizational Models of the Laboratory Network

The organization of laboratories in a homogeneous geographical area (corporate, district, provincial, regional, and even national) has undergone considerable changes over the years. Simplifying, we have passed from a model (typical in the 70s) of extreme parceling out of laboratory diagnostics (facilities operating below a critical performance mass) to a model of progressive integration and consolidation, which has led to the unification or closure of numerous territorial realities. In Italy, this change was strongly supported by law number 296 of December 27, 2006, which states that "the Regions shall approve a reorganization plan of the network of public and private accredited facilities providing specialist services and laboratory diagnostics, in order to adjust the organizational and personnel standards consistent with the processes of increased efficiency made possible by the use of automated methods." The general principles that guide the reorganization process must be:

- Close interrelation between the type of hospital and type of hospital laboratory
- Hospital territory continuity
- Proximity to the needs of the patient and the clinician
- The role of continuing education and research
- The role of information systems and (health) technology assessment
- The role of external quality assurance
- Centrality of the promotion and control of appropriateness
- The role of a targeted reporting system on laboratory activities

The paradigm that has inspired most reorganization processes is the so-called hub-and-spoke model. This model was developed in the United States following the deregulation of commercial civil aviation and was originally introduced by the airline Delta in 1955, with the identification of Atlanta airport as the nodal point on which to concentrate most long-distance flights. In particular, the channeling of long-distance connections to a reference pole, called a hub, and the creation of a series of connecting flights from peripheral airports, called spoke, allowed the company to make much more efficient use of resources without penalizing passengers too much. In essence, instead of starting a daily direct flight to London from Washington, Miami, Baltimore, Indianapolis, and Atlanta, the daily number of flights from the Atlanta hub was increased (thus increasing the offer of departure times), eliminating those from spoke airports and transporting passengers by connecting flights from peripheral airports to that of Atlanta.

In general, the hub-and-spoke model has also progressively spread to the organization of health-care structures and has created an impact in terms of services provided. In this case, borrowing the model of aviation, there has been, within homogeneous geographical areas, a progressive reorganization of laboratory facilities, also divided into spoke (peripheral laboratories) and hub (central laboratories). The former has, therefore, been assigned a much lower degree of complexity (in relation to the types and number of tests performed) than has the latter. In this perspective, instead of transporting passengers from a peripheral airport to the central one, it was planned to transport patients' test tubes from a peripheral laboratory to the central one. In the economy of scale, this allowed considerable savings related to the abolition of duplicate analyses between two laboratories in proximity, thus saving on the number of instruments, reagents, and personnel. This model functionally operates according to a network in which each laboratory has a well-defined task and function (Fig. 3.3) and is qualitatively and economically sustainable after an accurate feasibility study. Briefly, it is necessary to create a balance between two opposing models, characterized by an unjustified multiplication of laboratories in the same area and by an irrational concentration of exams in laboratory "mega-structures" that, because of distance and volume, risk losing proximity to the clinician and the patient. In concrete terms, the facilities operating within a network should be characterized by a proper balance between general laboratories, also divided into specialized sectors, and specialized laboratories with organizational autonomy. This can be achieved in relation to the discipline and the type of activity involved, even for over-company catchment areas, and in explicit compliance with all the criteria provided in the reorganization process. Therefore, effective departmental inte-

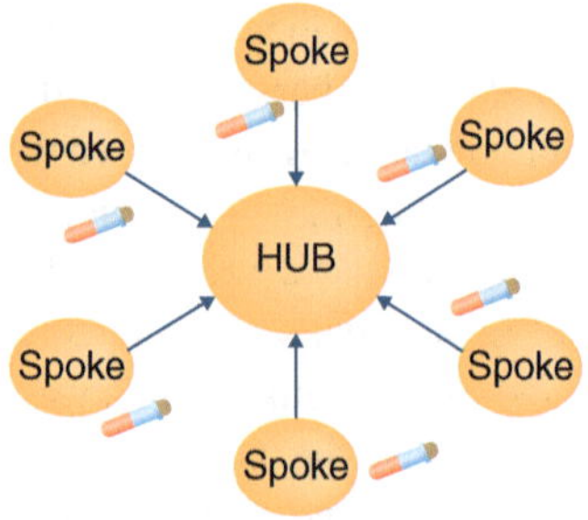

Fig. 3.3 Organization of a network of laboratories according to the hub-and-spoke model. (Copyright EDISES 2021. Reproduced with permission)

gration of the structures in the corporate, supra-provincial, and regional network must be developed, through a strong level of coordination, sharing of management processes, quality policies, and continuous staff training. Another critical aspect concerns the transport of samples, especially when the distance between the central and peripheral laboratories is considerable and transport is further complicated by complex orography (mountains, busy roads). The use of systems that allow adequate storage of samples and monitoring of transport conditions is therefore essential to ensure the integrity of samples and the quality of the results. In this highly complex scenario, there is also the problem related to the so-called decentralized diagnostics, carried out using point-of-care instrumentation, which is not dealt with in this chapter.

Roles of the Laboratory Staff

In the description of the different professional figures working in a laboratory, it is important to underline how the activity is always, in any case, carried out as a team, in a harmonious equilibrium prodromal to the production of the results in the form of the final report. For the sake of simplicity, we will describe in this part of the chapter the roles and functions of the director, the health manager, and the laboratory technicians, since only for them, in Italy, there is a professional profile well-defined by law regarding the functions to be performed in a laboratory.

The figure of a laboratory director requires a length of service of no less than 7 years in the discipline, a specialization diploma in the discipline, or a length of service of 10 years in the discipline, participation in a management training course, and a suitable professional case history. It should be noted that law number 229 of June 19, 1999, identifies only one level of management for access to the apical position of director of complex operational units (UOCs) of laboratory analysis and/or clinical pathology and/or clinical biochemistry (or other similar denominations). They are, therefore, provided for single competition procedures for graduated doctors, biologists, and chemists, provided they have a specialization, thus emphasizing the subordination of the degree to the course of study and professional. In the event of an absence or an impediment, the director delegates his/her functions to a collaborating manager. In summary, the director in charge of a biomedical laboratory performs a series of complex tasks that include:

- Negotiation of the budget of the laboratory for which he/she is responsible and transmission ("cascading") of the outcome of the negotiations
- Choice and approval of analytical methods, being personally responsible for the reliability of test results
- Formulation of proposals for the acquisition of diagnostic systems, taking into account the state of the art of technology, consistent with predefined financial sustainability
- Organization of services and quality control, being personally responsible for the suitability of equipment and facilities
- Signature of the results of analyses and/or diagnostic judgments
- Recording and archiving of test results
- Application of the internal regulations (hygienic state of the premises, good functionality of the installations, and materials used)
- Response to warnings and complaints
- Application of the rules protecting operators against the risks arising from the specific activity
- Formulation and management of proposals for the updating and continuous training of personnel health in light of the company's mission and the resources actually available

The figure of a health manager (doctor, biologist, chemist, or other qualifying degrees), who works in a laboratory, is purely managerial, requiring specific skills in the organization of activities for maintaining and improving the quality of the results, and is also expressed through an activity of advice to clinicians on the appropriateness of the request, interpretation of the results, and possible continuation of the diagnostic process. A manager, however, enjoys a relative professional autonomy, as defined by his/her organizational or functional assignment, subordinate to the organization of the activities defined by the director. Within his/her sphere of responsibility, a manager organizes the activities of his/her sector in collaboration with the director, the technical coordinator, and the technical and auxiliary staff and participates in the solution of clinical questions using the diagnostic potential assigned. Access to health management in the laboratory requires a university degree, an exam qualifying the profession and its registration, and a postgraduate specialization in a qualifying discipline (clinical biochemistry, clinical pathology, microbiology, or equivalent). It follows, therefore, that the profile of a health manager is that of a professional who has in his/her curriculum vitae no less than 9–10 years of targeted studies (degree and specialization) that allow him/her to perform and/or coordinate analytical activities, both clinical and managerial. It is important to remember that, following the Italian Inter-ministerial Decree number 68 of February 4, 2015, on the reorganization of specialization schools in the health area, the specializations of clinical pathology and clinical biochemistry have been merged under the new name of specialization in clinical pathology and clinical biochemistry. In summary, a specialist in this discipline must:

- Develop theoretical, scientific, and professional knowledge (including the relative assistance activities) in the field of diagnostic clinical pathology and laboratory methodology in cytology, cytopathology, immunohematology, and genetic pathology and in the diagnostic application of cellular and molecular methodologies in human pathology
- Acquire skills in the diagnostic clinical aspects of reproductive medicine and in the laboratory of medicine of the sea and sports activities
- Acquire skills in the study of cellular pathology in the fields of oncology, immunology, and immunopathology, and genetic, ultrastructural, and molecular pathology
- Acquire theoretical, scientific, and professional knowledge for laboratory diagnostics on human samples related to the problems of hygiene and preventive medicine, control and prevention of human health in relation to the environment, occupational medicine, community medicine, forensic medicine, thermal medicine, and space medicine
- Develop theoretical, scientific, and professional knowledge in the study of biological and biochemical parameters in biological samples as well as in vivo, also in relation to the pathophysiological states and clinical biochemistry of nutrition and motor activities, at different levels of structural organization, from single molecules to cells, tissues, and organs, up to the whole organism, both in humans and in animals
- Acquire the necessary skills to study the indicators of alterations underlying hereditary and acquired genetic diseases
- Acquire the necessary skills for the development, use, and quality control of (1) clinical molecular biology, molecular diagnostics, and recombinant biotechnology methodologies, including for the diagnosis and assessment of disease susceptibility; (2) instrumental technologies, including automated ones, which allow a quantitative and qualitative analysis of the above parameters at high levels of sensitivity and specificity; and (3) biochemical molecular technologies linked to human and/or veterinary clinical diagnostics and to environmental diagnostics relating to xenobiotics, residues, and additives, including in food.

The professional profile of a laboratory technician is well-defined by law in Italian decree number 745 of September 26, 1994 (and subsequent law number 42 of February 26, 1999, and law number 251 of August 10, 2000), which has as its objective the "regulation concerning the identification of the figure and the relative professional profile of the biomedical laboratory technician." Specifically, a laboratory technician is a professional figure in the health sector in possession of a qualifying university diploma, with duties aimed at analysis and research related to biomedical and biotechnological analysis, in particular biochemistry, microbiology and virology, drug toxicology, immunology, clinical pathology, hematology, cytology, and histopathology. Genetics and molecular biology have obviously been recently added to these disciplines. The activities are carried out (in public and private facilities, authorized according to the current regulations) in technical and professional autonomy, in direct collaboration with the laboratory manager in charge of the various operational responsibilities. A laboratory technician is also responsible for the correct fulfillment of the analytical procedures and his/her own work within the scope of the functions in the application of the work protocols defined by the responsible managers. He/she also:

- Verifies the correspondence of the services provided to the indicators and the standards predefined by the head of the structure
- Checks and verifies the correct functioning of the equipment used
- Provides for routine maintenance and the possible elimination of minor inconveniences
- Participates in the planning and organization of work
- Contributes to the training of support staff and directly contributes to updating their professional profile and research

In his/her activity, a laboratory technician must be able to autonomously assume responsibility for processes and decisions to implement interdisciplinary and interprofessional work in the complex care contexts in which the user expresses his/her health needs. In addition to specific professional skills, the figure of a laboratory technician is equipped with transversal skills, not specific to particular roles but related to knowing how to act in different situations. Law number 43 of February 1, 2006, regarding provisions on health professions, states that as a result of the new university curricula, the graduate staff belonging to health professions are classified into four categories, which contemplate the functions of health-care coordination, specialist coordination, and management, reserved to holders of a specialist degree.

The technical staff is therefore subordinate to the figure of the technical coordinator, the person in most direct contact with the director of the unit in the organization of laboratory activities. The coordinator coordinates and organizes the professional and economic resources assigned, thus full time carrying out many management tasks within a laboratory in which he/she operates. His/her role is crucial because, through an analysis of the needs of the laboratory, workloads, and availability of staff, he/she has the task of planning and managing the work, even in the event of unpredictable and unforeseen events (absences, malfunctions of technical and/or instrumental aids).

The Role of Statistics in Laboratory Medicine

4

Matteo Vidali

Introduction

Statistics plays a crucial role in many areas of laboratory medicine, from validation of new methods, to verification of analytical and diagnostic performances, definition of reference intervals, and design and conduction of experiments. The knowledge and the correct application of statistical methods allow to deal with the variability of laboratory data, to organize and synthesize information in order to make objective clinical decisions.

Moreover, methodological errors in quality monitoring, in terms of both data analysis and experimental design of procedures, can determine a serious clinical risk for the patient and an excessive consumption of economic resources. Finally, the introduction of new and complex statistical methods in the clinical laboratory, their increasing diffusion in scientific publications in the field and their implementation in easily available software packages, have led to consider as basic statistical tools those that until a few years ago were advanced statistical techniques.

In this chapter, together with introductory statistical concepts, the main statistical methodologies used in the most common scenarios of laboratory medicine are explained. However, these data analysis techniques are not presented as a mere list of tools, but organized into possible experiments in order to make the reader to understand their use and impact and to provide correct and practical solutions to the main problems encountered in the activity of a laboratory.

M. Vidali (✉)
Clinical Pathology Unit, Fondazione IRCCS Ca' Granda Ospedale Maggiore Policlinico, Milan, Italy

Basic Statistical Tools

Samples and Populations

In statistics, population means the finite or infinite set of all possible elementary units, homogeneous for a given characteristic, to which the statistical investigation refers. Rarely, the experimenter has access to all the units of the population (too costly in terms of money or time, or because the population is as infinite as the infinite number of measurements that can be made on a sample under specific conditions). Usually, the experimenter extracts a representative subset (sample) of the population (sampling process), calculates the desired sample statistics, and uses this information to estimate the true parameters of the population (inference process). Our interest is therefore not limited to the sample but to what the latter can tell us about the population. However, the sample estimate is affected by uncertainty (a different sample would have provided a different estimate) and must be interpreted in probabilistic terms.

We can therefore distinguish between descriptive statistics, which uses graphical and numerical methods to describe, summarize, and present data, and inferential statistics, which uses the information obtained from the sample to make more general statements that are valid and referable to a broader context than the data from that single experiment.

Descriptive Statistics

The statistical analysis of the results of an experiment should always be preceded by their visualization; for this purpose, a valid tool is the histogram which is the graphical representation of the frequency distribution of a quantitative variable. The frequency distribution is obtained by dividing the measurement scale into intervals or classes of equal amplitude and counting the number of observations that fall within each class (absolute frequency). The histogram is made up of many

M. Ciaccio (ed.), *Clinical and Laboratory Medicine Textbook*, https://doi.org/10.1007/978-3-031-24958-7_4

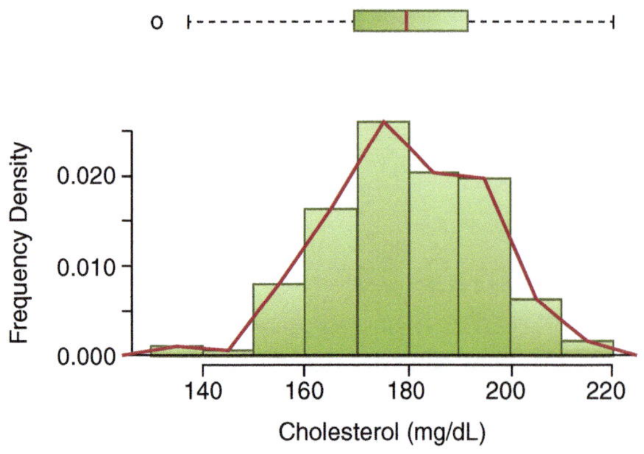

Fig. 4.1 Distribution (relative frequency density) of 300 cholesterol observations, displayed via histogram, frequency polygon, and boxplot

adjacent rectangles whose base is the class and whose height is the frequency. In order to avoid loss of information (when the number of classes is too small), or efficiency in the synthesis (when the number of classes is high), it is advisable to choose a number of classes between 5 and 20 and of equal amplitude, or the use of formulas such as Sturges' one $k = 1 + (10/3)\log_{10}N$, where k and N represent the number of classes and observations, respectively. As an alternative to absolute frequency, it is possible to use relative frequency (ratio of absolute frequency to the total number of observations) or relative frequency density (ratio of relative frequency to class amplitude) (Fig. 4.1). Note that in the case of relative frequency density, the area of each bar of the histogram is equal to the relative frequency for that class, so the area of the entire histogram is equal to 1 (sum of all relative frequencies). The histogram makes easy to check how the data are distributed, that is, where the observations are concentrated, the possible presence of extreme observations and/or asymmetry. From the histogram, it is possible to construct the frequency polygon by joining the midpoints of the upper bases of each rectangle and closing the polygon by joining the ends of the line thus obtained with the x-axis (Fig. 4.1). The frequency polygon, unlike the histogram, allows us to compare different distributions in the same graph.

Although the frequency distribution is a useful tool to illustrate a quantitative variable, it is sometimes necessary to summarize the data with a few values: position (or central tendency) and dispersion measures are used for this purpose. One must remember the conceptual difference between population parameters (denoted by letters of the Greek alphabet), which are generally unobservable but estimable quantities, and estimates of these parameters or statistics (denoted by letters of the Latin alphabet) calculated from sample data. The two most widely used measures of position are the arithmetic mean and the median. The former, calculated as the ratio of the sum of all observations to their number ($\mu = \frac{\Sigma x_i}{N}$ for the population, $\bar{x} = \frac{\Sigma x_i}{N}$ for the sample), has the advantage of synthesis and ease of calculation, but the disadvantage of being influenced by extreme values. The second one, less dependent on extreme values, represents the value that divides an ordered set of data into two equal parts, so that half of the observations have a value lower than the median and half a value higher than the median; in other words, in an ordered set of observations, the median is the value corresponding to the $[(N + 1)/2]$-th observation (if the number of observations is odd) or the value corresponding to the arithmetic mean of the values of the two central observations $(N/2)$ and $[(N/2) + 1]$ (if the number of observations is even).

Measures of dispersion are useful in assessing the variability of data around their central tendency. The most widely used are the variance ($\tilde{A}^2 = \frac{\Sigma(x_i - \mu)^2}{N}$ for the population, $s^2 = \frac{\Sigma(x_i - \bar{x})^2}{(N-1)}$ for the sample) and its square root, that is, the standard deviation ($\sigma = \sqrt{\frac{\Sigma(x_i - \mu)^2}{N}}$ for the population, $s = \sqrt{\frac{\Sigma(x_i - \bar{x})^2}{(N-1)}}$ for the sample). The advantage of the standard deviation, compared to the variance, is that it is expressed in the same units as the observations. Note that in the denominator of the variance and of the standard deviation does not appear N but $(N - 1)$, that is, the degrees of freedom. In general, the degrees of freedom (df) are equal to the difference between the number of data and the number of estimated parameters (in the standard deviation, we have $N - 1$ degrees of freedom because the mean was estimated). When we want to compare the variability of two or more sets of data, in particular when the observations are expressed with different units or orders of magnitude, it is preferable to use the coefficient of variation, which expresses the standard deviation as a percentage of the mean $\left(\%CV = \frac{s}{\bar{x}} \times 100\right)$.

Other measures often used to summarize a distribution are quantiles. A quantile q is a value such that, in an ordered set, a proportion q of data is less than the value corresponding to q and a proportion $1 - q$ of data is greater than the value corresponding to q. In particular, a quantile q in an ordered set is the value corresponding to the observation $i = q(N + 1)$ (if i is not an integer, the quantile is found by interpolation). Particularly useful quantiles are the quartiles (Q) and the percentiles (P) that correspond to those values that divide an ordered set into 4 and 100 parts of equal size, respectively. The median corresponds to the second quartile (Q2) and the 50th percentile (P50). From the definition of quartiles, it is clear that 25% of a data set will have a value lower than Q1 (P25), 25% higher than Q3 (P75), and the remaining 50% will have a value within the interquartile range (Q3-Q1 or IQR). Similarly, the interval between the 2.5 percentile (P2.5) and the 97.5 percentile (P97.5) includes 95% of the observations: in fact, 2.5% of the observations will be lower than P2.5, 95% between P2.5 and P97.5 and the remaining 2.5 will be greater than P97.5. The combination of five numbers, Q1, Q2, Q3, min, and max, is a convenient and often used way in the literature to summarize a distribution. This information can be visualized in the box and whisker plot or boxplot, in which the box, bisected by the median Q2, has Q1 and Q3 as its ends, with whiskers extending from Q1 and Q3, respectively, to the minimum and maximum values equal to or less than 1.5 times the IQR interval. The extreme observations, not reached by the whiskers, are represented as points (Fig. 4.1).

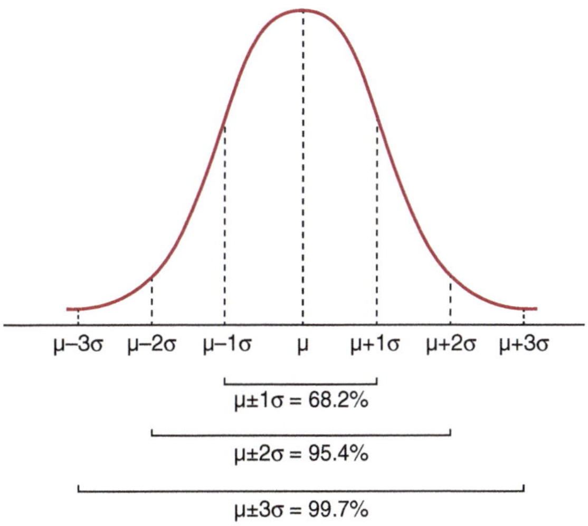

Fig. 4.2 Gaussian probability distribution

Probability Distributions and the Normal Distribution

As the sample size (N) increases, the class size (A) of the histogram may be decreased; in general, as N tends to infinity, A tends to zero and the histogram (or frequency polygon) is approximated by a smooth curve. When N tends to population size (i.e., with N very large), this smooth curve becomes the relative frequency density of the population. Just as the total area of the histogram is equal to 1, the area under the curve is also equal to 1. The proportion of observations that falls between two fixed limits is equal to the area under the curve between those two limits. Thus, since the probability of a randomly chosen observation falling between two limits is equal to the proportion of observations falling between these two limits, the population frequency distribution is called the probability distribution or probability density function.

The most common probability distribution in statistics is the normal or Gaussian distribution (or bell curve) (Fig. 4.2). Its probability density has equation: $f(x) = \frac{1}{\sigma\sqrt{2\pi}} e^{\left(-z^2/2\right)}$ with $z = \frac{x-\mu}{\sigma}$.

The Gaussian distribution is completely defined by mean and standard deviation. It is a symmetrical distribution in which the mean, median, and mode (most frequent value) coincide. This distribution is particularly important for at least two reasons: many biological variables of interest to the laboratory, including measurement errors, are approximately normally distributed and, in addition, many statistical methods are based on the assumption of normal distribution of the data. Knowing that a variable is normally distributed and knowing the mean and standard deviation, we can know the probability that a randomly selected individual will exhibit a value greater than, less than, or within 2 fixed limits. For example, knowing that the variable glycemia has $\mu = 90$ mg/dl and $\sigma = 15$ mg/dl, we might want to know the probability that a randomly selected individual has a glucose level greater than 120, or less than 75, or between 85 and 100 mg/dl. To answer this question, we must determine the proportion of the area under the curve to the right of $x_1 = 120$, or to the left of $x_2 = 75$, or between $x_3 = 85$ and $x_4 = 100$ mg/dl, respectively. These areas have been calculated and tabulated and can be found in statistical texts, literature, or statistical software. However, it would not be possible to tabulate values for all the infinite normal curves given by the combination of the different means and standard deviations. In fact, only one particular curve was tabulated, called the standardized normal distribution, with $\mu = 0$ and $\sigma = 1$. To use the tabulated area table in order to estimate the probability associated with a normal variable X, it is necessary to transform the normal variable X into a standardized normal variable Z with the operation $Z = \frac{X-\mu}{\sigma}$ (standardization) and then look for the relative probability value in the table. In our example, $z_1 = \frac{120-90}{15} = 2$; $z_2 = \frac{75-90}{15} = -1$; $z_3 = \frac{85-90}{15} = -0.333$; $z_4 = \frac{100-90}{15} = 0.667$. The area under the curve to the right of $z = 2$ is equal to 0.023; to the left of $z = -1$ is 0.159; included between $z = -0.333$ and $z = 0.667$ is 0.378; this means that the probability that a subject randomly selected from this population presents a glycemia value higher than 120 $P(X > 120)$, or lower than 75 $P(X < 75)$, or included between 85 and 100 mg/dl $P(85 < X < 100)$ is, respectively, equal to 2.3%, 15.9%, and 37.8% (Fig. 4.3). It is easy to verify that in a Gaussian distribution 68.4% of the

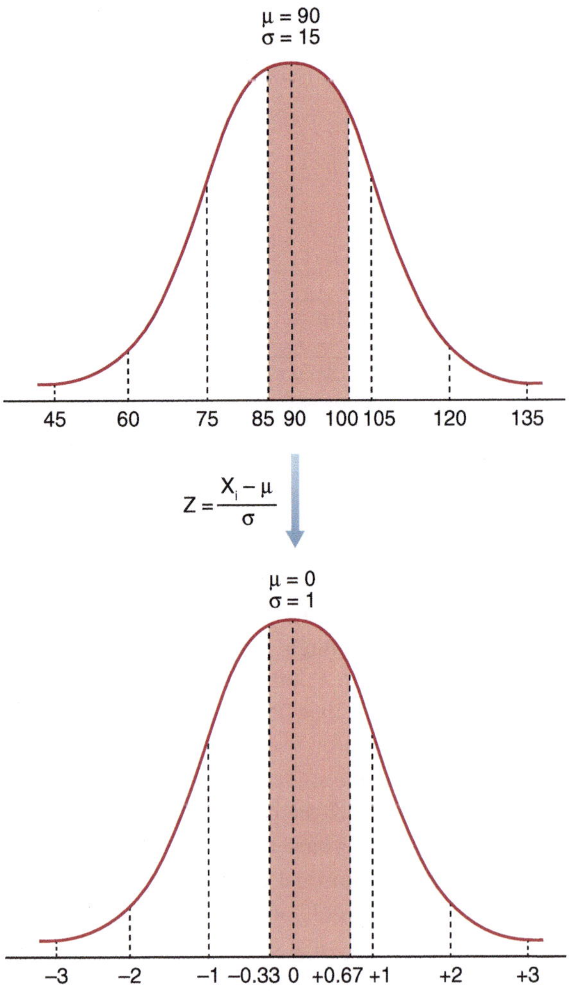

Fig. 4.3 Standardization procedure: from normal curve to standardized normal curve with mean = 0 and standard deviation = 1

area under the curve is between mean ±1 standard deviation (in the standardized normal between $z = -1$ and $z = 1$), 95.4% between mean ±2 standard deviations (between $z = -2$ and $z = 2$), and 99.7% between mean ±3 standard deviations (between $z = -3$ and $z = 3$) (Figs. 4.2 and 4.3).

Since many statistical methods can only be used if the observations are normally distributed, it is necessary to verify this assumption. Graphical and formal statistical methods can be used for this purpose. The visualization of the data through histogram and boxplot and the comparison of the position indices (mean and median) allow to easily verify the presence of deviations from normality: since, in fact, the mean is affected by extreme values, in distributions with positive asymmetry (right long tail), the mean will be higher than the median, and, in distributions with negative asymmetry (left long tail), the mean will be lower than the median. A reliable tool, also for small samples, is represented by the normal probability plot, a graphical method in which the observed quantiles (y-axis) are graphed toward the theoretical quantiles (x-axis). The presence of pronounced deviations from the theoretical quantile-quantile line suggests the

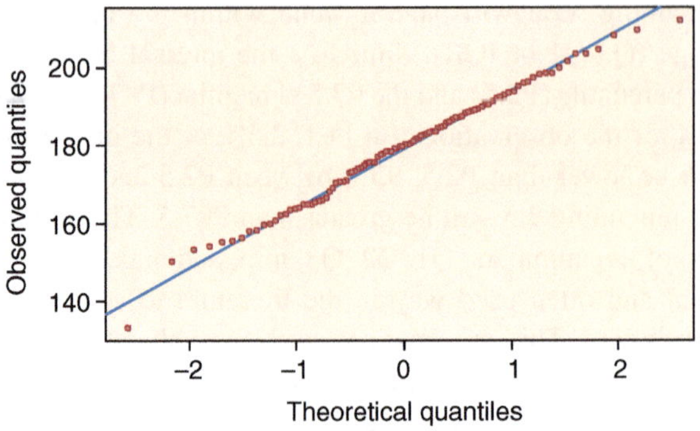

Fig. 4.4 Normal probability plot to verify the normal distribution of the data. The points are well aligned along the theoretical line that passes through the first and third quartiles. The graph shows only a slight curvature to the left (bottom of the curve) with a point identified as aberrant by Horn's method

assumption of non-normality in the data (Fig. 4.4). The assumption of normality can also be verified with formal statistical tests such as the Kolmogorov-Smirnov test, the Anderson-Darling test, and the more popular Shapiro-Wilk test (not usable, however, for large samples).

In the presence of significant deviations, it is possible to attempt mathematical transformations of the data in order to proceed subsequently to a new normality check. Several transformations can be applied to the data (including logarithm, reciprocal, square root, power-elevation), and it is often necessary to proceed by trial and error. Instead of using arbitrarily chosen transformations, it is possible to use the transformation proposed by Box and Cox, a technique that allows to determine which is the best transformation to apply to the variable y. The transformation is defined by $W = \begin{cases} (y^{\lambda} - 1)/\lambda, & \text{if } \lambda \neq 0 \\ \log(y), & \text{if } \lambda = 0 \end{cases}$ Where W is the transformed variable, y the original variable, and λ the parameter defining the transformation estimated using the maximum likelihood criterion.

Inferential Statistics: Confidence Intervals and Hypothesis Tests

Imagine extracting all possible samples of size n from the population and calculating the mean of each of them. The distribution of all possible averages is called the sampling distribution of the mean. More generally, we speak of sampling distributions to denote the distribution of the values of different statistics computed over all possible samples.

The sampling distribution of the mean will have its own mean and standard deviation. It can be shown that: (1) the mean of the sampling distribution (i.e., the mean of the averages) is equal to the population mean μ; (2) the standard deviation of the sampling distribution of the mean, or standard error (es), is equal to $\sigma / \sqrt{n}$; (3) if n is sufficiently large, the sampling distribution is approximately normal (the more the population deviates from normality, the larger n must be).

Confidence Intervals

Just as 95% of the observations in a normal distribution are within ±2 standard deviations (more precisely 1.96 sd) from the mean (Fig. 4.2), so, if n is sufficiently large, 95% of the averages in the sampling distribution of the mean are within ±1.96 standard errors from μ (i.e., $\bar{x}$ $\mu - 1.96\,\text{Ã}/\sqrt{n} < \bar{x} < \mu + 1.96\,\text{Ã}/\sqrt{n}$), which means that there is a 95% probability that the average $\bar{x}_i$ of a particular sample is within the range $\mu \pm 1.96\sigma/\sqrt{n}$); this, conversely, is equivalent to saying that μ is within the range $\bar{x} \pm 1.96\sigma/\sqrt{n}$ or that there is a 95% probability that the limits of the interval delimited by $\bar{x} \pm 1.96\sigma/\sqrt{n}$ contain the true population mean μ. This interval is called the 95% confidence interval, and the extremes of the interval $\bar{x} - 1.96\sigma/\sqrt{n}$ and $\bar{x} + 1.96\sigma/\sqrt{n}$ are called the 95% confidence limits. In real life, we do not have all possible samples of size n extracted from a population, and σ is unknown; however, if n is sufficiently large ($n > 60$), not only does the sampling distribution approximate well to a normal distribution but s (the sample standard deviation) is a good estimate of σ (the unknown standard deviation of the population) and $\text{se} = s/\sqrt{n}$ is a good estimate of $\sigma/\sqrt{n}$. This result allows us to associate a statistics calculated on a sample (point estimate) with an interval in which we have some level of confidence that it contains the true population parameter (interval estimate). The confidence interval $\left(\pm z\sigma/\sqrt{n}\right)$ increases as the selected confidence level increases ($z = 1.64$ for 90%, $z = 1.96$ for 95%, $z = 2.58$ for 99%) and decreases as n increases (the estimate becomes less uncertain). With small samples ($n < 60$), s may not be a reliable estimate of σ, because s can vary greatly from sample to sample, and the ratio $(\bar{x} - \mu)/\sqrt{s^2/n}$ does not follow a standardized normal distribution. However, for small samples, it can be shown, that if the observations come from a normal distribution, the ratio $(\bar{x} - \mu)/\sqrt{s^2/n}$ follows a Student's t distribution with $(n - 1)$ degrees of freedom. The t distribution is flatter and has thicker tails than the normal distribution. It is actually a family of distributions whose shape depends on the number of degrees of freedom. In general, for the same proportion of observations under the curve or the same probability, t has a larger value than z, resulting in wider confidence intervals and hence greater uncertainty in the estimate. As n, and hence degrees of freedom, increase, t tends to z.

Hypothesis Testing

A second type of inference is represented by statistical hypothesis testing. Hypothesis testing checks the validity of a hypothesis and involves first formulating a hypothesis relative to a characteristic of the population and then evaluating the probability of obtaining the observed sample (and therefore the statistics calculated on the sample) if the hypothesis is true. The hypothesis to be tested is called null hypothesis and is indicated with H_0, while the alternative hypothesis, complementary to the null hypothesis, is indicated with H_1. Since hypothesis testing is a procedure based on probability, two errors may occur: an error of the first type, or alpha error (α), which is the one committed if we reject H_0 when it is true, and an error of the second type, or beta error (β), which is committed if we accept H_0 when it is false.

In general, we proceed as follows: (1) we fix H_0 and H_1; (2) we calculate the desired sample statistic and the value of the statistical test; (3) using the tables of the appropriate distribution, we calculate the probability p of obtaining a statistic like the observed one, or more extreme, under the hypothesis that H_0 is true; (4) we compare p with α (fixed in general equal to 0.05): if p is less than α, we consider the test statistically significant and reject the null hypothesis (the null hypothesis is not consistent with the observed results) otherwise we accept it.

Example: To test the hypothesis that smoking affects blood pressure, we design a study in which the mean blood pressure of 20 heavy smokers is measured. The working hypothesis or alternative H_1 is "heavy smokers have a different mean blood pressure than the non-smoking population," while the null hypothesis H_0 is "heavy smokers have the same mean blood pressure as the non-smoking population." The value of the mean arterial pressure of the population (obtained from previous studies) is μ = 148 mmHg; we set $\alpha = 0.05$; H_0:μ = 148 mmHg; H_1:$\mu \neq$ 148 mmHg; the sample of subjects studied (n = 20) has mean $\bar{x}$ = 152 mmHg and standard deviation s = 8 mmHg; we calculate the value of the statisticaltestthatinthiscaseis$t_{n-1} = \dfrac{\bar{x} - \mu}{s/\sqrt{n}} = \dfrac{152 - 148}{8/\sqrt{20}} = 2.24$; using the distribution t (non-large sample) with ($n - 1 = 19$) degrees of freedom, we look for the probability value corresponding to $t_{\alpha/2;n-1} = 2.24$, which is $p = 0.037$. This means that, if the null hypothesis is true (μ = 148 mmHg), the probability of obtaining by chance a result like the observed ($\bar{x}$ = 152 mmHg), or even more extreme, is 3.7%. Therefore, since $p < \alpha$, we have sufficient evidence to reject H_0 and conclude that the mean arterial pressure of smokers is different from the pressure of non-smokers. With the data from the previous example, we can also calculate the 95% confidence interval of the mean: $\bar{x} \pm t_{\alpha/2;n-1} s/\sqrt{n}$, that is, $152 \pm t_{0.025;19} \times 8/\sqrt{20} = 152 \pm 2.093 \times 1.79 = 148.3 - 155.7$; confirming what previously concluded, the confidence interval does not contain the value 148 mmHg (we are 95% confident that the observed sample does not come from the population with mean μ = 148 mmHg).

In general, from a sample statistic (a mean, a difference between means, or some other estimator), it is possible to construct a confidence interval and to calculate the test statistic:

$$95\%\text{CI} = \text{Estimated parameter} \pm k \times \text{se}$$

$$\text{Test} = \text{Estimated parameter} / \text{se}$$

with the standard error (se) calculated in different ways depending on the estimator and k depending on the distribution considered. Let us see some examples (z can substitute t for large n), where the statistical test, calculations for CI and the specific statistics are reported, together with H_0 and H_1:

1. Test for two paired samples (e.g., subjects measured before and after a treatment): t-test for paired samples with $(n - 1)$ degrees of freedom, where n is the number of pairs of observations

$$\text{CI} = \bar{d} \pm t_{\alpha/2;n-1} s_d / n; \quad t = \frac{\bar{d} - \delta}{s_d / \sqrt{n}}$$

$$\text{with} \quad H_0 : \delta = \mu_1 - \mu_2 = 0 \quad \text{and} \quad H_1 : \delta \neq 0$$

with $\bar{d}$ e s_d mean and standard deviation of the differences, respectively, and δ the true difference between population means

2. Test for two independent samples (n_1 and n_2): Student's t-test with $(n_1 + n_2 - 2)$ degrees of freedom
   (a) With variances s_1^2 and s_2^2 equal (hypothesis of homogeneity or homoscedasticity of variances verifiable by tests for variance $F = s_1^2 / s_2^2$ with $s_1^2 > s_2^2$ and F following a Fisher distribution with $(n_1 - 1)$ and $(n_2 - 1)$ degrees of freedom (other tests that can be used for the same purpose are Bartlett's test and Levene's test):

$$\text{CI} = (\bar{x}_1 - \bar{x}_2) \pm t_{\alpha/2;n_1+n_2-2} s_p \sqrt{\frac{1}{n_1} + \frac{1}{n_2}}$$

$$t = \frac{(\bar{x}_1 - \bar{x}_2) - (\mu_1 - \mu_2)}{s_p \sqrt{1/n_1 + 1/n_2}} \quad \text{with} \quad H_0 : \mu_1 = \mu_2$$
$$\text{and} \quad H_1 : \mu_1 \neq \mu_2$$

   where μ_1 and μ_2 are the population means and s^2_p is the common or pooled variance: $s^2_p = \frac{(n_1 - 1)s_1^2 + (n_2 - 1)s_2^2}{n_1 + n_2 - 2}$

   (b) With variances s_1^2 e s_2^2 non-homogeneous (heteroscedasticity): other tests must be used (e.g., Welch test)

3. Test for two proportions (p_1 and p_2 calculated in two samples of size n_1 and n_2): test z

$$\text{CI} = (p_1 - p_2) \pm z_{\alpha/2} \sqrt{\frac{(p_1(1-p_1)}{n_1} + \frac{(p_2(1-p_2)}{n_2}};$$

$$z = \frac{(p_1 - p_2) - (\pi_1 - \pi_2)}{\sqrt{(p(1-p)\left(\frac{1}{n_1} + \frac{1}{n_2}\right)}} \quad \text{with} \quad p = \frac{n_1 p_1 + n_2 p_2}{n_1 + n_2},$$

$H_0 : \pi_1 = \pi_2$ and $H_1 : \pi_1 \neq \pi_2$, where π_1 and π_2 are the population proportions, assuming that, n_1p, $n_1(1-p)$, n_2p and $n_2(1-p)$ are all ≥ 5.

4. Tests for three or more independent samples: ANOVA test or analysis of variance

The null hypothesis H_0 in this case is that the samples are from the same population (same means). One approach would be to perform as many two-way comparisons with as many Student's t-tests. However, with m groups, the possible two-way comparisons are $m(m - 1)/2$ and, as the number of groups, and hence comparisons, increases, the probability that at least one comparison is statistically significant increases even if H_0 is true (i.e. greater probability of committing a Type I error: if for one comparison the probability of not rejecting H_0, when true, is 0.95, for k comparisons it is $(0.95)^k$ and so the probability of rejecting H_0 in at least one comparison, if H_0 is true, is $1–0.95^k$; for example, if $m = 5$ groups, $k = 10$ comparisons and the probability of rejecting H_0, when true, in at least one comparison is $1–0.95^{10} = 0.40$ much larger than α fixed).

The ANOVA test instead of considering averages considers variances. It is in fact based on the reasoning that when dealing with several populations (e.g., different treatments), the total variability depends on the variability of individual values with respect to the mean of their population, or within-group variance (due to measurement error, individual characteristics, or uncontrollable factors), and on the variability of their population averages with respect to the overall mean, or between-group variance. If the variability within populations (within-group variance) is small compared to the variability between population averages (between-group variance), we conclude that the averages are different. With m groups, we proceed in this way by calculating:

- Deviance within groups: $SS_{within} = \sum_{m}^{j=1} \sum_{n_j}^{i=1} (x_{ij} - \bar{x}_j)^2$, with $\sum_{m}^{j=1} (n_j - 1)$ degrees of freedom, which measures the variability of the data around the group mean
- Deviance between groups: $SS_{between} = \sum_{m}^{j=1} n_j (\bar{x}_j - \bar{\bar{x}})^2$, with $(m - 1)$ degrees of freedom, which measures the variability of group averages around the overall mean
- Total failure: $SS_{total} = SS_{within} + SS_{between}$, with $df_{total} = df_{within} + df_{between}$
- Variance within groups: $MS_{within} = \frac{SS_{within}}{df_{within}}$
- Variance between groups: $MS_{between} = \frac{SS_{between}}{df_{between}}$
- Fisher's test: $F = \frac{MS_{between}}{MS_{within}}$

if $F > F_{\alpha;df_{within};df_{between}}$, then $p < \alpha$ and we reject H_0: at least one of the m groups is significantly differenttinct from the others. A significant F-test indicates that not all averages are equal but does not allow us to know which and how many are different from each other. To answer this question, we need to run tests for multiple comparisons, that is, comparisons for each pair of averages. To avoid, as shown before, that increasing the number of comparisons leads to an increase in the probability of committing a type I error, various approaches or various corrections can be used. For example, Bonferroni's correction suggests fixing the type I error by calculating a modified α as $\alpha^* = \frac{\alpha}{\text{number of comparisons}} = \frac{\alpha}{m(m-1)/2}$, where m is the number of groups. The main drawback of the Bonferroni correction is the decrease in statistical power as the number of comparisons increases (i.e., the probability of rejecting H_0, when it is false, decreases). Several other tests for multiple comparisons are described in the literature including the Student-Newman-Keuls test, Tukey's method, and the Holm-Bonferroni method.

Like the t-test, two assumptions must be met for the ANOVA test: (1) the variable is normally distributed; (2) the variances of the groups are equal (homoscedasticity). Both assumptions can be verified through graphical methods (histogram, normal probability plot, scatter plot) and formal statistical methods (Shapiro-Wilk test for normality and Levene's test for homoscedasticity). While moderate deviations from normality can be ignored (especially for groups of equal numerosity), non-homogeneity of variances can greatly influence the results. In the presence of violations of these assumptions, it is advisable to try a mathematical transformation (which can both normalize the distribution and make the variance similar across groups) or to use non-parametric methods, that is, methods that do not depend on the shape of the distribution and do not involve the estimation of statistical parameters: the Mann-Whitney test for two independent samples, the Wilcoxon test for two paired samples, and the Kruskal-Wallis test for three or more samples. Although non-parametric tests do not require the assumptions underlying parametric tests, it should be remembered that, if these assumptions are met, non-parametric tests have less statistical power.

Correlation and Linear Regression

Sometimes we are interested in evaluating the relationship between two quantitative variables. In this situation, it is necessary to initially display the data in a scatter plot, that is, a dot plot whose coordinates are represented by the two variables being studied. Linear regression allows us to estimate the mathematical relationship between an independent variable (x-axis) and a dependent variable (y-axis). Correlation, on the other hand, allows us to estimate the strength of the association between the two variables.

The regression provides the equation of the line that best describes the dependent variable as a function of the independent variable: $Y = a + bX + E$, where a and b are, respectively, the intercept and the slope coefficient (or regression coefficient) of the line and E is the error, that is, the part of the variability of Y not explained by the line. The parameters of the line are estimated through the method of least squares, that is, looking for the values of the parameters that minimize the sum of the squares of the vertical distances of the points from the line. The following formulas are used:

$$b = \frac{\sum_{i=1}^{n}(x_i - \bar{x})(y_i - \bar{y})}{\sum_{i=1}^{n}(x_i - \bar{x})^2}; \quad a = \bar{y} - b\bar{x}$$

The parameters can then also be used to make predictions by the equation of the line.

As with other estimators, we can calculate confidence intervals and conduct hypothesis tests for a and b.

The confidence intervals are as follows:

for a: $\text{CI} = a \pm t_{\alpha/2;\,n-2}\text{SE}(a)$ with $\text{SE}(a) = s\sqrt{\frac{1}{n} + \frac{\bar{x}^2}{\sum_{i=1}^{n}(x_i - \bar{x})^2}}$

for b: $\text{CI} = b \pm t_{\alpha/2;\,n-2}\text{SE}(b)$ with $\text{SE}(b) = \frac{s}{\sqrt{\sum_{i=1}^{n}(x_i - \bar{x})^2}}$

with $s = \sqrt{\frac{\sum_{i=1}^{n}(y_i - \hat{y}_i)^2}{n-2}}$, where $\hat{y}_i$ are the corresponding values of y_i predicted by the regression line. The degrees of freedom are $(n - 2)$ because we estimated two parameters a and b.

It is also possible to test the null hypothesis that the population regression coefficient is 0, that is, that there is no linear relationship between X and Y.

So for H0:$\beta = 0$ we have $t = \frac{b}{\text{SE}(b)}$.

The use of the least squares method and the t distribution for hypothesis testing and confidence intervals are based on the assumption that the residuals $(y_i - \hat{y}_i)$ are normally distributed and have equal variance. These assumptions can be verified by a normal probability plot of the residuals and by graphing the predicted values $\hat{y}_i$ toward the residuals $(y_i - \hat{y}_i)$. Analysis of the residuals can also reveal extreme points that are highly influential on the regression line.

Instead, the strength of the association between two quantitative variables is expressed by the linear correlation coefficient r estimated by:

$$r = \frac{\sum_{i=1}^{n}(x_i - \bar{x})(y_i - \bar{y})}{\sqrt{\left(\sum_{i=1}^{n}(x_i - \bar{x})^2\right)\left(\sum_{i=1}^{n}(y_i - \bar{y})^2\right)}}$$

The coefficient r ranges from -1 (perfect association, as one variable increases, the other decreases) to $+1$ (perfect association, as one variable increases, the other also increases). If $r = 0$, there is no relationship between the variables. The correlation coefficient measures the strength of association of two linearly correlated variables. If they are linked by a strong non-linear relationship (e.g., quadratic), the correlation coefficient may be falsely low. This underscores, as noted earlier, the importance of always viewing data before statistical analysis.

The goodness of fit of the regression model to our data can be assessed through the coefficient of determination R^2, which is the square of the correlation coefficient: $R^2 = r^2$. Since r varies from −1 to +1, R^2 varies from 0 to 1 and indicates the proportion of variability of Y explained by the regression line.

Evaluation of Analytical Performance

Accuracy

The precision of a method can be defined as the degree of agreement between different independent measurements under defined experimental conditions. Precision can be expressed in terms of imprecision as standard deviation, variance and relative standard deviation (RSD), or coefficient of variation (CV). Several types of precision are defined:

- Repeatability (or within-run precision): precision under experimental conditions that minimize other sources of variability (same laboratory, same method, same instrument, same operator, same operating conditions, with measurements made within a short time interval).
- Intermediate precision (formerly total precision, recently within-laboratory precision): precision under well-defined experimental conditions (same laboratory, same instrument, same method, with measurements made over an extended time interval, with different calibrators, operators, and/or lots of reagents).
- Reproducibility: accuracy obtained under different conditions (laboratory, operator, instrument). Reproducibility can only be assessed by inter-laboratory studies.

Verification of repeatability with the protocol 5 days × 5 replicates.

Before using a method validated by a third party (e.g., a manufacturing company), the laboratory must perform a rapid verification protocol. To this aim, various experimental designs are reported in the literature, including the 3 × 5 scheme (3 replicates for 5 days); however, the most recent literature suggests increasing the number of replicates and/or days in order to obtain a more realistic estimate of repeatability. It is therefore advisable to use a 5 × 5 scheme (5 replicates for 5 days) or one in which the product of replicates × days is at least ≥20. For methods that are not validated by third parties but developed on their own, a more extensive protocol should be adopted (the literature suggests a 2 × 20 scheme, i.e., 2 replicates for 20 days).

The repeatability verification protocol includes the following steps:

- Sample selection.
- Sample analysis (5 replicates × 5 days) using the method to be verified.
- Processing of any aberrant data and possible integration with new measurements until the desired size is obtained.
- Calculation of repeatability.
- Verification of what has been declared by the producer.

The verification should be performed using at least two controls, or pools of patients, with concentrations close to both any decision values (and/or cut-off), and to those used by the manufaturer in the validation phase and reported in the data sheet.

If one or more observations from the collected data deviate to a certain extent from all the others (extreme or aberrant values), it is necessary to perform a statistical test for aberrant data, after checking that these values are not the result of analytical or gross errors. A number of statistical tests useful in identifying extreme values are described in the literature, including the Dixon-Reed test, Grubbs' test, Huber's test, and Horn's method. In the Dixon-Reed test, the ratio D/R is calculated, where D is the absolute difference between the aberrant and the least extreme value immediately preceding it and R is the maximum-minimum range of all the data; if the ratio is greater than 1/3, the observation is considered aberrant. In the Grubbs' test, instead, we calculate the ratio $G = \frac{\text{max} - \text{mean}}{s}$ or $G = \frac{\text{mean} - \text{min}}{s}$, where min and max are, respectively, the presumed minimum or maximum aberrant value, mean and s are, respectively, the mean and standard deviation of the sample. If the calculated G is greater than a certain critical value, which can be found in the literature and depends on the sample size, the data is considered an aberrant. Huber's test is based instead on the median absolute deviation (MAD), that is, the median of the absolute differences (i.e., without sign) between the observations and the median. The procedure is as follows: the median is calculated; then the absolute differences $D_i = \left| x_i - \tilde{x} \right|$ between the median and the individual observations are calculated and then the MAD, that is, the median of all these absolute differences, multiplied by an appropriate constant $b = 1.4826$ (when data follow a gaussian distribution), is obrained. Thus, $\text{MAD} = b \times \text{median}\left(\left| x_i - \tilde{x} \right|\right)$, where $\tilde{x}$ indicates the median of the observations. Finally, we calculate the limit value as $D_{\text{crit}} = \text{MAD} \times 3$ (very conservative; or 2.5 moderately conservative). If an observation has an absolute difference from the median greater than this critical value ($D_i > D_{crit}$), it is considered aberrant. Finally, in the Horn's method, we first apply the Box-Cox transformation, to obtain a more Gaussian distribution of the data, and then Tukey's robust method, calculating the interval between Q1–1.5 × IQR and Q3 + 1.5 × IQR, where Q1, Q3, and IQR are the first quartile, the third quartile, and the interquartile range (Q3-Q1), respectively: all data outside this interval are considered aberrant.

Example: for the 25 replicates (5 × 5 design), the following data were obtained: mean = 101.2; s = 14.0; min = 76.4; max = 150.0. The graphical analysis seems to suggest that the value 150.0 is an aberrant. We test this hypothesis with the four tests illustrated.

(a) Dixon-Reed test:
Given less extreme than the max: 113.2 then

$$\frac{D}{R} = \frac{150.0 - 113.2}{150 - 76.4} = 0.50 > 1/3 \text{ aberrant.}$$

(b) Grubbs' test: for $N = 25$ $G_{critical} = 3.14$

$$G = \frac{150.0 - 101.2}{14.0} = 3.49$$ since $G > G_{critical}$ we conclude that 150.0 is aberrant.

(c) Horn's method: after having found through statistical software $\lambda = -0.8511372$ and having transformed the data according to Box-Cox, that is, with $\frac{x^{\lambda} - 1}{\lambda}$ the limits of the interval Q1–1.5 × IQR and Q3 + 1.5 × IQR (1.147–1.156) are calculated. The suspected aberrant 150.0, after transformation, is equal to 1.158, which is outside the interval, confirming what was found with the previous tests.

(d) Huber's method: the median of the observations is equal to 100.8; the absolute differences between this median and all the observations are calculated and then the MAD, which results to be 9.17 (already corrected for the constant b). The critical value is $D_{crit} = 9.17 \times 3 = 27.5$. The observation 150.0, since it presents an absolute difference from the median equal to 49.2 (since $D_i > D_{crit}$; 49.2 > 27.5), is considered aberrant.

The most critical issues of these tests are the assumption of normality of the data distribution and/or the failure to identify the aberrant data in the presence of more than one extreme data. In fact, the reliability of the identification of an observation as aberrant depends on the distribution of the data (in particular for the Grubbs' test and the Dixon-Reed test). Before applying these tests, it is therefore necessary to check the Gaussian distribution and, if necessary, to transform the data. If a non-normal distribution is assumed for the data, the multiplicative constant for Huber's test changes from b = 1.4826 to b = 1/Q3, where Q3 is the third quartile. Graphical analysis of the data also allows to check for multiple aberrant data and thus choose a suitable test. In this regard, it should be noted that the Dixon-Reed test may not identify the aberrant data in the presence of more than one aberrant. In addition, while Tukey's test can identify multiple aberrants, Grubbs' test identifies one aberrant at a time, and it is therefore necessary to apply the test to the data set several times after removing the suspected aberrant (a minimum data size of >6 is recommended for Grubbs' test). Finally, it is advisable not to use Tukey's method for small data sets.

After identifying any aberrant data, we proceed to calculate total or within-laboratory repeatability through analysis of variance (ANOVA), as shown in the introductory section on basic statistical methods:

$$\text{DEV}_{\text{between}} = \sum_{j=1}^{d} n_j \left(\bar{x}_j - \bar{\bar{x}}\right)^2; \quad \text{DEV}_{\text{within}} = \sum_{j=1}^{d}\sum_{i=1}^{r} \left(x_{ij} - \bar{x}_j\right)^2;$$

$$\text{DEV}_{\text{tot}} = \text{DEV}_{\text{between}} + \text{DEV}_{\text{within}}$$

$$\text{DF}_{\text{between}} = d - 1; \quad \text{DF}_{\text{within}} = d(r-1); \quad \text{DF}_{\text{tot}} = \text{DF}_{\text{between}} + \text{DF}_{\text{within}};$$

$$\text{VAR}_{\text{between}} = \text{DEV}_{\text{between}} / \text{DF}_{\text{between}}; \quad \text{VAR}_{\text{within}} = \text{DEV}_{\text{within}} / \text{DF}_{\text{within}},$$

with d *being* the number of days, r the number of replicates, x_{ij} the i-th replicate of the j-th day, $\bar{x}_j$ the average of the j-th day, and $\bar{\bar{x}}$ the average of all replicates.

From the above data, we calculate the following:

The variance of repeatability (within-run or within-run): $V_W = \text{VAR}_{\text{within}}$.
The repeatability (within-run or within-run): $S_r = \sqrt{V_W}$ and $\%\text{CV}_r = \left(S_r / \bar{\bar{x}}\right)100$.
The "pure" between-run repeatability variance (i.e., corrected for the contribution of V_W): $V_B = (\text{VAR}_{\text{between}} - \text{VAR}_{\text{within}})/n_0$, where n_0 represents the average number of replicates of the runs (in the 5 × 5 design $n_0 = 5$). If $\text{VAR}_{\text{between}} \leq \text{VAR}_{\text{within}}$, then $V_B = 0$.
The "pure" (between-run) repeatability: $S_B = \sqrt{V_B}$ e $\%\text{CV}_B = \left(S_B / \bar{\bar{x}}\right)100$.
Total repeatability or within the laboratory: $S_{WL} = \sqrt{V_W + V_B}$ and $\%\text{CV}_{WL} = \left(S_{WL} / \bar{\bar{x}}\right)100$.

Example: we want to evaluate the repeatability of a method for the measurement of cholesterol (5 × 5 design). From the analysis of the data, we obtain:

$DEV_{between}$: 132.56; $DF_{between}$: 4; $VAR_{between}$: 33.14
DEV_{within}: 150; DF_{within}: 20; VAR_{within}: 7.50
V_w = 7.50; since $VAR_{between}$ (33.14) > VAR_{within} (7.50) then
$V_B = (33.14 - 7,50)/5 = 5,13$

$$S_r = \sqrt{V_W} = \sqrt{7.50} = 2.74\text{mg/dl}$$
$$S_B = \sqrt{V_B} = \sqrt{5.13} = 2.26\text{mg/dl}$$
$$S_{WL} = \sqrt{V_W + V_B} = \sqrt{7.50 + 5.13} = 3.55\text{mg/dl}$$

Since the average of all 25 replicates is $\bar{\bar{x}} = 203.76$ mg / dl, then:

$$\%CV_r = (2.74/203.76) \times 100 = 1.34\%$$
$$\%CV_B = (2.26/203.76) \times 100 = 1.11\%$$
$$\%CV_{WL} = (3.55/203.76) \times 100 = 1.74\%$$

The calculated total repeatability can be compared with that stated by the manufacturer. The manufacturer can provide the total repeatability either as a standard deviation (S_p) or as $\%CV_p$. In the latter case, to obtain the S_p, it is sufficient to apply the formula $S_p = (\%\ CV_p / 100)\bar{\bar{x}}$.

In the cholesterol example above, the manufacturer provided an S_p = 3.20. If he had provided, for example, a $\%CV_p$ = 1.6%, then S_p = 1.6%/100 x 203.76 = 3.26 would have been obtained.

It is then calculated: T (effective degrees of freedom):

$$T = \frac{\left[(r-1)\cdot S_r^2 + \left(r \cdot S_b^2\right)\right]^2}{\left(\frac{r-1}{d}\right)\cdot\left(S_r^2\right)^2 + \left(\frac{r^2 \cdot \left(S_b^2\right)^2}{d-1}\right)}$$, where S_b^2 is the "non-pure" between-run variance, that is, not corrected for the contribution of the within-run variance: $S_b^2 = V_B + V_W / r$.

In our example: r = 5; d = 5; $S_r^2 = V_w = 7.50$; $S_b^2 = V_B + V_W / r = 5.13 + 7.5/5 = 6.63$; T = 12.475. Using the calculated parameter T and the statistical table of the chi-square distribution, we find the critical value $C = 23.337$. Finally, we calculate a limit value (or verification value) $V = S_P\sqrt{C}/\sqrt{T}$, where S_p represents the total repeatability declared by the manufacturer. In our example, S_p = 3.20; V = 4.38. Since the total repeatability obtained experimentally (3.55) is lower than the limit value (4.38), the total repeatability declared by the manufacturer is verified.

If substantial changes are made to the method in use, it is necessary to revalidate the method; however, even in the absence of such changes, it is good laboratory practice to periodically verify the repeatability demonstrated at the time the method was introduced. This verification can be easily conducted through the so-called duplicate test experiment: two independent replicates of the same sample are analyzed under repeatability conditions (x_1 and x_2). For the verification to be positive, it must be $|x_1 - x_2| < \sqrt{2} \times t_{a/2;\text{df}} S_r$, where S_r is the documented repeatability at the introduction of the method.

Example: S_r = 2.74 mg/dl (see previous example); degrees of freedom (df) 20. S_r had been determined in a 5 day × 5 replicate design, with df = days × (replicates − 1) = 5 × (5–1) = 20; if repeatability had been determined in a single run of N replicates, the df would have been (N − 1)); $t_{(a/2;df = 20)}$ = 2.086. If x_1 = 202 and x_2 = 196, $x_1 - x_2 = 6$, we get: $6 < \sqrt{2} \times 2.086 \times 2.74 = 8.08$ and thus it can be assumed that the laboratory is operating under the established repeatability conditions.

Trueness and Bias

Trueness can be defined as the closeness of agreement between the mean value, obtained from a large number of determinations of a sample, and the true value. Note that the true value is defined as an accepted or assigned reference value. Trueness is expressed as bias, or systematic error, which can be calculated as the difference between the mean value of the replicates and the true value or as the ratio of this difference to the true value. Thus, $B = \bar{x} - x_0$ or $\%B = \frac{\bar{x} - x_0}{x_0}100$, where $\bar{x}$ represents the average of the values obtained by repeatedly analyzing the sample and x_0 the assigned value. Bias is often reported as recovery and expressed as $\%R = \frac{\bar{x}}{x_0}100$, where the relation $\%R = 100 + \%B$. Recovery is frequently evaluated by making repeated determinations of a sample before and after the addition of a known amount of analyte; we have that $\%R = \frac{C_2 - C_1}{C_a}100$, where C_1 and C_2 are, respectively, the mean value of the replicates of the sample before and after addition of the analyte, while C_a is the expected increase in the concentration of the analyte in the sample after addition. C_a can be obtained from $C_a = C_{st}(V_{st}/V_f)$, where C_{st} is the concentration of the standard used for addition, V_{st} is the added volume of the standard solution, and V_f is the final volume of the sample (initial volume + added volume of the standard solution).

Example: 0.2 ml (V_{st}) of a standard calcium solution 21.0 mg/ml (C_{st}) is added to a serum sample of 2.0 ml volume. Four determinations are carried out before and after the addition.

$$\text{Before addition}: C_1 = (9.3 + 9.6 + 9.5 + 9.2)/4 = 9.4\text{mg/dl}$$

$$\text{After addition}: C_2 = (11.2 + 11.0 + 10.9 + 11.3)/4 = 11.1\text{mg/dl};$$

$$C_a = 21\times(0.2/2.2) = 1.91\,\text{mg/dl}$$

$$\%R = \frac{C_2 - C_1}{C_a}100 = \frac{11.1-9.4}{1.91}100 = 89.0\%$$

The recovery experiment allows to estimate the proportional systematic error, that is, a type of error that increases as the concentration of the analyte increases, but not the constant systematic error; in fact, even in the presence of a recovery close to 100%, it is not possible to exclude the presence of bias due, for example, to interferents or matrix components.

Bias can also be assessed: (1) by measuring certified reference materials; (2) through a method comparison study.

Bias Assessment with a Certified Reference Material

A certified reference material (CRM), with an assigned concentration value (and associated measurement uncertainty) and with a composition as close as possible to that of the test samples, is analyzed N times. It is obtained that $t_{\text{calc}} = \frac{|\bar{x} - \text{known value}|}{s/\sqrt{N}}$, where $\bar{x}$ and s are, respectively, the mean and the standard deviation of the replicates. If t_{calc} is greater than a tabulated t_{critical}, which depends on N and the chosen significance level, then the bias is considered statistically significant. The confidence intervals of the mean value of the N replicates and the bias are, respectively, $\bar{x} \pm t_{a/2;N-1} s/\sqrt{N}$ and $B \pm t_{a/2;N-1} s/\sqrt{N}$.

Example: The assigned value for a CRM for cholesterol is 250 mg/dL. Ten measurements are made obtaining a mean of 236.8 mg/dl and a standard deviation of 10.2 mg/dl. You can then calculate:

$$B = 236.8 - 250 = -13.2\,\text{mg/dl}$$

$$\%B = (236.8 - 250)/250\times100 = -5.3\%.$$

$$\%R = 236.8/250\times100 = 94.7\% \text{ or}$$

$$\%R = 100 + \%B = 100 - 5.3 = 94.7\%.$$

$$t_{\text{calc}} = \frac{|236.8 - 250|}{10.2/\sqrt{10}} = 4.09$$

For (10 − 1) = 9 degrees of freedom and a 95% significance level, we have that $t_{a/2;9}$ = 2.262; since $t_{\text{calc}} > t_{\text{crit}}$ (4.09 > 2.262) we have that the bias is statistically significant, that is, the mean measured value is different from the known or assigned value (with a 95% confidence level).

To confirm this, the 95% confidence interval (95%CI) of the mean value of the N determinations is $236.8 \pm 2.262\times10.2/\sqrt{10} = 229.5 - 244.1$, which does not include the assigned value of 250 mg/dl. Similarly, the 95% confidence interval of the bias does not include zero (in fact: $13.2 \pm 2.262\times10.2/\sqrt{10} = 5.9 - 20.5$).

A bias, although statistically significant, could still be acceptable if it is less than an error established on the basis of different criteria (expert consensus, biological variability).

When assessing bias by repeated measurements of a certified reference material, the investigator should consider both the uncertainty of the method and that of the CRM itself. In such a case, it follows that: $t_{\text{calc}} = \frac{|\bar{x} - \text{known value}|}{\sqrt{\frac{s^2}{N} + u_{\text{CRM}}^2}}$ with the degrees of freedom given by $\text{df} = \frac{\left(\frac{s^2}{N} + u_{\text{CRM}}^2\right)^2}{\left(\frac{s^2}{N}\right)^2}(N-1)$, where u_{CRM}^2 represents the uncertainty of the reference material. Note that, with u_{CRM}^2 very small, the degrees of freedom approximate to (N − 1). The confidence interval, including the contribution of the CRM uncertainty, becomes $\pm t_{\text{critical}} \times \sqrt{\frac{s^2}{N} + u_{\text{CRM}}^2}$.

In the previous example, the CRM has an uncertainty equal to u_{CRM} = 1.4 mg/dl; it is then that $t_{\text{calc}} = \frac{|236.8 - 250|}{\sqrt{\frac{10.2^2}{10} + 1.4^2}} = 3.75$ with

$$\text{df} = \frac{\left(\frac{10.2^2}{10} + 1.4^2\right)^2}{\left(\frac{10.2^2}{10}\right)^2}\times(10-1) = 13 \text{ and } t_{\text{crit}} = 2.160.$$

Also in this situation, with $t_{\text{calc}} > t_{\text{crit}}$ (3.75 > 2.160), we have that the bias is statistically significant. The 95% confidence intervals for the mean value and the bias will be, respectively, $236.8 \pm 2.160\times\sqrt{\frac{10.2^2}{10} + 1.4^2} = 229.2 - 244.4$ and $13.2 \pm 2.160\times\sqrt{\frac{10.2^2}{10} + 1.4^2} = 5.6 - 20.8$. As can be seen, the confidence intervals, calculated including the contribution of the CRM uncertainty, are only slightly wider than the previous ones but the conclusion remains the same.

The Experiment of Comparison of Two Analytical Methods

Systematic error can also be assessed through a method comparison experiment, where patient samples are measured with two different methods or instruments. This approach allows both calculation of the error at decision levels and assessment of proportional (slope of the regression line) and constant (intercept of the regression line) systematic error. The comparison experiment between analytical methods consists of these steps:

1. Familiarization with the new method and verification of the analytical performance of the method under test.
2. Definition of acceptability levels for maximum allowable error:

    - Based on the combined imprecision of the methods: if the two methods do not differ significantly, their differences will be within the range: $\pm 1.96\sqrt{CV^2_{method1} + CV^2_{method2}}$, where CV represents the total repeatability of the method.
    - Based on quality specifications reported in the literature (e.g., $TE_{max} = B_{max} + 1.65 \times I_{max}$, where B_{max} and I_{max} are the maximum imprecision and inaccuracy, respectively).

3. Sample selection and analysis: 40–60 patient samples distributed over the entire reportable range (range in which analyte concentrations can be measured with acceptable accuracy and precision); a greater proportion of samples should be selected around any clinical decision values and/or cut-offs.
4. Plotting and statistical analysis of the data: the most used tools are the regression model and the Bland-Altman difference graph (Fig. 4.5).

    To conduct the regression analysis, the data obtained with the reference or in-use method (x-axis) are plotted against the data obtained with the method under test (y-axis) together with the identity line ($y = x$) and the regression line. Different regression models can be used to find the regression line. The widely used simple (or ordinary) linear regression model has the merit of computational simplicity. However, it is not always correct to use it. In fact, simple linear regression assumes that the method in use (the one on the x-axis) is error-free and that the error of the method under test (y-axis) is normally distributed and is constant throughout the range of concentrations studied (homoscedasticity). When the assumptions underlying the linear regression model are not met, it is necessary to use another regression model (Deming, weighted Deming, or the non-parametric Passing-Bablok model). The non-parametric Passing-Bablok model requires the least number of assumptions. In fact, in Passing-Bablok regression, extreme values (outliers) can be included, imprecision is allowed in both methods, and it is not necessary either that the error be normally distributed or that it be constant over the concentration range. Regression analysis allows the assessment of the systematic constant (intercept) and

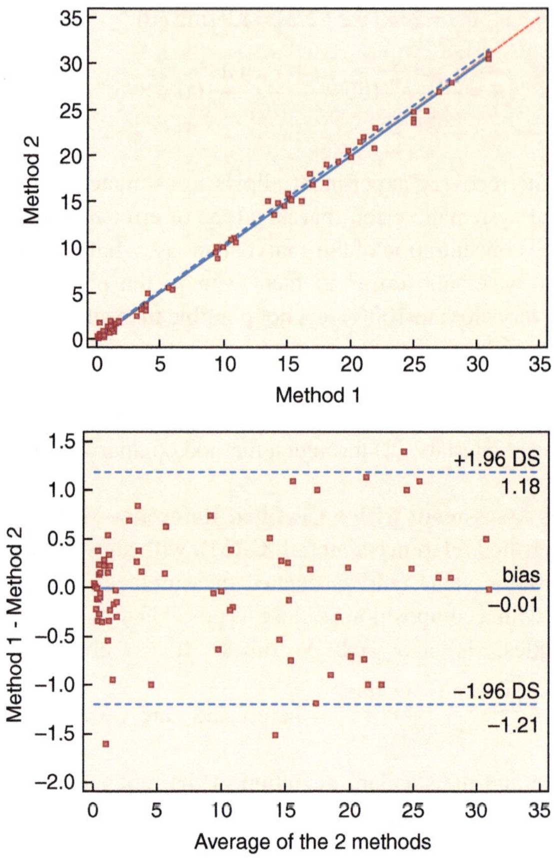

Fig. 4.5 Passing-Bablok regression analysis and Bland-Altman graph for the evaluation of systematic error in the comparison experiment between methods. In the Bland-Altman graph the line corresponding to the bias and the two lines delimiting the interval which includes 95% of the differences between the methods are shown

proportional (slope) error. It is always necessary to report not only these two parameters, point estimates of the true intercept and the true slope, but also their 95% confidence intervals (interval estimation), that is, the intervals in which we have some confidence (usually 95%) that the true intercept and slope can be found. In particular, to demonstrate the absence of constant, proportional systematic error, the relevant confidence intervals must include 0 (zero) for the intercept and 1 (one) for the slope. Small proportional or constant systematic errors identified by regression analysis, while statistically significant, do not preclude the use of a particular method if the error is less than the established maximum acceptable error. It is also important to note that the correlation coefficient should not be used to evaluate the agreement between two methods; in fact, it allows us to evaluate only the association between two variables but not the agreement (Fig. 4.5).

The Bland-Altman plot, on the other hand, allows calculation of the bias between the two methods, expressed as the mean of the differences. In this graph, the analyte concentrations on the x-axis (expressed as the mean of the two methods) are graphed toward the differences between the two methods on the y-axis. Alternatively, the mean of the two methods (x-axis) can be graphed toward the differences of the two methods expressed as a percentage of the mean, that is, (method1 – method2)/mean × 100 (y-axis). This method is especially useful when the concentration range is wide.

The bias is then calculated as the mean of all differences and expressed along with the associated 95% confidence interval. Significant bias occurs when the value 0 (zero) is not within the relevant 95% confidence interval. In the absence of bias (bias = 0), the points corresponding to the differences between the two methods should accumulate randomly around the zero line, that is, positive and negative differences close to zero should be observed for each concentration level. It is also possible to observe the presence of particular trends in the differences and/or the widening of the agreement range with increasing concentrations, a sign of the dependence of the differences on the concentration level. In addition to the mean difference (and the relative 95% confidence interval), it is also necessary to calculate the interval including 95% of such differences defined by mean ± 1.96 × SD, where mean and SD are, respectively, the mean (bias) and the standard deviation of the differences found.

On the other hand, this interval, which is often overlooked and not reported, is of considerable importance in the correct interpretation of the comparison experiment, because it allows us to identify the magnitude of most (95%) of the differences found. Alternatively, the interval comprising 95% of the differences can also be calculated using non-parametric methods, for example, as the interval between the percentiles 2.5 and 97.5 (Fig. 4.5).

5. Evaluation of the acceptability of the new method: the bias identified is compared with the maximum acceptable error defined at the beginning of the experiment.

For acceptability based on the combined imprecision of the two methods, we proceed as follows:

1. For each concentration level, the range defined by 0 ± 1.96 × CV% × C shall be calculated, where CV% is the combined inaccuracy of the two methods and C is the concentration. Since repeatability frequently varies over the concentration range, different CV% must be used for the two methods over the concentration range and different combined inaccuracies must be calculated.
2. Count how many method1-method2 differences fall outside the range. If the two methods are identical within the combined imprecision, their differences should be symmetrically distributed around zero and 95% of them should be within the interval.

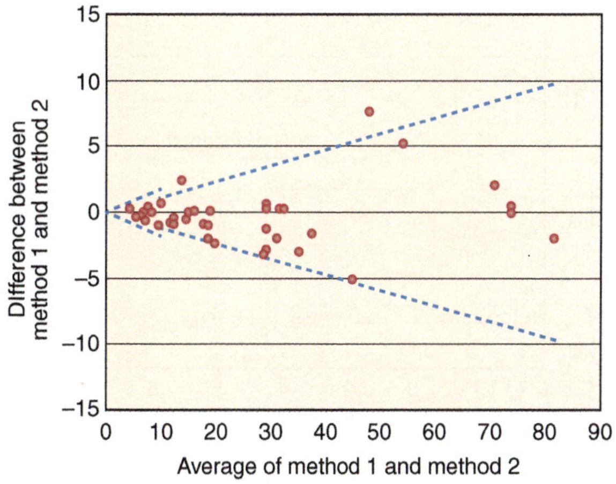

Fig. 4.6 Acceptability assessment based on combined imprecision

For a better visualization it is possible, in the same difference graph, to add the two lines representing the range 0 ± 1.96 × CV% × C (Fig. 4.6).

To assess acceptability based on predefined quality specifications, the previously defined maximum total allowable error is identified and a diagram (MEDx chart) is constructed, plotting the imprecision (x-axis) against the systematic deviation (bias) (y-axis) and representing the performance of the test under evaluation as a point. The coordinates of the point are obtained using for the abscissa the imprecision (calculated as repeatability within the laboratory) and for the ordinate the bias (using the equation of the regression line applied at a certain chosen concentration level, generally at clinical decision levels or cut-offs). The chart also shows four zones corresponding to different quality specifications (not acceptable, marginal, good, excellent). For the construction of the MEDx chart, we proceed as follows:

1. The point corresponding to the maximum allowable error is fixed on the y-axis.
2. On the abscissa axis, we fix three points corresponding to the various quality criteria: (max/2 error), (max/3 error), (max/4 error).
3. Three lines are drawn, starting from the point on the ordinate axis to the three points on the abscissa axis, which divide the graph space into four zones corresponding to the different quality specifications;
4. A point is represented using the performance of the method under test as coordinates (abscissa = within-laboratory repeatability; ordinate = bias as assessed by regression analysis). If $y = a + bx$ is the regression model, at a given concentration the bias will be given by $\%\text{bias} = \frac{|C-(C\times b+a)|}{C}\times 100$, where C is the expected concentration, a and b, respectively, the intercept and

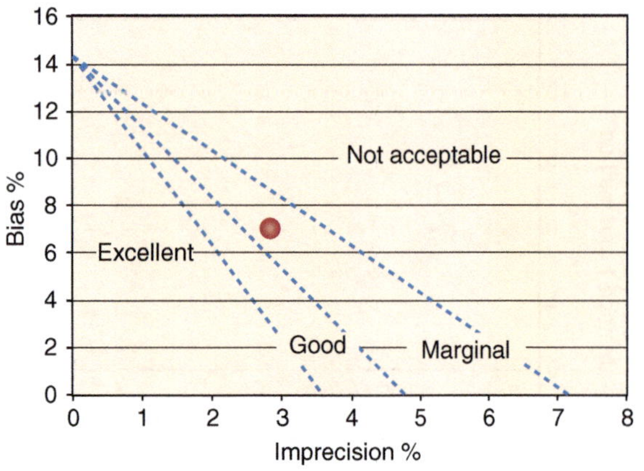

Fig. 4.7 MEDx chart for the assessment of acceptability based on quality specifications

slope obtained with the regression model. Finally, we evaluate in which region of the graph the point falls (Fig. 4.7).

Limit of Blank (LOB), Limit of Detection (LOD), Limit of Quantification (LOQ)

Determining the lower limit at which the analytical method under study is able to accurately identify and quantify the analyte is extremely important in many laboratory settings, including toxicology, drug monitoring, assessment of biochemical recurrence in cancer patients after surgery, and the assay of certain analytes (e.g., TSH, PSA, troponin).

The most recent international literature suggests adopting a common terminology. It is denoted by the following:

- Limit of Blank (LOB): the largest result that is likely (with a given probability) to be observed in repeated measurements of a blank sample, that is, not containing the analyte of interest.
- Limit of Detection or Sensitivity (LOD): the smallest amount of analyte in a sample that can be detected, but not quantified, with a given probability.
- Limit of Quantitation (LOQ): the smallest amount of analyte that can be quantified with a given precision.

Calculation of LOB and LOD

There are several protocols to determine these parameters, based on the signal-to-noise ratio (used for example in chromatography), on multiples of the standard deviation of replicates of white samples (LOD = $\text{mean}_B + 3 \times SD_B$; LOQ = $\text{mean}_B + 10 \times SD_B$), on the regression parameters of the calibration line (LOD = $3S_a/b$; LOQ = $10S_a/b$; where S_a is the standard deviation of the y residuals and b the slope).

The traditional approach for the determination of LOB and LOD, suggested by the CLSI protocol, involves the following steps:

- Analysis of blank samples, that is, samples without the analyte (samples that do not contain the drug or substance of abuse, samples pre-treated with resins or precipitating agents, samples from pathological subjects): K blank samples are analyzed r times with $N = K \times r$.
- Assessment of the presence of aberrants.
- Calculation of LOB = $\text{mean}_B + C_p \times S_B$, where mean_B and S_B are the mean and standard deviation of the replicates, respectively. C_p is a multiplicative factor related to the type I error at 5%, that is, the error that is made if a blank sample result exceeds the LOB and the investigator mistakenly concludes that the sample contains the analyte. C_p is calculated as: $C_p = \frac{1.645}{1-\left(\frac{1}{4\times(N-K)}\right)}$, where N is the total number of replicates and K is the number of blank samples. Since the results of the blank samples are often not distributed according to a Gaussian pattern, it is preferable to estimate the LOB with a non-parametric method: (1) the N results are ordered from the smallest to the largest; (2) the rank = $0.5 + N \times 0.95$ is determined and the LOB is equal to the value of the observation corresponding to the calculated rank (if the rank is not integer, interpolation is performed).

Example: $N = 45$; rank = $0.5 + N \times 0.95 = 43.25$. Since it is non-integer, we proceed by interpolation with neighboring observations 43 and 44. If $N_{43} = 2.6$ and $N_{44} = 2.9$ then LOB = $N_{43} + 0.25 \times (N_{44} - N_{43}) = 2.7$.

- Analysis of samples containing low concentrations of analyte (analyte concentration from 1 to 5 times the estimated LOB).
- Calculation of the standard deviation for the replicates of the J samples: S_i (standard deviation of the i-th sample) and calculation of the pooled standard deviation: $S_{\text{pooled}} = \sqrt{\frac{\sum_{i=1}^{J}(n_i-1)\times S_i^2}{\sum_{i=1}^{J}(n_i-1)}}$.

Example: Four blank samples are analyzed (10 replicates × sample). The standard deviations of the 10 replicates of the 4 samples are as follows: $S_1 = 3.20$; $S_2 = 2.97$; $S_3 = 3.54$; $S_4 = 3.75$. The pooled standard deviation will be $S_{\text{pooled}} = \sqrt{\frac{9\times3.20^2+9\times2.97^2+9\times3.54^2+9\times3.75^2}{9+9+9+9}} = 3.38$.

- Calculation of the LOD = LOB + $C_p \times S_{\text{pooled}}$, where C_p is a multiplicative factor related to the type II error at 5%, that is, the error that is made if a result from a sample containing the analyte is less than the LOB and the experimenter mistakenly concludes that the sample does not contain the analyte. C_p is calculated as $C_p = \frac{1.645}{1-\left(\frac{1}{4\times(L-J)}\right)}$, where L is the total number of replicates and J is the number of samples containing the analyte at low concentrations.

LOQ Calculation

Unlike LOB and LOD, where the calculation is based on statistical reasoning (in terms of type I and type II error), the determination of LOQ depends on operator-defined quality specifications (e.g., maximum allowable error, maximum imprecision). Two, among the many approaches suggested in the literature, are LOQ as functional sensitivity (accuracy profile) (Fig. 4.8) or as total error. In the first approach, the following steps can be identified:

1. The quality specification is set as total or within-laboratory repeatability (e.g., 10%CV_{WL}).
2. The protocol for determining total repeatability (design 5 days × 5 replicates or 20 days × 2 replicates) for 6–9 samples of known concentration distributed along the lower part of the concentration range) is applied.
3. The presence of any aberrants is checked.
4. For each sample (of the 6–9 tested), the mean concentration of all replicates is calculated $(\overline{x}_i)$ and, following the protocol described above, the relative repeatability within the laboratory (S_{WL} and %CV_{WL}).
5. The precision profile is drawn, plotting the average concentration $(\overline{x}_i)$ on the x-axis and the repeatability within the laboratory (%CV_{WL}) on the y-axis.
6. The best power model explaining the data is found: $\%CV = C_0 X^{C_1}$, where C_0 and C_1 are the two parameters of the model (Fig. 4.8).
7. Having identified C_0 and C_1, we find the concentration of the analyte (X) at which the repeatability within the Laboratory %CV = 10% (the chosen quality target): $LOQ = X = \left(\frac{Y}{C_0}\right)^{1/C_1}$, where $Y = 10\%$.

The experiment for determining LOQ based on total desirable error (TE) consists of these steps:

1. The quality specification, that is, the total desirable error, is set, based on biological variability, expert consensus, or other criteria.
2. Four to five CRMs are analyzed at assigned low concentrations: a 3 × 5 × (2 reagents) or 5 × 5 × (2 reagents) design is used for each CRM.

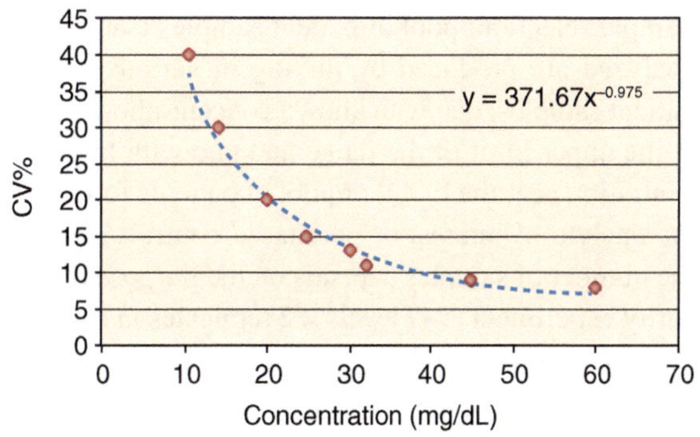

Fig. 4.8 Precision profile

3. The presence of any aberrants is checked.
4. For each sample/reagent pair, the mean of all replicates we calculate $(\overline{x}_i)$, the total repeatability (S_{WL}), and the bias as the difference between the mean $(\overline{x}_i)$ and the assigned value for that sample.
5. For each sample/reagent pair, the total error (TE) is calculated from the above data according to one of the following models:

$$TE = |\text{bias}| + 2S_{WL}\ (\text{Westgard model})$$

$$TE = \sqrt{S_{WL}^2 + \text{bias}^2}\ (\text{RMSmodel})$$

6. The TE of each pair/reagent is expressed as a percentage of the CRM assigned value: %TE = (TEx100)/(assigned value).
7. For each reagent, the sample with the lowest concentration that verifies the quality objective (i.e., with TE% < desirable total error) is selected; this concentration corresponds to the LOQ for that reagent,
8. The LOQ of the analytical method is set equal to the greater of the LOQs of the reagents.

If all of the samples used for one or more reagents do not meet the quality objective, the protocol for determining the LOQ must be rerun, using CRMs at higher concentrations.

Linearity

The linearity of an analytical method is its ability to give results that are directly proportional to the concentration of the analyte studied. Among the various methods that can be used for the determination or verification of linearity, the most recent literature suggests the robust method of polynomial regression in which regression models of different orders (first, second, and third order) are compared. The protocol consists of the following steps:

1. The maximum desirable error for non-linearity and repeatability is set.

2. Sample selection: pool of patient samples (samples to be analyzed are produced by mixing in various ratios two patient samples, one with known concentration of analyte at the upper limit of the range and one with known concentration near the LOQ) or pool of patients fortified with the analyte of interest or purchased controls/calibrators; the number of samples depends on the purpose of the linearity experiment (5–7 levels × 2 replicates in case of previously validated method, 9–11 levels × 2–4 replicates in case of new method to be validated).
3. Analysis of samples.
4. The results are graphed (results on the *y*-axis, theoretical concentrations on the *x*-axis) verifying the presence of evident non-linearity (in this case the experimental conditions should be re-evaluated).
5. The presence of possible aberrants is evaluated (in presence of more aberrant values the experimental conditions should be re-evaluated).
6. Using statistical software, first-, second-, and third-order regression models are obtained:
   First order: $y = b_0 + b_1x$ (linear model)
   Second order: $y = b_0 + b_1x + b_2x^2$ (non-linear model)
   Third order: $y = b_0 + b_1x + b_2x^2 + b_3x^3$ (non-linear model)
7. The possible significance for the coefficients b_2 b_3 is evaluated as the ratio between the non-linear coefficient (b_2 or b_3) and the relative standard error (provided by the software): $t = b/SE_{slope}$. t is compared with a critical value that depends on the chosen probability level (e.g., alpha = 0.05) and the degrees of freedom given by df = $C \times R - B$, where C is the number of samples, R the number of replicates for each sample, and B the number of coefficients of the regression model (first order: 2, second order: 3, third order: 4).
8. If neither b_2 or b_3 are significant, one concludes for linearity; otherwise, one selects among the significant non-linear models the one with the smallest standard error of regression (also provided by the statistical software). The difference between the value predicted by the non-linear model and the value predicted by the linear model is calculated for each level and expressed as a percentage of the concentration level (%diff = (100 × diff)/concentration). If all %differences are smaller than the fixed desirable error, the linear model is chosen anyway (the error, although statistically significant, is still within the desirable one). If, on the other hand, one or more of these differences% are greater than the fixed desirable error, one concludes for non-linearity and tries to identify any experimental conditions underlying this non-linearity; alternatively, if the major differences% are at the limit of the range, it is possible to run the statistical analysis again after removing the relevant concentration level (with obvious reduction of the linear range).
9. After verifying linearity, it is necessary to check that repeatability is also within the selected maximum error; to this end, compare S_r (obtained as in the repeatability experiment from the ANOVA table) or $\%CV_r$ (if repeatability is not constant along the interval) with the maximum error. If S_r or $\%CV_r$ is greater than the stated maximum error, the results of the linearity experiment are not reliable and the experimental conditions and the method itself must be verified.

When the repeatability varies a lot along the concentration interval (heteroscedasticity), in particular for concentration intervals that are very large at several orders of magnitude, it is preferable to use a weighted linear regression model rather than a simple linear regression model. A quick check for this purpose is obtained in the following way: (1) 10 replicates at the two extremes of the interval are analyzed (in total 10 × 2 = 20 replicates); (2) the variances at the two concentration levels are calculated; (3) an F-test for the variance is performed: $F_{calculated} = \frac{S_a^2}{S_b^2}$, where S_a^2 and S_b^2 are, respectively, the greatest and the least variance; (4) the $F_{calculated}$ is compared with the $F_{tabulated}$ (significance level = 0.05; $df_a = 10 - 1 = 9$; $df_b = 10 - 1 = 9$; $F_{tabulated} = 3.18$); (5) if $F_{calculated} > F_{tabulated}$, we conclude that the repeatability is not constant along the concentration interval and it is preferable to use the weighted linear regression model.

Finally, it should be noted that the correlation coefficient is not a measure of linearity but a measure of the degree of association between two variables.

Biological Variability

Several factors, acting on the concentrations of the different constituents of the body's biological fluids, can influence the analytical result. Among these sources of biological variability, age, sex, race, pregnancy, circadian, monthly or seasonal rhythms, and the presence of underlying medical conditions are of particular importance. Generally, these factors are considered controllable and included in the pre-analytical variability. The term "controllable" means that their effect can be controlled by appropriate measures, for example, by taking the sample under standardized conditions and comparing the subject with individuals who are comparable in as many characteristics as possible. Then there is an intrinsic variability of the organism, which cannot be controlled or further reduced, and which is the expression of random fluctuations around a homeostatic point. This random variability, different for the various components of the organism, is called intraindividual biological variability. The difference between the homeostatic points of different individuals is instead called interindividual biological variability and, with few exceptions, is generally greater than intraindividual variability. The knowledge of these sources of variability, measurable through the application of appropriate experimental protocols, is essential both for the definition of analytical goals and for the correct interpretation of a test result.

Different criteria can be used to interpret laboratory test results, such as comparing the current result with previous results from the same subject, or with a defined reference interval for the healthy population, with decision limits or following a probabilistic diagnostic approach.

Comparison of Current Results with Previous Results for the Same Subject: The Critical Difference

If, using standardized procedures, pre-analytical variability is minimized, the total variability associated with a laboratory result is $CV_T = \sqrt{CV_A^2 + CV_I^2}$, where CV_A and CV_I are, respectively, analytical variability and intraindividual biological variability, expressed as coefficients of variation (standard deviation/mean × 100%).

For two subsequent results obtained on the same individual and in the same laboratory, the total variability will be given by the sum of the individual variabilities and therefore equal to $\sqrt{2 \times CV_T^2} = \sqrt{2} \times CV_T$ from which the critical difference, that is, the maximum difference attributable to the effect of the total variability at a chosen probability level, can be derived:

$$\text{Critical difference}(\%) = \sqrt{2} \times Z \times \sqrt{CV_A^2 + CV_I^2}.$$

For a 95% probability level ($Z = 1.96$) the critical difference, expressed as a percentage, becomes $\text{critical difference}(\%) = 2.77 \times \sqrt{CV_A^2 + CV_I^2}$.

Differences greater than the critical difference may depend on variations in the subject's medical condition. Alternatively, from the above formula, it is possible to derive the value of Z and thus the probability that a certain percentage difference between two results from the same individual is significant: $Z = \frac{\text{difference}(\%)}{\sqrt{2} \times \sqrt{CV_A^2 + CV_I^2}}$.

The critical difference is simple to understand and easy to calculate from the values of CV_I available in the literature and of CV_A obtained in the laboratory with quality controls; however, its use presents some criticalities: (1) the intraindividual variabilities generally used are determined on healthy volunteers and are smaller than those presented by hospitalized or critical patients; (2) successive results obtained in a short time in the same individual tend to self-correlate and not to follow a Gaussian distribution; (3) different individuals present different intraindividual variabilities. In spite of these limitations, the critical difference is an important interpretative aid for the clinician, particularly in some situations (e.g., in serial analyses of tumor markers or in monitoring the effects or toxicity of a drug).

Definition and Use of Reference Intervals

Frequently, the result of a laboratory test is interpreted by comparison with a range of values determined in a reference population. The term "reference population" is preferred to the more ambiguous terms "normal population" or "healthy population" and indicates instead a set of reference individuals selected on the basis of specific criteria. The traditional approach to the definition of reference intervals foresees that from this population of reference individuals, a reference sample is extracted from which reference values are obtained, through measurements, which will present a certain distribution (reference distribution) and whose analysis, with appropriate statistical methods, will allow to estimate the reference limits and the relative reference interval. The most critical aspects of this process are the selection of the individuals, the standardization of the pre-analytical conditions, the analytical aspects, and the statistical processing. When the biological characteristics of the analyte are known, it is preferable to use the "a priori" selection approach, excluding and stratifying individuals before sampling according to specific criteria defined and well documented in the procedures (age, sex, genetic factors, physiological factors, risk factors, presence of pathologies). When, more rarely, the analyte is new and there is little information regarding biological variability, it is necessary to follow the "a posteriori" approach and therefore sampling a very large number of individuals and then excluding and stratifying on the basis of the measurements obtained. As an alternative to the previous direct methods, indirect methods use analytical results contained in existing databases and are based on the assumption that most of the data produced by a laboratory, even in hospitalized patients, are normal. From the mass of extracted data, the use of appropriate statistical methods allows to identify and exclude the abnormal values of the minority of "unhealthy" subjects and thus define the reference interval. Indirect methods are relatively simple and inexpensive and, in particular when it is difficult to collect samples of healthy subjects, they are the only methods that can be applied; however, they have important limitations: they are difficult to generalize, since the relevant pre-analytical, analytical, and clinical data are not available, and they strongly depend on the mathematical model used to define the interval, since the distribution of the values of the reference population is not known. These methods can therefore be considered to estimate or at most verify reference intervals determined by direct methods.

Pre-analytical conditions, particularly subject preparation, sample collection, processing, and transport, must be identified and carefully controlled to minimize their effect on the validity of results by introducing bias or increasing variability.

Analytical aspects shall also be carefully considered and described, including the method and instrumentation used, the analytical performance and, where possible, demonstration of "traceability" of the method used to higher metrological levels. In particular, this last aspect is crucial to enable the transfer of reference intervals between different laboratories.

In the definition of reference intervals, statistical methods are used to identify aberrant data, to stratify or partition reference individuals into homogeneous groups, and to calculate the limits of the reference interval and the associated confidence limits. Aberrant results can significantly affect the calculation of reference intervals. For their identification, and eventual elimination, it is necessary to observe the distribution of the collected data, to verify that the presence of apparently extreme data does not depend on pre-analytical, analytical, or gross errors and to apply appropriate statistical methods. The techniques used are the same as those described above. Several approaches have been described in the literature to assess the need to stratify or partition the reference population into homogeneous groups, including hierarchical ANOVA (or nested ANOVA), Sinton's method, Harrys and Boyd's method, and Lahti's method. According to the latter approach, we first calculate the limits that include 95% of all results (common distribution without subgroups) and then evaluate the number of observations in the subgroup distributions that fall outside these limits, that is, those that have values less than the lower limit or greater than the upper limit. Denoting with p the highest percentage of observations that in the distributions of the subgroups fall outside the limits (or in the right or left tail), if $p > 4.1\%$ it is necessary to stratify, if <3.2% it is not necessary to stratify. For intermediate percentages, it is necessary to decide on the basis of non-statistical criteria. Also for the calculation of the reference limits and the relative confidence intervals, there are numerous methods in the literature, including the non-parametric method and the robust method of Horn and Pesce. With the non-parametric method, traditionally preferred by the IFCC, the reference interval corresponds to the central portion of the reference distribution comprising 95% of the values, delimited by the quantiles 0.025 and 0.975. Their calculation requires at least 1/0.025 = 40 data. With such size ($n = 40$), the limits of the interval correspond to the values of the first and last observation, but it is not possible to estimate the relative confidence intervals and therefore the uncertainty. For 90% confidence intervals, a minimum number of 120 individuals is required: in this situation, the limits of the reference interval correspond to observations 3 and 118 and the respective confidence intervals to observations 1 and 7 and 114 and 120, respectively. For larger numbers, it is necessary to use tabulated values available in literature or statistical software. The main advantages of this method are the simplicity of calculation and the robustness; however, the high numerosity (even more if one considers that $n = 120$ is the minimum numerosity for each possible subgroup) often limits its applicability, in particular in small laboratories. In the robust method of Horn and Pesce, which can be used in samples of lower numerosity, the reference limits are estimated by iteratively applying robust indicators, after Box-Cox normalizing transformation. In the case of reference intervals determined by the robust method, the relative confidence intervals must be calculated by the bootstrap method, a statistical technique of sampling with re-entry; in short, from the sample of n reference individuals, N samples with replacement are extracted and of each one, by the Horn and Fish robust method, the two reference limits are calculated. Of the N estimates of the lower limit and N estimates of the upper limit of the reference interval, the percentiles 5 and 95 are calculated, which will constitute the limits of the two 90% confidence intervals of the reference interval.

An alternative to the fresh determination of reference intervals is the transfer of existing reference intervals. If the laboratory has already determined its own reference interval, and needs to change the analytical method, the new reference interval can be calculated from the previous one using its own samples and the equation of the regression model obtained by comparing the method currently in use with the new method (provided that the two methods have similar analytical performance). Alternatively, the laboratory, after verification, may implement a reference interval determined by a third party, provided that the pre-analytical and analytical conditions (method traceable to the same reference system) used to determine the interval are well documented and comparable with those of the laboratory and the two populations are similar. To verify the range, the laboratory selects 20 normal individuals (with characteristics that overlap with those of the reference population of the laboratory that determined the range), eliminates extreme values (by testing for aberrant values) and, if necessary, analyzes new samples until 20 results are obtained, and verifies how many results are outside the considered reference range: if 2 or less values fall outside the interval, it can be implemented; if 3 or 4, the verification experiment is redone; and if still more than 2 values fall outside the interval or if 5 or more in the first experiment, then the interval cannot be implemented and any pre-analytical or analytical differences or between the two populations must be checked.

Individuality Index

A problem with using reference intervals to interpret laboratory data is that of so-called individuality, that is, for some analytes the values of each individual occupy, or extend, only a small portion of the interval. In other words, the intra-individual biological variability (CV_I) is much smaller than the interindividual one (CV_G). This concept is well expressed by the individuality index: $II = \frac{\sqrt{CV_A^2 + CV_I^2}}{CV_G}$, which when $CV_A < CV_I$, can be simplified to $II = \frac{CV_I}{CV_G}$. A low individuality index (II < 0.6) means that the analyte has a marked individuality, while a high individuality index (II >1.4) indicates an analyte with little individuality. In the first case, reference values are not very useful and it is preferable to compare the current results with the previous results of the same subject, while in the second case, when the values of the individual extend to the whole range, the comparison with a reference range is preferable. The individuality index can be increased by stratification or partition, thus defining different reference intervals for different subgroups.

Decision Levels

Reference intervals are quite different from decision levels or limits. In fact, while the former are determined in apparently healthy subjects using statistical methods, and therefore the comparison with the result presented by the patient can at least suggest whether the latter is probably healthy or unhealthy, the latter are determined by expert consensus and serve to identify a risk threshold, a classification threshold for a given pathology or a threshold for a clinical decision or treatment.

Performance of a Diagnostic Test

Sometimes the clinician finds himself in the situation in which it is necessary to classify an individual on the basis of an analytical result expressed by a continuous quantitative variable. The simplest classification is represented by a dichotomous qualitative variable that can assume only two values: positive or negative, sick or healthy, or high or low risk. In order to classify individuals into two subgroups, the clinician uses a threshold value (cut-off). However, the choice of this cut-off value is not always easy as often the distribution of results of the healthy population is partly overlapped with that of the sick population. Hence, the attribution, on the basis of an analytical test, of an individual to a group or to another can be expressed only in probabilistic terms and the performance of the test in correctly classifying a subject is given by some indices such as clinical sensitivity, clinical specificity, positive predictive value, and negative predictive value, calculated by comparing in the so-called 2 × 2 contingency table the test whose performance is to be evaluated with a test of certain reliability (gold standard) (Table 4.1).

Table 4.1 Contingency table

		Gold standard		Total
		Sick	**Healthy**	
Tests under examination	**Pos**	True positives (TP)	False positives (FP)	Total positives
	Neg	False negatives (FN)	True negatives (TN)	Total negarives
Total		Total sick people	Total Healthy	

Clinical sensitivity (Se) is the probability of a test to give a positive result if the subject is sick [Se = P(pos | sick)] and answers the question "how many sick individuals tested positive?" Sensitivity can also be understood as the proportion of sick individuals who test positive. In other words, the sensitivity of a test is its ability to correctly identify sick individuals. Thus, $Se = \frac{TP}{sick} = \frac{TP}{TP + FN}$.

Clinical specificity (Sp) is the probability of a test to give a negative result if the subject is healthy [Sp = P(neg | healthy)] and answers the question "how many healthy subjects tested negative?" Specificity can also be understood as the proportion of healthy individuals who test negative. In other words, the specificity of a test is its ability to correctly identify healthy individuals. Thus, $= \frac{TN}{healthy} = \frac{TN}{TN + FP}$.

From the observation of the two ratios, it is clear that the sensitivity is greater the lower the number of false negatives; equally the specificity is greater the lower the number of false positives. This means that a test with high sensitivity will rarely fail to identify sick patients and can therefore be useful to exclude a diagnosis (rule out); vice versa, a test with high specificity will rarely classify a healthy individual as sick and can therefore be useful to confirm a diagnosis (rule in).

Since sensitivity and specificity are calculated on a sample, the respective confidence intervals must also be associated with the two statistics. For 95% confidence, the respective intervals are $CI95\% = \pm 1.96\sqrt{\frac{Se(1-Se)}{sick}}$ for sensitivity and $CI95\% = \pm 1.96\sqrt{\frac{Sp(1-Sp)}{healthy}}$ for specificity.

In real situations, where the distributions of healthy and sick populations are partially overlapping, as the selected threshold value varies, an increase in sensitivity of a test is associated with a decrease in specificity and vice versa. The preference for one or the other index depends on the particular situation: a high sensitivity test is preferable for rare pathologies, when the missed diagnosis may have serious consequences (fatal but potentially curable pathologies) or to exclude a diagnosis (rule out); a high specificity test is preferable for high prevalence pathologies in order to avoid waste of resources, and in situations where a false positive diagnosis may have serious consequences on the subject (economic, psychological, invasiveness of further tests).

Not all individuals who test positive or negative are sick or healthy, respectively. To answer this question, two other calculated indices are useful: the positive predictive value (PPV) and the negative predictive value (NPV).

The positive predictive value is the probability that an individual testing positive is ill [PPV = *P*(sick | pos)] and indicates the proportion of test-positive individuals who are ill. It is calculated as $PPV = \frac{TP}{positives} = \frac{TP}{TP+FP}$.

The negative predictive value is the probability that a test-negative individual is healthy [NPV = *P*(healthy | negative)] and indicates the proportion of test-negative individuals who are healthy. It is calculated as $NPV = \frac{TN}{negatives} = \frac{TN}{TN+FN}$.

Note that if PPV is the probability that a test-positive individual is ill, then 1-PPV is the probability that a test-positive individual is healthy. Equally, if NPV is the probability that a test-negative individual is healthy, then 1-NPV is the probability that a test-negative individual is sick.

Confidence intervals can also be calculated for the positive and negative predictive value:

$$CI95\% = \pm 1.96\sqrt{\frac{PPV(1-PPV)}{positives}} \text{ for the PPV, while}$$

$$CI95\% = \pm 1.96\sqrt{\frac{NPV(1-NPV)}{negativ}} \text{ for the NPV.}$$

One must pay close attention that the PPV and the NPV depend closely on the prevalence of the disease. This dependence is easily understood by applying Bayes' theorem:

$$PPV = \frac{prevalence \times Se}{(prevalence \times Se) + [(1-prevalence)(1-Sp)]}$$

$$NPV = \frac{(1-prevalence)Sp}{(1-prevalence)Sp + [prevalence(1-Se)]}$$

Thus, under conditions of low prevalence and specificity of a test, the positive predictive value drastically decreases and the negative predictive value increases.

From the formula of the PPV, it is also understood that, as the specificity tends to 0 (and therefore the sensitivity to 1), the PPV tends to the prevalence of the disease.

One approach to increasing PPV is obviously to operate in high prevalence situations, such as applying the test to high-risk subgroups.

Another index aimed at assessing the performance of a diagnostic test is the likelihood ratio (LR). The positive likelihood ratio (LR^+) indicates how many times a positive test is more likely in a sick individual than in a healthy one, that is, the ratio of the probability of having a positive result among those with the disease divided by the probability of having a positive result among those without the disease: $LR^+ = \frac{P(pos|\,sick)}{P(pos|\,healthy)} = \frac{Se}{1-Sp}$. Similarly, the negative likelihood ratio (LR^-) indicates how many times a negative test is more likely in a sick individual than in a healthy one, that is, the ratio of the probability of having a negative result among those with the disease divided by the probability of having a positive result among those without the disease: $LR^- = \frac{P(neg|\,sick)}{P(neg|\,healthy)} = \frac{1-Se}{Sp}$. For interpretation purposes, if LR > 1 the test supports the presence of the disease (particularly if LR > 10 the test is very effective in confirming), if LR < 1 the test excludes the presence of the disease (particularly if LR < 0.1 the test is very effective in excluding), if LR = 1 the usefulness of the test in the diagnostic process is null. Likelihood ratios are very useful to the clinician to understand how much the probability of a diagnosis hypothesized before the test increases in case of a positive or negative result. In other words, from the "a priori" or "pre-test" probability of a certain diagnosis (probability deduced from the literature for the general population, from the symptoms and signs presented by the subject or in any case from the available information), we pass to the "a posteriori" or "post-test" probability, that is, the probability of having a certain pathology after having obtained the results of a test. The calculation is much simplified if the odds are used instead of the pre- and post-test probabilities. In fact, in the case of odds, the effect of the likelihood ratio is multiplicative.

For this purpose, first we transform the pre-test probability into odds, $\text{odds}_{\text{pre-test}} = \frac{\text{prob}_{\text{pre-test}}}{1-\text{prob}_{\text{pre-test}}}$ (when the pre-test probability is equal to the prevalence of the pathology in the population or in that particular subgroup or class of individuals we will have $\text{odds}_{\text{pre-test}} = \frac{\text{prevalence}}{1-\text{prevalence}}$), the post-test odds are calculated $\text{odds}_{\text{post-test}} = \text{LR} \times \text{odds}_{\text{pre-test}}$ and, applying the inverse formula, the post-test probability is calculated $\text{prob}_{\text{post-test}} = \frac{\text{odds}_{\text{post-test}}}{1+\text{odds}_{\text{post-test}}}$. Alternatively, particularly in the clinical setting, it is possible to use Fagan's nomogram. The main advantages of using the likelihood ratio are represented by the fact that it is an easily understandable tool and that it can be used according to a sequential approach (the post-test probability of a given diagnostic test can be used as a pre-test probability for the next one). However, its use is limited in all those situations where a pre-test probability is not available.

ROC Curves

ROC (Receiver Operating Characteristic) curves are an important statistical tool to study the diagnostic performance of a test; in particular, they allow to evaluate how sensitivity and specificity change when the cut-off selected for a continuous quantitative result is changed.

In short, given a test with results described by a quantitative variable, various cut-offs are selected and for each of them the sensitivity and specificity of the test are calculated; in a Cartesian diagram, the curve given by the union of the points with the abscissa 1-specificity, that is, the proportion of false positives, and the ordinate sensitivity, that is, the proportion of true positives, is then drawn (Fig. 4.9). The slope of the line joining the origin of the axes with every single point of the ROC curve, that is, with every selected cut-off, is equal to the likelihood ratio at that point, that is, for that cut-off. The diagnostic power of the test, that is, its ability to correctly classify sick and healthy people, is expressed by the area under the ROC curve or AUC (Area Under the Curve) and is proportional to it. For AUC = 1, the test discriminates perfectly sick and healthy while for AUC = 0.5 the test is not informative. For intermediate values, between 0.5 and 1, the test will show poor, moderate, or high accuracy. The confidence interval for AUC can be calculated by different methods, including that of Hanley & McNeil, DeLong, or the bootstrap method, which are generally available in major statistical software. ROC curves, and their respective AUCs, can also be used to compare two or more different diagnostic tests in the same graph.

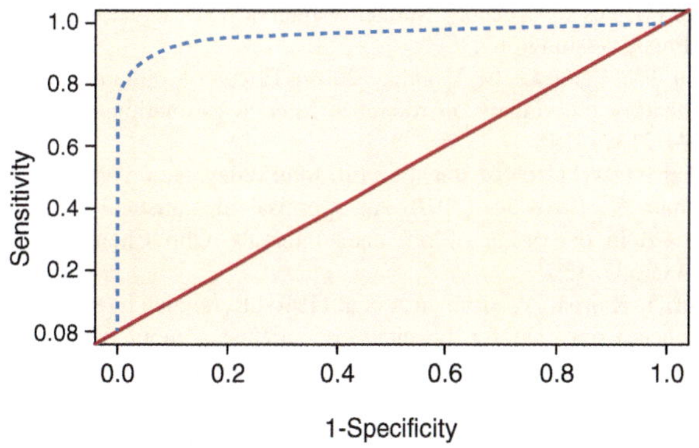

Fig. 4.9 ROC curve

Recommended Readings

Baadenhuijsen H, Smit JC (1985) Indirect estimation of clinical chemical reference intervals from total hospital patient data: application of a modified Bhattacharya procedure. J Clin Chem Clin Biochem 23:829–839

Bland JM, Altman DG (1999) Measuring agreement in method comparison studies. Stat Methods Med Res 8:135–160

Bock BJ, Dolan CT, Miller GC et al (2003) The data warehouse as a foundation for population-based reference intervals. Am J Clin Pathol 120:662–670

Box G, Cox D (1964) An analysis of transformations. J Royal Stat Soc B26:211–252

Ceriotti F (2007) Gli intervalli di riferimento nel nuovo millennio. Biochim Clin 31:254–266

Ceriotti F, Infusino I, Panteghini M (2012) Teoria e pratica degli intervalli di riferimento riferibili. Biochim Clin 36:171–176. Clinical and Laboratory Standards Institute. Defining, Establishing, and Verifying Reference Intervals in the Clinical Laboratory; Approved Guideline-Third Edition. CLSI document EP28-A3c. 2010

Clinical and Laboratory Standards Institute (2003) Evaluation of the linearity of quantitative measurement procedures: a statistical approach; Approved guideline. CLSI document EP06-A

Clinical and Laboratory Standards Institute (2005) User verification of performance for precision and trueness; Approved guideline-Second Edition. CLSI document EP15-A2

Clinical and Laboratory Standards Institute (2012) Evaluation of detection capability for clinical laboratory measurement procedures; Approved guideline-Second Edition. CLSI document EP17-A2

Clinical and Laboratory Standards Institute (2013) Measurement procedure comparison and bias estimation using patient samples; Approved guideline-Third Edition. CLSI document EP09-A3

Clinical and Laboratory Standards Institute (2014a) Evaluation of precision of quantitative measurement procedures; Approved guideline-Third Edition. CLSI document EP05-A3

Clinical and Laboratory Standards Institute (2014b) User verification of precision and estimation of bias; Approved guideline-Third Edition. CLSI document EP15-A3

Efron B (1982) The Jackknife, the bootstrap and other resampling plans. Society for Industrial and Applied Mathematics, Philadelphia

Ferre-Masferrer M, Fuentes-Arderiu X, Puchal-Ane R (1999) Indirect reference limits estimated from patients' results by three mathematical procedures. Clin Chim Acta 279:97–105

Harris EK, Boyd JC (1990) On dividing reference data into subgroups to produce separate reference ranges. Clin Chem 36:265–270

Hoffman RG (1963) Statistics in the practice of medicine. JAMA 185:864–873

Horn PS, Pesce AJ (2005) Reference intervals. A user's guide. AACC Press, Washington, DC
Horn PS, Feng L, Li Y et al (2001) Effect of outliers and non-healthy individuals on reference interval estimation. Clin Chem 47:2137–2145
Huber P (1981) Robust statistics. Ed. John Wiley, New York
Ichihara K, Boyd JC (2010) An appraisal of statistical procedures used in derivation of reference intervals. Clin Chem Lab Med 48:1537–1551
Kouri T, Kairisto V, Virtanen A et al (1994) Reference intervals developed from data for hospitalized patients: computerized method based on combination of laboratory and diagnostic data. Clin Chem 40:2209–2215
Lahti A (2004) Partitioning biochemical reference data into subgroups: comparison of existing methods. Clin Chem Lab Med 42:725–733
Lahti A, Petersen PH, Boyd JC et al (2004) Partitioning of nongaussian-distributed biochemical reference data into subgroups. Clin Chem 50:891–900
Leys C, Ley C, Kleina O et al (2013) Detecting outliers: do not use standard deviation around the mean, use absolute deviation around the median. J Exp Soc Psychol 49:764–766
Passing H, Bablok W (1983) A new biometrical procedure for testing the quality of measurements from two different analytical methods. Application of linear regression procedures for method comparison studies in clinical chemistry, Part I. J Clin Chem Clin Biochem 21:709–720
Petersen PH, Stockl D, Blaabjerg O et al (1997) Graphical interpretation of analytical data from comparison of a field method with a Reference Method by use of difference plots. Clin Chem 43:2039–2046
Sinton TJ, Crowley D, Bryant SJ (1986) Reference values for calcium, phosphate, and alkaline phosphatase as derived on the basis of multichannel-analyzer profiles. Clin Chem 32:76–79
Vidali M, Tronchin M, Dittadi R (2016) Protocollo per la comparazione di due metodi analitici di laboratorio. Biochim Clin 40(2):129–142
Westgard JO (1995) A method evaluation decision chart (MEDxchart) for judging method performance. Clin Lab Sci 8:277–283

Elements of Metrology

5

Orazio Ruzzenente and Claudio Brentegani

Introduction

Omnia in mensura et in numero et in pondere disposuisti (You have arranged everything by measure, number, and weight) (St. Augustine).

Measurement is the highest degree of knowledge of a thing or phenomenon.

Only the measurement of a "thing" or phenomenon allows us to know it in depth and make qualified decisions about it. Metrology is the science of measuring processes and is particularly important for laboratory professionals.

We provide below some definitions regarding metrology taken from "International Vocabulary of Metrology – Basic and General Concepts and Associated Terms":

- Measurement: a process to quantitavely describe the property (magnitude) of a thing or phenomenon
- Measure: the result of a measurement consisting of a set of three data elements (numerical value, unit of measure, measurement uncertainty)
- Magnitude: an attribute of a thing or phenomenon that can be expressed qualitatively or quantitatively
- Measurand: quantity intended to be measured
- Measurement unit: real scalar individual quantity, defined and adopted by convention, with which any other quantity of the same kind canbe compared to express the ratio of the two quantities as a number
- Measurement uncertainty: a parameter characterizing the dispersion of the values being attributed to a measurand, based on the information used

O. Ruzzenente
Section of Clinical Biochemistry, University of Verona, Verona, Italy

C. Brentegani (✉)
Section of Clinical Biochemistry, Department of Neurological, Biomedical and Movement Sciences, University Hospital of Verona, Verona, Italy
e-mail: claudio.brentegani@univr.it

The measurement process is characteristically carried out in successive steps. The first step is to identify the measurand and define its state during the measurement process since the latter can affect the characteristics of the measurand itself. The next step is to establish the reason for the measurement (appropriateness of the request) because its knowledge can influence the measurement procedure (choice of the measurement method or instrument) and the choice of acceptable limits of imprecision for that measurement.

Another step is to assess the suitability of the sample to be analyzed. The definition of a sample in biochemistry is different from that in metrology. In metrology, a sample is a medium that reproduces a known value of a quantity for use as a comparison. In biochemistry, a sample can be understood as a piece of material representing the whole. In this context, there is a fundamental part of clinical biochemistry that studies pre-analytical procedures that can also heavily influence the representativeness of the sample. In order to provide reliable measurements, it is fundamental to evaluate the representativeness of the material being measured, particularly in this historical moment, because completely automatic instruments carry out a large part of the analyses (measurements). Interferences due to incorrect preanalytical procedures or the presence of interfering substances cause a systematic, and often clinically significant, error in the determination of the concentration of an analyte.

Metrology in Laboratory

The choice of measurement method and instrumentation is the next step. Measurement methods can be direct or indirect. In biochemistry, the two direct methods still used are potentiometry, which measures the potential difference due to electrical charges carried by ions in solution (ions selective electrodes, blood gas analysis), and the lowering of the cryogenic point, which measures the osmolality of a sample. The direct methods usually allow for obtaining comparable results and are characterized by low measurement

M. Ciaccio (ed.), *Clinical and Laboratory Medicine Textbook*, https://doi.org/10.1007/978-3-031-24958-7_5

uncertainties. All other methods are indirect and often involve several steps to obtain the signal to be measured. It is, therefore, essential to know exactly the relationships among the products obtained in the various steps of the reactions during measuring procedures. Each step has a certain degree of uncertainty, and it is, therefore, important when studying a method to calculate the total inaccuracy by adding up the inaccuracies of the various steps. These methods are the most common way to determine the concentration of a substance. This quantity is not directly measurable but requires specific reagents to form with the substance under examination known products with a signal measurable by laboratory instruments.

The choice of instrumentation to perform a measurement must be weighted according to the measuring principle, and all the mechanical activities of the instrument must have a negligible influence (better not influence at all) on the result of the measurement. Once the signal has been detected, it must be transformed into a measurement by a logical or practical process known as calibration. The calibration of an instrument is a mechanical operation that prepares the instrument to provide measurement of a certain quantity equal to those of the reference material. Calibration, on the other hand, is a logical–practical operation that establishes the signal-to-size ratio by comparing, using indirect methods, known values of the same quantity of one or more reference materials.

In some cases, the association between signal and magnitude is established by chemical or physical relationships (e.g., Lambert–Beer law in spectrophotometry). However, the direct application is impossible because, in a routine laboratory, the measurement conditions cannot be perfectly controlled. The relationship is therefore obtained experimentally by constructing calibration lines or curves. It is preferable, during calibration, to use several reference samples and construct a calibration curve or straight line to minimize the inaccuracies related to the measurement of every single point of the curve. The choice of reference samples for the construction of the calibration curve shall consider the matrix effect. The samples must therefore be as similar as possible to those under analysis because all the substances present may influence the measurement even if they are not considered true interferents. Once all the steps described have been implemented, the expression of the measurement result is achieved. Three pieces of information should always be provided when communicating the results of a measurement: the numerical value, the unit of measurement, and the uncertainty of the measurement.

The unit of measure chosen to provide the result and the correct choice of the number of significant digits of the numerical value help to define the quality of a measurement. In recent years, there has been an innovative process whereby the focus of analysis is no longer the actual value of the measurand, but rather the quality of the measurement. The term measurement uncertainty is more appropriate than the term error and well describes the indeterminacy of any experimental result.

The evaluation of the measurement uncertainty is statistical and is therefore based on a calculation procedure codified in an international standard (UNI CEI ENV 13005, July 2007).

An indispensable prerequisite for considering a measurement as qualified information is the traceability of the measurements, that is, the existence of a chain of comparisons that enables each result obtained to be linked to internationally recognized reference standards. In this case, an increase in uncertainty can be generated in the passages between one comparison and another, which must be accurately quantified.

To resume and widen the discussion, it is necessary to premise that it is quite easy in biochemistry to identify the measurand because it is almost always the concentration of a substance or analyte, even if sometimes it is not so easy to identify exactly (or completely) the substance itself whose concentration is to be measured. Furthermore, a straight line corresponding to the Lambert–Beer law often represents the marker-concentration relationship. Although it is difficult to apply the law directly in a routine laboratory, it is relatively easy to identify this relationship and the linearity limit, which also represents the concentration limit beyond which the relationship is no longer predictable. Even in the case of reactions involving enzymes, although the latter present some problems characteristic of enzymatic reactions (isoforms), they are oxidation–reduction reactions and therefore relatively easy to predict. For a more detailed discussion of this aspect, please refer to the chapter on enzymes. All the activities described above can therefore be summarized in the standardization of a method.

Standardization in Laboratory

The definition of standardization encompasses a set of specific rules that are the basis of guidelines aimed at achieving an optimal degree of uniformity in each discipline. The product of standardization must ensure the highest possible comparability in the standardization of measures, consistency, and quality in the various aspects of procedures, thus guaranteeing the producer and the consumer. Translating the definition into good laboratory practice, the product of standardization is a good test with optimal results that contributes to quality diagnostics, followed by better patient care and improved public health. Standardization in the laboratory can therefore be explained by:

- Production of reference standards
- Control in the production of calibrators
- Method validation protocols and their instrumental applicability

Production of Reference Standards

iIt is necessary to distinguish two important areas of the laboratory: the strictly chemical area and the immunochemical one. Concerning the first one, it is easy to find pure substances and obtain primary standards by weighing them (e.g., glucose quantified by gravimetric procedure). Regarding the second area, standardization is more complex since the procedure represents the becoming of a strictly biological process, including all its variables, hence the well-known problems of standardization of immunoassays. The Expert Committee on Biological Standardization (ECBS) of the World Health Organization (WHO) confirms international standards and other reference substances for biological fluids. For some hormones and tumor markers, the National Institute for Biological Standards and Control (NIB- SC) is the reference. Other institutions, such as International Federation of Clinical Chemistry (IFCC), College of American Pathologists (CAP), and Clinical Laboratory Standard Institute (CLSI), collaborate in specific areas. The primary reference standards, characterized by purified substances and harmonization procedures between biological activity and concentration, are International Standard (IS) and International Reference Preparation (IRP).

- IS (lot/year) is always identified by a lot and year of production and is approved after a long process. They are available in small quantities, freeze-dried in protein "medium," and can satisfy requests for a period of 10–20 years.
- IRP (batch/year) is proposed by WHO despite a route that has not been fully refined. These standards are generally used to define the concentrations of secondary standards,represented by sufficiently pure substances. They are generally used to perform calibration curves and are stored to control them over time.

Control in the Production of Calibrators

The above procedures give rise to the certification of the calibrator values accompanying the various diagnostic kits. A significant element in the calibration procedure is instrumentation (immunochemistry is affected by non-dedicated instrumentation). Calibrators for immunochemistry kits are provided either to generate multipoint calibration curves (in ELISA and RIA procedures) or, in the presence of dedicated instrumentation, to transfer the new calibration to the one obtained "in-house" using only two calibration points. The evaluation of the latter procedure cannot disregard the knowledge of mathematical elements that are at the basis of the study of the curves. The two-point calibration procedure is generated on the following mathematical basis:

- The "in-house" curve, generated on more than one point, is often characterized by a sigmoid, characteristic of all biological phenomena, from which, through mathematical processes (fitting study), is extrapolated the interval of greater linearity, which is represented by a straight line and, as such, characterized by two important parameters, i.e., the angular coefficient and the intercept.
- The adjustment curve obtained in the laboratory by two calibrators is a straight line easily comparable with the one obtained "in-house" using the same parameters. The best adjustment is obtained by well-defined calibrators concentrations, one of which should represent the part closest to the limit of detection (LoD) and the other a point close to 50–60% of the linearity range. The comparison of the parameters of the lines, whose knowledge allows for validation of the calibration, will transfer, together with specific algorithms, the obtained value to that of the curve obtained "in-house".

Method Validation Protocols and Their Instrumental Applicability

Instrumentation should never be thought of as an element to relieve the analyst's responsibility for the reliability of measurements. This aspect allows a double evaluation of the test limits and the instrumentation. The knowledge and verification of the analytical sensitivity and the measurement linearity range, performed on the instrumentation in use, can contribute to analytical choices in which the error can be reduced. Knowledge of these aspects leads to a better setting and interpretation of quality control.

Problems in Standardization of Immunochemistry

The complexity of standardization in immunochemistry is generated by numerous problems, including protein heterogeneity, matrix effects, cross-reactivity, the presence of heterophilic antibodies, and the need to measure concentrations close to the limit of detection. In addition, the measured immunoreactivity often represents a mixture of protein isoforms that differ in their degree of glycosylation, degradation, immune complex formation, and even activity.

For some proteins with purification difficulties, IS and IRP standards report the concentration in U/L standardized in terms of bioactivity. This aspect is difficult to quantify mathematically and, therefore, more suitable for harmonization of the results than for standardization.

Good protein purification allows the expression of the mass concentration that most clearly reflects what is being measured. However, even the determination of molar concentration presents difficulties since it can be accurately determined by amino acid analysis even in the presence of heterogeneity concerning the carbohydrate composition and the isoelectric point. In contrast, the gravimetric determination, especially for glycoproteins, is more inaccurate since the freeze-drying process fails to completely remove the water bound in the molecular structure.

Immunochemical reaction is based on an antigen, an antibody, and the "matrix" in which they are located, which exerts its action by maintaining their tertiary structure unaltered (immunoreactivity) through a suitable salt and protein concentration. The use of monoclonal antibodies has greatly improved specificity through the characterization of epitope mapping, knowledge of which would allow the immunoreactivity of dosage to be defined based on its similarity to the reference antibody.

Only validation protocols provide the means to understand their complexity and take action to improve their analytical performance.

Recommended Readings

Calcatelli A L'incertezza di misura. INRIM
Clinicals and Laboratory Standard Institute EP07-A2
Clinicals and Laboratory Standard Institute 2014
Plassa M, Le misure di grandezze fisiche, a cura di E. Arri e S. Sartori. Torino: Paravia (1984) Revisione di E. Amico di Meane (INRIM – Istituto Nazionale di Ricerche Metrologiche), 2009. dosaggio sulla base della somiglianza con l'anticorpo di riferimento. Solo i protocolli di validazione forniscono i mezzi per capire, comprendere la complessità di questi dosaggi e poter intervenire per migliorarne le performance analitiche
Price CP, Newman DJ (1997) Principles and practice of immunoassay, 2nd edn. Macmillan Reference, London
Sant'Agostino De Civitate Dei. XI, 21

The Pre-analytical Phase

6

Davide Giavarina

Introduction

The clinical laboratory produces information, which is a response to the knowledge needs of clinicians and patients, through physical and chemical measurements of constituents in biological fluids. Good laboratory information needs a good "analytical" phase of the measurement. However, all events that occur from the moment the doctor expresses the need for information to the moment the result of the measurements, properly interpreted, returns to the physician has importance in determining the quality and reliability of the information produced.

Thirty-five years ago, George Lundberg introduced for the first time the concept of the total testing process and described in his famous brain-to-brain loop, some of the essential events that make up the formation of information. Only one point of the loop represents the analytical phase, while all the others are part of the pre- and post-analytical phases (Fig. 6.1).

In the overall process leading to the production of useful information for patient care based on measurements of the subject's biological material, the pre-analytical phase includes all events and conditions that occur or are determined before the chemical, physical, or biological analysis of the sample. It includes the formulation of the clinical question; the request of the examination; the planning; the recognition of the patient; the execution of the sampling or the collection of the sample; the identification of the sample; the transport; the evaluation of the suitability in relation to the container, volumes, times, temperatures, internal alterations of the sample; and the intra-laboratory pre-analytical treatment (centrifugation, storage, etc.).

It is commonly believed that laboratory errors are most frequent in the analytical phase, the phase with the greatest technical complexity. In reality, developments in technology and the attention devoted over the decades by specialists to controlling this phase have greatly reduced variability. In the analytical phase, moreover, quality objectives have been defined and constantly monitored through intra- and inter-laboratory quality controls, which provide for continuous and immediate corrective actions. On the other hand, the greatest risk of error occurs precisely in the pre- and post-analytical phases, amounting to over 80% of all cases. In particular, more than 60% of all laboratory errors occur in the pre-analytical phase. Knowing and recognizing the most frequent errors in this phase allow laboratories to focus their efforts on improving quality and, consequently, patient care.

In recent years, it is more and more used to divide the pre-analytical phase into two parts: a pre-pre-analytical phase, which concerns all that happens before the analysis but outside the laboratory or outside the control of the laboratory, and the real pre-analytical phase, which is what happens before the analysis but under the direct responsibility of the laboratory. The pre-pre-analytical phase includes test request, patient identification, collection, and transport, while sample handling, aliquoting, secondary labeling, and centrifugation belong to pre-analytical phase.

For expository purposes, we will distinguish pre-analytic problems into patient-related variables and sample-related variables. The latter may occur before, during, or after sample collection.

Patient-Related Variables

We can distinguish "natural" variables, which are inevitable because they are determined by biological rhythms and by the continuous adaptation of the living organism to maintain its homeostasis, from deterministic variables, which are caused by or directly related to external actions or events, which can be modified. The first is defined as biological variability, while the second is pre-analytical variability related to the state and preparation of the patient.

D. Giavarina (✉)
Laboratory Medicine, St. Bortolo Hospital, Vicenza, Italy
e-mail: davide.giavarina@aulss8.veneto.it

M. Ciaccio (ed.), *Clinical and Laboratory Medicine Textbook*, https://doi.org/10.1007/978-3-031-24958-7_6

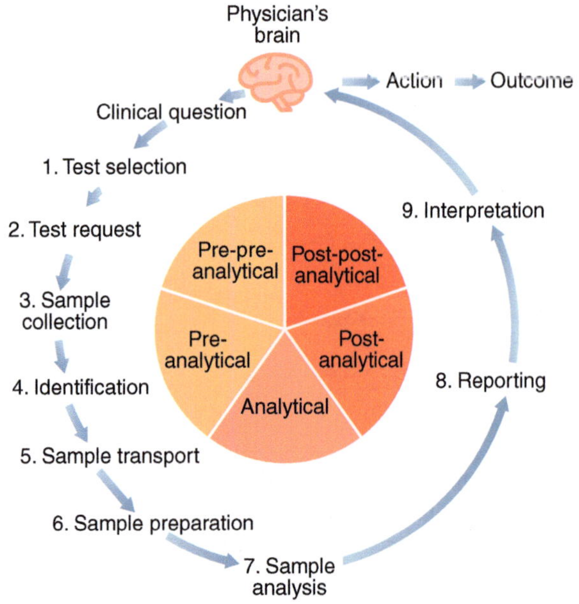

Fig. 6.1 Brain-to-brain loop. (Copyright EDISES 2021. Reproduced with permission)

Biological Variability

There are natural, biological variabilities that are somewhat predictable. Some of these define "diversity" among individuals but remain constant in the same individual, such as gender or ethnicity. However, it is important to consider that these variables significantly affect the interpretation of results. Different reference systems are often necessary for differences in gender, ethnicity, or for particular physiological states, such as pubertal growth and pregnancy.

Other variables, however, are also predictable but subject to change even in the same individual. For example, many analytes change over the course of life in everyone due to the natural processes of aging. Thus, plasma creatinine concentrations increase steadily throughout life with acceleration after age 65. Many other parameters have this kind of variation: among the most recently highlighted and discussed cases, D-dimer, which is higher after 60 years of age, as well as during pregnancy; ventricular natriuretic peptide, BNP; and troponin. Noteworthy, life is not the only "cycle" that influences our biorhythms. We have variations, also, of seasonal type, or monthly, or within a single day. It has been known for more than 40 years that cortisol is very variable during the day, with the highest concentrations in the morning and the lowest around midnight. Variations within the day have been described for many other analytes, including several hormones, but also for differential counts of white blood cells, concentrations of serum and urinary electrolytes, hemoglobin, etc. Some metabolites have even more particular behaviors within the day. The growth hormone, for instance, is released in the growing subjects in a pulsatile way, during the night, with differences between males and females. There are also monthly rhythms, such as the well-known rhythm of the sexual hormones, estradiol and progesterone, and their stimulating hormones, FSH and LH, in women of childbearing age. Other rhythms may have seasonal cyclicity related to temperature and sun exposure. Conditions of higher plasma concentration due to sweating, greater physical activity, and more light and a different diet can be responsible for higher concentrations of enzymes, such as lactic dehydrogenase, total protein, vitamin D, and glycated hemoglobin.

All these natural variabilities must be considered when measurements of concentrations are compared with references to obtain judgments, information. For example, life course variability sometimes leads to the definition of reference intervals for age classes, which are essential for examinations in children but often also necessary for assessment in later life.

Seasonal variability is generally modest and usually not considered significant for the clinical interpretation of test results.

On the other hand, the monthly variability has greater relevance and requires precise control of the day on which samples are collected and sometimes the need for serial sampling during the month for consistent comparison of results with appropriate references.

Of considerable impact and importance are the variations within the day. Except for the most striking examples, such as cortisol, the fact that concentrations of blood constituents vary throughout the day is not generally known among clinicians and patients. In hospitals, samples for hematochemical and biological fluid tests are collected at all hours of the day and night, and data are frequently compared. Also, in outpatient activities, there is a tendency to extend the time for blood sampling throughout the day. All this can be an organizational necessity or of timeliness of the cures, but it is in contrast with the research to standardize the procedures (that we will see useful also for other aspects) also for the times of sampling and collection.

There is also random biological variability. The constituents of the blood are quite stable in their concentrations over time; however, these concentrations fluctuate randomly around a homeostatic point, with variability depending on the constituent. These random variabilities, considered individually and as population variability, determine that one cannot compare a measured value with a fixed concentration but must do so against a range of concentrations within which one ascribes the same clinical significance. Analysis of the impact of random biological variability on clinical practice is beyond the scope of this chapter.

Pre-analytical Variability Related to Preparation and Patient Status

Physiological or para-physiological activities and behaviors can alter the concentrations of the constituents of biological fluids, even to a very significant degree.

Diet, prolonged fasting or a recent meal can result in significant variations in the concentrations of many analytes. Prolonged fasting can be a common occurrence in preoperative hospital procedures; much greater are the variations caused by the assumption of various foods, which can be themselves part of blood constituents (e.g., lipids, sugars), or act by stimulation and induction (e.g., insulin, leukocytes, enzymes).

Coffee is often considered food as others and sometimes patients worry that the sweetening of this beverage may alter tests, and not so the caffeine itself. In reality, caffeine can be found in many everyday foods. Caffeine inhibits phosphodiesterase and, therefore, the degradation of cyclic AMP. Cyclic AMP in turn, promotes glycogenolysis, increasing plasma glucose concentrations. In addition, caffeine stimulates gluconeogenesis. Therefore, blood glucose increases, not so much due to the added sugar, but because of caffeine itself. It also acts on lipase, increasing the concentration of non-esterified fatty acids, which in turn can displace hormones related to transport proteins, including albumin, thus changing the free measurable concentrations. For example, increases in plasma renin activity and catecholamines have been found 3 hours after caffeine intake.

Smoking causes some acute and chronic changes in the concentration of many analytes. The chronic changes are rather modest. For example, it has been reported that the concentration of carcinoembryonic antigen (CEA) is higher in subjects not affected by neoplasms but smokers, to the point of suggesting differentiated cut-offs for this variable. Of some importance are, instead, the changes in the very short term, between 1 and 3 hours after smoking, with an increase in leukocytes, fibrinogen, decrease in ACE (Angiotensin Converting Enzyme), prolactin, etc. These modifications depend on the number of cigarettes smoked, on the type, on the modality (aspiration or not) and are also influenced by the subject's age and gender. They are, therefore, completely unpredictable a priori.

The consumption of ethyl alcohol has pre-analytical importance in the assumption of important and toxic quantities, as it can interfere in many metabolic processes. A decrease in the glycemia and an increase of the lactic acid are observed due to blockage of the gluconeogenesis, an increase of the uric acid, a state of metabolic acidosis; the concentrations of aldosterone increase, while they tend to decrease those of many other hormones, such as the osteocalcin, the prolactin, and the cortisol.

The effects of chronic alcohol intake can only partly be ascribed to pre-analytical variables since these are real changes in the concentrations of constituents in the body, signs of the chronic state of intake of an element toxic to the body. The increase in liver enzymes or blood count parameters is sign of liver damage and altered erythropoiesis, rather than pre-analytical variables. Similarly, many other changes, such as the increase in triglycerides, catecholamines, and cortisol levels, are due to chronic interference in various metabolisms.

Acute changes in analyte concentrations during exercise may be due to volume shift between the intravascular and interstitial compartments, volume loss caused by sweat, and changes in hormone concentrations (e.g., increased concentrations of catecholamines, glucagon, somatotropin, cortisol, ACTH, and decreased insulin). In addition, the effects of physical training on muscle mass must be remembered. The hypoxia-mediated increase in creatine kinase (CK) is dependent on training status and, therefore, shows a high degree of individual variability. The lower the individual's level of training, the greater the increase in CK. Many other analyte concentrations are similarly dependent on muscle mass and training level. Very vigorous exercise can, also, cause an increase in cardiac markers, or an increase in plasma creatinine; it can also cause the excretion of red blood cells or other blood cells in the urine. However, these exercise-induced changes normally disappear within a few days.

Some blood constituents show significant changes at high altitudes compared to the same at sea level. Significant increases with altitude are observed, for example, for C-reactive protein (CRP) (over 65% higher at 3600 m), serum β2-globulin (over 43% at 5400 m), hematocrit and hemoglobin (over 8% at 1400 m), and uric acid. Adaptation to altitude takes weeks, while return to sea levels takes a few days. A significant increase with increasing altitude is also seen for urinary creatinine, creatinine clearance, estriol (over 50% at 4200 m), serum osmolality, plasma renin, and transferrin.

The intake of drugs can induce important changes in the constituents measured by the laboratory. For chronic therapies, this is not a pre-analytical variable but the actual state of the subject. On the contrary, the "pre-analytic" is important in the therapeutic monitoring of drugs, when it is precisely the concentration of the drug to be measured to verify toxic levels, maintenance in therapeutic range, compliance with therapy, etc. In this case, the time elapsed between the assumption of the drug and the time of sampling for the determination of concentration is essential for the use of appropriate references.

The importance of psychological stress on laboratory results is frequently underestimated (anxiety before sample collection, preoperative stress, etc.). Indeed, it induces the increased secretion of several hormones (aldosterone, angiotensin, catecholamines, cortisol, prolactin, renin, somatotropic hormone, TSH, vasopressin) and other molecules, including albumin, fibrinogen, glucose, insulin, lactate, and cholesterol.

The posture of the subject at the time of collection and in the minutes preceding collection can result in important changes in the concentrations of the measured constituents. The effective capillary filtration pressure (the difference between the capillary pressure and the colloidal osmotic pressure in the plasma) increases in the lower extremities when changing from the supine to the upright position. Consequently, water moves from the intravascular compartment into the interstitium, leading to the reduction of the plasma volume by approximately 12% in normal individuals. Particles larger than 4 nm in diameter present in the blood are held by the membranes and cannot follow this fluid movement. A change from the upright to the supine position leads to a decrease in the effective filtration pressure and, thus, a shift in volume in the opposite direction. A change in plasma volume results in an apparent concentration of cells, macromolecules, and small molecules bound to proteins.

Effective capillary filtration pressure is also the basis for the variables determined by prolonged tourniquet maintenance (see below).

Sample Variables

We can distinguish variables that intervene before, during, and after sample collection/pickup (Table 6.1).

Table 6.1 Pre-analytical errors related to the sample

Time	Error
Before sample collection	Request for inappropriate examination or necessary test not required Patient identification error Sample identification error
During sample collection	Insufficient volume Incorrect anticoagulant Incorrect sample/anticoagulant ratio Coagulated sample Sample contamination Incorrect tube
After sample collection	Labeling error Inappropriate transportation Sample storage error (time and temperature) Centrifugation error (time, temperature, and speed)

Inappropriate Examination Request or Necessary Examination Not Requested (Appropriateness of Request)

This type of pre-pre-analytical error, determined by request for an examination that is incorrect for the suspected pathology or the failure to request it, carries the risk of a delay or failure to recognize a pathological condition and the consequent treatment. Today, appropriateness is considered a cornerstone of public and general health governance. In laboratory medicine, an examination is defined as appropriate when the result provides an answer to the clinical question and enables a decision to be made. There is much attention on over-prescribing, which may be a problem of wasted resources or information "noise," but the inappropriateness determines the greatest clinical risk. Measuring this type of error is not easy and basically requires being able to acquire the diagnostic question together with the test request. The laboratory can act on different levels, defining and spreading diagnostic paths and profiles by pathology, integrating exams with subsequent diagnostic levels (reflex test), proposing computer systems for requesting exams (order entry) structured by diagnostic problem and with systems of suggestion to the prescription, etc. The communication between the laboratory and the clinic is the fundamental element of continuous improvement for this pre-analytical problem.

Patient and Specimen Identification Error

There are many ways to run into this error, perhaps the most dangerous of all pre-analytical errors. One can confuse one patient for another because they have the same or similar names, or because they are neighbors in the hospital; one can take a document, label, or test tube labeled for one patient and use it for another; one can enter a result on another's file; and one can make a mistake in a patient's file. According to a recent report by ECRI, a non-profit care improvement consulting foundation in the United States, more than 72% of these errors occur at the point of the patient encounter, and as many as 36.5% are associated with diagnostic procedures. The correct identification of the patient is a fundamental element for "safe" care and is a primary objective of every healthcare organization, as also sanctioned by important standardization and control bodies, such as the Joint Commission, the 15189:2012 standard, and the LEPS (Laboratory Errors and Patient Safety) working group on laboratory errors and patient safety of the International Federation of Clinical Chemistry and Laboratory Medicine (WG-LEPS of the IFCC). Despite all this attention, this type of error continues to be present. Table 6.2 summarizes possible risk actions and best practices to counteract this type of event.

Table 6.2 Patient identification: actions to avoid and best practices

Dangerous actions to avoid	Good practices for correct patient identification
Use room or queue number, bed location, or diagnosis to identify a patient	Use two different identifiers, defined by the organization, to identify the patient at the time of the meeting
Ask the patient to confirm her name by asking: "Is your first name last name?"	Ask the patient to indicate their personal data by saying: "What is your name? What is your date of birth?"
Assuming that the patient will correct you when you use a wrong name (the patient may be confused, fearful, or feel they have not heard right)	Spread the importance of correct identification to patients
Place patients with similar names in the same room	Involve patients by explaining the importance of identification in each procedure
Label a container before obtaining the sample, away from the patient	Provide support for hearing impairments or language barriers, so that the patient can adequately confirm their identity
Bring multiple pre-printed labels for different patients	Take measures to avoid confusion when patients on the same operating unit have similar names
Later label series of samples taken from different patients	Confirm the identity of a patient before affixing a label to a container
Trust the identification made by another operator	Label the sample in the presence of the patient, one at a time
Failure to follow established procedures for correct patient identification	Reconfirm the identity in each hand change of the patient or his/her samples
	Apply patient identification techniques consistently, following the organization's policies
	Minimize interruptions and distractions during patient identification
	"Do not be silent," if deviations from standard patient identification procedures are observed

Container Type Error

While serum samples, or samples anticoagulated with heparin or EDTA, may be readily identified and possibly recognized as unfit for analysis when supplied to the laboratory in the primary tube, the error may not be easily detected when working with secondary tubes or worse if the samples have been mixed, transferred, or added from one type of primary tube to another. Training of personnel involved in the collection of biological specimens and dissemination of knowledge of the potential risk of mixing specimens from one tube to another is the best preventive action for this insidious pre-analytical risk.

Insufficient Sample

The insufficient sample represents an associated limit to the execution of the examinations required for the specific sample. The problem particularly affects pediatric samples, but it can be found in all other types of patients, both for difficulties in venous access and for technical difficulties in sampling. It is important that all the samplers and the prescribers known the criticality related to the volume of the sample, so that they can limit the examinations required and give priority to the execution. Laboratories should be organized to perform examinations in order of priority so that examinations that are not performed due to sample exhaustion are the least cogent for patient care. The use of alternative methods that consume fewer sample can sometimes help with this problem. For example, many POCT (Point Of Care Testing) instruments use smaller amounts of samples than laboratory instruments. The use of an alternative method should be reported with the test result.

In some cases, if the analytical methods have sufficient analytical sensitivity even for concentrations 50% or 75% lower than expected, the plasma or serum samples may be diluted with physiological saline or appropriate buffer (1:2 or 1:4) to increase volume, and then the final concentrations calculated by multiplying by the dilution. Be aware that this practice increases analytical imprecision due to possible dilution errors. The practice is not always applicable to all measurable constituents since matrix variations or changes in ratios to binding proteins (hormones, drugs, etc.) can cause gross errors.

It is also important to note that each primary tube is constructed with a defined ratio between the volume of sample to be collected and the amount of additive contained. It would always be recommended that the tubes are filled to the expected nominal volume to maintain the correct ratios between blood and additive/anticoagulant. In some cases, such as coagulation tests, this is absolutely mandatory.

Coagulated Sample

A coagulated sample is defined as a blood sample collected in a test tube with anticoagulants that has visible micro- or macro-clots. The coagulated sample represents an insurmountable problem in cell counts (blood count, lymphocyte typing, etc.) and in coagulation tests, where coagulation factors must be maintained in zymogen form. Inappropriate coagulation occurs essentially for two main reasons: difficult and prolonged collection over time, and failure to mix the sample inside the tube.

Hemolyzed Sample

Visible hemolysis in the sample after centrifugation is defined by the presence of free hemoglobin concentrations in serum or plasma >0.30 g/L. It is caused by the rupture of red

Table 6.3 Main causes of extravascular hemolysis

Sampling	Transport	Intra-laboratory pre-analytics	Storage
Traumatic sampling, difficult venous access, access point	Origin of the sample: obstetrics, first aid, intensive care unit	Time between collection and centrifugation	Storage temperature
Sampling from catheter/needle cannula	Transport by pneumatic mail	Centrifugation at extreme temperatures	Duration of storage
Capillary sampling	Pre-centrifugation and transport temperature	Spin speed	
Needle size	Transport by courier	Imperfect barrier of the separator (gel)	
Transfer from syringe	Duration of transport	Re-centrifugation	
Use of antiseptic for sampling			
Vigorous shaking			
Tube not completely filled			
Lack of agitation			

blood cells and the release of their contents into serum/plasma. The interference of hemolysis with laboratory measurements is dependent on the red blood cells spilling their intracellular contents into the plasma or on spectrophotometric interference in the absorbance readings of the reaction products.

Hemolysis is the leading cause of sample rejection in clinical laboratories. We can distinguish between intravascular hemolysis, which is a sign of hemolytic anemia, from extravascular hemolysis, which is due to problems occurring during and after venous sampling. The intravascular hemolysis represents less than 2% of all the detectable hemolysis. The main causes can be traced back to metabolic or systemic diseases (hepatic, oncohematologic, autoimmune diseases, etc.), chemical agents (drugs), physical agents (mechanical heart valves), infectious agents, etc.

Extravascular hemolysis, on the other hand, has numerous causes and concomitant causes. Table 6.3 summarizes the main ones, in the different pre- and post-analytical phases.

Interference can be detected "by eye," after centrifugation (pale pink to lacquered red serum-plasma), or be recognized by measuring the hemolysis index on analytical platforms (see below).

Lipemic Sample

After hemolysis, the lipemic, or rather turbid, the sample is the most frequent cause of non-idoneous samples in the laboratory. High lipid concentrations generate optical interferences in many analyses. Lipemia in samples is defined in terms of turbidity caused by a high concentration of lipoproteins visible to the naked eye or quantifiable at 660/700 nm. Generally, the cause of lipemia is too short an interval between the last meal or parenteral lipid infusion and sampling. The interference can be detected "by eye," after centrifugation (milk serum-plasma) or be recognized by measuring the lipemic index on analytical platforms. Excess lipid can be removed from the sample to allow measurements of the other constituents. However, ways to clarify the sample must be carefully chosen on a case-by-case basis depending on the analytes to be measured. For example, plasma lipid can be removed from a complete blood count (CBC) sample after centrifugation at low speed and replaced with an isotonic solution. This allows for a correct measurement of hemoglobin. However, cell counts, particularly of platelets, should be performed prior to this operation since the removal of lipemic plasma involves the removal of platelets that remain in suspension by centrifugation at low speed. Cell counts are not interfered with by lipemia. For many serum or plasma measurements, ultracentrifugation techniques can be used to separate lipids from other components of the sample. The limitation to these approaches is the availability of ultracentrifuges in clinical laboratories. Finally, polar solvents can be added to extract lipids from samples. However, recent work has shown that these methods cannot be used for all constituents measured by spectrophotometric or immunological methods, as the recovery is variable and sometimes significantly lower than the true value.

For analytes distributed in the lipid layer, methods that remove the lipid fraction are not acceptable. In such cases, measurement after dilution may be attempted. The sample should be diluted only enough to remove the interference caused by turbidity, but not too much, to ensure that the analyte concentration remains within the analytical sensitivity limits of the methods used (two or three times). This is probably the best approach for the measurement of therapeutic drugs in lipemic samples.

Jaundiced Sample

Jaundiced samples appear intensely yellow in color and contain high concentrations of bilirubin. Bilirubin can chemically interfere with analytical reactions, typically causing under-estimation of measurements (e.g., cholesterol, creatinine, and creatine kinase isoenzymes), or disturbing absorbance spectra in spectrophotometric measurements.

Measurement of Serum Indices in the Laboratory

The automatic measurement of interference for hemolysis, lipemia, and jaundice can be done "by eye," comparing with chromatic scales, or with automatic instruments based on the absorbance of light by hemoglobin, bilirubin, and "lipids" (so-called serum indices).

Hemoglobin is red and absorbs light between 340 and 440 nm and between 540 and 580 nm. Bilirubin has its absorbance peak at 460 nm. The apparent absorption of light by lipemia/turbidimetry is determined by the deflection of light operated by lipids and lipoproteins; it is greatest below 400 nm and gradually decreases along the visible spectrum. With appropriate choices of wavelengths and subtractions for areas of overlap, auto-analyzers provide an estimate of each individual interference that is more precise and repeatable than human assessment. Even with the need for harmonization between the different technologies, automatic measurement procedures are preferred and today essentially indispensable.

Laboratories should define in their operating procedures the levels of hemolysis, jaundice, and turbidity for which each specific test may be affected. Results that may have a significant bias should not be reported.

Sampling Order (Urban Legend?)

It was hypothesized that any entrainment of anticoagulants or different additives in the collection tubes could cause errors comparable to those resulting from mixing samples between different tubes. This led to a series of recommendations on the sequence to be followed during sampling, as follows: blood cultures, citrate, serum, heparin, EDTA, and in vitro glycolysis inhibitor. Recent observations have not confirmed this potential risk, recognizing a negligible effect. However, the European Study Group on Pre-Analytical Variability, recently reviewing all available literature, insists on recommending maintaining the order of the collection tubes.

Infusion Route Contamination

Collecting blood samples from peripheral venous catheters can frequently cause hemolysis and is therefore not recommended, although often unavoidable. If the infusion route is used to infuse saline, glucose, or other fluids, the sample may also be diluted or contaminated by the infused substances. The technique for infusion collection, when not avoidable, should include procedures for washing and discarding the first portion of the collected specimen. Washing with 5 mL of saline, discarding 2 mL of waste, and then withdrawing the actual sample (3 mL) will yield suitable samples in over 99% of cases. Alternatively, 5 mL of blood should be discarded prior to collection. With good sampling practice, the use of peripheral venous catheters is permissible; however, it should be noted that mild dilution and "spurious" contamination remain a potential error, which is difficult to detect in the laboratory.

Recommendations for Sampling

As described above, many pre-analytical problems are related to the sampling techniques, the procedures used, and the personnel involved. Good laboratory practices and good training programs for operators, dedicated to the collection of biological samples, and blood, in particular, can greatly improve this problem, significantly reducing errors and related clinical risks. Table 6.4 summarizes the recommendations made by the Intersociety Study Group on Extra-Analytic Variability of the two major Italian societies of laboratory medicine (SiBioC – SIPMeL).

Table 6.4 Recommendation for blood collection

Recommendation	Grading
Operator training	
For all professionals qualified for blood sampling, introduce training courses, based on frontal teaching, tutoring, and practice	A
Devices for the collection	
Use devices that provide for the integration of disposable needles, support systems (holders or shirts), and primary vacuum tubes (vacuum).	A
Syringes are a possible alternative in the following cases: Emergency situations when it is not possible to find the above devices Anatomical and/or physical situations make it impossible or inadvisable to use the above devices	B
Use disposable devices that provide for the elimination of all parts in direct contact with the patient's blood Use systems that do not allow to re-cap needles and any other possible sharp object used during the collection	A
If the holder is not contaminated with blood, it can be reused	B
If, on the contrary, there is even the suspicion of blood contamination, the holder must be:	
Sterilized	D
Eliminated	A

(continued)

Table 6.4 (continued)

Recommendation	Grading
Prefer traditional needles	A
Use butterfly in specific situations:	
Veins difficult to access by location or caliber with the traditional device	A
Express request by the patient	
Prefer needles of equal gauge of 20 or 21G	A
Reserve small caliber needles for sampling on very small veins	B
Do not use needle cannulas	A
Rules relating to the patient	
Identify the patient correctly, using at least two criteria, neither of which must be the patient's room number	A
Use only one set of tubes intended for one patient at a time	
Always collect only one patient at a time	
Check the patient's physical condition	
If the patient is not in a suitable condition for sampling, this must inevitably be deferred to another date	
Check the prescription, verifying that the number and type of tests coincide with those accepted	
Preferential sampling sites (in descending order): central veins of the forearm (cubital and cephalic), basilica vein, veins in the back of the arm, veins in the wrist and hand. The veins of the feet are the last resort	
Avoid blood sampling from:	
Extensive scars due to burns or surgery, ipsilateral arm resulting from mastectomy (test results may be altered due to the presence of lymphedema), sites adjacent to hematomas, thrombi, or edema	
Devices for intravenous (IV) therapy and/or blood transfusions	
When sampling from infusion sites, flow into the device should be stopped for at least 2 minutes and no less than 5 mL of blood should be eliminated	
To help the vein swell, you can:	
Briefly warm the sampling site with a warm cloth	B
Massage the site in the opposite direction to the venous flow	
Briefly warm the collection site with warm water	C
Hit the site	D
Do not apply the lace in the presence of:	A
Large, visible, and palpable veins	
Sampling for the determination of the venous pH	
If the lace is instead necessary:	
Place it about 10 cm above the chosen site	
Use sufficient pressure to generate venous stasis but not to cause pain, discomfort, or obstruct arterial circulation	
Do not keep it in place for more than 1 minute	
When more time is needed, release it and reapply it	
Sampling rules	
Wear gloves during collection	B
Use primary tubes with labels indicating the type of tube required and the volume of sample required	A
Label tubes before collection, never after	
Use automatic label production systems	
Use automatic tube labeling	B
Cleanse the skin with a cotton ball soaked in an appropriate product, always proceeding in the same direction and then dry the skin	A
Use a specific sequence for collecting tubes (order of draw)	B
For tube intended for coagulation tests, it is not necessary to collect and discard a previous tube	
Verify that the amount of blood aspirated from the primary tube is appropriate	A
Gently invert tubes containing anticoagulant four to six times	
Never open vacuum tubes, or transfer blood from one tube to another (except for the use of syringes for sampling)	
If there are errors, check for further specimen collection or contact the laboratory for clarification	
Release the lace before extracting the needle from the vein, immediately place a cotton ball on the sampling site, asking the patient to apply moderate pressure on it, keeping the arm extended	
Rules to be followed at the end of the collection	
Eliminate the contaminated material in special safety containers suitable for the recognition of the type of material	A
Do not re-cap, break, or crush the needle used directly	
Check the patient's state of health and the onset of any complications	
Other general rules	
Always observe an attitude of availability and courtesy	A
Avoid bothering with the needle inside the sampling site	
In case of failure on the first attempt:	
Carefully move the needle forward or backward	
Replace the tube	
Remove the needle and try again if the result is still negative	
Transfer the patient to a colleague after two failed attempts	

A, *B*, *C*, *D*, *E* strength of the recommendations, in accordance with the indications of the Istituto Superiore di Sanità (ISS), *A* strongly recommended, *B* registered mail, *C* uncertainty for or against the recommendation, *D* not recommended, *E* strongly not recommended

Recommendations for Sample Acceptability

Recognition and management of unsuitable specimens is an essential activity to counteract the clinical risk associated with pre-analytical errors. The study group on extra-analytical variability of the Italian Society of Clinical Biochemistry has produced recommendations that can help in standardizing behaviors and improving the quality of laboratory diagnostics. The key elements of these activities are the education, training, and empowerment of staff, the adoption of objective and standardized systems for the detection of non-conformities related to unsuitable samples, the introduction of systematic procedures for the detection and monitoring of non-conformities, as well as the management of these non-conformities, codified in precise operating procedures.

Transport Problems

The correct transport of biological samples guarantees the quality of the samples. Transport issues are related to time, mode and temperature. In general, serum or plasma samples should be centrifuged and separated within 2 hours after collection. However, some samples must be transported immediately after collection (e.g., arterial blood gas analysis or samples for ammonium determination). In some cases, agitation of the sample during transport may cause significant variations and, for example, transport by airmail may be discouraged. Transport should be done at an appropriate and controlled temperature, depending on the test required. Some samples should be protected from light, such as bilirubin. This pre-analytical phase needs urgent improvement initiatives, also because of the increasing trend toward the consolidation of laboratory facilities, with the consequent need for longer and longer transports.

Pre-analytical Processing Problems in the Laboratory

It would be a mistake to assume that all pre-analytical issues are outside the laboratory. Before analysis, the sample is also treated in the laboratory to prepare it for analysis or to preserve it. For instance, many biochemical parameters may be centrifuged indifferently with different forces, for example, at $2200 \times g$ for 10 minutes, or 5 minutes at $3000 \times g$, without major differences, but some constituents, like the enzyme lactate dehydrogenase, are more "sensitive" and need more attention. For example, the time elapsed between collection and centrifugation, as well as the storage temperature before centrifugation, has an influence on the measurement of adrenocorticotropic hormone (ACTH). Even the most sophisticated proteomic analysis techniques have recently been evaluated for this type of possible pre-analytical error. The study of the effects of pre-analytical manipulation on samples and measurements, but above all, the definition of standard procedures that always recreate the same operating conditions, as well as the systematic detection of alterations and non-conformities, are the countermeasures to counter these problems and their associated risk.

Pre-analytical Quality Indicators

The use of reliable pre-analytical quality indicators is crucial for the identification, evaluation, and monitoring of corrective actions for this major problem in laboratory analysis, as well as for the evaluation of the quality of services. The current lack of attention to extra-analytical problems with respect to analytical quality is in stark contrast to the amount of available evidence and especially to the multitude of errors that continue to occur in all parts of the world. Standard 15189:2012 defines the pre-analytical phase and recognizes the need to govern, monitor, and improve it. As for the analytical phase, where control is continuous, rigorous, and documented, the extra-analytical phases should also have the same attention in the interest of laboratory quality and, above all, patient care. The IFCC WG-LEPS has developed a model of reliable quality indicators that cover all aspects of this chapter, from patient identification to request for testing, specimen collection, transport, and laboratory acceptability. This is an essential step to ensure quality evidence in all procedures and processes of biological specimen analysis in order to reduce the risk of errors in clinical practice.

Recommended Readings

Adcock Funk DM, Lippi G, Favaloro EJ (2012) Quality standards for sample processing, transportation, and storage in hemostasis testing. Semin Thromb Hemost 38:576–585

Alexander S (1984) Physiologic and biochemical effects of exercise. Clin Biochem 17:126–131

Banfi G, Colombini A, Lombardi G et al (2012) Metabolic markers in sports medicine. Adv Clin Chem 56:1–54

Baron JM, Dighe AS (2014) The role of informatics and decision support in utilization management. Clin Chim Acta 427:196–201

Bjerner J, Høgetveit A, Wold Akselberg K et al (2008) Reference intervals for carcinoembryonic antigen (CEA), CA125, MUC1, Alfa-foeto-protein (AFP), neuron-specific enolase (NSE) and CA19.9 from the NORIP study. Scand J Clin Lab Invest 68:703–713

Carraro P, Plebani M (2007) Errors in a stat laboratory: types and frequencies 10 years later. Clin Chem 53:1338–1342

CLSI (2007) Procedures for the collection of diagnostic blood specimens by venipuncture; approved standard. CLSI document H3-A6, 6th edn. Clinical and Laboratory Standards Institute, Wayne

CLSI (2012) Hemolysis, icterus, and lipemia/turbidity indices as indicators of interference in clinical laboratory analysis; approved guideline. CLSI document C56-A. Clinical and Laboratory Standards Institute, Wayne

Cornes M, van Dongen-Lases E, Grankvist K, Working Group for Preanalytical Phase (WG-PRE), European Federation of Clinical

Chemistry and Laboratory Medicine (EFLM) et al (2017) Order of blood draw: opinion paper by the European Federation for Clinical Chemistry and Laboratory Medicine (EFLM) Working Group for the Preanalytical Phase (WG-PRE). Clin Chem Lab Med 55:27–31

Felding P, Tryding N, Hyltoft Petersen P et al (1980) Effects of posture on concentrations of blood constituents in healthy adults: practical application of blood specimen collection procedures recommended by the Scandinavian Committee on Reference Values. Scand J Clin Lab Invest 40:615–621

Fraser GF (2001) Biological variation: from principles to practice. AACC Press, Washington, DC

Giavarina D, Mezzena G, Dorizzi RM et al (2001) Reference interval of D-dimer in pregnant women. Clin Biochem 34:331–333

Guder WG, Narayanan S, Wisser H, Zawta B (2003) Samples: from the patient to the laboratory. The impact of preanalytical variables on the quality of laboratory results. Wiley-VCH Verlag GmbH & Co. KGaA, Weinheim

Halm MA, Gleaves M (2009) Obtaining blood samples from peripheral intravenous catheters: best practice? Am J Crit Care 18:474–478

Hassis ME, Niles RK, Braten MN et al (2015) Evaluating the effects of preanalytical variables on the stability of the human plasma proteome. Anal Biochem 478:14–22

https://www.ecri.org/Pages/Patient-Identification-Deep-Dive.aspx. Last visit August 2023

https://specimencare.com/resources-category/factors-affecting-hemolysis/. Last visit August 2023

Hyltoft Petersen P, Felding P, Hørder M et al (1980) Effects of posture on concentrations of serum proteins in healthy adults. Dependence on the molecular size of proteins. Scand J Clin Lab Invest 40:623–628

ISO 15189: 2012 Medical laboratories – requirements for quality and competence. http://www.iso.org/iso/catalogue_detail?csnumber=56115. Last visit August 2023

Keyzer JM, Hoffmann JJ, Ringoir L et al (2014) Age- and gender-specific brain natriuretic peptide (BNP) reference ranges in primary care. Clin Chem Lab Med 52:1341–1346

Lippi G, Salvagno GL, Montagnana M et al (2005a) Influence of short-term venous stasis on clinical chemistry testing. Clin Chem Lab Med 43:869–875

Lippi G, Salvagno GL, Montagnana M et al (2005b) Shortterm venous stasis influences routine coagulation testing. Blood Coagul Fibrinolysis 16:453–458

Lippi G, Salvagno GL et al (2006) The influence of the tourniquet time on hematological testing for antidoping purposes. Int J Sports Med 27:359–362

Lippi G, Banfi G, Buttarello M et al (2007) Recommendations for detection and management of unsuitable samples in clinical laboratories. Clin Chem Lab Med 45:728–736

Lippi G, Blanckaert N, Bonini P et al (2008a) Haemolysis: an overview of the leading cause of unsuitable specimens in clinical laboratories. Clin Chem Lab Med 46:764–772

Lippi G, Caputo M, Banfi G et al (2008b) Raccomandazioni per il prelievo di sangue venoso. Biochim Clin 32:569–577

Lippi G, Cervellin G, Mattiuzzi C (2013) Critical review and meta-analysis of spurious hemolysis in blood samples collected from intravenous catheters. Biochem Med (Zagreb) 23:193–200

Lippi G, Giavarina D, Gelati M et al (2014) Reference range of hemolysis index in serum and lithium-heparin plasma measured with two analytical platforms in a population of unselected outpatients. Clin Chim Acta 429:143–146

Lippi G, Brambilla M, Bonelli P et al (2015) Effectiveness of a computerized alert system based on re-testing intervals for limiting the inappropriateness of laboratory test requests. Clin Biochem 48:1174–1176

Lundberg GD (1981) Acting on significant laboratory results. JAMA 245:1762–1765

Miller M, Bachorik PS, Cloey TA (1992) Normal variation of plasma lipoproteins: postural effects on plasma concentrations of lipids, lipoproteins, and apolipoproteins. Clin Chem 38:569–574

Møller MF, Søndergaard TR, Kristensen HT et al (2017) Evaluation of a reduced centrifugation time and higher centrifugal force on various general chemistry and immunochemistry analytes in plasma and serum. Ann Clin Biochem 54:593–600

Müller EE, Locatelli V, Cocchi D (1999) Neuroendocrine control of growth hormone secretion. Physiol Rev 79:511–607

Nikolac N (2014) Lipemia: causes, interference mechanisms, detection and management. Biochem Med (Zagreb) 24:57–67

Olivieri F, Galeazzi R, Giavarina D et al (2012) Aged-related increase of high sensitive Troponin T and its implication in acute myocardial infarction diagnosis of elderly patients. Mech Ageing Dev 133:300–305

Ortells-Abuye N, Busquets-Puigdevall T, Díaz-Bergara M et al (2014) A cross-sectional study to compare two blood collection methods: direct venous puncture and peripheral venous catheter. BMJ Open 4:e004250

Peter J, Patole S, Fleming J et al (2011) Agreement between paired blood gas values in samples transported either by a pneumatic system or by human courier. Clin Chem Lab Med 49:1303–1309

Plebani M (2009) Exploring the iceberg of errors in laboratory medicine. Clin Chim Acta 404:16–23

Plebani M (2012) Quality indicators to detect pre-analytical errors in laboratory testing. Clin Biochem Rev 33:85–88

Plebani M, Carraro P (1997) Mistakes in a stat laboratory: types and frequency. Clin Chem 43:1348–1351

Plebani M, Sciacovelli L, Marinova M et al (2013) Quality indicators in laboratory medicine: a fundamental tool for quality and patient safety. Clin Biochem 46:1170–1174

Plebani M, Sciacovelli L, Aita A et al (2014) Quality indicators to detect pre-analytical errors in laboratory testing. Clin Chim Acta 432:44–48

Robertson D, FroÃàlich JC, Carr RK et al (1978) Effects of caffeine on plasma renin activity, catecholamines and blood pressure. N Engl J Med 298:181–186

Salvagno G, Lima-Oliveira G, Brocco G et al (2013) The order of draw: myth or science? Clin Chem Lab Med 51:2281–2285

Saracevic A, Nikolac N, Simundic AM (2014) The evaluation and comparison of consecutive high speed centrifugation and LipoClear® reagent for lipemia removal. Clin Biochem 47:309–314

Schouten HJ, Geersing GJ, Koek HL et al (2013) Diagnostic accuracy of conventional or age adjusted D-dimer cut-off values in older patients with suspected venous thromboembolism: systematic review and meta-analysis. BMJ 346:f2492

Steyn FJ, Tolle V, Chen C et al (2016) Neuroendocrine regulation of growth hormone secretion. Compr Physiol 6:687–735

Taghizadeganzadeh M, Yazdankhahfard M, Farzaneh M et al (2015) Blood samples of peripheral venous catheter or the usual way: do infusion fluid alters the biochemical test results? Glob J Health Sci 8:93–99

Weitzman ED, Fukushima D, Nogeire C et al (1971) Twenty-four hour pattern of the episodic secretion of cortisol in normal subjects. J Clin Endocrinol Metab 33(1):14–22

Wu ZQ, Xu HG (2017) Preanalytical stability of adrenocorticotropic hormone depends on both time to centrifugation and temperature. J Clin Lab Anal 31:e22081. https://doi.org/10.1002/jcla.22081

Young DS (2007) Effects of preanalytical variables on clinical laboratory tests, 3rd edn. AACC Press, Washington, DC

Zaninotto M, Tasinato A, Padoan A et al (2012) An integrated system for monitoring the quality of sample transportation. Clin Biochem 45:688–690

The Quality of Laboratory Results: Sources of Variability, Methods of Evaluation, and Estimation of Their Clinical Impact

7

Ferruccio Ceriotti and Mauro Panteghini

Introduction

This chapter examines the sources of result variability, how to define analytical performance specifications (APS), and their main uses. In particular, the main sources of analytical error are described, and measurement error and the concepts of imprecision, trueness, accuracy, and measurement uncertainty are discussed. Intraindividual and interindividual biological variabilities are described, with some hints about reference intervals and the concept of reference change value. APS, their meaning and use in the medical laboratory are then discussed. APS can be obtained based on the effect of the analytical error on the clinical outcome of the patient, on the biological variability of the measurand, or on the state of the art of the measurement.

Sources of Result Variability

The quantitative results (continuous variables) provided by laboratory measurements are affected by two main sources of variability: analytical variability, related to the way in which the measurement is performed, and biological variability, related to the physiological fluctuation of the concentrations of various components in body fluids.

Main Sources of Analytical Variability and Their Definition

The laboratory results can be influenced by aspects related to the preanalytical phase, which, if not properly controlled, can introduce bias, even extremely significant bias, in the results.

As far as the strictly analytical aspects are concerned, the sources of variability can be schematically classified as follows:

- Reagents: variability in how they are prepared (if not ready to use), improper storage or aging (pH change, degradation of components, loss of catalytic activity of enzymes), and variability between production lots.
- Calibrators/calibration: variability between batches of calibrators, how to store and possibly reconstitute the calibrators, and how to perform the calibration.
- Instrumentation used for measurements: volume measurements (pipettes, dispensers), temperature measurements (incubators, thermostatic systems), absorbance measurements, and mixing and washing system efficiency.
- Operators: inadequate handling of reagents, instruments, calibrators, or biological samples; the way of performing manual analysis.

The measurement errors that the variables listed above can introduce are of two types: random and systematic.

Random Error

Random error derives from the set of variables that can affect the measurement result and can be modified in a random way, both positively and negatively (e.g., volume measurement, temperature control, measurement of light intensity signal, electrical signal, radioactive emission, etc.). It can be reduced but never eliminated. The characteristic that expresses the entity of the random error is precision, defined by the International Vocabulary of Metrology as "closeness of agreement between **indications** or **measured quantity values** obtained by replicate **measurements** on the same or similar objects under specified conditions". "Specified

F. Ceriotti (✉)
IRCCS Ca' Granda – Ospedale Maggiore Policlinico, Milan, Italy
e-mail: ferruccio.ceriotti@policlinico.mi.it

M. Panteghini
Research Centre for Metrological Traceability in Laboratory Medicine (CIRME), University of Milan, Milan, Italy

M. Ciaccio (ed.), *Clinical and Laboratory Medicine Textbook*, https://doi.org/10.1007/978-3-031-24958-7_7

conditions" refers to the different ways in which precision can be estimated. These conditions are indicated as *repeatability* when we speak of the same measurement procedure, performed by the same operator with the same operating conditions, on the same or similar samples within a short interval of time. On the other hand, *intermediate precision* refers to repeated measurements extended over a wide interval of time (days or weeks) and may involve different operators, various calibrations, and the use of different lots of reagents. In this case, we also speak of *reproducibility*. Precision is a qualitative characteristic that can be defined as "good," "poor," etc. What can be quantified in numerical terms is the level of *imprecision*, which is expressed in terms of standard deviation (SD):

$$\mathrm{SD} = \sqrt{\frac{\sum_{i=1}^{n}\left(x_i - \bar{x}\right)^2}{n-1}}$$

or in terms of coefficient of variation (CV):

$$\mathrm{CV} = \frac{\mathrm{SD}}{\bar{x}} \times 100$$

Systematic Error

A systematic error is an error that remains constant or varies in a predictable way in repeated measurements. For example, it may result from a pipette that always dispenses a larger volume of sample than it should, a lower reaction temperature, an error in the value assigned to the calibrator, or the use of a different lot of reagents. The defining characteristic of systematic error is *trueness*, defined as "closeness of agreement between the average of an infinite number of replicated measured quantity values and a reference quantity value". Trueness is a qualitative characteristic, like precision, and for the quantification of systematic error, *bias* is used, which can be expressed in absolute value or in percentage as: bias = (measured value − reference value)/reference value × 100, where "measured value" is the average of several measurements of the same sample.

Unlike random error, systematic error can theoretically be eliminated. According to the above definitions, repeated measurements are required to identify the presence of random or systematic errors. When a single measurement is performed, the level of agreement between the measured value and the reference value is defined as *accuracy*. Good accuracy can only be obtained when both random and systematic errors are well controlled and result in as small as possible measurement uncertainity. *Measurement uncertainty* is defined as "a nonnegative parameter characterizing the dispersion of the quantity values being attributed to a measurand, based on the information used." It is based on the concept that the true value is not knowable, but it is possible to define, based on the measurement obtained, a range within which, with a defined probability, the "true" (unknown) value is contained. In other words, the uncertainty range that characterizes the measurement defines the range in which, with reasonable certainty (usually with a probability of 95%), the value falls. Accordingly, the result of any measurement should not be considered complete if its uncertainty is missing. The concepts expressed above are schematized in Table 7.1.

Table 7.1 Classification of the measurement error

Type of error	Feature of the clinical presentation	Expression
Random error	Precision	Standard deviation
Systematic error	Trueness	Bias
Measurement error	Accuracy	Uncertainty

The ISO 15189:2022 "Medical Laboratories. Requirements for Quality and Competence" at paragraph 7.3.4 states that the laboratory shall evaluate and regularly review the measurement uncertainty of measured quantity values, where relevant.

The measurement uncertainty concept assumes to eliminate (by an appropriate correction) the bias and to combine the variability due to the random error (u_{imp}) and that related to the elimination of the bias (u_{bias}) according to the error propagation formula:

$$u_{\mathrm{c}} = \sqrt{\left(u_{\mathrm{imp}}\right)^2 + \left(u_{\mathrm{bias}}\right)^2}$$

where u_{c} is defined as *combined standard uncertainty*. To pass to the so-called *extended combined measurement uncertainty* (U_{c}), which allows the definition of a 95% probability range, it is necessary to multiply this value by a "coverage factor" k. To obtain 95% coverage, the factor k is 1.96, but it is usually rounded to 2.

Information about the inaccuracy and measurement uncertainty of the laboratory methods can be obtained through internal quality control (IQC) programs and external quality assessment (EQA) schemes.

Biological Variation

The biological variation represents the physiological variability of the concentrations of a measurand in the biological fluid of interest, for example, serum, plasma, whole blood, urine, or others. Two types of biological variation are recognized: intraindividual (concentration variations within the same individual) and interindividual (dispersion of average concentration values among different individuals belonging to a population with defined characteristics).

Intraindividual Biological Variation (CV_I)

CV_I is the random fluctuation of a constituent of the organism, measured at different times in the same individual, around its homeostatic point. The extent of this variation depends on the conditions that underlie the control of the concentrations of a given analyte in the biological fluid of interest. If the analyte levels are critical for the proper functioning of the organism as a whole (e.g., sodium, whose variations in plasma lead to a shift of water from the intracellular to the extracellular compartment, with potential problems in the central nervous system), the homeostatic control will be very tight, and the concentrations of the analyte will fluctuate within a narrow range. Conversely, if the measurand level does not have specific functions in the evaluated biological fluid (e.g., the enzymes indicating cytolysis) or its variations in the fluid are related to tissue needs or dietary intake (e.g., the plasma concentrations of iron or triglycerides), the intraindividual variation will be higher. The CV_I is therefore characteristic of a specific measurand (analyte in a specific biological fluid), although there may be significant differences between individuals.

Interindividual Biological Variation (CV_G)

CV_G is defined as the difference in the average results of the same constituent obtained in different individuals, all under the same physiological conditions, due to the diversity of their homeostatic points. It represents the variability due to the different characteristics of the individuals belonging to a specific population. The extent of this variability is represented by the width of the range of values that a given measurand can assume in a population of appropriately selected subjects with criteria similar to those employed for defining a "reference population".

An example of the two types of biological variation is presented in Fig. 7.1, showing serum creatinine values obtained in 90 subjects without renal dysfunction, whose blood samples were drawn once a week for 10 consecutive weeks. It is noteworthy that, while the CV_I in different individuals is narrow, medians of the various subject results are dispersed over a wide range, with different values between males and females. The case of creatinine, presented in Fig. 7.1, is typical for a measurand that has sex-specific differences and a very low ratio between CV_I and CV_G, that is, an individuality index [II] equal to ~0.3. In this situation, when the CV_I is lower than the CV_G, the population-based reference intervals lose their sensitivity in identifying individuals who may already present variations indicating possible renal dysfunction. In Fig. 7.2, the individual with the lower mean creatinine (~65 µmol/L) will almost have to double his/her creatinine value to overcome the upper limit of the reference interval (104 µmol/L), which for this individual may already imply a relevant reduction of renal function.

Knowing the biological variation of the various measurands is important because, in addition to being one of the models for the definition of APS (see below), it allows one to estimate the reference change value (RCV), that is, the reference to statistically evaluate the significance of the variation between two consecutive measurements on the same individual beyond the total variation of the measurement. The RCV estimate allows one to establish if two consecutive measurements of the same measurand in the same subject can be considered different from the statistical point of view (with a given level of probability) from the variation only due to analytical and biological sources.

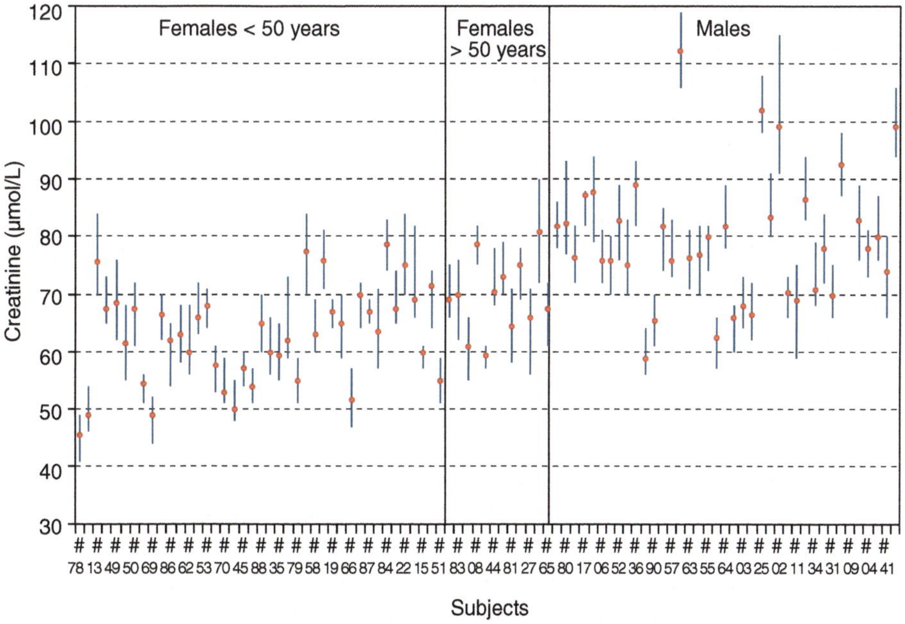

Fig. 7.1 Biological variability of creatinine in serum of 90 subjects with physiological renal function, subjected to sampling once a week for 10 consecutive weeks. The data are divided by gender and, within the same group, sorted by increasing age (from 20 to 69 years for women and from 20 to 59 years for men). The horizontal line indicates the median, and the vertical line indicates the range of dispersion of the values (maximum-minimum) of each subject

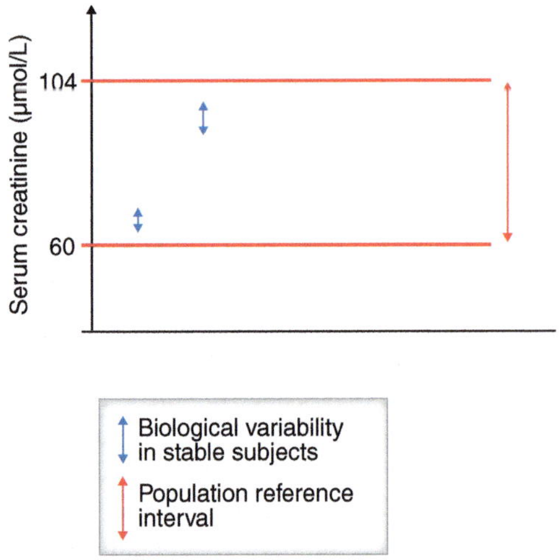

Fig. 7.2 Example of the impact of biological variability on the interpretation of the creatinine level. The intraindividual biological variability is much lower than the population reference range (in other words, the interindividual biological variability). Consequently, variations in the serum creatinine concentration in a subject, evaluated by the estimated critical difference (~12%), can signal a change in glomerular filtration even when the results are still within the reference range of the population

The RCV is calculated as follows:

$$\text{RCV} = 2.77 \times \sqrt{\text{CV}_\text{A}^2 + \text{CV}_\text{I}^2}$$

In the formula, the total variation (which derives from the square root of the sum of the analytical variance and the intra-individual biological variance) is multiplied by the factor 2.77, which is obtained from 1.96 (the statistical factor to estimate the 95% probability) multiplied $\sqrt{2}$ (because two measures are compared). From the formula, it appears that the smaller the analytical variation, the smaller the RCV is. It is also clear that if the intraindividual biological variation is high, the RCV value will be larger and, therefore, it will be more difficult to assess the actual clinical significance of a result variation. A limitation of the use of RCV is that the available biological variation data are usually obtained as a mean of a group of individuals, but the biological variation of the individual on which the measurements were performed and the RCV employed could be significantly different than the average of the data used to derive RCV.

Definition of APS

APS represent the level of analytical quality required for a given measurement to be suitable for clinical use. ISO 15189:2022, in paragraph 7.3.4, asks the medical laboratory to compare and document the measurement uncertainty of their measures against performance specifications. Therefore, the correct definition of the APS for each measurand is extremely important for several reasons: it can be the reference for the development of new analytical methods and technologies, for their evaluation, and for the assessment of the results obtained in IQC programs and in EQA schemes.

The Conference held in Milan in 2014 defined three models to be used in defining APS:

- A model based on the impact of analytical performance on clinical outcome
- A model based on biological variation of the measurand
- A model based on the state-of-the-art of the measurement

Model Based on the Impact on Clinical Outcome

This model should be based on defining the maximum permissible error of measurements that cannot affect the interpretation and, therefore, the clinical use of the laboratory data. It should be theoretically applied in all cases where there are defined decision thresholds, such as, for instance, in the case of serum cholesterol or blood glycated hemoglobin. The problem with this type of model is that studies to obtain the estimation of APS should allow the effect of measurement error to be quantified objectively. These types of studies are difficult to perform and, even when they employ simulations, are rarely available.

Model Based on Biological Variation of the Measurand

This model is based on the concept that the acceptable "noise" caused by analytical variation should depend on the biological variation of the analyte in question. As mentioned above, each measurement carried out in the medical laboratory is influenced by these two types of variability. To calculate their combined effect, it is necessary to obtain the square root of the sum of the two sources of variance according to the following formula:

$$\text{CV}_\text{T} = \sqrt{\text{CV}_\text{A}^2 + \text{CV}_\text{I}^2}$$

where CV_T is the total variation affecting the laboratory result. Considering that biological variation is an incompressible factor, the contribution of analytical variation should be minimized. Starting from the above formula, it is possible to calculate the effect of analytical variation on the total variation of the result: it can be easily verified that if $\text{CV}_\text{A} = 0.5\ \text{CV}_\text{I}$, CV_T increases by 11.8%; if $\text{CV}_\text{A} = 0.25\ \text{CV}_\text{I}$, CV_T increases by only 3%; and if $\text{CV}_\text{A} = 0.75\ \text{CV}_\text{I}$, CV_T

Table 7.2 Analytical quality levels and their effect on total variability

Quality level	*K* [$CV_A \leq K\ CV_I$]	Increase of CV_T
Optimal	0.25	+3%
Desirable	0.50	+12%
Minimum	0.75	+25%

CV coefficient of variation

increases by 25%. These ratios between CV_A and CV_I define three quality levels of APS, classified as "minimum," "desirable," and "optimal," respectively (Table 7.2).

The concept behind this approach is that if the CV_I is high, even the CV_T will be high, so it is not necessary to have a very low CV_A. On the contrary, if the CV_I of the measurand is low, the CV_A has to be very low to minimize its influence on the CV_T; otherwise, its contribution to increasing the total variation of the result will be such that it could potentially invalidate its correct clinical interpretation. Data about the biological variation of many analytes are available in the literature, and it is, therefore, possible to calculate APS based on this principle for almost all the most frequently performed tests. However, it must be kept in mind that for the definition of APS based on biological variation, it is necessary that the measurand in question have a certain level of stability in the biological fluid in which it is measured. For example, this does not happen for urinary components for which APS derived from biology are not relevant.

Using biological variation, it is also possible to define APS as the maximum acceptable bias. As mentioned above, when a nonnegligible bias is present, it is appropriate to eliminate it. However, in any case, the definition of negligible bias is useful to be able to define a maximum acceptable limit for it. The principle for its definition is based on the comparability between the obtained values and the reference interval adopted. Considering the value distribution of the reference population and the theory of the reference values that foresees considering the central 95% of the distribution in defining the reference limits, taking off the two tails (2.5% below and above the limits), a systematic deviation that leads to wrongly reclassifying a limited percentage of subjects is considered acceptable. Based on this principle, the following formula has been proposed:

$$\text{bias\%} \leq 0.25\sqrt{CV_I^2 + CV_G^2}$$

Model Based on State-of-the-Art of the Measurement

This model is not related to clinical needs or biological characteristics but depends only on the quality of available technology. However, it is very important to note that the definition of "*state-of-the-art*" corresponds to the highest level of quality obtainable at that given time and not, for instance, to the average of the performance provided by a group of laboratories. This model could lead to clinically inadequate APS if the quality of available technology is poor, or to stringent APS that is unnecessary for the clinical application of the test. However, when other models are not applicable or adequate data are not available, the quality achievable by, for example, the top 20% of laboratories participating in an EQA program may represent a good goal for all other participants.

Recommended Readings

Braga F, Panteghini M (2016) Generation of data on within-subject biological variation in laboratory medicine: an update. Crit Rev Clin Lab Sci 53:313–325

Braga F, Panteghini M (2020) The utility of measurement uncertainty in medical laboratories. Clin Chem Lab Med 58:1407–1413

Braga F, Panteghini M (2021) Performance specifications for measurement uncertainty of common biochemical measurands according to Milan models. Clin Chem Lab Med 59:1362–1368

Ceriotti F (2014) The role of external quality assessment schemes in monitoring and improving the standardization process. Clin Chim Acta 432:77–81

Ceriotti F, Fernandez-Calle P, Klee GG et al (2017) EFLM Task and Finish Group on Allocation of laboratory tests to different models for performance specifications (TFG-DM). Criteria for assigning laboratory measurands to models for analytical performance specifications defined in the 1st EFLM Strategic Conference. Clin Chem Lab Med 55:189–194

EFLM biological variation database. https://biologicalvariation.eu/

Panteghini M, Ceriotti F, Jones G, Oosterhuis W, Plebani M, Sandberg S, Task Force on Performance Specifications in Laboratory Medicine of the European Federation of Clinical Chemistry and Laboratory Medicine (EFLM) (2017) Strategies to define performance specifications in laboratory medicine: 3 years on from the Milan Strategic Conference. Clin Chem Lab Med 55:1849–1856

Sandberg S, Fraser CG, Horvath AR et al (2015) Defining analytical performance specifications: consensus statement from the 1st strategic conference of the European Federation of Clinical Chemistry and Laboratory Medicine. Clin Chem Lab Med 53:833–835

Principles of Immunochemistry

8

Antonio Fortunato

Introduction

The term immunochemistry identifies analytical methods characterized by high sensitivity and specificity. Identifying of the substance to be determined exploits the specific binding between molecules with complementary three-dimensional structures. These techniques are called ligand assay, and in most cases, they are based on the use of an antibody (Ab) to identify the analyte, therefore considered as antigen (Ag). The quantification of the analyte occurs by a tracer that, under certain conditions, allows detecting a signal proportional to the number of complexes between Ag and Ab formed in the reaction.

In addition to these characteristics, an immunochemical determination is based on several analytical steps:

- Incubation for the formation of Ag-Ab complexes
- Separation of Ag-Ab complexes from unbound Ag and Ab molecules
- Detection of the tracer analytical signal
- Data processing and extrapolation of analyte concentrations

The main classification of immunological methods, based on the architecture of the determination, distinguishes them into:

- Competitive (or in default of Ab/excess Ag)
- Non-competitive (or in excess of Ab/Ag)

In competitive methods, only one Ab identifies the Ag, and the analytical signal is inversely proportional to its concentration. In contrast, in non-competitive methods, the reaction involves two Abs, and the signal is directly proportional to the Ag concentration.

A. Fortunato (✉)
Clinical Pathology Laboratory, Ascoli Piceno, Italy
e-mail: antonio.fortunato@sanita.marche.it

The use of immunochemical methods began in the 1950s with the research of Rosalyn Yalow, who, in collaboration with Solomon Berson, developed the detection of insulin. Roger Ekins made a significant contribution to the development of methods for measuring molecules of various structures, to the introduction of non-competitive methods and systems on microchips.

The Immunochemical Assay

The fundamental components of the immunochemical assay, in addition to the analyte (Ag) to be determined, are the binding molecule (Ab), which allows the recognition of the Ag, and the tracer, which emits the analytical signal as a function of the bond between Ag and Ab.

The Analyte

In the immunochemical assay, the analytes are Ags toward which specific Abs are produced. Ags are generally macromolecules (proteins, glycoproteins, or polysaccharides) that can express many different epitopes. An antigenic determinant, or epitope, is defined the Ag region that is specifically recognized by the Ab. Ags capable of eliciting an immune response are called immunogens. Low-molecular-weight molecules (less than 2 kDa) can be antigenic, i.e., able to be recognized by an Ab, but not immunogenic, and are called haptens. To produce Ab directed toward a hapten, the latter must be conjugated to a carrier, consisting of a protein or a glucidic group, which makes it immunogenic. Immunochemical methods can be used to measure any analyte that can be recognized and bound by another molecule, so the Ab itself can be the Ag that is bound by an anti-Ab Ab (this is the case of the use of immunochemical methods to search for Ab produced during some viral infections or autoimmune diseases).

M. Ciaccio (ed.), *Clinical and Laboratory Medicine Textbook*, https://doi.org/10.1007/978-3-031-24958-7_8

The Binding Molecule

The binding molecules used in almost all immunochemical methods are Ab; only in some cases other molecules are used (e.g., receptors). The fundamental characteristics of an Ab are affinity (measured in chemical–physical terms by the equilibrium constant K of the reaction in which it binds the Ag) and specificity (ability to recognize the analyte, distinguishing it from molecules with a similar structure). The Abs most used for developing immunochemical methods are immunoglobulin G (IgG).

Ab used in immunochemical determinations can be either polyclonal or monoclonal. The production of polyclonal Ab can occur by immunization of an animal with an Ag from a different animal species. In this case, the Abs, produced by different clones of plasma cells, are heterogeneous and are directed against different Ag epitopes; their affinity for the Ag presents a high variability, but it can recognize different molecular forms of the same Ag. The characteristics of polyclonal produced Ab are specific for the response obtained in the individual animal, so the production of Ab with specificity is related, over time, to the survival of the animal itself.

Monoclonal Abs are produced by a single line of plasma cells. This phenomenon is obtained in vivo in certain myelomas in which tumor degeneration of a plasma cell clone leads to the hyperproduction of monoclonal immunoglobulin. The creation of hybrids between secretory B lymphocytes capable of producing specific Ab and plasmacytoma cells, with immortal characteristics, has made it possible to produce monoclonal Ab in adequate quantities to produce diagnostic reagents, reproducible over time (Table 8.1).

Therefore, the specificity of the immunochemical measurement is strictly dependent on the Ab used as a binding reagent and is defined as the ability of the Ab to recognize the Ag to be determined. The methods that use polyclonal Ab, show a lower specificity than those using monoclonal Ab, which constitutes a homogeneous population of Ab directed toward a single specific epitope. The degree of specificity can be expressed as a percentage of cross reaction relative to the ability of an Ab, typical for a given Ag, to recognize another Ag, called interferer.

Table 8.1 Characteristics of monoclonal antibodies

Advantages	Disadvantages
The characteristics of affinity and specificity are known and constant	Specificity may be excessive
A small amount of Ag is sufficient for production	It requires more sophisticated technologies for production
You can select the Ab with the desired specificity	They do not always react with protein A
Can be produced in unlimited quantities, and the process of purification is simple	They do not form a precipitate following binding with the Ag
They are characterized by high specificity	They are often characterized by a low affinity

The Tracer

The detection of the analytical signal in an immunochemical determination represents a critical step for the sensitivity of these methods used to measure concentrations of substances of the order of nanomoles and picomoles.

The term identifies the indicator of the analytical reaction. The substance to be measured is made detectable by introducing an appropriate marker into the molecule but without modifying its immunological behavior. The portion of Ag or Ab conjugated with the tracer in the immunochemical method is defined labeled.

The evolution of the tracers used has allowed the applicability of the immunochemical technique, even in laboratories not equipped for using radioactive tracers, which were initially the only ones used, and the development of automated instrumentation.

All tracers must possess certain general characteristics:

- Their chemical bond with the Ag or Ab molecule must have the least interference with the Ag-Ab complex.
- The analytical signal emitted shall have a high intensity per unit mass, which depends on the detection limit of the tracer type.
- The specificity of the signal shall be as high as possible in order to ensure a low background noise and, thus, contribute to improved detectability.
- The signal shall not be modified by the characteristics of the surrounding matrix.
- Whenever possible, the measurement should be repeated.
- The tracer must be detected over a wide range of concentrations.

The characteristics of the analytical signal depend on the nature of the signal itself and the instrumentation to measure it.

The tracers can be grouped into one of the following categories:

- Radioactive
- Enzymatic

Table 8.2 Tracers in immunochemical techniques

Tracers	Emitted signal	Instrumentation for the measurement	Mass (moles / tube)
Radiosotope	γ o β radiation emission	γ o β radiation counter	$0,1^{-10} \times 10^{-15}$
Fluorophore	Light radiation emission	Fluorimeter	$10^{-15} - 10^{-18}$
Enzyme	Absorption of light radiation	Photometer	$10^{-15} - 10^{-16}$
Enzyme	Light radiation emission	Luminometer	$10^{-16} - 10^{-18}$
Enzyme	Cyclic amplification	Luminometer	$10^{-20} - 10^{-21}$
Luminogenic molecule	Light radiation emission	Luminometer	$10^{-16} - 10^{-18}$

- Fluorescent
- Chemiluminescent

Each category of tracer is characterized by a different type of emitted signal and, consequently, by the type of detector used for its measurement (Table 8.2).

Some tracers provide an explicit signal (e.g., radioactivity, luminescence) that can be measured directly with the appropriate detection instruments; others carry an implicit signal (e.g., enzymatic activity, fluorescence) that requires an additional step before the measurement.

Radioactive Tracers

Radioactive tracers have long been the only tracers used in immunochemical determinations (isotopic methods).

Most natural elements are mixtures of various isotopes. Some isotopes, called radioactive isotopes, can be unstable and become stable after the emission of particles or electromagnetic radiation through disintegration. The two main radioelements used as markers in radioimmunochemistry are 125iodium (which emits γ radiation) and 3H (tritium) (which emits β radiation). The radioactive signal has the following advantages: it is a direct signal (emitted directly from the marker); it is positive (emission of electromagnetic or particle radiation); it is spontaneous (no need for an external energy source); and it is very specific (spurious radiation or contamination is exceptional). The emission of the signal is independent of the environment, and the activity (intensity of the physical signal) is proportional only to the number of labeled molecules. Signal detection is a simple numerical count and, thus, easily corrected for background noise.

Compared to other tracers, radioactive tracers have a reduced steric hindrance (in particular tritium), which slightly modifies the immunological behavior of the labeled concerning the analyte to be determined. In addition, the radioactive tracer emits a signal that is not destroyed during the measurement, and therefore, the measurement itself can be repeated several times. The limitations to the use of radioactive tracers are related to the regulations that govern the use of radioelements and define the methods of use (authorizations, radioprotective regulations) and, above all, to the difficulty in realizing fully automated instruments.

Enzyme Tracers

Immunochemical assays using an enzyme as a tracer (enzyme immunoassays) have been used since 1971 as an alternative to determinations using radioisotopes. In enzyme immunoassays, the label consists of an Ag, or an Ab, associated with an enzyme. The measurement of the enzyme activity allows for evaluating the amount of free or labeled bound in the immune complex. Quantification is made possible by the addition of the appropriate substrate that triggers the enzymatic reaction. Enzymatic markers do not emit a direct signal: the quantitatively measurable signal is proportional to the catalytic activity of the enzyme and, therefore, to concentration. The analytical signal emitted depends on the type of transformation induced on the substrate by the enzyme: it can be either the absorption of light radiation (absorbance) or the emission of light radiation (fluorescence or luminescence).

Absorption Signal

The light radiation absorption signal of an enzymatic reaction can be obtained by different methodologies described below.

Use of a Chromogenic Substrate

A chromogen is a molecule whose chemical transformation catalyzed by an enzyme (tracer) changes its absorption spectrum.

Example

$$\text{p-nitrophenylphosphate} \rightleftharpoons \text{p-nitrophenol} + \text{Pi}.$$

In this enzymatic reaction, alkaline phosphatase converts the chromogenic sub-layer (p-nitrophenylphosphate) to p-nitrophenol (whose peak light absorption is at a wavelength of 405 nm) and inorganic phosphate (Pi).

Use of Two Substrates, One of Which Is Chromogenic

Example

$$H_2O_2 + \text{o-phenyldiamine} \rightleftarrows \text{0-nitro-aniline} + 2H_2O$$

In this enzymatic reaction, horseradish peroxidase catalyzes the oxidation of o-phenyldiamine (reduced chromogen) to o-nitro-aniline and transfers electrons to hydrogen peroxide, which is reduced to water. The wavelength of the peak light absorption of o-nitro-aniline is at 492 nm.

Use of NAD+ and NADP+ (Cosubstrates That Absorb at 340 nm Only in the Reduced Form)

Example

$$\text{Glucose-6-phosphate} + \text{NADP}+ \rightleftarrows \text{6-phospho-D-glucono-}\delta\text{-lactone} + \text{NADPH} + \text{H}+$$

In this enzymatic reaction, the NADP+ cosubstrate, which does not absorb at 340 nm, is reduced to NADPH + H+, which exhibits an absorption peak at 340 nm.

Emission Signal

The light emission signal of an enzymatic reaction can be obtained by several methodologies described below.

Use of a Fluorogenic Substrate

A fluorogen is a molecule whose transformation gives rise to a fluorescent molecule.

Example

$$\text{4-methyl-umbelliferyl-}\beta\text{ galacto-pyranoside} \rightleftarrows \text{4-methyl-umbelliferone}$$

In this enzyme-mediated reaction, the substrate is hydrolyzed to 4-methyl-umbelliferone, which is fluorescent, whose maximum excitation wavelength is 364 nm, that of the emission peak at 448 nm.

Use of Two Substrates, One of Which Is Luminogenic

A luminogen is a molecule whose chemical transformation produces an emission of light (chemiluminescence reaction).

Example

$$H_2O_2 + \text{o-Phenyldiamine} \rightleftharpoons \text{Amino phthalate ion} + H_2O_2$$

In this reaction mediated by horseradish peroxidase, the luminogen is oxidized to aminophthalate ion, which is in an excited energy state. The spontaneous return of this ion to the fundamental state is associated with light emission (peak emission wavelength at 430 nm). Hydrogen peroxide functions as electron and proton acceptor.

Direct Fluorescent Tracers

Fluorescent tracers (fluorophores) re-emit, in most cases at a longer wavelength and, therefore, lower energy, the electromagnetic waves received after excitation by ultraviolet or visible wavelength radiation. In addition to the qualities common to all tracers (ease of marking, high marking efficiency, stability, low cost, etc.), fluorescent tracers should, if possible, emit an intense and specific signal. The intensity of a signal depends mainly on the power of the excitation source and the quantum yield of the fluorophore. Conventional fluorophores, such as fluorescein, emit an intense but unspecific signal because various interfering phenomena combine to increase the optical background noise in the detection passband. These interfering phenomena include Rayleigh and Raman scattering phenomena and other interfering fluorescence phenomena, such as the fluorescence of measuring cuvettes and the fluorescence of some natural plasma compounds (proteins, bilirubin, NADH, porphyrins, etc.) or reagents. Interfering diffusions increase the optical background noise when the Stokes shift (wavelength interval separating the maximum peaks of the excitation and emission spectra) is limited. Regarding spurious fluorescence, most frequently in blue and green, they contribute to the optical background noise when the fluorophore emits in the same wavelength region. In immunochemical determinations, in order to achieve an adequate limit of detection, a high signal-to-noise ratio is required, i.e., fluorophores that emit an intense, measurable signal with the highest specificity should be used. This measurement specificity can be improved by wavelength and time. When a fluorophore has a high emission wavelength (red), the contribution of interfering fluorescence phenomena is low at the level of the measurement passband. On the other hand, when the Stokes shift is important, scattering photons are not detected.

Direct Chemiluminescent Tracers

Chemiluminescent tracers are molecules whose excitation and subsequent emission of light radiation are achieved by a chemical energy input, usually resulting from the rapid oxidation of the molecule. Chemiluminescent compounds and marking methods that can be used in immunochemical assays must have the following characteristics: the ability to form covalent bonds with the molecule to be labeled; the chemical reaction that leads to the formation of this bond must ensure that the immunoreactivity of the complex formed remains unaltered; the quantum yield of luminescence must not decrease excessively after bonding. Some families of chemiluminescent compounds can be identified: phthalhydrazides, acridinium esters, and dioxyethanes.

Phthalhydrazide

Luminol, isoluminol, and isoluminol substitution derivatives are the most widely used. These molecules are transformed into excited species, which then return to the fundamental state with the emission of photons of energy hυ, in oxidation reactions in the presence of hydrogen peroxide and an enzymatic catalyst (peroxidase) according to the following scheme:

$$\text{Luminol} + H_2O_2 \underset{(\lambda_{max}=430\text{ nm})}{\overset{\text{peroxidase, OH}^-}{\rightarrow}} \text{Aminophthalate ion} + h\upsilon$$

These molecules can be used in direct and indirect chemiluminescence. In direct chemiluminescence, the tracer is the luminogenic substrate; at the end of the assay, the peroxidase is added together with the hydrogen peroxide. In indirect chemiluminescence, the tracer is the peroxidase, and the luminogenic substrate is added later together with hydrogen peroxide.

Acridinium Esters

The addition of hydrogen peroxide in an alkaline environment is sufficient to trigger the chemiluminescent reaction:

$$\text{Acridinium ester} + H_2O_2 \underset{(\lambda_{max}=430\text{ nm})}{\overset{\text{peroxidase, OH}^-}{\rightarrow}} \text{N-methyl-acridone} + h\nu$$

The forming compound (N-methyl-acridone) in the excited singlet state returns to the fundamental state with the emission of light radiation.

Dioxetans

Dioxyethanes are spontaneous chemiluminescent compounds. However, it is possible to stabilize them by conjugating them to a phosphate or galactose group. In this case, the light emission is obtained by the action of the enzyme corresponding to the stabilizing group (ALP or β-galactosidase). The most common substrate in commercially available systems is adamantyl-1,2-dioxyethane-phenyl-phosphate (AMPPD) ester.

Architecture of the Immunochemical Assay

The architecture of the immunochemical assay refers to the mode by which the concentration of analyte bound by the Ab is detected. It relies on the definition of the molecule (Ag or Ab) bound to the tracer, how many types of Ab are used to identify different epitopes of the Ag, and the relationship between the analytical signal measured and the Ag concentration. The sensitivity and specificity of immunochemical assays have greatly improved with the evolution of their architecture, replacing, where possible, the initial competitive method in the homogeneous phase with the non-competitive method in the heterogeneous phase.

Competitive Methods

Competitive methods are defined, in addition to Ab deficiency or Ag excess, also as displacement or saturation methods. The architecture of these methods is based on the competition of the analyte to be measured (Ag), in variable quantity in the sample, and of the labeled (Ag*) added as a reagent, in constant concentration, for the binding sites of the Ab in the reaction environment in constant concentration and in any case lower than the concentration of the same Ag*. Even in the absence of Ag, only a fraction of the total amount of Ag* added to the reaction environment can be bound by the Ab. The analytical signal represents the maximum binding capacity of the system. On the contrary, at increasing concentrations of Ag, at the end of the reaction, the amount of Ag* that will be bound in the Ab-Ag* complex will be lower because of competition.

Thus, the competitive reaction scheme involves the presence of the following components in the reaction environment:

1. The analyte to be analyzed contained in the sample (Ag)
2. The same Ag previously bound with the reaction reporter (labeled: Ag*)
3. An Ab for which both Ag and Ag* are immunologically indifferent, leading to the simultaneous formation of Ag-Ab and Ag*-Ab complexes according to the following reaction pattern:

$$\text{Ag} + \text{Ag}^* + \text{Ab} \leftrightarrow (\text{Ag-Ab}) + (\text{Ag}^*\text{-Ab})$$

If the concentrations of Ab and Ag* remain constant, an increase in the concentration of Ag in the biological sample to be analyzed leads to an increase in the concentration of the Ag-Ab complex at the expense of the formation of the Ag*-Ab complex (Fig. 8.1). If the nature of the tracer requires, before measuring the signal, the separation of the free and bound phases, it is essential to use a method that does not alter the balance of the concentrations of the Ag-Ab and Ag*-Ab complexes that have been formed. The measurement of the analytical signal emitted by the tracer used in the reaction makes it possible to determine the concentration of free Ag* (F*) and of the Ag*-Ab complex (B*) as the concentrations of Ag change. When equilibrium is reached in the formation of the complexes with Ab, the concentrations of F* and B* are present in the same ratio as the concentrations of the free (F) and bound (B) forms of the Ag being determined. For any level of Ag concentration, the following relationship is valid: B/F = B*/F*.

The main advantage of immunometric methods based on the competition principle is that they can be applied to all antigens, regardless of size. Moreover, they are the only method that can be used to determine haptens exposing only one epitope. However, some conditions are necessary for their application. In order to detect low concentrations of Ag, it is essential to use an Ab with a high-affinity constant and a tracer with high specific activity; in order to obtain adequate repeatability of the determinations, the presence of a constant number of Ab-binding sites must be guaranteed. This is

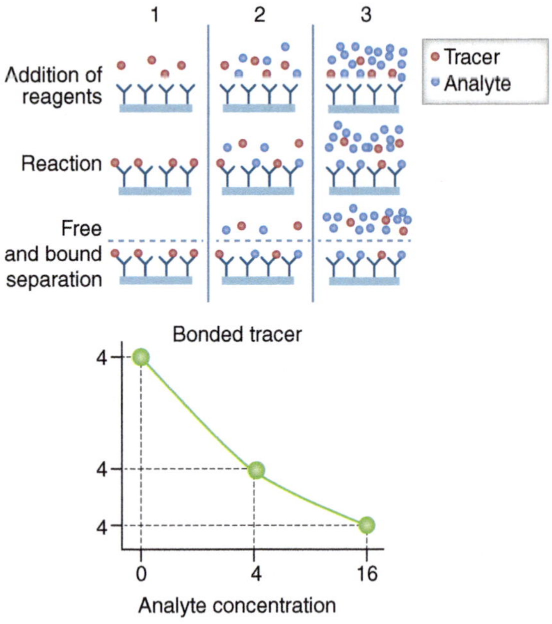

Fig. 8.1 Competitive methods. (Copyright EDISES 2021. Reproduced with permission)

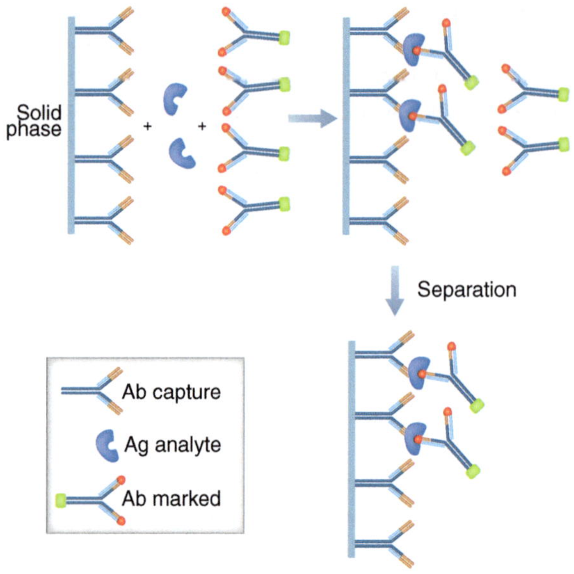

Fig. 8.2 Schematic representation of the non-competitive reaction (two-site immunometric methods or sandwich techniques in excess of Ab) in the heterogeneous phase. (Copyright EDISES 2021. Reproduced with permission)

particularly critical and is related to the fact that the concentration of Ab is much lower than the concentration of Ag + Ag*.

Since in the determination with competitive architecture, the Ag is recognized by the Ab through the binding of only one epitope, Ag fragments or metabolites containing this epitope are also measured as the intact Ag molecule. In this context, the Ag concentration is consequently overestimated, and the specificity of the method is reduced.

Non-competitive Methods

The non-competitive methods, or in excess of Ab or defect of Ag or sandwich, were introduced in the second half of the 1980s and have evolved significantly with the introduction of monoclonal Ab to replace most competitive methods in the determinations of high molecular weight molecules. These methods are characterized by two Abs, present in excess of the concentration of Ag to be measured, which recognize two different non-interfering epitopes of Ag. The second Ab is labeled with the tracer that allows the detection of the analytical signal (Fig. 8.2).

In the structure of these immunometric determinations in the heterogeneous phase, the first Ab (binding Ab or capture Ab) is fixed on solid support (tubes, marbles, different particles). The amount of immobilized Ab must guarantee several sites for Ag binding higher than the number of Ag molecules that may be present in the sample (calibrator or unknown sample). The Ag binds to the specific sites of the first Ab and then the Ab* binds to the same Ag (detection Ab). The second Ab can be added simultaneously with the other reaction components (one-step methods), or after an initial incubation and washing (two-step methods). The name sandwich given to this type of determination derives from the complex formed in the reaction in which the Ag is blocked between the two Abs. In these heterogeneous phase methods, a simple washing procedure separates the Ab-Ag-Ab* complexes from the free Ab* in excess of the Ag. In general, for the Ab* to bind to the Ag, already involved in the reaction with the first Ab, the two Abs must recognize different epitopes (if the Ag molecule exposes the same epitope in a repetitive manner, the same Ab can be used as capture Ab and as tracer Ab*). Capture Ab and tracer Ab* can be polyvalent with different specificity. This is the choice solution. Combinations of monoclonal Ab binding and polyclonal Ab tracing and the association of different monoclonal Ab can be used indifferently. The curves describing the relationship between the analytical signal deriving from the labeled complex (Ab-Ag-Ab*), shown in the ordinate, and the concentration of Ag in the abscissa have an increasing trend, contrary to what is obtained for the determinations by competitive methods in which this trend is decreasing.

The definition of the concentration of excess reagents must be carefully determined according to the analyte concentration range to be measured. Indeed, if the concentration of Ag approaches the concentration of Ab, both capture and detection, the relationship between concentration and signal tends to reach a level of saturation (plateau), tending to a situation of defect of Ab. On the contrary, an exaggeratedly high concentration of Ab* can be critical for the effectiveness of the removal system of the labeled not involved in the Ab-Ag-Ab* complex and, therefore, increase the background signal of the reaction.

The main advantages of the non-competitive architecture are:

- The high specificity (the Ag is recognized in at least two epitopes), which allows for distinguishing also macromolecules with a high homology, such as pituitary hormones
- The high analytical sensitivity due to the direct proportionality between Ag concentration and emitted analytical signal.

The limitations of non-competitive methods are:

- The fact that it cannot be used for the determination of small molecules such as haptens. Indeed, the Ag must have two different epitopes
- Possible underestimation of extremely high analyte concentrations due to the hook effect.

Heterogeneous Phase Methods

Heterogeneous-phase methods are defined as methods in which the assessment of the amount of Ag-Ab complexes formed, specifically Ag*-Ab complexes, should be performed only after physical separation because the tracer emitted signal is independent of whether Ag* is bound to free Ab or not. The process of free-ligand separation should be as complete as possible. It should not alter the balance between the forms to not decrease the sensitivity and precision of the determination. The simplicity and reproducibility of the free-ligand separation step are fundamental parameters for the quality of an immunometric method.

Free-Bound Separation Methods

The methods of separation of the free and bound phases have also undergone considerable evolution. Initially, in competitive methods, carbon was used, which, by adsorbing small molecules, allowed the separation of free Ag, both labeled and unlabeled, from those engaged in the complex with the Ab. Binding selectivity toward Ag of low molecular weight can be obtained by coating the carbon particles with dextran of appropriate molecular mass. The coated carbon then acts as a molecular sieve that adsorbs only molecules whose size is less than or equal to that of the dextran used for coating; all molecules with larger sizes are excluded. After centrifugation and removal of the supernatant, the carbon (or carbon-dextran) is recovered and the analytical signal emitted from the free form of the marker is measured.

Prior to the introduction of solid-phase separation methods, which are currently the most widely used, the following techniques were used for non-automated assays.

Non-specific Precipitation

Based on the difference between the molecular size of the free Ag fraction and the Ag-Ab complex, the latter can be separated by chemical precipitation. The addition of a critical concentration of the precipitating agent obtains the precipitation. After centrifugation, the Ag-Ab complexes become insoluble, then the free fraction of the labeled, remains in the supernatant. The most used precipitating agents are ammonium sulfate, ethanol, and PEG (polyethylene glycol) of molecular mass 6000 or 8000 kDa (PEG 6000 or PEG 8000).

According to Ab

They consist of precipitation of the Ag-Ab and Ag*-Ab complexes by an anti-Ig Ab in case of Ag determination. For example, if the first Ab is obtained by rabbit immunization, it is possible to use a second Ab from mutton or goat, directed against rabbit Ig. Binding with this additional Ab increases the difference in molecular size of the bound complex with respect to the free Ag fraction, facilitating its separation by centrifugation.

Protein A

It is a constituent of the wall of *Staphylococcus aureus*, and it possesses a particular affinity for the Fc fragment of human IgG, except IgG3, and forms insoluble complexes that allow free fractions to be separated after centrifugation.

The solid-phase separation technique is used, almost exclusively, for all immunochemical determinations, automated or manual, as it allows to distinguish more effectively the bound fractions from the free one, eliminating almost completely the non-specific bonds of the Ag with other molecules and minimizing the analytical background signal. In addition, solid-phase separation allows:

- To make the separation of free Ag and Ag-Ab complexes as simple as possible and fully automatable
- Facilitate and make homogeneous the chance of meeting between the molecules involved in the determination (Ag, Ag*, and Ab)

- To realize solid phase that allows the simultaneous determination of several parameters on the same support (biochips) or on different supports (microspheres with a size of a few microns)

The different types of solid phases used are very numerous and differ in size and in the technology to carry out the separation between free and bound fraction but especially in the kinetics of Ag-Ab complex formation. The use of solid phase consisting of smaller particles, which remain in suspension in the reaction environment, makes more homogeneous the possibility of encountering the molecules that must react. In contrast, in the other types of solid phases, the binding agent is immobilized on the wall of the container in which the reaction takes place. favoring the binding of the molecules near the edges of the container with respect to those further away. Chromium dioxide particles, latex microparticles, microparticles or paramagnetic particles, magnetic microparticles, cellulose foam, cones, and test tubes, microwells, marbles, etc can constitute solid supports. In technologies involving the determination of more than one analyte in the same sample reaction environment, the solid phases used are polystyrene microspheres and ceramic biochips.

Some methods employ "universal" solid phases that bind the captured Ab indirectly using the avidin-biotin system. In this case, the capture reagent is conjugated to biotin, and the solid support's surface is coated with avidin. The binding of the capture reagent to the solid phase is ensured by the extraordinary affinity of avidin for biotin (very high-affinity constant). In this case, the same solid phase can be used for all determinations in which Ab is conjugated with biotin.

Homogeneous-Phase Methods

The homogeneous-phase methods differ from the heterogeneous-phase methods (Table 8.3) in the detection of Ag-Ab complexes formed in the immunochemical reaction, which does not require the separation of free and bound fractions: all the molecular forms involved remain in the reaction environment until the analytical signal is measured. For the realization of these methods particular tracers, or more correctly tracer systems, which modify their ability to emit a detectable signal as a function of the involvement in the formation of the Ag-Ab complex must be used.

This technology was initially introduced in the 1980s for the determination of drugs with automated instruments and, subsequently, applied for the determination of other analytes, both haptens and proteins. The efficiency of this method is limited by the permanence in the reaction environment of all potentially interfering species and by the heterogeneous-phase methods that can be removed by separation procedures, and by the peculiar characteristics of the tracers:

Table 8.3 Characteristics of heterogeneous and homogeneous immunochemical methods

Heterogeneous methods	Homogeneous methods
More complex execution (need for separation of free and bound fractions) Can be automated only by dedicated systems High analytical sensitivity Wide range of measurable concentrations Reduced sensitivity to sample matrix interference	Quick and easy execution Ease of complete automation Higher detectable concentrations (low sensitivity) Narrow range of measurable concentrations High sensitivity to sample matrix interference

- In Ab defect determinations, the bond established between Ab and Ag must lead to the highest degree of inhibition of signal emission by the tracer. In the case of enzymatic tracers, e.g., EMIT® (Enzyme Multiplied Immunoassay Technique) and CEDIA® (Cloned Enzyme Donor Immunoassay) technologies, it is necessary to carry out the marking in such a way that the steric configuration assumed by the Ag-Ab complex can hinder the binding between the tracer enzyme and its specific substrate (Fig. 8.3);
- In Ab excess determinations, the binding between the two Abs and the Ag must occur in such a way that the interaction of the Ab-bound tracers is sufficiently close for them to interact effectively (Fig. 8.4).

Ag-Ab Complex Formation and Equilibrium Reaction

The immunochemical methods, based on the unique characteristics of selectivity and avidity of the bond between Ab and Ag, allow the realization of highly specific methods for quantifying of the analyte, even in complex matrices. The avidity of the Ag-Ab bond is measured by the affinity constant (K_{aff}), which is, to all intents and purposes, a thermodynamic constant of equilibrium (K_{eq}). An Ab, to be suitable for the development of immunochemical methods, must be characterized by a K_{eq} value greater than 10^9–10^{12} M^{-1}.

The binding reaction between Ag and Ab can be described by the following equilibrium:

$$\mathrm{Ab} + \mathrm{Ag} \underset{K_a}{\overset{K_d}{\rightleftharpoons}} \mathrm{AgAb}$$

$$K_{eq} = \frac{K_a}{K_d} = \frac{[\mathrm{AgAb}]_{eq}}{[\mathrm{Ag}]_{eq}[\mathrm{Ab}]_{eq}}$$

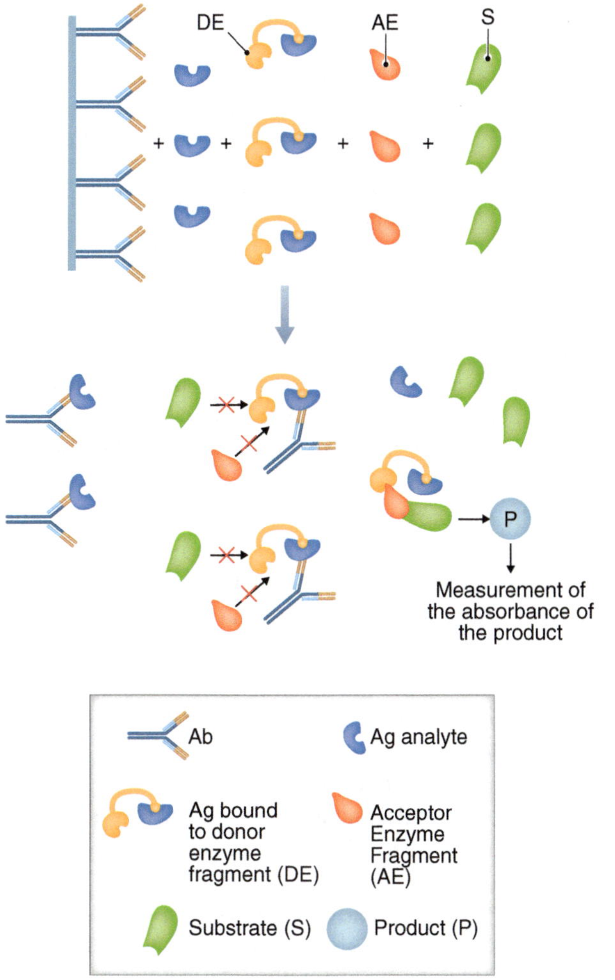

Fig. 8.3 Schematic representation of the principle underlying the CEDIA® technology: the binding of Ag * to Ab prevents the re-association of enzymatic fragments (donor enzyme and acceptor enzyme) and, therefore, the restoration of catalytic activity. The amount of reactivated enzyme, directly proportional to the measured signal, increases with increasing concentration of the Ag analyte. (Copyright EDISES 2021. Reproduced with permission)

where K_a and K_d are the constants of association and dissociation rates of the complex, and K_{eq} is the equilibrium constant of the process. The affinity of the antibody for the analyte can be determined graphically with the Scatchard diagram, which is obtained experimentally by incubating increasing amounts of the analyte with a constant amount of Ab and measuring, under conditions of equilibrium of the reaction of formation of the Ag-Ab complex, the concentration of free Ag and Ag bound to the Ab:

$$\frac{[AgAb]_{eq}}{[Ag]_{eq}} = K_{eq}Ab_t - K_{eq}[AgAb]_{eq}$$

where $Ab_t = [Ab]_{eq} + [AgAb]_{eq}$ is the total concentration of Ab. The Scatchard diagram is obtained by representing in abscissa the concentration of the Ag-Ab complex (bound or unbound) and in ordinate the ratio between the concentrations of free and bound Ag (bound/free) (Fig. 8.5).

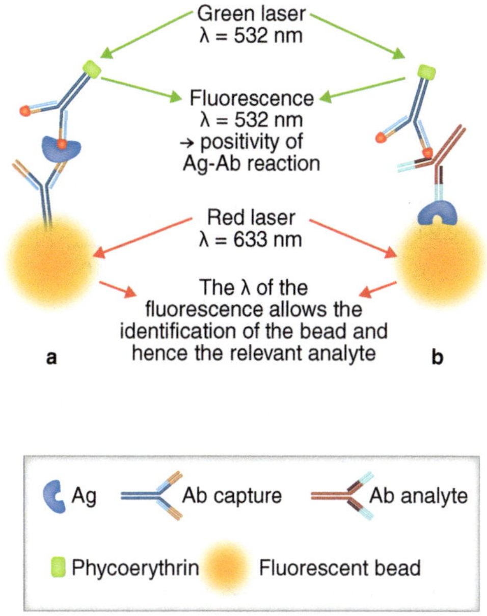

Fig. 8.4 Measurement of Ag or Ab by immunometric methods based on Luminex® technology. (Copyright EDISES 2021. Reproduced with permission)

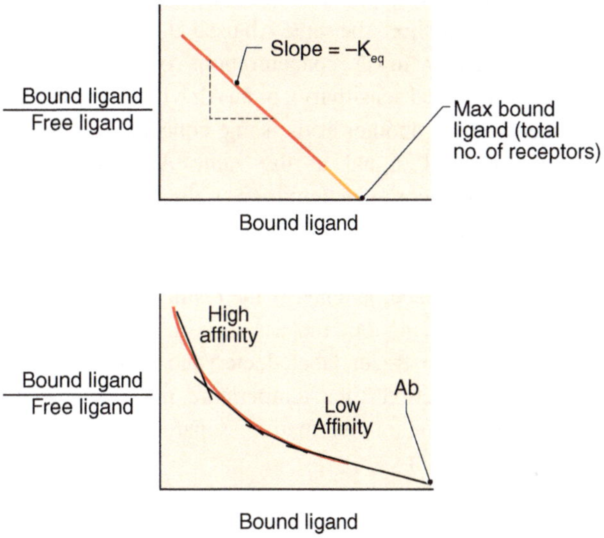

Fig. 8.5 The Scatchard diagram corresponding to a monoclonal antibody is a straight line where slope = $-K_{eq}$; intercept on the x axis = Ab. The Scatchard diagram corresponding to a polyclonal antibody, that is a heterogeneous class of antibodies with different K_{eq}, is curvilinear, as it derives from the superposition of different antigen-antibody association processes, each characterized by its own K_{eq}. An apparent K can be calculated for this system

The knowledge of the characteristics of the Ab used in an immunochemical determination is essential to assess the sensitivity of the method and, consequently, the range of

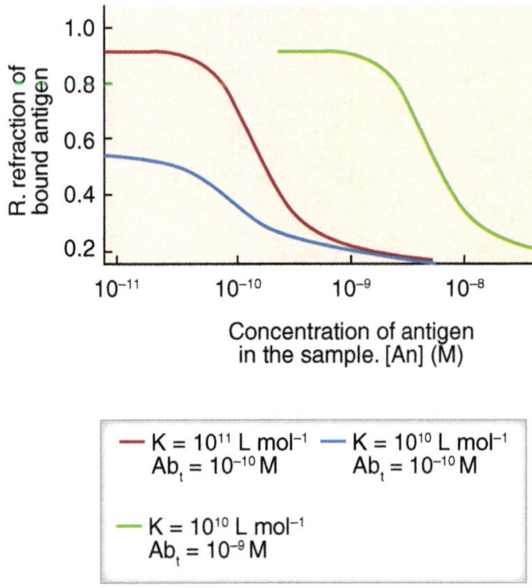

Fig. 8.6 Theoretical bond curves for a competitive immunoassay method. A reduction in Ab shifts the curve to lower analyte concentrations, but at the expense of sensitivity (comparison between green and blue curves), while an increase in K increases sensitivity (comparison between blue and red curves)

analyte concentrations that can be measured. For example, in a competitive method, the same Ab used at lower concentrations may allow lower concentrations of analyte to be detected (increased sensitivity), or an Ab with a greater affinity constant than another at the same concentration gives a higher analytical signal at the same Ag concentration (Fig. 8.6). The theoretical definition of the characteristics of an Ab is fundamental for predicting, in principle, the analytical performance of an immunochemical method. However, the real performance obtained is the result of all the factors involved in carrying out the assay, both in terms of the reagents used (Ab, tracer, labeled, etc.) and the physical conditions of the reaction (time, temperature, matrix), as well as instrumental (level of automation, separation of free and bound phases, etc.).

Immunochemical Methods Without Tracer

In label-free immunochemical determinations, the formation of the Ag-Ab complex can be detected directly, without resorting either to the separation of unbound Ag or Ab forms or to the measurement of the analytical signal emitted by a special tracer (Table 8.4).

In these methods, if the concentration is sufficiently high, the complexes formed as a result of the Ag-Ab binding are detectable by the naked eye, or lower concentrations can be quantified by exploiting their ability to deflect a light beam. The formation of the complex indicates that both molecular

Table 8.4 Classification of immunochemical methods that do not use a tracer

Analytical principle	Method
Agglutination	Hemagglutination Latex agglutination
Precipitation	Ouchterlony double immunodiffusion Rocket immunoelectrophoresis Immunoturbidimetry/nephelometry
Immunosensors	Resonant mirror biosensor (IAsysTM) Surface plasmon resonance (BIAcoreTM)

forms are present. Therefore, by using a constant concentration of Ab as a reagent, it is possible to detect a specific Ag and, vice versa, by using an Ag, it is possible to detect the presence of a specific Ab. If cells or very small particles are coated with the reagent (Ab or Ag), their binding with Ag or Ab in the sample can cause precipitation or agglutination, even at very low concentrations. Many methods still in use are based on the simple principle of precipitation. However, the most relevant for their application are turbidimetry and nephelometry. The main limitation of these methods is the reduced sensitivity and, in some cases. Additionally, an extremely high concentration of analyte can inhibit the formation of the complex.

Other label-free immunochemical methods are based on biosensors. Currently, and there are several commercial systems available that allow direct detection of Ab-Ag interaction.

Many of these are based on the generation of an evanescent wave from the surface of a sensor on which the ligand is immobilized, with the possibility of continuously monitoring the kinetics of Ag-Ab binding. These technologies, with the availability of specific low-cost instrumentation, will allow in the future a widespread application of immunochemical methods.

Turbidimetry and Nephelometry

Turbidimetry and nephelometry are the currently most widely used immunochemical techniques that do not require the use of a tracer.

These techniques are based on the evaluation of scattered light by a heterogeneous system, such as a colloidal solution or a suspension. Two limiting cases can be presented:

- The light is absorbed to a considerable extent so that the intensity of the emerging beam is particularly attenuated.
- The light is considerably diffused by the suspended particles by a series of phenomena of reflection and refraction of the suspended particles (Tyndall effect), originating the so-called opalescent light.

The two phenomena always coexist, but their intensity is always quite different. When absorption prevails over diffusion, it is preferred to evaluate the entity of the absorption by the dispersed phase: in this sense, a turbidimetric measurement is carried out. In turbidimetry, the intensity of light transmitted by a suspension is measured: normal colorimeters or UV-VIS spectrophotometers can be used to measure the transmittance or absorbance percentage.

When, on the other hand, the scattering phenomenon is much more intense than the previous one, in the case of excellent fine dispersed phase, then the evaluation of the light scattered by the suspension at 90°, with respect to that of the incident radiation, is carried out by performing a nephelometric measurement. Therefore, the detection system must be positioned at 90° to the beam entering the sample cell.

Turbidimetry is applied when the size of the particles causing turbidity is of the order of a micrometer or more, a condition in which absorption prevails over diffusion. When dealing with smaller particles, tens or hundreds of nanometers, the diffusive effect prevails and, therefore, nephelometry is used. Nephelometric measurements are very sensitive, and their accuracy is conditioned by many factors, such as the pH of the medium, and the presence of interfering substances or that can adsorb on colloidal particles of the suspension.

Relationship Between Analytical Signal and Concentration

For immunochemical determinations, the description of the relationship between analyte concentration and measured signal (calibration curve), achieved by measuring samples of known concentration (calibrators), is quite complex as it is not described by a linear function and, even if it does not involve analytical steps, it has a decisive influence on the precision and accuracy of the measurement.

The data processing process aims to define a mathematical function that can interpolate, all the points relating to the calibrators' measurements in best possible way. The expression of the analytical signal is first referred to as the measured labeled fraction: free (F) or bound (B). Or through transformations in direct ratios or referred to as the total signal (T) of the marker used in the assay: B/F, B/T, F/B, 1/B, 1/F, T/B, T/F. In other cases, indirect transformations are used, which refer to the measured signal for the fraction B to the signal obtained in the reaction system in the absence of unlabeled analyte (B0): B/B0, B0/B and logit(B/B0), where the following expression defines the logit function:

$$\text{logit}(Y) 5 \ln[Y/(1-Y)]$$

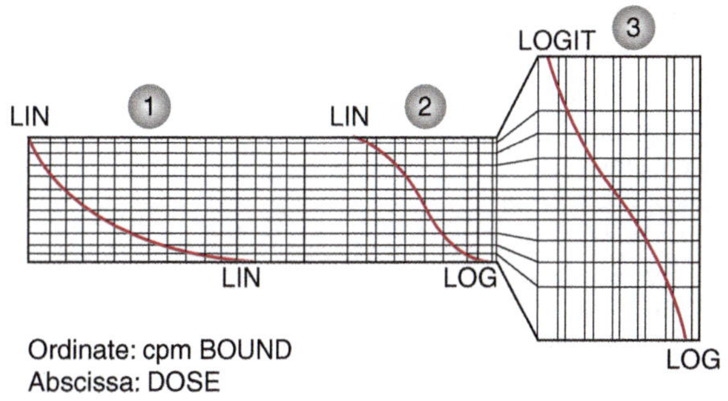

Fig. 8.7 Graphic description of the "concentration and analytical signal" relationship and modification of the graph obtained by transforming the scale: (1) linear-linear; (2) linear-logarithm; (3) logit-logarithm

According to the way of expression of the marker, the shape and the course of the curves describing the relation between concentration and analytical signal are modified. Given the curvilinear course of the function, which is obtained in any case, in order to "stretch" the curve more and make the function as linear as possible, at least in the central part, even the representation of the analyte concentration can be transformed from a linear scale to a logarithmic one (Fig. 8.7). This expedient proved to be indispensable until the early 1980s when, as electronic processors were not yet available, the only method available for calculating the results of an assay was manual graphic representation.

Interpolation and Regression Methods

The techniques used to search for mathematical functions that correctly describe the relationship between concentration and analytical signal, can be grouped into two categories:

- Interpolation
- Regression

With interpolation, the experimental points obtained by the measurement of the calibrators are joined with a broken line, being heavily affected by possible random errors. The regression technique tends to identify a mathematical model able to describe in a continuous way the relationship between concentration and signal, approximating the experimental points and mediating eventual measurement errors.

In the case of both interpolation and regression methods, several approaches have been developed, ranging from simple empirical observation of experimental data to models, which take into account the physicochemical aspects of the interactions between the species involved in the immunological reaction. According to these criteria, methods for describing the concentration–signal relationship can be grouped into the following categories (Table 8.5):

Table 8.5 Classification of methods for processing the relationship between concentration and analytical signal

Empirical methods
Manual graph
Linear interpolation
Polynomial expression
Approximate rational methods
Logit-log
4-parameter logistics
5 or more parameters logistics
Exact rational methods
Scatchard
Sips
Ekins
Feldman
Fernandez-Loeb
Rodbard
Wilkins

- Empirical methods
- Approximate rational methods
- Exact rational methods

The empirical methods include all the interpolation and regression techniques, more or less simple, which can describe the trend of the concentration–signal relation for several immunochemical assays. The choice among the various possibilities must be based on the relationship between the curve fit the experimental points and the "logical" trend of the relation, avoiding excessive degrees of curvature.

Rational methods tend to describe the experimental reality based essentially on the law of mass action that regulates the reaction between Ag and Ab. However, in order to make the data processing and the implementation of the calculation algorithms easier, some simplifications have been introduced that consistently move away from the experimental reality. In particular, in many of the applied models, it is considered that:

- The kinetics of the reaction are of first order.
- There is only one type of binding between Ag and Ab, i.e., there is no heterogeneity of epitopes and binding sites.
- The equilibrium constants for binding of the marker and analyte to the capture system are identical, i.e., have the same affinity.
- The signal measurement is carried out when thermodynamic equilibrium is reached.
- The separation of the signal from labeled bound is absolute in comparison to the free signal.
- In competitive assays, the labeled molecular weight shall be identical to that of the analyte.
- The mass of the tracer is known.
- The signal due to non-specific binding of the labeled is correctly measured.
- The signal due to non-labeled-specific binding is identical in each sample measured.

Hardly all these conditions are completely reached in an immunochemical determination. The exact rational methods, in addition to the achievement of a good definition of the concentration-signal curve, allow the determination of the main parameters of the immunological reaction, such as the equilibrium constant, the concentration of the binding sites, the concentration of the labeled, the magnitude of the signal due to non-specific binding, the degree of heterogeneity of the Ab, etc.

The use of electronic processors, facilitating the elaboration of complex algorithms, has allowed for intervention on fundamental aspects of the experimental by the introduction of statistic criteria, such as the attribution of different weight factors in function of the concentration because the properties of the analytical system confer variable precision characteristics in relation to the quantity of analyte measured and the possibility of exclusion from the procedure of extrapolation of the curve of points that suffer from random experimental errors (aberrant).

Since there is no absolute criterion to identify a data processing method with universal validity, the choice between empirical and rational models must be made according to the characteristics of the analyte and the properties of the binding system (Ab). This choice significantly influences the quality of the results obtained by immunochemical determination, as different algorithms may associate different analyte concentrations with the same signal value. The introduction of fully automated systems for these determinations has greatly limited or eliminated, the possibility for operators to choose the interpolation algorithm based on the characteristics of the reagents.

In almost all the completely automated systems for carrying out immunoassays, in addition to limiting the choice of the interpolation algorithm of the relationship between analytical signal and concentration, the direct measurement of all calibrators is not carried out, but the so-called re-calibration is applied: only two calibrators are measured in the laboratory (usually with low and high analyte concentration) to assess any variation of the signal obtained by the instrument in use compared to that of a full calibration curve, obtained from the manufacturer by averaging the signal of repeated measurements of an appropriate set of calibrators performed on several analyzers.

Interferences in Immunochemical Methods

For all the phases of an immunochemical determination, the analytical conditions must be carefully studied and monitored, starting from the collection of the sample up to the measurement of the analytical signal, to determine the concentration of the analyte to be measured.

Immunochemical methods are subject to interference both in forming the Ag-Ab complex and in the detection phase of the analytical signal. The leading causes are related to the presence of molecules in the sample that can bind nonspecifically the Ab (cross reaction), or interfere in the formation of the Ag-Ab complex (matrix effect), or that compete with the Ab for Ag binding (e.g., auto-Ab and heterophilic Ab).

Particular attention must be paid to the chemical–physical conditions under which the immunochemical reaction takes place. The architecture of these methods, particularly competitive ones, is based on the identical probability of Ag and Ag* being able to be bound by the Ab: any alteration of the conditions leading to a departure from the equilibrium leads to an alteration of the results.

Recommended Readings

Adison GM, Hales CN (1971) The immunoradiometric assay. In: Kirkham KE, Hunter WM (eds) Radioimmunoassay methods. Churchill Livingstone, Edinburgh, pp 447–461

Bazin H, Preaudat M, Trinquet E et al (2001) Homogeneous time resolved fluorescence resonance energy transfer using rare earth cryptates as a tool for probing molecular interactions in biology. Spectrochim Acta A 57:2197–2211

Berson SA, Yalow RS (2006) General principles of immunoassays. Clin Chim Acta 369:125–143

Boscato LM, Stuart MC (1986) Incidence and specificity of interference in two-site immunoassays. Clin Chem 32:1491–1495

Capolaghi B, Truchaud A (2004) L'Immunoanalyse automatisée au début du XXI siècle. Spectra Biol 141:55–65

Cohen R. Cours immunodosages, DES biologie medicale, module de biochimie, faculté de pharmacie. Université Claude Bernard, Lyon

Dudley RA, Edwards P, Eklns RP et al (1985) Guidelines for immunoassay data processing. Clin Chem 31:1264–1271

Ekins RP (1998) Ligand assays: from electrophoresis to miniaturized microarrays. Clin Chem 44:2015–2030

Ekins RP, Newman GN, O'riordan JLH (1970) Saturation assays. In: Mc Arthur JS, Colton T (eds) Statistics in endocrinology. MIT Press, pp 345–392

Gao Y, Yuan Z, Yu Y et al (2007) Mutual interference between serum thyroglobulin and antithyroglobulin antibody in an automated chemiluminescent assay. Clin Biochem 40:735–738

Gosling JP (2000) Immunoassays. Oxford University Press, New York

Henderson DR, Friedman SB, Harris JD et al (1986) CE-DIA®. A new homogeneous immunoassay system. Clin Chem 32:1637–1641

Kohler G, Milstein C (1975) Continuous cultures of fused cells secreting antibodies of predefined specificity. Nature 256:495–497

Le analisi immunometriche: basi teoriche e applicazioni cliniche. Padova: Piccin, 2014

Lequin RM (2005) Enzyme immunoassay (EIA)/enzyme linked immunosorbent assay (ELISA). Clin Chem 51:2415–2418

Long DA (2002) The Raman effect: a unified treatment of the theory of Raman scattering by molecules. Wiley, New York

Malvano R, Giraudi G (1998) L'immunochimica analitica. Ed. DiaSorin

Mathis G, Socquet F, Viguier M et al (1997) Homogeneous immunoassays using rare earth cryptates and lime resolved fluorescence: principles and specific advantages for tumor markers. Anticancer Res 17:3011–3014

McCreery RL (2000) Raman spectroscopy for chemical analysis. Ed. Wiley-Interscience, New York

Miles LEM (1977) Immunoradiometric assay (IRMA) and two-site IRMA systems. In: Abraham G (ed) Handbook of radioimmunoassay. Macel Dekker, New York, pp 131–177

Miles LEM, Hales CN (1968) Labelled antibodies and immunological assay systems. Nature 219:186–189

Miles LEM, Hales CN (1970) Immunoradiometric assay procedures: new developments. In: In vitro procedures in medicine. IAEA, Vienna, pp 483–489

Moalic V, Mercier B, Ferec C (2004) Technologie Luminex: principe, applications et perspectives. Immunoanal Biol Spec 19:181–187

Price CP, Newman DJ (1997) Principles and practice of immunoassay. Macmillan Press, London, pp 443–480

Purvis DR, Pollard-Knight D, Lowe CR (1997) Principles and practice of immunoassay. Macmillan Press, London, p 511

Robbins J, Rall JE (1960) Proteins associated with the thyroid hormones. Physiol Rev 40:415–489

Romano L, Marseglia S (1986) Aspetti termodinamici del dosaggio RIA. Ed. Radim

Romano L, Zofra S, Giglio G (1995) Guida pratica agli immunodosaggi. Ed. Poli industria chimica

Rubenstein KE, Schneider RS, Ullmann EF (1972) Homogeneous enzyme immunoassay. A new immunochemical technique. Biochem Biophys Res Commun 47:846–851

Ullman EF, Kirakossian H, Switchenko AC et al (1996) Luminescent oxygen channelling assay (LOCITM): sensitive, broadly applicable homogeneous immunoassay method. Clin Chem 42:1518–1526

Walker WHC, Keane PM (1977) TheoriÃÄtical aspects of radioimmunoassay. In: Abraham G (ed) Handbook of radioimmunoassay. Macel Dekker, New York, pp 87–130

Wide L (1983) Non competitive versus competitive binding assay. In: Odell WD, Franchimont P (eds) Principles of competitive protein-binding assay. Wiley, New York, pp 243–254

Wu JT (2000) Quantitative immunoassay: a practical guide for assay establishment, troubleshooting, and clinical application. AACC Press

General Information on Laboratory Tests and Biomarkers

9

Giuseppe Lippi, Elisa Danese, and Martina Montagnana

Introduction to Laboratory Tests and Biomarkers

The term "laboratory test" conventionally refers to a medical procedure consisting of the analysis of a biological fluid, such as blood, urine, or other biological matrices, as described in Table 9.1. Thus, the laboratory examination is not necessarily aimed at the measurement of a substance but rather at the quantification of a biological signal. Laboratory tests have become essential in the screening, diagnosis, prediction, and monitoring (including therapeutic) of most diseases.

According to the Biomarker Definition Working Group of the National Institute of Health (NIH) of the United States of America, a biomarker (or "biological marker") is instead a characteristic that can be directly determined (measured) and used as an indicator of (i) normal biological processes; (ii) pathological processes; (iii) response to any therapeutic intervention.

In general, although it is not obvious how the two definitions differ, it is preferred to use the term "laboratory test" to identify any analysis aimed at determining a biological signal in a matrix (regardless of the type of measurement). For example, the prothrombin time (a coagulation test) measures the overall clotting time of a plasma sample but does not provide direct information on the concentration of individual clotting factors that contribute to the coagulation process. The term "biomarker" is conventionally reserved to identify a directly quantifiable molecule in the biological sample (e.g., a biochemical parameter, a protein, and an enzyme). In a literal sense, prothrombin time is, therefore, a laboratory test, but it cannot be categorized as a biomarker. On the contrary, cardiac troponins are determined directly in serum or plasma (by monoclonal antibodies) to diagnose cardiac damage and are, therefore, to all intents and purposes, both laboratory tests and biomarkers.

Table 9.1 Biological matrices commonly analyzed in laboratory medicine

Blood (whole, plasma, or serum)
Urine
Feces
Saliva
Sweat
Tears
Nasal secretion
Sputum
Pus, exudates, transudate
Cavity fluids (physiologically or pathologically present)
Cerebrospinal fluid
Pleural fluid
Pericardial fluid
Synovial fluid
Bile
Stones
Liquids extracted from cysts and other neoformations
Skin and skin appendages

The number of potential biomarkers is now almost infinite, as evidenced by the evolution of their discovery, characterized by an almost exponential growth over the years. Nevertheless, of the several thousand biomarkers that are discovered each year, only a minority (an estimated ~1 out of 5000) enter clinical practice. This is due to a particularly long (it can last 5–10 years, or even longer) and complex process ("biomarker pipeline"), ranging from discovery to commercialization and requiring a series of intermediate steps, including verification, preclinical validation, clinical validation, development of commercial methods, clinical-analytical validation of commercial methods, introduction into guidelines or recommendations, and clinical use (Fig. 9.1).

This largely justifies the significant efforts (economic, human, and organizational) in preparation for the introduction of new biomarkers in clinical practice and the unfortunately very frequent finding of the disappearance of many

G. Lippi (✉) · E. Danese · M. Montagnana
Section of Clinical Biochemistry, University Hospital of Verona, Verona, Italy
e-mail: giuseppe.lippi@univr.it

M. Ciaccio (ed.), *Clinical and Laboratory Medicine Textbook*, https://doi.org/10.1007/978-3-031-24958-7_9

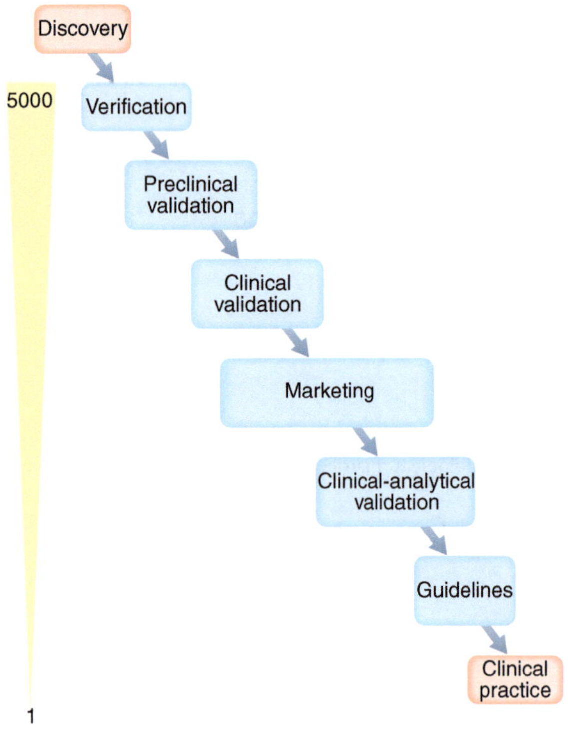

Fig. 9.1 Evolution of biomarkers, from discovery to clinical use. (Copyright EDISES 2021. Reproduced with permission)

biomarkers during the process of translation from the laboratory bench to the patient's bedside ("loss in translation between bench and bedside").

The Ideal Characteristics of a Biomarker

According to the NIH definition, the determination of a biomarker is nowadays essential for screening, diagnosis, prognosis, and monitoring of most (if not all) human diseases. Thus, it is evident that the biological characteristics and the analytical techniques used for its determination substantially influence the diagnostic performance in each of the potential fields.

In general, the ideal characteristics of a biomarker are summarized in Table 9.2.

The first essential feature is the presence of high concentrations in the target organ/tissue. This is related to the fact that if the biomarker is present in a low concentration in the target organ/tissue, its increase in circulation (or in biological fluids) following a pathology/organ damage will not be detectable by simple, rapid, and inexpensive techniques. The lower the concentration of the biomarker in the circulation (or biological fluids), the greater the analytical complications for its determination, resulting in the concept of low diagnostic "sensitivity" (many false negatives). In other words, if the amount of biomarker contained in the damaged tissue is small, its concentration in the biological fluid under

Table 9.2 Ideal characteristics of a biomarker

Present in high concentrations in the organ/tissue target
Absent in non-target organs/tissues
Absent in the biological fluids of healthy subjects
Rapidly released into the biological fluid as a result of pathology
Persist in the biological liquid for a period sufficient to ensure an adequate diagnostic window
Show increases in biological fluid related to the severity of the damage and prognosis
Measurable in the biological liquid simply, quickly, and economically

investigation will be equally modest, thus making its measurement quite complex.

The second ideal feature, a logical consequence of the first, is that the biomarker should be absent in non-target organs/tissues. The higher the concentration of the biomarker in other organs/tissues, the lower the possibility of correlating its increase to the damage to the target organ/tissue, which results in the concept of low diagnostic "specificity" (many false positives). Typical examples are the virtually ubiquitous enzymes, such as lactate dehydrogenase (LDH) and gamma-glutamyltransferase (GGT), whose blood concentrations increase following cellular damage in a myriad of organs and tissues. Their increase is certainly a sign of cellular damage but understanding their origin and cause (especially in the absence of specific clinical signs or symptoms) is virtually impossible.

The third ideal characteristic is also substantially consequential to the first and refers to the opportunity for the biomarker to be absent in the blood or other biological fluids of presumably healthy subjects. This aspect appears to be of substantial importance in optimizing the diagnostic sensitivity of the biomarker. An example is presented in Fig. 9.2 that shows the comparison between two hypothetical biomarkers of liver damage: biomarker "A," which is already physiologically present in the circulation under normal conditions, and biomarker "B," which is not normally present in the circulation under physiological conditions. In the first case, a reference range is available for the biomarker concentration (since biomarker "A" is present in the circulation also in healthy subjects), while in the second case, a diagnostic threshold is used corresponding to the analytical sensitivity limit of the method (since biomarker "B" is absent in the circulation and therefore usually not measurable in healthy subjects). Following organ damage, both biomarkers are released from the target tissue and enter the circulation. The concentration of biomarker "A" will progressively add to that already present in the circulation, and an unavoidable period must elapse before the concentration exceeds the upper limit of the reference range. In contrast, the concentration of biomarker "B" will almost immediately exceed the diagnostic threshold since its concentration in circulation is virtually undetectable in healthy subjects.

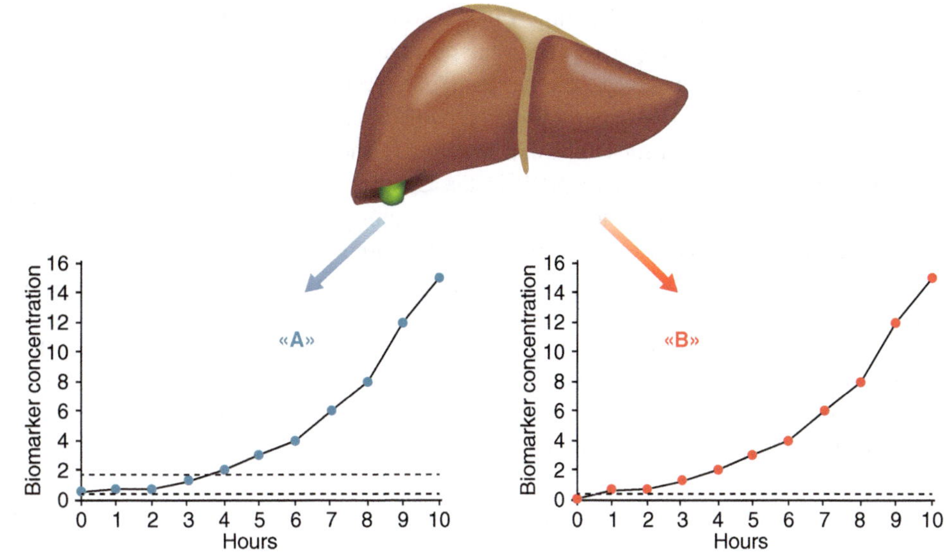

Fig. 9.2 Comparison of the diagnostic performance of biomarkers physiologically present in the circulation ("A"), or whose concentration is not detectable in the circulation in the absence of organ damage ("B"). (Copyright EDISES 2021. Reproduced with permission)

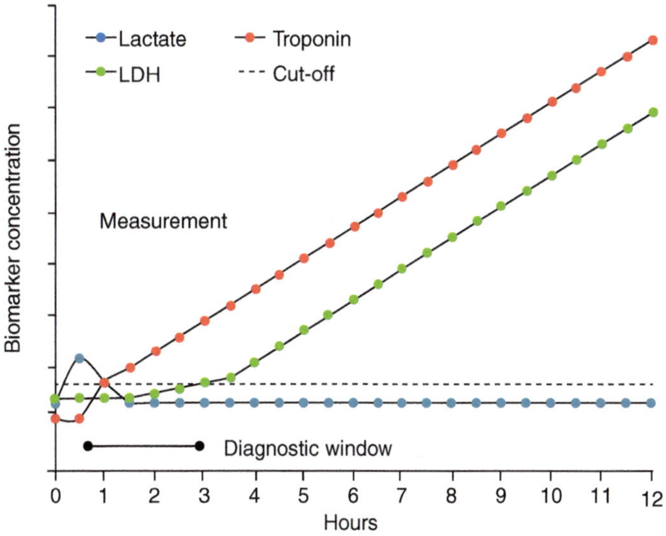

Fig. 9.3 Kinetics of lactic acid (lactate), cardiac troponin, and lactate dehydrogenase (LDH) following acute myocardial infarction in relation to the relative diagnostic window. (Copyright EDISES 2021. Reproduced with permission)

A fourth essential characteristic of the biomarker is its rapid release into the circulation (or biological fluids) following organ/tissue pathology. Obviously, this characteristic is always related to the timeliness of the request to reach a diagnosis and establish a therapy. As an example, the diagnostic timeliness translates into minutes (maximum hours) in pathologies in which an urgent intervention is required, as in the case of acute myocardial infarction or cerebral ischemia, in which the precocity of the therapeutic intervention cannot be delayed beyond 3–4 hours. In the case of the diagnosis of malignant neoplastic diseases, instead, the concept of precocity expands over days (or weeks) because the progression of the disease is much slower, and interventions are less radical. It seems, therefore, evident that the kinetics of biomarker release becomes the foundational element of diagnostic earliness, as long as the kinetics is not too rapid. This is illustrated in Fig. 9.3, which shows the kinetics of three potential biomarkers of cardiac damage in relation to an acute event (e.g., acute myocardial infarction) manifested at point 0. The effective recognition of this pathology (in terms of timely and efficient therapeutic management) requires a diagnostic window between 30 minutes and 3 hours from the event, depending on the fact that patients arrive at the emergency room after an average time of about 1 hour from the event, and the need to intervene to manage the most severe forms of infarction no later than 3 hours from the onset of symptoms. The concentration of the first biomarker (lactic acid) increases in the circulation immediately after the acute infarction, but it is often so early that when the patient arrives at the emergency room to perform diagnostic tests, its concentration may have already fallen within the reference interval (the half-life of lactic acid is about 10–15 minutes). The third test, LDH, demonstrates the opposite kinetics, with release into the circulation very late with respect to the acute event (not before 4–5 hours). This time window would not, therefore, allow an early diagnosis of cardiac damage since the concentration will not exceed the upper limit of the reference interval until 4–6 hours. In contrast to the two biomarkers described above, the kinetics of cardiac troponins I and T (which will be described in more detail in the Chapter on Cardiac Biomarkers) appear, instead, to be entirely appropriate to the diagnostic needs of acute myocardial infarction since their concentration begins to increase much earlier than LDH (usually between 30 minutes and 1 hour) but, unlike lactic acid, it remains elevated much longer. This results in kinetics that is entirely consistent with the patient's arrival in the emergency department and the subsequent diagnostic window defined by the need to institute rapid revascularization interventions.

A further important feature of biomarkers is related to the possible correlation between the magnitude of the increase, the severity of the damage, and/or prognosis of the pathology. Thus, the higher the biomarker concentration, the more severe (in terms of extension and functional impairment) the tissue/organ pathology should be and, therefore, the more inauspicious the clinical course. Although this concept may appear obvious, it is not always valid. Neoplastic biomarkers represent a paradigmatic example. Although it is true that the higher the concentration of a tumor marker in the circulation, the greater should be the mass of neoplastic tissue that has produced it, the release of the marker in the circulation also depends on a wide range of variables that affect the actual measurable value and its correlation with the clinic. In particular, for the same mass of neoplastic tissue, the circulating release of a tumor biomarker will also depend on the following: (i) degree of differentiation (frequently, the less differentiated the tumor, the lower the capacity to produce conventional biomarkers); (ii) vascularization (in less vascularized tumors, the release in the circulation is obviously lower); (iii) polarization (if the apical pole of the neoplastic cell is not facing the blood and/or lymphatic vessels, the ability to release molecules in the circulation will be lower); (iv) localization (tumors that have involved vital districts and/or organs are much more serious); and (v) response to therapy (in subjects undergoing chemo-/radiotherapy, the concentration of the biomarker may increase as a result of the therapy itself as well as cancer itself). It follows that forcibly generating parallelism between biomarker concentration and disease severity and/or prognosis can often be erroneous and misleading.

A final aspect concerns several pragmatic considerations related to the determination of the potential biomarker in the laboratory. As far as possible, this should use simple, rapid, and inexpensive techniques. Looking at each of these three aspects in detail, the possibility of using simple techniques (in relation to the volumes of tests that are required, of course) is always preferable. The determination of biomarkers with common techniques, such as enzymatic or immunochemical methods, almost always allows for automation, which translates into the possibility of not using sophisticated and dedicated instruments, avoiding the need for specialized personnel for analysis. Taking into account the example of acute myocardial infarction, even though a biomarker with absolute diagnostic efficiency (100%) is available, the diagnostic advantage would be completely nullified if its determination technique required complex instrumentation and very long times, more than 3–4 hours. One of the most important consequences of automation (partial or total) of the analysis is the greater simplicity and speed of execution. The cost/benefit ratio of the test represents a further and fundamental aspect. In a world of limited resources characterized by a strict expenditure ceiling for National Health Services (or Systems), the introduction of a new biomarker should always satisfy several essential requirements summarized in Table 9.3.

Table 9.3 Analysis necessary to evaluate the cost/benefit of a laboratory test

Diagnostic performance analysis (efficiency)
Clinical impact analysis (efficacy)
Incremental value compared to the state of the art
Cost/benefit analysis
Organizational impact analysis

The first two requirements (efficiency and effectiveness) are obviously essential for any other assessment. Not only the laboratory test (or the biomarker) must have diagnostic performances such as to allow an optimal diagnostic yield (efficiency), but it must also demonstrate that its determination allows changing the outcome of the disease (effectiveness). In short, even if a biomarker with a diagnostic efficiency of 100% (i.e., no false positives and no false negatives) is available, if its determination is not able to modify in any way the course (qualitative and/or quantitative) and the outcome of the disease, it becomes understandably superfluous. It is also evident that it is necessary to demonstrate that the determination of the biomarker can provide additional clinical information. If this cannot be demonstrated, its use would appear unjustified. If the biomarker has been proven to be efficient, effective, and generate new and important clinical information, a subsequent analysis must be performed to demonstrate that the expense of its determination is then completely reabsorbed within the patient's clinical management process. This is done by an analysis commonly called HTA (Health Technology Assessment) and defined according to the Ministry of Health as a "multidimensional and multidisciplinary approach for the analysis of medical, social, organizational, economic, ethical and legal implications of a technology through the evaluation of multiple dimensions such as effectiveness, safety, costs, social and organizational impact." In short, in laboratory medicine, the

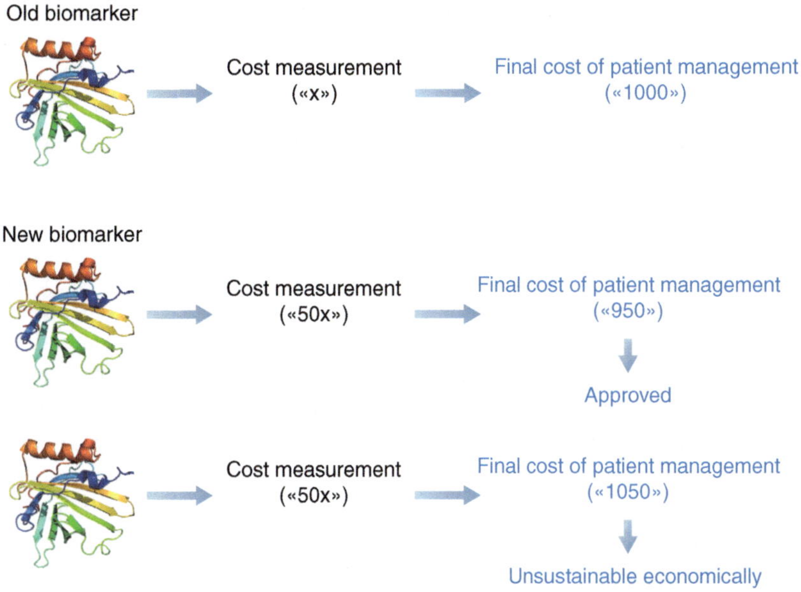

Fig. 9.4 Overall assessment of the cost/benefit ratio associated with the introduction of a new biomarker, according to the principle of Health Technology Assessment (HTA). (Copyright EDISES 2021. Reproduced with permission)

objective is to evaluate the actual and/or potential effects of a biomarker, as well as the consequences that its introduction or exclusion generates for the National Health Service, the economy and society. In other words, if the introduction of the determination of a biomarker generated a minimal increase in costs, completely reabsorbed by the consequent savings made possible by its determination, the biomarker could be introduced in clinical practice (Fig. 9.4). On the contrary, if the introduction of the determination of a biomarker generates an increase in costs of several thousand euros against a zero overall saving in terms of the overall management of the pathology, it could be said to be completely "unsustainable" by the National Health Service (Fig. 9.4). Finally, the decision on the possible introduction of a new biomarker is conditioned by other evaluations (ethical and clinical), which will not be dealt with in this chapter.

Possible Causes of Biomarkers Increase

It is necessary to introduce some concepts related to the possible reasons underlying the detection of increased concentrations (Table 9.4).

The increase due to a direct effect of the disease presupposes the existence of a causal link between the disease and the biomarker. To simplify, some tumor biomarkers increase in circulation because they are produced directly by tumor cells (e.g., carcinoembryonic antigen (CEA) is produced by colorectal cancer cells). Similarly, cardiac troponins I and T increase in the circulation because of cardiac injury following myocardial cell rupture. However, not all biomarkers that increase in the course of disease follow this logic, as changes in their concentration may occur because of other causes (Table 9.4). A frequent cause is the so-called dependent association from pathology. In short, the biomarker does not increase because it is produced directly by the cells of the tissue/organ affected by the disease, but rather due to the metabolic (local or systemic) consequences of the disease. A clarifying example is the frequent finding of a greatly increased concentration of ferritin in patients with cancer (e.g., colorectal). Ferritin is not directly produced by cancer cells, but the cancer itself generates a pro-inflammatory milieu, of which the increased concentration of ferritin is an epiphenomenon. A third hypothesis to explain the increase in biomarkers is their potential association with side effects and not with the pathology itself. Paradigmatic examples are the increase of creatinine in diabetic patients (due to renal damage resulting from diabetes and not because creatinine is produced by pancreatic cells), the increase of natriuretic peptides in patients with post-infarct heart failure (produced by cells other than those infarcted in the context of a decompensated heart), or the increase of calcemia in patients with bone metastases deriving from colon-rectal cancer (related to the erosion of bone tissue often vertebral consequent to the osteolytic effect of metastatic cancer). Finally, a possible random association between pathology and increased biomarker concentration should not be excluded. For example, patients with colorectal cancer in old age are often also diabetic (type 2 diabetes mellitus). Nevertheless, the increase in blood glucose in these patients is only coincidental with colorectal cancer, as

Table 9.4 Possible causes of increased biomarkers

Increase due to the direct effect of the disease (causation)
Association dependent on the disease
Direct effect of the complications of the pathology
Independent association (causal link)

there is no known causal link between colorectal cancer and increased blood glucose.

Diagnostic Efficiency and Effectiveness

A final aspect related to the introduction and possible use of biomarkers in clinical practice concerns the precise definition of the concepts of diagnostic efficiency and effectiveness mentioned above. Although the two concepts are often used as synonyms, they express two very different characteristics. Indeed, diagnostic efficiency refers to the specific diagnostic performance of a biomarker, identified based on a series of variables (diagnostic sensitivity and specificity, predictive value), which identify its ability to allow a precise diagnosis to be made. Diagnostic efficacy, on the other hand, refers to the possibility of modifying a clinical outcome following the determination of a biomarker. In other words, if a biomarker is characterized by a diagnostic efficiency close to 100%, but its determination is not also able to change the evolution of the disease, its real clinical significance could be questioned. A tangible example, still the subject of heated debate, is the determination of prostate-specific antigen (PSA).

There is now a large body of scientific evidence showing that PSA determination (in its different forms) has a very high diagnostic efficiency in identifying patients with prostate cancer. However, the characteristic of this cancer has led to the recent conclusion that the determination of PSA in prostate cancer screening may not determine any concrete advantage in terms of medium to the long-term survival of patients. Thus, in the recent revision of the guidelines in 2018, the determination of PSA is no longer recommended as a screening test for prostate cancer in certain categories of patients at risk.

Useful Glossary

In the remainder of this chapter, it is considered useful to provide a glossary to facilitate the use and interpretation of biomarkers in clinical practice:

- Limit of Blank (LOB): Signal generated by the measuring system (standard method or laboratory instrument) when a sample does not contain any molecule of the measurand (e.g., water or reaction buffer used by the instrument). In practice, it corresponds to the biomarker concentration that would be obtained by analyzing water or reaction buffer instead of a biological sample.
- Limit of Detection (LOD; "limit of measurement"): a signal generated by the measurement system (manual method or laboratory instrument) when analyzing a sample that contains the minimum measurable concentration of the measurand. In practice, it corresponds to the minimum concentration of biomarkers that can be quantified in a biological sample.
- Limit of Quantitation (LOQ): also known as "functional sensitivity," this is the biomarker value that can be quantified according to a predefined quality specification. In practice, it corresponds to the minimum biomarker concentration that can be quantified with inaccuracy below a certain limit, which varies from 10% to 20%, depending on the type of assay.

Recommended Readings

Armbruster DA, Pry T (2008) Limit of blank, limit of detection and limit of quantitation. Clin Biochem Rev 29 Suppl 1:S49–S52

Biomarkers Definitions Working Group (2001) Biomarkers and surrogate endpoints: preferred definitions and conceptual framework. Clin Pharmacol Ther 69:89–95

Lippi G, Plebani M, Guidi GC (2007) The paradox in translational medicine. Clin Chem 53:1555

Lippi G, Plebani M (2013) Biomarker research and leading causes of death worldwide: a rather feeble relationship. Clin Chem Lab Med 51:1691–1695

Lippi G, Mattiuzzi C (2015) The biomarker paradigm: between diagnostic efficiency and clinical efficacy. Pol Arch Med Wewn 125:282–288

Plebani M, Lippi G (2017) Uncertainty, quality, safety and accreditation in laboratory medicine. J Lab Precis Med 2:80

US Preventive Services Task Force (2018) Screening for prostate cancer: US preventive services task force recommendation statement. JAMA 319:1901–1915

Enzymes and Their Clinical Use

10

Ilenia Infusino, Ferruccio Ceriotti, and Mauro Panteghini

Introduction

Tissue injury leads to the release of cellular substances, which can represent biomarkers of tissue injury. A biomarker of injury that should be specific to an organ or tissue must derive primarily from the organ or tissue of interest. Many of the clinically useful markers of cellular injury are enzymes.

Enzyme determinations are used in medicine in two main ways. Enzymes are measured in serum and other body fluids to detect injury to an enzyme-producing tissue, and they are also measured, often within a tissue, to identify abnormalities or absence of the specific enzyme, which may underlie a disease. Some enzymes are found predominantly in specialized tissues (e.g., lipase in the pancreas); others, more widely distributed, have tissue-specific isoenzymes or isoforms (e.g., pancreatic α-amylase isoenzyme or bone isoform of alkaline phosphatase [ALP]), which can be assessed to improve tissue and organ specificity.

The timing of the enzyme's diagnostic window is another important aspect to consider when these markers are used to assess the acute injury. The diagnostic window of a marker of injury is the time interval following an acute episode of injury during which plasma concentrations of the marker have increased, thereby demonstrating that injury has occurred. Markers that rapidly enter the circulation (referred to as "early" markers) tend to have diagnostic windows that begin soon after the onset of the injury. In contrast, markers that are released slowly into the circulation or are cleared slowly from the circulation (called "late" markers) generally have diagnostic windows that begin later and last longer after the injury has occurred.

I. Infusino
Clinical Pathology Unit, ASST Fatebenefratelli-Sacco, Milan, Italy

F. Ceriotti (✉)
Fondazione IRCCS Ca' Granda Ospedale Maggiore Policlinico, Milan, Italy
e-mail: ferruccio.ceriotti@policlinico.mi.it

M. Panteghini
Research Centre for Metrological Traceability in Laboratory Medicine (CIRME), University of Milan, Milan, Italy

Factors Influencing the Plasma Concentration of Enzymes

The activity of an enzyme in the blood is the result not only of the total amount released from its cells of origin but also of the rate of its catabolism in the circulation, its passage to the extracellular compartment, and the rate at which it is eventually inactivated or removed.

Release of Enzymes from Cells

Enzymes are retained within their cells of origin by the cell membrane. The latter is a metabolically active component of the cell, and its integrity depends on the cellular production of adenosine triphosphate (ATP). Any process that penalizes ATP production by depriving the cell of oxidizable substrates or by reducing the efficiency of energy production through restriction of oxygen access (ischemia or anoxia) promotes deterioration of the cell membrane. The small molecules are the first to be released by the damaged or dying cell, followed by larger molecules such as enzymes and other proteins. Cytosolic proteins appear in plasma early after injury, followed later by mitochondrial and membrane-bound enzymes.

Because of the high intracellular concentrations of enzymes, thousands or even tens of thousands of times higher than concentrations in extracellular fluids, and because of extremely small amounts of enzyme can be detected by measurement of their catalytic activity, increased enzyme activities in plasma are an extremely sensitive indicator of cellular injury.

M. Ciaccio (ed.), *Clinical and Laboratory Medicine Textbook*, https://doi.org/10.1007/978-3-031-24958-7_10

Enzyme Efflux from Damaged Cells

Once the conditions for the release of enzymes from the cells have occurred, the ways in which the process is reflected in changes in the enzyme in the blood depend on many factors.

A first important factor modulating enzyme release is the concentration gradient that exists between the inside and outside of the cell, which is usually high. In addition, small molecules may appear in the extracellular fluid before larger ones. The way through which the released enzymes are transferred from the interstitial fluid to the blood varies from tissue to tissue; they may pass directly through the capillary wall or, more frequently, through the lymphatic system.

The intracellular localization of enzymes leaked following the injury influences the rate at which they appear in the circulation. As expected, the most sensitive indicators of cellular injury are molecules that are present in the soluble fraction of the cell. In contrast, the release of structurally bound membrane proteins requires cell membrane injury, which is a slower process. Enzymes associated with subcellular structures, such as mitochondria, are less readily released into the circulation and their appearance in the blood often indicates irreversible cellular injury.

The release of enzymes from damaged or dying cells and changes in the rate of enzyme production are the most important mechanisms by which changes in plasma enzyme activities are produced. However, there are other possibilities underlying some diagnostically important alterations. For example, much of the ALP activity in hepatocytes is localized on their outer surfaces. It is possible that ectoenzymes such as ALP may be eluted from cell surfaces, especially through the cleansing action of bile salts that accumulate in cholestasis. This process does not require cellular injury in the case of increased permeability of the cytoplasmic membrane, as evidenced by the poor correlation between increases in serum ALP and transaminase activities in different types of liver disease.

Impaired Enzyme Production

Small amounts of intracellular enzymes physiologically present in plasma may result from cell turnover or enzyme leakage from healthy cells. These amounts of enzymes in circulation may decrease because of a genetic deficit in enzyme production (e.g., ALP in hypophosphatasia or serum cholinesterase in individuals homozygous for the "silent" gene) or when enzyme production is depressed as a result of disease (e.g., cholinesterase in liver disease). However, cases in which enzyme production is increased are generally of greater interest in diagnostic enzymology. For example, an increase in the number and activity of osteoblasts is responsible for increased serum ALP concentrations in various types of bone diseases. The process of enzyme induction may also underlie an increase in enzyme production. An example is the increased serum γ-glutamyltransferase (GGT) activity that results from the administration of drugs such as barbiturates or phenytoin, or from alcohol intake.

Enzyme Clearance

Some enzymes are small enough to pass freely through the renal glomerular filter and be eliminated through this pathway. One example is pancreatic lipase (molecular weight, ~48 kDa). However, most of the enzymes present in plasma are rapidly removed primarily by the reticuloendothelial system by receptor-mediated endocytosis. For example, Kupffer cells in the liver have been shown to be able to pick up various tissue-derived enzymes, such as creatine kinase (CK) or aspartate aminotransferase (AST), by receptor-mediated endocytosis, having affinity for lysine residues on these enzymes.

Choice of Enzymatic Tests

The choice of which enzyme to measure for clinical purposes depends on several factors. One important factor is the distribution of the enzymes among the various tissues, as this affects their diagnostic specificity (Table 10.1). The amount of damaged or malfunctioning organ, together with the cell/blood gradient of the enzyme, obviously has a profound influence on the resulting increase in enzyme activity

Table 10.1 Main enzymes clinically important

Enzymes	Source of the enzyme	Main clinical applications
Alanine aminotransferase (ALT)	Liver	Parenchymal liver damage
Amylase (AMY)	Salivary glands, pancreas	Pancreatic diseases
Creatine kinase (CK)	Skeletal muscle, heart	Muscular diseases
Gamma-glutamyltransferase (GGT)	Liver, pancreas, kidney	Hepatocellular injury
Alkaline phosphatase (ALP)	Liver, bone, intestinal mucosa, placenta	Hepatobiliary injury, bone diseases
Lactate dehydrogenase (LDH)	Heart, erythrocytes, lymph nodes, skeletal muscle, liver	Hemolytic and megaloblastic anemias, leukemia and lymphomas, cancers
Pancreatic lipase	Pancreas	Pancreatic disease

in the blood. For example, it has been estimated that if even 1 in 750 liver cells is damaged, the increase in plasma alanine aminotransferase (ALT) concentration becomes detectable.

Knowledge of the intracellular localization of enzymes can help determine the nature and severity of a disease process if the appropriate enzymes are measured in the blood. For example, mild and reversible viral inflammatory injury of the liver is likely to increase only the permeability of the cell membrane, allowing cytoplasmic enzymes to leak into the bloodstream, whereas severe injury causing cellular necrosis can also destroy the mitochondrial membrane, allowing both cytoplasmic and mitochondrial enzymes to be detected in the blood.

To better clarify the clinical significance of the enzymes discussed in this chapter, individual enzymes will be discussed with respect to the related organ.

Muscle Enzymes

Creatine Kinase (CK)

CK is an 82 kDa enzyme that catalyzes the reversible phosphorylation of creatine by consuming ATP. Physiologically, when muscle contracts, ATP is converted to adenosine diphosphate (ADP) and CK catalyzes the rephosphorylation of ADP into ATP by using phosphocreatine as a phosphorylation reservoir. Its concentration is therefore higher in striated muscle and cardiac muscle.

CK is a dimer consisting of two subunits (B and M), each with a molecular weight of ~40 kDa. Since the active form of the enzyme is a dimer, only three different pairs of subunits can exist: BB, MB, and MM. All three of these isoenzyme forms are found in the cytoplasm of the muscle cell or are associated with myofibrillar structures. In skeletal muscle, MM represents the major isoform of CK (>99%), with a small proportion of MB. Cardiac tissue, on the other hand, contains the highest concentration of CK-MB, accounting for approximately 20% of cardiac CK. CK-BB is mainly and almost exclusively expressed in brain tissue (>90%). There is also a fourth isoenzyme (CK-Mt), which straddles mitochondrial membranes and, for example, in the heart, accounts for up to 15% of total CK activity. CK can also be found in macromolecular form, the so-called macro-CK, of which there are two types: type 1 and type 2. Type 1 is a complex consisting of CK, typically CK-BB, and an immunoglobulin, most often IgG. It has no clinical significance but can result in asymptomatic increases in CK, causing diagnostic confusion and leading to unnecessary further investigation. Its prevalence (80% of cases in women) has been estimated between 0.8% and 2.3%, but this depends on the study population. Type 2 macro-CK consists of oligomerized CK-Mt and has a prevalence between 0.5% and 2.6% in hospitalized patients. It is predominantly found in adults with malignant neoplasms or advanced liver disease and in children with significant tissue distress. The occurrence of this form in serum is usually associated with a poor prognosis.

Clinical Significance

Determination of CK is the laboratory test of choice when muscular injury is suspected. Serum CK concentrations are increased in almost all subjects following injury, inflammation, or necrosis of skeletal or cardiac muscle.

Increased serum CK activity may be the only sign of subclinical neuromuscular disorders. Serum CK activity is markedly elevated in all types of muscular dystrophy. In progressive muscular dystrophy, serum enzyme activity is highest in childhood and adolescence and may be increased long before the disease becomes clinically manifest. Serum CK activity decreases significantly with increasing age and as muscle mass declines with disease progression. Approximately 50–80% of asymptomatic female carriers of Duchenne dystrophy show CK activity three to six times higher than physiologic values.

High CK values (up to 50 times the upper reference limit [URL] in active disease) are found in viral myositis, polymyositis, and other inflammatory myopathies. In contrast, in neurogenic muscle diseases, such as myasthenia gravis, multiple sclerosis, poliomyelitis, and Parkinson's disease, serum enzyme activity is not increased. Very high CK activities are also found in malignant hyperthermia, a hereditary condition characterized by high fever and triggered by the administration of an anesthetic (usually halothane) in affected individuals.

In acute rhabdomyolysis due to crush injury, with severe muscle destruction, serum CK activity may be >200 times the URL. In this condition, very high serum CK concentrations, reflecting the marked myoglobinuria and the heme-induced mechanism of renal damage, have been associated with a high risk of developing acute renal failure. If CK remains <5000 U/L (~30 times the URL) during the first 3 days after the insult, the likelihood of developing renal failure appears to be low. Serum CK may also be moderately increased following other muscle trauma, such as a simple intramuscular injection and surgery. Finally, several drugs can increase serum CK activity. The drugs most frequently involved are statins, fibrates, antiretrovirals, and angiotensin II receptor antagonists. The clinical spectrum of statin-induced myotoxicity includes asymptomatic increases in serum CK activity, myalgia, myositis, and rhabdomyolysis (0.02%). Routine monitoring of CK in asymptomatic patients taking statins is not recommended; however, CK should be assessed in subjects with muscle pain and weakness, and statin treatment should be discontinued if values are >5 times the URL.

Changes in serum concentrations of CK and its isoenzyme MB following acute myocardial infarction have been the mainstay of diagnosis of this condition for many years. However, it is now more appropriate to use cardio-specific markers, such as cardiac troponin I or T.

Hypothyroidism is a common cause of endocrine myopathy. Approximately 60% of hypothyroid subjects shows an average increase in CK >5 times the URL.

During a physiological delivery, there is an average six-fold increase in maternal CK activity in serum. Surgical interventions during childbirth further increase this activity.

Reference Intervals

Serum CK activity undergoes several physiological variations. It is influenced by gender, age, ethnicity, muscle mass, and physical activity. Males have higher values than females and African Americans have higher values than Caucasians. Males, however, have a decrease in CK with aging.

In Caucasian subjects, the reference interval is 46–171 U/L for males and 34–145 U/L for females. Serum CK activity in healthy subjects is due almost exclusively to CK-MM activity (although small amounts of CK-MB may be present) and is the result of physiologic turnover of muscle tissue. Exercise, particularly if unusual, and muscle trauma can increase serum CK activity, which can rise above ten times the URL in the first 24 hours after activity. Infants generally have higher CK activity (up to 10 × adult URL) caused by muscle trauma sustained during birth. Concentrations return to the adult reference interval between 6 and 10 weeks of age.

Liver Enzymes

Enzymes included in this group are transaminases (ALT and AST), ALP, and GGT. These enzymes are often mistakenly referred to as "liver function tests" but they are markers of injury. The most common alterations underlying increases in liver enzymes encountered in clinical practice can be divided into two main pathophysiological subgroups: hepatocellular injury (increased transaminase activity) and cholestasis (increased ALP and GGT), although certain liver diseases may present a mixed biochemical picture.

Aminotransferases

The aminotransferases are a group of enzymes that catalyze the interconversion of amino acids to 2-oxy acids by transferring amino groups. While AST is found primarily in the heart, liver, muscle, and kidney, ALT is found primarily in the liver and kidney and, in smaller amounts, in the heart and skeletal muscle.

Clinical Significance

Hepatopathies are the most important cause of increased serum transaminase activity and thus represent the indication for requesting this test. Although the serum activity of both AST and ALT increases whenever a disease process affects the integrity of the liver cell, ALT is the most hepato-specific enzyme. Indeed, serum increases in ALT are rarely observed in conditions other than parenchymal liver disease. Furthermore, increases in ALT activity persist longer than those in AST. Therefore, it would be appropriate to abolish the determination of AST as a first level test, and laboratories that detect increased levels of ALT should offer the measurement of AST automatically as a reflex test on the same specimen, also calculating the AST/ALT ratio, since it can provide useful diagnostic and prognostic information.

In most liver diseases, ALT activity is higher than AST activity; exceptions may occur in alcoholic hepatitis, cirrhosis, and liver cancers. In viral hepatitis and other forms of liver disease associated with acute liver necrosis, serum transaminase activity increases before clinical signs and symptoms of the disease (such as jaundice) become apparent. The activity of both enzymes can reach values >100 times the URL, although 10- to 40-fold increases are more frequently detected. The most effective decisional cut-off for the diagnosis of acute hepatitis is 7 × URL (sensitivity and specificity >95%). In acute viral hepatitis, the peak of transaminase activity occurs between days 7 and 12 after disease onset; activity then gradually decreases, reaching physiological concentrations between weeks 3 and 5 if the disease progresses without complications. It is important to note that peak transaminase activity has no relation to prognosis and, indeed, concentrations may decrease as the patient's condition worsens, perhaps due to the loss of function of many hepatocytes.

Persistence of increased ALT levels for more than 6 months after an episode of acute hepatitis is used to diagnose chronic disease. Most patients with chronic hepatitis have a maximum ALT value <7 times the URL. However, ALT may be persistently normal in 15–50% of patients with chronic C virus hepatitis, although the likelihood of truly normal transaminases decreases with increasing frequency of measurements over time. It is therefore indicated that in patients with acute C virus hepatitis, ALT should be measured periodically over the next 1–2 years to determine its true normalization.

Nonalcoholic hepatic steatosis (NAFLD) is the most common cause of increased transaminases after viral and alcoholic hepatitis. NAFLD encompasses a spectrum of liver diseases, from simple steatosis to nonalcoholic steatohepatitis (NASH), in which inflammatory aspects and focal necrosis can progress to fibrosis, cirrhosis and liver failure. NAFLD is now considered an additional feature of metabolic syndrome. Indeed, increased transaminases in NAFLD are associated with the typical features of this syndrome, such as increased body mass index, fasting hypertriglyceridemia and hyperinsulinemia, and decreased levels of HDL cholesterol.

The transaminase activity observed during cirrhosis varies with the state of the cirrhotic process and can range from values around the URL to values four to five times higher, with an AST/ALT ratio >1, attributable to reduced ALT production in the damaged liver associated with reduced AST clearance as liver fibrosis progresses. An AST/ALT ratio ≥1 has a positive predictive value of ~90% in diagnosing the presence of advanced fibrosis in patients with chronic liver disease. In addition, the magnitude of the increase in the AST/ALT ratio may reflect the degree of fibrosis in these patients.

Two- to fivefold increases in both enzymes occur in patients with primary or metastatic liver cancer, with AST usually greater than ALT, but the activities are often normal in the early stages of malignant liver infiltration. Slight or moderate increases in AST and ALT activity have been observed after administration of various drugs, such as nonsteroidal anti-inflammatory drugs, antibiotics, antiepileptics, statins, or opioids. Over-the-counter medications and herbal preparations and homeopathic treatments may also be implicated. In subjects with increased transaminases, negative viral markers, and a negative history of drug or alcohol assumption, the diagnostic hypothesis should include the less common causes of chronic hepatopathy (hemochromatosis, Wilson's disease, autoimmune hepatitis, primary biliary cirrhosis, sclerosing cholangitis, celiac disease, α1-antitrypsin deficiency).

AST activity is also increased in progressive muscular dystrophy and dermatomyositis, reaching concentrations up to eight times the URL. Several authors have described the possible presence of circulating immunoglobulin-bound AST, or macro-AST. The typical picture of this condition includes a persistent increase in serum AST activity with physiological ALT concentrations in an asymptomatic subject, with the absence of any pathology affecting AST-rich organs. The increased AST activity in this condition reflects decreased clearance of the immune complex from plasma. Macro-AST has no clinical relevance. However, their identification is important to avoid unnecessary diagnostic procedures in these subjects.

Reference Intervals

The URL of AST in adults is 34 U/L, with no significant sex-related difference. In contrast, a clear difference between males and females was noted in ALT activity, whose corresponding URLs are 59 U/L and 41 U/L, respectively. ALT shows no age-related difference, whereas serum AST activity in infants and children aged <3 years is twice that of adults. Adult values are reached when the child enters puberty.

Alkaline Phosphatase

ALP catalyzes the alkaline hydrolysis of a wide variety of natural and synthetic substrates. This enzyme is present in many organs and is especially associated with cell surfaces located in the mucosa of the small intestine and in the proximal convoluted tubules of the kidney, in bone (osteoblasts), in the liver and in the placenta, anchored on the cell membrane by glycosylphosphatidylinositol (ectoenzyme). Although its exact metabolic function is not yet understood, it appears that ALP is associated with lipid transport in the intestine and with the calcification process in bone.

ALP exists in multiple forms (molecular weight 70–120 kDa), some of which are true isoenzymes, encoded by separate genetic loci. Bone, liver, and kidney forms of ALP share a common primary structure, encoded by the same genetic locus, but differ in carbohydrate content. The ALP activity present in the sera of apparently healthy adults comes primarily from the liver, with most of the remaining activity coming from the bone. The relative contributions of these two forms of the enzyme to total ALP activity are age dependent. Minimal amounts of intestinal ALP may also be present, particularly in the sera of individuals of blood group B or 0 (secretory subjects). Because intestinal ALP activity in serum may increase after a meal, ALP should preferably be measured in fasting state.

Clinical Significance

Increases in serum ALP activity commonly come from two major sources: liver and bone. Consequently, the measurement of serum ALP is of particular interest in the study of two groups of conditions: hepatobiliary diseases and bone diseases associated with the increased osteoblastic activity.

The liver's response to any form of biliary tree obstruction induces ALP synthesis by hepatocytes. The newly formed ectoenzyme is released from the cell membrane by the action of bile salts and enters the circulation by increasing enzyme activity in the serum. Increases tend to be greater (4 × URL) in extrahepatic obstruction (from gallstones or pancreatic head cancer) than in intrahepatic obstruction and are greater the more complete the obstruction. Enzyme activity can reach 10–12 times the URL and usually returns to baseline following surgical removal of the obstruction. Increased ALP (>2 × URL) can predict disease outcome (liver transplantation or death) in patients with primary biliary cirrhosis.

Liver diseases that primarily affect parenchymal cells, such as infectious hepatitis, typically show only moderately increased (<3 × URL) or even normal serum ALP activities. Increases could also be a side effect of drug therapy, and criteria based on the measurement of ALT and ALP have been recommended to discriminate the type of liver injury in drug-induced toxicity (Table 10.2). The intestinal isoenzyme of

Table 10.2 Criteria for drug-induced liver toxicity based on enzymatic data according to the Council of International Organizations of Medical Sciences (1990)[a]

Injury site	Alanine aminotransferase (ALT)	Alkaline phosphatase (ALP)	ALT/ALP ratio
Hepatocellular	>2 × URL	<URL	≥5
Cholestatic	<URL	>2 × URL	<2
Mixed	>2 × URL	>2 × URL	2–5

[a]Enzyme values are expressed as multiples of the upper reference limit (URL) in serum

ALP, an asialoglycoprotein normally picked up by specific hepatic receptors, is often increased in patients with liver cirrhosis.

An increase of up to two to three times the URL due to the placental-derived enzyme is observed in women in the third trimester of pregnancy. This makes ALP an unreliable marker of hepatobiliary disease in pregnancy. Benign familial increases in serum ALP levels due to increased intestinal ALP concentrations have also been described. Transient, benign increases in serum ALP may be observed in infants and children, with changes often >10 times the URL. These changes appear to reflect a reduction in blood ALP clearance caused by transient changes in glycosylation of the enzyme. Tissue-specific ALP gene mutations are associated with hypophosphatasia, a rare inherited disorder characterized by poor bone mineralization and low serum ALP concentrations.

Reference Intervals

Serum ALP activity varies with age and sex. Infants and children show higher ALP activity than healthy adults because of bone ALP production by osteoblasts during bone growth. However, activities in growing children are highly variable and it is known that the decrease in ALP activity to typical adult values differs from subject to subject, occurring on average 2 years earlier in females than in males.

In adult individuals, the following reference intervals have been established: 33–98 U/L for females of childbearing age and 43–115 U/L for males. A progressive increase in enzyme values is described for women after menopause.

γ-Glutamyltransferase (GGT)

GGT catalyzes the transfer of the γ-glutamyl group from peptides and other compounds to an acceptor. The enzyme acts only on peptides or peptide-like compounds containing a terminal glutamate residue joined to the rest of the compound via the (γ)carboxyl terminus.

GGT is present (in descending order of abundance) in the renal proximal tubule, liver, pancreas, and intestine. The enzyme is present in the cytoplasm (microsomes), but the largest amount is localized in the cell membrane and can transport amino acids and peptides into the cell across its membrane in the form of γ-glutamyl peptides. GGT activity is also important for the maintenance of adequate intracellular concentrations of reduced glutathione, an important antioxidant agent.

Serum GGT activity derives primarily from the liver, where it is found predominantly in the biliary pole of the hepatocyte. However, it is also found in the cytoplasm and smooth endoplasmic reticulum (where the enzyme is susceptible to induction).

Clinical Significance

Although renal tissue has the highest concentration of GGT, the enzyme present in serum comes primarily from the hepatobiliary system. GGT is a sensitive indicator of the presence of hepatobiliary disease, being increased in most individuals with liver disease regardless of cause, but its usefulness is limited by its lack of specificity. As in the case of ALP, it is higher in cases of intrahepatic or posthepatic biliary obstruction, reaching activities 5–30 times higher than URL. Marked increases in GGT are also observed in patients with primary or secondary (metastatic) cancers of the liver and other space-occupying liver lesions, presumably caused by intrahepatic obstruction. Moderate increases (two to five times the URL) occur in infectious hepatitis. Slight increases in GGT are observed in more than 50% of patients with NAFLD and similar but transient increases are noted in cases of drug intoxication. In acute and chronic pancreatitis and in some malignancies of the pancreas (especially when associated with hepatobiliary obstruction), enzyme activity may be 5–15 times higher than the URL.

Increased GGT activities are found in the sera of patients with alcoholic hepatitis and in most sera of heavy drinkers. GGT also increases with increasing body weight and in obese subjects. Increased concentrations of the enzyme are also found in the serum of subjects receiving anticonvulsant drugs, such as phenytoin and phenobarbital. Such an increase in serum GGT activity may reflect the induction of new enzymatic activity by alcohol and drugs or their toxic effect on the microsomal structures of the hepatocyte.

Unlike ALP, serum GGT is not increased in situations where osteoblastic activity is increased; therefore, measurement of the enzyme may be useful in differentiating the source of increased serum ALP activity, bone, or liver.

Epidemiological evidence has shown that serum GGT activity possesses an independent prognostic value for cardiovascular morbidity and mortality. This appears to be related to the ability of GGT to mediate redox/pro-oxidant reactions at the cellular level.

Reference Intervals

In adults, the URL of GGT in serum is 40 U/L for females and 68 U/L for males. In term infants, GGT activity at birth is ~6–7 times the adult reference interval. The activity then decreases, reaching adult values between 5 and 7 months of age.

Pancreatic Enzymes: Lipase

Lipases are enzymes that hydrolyze glycerol esters with long-chain fatty acids. Pancreatic lipase is a single-chain glycoprotein with a molecular weight of 48 kDa. The concentration of lipase in the pancreas is 5000-fold higher than in other tissues, and the concentration gradient between pancreas and serum is ~20,000-fold. Most of the lipase activity in serum comes from pancreatic acinar cells, but a small amount is also secreted by gastric, pulmonary, and intestinal mucosa. Lipase is a molecule small enough to be freely filtered through the glomerulus. It is totally reabsorbed by the renal tubules and is therefore not detectable in the urine.

Clinical Significance

Serum lipase measurement represents the recommended laboratory test for the diagnosis of acute pancreatitis. The clinical sensitivity is 80–100% depending on the selected decision level and the clinical specificity of 80–100% depending on the population to which the test is applied. After an attack of acute pancreatitis, serum lipase activity increases within 4–8 hours, peaks after ~24 hours, and values return to basal levels within 7–14 days. Increases between 2 and 50 times the URL have been reported, but it is important to remember that the magnitude of the increase in lipase activity is not necessarily proportional to the severity of the attack.

Acute pancreatitis is sometimes difficult to diagnose because it must be differentiated from other serious intra-abdominal disorders with similar clinical findings, such as perforated gastric or duodenal ulcer or intestinal obstruction. In the differential diagnosis, increased serum lipase concentrations >3 times the URL, in the absence of renal insufficiency, is a more specific diagnostic finding than increased serum α-amylase activity. In addition, lipase concentrations remain increased longer than α-amylase concentrations, which is another advantage in patients with delayed clinical presentation. Therefore, it is recommended that lipase should replace α-amylase determination as to the first diagnostic step for acute pancreatitis in a clinical emergency; on the other hand, simultaneous measurement of both enzymes is not warranted.

In subjects with reduced glomerular filtration rate, serum lipase activity is increased. Therefore, caution should be paid in the interpretation of increased lipase values in the presence of renal disease.

Reference Intervals

Among the most clinically essential enzymes, lipase is the only one whose measurement (and consequently the result) is method dependent. The reference intervals of lipase are, therefore, method dependent. For methods based on the use of diglycerides as a substrate, the suggested URL is 45 U/L, whereas the method using methylresorufin as a substrate has an URL of 64 U/L. There are no differences related to gender or age.

Amylase (Pancreatic)

The α-amylases are enzymes of the class of hydrolases that catalyze the hydrolysis of 1,4-α-glucosidic bonds present in polysaccharides (amylose, amylopectin, and glycogen). Amylases present physiologically in human plasma are small molecules with molecular weights ranging from 54 to 62 kDa. The enzyme is therefore small enough to pass freely through renal glomeruli, and amylase is the only plasma enzyme physiologically found in urine.

Amylases are present in numerous organs and tissues. The highest concentration is observed in the salivary glands, which secrete a potent amylase (the S form) to initiate the hydrolysis of starch when the food is still in the mouth and esophagus. In the pancreas, the enzyme (the P form) is synthesized by the acinar cells and is then secreted into the intestinal tract by the pancreatic ductal system. Amylase activity is also found in sperm, testes, ovaries, fallopian tubes, striated muscles, lungs, and adipose tissue. The enzyme physiologically present in serum and urine is predominantly of pancreatic and salivary origin. These isoenzymes are the products of two closely related loci present on chromosome 1. Amylase isoenzymes also undergo posttranslational modifications of deamidation, glycosylation, and deglycosylation to form several isoforms.

Clinical Significance

Plasma amylase activity is physiologically low and constant, and it increases markedly in acute pancreatitis and salivary gland inflammation. However, the specificity of amylase determination for the diagnosis of acute pancreatitis is low (20–60%, depending on the type of patient population studied), because increased values are also found in numerous other acute intra-abdominal disorders and in various extra-pancreatic conditions. The lack of specificity of total amylase measurement has shifted clinical interest to the direct measurement of pancreatic amylase. Applying the best decisional cut-off (an activity of three times the URL) and measuring the enzyme only under clinical conditions of diagnostic suspicion, the specificity of pancreatic amylase for the diagnosis of acute pancreatitis is >90%. However, it must be remembered that biliary tract diseases, such as acute cholecystitis, and various intra-abdominal events cause an increase in serum pancreatic amylase activity of up to four times the URL and sometimes even higher.

In renal insufficiency, serum amylase activity is increased in proportion to the degree of renal impairment. Hyperamylasemia also occurs in multiple myeloma, pulmonary neoplastic diseases, and serous and mixed ovarian carcinomas, with increases even >50 times the URL. In 1% of the population, the presence of macroamylases, which can cause hyperamylasemia, can be detected in the serum. These macroamylases are complexes composed of amylase (S-type amylase) and IgG or IgA. These macroamylases cannot be filtered by renal glomeruli because of their large size (>200 kDa) and are thus retained in plasma, where their presence, typically stable over time, can increase measurable amylase activity by two to eight times the URL. No clinical significance is associated with this biological manifestation.

Reference Intervals

In apparently healthy adults, pancreatic amylase accounts for ~40–50% of total serum amylase activity. The reference interval is 13–51 U/L. The activity of the pancreatic form is not demonstrable in the serum of most children <6 months of age, then slowly increasing to adult concentrations around 5 years of age; this reflects postnatal development of the exocrine pancreatic function. Consequently, the use of this enzyme for the diagnosis of suspected acute pancreatitis in young children should be avoided.

Bone Enzymes

Bone enzymes are direct products of activated osteoblasts (bone ALP [BAP]) and osteoclasts (tartrate-resistant acid phosphatase [TR-ACP]).

Alkaline Phosphatase (Bone Isoform)

The bone, liver, and kidney isoforms of ALP have posttranslational modifications of the product of the same gene and are differentiated according to their carbohydrate content. Bone alkaline phosphatase (BAP) is produced by osteoblasts: the enzyme is therefore an excellent indicator of global bone formation activity.

Clinical Significance

Advantages of using serum BAP concentrations as markers of bone formation in clinical practice include low biological variability and no influence by renal dysfunction. Among bone diseases, the highest BAP concentrations are found in Paget's disease (osteitis deformans) as a result of the action of osteoblastic cells because they attempt to rebuild bone that is being resorbed by uncontrolled osteoclast activity. Values of 10–25 times the URL are not unusual, and, in general terms, the magnitude of the increase reflects the extent of the disease. It is, however, worth commenting that, for a correct clinical setting, it is mostly sufficient to measure total ALP activity for the assessment of bone turnover and monitoring of Paget's disease.

In vitamin D deficiency (osteomalacia and rickets), BAP concentrations two to four times higher than the URL may be observed, slowly returning to baseline with treatment. Primary and secondary hyperparathyroidism is associated with mild to moderate increases in serum BAP, in which the degree of the increase reflects the presence and degree of skeletal involvement. BAP may be slightly increased in osteoporosis, but subjects with osteoporosis are not clearly distinguishable from healthy controls. Very high concentrations of BAP are present in patients with osteogenic bone cancer. Increased BAP indicates bone metastases in 70% of prostate cancers and its measurement has been recommended by the European Association of Urology guidelines for its staging. The Kidney Disease: Improving Global Outcomes (KDIGO) guidelines for the evaluation and management of chronic kidney disease recommend assessing associated mineral and bone disorders by measuring total ALP activity along with calcemia, phosphatemia, and parathyroid hormone (PTH) in adults with glomerular filtration rate (GFR) <45 mL/min/1.73 m^2. The level of evidence for recommending the specific measurement of BAP instead of total ALP was considered inadequate by the guideline.

Reference Intervals

The reference interval for adult men of serum BAP concentration does not change with age (7.5–26.1 μg/L), whereas menopause increases BAP concentrations in women (URL: 22.7 μg/L in women of childbearing age and 31.6 μg/L in postmenopausal women). Prepubertal BAP concentrations are six to seven times higher than in healthy adults.

Acid Phosphatase (Tartrate-Resistant Isoform 5b)

The term acid phosphatase (ACP) includes all phosphatases with optimal activity below a pH of 7.0. ACP is expressed in lysosomes, cell organelles present in every cell except erythrocytes; in addition, high concentrations of ACP are found in the prostate, bone (osteoclasts), spleen, platelets, and erythrocytes. Lysosomal and prostatic enzymes are strongly inhibited by tartrate ions, which, however, do not inhibit erythrocyte and bone isoenzymes. Most of the low activity of serum ACP (nonhemolysate) is tartrate-resistant (TR-ACP) and originates primarily from osteoclasts. The activity of this fraction is increased physiologically in growing children and pathologically under conditions of increased osteolysis and bone remodeling.

A gene on chromosome 19 encodes for TR-ACP expressed in osteoclasts and macrophages of other tissues, such as alveolar macrophages and Kupffer cells (ACP type 5). Isoenzyme 5 consists of two structurally related isoforms that differ in their carbohydrate content: TR-ACP 5a, which is derived primarily from macrophages and dendritic cells, and TR-ACP 5b, a more specific marker of osteoclastic activity.

Clinical Significance

TR-ACP 5b is a potentially useful marker in conditions characterized by a marked osteolytic component. Slight or moderate increases in TR-ACP 5b serum levels occur in Paget's disease, hyperparathyroidism with skeletal involvement, and in the presence of malignant bone invasion. TR-ACP 5b shows relatively small dynamic changes compared to other markers of bone resorption (e.g., those derived from type I collagen catabolism). This could be attributable to the fact that the enzyme is released into the sealed microenvironment of osteoclasts rather than directly into the circulation. However, unlike blood concentrations of bone resorption markers derived from mature type I collagen, TR-ACP 5b is not affected by renal dysfunction.

Reference Intervals

In the serum of apparently healthy adults, the reference interval of TR-ACP 5b activity is 1.5–4.5 U/L. Children show a higher activity (3.4–9.0 U/L).

Other Enzymes

Lactate Dehydrogenase

Lactate dehydrogenase (LDH) is an enzyme that, through hydrogen transfer, catalyzes the oxidation of L-lactate to pyruvate with the mediation of NAD+ as a hydrogen acceptor. The enzyme has a molecular weight of 134 kDa and is composed of four peptide chains of two types (M and H), each under separate genetic control. The structures of LDH-M and LDH-H are expressed, respectively, from loci localized on chromosomes 11 and 12. The subunit compositions of the five isoenzymes, in descending order of anodic mobility in an alkaline medium, are: LDH-1 (HHHH; H_4), LDH-2 (HHHM; H_3M), LDH-3 (HHMM; H_2M_2), LDH-4 (HMMM; HM_3), and LDH-5 (MMMM; M_4). A sixth different isoenzyme of LDH, LDH-X, composed of 4 X subunits, is present in postpubertal human testes. A seventh LDH, termed LDH-6, has been identified in the sera of critically ill patients.

LDH activity is present in many human cells and tissues. The concentration of enzymes in tissues ranges from 1500 to 5000 times that found physiologically in serum. Thus, the release of the enzyme from even a small amount of damaged tissue significantly increases the observed serum LDH activity. Different tissues show different isoenzyme compositions. In the heart, kidney, and erythrocytes, LDH-1 and LDH-2 isoenzymes predominate, while LDH-4 and LDH-5 isoenzymes predominate in the liver and skeletal muscle.

Clinical Significance

Because of its wide tissue distribution, increases in serum LDH occur in a variety of clinical conditions, including myocardial infarction, hepatitis, and hemolysis. However, the measurement of LDH in the serum is relevant only in hematology and oncology.

Hemolytic anemia significantly increases serum LDH concentrations. Marked increases in LDH activity, up to 50 times the URL, are observed in megaloblastic anemias. The latter usually result from a deficiency of folic acid or vitamin B12 and causes the weakening of the erythrocyte precursor cell in the bone marrow (ineffective erythropoiesis), leading to the release of large amounts of LDH. These increases quickly return to baseline values after appropriate treatment.

For monitoring purposes, LDH is relevant in predicting disease activity in leukemias and the probability of survival in Hodgkin's disease and non-Hodgkin's lymphomas.

Patients with malignant disease often show increases in serum LDH; up to 70% of patients with liver metastases and 20–60% of patients with other metastases (e.g., lymph nodes) show increased LDH activity. Increased LDH-1 is observed in germ cell tumors (~60% of cases), such as teratoma, testicular seminoma, and dysgerminoma of the ovary. The percentage of patients with increased LDH depends on the stage of the disease. LDH also appears to be a useful predictor of outcome in patients with nonseminomatous germ cell tumors of the testis, melanoma, and pulmonary microcytoma.

Increases in LDH are observed in hepatopathies, but their clinical use in a liver profile appears very limited because they do not add significant information to the determination of transaminases. Finally, measurement of LDH in pleural fluid (better if in combination with serum LDH) helps to distinguish between exudative (pleural fluid/LDH serum LDH >0.6) and transudative effusions (ratio <0.6).

The presence of macro-LDH due to the formation of an autoantibody–enzyme complex that leads to a persistent increase in the amount of circulating enzyme is estimated to occur in <1 person each 10,000. Its presence should be checked in suspected individuals to avoid further in-depth investigation or unnecessary treatment.

Reference Intervals

The reference interval of LDH in adult Caucasian subjects is 125–220 U/L, whereas Asian individuals show slightly higher values. The reference intervals are markedly higher in children, with a gradual decrease observed throughout the childhood period.

Cholinesterase (Pseudocholinesterase)

Two interrelated enzymes can hydrolyze acetylcholine. The first is acetylcholinesterase, also called true cholinesterase. This is found in erythrocytes, lungs and spleen, nerve endings and brain gray matter. It is responsible for the hydrolysis of acetylcholine released at nerve endings to mediate the transmission of nerve impulses across the synapse. The degradation of acetylcholine is necessary for the depolarization of the nerve so that it can be repolarized in the next impulse conduction event. The other enzyme is acetylcholine acetylhydrolase (CHE), also called pseudocholinesterase, serum cholinesterase, or butyrylcholinesterase, which is found in liver, pancreas, heart, and serum. Although CHE activity in the human body is ~3 times greater than acetylcholinesterase activity, its exact biological role is still unknown. Of clinical interest are atypical (genetic) variants of the enzyme characterized by decreased activity toward acetylcholine and other substrates, which are found in the serum of a small number of apparently healthy people.

The gene on chromosome 3 that controls CHE synthesis can exist in many allelic forms. Four of the most common forms are designated as E^u, E^a, E^f, and E^s. These four allelic genes can be combined into one normal and nine abnormal genotypes. The most common (normal) genotype is designated as E^uE^u or UU ("usual"). The E^a gene is designated as an atypical gene; sera from subjects homozygous for this gene (E^aE^a or AA) are weakly active toward most CHE substrates and show increased resistance to inhibition of enzymatic activity by dibucaine. The E^f (fluoride-resistant) gene gives rise to a weakly active enzyme but with increased resistance to inhibition by fluoride. The E^s (silent) gene is associated with the absence of the enzyme or the presence of a protein with minimal or no catalytic activity. Abnormal enzymes are fewer effective catalysts than the common form; the affinity of the enzymes for substrates and that for competitive inhibitors, such as dibucaine or fluoride, is reduced. This gives rise to the characteristic dibucaine- or fluoride-resistant properties of the genetic variants, which can be exploited in their characterization. The homozygous forms, AA and FF, are found in 0.3–0.5% of the white population.

Clinical Significance

Table 10.3 shows the indications for the measurement of serum CHE activity.

Table 10.3 Indications for measuring serum cholinesterase activity

To assess the liver function (e.g., in monitoring patients after liver transplantation)
To evaluate a possible acute poisoning by organophosphorus insecticides
To identify subjects with atypical forms of enzyme, who are at risk of prolonged responses to some muscle relaxants (e.g., succinylcholine) used in anesthesia

Measurement of serum CHE activity can serve as a sensitive indicator of liver synthesis capacity. In the absence of genetic causes or known inhibitors, any decrease in CHE activity reflects an impairment of hepatic synthesis of the enzyme.

Organophosphorus compounds that inhibit CHE activity include many insecticides, such as parathion, sarin, and tetraethylpyrophosphate. Agricultural and chemical workers may be subject to poisoning by inhalation of these substances or by direct contact. Both cholinesterases are inhibited, but the activity of the serum enzyme falls more rapidly than that of the erythrocyte enzyme.

Succinylcholine and mivacurium, myorelaxant drugs used in surgical procedures as an aid for endotracheal intubation, are hydrolyzed by CHE and their pharmacological effect normally persists enough to meet the needs of the surgical procedure. In subjects with low enzyme activities or in those with a weakly active variant, drug degradation does not occur rapidly enough, and the patient may enter a prolonged period of respiratory muscle paralysis (apnea), requiring mechanical ventilation until the effects of the drug gradually wear off. Preoperative screening of CHE activity has been proposed to identify individuals in whom succinylcholine administration may lead to complications; however, in many countries, this screening is not recommended, as all individuals undergoing surgery are considered potentially at risk. The phenotypes most susceptible to apnea after succinylcholine administration are AA, AS, FF, FS, SS, AF, and, to some extent, UA. Measurements of total CHE activity and determination of "dibucaine number" and "fluoride number" are necessary for the complete characterization of CHE variants. The dibucaine number and fluoride number indicate the percentage inhibition of enzyme activity in the presence of standard concentrations of these two compounds. Mutation genotyping can confirm alterations in the CHE gene.

Reference Intervals

The reference intervals for CHE are method dependent. At birth, CHE activity is below adult values by ~50%. It increases during the next 3–6 years until it exceeds adult values of ~30%. From the fifth year of life, the activity begins to decrease until it stabilizes at adult values with puberty. The significant decrease in CHE (~30%) during pregnancy

Table 10.4 Values of dibucaine (DN) and fluoride (FN) for various serum cholinesterase phenotypes

Phenotype	Interval DN	Interval FN
UU	≥77	≥55
UF	72–26	≥53
UA	48–72	≥44
AF	45–59	<44
FF	64–69	<44
AA	<35	[a]

[a]Redundant for the attribution of the phenotype AA

and puerperium is explained by the hemodilution inherent in this condition.

Based on the determination of dibucaine and fluoride numbers, the values characterizing the various CHE phenotypes were established (Table 10.4).

Recommended Readings

Beetham R (2000) Biochemical investigation of suspected rhabdomyolysis. Ann Clin Biochem 37:581–587

Bianchi ML (2015) Hypophosphatasia: an overview of the disease and its treatment. Osteoporos Int 26:2743–2757

Ceriotti F, Henny J, Queraltó J et al (2010) Common reference intervals for aspartate aminotransferase (AST), alanine aminotransferase (ALT) and γ-glutamyl transferase (GGT) in serum: results from an IFCC multicenter study. Clin Chem Lab Med 48:1593–1601

Chao T-Y, Wu Y-Y, Janckila AJ (2010) Tartrate-resistant acid phosphatase isoform 5b (TRACP 5b) as a serum marker for cancer with bone metastasis. Clin Chim Acta 411:1553–1564

Fahie-Wilson MN, Burrows S, Lawson GJ et al (2007) Prevalence of increased serum creatine kinase activity due to macrocreatine kinase and experience of screening programmes in district general hospitals. Ann Clin Biochem 44:377–383

Heiduk M, Päge I, Kliem C et al (2009) Pediatric reference intervals determined in ambulatory and hospitalized children and juveniles. Clin Chim Acta 406:156–161

Huijgen HJ, Sanders GT, Koster RW et al (1997) The clinical value of lactate dehydrogenase in serum: a quantitative review. Eur J Clin Chem Clin Biochem 35:569–575

Infusino I, Frusciante E, Braga F et al (2017) Progress and impact of enzyme measurement standardization. Clin Chem Lab Med 55:334–340

Kidney Disease: Improving Global Outcomes (KDIGO) CKD Work Group (2013) KDIGO 2012 clinical practice guideline for the evaluation and management of chronic kidney disease. Kidney Int Suppl 3:1–150

Kim WR, Flamm SL, Di Bisceglie AM et al (2008) Serum activity of alanine aminotransferase (ALT) as an indicator of health and disease. Hepatology 47:1363–1370

Lilford RJ, Bentham LM, Armstrong MJ et al (2013) What is the best strategy for investigating abnormal liver function tests in primary care? Implications from a prospective study. BMJ Open 3(6):e003099

McQueen MJ (1995) Clinical and analytical considerations in the utilization of cholinesterase measurements. Clin Chim Acta 237:91–105

Michelsen J, Wallaschofski H, Friedrich N et al (2013) Reference intervals for serum concentrations of three bone turnover markers for men and women. Bone 57:399–404

Moss DW (1992a) Changes in enzyme expression related to differentiation and regulatory factors: the acid phosphatase of osteoclasts and other macrophages. Clin Chim Acta 209:131–138

Moss DW (1992b) Perspectives in alkaline phosphatase research. Clin Chem 38:2486–2492

Panteghini M (2009) Lactate dehydrogenase and lactate dehydrogenase isoenzymes. In: Hickman PE, Koerbin G (eds) Methods in clinical chemistry. Pesce Kaplan Publishers, pp 793–796

Panteghini M (2010) Laboratory evaluation of pancreatic diseases. Biochim Clin 34:19–25

Panteghini M (2012) The never-ending search of an acceptable compromise for pancreatic lipase standardisation. Clin Chem Lab Med 50:419–421

Panteghini M, Pagani F, Bonora R (1993) Clinical and analytical evaluation of a continuous enzymatic method for measuring pancreatic lipase activity. Clin Chem 39:304–308

Phillip V, Steiner JM, Algül H (2014) Early phase of acute pancreatitis: assessment and management. World J Gastrointest Pathophysiol 5:158–168

Pratt DS, Kaplan MM (2000) Evaluation of abnormal liver-enzyme results in asymptomatic patients. N Engl J Med 342:1266–1271

Sardiwal S, Magnusson P, Goldsmith DJA et al (2013) Bone alkaline phosphatase in CKD-mineral bone disorder. Am J Kidney Dis 62:810–822

Sathasivam S (2012) Statin induced myotoxicity. Eur J Intern Med 23:317–324

Stein P, Rosalki SB, Foo AY et al (1987) Transient hyperphosphatasemia of infancy and early childhood: clinical and biochemical features of 21 cases and literature review. Clin Chem 33(2 Pt 1):313–318

Tenner S, Baillie J, DeWitt J et al (2013) American College of Gastroenterology guideline: management of acute pancreatitis. Am J Gastroenterol 108:1400–1415

Von Eyben FE (2001) A systematic review of lactate dehydrogenase isoenzyme 1 and germ cell tumors. Clin Biochem 34:441–454

Liver: From Biochemistry to Clinical Biochemistry

11

Marcello Ciaccio, Luisa Agnello, and Anna Maria Ciaccio

Introduction

Generally, liver function is studied by evaluating parameters that allow the discrimination of the main clinical pictures, which will then be investigated by specialized second- and third-level laboratory tests and imaging investigations. This chapter describes the main clinical pictures that can affect the liver and the relative laboratory tests that allow their correct classification.

Jaundice

The term jaundice, from the Greek ίκτερος (yellow), refers to the yellowish discoloration of the skin, sclerae, and mucous membranes, caused by the deposition of bilirubin. The latter is the product of the catabolism of heme, the prosthetic group, i.e., the nonprotein part of various proteins, including hemoglobin, myoglobin, and enzymes belonging to the cytochrome P450 family, catalases, and peroxidases. In this chapter, the metabolism of bilirubin is initially described to allow a better understanding of the classification of jaundice. Subsequently, the various classifications of jaundice are presented with a description of the main causes.

M. Ciaccio (✉) · L. Agnello
Department of Biomedicine, Neurosciences and Advanced Diagnostics, Institute of Clinical Biochemistry, Clinical Molecular Medicine and Clinical Laboratory Medicine, and Department of Laboratory Medicine, University Hospital "P. Giaccone", Palermo, Italy
e-mail: marcello.ciaccio@unipa.it

A. M. Ciaccio
Department of Health Promotion, Mother and Child Care, Internal Medicine and Medical Specialties (ProMISE) "G. D'Alessandro", University of Palermo, Palermo, Italy

Bilirubin Metabolism

Every day an adult produces about 250–400 mg of bilirubin, most of which (>80%) comes from the catabolism of hemoglobin and a small amount from the hepatic catabolism of other hemoproteins (proteins containing the heme group). In particular, the bilirubin produced by the catabolism of hemoglobin can be divided into a hemo-cateretic portion (70%), which comes from the physiological destruction of red blood cells in the reticuloendothelial system of the spleen, and an erythropoietic portion (30%) from ineffective erythropoiesis in the bone marrow (catabolism of heme molecules not used in erythropoiesis and destruction of erythroblasts, reticulocytes, and newly formed red blood cells).

Bilirubin is formed by the sequential catalytic degradation of heme mediated by two enzymes, heme oxygenase and biliverdin reductase. Heme oxygenase initiates the opening of the porphyrin ring of heme, leading to the formation of the green pigment, known as biliverdin, which, subsequently, will be reduced by biliverdin reductase to a yellow-orange pigment, namely the bilirubin (Fig. 11.1).

The bilirubin synthesized, called indirect or unconjugated, is poorly soluble in water at physiological pH and, being liposoluble, is responsible for the toxic effects of bilirubin. This form of bilirubin circulates in the plasma bound to proteins, mainly albumin and, to a much lesser extent, to high-density lipoproteins; the binding to lipoproteins becomes significant in severe hypoalbuminemia. Binding to albumin keeps bilirubin in the vascular space, thus preventing its deposition in extra-hepatic tissues, including the brain, and minimizing its filtration at the glomerular level. Albumin is also vital in transporting indirect bilirubin to the sinusoidal surface of hepatocytes, where the pigment dissociates from albumin and enters the hepatocytes by facilitated diffusion. Defects in genes encoding for bilirubin transporters may result in hyperbilirubinemia. The passage of bilirubin across the membrane of the sinusoidal surface of

M. Ciaccio (ed.), *Clinical and Laboratory Medicine Textbook*, https://doi.org/10.1007/978-3-031-24958-7_11

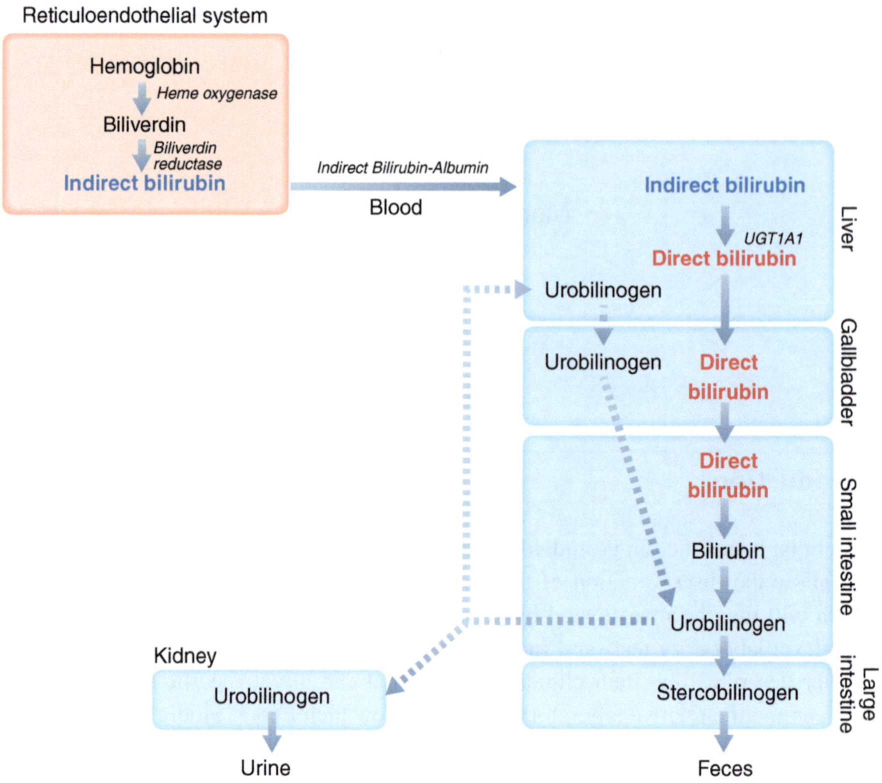

Fig. 11.1 Bilirubin metabolism. (Copyright EDISES 2021. Reproduced with permission)

hepatocytes is bidirectional. In order to prevent the escape of bilirubin from hepatocytes, bilirubin, as well as other organic anions, is bound to the cytosolic protein glutathione S-transferase (GST), also known as ligandin, which mediates its transport to the endoplasmic reticulum (ER). In the ER, indirect bilirubin undergoes a process of glucuronidation that converts it to a water-soluble form, the direct bilirubin (Fig. 11.1). The glucuronidation of bilirubin, as well as of various compounds, both endogenous (steroid hormones, thyroid hormones, and catecholamines) and exogenous (drugs, toxins, carcinogens), is mediated by a family of enzymes known as uridine diphosphate glucuronosyltransferases (UGT). The resulting glucuronides are water-soluble and, therefore, readily excreted in bile and urine. Glucuronidation is one of the most important detoxification mechanisms in our body. Among isoforms of the UGT family, only UGT1A1 is physiologically crucial in bilirubin metabolism. Following glucuronidation, mono- and diglucuronide bilirubin molecules are formed and directed toward the apical canalicular membrane and then transported into the bile canaliculus. Conjugated bilirubin, also called direct bilirubin, is secreted by active transport against a concentration gradient mediated by an ATP-dependent pump, known as MRP-2 (Multidrug Resistance Protein) or ABC-C2, expressed on the membrane of the biliary canaliculi of hepatocytes. Most of the bilirubin (approximately 80%) excreted in bile is present in the form of diglucuronide. In subjects with reduced glucuronidation activity, the bilirubin diglucuronide rate decreases, and the bilirubin monoglucuronide rate increases. Inhibitors of hepatic UGT1A1 may be present in breast milk, leading to neonatal jaundice, or in maternal plasma and cross the placental barrier, reaching the fetus (Lucey Discroll syndrome). A UGT1A1 deficiency can be found in newborns, in patients with chronic hepatitis, and in patients with some hereditary disorders, such as Gilbert's syndrome or Crigler–Najjar syndrome. Biliary excretion of conjugated bilirubin may be impaired in several acquired conditions, such as viral or alcoholic hepatitis and cholestasis in pregnancy, or congenital conditions, such as Dubin–Johnson syndrome and Rotor syndrome.

Conjugated bilirubin, produced at the hepatic level, after a brief stay in the gallbladder, reaches through the bile in the small intestine where, being water-soluble, it is not absorbed through the membrane of the epithelial cells. In the intestine, beta-glucuronidase enzymes mediate a process inverse to the hepatic one, removing glucuronic acid from bilirubin. Bilirubin is further metabolized by enzymes produced by the bacterial flora that mediate its reduction to urobilinogen (Fig. 11.1). A small fraction of urobilinogen is reabsorbed and returns to the circulation to be subsequently picked up in the liver or eliminated in urine, where urobilinogen undergoes further oxidation to urobilin. Urobilin is the molecule that gives urine its characteristic golden-yellow or straw-yellow color. Most of the urobilinogen (80%) in the colon is further processed by the bacterial flora into stercobilinogen and eliminated in the feces, where it undergoes further oxi-

Table 11.1 Reference range of bilirubin

Total bilirubin	<1 mg/dL
Non conjugated bilirubin	<0.8 mg/dL
Conjugated bilirubin	<0.3 mg/dL

dation to stercobilin. Stercobilin gives the characteristic color to the feces.

Classification of Jaundice

Jaundice is the clinical sign of hyperbilirubinemia. Normal values of total bilirubinemia are <1mg/dL, of which the conjugated form accounts for less than 5% (Table 11.1).

Total bilirubinemia values between 2 and 3 mg/dL result in sub-jaundice, characterized by a yellowish discoloration of the sclerae; values >3 mg/dL result in a clinically manifest jaundice condition characterized by a yellowish discoloration of the sclerae, skin, and mucous membranes. Jaundice and asymptomatic hyperbilirubinemia represent common clinical conditions that may be caused by a variety of disorders, which include increased bilirubin synthesis, impaired bilirubin conjugation, and biliary obstruction, and hepatic inflammation. Jaundice may represent the first and only sign of liver disease; therefore, its evaluation is of critical importance.

Various classifications of jaundice have been proposed. The following are the main ones.

According to traditional classification, jaundice can be distinguished based on the site of the cause into:

- Prehepatic: it results from the hyperproduction of bilirubin and it is characterized by an increase in the indirect bilirubin form
- Hepatocellular: it results from parenchymal lesions of the liver that impair the hepatic metabolism of bilirubin, leading to an increase in total bilirubinemia
- Posthepatic: it results from intra- and/or extra-hepatic biliary obstruction, with an increase in the direct form

According to a clinical biochemistry classification based on the form of altered bilirubin, jaundice is divided into three main categories:

- Jaundice characterized by a predominant increase in unconjugated bilirubin, which may be due to three pathophysiological mechanisms: increased production of bilirubin; altered uptake; or altered conjugation of bilirubin in the liver. The physiological jaundice of the newborn is a classic example in which all three pathophysiological mechanisms are present.
- Jaundice characterized by increased circulating levels of conjugated bilirubin, mainly due to biliary obstruction.
- Jaundice characterized by increased circulating levels of both conjugated and unconjugated bilirubin due to hepatocellular disease.

Finally, according to a pathophysiological classification, jaundice is distinguished as:

- Increased production of bilirubin
- Bilirubin uptake defect
- Bilirubin conjugation defect
- Defective bilirubin excretion
- Mechanical obstruction of the intra- and/or extra-hepatic biliary tract

Jaundice due to increased Bilirubin production

Overproduction of bilirubin is the result of excessive catabolism of heme derived from hemoglobin that may be due to hemolytic anemia, extravasation of blood into tissues, or dyserythropoiesis (characterized by altered incorporation of hemoglobin in erythrocytes resulting in the degradation of a large fraction of unincorporated hemoglobin).

These forms of jaundice are characterized by an increase in indirect bilirubinemia due to the inability of the liver to metabolize excess bilirubin.

Jaundice due to Bilirubin Uptake and Conjugation Defect

These forms of jaundice are characterized by a normal production of bilirubin, but the liver is unable to fully metabolize, leading to an increase in indirect bilirubinemia. Alterations in the release of bilirubin to the liver and in bilirubin internalization in hepatocytes result in reduced hepatic uptake. Congestive heart failure or portosystemic shunt (spontaneous collaterals in subjects with cirrhosis or surgical shunts) reduce hepatic blood flow and bilirubin release to hepatocytes, resulting in mainly indirect hyperbilirubinemia. In some patients with cirrhosis, direct plasma contact with hepatocytes may be impaired by capillarization of sinusoidal endothelial cells (loss of fenestrae), resulting in further reduced bilirubin uptake. Other causes of altered bilirubin uptake include Gilbert's syndrome and administration of various drugs, such as rifamycin, flavaspidic acid, and cholecystogram agents; drug-induced defects usually resolve within 48 hours of discontinuation of therapy.

Congenital Defects

Reduced or absent UGT1A1 enzyme activity is a major cause of jaundice due to bilirubin conjugation defects and can be detected in several acquired or inherited disorders, such as Gilbert's syndrome and Crigler–Najjar's syndrome types I and II. UGT1A1 activity is, moreover, modulated by various hormones; for example, thyroid hormones and ethinylestradiol inhibit bilirubin glucuronidation. Conversely, the combination of progestin and estrogenic steroids results in an increase in enzyme activity. Bilirubin glucuronidation may also be inhibited by some antibiotics (e.g., novobiocin or gentamicin at serum concentrations above therapeutic levels) and in chronic persistent hepatitis, advanced cirrhosis, and Wilson's disease.

Gilbert's syndrome is a benign hereditary disorder transmitted in an autosomal dominant trait, characterized by a defect in bilirubin uptake associated with a glucuronidation deficit due to alterations in the gene encoding for the enzyme UGT1A1. From an epidemiological point of view, Gilbert's syndrome is quite common, with a prevalence that varies between 4% and 16% in various populations. Patients with Gilbert's syndrome have moderately elevated levels of unconjugated bilirubin, with total bilirubin levels generally <3 mg/dL. Some associated diseases or physiological or paraphysiological events, such as stress, fasting, menstrual cycle, or asthenia, may determine an increase in total bilirubinemia, generally not exceeding 6 mg/dL. Other than jaundice, patients are typically asymptomatic, and no treatment is required.

Crigler-Najjar syndrome, also known as "congenital nonhemolytic jaundice with glucuronyltransferase deficiency" is a rare autosomal recessive disorder of bilirubin metabolism. Depending on the severity of the disease, two forms can be distinguished.

- Type I: It results from mutations (deletions, insertions, missense mutations, etc.) in one of the five exons of the gene encoding for the enzyme UGT1A1, which are associated with an altered protein characterized by a complete loss or a strong reduction in catalytic activity. Affected patients have severe jaundice and neurological impairment due to bilirubin encephalopathy, which can result in permanent neurological sequelae. Frequently, subjects affected by type I die in the neonatal period due to kernicterus (bilirubin encephalopathy). The hallmark of type I syndrome is the presence of pure unconjugated hyperbilirubinemia, with levels ranging from 20 to 25 mg/dL up to 50 mg/dL.
- Type II: It is known as "Arias syndrome" and it is phenotypically similar to type I. However, the unconjugated hyperbilirubinemia is usually lower than type I, with a concentration ranging from 8 to 18 mg/dL. Patients commonly survive to adulthood without neurological alterations. The disease is due to mutations in any of the five exons of the gene encoding for UGT1A1 but compared with the type I syndrome, which can be caused by a wide variety of alterations, the type II syndrome is due only to point mutations resulting in the substitution of single amino acid, with consequent reduction, but not abrogation, of the catalytic activity of the enzyme. The residual catalytic activity is responsible for the partial conjugation of bilirubin and the reduced severity of hyperbilirubinemia.

The differential diagnosis between Crigler–Najjar syndrome type I and type II is based on hyperbilirubinemia levels (lower in type II) and age of onset (late for type II). In addition, administration of phenobarbital may support the differential diagnosis because in most type II patients, but not type I, phenobarbital (60–120 mg for 14 days) can reduce bilirubinemia levels by 25%. Table 11.2 summarizes the main features of Gilbert's syndrome and Crigler–Najjar's syndrome types I and II.

Jaundice due to Defective Excretion of Bilirubin or Mechanical Obstruction of the Intra- and/or Extra-hepatic Biliary Tracts

These forms of jaundice are characterized by a normal production of bilirubin, which is taken up and conjugated in the liver but is not regularly excreted, leading to an accumulation of circulating direct bilirubin. Since direct bilirubin is water-soluble, it is filtered through glomerulus and eliminated in urine. In these forms of jaundice, therefore, there is direct hyperbilirubinemia and bilirubinuria. The altered excretion may be due to intra- and/or extra-hepatic cholestasis, the causes of which are summarized in Tables 11.3 and 11.4.

Table 11.2 Characteristics of Gilbert syndrome and Crigler–Najjar syndrome type I and II

	Crigler–Najjar syndrome type I	Crigler–Najjar syndrome type II	Gilbert syndrome
Hepatic histology	Normal	Normal	Normal
Bilirubinemia	20–50 mg/dL	<20 mg/dL	<3 mg/dL
Liver function	Normal	Normal	Normal
Bile	Usually pale, it contains small amounts of unconjugated bilirubin	Increased proportion of bilirubin monoglucuronide	Increased proportion of bilirubin monoglucuronide
Hepatic activity of UGT1A1	Absent	Markedly reduced	Reduced
Effect of phenobarbital on bilirubinemia	None	Reduction	Reduction
Inheritance	Autosomal recessive	Autosomal recessive	Autosomal dominant
Prevalence	Rare	Rare	Common
Prognosis	Neurological alterations	Usually benign, neurologic impairment is rarely present	Benign

Table 11.3 Causes of jaundice from intra-hepatic cholestasis

Acute hepatocellular injury	Viral hepatitis Steatosis and/or alcoholic hepatitis Nonalcoholic steatohepatitis Drugs
Chronic hepatocellular injury	Primary sclerosing cholangitis Primary biliary cholangitis Drugs Hepatitis (viral, alcoholic, autoimmune) Cirrhosis due to any cause
Multifactorial	Total parental nutrition Systemic infection Postoperative Sickle cell anemia Organ transplant (rejection; veno-occlusive disease; graft-versus-host disease)
Various	Hypotension/hypoxemia/heart failure Budd–Chiari syndrome Parasitic infection (Clonorchis sinensis; Fasciola hepatica)
Hereditary/endocrine	Benign recurrent intra-hepatic cholestasis Benign familial intra-hepatic cholestasis Cholestasis associated with low phospholipid levels Dubin–Johnson syndrome Rotor syndrome Thyrotoxicosis Alagille syndrome Disorders of the metabolism of carbohydrates, lipids, or bile acids Caroli's disease Pregnancy Protoporphyria
Infiltrative/ granulomatous	Amyloidosis Lymphoma Sarcoidosis Tuberculosis

Table 11.4 Causes of jaundice from extra-hepatic cholestasis

Cholangiopathies and other disorders involving the bile ducts	Cholelithiasis Biliary stenosis after invasive procedure Cholangiocellular carcinoma Primary sclerosing cholangitis AIDS cholangiopathy Cytomegalovirus Cryptosporidium spp. HIV Cyst of the choledochus Dysfunction of the sphincter of Oddi Parasitic infections Ascaris lumbricoides Histiocytosis X
Extrinsic causes	Pancreatitis (acute and chronic) Pancreatic cancer Portal adenopathy (metastasis, tuberculosis) Periampullary carcinoma Periampullary diverticulum Mirizzi syndrome

Neonatal Jaundice

In newborns, jaundice is a relatively frequent condition due to several mechanisms. Almost all newborns have total bilirubinemia values >1 mg/dL. As bilirubinemia increases, neonatal jaundice is observed. First, it affects the face and then progresses cranio-caudally to the trunk and extremities. Approximately 60% of term infants and 80% of preterm infants have physiologic jaundice, which is the result of the different neonatal metabolism of bilirubin, characterized by:

- Increased production of bilirubin: hemocateretic (shorter erythrocyte half-life) and hepatic share.
- Insufficient metabolic capacity of the liver: reduced levels of ligandin, glucuronic acid and UGT1A1. Specifically, UGT1A1 activity in term infants, at 7 days after birth, is approximately 1% of that of adults and does not reach adult levels until 14 weeks of age.
- Insufficient development of intestinal bacterial flora, resulting in insufficient production of bilinogens.

All these conditions result in an increase in unconjugated bilirubinemia. In Caucasian and African-American term infants, the mean peak total plasma bilirubinemia is 7–9 mg/dL and is reached between 48 and 96 hours of life. East-Asian infants may have total bilirubinemia values up to 10–14 mg/dL between 72 and 120 hours of life.

Generally, neonatal jaundice resolves within the first 2 weeks after birth, depending on the maturation of the bilirubin clearance system. Infants with severe hyperbilirubinemia, defined as total bilirubinemia >25 mg/dL, are at high risk of developing bilirubin-induced neurologic dysfunction due to the passage of indirect bilirubin across the blood–brain barrier.

Infants with clinically relevant hyperbilirubinemia should undergo phototherapy, which exploits the ability of radiation in the visible spectrum to convert bilirubin into isomers that can be rapidly eliminated hepatically and renally.

Diagnosis

The diagnostic approach to a jaundiced patient is based on careful anamnesis, physical examination, and laboratory data. Although evaluation is usually not urgent, in some cases, jaundice may represent a medical emergency, such as ascending cholangitis or fulminant liver failure.

From a laboratory point of view, the initial assessment is based on the measurement of circulating levels of:

- Total bilirubin
- Alkaline phosphatase
- Aminotransferase
- Prothrombin time (PT)
- Albumin

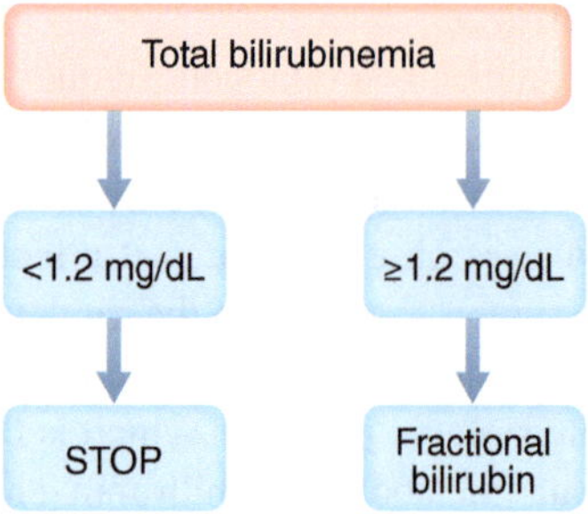

Fig. 11.2 Reflex test for the evaluation of bilirubin. (Copyright EDISES 2021. Reproduced with permission)

Aminotransferases (alanine aminotransferase [ALT] and aspartate aminotransferase [AST]), alkaline phosphatase, and bilirubin are biochemical markers of liver injury; albumin and PT are markers of hepatocellular function.

Some laboratories use a reflex test (Fig. 11.2) for bilirubinemia evaluation, where total bilirubinemia is determined first; if values are >1.2 mg/dL, direct bilirubinemia is determined. The value of indirect bilirubinemia is obtained with the following formula:

$$[\text{Indirect bilirubin}] = [\text{Total bilirubin}] - [\text{Direct bilirubin}].$$

The result of the laboratory investigations allows an initial differential diagnosis among the possible causes of jaundice (Table 11.5).

An increase in indirect bilirubinemia could be due to hemolytic anemia, medications that impair hepatic uptake of bilirubin, or inherited disorders, such as Gilbert's syndrome and Crigler–Najjar's syndrome. Direct hyperbilirubinemia, on the other hand, maybe due to biliary obstruction, intrahepatic cholestasis, hepatocellular injury, or hereditary disorders, such as Rotor's syndrome or Dubin–Johnson's syndrome. Notably, both syndromes are characterized by direct hyperbilirubinemia in the absence of other altered liver tests; normal levels of alkaline phosphatase support the differential diagnosis between these disorders and alterations associated with biliary obstruction.

Aminotransferases are generally normal in subjects with jaundice due to prehepatic causes, whereas levels increase in the case of jaundice due to posthepatic causes and, above all, to hepatic alterations. In addition, the pattern of aminotransferase alteration could point to a specific cause; for example, alcoholic hepatitis is characterized by a disproportionate increase in AST compared to ALT, with an AST/ALT ratio >2. Similar to aminotransferases, circulating levels of alkaline phosphate will be normal in subjects with jaundice due to prehepatic causes. However, they will be increased in subjects with jaundice due to hepatic and, above all, posthepatic alterations. In particular, an increase in alkaline phosphatase associated with normal or slightly increased aminotransferase values suggests biliary obstruction or intrahepatic cholestasis.

Table 11.5 Differential diagnosis of jaundice

	Prehepatic	Hepatic	Posthepatic
Total bilirubin	↑	↑	↑
Direct bilirubin (conjugated)	N	↑	↑↑
Indirect bilirubin (non-conjugated)	↑	↑	N/↑
Aminotransferase	N/↑	↑↑↑	↑
Alkaline phosphatase	N	N/↑	↑↑↑
Albumin	N	↓	N/↓

An increase in alkaline phosphatase could also be seen in granulomatous liver diseases, such as tuberculosis or sarcoidosis; these conditions could be characterized by jaundice. Increased alkaline phosphatase could result from increased extra-hepatic release, particularly in the bone. Extra-hepatic disorders do not cause jaundice, except in rare cases, such as bone tumors metastasizing to the liver. Measuring the serum canalicular enzyme γ-glutamyltransferase activity could help confirm the hepatic origin of alkaline phosphatase. Finally, an increased value of the prothrombin time that is corrected by the administration of vitamin K suggests an altered intestinal absorption of fat-soluble vitamins, compatible with obstructive jaundice (postepathic); on the other hand, an increase in the prothrombin time that is not corrected by vitamin K suggests the presence of moderate-severe hepatocellular disease with impaired synthesis function, especially if associated with hypoalbuminemia.

Based on the anamnesis, physical examination, and laboratory, an etiological diagnosis can be made, which can be confirmed with further investigations, such as liver imaging tests if biliary obstruction or intrahepatic cholestasis is suspected or serological tests if viral hepatitis or primary biliary cholangitis is suspected (Table 11.6).

Hepatitis

The term hepatitis refers to the inflammation of the liver. It can be due to various causes, mainly infection with hepatotropic viruses, alcohol abuse, drug intake, or alterations in the immune system. Hepatitis may occur in an acute or chronic form when it lasts for more than 6 months.

Viral Hepatitis

Viral hepatitis is an inflammatory process of the liver due to viral infection.

Although viral hepatitis is similar from clinical biochemistry and morphological point of view, it differs from an etiological (different viruses responsible for infection), epidemiological (different distribution and frequency of infection and disease), and immunopathogenic point of view.

Table 11.6 Common causes of liver laboratory test alterations

Disease	Characteristics and laboratory results
Alcoholic liver disease	History of alcohol abuse AST/ALT > 2
Chronic hepatitis C	ELISA test for the detection of anti-HCV antibodies PCR for the search for viral RNA, as a confirmation test
Primary biliary cholangitis	Anti-mitochondria antibodies Increased IgM
Primary sclerosing cholangitis	Strong association with inflammatory bowel diseases Cholangiography to establish the diagnosis Antinucleus, antismooth muscle and ANCA antibodies (not diagnostic)
Autoimmune hepatitis	Hypergammaglobulinemia Anti-core and anti-smooth muscle antibodies, and ANCA in type 1, anti-LKM-1 in type 2
Chronic hepatitis B	HBsAg, HBeAg, and, in some cases, HBV-DNA
Hereditary hemochromatosis	Family history of cirrhosis Transferrin saturation and ferritin levels should be measured, although they may be altered for liver disease per se Diagnosis established through genetic investigations or liver biopsy and calculation of the hepatic iron index
Wilson's disease	Family or personal history of cirrhosis at a young age Reduced circulating levels of ceruloplasmin Increased copper content on liver biopsy
Non-alcoholic steatohepatitis	History of diabetes mellitus or metabolic syndrome The diagnosis could be suspected from altered liver biochemical tests and liver imaging tests showing liver fat infiltration that could be confirmed with biopsy

ANCA anti-neutrophil cytoplasmic antibodies, *anti-LKM-1* anti-liver microsome type 1 antibodies, *ALT* alanine aminotransferase, *AST* aspartate aminotransferase, *ELISA* enzyme-linked immunosorbent assay, *HBeAg* antigen E of hepatitis B, *HBsAg* hepatitis B surface antigen, *HBV* hepatitis B virus, *HCV* hepatitis C virus, *PCR* polymerase chain reaction

Hepatitis viruses can be classified into major and minor. Major hepatitis viruses are hepatotropic agents, i.e., they have a marked tropism for the liver, which is, therefore, the primary site of their replication.

The main hepatotropic viruses that cause viral hepatitis are:

- Hepatitis A virus (HAV)
- Hepatitis B virus (HBV)
- Hepatitis C virus (HCV)
- Hepatitis D virus (HDV)
- Hepatitis E virus (HEV)

Table 11.7 shows the main characteristics of hepatotropic viruses.

Table 11.7 Characteristics of hepatotropic viruses

Virus	A	B	C	D	E
Genome	RNA	DNA	RNA	RNA	RNA
Transmission	Fecal-oral or sexual	Sexual, parental, or perinatal	Sexual, parental	Sexual, parental The presence of virus B is mandatory	Fecal-oral, perinatal
Chronic infection	No	Yes	Yes	Yes	No

On the other hand, minor hepatitic viruses do not have the liver as their primary target, but they can reach it during viremia, leading to hepatitis. Cytomegalovirus, Epstein–Barr virus, Coxsackie virus, yellow fever virus, and herpesvirus are tbe main minor hepatitic viruses.

Generally, hepatitic viruses do not have a direct cytopathic action on hepatocytes, but hepatic damage may be due to the immune response against viruses.

Hepatitis A

Hepatitis A is caused by an RNA virus that belongs to the Picornavirus family. HAV infection usually develops in an acute form.

Transmission occurs mainly via the fecal-oral route, consuming contaminated food (raw or undercooked seafood, berries, vegetables) or water, or through contact with infected persons. In recent years, sexual transmission has also been described. Travel to areas with high endemicity represents a critical risk factor.

From a pathogenetic point of view, viral replication occurs in the cytoplasm of hepatocytes. Liver injury is the result of the host's immune response to HAV.

The incubation period of hepatitis A infection averages 28 days (15–50 days). More than 70% of HAV-infected adults present with symptomatic illness, which begins with the sudden onset of nausea, vomiting, anorexia, fever, malaise, and abdominal pain. From a biochemical-clinical point of view, hepatitis A is characterized by increased serum ALT, serum bilirubin (typically ≤10 mg/dL), and alkaline phosphatase (up to 400 U/L) levels. Increases in serum ALT generally precede elevations in bilirubin. Serum aminotransferase peaks approximately 1 month after exposure to the virus and then progressively declines. Serum bilirubin concentration usually declines within 2 weeks of the peak. Other laboratory changes include increases in acute phase proteins and markers of inflammation.

Full biochemical-clinical recovery is observed within 2–3 months in 85% of patients and complete recovery within 6 months in almost all patients. HAV infection does not become chronic, and individuals cannot reinfect it.

The diagnosis of acute HAV infection is based on identifying serum anti-HAV IgM antibodies IgM is detectable from symptoms beginning, peaks during the convalescent phase of the disease, and remains detectable for 3–6 months. Detection of serum IgM antibodies without clinical symptoms may reflect a previous hepatitis A infection with prolonged IgM persistence, a false-positive result, or an asymptomatic infection (which is more common in children younger than 6 years of age than in older children or adults).

Serum IgG antibodies appear early in the convalescent phase of the disease, remain detectable for decades, and are associated with lifelong protective immunity. Detection of anti-HAV IgG in the absence of anti-HAV IgM indicates prior infection or vaccination.

Hepatitis B

Hepatitis B is caused by a DNA virus that belongs to the Hepadnavirus family. HBV infection can develop in an acute or chronic form. HBV represents a global public health problem. It is estimated that there are 248 million HBV carriers worldwide, of which approximately 600,000 die each year from HBV-related liver disease. The implementation of effective vaccination programs in many countries has resulted in a significant reduction in the incidence of new hepatitis B infections. However, HBV infection remains a significant cause of morbidity and mortality.

HBV can be transmitted sexually, parenterally, or perinatally.

The complete viral particle (virion) consists of:

- An outer envelope formed of the hepatitis B surface antigen (HBsAg) and components of host-derived lipids
- A core formed of the core hepatitis B antigen (HBcAg), the viral genome, and the polymerase

HBV also produces subviral particles in the form of filaments and spheres composed only of envelope proteins. These subviral particles do not contain the HBV genome and, therefore, are not infectious.

The HBV genome is a circular, partially double-stranded DNA molecule, with a longer L (−) chain consisting of 3200 nucleotides and a shorter S (+) chain varying in length from 1700 to 2800 nucleotides. Although its size is small, the genome can encode many different proteins (genetic economy), such as the S protein expressed on the surface, the functional protein DNA polymerase, which is a target of antiviral therapy, the core antigen, and the E antigen. The core antigen represents the structural protein of the capsid. The E antigen (HBeAg) is generated by a proteolytic cut of C protein. This protein, unlike the core protein, is soluble and, therefore, released in the serum.

HBV is generally not a cytopathic virus. The pathogenesis of HBV-related liver disease is mainly due to immune-mediated mechanisms.

Patients who progress to chronic HBV infection have an altered immune response.

The incubation period of HBV varies from 2 to 6 months.

The spectrum of clinical manifestations of HBV infection varies in both acute and chronic forms. During the acute phase, manifestations range from subclinical hepatitis (70% of cases) to jaundiced hepatitis and, in some cases, fulminant hepatitis (30% of cases). During the chronic phase, manifestations range from an asymptomatic carrier state to chronic hepatitis, cirrhosis, and hepatocellular carcinoma. The disease may be more severe in patients co-infected with other hepatitis viruses or underlying liver disease. Fulminant liver failure is rare (0.1–0.5% of patients) and is thought to be due to massive immune-mediated lysis of infected hepatocytes.

Complete eradication of HBV rarely occurs after recovery from acute infection, and T cells can control latent infection for decades after clinical recovery.

The natural course of chronic HBV infection is determined by the interaction between viral replication and the host immune response. Other factors that may play a role in the progression of HBV-related liver disease include sex, alcohol consumption, and concomitant infection with other hepatitis viruses. It is important to note that not all patients with chronic HBV infection develop chronic hepatitis. In 20% of cases, chronic hepatitis can progress to cirrhosis within approximately 5 years. Patients with cirrhosis have a high risk of developing hepatocarcinoma.

The evaluation of the different available markers, serological or genetics, depends on the clinical presenting characteristics of the patient, or if screening is to be performed in an asymptomatic person.

Regarding serological markers, HBV infection is characterized by particular changes in serum levels of HBV antigens and anti-HBV antibodies. Therefore, these markers are used to define different clinical states (Table 11.8).

HBsAg and Anti-HBs

HBsAg surface antigen is the serological sign of HBV infection and the first marker to appear. Serum HBsAg can be detected 1–10 weeks after acute HBV exposure, before the onset of clinical symptoms or ALT increase. In most cases, HBsAg is no longer detectable after 2–4 months. The persistence of HBsAg for more than 6 months indicates chronic infection. The disappearance of HBsAg is followed by the appearance of serum antibodies against the surface antigen (anti-HBs). In most patients, anti-HBs persist for life, thus conferring long-term immunity. In some patients, however, anti-HBs may not be detectable until several weeks or months, during which neither HBsAg nor anti-HBs can be detected. In this case, serological diagnosis can be performed by detecting IgM antibodies against the core antigen (IgM anti-HBc), which appear as early as 1–2 weeks after the appearance of HBsAg.

Table 11.8 Serological diagnosis of HBV infection

HbsAg	HBeAg	IgM-anti-HBc	IgM and IgG anti-HBc	Anti-HBs	Anti-HBe	HBV-DNA	ALT	Interpretation
Acute HBV infection								
+	+	+				+++	Elevated	Early phase
		+				+	Elevated	Window period
			+	+	+	+/−	Normal	Recovery phase
Chronic HBV infection (HBsAg positive >6 months)								
+	+		+	−	−	+++	Normal or modestly elevated	Immunotolerance phase
+	+		+	−	−	+++	Persistently elevated	Immunoactive phase, HBeAg positive
+	−		+	−	+	++	Elevated	Immunoactive phase, HBeAg negative
−	−		+/−	+/−	+/−	+ liver +/− serum	Normal	Occult HBV

HBcAg and Anti-HBc

The core antigen, HBcAg, is expressed intracellularly in infected hepatocytes; it is not detectable in serum. Antibodies against the core antigen (anti-HBc) can be detected during HBV infection. Specifically, anti-HBc is predominantly of the IgM class in acute infection. Anti-HBc IgM is the only detectable marker of HBV infection during the window period between the disappearance of HBsAg and the appearance of anti-HBs. The presence of anti-HBc IgM is usually considered indicative of acute HBV infection. However, anti-HBc IgM may remain detectable up to 2 years after acute infection. In addition, anti-HBc IgM titer may increase to detectable levels during exacerbations of chronic hepatitis B. Common causes of acute exacerbation of chronic hepatitis B are superinfection with hepatitis D virus or hepatitis C virus. Anti-HBc IgG persists along with anti-HBs in patients recovering from acute hepatitis B and along with HBsAg in patients progressing to chronic HBV infection.

HBeAg and Anti-HBe

HBeAg is a secretory protein generally considered a marker of HBV replication and infectivity. HBeAg is usually associated with elevated serum HBV-DNA levels and higher transmission rate of HBV infection. HBeAg to anti-HBe seroconversion occurs early in patients with acute infection, before HBsAg to anti-HBsAg seroconversion. However, HBeAg seroconversion may be delayed for years or decades in patients with chronic HBV infection. In such patients, the presence of HBeAg is usually associated with high serum HBV-DNA levels and active liver disease.

Seroconversion from HBeAg to anti-HBe is only associated with a reduction in serum HBV-DNA and remission of liver disease.

HBV-DNA

Qualitative and quantitative assays have been developed to measure serum HBV-DNA to assess HBV replication. The sensitivity limit of these assays depends on the techniques used. Currently, most assays for the assessment of HBV-DNA are based on real-time polymerase chain reaction (PCR).

Recovery from acute hepatitis B is usually accompanied by the disappearance of serum HBV-DNA, as determined by hybridization analysis. However, HBV-DNA may remain detectable in serum for many years when tested by PCR. This observation suggests that the virus persists after "recovery" but is controlled by the immune system. HBV-DNA levels are also detectable in patients with HBeAg-negative chronic hepatitis, although the levels are generally lower than in patients with HBeAg-positive chronic hepatitis.

The main clinical utility of HBV DNA testing in patients with chronic HBV infection is the assessment of HBV replication and eligibility for antiviral therapy. Indeed, the indications for HBV treatment are based on active liver disease and high HBV-DNA levels. HBV-DNA suppression is also used to assess response to antiviral treatment.

Table 11.9 shows the characteristics of HBV serological markers.

The diagnosis of acute hepatitis B is based on identifying HBsAg and anti-HBc IgM. During the early phase of infection, HBV replication markers, such as HBeAg and HBV-DNA, are also detectable. Recovery is accompanied by the disappearance of HBV -DNA, seroconversion from HBeAg to anti-HBe, and then seroconversion from HBsAg to anti-HBs. Rarely do patients present during the window period when HBsAg has become negative, but anti-HBs are not yet positive. This condition is more common in patients with fulminant hepatitis B because virus clearance tends to be more rapid and anti-HBc IgM is the only marker of acute HBV infection.

Prior HBV infection is characterized by anti-HBs and anti-HBc IgG. Immunity to HBV infection after vaccination is indicated only by the presence of anti-HBs.

The diagnosis of chronic HBV infection is based on the persistence of HBsAg for more than 6 months. Additional tests to assess HBV replication, such as HBeAg and serum

Table 11.9 Main characteristics of serological markers of HBV

Marker	Acronyms	Appearance in serum	Clinical usefulness
Surface antigen of the hepatitis B virus	HBsAg	It is the first marker to be detectable during the incubation period	First diagnostic marker. Its disappearance does not always indicate recovery.
Antigen E of the hepatitis B virus	HBeAg	It is the second marker to appear very early after HBsAg	Levels peak at the same time as HBsAg but disappear earlier. It indicates viral replication, and the patient is highly contagious.
Antibodies to the E antigen of the hepatitis B virus	Anti-Hbe	Low levels can be detected even before HBeAg disappears	Seroconversion from HBeAg to anti-HBe is an early indicator of recovery. The patient is less contagious.
IgM to the core antigen of the hepatitis B virus	Anti-HBc IgM	Usually 2 weeks after the onset of HBsAg and up to 6 months before the onset of symptoms	Diagnostic marker of acute HBV infection. It has no immunoprotective function.
IgG to the core antigen of the hepatitis B virus	Anti-HBc IgG	Levels rise early after the onset of anti-HBc IgM	It remains detectable throughout life. It cannot be used to distinguish between acute and chronic infection.
Antibodies to the surface antigen of the hepatitis B virus	Anti-HBs	It is the last marker to appear in the serum following the disappearance of HBsAg	They are the only antibodies that neutralize HBV and thus indicate immunity. The WHO has established that 10 mUI/L is the threshold value to define a state of immunity following natural exposure or vaccination.
HBV-DNA (detected by PCR)		Indicative of the presence of the virion	Indicative of active viral replication. Frequent false positives

HBV-DNA, should be performed to determine if the patient is eligible for antiviral therapy. All patients with chronic HBV infection should be monitored regularly because HBV-DNA and ALT levels vary throughout the infection.

HBeAg-negative patients who have normal or low ALT levels or undetectable HBV-DNA are in an inactive carrier status. These patients generally have a good prognosis, and antiviral treatment is not indicated. However, they should undergo evaluation of ALT and HBV-DNA levels at 3-month intervals during the first year.

Hepatitis C

Hepatitis C is caused by a single-stranded RNA virus belonging to the Flaviviridae family. Chronic HCV infection is one of the most common chronic liver diseases.

According to the WHO, approximately 71 million people worldwide are chronic carriers of the hepatitis C virus, and 399,000 people die each year from hepatitis C–related liver disease. HCV is mainly transmitted by exposure to contaminated biological fluids such as blood, vaginal secretions, or semen. The infection could occur through needles and syringes, blood transfusions, piercings, acupuncture, dental procedures, endoscopy, and unprotected sexual intercourse. In addition, it can be transmitted perinatally when passing through the birth canal. Another transmission mode is through parenteral drug use with an exchange of infected syringes. The risk of infection is increased if there is co-infection with HIV.

HCV is not a cytopathic virus, but liver damage is immune-mediated.

In most cases (80–90%), acute HCV infection is asymptomatic, whereas, in the remaining 10–20% of cases, the patient has an acute infection with overt clinical symptoms that are generally nonspecific (jaundice, nausea, dark urine, and right upper quadrant pain). Patients with acute HCV infection typically present with moderate-to-high ALT elevations. Approximately 50–80% of patients with acute hepatitis will develop chronic infection, while 20–50% will recover. The mechanisms responsible for the high prevalence of viral persistence, and therefore chronic infection, are not fully understood. However, it is likely that both viral and host factors, such as the age of acquisition of infection (acquisition at age > 50 years is associated with a higher risk of infection), co-infections with other liver viruses or HIV, alterations in immune status, sex (women are at lower risk), ethnicity (Caucasians are at higher risk), and the presence of other causes of liver disease such as alcohol consumption may contribute.

In 20% of cases, chronic hepatitis progresses to liver cirrhosis, characterized by a continuous deposition of connective tissue, which alters the normal architecture of the organ, leading to a progressive loss of function to death; in 1–4% of cases, there is the development of hepatocarcinoma. Figure 11.3 shows the natural history of HCV infection.

By convention, acute HCV infection refers to the first 6 months of infection following presumed exposure to the virus. HCV infection is estimated to account for 15% of symptomatic cases of acute hepatitis, whereas most patients with acute HCV are undetected. This is mainly because patients with acute HCV infection are typically asymptomatic.

ALT levels are often greater than 10–20 times the upper reference limit in patients with acute HCV infection but exhibit high variability. Furthermore, during

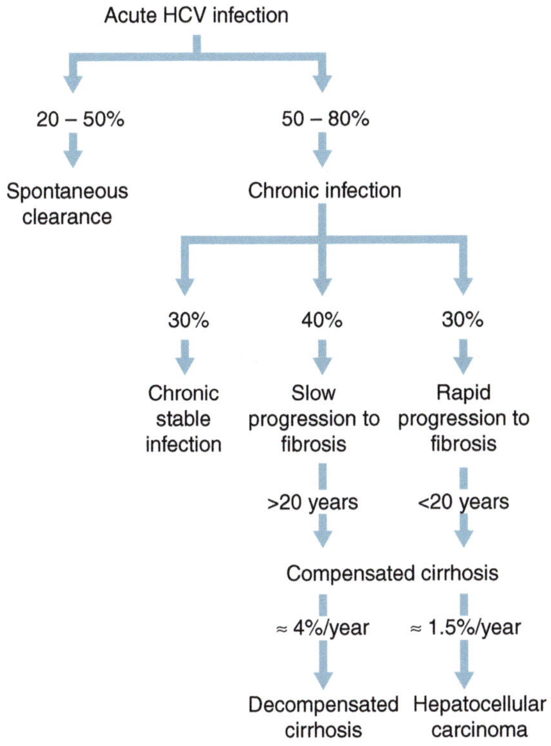

Fig. 11.3 Natural history of HCV infection. (Copyright EDISES 2021. Reproduced with permission)

acute infection, ALT levels can vary widely within short intervals, in contrast to a chronic infection, during which they are often elevated but relatively stable over time. Among patients who develop symptoms, aminotransferases rise shortly before the onset of clinical symptoms and usually before anti-HCV antibodies are detectable. However, because levels often fluctuate and may even normalize, not all patients will have elevated aminotransferase levels at the presentation time. Patients with acute HCV infection may also have elevated total bilirubin levels (>3 mg/dL).

The detection of HCV-RNA by PCR in subjects with undetectable HCV antibodies, which subsequently become detectable within 12 weeks, is generally indicative of acute HCV infection. Alternatively, the finding of HCV antibodies and HCV-RNA in a patient with negative tests in the preceding 6 months is indicative of acute HCV infection. In the absence of such documentation, the distinction between acute HCV infection and newly discovered chronic infection is not straightforward, as patients in both settings may have detectable HCV-RNA, anti-HCV antibodies, and elevated serum ALT levels.

Serum HCV-RNA is detectable by PCR within a few days to 8 weeks after exposure to the virus.

Enzyme-linked immunosorbent assays (ELISAs) that detect HCV antibodies become positive as early as 8 weeks after exposure, and most patients seroconvert 2–6 months after exposure. HCV antibody positivity does not allow differential diagnosis between acute and chronic infection.

Figure 11.4 illustrates the algorithm for the valuation of acute HCV infection in a patient with acute hepatitis (elevated ALT and/or jaundice). For a patient presenting following known exposure to HCV, first tests for HCV-RNA, anti-HCV antibodies, and ALT are performed to establish baseline HCV status (Fig. 11.4). Within 1–2 days after an exposure, a patient without a previous infection should test negative for HCV-RNA and anti-HCV antibodies, with nominal ALT values. Figure 11.5 shows the algorithm for assessing acute HCV infection in a patient with recent exposure to the virus.

Acute HCV infection can resolve but, in most cases, will become chronic. Chronic HCV infection progresses slowly and may not result in clinically evident liver disease. Approximately 5–30% of chronically infected persons develop cirrhosis within 20–30 years. Patients who develop cirrhosis are at risk for complications, such as varicose bleeding, ascites and encephalopathy, and hepatocellular carcinoma, although many patients with compensated cirrhosis remain stable for years. Patients with chronic infection have wide variability in serum ALT levels over time. Up to one-third of patients have normal ALT levels; mild increases are usually observed in about two-thirds of patients, and only about 25% have a serum ALT concentration greater than two times the upper reference limit. Generally, however, there is little correlation between aminotransferase levels and liver histology. During chronic HCV infection (i.e., after the acute phase), HCV viral levels remain constant, although significant fluctuations may occur. Useful laboratory tests to identify cirrhosis in HCV-infected patients include serum bilirubin concentration (increased), hypoalbuminemia, or platelet count (decreased). Serum alpha-fetoprotein (AFP) concentration may be slightly increased in chronic HCV infection and does not necessarily imply the presence of hepatocellular carcinoma or cirrhosis; up to 43% of patients with cirrhosis without hepatocellular carcinoma have a serum AFP between 10 and 100 ng/mL. However, an elevated serum AFP concentration requires liver imaging to rule out hepatocellular carcinoma.

Hepatitis D

Hepatitis D is caused by a defective RNA virus, the hepatitis D virus (HDV), also called delta virus; it is defined as defective because, although it can replicate on its own, it requires the simultaneous presence of HBV for the assembly and secretion of complete virions. Thus, HDV cannot infect a cell on its own but requires the simultaneous presence of HBV infection. Consequently, people with hepatitis D are always infected with HDV and HBV. Due to incompletely understood interference mechanisms, HBV replication is suppressed in most HDV-infected individuals.

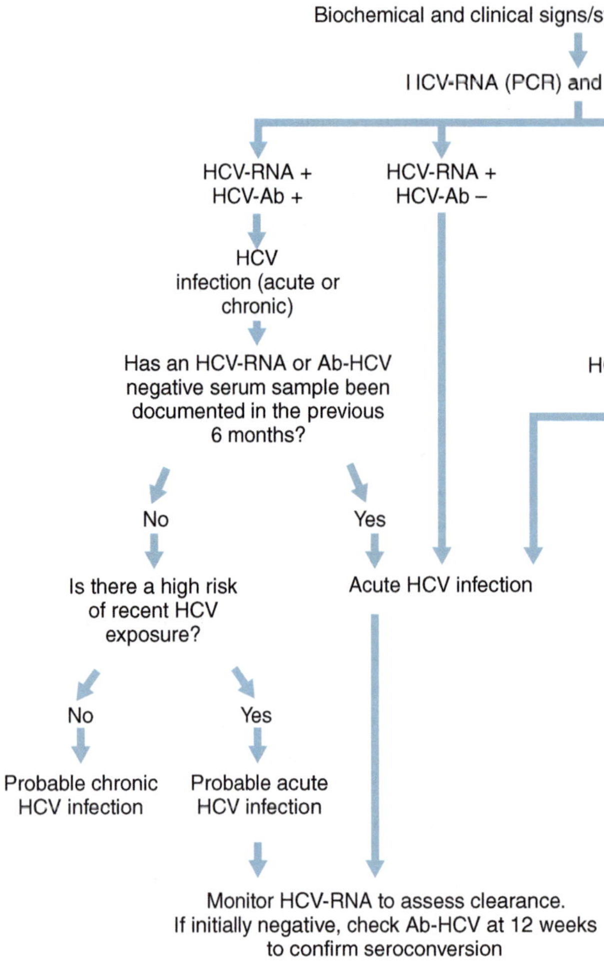

Fig. 11.4 Diagnostic algorithm for acute HCV infection in a patient with clinical symptoms. (Copyright EDISES 2021. Reproduced with permission)

Vaccination against hepatitis B has contributed to the progressive decrease in the incidence of HBV over the past 30 years.

HDV has a very high replication rate in hepatocytes.

This virus can give rise to two types of infection.

- Acute HDV superinfection: occurs in a patient who was already infected with HBV and became overinfected with the hepatitis D virus. HDV superinfection of a chronic HBsAg carrier may occur as severe acute hepatitis in a previously unrecognized HBV carrier or as an exacerbation of preexisting chronic hepatitis B. Progression to chronic HDV infection occurs in almost all patients. However, HBV replication is usually suppressed by HDV.
- Acute HBV/HDV co-infection: occurs in a patient who is simultaneously infected with both viruses. Co-infection of HBV and HDV causes acute hepatitis B + D, which is clinically indistinguishable from classical acute hepatitis B and is usually transient and self-limited; the risk of chronicity is low.

The detailed mechanisms by which HDV induces liver damage are unknown. However, the pathogenesis of hepatitis D–related liver disease appears to depend on the interaction between host-associated factors, such as the immune response, and virus-associated factors, such as HBV genotype and HBV replication level.

HDV is thought to cause direct cytopathic injury during acute infection, while immune-mediated damage predominates during chronic infection.

Hepatitis D can only be diagnosed in HBsAg-positive patients.

In HDV-infected individuals, the timing of onset and circulating levels of HDV-RNA, HDAg, and anti-HDV allows discrimination between three different clinical conditions:

- Acute HBV/HDV co-infection
- Acute HDV superinfection in a chronic HBV carrier
- Chronic HDV infection.

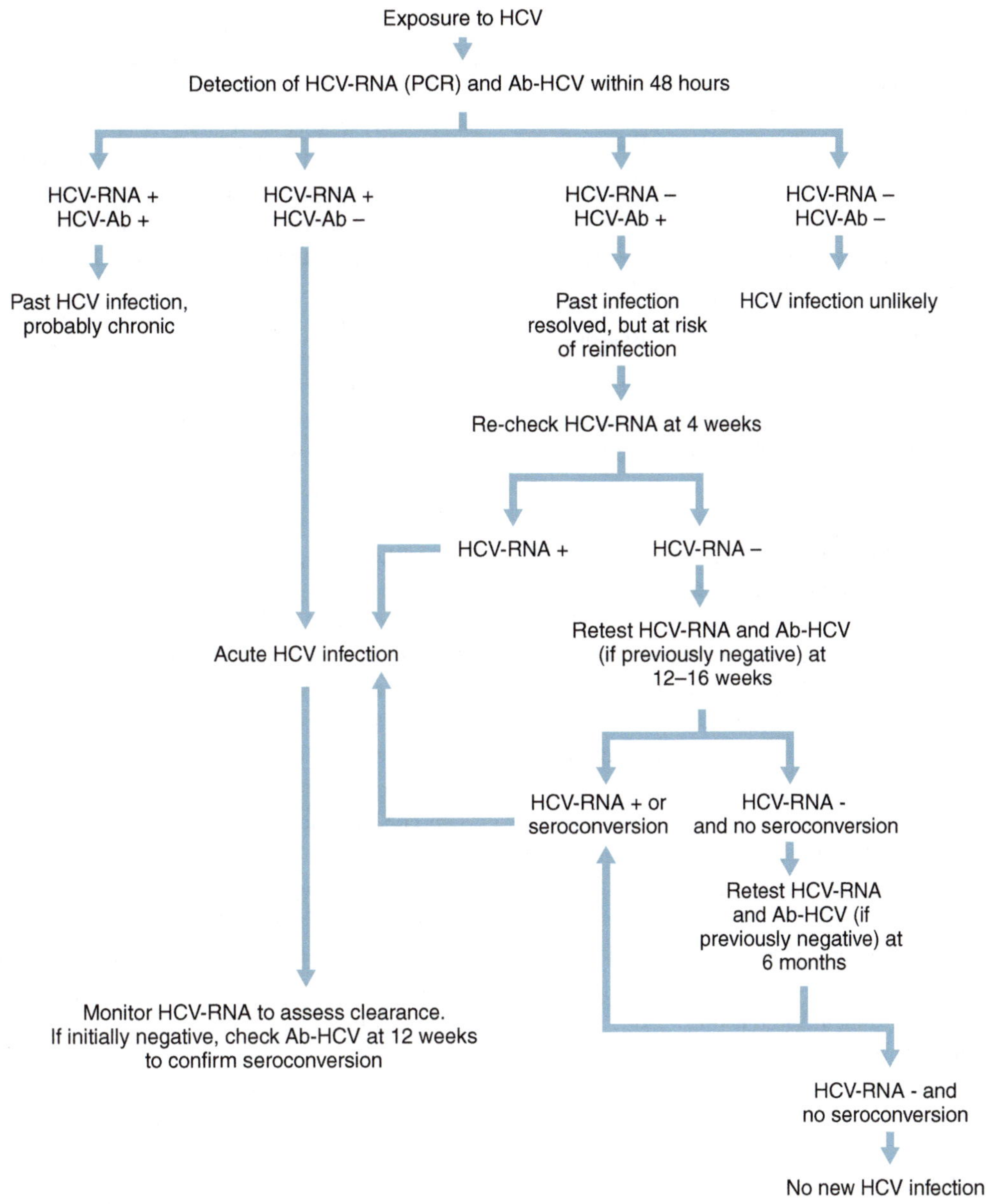

Fig. 11.5 Diagnostic algorithm for acute HCV infection in a patient with recent exposure to the virus. (Copyright EDISES 2021. Reproduced with permission)

Due to the dependence of HDV on HBV, the presence of HBsAg is required to diagnose HDV infection. The additional presence of anti-HBc IgM is required acute HBV/HDV co-infection (Table 11.10).

Serum HDAg can be detected by enzyme-linked immunosorbent assay (EIA) or radioimmunoassay (RIA) on microplates.

In acute HDV infection, serum HDAg appears early but has a very short half-life and may escape detection if repeated testing is not performed. In chronic HDV infection, anti-HDV is present at a high titer.

HDV RNA can be detected in serum by Reverse Transcriptase-PCR (RT-PCR) (Table 11.10).

Total HDV antibodies (IgM and IgG) usually appear after 4 weeks of acute HDV infection. Consequently, their clinical value is limited unless repeat testing is performed. However, a well-documented anti-HDV seroconversion may be the only way to diagnose acute HDV infection in the absence of other markers of HDV infection.

High titer anti-HDV IgG is present in chronic HDV infection and correlates well with ongoing HDV replication. It may help differentiate between recent and past HDV infection. Anti-HDV IgM is transient and delayed if the course of acute hepatitis D is self-limited but maybe the only serum marker of acute HDV infection. The differential diagnosis between HBV/HDV co-infection and HDV superinfection in

Table 11.10 Diagnostic marker of HDV hepatitis

Diagnostic marker	Acute co-infection HBV/HDV	HDV superinfection	HDV chronic infection
HBsAg	Positive	Positive	Positive
Anti-HBc IgM	Positive	Negative	Negative
HDAg	Early and short-lived	Early and transient	Undetectable
HDV-RNA	Early, transient but lasts longer than HDAg	Early and persistent	Usually, positive
Total anti-HDV	Late, low title	Titer progressively increases	High titer
Anti-HDV IgM	Transient but it could be the only marker	Titer rises rapidly and persists	Variable titer, usually high

an HBV carrier is mainly based on the detection of high titer anti-HBc IgM in patients with co-infection.

Anti-HDV IgM is present at high titers during chronic HDV infection, and titers correlate with the HDV replication rate and the severity of the liver disease. Anti-HDV IgM gradually disappears from serum in patients who have remission after interferon therapy and after liver transplantation.

Hepatitis E

Hepatitis E is caused by the hepatitis E virus, which belongs to the family Hepeviridae and consists of a linear single helix RNA genome.

Transmission of HEV can occur through contaminated food and water, blood transfusions, and via the perinatal route. Person-to-person transmission is quite rare.

The pathogenesis of hepatitis E is poorly understood. The virus is not cytopathic and, therefore, the liver injury could be immune-mediated by cytotoxic T cells and natural killer (NK) cells.

HEV generally causes acute self-limited infection, although acute liver failure may develop in a small percentage of patients. The incubation period of HEV infection ranges from 15 to 60 days. Most patients with acute HEV infection are asymptomatic or mildly symptomatic. In symptomatic patients, jaundice is usually accompanied by malaise, anorexia, nausea, vomiting, abdominal pain, fever, and hepatomegaly. Other less common features include diarrhea, arthralgia, pruritus, and urticarial rash.

Laboratory findings include elevated serum concentrations of bilirubin and ALT. Symptoms coincide with an abrupt increase in serum ALT levels. In immunocompromised patients, the development of chronic hepatitis leads to inflammation and progressive fibrosis, which can lead to cirrhosis. Most patients who acquire HEV spontaneously eliminate the virus. However, patients may develop complications, such as acute liver failure, cholestatic hepatitis, or chronic HEV infection. Chronic HEV infection occurs almost exclusively in immunosuppressed patients (e.g., patients with HIV infection after solid organ or bone marrow transplantation).

The diagnosis of HEV hepatitis should be considered in patients presenting with acute or chronic hepatitis that other causes cannot explain. The lack of a standardized assay complicates the diagnosis of acute HEV infection. Several EIA kits have been developed that exhibit high variability in test performance; both false positives and negatives are common with available tests.

The timing of the appearance of HEV markers is important for interpreting serologic test results in the context of acute hepatitis.

Anti-HeV IgM appears during the early phase of clinical disease and disappears rapidly within 4–5 months. Anti-HeV IgG appears shortly after IgM, and the titer increases throughout the acute phase until the convalescent phase. It is unclear how long anti-HEV IgG antibodies persist. HEV-RNA can be detected in the stool approximately 1 week before the onset of illness and may persist up to 2 weeks later. In serum, HEV-RNA can be detected 2–6 weeks after infection and may persist for 2–4 weeks in those who resolve the acute infection. Although HEV viremia is short-lived in most patients with acute infection, it may persist for years in those who develop chronic infection.

In general, the first test to evaluate the presence of acute HEV infection should be the detection of anti-HEV IgM. Confirmatory testing should be performed if this test is positive, as no single EIA method achieves high specificity. Confirmatory testing may include evaluating anti-HEV IgM by a different method, evaluating an increased anti-HEV IgG titer (>5 times over 2 weeks), or detecting HEV-RNA in serum or stool. If the initial test is negative and there is still a high suspicion of HEV infection, the test should be repeated, preferably in conjunction with the detection of HEV-RNA.

In patients with suspected chronic HEV infection, detecting serum HEV-RNA is the cornerstone of diagnosis. Chronic HEV infection is defined by identifying HEV-RNA in serum or stool over more than 6 months. The HEV IgG antibody test is of limited utility in the diagnosis of chronic HEV infection. Indeed, anti-HEV IgG is a marker of HEV exposure, whether recent or previous. In addition, the reduction of anti-HEV IgG titer over time could adversely affect its sensitivity to detect past infection.

Iatrogenic Hepatitis

Many drugs, both prescription and over-the-counter, as well as herbal products, can have hepatotoxic effects through various mechanisms. Drug-induced liver injury accounts for approximately 10% of all causes of acute hepatitis and is one of the most frequent reasons for drug withdrawals.

Drug-induced liver damage has an estimated annual incidence of about 10–15 cases per 10,000–100,000 people on prescription.

Several risk factors have been associated with the development of hepatic drug injury. In general, adults are at higher risk than children (with the notable exception of valproic acid damage, which is more common in children); women may be more susceptible than men, which may be due in part to their generally smaller body size; alcohol abuse and poor nutrition in some cases are predisposing factors, as in the case of acetaminophen.

More than 1000 drugs and herbal products are known to induce liver damage, and the list continues to grow. The National Institutes of Health maintains a research database of drugs, medications, herbal products, and dietary supplements associated with liver damage. The drug most implicated in acute liver injury is acetaminophen, followed by antibiotics; in particular, amoxicillin-clavulanate is one of the most reported causes of hepatopathy.

The liver is responsible for the selective absorption, concentration, metabolism, and excretion of most drugs and toxins introduced into the body. While some drugs can directly cause hepatotoxicity, their metabolites generally result in liver damage. Drugs are processed by a wide variety of enzymes, and each has its own specific biotransformation pathway involving one or more enzyme systems.

In general, factors that alter the activity of an enzyme have the potential to increase the toxicity of a compound (by reducing its conversion to nontoxic metabolites or increasing its conversion to toxic metabolites) or decrease its therapeutic efficacy (e.g., by increasing the rate of inactivation of the active compound).

Several factors can alter the activity of drug-metabolizing enzymes and, consequently, affect drug metabolism, contributing to hepatotoxicity:

- Genetics: Several polymorphisms are known in the genes encoding for enzymes involved in drug metabolism that may contribute to a reduction or excess of metabolization of a compound.
- Alcohol intake: Chronic alcohol ingestion increases the activity of certain enzymes.
- Nutrition: Some foods can alter the activity of specific enzymes.
- Presence of other drugs: The concomitant use of two or more drugs is one of the most important factors affecting the components of the cytochrome system and, consequently, drug metabolism.
- Patient demographics: It has been reported that a general decrease in cytochrome activity develops with increasing age. Infants, on the other hand, present considerable immaturity in the enzymatic kit, which develops over time. Finally, women are more susceptible to hepatic drug injury than men. Ethnic differences in the presentation and prognosis of drug-related liver injury have also been reported. African-American subjects are at greater risk for severe skin reactions and severe hepatic injury than Caucasian subjects.
- Underlying liver disease: Both acute and chronic liver disease have a variable effect on the metabolism of many drugs. Enzyme activities are generally reduced with increasing disease severity.
- Dose: Hepatic injury from intrinsic hepatotoxins is clearly dose-related.

Drug-induced liver injury is a complex process involving the drug, its metabolites, and the host immune system. Most of the effects of hepatotoxic drugs result in necrosis or apoptosis of hepatocytes. However, some drugs predominantly damage bile ducts, export proteins, and canaliculi (cholestasis), vascular endothelial cells (sinusoidal obstruction syndrome), or stellate cells. Thus, drug-induced liver damage may have the features of hepatitis, cholestasis, or mixed; the differential diagnosis is made based on the biochemical-clinical picture.

The acute presentation of drug-induced liver damage varies from a mild alteration of the hepatic biochemical-clinical profile without symptoms or with nonspecific symptoms such as malaise, mild fever, anorexia, nausea, vomiting, right upper quadrant pain, jaundice, or dark urine, to cholestasis with pruritus, acute hepatitis with jaundice reminiscent of viral hepatitis, and acute liver failure. Chronic liver damage manifests like other chronic liver diseases, such as autoimmune hepatitis, primary biliary cirrhosis, or sclerosing cholangitis. In some patients, chronic drug-induced damage progresses to cirrhosis.

In most cases, patients are asymptomatic, and laboratory tests reveal the injury. Diagnosis is based on obtaining a thorough history and laboratory investigations to rule out other potential causes of liver damage. Patients with hepatocellular damage will have a marked increase in aminotransferase levels (>25 times the upper limit of normal). In contrast, patients with cholestatic damage will predominantly have an increase in alkaline phosphatase. Serum bilirubin may be elevated in both cases.

If there is evidence of cholestasis, imaging to rule out biliary obstruction is also indicated. If testing for alternative causes of liver damage is negative and the patient has been exposed to a drug associated with liver damage, a liver biopsy is generally not performed. However, a liver biopsy should be performed if the diagnosis remains uncertain (especially in the context of acute liver failure) or if there is clinical evidence of chronic hepatopathy.

There are no specific circulating biomarkers that can reliably identify a drug as a cause of liver injury.

Key elements for attributing liver injury to a drug include:

- Exposure to the drug prior to the onset of liver injury (although the latency period is highly variable)

- Exclusion of underlying liver disease
- Improvement in liver injury after discontinuation of the drug
- Rapid and severe relapse with repeated exposure to the drug
- Assumption of a drug already associated with possible liver damage.

Steatosis, Fibrosis, and Cirrhosis

Nonalcoholic Hepatic Steatosis

Nonalcoholic fatty liver steatosis (NAFL) is defined by the accumulation of lipids in the hepatocyte greater than 5% in subjects in whom chronic alcohol abuse or other conditions that may generate hepatic steatosis are excluded (Table 11.11).

Steatosis, with or without inflammation, is often associated with fibrosis and may progress to cirrhosis. Generally, benign NAFL can be distinguished from Nonalcoholic Steatohepatits (NASH) with or without fibrosis, which may progress to cirrhosis, liver failure, and hepatocarcinoma. There is no significant evidence of inflammation in hepatic steatosis, whereas, in steatohepatitis, steatosis is associated with hepatic inflammation, which may be indistinguishable from alcoholic steatohepatitis on histological examination.

Nonalcoholic Fatty Liver Disease (NAFLD), which includes the entire spectrum of nonalcoholic liver diseases ranging from steatosis to steatohepatitis and cirrhosis, is the most frequent chronic liver disease in Western countries, where its main risk factors – central obesity, type 2 diabetes mellitus, dyslipidemia, and metabolic syndrome – show an increasing prevalence. In Europe, the incidence of NAFLD in the adult population is 20–30% but reaches up to 70% when considering diabetic subjects.

Table 11.11 Main causes of steatosis

Macrovesicular steatosis	Alcoholism Hepatitis C (genotype 3) Wilson's disease Lipodystrophy Undernutrition Parenteral nutrition Abetalipoproteinemia Use of drugs such as mipomersen, lomitapide, amiodarone, methotrexate, tamoxifen, corticosteroids
Microvesicular steatosis	Reye's syndrome Use of drugs such as valproate, and anti-retroviral agents Acute fatty liver disease of pregnancy HELLP syndrome Congenital defects of metabolism (lecithin-cholesterol deficiency acyltransferase, cholesterol ester storage disease, Wolman syndrome)

NAFLD is most frequently diagnosed during the fifth or sixth decade of life, while the association with sex has not yet been clarified since some studies report it to be more frequent in men and others in women. Patients with NAFLD, especially the subgroup with NASH, frequently present one or more components of the metabolic syndrome, particularly central obesity, arterial hypertension, dyslipidemia, and insulin resistance. However, the association between NAFLD and cardiovascular disease appears independent of the metabolic syndrome. Other conditions associated with NAFLD are cystic ovary syndrome, hypothyroidism, obstructive sleep apnea, hypopituitarism, and hypogonadism.

The pathogenesis of NAFLD has not yet been elucidated in many aspects, but insulin resistance would seem to play an important role. It has also been proposed that a second insult, such as oxidative damage, is required for the inflammatory necrosis typical of steatohepatitis. Intrahepatic iron accumulation, leptin, deficiency of antioxidant agents, and an imbalance in the intestinal microbiota have been proposed as potential pro-oxidant elements against the hepatocyte.

Most patients with NAFLD are asymptomatic. Fatigue, generalized malaise, and diffuse right upper abdominal pain may occasionally be observed in patients with NASH. Patients often come to the physician's attention for the occasional finding of increased serum transaminase values or steatohepatitis on abdominal imaging. On physical examination, patients with NAFLD may present with hepatomegaly caused by fatty liver infiltration, which is sometimes the first sign. However, the presence of hepatomegaly is highly variable in these patients.

From a biochemical-clinical point of view, patients with NAFLD have a slight-to-moderate increase in ALT. However, normal transaminases should not exclude NAFLD. When elevated, transaminases are two to five times the upper reference limit, with an AST/ALT ratio of less than 1. The increase in transaminases does not predict the severity of inflammation or fibrosis, just as transaminase values in the normal range do not rule out clinically relevant diseases. Alkaline phosphatase may exceed two to three times the upper reference limit. Albuminemia and bilirubinemia are typically in the normal range but may show variation in patients in whom the disease has progressed to cirrhosis. Similarly, prolonged prothrombin time, thrombocytopenia, and neutropenia are observed in patients with cirrhosis.

Patients with NAFLD may present with increased serum ferritin or transferrin saturation. Ferritin values greater than 1.5 times the upper reference limit have been shown to be associated with greater disease severity and more advanced liver fibrosis. Patients with NAFLD may also have positivity for antinuclear autoantibodies (ANA) and antismooth muscle autoantibodies (ASMA); however, the nature of this association remains controversial.

A diagnosis of NAFLD is made when all the following conditions are met:

- Evidence of steatosis on liver biopsy or imaging
- Exclusion of chronic alcohol abuse
- Exclusion of other causes of steatosis
- Absence of concomitant chronic liver disease.

Most often, imaging, along with a detailed history, is sufficient to diagnose NAFLD. Although not recommended for most patients, biopsy examination is indicated in doubtful cases or to determine the degree of hepatocellular injury. In addition, liver biopsy is currently the only diagnostic tool to differentiate NAFLD from NASH. Although often altered in patients with NAFLD, laboratory tests are not useful for diagnostic purposes but are essential to evaluate other conditions in the differential diagnosis. To this end, serologic tests for HCV, HAV, and HBV infection should be performed, and other chronic liver diseases, such as autoimmune hepatitis and hemochromatosis, should be excluded (Table 11.12).

Based on the patient's symptomatology and personal and family history, Wilson's disease, hypo- or hyperthyroidism, celiac disease, α1-antitrypsin deficiency, HELLP syndrome during pregnancy, and Budd–Chiari syndrome should be ruled out.

Radiological diagnosis usually requires an abdominal ultrasound. Although steatosis can also be detected by computed tomography or magnetic resonance imaging, none of these investigations allows differentiation of NAFL from NASH. The radiographic diagnosis should include radiographic findings of fatty liver infiltration, exclusion of other causes of steatosis, absence of signs or symptoms of cirrhosis, and low risk of advanced fibrosis. If these criteria are not met, a liver biopsy should be performed for diagnostic confirmation and to ascertain the degree of severity of the disease.

Regarding the opportunity to screen individuals at risk of NAFLD, such as obese or diabetic individuals, several scientific societies, such as the American Association for the Study of Liver Diseases, have expressed some skepticism toward this approach, given the uncertainty regarding which diagnostic tests to use and how to treat affected individuals.

Table 11.12 Main laboratory tests for the differential diagnosis of NAFLD

Anti-HCV antibodies
Anti-HAV antibodies (IgG)
HBsAg, HBeAg
Sideremia, ferritin, transferrin saturation
ANA, ASMA, anti-LKM-1

ANA anti-nuclear antibodies, *anti-LKM-1* anti-liver microsome type 1 antibodies, *ASMA* anti-smooth muscle antibodies, *HBeAg* hepatitis B antigen E, *HBsAg* hepatitis B surface antigen

Alcoholic Hepatopathy

Chronic alcohol abuse is associated with various liver manifestations, including steatosis (with or without steatohepatitis), alcoholic hepatitis, cirrhosis, and hepatocarcinoma. Chronic alcohol abuse is defined by an intake >210 g/week in men or >140 g/week in women over at least 2 years, with particular attention to weekly and daily intake patterns. Alcohol abuse is widespread worldwide. A prevalence of 18% in the adult population in the United States has been documented.

Hepatic steatosis is present in 90% of subjects who abuse alcohol and is typically macrovesicular. It can occur as early as after a couple of weeks of regular alcohol ingestion and can resolve spontaneously by abstaining from alcohol. Approximately one-third of patients with steatosis develop liver inflammation (steatohepatitis) if alcohol intake continues over time. In about 10% of patients, steatosis progresses to cirrhosis. Steatohepatitis represents a condition with a higher risk of cirrhosis than steatosis alone.

The clinical manifestations of alcoholic hepatopathy vary depending on the severity of the liver injury. On physical examination, patients with steatosis alone are almost always asymptomatic and often present with hepatomegaly on physical examination. Patients with alcoholic hepatitis typically present with jaundice. Alcoholic cirrhosis is associated with jaundice, asthenia, peripheral edema, symptoms of gastrointestinal bleeding, palmar erythema, ascites, and confusional states of hepatic encephalopathy of the more advanced forms characterized by hepatic insufficiency.

From a clinical biochemistry point of view, patients with alcoholic hepatopathy present various alterations in the haematochemical picture, although none of them is diagnostic. The classic picture includes an increase in serum transaminases, with an AST/ALT ratio >1, often even >2, unlike what happens in other forms of hepatopathy in which the most marked increase is in ALT compared to AST, and therefore, the AST/ALT ratio remains below 1. The increase in AST is usually over eight times the upper reference limit, while that in ALT usually does not exceed five times the upper reference limit. The fact that the relative increase in ALT is smaller than that of AST has been attributed, at least in part, to the deficiency of pyridoxal-phosphate, a cofactor of ALT. According to this hypothesis, the altered AST/ALT ratio would reflect an inappropriate increase in ALT rather than a disproportionate increase in AST.

The extent of the increase in transaminases does not reflect the degree of severity of the liver disease. Normal or only moderately elevated transaminases are not uncommon.

An AST/ALT ratio >1 may be observed in patients with steatohepatitis and, more frequently, in patients with nonalcoholic cirrhosis. However, if the AST/ALT ratio is >2, the etiology will be alcoholic since such values are very rarely seen in other forms of hepatopathy.

Serum γ-glutamyltransferase (GGT) is often increased in alcoholic liver disease and remains so even after several weeks of abstinence from alcohol intake. GGT is not specific for alcoholic liver disease because it may be increased in all forms of cholestatic jaundice or following the intake of certain drugs, such as barbiturates or phenytoin.

Increased total bilirubin is common in decompensated cirrhosis, regardless of the cause.

Cirrhotic or malnourished patients have hypoalbuminemia.

Haematological abnormalities in patients with alcoholic hepatopathy include thrombocytopenia, anaemia, increased mean corpuscular volume (MCV), neutropenia, increased erythrocyte sedimentation rate (ESR), and prolonged INR (International Normalized Ratio). Macrocytosis is an expression of disease that persists over time and results from folate and vitamin B12 deficiency, direct alcohol toxicity, and lipid accumulation in the erythrocyte membranes. Thrombocytopenia may result from primary bone marrow hypoplasia or splenic sequestration caused by portal hypertension and splenomegaly.

In patients in whom cirrhosis is accompanied by hepatorenal syndrome, hyponatremia, and increased serum creatinine may also be observed.

Alcoholic hepatopathy is diagnosed based on a compatible clinical history, finding of hypertransaminasemia, hepatic steatosis on the radiographic investigation, or liver biopsy. It is a diagnosis of exclusion from other causes of hepatic steatosis listed in Table 11.11.

The differential diagnosis of alcoholic hepatopathy should include chronic viral hepatitis, hemochromatosis, primary biliary cirrhosis, primary sclerosing cholangitis, and autoimmune hepatitis. In steatosis and cirrhosis, the alcoholic etiology should be established based on clinical history, physical examination, and clinical biochemistry evaluations.

Evaluation of the patient with suspected alcoholic hepatopathy should include:

- Accurate assessment of daily and weekly alcohol intake patterns
- Physical examination to identify the typical signs of chronic hepatopathies, such as hepatomegaly, spider nevi, ascites, splenomegaly, gynecomastia, jaundice, and palmar erythema
- Clinical biochemistry evaluation of liver function by measuring AST, ALT, bilirubin, alkaline phosphatase, GGT, blood cell count, albuminemia, PT, and INR. None of these tests is diagnostic of alcoholic liver disease, but an AST/ALT ratio >2 is strongly indicative of alcoholic etiology
- Biochemical and clinical evaluation of other potential causes of chronic liver disease by measuring HBsAg, HBeAg, and anti-HCV antibodies to exclude viral hepatitis
- Ferritin and transferrin saturation to rule out hemochromatosis
- Evaluation of total IgG, ANA, ASMA, and anti-LKM-1 to rule out autoimmune hepatitis.

It should be noted that ferritinemia increases even in the absence of martial overload in some conditions, including alcoholic hepatitis and acute or subacute hepatitis. It follows that in the patient with acute or subacute hepatitis, it is necessary to wait until the acute event has resolved to assess iron metabolism properly. Moreover, transferrin saturation in alcoholic hepatitis can reach values of 60% or more due to inhibitory action of alcohol on transferrin synthesis. For these reasons, specialist diagnostic tests should be performed if hemochromatosis is suspected.

Other laboratory investigations may be required if, based on the patient's medical history, α1-antitrypsin deficiency, hyperthyroidism, celiac disease, primary biliary cirrhosis, or primary sclerosing cholangitis is suspected.

Imaging provides evidence of steatosis or cirrhosis. However, it does not allow to establish the etiology, which instead can be determined based on the patient's clinical history and clinical biochemistry evaluation. The liver biopsy is required when the diagnosis remains uncertain after noninvasive investigations, for example, the clinical and biochemical picture is only partially compatible with alcoholic liver disease (moderate alcohol intake and AST/ALT ratio >1) and other causes cannot be excluded.

Fibrosis

Hepatic fibrosis is a scarring process in which the extracellular matrix encapsulates the region of the liver parenchyma that has suffered an inflammatory insult. Fibrosis develops in almost all patients with chronic hepatopathy, although varying degrees depending on the type of stimulus that generated it and factors related to the host. Fibrosis is a dynamic pathological event, which is reversible in the early stages.

The composition of scar tissue is independent of the type of insult that led to fibrosis and involves the presence of macromolecules present in the normal extracellular matrix, including collagen of type I, III, V and IX, fibronectin, laminin, and elastin. The transition to a fibrotic tissue involves a significant change in matrix composition, with at least a three to tenfold increase in collagen, glycoproteins, proteoglycans, and glycosaminoglycans. These changes in the composition of the extracellular matrix result in the replacement of the normal low-density matrix of the subendothelial space by the interstitial matrix, with important consequences for the function of hepatocytes, stellate cells, and endothelial

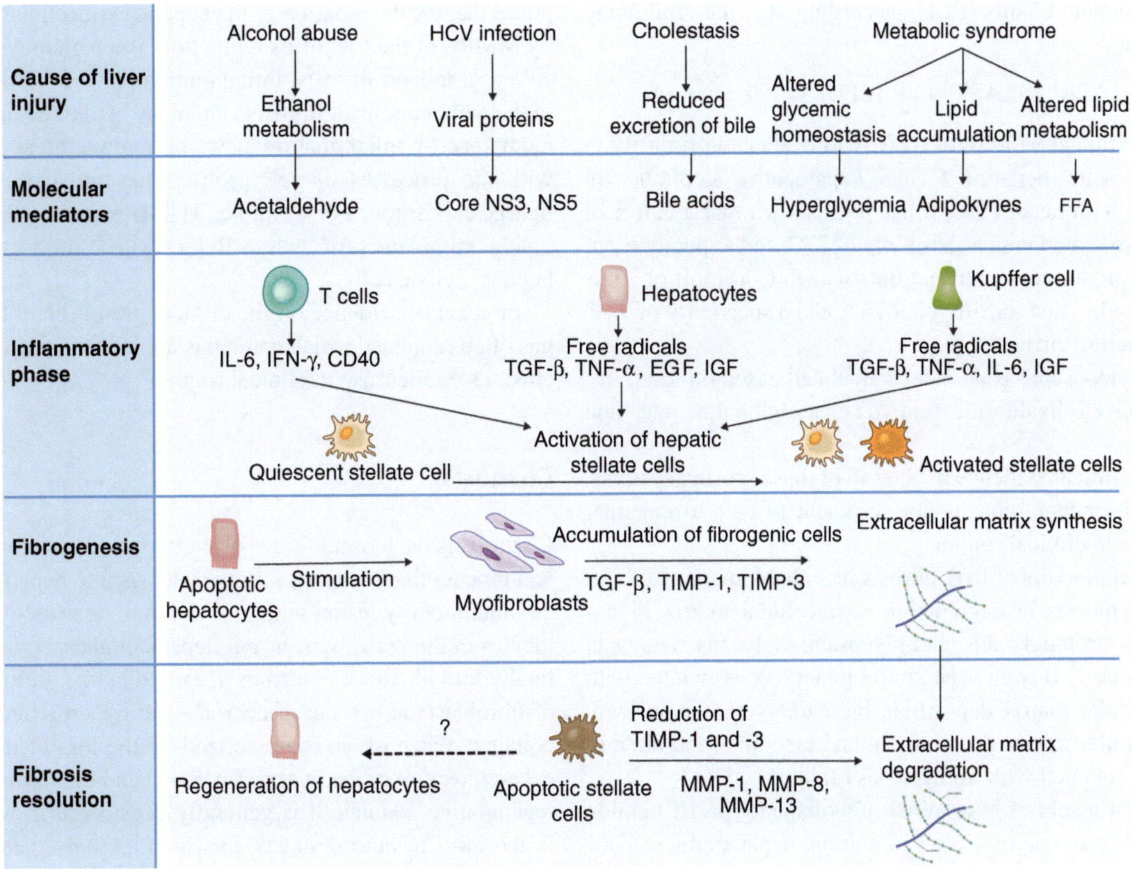

Fig. 11.6 Schematic representation of the mechanisms leading to hepatic fibrosis. (Copyright EDISES 2021. Reproduced with permission)

cells. When the processes of cell proliferation, deposition of the extracellular matrix, regeneration of the hepatic parenchyma, and inflammation occur in a contextual and, above all, uncontrolled manner, a state of fibrosis is established that can progress into cirrhosis (Fig. 11.6). Recently, the knowledge of the molecular mechanisms leading to fibrosis has aroused great scientific interest due to the potential of developing therapies aimed at stopping the progression of fibrosis.

Diagnostic Investigations

The diagnostic approach to liver fibrosis includes biochemical, imaging, and histopathological evaluations, which are of little help if considered in isolation. Histopathological investigations represent the gold standard for the diagnosis and classification of fibrosis.

Noninvasive, serological, and imaging investigations (mainly hepatic elastometry) are performed to monitor any fibrosis progression.

Many biomarkers have been proposed for predicting fibrosis severity, with variable results. Generally, biomarkers of hepatic fibrosis reflect matrix turnover, but not the extent of deposition of its components. Therefore, they may increase significantly when a relevant inflammatory state occurs, regardless of the degree of extracellular matrix deposition. None of the biomarkers available today are specific for hepatic fibrosis, and inflammatory states at other sites may contribute to increased circulating levels.

Serological markers can be distinguished into:

- Indirect markers reflecting changes in liver function.
- Direct markers of fibrosis, or biomarkers reflecting the turnover of the extracellular matrix. Indirect markers include all serological tests, which are indicative of an alteration in liver function, namely AST, ALT, platelet count, basic coagulation parameters (PT and INR), GGT, total bilirubin, α2-macroglobulin, and α2-globulins (mainly haptoglobin). Since these tests, individually, provide rather limited clinical information about the presence or absence of fibrosis, several panels have been proposed that combine these tests in various ways to increase their diagnostic accuracy. The APRI (AST to Platelet Ratio Index) and the Hepascore are the best known.

The APRI is calculated using the serum AST concentration, its upper reference limit (URL) used in the laboratory, and platelet count (PLT) according to the following formula:

$$\text{APRI} = \lfloor (\text{AST} / \text{LSR}) / \text{PLT} \rfloor \times 100$$

The clinical validity of APRI was evaluated primarily in patients with hepatitis C and hepatopathy alcoholic. In patients with hepatitis C, it has been shown that a cutoff of 0.7 is associated with a sensitivity of 77% and a specificity of 72% in predicting significant fibrosis, while a cutoff of 1.0 is associated with a sensitivity of 76% and a specificity of 72% in predicting cirrhosis.

The Hepascore is an index calculated based on total bilirubin, GGT, hyaluronic acid, α2-macroglobulin, age, and sex.

The clinical evidence on the use of these scoring systems, although encouraging, is not sufficient to date to consider their use in clinical routine.

Direct markers of liver fibrosis include biomarkers of collagen synthesis or degradation, extracellular matrix glycoproteins, proteoglycans, and glycosaminoglycans. They can be summarily distinguished into biomarkers associated with extracellular matrix deposition, biomarkers associated with extracellular matrix degradation, and cytokines and chemokines associated with fibrogenesis (Table 11.13).

Serum levels of N-terminal procollagen type III peptide (PIIINP) increase in acute and chronic hepatopathies. They correlate with transaminase and bilirubin levels in cirrhotic patients and histological degree of fibrosis and inflammation in patients with alcoholic hepatopathy, viral hepatitis, and primary biliary cirrhosis. N-terminal procollagen type I peptide (PINP) levels increase in patients with cirrhosis; however, this marker is less accurate than procollagen III-derived peptides in predicting the severity of fibrosis and the presence of hepatitis. Extracellular matrix degradation is primarily mediated by metalloproteinases (MMPs). These enzymes are synthesized intracellularly and secreted as proenzymes. Their activation requires proteolytic cutting mediated by cell surface enzymes. Finally, their action is inhibited by tissue inhibitors of metalloproteinases (TIMPs). It has been proposed that the degradation of the hepatic extracellular matrix by MMPs, or the loss of its regulation, is a pathophysiological event in liver fibrosis. Inflammation plays a relevant role in the pathogenesis of fibrosis, and many cytokines and other mediators of inflammation have been directly associated with the processes of cell proliferation and extracellular matrix deposition. For example, TGF-β is among the most potent stimulators of extracellular matrix production by hepatic stellate cells.

In general, evidence on the clinical use of direct labelers has often remained isolated and has not led to definitive conclusions on their use in clinical routine.

Table 11.13 Main biomarkers of fibrogenesis and hepatic fibrinolysis

Deposition of the matrix
N-terminal peptide of type I procollagen (PINP)
N-terminal peptide of type III procollagen (PIIINP)
C-terminal peptide of type III procollagen (PIIICP)
YKL-40
Degradation of the matrix
MMP-2
Tissue inhibitor of metalloproteinases (TIMP-1, -2)
Cytokines
TGF-β
TNF-α
PDGF

Cirrhosis

Cirrhosis is the terminal stage of the chronic pathophysiological process that progresses from steatosis into hepatitis, i.e., an inflammatory lesion with infiltration of neutrophil granulocytes in the portal spaces and hepatocellular necrosis, and finally into fibrosis and cirrhosis, caused by the proliferation of fibroblasts at the site of necrosis and hyperproduction of collagen. Cirrhosis is characterized by the loss of the typical architecture of the liver parenchyma and the formation of regenerative nodules. It is generally considered irreversible in the most advanced stages, in which the only therapeutic possibility is liver transplantation. Cirrhotic patients are at high risk of severe complications that limit their life expectancy. Several pathological conditions can lead to cirrhosis, the most frequent of which are certainly hepatitis C, alcoholic hepatopathy, and nonalcoholic hepatic steatosis, which together account for 80% of cases in the United States (Table 11.14).

Severe complications may be observed in cirrhotic patients, whose onset marks the passage from a compensated

Table 11.14 Causes of liver cirrhosis

Main causes
Chronic viral hepatitis
Alcoholic liver disease
Hemochromatosis
Non-alcoholic fatty liver disease (NAFLD)
Less frequent causes
Autoimmune hepatitis
Primary and secondary biliary cirrhosis
Primary sclerosing cholangitis
Use of methotrexate
Wilson's disease
α1-antitrypsin deficiency
Celiac disease
Hepatic granulomas
Hepatic polycystosis
Infections (brucellosis, syphilis, echinococcosis)
Hereditary haemorrhagic telangiectasia
Portal thrombosis

Table 11.15 Calculation of the Child–Pugh score

	Score		
	1	2	3
Encephalopathy	Absent	Grade 1–2	Grade 3–4
Ascites	Absent	Slight	Mild
Bilirubin (mg/dL)	<2	2–3	>3
Albumin (g/L)	>35	28–35	<28
INR	<1.7	1.7–2.2	>2.2

Table 11.16 Interpretation of the Child–Pugh score

Child-Pugh class	Score	Interpretation
A	5–6	Well compensated cirrhosis
B	7–9	Significant functional impairment
C	10–15	Decompensated cirrhosis

stage to decompensated cirrhosis. These include ascites, esophageal varices with bleeding, spontaneous bacterial peritonitis, hepatic encephalopathy, hepatocarcinoma, hepatorenal syndrome, and hepatopulmonary syndrome. Many of these conditions are consequences of portal hypertension. The most frequent complication of cirrhosis is ascites, a direct consequence of portal hypertension. While in patients with compensated cirrhosis, the average survival is about 12 years, the transition to a stage of decompensation determines a worse prognosis and a significant increase in the mortality risk.

In this regard, several predictive models have been formulated to be applied in patients with chronic hepatopathy for prognostic purposes, among which the most popular is the Child–Pugh classification and the MELD score.

The Child–Pugh classification is based on the presence of encephalopathy, ascites, bilirubinemia, albuminemia, prothrombin time and, by a scoring system ranging from 5 to 15, classifies the patient into three categories: Child A, well-compensated cirrhosis (5–6 points); Child B, significant functional impairment (7–9 points); Child C, decompensated cirrhosis (10–15 points) (Tables 11.15 and 11.16).

Another model to predict the prognosis of cirrhotic patients is the MELD score (Model for End-Stage Liver Disease), which is based on bilirubinemia, creatininemia, natremia, and INR. This score was initially developed to establish the priority of access to liver transplantation, but it was later extended to different settings. The predictive model underlying the MELD score has undergone various revisions over time, including the insertion of the variable sodium (Na), for which various versions are now available, as summarized in Table 11.17. A MELD value >15 represents the degree of severity of the disease below which liver transplantation would not lead to substantial benefit. An exception is hepatocarcinoma, for which access to transplant is independent of the MELD score.

Clinical manifestations of cirrhosis include both nonspecific symptoms (anorexia, weight loss, asthenia, fatigue) and

Table 11.17 Formulas for calculating the MELD score

MELD	9.57 ln [creatinin (mg/dL)] + 3.78 ln [bilirubin (mg/dL)] + 11.2 ln (INR) + 6.43
MELD-Na	MELD + 1.59 × [135 − Na (mmol/L)]
MESO	[MELD/Na (mmol/L)] × 100
MELD	MELD + [età (anni) × 0.3] − [0.7 × Na (mmol/L)] + 100

MELD-Na MELD sodium, *MONTH* MELD/sodium ratio, *iMELD* integrated MELD

signs and symptoms of hepatic decompensation (jaundice, pruritus, signs of gastrointestinal bleeding, such as hematemesis, melena, hematochexia, abdominal distension following ascitis, confusional states due to hepatic encephalopathy). The progression of cirrhosis is accompanied by a decrease in mean arterial pressure, which may contribute to the establishment of hepatorenal syndrome and is recognized as a significant predictor of mortality in these patients.

Several hematochemical parameters are altered during cirrhosis. Often, the occasional finding of such alterations leads to the suspicion of cirrhosis. The most common abnormalities include increased bilirubin, transaminases, alkaline phosphatase, GGT, INR prolongation, hyponatremia, and thrombocytopenia.

Transaminases are usually moderately elevated in cirrhotic patients. The increase in AST is more significant than that of ALT. However, the finding of normal transaminases does not ezclude the diagnosis of cirrhosis. In many forms of chronic hepatitis, except for alcoholic hepatitis, the AST/ALT ratio is <1. However, as hepatitis progresses to cirrhosis, the AST/ALT ratio may reverse.

Alkaline phosphatase usually increases in cirrhosis up to three times the upper reference limit. Higher values are an expression of cholestatic diseases, such as primary sclerosing cholangitis or primary biliary cirrhosis.

GGT correlates with alkaline phosphatase in liver disease but lacks any specificity. GGT increases more in chronic alcoholic hepatopathy than other hepatopathies of different etiology. Alcohol induces microsomal GGT expression and GGT release from the hepatocyte.

Bilirubin may be normal in well-compensated cirrhosis but increases as the degree of hepatic decompensation progresses. Increased bilirubin is an unfavorable prognostic sign in patients with primary biliary cirrhosis.

Albumin is synthesized exclusively by the liver. Its serum levels decrease as the biosynthetic capacity of the liver declines as cirrhosis progresses. For this reason, albuminemia is helpful in determining the severity of cirrhosis. Hypoalbuminemia is not specific to liver disease but can also be observed in nephrotic syndrome, protideless enteropathies, and malnutrition.

Most of the proteins involved in the hemostatic process are synthesized by the liver. For this reason, the prothrombin time and INR reflect the degree of hepatic dysfunction. A

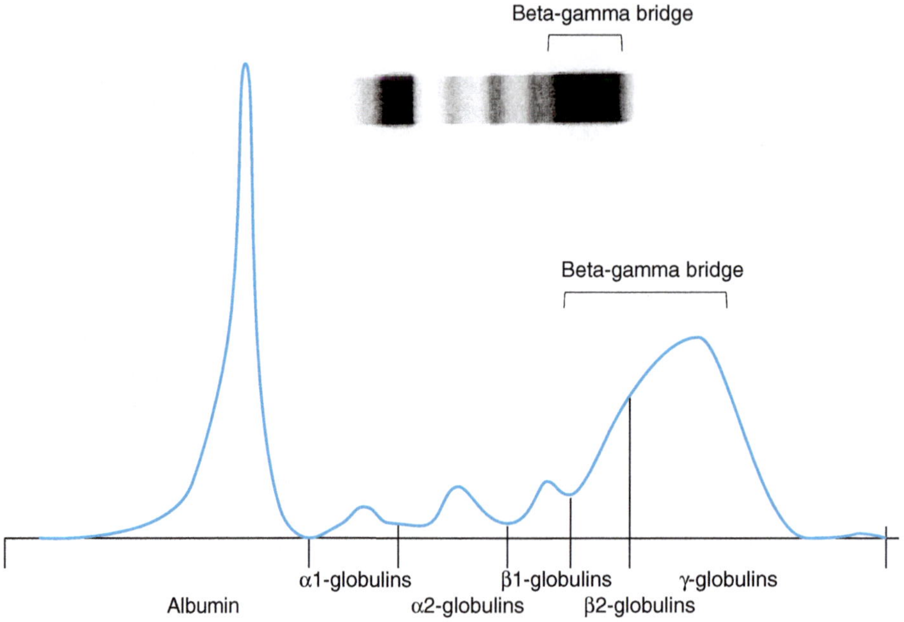

Fig. 11.7 Electrophoresis pattern of a cirrhotic patient in which the β2-globulin band overlaps the γ-globulin band, forming the typical beta-gamma bridge. The values of the single fractions were albumin, 38.4% (IR: 54.5–65); alpha-1, 2.5% (IR: 3.1–6); alpha-2, 5.1% (IR: 7.1–11.8); beta-1, 4.2 (IR: 5–7.2); beta-2, 9.9 (IR: 3.2–6.5); gamma, 39.9 (10.5–18.8). The patient had the following biochemical-clinical values: albumin, 27.3 g/L (IR: 35–52); protidemia, 71.2 g/L (IR: 66–87); ferritin, 1977 ng/mL (IR: 30–400); transfer, 158 mg/dL (IR: 200–360); AST, 47 U/L (IR: 0–37); ALT, 15 U/L (IR: 0–41); total bilirubin, 12 mg/dL (<1.2); alpha-fetus-protein, 2.99 μg/L (IR: 0–7); INR 2.04; APTT, 41 seconds (IR: 24–36)

prolongation of the INR is considered an unfavorable prognostic index in cirrhotic patients.

Hyponatremia is common in patients with cirrhosis and ascites and results from reduced water excretion caused by increased antidiuretic hormone (ADH) release.

Patients with cirrhosis often present with different haematological anomalies, including varying degrees of cytopenia. Thrombocytopenia is the most frequent abnormality; leukopenia and anemia occur in more advanced stages. Thrombocytopenia is primarily caused by portal hypertension related to splenomegaly. An enlarged spleen may result from the sequestration of approximately 90% of circulating platelets. However, it is rare for platelet counts to be <50,000/mL and, unless complicated by coagulopathy, is not a clinical problem.

Anemia has a multifactorial origin. Probable causes include acute and chronic blood loss, folate deficiency, direct alcohol toxicity to erythrocyte membranes, splenomegaly, bone marrow suppression (hepatitis-associated aplastic anemia), inflammation-associated anemia, and hemolysis.

Leukopenia and neutropenia result from splenic sequestration.

On serum protein electrophoresis, an increase in globulins is frequently found with the typical beta-gamma bridge (Fig. 11.7).

Patients with suspected cirrhosis undergo ultrasonography of the abdomen and the abovementioned laboratory tests, which are often sufficient, together with a complete history and physical examination, to establish the diagnosis. However, diagnosis must be confirmed by liver biopsy in doubtful cases.

Alcoholism

Introduction

Ethyl alcohol, or ethanol (CH_3CH_2OH), is a small water-soluble molecule rapidly and completely absorbed from the gastrointestinal tract. The lungs can also absorb ethanol vapors.

Ethanol is mainly present in alcoholic beverages; small quantities are produced by the intestinal bacterial flora (0.1–2 mg/100 g). In addition, in vitro studies have demonstrated its production in various body tissues, including the brain.

Ethanol has a high energy content, 7 kcal/g, an intermediate value between carbohydrates and lipids, but unlike the latter that are accumulated, ethanol is metabolized or eliminated through urine and exhaled air.

The chronic and/or excessive ingestion of ethanol causes alcoholism, a metabolic disease that mainly affects the liver, the digestive system (impaired digestion and absorption), and the respiratory system, with alterations in respiratory (ventilation, diffusion, vascularization, and surfactant synthesis) and nonrespiratory (ciliary apparatus, macrophages, lymphocytes) function, and the central and peripheral nervous system (the main effects of ethanol are on the central nervous system, where the depressive actions are similar to those of volatile anesthetics).

Ethanol Metabolism

Ingested ethanol is rapidly absorbed into the digestive system by passive diffusion. Ethanol is measurable in the bloodstream as early as 5 minutes after ingestion, and the peak concentration is reached after 30–90 minutes.

After absorption, ethanol is predominantly metabolized oxidatively (90–98%), at a constant rate over time (order 0 kinetics) of 100 mg/kg/hour. Most of the ethanol intake (>80%) is metabolized in the liver; the remaining part is metabolized in the kidney, lung, stomach, intestines, brain, and muscles.

Elimination occurs:

- Through the liver, where it is converted to CO_2 and H_2O (>90%)
- Through the respiratory route (1–4%)
- Renal (1–2%) (ethanol stimulates diuresis)
- In the milk
- In sweat

Ethanol is mainly metabolized through two oxidation reactions: the first converts it to acetaldehyde, and the second converts acetaldehyde to acetate.

In the hepatocyte, at least three alternative enzymes catalyze the first oxidation reaction (Fig. 11.8):

- Alcohol dehydrogenase (ADH)
- Microsomal oxidant ethanol system (MEOS)
- Catalase

ADH is the most important enzyme in ethanol metabolism. It is localized in the cytosol and is dependent on NAD+ and zinc.

ADH is encoded by seven gene loci (ADH1–ADH7), located on the long arm of chromosome 4, whose allele frequency differs in different ethnicities. In addition, gene variants associated with altered (accelerated or reduced) ethanol oxidation have been identified that may explain different alcohol tolerance; for example, Eastern peoples (Japanese and Chinese) have poor alcohol tolerance due to the high frequency of a polymorphic variant associated with reduced ADH activity.

MEOS is a mixed-function oxygenase associated with the smooth endoplasmic reticulum (SER) of the hepatocyte; it is cytochrome P450 and NADPH-dependent. It functions at high ethanol concentrations (chronic intoxication), following hypertrophy of the SER (hypertrophy is an adaptive phenomenon that improves the liver capacity to metabolize ethanol) and is induced by ethanol itself. This enzyme can also oxidize other substances, such as drugs (paracetamol); therefore, excessive stimulation of this system leads to important changes in the metabolism of these substances. Alcoholic patients are more sensitive to these types of substances to the point of suffering serious hepatic injury even at low drug doses.

Catalase is an enzyme in peroxisomes; it has a limited effect on ethanol metabolism due to the limited availability of hydrogen peroxide in the hepatocyte.

Under low ethanol intake, ADH is the only enzyme responsible for the ethanol oxidation to acetaldehyde; however, MEOS and catalase also function at high ethanol concentrations.

Most of the acetaldehyde produced in the hepatocyte is normally oxidized to acetate by the mitochondrial NAD-dependent acetaldehyde dehydrogenase, which has a high affinity for acetaldehyde and is highly specific. The remaining acetaldehyde is oxidized by cytosolic acetaldehyde dehydrogenase and partly by aldehyde oxidase and xanthine

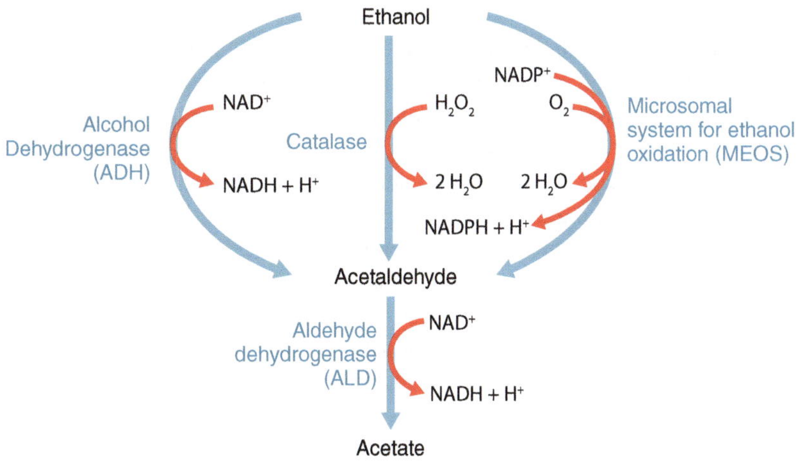

Fig. 11.8 Ethanol metabolism. (Copyright EDISES 2021. Reproduced with permission)

oxidase. At very high concentrations of ethanol, not all of the acetaldehyde can be converted to acetate, and therefore, this passes into the circulation, with consequent lesions at various levels.

Most of the acetate synthesized in the liver is released into the circulation. It reaches the extra-hepatic tissues, where it is rapidly converted to acetyl-CoA and oxidized to carbon dioxide (CO_2) in the Krebs cycle. A small amount of acetate, however, is converted hepatically to acetyl-CoA by acetyl-CoA synthetase and then oxidized to CO_2 or used for ketone bodies or fatty acids synthesis.

Very small amounts of ethanol are eliminated via the urinary or biliary tract following conjugation with glucuronic acid or sulfuric acid. Finally, a tiny proportion of ethanol (<2%) may be excreted in an unmodified form via the kidneys and lungs; this proportion increases with massive ethanol ingestion.

Ethanol can induce various biochemical alterations, including increased production of NADH, with increased $NADH/NAD^+$ ratio and, therefore, reducing power:

- In fasting, the change in the balance of lactate dehydrogenase to lactate rather than pyruvate, with a tendency to lactic acidosis and hypoglycemia (from impaired gluconeogenesis and reduced glucose uptake).
- In the postprandial, transient hyperglycemia due to the inhibition of glycolysis at the level of glyceraldehyde-3-phosphate dehydrogenase.
- The inhibition of β-oxidation of fatty acids and the consequent increase in the liver of fatty acids, which are converted into triglycerides; is reflected in the increase in the hepatic synthesis of very low-density lipoproteins (VLDL) and defines their deposition in the liver (hepatic steatosis), as well as the increase in their plasma concentration. In addition, the increased arrival of acetyl-CoA to the liver results in increased ketogenesis (due to oxalate deficiency following reduced pyruvate availability).

The high production of acetaldehyde, on the other hand, leads to an increased release of vascular prostacyclin (vasodilatory and antiplatelet action – protective effect of small doses of ethanol against vascular disease); acetaldehyde-tubulin binding, which determines a deficit in the polymerization of microtubules with the consequent reduced hepatic secretion of proteins and an increase in intracellular oncotic pressure and swelling of the hepatocytes; an increase in membrane lipid peroxidation; the binding with various proteins leading to the activation of liver the immune response against the acetaldehyde–protein complex (Fig. 11.9).

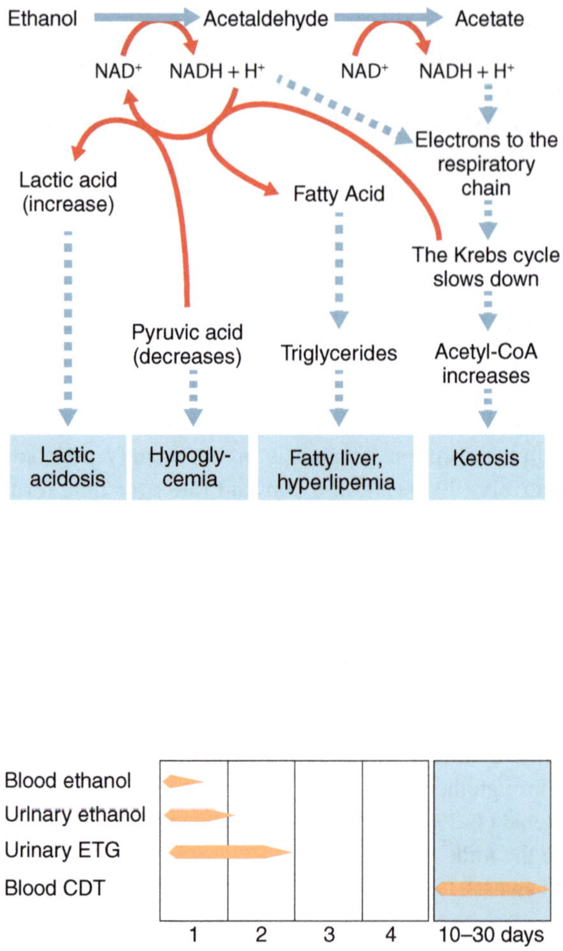

Fig. 11.9 Metabolic effects of ethanol. (Copyright EDISES 2021. Reproduced with permission)

Alcohol-Related Clinical Alterations

Excessive voluntary ingestion of ethyl alcohol is called alcoholism or ethylism and can appear in an acute or chronic form. A state of drunkenness characterizes the acute form. The chronic form, on the other hand, represents a true addiction and can lead to early death. Alcoholism must be considered one of the most significant issues for public health, which involves, in addition to the state of physical and mental health, also the social relationships of the individual.

The main alcohol-related clinical alterations are:

- Alcoholic gastritis
- Increased incidence of peptic ulcer
- Gastrointestinal bleeding
- Pancreatitis
- Cardiomyopathy
- Cardiac arrhythmias

Table 11.18 Biomarkers of alcohol use

Biomarkers	Acronymous	Biological sample	Characteristics
Ethanol	EtOH	Blood Urine	Limited to the conditions in which ethanol is still in circulation.
Ethylglucuronide	EtG	Urine Serum Cerebrospinal fluid Hair Nails	Minor metabolite of ethanol. It remains positive in the urine for 2–5 days after taking ethanol
Carbohydrate-deficient transferrin	CTD	Serum Cerebrospinal fluid	Specific marker of chronic alcohol use.
Gamma-glutamyltransferase	GGT	Serum/plasma	Sensitive marker of alcohol consumption, liver dysfunction and oxidative stress. It normalizes after 2–3 weeks.
Median corpuscular volume	MCV	Blood	Typically increased in alcoholics. It normalizes after 2–4 months.
Alanine and aspartate aminotransferase	ALT and AST	Serum/plasma	Used for the screening of liver dysfunction in subjects consuming alcohol. The AST/ALT ratio increases in alcoholic liver disease.

- Fetal alcohol syndrome
- Wernicke–Korsakoff syndrome
- Alcoholic steatosis
- Alcoholic hepatitis
- Cirrhosis of the liver

Biomarkers of Alcoholism

Several biochemical markers of alcohol abuse, both acute and chronic, have been proposed and can be measured in urine and/or blood. They represent a tool to assess the amount and mode (chronic or acute) of alcohol consumption, and the damage induced on the body by alcoholism. Biomarkers can be classified as direct, i.e., derived directly from the metabolism of ethanol, and indirect, which are released due to alterations induced by ethanol. The latter is characterized by low diagnostic sensitivity and specificity because it increases significantly only after high and regular intake of alcohol, and its levels may vary in response to factors unrelated to ethanol.

Alongside established biomarkers, such as aspartate aminotransferase, alanine aminotransferase, γ-glutamyltransferase, and mean corpuscular volume of erythrocytes, which are positive only in advanced states of alcohol-related diseases, new, earlier, and more specific markers have been identified (Table 11.18), the main ones of which are (Fig. 11.10):

- Blood ethanol (EtOH)
- Urinary ethanol (EtOH)
- EtG (ethylglucuronide)
- CDT (carbohydrate transferrin)

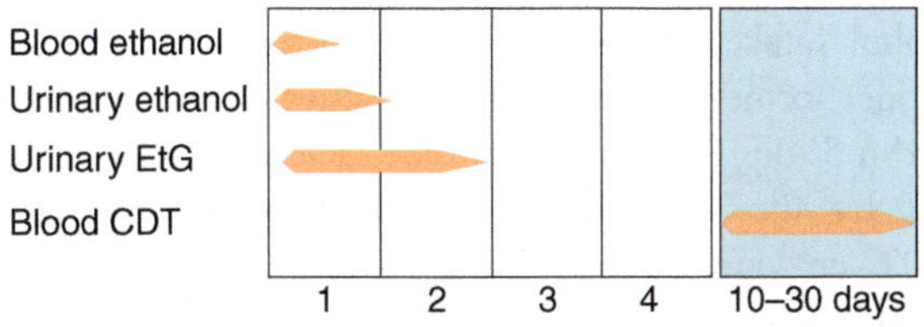

Fig. 11.10 Detection times of the biomarkers of alcoholism. (Copyright EDISES 2021. Reproduced with permission)

Table 11.19 Interpretation of blood ethanol concentration

Blood ethanol (mg/dL)	Interpretation
≥150	Indicative of excessive alcohol consumption
>100	Indicative of alcoholism
≥150	Indicative of risky drinking, tolerance, and possible addiction[a]

[a]If it can be safely excluded that such high concentrations are not due to acute ethanol intoxication

Blood Ethanol

Circulating ethanol levels are a direct and reliable marker of recent alcohol intake. In addition, assessment of blood ethanol values can give insight into long-term ethanol consumption patterns (Table 11.19).

Ethanol has a short half-life, about 30 minutes, and therefore, its presence can only be detected for a short period.

Urinary Ethanol

Ethanol peaks in the urine 45–60 minutes after ingestion. Ethanol levels in the urine are generally higher than the corresponding blood levels; this is true in the elimination phase, after ethanol has reached its peak in the blood and, therefore, its circulating levels begin to decline. In addition, alcohol in urine can be detected for a more extended period (up to

1–2 hours longer) than in blood because the urine remains in the bladder for a period. Urinary ethanol is characterized by marked interindividual fluctuations.

The presence of alcohol in the urine indicates recent previous intake but may not be related to the degree of intoxication observed at the time of sample collection. The level of ethanol in the sample refers only to the average concentration of alcohol in the blood during the time it takes for the urine sample to accumulate in the bladder, not to the concentration of alcohol in the blood at the time of collection.

False-negative results may be caused by the volatility of the alcohol. Urine alcohol concentrations may decrease by 10–25% during each hour before the urine sample is analyzed.

Ethylglucuronide (EtD)

Ethylglucuronide is a direct minor metabolite of ethanol and is considered highly specific for the assessment of recent alcohol intake. In particular, it is an ethanol conjugation product formed by reaction with activated glucuronic acid (uridin-5′-diphospho-β-glucuronic acid [UDPGA]) mediated by the uridine-diphosphate glucuronosyltransferase (UGT) enzyme.

EtG is in various biological fluids, tissues, and hair. However, it is generally measured in urine. EtG can remain positive for several days after the discontinuation of ethanol intake and, therefore, may provide added value in assessing recent alcohol consumption. In particular, in urine remains for 40–60 hours after alcoholic intake. In the blood, instead, it is detectable up to 14 hours after the intake; an occasional intake of alcohol is enough to exceed the threshold of 0.5 mg/L.

Some studies have indicated diagnostic applications of EtG in postmortem assessments of alcohol consumption and fetal alcohol exposure or in patients awaiting liver transplantation.

Polymorphic variants in genes encoding UGT may have a significant impact on the ability of humans to synthesize EtG and may, therefore, explain interindividual differences in EtG levels after alcohol consumption.

Carbohydrate Transferrin (CDT)

Carbohydrate transferrin is a specific marker of chronic alcohol abuse.

Transferrin is a serum protein that transports iron in the circulation to the target organs (bone marrow, liver, and spleen). From a structural point of view, transferrin is a glycoprotein consisting of a polypeptide with two polysaccharide chains linked with sialic acid residues. The addition of sialic acid occurs through a posttranslational reaction mediated by an enzymatic system with ethanol-dependent glucosyltransferase activity; ethanol and its metabolite, acetaldehyde, reduce the enzymatic activity of this system.

There are several forms of transferrin, which differ in the number of iron atoms carried (0–2), the primary structure (genetically determined), and the number of sialic acid residues (0–8). Most circulating transferrin (>80%) contains four sialic acid residues (tetrasialotransferrin, TeST); the form with two residues (disialotransferrin, DST) is physiologically present in amounts <2%, while the desialylated form (asialotransferrin, aST) is almost absent. In the case of massive and continuous ethanol intake, the desialylated fraction increases. The term CDT refers to the set of low sialic acid transferrins and, particularly to the aST and DST forms. CDT has a longer half-life (10 days) than transferrin in the circulation (7 days).

CDT elevation requires the consumption of at least 50–80 g of ethanol per day for several weeks and, therefore, has low sensitivity when used as a screening tool in the general population. However, in alcohol-dependent patients, it is sufficiently sensitive to detect relapse and monitor sobriety.

Liver Cancers

There are numerous types of primary liver cancers, both benign and malignant (Table 11.20). Most of them arise from the main cells constituting the organ, such as hepatocytes (hepatocarcinoma), biliary epithelial cells (cholangiocarcinoma and biliary cystadenocarcinoma), endothelial cells (angiosarcoma, epithelioid hemangioendothelioma), or combinations of these cells with various mesenchymal cells (e.g., hepatoblastoma). These lesions generally do not spread to other organs but within the liver itself, giving rise to multifocal tumors.

In addition, due to its high vascularity, the liver has a high incidence of secondary cancers, i.e., metastases from cancers originating in other organs, especially the gastrointestinal tract. Epidemiological data show that up to 50% of patients with colorectal cancer are diagnosed with or develop distant metastases, mainly hepatic, in the years following resection of the primary cancer.

Benign cancers are quite frequent in the population (up to 20%) and are usually discovered incidentally during investi-

Table 11.20 Benign and malignant liver lesions

Benign lesions	Malignant lesions

- Hepatic hemangioma
- Hepatic adenoma
- Focal nodular hyperplasia
- Idiopathic portal hypertension
- Regenerative nodules
- Inflammatory pseudotumor (rare benign liver tumor consisting of proliferating fibrous tissue infiltrated by inflammatory cells)
	- Hepatocarcinoma
	- Cholangiocarcinoma
	- Fibrolamellar carcinoma
	- Hepatoblastoma
	- Mesenchymal tumors:
	  - Epithelioid hemangioendothelioma
	  - Angiosarcoma

gations for other conditions in patients with noncirrhotic liver.

Hepatocarcinoma (HCC), on the other hand, is the most frequent hepatic malignancy.

Diagnosis is based on evaluating clinical history, risk factors, laboratory test results, imaging studies, and, in some cases, histopathologic examination of the lesion.

Regarding laboratory tests, in addition to the evaluation of liver parameters, which may be normal in a patient with hepatic neoplasia, the serum concentration of alpha-fetoprotein (AFP) should be evaluated.

AFP is a glycoprotein normally produced by the fetal liver and the vitreous sac during gestation. Its concentrations in amniotic fluid and maternal blood increase progressively during pregnancy, reaching a peak between the 12th and 16th week of gestation and then decreasing until delivery. It is the most abundant protein in fetal circulation; it is like albumin in molecular weight, amino acid sequence, immunological characteristics, and biological functions.

In adults, its levels are generally very low, almost undetectable. However, AFP levels can increase significantly in specific disease conditions, such as HCC. In particular, serum AFP levels are typically higher in advanced HCC than in early HCC, but overall, the levels do not correlate well with clinical features of HCC, such as tumor size or vascular invasion. In addition, not all tumors secrete AFP, and serum concentrations are normal in approximately 40% of individuals with small HCC. High levels of AFP have good specificity for HCC at the expense of low sensitivity. Indeed, it has been generally accepted that serum AFP levels >400 ng/mL (normal value is generally 10–20 ng/mL) in a high-risk patient are diagnostic of HCC, with >95% specificity; however, less than one-fifth of patients with HCC have such elevated AFP levels. Furthermore, AFP levels may increase in different pathological conditions. In particular, the differential diagnosis of elevated circulating AFP levels includes:

- Chronic liver disease without hepatocellular carcinoma
- Pregnancy
- Cancers of gonadal origin (germ and nongerm cell)
- Other malignant cancers, of which gastric carcinoma is the most common.

Given the sensitivity and specificity issues, serum AFP assessment has been removed from some guidelines as a diagnostic test for HCC. However, the AFP test may be helpful, in combination with the results of other investigations, to guide the management of patients in whom a diagnosis of HCC is suspected. An elevated AFP level in combination with suspicious but nondiagnostic imaging findings may have a positive predictive value in the absence of biopsy.

All patients with a solid liver lesion should have their serum AFP levels checked. If elevated, a diagnosis of HCC becomes more likely, although imaging studies and possibly biopsy are necessary to confirm the diagnosis.

AFP levels are generally normal in patients with benign liver tumors, such as hemangioma, adenoma, and focal nodular hyperplasia.

In the case of cholangiocarcinoma, although nonspecific, some serum tumor markers may be of diagnostic value. Carbohydrate antigen 19.9 (CA19.9) and carcinoembryonic antigen (CEA) are the two best-studied markers, although their diagnostic utility is limited due to low specificity and sensitivity. Serum CA19.9 levels are widely used to detect cholangiocarcinoma in patients with primary sclerosing cholangitis (PSC). For patients with PSC, periodic CA19.9 measurement is used for cholangiocarcinoma surveillance. In addition, elevated CA19.9 levels before treatment are associated with a worse prognosis, and CA19.9 concentrations >1000 units/mL are consistent with advanced disease, often involving the peritoneum. If initially elevated, serum CA19.9 levels may be useful in monitoring response to treatment and detecting disease recurrence. However, there are limitations to using CA19.9 as a diagnostic marker for cholangiocarcinoma because it is frequently elevated in patients with various benign pancreatico-biliary disorders, including cholangitis, and with other malignancies, including pancreatic carcinoma.

Serum CEA levels may be elevated in cholangiocarcinoma. However, serum CEA is neither sufficiently sensitive nor specific to diagnose cholangiocarcinoma. Many conditions other than cholangiocarcinoma may induce increased serum CEA levels, such as many primary cancers of the gastrointestinal tract, breast cancer, as well as extra-neoplastic causes such as gastritis, peptic ulcer disease, diverticulitis, liver disease, chronic obstructive pulmonary disease, diabetes, and any acute or chronic inflammatory state. However, if levels are elevated, they may be helpful for monitoring therapy.

In contrast, AFP differentiates intrahepatic cholangiocarcinoma from hepatocellular carcinoma. Even in the case of fibrolamellar carcinoma, AFP is not useful for diagnosing and monitoring disease progression. Only 7–11% of patients have an elevated serum level of AFP, with values of 100–200 ng/L.

Recommended Readings

Advisory Committee on Immunization Practices (ACIP), Fiore AE, Wasley A, Bell BP (2006) Prevention of hepatitis A through active or passive immunization: recommendations of the Advisory Committee on Immunization Practices (ACIP). MMWR Recomm Rep 55:1

Atkinson M, Budge H (2011) Review of the NICE guidance on neonatal jaundice. Arch Dis Child Educ Pract Ed 96:136–140

Bianchi V (2015) Transferrina carboidrato-carente e abuso alcolico. Biochim Clin 39(1)

Bianchi V, Raspagni A, Alfini C (2009) Etilglucuronide: un vecchio e nuovo indicatore di uso recente di alcol. Biochim Clin 33(3)

Blachier M, Leleu H, Peck-Radosavljevic M, Valla DC, Roudot-Thoraval F (2013) The burden of liver disease in Europe: a review of available epidemiological data. J Hepatol 58(3):593–608

Bohm SR, Berger KW, Hackert PB et al (2015) Hepatitis A outbreak among adults with developmental disabilities in group homes--Michigan, 2013. MMWR Morb Mortal Wkly Rep 64:148

Centers for Disease Control and Prevention (2016) Viral hepatitis surveillance: United States, 2013. US Department of Health and Human Services, Atlanta. https://www.cdc.gov/hepatitis/statistics/2014surveillance/pdfs/2014HepSurveillanceRpt_Rev2016-09-26.pdf. Accessed on September 11, 2017

Chalasani N, Younossi Z, Lavine JE et al (2018) The diagnosis and management of nonalcoholic fatty liver disease: practice guidance from the American Association for the Study of Liver Diseases. Hepatology 67:328

Chan ES, Yeh MM (2010) The use of immunohistochemistry in liver tumors. Clin Liver Dis 14:687

Christensen RD, Yaish HM (2015) Hemolytic disorders causing severe neonatal hyperbilirubinemia. Clin Perinatol 42(3):515–527

Czauderna P, Lopez-Terrada D, Hiyama E et al (2014) Hepatoblastoma state of the art: pathology, genetics, risk stratification, and chemotherapy. Curr Opin Pediatr 26:19

Erlinger S, Arias IM, Dhumeaux D (2014) Inherited disorders of bilirubin transport and conjugation: new insights into molecular mechanisms and consequences. Gastroenterology 146:1625

Fargo MV, Grogan SP, Saguil A (2017) Evaluation of jaundice in adults. Am Fam Physician 95(3):164–168

Forlani G, Giorda C, Manti R et al (2016) The burden of NAFLD and its characteristics in a nationwide population with type 2 diabetes. J Diabetes Res 2016:2931985

Goodman ZD (2007) Neoplasms of the liver. Mod Pathol 20(Suppl 1):S49

Graham RP, Jin L, Knutson DL et al (2015) DNAJB1-PRKACA is specific for fibrolamellar carcinoma. Mod Pathol 28:822

Gutierrez HL, Hund L, Shrestha S, Rayburn WF, Leeman L, Savage DD, Bakhireva LN (2015) Ethylglucuronide in maternal hair as a biomarker of prenatal alcohol exposure. Alcohol 49:617–623

Haque T, Sasatomi E, Hayashi PH (2016) Drug-induced liver injury: pattern recognition and future directions. Gut Liver 10:27

https://www.cdph.ca.gov/Programs/CID/DCDC/Pages/Immunization/Hepatitis-A-Outbreak.aspx. Accessed on December 7, 2017

https://www.who.int/substance_abuse/publications/global_alcohol_report/gsr_2018/en/

Huang MJ, Kua KE, Teng HC, Tang KS, Weng HW, Huang CS (2004) Risk factors for severe hyperbilirubinemia in neonates. Pediatr Res 56(5):682–689

Huang Y, Adams LA, Joseph J, Bulsara MK, Jeffrey GP (2017) The ability of Hepascore to predict liver fibrosis in chronic liver disease: a meta-analysis. Liver Int 37:121–131

Kazemi-Shirazi L, Veloso MP, Frommlet F et al (2008) Differentiation of nonalcoholic from alcoholic steatohepatitis: are routine laboratory markers useful? Wien Klin Wochenschr 120:25

Kim WR, Biggins SW, Kremers WK (2008) Hyponatremia and mortality among patients on the liver-transplant waiting list. N Engl J Med 359:1018–1026

Klevens RM, Miller JT, Iqbal K et al (2010) The evolving epidemiology of hepatitis a in the United States: incidence and molecular epidemiology from population-based surveillance, 2005-2007. Arch Intern Med 170:1811

Kushel M (2018) Hepatitis A outbreak in California – addressing the root cause. N Engl J Med 378:211

Kwo PY, Cohen SM, Lim JK (2017) ACG clinical guideline: evaluation of abnormal liver chemistries. Am J Gastroenterol 112(1):18–35. https://doi.org/10.1038/ajg.2016.517

Latash J, Dorsinville M, Del Rosso P et al (2017) Notes from the field: increase in reported hepatitis A infections among men who have sex with men – New York City, January-August 2017. MMWR Morb Mortal Wkly Rep 66:999

Lee YA, Wallace MC, Friedman SL (2015) Pathobiology of liver fibrosis: a translational success story. Gut 64:830

Lin ZH, Xin YN, Dong QJ et al (2011) Performance of the aspartate aminotransferase-to-platelet ratio index for the staging of hepatitis C-related fibrosis: an updated meta-analysis. Hepatology 53:726

Lurie Y, Webb M, Cytter-Kuint R, Shteingart S, Lederkremer GZ (2015) Non-invasive diagnosis of liver fibrosis and cirrhosis. World J Gastroenterol 21(41):11567–11583

Madan A, MacMahon JR, Stevenson DK (2005) Neonatal hyperbilirubinemia. In: Taeusch HW, Ballard RA, Gleason CA (eds) Avery's disease of the newborn. Elsevier Saunders, pp 1226–1256

Murphy TV, Denniston MM, Hill HA et al (2016) Progress toward eliminating hepatitis A disease in the United States. MMWR Suppl 65:29

Mutsch M, Spicher VM, Gut C, Steffen R (2006) Hepatitis A virus infections in travelers, 1988-2004. Clin Infect Dis 42:490

National Institute for Health and Clinical Excellence (2011). http://pathways.nice.org.uk/pathways/neonataljaundice

Nguyen T, Phillips D, Jain D et al (2015) Comparison of 5 immunohistochemical markers of hepatocellular differentiation for the diagnosis of hepatocellular carcinoma. Arch Pathol Lab Med 139:1028

Niemela. (2016) Biomarker-based approaches for assessing alcohol use disorders. Int J Environ Res Public Health 13:166. https://doi.org/10.3390/ijerph13020166

Pavlov CS, Casazza G, Nikolova D et al (2015) Transient elastography for diagnosis of stages of hepatic fibrosis and cirrhosis in people with alcoholic liver disease. Cochrane Database Syst Rev 22(1):CD010542

Rainio J, Ahola S, Kangastupa P, Kultti J, Tuomi H, Karhunen PJ, Helander A, Niemelä O (2014) Comparison of ethyl glucuronide and carbohydrate-deficient transferrin in different body fluids for post-mortem identification of alcohol use. Alcohol Alcohol 49:55–59

Rehm J, Samokhvalov AV, Shield KD (2013) Global burden of alcoholic liver diseases. J Hepatol 59:160

Richardson M, Elliman D, Maguire H et al (2001) Evidence base of incubation periods, periods of infectiousness and exclusion policies for the control of communicable diseases in schools and preschools. Pediatr Infect Dis J 20:380

Roche SP, Kobos R (2004) Jaundice in the adult patient. Am Fam Physician 69(2):299–304

Singal AK, Kodali S, Vucovich LA et al (2016) Diagnosis and treatment of alcoholic hepatitis: a systematic review. Alcohol Clin Exp Res 40:1390

Società Italiana di Neonatologia (2013) Task force per l'iperbilirubinemia neonatale raccomandazioni per il trat-tamento dell'iperbilirubinemia neonatale. https://www.neonatologia.it/upload/1216_Raccomandazioni%202013_interattivo.pdf

Staufer K, Andresen H, Vettorazzi E, Tobias N, Nashan B, Sterneck M (2011) Urinary ethyl glucuronide as a novel screening tool in patients pre- and post-liver transplantation improves detection of alcohol consumption. Hepatology 54:1640–1649

Taylor RM, Davern T, Munoz S et al (2006) Fulminant hepatitis A virus infection in the United States: incidence, prognosis, and outcomes. Hepatology 44:1589

Terrault NA, Lok ASF, McMahon BJ et al (2018) Update on prevention, diagnosis, and treatment of chronic hepatitis B: AASLD 2018 hepatitis B guidance. Hepatology 67:1560

Vento S, Garofano T, Renzini C et al (1998) Fulminant hepatitis associated with hepatitis A virus superinfection in patients with chronic hepatitis C. N Engl J Med 338:286

Wang Z, Chen Y, Xie S, Lv H (2016) Changing epidemiological characteristics of hepatitis A in Zhejiang Province, China: increased susceptibility in adults. PLoS One 11:e0153804

Watchko JF, Tiribelli C (2013) Bilirubin-induced neurologic damage--mechanisms and management approaches. N Engl J Med 369:2021

WHO (2016) Guidelines for the screening, care and treatment of persons with chronic hepatitis C infection, updated version. WHO, Geneva

WHO (2017) Guidelines on hepatitis B and C testing. WHO, Geneva

Wickremasinghe AC, Karon BS, Saenger AK, Cook WJ (2012) Effect of universal neonatal transcutaneous bilirubin screening on blood draws for bilirubin analysis and phototherapy usage. J Perinatol 32:851–855

Wiesner RH, McDiarmid SV, Kamath PS (2001) MELD and PELD: application of survival models to liver allocation. Liver Transpl 7:567–580

Wiseman R, Weil LM, Lozano C et al (2016) Notes from the field: health care-associated hepatitis A outbreak – Texas, 2015. MMWR Morb Mortal Wkly Rep 65:425

Woodgate P, Jardine LA (2015) Neonatal jaundice: phototherapy. BMJ Clin Evid 2015:0319

Workowski KA, Bolan GA, Centers for Disease Control and Prevention (2015) Sexually transmitted diseases treatment guidelines, 2015. MMWR Recomm Rep 64:1

World Health Organization (2016) Hepatitis E fact sheet. Updated July 2016. http://www.who.int/mediacentre/factsheets/fs280/en/. Accessed on June 20, 2017

Younossi ZM, Stepanova M, Afendy M et al (2011) Changes in the prevalence of the most common causes of chronic liver diseases in the United States from 1988 to 2008. Clin Gastroenterol Hepatol 9:524

Protein Diagnostics

12

Maria Stella Graziani and Anna Caldini

Introduction

Protein diagnostics mainly deals with monoclonal gammopathies, conditions characterized by the presence of a monoclonal immunoglobulin (or part of it) in the blood or in the urine. This chapter first examines the clinical relevance of the condition, which ranges from clinically occult forms, represented by monoclonal gammopathy of undetermined significance, to forms in which the clinical condition is determined by effects due to the peculiar characteristics of monoclonal immunoglobulin, to true neoplasms, such as multiple myeloma. Then the laboratory techniques that allow the management of patients with monoclonal gammopathy are described: protein electrophoresis (serum and urine), for the detection and quantification of the monoclonal component; immunological typing, for the definition of the class and type of immunoglobulin involved; measurement of serum-free light chains and of the immunoglobulins not involved in monoclonality. In the second part, the other field of protein diagnostics is introduced, i.e., the measurement of some clinically relevant serum proteins. Those taken into consideration are albumin, α1-antitrypsin, haptoglobin, β2-microglobulin, ceruloplasmin, complement, immunoglobulins, C-reactive protein, transthyretin. For each of these, a summary of the physicochemical characteristics and biological function is presented as well as indications for their appropriate clinical use.

M. S. Graziani (✉)
Section of Clinical Biochemistry, University of Verona, Verona, Italy
e-mail: mariastella@graziani.eu

A. Caldini
General Laboratory, Careggi University Hospital, Florence, Italy

Plasma Cell Dyscrasias

Clinical Relevance

The plasma cell dyscrasias (or monoclonal gammopathies, MG) are proliferative plasma cell disorders characterized by the production and secretion of a monoclonal component (MC) constituted by the whole immunoglobulin (Ig) or of a part of it. Indeed, when an expanding plasma cell clone is present, the single Ig produced may be in such a quantity that it reaches a sufficiently high concentration in serum or urine to be detectable with appropriate techniques, usually electrophoretic (Fig. 12.1).

Therefore, MC can be used as a serological marker of the clinical condition, for diagnostic purposes to detect the plasma cell clone, as well as during patient monitoring to assess the evolution of the disease. Unlike other tumor biomarkers, MCs present an extreme biochemical variability, because each single monoclonal Ig has a unique sequence in the variable region and can be constituted by the whole Ig or by a part of it. This means that the molecular weight range of an MC can vary from 24 kDa in the case of monomeric free light chains up to 900 kDa for pentameric IgM. Furthermore, although some plasma cell dyscrasias occur with serum concentrations of MC in the order of g/L, under other clinical conditions MC is present at very low concentrations or is even virtually absent. Thus, while the presence of an MC in serum or urine defines the clinical condition of MG, its absence on electrophoretic tracing cannot exclude it, because there are oligosecretory/nonsecretory plasma cell clones that do not produce sufficient MC to be visualized; or visualization is not possible due to technical problems, such as when MC is masked by other proteins comigrating in the same electrophoretic position. The clinical conditions that can be associated with the presence of MC are numerous and range

M. Ciaccio (ed.), *Clinical and Laboratory Medicine Textbook*, https://doi.org/10.1007/978-3-031-24958-7_12

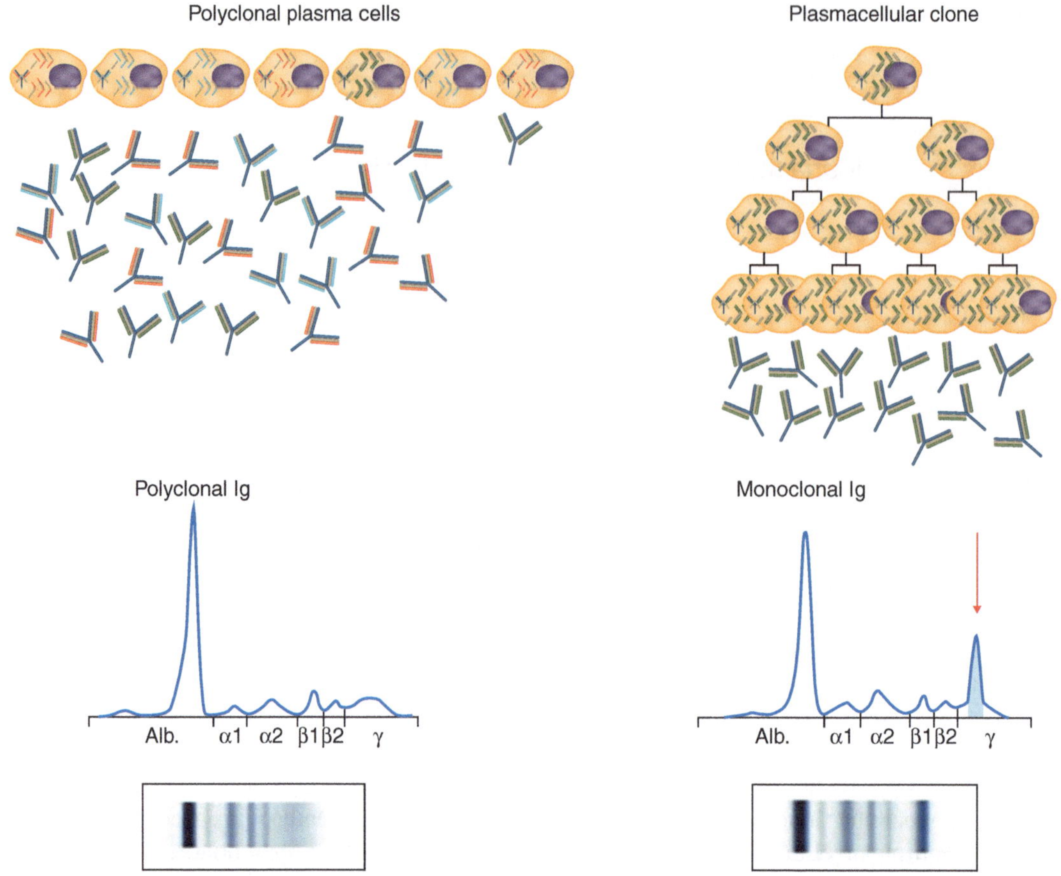

Fig. 12.1 Under normal conditions, only polyclonal plasma cells are present in the bone marrow, each of which secretes its particular immunoglobulin. The consequent great variety of molecular forms of immunoglobulins will lead to a Gaussian distribution of the γ zone of the electrophoretic trace shown in the lower left. On the other hand, when a single clone reproduces at a much higher rate than the rest of the plasma cell population, the concentration of the immunoglobulin secreted by it may be such as to be clearly visible as a narrow migration band such as that highlighted on the electrophoretic trace at the bottom right

from frankly neoplastic forms (such as multiple myeloma – MM) to clinically silent forms, up to clinical manifestations of organ damage due to the toxicity of the specific protein secreted by the plasma cell clone. For the latter group, the term Monoclonal Gammopathies of Clinical Significance (MGCS) has been coined.

Based on the type of clinical manifestations, MGs can be classified as shown in Table 12.1.

Among the clinically occult forms, i.e., those conditions in which MG is not associated with clinical manifestations, the most frequently encountered is Monoclonal Gammopathy of Undetermined Significance (MGUS), a term coined by Robert Kyle in 1978. MGUS is defined as the presence of a MC in serum or urine without clinical evidence of multiple myeloma, AL amyloidosis, Waldenström's disease, or other related disorders. Epidemiologically, MGUS is far from to be a rare condition, its prevalence increases with age, is higher in males and in Africans and African-Americans than in Caucasians. In an important population-based study in the United States, the prevalence of this condition in the general population over 50 years of age was 3.2%, a figure quite similar to that found in studies conducted in Italy. Smoldering or asymptomatic myeloma (SMM) is considered an intermediate condition between MGUS and MM, as it has biochemical and cellular characteristics intermediate between MGUS and MM but without presenting the clinical symptoms of the latter. Although asymptomatic, MGUS and SMM are considered "premalignant" conditions, as it has been shown that almost all cases of MM are preceded by either condition. While MGUS is associated with a risk of progression of approximately 1% per year, the risk of progression of SMM is 10% in the first 5 years after diagnosis and then gradually decreases to 1%, like MGUS, within 15 years. Many of those with MGUS and all those with SMM require continuous monitoring, to prevent organ damage, such as bone injury or renal insufficiency. The most common proliferating plasma cell neoplasm is MM, second among hematologic neoplasms only to non-Hodgkin's lymphoma. In 2015, 18,545 cases were observed in Italy, 5643 of which were newly diagnosed; the number is substantially stable, with 5759 new cases in

Table 12.1 Classification of the monoclonal gammopathy according to clinical manifestations

Clinically manifest forms due to the proliferation of the neoplastic clone
Multiple myeloma and its variants
Smoldering myeloma
Nonsecreting myeloma
Light-chain myeloma
IgM myeloma
Lymphoproliferative diseases
Waldenström macroglobulinemia
Heavy-chain disease
Non-Hodgkin's lymphoma
Chronic lymphocytic leukemia
Other plasma cell dyscrasias
Plasma cell leukemia
Medullary and extramedullary solitary plasmacytoma
Clinically manifest forms due to pathological effects of MC
Light chain amyloidosis (AL)
Type I and II Cryoglobulinemia
Monoclonal immunoglobulin deposition disease
Chronic cryoagglutinin disease
Light-chain deposition disease
Adult-acquired Fanconi syndrome
Clinically occult forms
Monoclonal gammopathies of undetermined significance (MGUS)
Transient monoclonal gammopathies

2020. MM predominantly affects the elderly population with a mean age at diagnosis of 69 years, while it occurs very rarely under 40 years. Myelomatous cells reside in the bone marrow, where they often prevail over other cell types causing typical lytic lesions in bone tissue. In these neoplastic forms with large expanding clones of secreting plasma cells, the clinic is dominated by systemic effects caused by the expansion of the malignant clone, such as anemia, bone lesions, hypercalcemia, renal damage, and infections. Despite the significant increase in 5-year survival from 25% between 1975 and 1977, to 43% between 2002 and 2008, MM is still considered a poor prognosis disease. In the non-neoplastic forms with small or very small clones, the clinical manifestations may be due to biological effects caused by the biochemical characteristics of the MC. These can range from vasculitis and neuropathy, due to the specific antibody activity of MC, to cardiac or renal functional deficit due to the deposition of MC in organs or tissues. Among these, the most frequently encountered pathology is immunoglobulin light-chain amyloidosis (AL amyloidosis), characterized by a plasma cell clone producing light chains with conformational abnormalities that cause systemic proteotoxicity with rapid deterioration of the function of vital organs where amyloid fibrils are deposited. In this group of diseases, cryoglobulinemias can also be encountered. Cryoglobulins are Ig that in vitro precipitate if they are subjected to temperatures below 37 °C, but they redissolve when brought to 37 °C again. In Brouet's 1974 classification, which remains the most widely used, three types of cryoglobulinemia are distinguished based on clonality and immunoglobulin class. Type I consists of monoclonal Ig only, usually IgG or IgM. Type II is a mixture of polyclonal IgG and a monoclonal IgM, with rheumatoid activity. Finally, in Type III, both polyclonal IgG and IgM are present. Type II and Type III are called mixed cryoglobulinemias. Cryoglobulins are associated with a clonal expansion of B cells, either in the context of lymphoproliferative disorders or in the presence of a persistent stimulation of the immune system triggered by a chronic infection or an autoimmune disease. The term cryoglobulinemia refers to the presence of cryoglobulins in the serum, whereas the term cryoglobulinemic disease or vasculitis is used to describe conditions in which there are symptoms related to the presence of cryoglobulins, since many subjects with cryoglobulinemia remain asymptomatic. The two main pathogenetic mechanisms are the precipitation of cryoglobulins in the micro vessels and the inflammatory process mediated by the presence of immune complexes.

Table 12.2 Prevalence of different monoclonal gammopathies[a]

Disease	No. of cases	%
Monoclonal gammopathies of undetermined significance (MGUS)	23,629	59.1
Multiple myeloma	6974	17.5
Light-chain amyloidosis (AL)	3781	9.5
Smoldering myeloma	1494	3.7
Lymphoproliferative disorders	1296	3.3
Waldenström macroglobulinemia	940	2.4
Plasmacytoma	774	1.9
Cryoglobulinemia	379	0.9
POEMS syndrome	217	0.5
Other clinical conditions with prevalence <0.5%	443	1.2

[a]Mayo Clinic case series 1960–2008

The prevalence of the main plasma cell dyscrasias, obtained from an impressive case series from the Mayo Clinic, is reported in Table 12.2.

Separation Technology

Electrophoretic techniques are at the heart of protein diagnostics, as they can detect and characterize serum and urine MCs.

Protein Electrophoresis

Electrophoresis (EF) is a separative technique based on the different rates of migration of electrically charged particles through a solution and under the influence of an applied electric field. The rate of migration of a protein depends not only on several independent factors such as the nature of the medium and the strength of the applied electric field but also on the mass, size, shape and charge of the particle itself, i.e., its electrophoretic mobility, defined by the following equation:

$$\mu = Q / 6\pi\eta\gamma$$

where η is viscosity of the medium; Q is the number of charges of the particle; γ is the ionic radius.

A very common electrophoretic technique in protein diagnostics is zonal electrophoresis on a solid support, typically agarose gel (AGE), schematically represented in Fig. 12.2a. The solid support, soaked in the electrolytic solution, is connected to the reservoirs of the latter, which are in turn in contact with the electrodes. The sample to be analyzed is deposited on the surface of the gel, and then a potential difference between the electrodes is applied by means of an electric current generator to make the charged particles present in the sample migrate. Proteins are amphoteric substances, and therefore, their net electric charge varies according to the pH of the solution in which they are found. At the alkaline pH, usually used in this type of separation, most proteins have a negative net electric charge and therefore migrate, with a speed proportional to the charge/radius ratio, from the cathode toward the anode. At the end of migration, the gel is placed in a staining solution also containing a fixative that prevents the diffusion of proteins. The excess of the dye is then removed from the support, but not from the fixed proteins, which will now appear as distinct bands. The gel is finally dried and subjected to densitometric scanning. The result is a graph consisting of a series of peaks corresponding to the separated protein fractions. The area delimited by each peak is proportional to the concentration of each fraction and is expressed as a percentage of the total protein concentration (Fig. 12.3a).

Widely used in the clinical laboratory for its high resolution and the possibility of complete automation of the process is capillary electrophoresis (CE), a separative technology in which the electrophoretic run takes place in liquid phase inside a long, thin capillary of fused silica filled with electrolyte solution. The size of the capillaries can vary from 20 to 100 cm in length and from 20 to 80 μm in internal diameter: therefore, for example, a capillary 50 cm long and with an internal diameter of 50 μm will have a volume of only 1 μL. Separation takes place in a few minutes by applying a high voltage (8–15 kV) under a controlled temperature. A schematic representation of the CE is shown in Fig. 12.2b. In contrast to AGE, electrophoretic mobility is not the prevailing force in CE, where instead electroendosmotic flow (EOF) predominates. The EOF (Fig. 12.4) is the result of the interaction between the negative groups present on the inner surface of the capillary and the positive ions in the electrolyte buffer. The polar molecules of water surround the positive ions, and when an electric current is applied, they are dragged toward the cathode, thus generating a powerful flow (called electroendosmosis) of solvent from the anode to the cathode (in the opposite direction to the electrophoretic migration). The proteins present in the sample, which according to their negative electric charge should move toward the anode if separating according to their electrophoretic mobility, are dragged by the EOF toward the cathode. Detection occurs without the aid of dyes, by measurement of absorbance around 210 nm by a UV detector placed at the end of the capillary, generating an electropherogram similar to that seen for AGE (Fig. 12.3b). The migration position of the

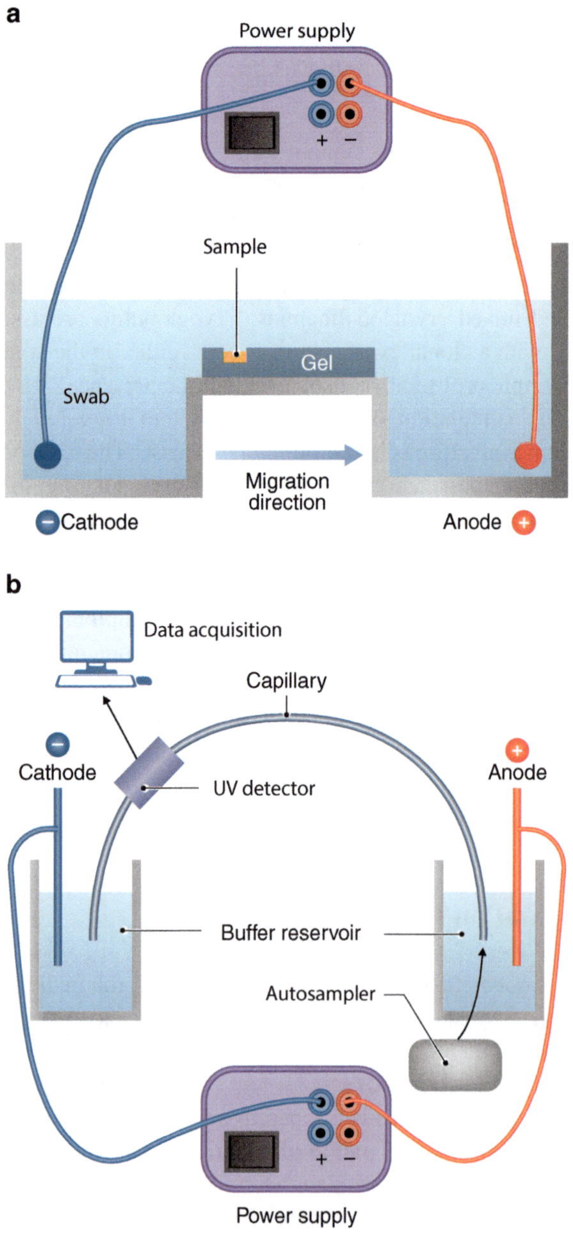

Fig. 12.2 Schematic representation of agarose gel electrophoresis system (**a**) and capillary electrophoresis instrument (**b**)

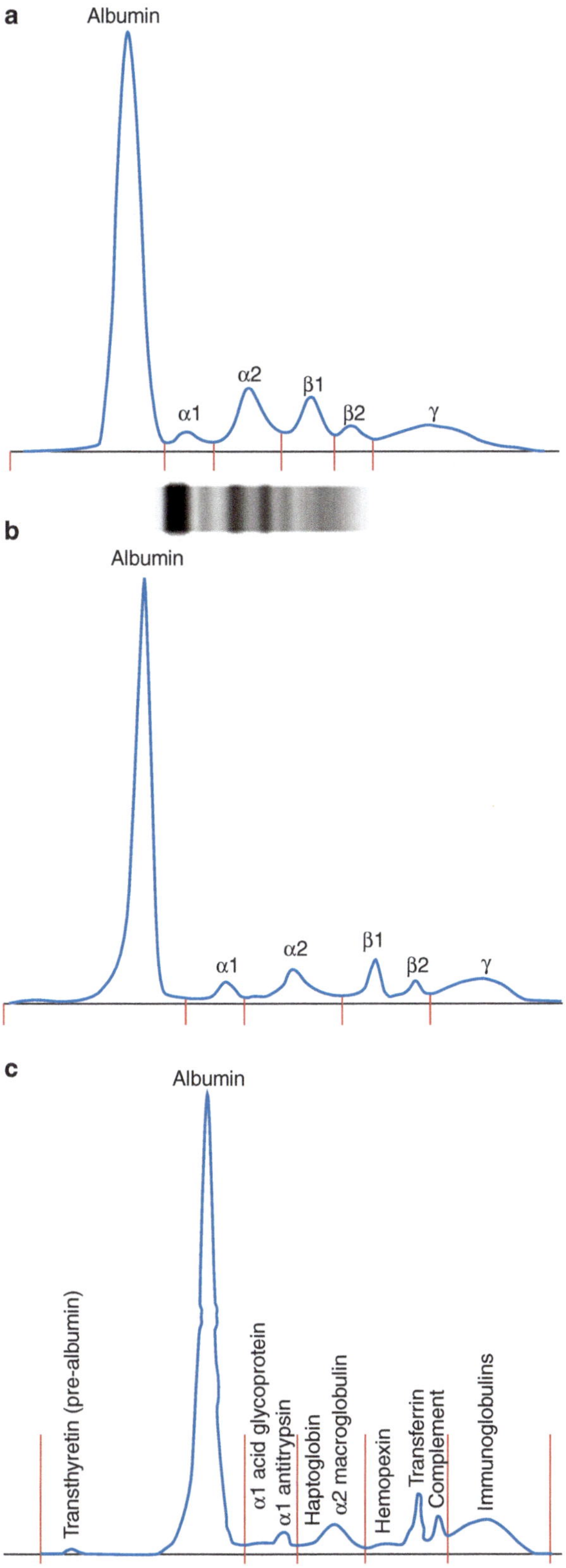

Fig. 12.3 Serum protein path performed on agarose gel (**a**) and capillary electrophoresis (**b**). In (**c**) the migration positions of the main serum proteins are indicated

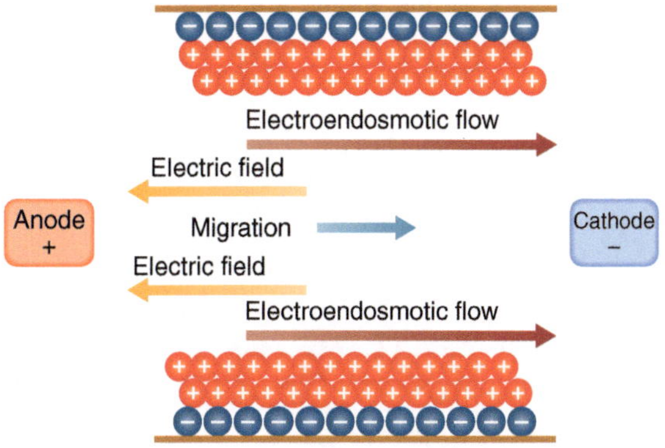

Fig. 12.4 Illustration of the principle behind the electroendosmotic flow. The negative charges present on the internal surface of the capillary attract the positive ions of the electrolyte solution, which in turn are surrounded by the polar molecules of the solvent. Under the influence of the strong electric field, the cations drag the water molecules with them, generating a powerful flow that opposes the normal migration of proteins in the anodic direction

main serum proteins with both techniques is shown in Fig. 12.3c.

Immunological Typing

Definitive confirmation of the presence of a MC in serum or urine and its immunological typing are performed by electrophoretic immunofixation (IFE), typically on agarose gel. The first step of this technique consists of an electrophoretic run in multiple lanes, usually six, for each sample to be tested. Subsequently, a fixative is layered on the first lane, while antisera directed against the Ig heavy (anti-γ, -α, -μ) and light (-κ, -λ) chains are layered on the others. If there is a reaction with one of the two light chains without association with one of the three heavy chains, it will be necessary to also test the sample with anti δ and ε antisera to exclude the presence of a rare IgD or IgE MC. The reaction of the antisera with the Ig present in the sample leads to the formation of an immunoprecipitate, which is trapped into the gel. The gel is then washed to remove excess antiserum and sample proteins that did not react with the antiserum and finally stained. This way, the clonality of the MC is defined with certainty, and it will appear as a distinct band, defining at the same time also its class and type. If, on the other hand, we are in the presence of polyclonal Ig, the immunoprecipitate will be visualized as a broad and shaded zone (Fig. 12.5).

Similarly to electrophoresis, CE can also be used for immunological typing of MCs, with a technique called immunosubtraction or immunotyping (ISE). In ISE, the sample is subjected to electrophoresis in the absence and in the presence of antisera (anti-γ, -α, -μ, -κ, -λ). The immunocomplexes that are thus formed migrate in the capillary at a very different

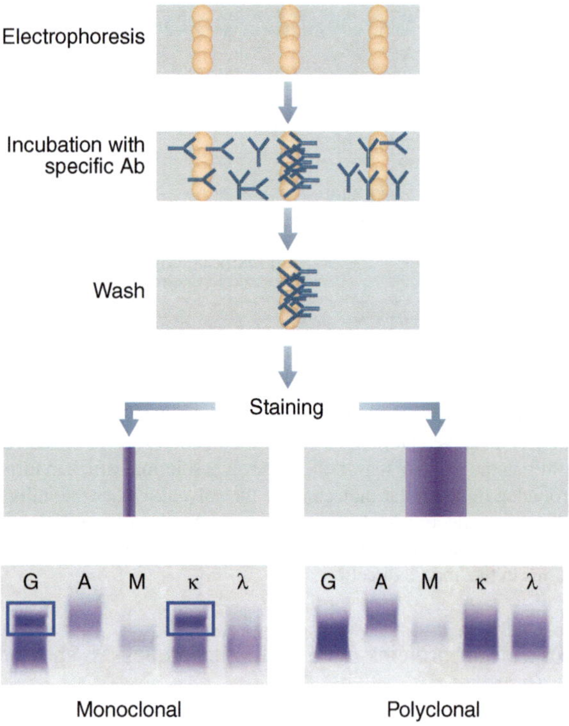

Fig. 12.5 Principle of the electrophoretic immunofixation technique. The sample is first subjected to electrophoresis and then reacted with specific antisera for the heavy and light chains of immunoglobulins. After washing, only the immune complexes formed in correspondence with the migration band of the specific immunoglobulin remain on the gel. After coloring the gel, in the absence of monoclonal components, only broad and blurred migration bands are observed as shown in the example below on the right. On the other hand, when a monoclonal component is present, this is highlighted as a narrow band of migration in correspondence with the heavy and light chain from which it consists. This allows not only to confirm its presence, but also to define its class and type, as in the example below on the left, in which a monoclonal component of the IgG κ type is highlighted

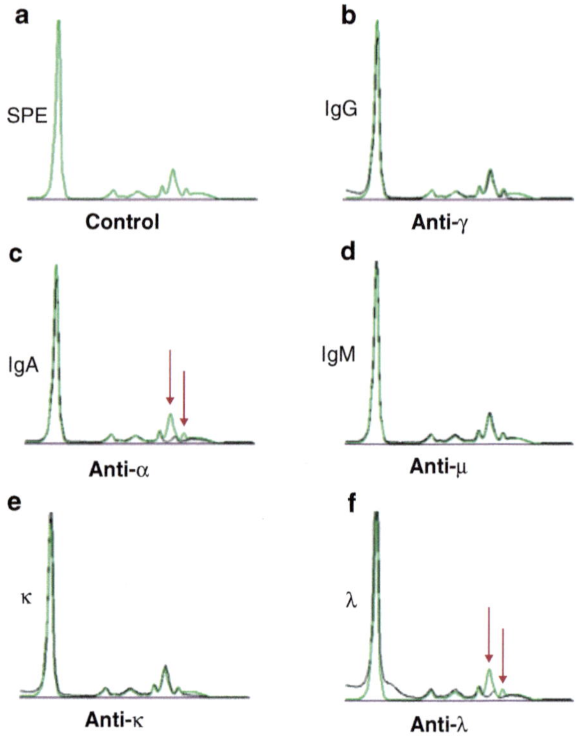

Fig. 12.6 Example of immunological typing performed in capillary electrophoresis. In (**a**) the reference trace is reported, ie carried out in the absence of antisera. In (**b**–**f**) the electrophoretic runs performed after incubation with the anti-γ, -α, -μ, -κ and -λ antisera, respectively, are reported. Overlapping the individual plots with the reference plot allows for easy interpretation of the results. In fact, in (**c** and **f**) two peaks can be seen which disappear after incubation with the anti-α and anti-λ antisera, respectively. In this case, immunotyping allows the typing of a monoclonal IgA λ component present in both monomeric and dimeric form

speed compared to the untreated sample, resulting in a reduction of the signal corresponding to the Ig under examination. By comparing the results of the different runs with the reference electropherogram, i.e., the run carried out in the absence of antiserum, it is possible to confirm the presence of a clearly visible MC in the CE and to type it (Fig. 12.6).

Laboratory Diagnostics

Serum Protein Electrophoresis

Serum electrophoresis (S-EF) is the test of choice for the detection and quantification of serum MCs, as it is able to detect the molecular homogeneity of the protein. MCs appear as a sharp band in the electrophoretic pattern and can migrate to different areas of the trace; some examples are shown in Fig. 12.7. Whichever technique is chosen (AGE or CE), it is important that the employed method shows a high resolution, reaching a sensitivity of <1 g/L (1.16) and that the personnel in charge of the visual inspection of the traces are specifically trained and experienced. S-EF is performed both for screening purposes to detect the presence of MC and during the subsequent diagnostic framing and monitoring of patients with MG. It is also used in the quantification of MC as well as to highlight qualitative alterations of MC compared to previous findings. The detection of an MC is not a rare event in clinical laboratory practice, thus posing the need for proper patient management. In this context, it should be emphasized that the choice of tests to be performed depends on the reasons of the request. When an S-EF is requested for subjects for whom there is no clinical suspicion of MG, it is reasonable that in the absence of alterations of the serum electrophoretic pattern, the investigation is not continued. In the latter case, however, a specific comment should be included in the report, clearly indicating that the S-EF does not indicate the presence of an MG. If instead S-EF detects an alteration, this must be indicated in the report and investigated as specified below.

Monoclonal Component Typing

Any alteration in the electrophoretic pattern (supranumerary band, monoclonal peak, morphological alteration) must

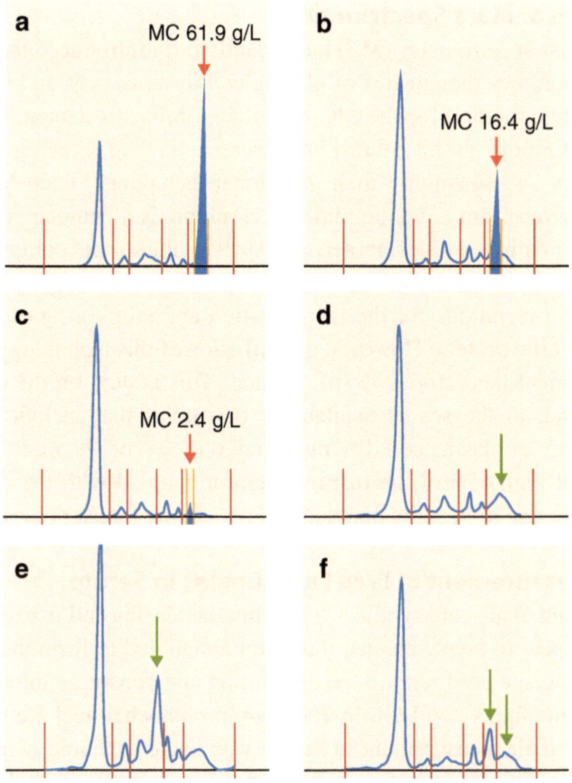

Fig. 12.7 Examples of electrophoretic tracings showing the extreme variability of electrophoretic mobility of the mono-clonal components. In (**a**), in (**b**) and in (**c**) three monoclonal components are shown in a wide concentration range, which migrate in the γ zone and which are all quantifiable by densitometric scanning in relation to the total serum proteins. In (**d**) is reported a monoclonal component in zone γ of slight entity, which cannot be quantified as it is poorly distinguished from the remaining very well represented polyclonal immunoglobulins. In (**e**) a rare monoclonal component is shown that migrates in the α2 area, while in (**f**) a monoclonal component is shown that migrates in the β2 area accompanied by another small band in the γ area. In the latter two cases, in (**e**) and in (**f**), quantification by densitometric scanning is not recommended as the monoclonal components migrate together with other serum proteins and are not distinguishable from them

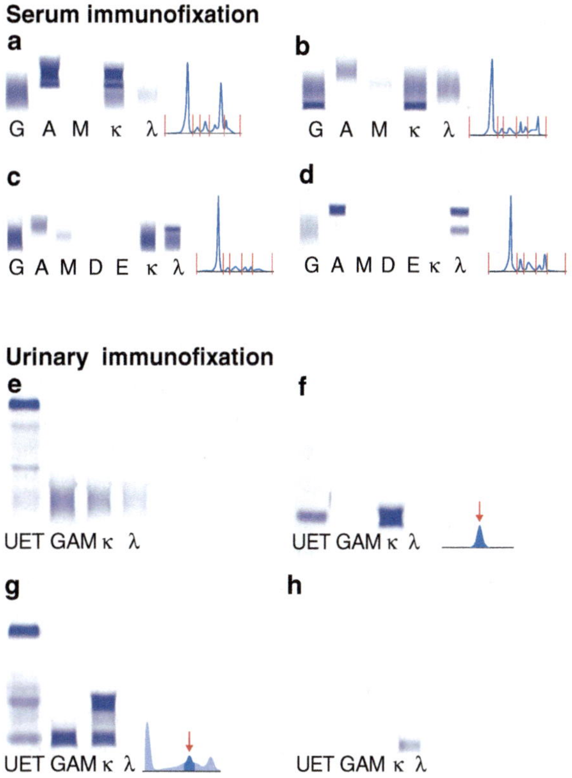

Fig. 12.8 Examples of serum immunofixations presented with the corresponding serum electrophoresis (**a**–**d**). In (**a**), we can observe a monoclonal IgA κ component consisting of the monomer migrating in the β2 position and the dimer migrating in the γ zone. In (**b**), an IgG κ monoclonal component in the γ position is reported. In (**c**), a monoclonal component consisting of only light chains λ is presented, which is difficult to identify on the electrophoretic trace, which well illustrates the greater sensitivity of immunofixation in detecting monoclonal components. Finally, in (**d**), a monoclonal component IgA λ in zone β2 is represented accompanied by another monoclonal component consisting of free light chains λ in zone γ. Examples of urinary immunofixations (**e**–**h**) with next to the urinary electrophoretic trace (ELP) in which a trivalent antiserum for the heavy chains γ, α, and μ (GAM) was used. The sample in (**e**) is negative for the presence of Bence Jones protein (PBJ), as only polyclonal immunoglobulins are present in a context of glomerular proteinuria. In (**f**) there is an example of κ-type PBJ in the absence of other proteins, while in (**g**) there is a very distinct κ-type PBJ from whole monoclonal immunoglobulin in a picture of frank glomerular proteinuria. In both cases, the PBJ is quantifiable densitometrically as evidenced by the corresponding urinary densitometric tracings. Finally, in (**h**), a sample is shown in which the PBJ is highlighted only in the presence of the anti-λ antiserum and not in the electrophoretic trace, which further demonstrates the greater sensitivity of urinary immunofixation. In this case, quantification is not possible

be investigated to highlight the nature (immunoglobulin or not) of the abnormality found. The typing (or immunological characterization) of serum or urine MC carried out by IFE or ISE has the aim of confirming the immunoglobulin nature and the monoclonality of the band highlighted by S-EF. It also allows the attribution of the heavy and light chain of the Ig involved. It is also capable of highlighting MCs that cannot be detected by the S-EF, because they are of slight entity or comigrating with other proteins present physiologically. This ability is more attributable to IFE, being a technique with higher sensitivity than ISE. IFE must be performed at the first electrophoretic finding of an MC (but also in the clinical or laboratory suspicion of the presence of an MC), as the type of Ig involved provides useful elements for both diagnostic and prognostic purposes. The typing must also be performed during the monitoring of the patient whenever the S-EF pattern shows qualitative alterations in the morphology of the MC compared to the previous ones and to confirm the disappearance of the MC for the definition of the complete response after treatment in MM. It is important that the method adopted for S-IFE has a high resolution and that the personnel in charge of its interpretation are adequately and specifically trained. Some examples of typing are presented in Fig. 12.8 (panels a–d).

Quantification of Serum Monoclonal Component

The most correct way to measure serum MC is by direct proportion with total serum protein concentration after delineation of the monoclonal peak in the electrophoretic pattern. However, the accuracy of this method is not optimal as it is influenced in a non-negligible way by the operator. In addition, the two separative techniques available (AGE and CE) present non-negligible differences. The immunochemical measurement of MC is inaccurate and therefore not advisable as it suffers from some limitations such as: the simultaneous measurement of polyclonal together with monoclonal Ig, which can be significantly represented in the sample; the absence of parallelism in the immunoassay between the polyclonal standard used for calibration and monoclonal Ig and finally MC can present little or poorly recognized antigenic specificity for the polyclonal antiserum. These problems can lead to both non-negligible over- and under-estimates. It is therefore important that the individual patient is monitored, as far as possible, always with the same method and that this is clearly indicated on the report.

Multicenter studies have evaluated the variability related to the quantification of MC and have confirmed that its accuracy and precision depend on a not small number of variables: operator experience, amount of MC, presence of a polyclonal Ig background, its position within the electrophoretic pattern. The quantification of the MCs is a key parameter for the management of the patient with GM because in all secretory diseases it is an index of the tumor mass. Quantification of MC is necessary in differential diagnostics, risk stratification, and assessment of response to therapy. Some examples of MC quantification are presented in Fig. 12.7.

Determination of Bence Jones Protein

Bence Jones protein (BJP) consists of monoclonal free light chains produced in excess by the plasma cell clone and are excreted in urine in the urine. Urine immunofixation (U-IFE) is used to detect BJP, as this is currently the only method that can ascertain the two characteristics of BJP, i.e., monoclonality and absence of the heavy chain. Once its presence has been ascertained, the quantification of BJP must be performed on a timed (24-hour) urine sample by densitometric scanning of the monoclonal peak of the urine electrophoretic pattern against total urine protein concentration. Also, for this investigation, it is important that the method adopted has a high resolution and sensitivity and that the personnel in charge of its interpretation have adequate experience. The determination of BJP is essential in screening for plasma cell dyscrasias when AL amyloidosis is suspected or when clinical suspicion of MG persists and serum tests are negative; moreover, it is one of the tests to be performed at the diagnosis of MG and in verifying the response to therapy in MM and AL amyloidosis. Some examples of BJP determination are presented in Fig. 12.8 (panels e–h).

Use of Mass Spectrometry

Mass spectrometry (MS) has recently been introduced in the laboratory diagnostics of plasma cell dyscrasias as an alternative method for the MC identification, measurement, and definition of minimal residual disease, showing greater accuracy and sensitivity than traditional techniques. Each MC, derived from a unique plasma cell clone, is a "unique" protein with its own characteristics such as amino acid composition and molecular mass. MS is therefore theoretically the ideal technique for the identification and monitoring of the specific protein. However, the diffusion of this technology in clinical laboratories is still limited. This is due, on the one hand, to the scarce availability of studies that include all types of plasma cell dyscrasia and, on the other hand, to the still high cost of the instrumentation coupled with the specific and high skills that dedicated personnel must possess.

Measurement of Free Light Chains in Serum

Light chains are synthesized within the plasma cell in excess respect to heavy chains; those not assembled to form the Ig molecule are found in the circulation and constitute the free light chains (S-FLC). In 2001, an immunochemical method for their measurement in serum was made available, which uses antisera directed against epitopes that remain hidden when the protein is bound to the heavy chain. This measurement has made it possible, among other things, to obtain a quantitative parameter useful both for diagnosis and for monitoring conditions in which MC is difficult to detect and measure (AL amyloidosis, nonsecreting/oligosecreting MM). The measurement of S-FLC is used in screening of MG, differential diagnosis, risk stratification, and evaluation of response to therapy and is a parameter that has become increasingly important over the years in the management of patients with MM. International recommendations for its clinical use are currently available as well as alternative methods for measuring FLC, which however present differences in their analytical and diagnostic performance.

Measurement of Immunoglobulins in Serum

The measurement of Ig not involved in clonal expansion is aimed at verifying the presence or absence of an immunoparesis. The decrease in the serum concentration of polyclonal Ig indicates that the expanding clone within the bone marrow is predominant compared to the other clones synthesizing Ig. The measurement should be performed at the first detection of MC and for diagnostic framing. In case of MC overlapping with the band of other proteins, as it frequently happens in case of IgA MC often overlapping with transferrin or C3, its quantification can be performed with the immunochemical measurement of monoclonal Ig. The methods used to measure Ig are immunochemical methods (nephelometric or turbidimetric), which are described in another chapter of this volume.

Concluding Remarks

The first finding of an MC is a typical situation where the laboratory could (should) proceed with cascade examinations (reflex testing) for the correct classification of the patient. If the laboratory, for administrative and/or regulatory reasons, cannot perform the necessary tests on its own, these should be suggested in the report to the clinician, who will then evaluate the continuation of the investigations. Also, for other situations (differential diagnosis, risk stratification, monitoring of the condition, and evaluation of response to therapy), a close interaction between laboratory and clinician is necessary, as it is always desirable in other fields of laboratory diagnostics.

In the suspicion of some pathologies, such as amyloidosis AL or monoclonal Ig deposition disease, it is necessary to associate to S-EF also the measurement of free light chains and S-IFE and U-IFE.

For the management of patients with plasma cell dyscrasia, national and international guidelines and recommendations have been produced and are continuously updated, containing indications for the correct clinical-laboratory pathway to be followed to ensure the best patient outcome. It is highly advisable for laboratory professionals to keep abreast of developments in this field.

Main Serum Proteins of Clinical Relevance

Albumin

Albumin (ALB) is a nonglycosylated protein with an MW of about 66 kDa, synthesized by hepatocytes and with a half-life of about 20 days. The main functions of ALB are the maintenance of the colloidosmotic pressure, the binding and transport of numerous endogenous and exogenous substances (metal ions, bilirubin, free fatty acids, amino acids, steroid and thyroid hormones, drugs), constituting a reserve of amino acids for protein synthesis; it also exerts an antioxidant action. ALB synthesis may decrease due to increased oncotic pressure in the liver extracellular fluid or decreased amino acid availability. In addition, ALB is a negative acute-phase protein, as its synthesis is downregulated by interleukin-6 (IL-6). The reference values in plasma for adults, according to the IFCC international standardization (CRM 470), are 35–52 g/L.

Measurement of serum ALB is used in the evaluation of dysprotidemic states, but it should be kept in mind that the only clinically significant change is its decrease; in fact, its plasma concentration can increase only by hemoconcentration. Hypoalbuminemia may be due to the following causes:

- Decreased synthesis (liver dysfunction or protein-poor diet)
- Altered distribution between the blood compartment and extravascular spaces due to increased capillary permeability as in the case of septic shock
- Leakage into the "third space" as a result of edema or ascites
- Outward leakage, as in nephrotic syndrome, burns, or exudative enteropathy
- Inflammatory process, where ALB synthesis is decreased in favor of acute phase proteins
- Pregnancy, due to increased plasma volume
- In rare congenital disorders of albumin synthesis

In clinical practice, the measurement of serum ALB is widely used for the evaluation of liver function and situations of protein loss. The reasons for the request with the best evidences are the use in hemodialysis patients as a marker of therapeutic adequacy; in patients with MM for the staging of the disease; in patients who are candidates for human ALB replacement therapy for the calculation of the ALB dose to be administered and for the monitoring of the therapy; as a risk factor for acute renal failure (AKI) and poor prognosis after AKI.

α1-Antitrypsin

α1-antitrypsin (AAT) is a monomeric glycoprotein that belongs to the serpin family ("SERine Protease INhibitor"). The encoding gene (SERPINA1 or PI-"Protease Inhibitor") is located on chromosome 14. AAT inhibits several serine proteases, but the main target is neutrophil elastase (NE), an enzyme released by neutrophils that is responsible for the proteolysis of many components of the extracellular matrix, including elastin. When neutrophils are activated, NE is released into lung tissue, where, if not inhibited, it manifests its proteolytic activity. The reference values in plasma for adults, according to the IFCC international standardization (CRM 470), are 0.9–2.0 g/L. AAT deficiency is an autosomal recessive disorder, which increases the risk of developing some diseases, among which the most important for its social relevance are lung diseases (chronic obstructive pulmonary disease, emphysema) and liver diseases (liver fibrosis, cirrhosis, liver carcinoma). The diagnosis of the genetic defect is prerogative of specialized centers. In the general population, the occasional finding of a plasma concentration of AAT <1 g/L by immunometric measurement makes it necessary to confirm the data through quantitative and qualitative tests in specialized laboratories. AAT migrates in the α1 zone of the S-EF (Fig. 12.3c); the occasional finding of two separate bands in this zone (heterozygosity of AAT) or of the absence of the α1-globulin peak or of a percentage of the α1 zone lower than the reference values must be followed by the immunometric measurement of AAT on the sample. If the plasma concentration of AAT is below the lower reference limit, an in-depth diagnostic examination with qualitative and genetic tests at specialized laboratories is necessary. The main indications for the request are early emphysema, bronchiectasis without obvious etiology, hepatopathy not otherwise explainable, first-degree relatives of patients with ascertained deficiency. Since AAT is an acute-phase protein, it is advisable to simultaneously measure an acute-phase marker, such as C-reactive protein, to exclude an increase in AAT linked to an ongoing inflammatory phenomenon, which could mask the presence of a possible deficiency.

Haptoglobin

Haptoglobin (HPT) is a glycoprotein synthesized in the liver and composed of four polypeptide chains: two light α chains and two heavy β chains. HPT has an antioxidant and scavenger action for free hemoglobin, released following physiological erythrocyte turnover or hemolytic processes. The reference values in plasma for adults, according to the IFCC international standardization (CRM 470), are 0.3–2.0 g/L. The plasma concentration of HPT is rapidly reduced in the case of intravascular hemolysis, since the half-life of the HPT-hemoglobin complex is only 8 minutes. HPT is an acute-phase protein: its synthesis in liver cells increases due to stimulation of proinflammatory cytokines, consequently. Although decreased plasma concentrations of HPT possess a high predictive power of intravascular hemolysis, elevated values may be found during an acute-phase response. For a proper evaluation, its measurement should therefore be accompanied by that of C-reactive protein. Since even mild hemolysis is sufficient to saturate physiologic HPT concentrations and compensate for any increase due to inflammation, lactate dehydrogenase (LDH) and bilirubin should be measured when clinical suspicion exists. In the monitoring of intravascular hemolysis, considering the short half-life of the HPT-hemoglobin complex, the measurement of HPT is not very useful. Under conditions of chronic hemolysis, hemopexin measurement may play a role. If HPT concentrations are not within the measurement range, the extent of hemolysis is better assessed by the determination of LDH and bilirubin.

β2-Microglobulin

β2 microglobulin (B2M) is a low-MW protein (12 kDa); it belongs to the major histocompatibility class I system (MHC I or class I antigen), which consists of glycoproteins expressed on the surface of most nucleated cells. B lymphocytes have high concentrations of MHC I; the soluble form of the molecule, present in the circulation, represents the turnover of cells expressing MHC I. The plasma concentration of B2M depends on both the turnover of B lymphocytes and the glomerular filtration rate, since the catabolism of the protein is almost entirely renal, as B2M freely passes the glomerular filter and is more than 99% reabsorbed by the proximal convoluted tubule. Its plasma concentration is 1.2–2.5 mg/L. In uremic patients with end-stage renal disease, B2M accumulates in the circulation and tends to form deposits of fibrillar substance (amyloid) in the tissues, after ~8 years of dialysis treatment. The importance of the parameter in plasma cell dyscrasias is since B2M reflects both tumor mass and renal function. B2M is a recommended parameter in the initial evaluation of patients with MG as it is one of the parameters used in the staging of MM (together with ALB). Measurement of B2M in dialyzed patients is useful in assessing the efficiency of hemodialysis in removing medium MW molecules. Determination of B2M in urine is strongly discouraged due to the rapid degradation of the protein at acidic pH both in vivo (in the bladder) and in vitro (in the collection tube).

Ceruloplasmin

Ceruloplasmin (CER) is a glycoprotein composed of a single polypeptide chain with an MW of ~132 kDa. CER is a copper-oxidizing enzyme, synthesized in the liver; it binds about 95% of circulating copper and transports it to tissues. CER acts on iron (Fe) regulation and other metal ion status by exerting an antioxidant action as it inhibits metal ion–catalyzed oxidation on membrane lipids. The reference values in plasma for adults, according to the IFCC international standardization (CRM 470), are 0.2–6.0 g/L. Clinically significant decreases in plasma concentration of CER are present in hereditary synthesis defect, which is very rare, or in secondary deficits, such as those found in Wilson's disease (an autosomal recessive defect of copper metabolism). Since CER is an acute-phase protein, physiological concentrations of CER can be observed in conjunction with inflammatory states. In addition, the synthesis of the protein is estrogen-dependent and, therefore, high concentrations of CER are found in pregnancy and during estrogen administration in menopause, making the diagnostic specificity of this test rather low and not allowing to use the measurement of the protein as a screening test for Wilson's disease. CER values <0.05 g/L are strongly indicative for the diagnosis of disease; however, it should be noted that concentrations within the reference range do not allow to exclude the presence of Wilson's disease for the reasons stated above. The diagnostic value of low plasma concentrations of CER increases significantly when associated with the presence of Kaiser–Fleischer corneal rings. In the absence of this sign, as commonly occurs in the hepatic manifestation of the disease, on the one hand, low concentrations of CER cannot be considered diagnostic because they could be due to other pathologies (autoimmune hepatitis, celiac disease), on the other hand, being

CER an acute phase positive protein, during inflammation its concentrations could fall within physiological limits. In view of the low predictive value of CER, it should be associated with the measurement of serum and urine copper for the diagnosis of Wilson's disease.

Complement

The complement system is made up of complement factors, which circulate in the blood in an inactive form, and of regulatory proteins (inhibitors, activators, and receptors), by the expression of which the system is highly regulated. The complement system is a multienzyme cascade and is the major component of innate immunity; it plays a crucial role in the microbial killing, controlling the formation of immune complexes, in the clearance of apoptotic cells, and modulating the acquired immunity. The evidence of its role in acquired immunity is now more solid.

The pathways of complement activation are as follows:

- The classical pathway (C1q-C1r-C1s, C4, C2 factors), mainly triggered by the antigen–antibody complex.
- The alternative pathway (factors C3, B, D, P, H), triggered by lipopolysaccharides and surface polysaccharides of viruses, bacteria, fungi, parasites, by altered self-antigens of tumor cells.
- The lectin pathway (factors C4, C2), similar to the classical pathway, but triggered by mannose-binding lectin, which specifically recognizes residues of mannose and other sugars present on bacterial and other pathogenic surfaces. All pathways of complement activation converge through factor C3 into the common terminal pathway (factors C3, C5, C6, C7, C8, C9), which culminates in the formation of the lytic attack complex at the cell membrane. The decrease in complement proteins may be due to a genetically determined synthesis deficit or to consumption by activation of the system, whereas the increase in their concentration is of no clinical importance per se. Since C3 is the central point where the three activation pathways converge and C4 is a fundamental factor of the classical pathway and the lectin pathway, the measurement of these two proteins provides with good approximation an evaluation of the activation status of the complement system. The reference values in plasma for adults, according to the IFCC international standardization (CRM 470), are 0.9–1.8 g/L for C3 and 0.1–0.4 g/L for C4.

The request for the measurement of complement proteins is therefore appropriate in patients with severe, recurrent, atypical or not easily resolvable infections or with important allergic manifestations (congenital immunodeficiency suspected). It is also appropriate in the diagnostic and monitoring pathway of patients with diseases sustained by immune complexes (such as systemic lupus erythematosus, cryoglobulinemias, and vasculitis in general). It should be noted that the sole immunochemical measurement of complement proteins is not appropriate to define the problem: only the functional evaluation of the complement system (CH50 and APCH50) is able to identify the actual deficiency of activity of the classical, alternative, or common pathway.

Immunoglobulins

Immunoglobulins (Igs) represent a rather heterogeneous group of proteins with antibody function, synthesized by plasma cells. They are tetrameric glycoproteins consisting of two heavy chains of ~440 amino acids and two light chains of ~220 amino acids. The heavy H (γ, α, μ, δ, ε) and light L (κ, λ) chains are held together by intercatenary disulfide bridges. Each chain is divided into a constant and a variable (amino-terminal) region; on the latter are located the amino acid sequences that contribute to the formation of the antigen-binding site. Some structural differences in the constant region allow further differentiation of Ig classes into subclasses. IgGs ($\gamma 2\kappa 2/\gamma 2\lambda 2$), with a molecular weight of ~150 kDa, are monomers present in the circulation with the highest concentration and are the circulating antibodies of the secondary response; IgAs, ($\alpha 2\kappa 2/\alpha 2\lambda 2$), with a molecular weight of ~160 kDa, are also monomers; these antibodies are present as dimers on the surface of the mucous membranes, especially of the bronchial and intestinal tracts. Finally, IgMs ($(\mu 2\kappa 2/\mu 2\lambda 2)5$) are high-molecular-weight pentamers (~970 kDa); they are the circulating antibodies of the primary response. The reference values in plasma for adults, according to the IFCC international standardization (CRM 470), are: for IgA 0.7–4.0 g/L; for IgG 7.0–16.0 g/L and for IgM 0.4–2.3 g/L.

The role of Ig measurement in the diagnostic pathway of plasma cell dyscrasias is described in the dedicated paragraph. The increase in polyclonal Ig synthesis can be caused by infectious, inflammatory, autoimmune, or neoplastic processes. Increased plasma concentrations can therefore be associated with a wide variety of diseases, but in most cases these are nonspecific responses that add little or nothing to the diagnosis or management of the patient. The most useful measurement of Ig is linked to the diagnostic pathway of immunodeficiencies, within which the measurement of serum Ig is one of the key parameters. The establishment of a suspected immunodeficiency in patients with severe, recurrent, or atypical infections therefore typically begins with the measurement of serum Ig. The most frequent is selective IgA deficiency (approximately 1/400 in Europe), infrequently linked to clinical problems. Measurement of IgA is critical within the diagnostic pathway for celiac disease because the specific antibodies are IgA antibodies; anaphylactic reactions may occur if products containing traces of IgA are administered to patients with total IgA deficiency, as may occur during therapeutic IgG administration, so measurement of IgA in these patients is appropriate. The IgG4 subclass is related to an infrequent specific clinical situation (IgG4-related disease) that requires its measurement in the laboratory.

European guidelines on the management of cholestatic diseases indicate that treatment of primary biliary cirrhosis with ursodeoxycholic acid induces a marked decrease, in addition to liver enzymes, of IgM, one of the typical signs of the disease; the measurement of IgM is therefore indicated in the diagnosis and monitoring of therapy.

C-Reactive Protein

C-reactive protein (CRP) has an MW of 118 kD and is a member of the pentraxin family, its molecule being composed of five protomers containing 206 amino acids each, arranged with cyclic symmetry. The PCR gene has been mapped to chromosome 1, but no genetic defects are known, which supports the hypothesis that it is essential for life. Its synthesis by hepatocytes is under the control of proinflammatory cytokines, mainly IL-6. When stimulation by IL-6 ceases, CRP production by hepatocytes normalizes within 2–4 hours; the half-life of the protein in blood is ~24 hours. Functionally, CRP can bind a wide group of exogenous and endogenous substances, facilitating their elimination from the bloodstream. Removal of the CRP/ligand complex occurs through the activation of several biological systems, such as activation of the classical complement pathway, binding to phagocyte receptors, and binding to the lymphocyte receptor of the Fc fragment of IgG. CRP is considered the marker of inflammation par excellence, as its concentration increases during both acute and chronic inflammation. A value of <5 mg/L is considered normal; however, this value should be considered indicative as the parameter has not yet been included in international standardization. Its plasma concentration reflects the degree of inflammation and the mass of tissue involved, particularly in acute situations. However, plasma concentrations of CRP within the reference range do not allow the exclusion of a mild or localized inflammatory state in which the magnitude of the acute-phase response is modest. Serial determinations of CRP allow monitoring of disease progression and response to therapy, whereas a single determination may be uninformative due to the high intraindividual biological variability of the protein (CV 26%), and its rapid kinetics. PCR measurement is indicated in rheumatoid arthritis, at first detection, to identify patients at higher risk of progression and for monitoring therapy. In chronic inflammatory bowel disease (e.g., Crohn's disease), the demonstration of an inflammatory picture is a primary criterion for differential diagnosis from other noninflammatory diseases (such as irritable bowel disease), evaluation of disease activity, and monitoring of therapy. PCR must also be measured when the existence of a possible inflammatory state must be taken into account for correct interpretation of results of other laboratory tests (e.g., in the measurement of AAT, CER, HPT).

It has long been known that chronic inflammation plays a fundamental role in the development of cardiovascular disease, and that the persistence of the inflammatory phenomenon represents an important prognostic and risk stratification factor. CRP measurement with methods characterized by high analytical sensitivity (detection limit, ~0.3 mg/L) is the marker that best meets the requirements of practicability, due to the adaptability of measurement in automation. Therefore, PCR can currently be considered the parameter of choice in this diagnostic field.

Transthyretin (Prealbumin)

Transthyretin (TTR) (or prealbumin) is an unglycosylated globular protein with an MW of ~55 kDa. TTR is a tetramer composed of four identical subunits. Each monomer of TTR has a binding site for retinol-binding protein (RBP) and thyroid hormones; the TTR-RBP4 complex binds approximately 20% of circulating thyroid hormones. Circulating TTR is synthesized by liver parenchymal cells; small amounts of TTR are also produced by the choroid plexus, pancreas, and retina. The reference values in plasma for adults, according to the IFCC international standardization (CRM 470), are 0.2–0.4 g/L.

TTR is an acute-phase negative protein downregulated by proinflammatory cytokines; its blood concentration decreases during inflammation similarly to that of albumin. Its half-life is approximately 2.5 days. The interest in measuring TTR lies mainly in the assessment of nutritional status, as a marker of malnutrition. However, the decrease of its plasma concentration seems to be more determined by the inflammatory response that is often associated with these conditions than by the actual nutritional status. Consequently, sensitivity and specificity of TTR measurement for the diagnosis of malnutrition are low. Often, in malnutrition, the TTR/PCR ratio is used as a prognostic marker. However, consensus documents on the use of protein measurement in the management of malnutrition are still lacking.

Recommended Readings

Aguzzi F, Poggi N (1977) Immunosubtraction electrophoresis: a simple method for identifying specific proteins producing the cellulose acetate electrophoretograms. Boll Ist Sieroter Milan 56:212–216

AIRTUM Working Group (2016) Rapporto AIRTUM 2015. The burden of rare cancers in Italy. Epidemiol Prev 40(Suppl 2):1–120

Aletaha D, Smolen JS (2018) Diagnosis and management of rheumatoid arthritis: a review. JAMA 320:1360–1372

Aringer M, Costenbader K, Daikh D et al (2019) 2019 European League Against Rheumatism/American College of Rheumatology classification criteria for systemic lupus erythematosus. Ann Rheum Dis 78:1151–1159

Berggård B, Björck L, Cigén R (1980) Beta-2 microglobulin. Scand J Clin Lab Invest Suppl 154:13–25

Bradwell AR, Carr-Smith HD, Mead GP et al (2001) Highly sensitive, automated immunoassay for immunoglobulin free light chains in serum and urine. Clin Chem 47:673–680

Caraceni P, Angeli P, Prati D et al (2016) AISF-SIMTI position paper: the appropriate use of albumin in patients with liver cirrhosis. Blood Transfus 14:8–22

Dati F, Schuman G, Thomas L et al (1996) Consensus of a group of professional societies and diagnostic companies on guidelines for interim reference ranges for 14 proteins in serum based on the standardisation against IFCC/BCR/CAP reference material (CRM 470). Eur J Clin Chem Clin Biochem 34:517–520

European Association for the Study of the Liver (2009) EASL Clinical Practice Guidelines: management of cholestatic liver diseases. J Hepatol 51:237–267

European Association for the Study of the Liver (2012) EASL Clinical Practice Guidelines: Wilson's disease. J Hepatol 56:671–685

Fermand JP, Bridoux F, Dispenzieri A et al (2018) Monoclonal gammopathies of clinical significance: a novel concept with therapeutic implications. Blood 132:1478–1485

Gomollón F, Dignass A, Annese V et al (2017) 3rd European evidence-based consensus on the diagnosis and management of Crohn's disease 2016: part 1: diagnosis and medical management. J Crohns Colitis 11:3–25

Graziani MS (2016) Measurement of free light chains: pros and cons of current methods. Clin Chem Lab Med 54:1015–1020

Graziani MS, Merlini GP (2013) Recommendations for appropriate serum electrophoresis requests: the Italian approach. Clin Chem Lab Med 51:e117–e118

Graziani MS, Merlini G, Petrini C, for the IFCC Committee on Plasma Proteins and the SIBioC Study Group on Proteins (2003) Guidelines for the analysis of Bence Jones protein. Clin Chem Lab Med 41:38–46

Hellman N, Gitlinn J (2002) Ceruloplasmin metabolism and function. Annu Rev Nutr 22:439–458

Jacobs JFM, Turner KA, Graziani MS et al (2020) An international multi-center serum protein electrophoresis accuracy and M-protein isotyping study. Part II: limit of detection and follow-up of patients with small M-proteins. Clin Chem Lab Med 58:547–593

Johansson BG (1972) Agarose gel electrophoresis. Scand J Clin Lab Invest 29:7–21

Katzmann JA (2009) Screening panels for monoclonal gammopathies: time to change. Clin Biochem Rev 30:105–111

Katzmann JA, Kyle RA, Benson J et al (2009) Screening panels for detection of monoclonal gammopathies. Clin Chem 55:1517–1522

Kyle RA (1978) Monoclonal gammopathy of undetermined significance. Natural history in 241 cases. Am J Med 64:814–826

Landers JP (1995) Clinical capillary electrophoresis. Clin Chem 41:495–509

Lanzillotta M, Mancuso G, Della-Torre E (2020) Advances in the diagnosis and management of IgG4 related disease. BMJ 369:m1067

Merlini G, Stone MJ (2006) Dangerous small B-cell clones. Blood 108:2520–2530

Miravitlles M, Asger Dirksen A, Ferrarotti I et al (2017) European Respiratory Society statement: diagnosis and treatment of pulmonary disease in α1-antitrypsin deficiency. Eur Respir J 50:1700610

Murray DL, Puig N, Kristinsson S et al (2021) Mass spectrometry for the evaluation of monoclonal proteins in multiple myeloma and related disorders: an International Myeloma Working Group Mass Spectrometry Committee Report. Blood Cancer J 11:24. https://doi.org/10.1038/s41408-021-00408-4

Napodano C, Gulli F, Ludovico G et al (2021) Cryoglobulins: identification, classification, and novel biomarkers of mysterious proteins. Adv Clin Chem 104:299–340

National Kidney Foundation. KDOQI clinical practice guideline for hemodialysis: 2015 update. https://www.kidney.org/professionals/guidelines/hemodialysis2015. Last accessed: March 2020

Palumbo A, Avet-Loiseau H, Oliva S et al (2015) Revised international staging system for multiple myeloma: a report from International Myeloma Working Group. J Clin Oncol 33:2863–2869

Pepys MB, Hirschfield GM (2003) C-reactive protein: a critical update. J Clin Invest 111:1805–1812

Picard C, Al-Herz W, Bousfiha A et al (2015) Primary immunodeficiency diseases: an update from the International Union of Immunological Societies Primary Immunodeficiency Diseases Classification Committee. J Allergy Clin Immunol 35:696–726

Rajkumar SV, Dimopoulos MA, Palumbo A et al (2014) International Myeloma Working Group updated criteria for the diagnosis of multiple myeloma. Lancet Oncol 15:e538–e548

Ranasinghe RN, Biswas M, Vincent RP (2022) Prealbumin: the clinical utility and analytical methodologies. Ann Clin Biochem 59:7–14

Rapporto AIOM-AIRTOM (2021) I numeri del cancro in Italia 2020. https://www.aiom.it/wp-content/uploads/2020/10/2020_Numeri_Cancro-operatori_web.pdf. Last accessed: March 2022

Rishi K, Wadhera RK, Phil M et al (2010) Prevalence of monoclonal gammopathy of undetermined significance: a systematic review. Mayo Clin Proc 85:933–942

Ritchie RF, Smith R (1976) Immunofixation. I. General principles and application to agarose gel electrophoresis. Clin Chem 22:497–499

Rothschild MA, Oratz M, Schreiber SS (1972) Albumin synthesis. N Eng J Med 286:816–821

Shih AWY, McFarlane A, Verhovsek M (2014) Haptoglobin testing in hemolysis: measurement and interpretation. Am J Hematol 89:443–447

Silverman EK, Sandhaus RA (2009) α1-antitrypsin deficiency. N Engl J Med 360:2749–2757

Sun L, Pennells L, Kaptoge S et al (2021) Polygenic risk scores in cardiovascular risk prediction: a cohort study and modelling analyses. PLoS Med 18:e1003498

Swain S, Selmi C, Gershwin ME et al (2019) The clinical implications of selective IgA deficiency. J Transl Autoimmun 2:100025

Tate J, Caldwell G, Daly J et al (2012) Recommendations for standardized reporting of protein electrophoresis in Australia and New Zealand. Ann Clin Biochem 49:242–256

Touzeau C, Moreau P, Dumontet C (2017) Monoclonal antibody therapy in multiple myeloma. Leukemia 31:1039–1047

Turner KA, Frinack JL, Ettore MW et al (2020) An international multicenter serum protein electrophoresis accuracy and M-protein isotyping study. Part I: factors impacting limit of quantitation of serum protein electrophoresis. Clin Chem Lab Med 58:533–546

Vidya Sarma J, Peter A, Ward PA (2011) The complement system. Cell Tissue Res 343:227–235

Watson J, Round A, Hamilton W (2012) Raised inflammatory markers. BMJ 344:e454

Weiss BM, Abadie J, Verma P et al (2009) A monoclonal gammopathy precedes multiple myeloma in most patients. Blood 113:5418–5422

Wiedermann CJ, Wiedermann W, Michael Joannidis M (2017) Causal relationship between hypoalbuminemia and acute kidney injury. World J Nephrol 6:176–187

Willerson JT, Ridker PM (2004) Inflammation as a cardiovascular risk factor. Circulation 109(21 Suppl 1):II2–I10

Willrich MAV, Katzmann JA (2016) Laboratory testing requirements for diagnosis and follow-up of multiple myeloma and related plasma cell dyscrasias. Clin Chem Lab Med 54:907–919

Zajec M, Langerhorst P, VanDuijn MM et al (2020) Mass spectrometry for identification, monitoring, and minimal residual disease detection of M-proteins. Clin Chem 66:421–433

Biomarkers of Nutritional Status

13

Marcello Ciaccio, Luisa Agnello, Rosaria Vincenza Giglio, and Anna Maria Ciaccio

Introduction

Nutritional status is traditionally defined as the condition resulting from introducing, absorbing, and utilizing nutrients in our bodies.

The nutritional state is closely related to the state of health. Indeed, it plays a fundamental role in maintaining structural (body composition) and functional (body function) integrity through the exchange of energy (energy balance) and matter (energy and nonenergy nutrients) with the environment.

An altered nutritional status is associated with an increased risk of disease in the general population and a worse prognosis in patients. Good nutritional status is, however, a preventive strategy, for many diseases, including cancer.

An altered state of health, in turn, represents an important risk factor for the onset of alterations in nutritional status. Indeed, it has been estimated that about 20–50% of hospitalized patients have an inadequate nutritional status, which favors complications, adversely affects the outcomes of treatment, reduces the immune response and predisposes to infection, delays wound healing, compromises the functioning of organs and systems, reduces muscle mass and strength, and increases the risk of sarcopenia. It has been estimated that malnutrition in hospitalized patients is associated with an increased hospital stay and mortality (3 times higher, 12% compared to 4%).

M. Ciaccio (✉) · L. Agnello · R. V. Giglio
Department of Biomedicine, Neurosciences and Advanced Diagnostics, Institute of Clinical Biochemistry, Clinical Molecular Medicine and Clinical Laboratory Medicine, and Department of Laboratory Medicine, University Hospital "P. Giaccone", Palermo, Italy
e-mail: marcello.ciaccio@unipa.it

A. M. Ciaccio
Department of Health Promotion, Mother and Child Care, Internal Medicine and Medical Specialties (ProMISE) "G. D'Alessandro", University Hospital "P. Giaccone", Palermo, Italy

The following parameters influence nutritional status :

- Diet
- Physical activity
- Inflammatory state (can induce a condition of hypercatabolism, especially at the level of the lean mass, metabolically active)

Malnutrition

Malnutrition is defined by the Council on Food and Nutrition of the American Medical Association as a state of functional, structural, and developmental alteration of the organism resulting from a discrepancy between specific nutritional requirements and the intake and use of nutrients. According to a broader definition proposed by Stratton in 2003, malnutrition is when a deficit or excess of energy, protein, and other nutrients causes effects on body composition and/or on the functionality of organs and/or tissues.

Malnutrition can be due to:

- Increased food requirements
- Inadequate use of food

In malnutrition, the balance between intake and demand for a nutrient can be skewed either too high (overnutrition) or too low (undernutrition). However, in general terms, malnutrition refers to a lack of nutrients.

High-risk conditions for malnutrition are:

- Childhood and adolescence (high demand for energy and essential nutrients)
- Pregnancy and lactation (increased need for all nutrients)
- Advanced age (decreased physical activity, psycho-social problems, etc.)
- Chronic diseases
- Vegetarian diets (iron deficiency)

M. Ciaccio (ed.), *Clinical and Laboratory Medicine Textbook*, https://doi.org/10.1007/978-3-031-24958-7_13

- Fad diets (deficiency of vitamins, minerals, and proteins)
- Alcohol or drug addiction (altered lifestyle, high absorption and metabolism of nutrients, etc.)

Undernutrition may be due to inadequate intake, malabsorption, systemic loss of nutrients due to diarrhea, excessive sweating, hemorrhage, severe burns, kidney failure, infection, and drug use. Risk factors for undernutrition are:

- Severe underweight: body mass index (BMI) <80% of the standard value
- Loss >10% of usual body weight during a 3-month period
- Alcohol intake >170 ml per day
- No oral intake for >10 days
- Prolonged loss of nutrients due to malabsorption syndromes, short bowel syndromes, fistulas, diabetes, renal dialysis, secreting abscesses, or wounds
- Increased metabolic demand due to extensive burns, infection, trauma, prolonged fever, or hyperthyroidism
- Intake of drugs with antinutritive or catabolic properties (e.g., appetite depressants, corticosteroids, immunosuppressants, and anticancer drugs)

Overnutrition, on the other hand, may be due to overeating, poor exercise, parenteral nutrition, vitamin overdose (vitamin B6, niacin, A, and D), or mineral trace element overdose. The main risk factors are as follows:

- Good appetite combined with a lack of exercise and weight gain
- High-fat, high-salt diet
- High doses of nicotinic acid for hypercholesterolemia
- High doses of pyridoxine for premenstrual syndrome
- High doses of vitamin A for skin disorders
- High doses of iron and other mineral trace elements without a prescription

From a clinical point of view, the most important problems are associated with protein-energy malnutrition (PEM), an energy deficit due to a deficiency of all macronutrients, which can manifest itself in an acute or chronic form.

Malnutrition initially causes the breakdown of adipose tissue for energy, followed by protein depletion or imbalance. Since proteins are involved in many physiological processes, PEM directly compromises the organs' function and negatively impacts several systems and apparatuses, including the immune system. Visceral organs and muscles are also destroyed, resulting in weight loss. Organ weight loss is greatest in the liver and intestines, intermediate in the heart and kidneys, and minimal in the nervous system. The increasing number of individuals of advanced age and patients predisposed to malnutrition determines the existence of a population with marked chronic PEM.

PEM can be primary, caused by inadequate nutrient intake, or secondary to diseases or medications that interfere with nutrient utilization. Primary PEM generally affects children and the elderly lacking access to nutrients, although depression is a frequent cause in the elderly. Primary PEM can also result from fasting or anorexia nervosa.

Secondary PEM, on the other hand, may be due to the:

- Changes in gastrointestinal function, such as impaired digestion (pancreatic insufficiency), impaired absorption (enteropathies, inflammatory bowel disease), or impaired lymphatic transport of nutrients (retroperitoneal fibrosis)
- Pathological conditions that cause wasting, such as neoplasms or hemodialysis
- Pathological conditions that increase metabolic requirements, such as infections, hyperthyroidism, pheochromocytoma, other endocrine disorders, burns, trauma, surgery, and other critical illnesses

Malnutrition is a rather frequent condition in cancer patients, with a prevalence between 25% and 70% in various European and non-European countries. Cancer patients show alterations in nutritional status even in the extremely early stages of the disease, such as immediately after radical surgery and, therefore, in the absence of metastases. The consequences of malnutrition in these patients include an increased risk of infections, fatigue, and reduced muscle function; about 20–30% of cancer patients die from the direct and indirect consequences of malnutrition. The risk of malnutrition and its severity depends on the type of cancer, its stage, and the therapy employed. In addition, malnutrition has a negative impact on prognosis, response and tolerance to therapy, and quality of life. Indeed, malnutrition is a truly independent predictor of increased morbidity and mortality, and loss of body weight and muscle mass induces an increased risk of chemotherapy toxicity.

Malnutrition is also a common problem in patients with chronic kidney disease, especially those undergoing hemodialysis treatment. It has been estimated that about 40% of hemodialysis patients present with varying degrees of malnutrition, and about 10% present with severe malnutrition. Many factors are related to malnutrition in dialysis patients, including the dialysis procedures themselves. The uremic state is the most important factor. It is associated with inflammation and endocrine-metabolic alterations that induce catabolic processes by inhibiting anabolic ones and reducing appetite. The survival of adequately dialyzed patients depends mainly on their age and nutritional status.

Obesity is the most common form of overnutrition. According to the World Health Organization (WHO), obesity is one of the main public health problems in the

Table 13.1 Cause of obesity

Primary causes	Genetic factors
	Metabolic factors
	Nutritional factors
	Social and cultural factors
	Decreased metabolism
Secondary causes	Hereditary (Prader–Willi syndrome, Laurence–Moon–Biedl syndrome, Allstrom syndrome, Cohen syndrome, Carpenter syndrome)
	Neuroendocrine alterations (tumors, hemorrhages, inflammation, surgery)
	Endocrine alterations (Cushing's syndrome, hypothyroidism, hypogonadism, polycystic ovarian syndrome, growth hormone (GH), and insulinoma)
	Drug therapies (antidepressants, lithium, antipsychotics, benzodiazepine, antiepileptics, antihistamines, cortisones, insulin, estroprogestics, and oral antidiabetics)

Table 13.2 Body weight classification based on BMI

BMI, kg/m^2	Classification
<16	Severe thinness
16–18.49	Underweight
18.5–24.99	Normal weight
25–29.99	Overweight
30–34.99	Obesity class 1
35–39.99	Obesity class 2
>40	Obesity class 3

BMI body mass index

world, representing a truly global epidemic, with a prevalence that is on the rise and growing at a worrying rate, not only in Western countries but also in low-income countries. In addition, obesity is a major risk factor for various chronic diseases, such as type 2 diabetes mellitus, cardiovascular disease, and cancer. It is estimated that 44% of cases of type 2 diabetes, 23% of cases of ischemic heart disease, and up to 41% of some cancers are attributable to obesity. Overall, obesity is the fifth most important risk factor for global mortality, and deaths attributable to obesity are at least 2.8 million per year worldwide. WHO has estimated that the prevalence of obesity globally has doubled from 1980 to 2018, affecting the youngest segments of the population. It is estimated that there were over 40 million overweight children under 5 years of age worldwide in 2011. Table 13.1 summarizes the main causes of primary and secondary obesity.

According to the WHO, obesity is defined as a BMI ≥30 kg/m^2.

BMI is calculated according to the following formula:

$$\text{BMI} = \text{weight}(\text{kg}) / \text{height}^2(\text{m}).$$

Based on BMI, various categories are distinguished (Table 13.2).

Assessment of Nutritional Status

Assessment of nutritional status is important to:

- Identify patients with nutritional problems (already malnourished or at risk of caloric-energetic malnutrition or with depletion of specific nutrients) who require specific therapeutic intervention
- Frame and manage a patient
- Monitor the adequacy of any nutritional support

There is no gold standard for the assessment of nutritional status. It is currently based on the integration of anamnestic data (physiological, pathological, and in-depth nutritional history), clinical evaluations (accurate and objective examination), anthropometric measurements (body weight, BMI, plicometry, cross-sectional area of the arm), and biochemical parameters.

The nutritional status should be assessed in the following cases:

- High-risk patients
- Subjects with recent significant weight loss; a significant weight loss is defined as an unintentional weight loss in the last 6 months >10% compared to the usual weight or greater than 5% in 1 month. In the absence of usual weight, body weight 20% below ideal weight may be considered indicative of malnutrition
- Subjects with loss of appetite for 2–3 weeks
- Subjects on enteral or parenteral nutrition

Biomarkers of Nutritional Status

The ideal biomarker for the assessment of nutritional status should have the following characteristics:

- Short half-life
- Sensitivity
- Constant catabolism
- Rapid response to nutritional support
- Specificity
- Nonsusceptibility to nonnutritional factors

None of the biomarkers identified over the years has sufficient sensitivity and specificity; indeed, they are usually influenced by nonnutritional factors, are poorly reproducible, and are not very sensitive to nutritional therapy. Moreover, they have a poor predictive value for complications attributable to malnutrition. However, several laboratory parameters can provide valuable information on nutritional status: total lymphocyte count for the assessment of malnutrition-related cell-mediated immune response, albumin, prealbumin, retinol-binding protein (RBP), transferrin, and urinary creatinine.

Albumin

Albumin is a serum protein synthesized in the liver, with a molecular weight of about 66 kDa and a half-life of 15–20 days. Its normal values are 35–52 g/L. Its main functions are the transport of liposoluble molecules, such as hormones, drugs, and free fatty acids; the maintenance of colloid-osmotic pressure; and the reserve function of amino acids. Moreover, it is an acute-phase negative protein, so its levels are reduced in cases of chronic inflammation.

Albumin reflects well the nutritional status of stable patients but has several disadvantages, including a long half-life, which makes it insensitive to short-term fluctuations. In addition, causes unrelated to nutritional status may lead to hypoalbuminemia, such as hemodilution, maldistribution between intravascular and extravascular compartments, reduced hepatic synthesis, or sepsis. Finally, albumin is often administered as therapy.

Prealbumin

Prealbumin is a serum protein rich in essential amino acids, with a molecular weight of about 55 kDa and a half-life of 2 days. Its normal values are 20–40 mg/dL. Its main function is the transport of vitamin A and thyroid hormones in circulation. Like albumin, prealbumin is an acute-phase negative protein.

Prealbumin is a more reliable marker of acute changes in a patient's nutritional status than albumin because it has a shorter half-life and its total body reserve is considerably lower. However, prealbumin is degraded renally, and consequently, any renal dysfunction results in an increase in its serum levels. In addition, one of the functions of prealbumin is the transport of thyroxine. Therefore, under conditions of hyperthyroidism, prealbumin molecules are saturated with thyroxine, and, therefore, serum levels are low. Similarly, prealbumin levels increase in hypothyroidism.

Transferrin

Transferrin is a serum protein of about 80 kDa, synthesized in the liver, with a half-life of 8–10 days. Its normal values are 2–3 g/L. It has the function of transporting iron ions in the plasma. Similar to albumin and prealbumin, transferrin is an acute-phase negative protein. Transferrin, with its short half-life and relatively small body pool, reflects short-term protein loss and recovery.

However, it has several disadvantages that make it an unreliable marker of nutritional status. Its levels decrease in cases of hepatic alterations (due to reduced synthesis) or in cases of increased protein loss due to renal damage; high levels, instead, can be found in cases of severe iron deficiency. Therefore, transferrin is less accurate than other plasma proteins in assessing protein status because it may be falsely normal during protein malnutrition combined with iron deficiency.

Retinol Binding Protein

RBP is a serum protein of approximately 22 kDa, synthesized in the liver, with a half-life of 12 h. Its normal value is 30–60 mg/L. Its function is to transport retinol into circulation. Its increase could indicate an acute change in the patient's nutritional status. Indeed, this protein responds readily to nutritional depletion because of its small body pool and very short half-life. However, RBP levels are reduced due to protein restriction and deficiency of vitamin A and zinc, which are essential for its proper functioning. In addition, RBP is degraded in the kidney, and consequently, its levels are increased in patients with renal impairment.

In summary, prealbumin and RBP represent the most reliable biomarkers for assessing the short-term effects of nutritional changes because of their rapid turnover. Albumin, on the other hand, has a greater body reserve and a longer half-life; consequently, it represents an indicator of long-term changes in protein status. Table 13.3 summarizes the characteristics of albumin, transferrin, and prealbumin.

Table 13.3 Characteristics of protein biomarkers

Biomarker	Half-life	Normal range	Advantages	Disadvantages
Albumin	≈ 20 days	35–52 g/L	Easy to measure Low cost Reproducibility Marker of chronicity and severity of the nutritional status	Long half-life Reduced levels in infections, burns, fluid overload, liver failure, cancer, and nephrotic syndrome
Transferrin	≈8 days	2–3 g/L	Short half-life (8–10 days) Fast response to changes in protein status	Influenced by several factors, including liver disease and stress Unreliable in the evaluation of mild malnutrition and response to nutritional intervention Costly
Prealbumin	≈2 days	20–40 mg/dL	Shorter half-life than albumin (2–3 days) Easily available Not affected by the state of hydration Earliest marker of nutritional status	Levels may increase in renal dysfunction and corticosteroid therapy Levels can be reduced in case of physiological stress, infection and liver dysfunction

Urinary Creatinine

Urinary creatinine, commonly used to assess kidney function, can give useful information about an individual's muscle mass. Creatine is almost completely present in muscle (in the form of creatine phosphate) and is converted to creatinine at a relatively constant rate; therefore, excretion of creatinine in the urine may be a reliable indicator of muscle mass. The most widely used method for assessing urinary creatinine is to calculate the creatinine/height index, which is the ratio of the subject's creatinuria to the ideal creatinuria of a reference subject of the same sex and height.

$$c/a \text{ index} = 24\text{-hour urinary excretion} \;/ 24\text{-hour urinary excretion ideal} \times 100$$

Its reduction indicates depletion of the muscle compartment.

Recommended Readings

AMA Council on Food and Nutrition (1970) Malnutrition and hunger in the United States. JAMA 213:272

Bharadwaj S, Ginoya S, Tandon P, Gohel TD, Guirguis J, Vallabh H, Jevenn A, Hanouneh I (2016) Malnutrition: laboratory markers vs nutritional assessment. Gastroenterol Rep (Oxf) 4(4):272–280. Epub 2016 May 11

Council of Europe, Public Health Committee, Committee of Expert on Nutrition, Food Safety and Consumer Health (2002) Ad hoc group nutrition programmes in hospitals. Food and nutritional care in hospitals: how to prevent undernutrition. Reports and Guidelines Strasbourg

Stratton RJ, Green CJ, Elia M (2003) Disease-related malnutrition: an evidence-based approach to treatment. CABI Publishing, Wallingford

Dyslipidemias

14

Marcello Ciaccio, Luisa Agnello, Bruna Lo Sasso, Rosaria Vincenza Giglio, and Anna Maria Ciaccio

Introduction

Dyslipidemia identifies a broad spectrum of pathological conditions associated with abnormalities of lipid metabolism involving a significant alteration in circulating levels of lipoproteins and associated lipids, particularly cholesterol and triglycerides.

Dyslipidemia represents an important cardiovascular risk factor, constantly increasing incidence and prevalence in the general population. In particular, the alterations that are of greatest interest from the point of view of cardiovascular risk are hypercholesterolemia, hypertriglyceridemia, and low HDL levels, which can each occur in isolated form or in association as a result of genetic alterations or secondary to other diseases or drug intake.

Lipoproteins

Lipoproteins are macromolecular complexes consisting of a hydrophobic core formed by nonpolar lipids, mainly cholesterol esters and triglycerides, surrounded by a hydrophilic membrane consisting of phospholipids, free cholesterol, and apolipoproteins (or apoproteins).

Based on the size, lipid, and apoprotein composition, lipoproteins are divided into seven classes (Table 14.1 and Fig. 14.1):

- Chylomicrons
- Chylomicron remnants
- Very low-density lipoprotein (VLDL)
- Intermediate-density lipoprotein (IDL)
- Low-density lipoprotein (LDL)
- High-density lipoprotein (HDL)
- Lipoprotein(a) [Lp(a)]

Chylomicron remnants, VLDLs, IDLs, LDLs, and Lp(a), are proatherogenic lipoproteins, whereas HDLs are antiatherogenic.

Chylomicrons and Chylomicron Remnants

Chylomicrons are the largest lipoproteins synthesized at the intestinal level, with the primary role of transporting fats of exogenous origin, mainly triglycerides, absorbed at the enterocyte level, to the liver and peripheral tissues.

Among the apolipoproteins constituting chylomicrons, ApoB-48 represents the most important isoform, having not only a structural but also a functional role as it mediates the exocytosis of the particle from the enterocyte. A defect in ApoB-48, both qualitatively and quantitatively, limits the release of chylomicrons leading to enterocyte thesaurismosis. The size of chylomicrons varies according to the amount of ingested fat. A high-fat meal leads to large chylomicrons, whereas the fasting state results in small particles. The neosynthesized chylomicrons are secreted into the lymphatic circulation and then released through the thoracic duct into the systemic circulation. In this way, chylomicrons do not reach the liver directly via the portal circulation, thus avoiding an overload of fat that could lead to hepatic steatosis. In circulation, the triglycerides removal from chylomicrons by endothelial lipases leads to smaller particles, known as chylomicron remnants. Compared to chylomicrons, remnants are enriched in cholesterol and are proatherogenic.

M. Ciaccio (✉) · L. Agnello · B. Lo Sasso · R. V. Giglio
Department of Biomedicine, Neurosciences and Advanced Diagnostics, Institute of Clinical Biochemistry, Clinical Molecular Medicine and Clinical Laboratory Medicine, and Department of Laboratory Medicine, University Hospital "P. Giaccone", Palermo, Italy
e-mail: marcello.ciaccio@unipa.it

A. M. Ciaccio
Department of Health Promotion, Mother and Child Care, Internal Medicine and Medical Specialties (ProMISE) "G. D'Alessandro", University Hospital "P. Giaccone", Palermo, Italy

M. Ciaccio (ed.), *Clinical and Laboratory Medicine Textbook*, https://doi.org/10.1007/978-3-031-24958-7_14

Table 14.1 Characteristics of lipoproteins

Lipoprotein	Density (g/mL)	Diameter (nm)	Lipids	Apolipoproteins
Chylomicrons	<0.930	75–1200	Triglycerides	B-48, C, E, A-I, A-II, A-IV
Chylomicrons *remnants*	0.930–1.006	30–80	Triglycerides Cholesterol	B-48, E
VLDL	0.930–1.006	30–80	Triglycerides	B-100, E, C
IDL	1.006–1.019	25–35	Triglycerides Cholesterol	B-100, E, C
LDL	1.019–1.063	18–25	Cholesterol	B-100
HDL	1.063–1.210	5–12	Cholesterol	AI, A-II, C, E
Lp (a)	1.055–1.085	~30	Cholesterol	B-100, (a)

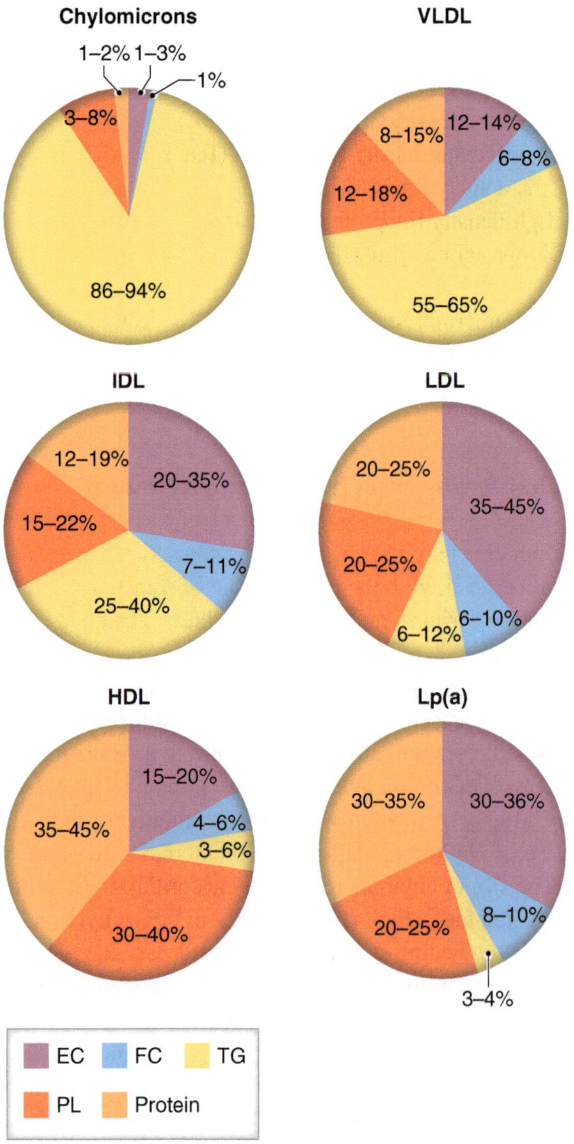

Fig. 14.1 Composition of plasma lipoproteins. HDL, high-density lipoprotein; IDL, intermediate-density lipoprotein; LDL, low-density lipoprotein; Lp (a), lipoprotein (a); VLDL, very low-density lipoprotein. EC, esterified cholesterol; FC, free cholesterol; TG, triglycerides; PL, phospholipids (Copyright EDISES 2021. Reproduced with permission)

VLDL and IDL

VLDLs are the very low-density lipoproteins synthesized in the liver. They represent the main plasma transport vehicle of endogenous triglycerides. The most represented apolipoprotein is ApoB-100. Similar to chylomicrons, the size of VLDL can vary depending on the number of triglycerides transported by the particle.

As soon as they are synthesized in the hepatocytes, VLDLs are released into the circulation, where they interact with lipoprotein lipases which, through the hydrolysis of triglycerides, cause their conversion first into IDL and then into LDL.

LDL

LDLs, the low-density lipoprotein, originate from IDL and have a high cholesterol content. Indeed, their main function is the transport of cholesterol in the plasma, both in free and esterified form.

Like VLDL, the major apolipoprotein in LDL is ApoB-100, which plays a key role as it mediates the interaction of the lipoprotein with its receptor expressed on the cell membrane.

When LDLs are in plasma at high concentrations, they are potentially atherogenic. Indeed, they can passively diffuse through the junctions of endothelial cells and accumulate in the intima, triggering an inflammatory response. Under these conditions, LDL undergoes a process of oxidation; oxidized LDL is phagocytized by macrophages, which progressively turn into "foamy cells," typical of both early and late atherosclerotic lesions. The accumulation of intracellular cholesterol does not negatively regulate macrophages; therefore, they continue to internalize oxidized LDL until they undergo apoptosis or necrosis, thus contributing to the formation, within the plaque, of a soft and destabilizing necrotic core rich in lipids.

LDL consists of a spectrum of particles that vary in size and density. Small and dense LDL is considered the most atherogenic for several reasons. First, small and dense LDL has a lower affinity for the LDL receptor prolonging the lipoprotein's stay in the bloodstream. In addition, they penetrate the artery wall more easily, where they bind more avidly to proteoglycans and become trapped. Finally, they are more susceptible to oxidation resulting in greater uptake by macrophages.

HDL

HDLs, the high-density lipoprotein, acts as a scavenger, mediating the cholesterol transport, from peripheral tissues to the liver. Nascent HDLs, synthesized in the liver and intestine, known as pre-beta HDLs, pick up free cholesterol at the periphery through interaction with the ABCA1 (ATP-Binding Cassette transporter A1) enzyme, and esterify it through the membrane-associated LCAT (lecithin-cholesterol acyltransferase) enzyme. The main inducer of the LCAT enzyme is ApoA-I apolipoprotein, which represents the most abundant protein component of HDL. The continuous accumulation of esterified cholesterol in the lipoprotein core transforms the nascent HDL from discoidal particles to spherical parts, called HDL-3. Through the esterified cholesterol transfer protein (CETP), HDL gives up esterified cholesterol, in exchange for triglycerides, to VLDL, IDL, and chylomicrons and is transformed into HDL-2. The latter, enriched in triglycerides, can be metabolized by hepatic and lipoprotein lipase and, thus, return to the nascent HDL stage, or be directly captured and degraded by the liver. The accelerated catabolism of triglyceride-rich HDL-2 explains the common finding of low HDL cholesterol levels during hypertriglyceridemia.

Lp(a)

Lipoprotein(a) is synthesized in the liver and it is structurally similar to LDL, consisting of an ApoB-100 molecule covalently linked, by disulfide bridges, to apolipoprotein(a). Apo(a) has a molecular weight related to the number of repeated sequences (kringles), which varies widely in the population between 300 and 800 kDa. This variability is genetically determined and influences Lp(a) levels: low molecular weight isoforms are associated with high plasma Lp(a) levels and viceversa. In addition, the Apo(a) structure has a high homology with plasminogen.

The physiological role of Lp(a) is not yet completely clear, but levels >30 mg/dL are associated with an increased atherosclerotic risk. It is not known whether Lp(a) participates directly or indirectly in the formation of atheroma. However, due to its homology with plasminogen, Apo(a) may inhibit fibrinolysis and promote thrombosis, thus interacting with the physiological mechanism of coagulation.

Table 14.2 Classification of apolipoproteins

Apoprotein	Site of synthesis	Associated lipoprotein	Function
ApoA-I	Intestine, liver	Chylomicrons, HDL	Activate LCAT Structural protein for HDL
ApoA-II	Intestine, liver	Chylomicrons, HDL	Activates hepatic lipase Structural protein for HDL
ApoA-IV	Intestine	Chylomicrons, HDL	Facilitates the outflow of cholesterol from cells
ApoA-V	Liver	VLDL, chylomicrons, HDL	Promotes LPL-mediated lipolysis of triglycerides
Apo (a)	Liver	Lp (a)	Inhibits plasminogen activation
ApoB-48	Intestine	Chylomicrons	Structural protein
ApoB-100	Liver	VLDL, IDL, LDL, Lp (a)	Structural protein Ligand for the LDL receptor
ApoC-I	Liver	Chylomicrons, VLDL, HDL	Activate LCAT (less important than AI)
ApoC-II	Liver	Chylomicrons, VLDL, HDL	Cofactor for LPL
ApoC-III	Liver	Chylomicrons, VLDL, HDL	Inhibits LPL
ApoE	Liver	Chylomicrons *remnants*, VLDL, IDL, HDL	Ligand for the LDL receptor

Apolipoproteins

Apolipoproteins have four main functions: (1) structural role; (2) binding of lipoprotein receptors; (3) guidance in lipoprotein formation; and (4) activation or inhibition of enzymes involved in lipoprotein metabolism (Table 14.2). Thus, apolipoproteins play a crucial role in lipoprotein metabolism.

Molecules involved in Lipoprotein metabolism

LDL Receptor

The LDL receptor (R-LDL) is expressed on the membrane of cells of the liver and most tissues, where it mediates endocytosis of LDL, remnant chylomicrons, and IDL through interaction with ApoB-100 and ApoE. Following internalization of the R-LDL/LDL complex, the cleavage of the receptor–lipoprotein bond occurs; the receptor is recycled and transported to the membrane, while the lipoprotein undergoes lysosomal degradation, which leads to the release of its protein and lipid components. Cholesterol inhibits the expression of the gene encoding for hydroxymethylglutaryl coenzyme A (HMG-CoA) reductase, a key enzyme in cholesterol biosynthesis, and the gene encoding for the LDL receptor. In addition, cellular cholesterol is oxidized, and oxidized sterols activate LXR, a nuclear hormone receptor that acts as a transcription factor and stimulates transcription of the ubiquitin E3 ligase gene, which mediates ubiquitination and degradation of the LDL receptor. Finally, the LDL receptor interacts with PCSK9 (proprotein convertase subtilisin/kexin type 9), a protein that promotes its early lysosomal degradation, preventing its recycling to the plasma membrane. PCSK9 gene mutations with loss of function are associated with increased LDL receptor activity and reduced LDL levels. Conversely, mutations resulting in increased PCSK9 protein activity are associated with reduced LDL receptor activity, resulting in increased LDL levels that can lead to severe familial hypercholesterolemia.

Class B Scavenger Receptor Type 1 (SR-B1)

The receptor belongs to the scavenger receptor super family and is expressed in the liver, adrenal glands, ovaries, testes, macrophages, and other cell types. In liver and steroid-producing cells, it mediates the selective uptake of cholesterol esters from HDL.

ATP-Binding Membrane Cassette Transporter A1 (ABCA1)

The ABCA1 is expressed in most cells of the body, including hepatocytes, enterocytes, and macrophages, where it mediates the transport of cholesterol and phospholipids from cells to lipid-poor HDLs (pre-beta HDLs). Therefore, it plays a key role in the initial lipidation processes of nascent HDLs.

ATP-Binding Cassette Transporter G1 (ABCG1)

The ABCG1 is expressed in many cell types and mediates cholesterol efflux from cells to mature HDLs.

ATP-Binding Cassette Transporters G5 and G8 (ABCG5/ABCG8)

The ABCG5/ABCG8 are expressed in the liver and intestine as heterodimers. In the intestine, these transporters mediate the flow of cholesterol from the enterocyte to the intestinal lumen, thereby reducing its absorption. In the liver, however, they promote the transport of cholesterol into the bile.

Acylcoa-Cholesterol Acyltransferase (ACAT)

The ACAT is an enzyme that mediates the esterification of cholesterol with a fatty acid molecule. Esterification makes cholesterol even more lipophilic and suitable for storage or transport via plasma lipoproteins.

Lipoprotein Lipase (LPL)

LPL is an enzyme synthesized in numerous tissues, including muscle and adipose tissue, secreted and anchored to heparan sulfate on the luminal surface of endothelial cells.

LPL catalyzes the hydrolysis of triglycerides, chylomicrons, and VLDL, releasing glycerol and fatty acids, which, once released, are picked up by cells. Triglyceride catabolism leads to the conversion of chylomicrons into chylomicron remnants and VLDL into IDL. LPL requires ApoC-II as a cofactor. In addition, ApoA-V also plays a key role in the activation of the enzyme. In contrast, ApoC-III and ApoA-II inhibit its activity. Insulin stimulates the expression of LPL, whereas the activity of the enzyme is reduced in patients with poorly controlled diabetes, and this could impair the metabolism of triglyceride-rich lipoproteins, leading to hypertriglyceridemia.

Hepatic Lipase

Hepatic lipase is located on the sinusoidal surface of hepatocytes, where it mediates the hydrolysis of triglycerides and phospholipids of IDL and LDL, leading to the formation of smaller particles (IDLs are converted into LDLs, and large LDLs are converted into smaller particles).

Lecithin Cholesterol Acyltransferase (LCAT)

LCAT is an enzyme synthesized in the liver and secreted into plasma, where it circulates mostly associated with HDLs. On HDLs, it catalyzes the synthesis of cholesterol esters, facilitating the transfer of fatty acid from the 2-position of lecithin to the hydroxyl group in 3-position of cholesterol. ApoA-I serves as a cofactor for LCAT.

The free cholesterol on the surface of HDL, once esterified, is transferred within the particle. In this way, HDL promotes the uptake of free cholesterol through the reduction of cholesterol on the surface of HDL.

Cholesteryl Ester Transfer Protein (CETP)

It is a glycoprotein synthesized in the liver and released into plasma, where it mediates the transfer of cholesteryl esters from HDL to VLDL, chylomicrons, and LDL, and the transfer of triglycerides from VLDL and chylomicrons to HDL. Inhibition of CETP activity leads to an increase in HDL cholesterol and a decrease in LDL cholesterol.

Lipoprotein and Lipid Metabolism

The metabolism of lipoproteins can be divided into an exogenous and an endogenous pathway, which are connected through the liver (Fig. 14.2).

Exogenous Pathway of Lipid Metabolism

The exogenous pathway of lipoprotein metabolism begins in the intestine, where dietary triglycerides (about 100 g/day) are hydrolyzed into fatty acids and monoacylglycerol by intestinal lipases and emulsified with bile acids, cholesterol, and fat-soluble vitamins to form micelles. The latter are then transported into the enterocytes, where cholesterol is esterified by the enzyme ACAT, while fatty acids and monoacylglycerol are converted into triglycerides. In the endoplasmic reticulum, triglycerides and cholesterol esters are "packed" into chylomicrons to be secreted into the lymphatic circulation and, later released, through the thoracic duct, into the systemic circulation. In the capillaries of the adipose and muscular tissue, LPL hydrolyses the triglycerides carried by the chylomicrons into glycerol and fatty acids. The latter are picked up by skeletal and cardiac muscle cells for energy production (beta-oxidation) and by adipocytes, where they are used as substrates for triglyceride synthesis and subsequent accumulation as an energy reserve.

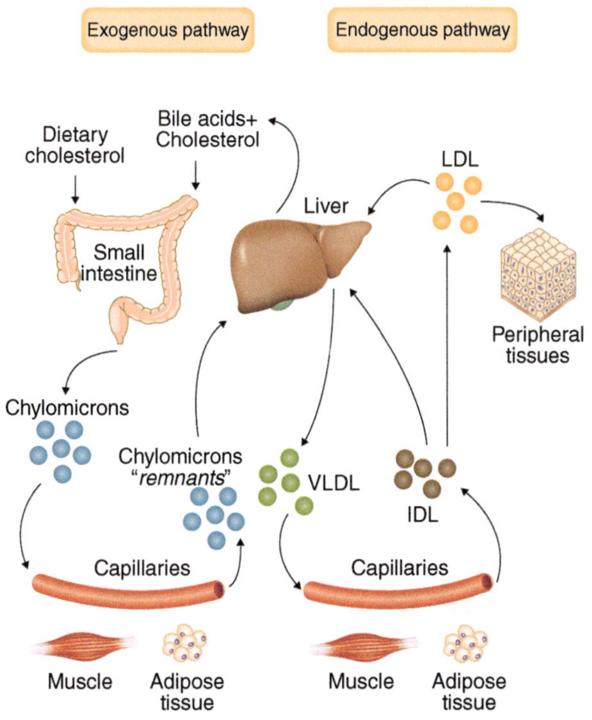

Fig. 14.2 Lipoprotein metabolism. IDL, intermediate-density lipoprotein; LDL, low-density lipoprotein; VLDL, very low-density lipoprotein (Copyright EDISES 2021. Reproduced with permission)

Following the lipolytic activity of the LPL, the core of the chylomicrons, which is rich in triglycerides, condenses, resulting in excess of components of the outer structure (nonesterified cholesterol, phospholipids, ApoA-I, and ApoA-II), which are transferred to the nascent HDLs, from which they receive ApoE and ApoC. As a result of these modifications and the reduction in triglyceride content, chylomicrons transform into chylomicron remnants. ApoE plays a key role in mediating the interaction of chylomicron remnants with the LDL receptor expressed on hepatocytes, where these particles will be degraded. Mutations in the ApoE gene (e.g., the ApoE2 isoform) may result in reduced clearance of chylomicrons and increased plasma levels of cholesterol and triglycerides, a condition known as familial dysbetalipoproteinemia. In the liver, cholesterol can be used for the VLDL synthesis to it can be excreted in the bile as such or after transformation into bile acids and then released in the intestine.

Endogenous Pathway of Lipid Metabolism

In the liver, triglycerides and cholesterol are released into the systemic circulation as VLDLs. VLDL assembly is initiated by microsomal transfer protein (MTP), which transports lipids to ApoB-100. MTP mobilizes esterified cholesterol, triglycerides, and phospholipids from the cytosolic pool to lipoproteins forming in the endoplasmic reticulum. Mutations involving loss of function of ApoB-100 fall to produce VLDL and a marked reduction in circulating cholesterol and triglycerides (familial hypobetalipoproteinemia or abetalipoproteinemia).

Similar to chylomicrons, circulating VLDLs are exposed to the action of the LPL of adipose and muscle tissue capillaries which, by hydrolyzing the triglyceride-rich core, increases the esterified cholesterol content and transforms the particles making them smaller and denser; the excess surface components (nonesterified cholesterol, phospholipids, apolipoproteins) are transferred to HDLs. In this way, VLDL is transformed into IDL.

A portion of IDL (about half) is reuptake by the liver through the interaction between ApoE (expressed on IDL) and the LDL receptor expressed on hepatocytes; the remaining portion undergoes further triglyceride hydrolysis by hepatic lipase, leading to the formation of LDL, consisting mainly of esterified cholesterol and ApoB-100.

LDLs have the role of transporting cholesterol to peripheral tissues, where they regulate both *de novo* synthesis and cholesterol uptake. Approximately two-thirds of LDL are removed from the circulation by such mechanism, which involves the internalization of particles following their interaction with R-LDL; the remaining portion is incorporated within the vascular wall where, by binding to proteoglycans, it becomes susceptible to modifications, such as oxidation. Oxidized LDLs are recognized and phagocytized by macrophages through interaction with scavenger receptors. This mechanism is the basis of the genesis of atheromatous plaques (Fig. 14.3).

Reverse Cholesterol Transport

Peripheral tissue cells accumulate cholesterol through the uptake of circulating lipoproteins or by *de novo* synthesis, but most do not possess a mechanism for the degradation of excess cholesterol. Cells synthesizing steroid hormones can convert cholesterol to glucocorticoids, estrogen, testosterone, etc.; intestinal cells can secrete cholesterol into the intestinal lumen. Most cells reduce their cholesterol content through reverse transport, mediated by HDLs, which return cholesterol to the liver. Specifically, HDLs pick up cholesterol in free form from peripheral tissue cells and release it in

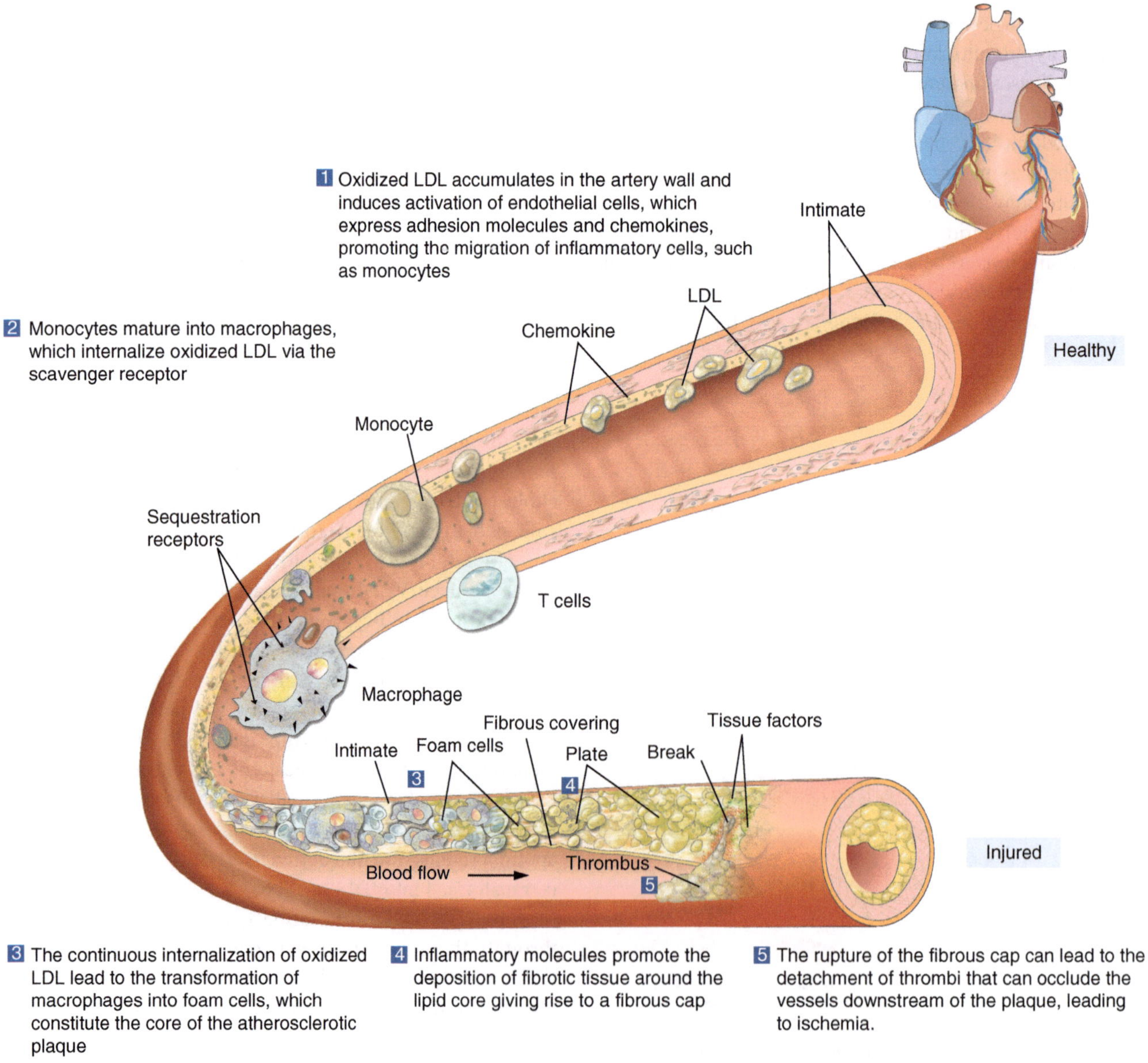

Fig. 14.3 Mechanism of atherosclerotic plaque formation (Copyright EDISES 2021. Reproduced with permission)

the esterified form to the liver, where it can be recycled to synthesize new lipoproteins, excreted in bile as free cholesterol, or eliminated as bile acids.

The first step (Fig. 14.4) in the reverse transport of cholesterol is the secretion by the gut and liver of pre-beta HDLs, consisting mainly of ApoA-I and small amounts of cholesterol and phospholipids (about 10%). Pre-beta HDLs represent initial acceptors of free cholesterol from cells. The membrane transporter ABCA1 plays a key role in mediating the efflux of free cholesterol from peripheral tissue cells to native HDLs, where the plasma enzyme LCAT mediates the esterification of the incorporated cholesterol, thereby transforming pre-beta HDLs into mature HDLs. The efficiency of reverse cholesterol transport depends primarily on the ability of ApoA-I to promote cholesterol efflux through interaction with ABCA1 and activate the LCAT enzyme.

Mature HDL releases esterified cholesterol to the liver primarily through two pathways, direct and indirect.

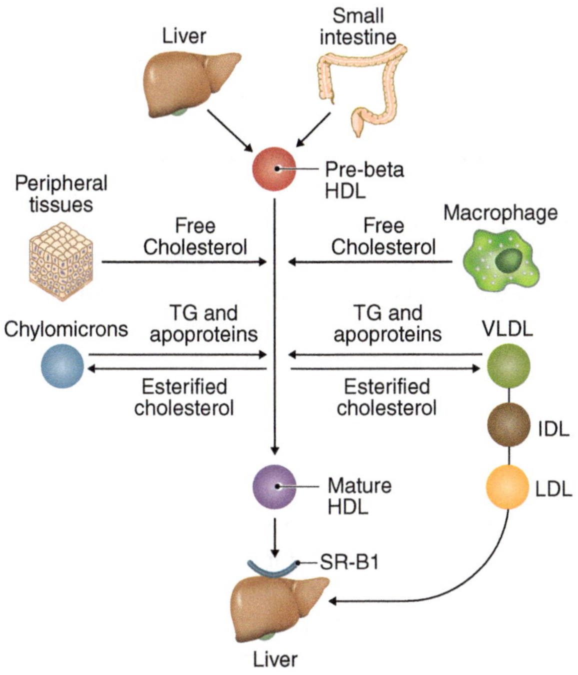

Fig. 14.4 Reverse transport of cholesterol. HDLs are released by the liver and intestine in an immature form, known as pre-beta HDL, or nascent HDL, consisting of ApoA-I, free cholesterol, and phospholipids. Peripheral tissues and macrophages yield free cholesterol to nascent HDL while VLDL and chylomicrons yield triglycerides and apoproteins (ApoC-II, ApoE). Free cholesterol is esterified by the LCAT enzyme associated with HDL, thus leading to the formation of mature HDL. The latter can release the esterified cholesterol directly to the liver, through the interaction with SR-B1, or indirectly, through the action of CETP (cholesterol ester transfer protein), giving up the esterified cholesterol, in exchange of triglycerides, VLDLs and chylomicrons. HDL, high-density lipoprotein; IDL, intermediate-density lipoprotein; LDL, low-density lipoprotein; SR-B1, class B type 1 scavenger receptor; TG, triglycerides; VLDL, very low-density lipoprotein (Copyright EDISES 2021. Reproduced with permission)

In the direct pathway, SR-B1 receptor, expressed on the surface of hepatocytes, mediates the HDL interaction and modification by selectively removing cholesterol from mature HDL, and thus transforming them into lipid-poor HDL remnants, ready for a new cycle.

In the indirect pathway, circulating HDL, through the action of the CETP enzyme, releases cholesterol to ApoB-containing lipoproteins (VLDL and LDL), which are subsequently captured and degraded in the liver.

Classification of Dyslipidemias

From a nosographic point of view, there are different classifications of dyslipidemias, depending on the lipid alteration (hypercholesterolemia, hypertriglyceridemia, or mixed forms), the phenotypic aspect (Fredrickson classification), the biochemical-genetic defect (mono and polygenic alterations) or the pathogenic mechanism (primary and secondary dyslipidemias). Each of these classifications has strengths and limitations.

Fredrickson Classification

In 1967, Fredrickson proposed a classification of the dyslipidemias according to the phenotype. At the beginning of the 1970s, the Fredrickson classification was adopted in a modified form by the World Health Organization (WHO). According to this classification, dyslipidemias are distinguished into six phenotypes based on the altered lipoprotein and lipid profile (Table 14.3).

Type I

The lipoprotein alteration is represented by increased chylomicrons associated with severe hypertriglyceridemia (>1000 mg/dL) and low HDL levels. This phenotype is more frequently found in individuals with primary dyslipidemias due to a familial defect of the lipoprotein lipase enzyme, with the absence or reduction of function, or to the absence of ApoC-II, an activator of lipase, or in individuals affected by other diseases, such as diabetes (insulin is an activator of lipase). Individuals with this phenotype frequently experience pancreatitis due to elevated triglyceride levels.

Table 14.3 Fredrickson's classification of dyslipidemias

Type	Lipoprotein increased	Cholesterolemia	Triglyceridemia
I	Chylomicrons	Normal or ↑	↑↑↑↑
IIa	LDL	↑↑	Normal
IIb	LDL and VLDL	↑↑	↑↑
III	IDL	↑↑	↑↑↑
IV	VLDL	Normal or ↑	↑↑
V.	VLDL and chylomicrons	↑ or ↑↑	↑↑↑↑

Table 14.4 Classification of dyslipidemias

Disorder	Cholesterol ↑	Triglycerides ↑	Cholesterol and triglycerides ↑
Primary	Familial hypercholesterolemia Familial deficit by ApoB-100 Polygenic hypercholesterolemia	Familiar hypertriglyceridemia Familial deficit by LPL Familial deficiency of ApoC-II	Familial combined hyperlipidemia Hyperlipoproteinemia type III
Secondary	Hypothyroidism Nephrotic syndrome	Diabetes mellitus Alcoholic hyperlipemia Chronic kidney failure Monoclonal gammopathies Lymphomas Drugs: estrogen, progestins, glucocorticoids, and androgens	Drugs: thiazide diuretics, beta-blocker, and cyclosporine

Type IIa

The characteristic lipoprotein alteration is increased LDL with pure hypercholesterolemia and normal triglyceridemia. This phenotype is quite common and is found in several forms, primary, such as familial hypercholesterolemia, and secondary, such as hypothyroidism and nephrotic syndrome.

Type IIb

This phenotype is characterized by increased LDL and VLDL with mixed dyslipidemia (hypercholesterolemia and hypertriglyceridemia). It is found in primary forms, such as familial combined hyperlipemia, and secondary to diseases, such as hypothyroidism, diabetes, and kidney disease, or secondary to the use of drugs, such as diuretics.

Type III

It is characterized by mixed dyslipidemia with significant hypertriglyceridemia and IDL accumulation. This phenotype is characteristic of two rare primary disorders, familial dysbetalipoproteinemia and hepatic lipase deficiency.

Type IV

It is one of the most frequent forms, characterized by VLDL accumulation with hypertriglyceridemia, normal or slightly increased cholesterolemia, and low HDL. This phenotype can be found in subjects affected by a primary disorder, such as familial hypertriglyceridemia, and in subjects affected by other diseases, such as diabetes, or in subjects who abuse alcohol or drugs.

Type V

It is characterized by a significant accumulation of VLDL and chylomicrons with important hypertriglyceridemia, a modest increase in cholesterolemia, and low HDL. Indeed, this phenotype represents a complication of phenotype I and phenotype IV. Patients experience pancreatitis.

Pathogenetic Classification

Fredrickson's classification, which has been adopted for several years and is still considered a reference system for indicating patterns of lipoprotein alterations, is helpful but insufficient for a correct pathogenetic classification of dyslipidemia. Indeed, it does not consider the different etiologias of the alterations.

Based on the etiopathogenesis, it is possible to classify dyslipidemias into primary and secondary forms (Table 14.4).

Primary dyslipidemias are genetic forms due to mutations in genes that encode molecules involved in lipoprotein metabolism. Secondary dyslipidemias, instead, are consequent to other pathologies or conditions. In this case, the therapeutic intervention is aimed at removing the primary cause and not directly dyslipidemia; for example, in the case of dyslipidemia secondary to hypothyroidism, statins are not administered, but hormone replacement therapy for hypothyroidism will be sufficient. Table 14.4 shows the main forms of primary and secondary dyslipidemias, subdivided according to the type of lipid alteration: pure hypercholesterolemia, pure hypertriglyceridemia, or mixed forms.

Primary Dyslipidemia

Primary dyslipidemias are associated with severe and early onset of cardiovascular events.

For some forms, the involved gene and the transmission mode have been identified; for most of them, instead, the pathogenesis is still to be clarified, and it is believed to be due to multiple gene defects that, combining, can produce important effects on the plasma lipid and lipoprotein concentrations. Table 14.5 shows the characteristics of the main primary dyslipidemias.

Early identification of these dyslipidemias is critical for the initiation of effective treatment in preventing/limiting injury due to prolonged exposure to elevated plasma lipid levels.

Combined Familial Hyperlipidemia

It is the most frequent familial form in the general population, although it is often misrecognized. The phenotypic manifestation results from multiple genetic mutations leading to altered VLDL and LDL metabolism and is characterized by variability, both among members of the same family and within the same individual over time. From a biochemi-

Table 14.5 Primary dyslipidemias

Disorder	Mutated gene	Transmission	Frequency	Phenotype Frederickson	Cardiovascular risk
Familial combined hyperlipidemia	Unknown	Autosomal dominant	1/100	IIa, IIb, IV or V	+
Familial hypercholesterolemia	*R-LDL; PCSK9; APOB*	Autosomal dominant	1/250 heterozygous form; $1/10^6$ homozygous form	IIa or IIb	+
Familial LPL deficiency	*LPL*	Autosomal recessive	$1/10^6$	I, V	
Familial ApoC-II deficiency	*APOC-II*	Autosomal recessive	$1/10^6$	I, V	
Familial ApoB-100 deficiency	*APOB*	Autosomal recessive	1/1000	IIa	+
Familial dysbetalipoproteinemia	*APOE*	Autosomal recessive	1/10,000	III	+
Familial hypertriglyceridemia	Unknown	Autosomal dominant	Unknown	IV or V	+ (?)

cal point of view, it is characterized by increased plasma levels of triglycerides and/or C-LDL. The metabolic defect consists of VLDL-ApoB increased hepatic synthesis, which is associated with the prevalent reduction of LPL activity or postprandial catabolism of VLDL, which could explain the phenotypic variability. Patients are often asymptomatic until adolescence and present a high predisposition to early Coronary Artery Disease (CAD).

Familial Hypercholesterolemia

Familial hypercholesterolemia (FH) is an autosomal dominant monogenic disease due to mutations in genes encoding for key proteins of the LDL metabolism, leading to an increase in circulating LDL cholesterol. Mutations in R-LDL gene, with loss of function; the APOB gene, with alterations in the binding domain of ApoB with R-LDL; and the PCSK9 gene, with a gain of function. Mutations in the R-LDL gene are the most frequent. To date, more than 1000 mutations in the R-LDL gene associated with different functional alterations of the receptor have been described: from a lack of synthesis to a lack of transport to the cell surface, failure to internalize the receptor-LDL complex, or failure to recycle the receptor after internalization.

FH can occur in heterozygous (HeFH) or homozygous (HoFH) forms. HeFH has a relatively high frequency in the population, with an incidence of 1/250 individuals worldwide, and is characterized by an increase in LDL cholesterol of about 2–3 times. From a clinical point of view, subjects who present the heterozygous form, often, before developing an acute event, such as acute myocardial infarction or stroke, do not show any discomfort objectable by the physician. Usually, the finding occurs incidentally during routine blood laboratory tests. Early diagnosis and adequate treatment allow for maintaining a cardiovascular risk similar to that observed in the general population.

HoFH is a rarer and more severe condition characterized by increased LDL cholesterol of about 6–8 times. From a clinical point of view, subjects with the homozygous form develop cardiovascular events very early (18–20 years).

Both forms (both homozygous and heterozygous) are characterized by the presence of xanthelasmas, skin, and tendon xanthomas due to the accumulation of cholesterol in tissue macrophages that infiltrate the skin and tendons. Homozygous individuals may present with xanthomas before 10 years of age, whereas in heterozygotes, the incidence of xanthomas increases with each decade. In some cases, especially in women, inflammation of the Achilles tendon is recurrent.

There is also a form of familial autosomal dominant hypercholesterolemia due to mutation in the LDLRAP1 gene, which encodes for the R-LDL adaptor protein 1, which has a fundamental role in the correct functioning of the LDL receptor. It is a very rare and severe familial dyslipidemia.

Familial LPL or ApoC-II Deficiency

These are very rare dyslipidemias, characterized by a severe deficit in the metabolism of triglyceride-rich lipoproteins (VLDL and chylomicrons) due to mutations in the gene encoding for LPL or ApoC-II. Triglycerides are particularly elevated (≈10 mmol/L, 880 mg/dL), and the serum shows a milky appearance. The possible association of triglycerides with pancreatitis requires prompt and appropriate treatment.

Familial ApoB-100 Deficiency

It is a very rare condition, with recessive autosomal transmission caused by mutations, usually missense, in the gene encoding for ApoB, which make the protein poorly or not recognizable by the LDL receptor, resulting in a reduction in the rate at which LDL is removed from the circulation and catabolized.

LDL levels are lower than in familial hypercholesterolemia, and xanthomas are rare.

Familial Dysbetalipoproteinemia

This condition, also known as familial hyperlipemia type III, is a rare form of dyslipidemia characterized by the circulating accumulation of VLDL migrating in the β region, associated with pathognomonic flat, tuberous-eruptive xanthomas, and a significant predisposition to early severe atherosclerosis. It is due to mutations in the ApoE gene that alter its ability to bind to hepatic lipoprotein receptors. In particular, the APOE gene has a polymorphic sequence that leads to the expression of three common isoforms: ApoE3, ApoE2, and ApoE4. The product of the APOE2 allele has reduced affinity for the LDL receptor, and consequently, lipoproteins presenting this isoform are removed more slowly from the circulation. The presence of the APOE2 allele in homozygosity predisposes to the risk. However, the onset of the disease is due to the coexistence of other conditions, such as high-calorie diet, diabetes, obesity, hypothyroidism, alcohol intake, etc.

Clinically, patients have high levels of otal cholesterol and triglycerides and low levels of HDL cholesterol. The disease onset in adulthood, but not all subjects with E2/E2 genotype develop it.

Familial Hypertriglyceridemia

It is a disorder characterized by an isolated increase in triglycerides of endogenous origin carried by VLDL. It manifests itself, usually in adulthood, with increased triglyceride levels (200–600 mg/dL) and usually normal or slightly increased LDL cholesterol. The increase in cholesterol is because VLDL also carries a small amount of cholesterol. Characteristic clinical signs are eruptive cutaneous xanthomatosis and lipemia retinalis. Individuals with this disease have a high risk of acute pancreatitis.

Secondary Dyslipidemia

Lipoprotein alterations are often the consequence of endocrine-metabolic and nonendocrine diseases, or of drug assumption. Secondary dyslipidemias can coexist with primary dyslipidemias. The leading causes of secondary dyslipidemias are reported in Tables 14.6 and 14.7.

Table 14.6 Main secondary dyslipidemias classified according to lipid alteration

Lipid alteration	Clinical condition
Hypercholesterolemia	Cholestasis Nervous anorexia Hypothyroidism Obstructive liver disease Nephrotic syndrome Obstructive jaundice Therapy with progestins, cyclosporine, and thiazides
Hypertriglyceridemia	Alcoholism Obesity Type 2 diabetes mellitus Chronic kidney failure Pregnancy Systemic lupus erythematosus Monoclonal gammopathies: multiple myeloma and lymphomas Glycogenosis Sepsis Stress Cholelithiasis Pancreatitis Therapy with beta-blockers, glucocorticoids, diuretics, and estrogens
Combined hyperlipidemia	Acute hepatitis

Table 14.7 Dyslipidemias secondary to pathologies

Pathology	Fredrickson phenotype	Lipoproteins ↑	Pathogenetic mechanism
Diabetes mellitus	IV, V	VLDL Chylomicrons	↑ VLDL production ↓ VLDL catabolism
Hypothyroidism	IIa	LDL	↓ *Clearance* LDL
Hyperestrogenism	IV	VLDL	↑ VLDL production
Hypercortisolism	IIa, IIb	VLDL, LDL	↑ VLDL production with conversion to LDL
Hypopituitarism	IIb	VLDL, LDL	↑ VLDL production with conversion to LDL
Nervous anorexia	IIa	LDL	↓ Biliary cholesterol excretion and bile acids
Lipodystrophy	IV	VLDL	↑ VLDL production
Werner Syndrome	IIa	LDL	Unknown
Glycogenesis	IV	VLDL	↑ VLDL production ↓ VLDL catabolism
Alcoholism	IV	VLDL	↑ VLDL production
Nephrotic Syndrome	IIa, IIb	VLDL, LDL	↑ VLDL production
Uremia	IV	VLDL	↓ *Clearance* LDL
Cholestasis	–	LPX	↑ Entry of biliary cholesterol into the circulation
Hepatitis	IV	VLDL	↓ LCAT
Systemic lupus erythematosus	I	Chylomicrons	Heparin-binding antibodies ↓ LPL activity
Monoclonal gammopathies	IIa, III, IV	VLDL, IDL, LDL	Lipoprotein-binding antibodies that interfere with their metabolism

Dyslipidemia and Cardiovascular Risk

Lipid metabolism disorders represent a critical cardiovascular risk factor. Epidemiological data, *in vivo* and *in vitro* studies have contributed to delineating the role of lipids in atherosclerosis. In particular, lipids and the lipoproteins that mediate their transport are differently implicated in the atherogenic process (Table 14.8).

Chylomicrons are large molecules and, since they cannot penetrate the artery walls, they are not involved in the development of atherosclerosis. On the contrary, LDLs contribute to the development of atherosclerosis by damaging the endothelium, the first stage of the atherosclerotic process, and by accumulating in the intima of the artery, thus inducing the formation and growth of the atheromatic plaque.

The Lp(a), due to its peculiar intermediate composition between LDL and plasminogen, may contribute to the development of atherosclerosis and/or thrombosis. Indeed, like LDL, Lp(a) can accumulate in the arterial wall, promoting those initial processes of atherosclerosis, such as inflammation and the formation of foamy cells, which lead to the genesis of the atheromatic plaque. Moreover, since a part of the molecule possesses a sequence similar to plasminogen, it contributes to the formation of the clot through the inhibition of endogenous fibrinolysis, competing with plasminogen for the binding on the endothelial surface; the final result is an increased risk of thrombus formation which leads to the occlusion of the vessel.

HDLs, on the other hand, have a protective role against atherosclerosis, which they perform through multiple functions. Among these, the best known is the reverse transport of cholesterol and, in particular, the ability of HDL to mobilize excess cholesterol from macrophages in the arterial wall. In this way, HDLs reduce the accumulation of cholesterol-laden foam cells within the arterial wall, a phenomenon that constitutes one of the earliest stages in the pathogenesis of atherosclerosis. Moreover, HDLs exert their antiatherogenic role through other mechanisms, such as the prevention of LDL oxidation, anti-inflammatory effects, prevention of vascular endothelial cell apoptosis, profibrinolytic, and antithrombotic effects.

Table 14.8 Role of lipoproteins in atherosclerosis

Lipoprotein	Atherogenicity
Chylomicrons	0
VLDL	+
Chylomicrons and VLDL *remnants*	+ + +
LDL	+ + + +
Lp (a)	+ + + +
HDL	Protective

Diagnosis of Dyslipidemia

The diagnosis of dyslipidemia is based on a blood test aimed at assessing the basic lipid profile that includes the determination of triglycerides (TG), total cholesterol (TC), HDL cholesterol (C-HDL), LDL cholesterol (C-LDL), and their ratios. Therefore, the clinical laboratory has a critical role in dyslipidemia diagnosis.

Although, in most cases, dyslipidemia is diagnosed incidentally during routine examinations, screening is recommended in all male adults ≥40 years and female adults ≥50 years or postmenopausal, especially in the presence of other cardiovascular risk factors.

In addition, all individuals with evidence of atherosclerosis in any vascular bed or with type 2 diabetes, regardless of age, are considered to be at elevated risk; therefore, the evaluation of lipid profile is recommended for such individuals. Individuals with a family history of premature cardiovascular disease, children of patients with severe dyslipidemia (combined familial hyperlipidemia), patients with hypertension and central obesity, defined as circumference ≥94 cm in men and ≥80 cm in women, or with a BMI ≥25 kg/m^2, should be screened for dyslipidemia. Finally, chronic clinical conditions, such as autoimmune inflammatory diseases (lupus erythematosus and psoriasis) and renal failure, are associated with an increased risk of developing dyslipidemia and, therefore, affected patients should be screened. Evaluation of the lipid profile is also indicated in patients with peripheral artery disease or who have increased carotid intima-media thickness or carotid plaques.

Following the detection of elevated lipid levels, in at least two determinations carried out approximately 7 days apart, a period of correct diet and physical activity is recommended before making a diagnosis of dyslipidemia and possibly establishing therapy. If there is no improvement in the lipid profile at the end of this period, a familial form of dyslipidemia should be suspected.

Lipid Profile

Blood Collection

Proper patient preparation and sampling are essential for an accurate lipid profile evaluation.

Venous sampling for lipid parameters measurement should be performed under the following conditions:

- Habitual diet, including alcohol consumption, up to 24 h before collection.
- Whenever possible, discontinue therapies that affect lipemia (diuretics, beta-blockers, etc.) 3 weeks before performing blood collection.

- Do not test before 2–3 months after childbirth or trauma, such as surgery, or serious illness, such as myocardial infarction (after myocardial infarction, there is an increase in triglycerides and a decrease in cholesterol).
- Do not test until 2 weeks following an episode of inflammation (inflammation is associated with a reduction in total and LDL cholesterol, which is proportional to basal cholesterol concentration).
- Fasting is not required. Traditionally, fasting for 12–14 h was required for lipid analysis. Recently, it has been shown that analysis performed on samples obtained from fasting and nonfasting patients give overlapping results for TG, C-LDL, and C-HDL. TGs are affected by food, resulting in an increase of only about 27 mg/dL depending on composition and time since the meal.
- Maintain venous stasis only as long as necessary (maximum 1 min) and remove the tourniquet as soon as the blood begins to flow.
- The sample can be serum or plasma.
- Results should be confirmed on at least two samples collected 7 days apart.

Intraindividual Variability

There is considerable intraindividual variability in circulating lipid levels. In particular, variability of 5–10% for TC and >20% for TGs has been reported, especially in patients with elevated TGs.

These data are required to calculate significant differences between successive values of the same patient.

Intraindividual variability is due, in part, to analytical variability but also to environmental factors such as diet, physical activity, and seasonal variations; for example, TG and C-HDL levels increase in winter.

Measuring Methods

Enzymatic colorimetric methods are commonly used for lipid assay.

Total Cholesterol

The CHOD/POD/Trinder enzymatic method is uniformly used in laboratories for the TC assay. According to the National Cholesterol Education Program (NCEP) guidelines, to ensure an adequate quality of the assay, the values obtained by the individual laboratory should not deviate from the value of the reference method by more than 3% and the measurement methods should have an inaccuracy (expressed as a coefficient of variation) of less than 3%. The total error must be <8.9%.

The desirable value of TC is ≤5.00 mmol/L (≤190 mg/dL; Table 14.9).

LDL-Cholesterol

In most laboratories, the C-LDL is calculated by the Friedewald formula:

in mmol/L: C-LDL = TC - C-HDL - (TG/2.2)

in mg/dL: C-LDL = TC - C-HDL - (TG/5)

The value of C-LDL calculated according to this formula must consider the following assumptions:

- Methodological errors that may accumulate during the analysis of each of the three parameters included in the formula (CT, TG, and C-HDL).
- The formula can be applied assuming that there is a constant cholesterol/triglyceride ratio in VLDL. In the presence of elevated TG values (>4.5 mmol/L or >400 mg/dL), the formula cannot be applied because of the possible presence of circulating chylomicrons and VLDL remnants.
- The formula may not be applicable when blood is collected under nonfasting conditions. In that case, non-HDL cholesterol should be determined.

Recently, direct methods (immunological or ultracentrifuge) have been developed for the determination of C-LDL whose values are comparable with those obtained by the formula. The direct methods would seem to have analytical performances superior to the Friedewald formula but, similarly to the formula, cannot be applied in the presence of high TG values.

According to NCEP guidelines, analytical performance should have an inaccuracy and bias <4% with a total error <12%.

In general, desirable values should be ≤3.00 mmol/L or ≤115 mg/dL (Table 14.9). Specifically, based on the subject's characteristics, the desirable values are:

- ≤3.00 mmol/L or ≤115 mg/dL in healthy subjects without cardiovascular disease and without other risk factors for cardiovascular disease
- ≤2.60 mmol/L or ≤100 mg/dL in subjects with other risk factors for cardiovascular disease but without current or previous cardiovascular disease
- ≤1.8 mmol/L or ≤70 mg/dL in subjects with prior or ongoing cardiovascular disease

Table 14.9 Desirable plasma values of lipids, lipoproteins, and apolipoproteins

	Value		
	mmol/L	mg/dL	g/L
Total cholesterol	≤5.00	≤190	
C-LDL	≤3.00	≤115	
Non-HDL cholesterol	≤3.80	≤145	
C-HDL	≥1.00 (males) ≥1.20 (females)	≥40 (males) ≥45 (females)	
Triglycerides	≤1.70	≤150	
ApoA-I		≥125	≥1.25
ApoB		≤100	≤1.00

HDL Cholesterol

In the past, C-HDL was measured by methods that involved initial precipitation of ApoB-containing lipoproteins (VLDL and LDL) and subsequent determination of the cholesterol present in the supernatant (C-HDL) by enzymatic methods. Currently, these methods have been replaced by "direct" methods that allow measuring C-HDL in a homogeneous phase without the need for prior separation of HDL from other lipoproteins.

According to NCEP guidance, analytical performance should have an inaccuracy <4% and bias <5% with a total error <13%.

Desirable values should be ≥1.00 mmol/L or ≥40 mg/dL for males and ≥1.20 mmol/L or ≥45 mg/dL for females (Table 14.9).

For the correct evaluation of cardiovascular risk, the risk index, i.e., the CT/C-HDL ratio and the C-LDL/C-HDL ratio, must be considered. The CT/C-HDL ratio in healthy subjects must be <5 for males and <4.5 for females. Regarding the C-LDL/C-HDL ratio, the risk is defined as low when the ratio is 1.47, medium 3.22, moderate 5.0, and high 6.14.

Triglycerides

TGs are commonly assessed by enzymatic methods that measure the free glycerol produced by TGs hydrolysis. Although these methods also measure free glycerol obtained from lipase-mediated hydrolysis of circulating TGs, its concentration is usually negligible. In rare conditions, such as patients with severe multiorgan diseases, a hyperglycerolemic condition may occur, resulting in excessively high false TG values. In these cases, it would be appropriate to use methods for measuring true TG with subtraction of free glycerol.

Triglycerides have a particularly high intraindividual biological variability. In particular, smoking, physical exercise, and alcohol can significantly modify TG levels and, therefore, before collecting the sample, it is important to take precautions. In addition, the time of sampling is important because, regardless of food intake, TG concentration varies significantly throughout the day. Therefore, collecting the sample in the early morning would be preferable. TGs can be assayed under nonfasting conditions for general screening and cardiovascular risk assessment.

According to NCEP guidelines, analytical performance should have an imprecision and bias <5% with a total error <15%.

Desirable values should be ≤1.70 mmol/L or ≤150 mg/dL (Table 14.9).

Non-HDL Cholesterol

Non-HDL cholesterol is helpful for quantifying atherogenic lipoproteins. It is obtained by subtracting the C-HDL value (the only nonatherogenic cholesterol) from the TC value. It is advisable to evaluate this parameter when TGs are ≥150 mg/dL because TG-rich lipoproteins also carry an atherogenic potential.

Apolipoproteins

The ApoA-I and ApoB apolipoproteins are not evaluated in clinical laboratory practice; their evaluation represents a second level test to be performed in selected cases. In particular, ApoB, being the main apolipoprotein of VLDL, IDL, and LDL, provides information on atherogenic lipoproteins overlapping those provided by non-HDL cholesterol, while ApoA-I provides information on nonatherogenic lipoproteins, overlapping those provided by C-HDL.

There are no standard methods for the apolipoprotein measurement. However, it is recommended that commercial kits be used whose calibrator values are assigned based on internationally recognized preparation.

The reference values for ApoB are ≤100 mg/dL or ≤1.00 g/L and for ApoA-I are ≥125 mg/dL or 1.25 g/L (Table 14.9).

ApoC-III has recently been identified as a potentially important risk factor. It is a key regulator of TG metabolism and high plasma levels are associated with high plasma levels of TG and VLDL. In addition, mutations that cause loss of function are associated with low TG levels, and reduced cardiovascular risk. However, the role of ApoC-III in clinical practice is not yet known.

Lp(a)

Several immunometric methods for Lp(a) measurement are available but are not yet standardized. Currently, Lp(a) assay is not recommended for cardiovascular risk assessment in the general population. However, it should be considered in subjects with high risk (familial, high C-LDL, and low C-HDL levels) or with a strong family history of premature atherothrombotic disease. Risk is considered significant when Lp(a) levels are >30 mg/mL.

Diagnosis of Familial Dyslipidemia

The diagnosis of familial dyslipidemia is based on diagnostic algorithms that include biochemical, clinical, and anamnestic criteria.

First, since familial forms are characterized by elevated lipid levels, in adults, the finding, at least once, of a total cholesterol value >250 mg/dL and/or triglycerides >250 mg/dL may lead to the diagnostic suspicion of familial dyslipidemia (Table 14.10). It is, therefore, necessary to evaluate the complete lipid profile following a period of adequate diet, reduction of any excess weight, and regular physical activity. In addition, it is necessary to evaluate family history

Table 14.10 Main primary dyslipidemias classified according to the lipid alteration

Dyslipidemia	Altered lipids mmol/L (mg/dL)	Clinical signs
Pure hypercholesterolemia		
Common polygenic hypercholesterolemia	Total cholesterol 6.5–9.0 (250–350)	Very common form; generally asymptomatic up to the onset of cardiovascular events
Familial hypercholesterolemia		
Heterozygous form	Total cholesterol 7–13 (275–500)	Prevalence 1/500; high risk of ischemic heart disease and presence of xanthomas in adults
Homozygous form	Total cholesterol >13 (>500)	Prevalence 1/1000,000; high cardiovascular risk and presence of xanthomas from childhood
Pure hypertriglyceridemia		
Familial hypertriglyceridemia	Triglycerides 2.8–8.5 (250–750)	Prevalence 1/1000; increased risk of vascular disease and pancreatitis
Familial lipoprotein lipase or ApoC-II deficiency	Triglycerides >8.5 (>750) (milky plasma)	Prevalence 1–2/1000,0000; possible association with pancreatitis and hepatosplenomegaly
Hypercholesterolemia + Hypertriglyceridemia		
Familial combined hyperlipidemia	Triglycerides 2.8–8.5 (250–750) Total cholesterol 6.5–13.0 (250–500)	Prevalence 1/100; high risk of premature ischemic heart disease
Familial dysbetalipoproteinemia	Triglycerides 4.5–5.6 (400–500) Total cholesterol 10.4–13 (400–500)	Prevalence 1/10,000; high risk of early atherosclerosis and the presence of xanthomas

(early cardiovascular events and dyslipidemia), complete physical examination aimed at searching for xanthomas, xanthelasmas, corneal arch and retinalis lipemia, and the exclusion of secondary forms by appropriate laboratory investigations, which include the evaluation of TSH, renal and hepatic function, and fasting blood glucose. Then, it will be possible to evaluate the diagnosis of a familial form of dyslipidemia.

Familial Hypercholesterolemia

Heterozygous Familial Hypercholesterolemia

The diagnostic pathway is quite complex, and it is generally based on the criteria of the Dutch Lipid Clinic Network (Table 14.11).

Homozygous Familial Hypercholesterolemia

It is a much rare and severe condition. The diagnosis is made based on the finding of:

- TC ≥ 13 mmol/L (≥500 mg/dL)
- Early clinical manifestations (around 10 years of age), which include xanthomas in the tendons of the hands and Achilles tendons
- Premature cardiovascular disease, as early as the first 10 years of age. The diagnosis is usually made in childhood. Early identification of affected children and immediate referral to a specialized center are crucial. Patients should be treated with cholesterol-lowering drugs and, when possible, lipoprotein apheresis.

Table 14.11 Criteria of Dutch Lipid Clinic Network for heterozygous familial hypercholesterolemia diagnosis

Criteria	Points
Family history	
First-degree relatives with premature coronary heart disease (CHD) (<55 years in men, <60 years in women)	1
First-degree relatives with cholesterol >8 mmol/L (310 mg/dL) or >95th percentile of the origin country	1
First-degree relatives with tendon xanthomas and/or corneal arch	1
Children <18 years with cholesterol >6 mmol/L (≥230 mg/dL) or >95th percentile of the origin country	2
Clinical history	
Subject with premature CHD (<55 years in men, <60 years in women)	2
Subject with premature cerebral or peripheral vascular disease (<55 years in men, <60 years in women)	1
Physical examination	
Tendon xanthoma	6
Corneal arch before 45 years	4
Biochemical results	
C-LDL > 8.5 mmol/L (325 mg/dL)	8
C-LDL 6.5–8.4 mmol/L (251–325 mg/dL)	5
C-LDL 5.0–6.4 mmol/L (191–250 mg/dL)	3
C-LDL 4.0–4.9 mmol/L (155–190 mg/dL)	1
DNA analysis	
Known causative mutation in *R-LDL, APOB,* or *PCSK9 genes*	8
Choose only one point per group; the diagnosis is based on the total sum of the points obtained "Certain" diagnosis with score > 8 "Probable" diagnosis with score between 6 and 8; "Possible" diagnosis with score between 3 and 5; "Unlikely" diagnosis with score between 0 and 2	

Combined Familial Hyperlipidemia

It represents an important cause of premature coronary artery disease. It is a complex disease whose phenotype is determined by the interaction among several susceptibility genes and environmental factors. It is characterized by a high phenotypic variability, both inter- and intraindividual based on lipid values (TG, C-LDL, C-HDL, and ApoB). Consequently, it is often misrecognized in clinical practice.

The diagnosis is made following the finding of:

- C-LDL ≥4.15 mmol/L (≥160 mg/dL) and/or triglyceridemia ≥2.25 mmol/L (≥200 mg/dL)
- Documentation of hypercholesterolemia and/or hypertriglyceridemia (multiple phenotypes) in the proband's first- and second-degree relatives, often with phenotypic variability over time (transition from hypercholesterolemia to hypertriglyceridemia, or to mixed forms).

In the absence of documentation on family members, dyslipidemia is strongly suspected in the presence of an anamnestic or clinical or instrumental diagnosis of early atherosclerosis.

It is essential to exclude families in which only hypercholesterolemia or hypertriglyceridemia is present.

Familial Dysbetalipoproteinemia

It is a very rare disease occurring in homozygous for the E2 isoform of ApoE. However, not all subjects with E2/E2 genotype manifest the disease.

The diagnosis is based on the following criteria:

- Cholesterol and triglyceridemia around 400–500 mg/dL
- Presence on electrophoresis of the broad beta band due to the fusion of VLDL and LDL
- The presence of any of the following factors increases the validity of the diagnosis:
  - Tuberous xanthomas
  - Palmar striated xanthomas (yellowish striae in the interdigital folds or on the palmar surface of the hands, to be considered very specific)

Familial Hypertriglyceridemia

The diagnosis is based on the detection of elevated plasma triglyceride levels (TG > 400 mg/dL) in the patient and in at least one first-degree relative, and the exclusion of secondary forms.

Treatment and Therapy

The main goal of treating dyslipidemia is to reduce C-LDL levels.

Treatment involves the correction of eating habits and lifestyle. Only after 3 months of diet and lifestyle modification, the initiation of drug therapy can be evaluated, and it must not be a substitute for but a complement to diet and exercise.

Before drug therapy, it isessential to define the individual global cardiovascular risk. The guidelines recommend calculating the cardiovascular risk using the SCORE (Systemic COronary Risk Estimation) project risk maps.

According to the SCORE system, the risk is defined as the probability of developing a fatal cardiovascular event at 10 years. This probability is estimated based on the following variables: sex, age, systolic blood pressure (PAS), and total cholesterol. The combination of the above variables defines four risk classes: low, moderate, high, and very high.

Subjects with high and very high risk fall into the high and very high-risk classes:

- Documented cardiovascular pathology (previous myocardial infarction, unstable angina, stable angina, patients undergoing bypass surgery, etc.)
- Type I or type II diabetes
- High levels of individual risk factors
- Chronic renal failure

Desirable C-LDL levels, cutoffs for initiating treatment, and type of treatment, lifestyle modification, or drug therapy are established for each risk class (Table 14.12).

C-HDL and TG are not treatment targets, although an increase in C-HDL predicts regression of atherosclerosis, while low C-HDL levels are associated with increased mortality. The desirable level of TG is ≤1.7 mmol/L (150 mg/dL). In contrast, non-HDL cholesterol and ApoB represent secondary treatment targets. Specifically, when available, it is recommended to achieve levels of:

- Non-HDL cholesterol <2.6 mmol/L (100 mg/dL) and <3.4 mmol/L (130 mg/dL) in very high and high cardiovascular risk individuals, respectively
- ApoB <80 mg/dL and <100 mg/dL in very high and high risk individuals, respectively

Lifestyle Modifications to Improve Lipid Profile

Diet is crucial in modifying an individual's lipid profile. In particular, the guidelines recommend a Mediterranean diet, which includes high consumption of fruit, vegetables, and whole grains, a frequent intake of legumes, fish, and low-fat dairy products, while the consumption of sweets, sweetened drinks, and red meat is limited. A strength of the Mediterranean diet is the use of extra virgin olive oil, which has been widely shown to reduce the incidence of major car-

Table 14.12 Therapeutic goals and recommended treatments based on C-LDL levels

Patient's risk class	C-LDL levels mmol/L (mg/dL) to start with: Lifestyle intervention	Pharmacological therapy	Therapeutic goals C-LDL mmol/L (mg/dL)
Low	≥4.9 (≥190)	≥4.9 (≥190) if lifestyle intervention is insufficient	<3.0 (<115)
Moderate	2.6 (≥100)	2.6 (≥100) if lifestyle intervention is insufficient	<3.0 (<115)
High	≥1.8 (70)	1.8–2.6 (70–100) if lifestyle intervention is insufficient ≥2.6 (≥100) always	<2.6 (<100) or a reduction of at least 50% if the baseline is between 2.6 and 5.2 (100–200)
Very high	Always recommended for any C-LDL value	≥1.8 (70) always	<1.8 (<70) or a reduction of at least 50% if the baseline is between 1.8 and 3.5 (70–135)

diovascular events. Fats are an important element of the diet, but it is essential to distinguish between saturated and unsaturated fatty acids. Saturated fatty acids are solid at room temperature (butter, lard, bacon, margarine) and significantly impact C-LDL levels, so their consumption should be limited or avoided.

Unsaturated fatty acids, also called essential fatty acids because they are not synthesized in our bodies, are liquid (oil) and are divided into monounsaturated, such as oleic acid contained in olive and peanut oil, which promote the formation of C-HDL, and polyunsaturated omega-6 series (linoleic acid), contained in sunflower oil, corn, grape seeds and sensitive to heat (so always use raw), polyunsaturated omega-3 series (linolenic acid), found in dried fruit, anchovies, mackerel, herring, and salmon.

Daily salt intake should be <5 g, limiting its use in food preparation and choosing fresh or frozen unsalted foods; many prepackaged foods have a high salt content.

Weight loss and regular physical activity (at least 30 min a day, every day), although to a lesser extent than diet, have positive effects on C-LDL levels. Finally, alcohol consumption should be limited (up to 20 g per day for men and 10 g per day for women) and smoking should be avoided.

Drug Therapy

Statins, inhibitors of HMG-CoA reductase, the key enzyme in the cholesterol biosynthetic pathway, are the drugs of choice for the treatment of hypercholesterolemia. Statins can lower C-LDL levels from 20% to 55% depending on the dose and type of statin administered, with a linear dose-response effect. A significant proportion of patients with elevated C-LDL levels require additional treatment. Moreover, some patients may be nontolerant to statins or may not tolerate high doses. In these cases, polytherapy should be considered, combining other drugs such as bile acid sequestration resins, ezetimibe (cholesterol absorption inhibitor), and PCSK9 inhibitors.

Treatment should lead to a 30-40% reduction in C-LDL levels .

In the treatment of hypertriglyceridemia, the first step is to consider the possible causes of hypertriglyceridemia and assess total cardiovascular risk. Although the cardiovascular risk increases with fasting TG levels >1.7 mmol/L (150 mg/dL), the use of drugs lowering TG levels is recommended only in high-risk subjects, i.e., when TGs are >2.3 mmol/L (200 mg/dL) and cannot be reduced by lifestyle modification alone. Pharmacological treatments include the use of statins, fibrates, PCSK9 inhibitors, and omega-3.

Recommended Readings

Averna M, Brignoli O, Bucci M et al (2013) Linee Guida cliniche per la prevenzione della cardiopatia ischemica nella ipercolesterolemia familiare. Giorn It Ateroscl Suppl 1

Bachorik PS, Ross JW (1995) National Cholesterol Education Program recommendations for measurement of low-density lipoprotein cholesterol: executive summary. Clin Chem 41:1414–1420

Catapano AL, Gaddi A, Graziani MS et al (1998) Linee Guida per la Diagnostica di Laboratorio delle Dislipidemie. Biochim Clin 22(3):145–153

Catapano AL, Graham I, De Backer G et al (2016) 2016 ESC/EAS Guidelines for the Management of Dyslipidaemias: The Task Force for the Management of Dyslipidaemias of the European Society of Cardiology (ESC) and European Atherosclerosis Society (EAS) Developed with the special contribution of the European Association for Cardiovascular Prevention & Rehabilitation (EACPR). Atherosclerosis 253:281–344. https://doi.org/10.1016/j.atherosclerosis.2016.08.018

Deb A, Caplice NM (2004) Lipoprotein(a): new insights into mechanisms of atherogenesis and thrombosis. Clin Cardiol 27:258–264

Examination Survey Research Group (2015) Cardiovascular health in Italy. Ten-year surveillance of cardiovascular diseases and risk factors: Osservatorio Epidemiologico Cardiovascolare/Health Examination Survey 1998–2012. Eur J Prev Cardiol 22(2 Suppl):9–37. https://doi.org/10.1177/2047487315589011

Feingold KR, Grunfeld C (2000–2015, June 10) Introduction to lipids and lipoproteins. In: De Groot LJ, Beck-Peccoz P, Chrousos G (eds) MDText.com, Inc., South Dartmouth (MA)

Filippi A. La nota 13 e la diagnosi delle dislipidemie familiari. http://www.sisalombardia.it/sisa_novitaletteratura_200903/PTA4_3_La_nota13_MMG.pdf

Graziani MS, Catapano A, Frascatore S (1993) Trigliceridi. Giorn It Chim Clin 15:1448–1455

Graziani MS, Ceriotti F, Zaninotto M et al (2016) La diagnostica di laboratorio delle dislipidemie. Documento di Consenso di: SocietaÃÄ Italiana di Biochimica Clinica e Biologia Molecolare Clinica-Medicina di Laboratorio (SIBioC), SocietaÃÄ Italiana per lo Studio dell'Aterosclerosi (SISA), SocietaÃÄ Italiana di Medicina

Generale e delle Cure Primarie (SIMG), Associazione Nazionale Medici Cardiologi Ospedalieri (ANMCO). Bioch Clin 40:338–346
Gulizia MM, Colivicchi F, Ricciardi G et al (2016) Colesterolo e Rischio Cardiovascolare: percorso diagnostico-terapeutico in Italia. Documento di consenso intersocietario. Ital Cardiol 17:24–25
Ramasamy I (2014) Recent advances in physiological lipoprotein metabolism. Clin Chem Lab Med 52(12):1695–1727. https://doi.org/10.1515/cclm-2013-0358
Rosenson RS, Brewer HB Jr, Ansell BJ et al (2016) Dysfunctional HDL and atherosclerotic cardiovascular disease. Nat Rev Cardiol 13(1):48–60. https://doi.org/10.1038/nrcardio.2015.124
Usifo E, Leigh SE, Whittall RA et al (2012) Low-density lipoprotein receptor gene familial hypercholesterolemia variant database: update and pathological assessment. Ann Hum Genet 76:387–401

Hematological Diagnostics

15

Gian Cesare Guidi

Definition of Anemia

Anemia is a blood disorder characterized by a reduced ability of the blood to carry oxygen due to a decrease in the mass of erythrocytes and, therefore, in the oxygen transporter they contain, hemoglobin. Symptoms accompanying anemia are generally dependent on the extent of the reduction of erythrocyte mass, and consequently of hemoglobin, and on the time within which it occurred. Specifically, a slow progression toward anemia is compensated and tolerated up to limits of erythrocyte mass and hemoglobin that would not be tolerated if they were reached in a short time, for example, during massive hemorrhages. In such a compensatory effect, immediate and delayed mechanisms are of great importance. These include the increase in heart rate, peripheral vasoconstriction, and the increase in 2,3-diphosphoglycerate, which is accompanied by the shift of the oxygenation curve of hemoglobin to the right, on the one hand, and by the secretion of erythropoietin, in turn, stimulated by the combined action of the hypoxia-inducible factor (HIF-1, -2), on the other hand.

It is, therefore, a complex picture in which the laboratory plays a fundamental diagnostic role in the identification of the primary causes of anemia (congenital or acquired), in the evaluation of the current status and severity of the anemia, and in the therapeutic follow-up of the anemic patient.

The World Health Organization (WHO), in view of the high incidence of anemia, defined as "low hemoglobin concentration in the blood" according to the values in Table 15.1, has implemented a series of actions aimed at reducing its global incidence.

Indeed, anemia has been shown to be a public health problem that affects low-, middle-, and high-income countries and has significant adverse health consequences as well as negative effects on social and economic development.

G. C. Guidi (✉)
School of Medicine e Surgery, University of Verona, Verona, Italy
e-mail: giancesare.guidi@univr.it

Table 15.1 Hemoglobin values (g/L) for the diagnosis of sea level anemia

		Anemia		
Population	Nonanemia	Mild[a]	Moderate	Serious
Children from 6 months to 5 years	≥110	100–109	70–99	≤70
Children 5–11 years old	≥115	110–114	80–109	≤80
Children 12–14 years	≥120	110–119	80–109	≤80
Nonpregnant women ≥15 years	≥120	110–119	80–109	≤80
Pregnant women	≥110	100–109	70–99	≤70
Men ≥15 years old	≥130	110–129	80–109	≤80

Modified from: WHO/WMNIS (2016)
[a]It is useful to keep in mind that even in the case of mild anemia, an iron deficiency can already be severe and present for some time

Although the most reliable indicator of anemia is the concentration of hemoglobin in the blood, its measurement alone cannot indicate the cause. Anemia can result from several causes, and iron deficiency is the most important. About 50% of cases of anemia are due to iron deficiency, but the percentage probably varies among population groups in different areas depending on the local conditions, primarily the quality of the diet. Other causes of anemia include other micronutrient deficiencies (e.g., folate, riboflavin, vitamin A, and vitamin B12), acute and chronic infections (e.g., malaria, cancer, tuberculosis, and HIV), and inherited or acquired diseases affecting hemoglobin synthesis, red blood cell production, or red blood cell survival (e.g., hemoglobinopathies).

Complete Blood Count (CBC)

Anemia can be identified by the CBC exam, which consists of a group of tests that evaluate the cellular components of the peripheral blood: red blood cells (or erythrocytes), white blood cells (or leukocytes), and platelets.

M. Ciaccio (ed.), *Clinical and Laboratory Medicine Textbook*, https://doi.org/10.1007/978-3-031-24958-7_15

Hemoglobin Measurement

The reference method for measuring hemoglobin is the well-known method based on the conversion of hemoglobin to cyanmethemoglobin and subsequent reading of the stable derivative by spectrophotometry. The blood sample must be collected in a tube containing EDTA-K_2 as the anticoagulant of choice.

Hematocrit Measurement

According to the recommendation of the International Council for Standardization in Hematology (ICSH), the measurement of hematocrit (Hct) is a "surrogate" reference method, traceable to the reference method. Indeed, the recommendation still refers to packed cell volume (PCV), according to the previous reference method, which was based on the centrifugation of a blood sample in a borosilicate capillary and subsequent measurement of the ratio of cell mass to total blood volume in the capillary. The "surrogate" method makes the evaluation of hematocrit simpler than the previous method. It provides two measurements of hemoglobin, one on the whole blood sample and the other on the red blood cells of the same sample packed after centrifugation. The ratio between the two measurements of hemoglobin provides the reference value of hematocrit. The reference standard thus identified is applied to the manufacture and certification of capillary tubes for microhematocrit and to the calibration of automatic instruments for hematology laboratory tests. Current hematology instruments actually measure hematocrit on the basis of other direct parameters, for example, cell volume or differential absorption characteristics of blood in the infrared.

Reference ranges for hematocrit are reported in Table 15.2.

Cell Counts

Nowadays, all laboratories use electronic counting instruments based on different principles developed in research and applied by various manufacturers.

The degree of complexity of the different instruments is a function of the use for which each instrument is intended. It is possible to distinguish basic analyzers, which provide the following basic parameters: red blood cells, leukocytes, and platelets; measurement of hematocrit; measurement of the main erythrocyte parameters (MCV, MCH, MCHC, RDW); and sometimes of platelet parameters (PCV, PDW). Additionally, basic instruments can provide differential leukocyte counts over 3–5 classes. These instruments can be used for urgent tests and can analyze 60–90 samples per hour. The technological advancement led to the development of new-generation hematological instruments that can provide increasingly detailed and specific measurements of numerical, qualitative-quantitative, and morphological alterations observable in many hematological pathologies, both of the erythrocyte series and of the leukocyte and platelet series. To better understand the possible fields of application of modern instrumentation, it is necessary to provide some basic information on the principles of operation of the three major technologies: (1) electrical impedance; (2) flow cytometry; and (3) fluorescent flow cytometry. The combination of the three technologies and the features chosen by the different manufacturers characterizes the various instruments. It should also be borne in mind that the instruments use reagents capable of modifying certain cellular characteristics in order to better detect their distinctive features (e.g., cytoplasmic lysers to better differentiate leukocyte nuclei).

Table 15.2 Hematocrit: reference ranges

Population	IS unit L/L	Traditional units %
Children	0.53–0.69	53–69
Women	0.36–0.46	36–46
Men	0.40–0.54	40–54

Electrical Impedance

The principle of electrical impedance was discovered by W.H. Coulter in 1953 and applied to the counting of microscopic particles suspended in liquids. A blood sample, appropriately diluted in an isotonic electrolyte solution, is aspirated through a small capillary hole (diameter ~100 μm) into a tube immersed in the suspension of cells. Two electrodes are also immersed in the same isotonic solution. Specifically, one is internal, and one is external to the suction tube, and a potential difference is applied to them to generate a weak electric current conducted by the ions of the electrolyte solution that pass through the capillary slit. When a cell passes through the same hole, there is a reduction in the intensity of the current, which can be detected and whose magnitude depends on the volume of the cell. Measurement of the number and magnitude of current intensity reductions provides the number and size of cellular elements present in the solution (Fig. 15.1). Impedance analysis, therefore, provides the erythrocyte cell count and can group the leukocytes according to size into three classes: granulocytes, lymphocytes, and monocytes. The impedance-based technique allows the counting of about 10,000 elements per minute (with variations depending on the instruments). Since a typical counting period ranges from 40–45 s to 1 min, it is clear that the large final number of elements counted is the guarantee of accuracy that is inherent in impedance-based counting. The only correction that is made automatically by the instruments concerns the statistical coincidence of the passage of two

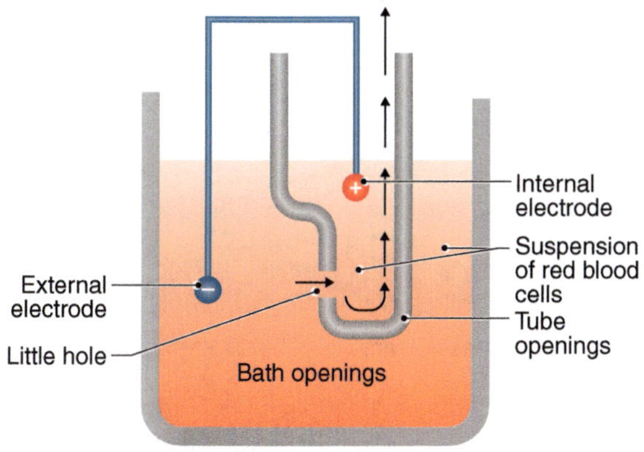

Fig. 15.1 Diagram of the electrical impedance cell counter (section). The cell suspension is drawn (arrows directed upward) from the outer container (bath apertures) to the inner container (tube apertures) through a capillary bore (apertures). The current flow that is maintained between the inner (+) and outer (−) electrodes passes only through the small hole; it is constant until a cell also passes through it. Since the current is weak and the cell has dielectric properties this results in a momentary reduction in charge flow, the magnitude of which is proportional to the volume of the cell. (Copyright EDISES 2021. Reproduced with permission)

elements at the same time, which increases as the number of cells in the sample increases (e.g., a sample with two million erythrocytes has a lower coincidence than one with four million erythrocytes). A special statistical algorithm corrects the count for this predictable phenomenon, which would otherwise result in a default assessment of the number of cells.

Flow Cytometry

Flow cytometry is based on the detection of the absorption and diffraction of light at certain angles and the differential evaluation of the signals received. Current instruments operate with laser lights at certain wavelengths, which can guarantee the constancy of the light source. Each stream of cells suspended in solutions in which reagents chosen to identify distinct characteristics are premixed is passed through a thin optical glass cell so that the cells are aligned in a single row during the transit. The beam of light passing through each cell is partly absorbed (photons passing through the cell) and partly deflected (scattering photons impinging at an angle on the cell or being deflected by structures within the cell, such as the nucleus and granules) (Fig. 15.2). The detection of the magnitude of absorption and the signal from different angular degrees of deflection are useful elements to diversify individual classes of cells by number, size, and peculiar characteristics (Fig. 15.3). Flow cytometry is widely used in modern hematological instruments in association with electrical impedance so as to guarantee high counting accuracy and allow correct detection of all the cellular elements of the blood, including both immature elements and elements found in hematological disorders (Fig. 15.4).

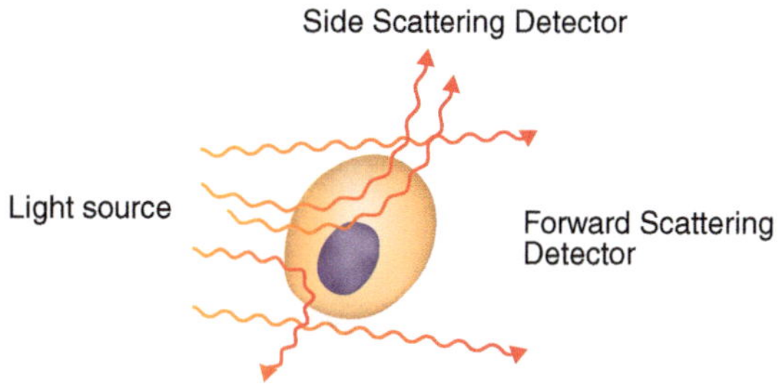

Fig. 15.2 Light scattering is observed when a beam of light impinges on a cell without being absorbed. Part of the photons are deflected with a low angular degree (forward scattering) in relation to the size of the cell, part is deflected with a high angular degree (side scattering) in relation to the contents inside the cell. Appropriate detectors collect the signal obtained from the cell flow. (Copyright EDISES 2021. Reproduced with permission)

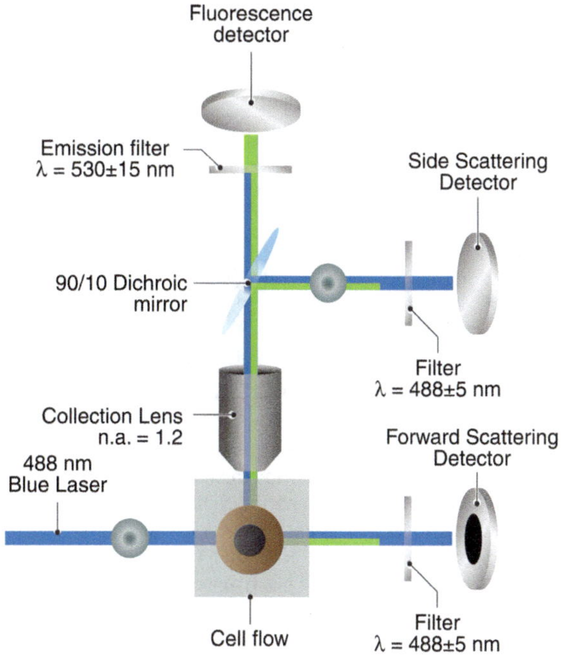

Fig. 15.3 Flow cytometer scheme. Each cell passing through the laser light beam generates different signals that are collected by appropriate detectors. The combination of the signals, through a special algorithm, allows each cell to be assigned to its own distinctive class. (Copyright EDISES 2021. Reproduced with permission)

Fluorescence Cytometry

An extension of the principle of flow cytometry is present in fluorescence cytometry, in which the flow of cells is suspended in solutions containing fluorochromes that bind specifically to nuclear structures or contain nucleic acids,

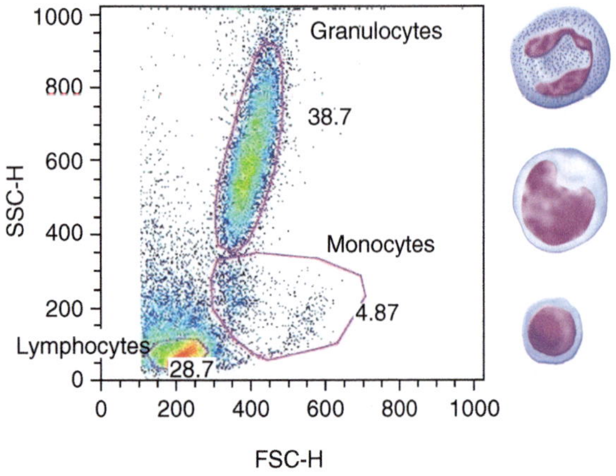

Fig. 15.4 Scattering graph. Each cell class is reported in the graph to the area identified by the intersection of signals from FSC-H (forward scattering) and SSC-H (side scattering). The graduation of colors within each area is in relation to the cell density

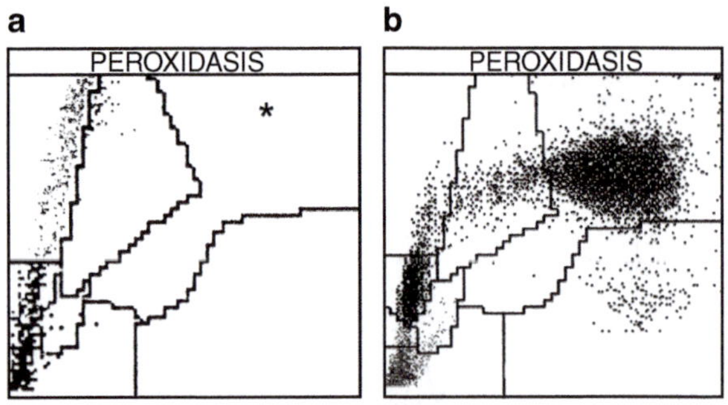

Fig. 15.5 Myeloperoxidase activity graph. In abscissa, the peroxidase activity increases toward the right, and the number of cellular elements increases toward the top. Shown on the right is a normal specimen in which cellular activity in neutrophil granulocytes is evident in the upper-right portion; lower-right is eosinophils; and middle-upper is monocytes. The middle-left band has lymphocytes, which lack myeloperoxidase activity. The bottom left are platelets lacking activity. On the left is a sample with myeloperoxidase deficiency

allowing further elements of characterization of certain classes of cells (e.g., reticulocytes, platelets).

The manufacturers personalize their instruments by varying the combinations of the counting principles or by modifying the performance with the intention of introducing improvements to the diagnostic possibilities. Mention should be made of the use of reagents that detect the peroxidase activity typical of cells of myeloid origin that allows, for example, the distinction between polymorphonucleates and monocytes (Fig. 15.5). Others have made the measurement of hemoglobin more sensitive and reduced analytical interference, and still others have employed flow cytometry systems with multicolor fluorescence and multiangle scattering detection to increase the specificity of detection of different leukocyte classes (Fig. 15.6).

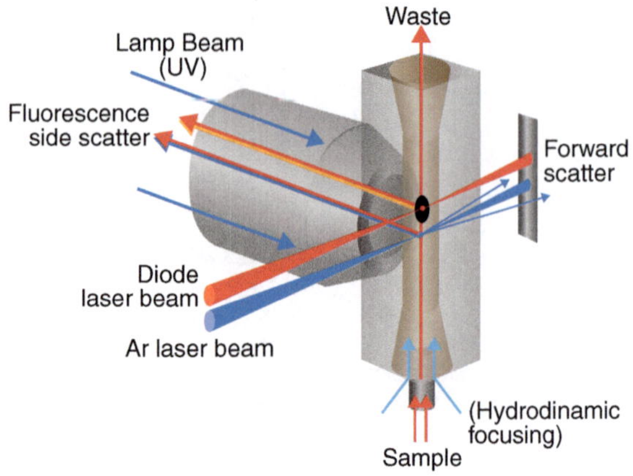

Fig. 15.6 Graph of dual laser beam flow cytometric cell with multicolor fluorescent signal emission. (Copyright EDISES 2021. Reproduced with permission)

Erythrocyte Indices and Distribution Measurements

The erythrocyte indices are derived from the measurements described above. They are mainly used as first-line diagnostic indicators for different types of anemia.

Mean Corpuscular Volume

The mean corpuscular volume (MCV) is the value obtained from the ratio of the Hct to the number of red blood cells:

$$MCV = Hct / RBC \times 10$$

Values are normally expressed in femtoliters (1 fL = 1 × 10^{-15} L), according to the International System (IS), or in microliters (μL), according to traditional units. The reference range is between 80 fl and 100 fL, with variations depending on age and status (e.g., pregnancy). Anemic conditions are defined according to the measured MCV value as microcytic if MCV <80 fL, normocytic if MCV is between 80 fl and 100 fL, and macrocytic if MCV >100 fL (Table 15.3).

Mean Corpuscular Hemoglobin

The mean corpuscular hemoglobin (MCH) is the value obtained from the ratio of hemoglobin to the number of red blood cells and expresses the hemoglobin content as mass-weight.

$$MCH = Hb / RBC \times 10$$

Values are normally expressed in picograms (1 pg = 1 × 10–12 g) according to IS. The reference range is 27–31 pg, with variations depending on age and status (e.g., pregnancy). Anemic conditions are defined according to the measured MCH value as hypochromic if MCH <27 pg, normochromic if MCH is between 27 pg and 31 pg, and hyperchromic if MCH >31 pg.

Table 15.3 Classification of anemia according to mean corpuscular volume (MCV) and morphology

Microcytic (MCV < 80 fL)	Normocytic (MCV 80–100 fL)	Macrocytic (MCV > 100 fL)
Generally microcytic Sideropenia (iron deficiency) Thalassemias Hereditary sideroblastosis Occasionally microcytes Chronic conditions Hemoglobinopathies	Generally normocytic Hypoproliferative anemia Anemia secondary to malignancy Refractory myelodysplastic anemia Hemolytic anemia Hemoglobinopathies Post-hemorrhagic anemia Chronic conditions Acquired sideroblastosis Occasionally normocytic Early stage iron deficiency	Generally macrocytic Folic acid deficiency Vitamin B12 deficiency Liver diseases Hemolytic anemia Post-hemorrhagic anemia Occasionally macrocytic Hypoproliferative anemia Refractory anemia

Mean Corpuscular Hemoglobin Concentration

Mean corpuscular hemoglobin concentration (MCHC) is the value obtained from the ratio of hemoglobin to hematocrit and expresses the concentration of hemoglobin compared to the total erythrocyte volume.

$$\mathrm{MCHC} = \mathrm{Hb} / \mathrm{Ht} \times 100$$

The values are normally expressed in g/L according to the IS or in percentage according to the traditional system. The reference range is 320–360 g/L (IS) or 32–36 g/dL or % (traditional system). Its use as a diagnostic index is limited to conditions affecting the erythrocyte membrane, such as an increase in its outer surface due to the adhesion of cold agglutinins to the membrane or a decrease in membrane thickness, which can be found in some clinical conditions, such as hereditary spherocytosis (HS) and elliptocytosis. In addition, increased MCHC is found in pathological conditions that cause the concentration of erythrocyte content to increase due to dehydration, such as some hemoglobinopathies (mainly HbS and HbC).

Red Cell Distribution

Red cell distribution width (RDW) is a measure of the range of variation in erythrocyte volumes in terms of a coefficient of variation (CV) in percent. RDW can be reported statistically as a coefficient of variation (CV) and/or standard deviation (SD).

- RDW-CV, which represents the coefficient of variation in percent with respect to MCV, is expressed by the formula: RDW = $SD_{(MCV)}$/MCV × 100.
- RDW-SD, which reports in femtoliters the true value of the standard deviation obtained from the distribution of erythrocyte volumes measured at the height of 20% on the curve itself (Fig. 15.7).

The reference range is between 12% and 15% (RDW-CV) and between 40 fl and 55 fL (RDW-SD), with variations depending on age and status (e.g., pregnancy). In evaluating this index, two factors that influence the variation of erythro-

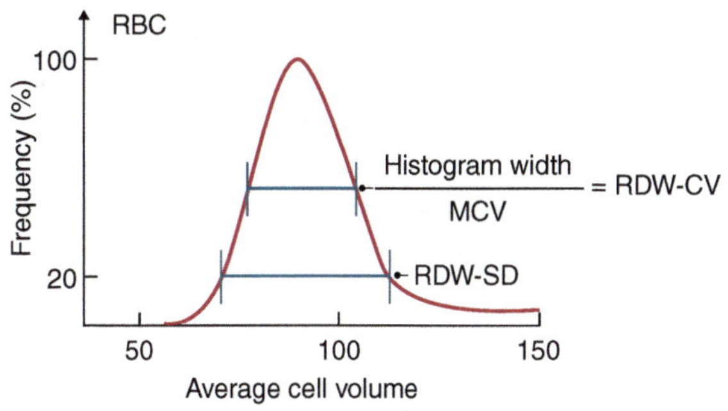

Fig. 15.7 Graph representing the distribution of the erythrocyte population with RDW-CV and RDW-SD. (Copyright EDISES 2021. Reproduced with permission)

cyte volumes within physiological ranges should be considered:

- Variability of production by the bone marrow, which is kept within narrow limits and controlled at the site by eliminating, through the medullary reticuloendothelial system, cells that cannot be released into circulation.
- The progressive aging of the circulating erythrocyte population is accompanied by a reduction in cell volume due to dehydration. Young red blood cells, since the reticulocyte state, are significantly more voluminous than red blood cells that have reached the end of their useful life, and, passing from one extreme to the other, a progressive and regular reduction in volume is observed. In many conditions of anemia, often high values of RDW are observed, so this index is used in association with MCV for a first differentiation among anemias that present similar characteristics; the most typical case is the differential diagnosis between two common microcytic anemias, the iron deficiency anemia and the β-thalassemic trait anemia, in which, besides a common reduced MCV, a high RDW value is more indicative of the first condition than the second one. Also, macrocytic anemia due to the deficiency of folate and vitamin B12 is characterized by an increase in RDW and an elevated MCV. However, it must be kept in mind that this is an index with relatively modest

specificity and that, therefore, more specific tests must be associated with it. In recent years, interest in RDW has increased because an association between RDW and cardiovascular disease has been described. Specifically, increased levels of RDW are a predictor of adverse outcomes in patients with cardiovascular diseases. Additionally, high RDW values have been described in several clinical conditions, including thromboembolic diseases, various cancers, diabetes, pulmonary diseases, kidney diseases, liver diseases, and other chronic disorders. The pathophysiological mechanisms underlying the RDW increase in these clinical conditions are not entirely understood, but erythrocyte eriptosis or apoptosis have been hypothesized. Eryptosis is a suicidal mechanism of the cell triggered by the entry of calcium ions, oxidative stress, alterations in the regulation of cytokines, and other mechanisms that occur in many chronic inflammatory diseases (e.g., heart and kidney failure), metabolic diseases (e.g., diabetes), and others such as neoplasms.

Platelet Indices and Platelet Count

Modern hematological instruments calculate several indices related to platelets, which provide useful information regarding the morphology and turnover of these cellular elements.

Mean Platelet Volume

The mean platelet volume (MPV) population, unlike erythrocytes, does not have a Gaussian distribution but is rather shifted to the left so that some graphs represent the curve after log transformation. The reference interval is between 7 fl and 12 fL. Newly produced platelets from the marrow are larger in size than older platelets, and greater MPV is observed when platelet production increases. MPV may be useful as an index to assess the extent of platelet production when accelerated destruction is suspected. In the latter case, increased MPV can be found in several inflammatory gastrointestinal diseases, thrombotic thrombocytopenic purpura (TTP), Bernard–Soulier syndrome, myeloproliferative diseases, preeclampsia, and some cancers. In many of these clinical conditions, an increase in MPV is accompanied by a significant decrease in the number of circulating platelets due, as mentioned, to early destruction in the circulation. On the contrary, a decrease in MPV can be observed in some genetic thrombocytopenias (Wiskott–Aldrich syndrome) and in bone marrow aplasia.

Platelet Distribution Width

The platelet distribution width (PDW) represents a measure of the range of variation of platelet volumes in terms of the percentage coefficient of variation. Similarly to RDW, PDW can be expressed both as a coefficient of variation(PDW-CV) or standard deviation (PDW-SD):

- PDW-CV = $SD_{(MPV)}$/MPV× 100
- PDW-SD, which reports in femtoliters the true value of the standard deviation obtained from the platelet volume distribution measured at the height of 20% on the curve itself.

The mode of expression depends on the type of instrumentation. The reference range for PDW-CV reported in the literature is 10.0–17.9%, with a mean value of 13.3%, depending on the instrumentation used. PDW generally increases in relation to an increase in the number of young circulating platelets, as seen in idiopathic thrombocytopenic purpura (or immune thrombocytopenia or immune thrombocytopenic purpura [ITP]) or macrothrombocytopenia. Although PDW is significantly increased in ITP compared to controls, its diagnostic power is limited when analyzed on its own. PDW should be considered a useful supporting parameter in association with MPV and immature platelet fraction, a parameter known as IPF. For example, PDW should be evaluated together with other platelet parameters for differentiating idiopathic thrombocytopenic purpura from Wiskott–Aldrich syndrome in patients with macrothrombocytopenia. In ITP, young platelets are produced with a larger size, and, thus, MPV is often elevated along with a high PDW. On the other hand, patients with Wiskott–Aldrich syndrome have uniformly large platelets, thus a high MPV but a normal PDW.

Immature Platelet Fraction

Immature platelet fraction (IPF) is a parameter recently introduced in automated analyzers. It provides the fraction of cross-linked platelets on the total number of platelets, that is, the youngest platelets, which, similar to reticulocytes, have a cytoplasmic reticular structure due to polyribosomal RNA. The reference range is between 1% and 8% of the total platelets, and the value increases when the production of platelets by the marrow increases. It is therefore indicative of the efficiency of the bone marrow and, in particular, the proportion of megakaryocytes produced. The parameter is useful in the diagnosis and treatment of thrombocytopenia, such as idiopathic thrombocytopenic purpura and thrombotic thrombocytopenic purpura, as well as in the differentiation between forms with destruction in the circulation and those due to suppression of bone marrow production.

Plateletcrit

The plateletcrit (PCT), similar to the hematocrit for erythrocytes, represents the portion of the volume occupied by platelets in the blood as a whole. The value of the PCT is expressed as a percentage and is calculated according to the following formula:

$$PCT = \text{platelet count} \times MPV / 10{,}000$$

The reference range for PCT is 0.22–0.24%. In normal subjects, the total mass of platelets is a constant ratio between the number of platelets and MPV, which therefore shows an inverse correlation between them. This ratio is altered in some conditions, including physical efforts, consumption of substances of abuse, such as alcohol and smoking.

Other Indexes

- Percentage of large platelets (>12 fL) over total platelets (platelet larger cell ratio, [P-LCR]), with normal values between 15% and 35%. Since the largest platelets are the youngest and most active, the index would provide an indirect estimate of platelet activity.
- The mean platelet component (MPC) indirectly provides an evaluation of the state of platelet activation through the scatter of incident light, whereby a lower deflection corresponds to lower granule content and, therefore, is indicative of the previous activation.
- Platelet component distribution width (PCDW) and mean platelet mass (MPM) are linked to the P-LCR and MPC.
- Immature platelet fraction (IPF) is a percentage index obtained by detecting the fluorescent signal emitted by platelets due to the presence of RNA within them. This is obtained in the same flow channel, which contains a special dye to detect the RNA, which is indicative of the percentage of maturation of reticulocytes. IPF increases with increased platelet production and decreases in thrombocytopenia due to reduced thrombocytopoiesis.

Reticulocytes and related Indices

Reticulocyte count is performed by microscopy or by electronic instruments. In both cases, the distinctive marker is polyribosomal RNA, which is well represented in reticulocytes. Preparation for microscopic counting is performed by supravital staining of a part of incoagulable blood suspended in an equal part of an isotonic solution of brilliant cresyl blue. Other supravital stainings are the new methylene blue and Azur B. Also, dyes for fluorescence microscopy, such as acridine orange and auramine, are commonly used. The observation is done by immersion (1000×) on a freshly swiped slide (Fig. 15.8). The reticulocytes observed are compared to 1000 erythrocytes, and their number is expressed as a percentage; a CV of 10% is acceptable, but this value must consider the variability among operators. Therefore, in practice, even higher values are reported to the point of invalidating their clinical-diagnostic significance. Although the advent of automatic instruments has displaced microscopic counting, it is still used in less-developed healthcare facilities.

Automated instruments allow not only obtaining a more accurate and precise reticulocyte count than microscopy-based methods but also measuring reticulocyte characteristics that cannot be captured by microscopic observation. In the automatic procedure, the blood is incubated with an isotonic solution of a dye, which crosses the cytoplasmic membrane and binds to RNA; the detection of reticulocytes is carried out by colorimetry or fluorescence through the use of supravital dyes, such as the new methylene blue, or fluorescent dyes, such as thiazole orange, osazine 750, and others.

The reticulocyte count allows us to evaluate the production rate and, indirectly, the replacement rate of the erythrocyte population. It can be reported as an absolute count or as a percentage of reticulocytes relative to erythrocytes. In this way, it is possible to distinguish between anemias of different origins, such as those resulting from hemolysis or other destructive processes, and those due to deficient production or deficiency of essential factors for erythropoiesis. The reticulocyte count allows determining not only if their production is qualitatively increased during anemia but also if the entity of increased production is adequately sustained and leads to the recovery of the state of normality. On the other hand, a decrease in the reticulocyte count can be a consequence of conditions such as deficiency states, bone marrow insufficiency, and cancer (Table 15.4).

The reference range for the adult percentile count is 0.5–1.5%. The absolute reticulocyte count is obtained by calculating its value over the total number of red blood cells; for example, in a subject with 4,800,000/fL erythrocytes and 1.5% reticulocytes: 4,800,000/(100 × 1.5) × 72,000/fL.

Hematological instruments automatically provide this calculation. If there is a state of anemia, the percentage of reticulocytes does not accurately reflect the true responsiveness of the bone marrow to the anemic state. It is necessary

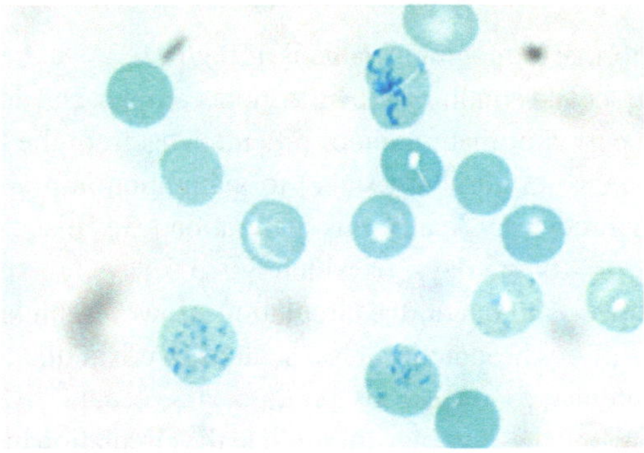

Fig. 15.8 Reticulocytes after supravital staining

Table 15.4 Conditions accompanying changes in the reticulocyte count

Increase in count from increased production	Decreased count from decreased production
Post-hemorrhage (e.g., trauma, gastrointestinal hemorrhages, obstetric-gynecological hemorrhages)	Iron deficiency, vitamin B12 deficiency, folic acid deficiency
Post-hemolysis (e.g., hemolytic anemia, neonatal hemolytic disease)	Decreased production of erythropoietin (e.g., in chronic kidney failure)
Response to ad hoc treatments (e.g., administration of iron, vitamin B12, folic acid)	Aplastic anemia and/or bone marrow failure syndromes
Recovery of bone marrow function after chemotherapy/irradiation	Chemotherapy and/or post-irradiation
Recovery of bone marrow function after transplantation	Bone marrow replacement from benign conditions (e.g., metabolic and *storage diseases,* infections, sarcoidosis) and/or malignant (e.g., leukemias, lymphomas, metastatic tumors)

to correct the percentage value according to the hematocrit value. While it appears evident that the absolute value does not need any correction as it expresses the actual number of reticulocytes, the percentage value needs correction for two main reasons: one intrinsic to the data itself and the other due to the period of maturation of the reticulocytes.

Reticulocyte Production Index

The first correction is performed by the reticulocyte production index (RPI) or corrected reticulocyte count using the following calculation:

$$\text{RPI} = \text{Reticulocyte Count} \times \text{patient hematocrit} / \text{Hematocrit } 45\%.$$

A hematocrit of 45% is taken as a reference. The RPI is normally between 0.5% and 2.5%. Alternatively, there are formulas based on hemoglobin, such as the following:

$$\text{RPI} = \text{Reticulocyte Count} \times \text{patient hematocrit} / \text{normal hemoglobin}$$

A normal hemoglobin value is 150 g/L.

The second condition is a little more complex and depends on the period of maturation of reticulocytes from the loss of the nucleus (erythroblast stage) to maturation into circulating erythrocytes. Usually, this maturation takes place over a period of 4.0–4.5 days, of which 3.0–3.5 days are spent in the marrow and one in the circulation. However, under conditions of erythropoietic stress, such as hypoxia due to anemia, premature release of reticulocytes occurs from the marrow, and the days of maturation in the circulation increase up to 2–3. This leads to increased reticulocyte counts that do not accurately reflect normal production and release kinetics.

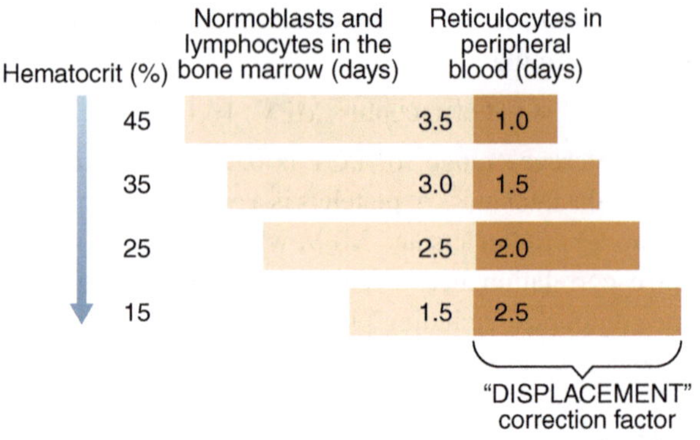

Fig. 15.9 The correction factor is the divisor to be applied to the percent count to modify the effect due to the lengthening of circulating maturation in case of anemia. (Copyright EDISES 2021. Reproduced with permission)

Graphs such as the one shown in Fig. 15.9 have been proposed to correct this effect.

Immature Reticulocyte Fraction

In the evaluation of anemia, the state of maturation of reticulocytes can introduce elements of complexity when a correct meaning must be attributed to their count. However, it provides further diagnostic opportunities that have been progressively perfected with automation. The microscopic observation gives the opportunity to detect different entities of staining in relation to the progressive reduction of RNA until its disappearance when the reticulocyte becomes an erythrocyte. To define this phenomenon, as evidenced very well by automatic instruments, an index called immature reticulocyte fraction (IRF) has been proposed and adopted internationally. IRF considers the reticulocytes subdivided into populations of different maturation degrees. Generally, instruments divide reticulocytes into three populations of different immaturities: low, medium, and high. The IRF index groups the fractions of medium and high immaturity. The reference range is 0.11–0.38% (percentage of medium and high immature reticulocytes in the total erythrocyte count), with possible variations related to the instrumentation used. The parameter represents an early and sensitive index of bone marrow erythropoietic activity, with the following main diagnostic features:

- After marrow transplantation, IRF may demonstrate successful engraftment earlier than other laboratory parameters, including absolute neutrophil granulocyte count. An increase in IRF of more than 20% over the post-transplant value is indicative of successful erythroid engraftment.
- IRF is a sensitive and early index of the recovery of bone marrow activity after chemotherapy treatment.
- IRF, measured in patients with anemia associated with chronic renal failure, neoplasms, infections, and chronic

Table 15.5 Clinical conditions using the combination of immature reticulocyte fraction (IRF) and reticulocyte count

Clinical condition	IRF	Absolute reticulocytes count
Medullary aplasia	↓	↓
Early erythropoietic response following anemia or after rooted bone marrow transplantation	↓/≈	↓
Response to erythropoietin or early post-hemorrhage recovery or post-radiation treatment	↑	↑
Hemolytic anemia or bleeding	↑	↑

conditions, is an early and reliable index of response to erythropoietin therapy.
- In surgical procedures with preservation of the patient's blood, it may be useful to monitor the response to erythropoietin.
- It can be used to monitor the effectiveness of treatments for deficient anemias, such as those caused by iron, vitamin B12, and folic acid deficiency.

Evaluation of IRF along with reticulocyte counts may provide additional insight, as shown in Table 15.5.

Reticulocyte Hemoglobin Content

Hemoglobin content in reticulocytes (reticulocyte hemoglobin content [CHr]) provides a measure of the efficiency of hemoglobin synthesis that the reticulocyte has implemented in the previous few (3–4) days. The reference range is approximately 24–32 pg. Values <24 pg are indicative of iron deficiency. Since hemoglobin synthesis depends on iron availability, CHr provides a measure of functional iron and, therefore, is an index for early assessment of iron deficiency. Accordingly, it could be useful in the following conditions:

- Pediatric age.
- Pregnancy.
- Complex clinical pictures, such as chronic inflammation and chronic kidney disease, as the ferritin level in these conditions, is elevated as an acute-phase reactant, despite the reduced iron stores.
- Nonmacrocytic anemias, which may be a better predictor of iron stores than traditional indices.

Reticulocyte Hemoglobin Equivalent

The hemoglobin equivalent in reticulocytes (Reticulocyte Hemoglobin equivalent [Ret-He]) is a parameter with a very similar meaning to the previous one. The reference range is approximately 28–35 pg. Values <28 pg are indicative of iron deficiency.

Differential Leukocyte Count (Leukocyte Formula)

The differential leukocyte count provides the relative percentage of each class of white blood cells. In addition to these values, hematological instruments generally report the absolute values of the various classes. The differential count also allows the detection of cells normally not present in the circulation, such as immature elements, or abnormal populations, such as, for example, leukemic cells. The differential count is obtained mainly by two methods: microscopic and automated. Leukocytes are classified into granulocytes, lymphocytes, and monocytes.

Granulocytes owe their name to the presence of distinct cytoplasmic granulations that define three classes: neutrophils (also called polymorphonuclear), eosinophils, and basophils. Neutrophil granulocytes are mature cells that originate from an initial morphologically identifiable precursor in the marrow, the myeloblast. As maturation progresses, the myeloblast becomes first a promyelocyte and then a myelocyte. In these proliferative stages, the number of cells increases geometrically. The metamyelocyte, the next cell, no longer has mitosis and then transforms into a granulocyte with a banded nucleus (band granulocyte). At this stage, the cell is released into the circulation, where it completes its maturation or enters medullary compartments to mature, from which it is later released into the circulation. About half of the intravascular polymorphonucleates circulate, maintaining a dynamic equilibrium with the other half, which is "marginalized" on the vascular endothelium. Therefore, only the circulating neutrophils are considered in the differential count. The half-life of mature circulating neutrophils is about 7 h. They irreversibly cross the vascular endothelium, migrate into the tissues to perform antibacterial surveillance, and die after 1 or 2 days. Eosinophilic and basophilic granulocytes have a similar development. After their release from the bone marrow, eosinophils rapidly leave the intravascular compartment and migrate into tissues, especially the gastrointestinal tract, lung, and skin, and never return to circulation. The function of these cells is probably to defend against multicellular parasites and to control the mechanisms associated with allergies and asthma, as well as inflammatory phenomena. Basophil granulocytes constitute about 1–2% of circulating leukocytes. Their physiological role is not yet completely clarified. They are present in inflammatory reactions during the immune response, as well as in allergic manifestations. Their granules contain heparin and histamine.

Lymphocytes and monocyte-macrophages are known as mononuclear leukocytes. Both play an important role in cellular and humoral immunity. These cells can get in and out of circulation, maintaining their function, and can spend a long time in tissues or lymph nodes.

Lymphocytes are the key immune cells for cellular and humoral immunity. In the blood, they represent 20–45% of leukocytes. Morphologically indistinguishable, they belong to two distinct systems called T (thymus) and B (bursa or bone marrow), having different functions. B lymphocytes are responsible for the synthesis of antibodies. When a lymphocyte of the B system (humoral system) is properly stimulated, it first proliferates and then turns into a plasma cell, the effector part of the immune response. Each B lymphocyte can synthesize only one type of antibody. The T lymphocyte system constitutes the cellular immune system and regulates the entire immune system. Different subsets of T cells can be identified by monoclonal antibodies specific against different membrane antigens. For example, T helper lymphocytes promote the function of B lymphocytes, while T suppressor lymphocytes inhibit them. Some T cells are responsible for cell-mediated cytotoxicity; natural killer (NK) lymphocytes are responsible for the nonspecific lysis of certain cells, for example, cancer cells. In peripheral blood, about 15–25% of lymphocytes are B cells, and about 40–75% are T cells.

Monocyte-macrophages originate in the bone marrow, from where they are rapidly released into the circulation, where they account for ~5% of leukocytes. Once they migrate into tissues, they become macrophages. Monocyte-macrophages play phagocytic functions on bacteria and particulate material, participate in inflammatory reactions, and are important to the immune system as they process antigenic material and communicate related information to T lymphocytes through a process of cell–cell interaction. Monocytes can also secrete interleukins, which boost B and T lymphocytes. Additionally, they participate in fibrinolysis by secreting plasminogen activators.

Microscope Differential Count

It is performed under the microscope after staining a smear with the May-Grünewald-Giemsa technique (more used in Europe) or with Wright's technique (in the USA and Anglo-Saxon countries). The 100× immersion objective is used with a 10× eyepiece, and then the final magnification is 1000×. Running the slide in the tail of the smear, according to a Greek profile that includes in part the cells on edge and in part those more centrally located, a total of 100 leucocytes are counted, classifying them in percentages according to the class to which they belong (Fig. 15.10). Greater accuracy is obtained by counting 200 cells.

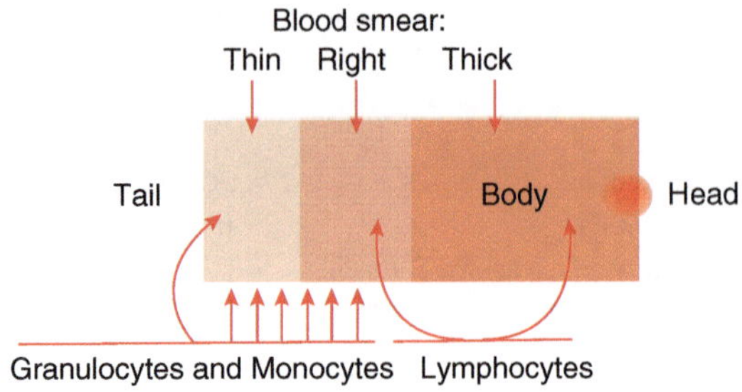

Fig. 15.10 Correct procedure for differential leukocyte count. (Copyright EDISES 2021. Reproduced with permission)

Automated Differential Count

Multiparametric flow cytometry is currently the most popular (and preferred) alternative to microscopic differential counting since it allows analyzing thousands of cells in a standardized way in a short time, with the possibility of detecting not only the populations usually recognized and classified in a blood smear, that is, neutrophils, eosinophils, basophils, lymphocytes, monocytes, immature granulocytes, and blasts, but also other subsets of cells, including various types and subpopulations of blasts and lymphocytes. In addition, recent studies have demonstrated the feasibility of flow cytometry using combinations of monoclonal antibodies that allow the detection and quantification of up to 16 cell subsets. The instruments most commonly used in laboratories are based on two technologies, either alternative or combined, depending on the complexity of the instrumentation and the manufacturer, namely electrical impedance and cytochemistry.

The differential leukocyte count is primarily necessary for the following reasons:

- To search for quantitative anomalies in a population of morphologically normal leukocytes, such as in infectious or allergic diseases.
- To therapeutic monitoring of cytotoxic or myelotoxic drugs. Of course, this requires a high level of precision and accuracy.
- To detect morphological abnormalities of white blood cells (e.g., when abnormal populations of white blood cells are circulating or when the presence of immature or atypical cells is suspected). This requires a high level of clinical sensitivity, that is, the ability to identify all patients with abnormal circulating leukocytes.

Automated differential counting provides a high level of accuracy and precision in quantifying and identifying normal white blood cells. However, the method is not always sufficiently sensitive in identifying abnormal or immature cells, and it cannot identify and classify all types of white blood cells accurately. To overcome this problem, many automated analyzers flag the presence of possible abnormal white blood cell populations, thereby indicating the need for a peripheral smear examination by personnel capable of identifying abnormal cells. Monocytes and basophils are the most difficult populations to count, and as a result, counting them has a low level of precision and accuracy. In addition, automated analyzers tend to underestimate basophil counts in true basophilic conditions. It should also be noted that both automated and manual methods are unable to detect a small number of abnormal cells. The false-negative rate reported in the detection of abnormal cells varies from 1% to 20%, depending on the instrument and the required detection limit. The identification of lymphoma cells and reactive lymphocytes is the most difficult condition for both automated instruments and microscopic examination. The value of reporting the presence of banded core granulocytes is questionable. However, the measurement of immature cells in myeloid lines, particularly bands, has been considered clinically useful in the diagnosis of infections, particularly neonatal sepsis. However, it should be considered that neutrophil granulocytes with banded nuclei cannot be enumerated by automated analyzers, as they are reported together with segmented neutrophils as absolute neutrophil counts. Identification of neutrophils with banded nuclei by microscopic examination is neither precise nor consistent due to the high variability in morphological classification since quantification of banded neutrophils reflects interexaminer variability.

The Primary Forms of Anemia

Microcytic Anemias

As reported in Table 15.3, there are three forms of microcytic anemia: sideropenic, thalassemias, and hereditary sideroblastosis. The first two have a higher incidence, even if sideropenic anemia (from iron deficiency) is the most prevalent worldwide. Thalassemia is more frequent in regions (central-northern Africa, southern Asia, and the Mediterranean area) in which the contemporary presence of favorable factors, in particular the malarial disease, has increased its diffusion.

Sideropenic and Assimilated Anemia

These are disorders in which the production of red blood cells is decreased due to low iron stores in the body and/or reduced iron utilization of the stores. The former is the most common nutritional disorder worldwide and accounts for about half of all cases of anemia. The development of anemia can result from insufficient iron intake, decreased iron absorption, increased demand for iron by the body, and

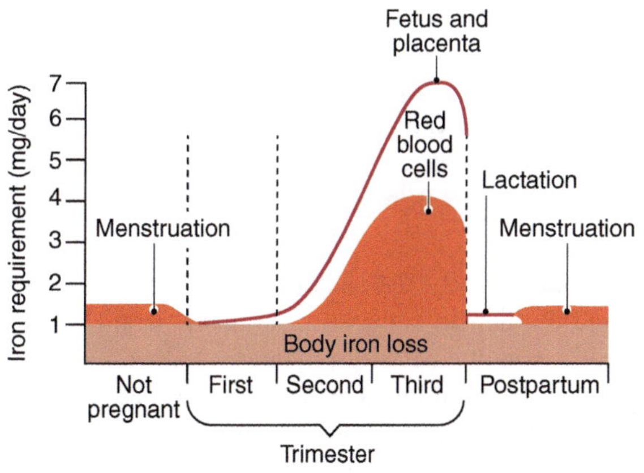

Fig. 15.11 Iron demands in different physiological states in women. (Copyright EDISES 2021. Reproduced with permission)

increased iron losses. Correct identification of the underlying etiology and administration of appropriate therapy are the keys to the evaluation and management of this condition. As mentioned at the beginning, sideropenic anemia should be differentiated first from the anemia of chronic conditions, which generally occurs in subjects affected by chronic inflammatory diseases and presents many clinical and laboratory features similar to sideropenic anemia, even if it differs in evolution and treatment.

The regulation of the body's iron balance is based on the meticulous control of intestinal absorption, as there is no excretory regulation. Daily losses, which in a normal male subject are between 1 mg and 2 mg and in the female are greater but variable depending on the physiological state (as shown in Fig. 15.11, they range from values similar to the male in childhood and menopause to values 30–100% greater in childbearing age, to values more than 3.5 times greater in pregnancy), are then rebalanced by absorption. In turn, absorption depends on the iron content of the diet. An average European diet contains 6 mg of iron per 1000 kcal; this corresponds to total values of 10–30 mg/day (Table 15.6). It should be borne in mind that, under conditions of normal balance regulation, the iron absorbed corresponds to 5–10% of the total content, and that under deficit conditions, this percentage may increase by 3–5 times. The localization of absorption begins in the stomach, proceeds to a maximum in the central part of the proximal jejunum, and then decreases caudally. Absorption is influenced by pH and redox potential. Of the various forms of iron presented for absorption, hemic iron (Fe^{2+}) derived from hemoglobin, myoglobin, and hemic enzymes from animal-derived foods is the most easily absorbed. Proteases and gastric acid juice release heme from the apoproteins to which it is bound. The iron in heme is oxidized to hemin (Fe^{3+}), and the molecule enters the enterocyte intact, from which it will then pass into the blood. European nonvegetarian diets contain

Table 15.6 Recommended daily intake of iron in food

Sex/eating habits	mg/day
Nonvegetarian woman in premenopause	18
Postmenopausal nonvegetarian man/woman	8
Vegetarian man/woman	♂14 and ♀33
Maximum iron intake in the diet	45

approximately 10–15% heme iron. The latter is not affected by diet, in contrast to inorganic iron (Fe^{3+}). It is generally complexed with phytates, oxalate, citrate, lactate, sugars, and amino acids and must be reduced to Fe^{2+} to be absorbed. Other foods and substances present in the diet can favor its absorption (ascorbates and animal meats through low molecular weight factors due to proteolysis that carry inorganic Fe−, keto sugars, amino acids, organic acids, and human milk) or slow down or even hinder its absorption (vegetable proteins, soy, egg white, cow's milk, phytates, bran and vegetable fibers, polyphenols, phosphates and phosphoproteins, tannin, and other metals that compete for absorption, such as Zn, Cd, and Ca). The recommended amount of iron in the diet also depends on dietary habits; subjects with a predominantly or totally plant-based diet should include higher amounts of iron in the diet. Once entering the enterocyte cytoplasm through the apical membrane, absorbed Fe^{3+} is reduced to Fe^{2+} by a ferro-reductase and internalized via the divalent metal transporter (Fig. 15.12). In contrast, heme iron (Fe^{2+}), transported internally by heme carrier protein 1, is released from the heme by heme oxygenase. Fe^{2+} in the cytoplasm can now take two routes: (1) the pathway by which iron is released into the circulation for use in erythropoiesis, which is mediated by ferroportin, a molecule located on the basolateral membrane that releases Fe^{2+}, which is simultaneously oxidized to Fe^{3+} by the ferroxidase hephaestin and immediately bound to circulating transferrin; and (2) the intracellular deposition pathway, whereby iron is incorporated into the cell's ferritin molecules when the cell receives the signal that no more iron is needed for erythropoiesis. Since the enterocyte is exfoliated after a few days, the iron inside it is eliminated with feces. A molecule, hepcidin, interferes with the mechanisms of iron entry into the circulation. It performs its function by inducing the lysosomal degradation of ferroportin. Hepcidin is expressed in the liver and is regulated by the BMP6 protein (bone morphogenic protein 6), together with its co-receptor hemojuveniline, and thus represents a mechanism for regulating iron absorption according to the needs and deposits existing mainly in the liver. Its action also occurs at the level of the reticuloendothelial macrophages.

Since hepcidin behaves as an acute-phase protein, it increases during inflammation upon induction by interleukin-6 and inhibits iron absorption. In chronic inflammation, this prolonged effect results in anemia with features similar

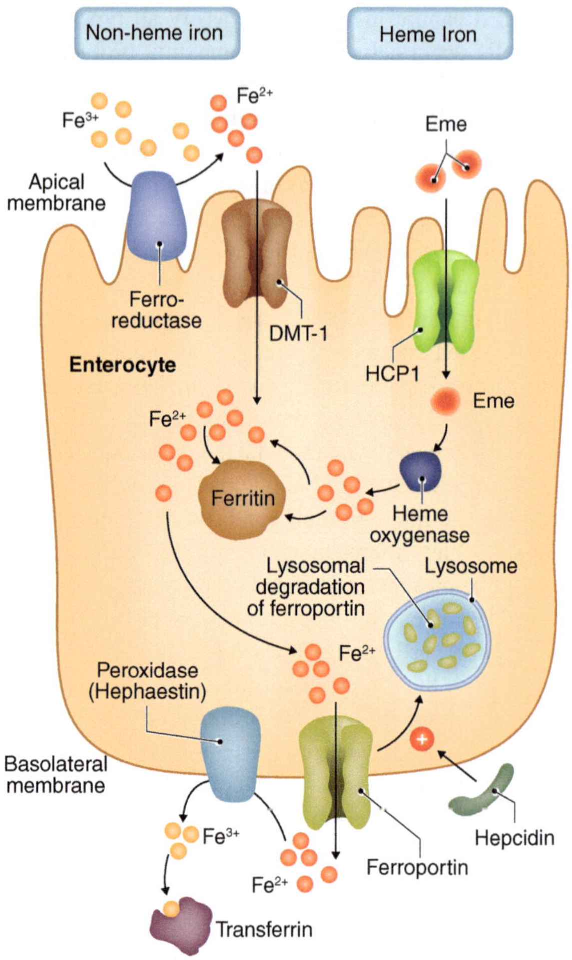

Fig. 15.12 Molecules contributing to the iron cycle in the enterocyte. (Copyright EDISES 2021. Reproduced with permission)

to those of iron deficiency anemia, from which, however, it must be differentiated. Indeed, anemia due to chronic inflammation results from the inability to allocate sufficient iron to erythropoiesis as it remains sequestered in the reticuloendothelial system.

Genetic mutations involving hepcidin are also known to cause juvenile hemochromatosis. This pathology can also be due to mutations in the hemojuvenilin regulator, which are the most frequent causes. Other important genes involved in iron regulation are transferrin receptor 2 (TfR2) and *HFE*. Some mutations of the latter, mainly C282Y and H63D, are responsible for hereditary hemochromatosis in adults, which occurs around 40 years of age. The laboratory analysis useful for the diagnosis of hemochromatosis is shown in Table 15.7.

The measurement of serum or plasma iron (sideremia) reflects the amount of iron bound to transferrin, which is continuously variable throughout the day as a function of diet and activity. Generally, sideremia values tend to be higher in the morning than in the evening and higher after meals than during fasting. For these reasons, sideremia is not reliable for assessing the iron status of the body.

Table 15.7 Investigations for the diagnosis of hemochromatosis

Test	Reference Range/ Methods	Comment
Sideremia (Fe)	12–32 μmol/L M 10–30 μmol/L F	Not singularly diagnostic, it must be interpreted in association with transferrin saturation
Transferrin saturation	18–40%	Values above 45–50% are indicative of iron overload
Ferritinemia	18–250 μg/L M 14–150 μg/L F	Values above the limits indicate the risk of iron accumulation in the deposits Values above 1000 are diagnostic Note that ferritin behaves as an acute phase reactant
Liver biopsy	By specific coloring Iron accumulations and consequent fibrosis are normally absent	In case of iron accumulation, semiquantitative measures of the concentration and distribution of iron in the liver lobules can be detected. The extent of fibrosis is assessed
Hepatic SQUID/ MRI	Noninvasive methods of image	SQUID is used only in very few centers; considered superimposable to the biopsy RM is like SQUID, albeit less accurate
Mutations in the gene *HFE* and other genes involved in iron regulation	Search for C282Y, H63D, S65C (E168X, W169X)	Second level test The choice depends on the frequency and geographic distribution

The transport of iron in the circulation is mediated by transferrin, a molecule circulating in four main forms in various proportions:

- Apotransferrin, an iron-free protein
- Monoferric transferrin with Fe bound to the carboxyl terminus
- Monoferric transferrin with Fe bound to the amino-terminal
- Holotransferrin, with two bound Fe atoms

Transferrin can be measured by two different methods. Specifically, it can be assessed by an immunometric assay of transferrin protein, whose values are expressed in g/L (IR 1.9–3.7), or by a functional method, which evaluates the total iron-binding capacity (TIBC) and is obtained after saturation of serum or plasma with Fe^{3+} and measurement of bound iron. The latter provides diagnostic information for microcytic anemias, such as the latent iron-binding capacity (LIBC) and the percentage of saturation of transferrin (TSat %), through the following calculations:

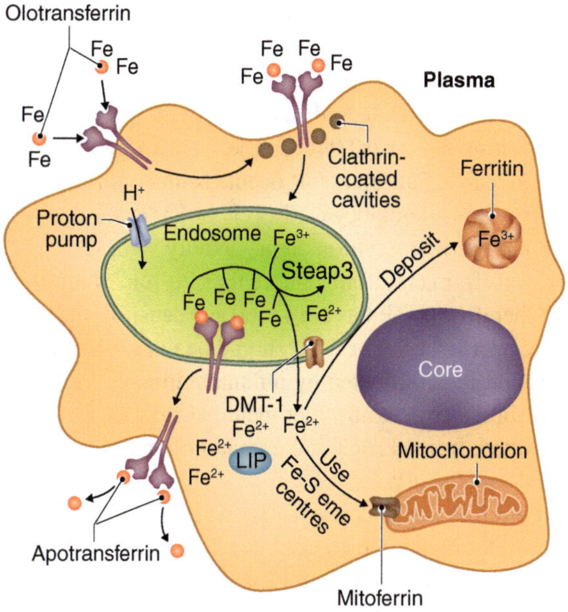

Fig. 15.13 Iron metabolism. Holotransferrin binds to TfR1 on the cell surface and the complex undergoes endocytosis in the clathrin-coated vesicle. The proton pump acidifies the endosome and causes Fe^{3+} release. Step 3 reduces Fe^{3+} to Fe^{2+}, which exits the endosome via divalent metal transporter (DMT-1). Iron is utilized in mitochondria via mitoferrin or deposited (nonerythropoietic cells) as ferritin. A fraction of intracellular Fe (redox) remains to constitute the Labile Iron Pool (LIP). Apotransferrin and TfR1 go to the membrane and dissociate at neutral pH. The apotransferrin cycle resumes. (Copyright EDISES 2021. Reproduced with permission)

- TIBC μmol/L S-Fe μmol/L = LIBC μmol/L
- TSat % = Fe/TIBC × 100

The reference values are, respectively:

- S-Fe: 11.6–31.7 μmol/L (M) and 10.0–30.4 μmol/L (F)
- TIBC: 45–66 μmol/L
- Transferrin: 1.9–3.7 g/L
- TSat: 20–50% (M) and 15–50% (F)
- Absolute ratio of TIBC μmol/L over transferrin (g/L) ≈ 25.0

Unlike sideremia, the TIBC/transferrin ratio does not vary continuously in the blood; however, the diagnostic information of TIBC/transferrin in sideropenic anemia is delayed as it increases when iron stores are depleted.

Transferrin is mostly in the apo form (approximately 65–80%). Its function is to bind Fe^{3+}, making it soluble to dampen its reactivity, and deliver it to cells via special transferrin receptors (TfRs). Figure 15.13 shows the mechanism through which iron is incorporated into cells. Cells receive Fe via TfRs, which are located on the cell membrane and are particularly represented in erythropoietic cells. Their expression varies according to the intracellular Fe concentration. Two types of TfRs are known, 1 and 2. TfR1 shows a much greater affinity for Fe and is ubiquitous in cells, while TfR2 has been shown to perform different functions, not yet fully elucidated. Once its function has ceased, the extracellular portion of TfR is released into the circulation as a soluble truncated monomer (sTfR) complexed with transferrin. The major release of plasma sTfR occurs from erythroblasts, secondarily from reticulocytes. Levels of sTfR are determined by bone marrow erythropoietic activity, which can cause variations from eightfold to more than 20-fold of normal values. Therefore, sTfR has been used as a sensitive index of the recovery of erythropoiesis. It should be noted that sTfR concentration is not affected by inflammation and thus may be useful in differentiating sideropenic anemia, in which sTfR is elevated, from chronic inflammatory conditions, in which sTfR does not vary.

The reference values of sTfR are 2.2–5.0 mg/L in males and 1.9–4.4 mg/L in females.

Ferritin is the main form of iron storage in the body. Its globular conformation with a high MW (474 kDa) and 24 units of two types of chains (H and L) enclose within the shell over 4000 atoms of $Fe3+$ complexed with hydroxy phosphate in the form of ferrhydrite, so as to minimize any reactivity due to contact. Most organs have ferritin deposits, which vary greatly according to the general state of iron. The liver, spleen, and bone marrow are the organs with the largest iron deposits in the form of ferritin. The values of circulating ferritin are in constant equilibrium with those of the stores, with 1 μg of circulating ferritin approximately corresponding to 7.5 mg of iron in the stores. The exceptions are mainly due to inflammatory and neoplastic diseases, as well as hepatopathies. In such conditions, ferritinemia values are unpredictably abnormally elevated and thus not reliable as biomarkers of the status of iron stores, even if the values are apparently in the normal range.

The reference values of ferritin are 18–250 μg/L in males and 14–150 μg/L in females.

Zinc protoporphyrin (ZnPP) is a surrogate biomarker of iron deficiency. Zinc fits into the protoporphyrin ring in the absence of iron, and, therefore, its measurement in red blood cells can give an assessment of the extent of erythropoietic iron deficiency. Normally, the Fe/Zn concentration ratio is 30,000/l. In the case of iron deficiency, a decrease in the ratio is observed. The test is not widely used but may be useful in some diagnostic conditions.

The differential diagnosis of iron deficiency anemia and similar conditions is based on a reasoned combination of the tests described above. In Table 15.8, the diagnostic criteria that should be mainly considered in the conditions of microcytic anemia resulting from iron deficiency, chronic inflammatory conditions, and the combination of the two conditions, to which sometimes difficulties in making a correct diagnosis are added, are presented. The iron overload condition is also reported in Table 15.8.

Anemia Due to Thalassemia Syndromes

This is a very heterogeneous group of conditions that have a common feature: they are caused by mutations/deletions affecting the gene complexes responsible for the synthesis of globin chains. Due to its complexity, it is important to briefly describe genetic-molecular and structural information concerning the three types of normal human hemoglobin, namely HbA ($\alpha2\beta2$), HbF ($\alpha2\gamma2$), and HbA2 ($\alpha2\delta2$), and the constituent globin chains, namely α, β, γ (in turn distinguished in Aγ and Gγ), and δ globins. In addition to these, two other globin chains, ε and ζ, are normally expressed in the embryonic period and, together with the α-globin, constitute the Portland hemoglobins (Hb), Gower 1 and Gower 2.

Gene Complexes, Molecular Expression, Globin Chains

Genes encoding for globin chains are organized into two gene complexes (clusters): the β complex, located on chromosome 11, and the α complex, located on chromosome 16 (Fig. 15.14). At the cis position with respect to the β complex is the Locus Control Region (LCR), which regulates the expression of genes in the β complex together with other proximal gene elements (promoters, silencers, and enhancers). The LCR and proximal region are important as some forms of β-thalassemia are due to abnormalities involving them.

Table 15.8 Differential diagnosis of iron deficiency anemia, chronic inflammatory condition, and iron overload

	Iron deficiency anemia	Combined anemia (iron deficiency and chronic-inflammatory)	Anemia of chronic-inflammatory conditions	Iron overload
S-Iron	Decrease	Decrease	Decrease	Increase
TIBC/Transferrin	Increase	Decrease/normal	Decrease/normal	Decrease
Transferrin saturation	Decrease	Decrease/normal	Decrease	
Ferritinemia	Decrease (diagnosis if <12 μg/L)	Normal (typically >100 μg/L)	Normal/increase	Increase
sTfR	Increase	Increase	Normal	Decrease
ZnPP	Increase	Increase	Increase	Decrease

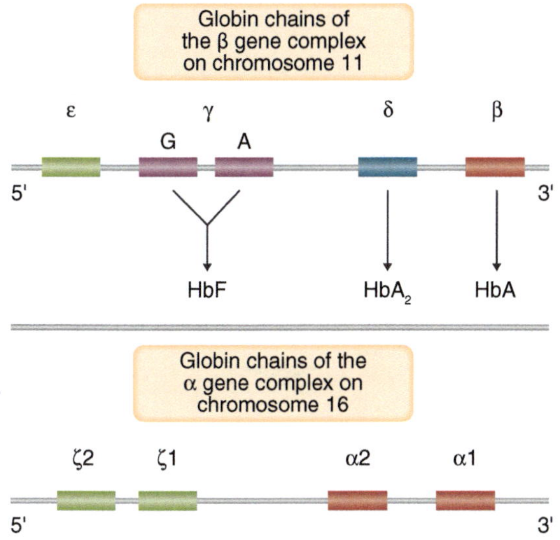

Fig. 15.14 Schematic diagram of the β and α gene complexes. In the β complex in 5′ → 3′ sequence, there are the ε, Gγ, and Aγ, δ, β genes. In the α complex in sequence, there are the genes ζ2, ζ1, α2, and α1. (Copyright EDISES 2021. Reproduced with permission)

Embryonic Hemoglobins, Fetal Hemoglobin (HbF), HbF/HbA Switches

As stated above, during the early part of the first trimester of embryonic development, there is a consistent expression of an embryonic form of globin of the β gene complex, known as globin ε, in the erythropoietic cells of the yolk sac. Shortly thereafter, when production of the first erythrocytes by erythropoietic cells of the fetal liver begins, the expression of globin ε ceases and globin γ becomes predominant. Therefore, HbF (α2γ2) appears in the 5th week of gestation and rapidly becomes the major Hb of the fetus (Fig. 15.15). After the 10th week, HbF is approximately 90% Hb and remains so until 32–36 weeks; then, adult HbA synthesis begins. At birth, the β/γ ratio is approximately 1.0. During fetal life, the Gγ/Aγ ratio is ≈ 3/1 and, starting at birth (HbF/HbA switch), it changes to 2/3 due to the progressive silencing of the two Gγ genes along with the change in erythropoiesis sites (yolk sac → liver → bone marrow), the decrease in erythrocyte volume from 120 fL to 80 fL, the change in RBC membrane antigens (i → I), and some changes in the isoenzyme profile. Table 15.9 shows the normal hemoglobins and their respective values at birth and in adults.

Genetics of Thalassemias

α- and β-thalassemias are mostly inherited by Mendelian modality as autosomal recessive forms; therefore, the complete manifestation of the thalassemia is observed in children of both parents' carriers. Given the heterogeneity of the conditions, thalassemias with double heterozygosity, in which the phenotype presents the combined expression of the two altered genes, are not uncommon. Rare, dominant modes of transmission are also described.

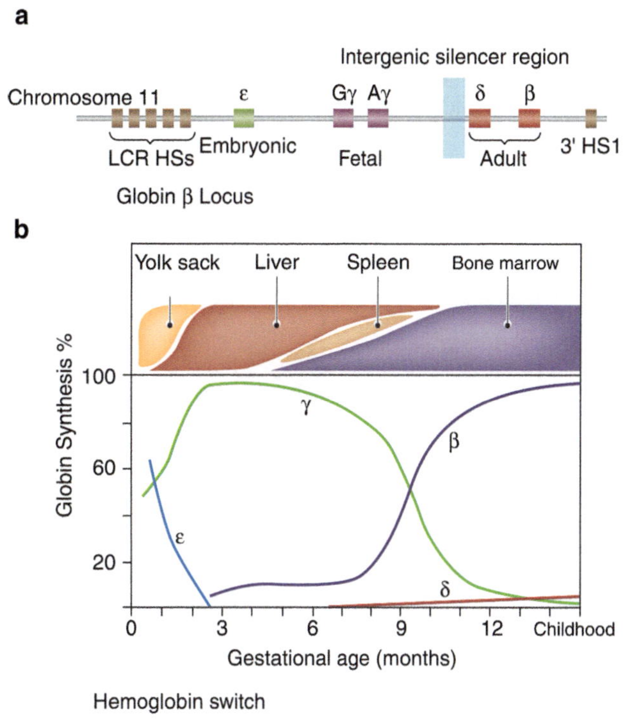

Fig. 15.15 Hemoglobin (Hb) synthesis during pre and postnatal development. Shown are the sites of erythropoiesis and the embryonic fetal and adult hemoglobins and their epochs of appearance and disappearance. As noted, the switch occurs shortly after birth at the equivalence of HbF and HbA. (Copyright EDISES 2021. Reproduced with permission)

Table 15.9 Types of hemoglobin and normal variants at birth and in adulthood

Type of hemoglobin	Globine components	Values at birth	Values in the adult	Note
A	$\alpha_2\beta_2$	20–25%	97%	Values of the adult at 1 year of age
F	$\alpha_2\gamma_2$	75–80%	<1%	Values for adults at 1 year of age If it remains high, suspicion of thalassemia
A_2	$\alpha_2\delta_2$	0.5%	2.5%	High in *trait* β-thalassemia, of which it is a biomarker

Molecular Alterations in Thalassemias

Under normal expression conditions, the synthesis of the two major globin chains is balanced. Measurement of the α mRNA/β mRNA ratio shows that in α-thalassemias, this ratio correlates with the number of functional α genes and that in β-thalassemias, it is an indicator of disease severity. In thalassemias, there is an unbalanced expression of one or more chains, resulting in a reduction of intraerythrocytic cells and a decrease in MCH (Fig. 15.16). When MCH

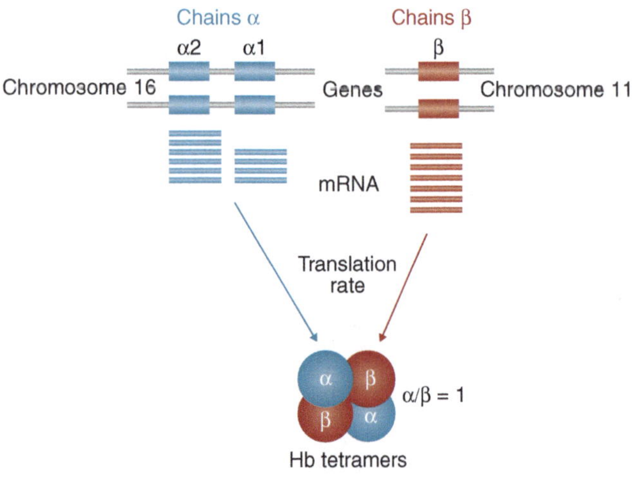

Fig. 15.16 Translation of genes for α and β globin chains and their expression under normal conditions. In α-thalassemias, there is usually deletion of one or more genes, whereas in β-thalassemias, there is usually reduced/altered transcription of the related gene, affecting translation and expression. (Copyright EDISES 2021. Reproduced with permission)

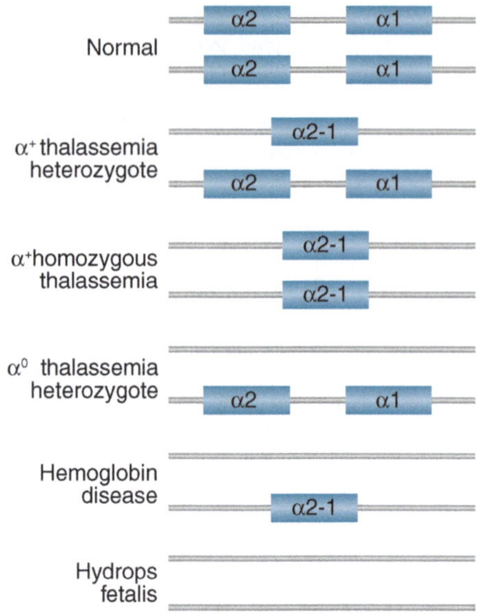

Fig. 15.17 Genotypes of α-thalassemias. (Copyright EDISES 2021. Reproduced with permission)

<27 pg and after excluding iron deficiency, it is appropriate to ascertain the existence of possible thalassemia. From a clinical point of view, the consequences for the patient are due to the excess presence of one or the other type of free globin chain.

Pathophysiological and General Clinical Aspects of α-Thalassemias

Most of the clinical consequences of α-thalassemias are due to excess β globin chains in the adult and γ in the newborn due to the deletion of 1–4 α genes. Indeed, excess β chains tend to aggregate into hemoglobin H (HbH) tetramers, in the case of β tetramers, or hemoglobin Bart (Hb Bart) tetramers, in the case of γ tetramers. Both HbH and Hb Bart tetramers show a high affinity for oxygen, are unstable, and tend to precipitate inside the red blood cell, leading to the formation of visible Heinz bodies. As they adhere to the cell membrane, they cause a loss of the cell's plasticity characteristics and make it difficult to pass through the microcirculation. Eventually, this causes the early elimination of the erythrocytes by the reticuloendothelial system.

Clinical Forms of α-Thalassemia

The α-thalassemias are responsible for very different clinical forms among them, ranging from apparently normal health conditions up to intense hemolytic anemia, which could be lethal, such as the fetal hydrops by Hb Bart (γ4) due to deletion of all four α globin genes.

The heterogeneity of the syndrome is explained by the fact that on the pair of chromosomes 16 of everyone there are a total of 4 α genes, and, therefore, the deletion can affect from 1 to 4 genes, with results that vary according to the genotype, as illustrated in Fig. 15.17. HbH disease is due to the deletion of 3 α genes and, consequently, to the excess of β chains forming the β4 tetramers (HbH). In HbH disease, the anemia is due both to a reduced amount of Hb and to a reduced functionality of the residual Hb because the presence of HbH with an increased affinity for oxygen further reduces its functional adaptation capacity. It should be borne in mind that hemoglobin with increased affinity for oxygen is always disadvantageous for homeostasis because it yields smaller volumes of bound oxygen at the same gradient compared to normal hemoglobin.

α-Thalassemia Due to −3.7 Deletion

In these forms, an α-thalassemic allele corresponds to the presence of only one gene on chromosome 16, while the other one has undergone a deletion of −3.7 kb due to unequal crossing-over between the two genes on homologous chromosomes. Forms with a −4.2 kb deletion are also known.

α-Thalassemias Due to Nondeletion

These are molecular forms whose clinical features do not differ or differ little from the deletion forms. They are due to mutations in the termination codon. The best known is the Hb Constant Spring (HbCS), in which the TAA → CAA mutation at position 142 (the usual position of the termination codon of the gene) causes an elongation of the expressed α globin by an additional 31 amino acids (from 142 to 173). As the related mRNA is less stable and degrades, a reduction in expression also occurs. In addition, HbCS is unstable and tends to precipitate on the membrane of not only the erythrocyte but also the bone marrow precursors. This causes both hemolytic anemia (circulating cells) and ineffective erythropoiesis (bone marrow precursors). Other forms of α-thalassemia due to nondeletion are Hb Pakse, Hb Icaria, Hb Koya Dora, and Hb Seal Rock. All related mutations elongate the expressed α globin.

Pathophysiological and General Clinical Aspects of β-thalassemias

From a pathophysiological point of view, the clinical consequences of β-thalassemia are due to an excess of α-chains, which are unable to associate and are highly unstable. In normal erythropoiesis, α-globins are associated with β-globins via AHSP (α-chain stabilization protein). Free α chains behave as active oxidants capable of causing apoptosis and thus inducing ineffective erythropoiesis. The main difference between α and β-thalassemias is that α-thalassemias are hemolytic diseases, while β-thalassemias are dyserythropoietic diseases. Furthermore, β-thalassemias are predominantly due to point mutations or short insertions/deletions of a few nucleotides, not extensive deletions like the α-thalassemic forms. From a translational point of view, the β-thalassemias are distinguished in β+ forms and β0 forms. In β+-thalassemias, mutated alleles encode for reduced amounts of mRNA and thus β-globin, with wide variations depending on the type of mutation and with normal or abnormal chain (e.g., HbE, Hb Knossos). In β0-thalassemias, the mutation completely prevents the production of mRNA, and therefore there is an absence of β globin.

The main molecular mechanisms underlying β-thalassemias involve mutations affecting:

- Promoters, so the transcription efficiency is decreased (the consequence is a β+-thalassemia).
- Splicing, which is completely abolished due to abnormalities of the exon–intron junction leading to β0-thalassemia.
- Splicing, in which the incorrect position of the splicing itself results in structural anomalies leading to β+-thalassemia; the typical example of this anomaly is HbE.
- Splicing, in which an alternative splicing sequence results in a premature stop codon leading to β+-thalassemia with expressed protein rapidly eliminated by proteolysis; the typical example of this anomaly is Hb Knossos.
- Introns (e.g., intron-1 at 110 G → A), with the creation of a new splicing acceptor site leading to β+-thalassemia. Another mutation creates a GT donor site that, interacting with the normal acceptor, unmasks a new acceptor that interacts with the normal donor, leading to the division of intron-2 into two separate introns with a reduction of chain production; also, in this case, β+-thalassemia is observed.
- Polyadenylation site, where there is inefficient cutting and defective polyadenylation at the normal site, resulting in long and unstable mRNA and reduced production of normal mRNA leading to β+-thalassemia is observed.
- Translation, in which the cause is the formation of a nonsense codon; for example, at codon 36, the CAG → TAG mutation is typically present in Sardinia and is frequent in the Mediterranean; at codon 17, the AAG → TAG mutation is frequent in Asians and Chinese (such mutations do not allow the expression of globin, and therefore we are dealing with β0-thalassemias).
- Exons in which microdeletions or small insertions cause a frameshift and premature stop signal, with the expression of highly unstable globin that is oxidizing and toxic for the erythroid cell (the condition causes severe hemolytic anemia also in the heterozygote and is known as dominant thalassemia).

Among the about 200 defects causing β-thalassemia, 20 are due to deletion. These variants are quite rare, except for the 619 bp one that is frequently found in south-eastern Asia. Among them, there are also two forms present with a certain frequency in the Mediterranean area:

- Hereditary persistence of fetal hemoglobin (HPFH), a deletion syndrome characterized by decreased/absent production of β-globin and variable compensatory increase of γ-globin, is present in Mediterranean and African subjects, sometimes with double HPFH/HbS heterozygosity. Some HPFH deletions are very large and can reach up to 100 kb. There are also HPFH forms due to nondeletion that are classified as forms with only increased globin Aγ and forms with only increased globin Gγ.
- δβ-thalassemia: in the heterozygote state, it is very similar to the β-thalassemia, with the difference of increased levels of HbF included between 5% and 15% and a characteristic cellular distribution (F cells); the homozygotes and the double heterozygotes δβ/β-thalassaemia have a picture similar to thalassemia intermedia or major. The δβ-thalassemia is subdivided into two groups, within which there are further differences in the deletions:
  - δβ-thalassemias with the production of Aγ and Gγ chains, called AγGγ (δβ)0-thalassaemias; in them are included the Hb Lepore, hemoglobins whose globin chains are the result of unequal crossing-over δβ.
  - δβ-thalassemias with the production of Gγ only, called Gγ $(A\gamma\delta\beta)^0$-thalassemias.

HPFH and δβ-thalassemia are strictly related from a molecular point of view, and the distinction has mainly clinical value; however, some forms of HPFH are due to punctiform mutations, which denotes their heterogeneity with respect to δβ-thalassemia. The heterozygote for HPFH has a normal hematological picture with normal indices, normal HbA2, and HbF between 15% and 30% with homogeneous pan cellular distribution. The homozygote for HPFH is hematologically normal, except for a reduction of the MCV and MCH indexes, and so also appears the double heterozygote HPFH/β-thalassaemia.

Clinical Forms of β-thalassemia

The subjects with the minor forms of β-thalassemia (or β-thalassemic trait) do not generally present clinical consequences but mainly of the laboratory:

- Low MCH.
- Low MCV.
- Minor morphological changes.
- Modest increase in HbA2.
- Sometimes a slight reduction in hemoglobin.
- Often an altered globin α/β ratio.

Patients with the major forms of β-thalassemia (homozygous or double heterozygous) always present, with different degrees:

- Severe hemolysis
- Ineffective erythropoiesis
- Iron overload

The intermediate forms of β-thalassemia represent a heterogeneous group with clinical manifestations intermediate between the two previous forms, in particular:

- Hemoglobin is between 60 g/L and 90 g/L.
- Transfusions are sometimes necessary.
- From a hereditary point of view, homozygotes for mild forms or double heterozygotes for forms of different severity can be found.
- These forms could be present in subjects with a combination of other types of hemoglobinopathy (e.g., association with unstable hemoglobin).

Diagnosis of Thalassemia

Molecular Diagnosis

In the case of α-thalassemias, the most common deletion forms are investigated by Gap-polymerase chain reaction (PCR). The rarer nondeletion forms are investigated by specific hybridization methods or using restriction enzymes, followed by sequencing.

In the case of β-thalassemia, two PCR-based techniques are preferably used:

- Hybridization with allele-specific oligonucleotide (ASO)
- Allele-specific priming

Larger deletions should be investigated by Gap-PCR. If PCR-based methods are not sufficient, DHPLC (to locate the mutation site region) and/or sequencing (to characterize the mutation) will be used in the presence of a suspected mutation.

Laboratory Diagnosis (Phenotypic) of Thalassemias

Screening methods based on the measurement of erythrocyte osmotic resistance in progressively hypoosmotic saline solutions have been abandoned as they are nonspecific and have poor sensitivity. In some less equipped realities, techniques based on the hemolysis time in isosmotic glycerol solutions, the so-called GLT50, and its commercial variants (Green test, Osmored A, and others), are still used.

The diagnostic process for thalassemia suspicion is based on the use of hematological methods, such as the CBC and the reticulocyte count, electrophoretic methods, and HPLC (with the use of dedicated instruments, which are today more common than electrophoretic systems), with eventual mass spectrometry in particular cases.

In some situations, the use of molecular biology techniques based on nucleic acids is appropriate. In the α-thalassemic trait, the erythrocyte indices show alterations characteristic of microcytic anemia, that is, decreased total hemoglobin, MCH <27 pg with a progressive decrease in relation to the number of deletions of α genes, MCV <80 fL, erythrocyte count mostly increased, normal or little increased reticulocytes, and normal or decreased HbF. Indeed, there is no HbA2 marker as in the case of the β-thalassemia trait because it is also affected by α gene deletion. Moreover, the observable effects depend on the number of nonfunctioning genes and on the techniques used. For example, the presence of HbH (or rarely of Hb Bart) can already be evident in subjects with two deletions when sensitive methods like HPLC, and eventually mass spectrometry, are used, which is more difficult with electrophoresis. HbH is instead well evident on electrophoresis and HPLC in subjects with three nonfunctioning genes, together with HbH intraerythrocytic inclusion bodies.

In the β-thalassemic trait, the erythrocyte indices showing variations are as follows:

- Decreased hemoglobin (9–10 g/L)
- Decreased MCH (<27 pg)
- Decreased MCV decreased (<80 fL)
- Increased erythrocyte count (>5.0 T/L)
- Normal or slightly increased RDW, in contrast to sideropenic anemia
- Normal or increased HbF
- Increased HbA_2 due to increased relative to HbA
- Slightly increased reticulocytes

Stained smears may highlight morphological features, such as microcytes, target cells, F cells (rich in HbF), and included bodies. These elements are not determinative per se but contribute to the diagnosis.

Electrophoresis at alkaline pH shows the relative increase in HbA2 and, where increased, HbF.

HPLC shows the relative increase in HbA2 and, where increased, HbF.

HPFH and δβ-thalassemia both present high HbF values. The distinction between the two forms lies in the distribution of intraerythrocytic HbF, homogeneously diffused in HPFH and concentrated in erythrocytes particularly rich in HbF (F cells) in δβ-thalassemia. The microscopic observation after staining with Kleihauer's method or a more recent method with specific antibodies in flow cytometry can provide differential diagnostic elements.

Normocytic Anemias

Normocytic anemias include a series of manifestations with different etiologies. Some of them can be considered primitive because they are due to direct action in and/or on the erythrocyte, while others are secondary to concomitant pathologies and therefore depend on the evolution of such causal pathologies. Among the primitive forms, there are qualitative hemoglobinopathies that, with some exceptions (e.g., HbC), are to be considered normocytic anemias, even if sometimes accompanied by morphological alterations. Qualitative hemoglobinopathies involve molecular modifications that affect the structural and functional characteristics of the hemoglobin involved.

This chapter is limited to sickle cell anemia and hemolytic anemias.

Sickle Cell Anemia (Hemoglobinopathy S, Sickle Cell Disease)

The term sickle cell anemia refers to a collection of autosomal recessive conditions with Mendelian transmission characterized by the presence of the HbS mutation, also in combination with other Hb variants. Mixed heterozygotes are defined as individuals who have one copy of HbS and one copy of another variant, for example, HbD^{Punjab}, or β-thalassemia. They usually present modest clinical manifestations, especially those with β-thalassemia, but there are mixed forms leading to sickle cell syndrome-like conditions, thus with severe clinical manifestations. Carriers have a single copy of HbS and a normal β gene (HbAS) or a variant that does not cause sickling (e.g., HbS/Hb Korle Bu); even in carriers, the clinical manifestations are usually mild or absent. The condition is referred to as sickle cell trait. In sickle cell anemia (HbSS genotype), the affected individual has two copies of the variant Hb, and this determines the greater severity of the condition. In circulation, the main form of hemoglobin is HbS, with variable levels of HbF and normal levels of HbA2. This is a qualitative hemoglobinopathy caused by a single missense substitution (GAG → GTG) that results in the insertion of a valine instead of a glutamic acid at position 6 of the β globin chain. The structural consequence of the substitution is the propensity of the molecule to form polymers under conditions of reduced oxygen pressure with a characteristic elongated and curved shape (sickle). This causes deformation of the erythrocytes that contain the polymerizing molecules, a phenomenon that occurs above all in the microcirculation, slowing down to the point of stopping the circulation (Fig. 15.18). The main clinical features are as follows:

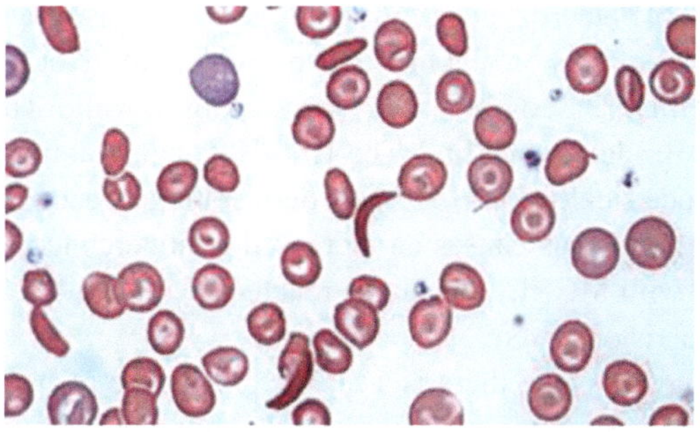

Fig. 15.18 Blood smear showing the presence of sickle cell erythrocytes

- Reduced erythrocyte survival
- Vaso-occlusion phenomena, which occur at the level of the cerebral vessels can cause strokes
- Painful seizures, respiratory complications, and organ damage
- Anemia
- Episodic hemolytic seizures
- Blood sequestration in organs like the spleen, lungs, and liver
- Tendency to infections of varying severity

From a clinical point of view, it is also important to determine the haplotype of the patient affected by sickle cell anemia. Indeed, some haplotypes favor the contextual expression of HbF, whose presence within the erythrocyte has an inhibiting effect on the polymerization of HbS and, therefore, on the sickling, with a consequently better prognosis. There are five haplotypes of HbS, named according to the regions/populations in which they have been observed (Senegal, Benin, Cameroon, Bantu, and Arab-Indians). The Senegal and Arab-Indian haplotypes favor a higher expression of HbF, while the Cameroon haplotype favors a low expression of HbF and is therefore associated with a less favorable prognosis than the previous one.

Double heterozygosity for other hemoglobin variants has different effects on sickling. For example, HbS/HbO^{Arab} and HbS/HbD^{Punjab} combinations are sickling, while HbS/Hb Korle Bu delays sickling. The HbS/HbC combination tends to reduce sickling as HbC has a dehydrating effect on erythrocytes and thus causes an increased relative concentration of Hb with MCHC, which can reach 40%, and at that point, sickling is observed.

In areas where there is a higher prevalence of hemoglobinopathy S, infant mortality is high, with a peak between 1 and 3 years, mainly due to infections. In the most developed areas, there is better survival, guaranteed by neonatal screening programs, early vaccinations, prophylaxis with oral penicillin, and, in general, better treatment of infections. In the United States, for example, the median life expectancy for individuals with HbS is 42 years for males and 48 years for females.

The laboratory diagnosis of HbS is based first on the identification of the carrier (HbA/HbS), usually silent, by the following:

- Separation techniques:
  - Cellulose acetate electrophoresis at alkaline pH and at acid pH
  - Capillary electrophoresis
  - Isoelectrophoresis, if other hemoglobin fractions are present
  - HPLC
- Confirmatory function tests:
  - Sick cell test
  - Solubility test for HbS
- Identification of the mutation by molecular biology tests

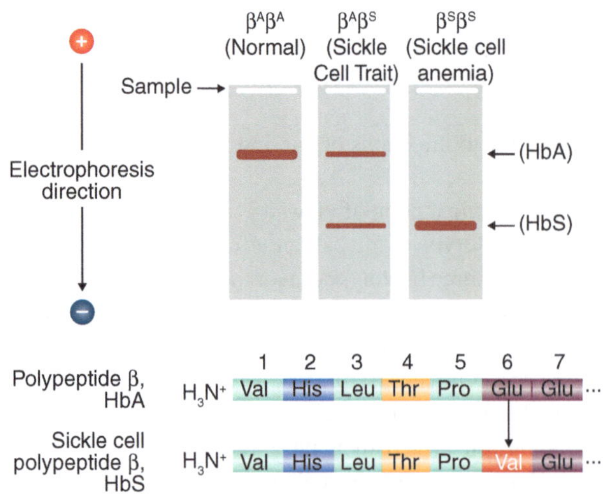

Fig. 15.19 Electrophoresis at alkaline pH of Hb lysates belonging to samples from subjects with HbAA (left), HbAS (center), and HbSS (left). (Copyright EDISES 2021. Reproduced with permission)

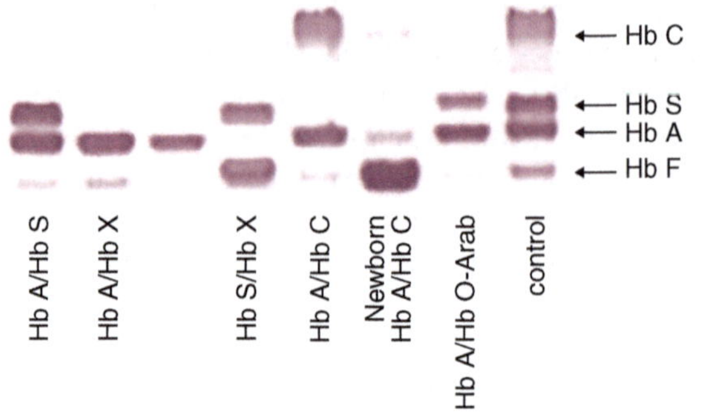

Fig. 15.20 Electrophoresis at acid pH of lysates from samples with different hemoglobin variants

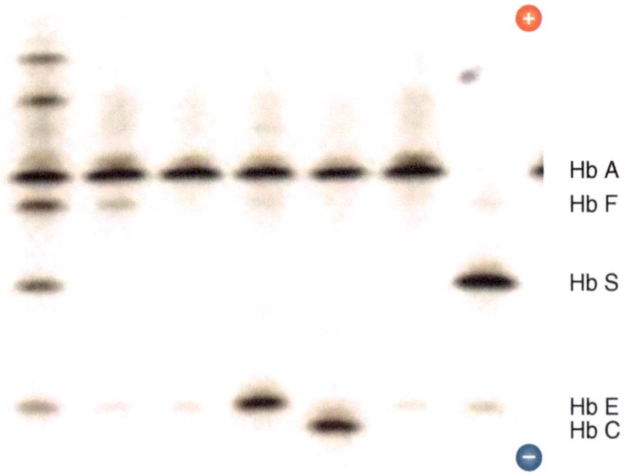

Fig. 15.21 Isoelectrophoresis pattern at pH range of 6.0–9.0 for lysates from samples with different hemoglobin variants

During electrophoresis at pH 8.5, HbS migrates more slowly than HbA (Fig. 15.19). Noteworthy, other Hbs also migrate to the same position, such as HbD^{Punjab}, which can coexist in heterozygosis with HbS (see above).

Electrophoresis at an acidic pH (5.2) in agarose gels can provide additional diagnostic criteria to identify HbS and differentiate it from other hemoglobins (Fig. 15.20).

Isoelectrophoresis can better define the different fractions of hemoglobin (Fig. 15.21).

The use of HPLC is favored by the fact that it is an automated method for measuring HbA1c, making it widespread in laboratories and useful for separating and quantifying the remaining hemoglobin fractions as well. Presumptive identification of variants is made based on retention time (Fig. 15.22).

The in vitro sickle cell test is performed by placing a drop of the patient's blood on a slide and covering it with a coverslip. In a short time, the lowering of the oxygen level in the preparation causes the appearance of sickles visible under the microscope (Fig. 15.18). The test is made more rapid and sensitive by adding a small amount of sodium dithionite or tetrathionate powder.

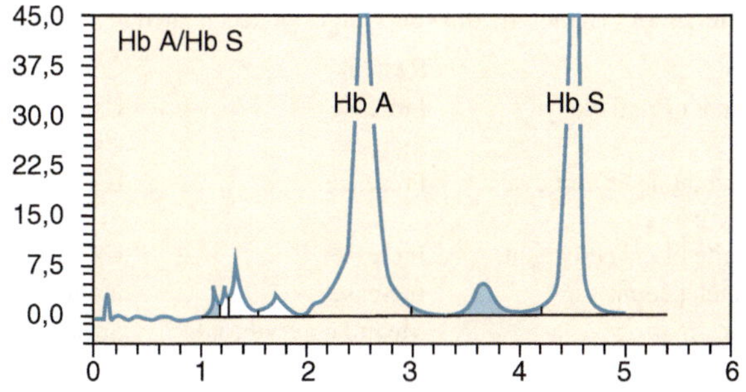

Fig. 15.22 HPLC pattern of lysate from sample with HbA/HbS genotype

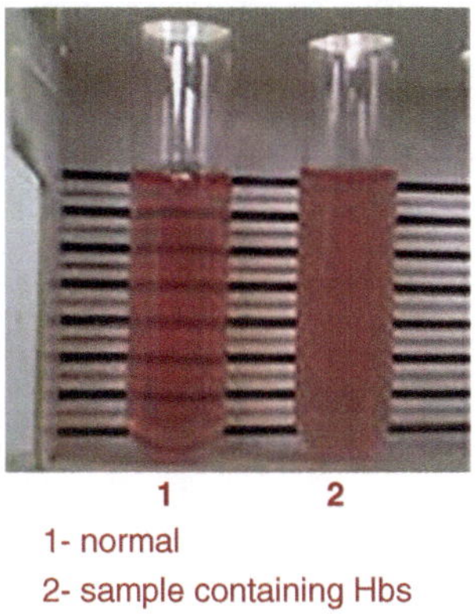

Fig. 15.23 Solubility test for HbS. Evident turbidity in the sample on the right. (1) Normal and (2) sample containing HbS

The solubility test for HbS is done by adding a drop of blood to phosphate buffer pH 6.8–2.8 mol/L, deoxygenated with Na dithionite. A precipitate is observed in the presence of HbS alone (Fig. 15.23). The test is specific but does not distinguish heterozygous from homozygous status.

Molecular biology tests involve amplification by polymerase chain reaction (PCR) of the DNA tract with the first exon of the β-chain, the use of restriction enzymes, electrophoresis, and restriction fragment length polymorphism (RFLP) analysis. The A → T mutation eliminates a restriction site for various enzymes (e.g., MstII). Note that the method is not specific to HbS, as there may be other site mutations. PCR-ARMS (amplification refractory mutation system methods with allele-specific probes can also be used.

Hemolytic Anemias

Common features of hemolytic anemias are as follows:

- Increased rate of red blood cell destruction
- Variable degree of intensity of hemolysis
- Transient marrow response, increasing up to six- to eight-fold on baseline erythropoiesis
- Progressive decrease in the marrow's ability to respond to demands

Table 15.10 Etiopathogenetic classification of hemolytic anemias

Forms of hemolytic anemia	
Hereditary	Acquired
Membrane defects Defects of enzymes of the glycolytic pathway Abnormality of nucleotide metabolism Defects of the pentose phosphate and glutathione cycle Defects of structure and synthesis of globins	Immunohemolytic diseases Traumatic and microangiopathic anemias Anemia due to infectious agents Anemias from drugs, chemicals, poisons Physical agent anemias Hypophosphatemia anemia Paroxysmal nocturnal hemoglobinuria Anemias of liver disease (with *spur cells*) Vitamin E deficiency anemia in infants

- Reduction in erythrocyte survival up to 15–20 days without anemia (compensated state, limited in time and variable in magnitude)

The wide variety of hemolytic conditions makes it necessary to proceed systematically, subdividing hereditary and acquired disorders on an etiopathogenetic basis (Table 15.10). Membrane defects represent the most numerous groups of hereditary hemolytic conditions, and among them, hereditary spherocytosis is relatively more frequent. The main clinical features to be detected and evaluated in the hereditary forms are listed below.

- Degree of anemia:
  - Absent, sometimes in crisis
  - Variable, moderate to mild (compensation)
  - Severe with cardiovascular syndrome
- Presence of jaundice:
  - Neonatal
  - Persistent, episodic, absent
  - Acholuric, no itching
- Episodes of aplastic crisis:
  - From human parvovirus type B19 (HPV) → inhibition CFU-Es
  - Hb drop of 20–60 g/L
  - Erythroid hypo/aplasia (folate deficiency)
- Splenomegaly
- Cholelithiasis ➔ it is often the first symptom
- Lower limb ulcers, especially in cases of spherocytosis and sickle cell trait
- Skeletal changes due to erythroid expansion in the marrow

Hemolytic crises can be characterized by acute onset (e.g., after a noncompatible transfusion) similar to those of acute febrile illness with pain in the back, abdomen, head-

Table 15.11 Hemolytic disorders: signs of increased red blood cell destruction

Test	Results	Comments
RBC survival (^{51}Cr)	Up to <30 days	Test performed in specialized centers
Endogenous CO	>1.5% to 10–12% (infants)	Little/not used
Quantitative fecal urobilinogen	Increased	Little/not used
Hemolytic index (urobilinogen/Hb mass)	32–792 mg/day/100 g Hb (IR 11–21)	Little/not used
LDH-2 in plasma	Increased	Not very specific
Haptoglobin	Hemolysis for values <250 mg/L (sensitivity 83% and specificity 96%)	Useful

Table 15.12 Hemolytic disorders: signs of intravascular hemolysis

Test	Results	Comments
Hemoglobinemia	IR < 10 mg/L	Test requiring technical precautions
Hemoglobinuria	Search for oxidized dimers of Hb	Rarely used
HbA1c	Reduced	Little/not used
Sideruria with Prussian blue	Hemosiderinuria in tubular cells	Little/not used
Methemalbumin, hemopexin in plasma	Presence of spectral bands at 620–630 nm that do not disappear with H_2O_2	Little/not used

Table 15.13 Hemolytic disorders: signs of accelerated erythropoiesis

Test	Results	Comments
Count of reticulocytes	Increase	Used
MCV	Increase	Used
Circulating erythrocyte precursors	Presence	Used
White blood cell count	Increase	Used
Platelet count	Increase	Used
Bone marrow	Medullary erythroid hyperplasia	Specialist
Fe plasma/Fe erythrocyte turnover	Increase	Little/not used
Erythrocyte creatine	Increase (IR 14–90 mg/L)	Little/not used
Erythrocyte enzymes (Hx, GOT, etc.)	Increase	Little/not used

ache, vomiting, oligo/anuria, pallor, tachycardia, or subacute onset with gradual and insidious progression, with few symptoms, good compensation at least initially, jaundice, pallor, then crises with worsening. From the point of view of the clinical laboratory, the diagnostic indicators should be sought in different directions, as reported in Tables 15.11, 15.12, and 15.13.

In hemolytic conditions, direct and indirect antiglobulin tests (the Coombs test) should always be performed. However, it should be considered that 2–5% of patients with anemia of immunohemolytic origin are falsely negative due to the reduced sensitivity of the test.

The evaluation of erythrocyte osmotic fragility is based on tests that are parallel to those described for the screening of thalassemias; however, the mechanism and meaning are completely different, which makes these fragility tests useful for screening and therefore sensitive but little specific. In these tests, the principle is always that of hemolysis for osmotic stress, obtained either with progressively hypoosmotic solutions of NaCl or with solutions of glycerol in the acid buffer. Hemolysis begins when the erythrocyte surface/volume ratio reaches a critical volume, which in a normal subject is ~165%, and occurs at a NaCl concentration of 0.45–0.50% to reach a 50% hemolysis value at 0.40–0.45%

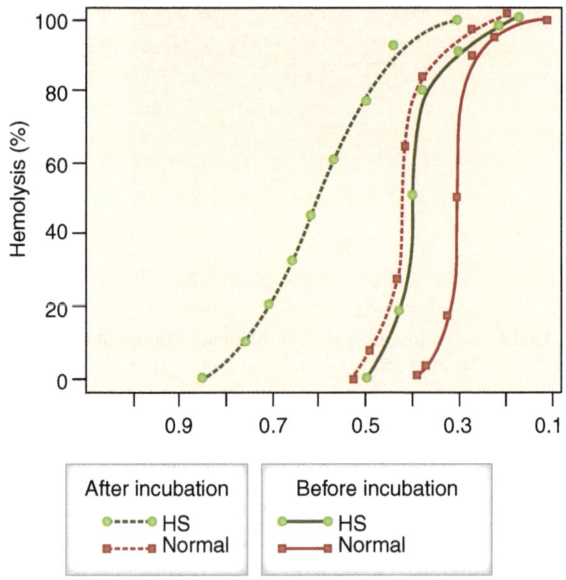

Fig. 15.24 Osmotic fragility test in progressively hyposmotic NaCl solutions of a sample from a normal subject and one from a subject with hereditary spherocytosis before and after incubation. (Copyright EDISES 2021. Reproduced with permission)

NaCl. If hemolysis is observed at higher NaCl concentrations, this is referred to as increased fragility due to the shift of the hemolysis curve to the left (Fig. 15.24). A normal erythrocyte swells when placed in 0.6% NaCl; hemolysis begins at 0.42% NaCl when the erythrocyte volume rises to ~145% of normal; in 0.35% NaCl, hemolysis is complete. In borderline cases, the test can be made more sensitive by performing it on a blood sample preincubated at 37 °C for 24 h under sterile conditions. The test is positive for conditions associated with hereditary spherocytosis. Evaluation of erythrocyte osmotic fragility is also performed with variants of the glycerol test, previously described for thalassemias, using an acidified glycerol solution. Additional commercial tests are the Pink Test and the Osmored B.

The self-hemolysis test, once widely used for the evaluation of erythrocyte fragility but now less practiced, is based on the evidence of spontaneous hemolysis after incubation of sterile defibrinated blood at 37 °C for 48 h in the presence of two parallel tubes containing glucose and ATP. In spherocytosis and various enzyme deficiency conditions, hemolysis is corrected by both additions; however, the test is not very sensitive and specific.

Various hemolytic conditions show the presence of Heinz bodies. These are bodies of denatured hemoglobin adherent to the erythrocyte membrane. Causal conditions include glucose-6-phosphate dehydrogenase (G6PD) deficiency, the presence of unstable hemoglobins, some hemolytic thalassemias, and some chemical agents. Heinz bodies are visible microscopically after staining with supravital staining (e.g., methyl violet). In enzymatic deficiency conditions, their presence is accentuated by incubation with in vitro oxidizing agents (e.g., acetylphenyl hydrazine).

The glutathione stability test evaluates the residual reducing capacity of the erythrocyte under conditions of oxidative stress. Glutathione is mainly present in the erythrocyte in its reduced form (GSH), at concentrations in the reference range of 5.3–6.8 μmol/g Hb. Under normal conditions, these values vary little, even under oxidative stress (e.g., with acetylphenyl hydrazine). In enzyme deficiencies, particularly G6PD deficiency, GSH falls to values near zero.

Other laboratory tests generally used in hemolytic conditions include measurement of intrinsic enzyme activities of erythrocytes, assessment of hemoglobin balance when unstable hemoglobins are suspected, and Ham's test for paroxysmal nocturnal hemoglobinuria (PNH), a membrane structure disorder, performed to reveal increased sensitivity of cells to complement lysis.

The diagnostic approach to hemolytic anemias involves the establishment of two conditions, the first of which is preliminary: (1) demonstration of the presence of the hemolytic state and (2) determination of the specific cause of hemolysis. In the differential diagnosis of hemolytic conditions, it is important to consider conditions that may mimic the presence of hemolysis (Table 15.14).

Once the presence of hemolysis has been demonstrated, the process of identifying the specific cause must begin with an accurate medical history, a microscopic examination of the blood smear, and antiglobulin testing. Based on the data obtained, patients can be distinguished into five groups:

1. Patients with exposure to hemolytic agents.
2. Patients with positive antiglobulin tests.
3. Patients with a negative antiglobulin test are likely to have spherocytosis.
4. Patients with morphological abnormalities (ellipsocytes, schistocytes, ovalocytes, etc.).
5. Patients with a negative antiglobulin test and normal morphology, on which to proceed with investigations on erythrocyte enzymes, hemoglobin balance, PNH tests, etc.

Hereditary Spherocytosis

Hereditary spherocytosis (HS) is a form of hereditary hemolytic anemia common in Northern Europe and is inherited in an autosomal dominant pattern due to alterations of a gene on chromosome 8, in which about a quarter of the cases are clinically normal at diagnosis and of which sporadic cases due to new mutations are described. The disease is due to defects in the expression of erythrocyte membrane proteins that cause structural changes. The mechanism of sphericization in spherocytosis is caused by a spectrin deficiency, either primary or secondary to an ankyrin deficiency. This results in reduced cytoskeleton density, failure of the cytoskeleton to bind to band 3 (structural membrane protein), and subsequent instability of the lipid bilayer. Thus, lipid loss in the form of vesicles is observed both *in vitro* and *in vivo*. A similar effect is caused by the reduction of band 3. Further causes of spherocytosis are due to protein 4.2 deficiency, in which altered binding of spectrin via ankyrin and band 3 to the membrane protein CD47 is observed, and defective spectrin-binding protein 4.1. The various defects cause alterations in membrane properties that are reflected in erythrocyte metabolism (Table 15.15). Increased erythrocyte fragility is

Table 15.14 Conditions mistaken for hemolytic state

Associated with anemia and reticulocytosis Hemorrhage Deficiency of Fe, folate, or vitamin B12 in the recovery phase Recovery after bone marrow deficiency (e.g., poisoning)
Associated with anemia and acholuric jaundice Ineffective erythropoiesis (intramedullary hemolysis) Intracavitary or intratissutal hemorrhages
Acholuric jaundice without anemia
Bone marrow invasion (myelofibrosis, metastasis)
Myoglobinuria

Table 15.15 Erythrocyte abnormalities inherent in hereditary spherocytosis

Defect of erythrocyte membrane skeleton	Spectrin deficiency Protein 4.2 deficiency Spectrin–protein 4.1 bond defective
Altered properties of membrane	Loss of surface area Lipid alterations Altered calcium content Altered membrane proteins Increase of membrane catalase and/or Hb Altered protein phosphorylation Protein aggregation defective
Altered metabolism cell phone	Increased sodium permeability Increased glycolysis Increased turnover of ATP Decreased transport of phosphoenolpyruvate (PEP)

directly related to spectrin deficiency, which can range from 30 to 80% of normal, and the severity of hemolysis. The chronic hemolysis that usually accompanies spherocytosis requires splenectomy in the most serious cases. The efficacy of the treatment appears to be related to the extent of spectrin deficiency so that with a deficiency >70%, a practically complete reduction of chronic hemolysis is usually observed. The laboratory data characteristics of hereditary spherocytosis are shown in Table 15.16. The diagnosis and judgment of the severity of hereditary spherocytosis are based on the integration of clinical and laboratory data. Sometimes anamnestic and family reports, as well as splenectomy, are useful (Table 15.17). A consistent presence of spherocytes is not always detectable; indeed sometimes, they are scarce and therefore do not represent a safe indicator. In addition, there are other conditions that can cause spherocyte formation, such as immunohemolytic anemias, certain enzymatic deficits of the erythrocyte and trauma. Although not very sensitive, an increase in self-hemolysis and its correction with glucose and ATP are indices with good diagnostic value. To have diagnostic certainty, it is necessary to analyze the membrane protein content by separation procedures (SDS-PAGE electrophoresis, HPLC+MS). An alternative procedure is flow cytometry of intact red blood cells labeled with eosin-5′-maleimide, which binds to band 3.

Table 15.16 Tests and laboratory data in pre and postsplenectomy hereditary spherocytosis

Presplenectomy	
Modest anemia often compensated	Erythrocyte osmotic fragility curve with hemolysis starting from 0.5 to 0.7 g/dL of NaCl with tails
MCHC increased due to erythrocyte dehydration	Increase in fragility after incubation at 37 °C for 48 h
Normal/variable MCV and MCH	Corrected hemolysis by glucose and ATP
RDW mostly increased	Increase of unconjugated bilirubin
Reticulocytosis (5–20%)	Decrease of haptoglobin
Polychromatophilia and anisocytosis	Coomb's test negative
Spherocytes (Ø6.2–7.0 μm, thickness 2.2–3.4 μm) in number from 1% to 30%	Modest neutrophilia-platelet disease
Normoblasts and normal reticulocytes	Medullary hyperplasia (megaloblastosis)
Postsplenectomy	
Hb increases	Hb increased to normal-high values
RDW returns to normal	Normal 2,3-DPG
Reduced microcytes	P_{50} normal-high
Hyperchromic cells increased	Reticulocytosis remains moderate
	Normal-high bilirubin
	Erythrocyte osmotic fragility curve loses tails

Table 15.17 Severity classification of hereditary spherocytosis

	Mild	Moderate	Serious
Hemoglobin (g/L)	110–150	80–120	60–80
Reticulocytes (%)	3–6	>6	>10
Bilirubin (mg/L)	17–34	>34	>51
Splenectomy	Usually not necessary	Indicated during school age, before of puberty	Necessary (delay up to 6 years of age if possible)

Rare forms of morphological abnormalities are represented by hereditary elliptocytosis and ovalocytosis. These are a heterogeneous group of conditions in which the suggested, but not exclusive, diagnostic criterion is the presence of at least 25% abnormal (elliptical/oval) erythrocytes under the microscope. These forms are also due to genetic mutations affecting the expression of membrane cytoskeleton proteins.

Glucose-6-Phosphate Dehydrogenase (G6PD) Deficiency

G6PD is an enzyme widely distributed in cells of all organs. In erythrocytes, it catalyzes the first reaction of the pentose phosphate pathway, which performs an essential function in oxidoreductive balance, as it is the only source of hydride ions (one proton and two electrons) for the reduction of NADP+ to NADPH and thus for the reduction of glutathione and the elimination of hydrogen peroxide (H_2O_2) (Fig. 15.25). The gene encoding the enzyme is in the telomeric region of the long arm of the X chromosome. More than 300 variants of the enzyme are known, of which four are the most common, with different properties (Table 15.18). Since the hereditary transmission is linked to the X chromosome, males are hemizygous for the relevant gene, so in them, there are two possibilities of expression, normal or deficient G6PD. In females, since there are two X chromosomes, there are three possibilities of expression: normal, heterozygous, and homozygous (in populations with a high frequency of deficiency). Heterozygous females have mosaicism due to random inactivation of the X chromosome. Therefore, there can also be conditions of complete inactivation and deficit similar to that of the male, but these conditions are rarer as usually, nonheterozygous females do not manifest severe deficits as the male. G6PD variants are classified according to their activity values compared to those of the normal contextual population. Specifically, class I variants cause congenital nonspherocytic hemolytic anemia (activity <10% of normal); class II variants exhibit severe deficiency but not hemolytic anemia (activity <10% of normal); class III variants exhibit mild to moderate deficiency (activity 10–60% of normal); class IV variants exhibit very mild deficiency (activity 60–100% of normal); and class V variants are very rare and exhibit increased enzyme activity compared to normal, over 150%. Although many individuals with deficiency are asymptomatic throughout life, acute hemolytic crises can be observed unpredictably because of oxidative stress, which

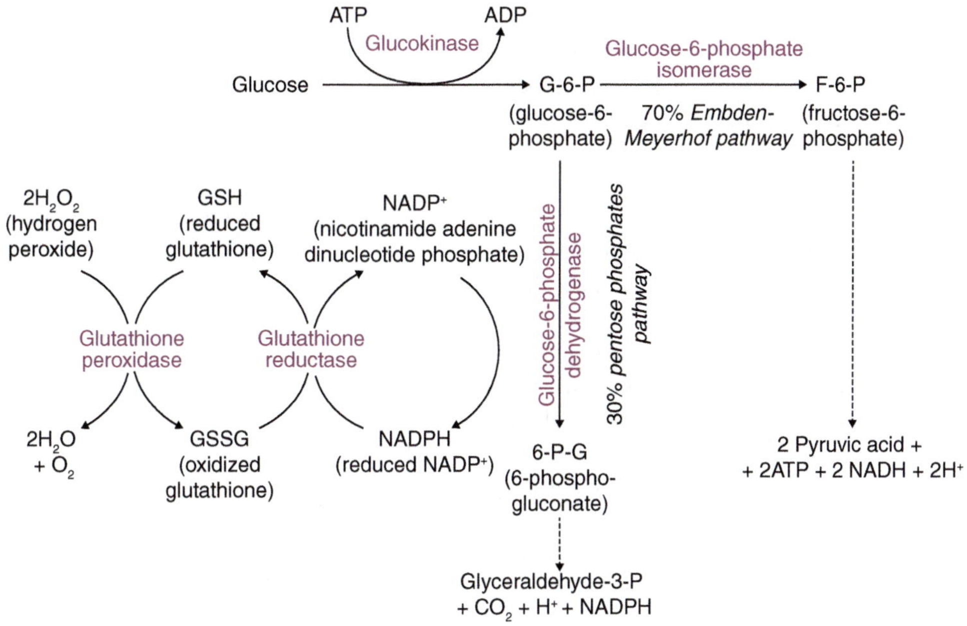

Fig. 15.25 Metabolism of the red blood cell with the cycle based on G6PD, 6PGD and glutathione oxidation-reduction by NADP+/NADPH and glutathione-peroxidase/glutathione-reductase highlighted, with a final reduction of H_2O_2 to H_2O. (Copyright EDISES 2021. Reproduced with permission)

Table 15.18 Common variants of glucose-6-phosphate dehydrogenase

Variant	Activities, characteristics, and structure	Diffusion
Gd^B	Normal activity Half-life 62 days	All world populations
Gd^A	Normal activity Half-life 13 days Aspartic acid → Asparagine[126] G → A[376]	Africa (the most common)
Gd^{A-}	8–20% of normal activity Methionine → Valina[67] Aspartic acid → Asparagine[126] A → G[202] G. → A[376]	Africa
Gd^{Med}	Activity <5% of normal Half-life 13 days Highly unstable Phenylalanine → Serine[188] T → C[563]	Mediterranean (Greece, Sardinia, etc.), Iran, Iraq, India, Pakistan

can be caused by drugs, infections, ingestion, or simple contact with broad beans, especially in Mediterranean forms. The precise mechanism by which the crises are triggered is not known, and they appear to be due to increased sensitivity to oxidative damage. Lists of substances that are capable of triggering seizures have been prepared and can be found on several websites, which are continuously updated: http://www.cych.org.tw/pharm/MIMS%20Summary%20TableG6PD.pdfandhttp://www.g6pd.org/it/G6PDDeficiency-it/SafeUnsafe-it/DaEvitare_ISS-it.aspx.

During a crisis, fatigue, back pain, anemia, jaundice with increased unconjugated bilirubin, reticulocytosis, increased LDH, the appearance of Heinz bodies, hemoglobinemia, hemoglobinuria, and other symptoms common to acute hemolytic manifestations are generally observed. A G6PD deficiency may be revealed at birth in the form of neonatal hemolytic jaundice, which should be considered in the differential diagnosis. To prevent such occurrences, in many countries, screening for G6PD deficiency is part of neonatal prevention programs. Seizures are more characteristic of Gd^{Med} enzyme variants, while chronic hemolysis with chronic jaundice and cholelithiasis are observed more in Gd^A variants. In Africa and the Mediterranean, the selection of the deficient gene has been operated by malaria, particularly the one due to Plasmodium falciparum, as the parasite entering the erythrocytes of a subject with deficiency is in an unfavorable metabolic environment, with limited ability to replicate.

The laboratory diagnosis of G6PD deficiency is based on the evaluation of enzymatic activity by quantitative spectrophotometric analysis of NADPH produced by NADP+ in a physiological pH system having as substrate glucose-6-phosphate to which is added hemolysate of the sample to be examined. There are standardized commercial methods based on a procedure initially recommended by the World Health Organization (WHO). Screening tests are also available, which can make use of samples collected on bibulous paper. Since the activity of the enzyme in reticulocytes is much higher than the average of red blood cells (the half-life of the enzyme is 62 days), it is necessary to interpret very carefully the results obtained from subjects with reticulocytosis, as can happen at the end of a hemolytic crisis because the values obtained could simulate a normal condition.

Unstable Hemoglobins

Historically, the name refers to the property of some Hb to precipitate when incubated at 50 °C for 1 h, unlike HbA, which remained stable. Subsequently, new tests were developed to assess stability, such as incubation at higher temperatures (up to 65 °C) and kinetic assessment of stability. One test that has proven to be very valid for assessing the stability of Hb is the isopropanol test, which, however, should be considered parallel to and not a substitute for traditional heat tests.

The instability is caused by structural changes that allow water (H_2O) access to the heme (hydrophobic) pocket, resulting in the development of free radicals, denaturation, and precipitation of the molecule.

Typically, it is the introduction of proline or glycine into an Hb helix that disrupts the secondary structure and causes instability and precipitation with the detachment of the heme. Structural abnormalities can be close to the heme or in the contact area between globin and heme or be due to insertion or deletion (Fig. 15.26). The α, β, and γ globins are affected.

Of the approximately 250 mutations that affect globin chains and can reduce the stability of the molecule, roughly half are important because of the clinical consequences of the variant. Furthermore, since instability is caused in only a portion of the variant hemoglobin molecules, few Hbs are detectable by traditional methods. Mutations can cause four conditions:

- Instability detectable only in vitro, without clinical-hematological consequences.
- The destruction of the most unstable Hb within minutes/hours of synthesis causes ineffective erythropoiesis and thalassemia-like syndrome. They should be identified by molecular biology methods.
- The intermediate condition between the two previous ones, with a variant present in erythrocytes where it precipitates under conditions of oxidative stress.
- Dominant phenotypic characteristics due to de novo mutations.

From a clinical point of view, they are present at varying degrees depending on the type of variant:

- Congenital nonspherocytic hemolytic anemia with splenomegaly and cholelithiasis (bilirubin stones)
- Hemolytic anemia with Heinz bodies, with sensitivity to oxidizing drugs such as sulfonamides

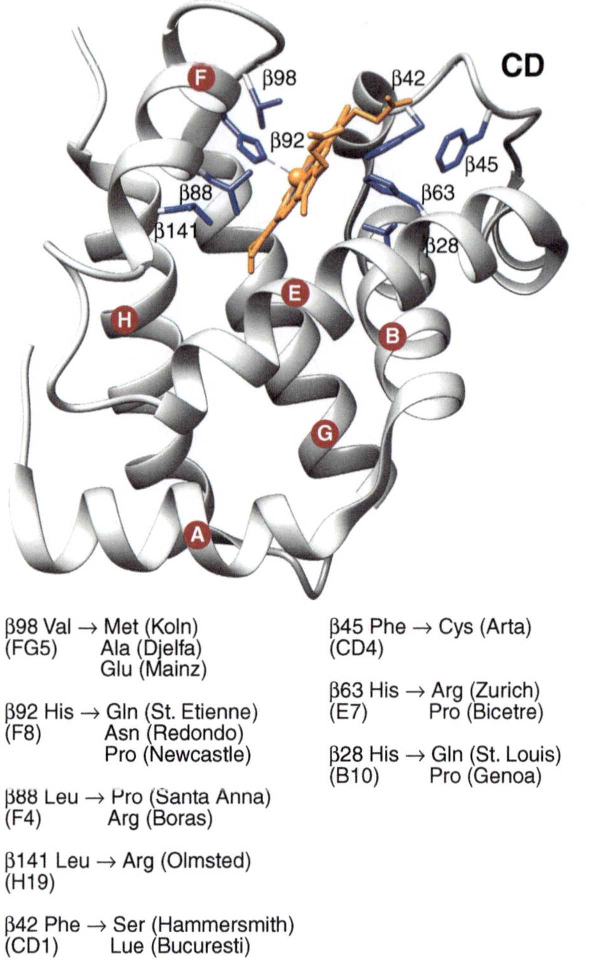

Fig. 15.26 Molecular abnormalities causing hemoglobin instability. Next to each type of abnormality are the names of some better-known unstable hemoglobins. (Copyright EDISES 2021. Reproduced with permission)

- Modest/slight anemia with reticulocytosis disproportionate to the level of circulating hemoglobin
- Peripheral hematological picture similar to thalassemia with hypochromic erythrocytes
- Increased methemoglobin formation

Among the most known variants, Hb Koln (β 98 Val → Met) has been described in different groups and geographical locations; Hb Hasharon (α 47 Asp → His) is predominantly described among Ashkenazi Jews where it causes neonatal hemolysis; unstable γ variants, such as Hb Poole (γ 130 Trp → Gly), are associated with hemolysis in the first months of life and then disappear as the synthesis of γ chains decreases within the first year of life; on the contrary, children with unstable β variants appear normal at birth, with progressive hemolysis developing during the first year of life as a result of increased synthesis of β chains. Unstable α variants differ depending on whether the affected gene is α1 or

Table 15.19 Main unstable hemoglobins

First name	Mutation	Mechanism	Clinical expression	Observation
Hb Koln	β98 (FG5) Val → Met	Heme contact, heme detachment, increased O_2 affinity	Medium to severe chronic hemolytic anemia	Frequent all over the world
Hb Hammersmith	β42 (CD1) Phe → Ser	Heme contact, heme shift to the deoxy form, decreased O_2 affinity, hemicrome formation	Very severe hemolytic anemia with Heinz bodies	Very rare
Hb Genoa	β28 (B10) Leu → Pro	Internal, α-helix disruption	Very severe hemolytic anemia with Heinz bodies	Rare, in a few families
Hb Medicine Lake	β98 (FG5) Val → Met + β32 (B14) Leu → Gln	Two mutations are associated in the same gene, each causing instability. The hyper-unstable β-chain variant is destroyed in erythroid precursors	Thalassemia major-like syndrome	Single case described
Hb Aghia Sophia	α62 (E11) Val → 0 (α1)	Deletion in frame of a residue in the middle of a helix with alteration of the heme pocket structure	Determine α+-thalassemia	Interaction with a deletion – MED in a patient with HbH disease
Hb Constant Spring	α142 Term → Gln (+31 residues) (α2)	Elongated unstable C-terminal α-chain	Determine α+- thalassemia, severe hemolytic anemia, and HbH disease when associated with α1 thalassemia (with deletion –)	Frequent in southeast Asia

α2, which encodes for higher levels of mRNA. Consequently, a mutation in the α2 gene has a stronger clinical impact than that in α1, resulting in classic unstable Hb syndrome and syndrome with thalassemia-like phenotype with intramedullary Hb destruction. Some forms of HbH (β4) are due to the unstable α gene. These forms, not being of thalassemic nature, should be considered during genetic counseling. The main unstable hemoglobins are shown in Table 15.19.

The diagnosis of unstable hemoglobin should be made based on the following:

- Regenerative hemolytic anemia with increased reticulocytosis.
- Heinz bodies (not always pathognomonic as they are also present in erythroenzymopathies and following the intake of certain drugs).
- Anisocytosis, basophilic punctuation, and macrocytosis with reduced MCH.
- Separation procedures (electrophoresis, HPLC). However, they do not always allow identifying variants unless they are mutations causing altered HPLC mobility and/or electrophoretic migration.
- The presence of hemichromes, semi-Hb, or free α-chains, may indicate the presence of unstable β-chain.
- Combinations of methods and use of HPLC combined with MS.

Macrocytic Anemias

Two types can be distinguished according to the causes:

- Vitamin B12 (cobalamin) or folate deficiency causes macrocytic anemia, in which the marrow is megaloblastic. Folate and vitamin B12 are closely involved in DNA synthesis through the methylation of deoxyuridine monophosphate to deoxythymidine monophosphate.
- Anemias from other causes, in which the marrow is normoblastic.

Hyporegenerative Macrocytic Anemias/ Megaloblastic Anemias

They are due to vitamin B12 and folate deficiency. Absorption of vitamin B12 occurs through the ileum, facilitated by Castle intrinsic factor (IF), which is secreted by the parietal cells of the stomach; underlying the anemia is an autoimmune mechanism that leads to loss of intrinsic factor secretory capacity. The causes of megaloblastic anemia are as follows:

1. Folate deficiency:
   - Diet (elderly patients and alcoholics)
   - Malabsorption (ethylism, coeliac disease, tropical sprue, Crohn's disease, and jejunal resection)
   - Increased demand (pregnancy, prematurity, chronic hemolytic anemia, inflammatory diseases, and neoplasms)
   - Drug interference (antiepileptics, folate antagonists such as methotrexate, trimethoprim, sulfasalazine, and oral contraceptives)
   - Renal loss (congestive heart failure and dialysis)
   - Hereditary abnormalities of folate absorption and metabolism (absence of enteric folate receptors)
2. Vitamin B12 deficiency:
   - Reduced intake and release of vitamin B12 by diet
   - Inadequate secretion of intrinsic factors, which can be distinguished in:
     - Classic pernicious anemia, an autoimmune condition with antibodies directed against the parietal cells of the stomach; in the affected patient there is

autoimmune gastritis or atrophy of the parietal cells of the gastric mucosa, with achlorhydria and reduced secretion or absence of intrinsic factor. This deficiency limits vitamin B12 absorption. The lack of IF can be congenital (rare) or acquired on an autoimmune basis with the presence of highly specific pernicious anemia intrinsic factor antibodies (50–70% of cases). Anti-intrinsic factor antibodies have been found in some patients with Graves' disease and in insulin-dependent diabetes. They are of two types: type I antibodies, which block the binding of vitamin B12 to the IF, and type II antibodies, which block the binding of the B12-IF complex to the ileal receptor. Antiparietal cell antibodies have been reported in 50–90% of patients with pernicious anemia.
   - Total or partial gastrectomy.
   - Congenital intrinsic factor deficiency or abnormal intrinsic factor.
3. Diseases of the ileum:
   - Ileal resection
   - Crohn's disease
   - Chronic tropical sprue
   - Other acquired diseases
4. Different utilization of vitamin B12:
   - Abnormal intestinal flora (diverticula of the jejunum, excluded intestinal loop with stagnation of bacteria).
   - Parasitic infestation by *Diphyllobothrium latum* (botryocephalus).
5. Defective release of vitamin B12:
   - Pancreatic insufficiency due to lack of trypsin which detaches vitamin B12 from its complex.
   - Zollinger–Ellison syndrome: acidification of the small intestine delays the absorption of vitamin B12.
6. Hereditary abnormalities of vitamin B12 transport or its intracellular metabolism.

The causes of vitamin B12 deficiency listed at points 3–6 occur in the presence of a normal intrinsic factor.

Causes of Macrocytosis with Normoblastic Erythropoiesis

Some patients present with macrocytosis with MCV up to 110–112 fL independent of vitamin B12 and folate. Causes are as follows:

- Chronic alcohol abuse
- Myelodysplastic syndromes
- Hypothyroidism
- Chronic lung disease with hypoxia
- Heavy smokers
- Chronic liver diseases
- Chronic hemolytic anemia
- Therapy with antiepileptic drugs
- Myeloma (some cases)
- Hypoplastic anemia

In some physiological conditions, such as infants and women with physiological pregnancies, macrocytosis could be observed.

The laboratory diagnosis of megaloblastic anemia is based on the following findings:

- Normochromic macrocytic anemia (MCV >102 fL; ranging from 100 to 160 fL). Macrocytosis may not be as evident if iron deficiency anemia, thalassemic trait, or intercurrent infections coexist. It is possible that even in the presence of vitamin B12 deficiency, the hematocrit value may be normal; hemoglobin and erythrocyte count are both significantly reduced, and MCH is increased with normal MCHC. It should be kept in mind that MCHC represents the ratio of the mean mass of hemoglobin contained in each erythrocyte to the mean mass of the erythrocyte itself. This ratio tends to be rather conservative in many hematologic disorders, ranging between 32% and 35%. As mentioned above, the most well-known condition in which MCHC increases is hereditary spherocytosis. In this condition, MCHC is higher because the erythrocyte membrane mass decreases and erythrocytes accumulate excess H_2O due to altered extrusion mechanisms.
- Low reticulocyte count, unless a hemolytic component is present.
- Possible decrease of leukocytes and platelets.
- Hypersegmented neutrophils; a plurilobed nucleus with 5–6 lobes is always indicative of megaloblastic anemia.

Other data that may be altered are those related to possible ineffective bone marrow erythropoiesis and hemolysis, in particular:

- Indirect bilirubin: increased
- Sideremia: increased
- Ferritin: increased
- LDH: increased
- Haptoglobin: decreased or absent.

Erythrocytes show variations in shape (macro-ovalocytes) and volume. They contain within them basophilic inclusions and nuclear remnants (Cabot rings and Howell–Jolly bodies).

The bone marrow aspirate is rich in megaloblasts and hyperplastic cells, with a decreased ratio of myeloid and erythroid series. Obvious alterations are present in the precursors of all cell lines, but the erythroid line is especially affected. Widespread basophilia of the precursors gives the megaloblasts a blue May-Grünwald-Giemsa (blue marrow) coloration. The nuclear chromatin is scattered and appears poorly stained, with fenestrated aspects typical of megaloblastic anemia. The presence of ringed sideroblasts in the marrow, as well as hypersideremia and hyperferritinemia, indicates decreased iron utilization.

Platelet Alterations

The study of platelet alterations has benefited from considerable progress in recent decades, mainly due to important acquisitions in physiopathology, new knowledge in molecular biology, and the development of innovative technologies applied to instruments that have progressively spread in laboratories. The starting point, however, is the complete CBC with microscopic observation of the peripheral blood smear. From these initial tests, it is possible to obtain information on the diagnosis of two main platelet diseases: immune thrombocytopenic purpura, also known as Werlhof's disease, and thrombotic thrombocytopenic purpura (TTP). In addition, immunoglobulin measurements are sometimes required in pediatric patients to rule out common variable immunodeficiency as a cause of thrombotic thrombocytopenic purpura. Careful observation of the peripheral smear is essential in a patient with thrombocytopenia. It can rule out forms of pseudothrombocytopenia, such as those due to platelet satellites, in which platelets adhere to neutrophils, causing a false reduction in count on CBC (Fig. 15.27). Moreover, giant platelets could be found in patients with immune thrombocytopenic purpura; in this case, they are not specific markers of the condition but rather a sign of the stimulus exerted by thrombopoietin on the bone marrow due to thrombocytopenia (Fig. 15.28).

In thrombotic thrombocytopenic purpura, fragmentation of red blood cells (schistocytes) due to peripheral microangiopathy, as well as thrombocytopenia, could be observed. The minimum criteria for the diagnosis of thrombotic thrombocytopenic purpura are thrombocytopenia and microangiopathic hemolytic anemia without apparent etiology. In the complete form, neurological abnormalities, high fever, and acute renal failure occur. A smear examination shows thrombocytopenia and the presence of schistocytes. In addition, increased LDH and reticulocytosis are observed. Signs of intravascular coagulation are absent or only slightly increased in patients with TTP.

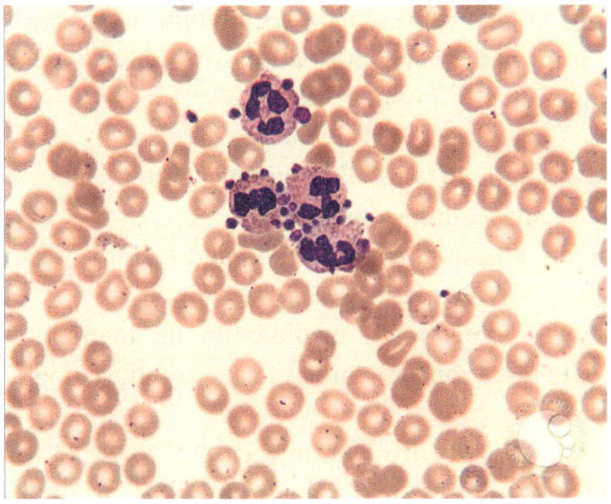

Fig. 15.27 Microscopic image of a blood smear from a patient with pseudothrombocytopenia from satellite

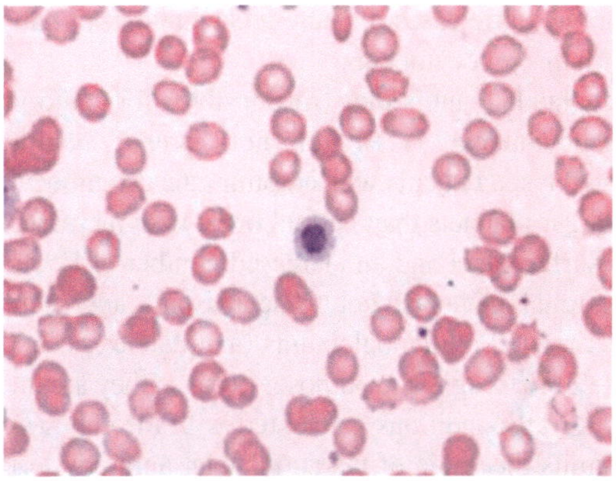

Fig. 15.28 Giant platelet in peripheral smear. Normal-sized platelets on the right are observed for comparison

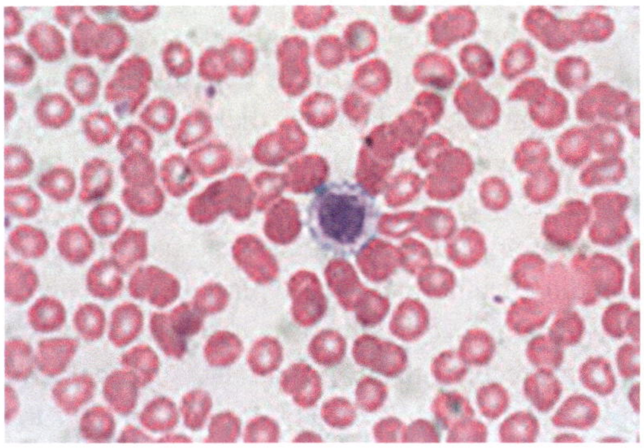

Fig. 15.29 Giant platelet in a patient with Bernard–Soulier syndrome

Rare congenital thrombocytopenia, known as Bernard–Soulier syndrome, which causes easy nose and gum bleeding as well as gastrointestinal bleeding, can also be suspected based on the examination of the peripheral smear (Fig. 15.29) when thrombocytopenia is accompanied by the presence of giant platelets. This syndrome is due to a molecular defect in which the glycoprotein complex of the platelet membrane, gp Ib-gp V-gp IX is missing, with a consequent inability to bind to von Willebrand factor (VWF) and therefore to adhere to the platelet. The genes involved are GP1BA, GP1BB, and GP9, which encode for gp Ibα, gp Ibβ, and gp IX subunits. The bleeding time is prolonged. A careful smear examination is essential to exclude thrombotic thrombocytopenic purpura and rare cases of acute leukemia that initially present with thrombocytopenia.

Antiplatelet antibodies responsible for auto-immune thrombocytopenia (ITP) do not induce complement-mediated lysis. Patients with autoimmune thrombocytopenia have larger platelets than normal ones, with increased membrane surface area and an increased number of Fc receptors than normal platelets. For these reasons, any increase in platelet-related immunoglobulins (antiplatelet antibodies) is not useful for diagnostic purposes because they are increased in almost all conditions associated with thrombocytopenia. This limits the value of the antiplatelet antibody test in the diagnosis of immune thrombocytopenic purpura.

One form of acquired thrombocytopenia that often comes first to the attention of the laboratory is heparin-induced thrombocytopenia. This is a particular alteration usually observed in samples from surgical (but also medical) patients who have received some form of heparin for prophylactic/therapeutic purposes, although the highest frequency is observed with unfractionated heparin and can affect up to 5% of patients. The cause is due to the exposure of a particular masked epitope that occurs in some individuals because of the binding that normally takes place between heparin and PF4. The generation of antibodies, in response to this epitope exposure, proceeds with the formation of circulating immune complexes that activate platelets through binding to the Fc receptor. Platelet activation continues with further release of PF4, and the phenomenon tends to amplify, causing thrombocytopenia due to the aggregation of additional platelets and hypercoagulability that can result in thrombosis. Heparin-induced thrombocytopenia appears approximately 5–10 days after the initiation of drug administration. A rapid-onset form of heparin thrombocytopenia is anamnestically observed in patients who had previously received heparin without developing thrombocytopenia. The immediate measure is to discontinue heparin administration. The risk is much lower with fractionated low molecular weight heparins. The risk is absent when using one of the direct anti-FXa anticoagulants (e.g., rivaroxaban) for the same indications.

A rare form of spurious thrombocytopenia that usually comes first to the attention of the laboratory is that resulting from samples collected in EDTA. This phenomenon is due to the presence of EDTA-dependent antiplatelet antibodies that are activated at low temperatures. They cause agglutination of platelets and, therefore, a lowering of count values in automated instruments. It is suspected when unexplained thrombocytopenia, without clinical signs (absence of petechiae, normal bleeding time), is observed in outpatients. In order to ascertain whether EDTA is the cause of thrombocytopenia, it is necessary to take a second sample with a different anticoagulant (e.g., sodium citrate) and perform the count again, which will give a normal value if positive.

Bleeding time is the most widely used test for assessing primary hemostasis. However, despite its theoretical usefulness, this test is highly operator-dependent and not recommended as a screening test in the diagnostic routine for disorders of hemostasis. Bleeding time is measured by measuring at 30-s intervals the time required to observe the cessation of bleeding following a 1-mm-deep, 1-cm-long superficial cut on the volar surface of the forearm made by a standard cutter. Under these conditions, cessation of bleeding results from the formation of a primary hemostatic plug. There is a fairly linear correlation between bleeding time and platelet count between 10,000/μL and 100,000/μL. Bleeding time is prolonged with platelet counts below 75,000/μL, although this does not provide an explanation as to why the platelet count is low. In patients with thrombocytopenia, bleeding time should not be performed as it has no diagnostic significance. In alterations of secondary hemostasis related to coagulation mechanisms (e.g., in patients with hemophilia A or B), the bleeding time is always normal.

Evaluation of Platelet Function In Vitro with PFA-100

The Platelet Function Analyzer 100 (PFA100) is an automated benchtop instrument that evaluates primary hemostasis in vitro in a shear-stressed blood sample. The PFA-100 uses disposable test cartridges, each containing a nitrocellulose membrane impregnated with collagen plus ADP (Col/ADP) or collagen plus epinephrine (Col/Epi). A sample of citrated blood (0.8 mL) is placed in a cup of the instrument and aspirated through the membrane opening. Shear stress occurs when platelets pass through Col/ADP or Col/Epi, which act as activation agonists, aggregate and progressively reduce until the flow through the membrane is stopped. This represents the endpoint, expressed as the time of closure. The formation of platelet aggregates depends on the following:

- The binding of von Willebrand factor to the collagen-coated nitrocellulose membrane
- Platelet adhesion to von Willebrand factor via activation of the gp Ib receptor
- Platelet aggregation is mediated by the interaction of gp IIb/IIIa with von Willebrand factor and fibrinogen

Normal closure times range from 77 to 133 s for the Col/ADP membrane and from 98 to 185 s for the Col/Epi membrane. The PFA-100 has been tested in patients with bleeding disorders. The closure time using the Col/Epi cartridge is abnormal in patients with congenital defects of platelet function and von Willebrand disease after aspirin intake, whereas the closure time using the Col/ADP cartridge is abnormal mainly in patients with von Willebrand disease and in congenital diseases. Aspirin prolongs the closing time in 94% of cases with the Col/Epi cartridge and only in 27% of cases with the Col/ADP cartridge. Glanzmann thrombasthenia, Bernard–Soulier syndrome, and most mild forms of von Willebrand disease are associated with prolonged closing time with both cartridges, whereas a storage pool defect and giant platelet thrombocytopathy have prolonged closing time only with the Col/Epi cartridge.

The advantages of this instrument include simplicity and reproducibility. The PFA-100 has been reported to have a coefficient of variation of less than 10%. It may be useful for determining global platelet function and assessing the efficacy of antiplatelet therapy.

Platelet aggregation is measured by turbidimetric methods. When platelets aggregate, the platelet-rich plasma suspension becomes clearer and allows more light to be transmitted. The degree of aggregation is determined by measuring the progression over time, usually for 5 min, of the increase in light transmission and by reporting the aggregation curve, which evaluates the reversibility/irreversibility of aggregation (primary vs. secondary), the shape change, that is, the change in platelet shape upon stimulation with agonists such as ADP and collagen, and the maximum amplitude of the aggregation wave.

Small doses of ADP (<1 µmol) induce a reversible form of platelet aggregation (primary wave), unaccompanied by thromboxane synthesis and intraplatelet ADP release. However, with increasing doses of ADP, sufficient stimulation of platelets occurs with intraplatelet ADP release and thromboxane A2 synthesis by rapid metabolism of arachidonic acid (released from platelet membrane phospholipids) by the cyclo-oxygenase enzyme; the final consequence is a more pronounced and irreversible aggregation wave (secondary wave). A secondary wave of aggregation is also caused by collagen and directly by arachidonic acid, which is used as an agonist. Ristocetin causes platelet aggregation by inducing von Willebrand factor to bind to the gp Ib protein complex on the platelet.

Platelet aggregation tests are useful in distinguishing various disorders of platelet function. They are also particularly used in the differential diagnosis of von Willebrand disease, in which ristocetin-induced platelet aggregation is defective.

Recommended Readings

Budak YU, Polat M, Huysal K (2016) The use of platelet indices, plateletcrit, mean platelet volume and platelet distribution width in emergency non-traumatic abdominal surgery: a systematic review. Biochem Med 26:178–193

Bull BS, Fujimoto K, Houwen B, ICSH Expert Panel on Cytometry et al (2003) International Council for Standardization in Haematology (ICSH) recommendations for "surrogate reference" method for the packed cell volume. Lab Hematol 9:1–9

Buttarello M, Plebani M (2008) Automated blood cell counts: state of the art. Am J Clin Pathol 130:104–116

Chaisue C, Kitcharoen S, Wilairat P et al (2007) alpha/beta-Globin mRNA ratio determination by multiplex quantitative real-time reverse transcription-polymerase chain reaction as an indicator of globin gene function. Clin Biochem 40:1373–1377

D'Angelo G (2013) Role of hepcidin in the pathophysiology and diagnosis of anemia. Blood Res 48:10–15

Davis BH, Jungerius B (2010) International Council for Standardization in Haematology technical report 1-2009: new reference material for haemiglobincyanide for use in standardization of blood haemoglobin measurements. Int J Lab Hematol 32:139–141

Farias MG, Schunck EG, Dal Bó S, de Castro SM (2010) Definition of reference ranges for the platelet distribution width (PDW): a local need. Clin Chem Lab Med 48:255–257

Felker GM, Allen LA, Pocock S, CHARM Investigators et al (2007) Red cell distribution width as a novel prognostic marker in heart failure: data from the CHARM Program and the Duke Databank for Cardiovascular Diseases. J Am Coll Cardiol 50:40–47

Harris JR (ed) (1990) Blood cell biochemistry. In: Erythroid cells, vol 1. Springer Science+Businness Media, New York

LaRue N, Kahn M, Murray M et al (2014) Comparison of quantitative and qualitative tests for glucose-6-phosphate dehydrogenase deficiency. Am J Trop Med Hyg 91:854–861

Lippi G, Plebani M (2012) EDTA-dependent pseudothrombocytopenia: further insights and recommendations for prevention of a clinically threatening artifact. Clin Chem Lab Med 50:1281–1285

Neunert C, Lim W, Crowther M, American Society of Hematology et al (2011) The American Society of Hematology 2011 evidence-based practice guideline for immune thrombocytopenia. Blood 117:4190–4207

Nierhaus A, Klatte S, Linssen J et al (2013) Revisiting the white blood cell count: immature granulocytes count as a diagnostic marker to discriminate between SIRS and sepsis-a prospective, observational study. BMC Immunol 12:14–18

Ryan DH (2010) Examination of blood cells. In: Lichtman MA, Kipps TJ, Seligsohn U et al (eds) Williams hematology, 8th edn. The McGraw-Hill Companies, Inc., New York. Chapter 2

Salvagno GL, Sanchis-Gomar F, Picanza A, Lippi G (2015) Red blood cell distribution width: a simple parameter with multiple clinical applications. Crit Rev Clin Lab Sci 52:86–105

Sankaran VG, Orkin SH (2013) The switch from fetal to adult hemoglobin. Cold Spring Harb Perspect Med 3:a011643

WHO/WMNIS. Hemoglobin concentration for the diagnosis of anemia and assessment of severity. http://www.who.int/vmnis/indicators/haemoglobin.pdf. Last accessed 8 Sept 2016

Won DI, Suh JS (2009) Flow cytometric detection of erythrocyte osmotic fragility. Cytometry B Clin Cytom 76B:135–141

World Health Organization (1967) Standardization of procedures for the study of glucose-6-phosphate dehydrogenase: report of a WHO Scientific Group. WHO Technical Report Series No. 366

World Health Organization (2015) The global prevalence of anaemia in 2011. WHO Library Cataloguing-in-Publication Data. ISBN 978 92 4 156496 0

Immunohematology

16

Gianluca Gessoni

Introduction

Immunohematology studies the antigens expressed on the surface of the red blood cells and the respective antibodies. It is a discipline that was born in 1900 with the description by Landsteiner and Wiener of the ABO system, from the 1940s comes the Rh system, since then the knowledge in the immune hematological field has expanded enormously, and at present, they are described in over 30 erythrocyte blood group systems comprising no less than 700 distinct antigens. This brief discussion aims to provide descriptive elements of the most relevant erythrocyte blood group systems in relation to genetic, biochemical and functional aspects. For each erythrocyte blood group system, an attempt was made to give indications on the Ig family and clinical significance of the relative antibodies. A brief illustration of the serological investigation methodologies for the search for antigens belonging to the erythrocyte blood group systems and related antibodies was included in the discussion. A schematic treatment of granulocyte and platelet antigens was introduced, as well as a mention of the application of genotyping techniques in the immunohematological field.

Historical Premise

Since the beginning of human history, blood has been recognized as a vital force, the essence of life. From the time of Hippocrates, disease was believed to be caused by an imbalance of the four humours: blood, phlegm, yellow bile, and black bile. Of these, blood was the most important, so bloodletting became a popular treatment for most diseases. In 1666, Richard Lower successfully transfused blood from one dog to another. In the seventeenth century, some unsuccessful attempts of animal-to-human transfusion were performed. After these attempts, transfusions were banned for more than 150 years, and interest in blood transfusions was revived during the nineteenth century. In August 1825, James Blundell, an obstetrician at Guy's Hospital in London, successfully transfused a woman dying from postpartum hemorrhage with blood from her husband.

Until the end of nineteenth century, the prevailing belief was that all human blood was the same. This paradigm changed in 1901 with Karl Landsteiner's discovery of ABO blood groups. Landsteiner, by incubating red cells from some individuals with serum from others (his coworkers in Wien School of Medicine), identified agglutination patterns, leading to the initial identification of three blood groups, A, B, and C (later renamed O) In 1902, Decastello and Sturli identified the fourth blood group, AB. The inheritance pattern of blood groups was proved by Felix Bernstein in 1924.

In 1939, Philip Levine observed a post transfusion hemolysis in a blood group O patient who received blood from her blood group O husband. Parallel experiments conducted by Landsteiner and Wiener in which antibodies produced by immunization of rabbits and with blood from rhesus monkeys caused red cell agglutination of 85% of humans tested. Those individuals whose red cells were agglutinated by these antibodies were classified as rhesus (Rh) positive. Later, it was appreciated that the Rh system is composed of numerous alleles. The current system of nomenclature—c, C, d, D, e, E—was proposed in 1944 by Ronald Fisher.

G. Gessoni (✉)
Transfusion Medicine, Department of Venice District, Dell'Angelo General Hospital, Mestre - Venice, Italy
e-mail: gianluca.gessoni@aulss3.veneto.it

M. Ciaccio (ed.), *Clinical and Laboratory Medicine Textbook*, https://doi.org/10.1007/978-3-031-24958-7_16

The use of the antiglobulin test, first described by Moreschi in 1908 and rediscovered in 1945 by Robin Coombs, Rob Race, and Arthur Mourant, allowed the identification of many other blood group antigens in the decades that followed. Currently, several hundred antigens are known to be present on the surface of erythrocyte, classified in 33 blood group systems, 7 blood group collections, 700 low-incidence antigens, 901 high-incidence antigens.

Blood Group Antigens

The antigenic determinants of the erythrocyte blood group systems are inherited, usually, following the dictates of formal genetics. During cell division (Mitosis), chromatin loses its homogeneous structure and the chromosomes in which the units of genetic information are present, called genes. The genes are distributed with a specific order along the chromosome, in a precise physical location called the locus. Alternative forms of genes, each of which can occupy a single locus on one of the homologous chromosomes, are termed alleles.

The ISBT terminology distinguishes between the alleles for blood group antigens and the antigens that they encode. For example, the major antigens of the ABO system are A, B, and O, yet the alleles are *A*, *B*, and *O*. In the Kell system, two alleles, *K* and *k*, determine the K and k antigens, respectively. In other cases, the name of the gene alleles is very different from that of the antigens they determine. For example, in the MNSs blood group system, the genes are named *GYPA* and *GYPB*. Almost all the gene alleles coding for the antigens of the erythrocyte blood group systems are in one of the 22 pairs of autosomes, only one allele coding for the Xg and Xk antigens has been located on the X chromosome.

Individuals who have identical alleles at a given locus on both chromosomes (i.e., *K/K*) are *homozygous*. In *heterozygous* condition, alleles present at the locus on each chromosome are nonidentical (i.e., *K/k*). Individuals who are homozygous for an allele, in some blood group systems, may have more antigen expressed on their red cells than persons who are heterozygous. For example, red cells from a person whose phenotype is Jk(a+b–) have a "double dose" of the *Jka* allele and, as a result, express more Jka antigen on the red cell surface than an individual whose phenotype is Jk(a+b+). The difference in amount of antigen expressed on the red cell membrane between a homozygous and a heterozygous phenotype sometime can be detected serologically (dosage effect).

Usually blood group antigens are expressed as codominant characters. Therefore, heterozygotes express the product of both alleles. For example, if an individual's erythrocytes are typed as K + and k +, a *K/k* genotype can be inferred.

In ABO system, the situation is more complicated. In fact, the genes of the ABO system do not code for membrane proteins, but control the production of enzymes: glycosyltransferases. These enzymes add sugars to a preexisting oligo-carbohydrate chain, modifying it and giving it the antigenic specificity A or B or leaving it unchanged (O). The antigens of the ABO system are therefore oligosaccharide chains. The determinants of the other erythrocyte blood group systems are usually the glycoproteins present on the erythrocyte membrane of which they are often integral constituents (intrinsic proteins).

Effectiveness of an antigen in evoking the immune response is due to numerous characteristics, the total effect of which is defined as immunogenicity. Some of these characteristics are the exposure of the antigen and the number of determinants present on the erythrocyte membrane, maturation of the antigen during fetal life, its spread in the population, and more. Some relevant data concerning blood groups antigens are reported in Table 16.1.

Table 16.1 Blood group systems with gene denomination and chromosome location

ISBT number	ISBT denomination	Conventional denomination	Genes denomination	Localization chromosome	Number of antigens	Clone denomination
001	ABO	ABO	*ABO*	9	4	–
002	MNS	MNSs	*GYPA, GYPB*	4	46	CD235
003	P	P	*P1*	22	1	CD77
004	Rh	Rh	*RHD, RHCE*	1	52	CD240
005	Lutheran	Lu	*LU*	19	20	CD239
006	Kell	K	*KELL*	7	14	CD238
007	Lewis	Le	*FUT3 (Le)*	19	6	–
008	Duffy	Fy	*FY*	1	5	CD234
009	Kidd	Jk	*JK*	18	5	–
010	Diego	Di	*AEI (SCL4AEI)*	19	22	CD233
011	Cartwright	Yt	*ACHE*	7	2	–
012	Xg	Xg	*XG*	X	2	CD99
013	Scianna	Sc	*SC*	1	7	–
014	Dombrock	Do	*DO*	12	8	CD297
015	Colton	Co	*AQP1*	7	4	–
016	Landsteiner-Wiener	LW	*LW*	19	3	CD242
017	Chido-Rodgers	Ch/Rg	*C4A, C4B*	6	9	–
017	Hh	H	*FUT1*	19	1	CD173
019	Kx	Kx	*XK*	X	1	–
020	Gerbich	Ge	*GYPC*	2	11	CD236
021	Cromer	Cromer	*DAF*	1	17	CD55
022	Knops	Kn	*CR1*	1	4	CD35
023	Indian	In	*CD44*	11	4	CD44
024	OK	OK	*CD147*	19	3	CD147
025	RAPH	MER2	*MER2*	11	1	CD151
026	JMH	JMH	*SEMA7A*	15	6	CD118
027	I	I	*CGNT2*	6	2	–
028	GLOB	P	*B3GALT3*	3	1	–
029	Gil	GIL	*AQP3*	9	1	–
030	Rh associated glyccoprotein	RHAG	*RHAC*	6	4	CD241
031	Forsman	FORS	*CBCT1*	9	1	–
032	JR	JR	*IJ, ABCC2*	4	1	–
033	LAN	LAN	*LAN, ABCB6*	2	1	CdW338
034	Se	Secretore	*FUT2*	19	2	

The first column shows the numerical nomenclature of the International Society of Blood Transfusion (ISBT), in the second the full name of the erythrocyte blood group system using the ISBT nomenclature, in the third column is reported the common name of the blood group systems. The fourth column shows the name of the genes and the fifth the location at chromosomal level. In the sixth the number of antigens and in the last column the clone denomination (CD)

Antibody Versus Blood Groups Antigens

Immunoglobulins (Ig) of prevailing immunohematological interest are IgM and IgG. IgMs are pentameric antibodies produced during the primary phase of the immune response, and which in immuno-hematology are classically identified with the “natural” antibodies of the ABO system. They are complete antibodies, being able to generate agglutination in physiological solution, and are extremely effective in activating complement according to the classical way. They are unable to cross the placenta, while they can react well at 4 °C and 22 °C.

IgGs are monomeric antibodies. In immunohematology, the IgG1 and IgG3 subclasses are particularly relevant, to which most of the allo-antibodies directed toward antigens belonging to the erythrocyte blood group systems belong. IgGs are produced during the secondary phase of the immune response, and in immunohematology, they are identified with immune antibodies. They are incomplete antibodies, not being able to generate agglutination in saline solution. Their presence is highlighted either by modifying the reaction conditions (low-ionic-strength solutions, polybrene, albumin) or with the antiglobulin test. IgG can cross the placenta. Their thermal optimum is from 22 to 37 °C.

In Transfusion Medicine, an antibody is defined clinically significant if associated, with a certain frequency, with hemolytic posttransfusion reaction (PTH) and/or foetal neonatal hemolytic disease (HDFN).

Antibody Antigen Reaction in Erythrocyte Immunohematology

Classical erythrocyte immunohematology is based on the detection of reaction between the antigens present on the surface of the red blood cells and the antibodies present in the antisera (or in patient serum). The antigen–antibody reaction can produce a variety of observable results. In blood group serology, the most observed reactions are agglutination, hemolysis, and precipitation.

Agglutination is the most widely used antigen–antibody reaction in erythrocyte immunohematology and consists of the aggregation, mediated by antibodies, of red blood cells that have the relative antigen on their surface. Agglutination of red blood cells occurs because the antibody molecules bind to the antigenic determinants of different but adjacent red blood cells, binding them together to form a visible aggregate.

Hemolysis consists of the rupture of red blood cells, resulting in the release of intracellular hemoglobin. In vitro hemolysis is a two-stage reaction: In the first stage, there is the binding of antibody to the blood group antigens; and in the second stage, there is activation of complement system leading to RBC destruction. Hemolysis is considered a positive result, because it demonstrates the antigen–antibody reaction with activation of the complementary cascade.

Precipitation consists of the formation of an insoluble complex, usually visible, which originates from the reaction of a soluble antibody with a soluble antigen. These insoluble antigen–antibody complexes are detectable in tube tests as an annular or button-shaped sediment or in agar tests as a precipitation line (Fig. 16.1).

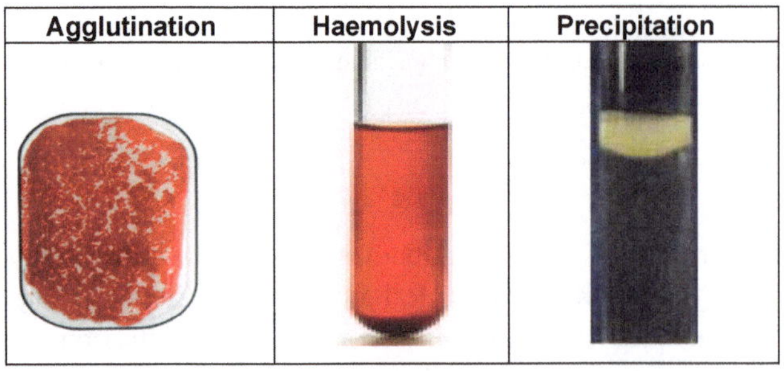

Fig. 16.1 Agglutination, hemolysis and precipitation

Erythrocyte Blood Group Antigenic Systems

ABO, H, and Lewis Blood Groups and Structurally Related Antigens

ABO blood group antigens, as well as those of the Hh, Lewis, and Ii systems, consist of structurally related polysaccharide molecules. The antigens are formed by specific glycosyltransferases, which add, in sequence, specific sugars on oligosaccharide chains that derive from a common precursor substance. The interactions of the products of the *ABO*, *Hh*, *Sese*, *Se/se* and *Lele* genes affect the expression of the ABO, H, and Lewis antigens, as well as the presence of substance A and B in body secretions.

The ABO system was discovered when Karl Landsteiner recorded the agglutination of human red cells by the sera of other individuals in 1901 and detailed the three patterns of reactivity called groups A, B, and O. He found that serum from group A individuals agglutinated the red cells from group B individuals, and conversely, the serum from group B individuals agglutinated group A. Red cells that were not agglutinated by the serum of either the group A or group B individuals were later called group O; the serum from group O individuals agglutinated the red cells from both group A and group B individuals. A year later a fourth group, named AB was described. Serum of group AB individuals did not agglutinate group A, B, and O red cells; conversely, the red cells from group AB individuals were agglutinate from group A, B, and O serum. From these observations derived two fundamental concepts in immunohematology: antibodies to A and B antigens are present when the corresponding antigen is missing and that these antibodies are almost always present in people who have had no exposure to human red cells (natural antibodies) (Fig. 16.2).

The loci of the genes H and Se (Secretor), identified, respectively, as *FUT1* and *FUT2*, are located, in close concatenation, on chromosome 19. Each locus can be occupied by two alleles, one of which, respectively, called h and se, is an amorphous gene. The active alleles H and Se encode two glycosyltransfers, which, by acting on a precursor, transform it into H antigen. The H gene encodes a transferase that acts at the cellular level to form the H antigen on red blood cells. The transferase encoded by the Se gene also produces the H antigen, but in secretions, such as saliva. The amorphous h

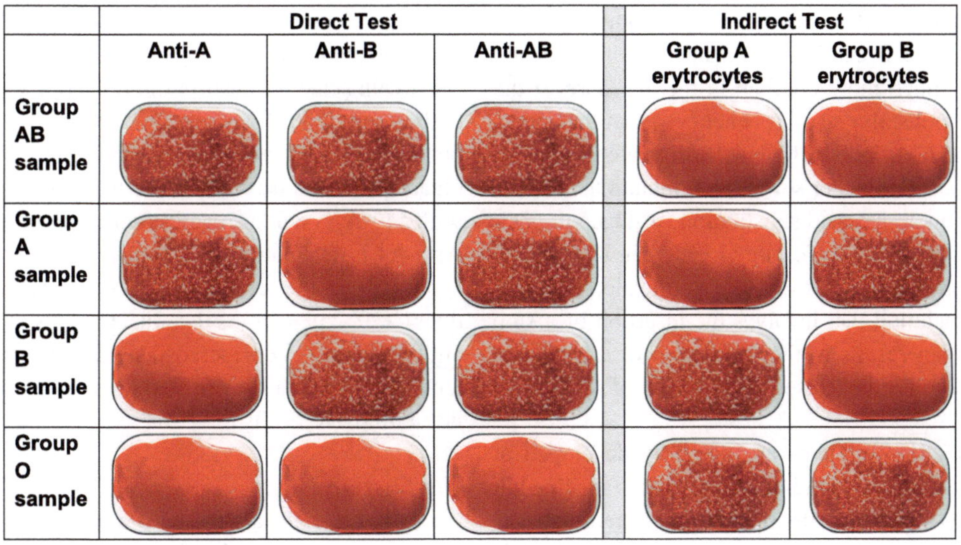

Fig. 16.2 The Karl Landsteiner experiment. Serum from group A individuals agglutinated the red cells from group B individuals, and, conversely, the serum from group B individuals agglutinated group A. Red cells that were not agglutinated by the serum of either the group A or group B individuals were later called group O; the serum from group O individuals agglutinated the red cells from both group A and group B individuals. Serum of group AB individuals did not agglutinate group A, B, and O RBC but red cells from group AB individuals were agglutinate from group A, B and O serum

genes and if they behave like recessive genes, the h gene is extremely rare.

The ABO locus is located on chromosome 9, and there are three common alleles: alleles *A* and *B* encode transferases, which produce antigens A and B, respectively; the *O* allele does not code for a functional enzyme. The red blood cells of individuals of group O are therefore devoid of antigens A and B, but have a high quantity of antigen H. The antigens of the ABO system are oligosaccharide chains, which can be joined to other macromolecules to form glycoproteins, glisfingolipids, glycolipids. The methods of conjugation affect the distribution of antigens. For example, ABO antigens present on the surface of erythrocytes are usually glycoproteins and glycosphingoplipids, in saliva they are present as glycoproteins, in milk they are present as free oligosaccharide.

Glycosyltransferases encoded by alleles *A, B, H*, and Se add a specific sugar to a preexisting oligosaccharide chain. This additional carbohydrate determines the antigenic specificity, which is, in fact, lost when it is removed (referred to as immunodominant). These reactions can only occur sequentially; therefore, the *H* gene encodes a fucosyl transferase that adds a molecule of fucose with α1–2 bond to the terminal galactose of the precursor oligosaccharide chain (therefore, fucose is the immunodominant sugar for the antigen H). The *A* allele encodes an N-acetyl-D-Galactosaminyltransferase, which adds, to the terminal Galactose of substance H, a molecule of N-acetyl-D-galactosamine with α1–3 bond (therefore, the N-acetyl-D-galactosamine is the immunodominant sugar for antigen A). The *B* allele encodes a galactosyltransferase, which adds a molecule of Galactose with a α1–3 bond to the terminal Galactose of substance H (therefore, Galactose is the immunodominant sugar for antigen B). Individuals of group AB possess both a gene *A* and a gene *B* and are therefore able to code for both transferases and express both antigens (A and B). The transformation of substance H into antigens A and B reduces the serological reactivity of antigen H. The subjects of group O are double-dose carriers of a nonfunctional recessive gene, which is unable to code for any glucosyltransferase. Substance H therefore remains unchanged on the surface of the red cells. The very rare subjects lacking the H gene are unable to modify the precursor oligopetide chain into H antigen. Consequently, the transcripts of genes A and B, although present and functioning, cannot carry out their action. These subjects are defined as Bombay phenotypes. If, on the other hand, the sese genotype is associated, these individuals do not present substance A and B even in the secretions, while substances A and B may be present in the secretions if a *Se* gene is present.

The polysaccharide chains that expose the immunodominant sugars H, A, or B can have a linear or multi-branched form, the latter much more efficient in the exposure of the antigen. At the molecular level, it is believed that the ancestral gene is gene A, from which gene B originated by mutation. The two genes differ for seven point mutations, four of which result in an amino acid substitution (residues in positions 176, 235, 266, and 268). The O gene can result from various mutations. Among these, the identification of a single nucleotide deletion that involves the insertion of a stop codon, which then results in the translation of a truncated protein devoid of transferase activity, is of particular importance.

ABO blood group antigens are not fully developed at birth, and their expression gradually increases to reach adult life levels after 2–4 years of life. It must also be remembered that ABO antigens are ubiquitously expressed on the body's cells.

Numerous subgroups have been described within the ABO system. The two main subgroups of A are called A1 and A2. On a molecular basis, they differ as the transferase in A2 subjects is 22 amino acids longer than that of A1 subjects and less active than this in the elaboration of highly repetitive and branched antigenic structures. Subjects A1 and A2 (as well as subjects A1B and A2B) can be differentiated by the use of lectins. They have been described under weak groups of A (A3, Ax, Am, Ael). Subgroups of B are much less frequent and of less practical importance.

Antibodies: Adult subjects normally have specific antibodies against antigen A or B, which is absent on its own erythrocytes. These are IgM class immunoglobulins with a wide thermal optimum (from 4 to 37 °C), able to fix the complement (therefore able to give hemolytic reaction after transfusion), reactive in saline solution. It is believed that the production of these antibodies (which are defined as "natural," since exposure to nonself-erythrocytes is not demonstrable) derives from the fact that the configurations that confer the antigenic specificities of determinants A and B also exist on the bacterial walls of the germs constituting the intestinal microbiota. At birth these antibodies are normally absent, while antibody production increases between the fifth and tenth year of life until it reaches adult levels and decreases in the latter part of life. IgM represents the predominant immunoglobulin class of anti-A produced by group B subjects and anti-B produced by group A subjects. However, modest amounts of IgG class antibodies may still be present in these subjects. IgG, on the other hand, is the dominant antibody class in subjects of group O. Since IgG is able to cross the placenta, infants of group A or B born to mothers of group O may develop neonatal hemolytic disease, usually of modest entity. The serum of the subjects of group O contains an antibody that is defined anti-A,B because it reacts with both the A and B red blood cells. The anti-A and anti-B reactivity of this antibody cannot be separated with selective absorption. It is not the sum of two antibodies (one anti-A specificity and one anti-B specificity), but an anti-A,B cross-reactive antibody. For example, an eluate prepared from group A red cells that have reacted with anti-A, B will be reactive with both A and B cells alone. Anti-A1 is detectable, as alloantibody, in a reduced percentage (<2%) in the serum of A2 subjects while it is more frequent (20–25%) in A2B subjects. This is usually a cold antibody and is considered insignificant. Dolichos biflorus lectin preparations are available, which react with A1 but not with A2 red cells.

H/h System

Group O red cells lack A and B antigens, and the membrane expresses unchanged H antigen. In fact, the red blood cells of group O subjects show a strong agglutination when tested with the anti-H lectin obtained from Ulex europaeus (Fig. 16.3). The reaction observed is in descending order: O> A2> B> A2B> A1> A1B. The rare subjects with the hh genotype whose red blood cells lack the H antigen (they do not react with the anti-H lectin) at ABO typing are classified as type O subjects but have, in addition to the anti-A and anti-B, an alloantibody anti-H (therefore able to agglutinate normal group O red blood cells). The term Oh Bombay is used to designate subjects with this phenotype. At the genotypic level, the Oh Bombay phenotype arises from the transmission of the hh alleles at the H locus and of sese at the Se locus. Since the If allele is necessary for the formation of the Leb antigen, the Oh red cells will, therefore, be Le (a + b–) or Le (a – b–).

I/i System

Antigens: Antigens I and i are expressed on the erythrocyte membrane on the same glycoproteins and glycosphingolipids that carry H, A, and B antigens and, in the secretions, on the same glycoproteins that carry H, A, B, Lea, and Leb. Nonetheless, antigens I and i are not antithetical, but are expressed in temporal succession. At birth, neonatal red blood cells are rich in i antigens, so for practical purposes, cord blood cells are considered I−i +. During the first 2 years of life, the expression of antigen I gradually increases, while that of antigen i decays. In adults, red blood cells are usually highly reactive with anti-I and are considered I + i−. There is a rare I−i + phenotype in adults. From a structural point of view, antigen i appears to consist of a linear chain with at least two repeating units Galactose-N-acetyl galactosamine, joined by a β1–4 glycosidic bond. To the surface of the red blood cells of adults, to give specificity I, these linear chains are modified by the addition of branched structures consisting of N acetyl galactosamine that join the linear chains by means of a β1–6 glycosidic bond.

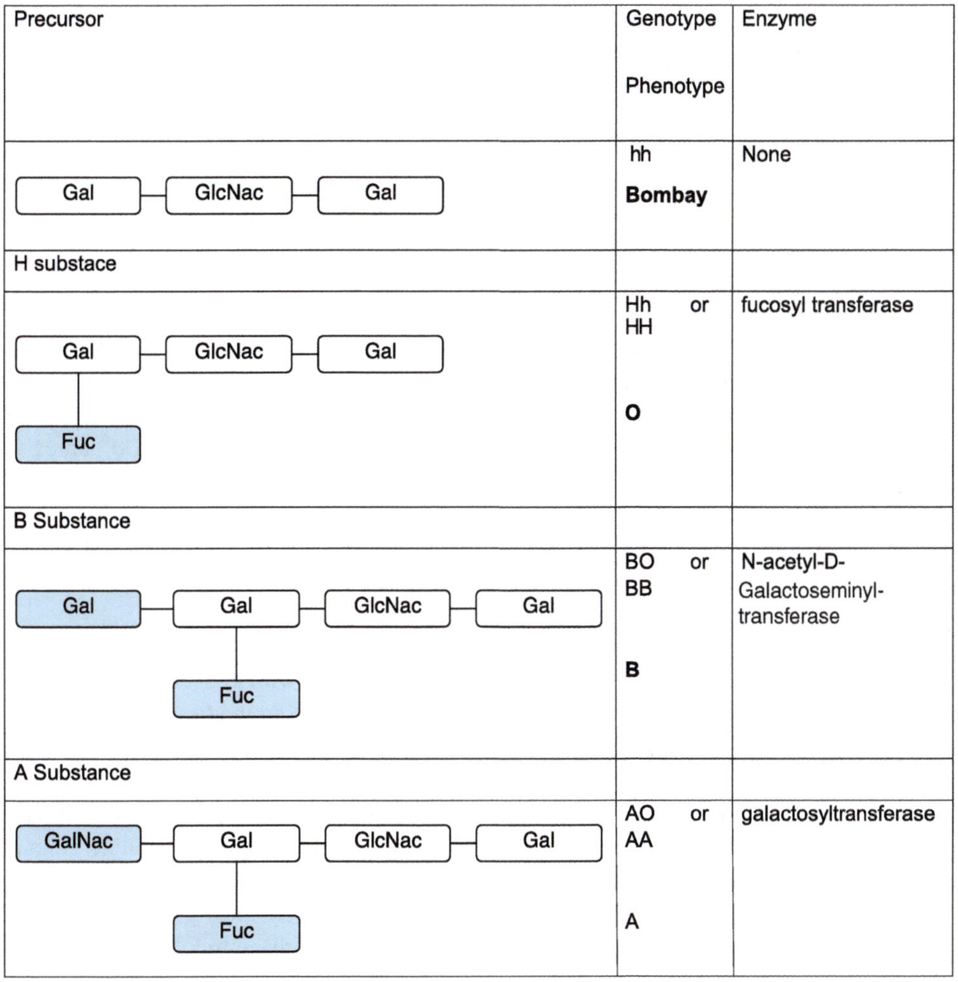

Fig. 16.3 ABO molecules. Alleles A, B, H encode for specific Glycosyltransferases that add a specific sugar to a preexisting oligosaccharide chain. This additional carbohydrate determines the antigenic specificity. These reactions occur sequentially: Therefore, the H gene encodes a fucosyl transferase that adds a molecule of fucose with α1–2 bond to the terminal galactose of the precursor oligosaccharide chain (therefore, fucose is the immunodominant sugar for the antigen H and for group O individuals). The A allele encodes an N-acetyl-D-Galactosaminyltransferase, which adds, to the terminal Galactose of substance H, a molecule of N-acetyl-D-galactosamine with α1–3 bond (therefore, the N-acetyl-D-galactosamine is the immunodominant sugar for antigen A). The B allele encodes a galactosyltransferase which adds a molecule of Galactose with a α1–3 bond to the terminal Galactose of substance H (therefore, Galactose is the immunodominant sugar for antigen B). Individuals of group AB possess both a gene A and a gene B and are therefore able to code for both transferases and express both antigens (A and B)

Antibodies: Anti-I and anti-i antibodies are often autoantibodies, are usually active in saline with thermal optimum at 4 °C, and are commonly identified with cold agglutinins, which can take on clinical significance if present at a titer greater than 1/64 and, with a wide thermal range, are able to fix the complement. Anti-I auto antibodies are often produced by patients with Mycoplasma pneumoniae pneumonia. These patients may have transient hemolytic episodes determined by the antibody. Patients with infectious mononucleosis often have anti-i antibodies.

Lewis System

Antigens: Antigens of the Lewis erythrocyte blood group system are called Lea and Leb originate from the activity of a glycosyltransferase encoded by the allele *Le* (or *FUT3*) located on chromosome 19. This glycosyltransferase adds a glucose residue to a precursor chain. The Lewis system appears associated with the Sese system. In fact, the Lea antigen is produced when the *Le* gene is inherited together *sese*, and Leb is produced when it is inherited together with *Sese* or *SeSe*. Therefore, Lea and Leb are not antithetical antigens produced by alleles but depend on the interaction of alleles inherited independently (*Le, Se, se*). The antigens of the Lewis system are also not intrinsic to the erythrocyte membrane, but are expressed on the Type 1 glycosphingolipid chains, which are adsorbed by the plasma on the red cell membrane. Also in this case, four different phenotypes are possible, which present very different frequencies in the various ethnic groups: Le (a + b−) is present in 20% of the Caucasian and African population, Le (a−b +) is present in 75% of Caucasians and 55% of black subjects, Le (a−b−) is present in 5% of Caucasians and 25% of black subjects, Le (a + b +) is rare in European populations and of African ancestry, while it is common in Asian populations.

From a biochemical point of view, the synthesis of Lewis antigens derives from the interaction of two different fucosyltransferases. One of them is produced by the *Se* locus and one by the *Le* locus. Both enzymes act on the same substrate. The fucosyltransferase encoded by the *Le* allele attaches a molecule of fucose with α1–4 bond to the oligosaccharide chain of the precursor. In the absence of the transferase encoded by the *Se* allele, this configuration gives rise to the Lea antigen. On the other hand, the Leb antigen is formed when the fucosyltransferase encoded by the *Se* gene and then, in sequence, the fucosyltransfease encoded by the *Le* gene acts on the oligosaccharide chain of the precursor. In this configuration, two fucose residues are added to the original oligosaccharide chain. The Leb antigen therefore reflects the presence of both the *Le* and *Se* alleles; while Lea antigen reflect the presence of *Le* without *Se* alleles. Lewis antigens are rapidly adsorbed and eluted by the erythrocyte membrane. The transfused red blood cells lose their Lewis antigens and assume the recipient's Lewis phenotype within a few days of being released into the circulation.

Antibodies: Antibodies to Lea (more frequently) or Leb (more rarely) are detectable, almost exclusively, in the sera of Le (a−b−) subjects, usually in the absence of antigenic stimulation evident from previous exposure to nonself red blood cells. self. These are generally IgM antibodies. Therefore, considering that Lewis antigens are also underdeveloped at birth, these antibodies are not associated with MEN. Lewis antibodies can bind complement and fresh sera containing anti-Lea can hemolyze incompatible red blood cells in vitro. Being IgM i, these antibodies react in saline solution, forming rather fragile agglutinates that can be easily dispersed during too vigorous manipulation.

MNSs System

Antigens: It is a complex system, consisting of over 40 glycoproteic antigens. The M, N, S, s antigens together with U (a high-frequency antigen) are the most relevant in Transfusion Medicine. The M and N antigens are localized on glycophorin A; S, s, and U are located on glycophorin B. The genes encoding these antigens exhibit marked linkage disequilibrium. As an example, the *Ns* gene complex is more frequent than *NS*. The MNS system also includes numerous low-frequency antigens, resulting from mutations involving amino acid substitutions and from modifications in glycation.

The genes that code for the antigens of the MNS system are located on chromosome 4. The gene that codes for glycophorin A is called *GYPA*, the gene that codes for glycophorin B is called *GYPB*. The two genes, which probably derive from a single ancestral progenitor, have a homology greater than 95%. Glycophorin A is a transmembrane glycoprotein present in approximately 1,000,000 copies for each erythrocyte. The M and N antigens are located in an extracellular portion consisting of 72 amino acid residues linked to side glucose chains. M specificity is characterized by a serine in position 1 and a glycine in position 5; the specificity N is characterized by a leucine in position 1 and a glutamate in position 5. Glycophorin B is smaller and is present in about 200,000 copies for each erythrocyte. The specificity S is characterized by a methionine in position 29, this position is occupied by a threonine to give the specificity s. Usually the M and N antigens are destroyed by treatment with proteolytic enzymes such as ficin or papain.

Antibodies: Anti-M is frequently detected as a cold reactive agglutinin with red cells suspended in physiological solution. These antibodies are usually of the IgM class (although the finding of IgG is not uncommon) and are frequently observed in the serum of subjects who have never been exposed to nonself human erythrocytes. These are antibodies that only rarely take on clinical significance, although they have been implicated in rare cases of PTH and HDFN. These were mostly IgG class antibodies able to react at 37 °C or with the antiglobulin test.

Anti-N is a relatively rare antibody, it is usually of the IgM class, and behaves like a cold agglutinin. They are not considered clinically significant antibodies. The "anti-S" and "anti-s" antibodies are usually immune antibodies of the IgG class, identifiable by the antiglobulin test. They are considered clinically significant, being implicated in PTH and HDFN. Anti-U is rare but should be considered if a reactive antibody against a high frequency antigen is identified in a subject of African origin previously exposed to nonself red blood cells.

P System

Antigens: The P group system historically included the antigens P, P1, Pk and Luke. However, further studies have clarified how two biosynthetic pathways and genes located in different loci are involved in the development and expression of these antigens. At present, according to the ISBT nomenclature, the P antigen is part of the GLOB system; P1 remains assigned to the P system, while Pk and Luke are included in the GLOB29 antigenic collection. The antigens of the P system, similarly to those of the ABO system, are synthesized sequentially by adding single glucose molecules to "precursor" oligosaccharide chains. These antigens are expressed on glycolipid type molecules. In the case of the P blood group system, the common precursor is lactosylceramide. This molecule, by means of two metabolic pathways, is converted to P1 antigens or to antigens of the globoside series (Pk, P or LKE). There are two common phenotypes associated with the P system (P1 and P2) and three rare phenotypes (p, P1k and P2k). The *P1* gene is located on chromosome 22, while the *P* gene is located on chromosome 3. The P antigen has been shown to be a receptor for erythrovirus (parvovirus) B19. Therefore, subjects with the p phenotype (lacking the globoside antigen) appear naturally resistant to infection.

Antibodies: The serum of P1 negative subjects often contains an anti-P1. It is, therefore, as a rule, a natural antibody, of the IgM class, which reacts optimally at +22 °C. It is not thus able to determine MEN and, being only rarely implicated in PTH, it is considered to be of little significance from a clinical point of view. The expression of the P1 antigen on the surface of the red cells can be very variable and tends to decrease with their conservation. This can lead to difficulties in identifying the antibody specificity. P1k and P2 subjects may develop an anti-P antibody. It is a natural antibody, consisting of a mixture of IgM and IgG, characterized by a wide thermal range (reacts from 4 to 37 °C), considered clinically significant in how much can be associated with PTH and, although more rarely, with HDFN. The anti-P autoantibody has also been associated with paroxysmal cold hemoglobinuria; it is a cold IgG autoantibody which is also described as biphasic hemolysis.

RhD RhCE System

As part of the numerous nomenclature and classification proposals of the Rh system antigens, in this paragraph the Fisher and Race nomenclature will be used, updated according to current genetic and biochemical knowledge. Historically, the Rh antigen (later identified with the D antigen) was described in 1939 by Levine in a woman whose infant was found to have HDFN, and she herself developed a hemolytic post-transfusion reaction after transfusion of red blood cells donated by her husband. The name Rh derives from the fact that in 1940, Landsteiner and Wiener described an antiserum obtained by immunizing guinea pigs with Macaques rhesus monkey red cells, capable of reacting with 85% of human red blood cell samples, and which behaved like the antisera obtained by women whose foetuses had suffered from MEN. Later, in the second half of the 1940s, four additional antigens named C, E, c and e were included in the Rh system, which currently consists of more than 50 antigens, thus becoming the much larger erythrocyte blood group system. Antigens of the Rh system begin to be represented on the surface of the red cells as early as the eighth gestational week, and are fully expressed in the term new-born. They are characteristic exclusively of red blood cells, not being expressed in other cells or tissues.

Genes: At present it is believed that the antigens of the Rh system (D, C/c, E/e) are encoded by two alleles called *RHD* and *RHCE*, located on chromosome 1 in two contiguous and closely linked loci. These alleles encode nonglycosylated polypeptides expressing Rh antigens. The *RHD* gene encodes a transmembrane protein expressing antigen specificity D, while a d antigen has never been described. In D-negative subjects of Caucasian origin the *RHD* gene is deleted, while in subjects of African or Asian origin the D-negative phenotype is frequently associated with an inactive or mutated *RHD* gene. The *RHCE* gene encodes a single nonglycosylated polypeptide, which expresses the certain antigenic C, c, E, and e (its alleles are *RHCe*, *RHCE*, *RHcE*, and *RHce*). From a biochemical point of view, both the *RHD* gene and the *RHCE* gene encode a transmembrane protein of 417 amino acid residues. Unlike what has been observed for other antigenic determinants of erythrocyte blood group systems, this polypeptide chain does not present glycosylation but bonds fatty acids. There is a high homology between the products of the different alleles of the *RHCE* gene. For example, the C and c antigens differ from each other only by four amino acids in positions 16, 60, 68 and 103. The presence of a proline or an alanine in position 226, on the other hand, differentiates the E allele from that e. Rh proteins (D and CE) are complexed with Rh-associated lipoprotein (RhAG), which has 37% homology with Rh polypeptides, and is encoded by the *RHAG* gene located on chromosome 6.

Rh phenotype and genotype: commercial anti-D, anti-C, anti-E, anti-c and anti-e antisera are used for the determination of the Rh phenotype. The set of antigens detected on a subject's red blood cells therefore constitutes its Rh phenotype. D negative subjects lack the RHD gene that encodes the D antigen or have a nonfunctional *RHD* gene. Most of the negative D subjects are homozygous for the *RHce* allele, which represents the gene encoding the antigens c and e. Less frequently they may have *RHCe* or *RHcE* alleles, which code for C and e or for c and E, respectively. The *RHCE* gene, which produces the C and E antigens, is very rare in D negative individuals. The D genotype of D positive persons cannot be determined with serological tests, as molecular biology techniques must be used.

In Rh system, the interaction between genes (position effect) is well described. If the interaction occurs between genes located on the same chromosome, it is referred to as the cis effect; if it occurs between genes placed on homologous chromosomes, it is indicated as a trans effect. In example (cis effect), the E antigen produced by DcE appears quantitatively less expressed than the E antigen produced by cE. It was also observed that both C and E were less expressed when they were encoded by the DCe/DcE genotype than the antigens encoded by DCe/ce or DcE/ce (trans effect).

Usually positive D red cells show, when tested with anti-D antiserum, strong agglutination, appreciable macroscopically, which allows rapid and safe classification. The classification of red blood cells that do not agglutinate after being tested with an anti-D is not so simple. Reactivity can in fact be reactive-dependent or can only be revealed by the antiglobulin test. These "weak" Ds (formerly called Du) are determined in most cases by an *RHD* gene that encodes a RhD protein with reduced quantitative expression of the D antigen on the erythrocyte membrane. This phenotype is quite common in populations of African origin. In example, erythrocytes of some subjects with the Dce/Ce genotype show a weak D antigenic expression due to a suppressor effect exerted by *RHC* in the trans position with respect to *RHD*. Moreover, it is well known that the D antigen consists of numerous distinct structural components; RBCs that lack some components of the D antigen are referred to as "partial D" and, especially those with the DVI phenotype, can react only with some anti-D monoclonal antisera and can produce anti-D antibodies. There are many other antigens belonging to the Rh erythrocyte blood group system. Only some of them, such as Cw (antithetic to C), the G antigen (characterized by a serine in position 103 of the Rh polypeptide), the f antigen (a ce fusion antigen), have practical importance. Very rare family groups are also described, whose red blood cells appear to be devoid of all Rh (Rh null) antigens. The Rh null phenotype can be of the "regulatory" type and therefore derive from a mutation of the *RHAG* gene, or of the "amorphous" type because of mutations of *RHCE* genes associated with deletion of the *RHD* gene.

Antibodies: Usually the antibodies directed towards antigens of the Rh system are of the immune type and derive from exposure to nonself red blood cells. It is therefore IgG with thermal optimum at 37 °C, which react with enhancing media or with the antiglobulin test. In some cases, the anti-Rh antibodies may be natural (for example, anti-E or anti-Cw). As regards the ability to induce the production of antibodies after stimulation, antigen D is undoubtedly the most powerful immunogen, followed in descending order by antigens c, E, C, and e. It must be remembered that individuals with partial D may develop an anti-D alloantibody. Rh antibodies do not generally fix complement. They therefore result in extra-vascular hemolysis in vivo. They are considered clinically significant, as they are potentially associated with even severe HDFN and post transfusion hemolysis (mainly extra-vascular).

Landsteiner–Wiener (LW) System

The Antigens: The Landsteiner–Wiener (LW) blood group system, although constituting an independent system, appears to be linked to the Rh system. In fact, it is the antigen identified by the original anti-Macaques rhesus serum used to define positive D subjects. Four antigens are known; to the original LW1 (present in the D + subjects) and LW2 (present in the D− subjects), further antigens named LW3 and LW4 were associated. The genes encoding the antigens of the LW system are located on chromosome 19.

Antibodies: The antibodies of the LW system are not considered clinically significant, as they are unable to generate MEN and PTH.

Lutheran System

Antigens: Lutheran erythrocyte blood group system antigens are named Lua and Lub. They are encoded by two codominant alleles in the *LU* gene, located on chromosome 19. Four phenotypes are therefore possible: Lu (a+b−) present in less than 1% of the population, Lu (a+ b+) present in 7.5%, Lu (a−b +) present in 92%, and Lu (a−b−) very rare. A series of high-frequency antigens (Lu4, Lu5, Lu6, Lu7, Lu8, Lu11, Lu12, Lu13, Lu16, Lu17, and Lu20) and two low frequency antigens (Lu9 and Lu14) were included in the Lutheran system. Antigens of the Lutheran system are carried by a glycoprotein that appears to play a role in cell adhesion. The antigens of the Lutheran system are poorly expressed on the surface of foetal red blood cells, and the expressiveness increases during life, while remaining all in all modest (from 500 to 4000 molecules per erythrocyte, depending on the phenotype).

Antibodies: These are antibodies that are rarely found. They are largely produced in response to pregnancy or transfusions but have also been detected in the absence of obvious exposure to nonself red blood cells. The anti-Lua antibody is not considered clinically significant, as it is not associated with PTH or HDFN. The anti-Lub antibody was associated with moderate HDFN and reduced survival of the transfused red cells, but not with PTH. Most anti-Lua and some anti-Lub antibodies directly agglutinate red blood cells in saline, producing a typical mixed-field reactivity pattern, with small agglutinates scattered among most free red blood cells.

Kell/Cellano System

Antigens: The major antigens of the system are the Kell (K) antigen present in 9% of Caucasians and 2% of blacks, and the cell (k) antigen present in over 99% of individuals. Other antigens of the Kell system, antithetical to each other, are Kpa/Kpb, and Kpc; Jsa/Jsb; K11/K17; K14/K24. An ethnic distribution is also observed for the Kpa and Jsa antigens, as Kpa is mainly observed in white subjects while Jsa is predominantly detected in subjects of African origin. Also described are subjects totally devoid of antigens referable to the Kell blood group system, defined as Ko or Knull. The antigens of the Kell system are located on a 93-kD transmembrane protein rich in cysteine and capable of forming sulfhydryl bonds. They are, therefore, conformational antigens, sensitive to the treatment of red blood cells with agents capable of altering the secondary structure of the molecule (mercaptoethanol, for example). It appears that this protein belongs to the family of neutral endopeptidases, which act as a bond for Zn; it also has a high similarity with the common antigen of lymphoblastic leukaemia (CALLA or CD10). Although the Kell system locus is located on chromosome 7 and the Kx (XK) locus is located on the X chromosome, it appears that the K and Kx proteins form a covalent complex on the surface of the red cells. The lack of Kx antigen is accompanied not only by a weak expression of the Kell antigens, but also by acanthocytosis and reduced erythrocyte survival. These erythrocyte abnormalities can be accompanied by an increase in the creatine kinase (CK) enzyme and neuromuscular abnormalities, constituting the so-called McLeod phenotype, probably linked to abnormalities of the XK locus.

Antibodies: The K antigen is highly immunogenic. Anti-K antibodies are, therefore, frequently detected in the serum of transfused kk patients, although anti-K have rarely been observed even in subjects never exposed to nonself erythrocytes. These are IgG class immunoglobulins, with an optimum temperature at 37 °C, reactive with the antiglobulin test. Being able to generate PTH and HDFN, they are considered clinically significant antibodies. The k antigen is also highly immunogenic, as it can generate the appearance of IgG-class allo-antibodies, with an optimum temperature of 37 °C, reactive with the antiglobulin test. Being able to generate PTH and HDFN, they are also considered clinically significant. Anti-Kpa, anti-Kpb, Anti-Jsa, and anti-Jsb antibodies are all much less frequent than anti-K, but have similar serological characteristics, and are therefore to be considered clinically significant.

Duffy System

Antigens: The antigens of the Duffy erythrocyte blood group system, called Fya and Fyb, are encoded by a pair of co-dominant alleles in the *FY* locus, located on chromosome 1. As these are co-dominant alleles, four phenotypes are possible: Fy (a+b−), Fy (a+b+), Fy (a–b+) and Fy (a–b–). In Caucasians, individuals with the Fy (a−b−) phenotype are extremely rare, while the frequency can be as high as 68% among individuals of African descent. From the biochemical point of view, the Fya and Fyb antigens are localized in the N-terminal fraction of transmembrane glycoproteins and are, therefore, eliminated by treatment with proteolytic enzymes. The above antigens have ubiquitous diffusion. In addition to the surface of the red cells, they are present in other tissues such as central nervous system, spleen, kidney, lung. Glycoprotein constitutes the receptor for the malarial parasite Plasmodium vivax; people with red blood cells lacking the Fya and Fyb antigens are therefore resistant to this infection.

Antibodies: Both anti-Fya and anti-Fyb antibodies are usually IgG class immunoglobulins, with thermal optimum at 37 °C reactive with the antiglobulin test. The anti-Fya antibody is quite common and is considered clinically significant as it is capable of generating PTH and HDFN. Anti-Fyb is a rather infrequent antibody, and only rarely implicated in PTH and HDFN. Weak anti-Fya and anti-Fyb often express a dose-dependent effect.

Kidd System

Antigens: The antigens of the Kidd erythrocyte blood group system, named Jka and Jkb, are encoded by two codominant alleles in the *HUT11* gene located on chromosome 18. Four phenotypes are therefore possible: Jk (a+b−), Jk (a+ b+), Jk (a−b+), and Jk (a−b−), the latter of which is extremely rare. From a biochemical point of view, the antigens of the Kidd system are expressed by an erythrocyte membrane protein involved in urea transport.

Antibodies: Both anti-Jka and anti-Jkb antibodies are usually IgG class immunoglobulins, with thermal optimum at 37 °C, reactive at the antiglobulin test. Both are considered clinically significant, being able to generate PTH and HDFN. Characteristic of anti-Jka and anti-Jkb antibodies is their ability to produce severe delayed post-transfusion hemolytic reactions. In these subjects, it is possible to highlight a previous exposure to the antigen, which resulted in immunization with a modest antibody titer and not detectable with routine use tests. Nonetheless, in response to a new exposure to the antigen, an "anamnestic" antibody response rapidly develops against antigens of transfused red blood cells, which are destroyed when they are still present in the circulation. Both anti-Jka and anti-Jkb are often weakly reactive, with a score that tends to decrease depending on the conservation of the sample. They are more easily detectable in the presence of complement if a polyspecific Coombs serum with anti-complementary activity is used. Alternatively, the identification of reactivity can be improved by using a two-stage antiglobulin test or enzyme treatment. Both antibodies are more easily detectable using homozygous red blood cells (dose-dependent effect).

Minor Blood Group Systems

- *Diego system.* Antigens: The Diego blood group system consists of two pairs of independent antithetic antigens, called Dia/Dib Wra/Wrb. The system also contains many low frequency antigens. The antigens of the Diego system are encoded by alleles located on chromosome 17 and are expressed on the AE-1 protein which appears to have the role of anion exchanger across the erythrocyte membrane. The Dia antigen is present almost exclusively in populations of Asian origin and in North and South American natives. Antibodies: These are usually immune antibodies (with the exception of anti-Wra), which can be detected even in the absence of exposure to nonself red blood cells) and can be clinically significant. The anti-Dia antibody has been associated with PTH and HDFN. The anti-Dib antibody is rare, but it too can generate HDFN. The anti-Wra antibody is quite common and, although rarely, can be associated with PTH and HDFN.
- *Cartwright system.* Antigens: The Yt (Cartwright) system consists of two antigens, Yta and Ytb, which are encoded by a pair of gene alleles located on chromosome 7. The Yt antigens are found on the acetylcholinesterase of red blood cells. Antibodies: These are immune allo-antibodies of modest clinical significance, as they have never been implicated in cases of PTH or HDFN, although rare cases have been reported in which an anti-Yta has generated an accelerated destruction of Yta positive red blood cells transfused.
- *Xg system.* Xga Antigen: Discovered in 1962, it is an antigen encoded by an on the X chromosome. Anti-Xga Antibody: It is a rare immune antibody, which reacts only to the antiglobulin test and is not considered clinically significant.
- *Scianna system.* Antigens: This erythrocyte blood group is composed of five antigens, named Sc1, Sc2, Sc3, Rd, and STAR, expressed on the cell adhesion protein ERMAP3. Sc1 is a high frequency antigen, while Sc2 occurs very rarely. The gene that encodes the Scianna antigens is located on chromosome 1. Antibodies: These are antibodies of rare finding and of modest clinical significance, not being associated with PTH and HDFN.
- *Dombrock system.* The Antigens: Initially, this system consisted of the Doa and Dob antigens. Currently three other named high-frequency antigens (Gya, Hy, and Joa) have been associated with this erythrocyte blood group system. Antibodies: Anti-Doa and anti-Dob are rare antibodies, sometimes identified in sera containing antibody mixtures. They are not associated with HDFN, but PTH sustained by these alloantibodies are described.
- *Chido/Rodgers system.* Antigens: The system is composed of two high frequency antigens called Chido (Ch) and Rodgers (Rg), which are not part of red cell membrane but are present on the C4 fraction of the complement and are therefore only subsequently adsorbed by the erythrocyte membrane. The gene is located on chromosome 6, in the region of the major histocompatibility complex (class III molecules). Antibodies: These are antibodies substantially devoid of clinical significance.
- *Gerbich system.* The Antigens: The Gerbich system includes eight antigens, three of which are high frequency (Ge2, Ge3 and Ge4) and five are low frequency (Wb, Lsa, Ana, Dha, and GEIS). These antigens are expressed on Glycophorin C and Glycophorin D. Antibodies: Only anti-Ge2 antibodies may have any clinical significance, having been associated with rare cases of HDFN.
- *Colton system.* The Antigens: The Colton system consists of a high frequency antigen (Coa) and a low frequency antigen (Cob), as well as an antigen considered to be the product of the Coa gene and the Cob gene, called Co3. The genes encoding the antigens of the Colton erythrocyte blood group system are located on chromosome 7 and expressed on an aquaporin. Antibodies: The anti-Coa antibody is considered clinically significant, since it can generate PTH and HDFN.
- *Cromer system.* Antigens: The Cromer system appears to consist of 10 high frequency antigens and three low frequency antigens. The antigens of this blood-group system are associated with the complement regulatory protein called DAF, (decay-accelerating factor), encoded by the DAF gene on chromosome 1. Antibodies: These are extremely rare immune antibodies of uncertain clinical significance.
- *Knops system.* Antigens: The antigens belonging to the Knops erythrocyte blood group system (Kna, Knb, McCa, McCb, Sla, Yka, Vil, and Sl3) are located on the receptor (CR1) of the C3b / C4b factors and encoded by a gene located on chromosome 1. Antibodies: Knops antibodies usually exhibit weak reactivity and varying intensity in serum antiglobulin tests. They are not considered to be clinically significant.
- *Indian system.* The antigens of the Indian system include a high-frequency antigen (Ina) and a low-frequency antigen (Inb). They are associated with a widely spread adhesion molecule (CD44).
- *System Ok.* The Ok system consists of a single high frequency antigen, called Oka. Ok (a–) subjects are extremely rare and are exclusively Japanese. It is possible to generate an anti-Oka antibody, which is able to decrease the survival of transfused red blood cells.
- *Raph system.* The Raph system consists of a single antigen, MER2, present in 90% of subjects.
- *John Milton Hagen system.* The John Milton Hagen (JMH) antigen is located on the CD108 glycoprotein.
- *Gill system.* In this system, a single high frequency antigen is represented, called GIL, localized on aquaporin 3. The antibody is not considered clinically significant.

- *Red blood cell collections*. In addition to the erythrocyte blood group systems already described previously, families or collections (defined collections) of antigens have been identified which, although having shared characteristics, do not meet the ISBT standards in order to be catalogued in an erythrocyte blood group system. These antigens were grouped in the Cost, Er and Vel collections. The Cost collection consists of two named antigens, Csa present in 98% of the population and Csb which has a frequency of about 34%. The Er collection consists of two antigens. It was present in 99% of the population and Erb which has a frequency of less than 1%. The VEL Collection includes two high frequency antigens called VEL and ABTI. Anti-VEL antibodies are frequently IgM, can fix complement, have been associated with PTH reactions.
- *"Public" antigens and "private" antigens*. With the term "public" antigens they are defined high frequency antigens (present in over 99.9% of the population), while with the term "private" antigens they are defined low frequency antigens (present in less than 0.5% of the population). Among the public antigens those named Lan, Jra, AnWj, Ata, and Sda should be mentioned, while among the private antigens those named: By, Lia, Bi, Milne, Bxa, Ola, Chra, Pta, HJK, RASM, etc.

Platelet and Granulocyte Antigens and Antibodies

On the surface of granulocytes and platelets, erythrocyte antigens, antigens of the HLA system and antigens of platelets and granulocytes can be expressed. This short dissertation will focus on this third type of antigens.

Platelet immunohematology From a clinical point of view, platelet antigens are important as they are implicated in neonatal alloimmune thrombocytopenia (FNAIT) and posttransfusion purpura (PTP). Almost always these clinical manifestations are related to the production of an alloantibody directed against the HPA-1a antigen. Forms of autoimmune thrombocytopenia are also described, in which autoantibodies can be directed against platelet-specific antigens.

Although numerous platelet-specific antigens have been defined, for some of them the platelet-specific denomination is not fully correct, being evidenced in many other cells, especially of endothelial derivation. Of the numerous glycoproteins expressed on the platelet membrane, only some (GP Ia, GP Ibα, GP Ibβ, GP IIb, GP IIIa, and CD109) show polymorphisms that make them immunogenic.

Granulocyte immunohematology Granulocyte antigens and related allo-antibodies can be implicated in several clinical syndromes, the two most relevant being neonatal immune neutropenia (FNAIN) and transfusion-related acute lung injury (TRALI). Antibodies directed to leukocyte antigens may also be the cause of the most common complication of blood component transfusion, namely the nonhemolytic posttransfusion febrile reaction. As reported in Table 16.2, seven proper granulocyte antigens have been described. The HNA-1a antigen and its antithetic HNA-1b antigen are expressed on CD16 (FcγRIIIb), both have been associated with TRALI and NAIN. The HNA-2a antigen is found on CD177, and is also associated with TRALI and NAIN. Further antigens have also been described (5b or HNA-3a, MARTa or HNA-4a and ONDa or HNA-5a) which, although present on granulocytes, are also identifiable on other cells. They are therefore not granulocyte specific. In neonatal autoimmune neutropenia, auto antibodies are directed towards HNA-1a or HNA-1b antigens in more than half of cases (Tables 16.3 and 16.4).

HLA System

The HLA system includes a complex family of genes and their protein products that contribute to the recognition of self and nonself, to the immune responses to antigenic stimuli, and to the coordination of cellular and humoral immunity. HLA major histocompatibility consist in a series of 35–40 closely linked genes physically grouped into three regions (class I, class II, and class III regions) located on the short arm of chromosome 6. Class I region encodes genes for the classic transplantation molecules HLA-A, HLA-B, and HLA-C. Class II region encodes genes for the molecules HLA-DR, HLA-DP, and HLA-DQ. Class III region encodes structurally and functionally diverse molecules, including complement factors and tumour necrosis factor.

The most important characteristic of HLA genes is that they are highly polymorphic, and several alleles exist at each locus. Each person has two alleles for each locus. Both alleles of a locus are expressed co-dominantly.

Class I molecules are found on the surface of platelets and of all nucleated cells of the body. Mature red cells surface usually lack HLA antigens, but nucleate immature erythroid cells express them. MHC Class II antigens are restricted to a few immunocompetent cell types such as B lymphocytes, macrophages, and dendritic cells.

Expression of different HLA antigens has significant repercussions in human pathology. It has been determined that HLA antigens are associated with disease susceptibility: i.e., narcolepsy is associate with DR2 allele and B27 allele is associate with ankylosis spondylitis. The HLA system is of great importance, second only to the ABO, in influencing the

Table 16.2 Principal RH gene and antigens encoded

Test results with the five principal Rh blood typing reagents.							
Anti-D	Anti-C	Anti-E	Anti-c	Anti-e	Antigens present	Conventional name	Probable phenotype
POS	POS	NEG	POS	POS	D, C, c, e	CcDee	R1r
POS	POS	NEG	NEG	POS	D, C, e	CCDee	R1R1
POS	POS	POS	POS	POS	D,C,c,E,e	CcDEe	R1R2
POS	NEG	NEG	POS	POS	D, c, e	ccDee	RoRo / Ror
POS	NEG	POS	POS	POS	D, c, E, e	ccDEe	R2r
POS	NEG	POS	POS	NEG	D, c, E	ccDEE	R2R2
POS	POS	POS	NEG	POS	D,C, E, e	CCDEe	R1Rz
POS	POS	POS	POS	NEG	D,C, c, E	CcDEE	R2Rz
POS	POS	POS	NEG	NEG	D, C, E	CCDDEE	RzRz
NEG	NEG	NEG	POS	POS	c, e	Ccdee	rr
NEG	POS	NEG	POS	POS	C, c, e	Ccdee	r'r
NEG	NEG	POS	POS	POS	c, E, e	ccdEe	r"r
NEG	POS	POS	POS	POS	C, c, E, e	CcdEe	r'r"

Commercial anti-D, anti-C, anti-E, anti-c and anti-e antisera are used for the determination of the Rh phenotype. The set of antigens detected on a subject's red blood cells therefore constitutes its Rh phenotype. In this table columns from one to five report the pattern of reactivity, column six report phenotype, column seven the common denomination and column eight the most probable genothype. D negative subjects lack the RHD gene that encodes the D antigen, or have a nonfunctional RHD gene. Most of the negative D subjects are homozygous for the RHce allele, which represents the gene encoding the antigens c and e. Less frequently they may have RHCe or RHcE alleles, which code for C and e or for c and E, respectively. The RHCE gene, which produces the C and E antigens, is very rare in D negative individuals. The D genotype of D positive persons cannot be determined with serological tests, as molecular biology techniques must be used

Table 16.3 Platelet antigens

System	Antigen	Frequency	Gene	Localization	Associated diseases
HPA-01	HPA-1a, HPA-1b	HPA-1 (a+b−) 72%, (a+b+) 28%, (a−b+) <1%	*ITGB3*	GPIIIa	NAIT, PTP
HPA-02	HPA-2a, HPA-2b	HPA-2 (a+b−) 85%, (a+b+) 15%, (a−b+) <1%	*GPIBA*	GP Ib	NAIT,
HPA-03	HPA-3a, HPA-3b	HPA-3 (a+b−) 37%, (a+b+) 48%, (a−b+) 15%	*ITGA2B*	GP IIb	NAIT, PTP
HPA-04	HPA-4a, HPA-4b	HPA-4 (a+b−) 99%, (a+b+) <1%, (a−b+) <1%	*ITGB3*	GP IIIa	NAIT, PTP
HPA-05	HPA-5a, HPA-5b	HPA-5 (a+b−) 80%, (a+b+) 19%, (a−b+) 1%	*ITGA2*	GP Ia	NAIT, PTP
HPA-06	HPA-6bw	<1%	*ITGB3*	GP IIIa	NAIT,
HPA-07	HPA-7bw	<1%	*ITGB3*	GP IIa	NAIT,
HPA-08	HPA-8bw	<1%	*ITGB3*	GP IIIa	NAIT,
HPA-09	HPA-9bw	<1%	*ITGA2B*	GP IIb	NAIT,
HPA-10	HPA-10bw	<1%	*ITGB3*	GP IIIa	NAIT,
HPA-11	HPA-11bw	<1%	*ITGB3*	GP III a	NAIT,
HPA-12	HPA-12bw	<1%	*GOIBB*	GP Ib	NAIT,
HPA-13	HPA-13bw	<1%	*ITGA2*	GP Ia	NAIT,
HPA-14	HPA-14bw	<1%	*ITGB3*	GP IIIa	NAIT,
HPA-15	HPA-15a, HPA-15b	HPA-15 (a+b−) 35%, (a+b+) 42%, (a−b+) 23%	*CD109*	CD109	NAIT, PTP
HPA-16	HPA-16bw	<1%	*ITGB3*	GP IIIa	NAIT,

This table indicates the name of the blood group system in the first column while the second column shows the name of the individual antigens, and their frequency is indicated in the third column. The name of the gene alleles is reported in the fourth column, while their chromosomal location is reported in the fifth column. The possible associations with the pathology are indicated in the last column., PTP: posttransfusion purpura, NAIT: neonatal immune thrombocytopenia

Table 16.4 Granulocytes antigens

System	Antigen	Frequency	Gene	Localization	Associated diseases
NA	NA1 NA2	HNA-1a 45% HNA-1b 85%	*NA1* *NA2*	CD16	TRALI, NAIN TRALI, NAIN
SH	HNA-1c	5% caucasici, 25% neri	*SH*	CD16	
NB	HNA-2a	97%	*NB1*	CD177	
5	HNA-3a	<1%	*5b*	GP 95kD	
MART	HNA-4a	99%	*MART*	CD11a	
OND	HNA-5a	99%	*OND*	CD11b	

This table indicates the name of the blood group system in the first column while the second column shows the name of the individual antigens, and their frequency is indicated in the third column. The name of the gene alleles is reported in the fourth column, while their chromosomal location is reported in the fifth column. The possible associations with the pathology are indicated in the last column. *TRALI* transfusion related Lung Injury, *NAIN* neonatal immune neutropenia long-term survival of transplanted solid organs (kidney, liver, lug, heart) and is of paramount significance in hematopoietic progenitor cell (HPC) transplantation. HLA antigens and antibodies are also important in such complications of transfusion therapy as platelet transfusion refractoriness, transfusion-related acute lung injury (TRALI), neonatal alloimmune neutropenia NAIN), neonatal alloimmune thrombocytopenia (NAIT), and post transplant and post transfusion graft-vs-host disease (GVHD).

Analytical Aspects of Erythrocyte Immunohematology

It is strongly recommended that an immunohematology laboratory can have access to different methods to face complex clinical cases overcoming the intrinsic limitations of a single method. For example, in our laboratory, which is in a provincial reference hospital, we routinely have three methods for the serological study of antigens and antibodies: tube tests, gel tests, and solid-phase tests. We also have a molecular biology laboratory using a micro-arrays-based assay for the genotyping of blood group antigens.

As previously reported, demonstration of red cell antigen-antibody reactions are key to immunohematology. The combination of antibody with antigen may produce a variety of observable results: agglutination, hemolysis, and precipitation. Agglutination is the antibody-mediated clumping of particles that express antigen on their surface, and it is the endpoint for most tests involving red cells and blood group antibodies. Agglutination is a reversible chemical reaction that occurs in two stages: (1) sensitization, the attachment of antibody to antigen on the red cell membrane; (2) formation of bridges between the sensitized red cells to form the agglutinate. Agglutination is also affected by physical and chemical conditions such as temperature, pH, ionic strength, relative antigen-to-antibody concentrations, time of incubation. Usually in routine immune hematology laboratory were in use some enhancement medium for antibody detection such as albumin, Polyethylene Glycol (PEG), low-ionic-strength solution (LISS). Proteolytic enzymes reduce the red cell negative surface charge cleaving sialic acid molecules from polysaccharide chains. Any mechanism that reduces the net charge should enhance red cell agglutination. Moreover, while enhancing agglutination by some antibodies, enzymes destroy certain red cell antigens, notably M, N, S, Fya, and Fyb. The proteolytic enzymes used most often in immunohematology laboratories are bromelin, ficin, papain, and trypsin. Figure 16.4 shows solid-phase tests.

Determination of the ABO Group

In the direct determination of the ABO group, the patient's RBCs are tested against known antisera. The presence of the antigens (agglutinogens) present on the surface of the red cells is detected by the agglutination that occurs after contact with the antisera. A positive reaction (agglutination) indicates

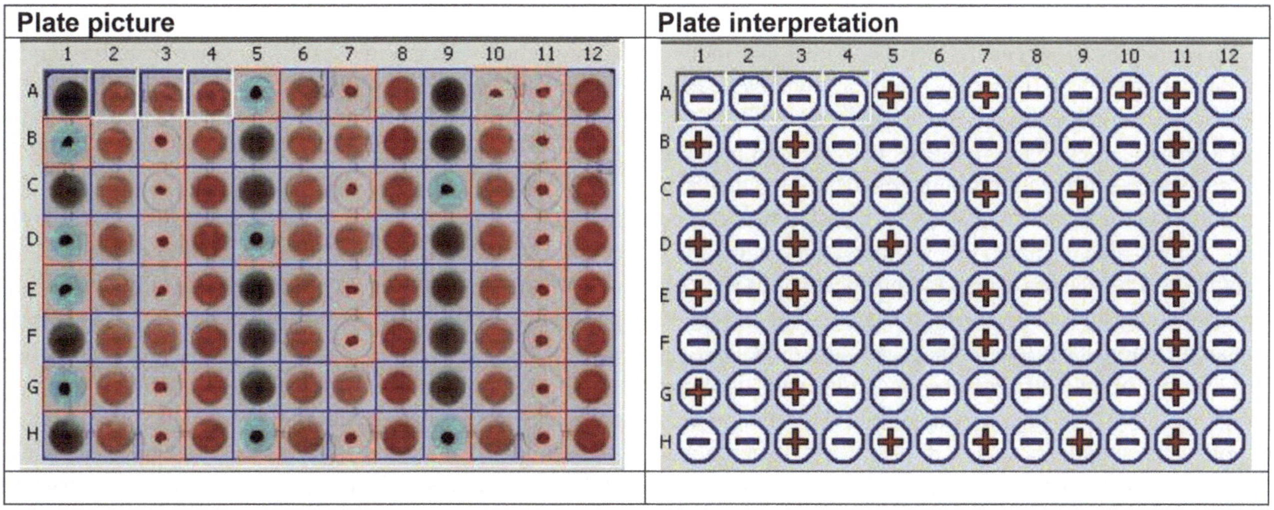

Fig. 16.4 Solid phase tests. In this figure are reported, on the left, the picture of a 96 wells microplate and, on the right, the results interpretation. A "button" in the bottom of the wells represent a positive reaction while a diffuse pattern represents a negative result

that the corresponding antigen is present on the surface of the red cells under examination. In the indirect determination of the ABO group, the patient's serum (or plasma) is tested against red blood cells of a known group. The patient's serum contains natural antibodies (agglutinins IgM) against antigens that are not present on the surface of their own red cells. A positive reaction (agglutination) indicates that antibodies are present in the serum, and therefore the relative antigen is missing on the surface of the red cells. Figure 16.5 shows direct and indirect ABO group analysis with RHD type determination.

In rare cases, subjects who do not react with anti-A, anti-B and anti-A, B sera strongly agglutinate the red cells of group O. If this agglutination is not due to the presence of cold agglutinins (in this case it disappears completely after 5 minutes at 37 °C), it could be the so-called Bombay phenotype (hh), an exceptional finding.

Weak variants of A and B are known. The best known A2 reacts well (4+) with anti-A, B sera and weaker (2+) with anti-A sera, does not react with anti-A lectin. A1 but reacts well with anti-H lectin. Natural anti-B antibodies are always present in the serum, and in 2–3% of cases anti-A antibodies may be present which react with group A1 red cells but not with group A2 ones. The variants called A3 and B3 have the characteristic of reacting very weakly with the anti-A and anti-B sera (respectively), while they react well with the anti-A, B serum. The expected natural antibodies are present in the serum of these subjects.

From these observations it follows that the ABO group must be determined using always and, on all samples, complete anti-A, anti-B and anti-A, B sera. The search for "natural" antibodies in the serum must be carried out using at least group A1 and B red blood cells.

There may be discrepancies between direct and indirect determination of the ABO group, mostly attributable to technical causes or anomalies present in the sample.

Technical causes: False negatives may be attributable to failure to add a reagent, errors in interpretation or registration (hemolysis) of the results, incorrect relationship between red blood cells and antisera, too low working temperature. False positives can be due to excessive centrifugation, use of contaminated reagents, misinterpretation or recording of results.

Sample anomalies: Interferences with the direct test can be observed in patients recently transfused with nonhomo group red blood cells (i.e., red blood cells O to a recipient A) or with weak subgroups of A or B, in case of poly agglutination of the red blood cells, presence of Wharton's gelatine in umbilical cord) or high concentrations of abnormal proteins (myeloma), cold agglutinins. Interference with the indirect test can be observed in specimens with clots, presence of allo antibodies to erythrocytes, auto-antibodies to erythrocytes, immunosuppressed patients, children under 6 months. Transplanted with ABO-incompatible allogenic marrow, recent transfusions with nonhomo group plasma. Figure 16.6 shows direct ABO determination with evidence of a double red cells population.

Search and titration of immune or natural anti-A and anti-B antibodies is done primarily to support ABO incompatible solid organ and hematopoietic stem cells transplant programs. The titration of anti-A and anti-B is carried out by diluting the serum to the doubling in saline and evaluating the reaction after immediate centrifugation. The last dilution of the serum in which hemolysis and / or agglutination is highlighted determines the titer of the hemolysins or anti-A and anti-B agglutinins. Subjects with high titer (>1/128) of anti-A and anti-B have antibodies of an immune nature alongside the natural ones. Search for anti-A and anti-B immune antibodies can be carried out with the same method after denaturation of the IgM with 2-mercaptoethanol. Alternatively, two doubling dilutions of the serum can be prepared, one incubated at 37 °C and the other at 4 °C. The difference in the titer (always greater in that incubated at 37 °C) will be attributable to the presence of IgG. Group O subjects naturally have IgG class anti-AB.

Determination of Type D and Rh Phenotype

In direct determination of Rh type and Rh phenotype, the patient's red cells are tested against known antisera. The presence of the antigens (agglutinogens) on the surface of the red cells is subtended by the agglutination that occurs after contact with the antisera. A positive reaction (agglutination) indicates that the antigen corresponding to the surface of the red cells under examination is present. As there are no natural antibodies, the indirect test is not used.

Search for weak D Pre-prepared columns are used for the execution of the Coombs test in Liss. A 1% suspension of the patient's red blood cells is prepared, and a drop of the suspension is placed in the incubation chamber of a microtube of the coupon. Then 25 μL of an IgG monoclonal anti-D serum or a serum blend in which an IgG component is present is added. It is incubated for 10–15 minutes at 37 °C, centrifuged for 10 minutes at 900 rpm on a dedicated centrifuge and visually assessed the presence of agglutination against a light source. In the latter circumstance, the patient will be classified as Du (Fig. 16.7).

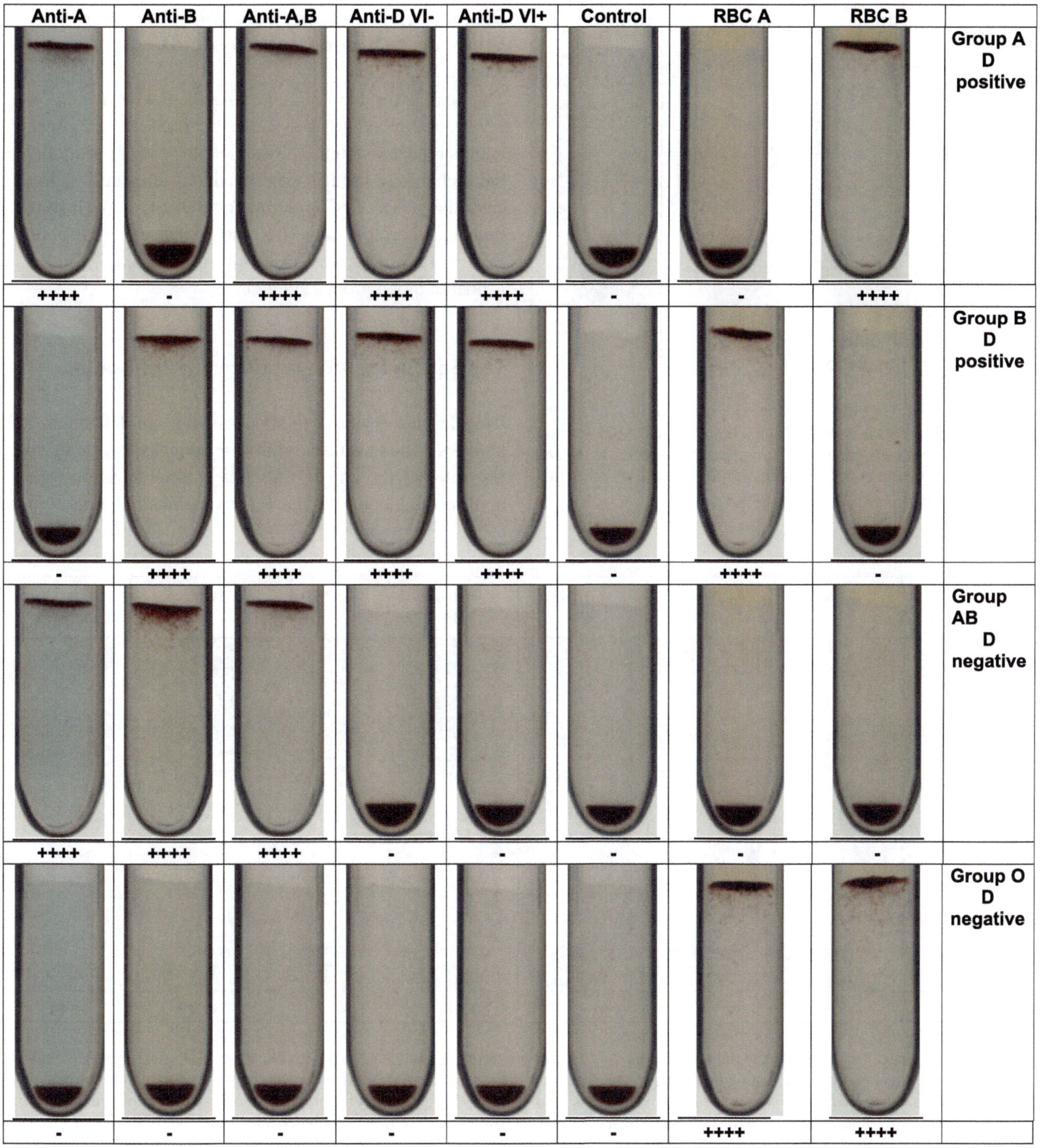

Fig. 16.5 Determination of the direct and ndirect ABO Group, with RHD type determination; some examples. The first sample reacts with the anti-A and anti-AB sera, does not react with the anti-B serum, it reacts with the anti-D sera (both Dvi− and Dvi+), plasma reacts with group B but not group A red blood cells, therefore the subject is A D positive. The second sample does not react with the anti-A serum, but reacts with the anti-B and anti-A sera, B, reacts with the anti-D sera (both Dvi− and Dvi+), plasma reacts with group A but not group B red blood cells. It is, therefore, a B D positive. The third sample reacts with the anti-A, anti-B and anti-A,B sera, does not reacts with the anti-D sera (both Dvi− and Dvi+), plasma does not react with group A and group B red blood cells. It is therefore a subject of AB D negative group. The fourth sample does not react with the anti-A, anti-B, anti-A,B sera, such as with the anti-D sera (both Dvi− and Dvi+), plasma reacts with group A and group B red blood cells. It is, therefore, a subject of group O D negative

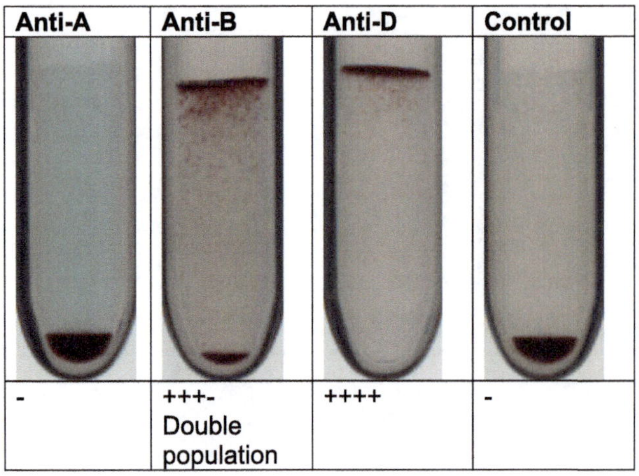

Fig. 16.6 Direct ABO determination with evidence of a double red cells population. This is a sample from a B group patient that in emergency room received two units of O group packed red cells. In the anti-B column are present two RBC population: the patient's B RBC (top) and the transfused O RBC (bottom) of the tube

Study of the Antigens Belonging to Other Erythrocyte Blood Group Systems

The search for the antigenic specificities of the different erythrocyte blood group systems is carried out using specific commercial antiserum. The methodology varies according to the technology used in the individual laboratories. In any case, the presence of agglutination indicates a positive reaction and therefore underlies the presence of the antigen. Figure 16.8 describes research for MN, Ss, Kidd, and Lutheran blood groups antigens.

The Search for Anti-erythrocyte Antibodies

Irregular anti erythrocyte alloantibodies are defined as antibodies directed against erythrocyte antigens, except for natural anti-A and anti-B antibodies. Immunization against erythrocyte blood group antigens can usually be traced back

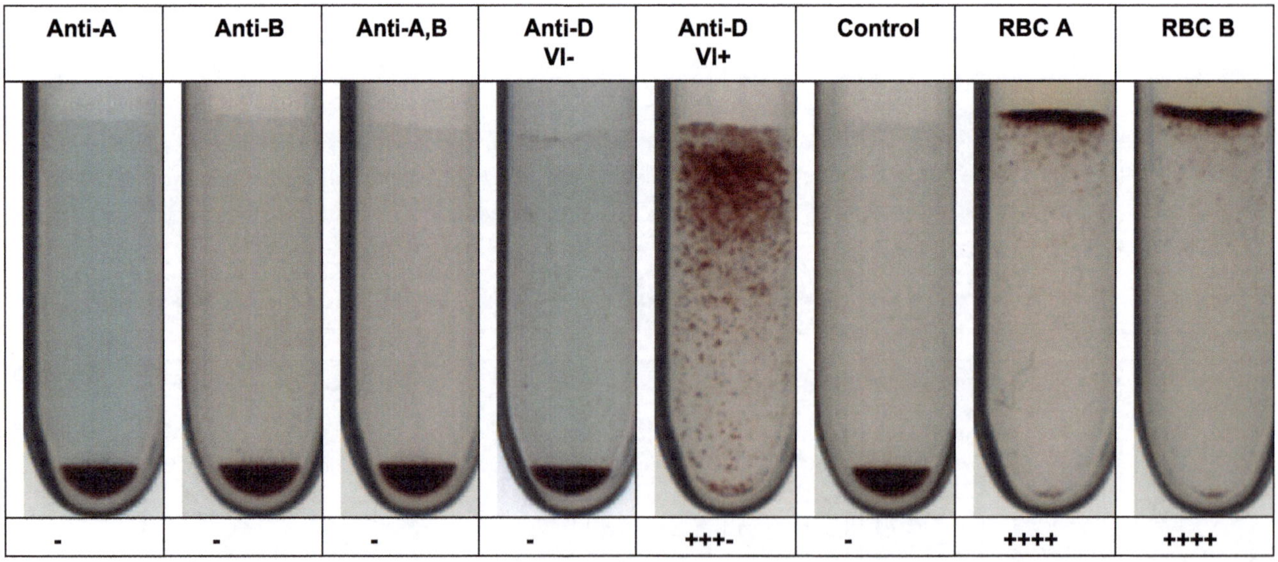

Fig. 16.7 Sample with a variant D. The image above shows the determination of the direct and indirect ABO group D type. The red blood cells of the sample under examination do not react with the anti-A, anti-B, and anti-AB sera, while the plasma of the test sample reacts strongly with red blood cell tests of groups A and B. The patient is therefore of group O. As regards the determination of D type, negativity with the first antiserum (DVI−) and a reactivity (+++−) with the second antiserum (DVI+), suggest the presence of a D variant. Further analysis using genotyping study are highly recommended

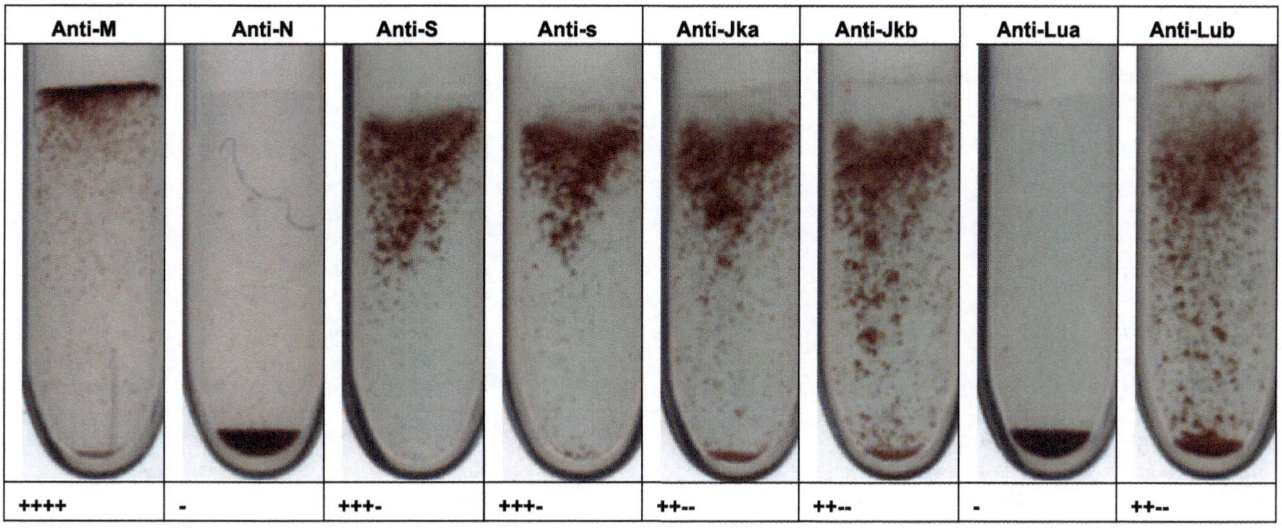

Fig. 16.8 Research for MN, Ss, Kidd and Lutheran blood groups antingens. This sample is positive for M (first column), negative for N (second column), positive for S and s (third and fourth column), positive for Jka and Jkb (fifth and sixth column), negative for Lua (seventh column), and positive for Lub (eighth column). Phenotype of this subjects was M+N−, S+s+, Jka+Jkab+, Lua-Lua+

Table 16.5 Serological characteristics of allo-antibodies directed towards erythrocyte blood groups antigens of ABO and RHD systems

Antibody	Ig class	Reactivity in NaCl at 22 °C	Reactivity al 37 °C	Reactivity in AHG	Enzymes	Complement binding	In Vitro hemolysis	PTH	NHD
Anti-H	IgM	Yes	Yes	–	No change	Yes	Yes	Yes	Mild
Anti-A	IgM (IgG*)	Yes	Yes	–	No change	Yes	Yes	Yes	Mild
Anti-B	IgM (IgG*)	Yes	Yes	–	No change	Yes	Yes	Yes	Mild
Anti-A,B	IgG (IgM*)	Yes	Yes	–	No change	Yes	Yes	Yes	Mild to moderate
Anti-A1	IgG	Yes	Rare	–	No change	Rare	Rare	No	No
Anti-D	IgG	Few	Yes	Yes	Enhance	No	No	Few	Severe
Anti-c	IgG	Few	Yes	Yes	Enhance	No	No	Few	Severe
Anti-C	IgG	Few	Yes	Yes	Enhance	No	No	Few	Severe
Anti-E	IgG (IgM*)	Few	Yes	Yes	Enhance	No	No	No	Mild to moderate
Anti-e	IgG	Few	Yes	Yes	Enhance	No	No	No	Mild to moderate
Anti-Cw	IgG (IgM*)	Few	Yes	Yes	Enhance	No	No	No	Mild to moderate

This Table reports the main characteristics of antibodies directed against erythrocyte blood groups antigen of ABO and RHD antigens. The first column report antibody specificity, the second the Ig class, the third the presence of reactivity in NaCl at +22 °C, the fourth the presence of reactivity al +37 °C, the fifth the presence of reactivity by using a Coombs antiserum, the sixth the sensitivity to enzymes, the seventh the ability in complement activation with in vitro hemolysis (eight column), the ninth and the eleventh columns report data about antibody's significance: post-transfusion hemolysis and newborn hemolytic disease

to exposure to nonself red blood cells, such as pregnancy or previous transfusions. In some rare cases, however, no exposure to nonself red blood cells is identifiable. If an anti-erythrocyte alloantibody is discovered, it is important to determine its specificity, titer, and to evaluate its clinical significance. An anti-erythrocyte alloantibody is defined clinically significant when it can generate a PTH (or is able to significantly reduce the survival of transfused red cells), or when it is implicated in cases of hemolytic disease of the foetus and neonate. Usually, the evaluation of the clinical significance of an anti-erythrocyte antibody is carried out on the basis of literature data, as reported in Tables 16.5 and 16.6.

Table 16.6 Serological characteristics of allo-antibodies directed towards erythrocyte blood groups antigens other than ABO and RHD

Antibody	Ig class	Reactivity in NaCl at 22 °C	Reactivity al 37 °C	Reactivity in AHG	Enzymes	Complement binding	In vitro hemolysis	PTH	HDFN
Anti-K	IgG, (IgM*)	Some	Some	Most	No change	Rare	No	Yes	Mild to severe
Anti-k	IgG	Few	Few	Most	No change	No	No	Yes	Mild
Anti-Kpa	IgG	Some	Some	Most	No change	No	No	Yes	Mild
Anti-Kpb	IgG	Few	Few	Most	No change	No	No	Yes	Moderate
Anti-Jsa	IgG	Few	Few	Most	No change	No	No	Yes	Moderate
Anti-Jsb	IgG	No	No	Most	No change	No	No	Yes	Mild
Anti-M	IgM, (IgG*)	Most	Few	Few	Destroy	No	No	Few	Mild to Severe
Anti-N	IgM	Most	Few	Few	Destroy	No	No	Rare	Moderate
Anti-S	IgG	Few	Some	Most	Variable	Some	No	Yes	Mild
Anti-s	IgG	Few	Few	Most	Variable	Rare	No	Yes	Mild to severe
Anti-U	IgG	Rare	Some	Most	No change	No	No	Yes	Mild to severe
Anti-Jka	IgG	Few	Few	Most	Enhance	Yes	Some	Yes	Mild
Anti-Jkb	IgG	Few	Few	Most	Enhance	Yes	Some	Yes	Mild
Anti-Fya	IgG	Rare	Rare	Most	Destroy	Rare	No	Yes	Mild to severe
Anti-Fyb	IgG	Rare	Rare	Most	Destroy	Rare	No	Yes	Mild
Anti-Lua	IgM, (IgG*)	Most	Few	Few	Variable	Some	No	No	Mild
Anti-Lub	IgG	Few	Few	Most	Variable	Some	No	No	Mild
Anti-Lea	IgM	Most	Few	Few	No change	Most	Few	No	No
Anti-Leb	IgM	Most	Few	Few	No change	Most	Few	No	No
Anti-I	IgM	Most	Few	Few	Enhance	Most	Few	Rare	No
Anti-i	IgM	Most	Few	Few	Enhance	Most	Few	No	Mild
Anti-P1	IgM	Most	Some	Rare	Enhance	Rare	Rare	No	No

This Table reports the main characteristics of antibodies directed against erythrocyte blood groups antigen other than ABO and RHD antigens. The first column report antibody specificity, the second the Ig class, the third the presence of reactivity in NaCl at +22 °C, the fourth the presence of reactivity al +37 °C, the fifth the presence of reactivity by using a Coombs antiserum, the sixth the sensitivity to enzymes, the seventh the ability in complement activation with in vitro hemolysis (eight column), the nineth and the eleventh columns report data about antibody's significance: posttransfusion hemolysis and newborn hemolytic disease

Indirect Antiglobulin Test (Indirect Coombs Test)

The indirect antiglobulin test (IAT) is used to detect the presence of serum antibodies directed to erythrocyte blood group antigens, there must be between 100 and 200 IgG or C3 molecules on the cell to obtain a positive reaction.

The presence of these antibodies is identified by reacting (in vitro) the test serum with suitably selected red blood cells. The reaction is revealed by using a Coombs serum. If the indirect Coombs' is positive, titration of the antibody and differentiation between IgG or IgM antibodies must always be carried out. It is also necessary to identify the erythrocyte blood group antigen towards which the antibody is directed. For the titration of the antibody a scalar dilution to doubling the serum in physiological is prepared. A IAT is then prepared from each dilution, using a polyspecific antiglobulin serum. The titer of the antibody is indicated by the highest dilution at which agglutination of the red cells that have given a positive reaction to the IAT still occurs.

To proceed with the identification of the specificity of an antibody, special RBC panels are used. If maternal-foetal incompatibility due to ABO incompatibility is studied, a panel consisting of 4 red blood cells of groups A1, A2, B and O is used. a panel consisting of group O polyantigenic red cells characterized by different phenotypes is used. Figure 16.9 shows screening for Indirect Antiglobulin Test using a three cells panel.

This screening is positive for cells 1 and 2. An antibody identification should be performed using an extended panel of fifteen group O polyantigenic red cells.

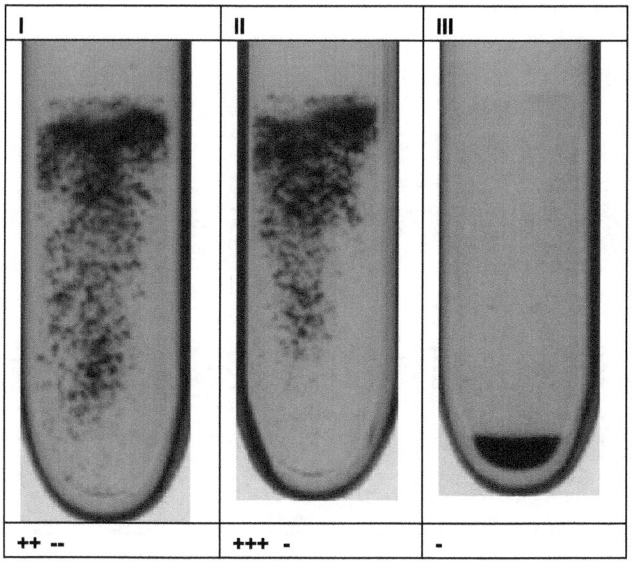

Fig. 16.9 Screening for Indirect antiglobulin test using a three cells panel. The figure above shows a screening search for anti-erythrocyte antibodies carried out using a panel of red blood cells test polyantigenic group O. This screening is positive for cells 1 and 2. Antibody specificity identification performed using a more extensive panel is reported in Fig. 16.10

Figure 16.10 describes characterization of an alloantibody mixing using an extended identification panel untreated and treated with enzymes and Fig. 16.11 antibody titration.

After reporting the reactivity pattern on the appropriate sheet, it is compared with the results reported on the accompanying sheet of the panel (identigram) (Fig. 16.12), excluding the antigens based on the negative results. It is good practice, after presumptively identifying an alloantibody directed towards a specific erythrocyte antigen, to demonstrate its absence on the patient's red cells.

Interpreting Results

Antibody screening results are interpreted as positive or negative based on the presence or absence of reactivity. Interpretation of panel results can be a more complex process because panel results generally will include a mix of both positive and negative results at different phases of testing, with different reactivity score. Moreover, patient's red cell phenotype also plays roles in the final interpretation and in differentiation between auto and allo antibodies.

Both positive and negative reactions are important in antibody identification: positive reactions (considering phase and strength of reactivity) can suggest certain specificities. Positive reactions also can be compared to the antigen patterns expressed by the panel cells to help assign specificity. Single alloantibodies usually show definite positive and negative reactions that create a clear-cut pattern with antigen-positive and -negative reagent red cell samples. Negative reactions are important in antibody identification because they allow tentative exclusion of antibodies to antigens expressed on the nonreactive cells. Exclusion (crossing-out) of antibodies is a widely used first approach to the interpretation of panel results. Once results have been recorded on the worksheet, the antigen profile of the first nonreactive cell is examined. If an antigen is *present* on the cell and the serum *did not react* with the cell, the presence of the corresponding antibody may be, at least tentatively ruled out. After all antigens present on that cell have been crossed off, interpretation proceeds with the other nonreactive cells and additional specificities are excluded. In most cases, this process will leave a group of antibodies that still have not been excluded. Next, the cells reactive with the serum are evaluated. The pattern of reactivity for each non excluded specificity is compared to the pattern of reactivity obtained with the test serum. If there is a pattern that matches exactly, that is most likely the specificity of the antibody in the serum. However, if there are remaining specificities that have not been excluded, additional testing may be needed to eliminate remaining possibilities and to confirm the specificity identified.

For example, patient reported in Figs. 16.9 and 16.10 is group O and has the CCDee Kk, control negative, IAT positive. Because negativity od autologous control, we can exclude the presence of antibodies present on the red blood cells in the patient: D, C, e, K, and k.

Considering the negative cells in the panel not treated with enzymes: cells 6, 11 and 12; we can provisionally exclude the antigens c, Cw, Kpa, Fya, Fyb, Jka, Jkb, N, S, s, Lua, Lub, Lea, P1, N, S, s, and Xg. On the basis of the analysis of this panel it is not possible to exclude antibodies directed to the antigens E, Jsa, Lea, M, lua, and Coa. It was therefore considered, also considering the differentiated reactivity scores, to evaluate a second extended panel of polyantigenic red blood cells of group O treated with enzymes. The treatment with enzymes eliminates the reactivity for cells 1, 2, 3, 7, 8, 10, 14, and 15 while the reactivity for cells 4, 5, and 13 remain. This differentiated reactivity allows us to hypothesize the presence of an antibody directed towards an antigen sensitive to the treatment with enzymes that according to the reactivity panel we can identify with an anti-M associated with an antibody directed towards an antigen resistant to the enzymatic treatment that we can identify with an anti-E.

The presence of these allo antibodies is in agreement with the patient's phenotyping.

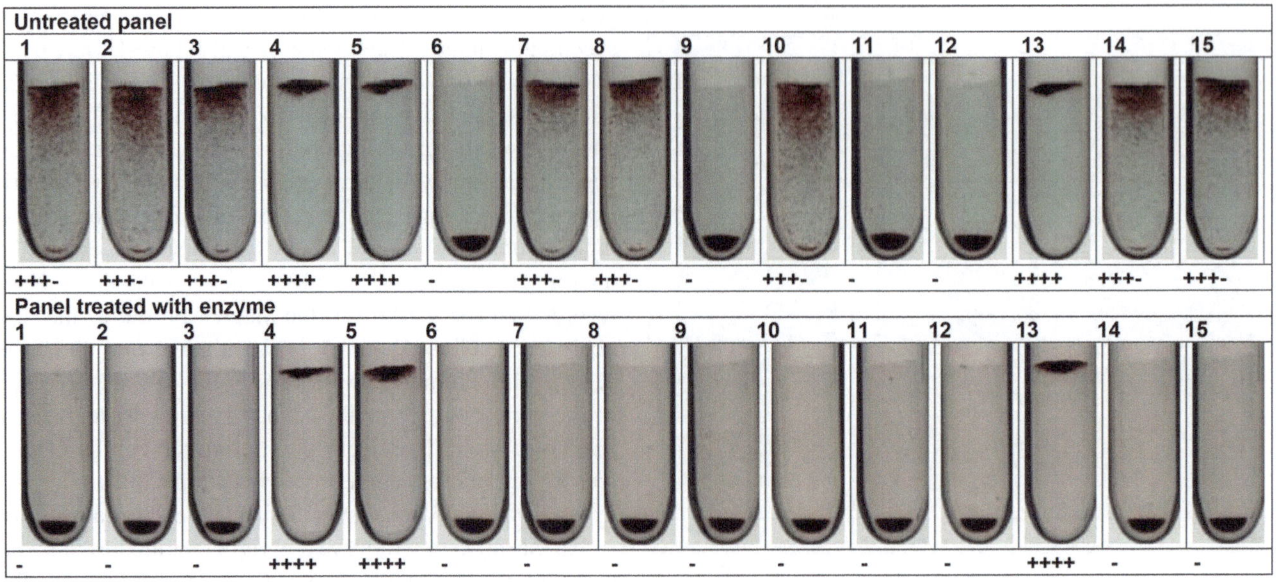

Fig. 16.10 Identification of allo-antibodies using an extended identification panel with and without enzyme treatment. This is the same sample already screened, as shown in Fig. 16.9. For identification we tested the sample against a panel of polyantigenic group O red blood cells not treated with enzymes, the test was performed in Liss-Combs. Some negative test cells 6, 11 and 12 were observed; such as some positive tests with different score: ++++ for columns 4, 5, and 13; and +++− for 1, 2, 3, 7, 8, 10, 14, and 15. after this series of tests, based on the results obtained we hypothesized the presence of an antibody mixture to solve which we decided to use a different panel and a different reaction method, as shown in the second series of images. We then tested the sample against the same panel of polyantigenic group O test red cells, but enzyme-treated, the test was performed in Liss. As can be seen, the negativities for cells 6, 11 and 12 remain as well as the positivity ++++ for columns 4,5 and 13. The positivity for cells 1, 2, 3, 7, 8, 10, 14 and 15 disappear; a sign that the antigens to which the antibody was directed was destroyed by the enzymatic treatment of the red blood cells. The combined evaluation of the results of the two tests made it possible to resolve the mixture by identifying an anti-E (specificity that remains even in the panel treated with enzymes), and an anti-M antibody whose reactivity is eliminated by the enzymatic treatment

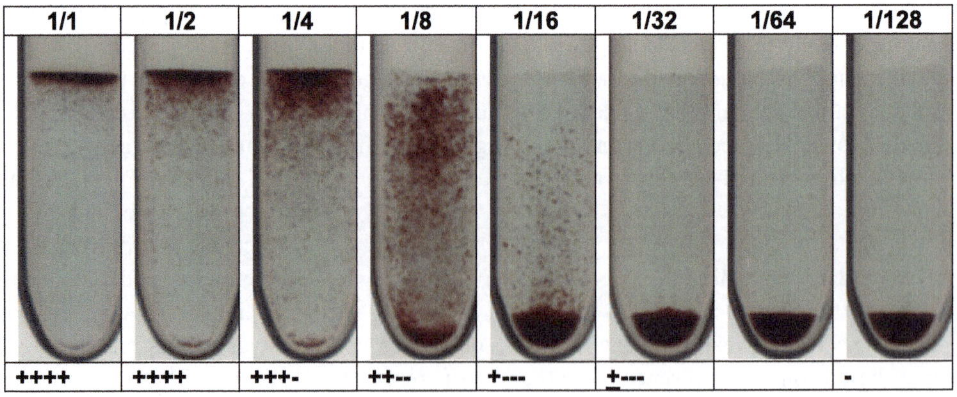

Fig. 16.11 Allo antibody titration. To carry out the titration of an antibody, serial factor two dilutions are prepared which are tested against a cell that has been tested reactive to IAT using the most appropriate phase. The dilution of the last positive column indicates the reciprocal of the antibody titer. Results: Titer 1/32

Direct Antiglobulin Test (Direct Coombs Test)

The direct antiglobulin test (DAT) is used to detect in vivo sensitization of erythrocytes by γglobulins or complement fractions. It is first performed using a polyspecific Coombs serum. DAT positivity occurs with the presence of at least 500 immunoglobulin molecules attached to the red blood cells, and can be traced back to one of the following factors:

- Auto antibodies directed to the patient's own erythrocyte antigens
- Allo antibodies present in a recently transfused patient that react with antigens present on the donor's red cells
- Passively acquired antibodies by administration of plasma or immunoglobulins or, in the case of maternal antibodies, that have crossed the placenta and react with fetal red blood cells (neonatal hemolytic disease)

	VIAL	Donor No.	Rh		Rh-hr						Kell				Duffy		Kidd		Lewis		P	MNS				Luth.	Colt.	Xg	SPECIAL TYPE	RESULTS
					D	C	E	c	e	C^w	K	k	Kp^a	Js^a	Fy^a	Fy^b	Jk^a	Jk^b	Le^a	Le^b	P_1	M	N	S	s	Lu^a	Co^a	Xg^a		
	1	2007303	CCDee	R_1R_1	+	+	0	0	+	0	+	+	0	nt	0	+	+	0	0	+	0	+	+	+	0	0	0	+		1
	2	2007304	Ccddee	r'r	0	+	0	+	+	0	0	+	0	nt	0	+	0	+	0	0	0	+	+	0	+	0	0	+		2
	3	2007319	ccDee	R_0r	+	0	0	+	+	0	0	+	0	nt	+	0	0	+	0	+	+	+	0	+	0	0	0	+		3
	4	2007306	ccddEe	r''r	0	0	+	+	+	0	0	+	0	nt	0	+	+	0	0	+	+	+	0	0	+	0	0	+		4
	5	2007307	ccDEE	R_2R_2	+	0	+	+	0	0	0	+	0	nt	+	0	0	+	0	0	+	0	+	0	+	0	0	+		5
	6	2006977	C^wCDee	$R_1^wR_1$	+	+	0	0	+	+	0	+	0	nt	+	+	+	+	0	0	0	0	+	+	+	0	0	0		6
	7	2003046	ccddee	rr	0	0	0	+	+	0	+	+	0	nt	+	0	+	+	0	+	+	+	0	+	0	0	0	+		7
	8	2007309	ccddee	rr	0	0	0	+	+	0	0	+	0	0	+	0	+	0	0	+	+	+	0	0	+	0	0	0		8
	9	2007310	ccddee	rr	0	0	0	+	+	0	0	+	0	nt	0	+	0	+	+	0	0	0	+	0	+	+	0	+*		9
	10	2007311	ccddee	rr	0	0	0	+	+	0	0	+	0	nt	0	+	+	0	+	0	+	+	+	0	+	0	0	+*		10
	11	2006109	CCDee	R_1R_1	+	+	0	0	+	0	0	+	+	nt	+	0	0	+	0	+	+	0	+	+	+	0	0	+		11
EXTEND	12	2007312	ccddee	rr	0	0	0	+	+	0	+	+	0	nt	0	+	+	0	0	+	+*	0	+	+	+	0	0	0		12
EXTEND	13	2007313	ccDEE	R_2R_2	+	0	+	+	0	0	0	+	0	nt	+	+	+	0	0	0	0	+	+	0	+	0	0	+		13
EXTEND	14	2007314	CCDee	R_1R_1	+	+	0	0	+	0	0	+	0	nt	+	0	+	0	+	0	0	+	0	+	+	0	0	+		14
EXTEND	15	2007060	CCDee	R_1R_1	+	+	0	0	+	0	0	+	0	nt	0	+	0	+	0	0	+	+	0	+	0	0	0	+		15
																														AC

Fig. 16.12 Identigram. The figure shows the data sheet showing the antigenic composition of the panel of polyantigenic red blood cells of group O used for the identification of the antibody mixture shown in Fig. 16.10. Each row corresponds to the typing of each of the 15 cells that make up the panel and for each column the positivity or negativity for each of the antigens is reported

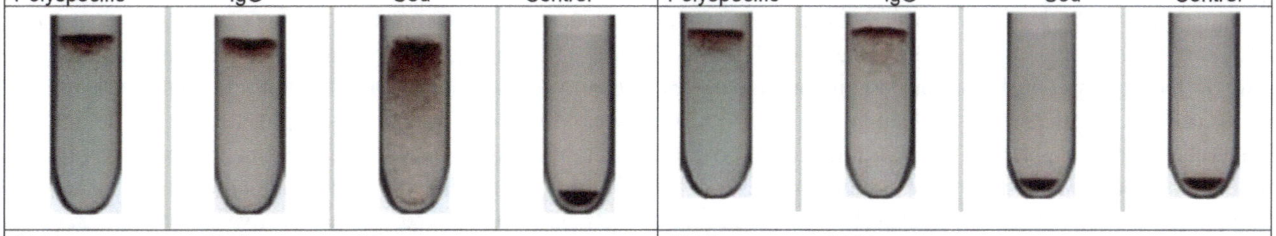

Fig. 16.13 Direct antiglobulin test. Patient shows a marked positivity (++++) with the polyspecific Coombs serum. The positivity is confirmed for IgG (++++) and for C3b (+++−), the auto-control is negative. The patient therefore presents both IgG and complement on the surface of the red cells. The second patient shows marked positivity (++++) with polyspecific Coombs serum. The positivity is confirmed for IgG (++++) but both the C3b and the auto-control are negative. The patient therefore presents exclusively IgG on the surface of the red cells

- Antibodies directed to drugs that bind to the erythrocyte membrane (such as penicillin)
- Nonspecific binding (mediated by drugs usually cephalosporins) of immunoglobulins to the erythrocyte membrane
- Presence of complement fractions on the surface of red blood cells with a mechanism mediated by drug-induced immune complexes
- Autoimmune diseases

Figure 16.13 shows two markedly positive samples for the direct antiglobulin test. The sample above is reactive with polyvalent Coombs serum (AHG.CD), with monospecific serum for IgG (IgG.IG) and C3d (C3d.IG); therefore, on the surface of these red cells there are IgG and complement fractions. The sample below is reactive with polyvalent Coombs serum and with monospecific serum for IgG but not for C3d; therefore, on the surface of these red cells there are IgG but not complement fractions.

In case of positive TCD with polyspecific Coombs serum, the antibody adhered to the red blood cells is then identified by TCD with monospecific sera and using elution methods. The direct antiglobulin test with monospecific sera is used to detect in vivo sensitization of erythrocytes by di γ-globulins or complement fractions. Secondly, monospecific Coombs sera (anti-IgG, anti-IgA, anti-IgM, anti-C3, anti-C3b, anti-C4) can be tested.

The elution methods are used to remove the antibodies adhered to the red blood cells, and to be able to proceed with their study using the eluate. Several methods of antibody elution based on physical methods are available: hot elution, particularly effective for IgM; by freezing and thawing, usable in cases of ABO related HNFN; or chemical methods: for example, elution with organic solvents or with acid glycine which is widely indicated in the elution of IgG antibodies. Once the antibodies have been eluted, it is also possible to try to identify their specificity.

Applications of Molecular Biology to Erythrocyte Immunohematology

Traditionally, erythrocyte immunohematology is based on the study of the phenotype, determined by the search for the presence on the surface of the red blood cells of antigens belonging to blood group systems using appropriate antisera. Direct genotype determination is now routinely applied in immunohematology.

DNA extraction usually takes place from whole blood collected in EDTA, using automated methods that can lighten the workload and at the same time ensure good quality of the extracted nucleic acids. After the extraction phase, we move on to the amplification phases of the target with a multiplex method, which allows the simultaneous and allele-specific amplification of multiple target sequences.

In the routine of our immunohematology laboratory, commercial gene-chips are usually used for the detection of amplified products, which allow to study polymorphisms of different blood group systems at the same time. The Blood Chip Reference system, in routine use in our laboratory, allows to determine the genotype of antigens related to ABO, RH, Kell, Duffy, Dombrock, Colton, Lutheran, Diego Kidd and MNS blood group systems. The system foresees that the amplicons, labelled with a fluorescent substrate, are detected by binding with allele-specific probes immobilized on a slide. The slide is then scanned by a laser, and for each probe the positivity for the single allele is assessed, based on the bound fluorescence, using specific software.

Figure 16.14 shows genotyping in immunohematology; the Blood Chip Reference system adopted in our Laboratory.

Another widespread system is the one called HEA BeadChip, capable of determining the genotype of antigens related to the blood group systems ABO, RHD, RHCE, Kell, Duffy, Dombrock, Lewis, Colton, Scianna, Lutheran, Diego, Kidd, and MNS. In this system, the amplified products are denatured and linked, through the use of specific allele probes, to color-coded microspheres. Upon successful bonding, the probe elongates and produces a fluorescent signal which is detected by an image capture system using a fluorescence microscope. The intensity of the signal and the probe related to it allow to trace the genotype.

At present, genotyping methods find routine applications in the extended typing of donors for the establishment of "banks of rare groups", which can only be computerized or provide for the storage of frozen erythrocyte concentrates. They are widely used in the study of transfusion-dependent subjects (i.e., patients with hemoglobinopathies), in which the determinable phenotype no longer corresponds to the genotype, as well as in the study of rare groups or carriers of weak antigenic variants. The applications in the field of prevention of MEN appear to be very interesting, with the possibility of carrying out foetal genotyping both from foetal cell material obtained with an invasive method and from free DNA in maternal plasma.

Fig. 16.14 Genotyping in immune hematology; the Blood Chip Reference system adopted in our laboratory

Recommended Readings

Anstee D (2009) Red cell genotyping and future pretransfusion testing. Blood 114:248–252

Anstee D (2011) The functional importance of blood group-active molecules in human red blood cells. Vox Sang 100:140–149

Blaney K, Howard P (2013) Basic and applied concepts of blood banking and transfusion practices, 3rd edn. Elsevier

Brown P (1994) Antibody identification. In: Rudman S (ed) Textbook of blood banking and transfusion medicine. Saunders & co, Philadelphia

Brown P, Resolving ABO (1994) Typing discrepancies and other typing problems. In: Rudman S (ed) Textbook of blood banking and transfusion medicine. Saunders & co Philadelphia

Cooling L (2016) Carbohydrate blood groups. In: Rossi's principles of transfusione medicine, 5th edn. Wiley Blackwell

Daniels G (2013) Human blood groups, 3rd edn. Wiley-Blackwell, Oxford

Fung MK, Grossman BJ, Hillyer CD, Westhoff CM (eds) (2014) Technical manual, 18th edn. AABB, Bethesda

Harmening D (2012) Modern blood banking and transfusion medicine, 6th edn. FA Davis edition, Philadelphia

Hiller S, Ness A, Robak. (2007) Blood banking and transfusion medicine basic principles & practice, 2nd edn. Churchill and Livingstone USA

Issitt PD, Anstee DJ (1998) Applied blood group serology. Montgomery Scientific Publications, Miami

Klein HG, Anstee DJ (2005) Mollison's blood transfusion in clinical medicine, 11th edn. Blackwell, Malden

McBean R, Hylard C, Flower R (2014) Approaches to determination of a full profile of blood group genotypes: single nucleotide variant mapping and massively parallele sequencing. Comp Struct Biotech J 11:147–151

McCullough J (1998) Transfusion medicine. Mc Grow Hill, New York

Rao G, Eastlund T, Jaganathan L (2007) Handbook of blood banking and transfusion medicine. Japee Medical Edition, New Delhi

Reali G (2009) Immunoematologia e Trasfusione. In: Trattato Italiano di Medicina di Laboratorio Volume VI. Seconda Edizione Piccin

Reid ME, Mohandas N (2004) Red blood cell blood group antigens: structure and function. Semin Hematol 41:93–117

Reid ME, Lomas-Francis C, Olsson ML (2012) The blood group antigen factsbook. Academic Press, London

Schenkel-Brunner H (2000) Human blood groups: chemical and biochemical basis of antigen specificity, 2nd edn. Spinger-Verlang, Vienna

Sorry J (2016) Other protein blood groups. In: Rossi's principles of transfusione medicine, 5° edizione edn. Wiley Blackwell

Storry JR, Castilho L, Daniels G et al (2014) International Society of Blood Transfusion Working Party on red cell immunogenetics and blood group terminology: Cancun report (2012). Vox Sang 107:90–96

Telen M (2000) Red cell antigen. In: Anderson K, Mess P (eds) Scientific basis of transfusion medicine, 2nd edn. WB Saunders co, Philadelphia

Turgeon M (1995) Fundamentals of innumohematology, 2nd edn. William & Wilkins, Baltimore

Westhoff C (2016) Red cell immunology and compatibility testing. In: Rossi's principles of transfusione medicine, 5° edizione edn. Wiley Blackwell

Westhoff C, Siegel D (2016) Rh and LW blood groups antigens. In: Rossi's principles of transfusion medicine, 5° edizione edn. Wiley Blackwell

Wilkinson S (1994a) The ABO anf H group systems. In: Rudman S (ed) Textbook of blood banking and trasfusion medicine. Saunders & co Philadelphia

Wilkinson S (1994b) Other blood group. In: Rudman S (ed) Textbook of blood banking and trasfusion medicine. Saunders & co Philadelphia

Wilkinson S (1994c) The Lewis, secretor, and soluble ABH antigents. In: Rudman S (ed) Textbook of blood banking and trasfusion medicine. Saunders & co Philadelphia

Wilkinson S (1994d) The Rh blood group system. In: Rudman S (ed) Textbook of blood banking and trasfusion medicine. Saunders & co Philadelphia

Hemostasis

17

Giuseppe Lippi, Gian Luca Salvagno,
and Massimo Franchini

Introduction

Hemostasis is a complex balance in which anticoagulant forces contribute to maintaining the fluidity of the blood within intact vessels and procoagulant forces act instead to avoid an excessive outflow of blood when the vessels lose their integrity (Fig. 17.1). Hemostasis can be divided into primary, in which platelets and the vascular system play a key role, and secondary, in which coagulation factors participate. The subsequent fibrinolytic process has the function of bringing the vessel back to patency when the lesion has been repaired. This chapter will describe the main pathophysiological mechanisms of hemostasis and the related laboratory tests useful in the diagnosis and therapeutic monitoring of the main coagulation disorders.

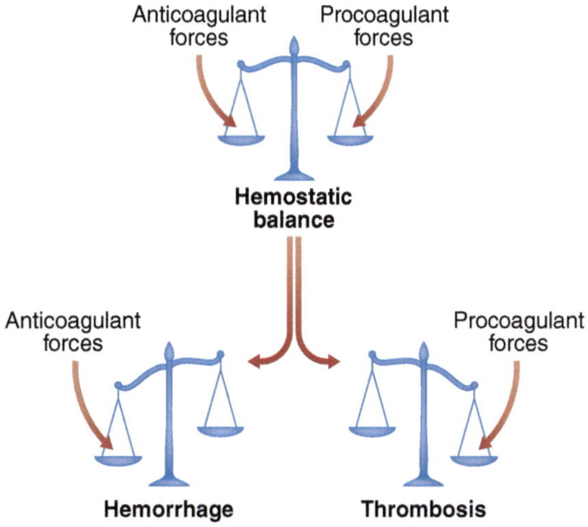

Fig. 17.1 Hemostatic balance. (Copyright EDISES 2021. Reproduced with permission)

Vascular System

The vascular system in mammals, including humans, is divided into four types, including arterial vessels, venous vessels, capillary vessel systems, and lymphatic systems, which are described below.

Arterial vessels (the arteries): Their function is to transport blood from the heart to the peripheral organs to carry oxygen, nutrients, and other molecules. Their function is essential to ensure the survival of the organs. They are, therefore, characterized by a very elastic wall (to allow the propagation of the sphygmic wave) and very high internal pressure (about 100–140 mmHg during systole and about 60–90 mmHg during diastole) to ensure efficient perfusion.

Venous vessels (veins): They have the function of conveying blood from the peripheral organs to the heart ("venous return"). They have much thinner walls than the arteries, and the flow of blood inside them is guaranteed by a residual arterial pressure (5–10 mmHg) and by the contractile action of the muscles that often surround them.

G. Lippi (✉)
Section of Clinical Biochemistry, University Hospital of Verona, Verona, Italy
e-mail: giuseppe.lippi@univr.it

G. L. Salvagno
Section of Clinical Biochemistry, Department of Neurological, Biomedical and Movement Sciences, University Hospital of Verona, Verona, Italy

M. Franchini
Department of Hematology and Transfusion Medicine, Carlo Poma Hospital, Mantua, Italy

M. Ciaccio (ed.), *Clinical and Laboratory Medicine Textbook*, https://doi.org/10.1007/978-3-031-24958-7_17

Capillary vessel system (capillary system): It represents the junction point between arteries and veins. Capillaries are essentially characterized by a small caliber (5–10 μm, sufficient for transit), very thin walls, low pressure (35–40 mmHg at the arterial end and 15–20 mmHg at the venous end), and low speed of blood flow (about 1 mm/second).

Lymphatic system (lymphatic vessels): It is a system of drainage vessels whose function is to carry the lymph from the interstitial space to the circulatory stream. It does not contain blood or even components of the hemostatic system but instead contains proteins and lipids.

Hemostatic Balance

The main hemostatic events, at least those of clinical relevance, occur in most cases in arterial and venous vessels. An imbalance in the hemostatic system can lead to two types of events: thrombosis, defined as a procoagulant tendency of the blood within an intact vascular system, and hemorrhage, defined as an inefficient coagulative capacity within a vascular system that has lost integrity.

A minimum degree of thrombosis can physiologically occur within intact vessels (e.g., as a result of a slight trauma), such, however, as never to exceed the effect of anticoagulant forces to prevent excessive amplification and transformation into a process of pathological thrombosis. Briefly, in a normal subject, thrombosis is effectively stemmed. Pathological thrombosis occurs, therefore, only when the procoagulant forces exceed the anticoagulant ones in circumstances where this is not necessary (intact vessel). It can be due to an excess of procoagulant forces (excess of prothrombotic factors) or a defect of anticoagulant forces (deficit of antithrombotic forces).

Depending on the extent of vascular damage, a variable degree of hemorrhage occurs physiologically after the rupture of a blood vessel (e.g., because of small cuts or wounds) and can involve a vein, an artery, or even the capillary system. Nevertheless, hemorrhage is usually counteracted by a series of mechanisms that will be defined in detail below. Briefly, in the presence of an efficient hemostatic system and compatible with the size of the damage, an abnormal (excessive or pathological) hemorrhage occurs when the anticoagulant forces overwhelm the procoagulant ones after the rupture of an intact vessel, and this is usually due to a defect of procoagulant forces (deficit of prothrombotic factors) or to an excess of anticoagulant forces (excess of antithrombotic factors). It is also easily comprehensible that the hemorrhagic phenomenon assumes different characteristics in relation to the type of vessel involved. In the veins, the pressure of the blood inside the vessels is such that the hemorrhage resulting from lesions of large caliber veins can be effectively counteracted by the combined action of the hemostatic system and external therapeutic interventions, such as compression or ligation of the vessel (obviously, potential "internal" hemorrhages that are not recognized are exceptions). On the contrary, the pressure and the flow rate in the arteries (especially in those of large caliber) are such that in the presence of a lesion in a femoral and/or humeral artery, for example, the hemostatic system cannot stop the bleeding and, at the same time, even external compression or "ligation" maneuvers can sometimes be ineffective, leading the patient to death for bleeding. Nevertheless, if the hemostatic balance is seriously altered in favor of an anticoagulant status, the loss of integrity of small arterial vessels and/or veins can lead to severe hemorrhages that cannot be found in subjects with normal hemostatic function.

Hemostasis

Hemostasis is conventionally divided into three phases: primary (also known as the vessel-platelet phase), secondary (also known as the coagulation cascade) (Table 17.1), and fibrinolysis. The first two phases contribute to the formation of a stable hemostatic plug (also known as thrombus) to prevent the leakage of blood from the damaged vessels, while fibrinolysis determines the lysis of the hemostatic plug when it is no longer necessary (*restitutio ad integrum* of the vascular wall).

Primary Hemostasis

Primary hemostasis is defined as the process that determines the formation of a first bulwark against hemorrhage to prevent the "acute" leakage of blood from the vessels. The main effectors of this phase are the vessels (endothelium and muscle cells) and platelets.

Vessel Participation

The participation of the vessels in the hemostatic process is often underestimated, but it is far from secondary. The wall of the blood vessels is composed of a concentric series of layers, including the tunica intima (composed of endothelium and subendothelial layers), the tunica media (with a prevalence of elastic and muscular fibers in the arteries and fibrous tissue in the veins), and the tunica adventitia (composed mainly of dense connective tissue). Injury to the vessel reflexively causes contraction of the elastic and muscular fibers of the middle layer, a process aimed at reducing the caliber of the vessel and its output. This phenomenon, also known as vasospasm, is mediated by thromboxane A2 (TXA2) and can also be activated secondary to neurogenic stimuli such as pain. The reduced blood flow within the contracting vessel is therefore reflected in a reduction in the amount of blood leaving the vessel following its rupture.

Table 17.1 Numbering of coagulation factors (F)

Number	First name	Origin	Function
FI	Fibrinogen	Liver	Precursor of fibrin; stabilizes the clot
FII	Prothrombin	Liver	Precursor of thrombin; activates fibrinogen, FV, FVIII, FXI, and FXIII and binds to thrombomodulin to activate protein C
FIII	Tissue factor (TF); tissue thromboplastin	Endothelium	Contained in perivascular fibroblasts and epithelial cells; it binds to FVII to form the extrinsicase complex which activates the coagulation cascade
FIV (Ca)	Calcium		Mediates the binding of the gamma-carboxylic end groups of the coagulation factors to the phospholipids of the platelet wall
PV	Proaccelerin; labile factor	Platelets, α granules	FXa cofactor in the prothrombinase complex that activates prothrombin
FVII	Proconvertine; stable factor	Liver	It binds the TF to form the extrinsicase complex which activates the coagulation cascade
FVIII	Anti-hemophilic A factor	Endothelium	FIXa cofactor in the tenase complex that activates FX
FIX	Anti-hemophilic B factor; Christmas factor	Liver	It binds to FIX to form the tenase complex; activate FX
FX	Stuart-Prower factor	Liver	It binds to the FV to form the prothrombinase complex; activates prothrombin
FXI	Anti-hemophilic C factor; antecedent plasma thromboplastin	Liver	Activate FIX
FXII	Hageman factor	Liver	Activates precallicrein and FXI; activator of the intrinsic pathway (mainly*in vitro*)
FXIII	Fibrin stabilizing factor	Liver	Activated by thrombin; catalyzes the formation of covalent bonds between two adjacent fibrin monomers, to make the fibrin and clot stable
PK	Precallicrein; Fletcher factor	Endothelium	Activate the FXII
HMWK	High molecular weight Kininogen; Fitzgerald factor	Liver and endothelium	Cofactor of the intrinsic pathway and mediator of inflammation; it has no intrinsic catalytic activity
VWF	Von Willebrand factor	Endothelium and megakaryocytes	It binds and stabilizes the FVIII; mediates platelet aggregation

Simplifying the concept, the less blood circulates within the vessel, the less chance it has of escaping from the point where the damage occurred.

It is easy to understand that vascular contraction is very efficient in the arterial district (since in the arteries the middle layer is much thicker), while its efficiency is obviously much lower in the venous district, in which the middle layer consists mostly of fibrous tissue. Nevertheless, this difference has a valid pathophysiological explanation: since the pressure in the arteries is much higher than in the veins, a much more effective vascular contraction is required in the arteries in order to prevent a considerable leakage of blood from the vessel. The functions of vasoconstriction can be summarized in three decisive aspects:

- Reduction of the flow rate of the vessel, thus limiting the amount of blood that can flow out of it
- Promotion of platelet margination at the site of endothelial damage and near the injured area
- Promotion of the accumulation, at the site of endothelial damage, of coagulation factors to enhance the activation and propagation of secondary hemostasis

The vascular endothelium takes part in the hemostatic processes with two prevalent activities: the liberation of tissue factor (Tissue Factor, TF), following the rupture of the cytoplasmic membrane, and the active production of Von Willebrand factor (VWF), contained in the so-called Weibel–Palade organelles, which will be discussed later in relation to secondary hemostasis. Accessory roles played by the endothelium are those related to the inhibition of hemostasis, especially when it is not necessary (intact vessel). In this context, the anti-hemostatic effect of the endothelium is mediated by a series of activities, including:

- Formation of an anatomical barrier to the subendothelium
- Coating with heparinoid substances, which help inhibit the coagulation cascade.
- Coating with thrombomodulin, which modulates the activation of the coagulation protein C and S complex
- Production of antiplatelet substances, such as prostacyclin (PGI-2) and nitric oxide (NO)
- Production of tissue plasminogen activator (t-PA), whose role is to activate fibrinolysis

Platelets

Platelets (also known as thrombocytes, from the Greek *θρόμβος, thròmbos*, "lump" and *κύτος, cytos*, "cell") are corpuscular elements of the blood without a nucleus. Platelets derive essentially from the lysis of the cytoplasm of bone marrow cells called megakaryocytes (each megakaryocyte produces about 4000 platelets), are round or oval in shape, with a diameter of about 1–3 μm, and circulate in the blood in a variable number, generally between 150–400 × 10^9/L, with a half-life of about 5–9 days and elimination mainly by splenic catabolism. Even if they are "terminal" elements, as they have no nucleus, these cellular fragments contain inside them three types of granulations, which represent the prerequisite for their functioning. Briefly, the platelet granules are subdivided into α, δ, and γ granules, as described below.

- Granules α (alpha): Slightly opaque, very numerous, and contain mainly fibrinogen, VWF, factors V and XIII, fibronectin, platelet factor 4 (FP4), β-thromboglobulin (β-TG), thrombospondin, and various growth factors, including platelet-derived growth factor (PDGF) and transforming growth factor (TGF)-β
- δ (delta) granules or "dense" granules: Contain mainly histamine, serotonin, calcium, ADP, and ATP
- γ (gamma) or lysosomal granules: Contain mainly lysosomal hydrolases and peroxisomes

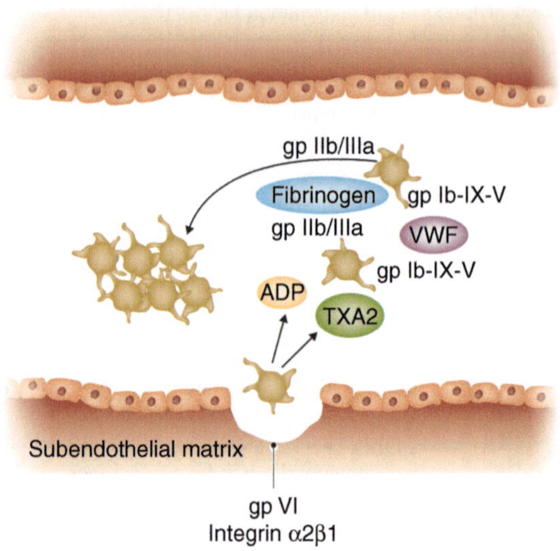

Fig. 17.2 Primary hemostasis. (Copyright EDISES 2021. Reproduced with permission)

As previously described, platelets play an essential role in physiological hemostasis, representing the main effectors of primary hemostasis. Briefly, the role of these small corpuscular elements is to form a first "plug" (called "platelet plug," formed by the "weak" association of platelets) or bulwark to the leakage of blood following the loss of integrity of the vessel.

To briefly summarize the cascade of events, following vascular damage, the platelets are exposed to the subendothelium (collagen, proteoglycans, fibronectin, and other glycoproteins), and this causes their activation. The platelet response is implemented through biochemical, structural, and morphological alterations. The response to a stimulus is due to the coordinated intervention of the membrane, granules, and cytoskeleton and can be divided into several phases, which may overlap (Fig. 17.2).

Phase I: Adhesion (to the Subendothelium)

The adhesion of platelets to the subendothelium occurs mainly by binding to the subendothelial collagen, which is exposed following the loss of integrity of the vascular wall (rupture of the endothelial layer). The binding between platelets and collagen is mediated not only by a receptor located on the platelet membrane, the glycoprotein Ia/IIa (gp Ia/IIa), but also by the bridging function exerted by VWF, which binds with a part of the molecule to the collagen and with another part to a receptor located on the platelet membrane, the glycoprotein Ib (gp Ib). The binding of the platelet to the subendothelium is also mediated by other platelet membrane receptors, such as the glycoprotein Ic/IIa (gp Ic/IIa), which binds laminin and fibronectin, and, most likely, by the glycoprotein IV (gp IV), which binds to collagen by means of a bridge-bond with thrombospondin. The activation of these platelet receptors is an essential prerequisite for the subsequent phases of the process, which consist of the change in the structure of the platelet (shape change) and its contextual activation. The binding of these first platelets to collagen (through the Ia/IIa receptor) and to VWF (which in turn acts as a bridge for further binding to collagen) causes a deformation of the three-dimensional structure of the platelets; this change facilitates further aggregation of the platelets among themselves and above all stimulates the so-called release reaction.

Phase II: Shape Change

The change in platelet shape (from roughly circular and/or ovaloid elements to a very irregular, almost starry shape, characterized by pseudopod generation) is sustained by the contraction of the platelet cytoskeleton and is a propaedeutic event to the granule release reaction. Of particular interest is the phenomenon known as the "flip-flop" of phosphatidylserines, in which FP4 is exposed on the platelet surface and represents the physiological binding site of coagulation factors during their activation process (described in detail below).

Phase III: Activation and Secretion of Granules (Release)

The activation of the platelet and the consequent release of internal granules (through the open canalicular system by direct exocytosis) are aimed at releasing the numerous mediators contained within the three different types of granules and at enhancing primary hemostasis. This process can be summarized with the term autocrine activation, a mechanism according to which a cellular element (the platelet) produces mediators (mainly ADP and serotonin) that have the function of activating other identical cellular elements (other platelets), thus amplifying the subsequent phenomenon, platelet aggregation. It is also important to remember that the release of TXA2 by the platelets contributes to further stimulating vasoconstriction and thus limiting the caliber of the vessel and the outflow of blood.

Phase IV: Aggregation (Between Them)

Following the contraction of the cytoskeleton (shape change), activation, and secretion of granules, platelets express on their membrane surface the glycoprotein IIb/IIIa (gp IIb/IIIa), which specifically binds fibrinogen. Since two contiguous activated platelets express gp IIb/IIIa at the same time, it is not difficult to understand how they can form bridges between them (gp IIb/IIIa-fibrinogen-gp IIb/IIIa), which, therefore, represents the prerequisite for the phenomenon known as platelet aggregation. Thrombospondin also seems to play a role in this process, increasing the size of the platelet aggregate by forming a bridge (cross-link) between platelets and fibrinogen or by positioning itself directly as a bridge between gp IIb and IIIa of two adjacent platelets. Interestingly, platelet aggregation occurs in two phases. Primary aggregation is promoted by a modest concentration of agonists (ADP, collagen, thrombin, and platelet-activating factor) and is essentially reversible, whereas secondary aggregation is determined by higher concentrations of agonists (ADP and TXA2).

The combination of all the phenomena described above, in the absence of quantitative or qualitative platelet abnormalities, effectively contributes to limiting the leakage of blood from the damaged vessel within a few minutes. The platelet plugs thus formed, however, do not have stability requirements to allow a permanent interruption of blood leakage, especially in vessels with higher blood flow and high pressure. The subsequent stabilization of the hemostatic plug will be guaranteed by the intervention of secondary hemostasis, described in detail in the next paragraph.

Platelet Disorders

Although a detailed description of platelet diseases is beyond the scope of this chapter, it may be useful to recall that, like many other human diseases, they are essentially subdivided into quantitative (decrease in the number of platelets, also known as thrombocytopenia) or qualitative (altered platelet function) diseases, which are in turn subdivided into congenital (or hereditary) and acquired diseases. The principal platelet pathologies are described in Table 17.2. It should be noted that thrombocytopenia with platelet counts greater than 50×10^9/L is usually asymptomatic, 30 to 50×10^9/L is usually accompanied by mild bleeding symptoms, and 10 to 30×10^9/L is associated with moderate bleeding symptoms,

Table 17.2 Main platelet pathologies

Quantitative (Thrombocytopenias)
Congenital
Wiskott–Aldrich syndrome
May–Hegglin syndrome
Bernard–Soulier syndrome
Acquired
Reduced/impaired bone marrow production of megakaryocytes
Bone marrow aplasia (due to cytotoxic drugs, radiation therapy, viruses, or other microorganisms)
Medullary neoplasms
Medullary neoplastic infiltration
Myelodysplastic syndrome
Deficiencies
Vitamin B12 deficiency
Folic acid deficiency
Increased destruction in circulation
Thrombotic thrombocytopenic purpura (Moschowitz syndrome)
Idiopathic thrombocytopenic purpura (Werlhof disease)
Hemolytic-uremic syndrome
Thrombocytopenic purpura associated with autoimmune diseases (e.g., autoimmune hemolytic anemia or Fisher–Evans syndrome, collagenopathies, and sarcoidosis)
Thrombocytopenic purpura associated with lymphoproliferative diseases (e.g., chronic lymphocytic leukemia, Hodgkin's and non-Hodgkin's lymphomas, and paraproteinemia)
Thrombocytopenic purpura from drugs (e.g. heparin, isopropylcarbamide, chlorothiazide, sulfonamides, para-aminosalicylic acid (PAS), quinine, and quinidine)
Thrombocytopenic purpura from isoantibodies (maternal–fetal immunization and transfusions)
Anti-phospholipid antibody syndrome
Eclampsia
Disseminated intravascular coagulation
Vascular prostheses
Extra-corporeal circulation
Kidnapping
Splenomegaly (hypersplenism)
Giant hemangiomas (e.g., Kasabach-–Merrit syndrome)
Pseudothrombocytopenia (spurious thrombocytopenia due to sampling issues or induced by ethylenediaminetetraacetic acid, cryoglobulins, platelet satellites, etc.)
Qualitative (platelet disorders)
Congenital
Bernard–Soulier syndrome (platelet adhesion defect)
Glanzmann's thrombasthenia (platelet aggregation defect)
Storage pool disease (defect in the release of granules)
Aspirin-like defects (defects in cyclo-oxygenase or thromboxane-synthetase enzymes)
Gray platelet syndrome (lack of α granules).
Acquired
Excess of fibrin breakdown products
Monoclonal gammopathies
Type I glycogenosis
Drugs (antiplatelet agents, aspirin, dipyridamole, indomethacin, phenylbutazone, chlorpromazine, dextran, and ethyl alcohol)
Uremic syndrome
Myeloproliferative syndrome

whereas a decrease in platelet counts to values <10 × 10^9/L frequently results in spontaneous mucosal skin bleeding, generalized purpura, and the risk of severe urinary, gastrointestinal, and central nervous system bleeding.

Secondary Hemostasis

The purpose of secondary hemostasis is to generate a fibrin network to "cement" the primary platelet plug.

Secondary hemostasis, also known as coagulation, consists of a complex series of events involving a large number of molecules (Table 17.1). The traditional model of secondary hemostasis is that of a "cascade," which describes coagulation as a sequential series of enzymatic or proteolytic reactions in each of which an inactive precursor of a proteolytic enzyme (termed zymogen) is converted to its active form (protease), which in turn activates the next zymogen in the series. Since the individual sequential reactions are catalyzed by an enzyme, and since each enzyme can catalyze the generation of a large number of other active molecules, it is easy to understand how this model is characterized by an enormous amplifying potential, hence the name "cascade." It is possible to summarize the main features of the cascade model in a few essential points.

- Each step in the cascade includes a serine protease and sometimes a cofactor (to form an enzyme "complex").
- Enzyme complexes consist of similar components with similar characteristics with respect to their assembly and function.
- The formation of the complex results in a considerable increase (about 100-fold) of the catalytic efficiency towards the substrate.
- To function effectively, individual zymogens and enzyme complexes require the binding of one of their terminal carboxyl groups to an appropriate surface consisting of negatively charged phospholipids (generally those present on the membrane of platelets), a binding mediated by the indispensable presence of calcium ions (Ca^{2+}) (Fig. 17.3). In the absence of the carboxylic group on the coagulation factors or as a result of the subtraction of calcium ions, the binding between the coagulation factor and the platelet occurs with much less efficiency, overall, also altering the effectiveness of the coagulation cascade.

During the last two decades, the model of the coagulation cascade has undergone considerable interpretative revisions, conditioned by the progressive discovery of new coagulation factors and new and complex biochemical interactions. The historical paradigm of coagulation is the one that, for years, was based on the existence of three different "pathways," also known as extrinsic, intrinsic, and common. Over the years, however, it has been demonstrated that this model, while maintaining stringent validity in the interpretation of laboratory abnormalities of the coagulation cascade, has no real feedback in vivo. In this case, it has been concluded that the "parallel" model, according to which extrinsic and intrinsic pathways concur in parallel to the activation of the common pathway, has no real biological correspondence since the physiological mechanism of activation of coagulation is represented only by the extrinsic pathway. According to the most current model, therefore, coagulation can be described in three sequential phases (Fig. 17.4):

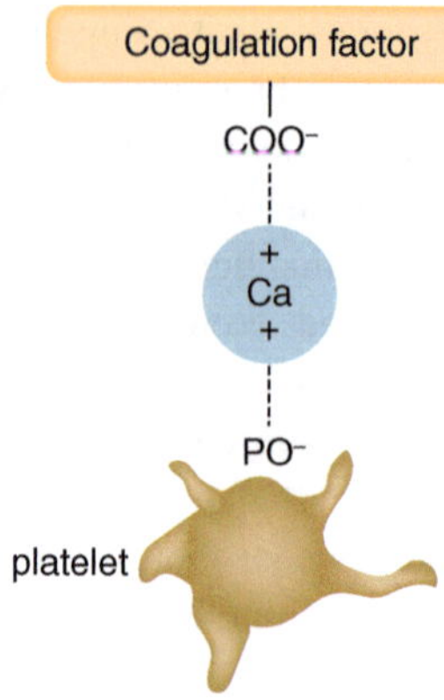

Fig. 17.3 Role of calcium in secondary hemostasis. (Copyright EDISES 2021. Reproduced with permission)

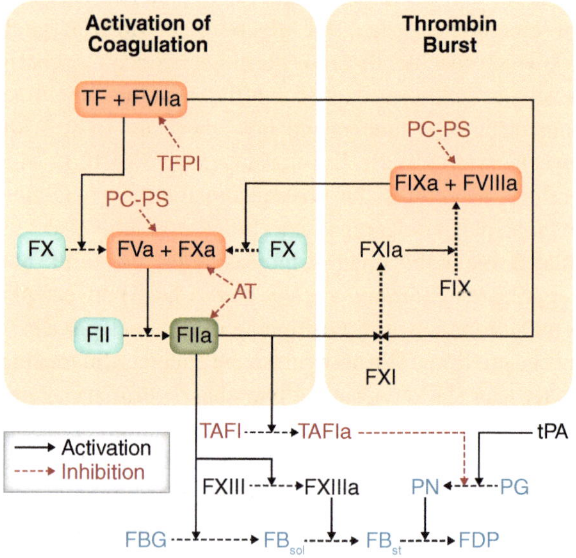

Fig. 17.4 Secondary hemostasis. The dotted lines define the critical processes of the thrombin burst. AT antithrombin, FB fibrin, FBG fibrinogen, FDP fibrin degradation products, FII prothrombin, FIIa thrombin, PC protein C, PG plasminogen, PN plasmin, PS protein S, TF tissue factor, tPA tissue plasminogen activator, TFPI inhibitor of the tissue factor pathway. (Copyright EDISES 2021. Reproduced with permission)

- Activation phase ("trigger") of the coagulation cascade: mediated by the extrinsic and common pathways. The exposure of the TF is the key event.
- Propagation phase (burst) of thrombin: mediated by an intrinsic and common pathway, it is characterized by the formation of enzymatic complexes aimed at the creation of a stable clot.
- Coagulation shutdown phase: supported by a discrete number of natural anticoagulants, aimed at circumscribing procoagulant events and preventing the formation of excess clots.

In the following paragraphs, we will try to fully describe the sequence of events; for the sake of synthesis, the term "coagulation factor" will be abbreviated with the letter "F" for all the different factors.

Coagulation Activation Phase

The physiological activation ("trigger") of coagulation is supported only and exclusively by the extrinsic pathway and, subsequently, by the common pathway. This evidence is supported not only by the physiological mechanism that will be described later but also by the clear and indisputable clinical evidence that total deficits of FXII (the factor upstream of the activation of the intrinsic pathway) are fully compatible with life and, indeed, do not cause any type of bleeding diathesis. As will be seen later, the task of the intrinsic pathway is not to constitute a mechanism of parallel activation of coagulation but rather to amplify the formation of thrombin and fibrin, generated in small quantities by the activation of the extrinsic pathway.

Activation of the extrinsic pathway thus originates from the rupture of the vascular wall, which directly results in damage to endothelial cells and exposure or release of the TF contained therein. Immediately after its release, TF binds with high affinity and specificity to FVII, thus triggering coagulation. Since TF is essentially a transmembrane protein, its propagation in circulation is very limited, except in the presence of considerable tissue damage. This is easily justified by the need for the coagulation process to be limited to the site where the endothelial damage has occurred without excessive propagation to other districts in which the coagulation process is not required (intact endothelium) and in which the excessive activation of coagulation could, on the other hand, generate the paradoxical phenomenon of thrombosis. It is interesting to note how the excessive propagation of TF is contrasted not only by the layer of platelets that have adhered to the collagen at the site of endothelial damage but also by the release, by the same endothelial cells damaged, of the inhibitor of the tissue factor pathway (TFPI), also known as EPI (extrinsic pathway inhibitor), actively released by endothelial cells and, probably, also by platelets. Unlike what was thought years ago, a small amount of TF is always present in the blood, even in the absence of endothelial damage. Although its origin remains controversial, it is legitimate to assume that it originates from cellular microparticles, tiny membrane vesicles released from stimulated cells, or cells undergoing apoptosis. The essential function of TF is to bind to FVII (zymogen) and activate it to FVIIa (active enzyme) by proteolytic cutting at the level of Arg152, which determines the conversion of FVII from a single chain to a double chain. Recent evidence shows that a modest amount of FVIIa (1–2% of total FVII) is nevertheless present in the circulation, despite its concentration being too modest to trigger the coagulation process, especially in the absence of TF (only the interaction between FVIIa and TF is able to "trigger" the active site of FVIIa, increasing its catalytic activity towards the next factor in the coagulation cascade, i.e., FX). Although the origins and function of this small amount of physiologically circulating FVIIa are still uncertain, it seems plausible to assume that the serine protease factor VII-activating protease (FSAP) can produce the physiological activation of a modest amount of FVII in order to make coagulation more ready and efficient when necessary.

The function of the TF-FVIIa complex (also known as the extrinsic complex) is primarily to activate FX to FXa and, to a small extent, also to activate FXI to FXIa. FXa, together with its essential cofactor FVa, forms the so-called tenase complex, which has the function of activating prothrombin (FII) to thrombin (FIIa). Thrombin, in turn, converts fibrinogen into fibrin, which, as it polymerizes, forms a "meshwork," or "reticulum," aimed at stabilizing the thrombus, according to a series of mechanisms described below. The quantities of thrombin and fibrin produced by the mechanisms now described are in the order of nanomoles, which are concretely too scarce (3–5% of the necessary amount) for a fibrin lattice large enough to stabilize the platelet plug (in practice, this is an "ineffective" clot). The main function of this small amount of thrombin produced is to activate a secondary "burst" (also known as the thrombin propagation phase), which is the most functional link between the extrinsic and intrinsic pathways.

Thrombin Propagation Phase

As repeatedly described, the initiation of coagulation by the extrinsic pathway produces small amounts of thrombin (3–5% of the total), which are nevertheless sufficient to accelerate the coagulation process by a series of mechanisms involving platelet activation, FXIII activation, and activation of FV and FVIII cofactors. Most thrombin (95–97%) is instead generated in the so-called propagation phase of coagulation (thrombin burst; Fig. 17.4).

The sequential series of events in the thrombin propagation phase involves the activation of FXI (to FXIa), which in turn catalyzes the conversion of FIX to FIXa. This latter factor, together with its essential cofactor FVIII (FVIIIa in activated form), forms a molecular complex called intrinsecase. The FIXa-FVIIIa complex can activate FX to FXa, with kinetics almost 100-fold higher than the single action of FIXa and 50- to 100-fold higher than the action of the TF-FVIIa complex. It is, therefore, evident how the maximum part of the activation of FX to FXa is sustained by the intrinsecase complex (FIXa-FVIIIa) and not by the extrinsic complex (TF-FVIIa), which justifies the onset of a severe hemorrhagic syndrome related to the deficiency of FVIII (hemophilia A) and FIX (hemophilia B).

In the final phase of coagulation, as described above, thrombin acts enzymatically on the fibrinogen molecule, determining the cutting of part of the α and β chains (releasing the corresponding fibrinopeptides A and B) and generating the so-called fibrin monomer. This peptide has a characteristic shape, consisting of a central domain (called E) and two side domains (called D). The fibrinopeptide A cleavage exposes a binding sequence on the central E domain that associates, by noncovalent bonding, with a complementary site on the lateral D region of an adjacent fibrin monomer. The overlap of fibrin molecules, where the central domain of one binds to the lateral domain of the other, results in the formation of double-stranded protofibrils. The release of fibrinopeptide B exposes an additional binding site on the central E domain, which in turn binds to a complementary site on the lateral D region of another adjacent fibrin monomer, thereby promoting the lateral association of the protofibrils and the formation of thicker fibers. The deposition of this first fibrin lattice on the platelet plug is essentially governed by weak forces, which are not sufficient to ensure the long-term stability of the clot. Its stabilization is, therefore, dependent on the intervention of FXIIIa. In summary, FXIIIa associates with fibrin to promote the formation of a series of stable bonds in the clot, initially between chains of two adjacent fibrin molecules and subsequently forming covalent bonds between two overlapping fibrin chains (Fig. 17.5). Once this process has been completed, the clot is sufficiently stable to provide an effective barrier against the leakage of blood from the injured vessel.

As a corollary to the description of the coagulation cascade, it is necessary to define, at least briefly, the role of VWF. This factor is a large adhesive protein synthesized mainly by endothelial cells and megakaryocytes and is therefore present in the α granules of platelets. It is present in the circulation in the form of multimers (aggregations of monomers), which, according to their extent, are classified as low, intermediate, and high molecular weight multimers. It performs two essential functions. The first, already described regarding primary hemostasis, is that of mediating platelet aggregation or adhesion. The second function, equally essential, is to act as a carrier of FVIII in the circulation, thus protecting it from early degradation. Deficiency of this factor generates Von Willebrand disease (VWD), whose clinical characterization is very complex (types 1 and 3 are due to partial or virtually complete VWF deficiency, while type 2 is due to a qualitative defect), depending on the nature of the deficiency. In general, however, it is a disease with highly variable symptoms; the dual function of VWF in primary and secondary hemostasis determines the appearance of "mixed" symptoms between the two conditions. The high molecular weight multimers of VWF are much more efficient in the process of platelet aggregation. The efficiency of the process is guaranteed by the catalytic action of the enzyme ADAMTS13, which acts by cutting large VWD into small fragments. Alterations in the action of ADAMTS13 lead to a polymerization deficit, as, for example, in thrombotic thrombocytopenic purpura, in which ADAMTS13 deficiency prevents the degradation of multimers that remain anchored to the endothelium, favoring the adhesion of platelets by means of gp Ib and gp IIb/IIIa and thus favoring the formation of potentially occlusive thrombi in terminal arterioles and capillaries of many organs (heart, pancreas, kidneys, adrenals, brain, spleen, and liver).

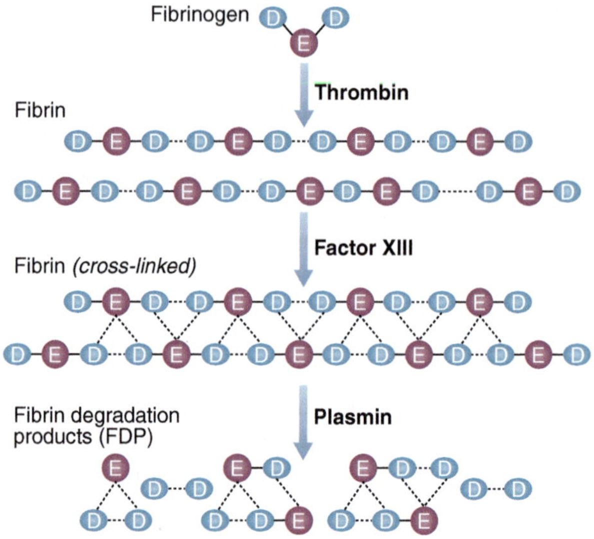

Fig. 17.5 Fibrinoformation and fibrinolysis. (Copyright EDISES 2021. Reproduced with permission)

Coagulation Shutdown Phase

The fine regulation of the hemostatic process is guaranteed by the intervention, at various levels and with different mechanisms, of a series of physiological inhibitors of coagulation, essentially represented by the thrombin activatable fibrinolysis inhibitor (TAFI), antithrombin (AT), and protein C (PC)–protein S (PS) complex.

Antithrombin

Antithrombin (AT) is by far the most potent and important inhibitor of coagulation. Present in the blood at a much higher concentration than any other coagulation factor generated by the TF pathway, it appears to be capable of inhibiting multiple coagulation factors, although its activity is directed primarily at inhibition of FXa and thrombin (Fig. 17.4). The inhibitory efficacy of antithrombin is strongly conditioned by the presence of physiological (heparan sulfates lining the inner wall of vessels) or pharmacological (heparin) heparin-like substances, in the presence of which its inhibitory activity on FXa and thrombin is increased almost 1000-fold. The importance of the inhibitory effect of antithrombin is reflected in the association between deficiency and venous thrombosis, from the incompatibility with life to the state of homozygous deficiency and the effectiveness of heparin therapy in preventing thrombotic events.

Protein C (PC) and Protein S (PS) System

The protein C (PC) and protein S (PS) system is the second most biologically and clinically important inhibitory mechanisms of coagulation. PC is a vitamin K-dependent protein whose activity is conditioned by binding to negatively charged surfaces (phospholipids), thrombomodulin, and the endothelial protein C receptor (EPCR). The activity of protein C is then strictly dependent on the presence of protein S, which therefore represents its essential cofactor, increasing the activity of protein C about 20-fold. Protein S is also vitamin K-dependent and circulates in the blood in a free form (30–40% of the total) or complexed with C4BP (60–70%), a complement regulatory protein. Only the free form of protein S can carry out its activity as a cofactor of protein C, hence the importance in the laboratory of effectively distinguishing the free circulating quantity (the "effective" one) from the total one. To better understand how the PC–PS system works, it is essential to briefly illustrate the structure of the two cofactors of coagulation (FV and FVIII) that represent the molecular target. The two proteins have considerable structural homology, which involves an arrangement into sequential domains (A1A2-B-A3-C1-C2). Thrombin activates both FV and FVIII by proteolytic cutting, causing the release of the B domain. The PC–PS system then inactivates FV by making an additional cut at the levels of Arg306, Arg506, and Arg679. Proteolytic cutting at the level of Arg506 results in the formation of an intermediate form of FVa that retains some of its procoagulant properties, whereas cutting at the level of Arg306 causes the total loss of the factor's procoagulant activity. The inactivation of FVIIIa by the PC–PS complex occurs instead by proteolytic cutting at the levels of Arg336 and Arg562, although only the latter (Arg562) determines the complete inactivation of the factor.

Fibrinolysis

Once the restitutio ad integrum of the injured vessel has been achieved, the clot no longer has any reason to exist and must, therefore, be removed to restore patency and canalization of the vessel. The process that determines the lysis of the clot is defined as fibrinolysis (Fig. 17.5). The major components of the fibrinolytic system are plasminogen, which is converted to plasmin by plasminogen activators known as t-PA and urokinase (u-PA). Plasmin is the main enzyme responsible for clot dissolution and works by degrading stabilized fibrin (but also fibrinogen and insoluble fibrin) into degradation products also known as fibrinogen/fibrin degradation products (FDP). The process of fibrin degradation begins at the α, β, and γ chains, generating C-terminal lysine residues that are in turn capable of promoting the conversion of further plasminogen molecules into plasmin. The progressive action of plasminogen releases, therefore, a wide and heterogeneous (both in terms of structure and molecular weight) series of degradation products, containing variable combinations of the central E and lateral D domains. The degradation of stabilized fibrin, but not of soluble fibrin or fibrinogen, determines the formation of FDP containing the so-called D-dimer, represented by degradation molecules of various structures and molecular weights that contain two lateral D units linked by a stable covalent bond. It is therefore evident that the determination of the D-dimer is useful in the laboratory to testify that the lysis of a stabilized clot has occurred (as it happens, e.g., in thrombotic processes), while the determination of the total FDP does not allow one to distinguish whether the action of plasmin has addressed only the stabilized fibrin or has also affected the fibrinogen and/or soluble fibrin (as it usually happens in primary fibrinogenolysis or in disseminated intravascular coagulation). This mechanism explains the higher diagnostic specificity of D-dimer compared to FDP for the diagnosis of venous thromboembolism.

As with the coagulation cascade, the progression of fibrinolysis is modulated by specific inhibitors. In the latter case, plasminogen activators are inhibited by PAI-1 (plasminogen activator inhibitor-1), α2-antiplasmin, and α2-macroglobulin.

TAFI (Thrombin Activatable Fibrinolysis Inhibitor), a metalloprotease of hepatic origin, acts as a bridge between the coagulation cascade and fibrinolysis. The activation of TAFI (bound to thrombomodulin) by thrombin promotes the degradation of carboxy-terminal residues of lysine and arginine on the fibrin surface, thus preventing the binding of fibrin itself to plasminogen. In summary, activated TAFI, in addition to inactivating partially degraded fibrin, also acts as a cofactor for plasmin generation, ultimately inhibiting fibrinolysis as a whole. Considering that thrombin is the activator of TAFI, it is therefore legitimate to conclude that the coagulation cascade regulates fibrinolysis as well. Therefore, a defect in thrombin generation is inevitably reflected in increased fibrinolytic power. The clinical consequences are easily predictable when the thrombin generation guaranteed by the TF pathway alone (the extrinsic pathway) is insufficient to activate such a quantity of TAFI as to inhibit fibrinolysis. A "useful" concentration of TAFI is obtained only when the thrombin burst is fully functional. Defects in the factors involved in this process (i.e., essentially FXI, FIX, and FVIII) not only lead to hemorrhagic diathesis due to insufficient fibrin generation but also to inefficient activation of TAFI. It could therefore be concluded that hemophiliacs bleed not only due to insufficient clot generation but also due to excessive lysis of the small clot that has formed.

Congenital Disorders of Coagulation Factors

Congenital coagulation factor disorders include a heterogeneous group of conditions characterized by quantitative (synthesis deficiency) or qualitative abnormalities (generation of nonfunctioning coagulation factors) or, rarely, by a combination of the two. They cause multiple bleeding problems of a spontaneous (in the most severe forms) or acquired nature. Although classification on a molecular basis often has a modest correlation with clinical severity, it has been unanimously agreed to classify the severity of the deficiency according to the residual activity of the coagulation factor, expressed in percentage terms (100% indicates a normal or reference concentration). It follows that:

- Severe forms: Factor activity between 0% and 1%
- Moderate forms: Factor activity between 1% and 5%
- Mild forms: Factor activity between 5% and 40%
- Deficiency (usually asymptomatic): Factor activity >40%

The above classification remains indicative since clinical severity depends largely on many factors, including (1) factor affected by the deficiency (patients with moderate FVIII deficiency usually have a worse bleeding diathesis than patients with moderate FV deficiency); (2) residual activity of the other coagulation factors (patients in whom the activity of one or more of the other factors is <80% usually have a worse hemorrhagic diathesis than patients whose activity of the other factors is close to 100%); (3) organic district involved in the hemorrhagic process (trauma, injury, or surgery in districts with high fibrinolytic potential, such as the tonsils or urogenital sphere, usually result in more severe hemorrhages than in districts with normal fibrinolytic potential). In epidemiological terms, VWD, FVIII deficiency (also known as hemophilia A), and FIX deficiency (also known as hemophilia B) represent about 80% of congenital bleeding disorders. Hemophilias A and B are inherited diseases with an incidence of 1:5000 (hemophilia A) and 1:30,000 (hemophilia B) in male infants. Both are transmitted as recessive characters, linked to the X chromosome. VWD is, however, the most common inherited hemorrhagic disease, with autosomal dominant transmission and an incidence that can reach 5–10:1000, even considering the mildest forms.

Laboratory Tests for the Study of Hemostasis

Laboratory tests for the evaluation of hemostatic abnormalities have essentially the role of directing the clinician towards the phase of hemostasis mainly responsible for the hemorrhagic diathesis (Fig. 17.6). A decisive aspect is therefore played by the anamnesis and clinical history. The main elements that can guide the clinical evaluation towards a specific defect are represented by the age of onset (it may reflect the presence of congenital or acquired anomalies, moderate or severe), familiarity (although not always indicative, a negative family history may suggest an acquired pathology), and by the characteristics of the hemorrhagic symptomatology (deep hemorrhages are more typical of abnormalities of secondary hemostasis, while superficial hemorrhages more commonly distinguish pathologies of primary hemostasis). Based on these elements, it seems, therefore, not strictly necessary to extend the potential analysis of coagulation defects to the general population but only to patients with an indicative hemorrhagic history and possibly to family members of patients with hereditary anomalies with a hemorrhagic imprint. Coagulation tests, especially level I tests, are also useful in defining bleeding risk in asymptomatic patients undergoing invasive procedures because bleeding diathesis in patients with mild forms may only begin following trauma or surgery.

In relation to these premises, the diagnostic approach foresees the distinction of coagulation tests in at least three levels:

- Level I examinations or screening examinations. They are basically subdivided into examinations that explore the primary hemostasis, or vasoplatelet phase (evaluation of the primary hemostatic function and/or platelet count), and examinations more specifically aimed at the coagulative phase, including activated partial thromboplastin time (aPTT), prothrombin time (PT), and determination of fibrinogen with functional methods.
- Level II or in-depth diagnostic examinations (mostly intended for patients with suspected hemorrhagic diathesis and an indicative family history, usually performed by specialized centers).
- Level III examinations (typically performed by highly specialized centers and essentially aimed at the complete typing, including molecular typing, of the defect or at the investigation of abnormalities not otherwise diagnosed with Level I and II examinations) (Table 17.3).

Although the participation of the intrinsic pathway in physiological hemostasis is actually irrelevant (patients with total FXII deficiency do not manifest significant hemorrhagic diathesis), abnormalities in the factors that compose it

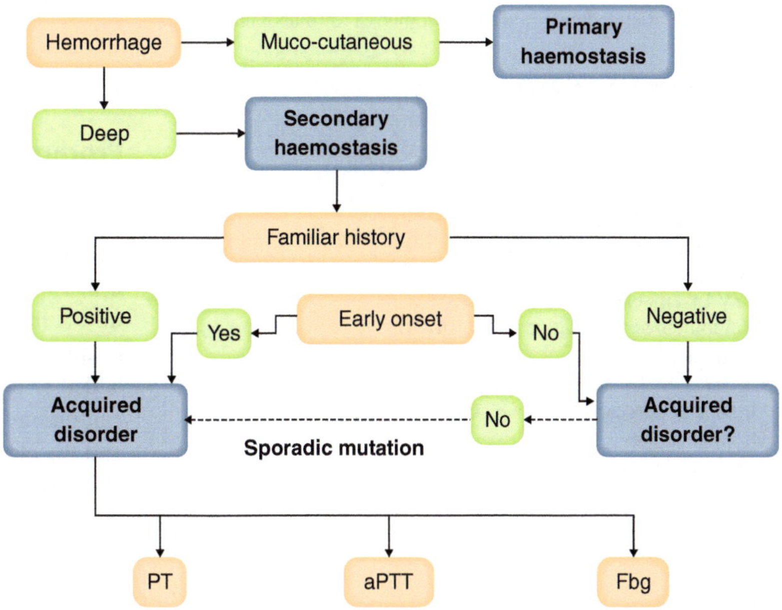

PT	aPTT	Fbg	Deficit
↑	N	N	FVII
N	↑	N	FVIII; FIX; FXI; FXII; precallicrein
↑	↑	N	FII; FX; FV; combined deficit
↑/N	↑/N	↓	Fbg
N	N	N	FXIII; Other?

Fig. 17.6 Diagnostic approach to hemostasis pathologies. aPTT activated partial thromboplastin time, Fbg fibrinogen, PT prothrombin time. (Copyright EDISES 2021. Reproduced with permission)

Table 17.3 Hemostasis test

I Level test (screening)
Activated partial thromboplastin time (aPTT) Prothrombin time (PT) Fibrinogen (functional method) Platelet count Platelet function test (e.g., PFA)
II Level test
Mix test Thrombin time Determination of coagulation factors by coagulation methods Antigenic Von Willebrand factor (VWF:Ag) Von Willebrand factor as a ristocetin cofactor (VWF:RCo) Platelet aggregation
III Level test
Determination of coagulation factors by immunological methods Fibrinogen (antigenic method) Von Willebrand factor as collagen binding activity (VWF:CB) Multimeric Von Willebrand factor α2-antiplasmin Molecular biology Flow cytometry and electron microscopy for the study of platelets

are validly reflected by prolongations of aPTT. On the other hand, abnormalities mainly affecting FVII result in variable PT prolongation, whereas abnormalities affecting common pathway factors usually result in variable combined PT and aPTT prolongation. In the next part of this chapter, the characteristics of the main coagulation tests and their potential clinical use will be briefly described. Besides these tests, there are many others that contribute to the definition of the so-called thrombophilic status, defined as the predisposition to develop venous thrombosis (Table 17.4).

Biological Sample

The quality of the biological sample used for the determination of coagulation tests plays an essential role. The principle behind the determination of coagulation tests is that the blood sample must be initially and temporarily rendered incoagulable and then be "re-coagulated" to assess whether the clotting time is "normal." To achieve this, an anticoagulant that only reversibly inhibits clotting should be used. Coagulation tubes therefore contain a buffered solution of sodium citrate at a concentration of 0.105–0.109 mmol/L (3.2%). The sodium citrate sequesters the calcium present in the blood sample, effectively making it impossible to activate the coagulation factors according to the mechanism described in the previous paragraphs. The advantage of using

Table 17.4 Risk factors for venous thrombosis (thrombophilia)

Congenital
Antithrombin deficiencyAT)
Coagulation protein C (PC) deficiency
Coagulation protein S (PS) deficiency
Congenital activated protein C resistance (aPCR)
Leiden FV mutation (Arg506Gln)
Haplotype HR2 (His199Arg)
FV Hong Kong (Arg306Gly)
FV Cambridge (Arg306Thr)
FV Liverpool (Ile359Thr)
Prothrombin gene mutation (G20210A polymorphism)
Acquired
Cancer
Orthopedic surgery/lower limb immobilization/paralysis
Hospitalization
Long-term flights
Major trauma
Pregnancy
Oral contraceptives and postmenopausal therapy
Obesity
Heart failure
Anti-phospholipid antibody syndrome
Acquired activated protein C resistance (aPCR)
Increased concentration of factor VIII

sodium citrate lies in its ability to sequester the calcium present in the sample in a manner directly dependent on the final concentration. In this case, the final volume between blood and anticoagulant inside the tube must be 9:1, a ratio that should ensure the sequestration of all the calcium physiologically present in the sample. However, this ratio may be slightly modified in circumstances where it is known a priori that calcemia is increased (very severe hypercalcemia) or decreased (e.g., in patients with a very high hematocrit value). The correct filling of the test tube is, therefore, essential to ensure the perfect success of coagulation tests, since an excess of anticoagulant compared to the blood (the so-called tube with "poor" filling) determines an increase in the final concentration of sodium citrate and, therefore, an inefficient subsequent recalcification (and necessary) for the activation of coagulation tests.

After collection, the tube must be centrifuged quite rapidly (usually within 2–4 h) to separate the corpuscular part (blood cells, including platelets) from the plasma. Indeed, most coagulation tests use citrated plasma as a biological matrix. The removal of platelets must be almost total (a residual value of platelets in citrated plasma of $<10 \times 10^9$/L is recommended) to standardize the performance of tests on samples from different patients, making the result independent from the platelet count of whole blood. This is conventionally achieved, according to current indications, by centrifugation of the tube at 1500 g for 15 minutes. The supernatant plasma is, therefore, defined as platelet-poor plasma (PPP) and virtually calcium-free. Since these two elements (platelet membrane phospholipids and calcium) are indispensable for the activation and development of the coagulation cascade, they are expected to be added back into common laboratory tests in a manner to be described later.

Prothrombin Time (PT)

The prothrombin time (PT) allows the evaluation of the extrinsic pathway of the coagulation cascade and is widely used for the monitoring of patients under treatment with vitamin K inhibitors. From the methodological point of view, it is quite simple: thromboplastin and calcium chloride are added to a plasma sample collected in sodium citrate (concentration 3.2%, corresponding to 0.105–0.109 mmol/L), and the clot formation time, expressed in seconds, is evaluated. The test can be performed either manually or using automated instruments. The PT is expressed as the ratio between the patient's PT (expressed in seconds) and an average time obtained from a sample of subjects without coagulation alterations or under therapy with anticoagulant drugs, according to the following relation:

$$\text{PT} = \text{PTpatient}(\text{seconds}) / \text{PTplasma pool.}$$

The test was first described by Quick in 1935. The central and defining element of the test is the use of thromboplastin, a mixture of tissue factor (animal extract or recombinant) and phospholipids. From the very first studies, it was evident that a standardization procedure was necessary in the preparation of thromboplastins since, depending on the preparation method, the results were extremely variable, making it difficult to evaluate patients over time. In 1977, the World Health Organization adopted the first thromboplastin calibration system, implementing an international reference preparation (IRP). This system was modified in 1982 with the introduction of the international sensitivity index (ISI) concept. Briefly, the calibration system was modified. In a group of subjects undergoing therapy with vitamin K antagonists, PTs performed simultaneously with both the reference preparation and the new commercially introduced thromboplastins were evaluated. The slope of the calibration line represents the ISI, which effectively establishes a relationship between the thromboplastin and the international reference preparation. Thus, the PT is expressed as an international normalized ratio (INR) according to the following relationship:

$$\text{PT}(\text{INR}) = \left[\text{PTpatient}(\text{s}) / \text{PTplasma pool}\right]^{\text{ISI}}$$

Currently, INR is the universally accepted method for expressing PT independent of thromboplastin and is used by individual laboratories.

In the past, PT was performed manually, but soon automatic analyzers (commonly called coagulometers) became available. These, by means of an optical system or a mechanical system, allow for improved performance in terms of reproducibility and reliability of the analysis.

The ISI depends not only on the type of thromboplastin used but also on the analytical characteristics of the coagulometer, so thromboplastin manufacturers provide information regarding the ISI characteristics of thromboplastins related to the instrumentation used.

Activated Partial Thromboplastin Time (aPTT)

It is commonly known that the period between 1935 and 1953 represents the so-called golden age of coagulation discoveries (e.g., 1936 discovery of FVIII, 1947 discovery of FV, 1951 discovery of FVII, 1936 discovery of FVIII, 1952 discovery of FIX, and 1953 discovery of FXI). In 1953, Langdell and collaborators described for the first time the partial thromboplastin time in a hemophilic patient. They modified the laboratory purification method to obtain thromboplastin using only a "partial" form that today is known to consist exclusively of phospholipids. Indeed, the PT was not altered in the plasma of hemophilic subjects. Instead, the use of a partial thromboplastin (without tissue factor) allowed us to easily distinguish hemophilic from nonhemophilic subjects. The aPTT is performed on a sample of citrated plasma, recalcified with calcium chloride, to which phospholipids and an activating agent are added. Indeed, to make aPTT more reproducible, in 1961, Rapaport introduced the use of kaolin (an activator) capable of activating FXII, which represents the starting point of the intrinsic pathway. Today, in addition to kaolin, ellagic acid, and silica crystals are also used.

The aPTT is expressed as the ratio between the time, expressed in seconds, of the patient and an average time obtained from a sample of subjects not affected by coagulation alteration or in therapy with anticoagulant drugs, according to the following relationship:

$$\text{aPTT} = \text{aPTTpatient}(\text{s}) / \text{aPTTplasma pool}$$

The aPTT allows evaluation of the most common factorial deficiencies of the intrinsic and common pathways (especially FXII, FVIII, and FIX). Moreover, the observations of Conley and Hartmann first and then of Bowie led to the evidence that the elongation of the aPTT was due to the presence of the so-called lupic anticoagulant (LAC). The different commercial reagents have different sensitivity in recognizing the presence of LAC, and, therefore, specific aPTT (aPTT-LA) have been developed to recognize the presence of these antibodies that can lead to thrombotic situations. The aPTT is, therefore, a very versatile test that has found wide use in the clinical laboratory for the assessment of both hemorrhagic and thrombotic risk.

Fibrinogen (Fbg)

Fibrinogen is the protein of the coagulation system most commonly present in plasma, with an average concentration between 1.5 and 4 g/L. During the last decades, many methods have been developed for the determination of plasma fibrinogen, using both coagulative tests and enzyme-linked immunoassay (ELISA), immuno-Radiometric Assay (IRMA), nephelometric techniques, etc. The methods that have become more established are:

- Clauss's method
- Methods derived from the PT
- Immunological methods in nephelometry

Clauss' method, described in 1957, is the most widespread, is fully automated, and allows a real evaluation of the active protein. Indeed, it results from a variation of PT. Briefly, a high concentration of thrombin (recombinant or animal-derived) is added to a sample of citrated plasma together with phospholipids and calcium, which represents the starting point (starter) of the reaction time. The presence of excess thrombin allows a rapid transformation of fibrinogen into fibrin. With a calibration curve previously performed on a plasma sample with a known concentration of fibrinogen, it is possible to derive the amount of protein present in the analyzed plasma.

Prothrombin time–based methods are fully automated and use the principle of the relationship between fibrinogen concentration and optical density variation. Briefly, PT is performed on a series of plasma samples of known and scaled fibrinogen concentrations, allowing the construction of a calibration curve.

Immunological methods (so-called antigenic methods) use an antibody that specifically recognizes the protein. Different detection systems based on ELISA or nephelometric techniques have been developed. Nephelometric methods are automatable and, therefore, have been more widely used in clinical laboratories. Compared to the Clauss' method and PT-based methods, which are called functional, immunological methods allow an assessment of dysfibrinogenemia when there is a discrepancy between protein levels obtained with functional and antigenic methods.

Thrombin Time (TT)

Thrombin time (TT) is a coagulation test that is used to evaluate congenital or acquired fibrinogen deficiency. The test is particularly sensitive to the presence of unfractionated heparin. A known amount of thrombin is added to a citrated plasma sample, which allows the conversion of fibrinogen to fibrin, forming a clot in the sample.

Mixture Test

In the evaluation of coagulation disorders, it is of particular interest, especially in emergency situations, to identify the causes that alter coagulation times, distinguishing elongations due to factor deficiencies (e.g., hemophilia A) from those due to the presence of coagulation inhibitors (e.g., both inhibitors of individual coagulation factors and the presence of LAC).

The mixture test is therefore a first-level procedure that is performed on citrated plasma samples. The mixture test is applied to both PT and aPTT, although it is more applicable in the latter case. Briefly, a PT or aPTT is performed on a sample with a 1:1 dilution ratio of normal plasma (i.e., without alteration of clotting times). The correction expressed in seconds is evaluated. If the clotting time performed on the mixture is close to that of normal plasma (the term "correction" is used), this is a factorial deficiency. If there is no correction, an inhibitor is suspected. It is useful to re-evaluate the test after 2 h of incubation at 37 °C to correctly recognize "slow" inhibitors, for example, those directed against FVIII, which require more time to exert their inhibitory activity. This test can therefore provide extremely useful information for the continuation of the diagnostic process.

Anticoagulant Therapy

Heparins and Fondaparinux

Heparins are glycosaminoglycans particularly rich in sulfuric groups with anticoagulant activity and used for the prevention and treatment of venous and arterial thromboembolic disease. Based on molecular weight, heparins are divided into two classes: unfractionated heparin (ENF), with a molecular weight between 3000 and 30,000 Da, and low molecular weight heparin (EBPM), with a molecular weight between 4000 and 6000 Da.

The anticoagulant activity of ENF is mediated by the pentasaccharide sequence that binds to antithrombin and, through this, exerts its inhibitory action, in an irreversible manner, on FXa and FIIa of the coagulation cascade (Fig. 17.7). ENF bound to antithrombin also inhibits, to a lesser extent, FIXa, FXIa, and FXIIa. To exert inhibitory action on thrombin (FIIa), heparin molecules must contain more than 18 saccharide units. Molecules with less than 18 saccharide units, such as EBPMs and the synthetic pentasaccharide fondaparinux, are not able to bind and inhibit FIIa, only FXa.

ENF is of extractive origin (from bovine lung and pig intestine) and is administered subcutaneously or intravenously, as it is not absorbed orally. Because of the wide variability in anticoagulant effects, ENF requires laboratory monitoring, via the aPTT, to ensure appropriate doses. The half-life of ENF varies from 30 to 150 minutes, depending on the doses administered. ENF is eliminated by two mechanisms: a saturable one, by hepatic macrophages and endothelial cells, and a nonsaturable one, slower, mainly renal. At therapeutic doses, a large percentage of heparin is eliminated through the rapid dose-dependent saturable pathway, and only for very high doses of ENF, when hepatic clearance is saturated, the elimination is renal. Therefore, ENF can also be used in patients with severe renal impairment. The antidote is protamine sulfate. A rare but serious complication of ENF use is the occurrence of heparin-induced thrombocytopenia (HIT) due to the formation of anti-FP4 antibodies.

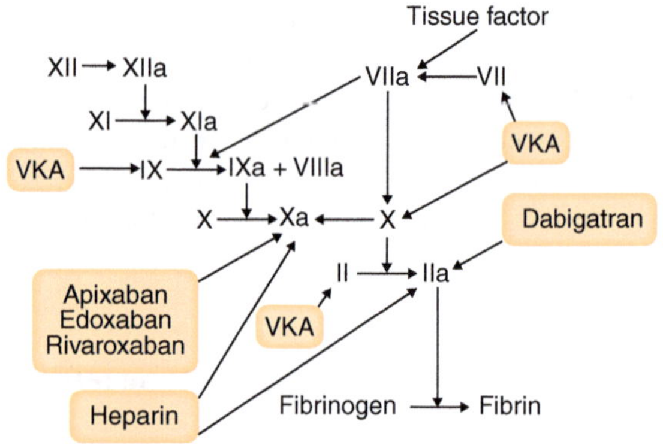

Fig. 17.7 Action of anticoagulant drugs. (Copyright EDISES 2021. Reproduced with permission)

EBPMs are derived from the fractionation, by chemical or enzymatic methods, of ENF. EBPM molecules contain less than 13 saccharide units (in addition to the pentasaccharide structure), so their inhibitory effect, mediated by binding with antithrombin, is mainly towards FXa. EBPMs have several advantages over ENF: longer half-life (3–6 h), which allows for single or dual daily subcutaneous administration; higher bioavailability (~90%); lower incidence of bleeding; greater predictability of the dose–response relationship; and lower incidence of HIT. They are administered at fixed dosages based on body weight. When used in therapeutic regimens, they do not require laboratory monitoring to evaluate the anticoagulant effect, except in some clinical conditions (renal insufficiency, obese patients, and pediatric patients). In these cases, laboratory monitoring is performed through the dosage of anti-FXa activity. On the other hand, weekly platelet count monitoring is necessary for the first 2 weeks of administration due to the possible, albeit very rare, occurrence of HIT. Elimination is mainly renal, so in patients with renal insufficiency, it is preferable to administer ENF or reduce the prophylactic or therapeutic dose of EBPM.

Fondaparinux is an entirely synthetic product consisting of five saccharide units (molecular weight of 1728 Da), which exerts its anticoagulant activity exclusively towards FXa through the strong link with antithrombin. The drug, administered subcutaneously, is highly bioavailable, has a plasma half-life of about 17 h, and is eliminated unmodified by the renal route (it is therefore contraindicated in patients with renal insufficiency). These characteristics allow for a single daily administration and a rapid onset of antithrombotic activity. Like EBPM, it does not require laboratory monitoring.

Vitamin K Antagonist (VKA)

Vitamin K antagonists (VKAs), used for more than 60 years for the prevention and treatment of numerous thromboembolic diseases, are low-molecular-weight compounds, derived from coumarin (dicumarol), rapidly absorbed orally. VKAs act by blocking the reduction of vitamin K-epoxide to vitamin K in hepatocytes through the competitive inhibition of specific epoxide reductases. Their anticoagulant activity is therefore exerted by preventing the γ-carboxylation (the enzyme γ-glutamylcarboxylase requires reduced vitamin K as a cofactor) of coagulation factors II, VII, IX, and X (Fig. 17.7), a phenomenon that is indispensable for their biological activity to take place. The most commonly used VKAs are warfarin sodium (half-life of about 36 h) and acenocoumarol (half-life of about 12 h, with about twice the potency of warfarin). VKAs interact with many foods and drugs, and their metabolism is genetically determined (polymorphisms of the genes of cytochromes VKORC1 and CYP2C, involved, respectively, in the vitamin K cycle and in the metabolism of warfarin). Due to their narrow therapeutic range, they require frequent laboratory monitoring (by INR) and therapeutic adjustments.

The VKA antidote is vitamin K.

Direct Inhibitors of Factors II and X

They are recently introduced anticoagulant molecules (also called direct oral coagulation inhibitors [[DOACs]) capable of binding directly to the target enzyme, blocking its interaction with the substrate. DOACs are absorbed orally, administered in fixed doses, and usually do not require laboratory monitoring. They also have fewer drug interactions than warfarin. They are currently registered for the treatment and prevention of venous thromboembolism and stroke in patients with nonvalvular atrial fibrillation. DOACs currently on the market include the direct thrombin inhibitor dabigatran etexilate (renal elimination 80%, half-life 14–17 h) and the direct FXa inhibitors apixaban (renal elimination 25%, half-life 8–14 h), edoxaban (renal elimination 35%, half-life 8–10 h), and rivaroxaban (renal elimination 33%, half-life 7–11 h). The only DOAC to have a specific inhibitor on the market at present is dabigatran (idarucizumab), while some inhibitors of anti-FXa DOACs (e.g., andexanet) are forthcoming. Figure 17.7 summarizes the mechanisms of action of anticoagulant drugs.

Antiplatelet Therapy

Antiplatelet agents are a class of drugs developed to reduce platelet function with the intent of preventing arterial thrombotic events. The antiplatelet agents currently in clinical use belong to four categories:

- Cyclo-oxygenase inhibitors
- P2Y12 receptor antagonists for ADP
- gp IIb/IIIa inhibitors
- Phosphodiesterase inhibitors

Acetylsalicylic acid, or aspirin, which has been known for more than 100 years, exerts its action by irreversibly blocking the platelet cyclo-oxygenase type 1 (COX1) enzyme by transferring its acetyl group to the ser529 residue of the enzyme's amino acid chain. Blockade of the platelet COX-1 enzyme prevents the production of TXA2 (an amplifier of platelet response to stimuli), resulting in a reduction in the extent of platelet activation in vivo.

Platelet P2Y12 receptor antagonists for ADP (which act by amplifying platelet activation) are represented by the thienopyridines, which include ticlopidine, clopidogrel, prasugrel, and ticagrelor. While the first three are prodrugs and require hepatic biotransformation to active metabolites, ticagrelor is already active on its own. Inhibitors of gp IIb/IIIa, which acts as a receptor for fibrinogen and other adhesive proteins, include abciximab, eptifibatide, and tirofiban. Phosphodiesterase inhibitors include dipyridamole and cilostazol.

18 Kidney

Michele Mussap

Introduction

Laboratory medicine has a role of primary importance in the diagnosis and follow-up of patients with renal disease. This is mainly due to two peculiar aspects: first, the deterioration of renal function remains asymptomatic for a long time, until the extension of parenchymal damage induces the appearance of evident but late signs and symptoms, such as oliguria or edema in the lower limbs. Therefore, in the period between the onset of the disease and the onset of symptoms, it is essential to measure biomarkers for early detection of the presence of clinically undetectable alterations. Second, the definition of the risk associated with the progression of the disease and the development of complications is essentially based on two biomarkers: creatinine and albuminuria. In particular, the determination of plasma creatinine plays a key role in calculating the estimate of glomerular filtrate using equations that include creatinine concentration. Creatinine and albuminuria are therefore two biomarkers that are certainly not new but of extreme importance in predicting the clinical outcome of patients with kidney disease.

Kidney Disease

"Kidney disease" is a generic term referring to the presence of organ damage. It replaces the term "renal failure," which was universally used prior to 2002, when the Kidney Disease Outcomes Quality Initiative (K/DOQI) guidelines on chronic kidney disease were published. The difference between renal disease and renal failure is not only semantic but also involves the clinical laboratory because the relationship between parenchymal damage and reduced renal function is not linear at all. The search for biomarkers capable of detecting tissue lesions of the organ before the functions of glomerular filtration of the plasma, reabsorption of substances from the proximal tubule, and hydroelectrolytic exchange in the loop of Henle and distal tubule compartments are impaired remains an open challenge for the early identification of the disease. The etiology of renal disease involves both primary diseases of the kidney (e.g., glomerulonephritis, interstitial nephritis, vascular nephropathy, neoplasia, etc.) and systemic diseases with involvement of the renal emunctory, including hemodynamic stress conditions, such as complex surgeries, extracorporeal circulation, etc. Pathological conditions affecting the structure and function of the kidneys can be considered acute or chronic, depending on their duration, identified by an arbitrary time threshold of 3 months. In particular, we distinguish:

- Chronic kidney disease (CKD)
- Acute kidney injury (AKI)
- Acute kidney disease (AKD)

AKI is part of AKD, and it can occur in the course of CKD or without concomitant CKD. Distinguishing renal disease into acute and chronic has a rationale that stems from fundamental differences between the two conditions, such as etiology, therapeutic approach, clinical outcome, and the choice of biomarkers. CKD and AKI have been defined by two separate working groups according to extremely different diagnostic and staging criteria; Table 18.1 summarizes the definitions of AKI, CKD, and AKD.

Chronic Kidney Disease

The K/DOQI CKD guidelines of 2002 were the first structured document to propose a clear and uniform definition of CKD associated with a disease staging system based on estimated glomerular filtration rate (eGFR) values. For the first time, three concepts were defined in a structured manner:

M. Mussap (✉)
Department of Surgical Sciences, University of Cagliari, Cagliari, Italy
e-mail: michele.mussap@unica.it

M. Ciaccio (ed.), *Clinical and Laboratory Medicine Textbook*, https://doi.org/10.1007/978-3-031-24958-7_18

Table 18.1 Definitions of acute kidney injury, chronic kidney disease, acute kidney disease, and no known kidney disease

	Functional criteria	Structural criteria
AKI	50% increase in SCr in 7 days	None
	or	
	Increase in SCr of 0.3 mg/dL (26.5 μmol/L) in 2 days	
	or	
	Oliguria	
CKD	GFR < 60 mL/min/1.73 m^2 for more than 3 months	Tissue damage >3 months
AKD	See AKI	Tissue damage <3 months
	or	
	GFR < 60 mL/min/1.73 m^2 for less than 3 months	
	or	
	Decreased GFR ≥ 35%	
	or	
	50% SCr increase for less than 3 months	
NKD	GFR ≥ 60 mL/min /1.73 m^2, SCr stable	No damage

AKD acute kidney disease, *AKI* acute kidney injury, *CKD* chronic kidney disease, *GFR* glomerular filtration rate, *NKD* no known kidney disease or absence of known kidney disease, *SCr* serum creatinine

- CKD is a global public health problem.
- The first stage of CKD includes low-risk subjects, with a GFR ≥90 mL/min/1.73 m^2, previously considered not at risk of CKD.
- Clinical laboratories should report the plasma creatinine result with the calculation of eGFR by using equations based on plasma creatinine.

In addition, the K/DOQI document highlights the importance of proteinuria and standard urine examination in the framing of a patient with CKD. This publication revolutionized not only the concept and management of CKD, generating a substantial boost in research in the following years, but also numerous controversies that stimulated lively debates in the scientific community and gave rise to various official documents. It has considerably influenced public health policy strategies and clinical laboratory routines. In 2013, new guidelines on CKD were published; they represent an extended update of the 2002 document based on new evidence and the result of the debates and controversies within the scientific community during the decade 2002–2012. In this new consensus document, CKD is defined as changes in kidney structure or function, which are detectable for a period of more than 3 months and have implications and repercussions on the patient's overall health status. The diagnostic criteria for CKD are summarized in Table 18.2. The risks associated with the onset of CKD are divided into: (1) sociodemographic and genetic factors and (2) predisposing factors to which an individual may be exposed. In most patients with CKD, renal excretory, endocrine, and metabolic functions decline only after morphological damage has already compromised a part of the organ. However, to date, eGFR is considered the best index of renal function and is therefore the most important parameter for assessing the stage of the disease. In contrast to what emerged in 2002, nowadays, albuminuria has been added to GFR, which is not only a functional index but also a biomarker of injury and a key prognostic index in defining the risk of death, end-stage kidney disease (ESKD), and cardiovascular disease. In addition, albuminuria represents a therapeutic target in patients with CKD. Therefore, the severity of CKD is classified according to three variables (strength of recommendation 1B) (Table 18.3):

- Etiology
- eGFR (six categories)
- Albuminuria (three categories)

Table 18.2 Diagnostic criteria for chronic kidney disease (each present for more than 3 months)

Biomarkers of kidney damage
Albuminuria: AER > 30 mg/24 h; ACR > 30 mg/g creatininuria
Abnormalities of the urinary sediments
Altered hydroelectrolytic or other balances related to tubule dysfunctions
Structural anomalies at tissue histology
Structural anomalies at diagnostic imaging
Organ transplant patient
Decrease in the glomerular filtration rate (GFR)
GFR < 60 mL/min/1.73 m^2 (categories G3A–G5)

Partially modified by KDIGO

ACR albumin-to-creatinine ratio, *AER* albumin excretion rate, *GFR* glomerular filtration rate

Table 18.3 Glomerular filtration rate and albuminuria categories for classifying chronic kidney disease

Category of GFR	*GFR (mL/min/1.73 m^2)*		*Assessment*
G1	≥90		Normal or increased
G2	60–89		Slight decrease
G3a	45–59		Mild-moderate decrease
G3b	30–44		Moderate–severe decrease
G4	15–29		Severe decrease
G5	<15		Absent (terminal stage)
Categories of albuminuria	*AER (mg/24 h)*	*ACR (mg/g of creatinine)*	*Assessment*
A1	<30	<30	Normal–slight increase
A2	30–300	30–300	Moderate increase
A3	>300	>300	Severe increase

ACR albumin-to-creatinine ratio, *AER* albumin excretion rate, *GFR* glomerular filtration rate

Table 18.4 Chronic kidney disease (CKD) staging and cumulative risk of clinical outcome by glomerular filtration rate (GFR) and albuminuria (ACR) categories

			Persistent albuminuria categories (ACR)[a]		
			A1	A2	A3
			Normal-slight increase	Moderate increase	Severe increase
Filtrate categories (GFR), mL/min/1.73 m^2			<30 mg/g (<3 mg/mmol)	30–300 mg/g (3–30 mg/mmol)	>300 mg/g (>30 mg/mmol)
G1	Normal or increased	≥90	Low	Medium	High
G2	Slight decrease	60–89	Low	Medium	High
G3a	Mild-to-moderate decrease	45–59	Medium	High	Very high
G3b	Moderate-to-severe decrease	30–44	High	Very high	Very high
G4	Severe decrease	15–29	Very high	Very high	Very high
G5	Kidney failure	<15	Very high	Very high	Very high

ACR albumin-to-creatinine ratio, *GFR* glomerular filtration rate
[a]The ACR values are reported in two different units of measurement: mg/g and mg/mmol

The combination of eGFR and albuminuria defines risk stratification and thus clinical outcome, according to the scheme shown in Table 18.4. The natural history of most patients with CKD is a progressive decline in GFR over time, leading to renal failure and the onset of complications. The onset of cardiovascular disease in patients with CKD is one of the main outcomes of the disease and is the most important cause of death in these patients. Other complications are drug nephrotoxicity, endocrine and metabolic complications, infections, and decline in cognitive function. Most causes of CKD are irreversible, and the evolution of the morpho-functional damage can only be slowed down before reaching the uremic end stage (ESKD), which can only be treated with replacement therapy (peritoneal dialysis, hemodialysis, or kidney transplantation). However, chronicity is not synonymous with irreversibility. In some cases, CKD can also regress spontaneously or following specific therapies, and, in other cases, therapeutic treatment can induce a partial regression of the morphological damage, improving physiological functions. A typical example is represented by the success of some immunosuppressive therapies in glomerulonephritis.

Acute Kidney Injury

The term acute kidney injury (AKI) is not an evolution of the traditional term acute renal failure (ARF), which defined the rapid deterioration (from a few hours to 1–2 days) of renal function with alteration of hydroelectrolytic and acid–base balances. The term "AKI" emphasizes the presence of a much broader morbid spectrum than that previously associated with acute renal failure; it was also coined because more than 35 definitions of acute renal failure were reported in the literature, which prevented systematic epidemiological and etiological studies and the progress of knowledge and scientific and clinical research for a long time. The proposal of a universal definition of acute renal failure and a classification system able to homogenize the diagnostic and therapeutic approach was the result of a collaborative effort between nephrologists and intensivists-resuscitators gathered in the working groups of ADQI (Acute Dialysis Quality Initiative) and AKIN (Acute Kidney Injury Network), which led not only to the coinage of the term "AKI" but also to the proposal of disease staging systems, such as RIFLE (Risk, Injury, Failure, Loss, End-stage renal failure) in 2004, AKIN in 2007, and the Kidney Disease: Improving Global Outcomes (KDIGO) in 2012. The etiopathogenesis of AKI is complex and is almost always the result of multiple interactions between various factors: renal diseases (e.g., acute interstitial nephritis, vasculitis and acute glomerulopathies, hemodynamic stresses such as ischemia followed by reperfusion, iatrogenic toxic nephropathies), extrarenal pathologies (e.g., hypoxia and anoxia, iatrogenic toxic nephropathies), and systemic pathologies (dehydration, hypovolemia, hypotension), among which sepsis accounts for approximately 50% of the causes of AKI. Each of these factors, and their combination, causes apoptosis and necrosis of tubular cells; at the same time, the concomitant development of inflammation is observed, with local recall of immunocompetent cells, cytokines, and acute-phase proteins. In turn, tubular damage, aggravated by the onset of the acute phase and the progressive impairment of the microcirculation, affects the functionality of the glomerulus, which results in the reduction of the GFR. With regard to exogenous substances that are potentially nephrotoxic and capable of contributing to the onset of AKI, a partial list is provided in Table 18.5.

AKI is defined by the presence of one of the following criteria:

- Increase within 24 h in plasma creatinine ≥0.3 mg/dL (≥26.5 μmol/L)
- ≥1.5-fold increase in plasma creatinine from a baseline value measured within the previous 7 days
- Oliguria, defined as a urinary volume < 0.5 mL/kg/h for the duration of 6 h

It is common opinion that oliguria identifies tubular damage; however, this is not always true because the relationships

Table 18.5 Potentially nephrotoxic exogenous substances involved in the pathogenesis of acute kidney injury

Acyclovir
Amphotericin
Aminoglycosides
β-Lactam antibiotics (interstitial nephropathy)
Nonsteroidal anti-inflammatory drugs
Cyclosporine
Cisplatin
Inhibitors of the enzyme that converts angiotensin
Angiotensin receptor inhibitors
Methotrexate
Contrast media
Sulfonamides
Tacrolimus

Table 18.6 Staging of acute kidney injury according to the KDIGO guidelines

Stage	Criteria based on plasma creatinine	Criteria based on urinary volume
1	SCr 1.5–1.9 higher than the baseline value or	<0.5 mL/kg/h for 6–12 h
	Increased SCr ≥0.3 mg/dL (26.5 μmol/L)	
2	SCr 2.0–2.9 higher than baseline	<0.5 mL/kg/h over 12 h
3	SCr 3.0 times higher than the baseline value;	<0.5 mL/kg/h over 12 h; or
	Increased SCr ≥4.0 mg/dL (353.6 μmol/L);	Anuria over 12 h
	Beginning of replacement therapy; or	
	In adult patients (>18 years), decrease in GFR (<35 mL/min/1.73 m^2)	

SCr serum creatinine

between oliguria, GFR, and tubular damage are highly complex. Severe oliguria can be observed in the absence of tubular damage in cases of severe hypovolemia and hypotension; these factors, by inducing the synthesis of the antidiuretic hormone (ADH), lead to the excretion of minimal volumes of urine with high osmolality (>500 mOsm/kg). This finding confirms the importance of determining, in specific cases, urinary osmolality by direct methods and to always take into account the urinary volume in the evaluation of results. On the contrary, in the presence of tubular damage, the ability to concentrate urine is compromised and urinary volume can also be normal (non-oliguric renal failure). The staging of AKI according to KDIGO is shown in Table 18.6.

Acute Contrast Media-Induced Nephropathy

Contrast Media-induced acute kidney injury, (CI-AKI), is a rapid deterioration of renal function within 48–72 h of intravascular administration of contrast and has been defined as an increase in plasma creatinine >0.5 mg/dL (≥44 μmol/L) or ≥25% from baseline. A 5% increase in creatinine at 12 h after contrast administration has been shown to be a predictive index of renal failure with a sensitivity of 75%, a specificity of 72%, and an odds ratio (OR) of 7.37. Patients with CKD have up to a 20-fold increased risk of contrast media-induced nephropathy compared to patients with GFR >90 mL/min/1.73 m^2. Particular attention should be paid to renal function when gadolinium is used because of the risk of development of systemic nephrogenic fibrosis in patients with GFR <60 mL/min/1.73 m^2. Other risk factors are age > 75 years, dehydration, diabetes, diabetes under metformin therapy, heart failure, hypotension, anemia, and allergies. The association of two or more factors (e.g., diabetes and chronic renal failure) exponentially increases the correlated risk. Finally, the use of iodinated contrast media is not recommended in patients with hyperthyroidism. There is no evidence that monoclonal gammopathies in general and, in particular, those of undetermined significance (MGUS), multiple myeloma, Waldeström macroglobulinemia, light chain immunoglobulin deposit diseases, and amyloidosis represent "per se" a risk factor for CI-AKI in patients who are well hydrated at the time of testing and with normal renal function. However, substantial deterioration of renal function very often coexists in patients with these conditions; therefore, they should be carefully evaluated before administration of the contrast because they fall into the categories of patients at risk of CI-AKI. Table 18.7 summarizes the recommendations that both the Italian Society of Medical and Interventional Radiology (SIRM) and the Italian Society of Clinical Biochemistry and Clinical Molecular Biology – Laboratory Medicine (SIBioC) have published on the correct use of laboratory tests for the evaluation of a patient to undergo imaging with contrast.

Table 18.7 Recommendations on the appropriate use of laboratory tests in assessing the risk of toxic contrast-induced nephropathy (contrast medium)

Preliminary assessment of the nephrotoxic risk	Determination of the basal creatinine by a method traceable to the international analyte measurement system; reporting must include calculation of glomerular filtrate (GFR) by the CKD-EPI equation
	The decentralized execution of the test, namely the Point-Of-Care Testing (POCT) is not recommended
Monitoring the nephrotoxic risk	Determination of creatinine for one or more times by methods similar to those reported above; it is recommended to integrate the result with the calculation of the critical difference (reference change, value, RCV), when possible, or, at least, the difference from the baseline value In the case of increased plasma creatinine >5%, remote control 48–72 h is recommended
Plasma cell dyscrasias and contrast media	Specific laboratory tests are not considered necessary to exclude the presence of monoclonal gammopathies, such as serum protein electrophoresis, the search for Bence Jones protein in urine, and the determination of plasma free light chains (FLCs); these tests alone cannot rule out a monoclonal gammopathy, and none of these tests are included in the panel of the stat clinical laboratories

Laboratory Diagnosis of Kidney Diseases

Laboratory diagnosis of kidney disease has historically been based on standard urine examination, determination of plasma creatinine, and determination of albuminuria or proteinuria. Since 2002, it has been recommended that all clinical laboratories combine creatinine determination with the calculation of estimated glomerular filtration using equations. Over the last 15 years, the scientific community has made a considerable effort and employed many resources in the search for biomarkers of renal parenchymal injury, considering the evidence that in kidney disease, morphological damage always precedes functional damage and that before an evident functional damage almost always the patient is asymptomatic. In the spring of 2004, a group of clinical professionals and the board of the American Society of Nephrology defined the widespread use of creatinine as an almost exclusive biomarker of AKI, an obstacle to the development of innovative therapies. Based on this observation, they determined that a priority for the immediate future should be the search for new biomarkers of AKI. During the subsequent 7 years, the scientific community identified an impressive number of candidate biomarkers for the management of patients with AKI; in clinical practice, the most interesting were neutrophil gelatinase-associated lipocalin (NGAL), kidney injury molecule-1 (KIM-1), interleukin-18 (IL-18), and liver-type fatty acid-binding protein (FABP-1). Unfortunately, there is no conclusive evidence to establish whether, and to what extent, the new biomarkers have improved the diagnosis and management of patients with kidney disease due to univocal interpretative difficulties arising from the lack of comparability of the case histories reported in the literature, the heterogeneity of the inclusion and exclusion criteria, and the analytical methods that are sometimes unreliable and not applicable in routine clinical practice. Finally, the most important problem is related to the evaluation of new biomarkers that is always performed considering creatinine as a reference (gold standard), and, therefore, the classification of patients into true positives and true negatives is affected by biological and methodological problems related to creatinine. For a long time, this problem hampered the definitive identification of cystatin C sensitivity and specificity, postponing the introduction of this biomarker in the routine of clinical laboratories.

Standard Urine Test

The first approach to the study of renal function is the standard urine examination, improperly called "complete urine examination." According to Angelo Burlina, the standard urine test is a set of qualitative data that is easy to obtain and of safe clinical interpretation. Indeed, it is an organ's metabolic profile that can provide some fundamental information both on the morpho-functional status of the kidneys and on the status of many metabolic pathways, including those involved in the hydroelectrolytic balance and in the metabolism of carbohydrates and proteins. Although the standard urine test is simple to perform and easy to interpret, it is necessary to properly manage the pre-analytical phase to avoid misleading results resulting from improper sample collection or from unsuitable transport and storage conditions. The standard urine examination allows detecting the presence of clinically relevant indices both in chemical–physical examination (hemoglobin, proteins, glucose, etc.) and in microscopic examination of the sediments (blood cells, parenchymal cells, microorganisms, etc.). In recent years, the application of cytofluorimetric techniques to the analysis of urinary sediments and the introduction of new automated techniques, such as optical flow microscopy with digitization and image analysis through neural networks, have allowed a reevaluation of urinary microscopic analysis, with undoubted advantages for the nephrologist, who has always assigned great clinical importance to this examination. The chemical–physical examination is carried out in all clinical laboratories by complete automation allowing semiquantitative results to be obtained from the photometric reading of dipsticks. Thanks to their wide diffusion and the simplicity of their use, the physical–chemical examination is also carried out by the patients themselves or by their families at home, especially in the monitoring of the therapy of chronic nephropathies such as nephrotic syndrome. However, physical–chemical examination has some limitations that must be considered when evaluating the results. For example, the presence of chromogenic catabolites in the urine or the presence of drugs may interfere with color formation in some dipstick reactions, such as the assessment of proteinuria, which represents one of the most clinically relevant data. The principle of a proteinuria dipstick is based on the protein error of pH indicators: the color of certain indicators at a given pH changes if they bind to proteins in the urine sample. In other words, it is a paper soaked in tetrabromophenol, a pH indicator, which changes its color when binding to proteins present in the sample. The presence of proteinuria evidenced by the standard urine test must be confirmed by quantitative and qualitative examinations, which allow assessing the concentration and composition of proteinuria. Even the absence of proteinuria on physical–chemical examination must be confirmed by further tests if there is a definite diagnostic suspicion (e.g., amyloidosis, multiple myeloma, etc.). Indeed, it is known that a dipstick is unable to detect the presence of Bence Jones proteinuria, even if present in significant amount and, therefore, the interpretation of the results must always take into account the clinical question underlying the request for standard urine test. Even hematuria, which can be detected both by physical–chemical and microscopic examination of the sediments, has a great clinical value, especially in cases of microhematuria, when no alteration is detectable by observation of the sample with the naked eye (e.g., the color). Once identified, hematuria must be clinically investigated to establish both the site of its origin (it can originate from any part of the urinary system) and the etiology. In CKD, some abnormalities of the urine sediments are indicative of parenchymal damage: microscopic hematuria with anisocytosis in glomerular basement membrane alterations, erythrocyte cylinders in proliferative glomerulonephritis, leukocyte cylinders in pyelonephritis and interstitial nephritis, hyaline cylinders with fat globule inclusions in renal diseases characterized by proteinuria, and granular cylinders with tubular epithelial cells in many conditions of renal parenchymal impairment. The National Institute for Health and Care Excellence (NICE) guideline on AKI recommends that urine chemistry and physical examination (dipstick) for blood, proteins, leukocytes, nitrites, and glucose should be performed as soon as possible in all patients with suspected or confirmed AKI.

Nitrogen (N)

The nitrogen (N) present in the human body can be schematically divided into protein nitrogen and nonprotein nitrogen. The term "nonprotein nitrogen" refers to all those non-peptidic substances, mostly of low molecular weight (MW), containing nitrogen, which are not precipitated by common deproteinizing agents. More than 50% of nonprotein nitrogen is derived from urea, 20–25% from creatinine, and 10–15% from uric acid, whereas negligible amounts may be derived from amino acids, ammonia, creatine, bile pigments, guanidine compounds, some hormones, and some vitamins. A small amount (about 50–150 mg/L) is finally located in the elements (residual nitrogen). The term azotemia indicates the amount of nonprotein nitrogen present in the plasma. The pathways of nitrogen elimination are mainly represented by the kidneys and the gastrointestinal system. The elimination of nitrogenous substances by the kidneys mainly occurs by glomerular filtration and tubular secretion; in the case of some substances, these mechanisms are combined. The balance between nitrogen introduced through diet and nitrogen eliminated through urine and feces is relatively constant in healthy adults, whereas it is positive (production higher than elimination) during the growth stage of the life cycle, namely infancy and adolescence. The determination of total nitrogen and nitrogen on biological samples was made possible already in the last century by the classical method of Kjeldahl in order to study renal function. Although it is now possible to measure nitrogen with accurate and precise analytical methods such as gas chromatography or chemiluminescence, this determination is no longer routinely carried out for diagnostic purposes in the clinical laboratory and can only be used for research purposes and in particular cases, for example, to assess the nitrogen balance in malnutrition, cirrhosis, and polytraumatized patients.

Urea

Oxidative deamination of amino acids results in the formation of amine groups that are released into the tissues in the form of NH_4^+ (ammonium ion). Urea, or carbonic acid diamide (CH_4N_2O), is a substance with a molecular weight (MW) of 60.06 Da that is formed in the liver through a cycle of reactions (urea cycle) to eliminate ammonia (NH_3), a toxic substance. In humans, ureogenesis is, therefore, an expression of protein catabolism. Urea diffuses freely between intracellular and extracellular fluids and is, therefore, uniformly distributed in body fluids. More than 90% of plasma urea is concentrated and excreted in the urine through the renal excretory system, constituting approximately 80–90% of urinary nitrogen. The concentration of urea in the serum and other biological fluids depends on two factors: production by the liver and elimination by the kidneys. In healthy adults, the reference range for urea is between 3.5 and 9.0 mmol/L (21–54 mg/dL); for those aged more than 60 years, it is slightly higher. At birth and in the first years of life, the reference range is between 1.0 and 5.0 mmol/L (6–30 mg/dL). Recently, pediatric and neonatal reference values have been divided by age groups and analytical methods. Most analytical methods for the determination of urea are based on the enzymatic reaction of hydrolysis of urea by urease (urea amidohydrolase; EC 3.5.1.5) leading to the formation of ammonium ions. There are also electrochemical methods that are based on conductivity measurements in cells containing the reaction mixture (sample + reagent) or on potentiometry measurements using ammonium ion-selective electrodes in contact with a membrane on which urease is distributed. The absolute method for the measurement of urea is mass spectrometry. To convert a urea value from international system of units (SI) (mmol/L) to traditional units (mg/dL), multiply the value by 6.02; conversely, to obtain a value expressed in SI units from a result expressed in milligrams per deciliter, multiply the value by 0.166. Increases in plasma urea concentration are observed during reduced renal function, whereas significant decreases can be observed only in the presence of severe hepatic damage with impaired hepatocyte function. Some extrarenal pathologies can lead to significant increases in blood urea (prerenal hyperazotemia), such as, for example, hypotension, states of dehydration and hypovolemic shock, cardiac insufficiency, sepsis, and hyperthyroidism. Moreover, urea nitrogen is a product of protein catabolism, especially of those of food origin, and, therefore, its plasma levels can be significantly influenced by high or low protein content diets.

The elimination rate of urea is a function of its plasma concentration, of the renal plasma flow (perfusion rate), and of the ability to remove this substance from the plasma by the kidneys (renal function). Indeed, the amount of urea excreted by urine depends not only on glomerular filtration but also on tubular reabsorption; it is estimated that about 40–50% of filtered urea is normally reabsorbed by the tubules. The value of the concentration of urea in the serum can be expressed by either considering the entire molecule or in terms of nitrogen content in the molecule; in this case, we refer to urea nitrogen (blood urea nitrogen (BUN)). Since nitrogen contributes 28/60 to the total weight of urea, starting from a given value of urea nitrogen concentration, urea concentration can be calculated by multiplying this value by 2.14.

Creatinine

Creatinine ($C_4H_7N_3O$), a cyclic amide of creatine, is the terminal product of the catabolism of creatine, a substance present above all in the muscular apparatus, where it participates, in phosphorylated form (phosphocreatine), in the storage of energy. During the reaction that allows the conversion of phosphocreatine to creatine, small amounts of creatine (1–2%), through a nonenzymatic process, are irreversibly transformed into creatinine, which passes from the muscle into the circulation, where it is finally removed through renal filtration due to its low MW, equal to 113.12 Da. In addition, a portion of creatinine (7–10%) is also eliminated through the mechanism of tubular secretion. Creatinine is a nontoxic substance, does not bind to proteins, is present in all biological fluids (sweat, bile, cerebrospinal fluid, etc.), and is physiologically inert. The amount of creatinine formed daily in the body is proportional to the amount of creatine, which in turn is a function of total muscle mass. Its production varies with age and sex, but in a healthy adult, the rate of formation is relatively constant and little affected by dehydration or other morbid conditions, such as sepsis, heart failure, etc. In a healthy adult, creatinine levels, obtained using a traceable enzymatic method, range from 64 to 104 μmol/L (0.72–1.18 mg/dL) in men and from 49 to 90 μmol/L (0.5–1 mg/dL) in women. In newborns, plasma creatinine concentration is elevated, primarily because of the proportion derived from maternal blood; it is also higher in low-birth-weight infants and is inversely related to gestational age and postnatal age. During the first week of life, there is a significant decrease in creatininemia due to the progressive elimination of the maternal rate.

To convert a creatinine value from SI units (μmol/L) to traditional units (mg/dL), multiply the value by 0.0113; conversely, to obtain a value in SI units from a result expressed in milligrams per deciliter, multiply the value by 88.4.

In renal disease, the progressive increase in creatinine is less than the reduction in filtrate because the creatinine formed is metabolized or excreted extrarenally (at normal plasma concentrations, the amount of creatinine eliminated by the intestine is negligible, but in uremia, it becomes significant). In addition, when a major reduction in renal function appears, the creatinine clearance/inulin clearance ratio increases in relation to an increase in tubular creatinine secretion. In the clinical laboratory, the determination of plasma creatinine is the most widely used renal function test due to the simplicity of execution, the availability of commercial methods adaptable to automation, low costs, and the possibility of performing the test in emergency. Less common, because it has been replaced by eGFR, is the performance of creatinine clearance, which requires collection of a timed urine sample (usually 24 h) to determine urinary creatinine in addition to plasma creatinine and to calculate the clearance value (mL/min) through diuresis (volume). The absolute and reference methods for creatinine determination are, the isotope dilution gas chromatography-mass spectrometry (IDGC-MS) and the high-performance liquid chromatography (HPLC), respectively. However, the most popular assay for the routine determination of creatinine is the colorimetric one, developed and described about 100 years ago by Jaffe. This method is based on the reaction of creatinine with alkaline picrate, resulting in the formation of an orange-red-colored complex. Even today, although with different and sometimes substantial modifications, the alkaline picrate is the most widely used method in clinical laboratories, although up to 20% of the color that develops from the colorimetric reaction may derive from substances other than creatinine, such as glucose, uric acid, bilirubin, pyruvic acid, and drugs. However, the need to express eGFR by equations using plasma creatinine concentration has promoted the acceleration of method standardization in order to minimize bias between methods and, in turn, between laboratories. Hence, the 2002 K/DOQI guidelines called for the urgent need to obtain traceable analytical methods, commercially available. Based on this requirement, a primary reference material (primary calibrator) was prepared, which allowed in vitro diagnostic (IVDs) companies to make available traceable methods for plasma creatinine. Clinical laboratories are, therefore, recommended to perform creatinine determination by traceable methods; moreover, the scientific community suggested a switch from colorimetric to enzymatic methods, based on the evidence of the lower analytical variability of the latter, in terms of imprecision, bias, and specificity. Unfortunately, even enzymatic methods are not completely free from analytical pitfalls: for example, high concentrations of ammonium significantly decrease the precision of enzymatic methods using creatinine deaminase, whereas high concentrations of catecholamines can give rise to important underestimates of the results.

Ultimately, although the spread of traceable methods has greatly improved the standardization of plasma creatinine measurement, critical issues remain regarding both the low number of laboratories using enzymatic methods and the differences between results obtained by colorimetric and enzymatic methods.

The validity of creatinine as a test of renal function is based on the observation that, in healthy subjects under steady-state conditions, the daily amount of creatinine released from the muscle mass is constant and is consistently removed from the kidneys. Despite the widespread use of plasma creatinine determinations, several critical pathophysiological issues make interpretation of the results problematic and limit the clinical significance of these tests. Creatinine concentration depends on extrarenal factors, the most important is muscle mass. Therefore, creatinine varies according to sex, age, ethnicity, and regular exercise (athletes). Other factors are diet, extrarenal creatinine elimination, and the mechanism of tubular secretion, which increases inversely with GFR and can be significantly affected by certain drugs, including cimetidine, cyclosporine, and trimethoprim. The tubular secretion of creatinine causes the overestimation of creatinine clearance, compared to the true value of GFR, by 10–40% in healthy subjects and by more than 40% in renal failure, and is the main cause of the poor sensitivity of the test (high number of false negatives); indeed, the GFR can decrease by up to 50% before the circulating concentration of creatinine significantly increases, just because the mechanism of tubular secretion counter balances the decrease in glomerular filtration. Depending on the clinical condition of the subject, further factors can induce the overestimation or underestimation of eGFR based on plasma creatinine, such as unstable metabolic status. This means that eGFR cannot be used in several unstable metabolic conditions, for example during pregnancy or in the early infancy. Finally, non-standardized water intake and incorrect or insufficient collection of timed urine samples (24-h, night urine, etc.) are additional causes of inaccurate creatinine clearance. In particular, the partial or inaccurate collection of the urine specimen during the 24-hours originates significant variations in clearance results. For these reasons, the K/DOQI guidelines advise against the determination of creatinine clearance and recommend the use of equations that employ plasma creatinine and other parameters in their calculation.

Estimation of the Glomerular Filtration Rate by Equations

The GFR represents the amount of plasma filtered by the kidneys in a unit of time; the GFR is universally considered a functional indicator, able to represent the function of the organ, and, for this reason, it has a high clinical value. In clinical practice, the direct measurement of GFR, that is the true GFR (mGFR), is unlikely and even the clearance of exogenous substances, such as inulin, ^{125}I-Iothalamate, ^{51}Cr-EDTA, ^{99}TC-DPTA, cannot be used in the routine, both for the complexity of the protocols to be applied and for the critical issues affecting methods for their measurement. Some authors have recently proposed the utilization of the iohexol clearance because of the availability of reliable and fast methods for the measurement of this compound. Hence, since 2002, the international scientific community has recommended to associate the measurement of creatinine with the estimation of GFR through the application of specific equations based on some physical parameters combined arithmetically with the plasma creatinine concentration, or other plasma biomarkers, and predefined constant factors.

A plethora of formulas/equations for eGFR have been developed over the past 50 years (Fig. 18.1 and Table 18.8); for a long time, the Cockcroft–Gault formula for adults and the Schwartz formula for children were widely used worldwide, mainly because of their simplicity. However, the Cockcroft–Gault formula did not allow the estimation of GFR but was simply the estimation of creatinine clearance based on plasma creatinine. In 1999, the MDRD (Modification of Diet in Renal Disease) equation was developed in a group of participants (1628) in a randomized trial studying the effects of dietary restriction and blood pressure control on the progression of CKD. In these patients, mGFR was measured by iothalamate clearance. The original equation had two shortcomings: a universal standard calibrator for creatinine determination was not available at that time (non-traceable method) and the equation was developed only in patients with CKD and, therefore, the results were inaccurate in healthy subjects with GFR >90 mL/min/1.73 m^2 (systematic bias of underestimating GFR for high values). The original 1999 equation included six variables: creatinine, age, gender, ethnicity, urea, and albumin. In 2000, the equation was simplified by removing urea and albumin as variables, and in 2006, it was reexpressed using a traceable method for creatinine. In 2009, the Chronic Kidney Disease Epidemiology Collaboration (CKD-EPI) equation was developed to overcome the problems associated with the MDRD equation. The new equation was derived from a meta-analysis study that included 8254 healthy adult subjects with CKD. Numerous clinical studies in various populations have demonstrated over time that the performance of the CKD-EPI equation is significantly better than that of the MDRD equation; for these reasons, clinical laboratories are still recommended to use the CKD-EPI equation for the calculation of GFR estimation. The CKD-EPI equation based on the determination of plasma creatinine is organized according to the scheme shown in Table 18.8. The original 2009 CKD-EPI equation

Fig. 18.1 Formulas for calculating the GFR estimate. (Copyright EDISES 2021. Reproduced with permission)

Cockroft-Gault formula

For women

$$\text{creatinine clearance (mL/min)} = 0{,}85 \times \frac{(140 - \text{age [years]}) - \text{body weight (kg)}}{72 \times \text{serum creatinine (mg/dL)}}$$

For Men

$$\text{GFR (mL/min)} = \frac{(140 - \text{age [years]}) - \text{body weight (kg)}}{72 \times \text{serum creatinine (mg/dL)}}$$

Schwartz formula

1976 Schwartz equation

$$\text{creatinine clearance} = \frac{\text{k} \times \text{length (cm)}}{\text{serum creatinine (mg/dL)}}$$

2009 Schwartz equation

$$\text{eGFR} = \frac{0.413 \times \text{height (cm)}}{\text{serum creatinine (mg/dL)}}$$

k is a constant that varies with age and creatinine assay method (Jaffe or IDMS)

MDRD formula

$$\text{GFR (mL/min)} = 170 \times \text{Serum creatinine}^{-0.999} \times \text{Age}^{-0.176} \times \text{Gender} \times \text{Race} \times \text{BUN}^{-0.170} \times \text{Albumin}^{0.318}$$

Simplified MDRD formula

$$\text{GFR (mL/min/1,73 m}^2) = 175 \times \{[\text{serum creatinine (mg/dL]} - 1.154 \times \{\text{age (years)} - 0.203\} \times 0.742$$
if female and x 1.21 if African-American

Table 18.8 CKD-EPI equation based on plasma creatinine

Ethnicity and sex	S-creatinine (mg/dL)	Equation
African American		
Woman	≤0.7	$eGFR = 166 \times (SCr/0.7)^{-0.329} \times (0.993)^{age}$
	>0.7	$eGFR = 166 \times (SCr/0.7)^{-1.209} \times (0.993)^{age}$
Man	≤0.9	$eGFR = 163 \times (SCr/0.9)^{-0.411} \times (0.993)^{age}$
	>0.9	$eGFR = 163 \times (SCr/0.9)^{-1.209} \times (0.993)^{age}$
Caucasian or other		
Woman	≤0.7	$eGFR = 144 \times (SCr/0.7)^{-0.329} \times (0.993)^{age}$
	>0.7	$eGFR = 144 \times (SCr/0.7)^{-1.209} \times (0.993)^{age}$
Man	≤0.9	$eGFR = 141 \times (SCr/0.9)^{-0.411} \times (0.993)^{age}$
	>0.9	$eGFR = 141 \times (SCr/0.9)^{-1.209} \times (0.993)^{age}$

eGFR estimated glomerular filtration rate, *SCr* serum creatinine

has been modified in various subsequent studies, with the aim to minimize the bias between mGFR and eGFR in different ethnic groups, especially African Americans. A recent position statement recommended the utilization of the new European Kidney Function Consortium (EKFC) equation or, alternatively, the 2009 CKD-EPI equation based on serum creatinine without applying the race correction factor. Further modifications of the CKD-EPI equation, including cystatin C, are discussed below. The CKD-EPI equation is inaccurate in people over the age of 70 years; therefore, two new equations designed specifically for populations over 70 years of age were proposed in 2012: the Berlin Initiative

Table 18.9 Recommendations for laboratory evaluation of renal function

All clinical laboratories must perform the determination of plasma creatinine using methods referred to the absolute method in mass spectrometry and express the result in mg/dL.
All clinical laboratories must participate in an external quality assurance service (EQAS), including plasma creatinine.
All clinical laboratories must automatically generate the glomerular filtrate calculation based on plasma creatinine using the CKD-EPI equation and provide the result together with creatinine. This recommendation is valid for the ages between 12 and 70 years; under 14 years, it is preferable to use the Schwartz formula, whereas above the age of 70 years, it is preferable to use the EKFC equation.
The laboratory report must show the numerical value of eGFR expressed in mL/min/1.73 m^2 of the body surface.
The calculation of eGFR must be interpreted with caution in all conditions of hemodynamic and metabolic instability, such as postoperative course, follow-up organ transplantation, pregnancy (especially in the third trimester), obesity, malnutrition, diabetes, competitive sports, acute kidney injury, sepsis, septic shock, and systemic infection, and therapy with nonsteroidal anti-inflammatory drugs (NSAIDs).
The calculation of eGFR does not hamper any eventual requests for creatinine clearance (blood sampling and 24-h urine collection), especially in certain clinical conditions mentioned above.
The determination of creatinine in whole blood in POCT (decentralized bedside analysis) cannot be accompanied by the calculation of the eGFR.
Several websites allow calculating the GFR by the plasma creatinine value; the official site that everyone can access for any checks is http://www.nephron.com/MDRD_GFR.cgi

Study 1 (BIS1), which incorporates creatinine into the calculation, and the Berlin Initiative Study 2 (BIS2), which incorporates creatinine and cystatin C. Based on the evidence that the 1976 Schwartz equation overestimated the mGFR by approximately 20% (mGFR assessed by the iohexol clearance), in 2009 the original equation was changed by using a traceable, enzymatic method for plasma creatinine.

Equation for eGFR show at least two major advantages: (1) decreased influence of extrarenal factors that influence the creatinine result and (2) increased focus by the clinicians on organ function, expressed as a dynamic parameter (mL/min/1.73 m^2) rather than as a numerical value (mg/dL) of creatinine. These advantages are achievable only if the measurement of the biomarkers included in the equation (creatinine, cystatin C, or others) is accurate and standardized. Indeed, the relationship between the GFR and biomarkers, such as creatinine or cystatin C, is of the exponential type, which means that even small variations in biomarker concentration due to imprecision and bias are reflected by significant variations in the eGFR. This explains the importance of using traceable analytical methods with excellent quality specifications. Regardless of the type of equation, the calculation of estimated GFR should be interpreted with caution in all conditions of hemodynamic and metabolic instability, such as postoperative course, organ transplant follow-up, pregnancy (especially in the third trimester), obesity, malnutrition, diabetes, competitive sports activity, acute kidney injury, therapeutic treatment with nonsteroidal anti-inflammatory drugs (NSAIDs), sepsis, septic shock, and systemic infection. Some recommendations on laboratory assessment of renal function are described in Table 18.9.

Low-Molecular-Weight Plasma Proteins

Proteins of MW not exceeding 40–50 kDa, often called microglobulins, are freely filtered by the glomerulus and then almost completely reabsorbed (up to 97–99%) by the proximal convoluted tubule. Although they are reabsorbed by the tubule, these proteins do not return to the circulation because within the tubule itself they are degraded into lysosomes, where specific proteolytic enzymes break down the polypeptide chains into single amino acids. Consequently, if the catabolism of a protein of low MW is renal, its plasma concentration can be directly correlated to the amount of renal filtrate because the lower the GFR, the more it increases in the circulation. Another consequence is that the determination of the clearance of a microglobulin has no clinical or pathophysiological significance. Unfortunately, it has not been easy to identify a protein of low MW that in some way could be proposed as a biochemical marker of GFR because it should be eliminated exclusively renally and its concentration should not be influenced by extrarenal factors. Moreover, this protein should be able to be measured in the laboratory by simple, cheap, and reliable methods, easily optimizable in the clinical routine. For these reasons, some proteins, such as β2-microglobulin, lysozyme, and α1-microglobulin, have not been used as possible plasma biomarkers of GFR, whereas other proteins, such as cystatin C and, more recently, β-trace proteins, have been studied in detail.

Cystatin C

Cystatin C, also called γ-trace protein or post-γ protein because of its electrophoretic mobility, is a basic protein of MW of 15,359 Da containing 120 amino acids. It belongs to the cystatin superfamily together with about a dozen other cysteine protease inhibitors. Cystatin C is expressed by most tissues and many cells, including skin fibroblasts and macrophages, neurons, embryonic tissues, smooth muscle cells, and neuroendocrine cells. The mature form consists of a single non-glycosylated polypeptide chain with two intramolecular disulfide bridges. The only posttranslational modification demonstrated to date is the 50% hydroxylation of proline at position 3. The monomeric form secreted by cells represents the active form of the protein. At physiological pH, cystatin C is a cationic protein and has an isoelectric point (IP) of 9.3. In humans, the cystatin C gene is located on chromosome 20p11.2 and belongs to the so-called class of housekeeping genes, i.e., "conservative" genes, constantly expressed in all the cells of the body. The kidney is the main catabolic sites of cystatin C. Due to its low MW, the protein is freely filtered by the glomerulus and almost completely reabsorbed and catabolized in the cells of the proximal convoluted tubule. Under physiological conditions, it is present in the urine only in trace amounts. The plasma concentration of cystatin C depends mainly on GFR because its extrarenal elimination rate is negligible and it is not affected by extrarenal factors, such as muscle mass, diet, binding to other plasma proteins, etc. Consequently, cystatin C increases in the circulation as the GFR decreases. Additionally, it does not cross the placental barrier and, therefore, the plasma concentration at birth and in the first week of life is not affected by maternal concentration, unlike creatinine. Since 1994, various immunological methods have been developed for the routine determination of cystatin C; in particular, nephelometric methods are limited to a few platforms, whereas turbidimetric methods can be optimized on many analytical platforms. More than 20 years after the release of the first routine method, cystatin C has entered the clinical evaluation of a patient affected by renal disease. Efforts have also been made to standardize methods for cystatin C: in 2010, the International Federation for Clinical Chemistry and Laboratory Medicine (IFCC) working group made available a primary calibrator standard (ERM-DA471/IFCC); the availability of an absolute method in liquid chromatography combined with mass spectrometry now allows certification of the exact content of the protein in biological fluids and calibration materials. Despite these advances, the reduction in variability among methods resulting from standardization has been described in only a few studies but not in others, confirming the presence of controversy over whether or not the analytical problems plaguing the determination of cystatin C are resolved. In the last 15 years, all of the most important clinical studies have shown that cystatin C has a high sensitivity and a high positive predictive value for renal disease. In other words, data from the literature confirm that cystatin C is an early index of even mild reduction in GFR. Moreover, cystatin C is a better prognostic index of death and cardiovascular events than creatinine. Cystatin C has been incorporated into equations for estimating filtrate alone or in combination with creatinine, proving to make GFR estimation more accurate, especially considering the standardization of analytical methods. In particular, the CKD-EPI equation, which includes both creatinine and cystatin C in the calculation, provides better performance than either the creatinine-only CKD-EPI or the cystatin C-only CKD-EPI, although not in all clinical conditions, such as in an organ transplant patient. It is desirable that the use of cystatin C becomes part of routine laboratory practice, especially in the evaluation of patients in whom filtrate reduction is mild but not without the risk of renal disease progression. Conversely, the determination of cystatin C in patients with GFR <30 mL/min/1.73 m^2 is inappropriate because at those filtrate levels creatinine is more than sufficient to assess disease progression and estimate GFR.

Biomarkers of Tubulo-Interstitial Injury

Morpho-functional damage of the tubulo-interstitial compartment is almost always found in the early stages of renal disease, even in nephropathies characterized by prevalent glomerular involvement, and its recognition is of great importance in acute diseases such as AKI. Indeed, the high perfusion of the tubulointerstitial compartment makes this district extremely sensitive to ischemia and hypoxia, which are often the first consequence of a pathological process. In conditions of ischemia followed by rapid reperfusion, which are frequent during complex surgery and accompanied by periods of extracorporeal circulation, tubulointerstitial damage is acute and extensive and identifies the onset of AKI. Moreover, in this area, there is a considerable concentration of solutes, including any exogenous and/or endogenous toxic substances (e.g., drugs, bilirubin, etc.) capable of causing parenchymal damage. Particularly vulnerable to the toxic insult of many drugs are the cells of the proximal tubule, which, as is known, play a fundamental role in the mechanisms of reabsorption and tubular secretion. However, tubular cells have a marked tendency to "repair" the morpho-functional damage and, in the presence of remission of "noxae," such as hypoxia, ischemia, iatrogenic toxicity, etc., they demonstrate a rapid anatomic functional recovery. In some cases, such as after antibiotic therapy with aminoglycosides, the speed of glomerular filtration does not decrease significantly in the first 5–7 days after the start of drug administration and this is due to the attempt to repair the toxic damage implemented by the tubular cells. Only when necrotic phenomena prevail over regenerative phenomena, the extension of tubular lesions affects the glomerulus and its filtration function. The evaluation of tubulointerstitial damage is mainly carried out through the study of urinary and plasma biomarkers. Historically, laboratory diagnostics have been based on the assessment of the hydroelectrolytic balance through the measurement of urinary volume, urinary density (specific gravity), osmolality, and acid–base balance. The determination of urinary electrolytes and the balance with plasma electrolytes is also part of the routine evaluation of the tubulointerstitial compartment. To date, there is no single organ-specific biomarker of injury, such as troponin I and T for the myocardium. The search for biomarkers of tubular damage has focused mainly on proteins. Since the second half of the 1970s, urinary biomarkers have been proposed, including some enzymes of renal origin and some low-molecular-weight proteins. Subsequently, tubular antigens, prostaglandins, renal-derived fibronectin, tubular adenosine deaminase-binding protein, and epidermal growth factor (EGF) have been evaluated. Many low-MW proteins, including myoglobin, immunoglobulin light chains, β2-microglobulin, cystatin C, and α1-microglobulin (heterogeneous in charge -HC- protein), have demonstrated high sensitivity but low specificity because they may increase in urine due to extrarenal physiological (exercise, posture, etc.) or pathological (proliferative diseases, muscle diseases, trauma, etc.) factors. Moreover, the lack of standardization of some methods on the urinary matrix and the instability of proteins in an "open" system such as urine represent very important barriers to the introduction of these methods in routine practice.

NGAL

NGAL, also called lipocalin 2 (LCN2), is the most evaluated candidate biomarker of kidney damage in the literature. It is a protein identified in the early 1990s, consisting of 178 amino acids and with a molecular weight of approximately 25 kDa. It belongs to the lipocalin family and is expressed not only by neutrophil granulocytes and their precursors but also by many tissues, including the liver, lungs, trachea, salivary glands, prostate, uterus, stomach, and, above all, renal tubular cells. All these tissues express NGAL in minimal amounts, but stimuli such as oxidative stress, acute phase with cytokine accumulation, infections, cancer, intoxication, and other conditions leading to cell necrosis, apoptosis, and death induce NGAL overexpression. Its main biological action is bacteriostatic through iron sequestration. During the early stages of AKI, NGAL is massively synthesized by cells in Henle's loop and in the distal and collector tubules (renal NGAL pool); at the same time, AKI induces an increase in NGAL synthesis also in organs such as the liver and lungs (systemic NGAL pool) and, for this reason, in the early stages of AKI, NGAL increases both in the circulation and in the urine. Both clinical and experimental studies in animal models and in human cells in vitro have unequivocally demonstrated that, just 3–6 h after the onset of ischemia or anoxia, tubule cells express amounts of NGAL increased by approximately 10-fold from the baseline. Unfortunately, NGAL is affected by some issues that have so far limited its use in clinical practice. First, NGAL is a protein expressed by various tissues and organs, as highlighted above, and, therefore, can increase in circulation even in many extrarenal pathologies. Clinical pictures such as systemic infections and sepsis are associated with increased NGAL due to hyperactivity of neutrophil granulocytes. Moreover, "noxae" of the tubule, such as ischemia, anoxia, oxidative stress, etc., cause not only the increased synthesis of NGAL by tubular cells but also the recall of inflammatory cells and, more specifically, of neutrophil polymorphonucleates, which, in turn, express NGAL. Therefore, the amount of NGAL detectable in the tubular interstitium, and, thus, also in the urine, derives from both the tubule and the neutrophils. There are three molecular isoforms of NGAL: monomers (MW ≅ 25 kDa), dimers formed by the binding of two monomers (MW ≅ 45 kDa), and heterodimers formed by the binding of a monomer to neutrophil matrix metalloproteinase-9 or gelatinase (NGAL-MMP-9, MW ≅ 135 kDa). Monomers and dimers are expressed and released by neutrophils, whereas monomers and heterodimers are isoforms expressed and released by renal tubule cells. It follows that a heterodimer can be considered tubule-specific, and its presence in the urine or in the circulation identifies release by tubular cells, excluding the share of NGAL from neutrophils or other cell types or tissues. Since 2010, automated methods for the determination of NGAL also in urine have been developed, which allow the test to be introduced in clinical laboratories on an urgent/emergency basis. Unfortunately, these methods use mixtures of antibodies that recognize the epitopes common to the three isoforms, and this prevents the differentiation of the isoform corresponding to the heterodimer. In a restricted population of healthy adult subjects ($n = 174$), the 95th percentile urinary NGAL concentration was 107 μg/L; distinguishing the population into men ($n = 100$) and women ($n = 74$), the 95th percentile was 91 and 129 μg/L, respectively. To reduce intra- and interindividual variability, it is recommended to express NGAL in urine in relation to creatininuria.

Proteinuria

An organic nephropathy is almost always characterized by the presence of proteinuria, which, in many cases, is the earliest sign of renal impairment. Most of the proteins present in the urine are of plasma origin; however, if proteinuria originates from renal parenchymal lesions, secondary to ischemic, hypoxic, or toxic insults, proteinuria may also be represented by tissue proteins of the kidney and urinary tract. Of particular interest is the Tamm–Horsfall protein (THP), or uromodulin, a glycoprotein with immunosuppressive action secreted by the ascending tract of Henle's loop: its average urinary excretion value is around 20–200 mg/day. THP has a MW of 85 kDa and is found in the urine in polymerized form, in aggregated subunits; it has an important role in the formation of urinary cylinders and shows antiviral activities, but, due to the high content in carbohydrates, it escapes the common methods used for the measurement of total urinary proteins. Recently, the role of THP in CKD and hypertension has been reevaluated, especially as a risk factor. It has also been shown that high THP concentrations reflect good renal reserve and are inversely correlated with the risk of GFR decay and AKI. Other urinary proteins of renal origin are protein 1, or Clara cell protein (CC16; PM = 15.8 kDa), and tissue enzymes, such as urokinase, an enzyme with antifibrinolytic activity, as well as other enzymes, such as alanine aminopeptidase (AAP; E.C. 3.4.11.2) and *N*-acetyl-β-D-glucosaminidase (NAG; E.C. 3.2.1.30), whose increase in urine is of particular interest in the diagnosis of toxic nephropathy and, more generally, of tubulointerstitial nephritis. From a general point of view, proteinuria can be classified into physiological, transitory (intermittent or functional), and associated with nephropathies. There is a general agreement in defining "physiological" a content of urinary proteins in healthy adults not higher than 150–200 mg/die/1.73 m^2 of body surface (about 100–150 mg/L). Transient proteinuria is the nonconstant presence of protein in the urine. This type of proteinuria is also defined as functional, and it is essentially related to transient hemodynamic alterations (renal blood flow), which affect the reversible increase in glomerular permeability both in certain physiological situations, such as prolonged physical activity or heavy exertion, pregnancy, and intense cold, and in pathologies involving extrarenal organs and apparatuses, such as fever and hyperthermia, emotional stress, noradrenaline infusion, heart failure, and prolonged hyperlordotic posture, typically characteristic of orthostatic proteinuria. When the concentration of proteinuria stably exceed (at least three successive determinations) the threshold of physiological proteinuria, we talk about pathological or overt proteinuria. In the presence of proteinuria, it is absolutely necessary to establish its entity and composition. According to its composition, proteinuria can be classified into glomerular, tubular, mixed, and overflow proteinuria.

Glomerular proteinuria consists mainly of plasma proteins of PM >60 kDa. The term "selective proteinuria" indicates a glomerular proteinuria consisting almost exclusively of proteins with a MW between 60 and 150 kDa. The selectivity of glomerular proteinuria positively correlates with the ability of the glomerular filter to retain proteins with high and very high MWs (>150 kDa), distinguishing them from those with lower MWs. "Non-nephrotic glomerular proteinuria" refers to the excretion that does not exceed 3.5 g/day; above this limit, we talk about "nephrotic proteinuria," which is usually accompanied by the clinical and humoral manifestations of the nephrotic syndrome: edemas, hypoalbuminemia, and increased globulins with the simultaneous presence of hypogammaglobulinemia. In these cases, protein loss can be marked (5–10 g/day), even reaching massive levels (more than 40 g/day). Tubular proteinuria consists of plasma proteins of low MW or microglobulins (MW ≤50 kDa). Mixed proteinuria results from the presence of high- and low-MW plasma proteins and proteins and enzymes of renal origin. It is predominantly found in conditions of impaired glomerular and tubular function. Finally, proteinuria due to overload, also called extrarenal proteinuria, is constituted by low-MW plasma proteins that are not normally present in the circulation or at most are present in low concentrations. In this case, the presence of extrarenal pathologies characterized by specific pathological mechanisms affecting organs and tissues (proliferation, cytolysis, etc.) causes an abnormal increase of proteins in the plasma and in the ultrafiltrate; this increase quickly saturates the tubular reabsorption mechanisms, causing their appearance in the urine. Typical examples include myoglobin, and immunoglobulin free light chains.

The approach to the study of proteinuria is developed through three phases: research, measurement, and characterization of the composition of proteinuria. The presence of proteinuria must be ascertained in all cases of suspected nephropathy or, in the absence of signs and/or symptoms, in all cases of hematuria, urinary tract infection, etc. If the presence of proteinuria is ascertained, it should always be measured to ascertain its extent. Finally, the characterization of the composition of proteinuria is an important evaluation criterion for the diagnosis and clinical classification of renal pathology. For these reasons, the methods can be schematically divided into qualitative and quantitative methods. Qualitative methods, such as dry chemistry methods (dipstick, see the section "Standard Urine Test") and electrophoretic separative methods (solid support or liquid phase, namely capillary), allow the detection and characterization of proteinuria. Separative electrophoretic methods are the only ones able to highlight the morphology of the free light chains, allowing the diagnosis of Bence Jones proteinuria in the case of monoclonality. Quantitative methods allow determining the concentration of total proteins and of the main urinary specific proteins. In particular, the determination of total proteins in urine has several critical issues that have not yet been resolved, mainly due to the poor standardization of methods and numerous analytical interferences. The direct colorimetric methods constantly show, even if in a pronounced form, a limitation given by the different affinity of the dye for the different protein structures. It follows that in a urine sample, the prevalence of a certain type of protein structure rather than another (globulins rather than albumin, or vice versa, etc.) inevitably leads to overestimates or underestimates of the real protein content, with poor accuracy of the determination. To overcome this serious problem, the use of sodium dodecyl sulfate (SDS) was introduced, to be added in a defined amount to the reaction mixture containing the dye. The use of SDS increases the linearity range of the colorimetric method. To date, there is no definitive consensus on the choice of reference method for the determination of total urinary proteins: one of the few recommended methods for urinary proteins is the biuret colorimetric method after gel filtration, identified as the selected method. However, other recommended methods are affected, to a greater or lesser extent, by pitfalls related to interferences, protein pattern composition, etc. However, it is desirable that a good method can achieve wide ranges of linearity, allowing accurate measurements of protein concentrations between about 0.025 and 4.0 g/L.

Albuminuria

Plasma albumin is normally filtered in very low amount by the glomerulus, and then more than 99% of it is reabsorbed by the cells of the proximal convoluted tubule. The term albuminuria refers to the loss of albumin by urine, regardless of the extent of the loss. Under stable physiological conditions and at rest, albuminuria is present only in trace amounts, i.e., in concentrations ≤10 mg/L. The most important physiological factors that can cause significant changes in albuminuria are sex, age, body mass index (BMI), and a high-protein diet. Other factors that cause albuminuria, independently of kidney function, are physical exercise over the 24 h prior to urine collection, fever, congestive heart failure, marked hyperglycemia, urinary tract infections, severe hypertension, menstrual period, and certain medications. Albuminuria is a prognostic index of progression of diabetic nephropathy in both type 1 and type 2 diabetes, according to the scheme shown in Table 18.10, broken down by various modes of expression of the outcome. In 80% of patients with type 1 diabetes and albuminuria, urinary albumin excretion increases at an annual rate of 10–20%, with development of clinical proteinuria within 10–15 years. After the onset of clinical proteinuria, most of these patients (>80%) manifest reductions in GFR until the development, in a relatively short time, of ESKD. 20–40% of patients with type 2 diabetes and albuminuria evolve to clinical proteinuria, but, in the 20 years following the onset of clinical proteinuria, only 20% of these patients evolve to ESKD. Finally, in patients with type 1 or 2 diabetes and albuminuria between 30 and 300 mg/g creatinine, the risk of cardiovascular disease is significantly increased: meta-analysis studies have shown that in type 2 diabetes, albuminuria doubles cardiovascular morbidity and

Table 18.10 Levels of excretion of albuminuria referred to as impromptu and timed collection

Category		Extemporaneous collection (sample spot)		Timed collection	
		mg/mmol (creatinine)	µg/mg (creatinine)	mg/24 h	µg/min
Normoalbuminuria	Men	<2.5	<25	<30	<20
	Women	<3.5	<35		
Microalbuminuria	Men	2.5–30	<25–299	30–299	20–199
	Women	3.5–30	<35–299		
Clinical albuminuria (overt nephropathy)		≥30	≥300	≥300	≥200

Modified from Karalliedde and Viberti (2004)

mortality (OR 2; 95% CI 1.4–2.7) and more than doubles all other causes of death (OR 2.4; 95% CI 95%). According to the new recommendations of the American Diabetes Association (ADA) on the standards of care for diabetic patients, a significant albuminuria value must be found in at least two out of three samples over 3–6 months before a patient can be considered albuminuric. The threshold values defined by the scheme reported in Table 18.10 are very useful for defining the onset and the extent of progression of diabetic nephropathy; regarding CKD, it is defined by a constant value of albuminuria >30 mg/24 h or >30 mg/g of urinary creatinine in the extemporaneous sample, for more than 3 months. The choice of this threshold value is based, among other things, on the evidence that 30 mg corresponds to a value three times higher than the average value found in a healthy young adult (10 mg). Recent studies have shown that, in healthy adult populations, changes in urinary albumin excretion expressed in relation to urinary creatinine (albumin-to--creatinine ratio (ACR)) are significantly associated with the risk of ESKD and death. However, nowadays, albuminuria has assumed a more extensive clinical significance as a marker of cardiac and extracardiac organ damage and as an independent cardiovascular risk factor, and values <30 mg/24 h or <30 mg/g urinary creatinine should be clinically evaluated. Albuminuria is a marker of endothelial damage: factors such as hyperglycemia and hypertension, as mentioned above, and dyslipidemia induce an increase in endothelial permeability, both in the renal and systemic compartments, mediated by different metabolic pathways, with passage of molecules from the vascular to the interstitial compartment, including albumin. This passage triggers the acute phase, and the consequent recall of immunocompetent cells and accumulation of cytokines and acute-phase proteins, and this process maintains and increases endothelial damage. The prevalence of albuminuria in essential hypertension varies between 10% and 30% and increases with age and disease duration. Albuminuria is associated with signs of extracardiac vascular damage, such as increased carotid intima-media thickness, an expression of atherosclerosis, and increased pulse wave propagation velocity, an expression of arterial stiffness. In addition, albuminuria is associated with left ventricular hypertrophy and increased intrarenal vascular resistance. For these reasons, concentrations below 30 mg/day are significant, since they define the risk, which is a continuous variable and, therefore, changes considerably, for example, between an albuminuria value of 15 mg and one of 25 mg. As described above, in a CKD patient, albuminuria and eGFR define the cumulative risk of adverse outcome and progression to ESKD, as reported in Table 18.4. The determination of albuminuria requires great attention from the clinical laboratory, which is called to choose analytical methods with excellent quality specifications, especially those able to guarantee reliable measurements even in the range of concentrations below 30 mg. The choice of the type of sample is very important and is closely related to the expression of the results. Spot urine sample allows expressing albuminuria as albumin-to-creatinine ratio (ACR) (mg/g or mg/mmol of urinary creatinine); the timed collection (albumin excretion rate (AER)) allows expressing albuminuria as a share of daily excretion or in any case referred to as a time frame (mg for 24, 12, 6, or 3 h). There is a wide consensus on the recommendation to avoid expressing the results in relation to the urinary volume (mg/L) because too many variables contribute in this case to make the results not very reproducible. The "normalization" of results for urinary creatinine decreases the variability of albuminuria. In a 1989 study, it was shown that the intraindividual biological variability (coefficients of variation (CVs)) is lower in the first morning sample (36%) and further decreases (31%) when the result is expressed as an albuminuria/creatinuria ratio. In a more recent study, these results have been confirmed: the total variability of albuminuria is about 26% on the second morning urination when related to volume (mg/L) and drops to about 13% when related to urinary creatinine (mg/g creatinine). Several studies have confirmed that the best accuracy and reproducibility in the measurement of albuminuria is achieved by using the spot urine sample and expressing the result in relation to urinary creatinine, and, therefore, it is recommended that all laboratories follow this recommendation. Some recommendations for the determination and reporting of albuminuria are summarized in Table 18.11. However, factors related to reduced muscle mass, such as older age, female sex, and low body weight, result in reduced urinary excretion of creatinine,

Table 18.11 Recommendations for the laboratory evaluation of albuminuria

The use of the terms "microalbuminuria" and "microalbumin" is not recommended because they are misleading. The use of the term "albuminuria" is recommended to indicate the presence of albumin in urine, regardless of concentration.
For screening and outpatient monitoring, the sample of choice is the midstream urine collected upon awakening or, alternatively, the midstream urine from a sample spot or random.
The determination of albuminuria on timed samples (24 h, 12 h, etc.) is foreseen in particular cases (follow-up kidney transplant, nephrotic syndrome, highly developed muscle mass as in competitive athletes, malnutrition, etc.); in these cases, the result refers to diuresis in 24 h (die).
Sample collection must be performed in a sterile container. Since albumin is relatively stable in urine at pH >5.0, the sample can be stored for approximately 8 h after urination at 6–20 °C.
Samples collected during a feverish episode or after even modest physical exertion (walking, etc.) or in the presence of bacteriuria give rise to poorly reliable results from the clinical point of view and must be confirmed in subsequent checks.
To reduce the variability of the result, it is necessary to "normalize" the albuminuria concentration values (mg/L) in relation to the creatininuria concentration and report the albuminuria/creatininuria ratio expressed in mg/g creatinine (ACR) in the laboratory report. Therefore, each request for albuminuria involves the determination of two parameters (albumin and creatinine) and the calculation of the ratio.
The method of choice for the determination of albuminuria must be endowed with high sensitivity and good analytical performance even for ranges of concentration < 10 mg/g of creatinuria.
The diagnosis of albuminuria requires at least two determinations of albuminuria over a period of at least 2–3 months.

which, in turn, affects the ACR calculation at least partially, thus resulting in falsely increased results. Various remedies have been proposed to overcome this problem, including the development of equations to estimate urinary creatinine and their use as a correction factor to be multiplied by ACR. Finally, about the time of sample collection, it has been found that ACR measured in the extemporaneous random sample leads to an overestimation of the prevalence of albuminuria compared with ACR measured in the first morning urination. The clinical importance of albuminuria determination requires an effort to standardize analytical methods commonly used in routine practice. Although a candidate method has been developed as a reference method, all the elements of a "reference system" to standardize the determination of albuminuria, such as a reference material with a "target" value assigned by a reference method, are not yet available. Analytical issues related to the determination of albuminuria involve conformational differences of the molecule, which include the amount and types of bound substances and the degree of glycation. In urine, in addition, factors such as pH, ionic strength, and high concentrations of urea, glucose, and ascorbic acid contribute to albumin changes. Most of the routine methods for the determination of albuminuria are represented by immunoturbidimetric ones, but a further factor of variability among methods is just the type or mixture of antibodies used (polyclonal, monoclonal, mixed). This variability makes the methods nonhomogeneous, assigning different specifications and performance to each. Based on the data of biological variability, the analytical goal of inaccuracy, expressed as a coefficient of variation (CV, in percentage), should be <15%, a result confirmed by the external quality control schemes managed by various international bodies. There is also sufficient evidence to define stable albumin in urine samples up to 7 days between +2 and +8 °C, even if it is preferable to measure it on fresh samples or stored at −80 °C. Ultimately, the clinical importance of microalbuminuria requires the utmost commitment of the clinical laboratory in providing precise and accurate results, in the choice of sample type and mode of expression of results, and in the constant monitoring of analytical quality. These requirements are essential not only in the screening but also in the follow-up of patients with CKD, cardiovascular diseases, diabetes, and patients under therapeutic treatment, to improve the analytical reliability of a test so important in the definition of cardiovascular risk and mortality.

Recommended Readings

Abdelmalek JA, Gansevoort RT, Lambers Heerspink HJ et al (2014) Estimated albumin excretion rate versus urine albumin-creatinine ratio for the assessment of albuminuria: a diagnostic test study from the Prevention of Renal and Vascular Endstage Disease (PREVEND) Study. Am J Kidney Dis 63:415–421

American Diabetes Association (2017) Standards of medical care in diabetes – 2017. Diabetes Care 40(Suppl 1):s1–135

Bachmann LM, Nilsson G, Bruns DE, et al (2014) State of the art for measurement of urine albumin: comparison of routine measurement procedures to isotope dilution tandem mass spectrometry. Clin Chem. 60:471–80

Bellomo R, Ronco C, Kellum JA, Acute Dialysis Quality Initiative Workgroup et al (2004) Acute renal failure – definition, outcome measures, animal models, fluid therapy and information technology needs: the Second International Consensus Conference of the Acute Dialysis Quality Initiative (ADQI) Group. Crit Care 8:R204–R212

Bellomo R, Kellum JA, Ronco C (2012) Acute kidney injury. Lancet 380(9843):756–766

Burtis CA, Ashwood ER, Bruns DE (eds) (2012) Tietz textbook of clinical chemistry and molecular diagnostics. Elsevier Saunders, St. Louis

Cai L, Rubin J, Han W, Venge P, Xu S (2010) The origin of multiple molecular forms in urine of HNL/NGAL. Clin J Am Soc Nephrol 5:2229–2235

Carrero JJ, Grams ME, Sang Y et al (2017) Albuminuria changes are associated with subsequent risk of end-stage renal disease and mortality. Kidney Int 91:244–251

Carter CE, Gansevoort RT, Scheven L et al (2012) Influence of urine creatinine on the relationship between the albumin-to-creatinine ratio and cardiovascular events. Clin J Am Soc Nephrol 7:595–603

Ceriotti F, Boyd JC, Klein G, IFCC Committee on Reference Intervals and Decision Limits (C-RIDL) et al (2008) Reference intervals for serum creatinine concentrations: assessment of available data for global application. Clin Chem 54:559–566

Cowland JB, Borregaard N (1997) Molecular characterization and pattern of tissue expression of the gene for neutrophil gelatinase-associated lipocalin from humans. Genomics 45:17–23

Delanaye P, Schaeffner E, Cozzolino M, et al (2022) The new, race-free, Chronic Kidney Disease Epidemiology Consortium (CKD-EPI) equation to estimate glomerular filtration rate: is it applicable in Europe? A position statement by the European Federation of Clinical Chemistry and Laboratory Medicine (EFLM). Clin Chem Lab Med. 61:44–47

Devuyst O, Bochud M (2015) Uromodulin, kidney function, cardiovascular disease, and mortality. Kidney Int 88:944–946

Dodder NG, Tai SS, Sniegoski LT et al (2007) Certification of creatinine in a human serum reference material by GC-MS and LC-MS. Clin Chem 53:1694–1699

Earley A, Miskulin D, Lamb EJ et al (2012) Estimating equations for glomerular filtration rate in the era of creatinine standardization: a systematic review. Ann Intern Med 156:785–795

Ebert N, Delanaye P, Shlipak M et al (2016) Cystatin C standardization decreases assay variation and improves assessment of glomerular filtration rate. Clin Chim Acta 456:115–121

Eckardt KU, Berns JS, Rocco MV, Kasiske BL (2009) Definition and classification of CKD: the debate should be about patient prognosis – a position statement from KDOQI and KDIGO. Am J Kidney Dis 53:915–920

Ferguson TW, Komenda P, Tangri N (2015) Cystatin C as a biomarker for estimating glomerular filtration rate. Curr Opin Nephrol Hypertens 24:295–300

Gansevoort RT, Matsushita K, van der Velde M, Chronic Kidney Disease Prognosis Consortium et al (2011) Lower estimated GFR and higher albuminuria are associated with adverse kidney outcomes. A collaborative meta-analysis of general and high-risk population cohorts. Kidney Int 80:93–104

Gentile G, Remuzzi G (2016) Novel biomarkers for renal diseases? None for the moment (but one). J Biomol Screen 21:655–670

Grams ME, Coresh J (2013) Assessing risk in chronic kidney disease: a methodological review. Nat Rev Nephrol 9:18–25

Greenberg N, Roberts WL, Bachmann LM et al (2012) Specificity characteristics of 7 commercial creatinine measurement procedures by enzymatic and Jaffe method principles. Clin Chem 58:391–401

Howey JE, Browning MC, Fraser CG (1989) Biologic variation of urinary albumin: consequences for analysis, specimen collection, interpretation of results, and screening programs. Am J Kidney Dis 13:35–37

Inker LA, Eneanya ND, Coresh J, et al (2021) New Creatinine- and Cystatin C-Based Equations to Estimate GFR without Race. N Engl J Med. 385:1737–49

Ix JH, Wassel CL, Stevens LA et al (2011) Equations to estimate creatinine excretion rate: the CKD Epidemiology Collaboration. Clin J Am Soc Nephrol 6:184–191

Karalliedde J, Viberti G (2004) Microalbuminuria and cardiovascular risk. Am J Hypertens 17:986–993

Kidney Disease: Improving Global Outcomes (KDIGO) Acute Kidney Injury Work Group (2012) KDIGO Clinical practice guideline for acute kidney injury. Kidney Int Suppl 2:1–138

Kidney Disease: Improving Global Outcomes (KDIGO) CKD Work Group (2013) KDIGO clinical practice guideline for the evaluation and management of chronic kidney disease. Kidney Int Suppl 3:1–150

Killeen AA, Horowitz GL (2022) New Equations for Estimating Glomerular Filtration Rate. Clin Chem. 68:491-493 Inker LA, Titan S (2021) Measurement and Estimation of GFR for Use in Clinical Practice: Core Curriculum 2021. Am J Kidney Dis. 78:736–749

Kilbride HS, Stevens PE, Eaglestone G et al (2013) Accuracy of the MDRD (Modification of Diet in Renal Disease) Study and CKD-EPI (CKD Epidemiology Collaboration) equations for estimation of GFR in the elderly. Am J Kidney Dis 61:57–66

Lambers Heerspink HJ, Gansevoort RT (2015) Albuminuria is an appropriate therapeutic target in patients with CKD: the pro view. Clin J Am Soc Nephrol 10:1079–1088

Levey AS, Greene T, Kusek J, Beck G (2000) A simplified equation to predict glomerular filtration rate from serum creatinine [abstract A0828]. J Am Soc Nephrol 11:155A

Levey AS, Coresh J, Balk E, National Kidney Foundation et al (2003) National Kidney Foundation practice guidelines for chronic kidney disease: evaluation, classification, and stratification. Ann Intern Med 139:137–147

Levey AS, Eckardt KU, Tsukamoto Y et al (2005) Definition and classification of chronic kidney disease: a position statement from Kidney Disease: Improving Global Outcomes (KDIGO). Kidney Int 67:2089–2100

Levey AS, Stevens LA, Schmid CH, CKD-EPI (Chronic Kidney Disease Epidemiology Collaboration) et al (2009) A new equation to estimate glomerular filtration rate. Ann Intern Med 150:604–612

Levey AS, Inker LA, Coresh J (2014) GFR estimation: from physiology to public health. Am J Kidney Dis 63:820–834

Levin A, Kellum JA, Mehta RL (2008) Acute kidney injury: toward an integrated understanding through development of a research agenda. Acute Kidney Injury Network (AKIN). Clin J Am Soc Nephrol 3:862–863

Martin H (2011) Laboratory measurement of urine albumin and urine total protein in screening for proteinuria in chronic kidney disease. Clin Biochem Rev 32:97–102

Miller WG, Bruns DE, Hortin GL, National Kidney Disease Education Program-IFCC Working Group on Standardization of Albumin in Urine et al (2009) Current issues in measurement and reporting of urinary albumin excretion. Clin Chem 55:24–38

Mishra J, Dent C, Tarabishi R et al (2005) Neutrophil gelatinase-associated lipocalin (NGAL) as a biomarker for acute renal injury after cardiac surgery. Lancet 365(9466):1231–1238

Mussap M, Plebani M (2004) Biochemistry and clinical role of human cystatin C. Crit Rev Clin Lab Sci 41:467–550

Mussap M, Ruzzante N, Varagnolo M, Plebani M (1998) Quantitative automated particle-enhanced immunonephelometric assay for the routinary measurement of human cystatin C. Clin Chem Lab Med 36:859–865

Mussap M, Noto A, Fravega M, Fanos V (2011) Soluble CD14 subtype presepsin (sCD14-ST) and lipopolysaccharide binding protein (LBP) in neonatal sepsis: new clinical and analytical perspectives for two old biomarkers. J Matern Fetal Neonatal Med 24(Suppl.2):12–14

Mussap M, Graziani MS, Caldini A et al (2014) Documento di consenso SIBioC e Società Italiana di Radiologia Medica (SIRM) sulla richiesta di esami di laboratorio per la valutazione del danno renale da mezzi di contrasto. Biochim Clin 38:140–142

Myers GL (2008) Standardization of serum creatinine measurement: theory and practice. Scand J Clin Lab Invest 241(Suppl):57–63

National Kidney Foundation (2002) K/DOQ1 Clinical practice guidelines for chronic kidney disease: evaluation, stratification and classification. Am J Kidney Dis 39:s1–s266

Panteghini M (2008) Enzymatic assays for creatinine: time for action. Clin Chem Lab Med 46:567–572

Rampoldi L, Scolari F, Amoroso A et al (2011) The redisco-very of uromodulin (Tamm-Horsfall protein): from tubulointerstitial nephropathy to chronic kidney disease. Kidney Int 80:338–347

Schwartz GJ, Brlon LP, Spitzer A (1987) The use of plasma creatinine concentration for estimating glomerular filtration rate in infants, children and adolescents. Pediatr Clin N Am 34:571–590

Stevens PE, Levin A, Kidney Disease: Improving Global Outcomes Chronic Kidney Disease Guideline Development Work Group Members (2013) Evaluation and management of chronic kidney disease: synopsis of the kidney disease: improving global outcomes 2012 clinical practice guideline. Ann Intern Med 158:825–830

19 Physical, Chemical, and Morphological Urine Examination

Fabio Manoni

Introduction

The physical, chemical, and morphological urine examination (ECMU) is a widely used test because of the wealth of information it can provide, the ease of sample collection, the possibility of performing it in any laboratory in a practical, accurate, and safe way, and the advantageous cost-effectiveness ratio. ECMU is a profile of tests aimed at evaluating different aspects of the urinary system, including indicators of injury (proteins, erythrocytes) and indicators of function (urinary concentration expressed as relative density, conductivity, or osmolality). Since ECMU, together with creatinine and estimated glomerular filtration rate (eGFR), is the first approach to diagnose lesions and/or dysfunction of the kidney and urinary system, it is particularly important that it is appropriately performed. Specifically, it can be divided into three phases: preanalytical, analytical, and postanalytical.

Urine is a fluid with very labile physical, chemical, and corpuscular characteristics. Thus, a sample collected or stored inappropriately should be eliminated. The attending physician should instruct the patient with the indications on how to collect and deliver the urine sample to the clinical laboratory.

The ECMU typically includes some or all the following investigations:

- Visual inspection: Color and appearance
- Physical analysis: Concentration (relative density/conductivity/osmolality)
- Chemical analysis: Protein, albumin, creatinine, hemoglobin, esterase, nitrite, pH, ascorbic acid, glucose, ketones, bilirubin, urobilinogen
- Counting and morphology of the corpuscular component: On automatic analyzers and/or by microscopy for erythrocytes, leukocytes, epithelial cells, cylinders, crystals, lipids, bacteria, mycetes, protozoa, parasites, contaminants, atypical cells.

Each laboratory defines the procedures to use and, in agreement with the clinicians, the test to perform considering different aspects, including guidelines, published studies, the prevalence of diseases in the population, and the type of patients studied. Indeed, the pretest probability in relation to the prevalence of renal or urological diseases should condition the use of highly specific procedures or, on the contrary, of highly sensitive tests. Specifically, if mainly nephrological/urological patients are evaluated, high-specificity methods should be preferred, while in mainly normal subjects, for example, in sports medicine, highly sensitive methods should be preferred.

Preanalytical Phase

The request for ECMU should be made based on a clinical question. However, ECMU is often requested to exclude the presence of pathology or as part of an overall holistic approach to frame a patient. The request for chemical and morphological examination of urine finds its rationale in the following conditions:

- Suspected urinary tract infection
- Suspicion or follow-up of kidney disease
- Suspicion or follow-up of noninfectious urinary tract disease, either primary or secondary to systemic diseases, such as rheumatic disease, hypertension, toxemia of pregnancy, or drug side effects
- Recurrent formation of urinary stones

The patient should be instructed to avoid strenuous exertion prior to urine specimen collection, as well as to avoid dietary overload. Women of childbearing age should avoid collection during the menstrual period. Sexual relations

F. Manoni (✉)
Department of Diagnostics and Care Service, Ospedali Riuniti Padova Sud "Madre Teresa di Calcutta", Padova, Italy
e-mail: fabio.manoni@aulss6.veneto.it

M. Ciaccio (ed.), *Clinical and Laboratory Medicine Textbook*, https://doi.org/10.1007/978-3-031-24958-7_19

should be avoided for at least 12 hours prior to collection. Urine obtained from the first urination in the morning should normally be collected on an empty stomach and before undertaking any physical activity.

Sample Collection

Urine from the first-morning urination or, alternatively, a urine specimen that has been in the bladder for at least 4 hours under resting conditions is recommended. The patient should be properly informed about how to prepare for the collection of a urine specimen. Specifically, the patient should be advised to: wash hands; precede urine collection by thoroughly cleansing the genitals using a nonantiseptic soap and plenty of water for rinsing, and then dry; discard the first urination, collecting the intermediate urination in a wide-mouthed, sterile container, avoiding contact between container and genitals; collect at least 15–30 mL of urine; close the container hermetically and send it to the laboratory as soon as possible; always report any pathologies or major urological interventions; in the presence of urine with abnormal pigments, report the intake of particular drugs or foods. The sample collected from intermediate urination is the most suitable for microbiological examination and for chemical-physical and morphological evaluation. Indeed, it is only minimally influenced by the presence of urethral secretions and mucus. The latter represents an important disturbing element in the chemical-morphological evaluation of urine as it can induce false positivity in the evaluation of proteins and, on automatic instrumentation, of cylindruria.

Sample Storage

The ECMU should be performed as soon as possible. If it is anticipated that the urine sample cannot be examined within 4 hours, refrigerated storage at +4/8 °C is recommended. Alternatively, and only if the alterations they induce are known, preserving agents may be used limited to the examination of certain parameters. The borate has a bacteriostatic action (it prevents bacterial replication during storage); higher alcohols are instead able to preserve the morphology of corpuscular elements of urine to allow the microscopic examination even after a few hours from sample collection.

Arrival at the Laboratory

The laboratory must always express an opinion on the suitability and acceptability of the sample received; depending on the type of nonconformity (NC) found, only analyses not affected by the NC may be carried out, or a new sample must be requested. In both cases, the reasons for this choice must be adequately explained in a note in the report.

Analytical Phase

Freshly emitted urine usually has the following characteristics:

- Relative density between 1007 and 1035 (individuals who have recently received iodinated contrast media may have a urinary relative density greater than 1035)
- pH between 4.5 and 7.5
- Temperature between 32.5 and 37.5 °C
- Creatinine, with a concentration 50–100 times higher than in other body fluids.

The concentration of urea, Na^+, K^+, and Cl^- in urine is significantly higher than in other body fluids. Negligible amounts of glucose and protein are usually found in the urine of healthy subjects, whereas they are present in high concentrations in plasma, amniotic fluid, and exudates (but not necessarily in exudates).

Visual Inspection

Volume

The volume of urine normally produced in 24 hours is between 600 and 1500 mL. This is important in the case of timed collections and in the calculation of clearance.

Color

Urine usually has its own color, which is more or less markedly yellow according to the different concentrations of urochromes, but it can assume different colors in case of systemic, renal, or urological diseases: dark red or cola-colored in case of hemoglobinuria, myoglobinuria, porphyria; brown in case of jaundice and alkaptonuria; blue in blue diaper syndrome due to the presence of indole in urine; flesh-colored in case of macroscopic hematuria. Different colors related to the intake of foods containing pigments and of drugs are not pathologically relevant (Table 19.1). In ECMU, the color is always evaluated, but it is expressed in the report only in case of abnormal coloration, with the obligation to comment.

Turbidity

Normal urine appears clear. Various degrees of turbidity are related to an increase in suspended corpuscles. This feature may not be stated in the report because the corpuscular elements that originate it are the subject of a specific evaluation.

Table 19.1 Main changes in urine color and their possible causes

Color	Pathology	Medicines and foods
Red	Hematuria, hemoglobinuria, myoglobinuria, porphyrinuria	Cascara, desferroxamine, doxorubicin, epirubicin, phenothiazines, phenytoin, ibuprofen, levodopa, rifampicin, senna (alkaline urine), sulfamethoxazole beets, blackberries, rhubarb
Orange	Dehydration	Fluorescin, rifampicin, sulfasalanzina (alkaline urine), warfarin chilli, rhubarb
Green-blue	Jaundice, *Pseudomonas* urinary infections, *blue diaper syndrome*	Amitriptyline, methylene blue, indomethacin, triamterene
Brown	Alkaptonuria, tyrosinosis Porphyrinuria, jaundice	Cascara, quinine, phenothiazines, phenytoin, iron, levodopa, metronidazole, methyldopa, nitrofurantoin, senna (alkaline urine)
Black	*Black water fever* (hemoglobinuric fever in course of some infectious diseases such as malaria, dengue, acute HBV+HDV coinfection) Malignant melanoma	Cascara, quinine, iron, methyldopa
Violet	Porphyrinuria Catheter syndrome bladder (*purple urine bag syndrome*)	Seine

Modified from Manoni et al. (2016)

Foam

The presence of abundant whitish foam is often linked to the presence of protein. Also, in this case, it is not necessary to evaluate and report this parameter because the protein constituents, which are the relevant element from a diagnostic point of view, are analyzed and reported.

Chemical-Physical Examination

Usually, in clinical laboratories, ECMU is performed using multipad test strips (dipsticks), whose sensitivity and specificity have been evaluated by several authors. Table 19.2 summarizes the characteristics of the main products on the market. However, dipsticks, if not stored correctly (humidity), can give false-positive results (Table 19.3).

Although to date no commercial methods are available in kits applicable on clinical chemistry analyzers for evaluation of the entire ECMU profile, an "upgrade" can be made by providing: (1) the reporting of only the clinically useful parameters (Table 19.4); (2) the liquid chemistry measurement of urinary proteins with more sensitive and specific methods (in medium-large laboratories and/or territorial reference laboratories); (3) the on-demand analysis by traditional chemistry methods of parameters useful only in particular clinical conditions (glucose, ketones, etc.).

Albumin/Protein

When proteinuria exceeds physiological limits, it almost always indicates the presence of impaired kidney function and integrity or systemic disease. The Kidney Disease: Improving Global Outcomes (KDIGO) guidelines state that the presence of protein in the urine is a negative prognostic factor for the development of chronic renal failure and for cardiovascular risk even at concentrations considered "physiological" (between 100–300 mg/L of albumin; or, using the albumin/creatinine ratio, between 10 and 30 mg/mmol), also when the eGFR is still normal. Consequently, early detection of albuminuria (and/or proteinuria) with sensitive and accurate methods can be the most effective weapon for an early diagnosis of renal disease and for cardiovascular prevention.

Classically, it is possible to distinguish three types of proteinuria: prerenal, renal, postrenal. In the prerenal form, we distinguish the functional form and the overload form. In the functional form, proteinuria, usually modest (always <1.5 g/day), appears after the subject has maintained an upright posture for some time (orthostatic proteinuria) or during febrile states, after physical activity, acute heart failure, while it is absent in the first morning urine. The overload form is due to an increase in ultrafiltrated plasma proteins. This increased load of low-molecular-weight proteins can occur, for example, in renal pathologies due to monoclonal gammopathies associated with light chain elimination, in severe septicemia, in acute (hemoglobin) or chronic (hemosiderin) hemolysis, in massive muscle traumas (myoglobin). Proteinuria of renal origin can be distinguished into tubular and glomerular. Glomerular proteinuria is the most severe and common form of proteinuria, often marked (>3.5 g/day, so-called proteinuria in the nephrotic range), which can be associated with hematuria in the case of glomerulonephritis or with lipiduria together with hypoalbuminemia and hyperlipemia in nephrotic syndrome. Usually, the protein most represented in the urine is albumin, but as the disease evolves, other protein species of higher molecular weight may appear. The presence of modest amounts of albumin in the urine (erroneously called microalbuminuria) has assumed an important prognostic value in some common diseases, such as diabetes and hypertension. Based on this, KDIGO considers albuminuria a marker of chronic kidney disease (CKD), if reconfirmed after at least 3 months.

Tubular proteinuria occurs when the normal tubular function of protein reabsorption fails; it is typically relatively modest proteinuria (<1.5 g/day), characterized by low-molecular-weight proteins (<35 kDa), such as lysozyme, retinol-binding globulin, α1-microglobulin. Mixed glomerular

Table 19.2 Comparative analytical sensitivity evaluation of some commercially available test strips for albumin, glucose, red blood cells/hemoglobin, esterases

	Albumin	Glucose	RBCs/hemoglobin	Esterase
AimStick	15 mg/dL	50 mg/dL	5 RBC/μL 0.3 mg/dL	5 WBC/μL
AutionSticks	15 mg/dL	50 mg/dL	20 RBC/μL 0.06 mg / dL	5 WBC/μL
Chemistrip	6 mg/dL	40 mg/dL	5 RBC/μL –	20 WBC/μL
CombiScreen plus	15 mg/dL	40 mg/dL	5 RBC/μL –	10 WBC/μL
DiaScreen	5 mg/dL	50 mg/dL	5 RBC/μL 0.02 mg/dL	20 WBC/μL
Dirui H Series	15 mg/dL	50 mg/dL	5 RBC/μL –	5 WBC/μL
MediTest C9	30 mg/dL	50 mg/dL	10 RBC/μL –	–
Mission	18 mg/dL	25 mg/dL	– 0.018 mg/dL	9 WBC/μL
Multistix	15 mg/dL	75 mg/dL	5 RBC/μL 0.015 mg/dL	5 WBC/μL
Sismex Meditape UC-11A	10 mg/dL	50 mg/dL	10 RBC/μL 0.03	25 WBC/μL
Uriflet S2	5 mg/dL	10 mg/dL	10 RBC/μL 0.03 mg/dL	20 WBC/μL
Uriscan	10 mg/dL	50 mg/dL	5 RBC/μL 0.015 mg/dL	2 WBC/μL
Uritest 13G	10 mg/dL	40 mg/dL	– 0.3 mg/dL	15 WBC/μL
Uro-Dip 10C	–	100 mg/dL	– 0.05 mg/dL	–
Uropaper alfa 3-9L	15 mg/dL	50 mg/dL	10 RBC/μL 0.3 mg/dL	25 WBC/μL
URS	15 mg/dL	100 mg/dL	5 RBC/μL 0.3 mg/dL	10 WBC/μL
vChem	15 mg/dL	45 mg/dL	5 RBC/μL 0.2 mg/dL	20 WBC/μL

Modified from Manoni et al. (2016)
RBC Red Blood Cells, *WBC* White Blood Cells, leukocytes

Table 19.3 Main interferers with dipstick

Parameter	Specificity/interferences
Relative density	Only ionic solutes Interference in reduction: Alkaline pH, glucose, and urea >1 g/L Increasing interference: Protein 500 mg/dL, ketoacidosis
pH	Interference in reduction: Formaldehyde
Blood/hemoglobin	*False positives:* Bacterial peroxidase, oxidizing agents, hydrochloric acid *False negatives:* Ascorbate, high relative density, reducing agents, formalin, nitrites, drugs
Leukocyte esterase	Only in granulocytes *False positives:* Hyperchromic urine, formalin, drugs, sodium azide detergents *False negatives:* Ascorbate, borate, glucose >3 g/dL, gg proteins >0.5 g/dL High relative density, oxidizing agents, soaps and detergents, drugs
Nitrites	*False positives:* Hyperchromic urine, drugs, poor sample storage *False negatives:* Non-nitrite-forming bacteria, diet low in nitrates, urine that has not stayed in the bladder, ascorbate
Proteins	*False positives:* Strongly alkaline urine, hyperchromic urine, drugs, quaternary ammonium, plasma expander *False negatives:* Presence of globulins, hyperchromic urine
Glucose	Specific method for glucose but interference from low temperature and/or high relative density *False positives:* Oxidizing agents, peroxides, hydrochloric acid *False negatives:* Ascorbate, poorly stored

Table 19.3 (continued)

Parameter	Specificity/interferences
Ketones	It does not show hydroxybutyric acid. *False positives:* Free sulfhydryl groups (N-acetyl cysteine), hyperchromic urine, metabolites of levodopa, phenolphthalein *False negatives:* Incorrect conservation
Bilirubin	*False positives:* Hyperchromic urine, chlorpromazine *False negatives:* Ascorbate, nitrites, poor storage, direct sunlight
Urobilinogen	*False positives:* Hyperchromic urine, sulfonamides, para-aminosalyl acid *False negatives:* Formalin, oxidizing agents, poor storage
Ascorbate	*False positives:* Free sulfhydryl groups (N-acetyl cysteine), reducing agents
Creatinine	*False positives:* Hemoglobin, myoglobin *False negatives:* EDTA

Table 19.4 Comparative analytical sensitivity of some commercially available strips for albumin, glucose, red blood cells/hemoglobin, esterase

Clinical utility	Parameter
High	Albumin/protein Urinary concentration
High in specific clinical conditions	Albumin/protein Urinary concentration
Good	Leukocyte esterase Nitrites Ascorbate
Poor	Biliary pigments

and tubular proteinuria occurs when glomerular and tubular damage is associated; it is nonselective proteinuria characterized by proteins with different molecular weights.

Urinary protein determination should be expressed as a ratio to urinary creatinine as an indicator of urinary concentration: PCR (protein creatinine ratio) or ACR (albumin creatinine ratio).

The postrenal form is characteristic of all phlogistic or erosive pathologies of the urothelial mucosae.

The determination of urinary proteins must be performed on the intermediate urination of the first-morning sample by sensitive methods in liquid chemistry, and the result must be expressed in relation to the concentration of urine. ACR (first choice test in adults) and PCR (first choice test in children) are suggested. The lower sensitivity of ACR in pediatric subjects is determined by the higher frequency, in this age group, of tubular pathologies than the glomerular ones.

The analytical method used in the ECMU is very relevant. Indeed, immunoturbidimetry on automatic analyzers is the most reliable but also the least used at present; this is followed by dry chemistry methods on dipsticks by specific dyes for albumin and expression in relation to urinary creatinine; methods that assess protein concentration (almost exclusively albumin) on the basis of the variation of a pH-sensitive indicator with marked interferences in case of alkaline pH are not recommended. The accuracy (and harmonization) of urinary albumin and creatinine measurement is still an open issue. The International Federation of Clinical Chemistry (IFCC) is currently working to overcome it. In any case, the choice of the method must ensure the detection, at an adequate sensitivity, of both albumin and globulins. The term microalbuminuria, coined to identify the measurement of albumin at low concentrations, is wrong and should be avoided. When performing the test strip, we should not refer to protein but only to albumin. Indeed, the method used is sensitive to detect almost exclusively the presence of albumin and transferrin. Some multireactive strips use, besides the pad based on the above principle, a reactive area for the determination of albumin at low concentration (about 100 mg/L). Test strips capable of detecting the albumin/creatinine ratio have recently been marketed. Since it is a simple, inexpensive, and rapid method, dry chemistry on dipstick has spread to all laboratories and is currently still the most widely used method for protein determination.

Urinary Concentration

Urine is composed of 97–99% water and the remaining 1–3% of a mixture of solutes. The concentration of solutes in urine is an important index of the ability to concentrate urine by the kidney, as well as the state of hydration of the subject. It turns out, therefore, to be an indicator of considerable clinical value. It also has relevance in the analysis of urine sediment as the concentration of urine affects the preservation of cellular elements, which may undergo lysis, especially in low-concentrated urine, and morphological alterations, especially in strongly hypertonic urine. The expression of solutes in urine can be evaluated using different parameters, which differ in terms of meaning and type of solutes that will be detected: relative density, osmolality, and conductivity. Each of these parameters can be determined using different methods.

Relative Density

It measures the density of urine in g/L (mass/volume) and is usually performed in clinical practice. It is too often erroneously referred to as specific gravity, which is the ratio of weight to volume and is expressed in Newtons/L. Methods for the determination of relative density may be direct or indirect. Direct methods determine the relative density of urine regardless of the type of solute present as all solutes are detected and measured, both those physiologically present in urine, such as urea and electrolytes, and those indicative of pathologies, such as glucose and protein, and those of iatrogenic origin, such as radiological contrast agents. The relative density can be evaluated directly using the following methods: gravimetric, not used in routine diagnostics but still considered the reference method; urinometer and harmonic oscillation, obsolete; refractometric, not very suitable for routine use but still usable in particular cases (e.g., hyperchromic urine); test strips on automatic instruments.

Test Strip Determination

The method used in reactive dipstick pads is based on the determination of ions (mainly Na^+, Cl^-, K^+, $NH4^+$); the presence of other nonionic substances (glucose, proteins, radiological contrast media) is not detected. It is, therefore, a method capable of evaluating the ability of the kidney to manage the hydro-electrolytic balance through the selective reabsorption and elimination of water and ions. However, the method appears to be influenced by pH (overestimation at acidic pH and underestimation at alkaline pH) and by urine staining. In the case of strips with simultaneous detection of pH, it will be possible to make the correction for this interfering factor. It is an easily automated method, suitable for routine applications.

Osmolality

It is determined by exploiting methods capable of evaluating the colligative properties of solutes that interfere with changes in the state of the solution; these capabilities depend only on the number of particles present in the solution and not on their characteristics. Methods that evaluate the freezing temperature or the vapor pressure are used; in both cases, it is poorly automated and not suitable for routine applications. Also, in this case, the value is influenced by glycosuria and, therefore, it is not reliable in decompensated diabetics.

Conductivity

It is a parameter whose use in routine has been reproposed for availability on automatic instrumentation that makes it suitable for routine use. The conductivity depends on the concentration of electrolytes in the urine (Na^+, Cl^-, K^+, $NH4^+$, etc.) but not on the concentration of glucose and proteins; it is not influenced by the pH and the color of the urine and is, therefore, able to measure the result of the action of reabsorption and selective elimination of water and ions from the kidney even in diabetic subjects.

Creatinine

Measurement of urinary creatinine is relevant for several applications: identification of a liquid as urine, detection of adulteration, and normalization of proteinuria and ionuria.

Hemoglobin

The presence of hemoglobin and/or erythrocytes in the urine is, together with the presence of proteins, one of the most significant indicators of possible pathology of the urinary system; hematuria can be the only indicator of the presence of a glomerular or urological pathology, including those of neoplastic nature. The term hemoglobinuria indicates the presence of free hemoglobin, while the term hematuria defines the presence of erythrocytes in the urine. In low-concentrated urine (relative density <1010 g/L) or very alkaline urine (pH >8.0), blood cells can undergo lysis, releasing the hemoglobin they contain; usually in these cases, erythrocyte membranes (so-called ghost erythrocytes) remain, together with a variable number of erythrocytes, which have higher osmotic resistance. In addition to deficiencies of the preanalytical phase, hemoglobinuria is an expression of intra- and extra-erythrocytic hemolytic pathologies of different origin: metabolic, infectious, immune, and mechanical. In these diseases, the hemolytic crisis gives rise to a marked brown pigmentation of the urine due to the transformation into methemoglobin mediated by the acid pH of the urine.

Analytical Methods

All dipstick methods are based on the pseudoperoxidase activity of the complete tetrapyrrole ring of the central iron atom (protoporphyrin IX); the sensitivity is about 0.03 mg/dL corresponding to about 10 erythrocytes/μL. The test is reactive for both hemoglobin and myoglobin since both contain a tetrapyrrole ring. False negatives may be due to the interference of ascorbic acid, which, having strong reducing power, tends to bind the peroxide subtracting it from the reaction, while false positivity may be due to the activity of bacterial and/or leukocyte peroxidases or to contamination of the sample with substances having oxidizing-reducing activity (detergents, hypochlorite). The sensitivity does not exceed 80–90%; therefore, it is not possible to exclude the presence of hematuria by the use of the dipstick alone, but it must always be associated with microscopic and/or instrumental evaluation of the corpuscular fraction.

Hydrogenionic Concentration (pH)

Normally, the urine has a slightly acidic pH (between 5.0 and 6.0) because, under basal conditions, the endogenous production of acids is prevalent, and there is, therefore, the need to proceed to their elimination. It is an indispensable parameter from a clinical and laboratory point of view, with obvious limitations in measuring it by dry chemistry. Urinary pH can vary from 4.5 to 7.5. The determination of pH is very important for the laboratory because it allows for better interpretation of the other chemical reactions (albumin-proteins) and correctly evaluating crystalluria and eventual bacteriuria. For the clinician, it provides important information for urinary infections and for the evaluation of tubular function. The urinary pH can influence the conservation of the figured elements: for example, at alkaline pH, the lysis of the cells and the failure to form cylinders due to interaction with their protein matrix can occur.

Analytical Methods

Dipsticks with a range of 4.5–5.0 to 8.0–9.0 are used. The pH measured in this way (on an ordinal scale, with steps of 0.5 units) is not very accurate and is therefore of modest importance, if not indicative. If the measurement of pH is of clinical importance (stones, nephropathy, monitoring of acidifying/alkalizing therapies, etc.), it is advisable to make a specific request for urinary pH (other than ECMU) measured using a pH meter measurement.

Leukocyte Esterase

It is an enzyme present in granulocyte azurophilic granules but not in lymphocytes. Dipstick positivity occurs in the presence of esterase released from leukocytes during degeneration. In the presence of young, lysis-resistant leukocytes with no or minimal esterase release, leukocyturia can occur in the sediment with esterase negativity; conversely, leukocyte lysis (due to low sample concentration, exposure to very alkaline pH, or poor sample storage) can result in esterase positivity in the absence of detectable leukocytes. Therefore, the determination of leukocyte esterase should be combined with the detection of leukocytes by microscopy and/or automated instrumentation.

All commercially available dipsticks detect leukocyte esterase with a sensitivity equivalent to approximately 20–25 leukocytes/μL. This sensitivity is lower than that detected by automated analyzers of the corpusculate fraction (2–3 elements/μL) and at the reference limits for leukocytes (10–15/μL).

Nitrites

Certain bacteria (mainly Enterobacteriaceae) can convert nitrates normally present in urine into nitrites. Nitrates are introduced into the body by a diet rich in fresh vegetables, and their reduction to nitrites by bacterial metabolism requires a variable amount of time depending on the amount and type of bacteria involved. Therefore, in case of lack of intake of nitrates with the diet, reduced permanence of urine in the bladder, infections by bacteria not able to reduce nitrates to nitrites, the test for nitrites in urine may be negative even during urinary tract infections (UTI). Although the test has low sensitivity, it is useful because it has a high positive predictive value (PPV).

All commercially available dipsticks detect urinary nitrite, while no automated liquid chemistry kits are available.

Ascorbate

The presence of ascorbate in urine is rather frequent as it can originate both from the diet (citrus fruits, preservatives) and from the intake of drugs with vitamin C. In urine, the presence of ascorbic acid at a concentration of 100 mg/L can interfere with the determination of hemoglobin; at a concentration of 250 mg/L, it can interfere with the determination of nitrite and bilirubin; at a concentration of 500 mg/L, it can interfere with the determination of glucose. Thus, it is important to evaluate the possible presence of ascorbic acid.

Glucose

Glycosuria occurs when the amount of ultrafiltrated glucose exceeds the tubular reabsorption capacity. This can occur because of a decrease in the proximal tubule's ability to reabsorb glucose or because of increased ultrafiltration due to hyperglycemia. In the subject with intact kidneys, usually, glycosuria appears when glycemia exceeds 180 mg/dL; moreover, glycosuria with normal blood glucose concentration can occur in a congenital tubular pathology, known as normoglycemic glycosuria, and in nephrotic syndrome. It must be remembered that normoglycemic glycosuria can also be observed during therapy of diabetic disease with inhibitors of sodium-glucose cotransporters type 2, located in the proximal part of the convoluted tubule, which are responsible for about 90% of the reabsorption of ultrafiltrated glucose. Test strips can give false negatives in the presence of ascorbate and in case of urinary tract infection, while false positives are observed in the presence of oxidizing substances and acidic urine. The determination by liquid chemistry using a specific enzymatic method is useful in nephrological pathologies such as congenital or acquired tubulopathies (Fanconi syndrome) and in tubulo-interstitial diseases of various etiologies. Urinary glucose testing can be included, together with ketones, in a profile dedicated to pediatric subjects in early childhood due to the more frequent use of urine testing as the first test in clinical suspicion of diabetes.

Ketones

Ketones are a family of three compounds—acetone, acetoacetate, and β-hydroxybutyric acid—that derive from the metabolism (in glucose deficiency) of fatty acids. The presence of ketones in the urine is mostly related to fasting and is useful only in reference to specific patient populations and in specific clinical conditions (diabetes, hypothermia, fever, prolonged vomiting, fetal complications in the postterm) but rarely have real clinical utility, except for some situations in emergency medicine (diabetic ketoacidosis, alcohol abuse).

With a dipstick, false negatives are determined by the fact that β-hydroxybutyric acid is not detected, while false-positive results are observed in the presence of free sulfhydryl groups (e.g., drugs such as captopril, levodopa, cephalosporin) (Table 19.3).

Bile Pigments

Bilirubin and urobilinogen detected in urine have lost their clinical significance.

They are detected by common dipsticks; false negatives for bilirubin are observed in the presence of vitamin C and nitrite, false positives in the presence of chlorpromazine metabolites and for urobilinogen in the presence of some drugs (carbapenem and sulfanilamide) (Table 19.3).

Counting and Morphology of the Corpusculate Component

Analysis of the Corpusculate Fraction

The hierarchy of analytical processes recognizes four levels: (1) rapid tests; (2) routine methods; (3) qualified comparison methods; and (4) reference methods.

Level 1: Rapid Tests

These tests should ideally give a rapid and reliable answer for the individual patient; observation of a fresh preparation of native urine in a bright field by conventional coverslips and slides is performed, for example, in a physician's office.

Level 2: Routine Methods

These methods are usually applied for routine diagnostics in clinical laboratories. They are based on the microscopic evaluation of preparations obtained after centrifugation of the sample, aspiration of the supernatant, and resuspension of the base. The preparations are made in multicellular slides with predefined volumes equipped with a reticle for counting. The reading can be done in bright field or phase contrast.

Level 3: Comparison Methods

These are automated methods for the assessment of large series of samples and they require, for the appropriate use, trained personnel, and complex analytical equipment. A level 3 method for microscopic examination is the fresh evaluation of native (noncentrifuged) urine at 400 magnifications by two different operators using a phase-contrast cytometric chamber (e.g., Kovacs or Fuchs-Rosenthal). In some situations, in which it is important to search for elements of clinical importance (e.g., erythrocyte cylinders), it is advantageous to centrifuge the sample, even if this affects the correct quantification of the elements.

Level 4: Reference Methods

In urine microscopy diagnostics, since there is no reference test (level 4), level 3 methods are the highest level of diagnostic detail and can be used to evaluate the analytical performance of routine methods (level 2).

Principles of Hand Microscopy of Urinary Sediment

Evaluation of the corpusculate fraction of urine is performed using fresh urine sediment examination under bright field hand microscopy. The use of supravital staining is recommended only in pathological cases to improve the differentiation of cellular elements or cylinders. The use of phase contrast microscopy can improve the recognition and differentiation of corpuscular elements. Polarization microscopy is recommended for the detection and differentiation of crystals and lipids. It is appropriate the standardization of several aspects and preanalytical steps such as container, centrifugation, and preparation of microscopic slides. Finally, the recognition of the following elements is necessary:

- Blood cells: Erythrocytes and leukocytes
- Epithelial cells: Squamous, transitional (urothelial), tubular
- Cylinders: Hyaline, granular, waxy, lipidic, erythrocytic, leukocytic, epithelial (containing renal tubular cells), pigmented (by hemoglobin, myoglobin, bilirubin), including crystals or microorganisms, mixed
- Lipids
- Crystals: Calcium oxalate, uric acid, amorphous urates, amorphous phosphates, calcium phosphate, triple phosphate, cholesterol, cystine, 2,8 di-hydroxyadenine, drugs-derived
- Microorganisms: Bacteria, fungi, parasites, protozoa
- Other things: Mucus, sperm, contaminants.

The importance of a correct reading of the corpuscular fraction of urine and the considerable related clinical implications are exemplified in Table 19.5. The methods of observation of the urinary sediment under light microscopy, enunciated in the European guidelines of 2000, remain valid today in the general principles.

Morphological Evaluation of the Corpusculate Fraction by Light Microscopy

Identification and Quantification of Erythrocytes

Hematuria is defined as the presence of blood cells in the urine. It is called macrohematuria if the amount of blood is

Table 19.5 Corpuscular elements in urine and main clinical pictures

Main elements	Main clinical pictures
Isomorphic erythrocytes	Nonglomerular hematuria
Dysmorphic erythrocytes and acanthocytes	Glomerular hematuria
Polymorphonuclear leukocytes	Urinary infections, proliferative glomerulonephritis, acute interstitial nephritis Contamination from genital secretions
Squamous cells	Contamination from genital secretions
Renal tubular epithelial cells	Renal disorders associated with organic tubular damage (acute tubular necrosis)
Transitional superficial and/or deep cells	Pathologies associated with damage to the transitional epithelium
Lipids	Glomerular disorders associated with proteinuria of varying degrees, but especially of nephrotic entity Fabry disease (from lysosomal accumulation of glycosphingolipids)
Hyaline cylinders	They can be present in small numbers in healthy individuals They can be present in various types of kidney disease
Hyaline-granulosa cylinders	They can be present in small numbers in healthy individuals They can be present in various types of kidney disease
Grainy cylinders	They can be present in various types of kidney disease Acute tubular necrosis
Wax casts	Renal disease with significant loss of function
Lipid cylinders	Nephrotic syndrome
Erythrocyte cylinders	Glomerular hematuria, proliferative or necrotizing glomerulonephritis
Leukocyte cylinders	Acute interstitial nephritis, acute pyelonephritis, proliferative glomerulonephritis
Cell/epithelial cylinders (tubular epithelial cells)	Acute tubular necrosis, acute interstitial nephritis, nephrotic syndrome
Hemoglobin cylinders	Glomerular hematuria, proliferative or necrotizing glomerulonephritis, acute intravascular hemolysis
Myoglobin cylinders	Rhabdomyolysis
Bilirubin cylinders	Marked jaundice
Cylinders with included bacterial or fungal	Bacterial or fungal infections of the kidney
Cylinders with inclusion crystalline	Acute renal failure due to massive crystalluria

Modified from Manoni et al. (2016)

such as to alter the color of the urine. It is sufficient 2 mL of blood in a liter of urine to cause a visible change in color. In the case of macroscopic hematuria, the urine may have various colors, depending on the severity of the bleeding and on the timing during it occurs. For example, frank hematuria (red in color) indicates considerable bleeding in progress, "flesh-wash" hematuria indicates mild bleeding, "marsala" or "cola"-colored hematuria may indicate hemoglobinuria or previous bleeding. In the presence of red-colored urine, microscopic confirmation of the presence of hematuria in the sample is always necessary since some substances of food origin and some drugs can confer to urine a color like that determined by the presence of blood (Table 19.1). Microhematuria occurs when the quantity of blood is modest and not able to alter the aspect of the urine. There is no agreed-upon threshold for defining microhematuria. It is recommended that each laboratory defines its own reference values in relation to the population and the cases examined. One of the most widely accepted thresholds is that of the American Urological Association, which indicates as microhematuria the presence of three or more blood cells per microscopic field at 400×, equivalent to 10–12 erythrocytes/μL with automated analyzers.

Leukocyte Identification and Quantification

There is no agreed-upon threshold value for defining leukocyturia. It is recommended that each laboratory define its own reference values in relation to the population and case series tested. Leukocyturia is commonly defined as the presence of more than 3–5 leukocytes/microscopic fields (for high-resolution fields, HPF), equivalent to 10–20 white blood cells/μL of urine. Neutrophilic and eosinophilic granulocytes as well as lymphocytes and macrophages, can be found in the urine. In many inflammatory and infectious diseases sustained by bacteria – from urinary tract infection to glomerulonephritis – neutrophil granulocytes are a common finding. Eosinophilic granulocytes are present in several pathologies and have, therefore, lost their pathognomonic significance as a marker of acute interstitial nephritis. Lymphocytes are associated with conditions of chronic inflammation and viral diseases; they are present in urine during renal transplant rejection (8090% sensitivity) or during hematological pathologies (leukemia or lymphomas with kidney infiltration). Macrophages (histiocytes) can be present in various chronic inflammatory diseases, almost always associated with neutrophils and during marked proteinuria. They can assume various aspects: dendritic with pseudopodia, polygonal, like degenerating granulocytes, circular with inclusions, and evident nucleus.

Identification and Quantification of Cylinders

They are cylindrical elements with sometimes rounded and sometimes truncated ends, consisting of Tamm-Horsfall Protein (THP), which may be the only constituent (hyaline cylinders) or could be combined with cellular elements or elements of different origin. THP is the quantitatively most important component of physiological proteinuria and is produced at the level of the thick ascending tract of the loop of Henle. The formation of cylinders derives from the aggregation of THP fibrils, favored by acid pH, high osmolality, and ultrafiltrated proteins. In urine with alkaline pH, the finding of cylinders is quite rare due to the lack of aggregation of

THP fibrils. In normal subjects, the finding of hyaline cylinders is not uncommon. Based on their constitution, it is possible to distinguish the following types of cylinders: hyaline, granular (with small and large granules), leucocytic, erythrocytic, epithelial (renal tubular cells), lipidic, waxy, pigmented (by hemoglobin, myoglobin, and bilirubin), with bacterial or fungal inclusions, with crystalline inclusions, or mixed.

Hyaline Cylinders

They are made up only of THP scarcely visible by intense light (and are considered of recent formation), or more visible and with a more compact structure. A hyaline cylindruria can be observed even in normal subjects, more easily after physical exercise, dehydration, or exposure to cold, in acute heart failure, or in hyperpyrexia. A hyaline cylindrinuria may be present in all nephropathies, in which it is usually associated with cylinders of other types.

Granular Cylinders

The small granules are formed by conglutinates of ultrafiltrated proteins at the level of the glomerulus; the large granules are formed by the degeneration of cellular elements. They are not usually found in the urine of normal subjects although, even in the absence of renal pathology, small granule cylinders may be found after hyperpyrexia. Large granule cylinders are often found in many types of nephropathies, such as glomerulonephritis and diabetic nephropathy. In patients with acute renal failure, granular cylinders are considered a marker of organic tubular damage.

Leukocyte Cylinders

Their presence in the urine can be determined by all phlogistic pathologies of the kidney, such as lupus erythematosus, interstitial nephritis, and acute pyelonephritis.

Erythrocyte Cylinders

They can be observed in all nephropathies, causing hematuria for which they represent a marker of absolute specificity.

Epithelial Cylinders

They are an expression of acute organic tubular distress that can be detected, for example, in glomerular nephropathies, acute tubular necrosis, acute interstitial nephritis, and tubulopathies.

Lipid Cylinders

They are present in situations characterized by marked proteinuria, especially in nephrotic syndrome.

Wax Cylinders

They are the only cylinders in which the predominant protein matrix is not THP. They are compact and friable and only exceptionally present included elements because the long stay in the tubules leads them to complete degeneration. Waxy cylinders are therefore an expression of renal impairment. Their presence can be associated mainly with glomerulonephritis, diabetic nephropathy, and renal amyloidosis.

Pigmented Cylinders

They owe their coloration to the presence of chromogenic substances. Hemoglobin and myoglobin cylinders: reddish in color, they have the appearance of a granular cylinder; hemoglobin cylinders may derive from degenerated erythrocytes or from hemoglobinuria; myoglobin cylinders are found in acute renal failure associated with rhabdomyolysis of various origins; bilirubin cylinders. Finally, bilirubin gives the cylinder a dark orange color; they are observed in jaundiced patients with a high percentage of conjugated bilirubin.

Cylinders with Bacterial or Fungal Inclusions

The presence of cylinders with bacterial inclusions suggests a renal infection; the finding is of considerable importance because it is indicative of the presence of a particularly severe infection.

Cylinders with Crystalline Inclusions

The presence of crystalline inclusions indicates that crystals are present at the tubular level. It is very important clinically in crystalluric forms of acute renal failure, such as, acute uratic nephropathy.

Mixed Cylinders

These are pleiomorphic forms in which there may be different corpuscular elements (erythrocytes, leukocytes, cells, lipids, crystals, etc.) included in the THP matrix. Their clinical significance is like that of cylinders with single inclusions.

Identification and Quantification of Cells

The mucous membranes of the genito-urinary tract are surrounded by different types of epithelia. The urethra, in its first tract, is surrounded by a transitional epithelium in continuity with that of the bladder; in its anterior portion, instead, it is covered by a squamous pavement epithelium arranged in several layers up to the external urethral orifice. The squamous epithelial cells coming from the urethra and the bladder trigone (in women of fertile age) are large, leafy, with a small pycnotic nucleus. The bladder (except for the trigone) and ureters are lined by a multilayered transitional epithelium, called urothelium. At least three distinct cell morphologies can be recognized: the cells of the superficial layer, round or oval, umbrella-shaped, and large, with a small and central nucleus; the cells of the intermediate layer, generally smaller and more heterogeneous in shape (oval, or club-shaped, often binucleate); and the cells of the deep layer, cuboidal. The renal tubules are lined by the tubular epithelium. On a morphological basis, it is possible to distinguish the epithelium of distal and proximal tubules, monostratified with cubic or cylindrical cells, with a roundish central nucleus and short microvilli on the luminal side, from the epithelium of the collecting duct generally cubic, with an oval central nucleus and short microvilli. Epithelial cells may come from any portion of the genito-urinary tract and, therefore, are, by definition, extremely pleiomorphic.

The presence of squamous cells is a frequent event in the microscopic evaluation of urine and usually does not have pathological significance being usually an expression of genital contamination; therefore, it is most often an indicator of incorrect sample collection. The presence of elements of the urothelium (transitional cells) appears frequently related to inflammatory bladder disease, stones, invasive maneuvers (e.g., catheterization), and neoplastic disease. The presence of tubular cells always has a pathological significance and appears to be related to acute damage of the renal tubule, as observed in several acute diseases of the renal parenchyma.

Lipid Identification

From a morphological point of view, there are four categories of lipids: droplets (isolated or in aggregates), fat oval bodies, lipid cylinders, and cholesterol crystals. The identification of the first three categories is facilitated using polarized light, which shows the typical "Maltese crosses". These elements are associated with marked proteinuria.

Identification of Crystals

The presence of crystals in urine is significant only for large amounts and for certain types of crystals. In acid urine, crystals of uric acid and calcium oxalate can be found but also precipitates of amorphous urates. In alkaline urine, calcium phosphate crystals and precipitates of amorphous phosphates can be found. Some crystalluria is always considered pathological; these include the presence of triple phosphate crystals (genito-urinary tract infections), cystine (cystinuria), hydroxyadenine, tyrosine and leucine (hereditary diseases, hepatitis, leukemia), cholesterol (kidney diseases, nephrotic syndrome), bilirubin (clinically detectable jaundice), hemosiderin (severe hemolysis, hemolytic anemia, transfusion reactions). It should be borne in mind that some drugs can also give rise to the presence of precipitates in the urine. Most crystals are not strictly pathological precipitate in urine because of pre-analytical problems (e.g., refrigeration of the sample) or due to physiological or para-physiological factors, such as food, dehydration, etc.; in this case, their finding is of no clinical significance. In the evaluation of a subject with suspected calculotic diathesis, the examination of urine sediment should be performed exclusively on a urine sample just issued, examined "fresh": the finding of crystals not necessarily pathological (uric acid, calcium oxalate, etc..) but present in large quantities, with forms of medium and/or large and/or aggregate formation and/or persistent in repeated samples, should be further investigated by the metabolic study.

Microorganisms

The guidelines of the European Confederation of Medical Laboratories and the European Group on Urinalysis (ECLM/EUG) propose a classification of the etiological agents of UTI based on:

- Their potential uropathogenicity
- The integrity of the genito-urinary system
- The physiological conditions (e.g., pregnancy)
- The presence of systemic disease.

In addition, this classification considers the frequency at which different microorganisms are isolated from urinary specimens. Primary pathogens are defined as those bacteria that can frequently cause infection in healthy subjects without anatomical or functional abnormalities of the urinary tract (e.g., *Escherichia Coli* and *Staphylococcus Saprophyticus*) and secondary pathogens are those bacteria that are also found in healthy subjects but less frequently. These are often infections in institutionalized subjects or with functional/anatomical abnormalities of the urinary apparatus or with concomitant systemic pathology (e.g., *Enterococcus* spp., *Proteus* spp., *Pseudomonas aeruginosa*). Conditional pathogens are those bacteria that are not able to cause infection in healthy subjects but are pathologically significant in patients with functional or anatomical abnormalities of the urinary tract or with concomitant systemic pathology (e.g., mycetes, *Streptococcus agalactiae*). The fourth group is that of contaminating bacteria and, therefore, without pathological significance (e.g., *diphtheria* and lactobacilli).

The quantitative and qualitative evaluation of the bacterial flora in urine is not sufficient to make a diagnosis of UTI (which is a clinical diagnosis) but must be integrated with an assessment of the corpuscular component: leukocytes, erythrocytes, and epithelial cells. The presence of a high number of squamous epithelial cells suggests contamination, while the presence of bacteriuria without pyuria suggests colonization rather than infection.

Contaminants

All elements present in the sample under examination that do not originate from the urinary apparatus are considered contaminants. Specifically, contaminants should be distinguished according to the source as follows: from the subject from which the sample was collected, such as elements of genital origin including erythrocytes, leukocytes, squamous cells, bacteria, protozoa, mycetes, spermatozoa or cutaneous origin, including hair, parasites, talcum powder, creams, oils, aspersive powders or fecal origin, including fibers, cells, bacteria, parasites and their eggs; textile fibers from clothing, cellulose fibers from tampons, nappies, toilet paper; from the external environment during collection and storage (e.g. pollen, plant cells, sporefungals, fibers); from the laboratory during preparation and examination (glass fragments from microscopic slides, powder from gloves). It is essential to carry out the collection of the sample according to the recommended guidelines to reduce the risk of sample contamination. Specifically, accurate preliminary hygiene of the genitals, collection from an intermediate site by a disposable container and a vacuum tube device. The rare finding of protozoa of genital origin or helminths (or their eggs) of fecal origin in urine does not seem indicative of a urinary infestation but suggests contamination; however, since these elements are indicative of a genital or intestinal parasitic infestation, the data must be appropriately reported in the report for the relevant investigations and/or treatment.

Evaluation of the Corpusculate Fraction of Urine by Automated Instrumentation

Analyzers for automated evaluation and quantification of the corpusculate fraction of urine can be classified into three categories based on the principle of operation: automated microscopy, image capture, and cytofluorimetry.

Automated Microscopy

The most widely used analyzer for the analysis of the corpuscular fraction of urine by Cuvette-Based Microscopy (CBM) is the SediMAX analyzer (Menarini), which has recently been joined by a system called Cobas 6500 (Roche). The analysis requires a minimum sample volume of 2 mL. An aliquot of 200 μL is injected into a special cuvette which, once centrifuged, allows the formation of a thin liquid film on which the sample is read by a microscopic camera by means of high-power green LED illumination. For each sample, several microscopic fields can be analyzed at 400×, displayed through 15 photographs. A special neural network supported by a database containing thousands of images of the sediment identifies the elements based on their morphological characteristics. It is possible to highlight the identified elements by means of an acronym that appears on the element itself; anomalous, contiguous, and overlapping elements modify the perimeter morphological aspect and are, therefore, not recognized and, consequently, are excluded from the count. A new version of the SediMAX analyzer has recently been released, which uses phase contrast microscopy in addition to the bright field.

Image Capture

The first analyzer with image capture technology was the Iris iQ200 analyzer from Beckman. Other two systems are now available, namely FUS 100 and 200 (Dirui), which are based on the same principle of operation. These systems incorporate an automated microscope with focused optics on a laminar flow planar cell, in which the particles contained in the sample are focused hydrodynamically. The laminar flow allows the sample to be presented within the focal plane of the microscope objective, further orienting the asymmetric particles so that they are presented in an orthodromic position for better reading and classification. A stroboscopic lamp illuminates, at a frequency of 24 flashes per second, the sample passing through the flow cell, allowing a miniaturized digital camera to take, isolate and store a very high number of frames per sample. Every single image is subtracted from the background, which was previously captured and digitized, thus enhancing the morphology of the particle and its comparison with the liquid medium. The individual images of a particle are isolated within each frame. The particle recognition software analyzes each element by a neural network and compares it with over 26,000 unique images; then, considering the characteristics of size, shape, contrast, and internal content, it classifies.

Cytofluorimetry

The only automated analyzer of the corpuscular fraction of urine by cytofluorimetry is the Sysmex UF analyzer, of which the model 1000i is the latest development. The Sysmex UF-1000i (Dasit) combines impedance technology with cytofluorometry and uses a diode laser as the light source. The urine sample is aspirated and diluted with a buffer and then subjected to a staining process with two polymetin fluorochromes, which can bind to nucleic acids. After a process of hydrodynamic focusing, the sample is passed through two flow cells, one dedicated to the analysis of microorganisms and the other dedicated to the analysis of all other corpuscular elements. The passage of the single particles is recognized by the impedance method, which allows their accurate quantification and provides information about their size. In addition, the passage of particles suspended in the laminar flow deflects the laser light beam and generates a diffraction signal which is read by both frontal (forward, 45°) and lateral (side, 90°) scatter detectors, as well as by a fluorescence detector. The measured parameters are converted into electrical signals that, analyzed through mathematical algorithms, allow the identification of different elements present in the urine. The conductivity of the solution is also measured and indicates the concentration of electrolytes in the urine. A new cytofluorimetric system, UF-5000 Sysmex, which can be integrated with an image capture system, UD-10 Sysmex, is soon to be commercialized. The UF-5000 is the evolution of the UF-1000, while the UD-10 is a new system that, on the basis of grids selected by dedicated software, allows the evaluation of high-definition digital images to obtain a morphological picture to add to the cytofluorimetric count, where necessary.

All automated analyzers for the qualitative and quantitative evaluation of the corpusculate fraction of urine have in common some advantages and present some problems.

They examine native urine, thus eliminating centrifugation, aspiration of the supernatant, and preparation of the microscopic slides. These steps, present in traditional microscopy, introduce not only considerable analytical variability due to the substantial absence of standardization but also constitute organizational bottlenecks, with a critical delay in the analysis of the corpusculate fraction that can alter analytical outcomes as well as reduce precision and accuracy. This is more relevant in laboratories with high workloads, where process times can be critical due to the progressive degradation of the sample.

The quantification of corpusculate elements carried out by automated analyzers is much more repeatable than that carried out by a human observer under a light microscope. Moreover, quantitative results are expressed as the number of particles per unit volume, avoiding descriptive and subjective expressions such as "rare," "some," and "numerous". The possibility of quantifying the elements per unit volume examined allows for standardization of the analysis and constitutes an objective indicator for clinical evaluation.

The ability of the analyzers to correctly identify corpuscular elements of urine appears satisfactory for erythrocytes, leucocytes, squamous epithelial cells, bacteria, fungi, etc. On the other hand, the identification of cylinders appears to be less accurate and satisfactory, determining the risk of false positives due to the presence of mucus, cell aggregates, clusters of crystals; in this case, the morphological evaluation will be diriment. In this regard, it should be noted that the presence of mucus commonly indicates an incorrect collection of the sample performed by initial and not intermediate urination.

The need to define alarm thresholds, grids for reviewing results, and alarms for data and/or sample anomalies has made it possible to focus resources and professional skills on those cases that benefit from the appropriate investigation. Paradoxically, the introduction of automated analyzers for differential counting of the corpusculate fraction is increasingly enhancing the morphological expertise of operators, giving a significant added value to the ECMU. Moreover, these analyzers have easy internal quality control allowing the evaluation of the main corpuscular parameters.

These analyzers, now, are not able to recognize lipids, protozoa, and numerous crystals, signaling generic alarms, which must be caught in the review phase as well as they do not distinguish between tubular cells and transitional cells, limiting themselves generically to signal small round cells and large cells.

In conclusion, automated analyzers for the study of the corpusculate fraction of urine are an indispensable tool to ensure high analytical standards in laboratories, replacing the traditional morphological examination in cases of a simple definition, quantifying the corpusculate elements with precision and accuracy comparable to those of an experienced microscopist, using appropriate technology, methodology and with reduced workload and allowing an effective selection of cases for which must be deepened by traditional microscopy or other analytical methods.

Postanalytical Phase

For laboratory medicine, the report is the translation of the diagnostic tests performed into clinically useful information. The report must, therefore, provide clinically useful information for the diagnosis, therapy, monitoring and health promotion but also error-free results in time for the proper management of the patient. Given the clinical importance of the report, intended as a support to the clinician, it must be made to qualified personnel.

The Report of the Chemical-Physical and Morphological Examination of Urine (ECMU)

The report, or answer, in the urine test is usually expressed in a profile formulated on three types of tests: chemical-physical and morphological. The expression of the results of the parameters, both microscopic and chemical, must be quantitative and/or semi-quantitative. The chemical parameters (proteins, leucocyte esterase, pseudoperoxidase and nitrites) must, therefore, be quantified, just as the cells and other corpuscles present in the sample must be expressed as an average number per field or per microliter. Consequently, literal expressions, such as rare, some etc., should never be used.

Urine is an unstable biological fluid in which, at room temperature, several constituents undergo significant changes over time (pH, glucose, bilirubin, enzymes, leukocytes, and erythrocytes).

This makes the judgment of the suitability of the sample essential, which should be stated, when negative in the response.

In the report, the analytical methods used, and the areas of reference must be explained.

Any discrepancies (true or apparent) between parameters referring to the same source (e.g., nitrite/bacteria, pseudoperoxidase-hemoglobin/hematia, esterase/leukocytes) must be resolved in the report. The response should also be evaluated based on knowledge of possible interferences, false positives, and false negatives that dry chemistry methods may induce. However, the limitations of sediment preparation or automated examination of the corpusculate fraction are well known. In the case of discrepancies, the evaluation with confirmatory methods must be pursued and, in relation to the method used, may be the evaluation of graphical/imaging reports and alarms on automated systems, microscopic examination, liquid chemistry, or immunochemical execution. The clinical objective today is the exclusion, diagnosis, and monitoring of diseases of the kidney and urinary tract. Metabolic alterations of different apparatuses and organs can now be better evaluated by targeted serum and blood tests. In addition, it is possible to hypothesize pathways oriented to specific pathologies, with answers focused on the clinical question: morphology of hematuria, study of crystalluria, search for parasites, etc.

Recommended Readings

Aspevall O, Hallander H, Gant V et al (2001) European guidelines for urinalysis: a collaborative document produced by European clinical microbiologists and clinical chemists under ECLM in collaboration with ESCMID. Clin Microbiol Infect 7:173–178

Boven LA, Kemperman H, Demir A (2012) A comparative analysis of the Iris iQ200 with manual microscopy as a diagnostic tool for dysmorphic erythrocytes in urine. Clin Chem Lab Med 50:751–753

Brunzel N (2013) Fundamentals of urine & body fluid analysis, 3th edn. Elsevier

Burlina AB (1992) Medicina di laboratorio – Fondamenti di diagnostica. CG Edizioni Medico Scienti- fiche, Torino

Cappelletti P (2004) Il "referto" in Medicina di Laboratorio. Riv Med Lab – JLM 5(3)

Ceriotti F, Secchiero S, Plebani M et al (2011) Linee guida per la gestione dei Programmi di Valutazione Esterna di Qualità. Bioch Clin 35:107–126. Documento SIBioC, www.sibioc.it

Cho BS, Hahn WH, Cheong HI et al (2013) A nationwide study of mass urine screening tests on Korean school children and implications for chronic kidney disease management. Clin Exp Nephrol 17:205–210

CLSI GP-16 A3 (2009) Urinalysis and collection, transportation, and preservation of urine specimens; Approved guideline – 3rd ed. 29(4):4–21

College of American Professionists. http://www.cap.org CAP proficiency testing

Cosmacini G (2006) L'arte lunga, storia della medicina dall'antichità a oggi. Editori Laterza, Roma

Daudon M, Jungers P, Lacour B (2004) Clinical value of crystalluria study. Ann Biol Clin (Paris) 62:379–393

Du J, Xu J, Wang F et al (2015) Establishment and development of the personalized criteria for microscopic re- view following multiple automated routine urinalysis systems. Clin Chim Acta 444:221–228

ECLM (2000) European urinalysis guidelines. Scand J Clin Lab Invest 60:1–96

Fogazzi GB The urinary sediment an integrated view. 3rd edn. Masson

Fogazzi GB, Garigali G (2014) Urinalysis. In: Johnson RJ, Feehally J, Jurgen F (eds) Comprehensive clinical nephrology, 5th edn. Elsevier Saunders

Graziani M, Caldini A (2011) per il Gruppo di Studio IntersocietarioSIBioC-SIMeL Diabete Mellito. Indicazioni per la misura dell'albumina nelle urine per l'accerta- mento e il monitoraggio della nefropatia diabetica. Biochimica Clinica 35:127–130

Jiang T, Chen P, Ouyang J et al (2011) Urine particles analysis: performance evaluation of Sysmex UF-1000i and comparison among urine flow cytometer, dipstick, and visual microscopic examination. Scand J Clin Lab Invest 71:30–37

KDIGO 2012 (2013) Clinical practice guideline for the evaluation and management chronic Kidney disease. Kidney Int Suppl 3(1)

KDIGO clinical practice guideline for acute kidney injury (2012) Kidney Int Suppl 2(1)

Kilpatrick ES, Freedman DB (2011) National clinical bio-chemistry audit group. A national survey of interpre- tative reporting in the UK. Ann Clin Biochem 48:317–320

Kouri T, Laippala P, Kutter D et al (1999) Quality specifications for ordinal scale measurements with multi- property (multiple) urine test strips. Scand J Clin Lab Invest 59:523–526

Kouri T, Fogazzi G, Gant V et al (2000) European urinalysis guidelines. Scand J Clin Lab Invest 60(Suppl 231):39–47

Kuori T, Gyory A, Rowan M (2003) ISLH recommended reference procedure for the enumeration of particles in urine. Lab Hematol 9:58–63

Libeer JC (2001) Role of external quality assurance schemes in assessing and improving quality in medical laboratories. Clin Chim Acta 309:173–177

Linee Guida per la calcolosi delle vie urinarie (2007) AURO.it (Associazione Urologi Ospedalieri Italiani)

Manoni F, Fornasiero L, Ercolin M et al (2009) Laboratory diagnosis of renal failure: urine conductivity and tubular function. Minerva Urol Nefrol 61:17–20

Manoni F, Gessoni G, Alessio MG et al (2014) Gender's equality in evaluation of urine particles: results of a multicenter study of the Italian Urinalysis Group. Clin Chim Acta 427:1–5

Manoni F, Gessoni G, Fogazzi GB et al (2016) per il Gruppo Intersocietario Analisi delle Urine. Esame Chimico Fisico e Morfologico delle Urine: Proposta di Linee Guida per la Fase Analitica del Gruppo Intersocieta- rio Analisi delle Urine (GIAU). Biochimica Clinica 40:353–382

Manoni F, Caleffi A, Gessoni G et al (2011) L'esame delle urine chimico morfologico e colturale: proposta di linee guida per una procedura standardizzata della fase pre-analitica. Riv Ital Med Lab 7:25–35

Mc BL (1997) Textbook of urinalysis and body fluids: a clinical approach. Lippincott Williams & Wilkins, Philadelphia

McNulty CA, Thomas M, Bowen J et al (2008) Improving the appropriateness of laboratory submissions for urinalysis from general practice. Fam Pract 25:272–278

Mundt L, Shanahan K (2011) Graff's textbook of urinalysis and body fluids. Lippincott Williams & Wilkins, Philadelphia

National Health Service (2010 March) Evidence review. Automated urine screening systems. CAO 10030

Plebani M (2005) Il referto in medicina di Laboratorio. RI- MEL/ IJLaM (1 Suppl):57–59

Ross D, Neely A (1982) Textbook of urinalysis and body fluids. Appleton & Lange

Schürer-Maly C, Wood WG, Falbo R et al (2013) An educational web-based external quality assessment outcome and evaluation: first experiences with urinary sediment and hemostaseology. Clin Lab 59:1061–1069

Sciacovelli L, Secchiero S, Zardo L et al (2006) External quality assessment: an effective tool for clinical governance in laboratory medicine. Clin Clem Lab Med 44(6):740–749

Scottish Intercollegiate Guidelines Network (SIGN) (2006, July) Management of suspected bacterial urinary tract infection in adults. A National Clinical Guideline. Edinburgh (Scotland), Scottish Intercollegiate Gui- delines Network (SIGN). SIGN publication n. 88

Secchiero S, Fogazzi GB, Manoni F et al (2015) The Italian External Quality Assessment (EQA) program: results of the period 2012-2015. Clin Clem Lab Med 53(Suppl):S1495–S1502

Secchiero S, Epifani MG, Plebani M (2017) Verso l'armoniz- zazione dell'analisi delle urine con striscia reattiva: il ruolo di un programma di VEQ. Biochim Clin 41(3):248–257

Simerville J, Maxted W, Pahira J (2005) Urinalysis: a comprehensive review. Am Fam Physician 71:1153-morfologico. e colturale delle urine: proposta di linee 62

Takubo T, Tatsumi N (1999) Quality control in urinalysis. Southeast Asian J Trop Med Public Health 30(Suppl 3):136–148

Vasikaran S, Sikaris K, Klipatrick E et al (2016) On behalf of the IFCC WG Harmonization of quality assessment of interpretative comments. Assuring the quality of interpretative comments in clinical chemistry. Clin Chem Lab Med 54:1901–1911

Wood WG, Schwarz P, Illigen D, Fried R (2013) Experience with an alternative form of samples for external quality assessment of urinary sediment (visual sample EQA). Clin Lab 59:875–883

Zaman Z, Fogazzi GB, Garigali G et al (2010) Urine sediment analysis: analytical and diagnostic performance of sediMAX – a new automated microscopy image-based urine sediment analyser. Clin Chim Acta 411:147–154

Hydroelectrolytic Disorders

20

Marcello Ciaccio, Anna Maria Ciaccio, and Luisa Agnello

Introduction

Maintaining body fluid and electrolyte balance is of paramount importance for normal functioning. Any alteration in this balance falls under the definition of "hydroelectrolytic disorders."

This chapter describes the pathophysiology of the regulation of the hydroelectrolytic balance and the main causes water balance, sodium balance, and potassium balance alterations.

Pathophysiology

The human organism consists of 50–60% water; this percentage is slightly higher in children than in adults and men than in women. Water is distributed in two compartments, intracellular and extracellular, which differ in solute composition. The intracellular compartment contains about two-thirds of the total pool of water (intracellular fluid, ICF). In contrast, the extracellular compartment contains the remaining one-third of water (extracellular fluid, ECF) and can be further distinguished into extravascular or interstitial (connective, bone, serous cavities, and secretions) and intravascular (plasma) (Fig. 20.1). The intra- and extracellular compartments, although separated by cell membranes and capillary walls, are in dynamic equilibrium. The concentration of solutes in a liquid is termed "osmolality" and is expressed in milliosmoles per kilogram of water (mOsm/kg). Sodium (Na^+) and the anions Cl^- and HCO_3^- represent the primary solutes of ECF, whereas potassium (K^+) and organic phosphorus esters (adenosine triphosphate (ATP), creatine phosphate, and phospholipids) represent the primary solutes of ICF. The solutes in each compartment determine the effective osmolality of the compartment. The cell membrane and capillary wall are highly permeable barriers to water, which pass from one compartment to another to maintain osmotic balance. In particular, the water movement through the capillary wall depends on Starling's forces (hydrostatic pressure and colloid osmotic pressure), whereas that through the cell membranes depends on the difference in osmolality between ICF and ECF.

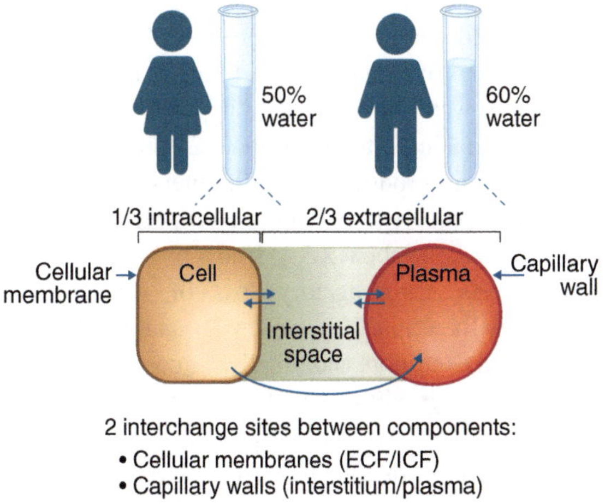

Fig. 20.1 Distribution of water in the body. (Copyright EDISES 2021. Reproduced with permission)

M. Ciaccio (✉)
Department of Biomedicine, Neurosciences and Advanced Diagnostics, Institute of Clinical Biochemistry, Clinical Molecular Medicine and Clinical Laboratory Medicine, University Hospital "P. Giaccone", Palermo, Italy

Department of Laboratory Medicine, University Hospital "P. Giaccone", Palermo, Italy
e-mail: marcello.ciaccio@unipa.it

A. M. Ciaccio
Department of Health Promotion, Mother and Child Care, Internal Medicine and Medical Specialties (ProMISE) "G. D'Alessandro", University Hospital "P. Giaccone", Palermo, Italy

L. Agnello
Department of Biomedicine, Neurosciences and Advanced Diagnostics, Institute of Clinical Biochemistry, Clinical Molecular Medicine and Clinical Laboratory Medicine, University Hospital "P. Giaccone", Palermo, Italy

M. Ciaccio (ed.), *Clinical and Laboratory Medicine Textbook*, https://doi.org/10.1007/978-3-031-24958-7_20

The composition of extracellular liquids can vary within narrow limits; water and the solutes dissolved play an essential role in cellular metabolic reactions, so excessive variations in extracellular liquids composition may lead to severe cell function alterations. It is, therefore, essential that the body's hydroelectrolytic content is kept constant according to the law of conservation of matter: everything that enters the body must leave it, if the body does not need it.

Daily, our organism acquires from the outside about 2 liters of water and food that contain NaCl, K^+, H^+, Ca_2^+, HCO_3^-, and phosphate ions. The body has several ways to excrete water and ions, mainly the kidneys; a small amount is also lost through stool, sweating, and respiration. In addition to physiological mechanisms, behavioral mechanisms, such as thirst and appetite for salt, also play an important role.

The maintenance of hydroelectrolytic balance is critical for the homeostasis. Water and Na^+ are critical for the volume and osmolality of ECF. Consequently, their alterations can result in cell volume changes, either shriveling or swelling, thus compromising cell function. Potassium (K^+) is, on the other hand, important for the membrane potential of excitable cells, and, its alterations can, therefore, compromise cardiac, muscular, and nervous functions.

K^+, which is more concentrated within the cell, tends to leave by passive diffusion. In contrast, Na^+ tends, by passive diffusion, to enter the cell; the Na^+/K^+ATP-dependent pump actively extracts Na^+ from the cell and reintroduces K^+, thus allowing the maintenance of the different concentration of these solutes between the two compartments. The regulation of hydroelectrolytic balance is highly complex. It involves not only renal and behavioral responses but also the cardiovascular and respiratory systems, which are under the control of the nervous system.

Water Balance

The introduced amount of water must be equal to the excreted amount of water excreted to maintain homeostasis. Alterations in water balance cause hypo- or hypernatremia. On average, an adult introduces about 2 liters of water daily through food and produces about 0.5 liters through cellular respiration (glucose + $O_2 \leftrightarrow CO_2 + H_2O$), resulting in a total water daily intake of about 2.5 L. Under physiological conditions, there is an obligatory loss of water of about 1.5 L through the urinary tract; a small quantity, about 0.1 L, is lost through excretion, and, finally, a quota, corresponding to about 0.9 L, is eliminated by *perspiratio insensibilis*, which consists of skin and respiratory tract evaporation (Fig. 20.2). In this way, the balance between the volume of incoming and outcoming water is maintained. Basically, body water volume is controlled by its intake, through the stimulus of thirst, and by its excretion at the renal level.

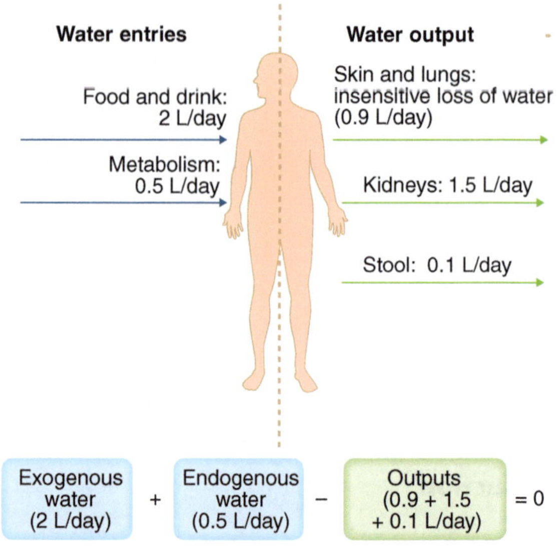

Fig. 20.2 Water balance. (Copyright EDISES 2021. Reproduced with permission)

If total body Na^+ is kept constant, a change in total body water alters osmolality. In other words, if there is an osmolality increase, the body's response is to improve water intake through the thirst sensation and the diuresis contraction. Viceversa, if there is an osmolality reduction, the organism determines a reduction in the thirst sense and a diuresis increase. The osmolality alteration is detected by osmoreceptors in the hypothalamus. In case of increased plasma osmolality, even by 12%, the osmoreceptors stimulate vasopressin synthesis, also known as the antidiuretic hormone (ADH), in supraoptic and paraventricular nuclei of the hypothalamus. Then, it is secreted by the posterior pituitary gland. The ADH exerts its action at the collector duct, making it permeable to water. Following the interaction of ADH and its receptor on the cellular membrane of the collector duct, the activation of a sequence of events leads to the insertion in the apical cellular membrane of channel proteins, known as aquaporins, allowing the water passage. The effect is the passive reabsorption of water according to an osmotic gradient, from the lumen of the collecting tubule to the hypertonic medullary interstitium, and the production of concentrated urine. The ADH also acts on the ascending tract of Henle's loop, increasing sodium transport into the medullary interstitium. The resulting increase in interstitial osmolality contribute to the osmotic gradient for water reabsorption. In the absence of the ADH, the collector duct is impermeable to water. The most powerful stimulus for ADH release is the increase of plasma osmolality; however, other non-osmotic factors regulate ADH secretion, including blood pressure changes and volemia, nausea, pain, stress, hypoglycemia, pregnancy, and various medications. Reductions in blood volume and blood pressure are detected, respectively, by voloreceptors, stretch receptors located in the cardiac

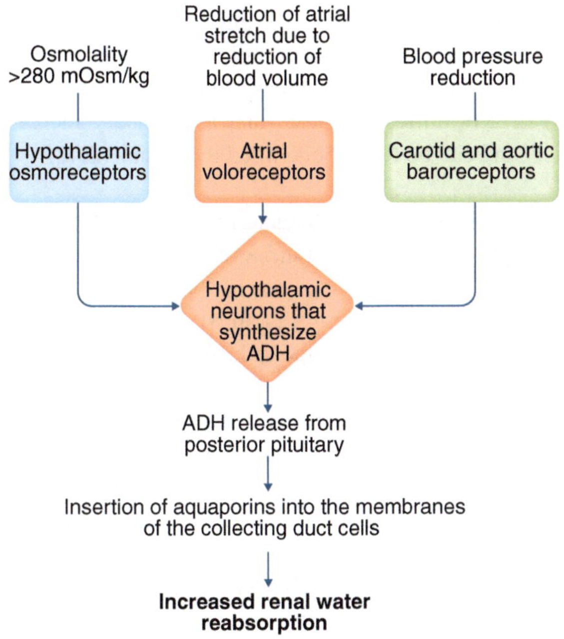

Fig. 20.3 Factors influencing ADH release. (Copyright EDISES 2021. Reproduced with permission)

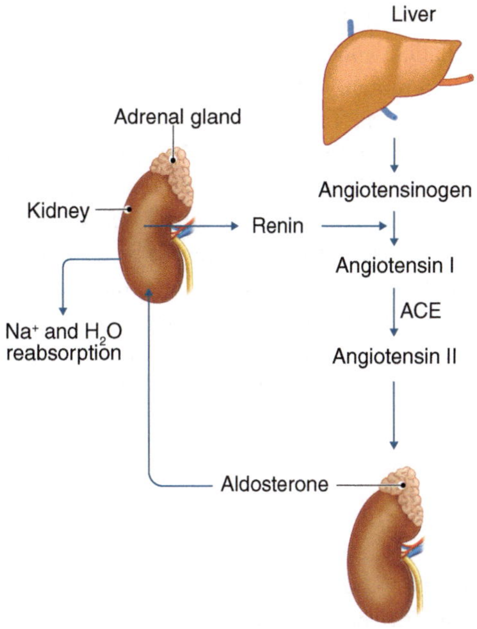

Fig. 20.4 Renin–angiotensin–aldosterone system. (Copyright EDISES 2021. Reproduced with permission)

atria, and baroreceptors in the carotid sinus, which stimulate hypothalamic ADH synthesis and release (Fig. 20.3). While the osmoreceptors detect even small changes in plasma osmolality (1%), the voloreceptors and baroreceptors are less sensitive, detecting a reduction of no less than 10%.

Sodium Balance

Sodium represents the primary electrolyte in extracellular fluids. Approximately 90% of Na^+ is in the extracellular compartment. Thus, the ECF volume reflects the body's Na^+ content. Na^+ is critical for controlling osmolality and maintening ECF volume and blood pressure.

Under physiological conditions, natremia (or sodiemia) has a value between 135 and 145 mmol/L. Our organism has regulatory mechanisms to maintain the balance between elimination and supply of Na^+. Indeed, as already mentioned above, everything that is introduced in excess into the body must be excreted. Since the Na^+ daily introduction by diet (200 mmol/L) exceeds our needs (20 mmol/L), renal excretion is necessary to keep sodium homeostasis. Pathological conditions, such as excessive sweating, bleeding, diarrhea, and vomiting are associated with Na^+ excretion through the kidneys.

Alterations in Na^+ intake and elimination balance can result in either an excess, which determines an increase in ECF volume, manifesting clinically with edema and hypertension, or a deficit, which results in ECF volume reduction, manifesting clinically with hypovolemia and hypotension. It is important to distinguish alterations between osmoregulation and volume regulation because osmolality and volemia are both independently regulated. Indeed, alterations in volume are perceived by detectors that stimulate compensatory mechanisms acting on the Na^+ balance regulation, whereas alterations in plasma osmolality are perceived by detectors that stimulate compensatory mechanisms acting on the water balance regulation.

Renal Na^+ excretion is fundamentally regulated by aldosterone, the primary mineralocorticoid hormone synthesized in the glomerular zone of the adrenal cortex, which controls Na^+ reabsorption in the distal tubule and collector duct. Increased aldosterone levels result in increased renal Na^+ reabsorption. Increased aldosterone-induced Na^+ reabsorption does not affect ECF Na^+ concentration because water reabsorption is simultaneously stimulated. The main stimulus for aldosterone secretion is angiotensin II, a protein of the renin–angiotensin–aldosterone system (RAAS) that is essential for the blood pressure maintenance. The RAAS system is activated in response to stimuli related, either directly or indirectly, to a reduction in blood pressure. In particular, it is activated when the juxtaglomerular cells of the afferent arterioles of the nephron secrete renin. This enzyme converts angiotensinogen, an inactive plasma protein, into angiotensin I, which is converted by the angiotensin-converting enzyme (ACE) into angiotensin II. The latter stimulates the adrenal gland to synthesize and release a ldosterone by (Fig. 20.4). Na^+ reabsorption results in an arterial pressure increase through indirect mechanisms, which include stimulation of water intake and expansion of blood volume.

Angiotensin II also plays a critical role in renal Na^+ excretion regulation, independently of aldosterone secretion. It acts, indeed, directly at the level of proximal tubule cells, stimulating the Na^+ reabsorption of Na^+.

As previous described, renal Na^+ reabsorption is also mediated by ADH at the Henle's loop. Finally, renal Na^+ reabsorption is regulated by atrial natriuretic peptide (ANP), a peptide hormone produced by specialized atrial myocardial cells. Following a volemia increase, atrial cells release ANP, which results in increased Na^+ renal excretion$^+$ through inhibition of renal renin secretion, a ldosterone production, and Na^+ reabsorption at the collector duct.

Hyponatremia

Hyponatremia, defined as plasma Na^+ concentration <135 mmol/L, represents the most common alteration of body fluids and electrolyte balance. From a clinical point of view, hyponatremia is characterized by a broad spectrum of signs and symptoms, from subtle to severe, and is associated with increased mortality and morbidity.

Hyponatremia is characterized by water excess over sodium content in ECF, which may result from water accumulation or Na^+ depletion.

Hyponatremia can be classified according to several criteria, such as Na^+ concentration (mild, moderate, severe), the timing of onset (acute or chronic), the severity of symptoms (moderate or severe), and serum osmolality (hypotonic, isotonic, and hypertonic) (Table 20.1).

Tonicity (or effective osmolality) is defined as the total osmolality capable of inducing the movement of water between compartments, and it is determined by all those solutes that cannot pass freely through cell membranes.

Table 20.1 Hyponatremia classification

Classification criterion	Description	Definition of hyponatremia
Na^+ serum concentration	130–135 mmol/L 125–129 mmol/L <125 mmol/L	Mild Moderate Severe
Serum osmolality	<280 mOsm/kg 280–295 mOsm/kg >295 mOsm/kg	Hypotonic Isotonic Hypertonic
Establishment times	For less than 48 hours For at least 48 hours	Acute Chronic
Symptoms	Moderate Severe	Moderate Severe

Hypotonic hyponatremia (or true hyponatremia), characterized by reduced plasma osmolality, is generally due to either a primitive water increase (and a secondary loss of Na^+) or a primitive Na^+ loss (and a secondary increase in water) and can be further classified into (Table 20.2):

- Hypovolemic: This is characterized by ECF depletion due to a primitive Na^+ loss (and a relative increase of water) that can be of renal or extrarenal origin. The contraction of ECF volume stimulates thirst and ADH secretion. Increased water intake and reduced renal excretion result in hyponatremia.
- Euvolemic: This is characterized by a modest ECF excess due to a primitive water increase and relative Na^+ deficiency. The syndrome of inappropriate ADH secretion (SIADH) represents the most common cause of euvolemic hyponatremia.
- Hypervolemic: This is characterized by an ECF excess due to a primitive Na^+ increase in Na^+ less than the secondary water increase. It is usually associated with edema. This form is due to pathologies, such as liver cirrhosis, which are associated with a reduction in effective circulating arterial volume resulting in increased ADH levels and a sense of thirst. Chronic and acute renal failure with oliguria may be associated with hyponatremia if water intake exceeds its elimination capacity.

Isotonic hyponatremia (or pseudohyponatremia), characterized by normal plasma osmolality, results from the accumulation of large amounts of isotonic substances lacking Na^+ that can freely cross membranes, such as urea, non-hypertonic mannitol, and ethanol, causing pseudohyponatremia (laboratory artifact).

Hypertonic hyponatremia, characterized by increased plasma osmolality, is the result of water transfer from cells to the extracellular compartment due to an excess of osmotic substances. It is usually due to hyperglycemia. In a patient with diabetes mellitus with poor metabolic control, glucose becomes an effective osmolyte capable of drawing water into the extracellular environment, resulting in hyponatremia; for each excess of 100 mg/dL in plasma glucose concentration, there is a 1.4 mmol/L reduction in plasma Na^+ concentration. In addition, intravenous administration of mannitol is another cause of hypertonic hyponatremia.

The clinical manifestations of hyponatremia are related to the water osmotic displacement, which results in an increased ICF volume. Neurons are the most susceptible to this alteration due to the inability of the cranial box to expand. Consequently, the symptoms of hyponatremia are predominantly neurological (hyponatremic encephalopathy) and their severity depends on the absolute Na^+ reduc-

Table 20.2 Causes of hypotonic hyponatremia classified by volume

Hypovolemia		Euvolemia	Hypervolemia
Kidney losses	*Extrarenal losses*	SIADH	Heart failure
Diuretics	Sweating	Excessive water intake, e.g., primitive polydipsia	Liver cirrhosis
Osmotic diuresis (glucose, mannitol)	Burns	Reduced food intake and urinary excretion of solutes (e.g., potomania or low protein diet)	Nephrotic syndrome Chronic and acute renal failure
Hypoaldosteronism	Vomiting	ADH secretion due to pain, nausea, drugs	
Sodium-dispersing nephropathy	Diarrhea	Glucocorticoid deficiency	
Post-obstructive diuresis	Drainage	Hypothyroidism	
Acute non-oliguric tubular necrosis	Fistulas	Chronic renal failure	
	Small bowel occlusion		
	Pancreatitis		
	Peritonitis		

SIADH syndrome of inappropriate, ADH secretion

Table 20.3 Symptoms of hyponatremia

Severity	Symptom
Mild	Asymptomatic
Moderate	Headache Nausea General malaise Mood alteration Difficulty concentrating Mental confusion
Serious	Vomiting Respiratory distress Lethargy Convulsions Coma (Glasgow Coma Scale ≤8)

tion in Na^+ and the development rapidity. Mild forms are generally asymptomatic. Nausea and general malaise are the first symptoms of moderate forms. The symptoms of hyponatremia, especially when it is mild or moderate, are nonspecific and often lead to a failure to diagnose it, with consequent therapeutic delay (Table 20.3). Seizures occur at extremely low natremia values, with a high risk of permanent brain damage. The guidelines suggest a threshold of 48 hours to distinguish acute from chronic hyponatremia because cerebral edema occurs more frequently when hyponatremia develops in less than 48 hours. Indeed, the brain needs about 48 hours to adapt to a hypotonic environment and, before adaptation, there is a high risk of edema. Adaptation involves the activation of compensatory mechanisms leading to the loss of Na^+ and K^+, followed by the loss of organic osmolytes, aimed at protecting cell volume through the displacement of water from ICF to ECF. However, once adaptation is complete, if Na^+ concentration increases too rapidly, neurons may undergo further damage. One of the most serious complications associated with correcting hyponatremia is the demyelinating osmotic syndrome, characterized by the breakdown of the myelin sheath that insulates neurons.

Diagnosis and Therapy

Hyponatremia represents the manifestation of numerous diseases. Therefore, the diagnostic approach to a patient with hyponatremia is first based on history and objective examination that includes the assessment of the ECF status and circulating volume. Gathering the patient's medical history may direct toward the cause of hyponatremia.

Clinical laboratory provides valuable information for the differential diagnosis of hyponatremia. In particular, the key tests include plasma and urinary osmolality, and urinary Na^+ concentration (Fig. 20.5). In addition, blood glucose and hormonal patterns (including the hypothalamic–pituitary–thyroid and hypothalamic–pituitary–adrenal axes) provides essential support for differential diagnosis.

Recently, an algorithm for the differential diagnosis of the causes of hyponatremia has been proposed and it is based on the determination of copeptin (Fig. 20.6).

SIADH is characterized by hypotonic hyponatremia with inappropriately concentrated urine (urinary osmolality >100 mOsm/kg). Patients usually are euvolemic and have a normal Na^+ balance.

The aim of hyponatremia therapy is, to correct the plasma Na^+ concentration and resolve the primary disease underlying altered Na^+ concentration. In an asymptomatic patient with mild hyponatremia, therapy consists of treating the underlying pathologies, removing the causal factors, such as the use of diuretics, the use of drugs that can cause SIADH, as well as stress, pain, and anxiety, which can stimulate the secretion of ADH. In a symptomatic patient, therapy is based on the correction of hyponatremia through the administration of 0.9% isotonic saline solution or 3% or 5% hypertonic saline solution by slow drip.

Hypernatremia

Hypernatremia is defined as a plasma Na^+ concentration >145 mmol/L. Since Na^+ is the main osmolyte of ECF, hypernatremia results in a state of hyperosmolality. Hypernatremia may be due to a primitive increase in Na^+ but,

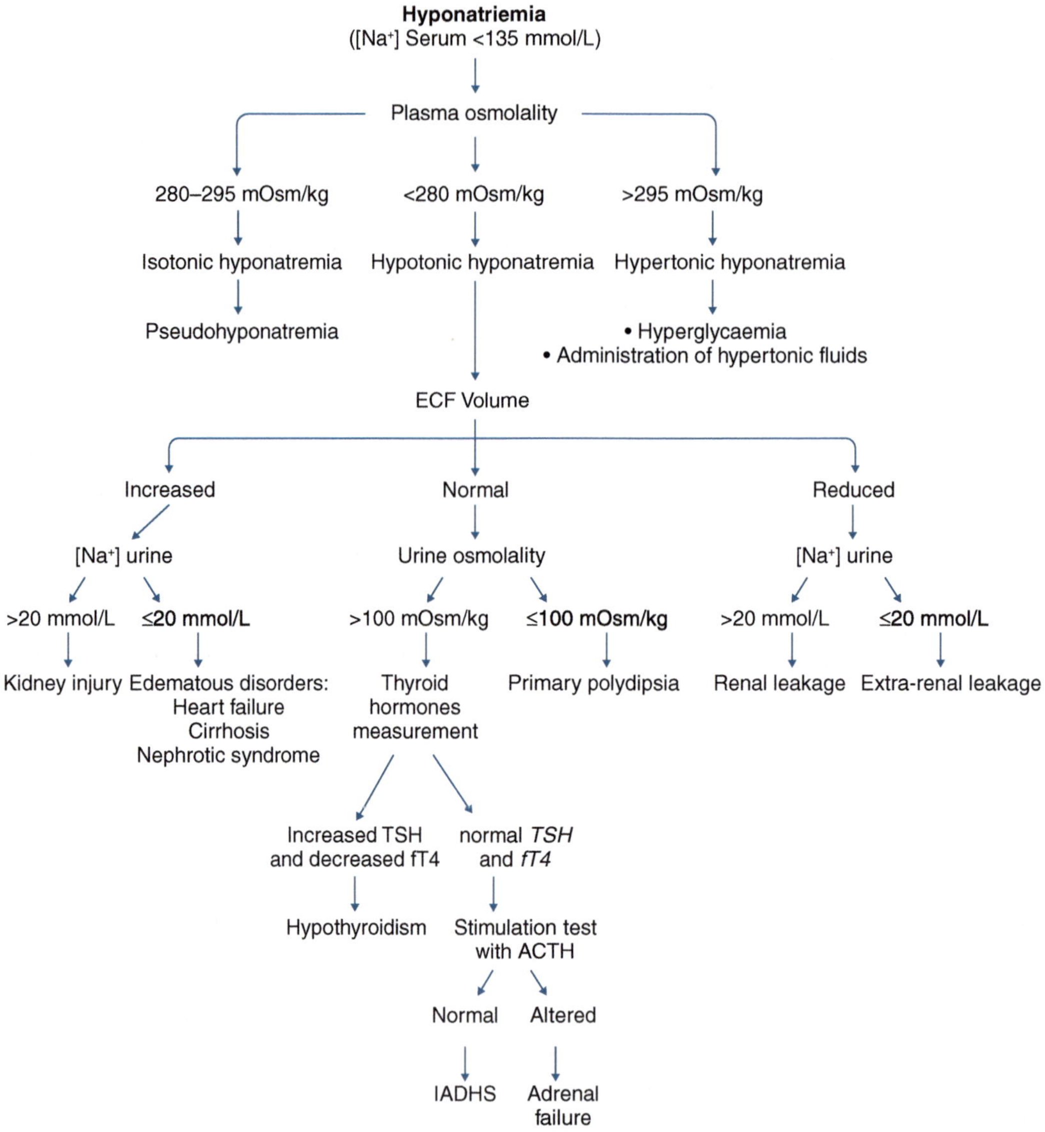

Fig. 20.5 Diagnostic algorithm of hyponatremia. (Copyright EDISES 2021. Reproduced with permission)

in most cases, is the result of water loss. Water loss may consist of:

- Loss of body water (with an associated increase in Na^+)
- Loss of hypotonic fluid, which can be of renal or extrarenal origin (the loss of body water is accompanied by a loss of Na^+, which is in lesser proportion and, therefore, there is a relative increase).

Table 20.4 shows the causes of hypernatremia.

Hyperosmolality due to hypernatremia results in the water movement to the extracellular environment, thus leading to the contraction of ICF volume. The cells of the central nervous system are the most affected. Therefore, the main symptoms of hypernatremia are neurological and include asthenia, neuromuscular irritability, confusion, coma, and convulsions. Patients may also present with polyuria, intense thirst, nausea, and vomiting. The severity of clinical manifestations depends on the rate and magnitude of increase in plasma Na^+ concentration. In chronic hypernatremia, brain cells activate an adaptive response consisting of initial electrolyte acquisition (early compensation) followed by production and accumulation of intracellular osmolytes, such as inositol. The patients at risk of developing hypernatremia are young children, elderly subjects, and neurological patients who have already been hospitalized.

Diagnosis and Therapy

A complete medical history, including a list of medications prior to and concurrent with the visit, and a physical examination, with the assessment of mental and neurological status, are the first investigations allowing the detection

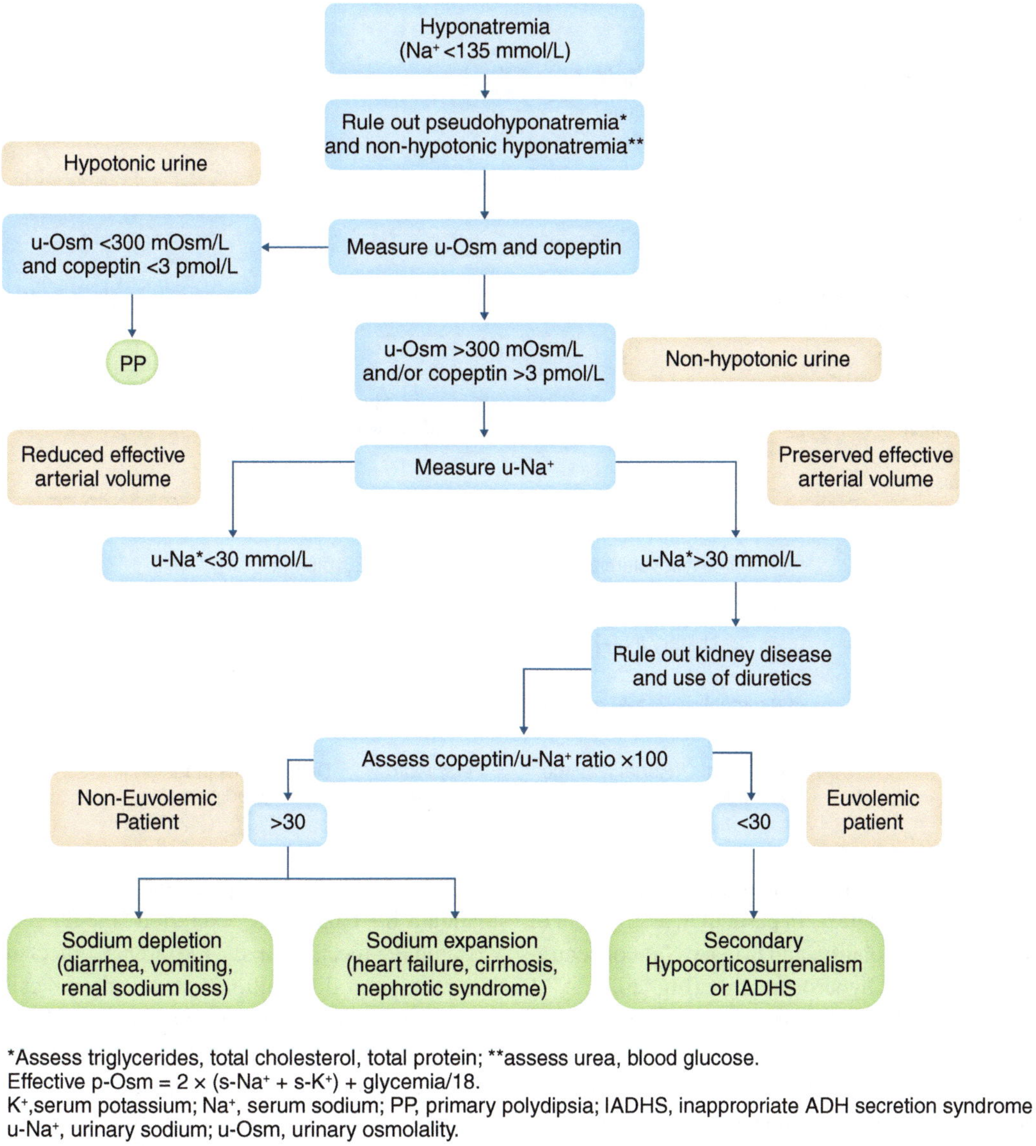

Fig. 20.6 Algorithm for the differential diagnosis of hyponatremia based on copeptin. (Copyright EDISES 2021. Reproduced with permission)

Table 20.4 Causes of hypernatremia

Water loss alone	– *Perspiratio insensibilis* (evaporation from the skin or respiratory tract) – Hypodipsia/adipsia (reduction/absence of thirst by insufficient water intake) – Neurogenic diabetes insipidus (excessive excretion of water by the kidneys due to impaired ADH production) – Nephrogenic diabetes insipidus (reduced sensitivity to ADH in the kidneys)
Hypotonic fluid loss	Kidney causes: – Loop diuretics – Osmotic diuresis (glucose, mannitol)
	Urinary obstruction
	Extrarenal causes: Vomiting Diarrhea Fever Profuse sweating Burns Nasogastric drainage

(continued)

Table 20.4 (continued)

Primitive Na$^+$ increase	Increased production: Infusion of hypertonic NaCl solution Introduction of sodium bicarbonate Emetics rich in sodium chloride Hypertonic saline enemas Enteral and parenteral nutrition Corticosteroid therapy Intake of improperly diluted infant formula Renal retention: Primitive hyperaldosteronism Cushing's syndrome

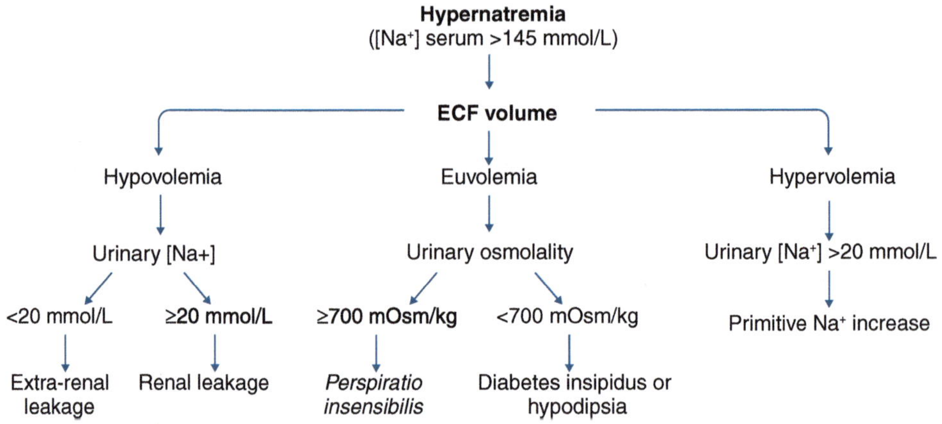

Fig. 20.7 Diagnostic algorithm of hypernatremia. (Copyright EDISES 2021. Reproduced with permission)

of hypernatremia cause. The clinical laboratory provide important support in this context. In particular, assessment of ECF volume, urinary osmolality, and urinary Na$^+$ concentration are essential to assess hyperosmolality (Fig. 20.7).

Therapy aims to treat the underlying disease and correct the water deficit. Patients with hypernatremia in less than 24 hours, natremia should be corrected rapidly, i.e., about 1–2 mmol/L per hour. Patients with hypernatremia for a more extended period, as early as a few days, the correction should be slower, i.e., 0.5 mmol/L per hour, because the cerebral adaptation has been activated and, therefore, a rapid correction could lead to cerebral edema, coma, convulsions, and death.

Potassium Balance

Potassium is the primary ICF electrolyte; only about 2% is found in ECF. The intracellular concentration is 140–150 mmol/L, whereas the extracellular concentration is 3.5–5 mmol/L. The difference in K$^+$ concentration between ICF and ECF is the primary determinant of the resting membrane potential, which, in turn, is critical for nerve impulse conduction and muscle cell contraction, including cardiac cells. K$^+$ can passively diffuse from the inside to the outside of the cell. The Na$^+$/K$^+$ATP-dependent pump actively transports it from the outside to the inside.

The dietary intake of K$^+$ is approximately 1–2 mmol/day per kg of body weight, of which 90% is absorbed in the gastrointestinal tract. An equal amount must be excreted to maintain K+ balance. K$^+$ excretion is regulated by renal and extrarenal mechanisms. Extrarenal mechanisms regulate potassium distribution (intra- and extracellular) by acting directly or indirectly on the Na$^+$/K$^+$ATP-dependent pump. The activity of this pump is influenced by various factors, such as insulin, catecholamines, aldosterone, acid–base balance, osmolality, and adrenergic drugs. In particular, beta-agonists promote the entry of K$^+$ into the cell, whereas beta-blockers and alpha-agonists promote the movement of K$^+$ into the extracellular environment. Insulin promotes the entry of K$^+$ inside the cells (especially at the hepatic and muscular level); insulin deficiency induces a shift of potassium from the intracellular to the extracellular compartment. Potassium balance is closely related to the acid–base balance. In particular, potassium and pH vary in opposite directions: a decrease in pH of 0.1 units results in an increase in potassium of 0.5 mmol/L and vice versa. In metabolic acidosis, there is the movement of K$^+$ toward the outside of the cell in exchange with protons, whose blood concentrations are reduced; vice versa, in metabolic alkalosis, there is the passage of K$^+$ inside the cell in exchange with protons, whose blood concentrations increase.

However, the main route of eliminating potassium excess is renal excretion at the distal nephron. The most important regulator of renal potassium excretion is aldosterone, which stimulates tubular potassium secretion by increasing the number of Na$^+$/K$^+$ATP-dependent pumps and sodium channels. Filtered potassium is almost completely reabsorbed in the proximal tubule and secreted into the distal convoluted tubule and the collector tubule. Net excretion is, therefore, the result of reabsorption and concomitant secretion.

Hyperpotassemia determines an increase in potassium excretion through basically two mechanisms: aldosterone secretion and tubular potassium secretion through the increase of the same in the tubular cell. Indeed, the Na^+/K^+ATP-dependent pump increases the K^+ intracellular concentration, which will be released into the tubular lumen. During hypokalemia, however, the renal excretion of potassium is decreased or absent.

Due to the important action of potassium on excitable tissues, it is essential to keep potassium values within the normal range; an alteration can lead to impaired functionality of the nervous and muscular tissues. In particular, low blood levels of potassium (hypokalemia) are associated with muscular weakness due to the greater difficulty of hyperpolarized neurons and muscles to trigger action potentials; hyperkalemia, on the other hand, is a more dangerous condition because the tissues depolarization initially makes them more excitable but subsequently, the cells are unable to repolarize entirely and, therefore, become less excitable. In this condition, they present action potentials smaller than normal or absent. Hyperkalemia can cause severe cardiac arrhythmias.

Hypokalemia

Hypokalemia is defined as a plasma potassium concentration <3.5 mmol/L. It is mild when serum potassium values are between 3 and 3.4 mmol/L; it is severe when serum potassium values are less than 3 mmol/L.

The causes of hypokalemia can be distinguished into three classes:

- Reduced dietary intake
- Passage of potassium into the intracellular compartment
- Increased renal or extrarenal leakage

Table 20.5 summarizes the main causes of hypokalemia.

Reduced dietary potassium intake is a rare cause because it can be compensated by reduced renal potassium excretion. Commonly, dietary potassium intake exceeds the amount excreted in the urine. Several causes of hypokalemia due to renal potassium loss lead to excess aldosterone (primary and secondary hyperaldosteronism) or, in general, excess mineralocorticoid and glucocorticoid hormones.

From a clinical point of view, the symptoms of hypokalemia rarely appear unless the potassium concentration is less than 3 mmol/L. Symptoms are related to alterations in muscular system (muscular asthenia, myalgia, muscle cramps), the cardiovascular system (palpitations, tachycardia, electrocardiographic changes), nervous system (mental confusion, depression), urinary system (polyuria, renal morphological damage), gastrointestinal system (dysphagia, constipation, nausea, vomiting), and metabolism (reduced protidosynthesis, reduced glucose tolerance, metabolic alkalosis).

Table 20.5 Causes of hypokalemia

Food intake reduction	Malnutrition due to: – Neoplasms – Nervous anorexia – Acquired immunodeficiency syndrome (AIDS) – Abuse of ethanol – Poverty Parenteral nutrition not integrated with potassium supplements	
Potassium passage into the intracellular compartment	Alterations of acid–base balance: – Metabolic or respiratory alkalosis Hormonal causes: – Insulin – Beta-2 adrenergic agonists (endogenous or exogenous) – Alpha-adrenergic antagonists Anabolic conditions: – Increased hematopoiesis following vitamin B_{12}, folate, and erythropoietin administration Other causes: – Accidental or induced hypothermia – Barium intoxication – Periodic hypokalemic paralysis	
Increased lost	Renal	Potassium-sparing diuretic therapy Excess of steroids: – Primary aldosteronism – Secondary aldosteronism – Cushing's syndrome – Adrenogenital syndromes – Steroid therapy – Apparent excess of mineralocorticoids due to abuse of licorice and chewing tobacco, containing glycyrrhizic acid Metabolic alkalosis Osmotic diuresis Abuse of ethanol, due to its diuretic effect Drugs (amphotericin B, cyclosporine, nonsteroidal anti-inflammatory drugs (NSAIDs)) Hypomagnesemia Renal tubular acidosis
	Extrarenal	Vomiting Nasogastric drainage Malabsorption Diarrhea Profuse sweating Extensive burns

Diagnosis and Therapy

Hypokalemia is often asymptomatic. Evaluation begins by looking for signs and symptoms that require urgent treatment, such as asthenia or palpitations, electrocardiogram

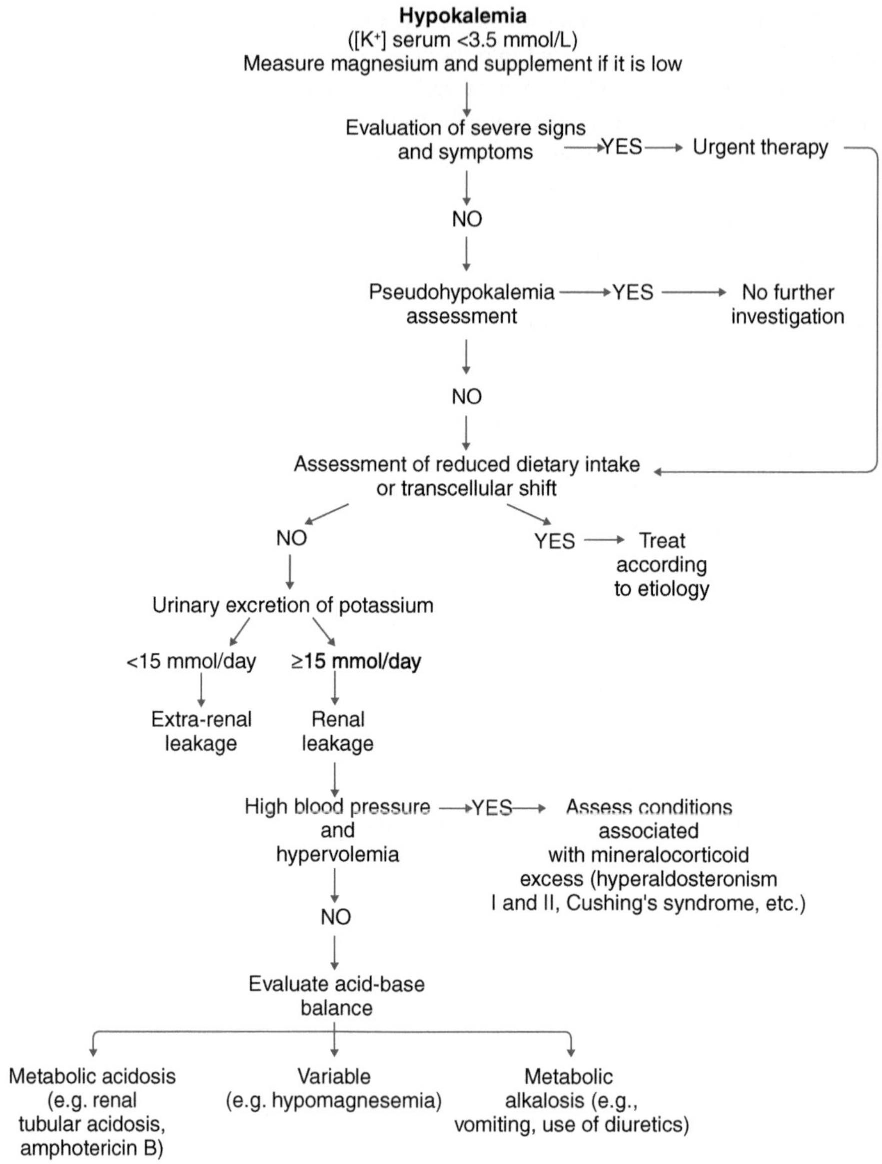

Fig. 20.8 Diagnostic algorithm of hypokalemia. (Copyright EDISES 2021. Reproduced with permission)

(ECG) changes, severe hypokalemia, rapid onset of hypokalemia, underlying heart disease, or cirrhosis. Identification and treatment of concurrent hypomagnesemia is critical because magnesium depletion often prevents potassium correction and may exacerbate hypokalemia-induced heart rhythm alterations.

In most cases, an accurate history allows to identify the K^+ depletion etiology. Diuretic or laxative abuse as well as induced vomiting, although difficult to ascertain, should be excluded. In patients with marked leukocytosis (e.g., in acute myeloid leukemia) and normokalemia, plasma potassium concentration may be reduced due to a laboratory artifact related to the ability of leukocytes to capture K^+ at room temperature. This pseudohypokalemia can be avoided by separating plasma (or serum) from cells immediately after collection.

A history of paralysis, hyperthyroidism, or the use of insulin or beta-agonists suggests a possible transcellular shift leading to redistribution-related hypokalemia (passage of K^+ within cells).

After ruling out reduced dietary intake or transcellular displacement of K^+, the evaluation of renal response can help detect the hypokalemia cause. Hypokalemia with minimal urinary potassium excretion <15 mmol/day suggests that potassium loss is due to extrarenal causes (through the skin or gastrointestinal tract). Assessment of acid–base balance by hemogasanalysis may provide helpful information about the potassium loss site. However, an increased urinary potassium excretion (>15 mmol/day) suggests the renal potassium loss. ECF volume status, blood pressure, and acid–base balance alterations may identify the cause of excessive renal potassium loss (Fig. 20.8).

Therapy aims to increase potassium levels, minimize ongoing losses, and treat the underlying disease to prevent complications. KCl administration (potassium chloride) is the choice therapy. KCl results in a rapid increase in plasma potassium concentration. In addition, the persistence of the Cl^- ion in ECF limits the initial potassium entry into the cells, thus promoting an increase in plasma potassium. Finally, chloride supports metabolic alkalosis, often associated with hypokalemia. In cases of mild hypokalemia, KCl is administered orally or intravenously, in more severe cases.

Hyperkalemia

Hyperkalemia, defined as plasma K^+ concentration >5 mmol/L, may be due to reduced renal K^+ excretion$^+$, increased cellular K^+ release, increased K^+ intake, and pseudohyperkalemia (Table 20.6). Increased dietary K^+ intake alone generally does not cause hyperkalemia because excess K^+ is rapidly excreted by the adaptive mechanism based on:

- Increased cellular potassium uptake, facilitated by insulin
- Increased beta-2 adrenergic tone
- Increased urinary potassium excretion

Therefore, hyperkalemia from increased intake can occur in the presence of other clinical conditions, such as renal impairment.

Pseudohyperkalemia represents an artificial increase in plasma potassium concentration due to potassium leakage from cells during venous sampling. Thrombocytosis causes hyperkalemia due to the potassium release by platelets during coagulation. Pseudohyperkalemia should be suspected in asymptomatic patients with no obvious underlying cause.

The etiology of hyperkalemia is often multifactorial; impaired renal function, medications, and hyperglycemia are the primary causes.

The symptoms of hyperkalemia are due to the difference in potassium concentration between ICF and ECF, which influence the resting membrane potential. Hyperkalemia results in partial membrane depolarization. The progression of the clinical picture is characterized by prolonged cellular depolarization , which reduces membrane excitability and is manifested by asthenia, which may progress to flaccid paralysis and hypoventilation. In addition, cardiovascular symptoms, such as bradycardia, hypotension, and conduction disturbances up to the cardiac arrest can also occur. Numerous electrocardiographic alterations may occur during hyperkalemia.

Table 20.6 Causes of hyperkalemia

Reduced renal excretion	Chronic renal failure Acute renal failure Drugs: – Inhibitors of the enzyme that converts angiotensinogen – Angiotensin receptor blockers – Nonsteroidal anti-inflammatory drugs – Potassium-sparing diuretics – Heparin – Lithium – Calcineurin inhibitors Reduced distal renal flow Hypoaldosteronism – Hyporeninemic hypoaldosteronism – Adrenal insufficiency – Adrenocorticotropic hormone deficiency Tubular diseases – Obstructive uropathy – Hereditary tubular defects – Amyloidosis
Increased potassium intracellular release	Insulin deficiency/resistance Metabolic and respiratory acidosis Hypertonicity – Hyperglycemia – Mannitol Drugs – Beta-blockers – Digitalis poisoning – Somatostatin – Arginine – Succinylcholine Increased tissue catabolism – Hemolysis – Widespread trauma – Rhabdomyolysis Hyperkalemic periodic paralysis Intense physical exercise
Increased intake	Potassium supplements Blood transfusion Foods
Pseudohyperkalemia	Hemolysis – Tourniquet kept for an excessive period – Excessive patient muscle contraction during sampling – Too rapid aspiration of blood by a fine needle Thrombocytosis (platelet count >106)

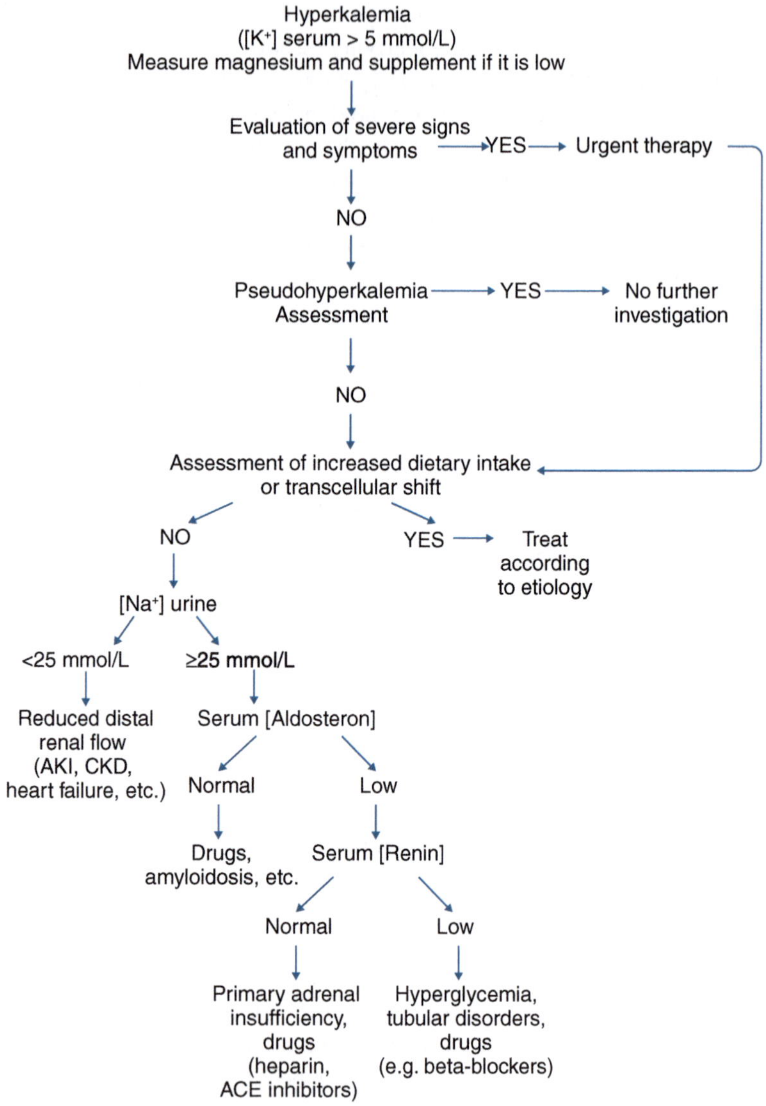

Fig. 20.9 Diagnostic algorithm of hyperkalemia. (Copyright EDISES 2021. Reproduced with permission)

Diagnosis and Therapy

Similar to hypokalemia, hyperkalemia alters cardiac conduction and muscle strength. Initial efforts should focus on determining the need for urgent intervention (Fig. 20.9). The absence of symptoms does not rule out severe hyperkalemia because it is often asymptomatic.

The hyperkalemia severity is assessed based on symptoms, plasma potassium concentration, and electrocardiographic alterations. A careful medical history is essential to identify the possible drugs intake that alter potassium excretion or induce its release from the cells. Urinary sodium, serum aldosterone and renin levels, and acid–base balance assessed by hemogasanalysis provide helpful information for the differential diagnosis (Fig. 20.8).

If the etiology is not initially evident and the patient is asymptomatic, the presence of pseudohyperkalemia should be evaluated.

Therapy depends on the degree of hyperkalemia, which is determined by potassium concentration, muscle signs, and electrocardiographic changes. Therapy aims to block the effects of hyperkalemia on cell membranes, increase potassium transfer within cells, and remove excess potassium and the cause of hyperkalemia. This is carried out using calcium gluconate, which reduces membrane excitability; insulin and glucose because insulin stimulates potassium entry into cells and glucose prevents hypoglycemia; alkalinizing treatment with sodium bicarbonate, especially in patients with severe hyperkalemia and metabolic acidosis, because it favors the movement of potassium inside the cells; and treatment with diuretics that favor the renal potassium excretion (the sodium excreted as a result of the diuretic must be reintegrated through the physiological solution). Finally, hemodialysis represents the quickest method to reduce plasma potassium concentration but it should be reserved for patients with renal insufficiency and for those with severe, potentially fatal, hyperkalemia who do not respond to other therapies.

Recommended Readings

Adrogue HJ, Madias NE (2000) Hyponatremia. N Engl J Med 342(21):1581–1589

Diercks DB, Shumaik GM, Harrigan RA (2004) Electro- cardiographic manifestations: electrolyte abnorma- lities. J Emerg Med 27:153–160

Elmi G, Faustini-Fustini M, Zaccaroni S, Zoni R (2011) Italian J Med 5(4):288

Gumz ML, Rabinowitz L, Wingo CS (2015) An integrated view of potassium homeostasis. N Engl J Med 373(1):60–72

Hoorn EJ, Zietse R (2017) Diagnosis and treatment of hyponatremia: compilation of the guidelines. J Am Soc Nephrol 28(5):1340–1349

Viera AJ, Wouk N (2015) Potassium disorders: hypokalemia and hyperkalemia. Am Fam Physician 92(6):487–495

Blood Gas Analysis

21

Paolo Carraro

Introduction

Blood gas analysis is the measurement of the pH of the blood and the partial pressures of oxygen (pO_2) and carbon dioxide (pCO_2). These laboratory variables are always associated with the determination of hemoglobin or hematocrit, the barometric pressure of local atmosphere, and several calculated variables.

With simplification, we could distinguish the two main purposes of this examination: the evaluation of the state of oxygenation and the acid–base balance.

Interpretation of the results requires adequate knowledge of pathophysiology and a precise picture of the clinical condition of the patient in terms of symptoms presented, previous pathological conditions, and treatment in progress, both pharmacological and ventilatory.

This profile of examinations is often also associated with other determinations that complete the information on the patient's status: oximetry (commonly co-oximetry) that verifies the fraction of oxygenated hemoglobin, reduced or inactivated by toxic situations, electrolytes, glucose and lactate metabolites, and sometimes other substrates such as urea, creatinine, and bilirubin.

It is well known that this topic is quite complex, but, here, in particular, we treat the aspects of competence and management by a clinical laboratory, whereas the clinical aspects related to the main pathological conditions will be visited in a more synthetic way.

Elements of Pathophysiology

As already mentioned, with blood gas analysis, we can actually obtain information on two particular pathophysiological balances, namely, the state of oxygenation and the acid–base balance, although these two functions are intrinsically connected.

The State of Oxygenation

Blood oxygenation varies with the barometric pressure, composition of the inhaled air, ventilatory volume, composition of the air at the alveolar level, state of the pulmonary alveoli, efficiency of blood circulation, and composition of the blood, in particular the amount of hemoglobin. To understand the gradient of gas exchanges, the composition of ambient air at sea level must be considered, as shown in Table 21.1. From this, the extent of the possible increase in oxygenation can be deduced by enriching the air with O_2. Indeed, if hypothetically pure humidified oxygen is administered (irrespective of however irritating to the mucous membranes), then it is possible to reach the partial pressure theoretically tending to 750 mmHg with the aim of forcing insufficient oxygenation for circulatory or parenchymal pulmonary disorders. Figure 21.1 shows the physiological pressure gradients that push oxygen from ambient air to the blood and then to the tissues. It is a schematic representation that has great limitations but can help us understand the physiology of oxygenation, at least under normal conditions, by breathing ambient air at sea level.

The variable of greatest clinical interest for this evaluation is represented by oxygen saturation (sO_2), normally provided by all blood gas analysis instruments. It is a parameter

Table 21.1 Relative composition of the main gases in ambient air and their typical partial pressure at sea level at a barometric pressure of 760 mmHg

	Percentage	Millimeters of mercury (mmHg)
Nitrogen	78	593
Oxygen	21	159
Argon and other gases	1	8
CO_2	0,03	0,2
H_2O vapor	0–7	

P. Carraro (✉)
Laboratory Medicine, Aziende ULSS3 Serenissima, Venice, Italy
e-mail: paolo.carraro@aulss3.veneto.it

M. Ciaccio (ed.), *Clinical and Laboratory Medicine Textbook*, https://doi.org/10.1007/978-3-031-24958-7_21

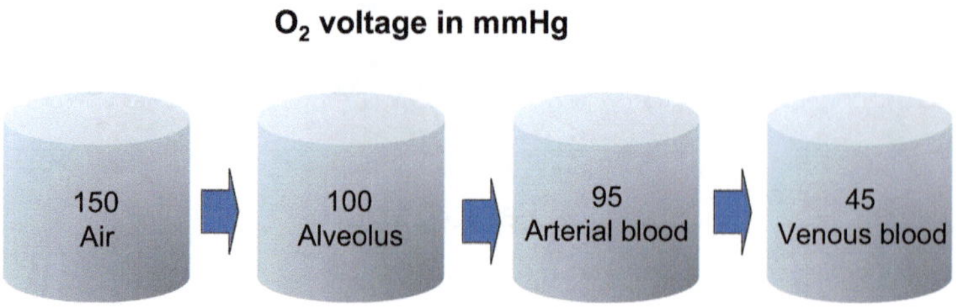

Fig. 21.1 A schematic representation of the partial pressure gradients of oxygen, from the inhaled air to the venous blood, in millimeters of mercury. This scheme refers to a healthy subject, at sea level, in the absence of therapeutic oxygen supplementation

that can be calculated from pO_2 and hemoglobin by means of an empirical equation of the hemoglobin dissociation curve, with correction factors related to pH, body temperature, and carbon dioxide (CO_2). The limitation of this estimate is not considering the effect of diphosphoglycerate (for example, in transfusions) or the presence of pathological hemoglobin. For these reasons, in the intensive field, it is preferred not to routinely use this type of calculation but to rely on the evaluation of an oximeter.

With this analyzer, fractions of hemoglobin are directly measured by the spectrophotometric method and more accurate measurements are obtained, through the formula: oxyhemoglobin/(oxyhemoglobin + reduced hemoglobin) %.

Values of sO_2 <95% are generally considered to indicate hypoxemia and the widespread, defined condition of "desaturation".

The Acid–Base Balance

The pH of the blood and, therefore, of the main metabolic environments of the body is close to 7.38, with an extremely low oscillation tolerance. In fact, we speak of acidemia already below 7.35 and alkalemia above 7.45, whereas those that exceed the limits of 7.10 and 7.70, respectively, are generally considered critical values. The buffer system of the blood most involved in its homeostatic maintenance is that of bicarbonate–carbonic acid, although there is buffering competition between hemoglobin, albumin, phosphates, and other substances. The physiological importance of bicarbonates, as shown in Fig. 21.2, is due to the intervention of organs, such as the lungs, which can modulate their effectiveness by intervening to compensate for any variations: variations in ventilation (volume of air displaced per minute) can either increase or decrease the content of carbon dioxide in the blood, considered the acid component of the buffer; even the kidneys carry out control function by acting on the urinary elimination of bicarbonates, the main alkaline component. All this happens under physiological conditions but can be accentuated in pathology, for example, if there is an increase in organic acids due to a metabolic disease or an introduction of acids or strong bases following an intoxication. However, we must also consider the limits of these compensations in the sense that the lungs, in case of metabolic acidosis, can certainly eliminate more CO_2 in an

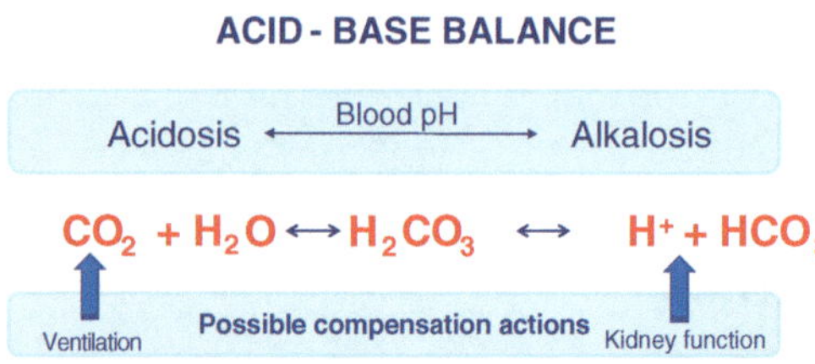

Fig. 21.2 The equilibrium equation of a bicarbonate buffer. Above are shown the effects of its displacement on pH; below the points of action of possible renal and ventilatory compensation

extremely short time but within a quantitative limit that normally stands at around 15 mmHg. Renal function in retaining or eliminating bicarbonates is also limited, in particular, by a certain slowness of effectiveness requiring 2 or 3 days to reach full capacity. It is precisely these limitations, however, that allow us to distinguish the alterations of balance in acute and chronic forms, whether compensated or not. Then, we must consider the pathologies that can affect the two control organs, namely, the kidneys and the lungs, producing imbalances that must be carefully interpreted. Finally, we must also keep in mind the interactions between the control of oxygenation and that of the acid–base balance: in particular, it is low oxygenation that induces, for example, an increase in ventilation as a compensatory attempt, determining an alkalotic tendency that the kidneys will tend to control by increasing the elimination of bicarbonate. It is therefore clear how these balances have rendered blood gas analysis the fundamental examination for understanding the state of vital functions, resulting in those who know how to correctly interpret the pathophysiological state of a patient, to reinforce this consideration and also to observe the mechanisms by which the body controls these functions. In practice, we are equipped with biological "sensors" that are highly similar to the electrodes used in the instruments, located at the brainstem and at the level of the carotids, which are able to detect even small variations in pH and pO_2 by rapidly activating, both nervously and hormonally, the corrective actions.

Instrumentation

A blood gas analyzer can be considered the paradigm of the instrument intended for point-of-care testing: since its debut, with the first models set up by Astrup in 1954 and by John

Severinghaus in 1957, the utmost attention was paid to simplicity of use. In highly synthetic terms, we can say that in the past the instruments were composed of four electrodes (one of which was a reference and three were selective) that required frequent replacement of the membranes and a fluidic system to move the sample of untreated blood into the measuring chamber; in addition buffered solutions to calibrate the pH, and two gas mixtures for the calibration of pO_2 and pCO_2 were required. Subsequent developments anticipated automation and miniaturization technologies that preceded laboratory analyzers by many years. In practice, these were the first so-called "point-of-care testing" devices suitable to be used even by operators with limited skills for laboratory diagnostics.

The heart of modern instruments is still based on the first one in 1957 that employed selective electrodes: the glass membrane electrode for pH, the Clarke electrode for oxygen, and the Astrup electrode for carbon dioxide. During the subsequent evolution, the measurement of hemoglobin soon appeared as a necessity to calculate oxygen saturation and other derived parameters and became available by conductivity or the colorimetric method. Between the 80s and 90s several other ion-selective electrodes for the most important species were also integrated, and later also biosensors based on the immobilization of enzymes for some substrates; finally colorimeters endowed of different wavelengths to measure the different states of hemoglobin on whole blood, after ultrasonic hemolysis, have been introduced. Based on this advancement, the analyzer has became a small laboratory, At the end of the 90s, was automatic, did not need manual calibration procedures, had reduced maintenance, and was able to analyze small volumes of blood.

The subsequent evolutions can be essentially summarized through four innovations:

- Abolition of the use of gas mixtures for the calibration of O_2 and CO_2 through the development of containers for calibration liquids capable of maintaining stable dissolved gas contents
- Prolongation of the life of individual electrodes, sometimes without the need to periodically replace the membranes
- Miniaturization of the electrodes and their integration into a single cartridge containing various solutions, fluidics, and the measuring cell (so-called "cartridge" instruments)
- Automation of quality control procedures, including operational decisions resulting from the results obtained
- Full connectivity for connection to the computer systems of the department, hospital, and laboratory

The different characteristics of the instruments currently on the market can be deduced from the greater or lesser ability to integrate these innovations into equipment that are progressively less demanding to manage.

The class of portable instruments has recently been enriched with some new models: they are based on a handheld (in some cases pocket) card containing fluidics, electrodes, and calibrators that allow maximum portability even in rescue vehicles and in the absence of gravity. Less miniaturized models are close in concept to bench tools but retain the logic of a disposable board. Their analytical reliability is precisely based on guaranteeing new materials to each test, whereas their criticisms concerns the sampling method and the small volume of blood used. For this reason, they require targeted training and methods of use strictly in accordance with the recommended procedures.

The Main Variables Measured and Calculated

pH: The concentration of H^+ ions is not directly measurable and is therefore deduced by the electrochemical method. This theoretical limitation justifies the use of the conventional scale that defines pH as $-\log [H^+]$.

pCO_2 and pO_2: Both of these are measurements of the partial pressure of the analyzed species, a fraction of the total pressure given by the atmospheric pressure under conditions in which the measurement is obtained. The traditional unit is millimeters of mercury (mmHg), and according to the International System (SI), it is kilopascals (KPa). For conversion, consider that 1 mmHg = 0.133 kPa.

Total CO_2 Concentration: This depends on the solubility constant of the gas and is therefore calculated with a formula starting from pCO_2, assuming a temperature of 37 °C.

Bicarbonates: They are estimated with an accurate formula starting from pCO_2 and pH, always at 37 °C.

Base Excess (BEecf): This is the estimated excess of bases for extracellular fluid compared to the standard theoretical conditions of pH of 7.40 and pCO_2 equal to 40 mmHg. It is also referred to as BE "in vivo."

Base Excess in the Blood or BE (B): This refers to the same standard theoretical conditions but also takes into account the buffering effect of hemoglobin, and it is also called excess of bases "in vitro." The estimates obtained with the calculation of the excess of bases have several limitations, not considering the variables of other blood buffers, proteins, and the volume of circulating and interstitial fluids. Nevertheless, they provide an immediate estimate of the possible dose of bicarbonates needed to bring the pH back into the normal range.

Total Hemoglobin: In blood gas analyzers, total hemoglobin can be determined with a simplified colorimeter, deducted by calculating the hematocrit measured by the conductivity method or by an oximeter at multiple wavelengths. The latter method provides accurate results.

Fractions of Hemoglobin: An oximeter allows analysis of hemolyzed blood without the addition of any reagent, by comparing the absorbance at different wavelengths, chosen on the basis of the behavior expected by different species: the percentages of carboxyhemoglobin (COHb), metahemoglobin, oxygenated hemoglobin, and reduced hemoglobin are thus determined. Obviously, interfering with this measure are conditions such as the presence of turbidity, abnormal stains, and sulfo-hemoglobin that cannot be estimated with this method.

Oxygen Saturation (sO_2): This is a fraction of oxyhemoglobin with respect to the overall ability of hemoglobin to bind oxygen. Clinically, it is one of the most important variables and is calculated from pO_2 and total hemoglobin, with corrections caused by variations in temperature, pH, and pCO_2. The calculation formula is deduced with empirical models on the basis of a normal oxygen dissociation curve and therefore cannot consider the content of diphosphoglycerate (DPG, low in transfused blood) and other factors. However, it remains a much more accurate measurement than the transcutaneous led pulse oximetric method that is often used in the clinic, in particular if a two-wavelength estimation model is used.

Oxygen Tension at 50% Hemoglobin Saturation (p50): This is the partial pressure of oxygen at which a hemoglobin saturation of 50% is obtained. It represents an estimate of the dissociation curve of hemoglobin evaluated at the critical point of 50%. It allows estimating the affinity of oxygen toward the patient's hemoglobin. It can be calculated only on venous blood.

The Pre-Analytical Phase

Patient Preparation

In emergency or intensive care conditions, it may be necessary to immediately carry out an examination. However, it must always be evaluated whether these are instantaneous determinations of variables that can change even in a few minutes. As a rule, therefore, it is required that the patient be under basal conditions, in particular that the patient should not be conditioned by changes in breathing, ventilation, and adjustments in the administration of oxygen. A measure after 20–30 minutes of stable conditions is therefore recommended, if possible.

An anxious state, frequently present in patients in the emergency room, or concerns about the pain of the puncture, can affect the outcome of the examination through moderate hyperventilation, thus resulting in a decrease in CO_2 and so with a tendency toward respiratory alkalosis.

An accurate measurement also requires that the request for the examination be accompanied by clinical information about body temperature, spontaneous or assisted ventilation, and oxygen flow to supplement the inhaled air [fraction of inspired oxygen (FIO_2)]. The instrument allows correcting the results based on this information.

Blood Sampling Materials

Blood gas analysis syringes must contain lyophilized heparin balanced for electrolytes normally measured at the same time (in particular calcium and possibly magnesium). Many commercial preparations allow a ready solubility and therefore a lower concentration. While using balanced heparins, it is not appropriate to obtain concentrations above 40 IU per ml, as interference would no longer be controllable; it is better to keep the concentration below 20 IU/ml. Two different types of syringes should be used: for arterial samples from a direct puncture, you can use a model with a vented rib with which the flow of blood leads to filling the syringe that is prepared with the piston already extended. In fact, when the blood reaches the rubber seal, the air passage is interrupted and the seal becomes airtight. The needle can now be locked by inserting it into a rubber block or it can be replaced, with normal safety procedures, with a cap, after ejection of each air bubble. Attention should be paid to the potential interference on some parameter due to the lubricant used, usually declared by the manufacturer.

After normal disinfection, the skin should be dry, particularly if small amounts of blood or capillary blood are taken. In fact, traces of benzalkonium chloride have a tensioactive action on membranes interfering with the determination of sodium, potassium, and calcium.

Site of Arterial Sampling

Several arteries can be used. The first choice is the radial artery, which is located on the side of the thumb of the wrist; due to its small size, the use of this artery requires a specific skill. Alternative sites for access are the brachial artery or femoral artery, but they have several disadvantages as they can be more difficult to detect because they are less superfi-

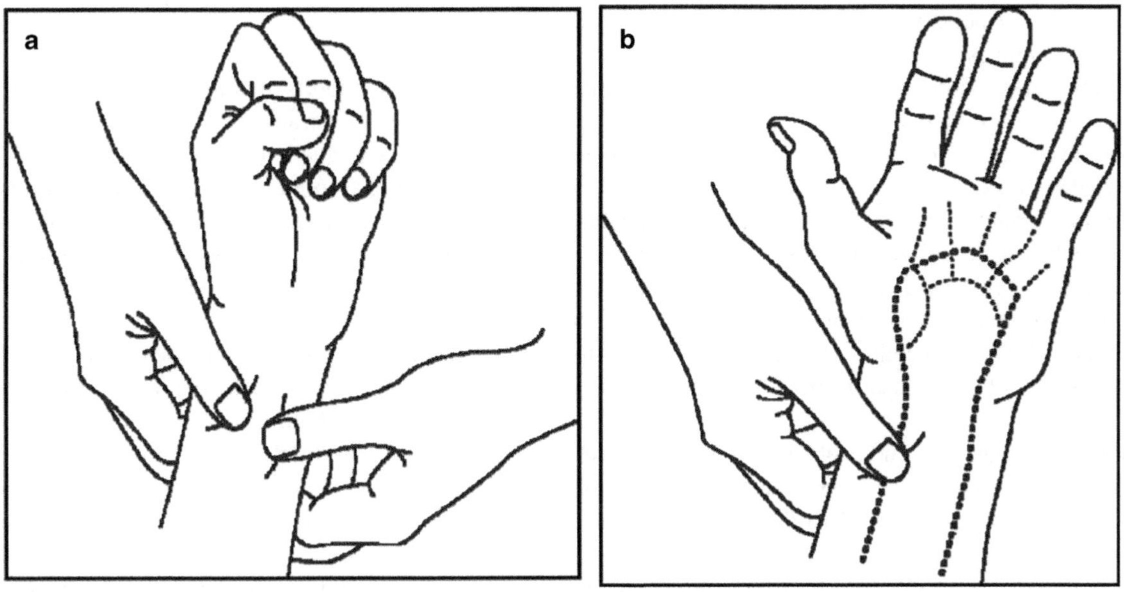

Fig. 21.3 The Allen test for the verification of the normal arterial flow of the radial and ulnar arteries. (**a**) Manual occlusion of both the arteries. (**b**) Resumption of regular flow while compressing only the radial artery

cial than the radial artery, may present collateral circulation problems, and are surrounded by structures that could be damaged by a defective technique.

Modified Allen Test

This is a test to assess collateral circulation of the palm of the hand, which must be performed before taking an arterial sample.

1. Ask the patient to clench his/her fist; if the patient is unable to do this, then personally close the patient's hand tightly.
2. With the thumbs, apply occlusive pressure on the ulnar artery and radial artery to hinder the flow of blood to the hand (Fig. 21.3a).
3. By applying this occlusive pressure to both the arteries, the patient should relax his/her hand and check whether the palm and fingers are "pale." If this does not occur, then the occlusion of the arteries exerted with the fingers is not complete, and, so, it is necessary to repeat the maneuver.
4. Release the occlusive pressure on the ulnar artery only to determine whether the test is positive or negative (Fig. 21.3b):

   Positive: The ulnar artery has good blood flow if blood flow to the hand is restored within 5–15 seconds.

   Negative: Ulnar circulation is inadequate or nonexistent if the blood flow to the hand is not restored within 5–15 seconds. In this situation, only the radial artery supplies arterial blood to that hand and does not have to be punctured.

Performing Radial Artery Sampling

1. Place the patient on his/her back, lying down or sitting in an armchair with armrests. Avoid hyperventilation by reassuring the patient. Keep your wrist in extension.
2. Locate the radial artery by performing the Allen test for collateral circulation.
3. If you cannot find the radial artery at the chosen site, then repeat the test on the other side.
4. Disinfect the sampling site on the patient with the induction products and allow it to dry. Pull the plunger of the syringe to the required level of filling (usually 2 ml).
5. With the index finger, locate the wrist, inform the patient of the imminent puncture, and, holding the syringe like a pen between the thumb, index, and middle fingers, insert the needle with an inclination of 30–45°, avoiding contaminating the area where the needle enters the skin.
6. Advance the needle into the radial artery until blood appears, and, then, wait for spontaneous filling. Do not pull back the plunger of the syringe. The gasket of the plunger has no hold until it is wet with blood, so you should not suck but wait for the filling due to blood pressure.
7. Take out the needle and syringe, place a cotton tablet over the site, and exert firm pressure long enough to stop

the bleeding. Check whether the bleeding has stopped after 2–3 minutes. Five minutes or more may be necessary for patients suffering from hypertension or clotting alterations or for those taking anticoagulants.

8. Cover the needle by closing the safety cover or pricking the rubber block.
9. With the syringe facing upward, expel the air bubbles, remove the needle and close with the cap, and then rotate the sample in your hands to mix the anticoagulant.
10. Immediately transport the sample to the laboratory in a refrigerated bag or directly to the decentralized instrument, if present in the department.

An illustrative video is also accessible on the site from which this information was obtained, subject to free registration (http://www.nejm.org/doi/full/10.1056/NEJMvcm0803851).

The vented rib syringe described here is absolutely unsuitable for venous or arterial sampling from the infusive line. In these cases, in fact, the piston prevents the suction of blood or even allows a retrograde flow of air to the blood sample. Finally, in the case of sampling from arterial infusive lines under pressure, must be carried out by personnel trained for this purpose to avoid accidents and possible dilution or contamination of the sample.

Capillary blood sampling is a valid alternative to the puncture of an artery to obtain a blood sample that is highly similar to an arterial one. Necessary conditions include correct peripheral perfusion as the subdermal blood may not be representative of the central one; this happens mainly in patients with severe hypotension, sepsis, and under conditions of peripheral vasoconstriction. So, you can proceed only if the skin is rosy and warm, heating the part to about 42 °C. Squeezing the skin should be absolutely avoided, and, therefore, the puncture must be deep enough to provide the necessary amount of blood.

Venous sampling does not replace arterial sampling. It can only be used to provide pH, pCO_2, electrolytes, metabolites, carboxyhemoglobin, and meta-hemoglobin. The degree of oxygenation of peripheral venous blood does not provide reliable information and should not appear in the report. It should also be considered that a venous sampling for blood gas analysis requires more attention than does a standard sampling, completely avoiding the application of the tourniquet or removing it, after the puncture, at least 1 minute before blood collection; this is to avoid spurious increases in pCO_2 and therefore lowering of pH.

Other particular types of sampling are that from a "shunt," usually used during dialysis sessions, and the sampling of mixed blood, obtained from a catheter in the pulmonary artery. The laboratory report must indicate these particular types of sampling, and the common reference values for arterial or venous blood should not be applied.

Transport and Storage

Some literature reports unanimously show a significant effect on the pO_2 of transport by the pneumatic tube system. Probably it is due to balancing of the gas with any air bubbles present, favored by shaking and pressure changes in the tube. These effects can be greatly reduced by systematically eliminating air bubbles after sampling and calibrating the pneumatic system to reduce acceleration and jolts.

Of great importance is the storage time before the examination. The main variables of the progressive decrease of CO_2 and O_2 (due to both loss and metabolic consumption) are as follows: the initial concentrations and therefore the gradient toward the environment, the amount of hemoglobin, the number of leukocytes, time and temperature, and the dissociation curve of hemoglobin. From another point of view, it should be considered that refrigeration also blocks the cellular pump of potassium, leading to an increase in its concentration. For these reasons, storage at room temperature should be preferred for the shortest possible time, no later than 30 minutes after collection. If waiting or transport does not allow this limit to be respected, then the sample must be refrigerated and analyzed within 2 hours. For this purpose, the aforementioned Clinical and Laboratory Standards Institute (CLSI) document recommends ice near the fusion, in order to obtain a temperature of just over 0 °C.

Analysis

All samples should be mixed for at least 1 minute by rotation between the hands and by means of a low-speed rotary agitator. Only in the case of freshly performed sampling is a shorter mixing time allowed, whereas a longer time is required in critical or dialysis subjects who normally have a high sedimentation rate.

Many analyzers suck up the sample and then introduce an air bubble into the syringe: this must be promptly removed if it is necessary to repeat the analysis.

The internal barometer should be periodically checked by comparing it, for example, with a professional meteorological barometer (airports, weather stations) located at the same altitude. Some Pmint of Care Testing (POCT) devices are no longer equipped with a barometer, and their use is guaranteed by calibrators at the current pressure.

There is no uniform behavior of instrument manufacturers regarding the expression of certain analytes: potassium can be corrected as serum potassium or expressed as plasma; glucose is sometimes corrected to blood glucose, other times as plasma. It is information that can be obtained from manufacturers.

The main analytical problem is the accidental aspiration of a clot. This can lead to not only the complete blockage of the instrument but also some interference on the electrodes if whips or protein residues stop near the measuring cells. In these cases, the instrument must be carefully sanitized before proceeding to new examinations.

pH and pCO_2 with selective electrodes are not susceptible to interfering substances. Other variables potentially intefererring are some gases used in anesthesia, benzalkonium of disinfectant solutions on electrolytes, salicylates and bromides on the chlorine electrode, electrolyte solutions can condition the hematocrit if measured by conductivity method and finally the presence of turbidity and lipemia that can produce overestimation of hemoglobin. If cord or newborn blood is measured, then a spurious increase in COHb can occur in the presence of fetal hemoglobin.

The Postanalytical Phase

Result Validation

Particular attention should be paid to critical pH values (generally if <7.1 or >7.6) and electrolytes. Even an internal inconsistency between some parameters can lead to suspect an analytical error: the CLSI suggests, for example, to evaluate as incongruent pH and normal bicarbonates with pCO_2 less than 25 mmHg or a sum of pO_2 and pCO_2 greater than 150 in a subject breathing ambient air.

Particularly altered values can be obtained from patients kept at low temperatures, as can happen in intensive care or during cardioplegia in the operating room: in these subjects, the results after correction should always be calculated.

Hemolysis, as a sampling artifact, is not normally recognized because it is not detectable by inspection but must be carefully considered. Ionized calcium should be reported with correction to pH 7.40 in cases of samples stored for more than 30 minutes.

Quality Control

Blood gas analysis has some peculiarities, usually having to substitute the use of materials other than fresh blood from donors. In practice, as quality control materials buffered aqueous solutions are often used, collected in glass ampoules or in plastic and metal bags strictly waterproof gas and without contact with ambient air. The main limitations of these materials are the abnormal surface tension and conductivity that can mask certain electrode drifts, with a matrix effect. Oxygen solubility is also limited, and this carries the risk of measurement artifacts if the liquid is exposed to air before or during sampling. This type of problem, in particular when different operators provide sampling using different procedures, may result in an apparent instability of the instrument in internal quality control or even erroneous situations of noncompliance in external quality assurance. Some commercial materials, to overcome these limitations, contain in solution albumin, free hemoglobin, or even erythrocytes. These materials can hardly be used in internal control when integrated into the instrument, according to the recent market trends. In an intensive environment, pO_2 values above 400 mmHg are often encountered, i.e., in a range of concentrations far from normal quality controls; for this reason, in these situations, it is recommended to also adopt materials with high oxygen values.

Table 21.2 Biological variability and related desirable analytical performance of some variables of blood gas analysis (CV is coefficient of variation)

	Biological variability (%)		Desirable performance (%)		
	CV% intra-individual	CV% inter-individual	Imprecision	Bias	Total error
pH	0.2	–	0.1	–	–
pCO_2	4.8	5.3	2.4	1.8	5.7
Bicarbonate	4.0	4.8	2.0	1.6	4.9

Alternative quality control procedures include those that are based on electronic checks of the state of the instrument and electrodes as well as data deriving from the signal obtained with the washing solutions, fluidic efficiency, and sample integrity. This methodology (called "alternative quality control") must in any case be supported by external materials that are periodically analyzed, even with a lower frequency.

The biological variability of blood gas analysis and the resulting analytical performance targets available in the literature are summarized in Table 21.2 (https://www.westgard.com/biodatabase1.htm).

The Report

There are some peculiarities regarding blood gas analysis that require specific information on the report. First is the type of blood taken. In the case of venous, shunt, or mixed blood, it is not appropriate to indicate the oxygenation parameters, which could be misleading or in any case do not have the possibility of associating them with reference values. If the patient is breathing a mixture of oxygen-enriched air, then the instrument allows you to enter the FiO_2 and consequently express the correct calculated parameters for this factor; the interpretation of the result will be correct only if this information is reported. A similar reasoning concerns the body temperature, in particular if extreme values are reached; this happens, for example, in hypothermic cardioplegia, where all the calculated parameters must be corrected for temperatures that can reach 32–25 °C in moderate hypothermia and 24–18 °C in deep hypothermia. This information should also be reported in the report when the deviation from the physiological values is important.

An interpretative suggestion can be introduced into the report through computer algorithms, which, however, must be shared with the clinicians who use the examination, in order to make them aware of the limitations offered by these systems. These algorithms should not be applied for extreme values of the variables considered.

Blood Gas in Various Situations

Emergency Room

This examination is the cornerstone of the initial evaluation of the critical patient or with a risk of organ deficiency; in particular, this is true where a subject is unable to report his health conditions, a traumatized, a chest pain, a severe dyspnea, a state of coma, an intoxicated. It is advisable that an extended panel of analytes is available in the emergency room, including electrolytes, metabolites, and oximetry.

Operating Room

In these situations, the clinical use mainly concerns ventilation (pCO_2) and oxygenation (sO_2). Due to the presence of anesthetic gases, it is also preferable that oxygen saturation be calculated with an oximeter.

Intensive Care and Coronary Units

In the context of intensive care, a further important parameter is lactic acid that allows the evaluation of peripheral perfusion in the state of shock and has a prognostic value in sepsis. The availability of blood sugar test deserves a special mention: in fact, tight blood glucose control protocols are frequently activated with repeated examinations and insulin infusion; under these conditions, the analytical performance of common portable glucometers is insufficient and the patient is exposed to the risk of severe hypoglycemia. For this reason the glucose concentration so must be measured with a blood gas analyzer, which exhibits sufficient analytical performance.

Pneumological Evaluation

Blood gas analysis is frequently an integral part of spirometry evaluation in the monitoring of pulmonopathic patients. For these types of subjects, an instrument equipped only with the classic parameters is normally sufficient. Given the great spread of chronic obstructive pulmonary disease, the measurement of pCO_2 under resting conditions is fundamental; decision-making levels indicative of disease severity are values >45 mmHg and >60 mmHg, often accompanied by pO_2 <60 mmHg. The current trend of care of these subjects involves pneumological visits at home, thus avoiding moving the patient; this solution necessarily requires the availability of a portable tool. Some pneumological structures are also equipped with semi-intensive units that therefore require instrumentation similar to other hospital intensive care units.

Dialysis Treatments

In subjects with severe renal insufficiency, in particular before and after dialysis treatments, the determination of blood gas analysis may be required, with particular regard to pH, bicarbonate, and potassium levels. This information is obviously related to the instability of the dialyzed subject with regard to electrolytes and pH. It should also be borne in mind that dialysis or hemofiltration can also be used as an acute treatment of hyperkalemia. For these specific purposes, the blood collection can be carried out at the level of the vascular shunt (a blood sample that is not completely arterialized) access that is more convenient in these cases. In the case of circulatory or respiratory imbalances induced by dialysis, arterial sampling will be needed, normally from the radial artery. Analyzers that also determine urea and creatinine can provide an overall picture of the patient that, together with hemoglobin, avoid the use of the laboratory for the completion of the biochemical profile.

Interpretation of Results

Classical Interpretation

This is based on the pathophysiological principles already described and can also be automated, albeit with the limitations already described (Table 21.3). The simple clinical conditions most often linked to these disorders are listed in Table 21.4. In this regard, dedicated software are also available. In this approach, the first step is the cross-evaluation of

Table 21.3 Interpretation of the acid–base balance in some simple clinical conditions

Disorder	pH	Main disturbance	Compensation
Metabolic acidosis	↓	↓ HCO_3^-	↓ pCO_2
Metabolic alkalosis	↑	↑ HCO_3^-	↑ pCO_2
Respiratory acidosis	↓	↑ pCO_2	↑HCO_3^-
Respiratory alkalosis	↑	↓ pCO_2	↓HCO_3^-

Table 21.4 Main causes of simple acid–base balance disorders

Respiratory acidosis	Airway obstruction, asthma, chronic obstructive pulmonary disease (COPD) Central nervous system depression Poor ventilation for neuromuscular or traumatic causes Hyperthermia
Respiratory alkalosis	Anxiety Central stimulation of breathing Hypoxemia Pneumonia, pulmonary edema, pneumothorax
Metabolic alkalosis	Acid loss, vomiting, gastric aspiration Hypochloremic hypovolemia Loss of H^+, diuretics
Metabolic acidosis	Renal failure Diabetic ketoacidosis Alcoholic intoxication Lactic acidosis

pH, pCO_2, and bicarbonates; the combination of these simple data provides an identification of the main disorder. An example of this mode of interpretation was suggested by G.S. Arbus with a specific nomogram (Fig. 21.4). It is estimated, however, that more than half of the cases that come for observation present a mixed or complex disorder. Table 21.5 summarizes the most frequent conditions of the combined disorders.

Stewart's Quantitative Approach

This is based on the ratio of strong ions to weak ions, the so-called strong ion difference (SID). The first category includes the Na, K, Ca, and Mg cations and the Cl anions and organic acids. It is therefore called SID, i.e., the difference between the first and the second. At the physiological pH for the principle of electrical neutrality, we will have SID + H+ = 0 or SID = -H+.

Weak acids are mostly represented by plasma proteins (A+). Starting from the concentrations of SID and A+, it will be possible to obtain an interpretation of acid–base disorders but only of metabolic or non-respiratory ones. This approach solves some limitations typical of classical interpretation but

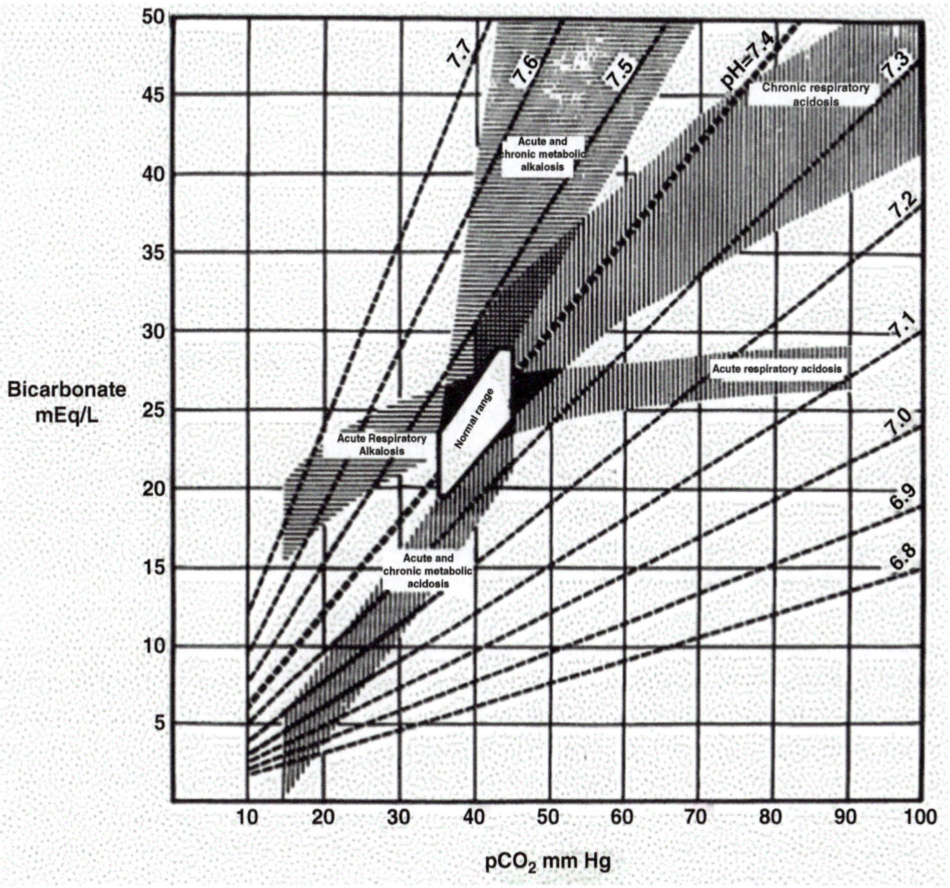

Fig. 21.4 The Arbus in vivo nomogram for an initial interpretation of acid–base simple disturbances

Table 21.5 Possible interpretation of some mixed imbalances

Disorder	Characteristics	Common conditions
Respiratory acidosis with metabolic acidosis	↓pH ↓ HCO_3 ↑ $PaCO_2$	Cardiac arrest Intoxication Multiple organ failure
Respiratory alkalosis with metabolic alkalosis	↑pH ↑ HCO_3^- ↓ $PaCO_2$	Hepatic insufficiency and diuretics Vomiting in pregnancy Hyperventilation in COPD
Respiratory acidosis with metabolic alkalosis	pH normal↑ pCO_2,↑HCO_3^-	COPD and diuretics Severe hypokalemia
Respiratory alkalosis with metabolic acidosis	pH normal ↓ pCO_2 ↓HCO_3	Sepsis Renal failure and hypoxia Severe hepatic impairment
Metabolic acidosis with metabolic alkalosis	pH normal HCO_3^- normal	Vomiting in renal failure and diuretics

has some critical calculations and requires the simultaneous concentration of several analytes, some of which can only be estimated. It is also understandable how this approach represents a refinement of the concept of the anion gap as an estimate of the amount of organic acids. In this sense, some interpretative formulas have been proposed that allow evaluating the contribution of proteins that represent a weak component but, at the same time, have a buffering effect, thus reducing the diagnostic inaccuracy of the anion gap (Palmer and Clegg 2023).

The Clinical Approach

Another highly pragmatic approach is based on clinical conditions. Leaving aside pathophysiology reasoning, we can in fact compare the clinical picture under consideration with that of the particular conditions listed in Table 21.6.

Table 21.6 Blood gas and electrolyte picture commonly associated with some pathological conditions

Pulmonary embolism	
	Moderate respiratory alkalosis
	Hypoxia related to severty
	More than 90 mmHg in ambient air excluding embolism
Acute pulmonary edema	
	Hypoxia
	CO_2 initially normal until the condition worsens
Asthma	
	Constant hypoxemia is proportionate to the severity of CO_2 that is normally decreased. If normal at 40 mmHg, then there is ventilator impediment and if 45 or more, then mechanical ventilation have to be considered
COPD	
	Only moderate hypoxia in "pink puffers"
	Increase in CO_2 in "blue bloaters," with the CO_2 severty index
	If normal pH, then consider a compensatory situation
Neurological, muscular, and traumatic pathologies	
	Acute respiratory acidosis even before hypoxemia
	If CO_2 tends to increase, then mechanical ventilation is needed
Sepsis	
	Initially, acute respiratory alkalosis
	Then, metabolic acidosis is associated with a mixed disorder: a normal pH with decreased HCO_3^-
	Critical condition if pH decreases, in particular in case of lactic acidosis
Diabetes	
	Metabolic acidosis, with variable phases during insulin treatment
	if decompensated HCO_3^- decreased, K elevated, anion gap increased
Diuretic therapy	
	Metabolic alkalosis
Nasogastric tube	
	Metabolic alkalosis
Severe diarrhea	
	Metabolic acidosis with decreased HCO_3^-
	Na and K are often deiminushed while Cl tending to increase

Integrated Management in the POCT

Blood gas analysis was one of the first decentralized laboratory diagnostics to be integrated into a laboratory management and control. This has been possible for many years due to the analyzers equipped with their own computer system for instrumental management, which lends itself well to connectivity. The management of the POCT is dealt with by specific national and international guidelines, but, here, we must remember some specific aspects mainly related to the mode of communication. In fact, the integration of this analysis into systems for data management is a priority for several reasons: the number and complexity of the results for each sample; the need to perform, in particular in intensive care; numerous analyses with the related problems of their historicization; and the need to monitor critical patients through computer systems that integrate different measurements and diagnostics.

Focusing more on the intensive context, it is clear that the architecture of the connections includes the following units: an analyzer, a concentrator for the bidirectional networking of the instruments and a remote control, a department management system (or an electronic medical record), and a laboratory computer system. In all, this must be added to the criticality of the safe identification of the patient, since these are operational contexts in which the analyses are normally assigned to nurses who also have to perform many other tasks and often face emergency situations with little time available. In practice, the identification can take place in two ways: the first requires a computer acceptance with the generation of a sampling label, similar to the other analyses for the laboratory, with which the syringe to be loaded on the instrument is identified; the second mode can allow simplified management, without pre-acceptance, and makes use of the reading of a personal identification code (for example, the bracelet) with the use of a portable reader connected to a bench instrument, and the automatic generation of the acceptance (and then a report) with the patient's personal details. It is understandable how, in any case, the secure identity of the patient in both ways can be well assured by requiring a second step: or with a confirmation with the name of the patient or with a manual validation of data.

Recommended Readings

http://www.nejm.org/doi/full/10.1056/NEJMvcm0803851

https://www.westgard.com/biodatabase1.htm

Palmer BF, Clegg DJ (2023) Respiratory Acidosis and Respiratory Alkalosis: Core Curriculum 2023. Am J Kidney Dis. 82(3);347–359. https://doi.org/10.1053/j.ajkd.2023.02.004

Cardiac Biomarkers

22

Aldo Clerico and Martina Zaninotto

Introduction

Cardiovascular diseases represent the first cause of morbidity, mortality, and hospital admission in industrialized countries and the most important item of expenditure of the National Health Service. For these reasons, the assessment of cardiovascular risk factors and the early diagnosis of heart disease are now universally considered primary objectives of the health organization. Cardiovascular biomarkers are essential for diagnosis, prognosis, and therapy monitoring. Thus, it is not surprising the growing interest in the research of new biomarkers and the development of suitable methods for their determination.

However, 30 years ago, the laboratory made available to the cardiologist only very few biomarkers that could support the diagnosis of myocardial necrosis, such as creatine phosphokinase (CK), lactate dehydrogenase (LDH), and some aminotransferases (transaminases) (Fig. 22.1). These biomarkers are insensitive and nonspecific for cardiac injury, as they are present not only in the myocardium but also in numerous other tissues. It was not until the 1980s that some more sensitive and specific methods for assessing myocardial injury were developed, such as immunometric methods for measuring CK-MB isoenzyme and myoglobin. The real revolution in cardiological diagnostics began only after the development of methods for measuring cardio-specific biomarkers, such as cardiac troponin I (cTnI) and T (cTnT) isoforms and cardiac natriuretic peptides.

At the same time, an increasing number of biomarkers have been proposed for assessing the risk of cardiovascular mortality together with the well-known and traditional markers of cardiovascular risk (Table 22.1). However, in recent years, biochemistry and pathophysiology studies led to the identification of molecules closely related to the pathophysiological processes underpinning cardiac remodeling and myocardial fibrosis. Some of these, especially galectin-3 and the soluble interleukin 33 receptor protein (IL-33), called sST2, have been intensively studied, and numerous clinical studies have shown their prognostic relevance. Indeed, some recent international guidelines have suggested their use in patients with heart failure.

Finally, in more recent years, the diffusion and broader availability of genetic analysis have also allowed their application to cardiovascular diseases. It is therefore hoped that new laboratory tests will soon be made available to clinicians, helpful both to ascertain the diagnosis of a cardiological disease and evaluate other possible risk factors to optimize pharmacological treatment (target therapy).

Based on the above considerations, laboratory tests, which can be used in clinical practice as an aid in the diagnosis, prognosis, and monitoring of patients with heart disease, can be classified into four groups: biomarkers of cardiac injury, biomarkers of cardiac function, biomarkers of myocardial remodeling and fibrosis and genetic biomarkers associated with cardiovascular disease. In truth, since some of the biomarkers may fall into more than one of the four groups, their differentiation into distinct sections serves more of a didactic than a conceptual function.

A. Clerico
Sant'Anna School of Advanced Studies Pisa, Pisa, Italy

M. Zaninotto (✉)
Department of Laboratory Medicine, University-Hospital of Padova, Padova, Italy
e-mail: martina.zaninotto@aopd.veneto.it

M. Ciaccio (ed.), *Clinical and Laboratory Medicine Textbook*, https://doi.org/10.1007/978-3-031-24958-7_22

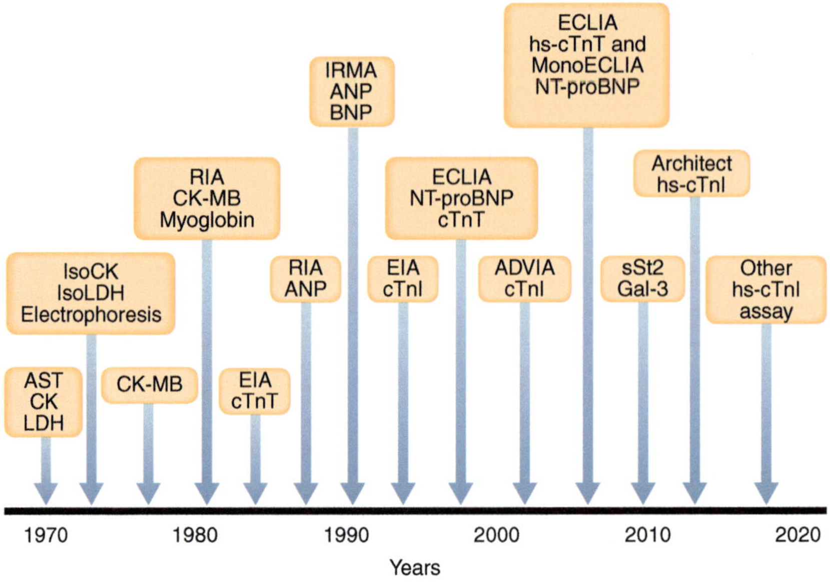

Fig. 22.1 History of the development of methods for determining cardiac biomarkers. Milestones in the development of methods for myocardial damage biomarkers (AST, LDH, CK, CK-MB, myoglobin, cTnI, and cTnT), function (ANP, BNP, and NT-proBNP), and remodeling (sST2 and Galectin-3). EIA immunoenzymatic method, IRMA immunoradiometric method, hs high sensitivity method, Iso isoform, Mono method using monoclonal antibodies, RIA radioimmunoassay method. (Copyright EDISES 2021. Reproduced with permission)

Table 22.1 Risk factors and biomarkers of cardiovascular diseases

Cardiovascular risk factors
Age
Sex
Positive family history for cardiovascular disease
Systemic arterial hypertension
Smoking habit
Dyslipidemia
Physical inactivity
Obesity
Diabetes
Prolonged states of psychophysical and/or psychological stress
Difficult economic conditions
Circulating biomarkers of cardiovascular risk
Natriuretic peptides (ANP, BNP, CNP)
Cardiac troponins (cTnI, cTnT)
Homocysteine
Fibrinogen
Lipoprotein (a)
Urinary albumin
γ-glutamyl transferase (GGT)
Renin–angiotensin–aldosterone system
Uric acid
Markers of coagulation and fibrinolytic function (tissue plasminogen activator, [TPA], its inhibitor PAI-1, D-dimer, and factor V Leiden)
Inflammation markers (C-reactive protein, adhesion molecules and proinflammatory cytokines, interleukin-6, tumor necrosis factor [TNF], sST2, galectin-3, and GDF-15)
Extracellular matrix renewal markers (MMP, TIMP, PICP, PIIINP, and ICTP)
Infectious agents (cytomegalovirus, herpes simplex virus, *Chlamydia pneumoniae*, and *Helicobacter pylori*)
MicroRNA

Biomarkers of Myocardial Injury

The first biomarkers of tissue injury used for the diagnosis and monitoring of acute myocardial infarction (AMI) were the enzymes of the amino-transferase group (in particular aspartate aminotransferase [AST]), which were then gradually replaced by the CK enzyme due to its better diagnostic sensitivity and specificity. Subsequently, the development of standardized and relatively fast chromatographic and electrophoretic procedures led to the development of methods for measuring CK and LDH isoenzymes to be implemented in clinical practice (Fig. 22.1). Finally, many immunoassays to measure proteins with structural and cytoplasmic localization, such as the CKMB isoenzyme (commonly referred to as "CK-MB mass" determination), myoglobin, myosin, and cTnI, and cTnT, have been developed in the last 25 years (Fig. 22.1).

According to some of the most recent national and international guidelines, the determination of CK-MB and myoglobin and, even more, AST and LDH enzymes, should be definitively abandoned and replaced, in patients with suspected ACS, by only cardiac troponins.

Biochemical Characteristics and Biological Function of Troponins

The sarcomere troponin complex is a structure that regulates muscle contraction and consists of three protein subunits: troponin C (TnC), troponin I (TnI), and troponin T (TnT). The cTnI and cTnT have a molecular weight of approximately 23,500 Da and 37,000 Da, respectively, and have slightly different peptide chains in the NH2-terminal part (the first 40 amino acids) than the musculoskeletal ones. TnC (18,000 Da) has three main domains: two terminal globular and one central α-helical. The C-terminal domain contains two high-affinity calcium sites. The other two sites (in the N-terminal region) have low affinity and bind Ca^{++} ions only after they have occupied the high-affinity sites. The low-affinity sites are the activators of the contraction process. The Ca^{++} ion at high concentrations occupies all the TnC sites-leading to conformational changes in the troponin complex that, through TnT, are transmitted to tropomyosin, which slides along the actin filament groove by removing the inhibitory effect of TnI on actin. Free actin can then interact with myosin by stimulating its ATPase function.

Clinical Relevance of Troponins Measurement

In recent years, the development of detection methods for cTnI and cTnT with progressively increasing analytical sensitivity, which can detect myocardial necrosis of microscopic size, has led to a revolution in the definition of acute myocardial infarction. According to the latest definition of myocardial infarction, the measurement of cardiac troponin plays a central and essential role in its diagnosis.

This new definition of AMI is having a significant clinical (but also social and economic) impact, as, since its introduction, diagnosed cases of infarction have increased by 20–40%, compared with when the World Health Organization (WHO) definition of AMI, published in 1973, was used. The WHO definition considered three distinct criteria for the diagnosis of myocardial infarction:

- The presence of typical ischemic symptoms
- The presence of typical electrocardiographic signs
- The typical upward/downward trend in the values of the so-called cardiac enzymes (AST, CK, and LDH).

However, only two of these three criteria were necessary for the diagnosis. According to this definition, in a patient with typical symptoms (anginal pain) and electrocardiographic (ECG) alterations (such as ST-segment elevation), the IMA diagnosis is established. In patients with ST-segment elevation myocardial infarction (STEMI), the cTnI or cTnT assay can only confirm the diagnosis and be used for monitoring and prognosis. The cardiac troponins measurement becomes essential for diagnosing patients with typical ischemic symptoms without ECG alterations; in these patients, if they have an altered troponins kinetics, a diagnosis of non-ST elevated myocardial infarction (NSTEMI) is made.

The most evident and relevant clinical advantage of the new methods for measuring cTnI and cTnT with higher analytical sensitivity is the reduction of the time required to diagnose AMI. Indeed, from diagnostic algorithms based on serial sampling on admission and after 12–24 hours by the first generation methods (1990–2002), there has been a gradual shift to more rapid diagnostic algorithms (6–12 hours in the years 2003–2007), which, by methods at high analytical sensitivity, require a second sampling 3 hours after admission (Fig. 22.2). Finally, in the European Society of Cardiology (ESC) guidelines published in 2015, and subsequently updated in 2018, an assessment of biomarker kinetics involving a second sampling (T1) after 1 hour is suggested.

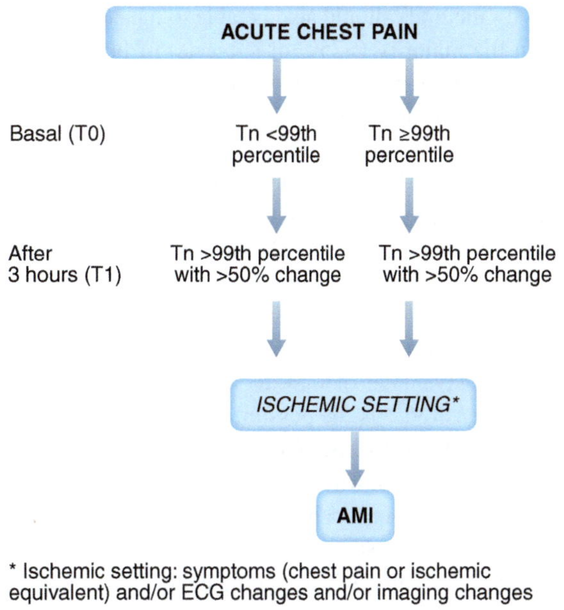

Fig. 22.2 Algorithm for diagnosing acute myocardial infarction (AMI) in accordance with national guidelines. 3-hour algorithm for the diagnosis of IMA NSTEMI, using a highly sensitive measurement method. Ischemic setting: patient with medium-high risk of the acute coronary syndrome, without a pathognomonic ECG for STEMI. The diagnosis of NSTEMI is ascertained when at least in a sample collected within 3 hours of admission, a troponin (Tn) value higher than the 99th percentile of the reference population is found and the change from the baseline value (admission to the Emergency Department) is greater than 50%. (Copyright EDISES 2021. Reproduced with permission)

Quality Specifications of Troponin Measurement Methods

All the most recent national and international guidelines published since 2000 recommend an increase in cTnI and cTnT concentrations above the 99th percentile of the distribution of values measured in a reference population of healthy individuals as a decision level for the diagnosis of AMI . The same guidelines recommend, at this concentration, an analytical imprecision ≤10% (expressed as coefficient of variation [CV]). However, only since 2007 methods with these quality specifications have been introduced commercially in Italy. Therefore, in accordance with national and international guidelines, only immunometric methods measuring the 99th percentile of the distribution of cTnI and cTnT concentrations in the reference population with a CV% ≤10% should be defined as new generation high sensitivity methods for troponin measurement. These high analytical sensitivity methods should also be able to measure cTnI and cTnT in the majority (>75% of subjects) of healthy adults. Newer generation methods, which have intermediate imprecision (CV between 10 and 20%) at the 99th percentile level, should be defined as clinically usable, i.e., they can be used in clinical practice but should not be defined as having higher sensitivity. Finally, methods (such as most POCT systems) that have an inaccuracy >20% should not be used to diagnose AMI but only for the initial screening of patients with suspected acute coronary syndrome, especially when higher sensitivity methods are not available.

Another advantage of introducing methods with higher analytical sensitivity is the possibility of measuring the biomarker in healthy subjects, including pediatric age. The most recent studies have shown that circulating cTnI and cTnT levels in healthy subjects depend on age and sex. From a clinical point of view, it is essential to emphasize that the 99th percentile varies greatly when calculated in a population of apparently healthy subjects over 65 compared to the value measured in blood donors between 20 and 55 years.

The immunometric methods for measuring cTnI currently available have very different reference values and provide different results since they use different antibodies and calibration materials. The method for the measurement of cTnT, which is governed by an international patent and uses control materials and reagents from one manufacturer, allows for more harmonized results, even though the reagents are applied to different platforms. Because cardiac troponins' decision values (99th percentile of the reference population) are both method- and population-dependent, national guidelines recommend that clinicians and laboratory medicine experts jointly agree on the most appropriate decision levels according to the clinical context in which the test is applied.

Clinical Interpretation of Results Obtained by High-Sensitivity Analytical Methods

An improvement in analytical sensitivity also leads to an improvement in clinical sensitivity, equating to an increase in the number of patients diagnosed with NSTEMI. This group of patients has less extensive myocardial ischemic lesions, which are not detectable by ECG, compared to patients with STEMI.

However, since an improvement in the clinical sensitivity of a laboratory test can be obtained only at the expense of specificity, the use of methods for measuring troponins with higher sensitivity results in an increase in the number of patients who, although not clinically having an AMI, have values of cTnI or cTnT above the decision level and, therefore, significantly detectable myocardial injury. Indeed, in recent years, there has been a progressive increase, over 30%, of patients diagnosed with NSTEMI and a parallel decrease in the number of patients diagnosed with unstable angina. The list of clinical conditions that can cause increased circulating levels of cTnI and cTnT are many and tend to increase with the age of patients. The most frequent causes of increased levels of cTnI and cTnT in conditions other than AMI are shown in Table 22.2.

It is important to emphasize that some of these clinical conditions may have a worse prognosis than AMI (e.g., dissecting aortic aneurysm, pulmonary embolism, and stroke), so an elevation of cardiac troponins must always be carefully evaluated. From a clinical point of view, two important aspects must be considered when an increase in cTnI and cTnT levels is observed. The first is that an increase in troponins always indicates cardiac tissue injury. However, a positive test cannot suggest the mechanism responsible for the cardiac injury, which may not even be ischemic. Without clinical evidence of myocardial ischemia, a biomarker value above the decision level should stimulate the clinician to thoroughly search for other causes of myocardial injury (Table 22.2). However, it is important to stress that, considering the close relationship between damaged tissue and released protein, higher circulating concentrations indicate larger areas of injury and, therefore, a higher cardiovascular risk. The second point that must be carefully considered is that there is no minimum threshold below which an increase in cTnI and cTnT is considered irrelevant from a prognostic

Table 22.2 Clinical conditions other than acute myocardial infarction in which circulating troponin levels above the 99th percentile of the reference population may be observed

Acute or chronic heart failure
Acute or chronic inflammatory myocarditis with myocardial infiltration (amyloidosis and sarcoidosis)
Tachycardia or bradycardia
Systemic arterial hypertension, especially if associated with ventricular hypertrophy
Heart surgery or ablation
Heart transplant rejection
Takotsubo myocardiopathy, stress heart
Pulmonary embolism or severe pulmonary hypertension
Respiratory distress syndrome
Dissecting aneurysm of the aorta
Stroke, subarachnoid hemorrhage
Shock
Sepsis
Renal dysfunction/failure
Hypothyroidism
Intense physical exercise (marathon, cycle race, and endurance tests)
Cardiac trauma
Cardiotoxic drugs
Critically ill patients admitted to intensive care
Chronic inflammatory musculoskeletal or congenital neuro-muscular diseases (cTnT only)
Severe anemia

AMI acute myocardial infarction

point of view, as demonstrated by numerous studies. Guidelines indicate cardiac troponins and natriuretic peptides as biomarkers of the first choice for stratifying cardiovascular risk in general population and patients with cardiovascular diseases.

Among the various clinical conditions that may present with increased circulating levels of cTnI and cTnT, it is essential to mention the intake of potentially cardiotoxic agents, particularly anticancer drugs. The development of symptomatic heart failure after anticancer therapy is quite frequent, with an incidence ranging from 30% to 40% ten years after treatment. It is important to emphasize that the troponin assay can highlight the onset of cardiac remodeling at a stage when the pathological process is still reversible, and symptoms and signs of cardiac dysfunction (including instrumental ECG and echocardiographic signs) are absent. At this early stage of cardiotoxic damage, an appropriate clinical intervention can prevent progression to symptomatic decompensation, as recent international guidelines recommend.

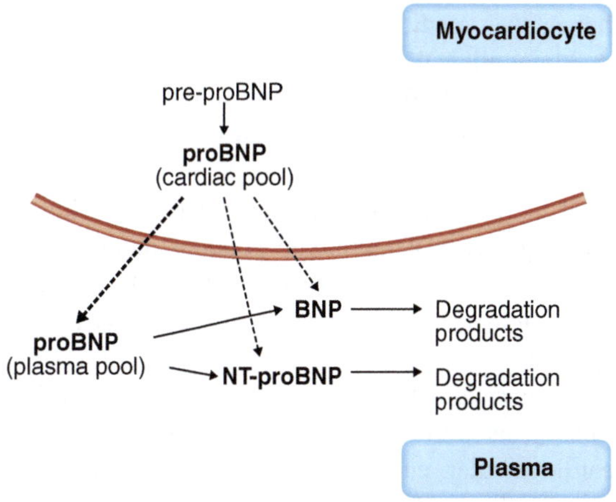

Fig. 22.3 The myocardiocyte's production, secretion, and peripheral degradation of natriuretic-type B peptides (proBNP, BNP, and NT-proBNP). The active hormone BNP derives from a 108 amino acid precursor, called proBNP. ProBNP can be transformed by specific enzymes (chorin and furin) in the cytoplasm of myocardiocytes (cardiac pool) and in the circulation (plasma pool) into two peptides, one longer NT-proBNP, inactive peptide, and one shorter BNP, the active hormone. These peptides, in turn, can be degraded by proteolytic enzymes into even shorter polypeptides. (Copyright EDISES 2021. Reproduced with permission)

Biomarkers of Cardiac Function

Biochemical Characteristics and Biological Function of Cardiac Natriuretic Peptides

It has been known since the first half of the twentieth century that atrial myocardiocytes contain secretory granules. However, it was demonstrated only in the years 1980/1990 that the heart also possesses an endocrine function because it synthesizes and secretes a family of peptide hormones, called cardiac natriuretic peptides, with a potent diuretic, natriuretic, and vasodilating action, as well as complex interactions with both the neurohormonal and immunological systems.

Natriuretic peptides constitute a complex family of interrelated peptides that include ANP, BNP, CNP, urodilatin, and DNP. However, only ANP and BNP peptides are produced and secreted in appreciable amounts by the heart and, thus, deserve the definition of cardiac natriuretic peptides. Cardiac natriuretic peptides are produced as prepro-hormones (prepro-ANP and prepro-BNP), containing a signal peptide at the NH_2-terminal level (Fig. 22.3). The respective pro-hormones (proANP and proBNP) are formed after the signal peptide's enzymatic removal and stored in the secretory granules of myocells. The pro-hormones are then cleaved into two fragments upon release from the myocardiocyte: a longer one, comprising the NH_2-terminal portion (termed NT-proANP and NT-proBNP), and a shorter COOH-terminal one, which constitutes the active hormone (ANP and BNP). The active hormones, ANP and BNP, have a faster plasma clearance than their respective propeptides, NT-proANP and NT-proBNP, and, consequently, lower circulating levels. Recent studies have also shown that proBNP is also present in the circulation (Fig. 22.3) and represents the predominant circulating peptide in patients with heart failure.

The contractile and neuroendocrine functions of cardiomyocytes are closely linked and influence each other through numerous complex feedback mechanisms, contributing together to determine cardiac function proper. Consequently, tests that assess contractile function (such as ECG and echocardiography) and those that measure neurohormonal levels provide different but complementary information about cardiac function. These aspects should be evaluated separately using appropriate methods to better understanding the role of the heart in some complex clinical conditions, especially in heart failure.

Clinical Relevance of Cardiac Natriuretic Peptide Measurement

BNP and related peptides (such as NT-proBNP) are now considered in all national and international guidelines as first-line biomarkers for the diagnosis of heart failure, acute and chronic, as well as for mortality/morbidity risk stratification, not only in patients with heart failure but also in those with the acute coronary syndrome and all other cardiovascular diseases. In particular, the BNP/ NT-proBNP assay is recommended in the differential diagnosis of acute and chronic heart failure, especially to exclude disease because of its high negative predictive value and favorable cost/benefit ratio. Many clinical studies, including meta-analyses, have also shown that BNP/NT-proBNP can reliably guide to personalized treatment of acute and chronic heart failure.

Interpreting BNP/NT-proBNP results requires knowledge of the pathophysiologic role of natriuretic hormones. Circulating levels of natriuretic peptides increase progressively during adult life, especially after 50 years and are higher in females than in males, especially during the fertile period. Moreover, some physiopathological conditions, such as physical activity, physical constitution (BMI), and pregnancy, as well as many pathological conditions, such as cardiovascular, endocrine, renal, hepatic, and neoplastic diseases, influence the circulating levels of these hormones. Finally, some drugs, often used in patients with cardiovascular diseases (digitalis, β-blockers, ACE-inhibitors, and diuretics), as well as neuro-hormonal factors (catecholamines, endothelin, vasopressin/adiuretin system, renin–angiotensin–aldosterone system, thyroid hormones, corticosteroid hormones, estro-progestinic, prostaglandinic, and cytokines), can interact with the natriuretic peptide system.

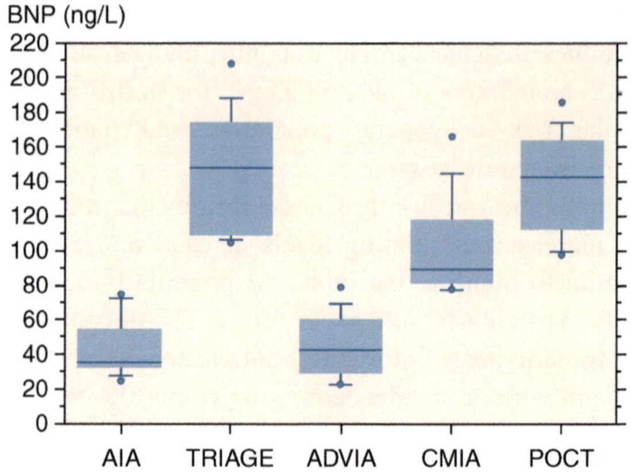

Fig. 22.4 Systematic differences between commercially available BNP assay methods. Results of external quality control CardioOrmocheck in the cycles from 2005 to 2008: 112 participating laboratories, 28 control samples, and 2354 determinations. The 10th, 25th, 50th (median), 75th, and 90th percentiles are shown in the box plots. The results obtained by all the methods show significant differences between them. In particular, the AIA and ADVIA methods measure BNP values that are approximately half of the TRIAGE Access and POCT methods. (Copyright EDISES 2021. Reproduced with permission)

Pathophysiological and Clinical Interpretations of Changes in circulating BNP/NTproBNP

Circulating BNP/NT-proBNP concentrations should not be considered an indicator of a specific cardiac disease (such as heart failure) nor, even less, a surrogate for echocardiographic examination. On the contrary, it is important to understand how the results of their determination must always be evaluated, keeping in mind the state of activation of the neurohormonal system and the general clinical state of the patient, especially in consideration of concomitant morbid conditions and the pharmacological therapy. BNP, a hormone produced and secreted by the heart, must be considered, above all, an index of the state of activation of the neuro-immune-hormonal system; this, among other, explains why circulating levels of BNP/NT-proBNP are influenced by therapy in heart failure. The most widely used and effective drugs (β-blockers, ACE inhibitors, specific angiotensin II receptor blockers, and diuretics) act by inhibiting the neuro-immune-hormonal system (including the adrenergic system, the renin–angiotensin–aldosterone system, the arginine-vasopressin system, the endothelin system, and the cytokine system). The response to drug treatment is clinically significant as it is the factor that most affects mortality.

Another essential aspect to consider when evaluating and, more importantly, comparing BNP and NT-proBNP values to each other is that the results are strictly method dependent. Indeed, there are differences between the circulating levels of the various peptides (BNP vs. NT-proBNP) and the different methods specific for a single peptide (Fig. 22.4). In particular, the methods for determining BNP currently on the market are sandwich-type solid-phase dual-antibody immunoassay systems, which use different calibrators and antibodies. Generally, one of the two antibodies is specific for the part of the peptide chain with the ring shape, which binds to specific biological receptors, while the other is specific for the NH_2 or COOH-terminal part. Since BNP is degraded *in vivo* and *in vitro* by various enzyme systems attacking the peptide chain at different positions, there are several degradation products and many potential interferent peptide molecules in plasma. Currently available commercial methods use different antibodies specific for different epitopes of the BNP peptide chain, as well as different calibration materials, resulting in different values and strictly method-dependent reference ranges and decision limits (Fig. 22.4).

As previously noted, from a strictly diagnostic point of view, determining BNP/NT-proBNP peptides should be used to exclude heart failure (HF). Many studies indicate that the most appropriate decision level for this differential diagnosis may correspond to the 97.5th (or 99th) percentile of the distribution of BNP/NT-proBNP values in the healthy population, preferably divided into groups concerning age and sex. Although this value is different for each measurement method, the distribution of peptide values in each method is relatively constant for the various populations studied. Thus, it seems reasonable to use in clinical practice the corresponding value indicated by scientific studies or by the manufacturing companies if calculated on a numerically significant population.

From a strictly clinical point of view, the information provided by ANP and BNP and their related peptides, MR-proANP or NT-proBNP, is not substantially different. From an analytical point of view, however, it is essential to note that active hormones (ANP and BNP) are less stable *in vitro* than their respective nonactive peptides, such as prohormones (proANP and proBNP) or N-terminal peptides (such as NT-proANP, MR-proANP, and NT-proBNP). Active hormones should only be measured in plasma-EDTA samples (which partially inhibits enzymes that degrade hormones in plasma), whereas NT-proANP, MR-proANP, and NT-proBNP peptides can be assayed in either plasma-EDTA or lithium-heparin plasma or serum.

Although BNP and NT-proBNP values are closely correlated in most clinical conditions, recent studies in patients with chronic heart failure treated with a new drug called LCZ696 (Entresto) have shown conflicting results. This pharmacological combination consists of two substances: a specific competitor of the angiotensin II receptor (valsartan) and an inhibitor of the enzyme neprilysin (sacubitril). A recent multicenter study, called PARADIGM-HF, showed that plasma BNP levels were higher in patients treated with LCZ696 than in those treated with enalapril. In contrast, circulating levels of NT-proBNP were lower during treatment with LCZ696 than with enalapril. Thus, for the first time, an opposed behavior between the levels of active BNP hormone and inactive NT-proBNP peptide in patients undergoing pharmacological treatment for heart failure has been described in the literature. These data, therefore, require an *ad hoc* physiopathological interpretation that takes into account the dual action of the drug: inhibition of the peripheral degradation of BNP (with a consequent increase in the circulating levels of the active hormone) and of the renin–angiotensin system with a decreased production of cardiac natriuretic peptides by the myocardial cells (Fig. 22.5). Consequently, the clinician's interpretation of natriuretic peptide changes during treatment with this pharmaceutical combination requires special attention. In particular, all cases of increased plasma levels of BNP not linked to the inhibition effects of the drug on the degradation of the hormone but secondary to a worsening of the patient's clinical conditions must be recognized.

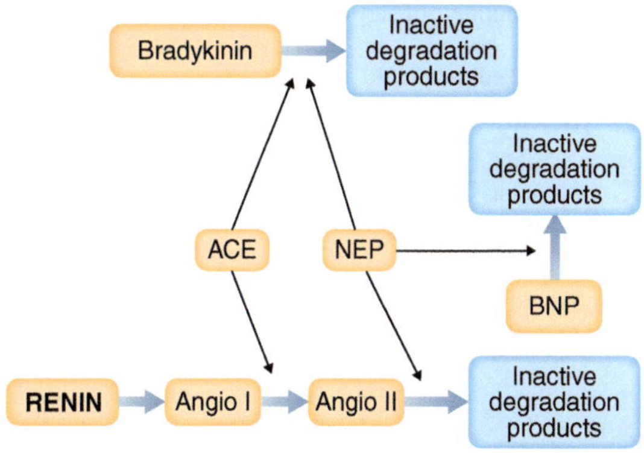

Fig. 22.5 Interrelationship between the renin-angiotensin system and the B-type natriuretic peptide system with plasma proteolytic enzymes ACE (angiotensin I converting enzyme) and NEP (neprilysin). The NEP enzyme degrades natriuretic peptides (including BNP), bradykinin, and angiotensins, transforming them into shorter and inactive peptides. The ACE enzyme acts on bradykinin, inactivating it, and on angiotensin I (angio I), transforming it into the much more active peptide angiotensin II (angio II). The inhibition of the NEP enzyme, therefore, causes high levels of bradykinin, angiotensin II, and natriuretic peptides (ANP, BNP, and CNP), which have contrasting effects on the cardiovascular system. (Copyright EDISES 2021. Reproduced with permission)

Biomarkers of Myocardial Remodeling and Fibrosis

Cardiac remodeling is generally considered the most important pathophysiological mechanism determining the progressive development of heart failure in patients with extensive ventricular myocardial infarction or long-standing myocardial disease. In the year 2000, a consensus conference defined cardiac remodeling as the result of changes in the expression of the cellular genome of myocardial tissue that induces changes, at the molecular level, of cellular structure and interstitial matrix that produce changes in the weight, shape, and function of the heart. These anatomopathological alterations are caused by hemodynamic overload and/or cardiac damage, and the subsequent progressive cardiac remodeling is influenced by hemodynamic alterations and activation of the neuro-immune-hormonal system. Heart failure is the final common pathway of all cardiovascular diseases (of which coronary ischemia covers more than half of the cases in Western countries), so patients with a wide variety of clinical conditions can be affected. For this reason, they show a different propensity to develop ventricular remodeling and fibrosis.

Patients with HF are distinguished into two groups concerning the value of left ventricular systolic ejection fraction (LVEF), usually assessed by echocardiographic examination, which can be reduced (≤40%, Heart Failure with Reduced Ejection Fraction [HFrEF]) or preserved (≥50%, Heart Failure with Preserved Ejection Fraction [HFpEF]) (Table 22.3). The epidemiology, etiology, pathophysiologic mechanisms, anatomopathological picture, and clinical disease progression that characterize these two groups of patients are substantially different. Patients with HFpEF are generally older, female, and have comorbidities (hypertension, obesity, diabetes mellitus, atherosclerosis, atrial fibrillation, and renal failure) and diastolic dysfunction on echocardiographic examination. The number of patients with HFpEF has increased in recent years, reaching that of patients with HFrEF, whose incidence has decreased. Thus,

Table 22.3 Definition of heart failure with preserved (HFpEF), mid-range (HFmrEF), and reduced (HFrEF) ventricular ejection fraction according to ESC guidelines 2016

Criteria	HFrEV	HFmrEV	HFpEF
1	Signs and symptoms	Signs and symptoms	Signs and symptoms
2	LVEF <40%	LVEF 40–49%	LVEF ≥50%
3	–	High concentrations of natriuretic peptides At least one of the following criteria: - Significant structural alterations (LVH and/or LAE) - Diastolic dysfunction	High concentrations of natriuretic peptides At least one of the following criteria: - Significant structural alterations (LVH and/or LAE). - Diastolic dysfunction

LAE left atrium enlargement, *LVH* left ventricular hypertrophy, *LVEF* left ventricular ejection fraction

the prevalence of the two different clinical conditions in the general population is estimated to be similar. Diagnosis and treatment are more complex in patients with HFpEF, for which the prognosis is more severe. In accordance with the 2016 ESC guidelines, the diagnosis of HFrEF is based only on the presence of clinical symptoms and the assessment of <40% LVEF reduction, whereas the diagnosis of heart failure with preserved or intermediate LVEF (LVEF between 40 and 50%) requires elevated natriuretic peptide values and structural or functional alterations in the ventricle, indicating the presence of diastolic dysfunction (Table 22.3).

Cardiac fibrosis, whose most evident aspect is ventricular wall thickening, occurs more frequently in patients with HFpEF than in those with HFrEF, representing the most important pathogenetic mechanism of diastolic dysfunction. This evidence has stimulated in recent years the research and clinical validation of reliable biomarkers of cardiac fibrosis for the diagnosis, risk stratification, and monitoring of patients with HFpEF. Considering the results obtained in these studies, the 2013 American College of Cardiology Foundation and American Heart Association guidelines included, for the first time, biomarkers of cardiac fibrosis, and in particular galectin-3 and the soluble IL-33 receptor, called sST2, among the biomarkers for risk stratification in patients with heart failure.

From a physiopathological point of view, these biomarkers have the critical limitation to not be cardio-specific since they can be produced by numerous immunocompetent cells, mature or immature fibroblasts, or even endothelial or epithelial cells, especially during local or even systemic inflammatory processes.

Thus, for the clinician, it is often impossible to link variations in the circulating levels of these biomarkers to possible pathophysiological processes specifically located in the myocardium (and not in other organs, such as kidney, liver, and lungs).

Cardiac Fibrosis

Fibroblasts account for about two-thirds of the myocardial cell population, whereas myocardiocytes account for about two-thirds of the myocardial mass. Myocardial remodeling in ischemic and nonischemic myocardial disease involves myocardiocytes, other myocardial tissue cells (especially fibroblasts and endothelial cells), and the extracellular matrix.

Specifically, collagen is secreted by fibroblasts as procollagen into the extracellular matrix, where proteases remove the carboxy-terminal amino acid propeptide, which will subsequently be degraded by matrix metalloproteases (MMPs), which in turn are regulated by the tissue inhibitor of metalloproteases (TIMP). Under pathological conditions, the cardiac interstitium may increase due to diffuse deposition of collagen fibers, edema (e.g., secondary to an inflammatory process), or pathological deposition of proteins that physiologically are not present in the cardiac extracellular matrix (such as amyloid). Recent studies have shown that activation of the renin–angiotensin–aldosterone system within cardiac tissue plays a central role in fibroblast activation and collagen deposition.

Cardiac fibrosis is generally defined as the proliferation of fibroblasts with increased deposition of cardiac muscle collagen fibers or (more rarely) fibrotic thickening of the heart valves. Fibrosis makes the heart muscle stiffer and less elastic, reducing the ability of the ventricles to dilate (ventricular diastolic dysfunction). In addition, fibrosis can affect the heart valves leading to valve dysfunction (stenosis and/or insufficiency).

Myocardial fibrosis can originate through two distinct pathophysiologic processes that result in two different phenotypes:

- Fibrosis can result from the loss of myocardial tissue (e.g., due to an extensive myocardial infarction), which must be considered as an actual scar (reparative or replacement fibrosis);
- In non-ischemic myocardiopathies, generally on a chronic inflammatory basis, an increase in the interstitial matrix is produced, which is generalized to the whole ventricle (or to a large part of it). This type of fibrosis is, therefore, called interstitial and is of reactive type and originates in the areas surrounding the blood capillaries from where it then radiates to the entire myocardial tissue.

The death of myocardiocytes and their replacement with fibrotic tissue cause important alterations in cardiac function. Both increased extracellular synthesis and decreased extracellular matrix can cause increased ventricular wall stiffness, the most important cause of ventricular diastolic dysfunction. In addition, extracellular matrix deposition between myocardiocytes can alter the propagation of electrical impulses through the myocardium, causing both abnormalities of contraction and cardiac arrhythmias, which can also be fatal. Finally, inflammatory edema and fibrotic tissue deposition around the perivascular areas, by slowing the flow of oxygen and nutrients to the myocardiocytes, trigger the vicious cycle that supports the progression of myocardial remodeling.

Biomarkers of Collagen Synthesis and Degradation

Extracellular matrix remodeling in the heart is an active and highly complex process orchestrated by matrix-degrading MMPs and their TIMP inhibitors. MMPs constitute a complex family of enzymes of the protease group, of which at least 28 members have been identified (indicated by progressive numbers from MMP-1 to MMP28), classified according to the nature of the most important functional groups in their catalytic site, which requires the presence of metal ions as cofactors (zinc or cobalt). Four members of the MMP inhibitor family (TIMP-1 to TIMP-4) were identified.

The fibrillar collagen in the heart is predominantly of type I (85%) and type III (11%); the latter provides elasticity to the myocardium, while the former contributes most to the strength and resistance of the myocardium to wall stress and deformation. Collagen undergoes turnover by the action of fibroblasts and myofibroblasts, which respond to mechanical stress and autocrine and paracrine factors (such as angiotensin II, aldosterone, cytokines, and growth factors), partly secreted by monocytes and macrophages. An increase in collagen deposition results from the increased proliferation of fibroblasts/myofibroblasts and their collagen synthesis and secretion rate. Fibrillar collagen is synthesized as pre-procollagen in the endoplasmic reticulum of fibroblasts and transformed into procollagen with a triple helix structure. Procollagen, secreted into the extracellular matrix, is then cut by proteases to form type I and III collagen fibrils that contribute to the extracellular matrix's structure and the myocardium's mechanical characteristics.

Although cardiac tissue biopsy is considered the gold standard for diagnosis, some markers of collagen turnover have been proposed for the noninvasive estimation of myocardial fibrosis. Although numerous markers of collagen turnover have been identified and described, the most robust evidence involves PICP (Procollagen type I C-terminal Propeptide), PIIINP (Procollagen type III N-terminal Propeptide), and ICTP (type I Collagen TeloPeptide).

Many studies, especially case-control, measured circulating biomarkers of extracellular matrix turnover in patients with heart failure. However, analytical difficulties limited their use in clinical practice. Additionally, the results of these studies are difficult to compare with each other because they differ in the number and type of biomarkers analyzed, the study design (case–control vs. cohort, prospective or retrospective study), and the number and demographic (especially age and sex) and clinical characteristics of the enrolled patients. Also, the methods used to measure the same biomarker in the different studies present very different analytical characteristics; generally, RIA or ELISA methods were used. In many studies, the main analytical characteristics and performance (sensitivity, reproducibility, and specificity of the antibodies used) of the methods used are not even specified, so it is impossible to compare the data obtained in different studies, even considering the same biomarker. Some authors have suggested that the PICP/ICPT ratio could better estimate type I collagen turnover than a single marker. However, it should be noted that the absolute differences between healthy subjects and patients with heart failure are generally much smaller than those observed in natriuretic peptides, which show differences of the order of 10-fold or more between the median value of a population of patients with heart failure compared to that observed in healthy subjects. This reduced differential between cases and controls reduces the diagnostic and prognostic accuracy of the biomarker, especially in identifying patients with early forms of the disease. This explains why these biomarkers have a much lower diagnostic and prognostic power than BNP and NT-proBNP. In conclusion, although there are many studies on the circulating levels of collagen and extracellular matrix turnover biomarkers, there is currently insufficient evidence to support their use in clinical practice for the diagnosis, prognosis, or treatment of patients with heart failure.

Biomarkers of Myocardial Fibrosis

Biomarkers of cardiac fibrosis, especially galectin-3 and the soluble cytokine receptor protein IL-33 (commonly referred to as sST2), have been increasingly studied since 2005. Considerable evidence, even if not definitive, has therefore been produced on their possible use as biomarkers of prognosis in patients with heart failure. Thus, the 2013 guidelines of the American College of Cardiology Foundation and American Heart Association scientific societies suggest their use for prognostic assessment in patients with acute HF.

Galectin-3

Galectin-3 is a lectin that binds compounds containing a β-galactosidic bond, characteristic of many glycans on the cell surface. Many experimental studies show that galectin-3 is involved in many functions at the cellular level, such as adhesion, activation, chemotactic attraction, growth, differentiation, and apoptosis. An increase in galectin-3 expression induces fibroblast proliferation and collagen production, thus contributing to the increase in cardiac fibrosis and subsequent remodeling.

Especially after the first decade of this century, numerous clinical studies on this biomarker have been published, favored by the commercialization of immunometric methods applied to automated platforms with excellent analytical performance. Galectin-3 was initially studied as a mediator of growth and progression of several types of tumors, and only later, its strong association with clinical conditions characterized by chronic inflammation and interstitial fibrosis was noted. Galectin-3 is present in the cytoplasm of different cell types and is particularly abundant in macrophages.

The rationale for the candidacy of galectin-3 as a marker in heart failure comes from studies in animal models published in the early years of the new century, which have shown that this leptin can regulate the hypertrophic response and functional cardiac damage following the administration of angiotensin II or constriction of the aorta. The results of these experimental studies paved the way for clinical trials that sought to evaluate whether galectin-3 could play a role as a marker of fibrosis, inflammation, and cardiac remodeling in patients with heart failure. Furthermore, in November 2010, the Food and Drug Administration (FDA) approved the determination of galectin-3 in conjunction with clinical evaluation to define the prognosis of patients with chronic heart failure. Subsequently, many studies have been published to assess the prognostic relevance of galectin-3 in patients with heart failure, especially HFpEF type. Unfortunately, these studies are primarily case–control. In 2015, a meta-analysis, considering 9 cohort studies, showed a significant association between mortality and biomarker levels; however, this study has, as an important limitation related to the significant heterogeneity of the data used for the meta-analysis. In addition, data derived from cohort studies are generally the result of *post hoc* analysis and, therefore, require confirmation with specific clinical protocols.

As previously pointed out for markers of inflammation and turnover of collagen and extracellular matrix, also for galectin-3 a modest difference is observed between the concentrations of healthy subjects and heart failure patients (on average from 15% to 40%). For this reason, in many statistical evaluations, the measurement of galectin-3 does not add significant prognostic information to other prognostic markers (especially that of natriuretic peptides).

It should be noted, however, that the biological variability of galectin-3 both intraindividually (about 5–8% in healthy subjects, as well as in patients with chronic stable heart failure) and interindividually (about 27% in healthy subjects and 40% in patients with chronic stable heart failure) is much lower than that of natriuretic peptides or troponins. The relatively low biological variability coupled with the excellent analytical performance of the currently available automated assay methods makes the biochemical information provided by galectin-3 very robust in monitoring individual subjects/patients.

Compared with the sST2 protein, galectin-3 possesses the advantage of being more closely related to chronic inflammatory processes resulting in interstitial fibrosis in patients with heart failure. If this hypothesis is confirmed, galectin-3 could be considered a biomarker of cardiac remodeling associated with myocardial interstitial fibrosis. Thus, Galectin-3 could be used to screen patients with HFpEF who are still asymptomatic or pauci-asymptomatic, to identify those who will need to undergo more expensive (such as gadolinium-enhanced MRI) or invasive (such as biopsy or arteriography) tests to confirm cardiac fibrosis, and then be targeted to specific and more aggressive therapies.

Soluble IL-33 Receptor (sST2)

The ST2 protein belongs to the interleukin-1 receptor family (IL-1 RL-1) and exists in two isoforms, transmembrane (ST2L) and soluble (sST2). The specific ligand of the STL2 receptor is IL-33, which plays an anti-inflammatory, antihypertrophic, and antifibrotic role in the myocardium. The soluble ST2 receptor is the ST2L receptor, which, through specific proteases, undergoes a cut at the transmembrane and cytoplasmic domains with subsequent release into the extracellular fluid. In the extracellular fluid and plasma, the sST2 protein can still bind its specific IL-33 high-affinity ligand, sequestering it and making it no longer available for binding to its membrane receptor. In this way, the sST2 receptor blocks the favorable effect of IL-33 on the myocardium by behaving as a decoy receptor.

High circulating levels of sST2 promote proinflammatory mechanisms leading to myocardial remodeling, cardiac fibrosis, and ventricular diastolic dysfunction. In accordance with this mechanism of action, many studies have shown that patients with heart failure have high circulating levels of sST2. However, as also pointed out for galectin-3, the measurement of sST2 does not always add significant prognostic information compared with other biomarkers, especially natriuretic peptides. Currently, it is impossible to measure sST2 by the most common automated platforms available in clinical laboratories, so its clinical use is limited to specialized centers, and, consequently, the results available in the literature are fewer than those related to galectin-3. sST2, however, provides better results than galectin-3 in evaluating patients with HFrEF than those with HFpEF. In addition, some authors have reported that sST2 can stratify the risk of progression of cardiac remodeling in patients with reduced left ventricular function (with or without symptoms of heart failure).

In conclusion, further prospective clinical cohort studies, with a sufficient number of heart failure patients, both with HFpEF and HFrEF, monitored for a significant number of years, are needed to evaluate and compare the possible prognostic relevance of galectin-3 and sST2 among themselves and concerning other biomarkers whose prognostic role is widely recognized, in particular natriuretic peptides and troponins, measured by methods with high analytical sensitivity.

Genetic Biomarkers Associated with Cardiovascular Diseases

The genomic analysis includes many procedures, very different from each other, which allow the identification of possible genes (candidate genes), whose altered function could be responsible for particular and rare forms of cardiovascular diseases (monogenic diseases). Additionally, some polymorphisms (gene variants) could be associated with an increased risk of cardiovascular events.

Malformations of the heart and great vessels account for a large proportion of congenital disabilities in about 1% of live births. Some of these forms may be caused by chromosomal alterations (such as deletion or translocation) or by mutations in a single gene. In addition, some of the idiopathic forms of dilated or hypertrophic myocardiopathy, especially familial forms, may be caused by the altered function of a single gene and be heritable. Finally, some familial forms of cardiac arrhythmias may also recognize a genetic basis because they are caused by an alteration in the function of a single or a few genes. In all these cases, the genetic test, if available, can not only confirm the diagnosis but also offer the opportunity for appropriate counseling in highly specialized departments and/or suggest new treatment protocols.

However, considering the cardiovascular system, the clinical utility of the analysis of possible candidate genes is theoretically limited by the fact that the most frequent disorders affecting this system recognize a multifactorial etiology, in which both environmental and behavioral factors interact dynamically with the function of different genes in determining the pathophysiological mechanisms that will lead to the establishment of the cardiovascular disorder. For this reason, it is impossible to associate single genes' to specific, persistent clinical conditions, such as acute coronary syndromes, hypertension, diabetes, and dyslipidemia, which recognize a complex multifactorial etiology. In these cases, a "genomic" or "proteomic" approach, rather than a genetic analysis based on a single gene locus or haplotype, has been more effectively suggested. The aim would be to define a profile or a fingerprint, recognized by molecular analysis or of the concomitant function of several genes or metabolites of the same biochemical pathway (gene clustering, expression patterns, proteomic fingerprint, or signature), associated with a specific cardiovascular disease. The limitations of this approach are that the analysis is expensive and available in few laboratories, and the results obtained are often not easily interpretable from a diagnostic and/or prognostic point of view, so this type of analysis is still not very usable in clinical practice.

Recently, the interest in studying the relationships between cardiovascular disease and gene expression has focused on the study of single-stranded RNAs, which may be present not only in the nucleus and cytoplasm of cells but also in the circulation. In particular, the focus has been on studying microRNAs (miRNAs), a class of small, noncoding RNAs that regulate the expression of complementary RNAs. Altered expression of intracellular miRNAs has been reported in many diseases, including cardiovascular diseases. In particular, many studies have recently highlighted the role of some miRNAs not only in some pathophysiological processes affecting the myocardium, such as fibrosis, hypertrophy, and angiogenesis, but also in some complex clinical conditions, such as dilated and hypertrophic myocardial diseases, myocardial infarction, and heart failure. Being circulating molecules, which can be assayed in blood samples by laboratory methods, miRNAs represent circulating biomarkers of disease and, therefore, do not present the limitations of the genomic or proteomic approach. A list of the most studied miRNAs as risk biomarkers in patients with heart disease is shown in Table 22.4.

Numerous meta-analyses have tried to evaluate the association of some of these circulating miRNAs with frequent cardiovascular disorders, such as stroke, ischemic heart disease, and heart failure. However, there are still some critical issues related to their use as biomarkers of cardiovascular diseases. First, it is not yet evident the relationship between the gene expression of these noncoding RNAs (e.g., in myocardiocytes) and their respective circulating levels. Moreover, the methods currently available for their measurement are not standardized, and there are no internationally agreed quality specifications, so results obtained in different laboratories and by different methods are not comparable. For these reasons, the identification and determination of these molecules are applicable only in the fields of pathophysiological and clinical research.

Table 22.4 List of microRNAs (miRNAs) most studied as risk biomarkers in patients with heart disease

MicroRNA
miR-1, miR-19, miR-21, miR-122, miR-126, miR-132, miR-133, miR-134, miR-140, miR-142, miR-145, miR-146, miR-150, miR-155, miR-186, miR-197, miR-208, miR-210, miR-223, miR-320, miR-328, miR-380, miR-451, miR-486, miR-499

Recommended Readings

Bao MH, Xiao Y, Zhang QS et al (2015) Meta-analysis of miR-146a Polymorphisms association with coronary artery diseases and ischemic stroke. Int J Mol Sci 16:14305–14317

Barison A, Grigoratos C, Todiere G, Aquaro GD (2015) Myocardial interstitial remodelling in non-ischaemic dilated cardiomyopathy: insights from cardiovascular magnetic resonance. Heart Fail Rev 20:731–749

Bettencourt P, Azevedo A, Pimenta J et al (2004) N-terminal-pro-brain natriuretic peptide predicts outcome after hospital discharge in heart failure patients. Circulation 110:2168–2174

Braunwald E (2015) The war against heart failure: the Lancet lecture. Lancet 385:812–824

Calvier L, Miana M, Reboul P et al (2013) Galectin-3 mediates aldosterone-induced vascular fibrosis. Arterioscler Thromb Vasc Biol 33:67–75

Cao W, Guo Q, Zhang T et al (2015) Prognostic value of microRNAs in acute myocardial infarction: a systematic review and meta-analysis. Int J Cardiol 189:79–84

Casagranda I, Cavazza M, Clerico A et al (2013) Proposal for the use in emergency departments of cardiac troponins measured with the latest generation methods in patients with suspected acute coronary syndrome without persistent ST-segment elevation. Clin Chem Lab Med 51:1727–1737

Caselli C, Cangemi G, Masotti S et al (2016) Plasma cardiac troponin I concentrations in healthy neonates, children and adolescents measured with a high sensitive immunoassay method: high sensitive troponin I in pediatric age. Clin Chim Acta 458:68–71

Chen A, Hou W, Zhang Y et al (2015) Prognostic value of serum galectin-3 in patients with heart failure: a meta-analysis. Int J Cardiol 182:168–170

Cheng C, Wang Q, You W et al (2014) MiRNAs as biomarkers of myocardial infarction: a meta-analysis. PLoS One 9:e88566

Clerico A, Giannoni A, Vittorini S, Passino C (2011) Thirty years of the heart as an endocrine organ: physiological role and clinical utility of cardiac natriuretic hormones. Am J Physiol Heart Circ Physiol 301:H12–H20

Clerico A, Passino C, Franzini M, Emdin M (2015a) Cardiac biomarker testing in the clinical laboratory: where do we stand? General overview of the methodology with special emphasis on natriuretic peptides. Clin Chim Acta 443:17–24

Clerico A, Ripoli A, Zucchelli GC, Plebani M (2015b) Harmonization protocols for thyroid stimulating hormone (TSH) immunoassays: different approaches based on the consensus mean value. Clin Chem Lab Med 53:377–382

Clerico A, Passino C, Franzini M, Emdin M (2016) Natriuretic peptides as biomarkers of cardiac endocrine function in heart failure: new challenges and perspectives. Futur Cardiol. in press

Cohn JN, Ferrari R, Sharpe N (2000) on behalf of an International Forum on Cardiac Remodeling. Cardiac remodeling-concepts and clinical implications: a consensus paper from an international forum on cardiac remodeling. J Am Coll Cardiol 35:569–582

Condorelli G, Latronico MVG, Cavarretta E (2014) MicroRNAs in cardiovascular diseases: current knowledge and the road ahead. J Am Coll Cardiol 63:2177–2187

Curigliano G, Cardinale D, Dent S et al (2016) Cardiotoxicity of anticancer treatments: epidemiology, detection, and management. CA Cancer J Clin 66:309–325

deFilippi C, Daniles LB, Bayes-Genis A (2015) Structural heart disease and ST2: cross-sectional and longitudinal associations with echocardiography. Am J Cardiol 115(7 Suppl):59B–63B

Dieplinger B, Mueller T (2015) Soluble ST2 in heart failure. Clin Chim Acta 443:57–70

Emdin E, Clerico A, Clemenza F et al (2005) Consensus document. Recommendations for the clinical use of cardiac natriuretic peptides. J Cardiovasc Med (Ital Heart J) 6:430–446

Fan D, Takawale A, Lee J, Kassiri Z (2012) Cardiac fibroblasts, fibrosis and extracellular matrix remodeling in heart disease. Fibrogenesis Tissue Repair 5:15

Franzini M, Lorenzoni V, Masotti S et al (2015) The calculation of the cardiac troponin T 99th percentile of the reference population is affected by age, gender, and population selection: a multicenter study in Italy. Clin Chim Acta 438:376–381

Friões F, Lourenço P, Laszczynska O et al (2015) Prognostic value of sST2 added to BNP in acute heart failure with preserved or reduced ejection fraction. Clin Res Cardiol 104:491–499

Biomarkers of Stroke

23

Marcello Ciaccio and Luisa Agnello

Introduction

Several biomarkers involved in the various stages of stroke pathogenesis (oxidative damage, inflammation, thrombus formation, cardiac function, and brain damage) have been identified. They could play a role in risk prediction, diagnosis, and differential diagnosis between ischemic and hemorrhagic stroke. However, further studies are needed to validate their use in clinical practice. This chapter describes the characteristics and potential usefulness of the primary stroke biomarkers.

Definition of Stroke

The term "stroke" refers to an acute vascular event n the brain. The World Health Organization (WHO) defines stroke as "a syndrome characterized by the sudden and rapid development of symptoms and signs referable to a focal deficit of brain function without any apparent cause other than vascular; the loss of brain function may be global (patients in a deep coma). The symptoms last more than 24 h or lead to death;" if the symptoms last <24 h, a transient ischemic attack (TIA) is diagnosed (Fig. 23.1). Two types of stroke are distinguished:

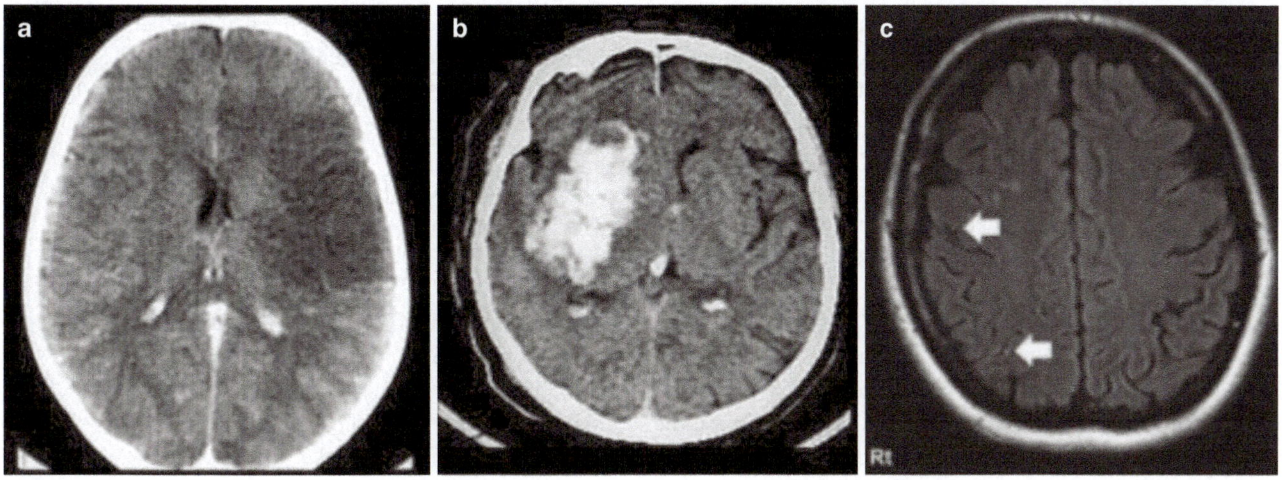

Fig. 23.1 CT finding of ischemic (**a**) and hemorrhagic (**b**) stroke. MR image of transient ischemic attack (**c**)

M. Ciaccio (✉) · L. Agnello
Department of Biomedicine, Neurosciences and Advanced Diagnostics, Institute of Clinical Biochemistry, Clinical Molecular Medicine and Clinical Laboratory Medicine, and Department of Laboratory Medicine, University Hospital "P. Giaccone", Palermo, Italy
e-mail: marcello.ciaccio@unipa.it

M. Ciaccio (ed.), *Clinical and Laboratory Medicine Textbook*, https://doi.org/10.1007/978-3-031-24958-7_23

Table 23.1 Pathophysiological classification of ischemic stroke subtypes

• Atherosclerosis of the great vessels (embolism/thrombosis)
• Cardioembolism
• Small vessel occlusion (lacuna)
• Stroke from various causes
• Stroke of undetermined causes – Two or more causes identified – Negative evaluation – Incomplete evaluation

- Ischemic stroke, the most frequent form (85%)
- Hemorrhagic stroke (15%)

Ischemic stroke is characterized by altered blood flow to a brain region, resulting from occlusion or spasm of a cerebral vessel. Table 23.1 shows the leading causes of ischemic stroke.

In hemorrhagic stroke, the lack of perfusion is due to intracerebral or subarachnoid hemorrhage. The rupture of deep or superficial parenchymal vessels due to arterial hypertension or amyloid angiopathy, respectively, represent the primary cause of intracerebral hemorrhage. Subarachnoid hemorrhage originates from the extracerebral vessels, and the blood pours into the cerebrospinal fluid (CSF); in 85% of cases, it is caused by an arterial aneurysm rupture.

Stroke is the third leading cause of death in Western countries, after coronary heart disease and cancer, and one of the leading causes of disability in adults. Mortality at 1 month is about 20–25%, at 1 year is 30–40%; at 1 year, about one-third of survivors have a high degree of disability. Every year, in Italy, 157,000 new stroke cases are expected, 196,000 if recurrences are also considered. The prevalence is around 800,000 cases, with an incidence that progressively increases with age, reaching a maximum value in subjects aged >65 years.

A complex and concatenated series of molecular processes (oxidative stress, excitotoxicity, endothelial damage, and blood–brain barrier alteration) follows temporary or permanent focal cerebral ischemia. Numerous factors increase the risk of stroke (Table 23.2), and their recognition constitutes the basis of both primary and secondary stroke prevention.

TIAs represent a strong risk factor for ischemic stroke, especially in the first hours or days after the event. The WHO defines TIA as "a sudden onset of signs and/or symptoms related to focal cerebral deficit attributable to the insufficient blood supply, lasting less than 24 h;" in most cases, TIA resolves within 1 h from the symptoms onset. The ABCD2 (Age, Blood pressure, Clinical features, Duration of symptoms, Diabetes) is a validated score predictive of the early risk of stroke in patients with TIA; it consists of the sum of points assigned to five clinical features independently associated with the risk of stroke:

Table 23.2 Risk factors for ischemic and hemorrhagic stroke

	Risk factors	
	Not modifiable	Modifiable
Ischemic stroke	Age Genetic Race	Hypertension Heart disease (patent foramen ovale) Atrial fibrillation Diabetes mellitus Cigarette smoking Hypercholesterolemia Reduced physical activity Excessive alcohol consumption Obesity Metabolic syndrome Use of oral contraceptives Anti-phospholipid antibodies, hemostasis factors, elevated Lp (a) values Drug use
Hemorrhagic stroke	Age Non-Caucasian race	Hypertension Excessive alcohol intake Smoking Therapy: thrombolytic and anticoagulant therapy in the acute phase and in the prevention of ischemic stroke; antiplatelet therapy only modestly increases the risk

- Age > 60 years (1 point)
- Systolic pressure > 140 mmHg or diastolic pressure > 90 mmHg (1 point)
- Clinical signs of TIA (unilateral hyposthenia, 2 points; aphasia without hyposthenia, 1 point)
- Duration of TIA (>60 min, 2 points; 10–59 min, 1 point)
- Diabetes (1 point)

In patients with TIA, the ABCD2 score classifies the 2-day stroke risk as:

- Low: score < 4
- Moderate: score 4–5
- High: score > 5

Hospitalization is indicated in patients with TIA at moderate-to-high risk of stroke.

Diagnosis and Therapy

The diagnosis of stroke and TIA is typically based on an accurate anamnesis and a scrupulous physical examination; in both cases, computed tomography (CT) or magnetic resonance imaging (MRI) is indicated for the differential diagnosis with other pathologies miming TIA or stroke. Moreover, a CT scan allows differential diagnosis between ischemic and hemorrhagic stroke; in the former case, it is negative in the acute phase.

In a subject with suspected stroke, the following laboratory tests are indicated, aimed at assessing the general condition and identifying the risk factors:

- Blood count with platelets
- Glycemia
- Cardiac biomarkers (troponin)
- Serum electrolytes (Na^+, K^+, Cl^-, Mg_2^+, Ca_2^+)
- Kidney function tests
- Coagulation tests (prothrombin time [PT], INR, and activated partial thromboplastin time [aPTT])
- Oxygen saturation

In addition, the following tests are indicated in selected patients:

- Chest X-ray
- Function profile and liver damage
- Arterial blood gas analysis
- Physical–chemical examination of the CSF
- Lipid profile
- Toxicological investigation
- β-hCG
- Alcohol
- Electroencephalogram

Stroke represents a medical emergency requiring immediate hospitalization. Stroke therapy must be promptly administred in the acute phase and will depend on the initial etiopathogenetic event. In the case of ischemic stroke, the therapy is thrombolytic and must be started within 4–5 h from the symptoms onset. In the case of hemorrhagic stroke, the patient starts a specific therapeutic process, including neurosurgical evaluation.

In the post-acute phase, physical rehabilitation is of paramount importance.

Biomarkers

For decades, research has been focused on identifying biomarkers that can improve the early stroke detection and positively modify the clinical, economic, and management outcomes. An ideal stroke biomarker should have the following characteristics:

- Diagnostic sensitivity and specificity
- The ability to differentiate between hemorrhagic and ischemic stroke
- Early and stable release after an acute event
- Predictable plasma clearance
- Potential to define stroke risk
- The ability to guide therapeutic choices
- The possibility of being quantitatively and rapidly measured using cost-effective techniques

The difficulties in identifying a biomarker to introduce in clinical practice are mainly related to the slow release of glial and neuronal proteins across the blood–brain barrier after stroke or traumatic injury and the reduced diagnostic specificity (they increase in many clinical situations simulating stroke).

Following cerebral ischemia, there is an increased production of oxygen free radicals, which triggers the inflammatory response activation that results in the recruitment into the damaged brain tissue of various immune cells, such as macrophages, neutrophils, and T lymphocytes. These inflammatory cells release proinflammatory cytokines, such as interleukin-6 (IL-6), which can cross the blood–brain barrier and reach circulation. In addition, biomarkers such as D-dimer and those that play a role in platelet function are involved in thrombus formation and propagation. The heart and large-caliber vessels, such as the aorta, are among the main sites of thrombus formation; the inability of the cardiovascular system to release sufficient blood to the brain may worsen the stroke outcome.

Several biomarkers have been identified in the various stages of stroke pathogenesis (oxidative damage, inflammation, thrombus formation, cardiac function, and brain damage) (Table 23.3).

Other biomarkers currently under investigation are:

- Lipoprotein-associated phospholipase A2 (Lp-PLA2)
- Asymmetrical dimethylarginine (ADMA)
- Matrix metalloproteinase-9 (MMP-9)
- S100-β protein
- *N*-methyl-D-aspartic acid (NMDA) receptor peptides and their antibodies
- Glial fibrillary acidic protein (GFAP)
- Parkinson disease protein 7 (PARK-7)
- Nucleotide diphosphate kinase A (NDKA)

Lp-PLA2

Lp-PLA2 is a calcium-dependent serine lipase that hydrolyzes oxidized phospholipids to release proinflammatory lysophosphatidylcholine and oxidized fatty acids. It circulates bound mainly to low-density lipoprotein (LDL) and partly to small, dense high-density lipoprotein (HDL) (anti-atherogenic effect). Lp-PLA2 is produced and expressed in macrophage-rich atherosclerotic lesions and represents an independent inflammatory marker of cardiovascular risk and a predictor of ischemic stroke.

Table 23.3 Biomarkers of stroke according to the pathogenesis

Mechanism and biomarker	Biological function
Inflammation	
IL-6	Inflammatory cytokine that acts as a messenger among leukocytes, vascular endothelium, and resident cells in the parenchyma
CRP	Acute-phase protein involved in the inflammatory response and innate immunity
VCAM-1	Transmembrane protein involved in endothelial cell–leukocyte signal transduction
MCP-1	Powerful mononuclear chemoattractant cell produced by endothelial and smooth muscle cells
Dyslipidemia/endothelial damage	
ApoC-I	Associated with LDL and VLDL; involved in plasma lipoproteins remodeling; inhibits CETP
ApoC-III	Associated with VLDL, HDL, and LDL; inhibits triglyceride hydrolysis by hepatic/lipoprotein lipase; interferes with physiological endothelial function
NT-proBNP	Neurohormone with natriuretic, diuretic, and vasodilatory functions
FABP	Cytoplasmic protein that modulates the cascade of signaling lipid; involved in the oxidation of fatty acids
Growth factors	
BDNF	Maintenance and survival of mature neurons
Endothelial damage	
MBP	Main proteolipid constituent of myelin, produced by oligodendroglia cells
NSE	Dimeric glycolytic isoenzyme in the cytoplasm of neurons and neuroendocrine cells
Coagulation/fibrinolysis	
D-dimer	Fibrin degradation product; it reflects a global activation of coagulation and fibrinolysis
Von Willebrand factor	Plasma glycoprotein involved in platelet adhesion

Apo apolipoprotein, *BDNF* brain-derived neutrophic factor, *BNP* type B natriuretic peptide, *CETP* cholesteryl ester transfer protein, *CRP* C-reactive protein, *FABP* fatty acid-binding protein, *HDL* high-density lipoprotein, *LDL* low-density lipoprotein, *MBP* myelin basic protein, *NSE* neuron-specific enolase, *MCP* monocyte chemoattractant protein, *VCAM* vascular cell adhesion molecule, *VLDL* very low-density lipoprotein

ADMA

The post-translational methylation of L-arginine produces the ADMA molecule. After proteolysis, it is released as free dimethylarginine along with symmetric (inactive) dimethylarginine. It is a potent nitric oxide synthase (NOS) inhibitor. Its levels are detectable in the blood, urine, and cerebrospinal fluid; the plasma form has been proposed as a predictive marker of stroke risk.

MMP-9

MMP-9 belongs to the family of zinc- and calcium-dependent endopeptidases responsible for the turnover and degradation of extracellular matrix proteins. It plays an essential role in several processes, including tissue remodeling, phlogosis, angiogenesis, and metastasis. In the brain, its expression is physiologically very low or undetectable, while its levels increase significantly early in the ischemic brain; in the acute phase, its concentrations correlate with the extent of ischemia, poor prognosis, and complications from hemorrhagic transformation. Several studies have highlighted the role of MMP-9 in stroke pathogenesis, including loss of blood–brain barrier integrity, neuronal death, and hemorrhage following stroke. In addition, MMP-9 plays a reparative role during brain regeneration and neurovascular remodeling in the subsequent phase of tissue repair.

S100-β

S100-β is a glial protein consisting of α- and β-subunits that combine in hetero- and homodimers (α–α, α–β, β–β); S100-β includes the β–β and α–β forms. It is present in melanocytes, adipocytes, and chondrocytes but is highly specific for the nerve tissue, localized in the cerebral astroglial compartment, and Schwann cells, which line peripheral nerve fibers. S100-β represents a not specific biomarker of blood–brain barrier dysfunction with a concentration in the cerebrospinal fluid significantly higher than in the serum (40:1). It has emerged as a biomarker of early ischemic stroke with a peak after 24 h. It correlates well with the extent of the infarct area. In addition, it allows differential diagnosis between ischemic and hemorrhagic stroke or stroke-mimicking diseases. One of the main disadvantages is its poor specificity, as it increases during other neurological pathologies.

NMDA Receptor Peptides

NMDA receptor peptides bind glutamate and are present on neurons throughout the encephalon. They consist of four subunits, 2 NR1 and 2 NR2. During ischemia, fragmentation of NR2 into NR2A and NR2B and production of anti-NR2 antibodies are observed. Therefore, NR2 fragments and their antibodies represent markers of ischemic stroke and TIA.

GFAP

GFAP is a monomeric filamentous protein specific to brain astrocytes. Its levels increase during ischemic stroke, peaking 2–4 days after the onset of symptoms. It also allows differential diagnosis between ischemic and hemorrhagic stroke.

PARK7

PARK7 has a reparative role in neurological damage in oxidative stress processes; its levels increase during stroke, with a peak between 30 min and 3 h after the symptoms onset. It does not allow making a differential diagnosis between ischemic and hemorrhagic stroke or TIA.

NDKA

NDKA is an enzymatic protein of neurons whose levels increase during stroke, with a peak between 30 min and 3 h after the symptoms onset. It does not allow the differential diagnosis between ischemic and hemorrhagic stroke, or TIA.

Table 23.4 shows the potential clinical applications of biomarkers at various stages of the disease, from risk prediction to diagnosis, differential diagnosis, and prognosis.

Given the high heterogeneity of stroke, a single biomarker may not be sufficient to capture the different aspects of stroke pathogenesis. Therefore, a multi-marker strategy that assesses the various stroke-related molecular alterations may be necessary. Translating stroke biomarkers into clinical practice is a goal of many researchers, clinicians, and pharmaceutical industries.

Table 23.4 Potential clinical application of stroke biomarkers

Type	Biomarker	Biological function
Risk	Lp-PLA2	Enzyme that catalyzes the degradation of platelet-activating factor by hydrolysis
	Oxidized LDL	Modified (oxidized) form of LDLs involved in the development of atherosclerotic plaque
	HDL	Lipoproteins with anti-atherosclerotic, antithrombotic, anti-inflammatory, and antioxidant activity
	MR-proADM	Peptide with vasodilatory activity
	ADMA	Endogenous inhibitor of nitric oxide synthetase; it is released by endothelial cells and causes endothelial dysfunction
	SDMA	Structural isomer of ADMA
	CRP	Acute-phase protein involved in the inflammatory response and innate immunity
	NT-proBNP	Neurohormone secreted by the myocardium with natriuretic, diuretic, and vasodilatory activity
	hsTnT	Biomarker of myocardial necrosis. It is a diagnostic biomarker of acute myocardial infarction and a prognostic diagnostic biomarker of ischemic stroke
	F2-isoprostanes	Oxidative degradation products of arachidonic acid
Diagnosis	Glutamate	Neurotransmitter amino acid that activates cell surface receptors on the membranes of neurons and glial cells
	S100-β	Homodimer glial protein that regulates intracellular calcium levels; biomarker of blood–brain barrier dysfunction
	VWF	Plasma glycoprotein involved in platelet adhesion
	MMP-9	Calcium-dependent proteolytic enzyme involved in the degradation of the basal lamina and the extracellular matrix, leading to the destruction of the blood–brain barrier
	MCP-1	Chemokine, which regulates the migration and infiltration of monocytes and macrophages
	VCAM-1	Transmembrane protein involved in endothelial cell–leukocyte signal transduction
	NT-proBNP	Neurohormone with natriuretic, diuretic, and vasodilatory activity
	CRP	Acute-phase protein involved in the inflammatory response and innate immunity
	D-dimer	Fibrin degradation product; it reflects a global activation of coagulation and fibrinolysis
Differential diagnosis: ischemic stroke	S100-β	Homodimer glial protein that regulates intracellular calcium levels; biomarker of blood–brain barrier dysfunction
	cFN	Main component of the extracellular matrix involved in the processes of healing and cell adhesion
	GFAP	Monomeric filamentous protein specific to astrocytes, whose levels increase during ischemic stroke
Differential diagnosis: hemorrhagic transformation	PAI-1	Major inhibitor of the fibrinolytic system
	TAFI	Enzyme that hydrolyzes the C-terminal bonds of fibrin to prevent fibrinolysis and inhibits the activation of fibrinogen
	cFN	Main component of the extracellular matrix involved in the processes of healing and cell adhesion
	MMP	Calcium-dependent proteolytic enzyme involved in the degradation of the basal lamina and the extracellular matrix, leading to the destruction of the blood–brain barrier

ADMA asymmetric dimethylarginine, *cFN* cellular fibronectin, *HDL* high-density lipoprotein, *hsTNT* high-sensitivity troponin T, *LDL* low-density lipoprotein, *Lp-PLA2* lipoprotein-associated phospholipase A2, *MCP-1* monocyte chemoattractant protein, *MMP* matrix metalloproteinase, *MRproADM* midregional proadrenomedullin, *PAI-1* type 1 inhibitor of plasminogen activator, *CRP* C-reactive protein, *SDMA* symmetric dimethylarginine, *TAFI* thrombin-activated fibrinolysis inhibitor, *VCAM-1* vascular cell adhesion protein, *VWF* von Willebrand factor

Recommended Readings

Adams HP Jr, Bendixen BH, Kappelle LJ et al (1993) Classification of subtype of acute ischemic stroke. Definitions for use in a multicenter clinical trial. TOAST. Trial of Org 10172 in Acute Stroke Treatment. Stroke 24(1):35–41

Johnston SC, Rothwell PM, Nguyen-Huynh MN et al (2007) Validation and refinement of scores to predict very early stroke risk after transient ischaemic attack. Lancet 369(9558):283–292

Maestrini I, Ducroquet A, Moulin S et al (2016) Blood biomarkers in the early stage of cerebral ischemia. Rev Neurol (Paris) 172(3):198–219. https://doi.org/10.1016/j.neurol.2016.02.003

Ng GJ, Quek AM, Cheung C et al (2017) Stroke biomarkers in clinical practice: a critical appraisal. Neurochem Int. https://doi.org/10.1016/j.neuint.2017.01.005. pii: S0197-0186(16)30460-0

SPREAD (2012) Ictus cerebrale: linee guida italiane di prevenzione e trattamento, VII ed

24 Thrombophilia

Marcello Ciaccio

Introduction

Thrombophilia screening is essential for recognizing patients at risk of venous thromboembolism and pulmonary embolism. This chapter describes the clinical conditions at high risk of thrombophilia and laboratory test for thrombophilia screening.

Hereditary and Acquired Causes of Thrombophilia

Thrombophilia is defined as the predisposition to develop thrombotic events in the venous and/or arterial district. It is typically characterized by the onset of clinical manifestations at a young age (<50 years). Although thrombophilia can be due to congenital and acquired causes, most thrombophilic alterations are congenital. The most frequent clinical manifestation of thrombophilia is deep vein thrombosis (DVT). This is a multifactorial condition resulting from the interaction of acquired and congenital risk factors (Table 24.1).

The simultaneous presence of congenital and acquired risk factors significantly increases the overall risk of DVT. Moreover, surgical interventions, pregnancy, puerperium, estroprogestinic therapy, and prolonged immobilization represent transient risk factors.

The prevalence of inherited disorders is variable and depends on the type of defect. For example, the prevalence of antithrombin III (ATIII) deficiency is around 0.2% in the general population, protein C deficiency around 0.5%, factor V of Leiden (FV Leiden) at 3–7%, and prothrombin G20210A (FII G20210A) at 5%. Overall, the incidence of venous thromboembolism remains high despite the primary and secondary prevention measures implemented in recent years. Pulmonary embolism, a dreaded complication of venous thromboembolism, is a high-mortality condition due to its nonspecific clinical presentation.

Thrombophilia can also be associated with thrombotic events of the arterial district. However, the usefulness of screening in these cases and, therefore, the potential therapeutic and clinical implications are still debated.

Table 24.1 Congenital and acquired causes of thrombophilia

Congenital	Acquired
ATIII deficiency	Pregnancy and puerperium
Protein C deficiency	Postoperative status
Protein S deficiency	Immobilization
Factor V Leiden	Trauma
Factor II G20210A mutations	Elderly
Dysfibrinogenemia	Estrogen and estroprogestinic therapy
Hyperhomocysteinemia	Anti-phospholipid antibody syndrome
	Hyperhomocysteinemia

Thrombophilia Screening

Thrombophilia screening, based on the search for conditions predisposing to thrombosis, is an essential diagnostic tool allowing to identify subjects at high risk of developing a thrombotic event and to define the most appropriate treatment for a patient with thrombosis.

The introduction of large-scale genetic testing and the spread of knowledge about the hereditary factors underlying thrombophilia led to an exponential increase in requests for thrombophilia screening in the early 1990s.

M. Ciaccio (✉)
Department of Biomedicine, Neurosciences and Advanced Diagnostics, Institute of Clinical Biochemistry, Clinical Molecular Medicine and Clinical Laboratory Medicine, University Hospital "P. Giaccone", Palermo, Italy

Department of Laboratory Medicine, University Hospital "P. Giaccone", Palermo, Italy
e-mail: marcello.ciaccio@unipa.it

M. Ciaccio (ed.), *Clinical and Laboratory Medicine Textbook*, https://doi.org/10.1007/978-3-031-24958-7_24

The laboratory plays a crucial role in thrombophilia screening by allowing the identification of the nature and causes of the coagulation disorder.

Indications for Performing Thrombophilia Screening

The main indication for thrombophilia screening is evaluating the risk of recurrence after the first thromboembolic episode or after recurrences. In addition to symptomatic subjects, thrombophilic screening should also be extended to asymptomatic subjects who are first-degree relatives of hereditary thrombophilia carriers or have a strong family history of thromboembolic diseases. Once the overall risk has been defined, the most appropriate anticoagulant therapy can be started. Immediately after the first thromboembolic event, the risk of recurrence is exceptionally high and remains higher than that of the general population for a long time. The estimated incidence of 5%/year supports prolonged secondary anticoagulant prophylaxis. However, the patient's compliance and the hemorrhagic risk are considered when choosing prolonged antithrombotic prophylaxis. It should also be considered that men have a higher risk of recurrence than women.

The indications for performing thrombophilia screening are:

- Young age of the thrombotic event (<50 years)
- Idiopathic thrombosis
- Recurrent thrombosis
- Thrombosis with atypical localization (thrombosis of the cerebral veins, mesenteric veins, and splanchnic circulation)
- Skin necrosis due to the use of oral anticoagulants
- Antiphospholipid antibody syndrome
- Neonatal fulminant purpura
- First-degree relatives of subjects with hereditary thrombophilia
- Poly-abortion

Laboratory Diagnosis

Thrombophilia screening is based on the following basic and specialist tests (Fig. 24.1):

- Prothrombin time (PT)
- Activated partial thromboplastin time (aPTT)
- ATIII (functional method)

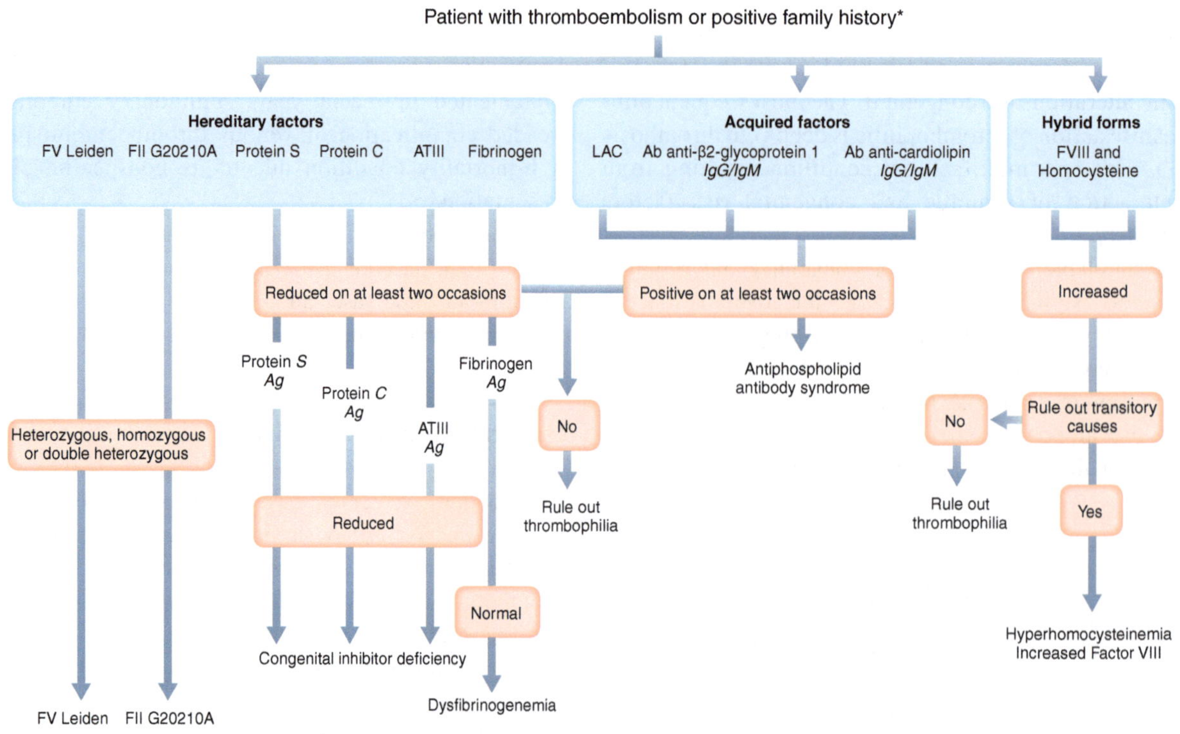

Fig. 24.1 Diagnostic algorithm of thrombophilia. Ag antigen, ATIII antithrombin III, LAC lupus anticoagulant. (Copyright EDISES 2021. Reproduced with permission)

- Protein C (functional test)
- S protein (functional test)
- Fibrinogen (Clauss method)
- Lupus anticoagulant (LAC)
- Anti-beta-2-glycoprotein-1 antibodies
- Anti-cardiolipin antibodies
- Factor VIII
- Homocysteinemia
- Genetic testing for FV Leiden and FII G20210A.

PT and aPTT are first-level tests that allow identifying a coagulative imbalance. In the case of positivity, the functional tests for ATIII, protein C, and protein S must be confirmed by the antigenic method based on the specific protein's measurement. In order to diagnose one of these defects, it is also essential to exclude clinical conditions clinical conditions associated with reduced levels of physiological inhibitors (e.g., hepatopathy, therapy with oral anticoagulants or heparin, intake of estroprogestin, pregnancy). Thus, it is advisable to also perform PT and aPTT as indicators of possible hepatic functional alteration and vitamin K-dependent factor deficiency. The fibrinogen measurement performed by the traditional method (Clauss method) allows for the identification of dysfibrinogenemia, i.e., a condition characterized by normal fibrinogen levels with an altered biological activity. LAC tests, anti-beta-2-glycoprotein-1 antibody assay, and anti-cardiolipin antibody assay are the three diagnostic criteria for antiphospholipid antibody syndrome, and their positivity should be confirmed after 12 weeks. FV Leiden, unlike the wild-type form, has an amino acid substitution making it insensitive to degradation by the activated protein complex C. FII G20210A is associated with increased gene transcription and increased circulating levels of prothrombin. Genetic tests for C677T and A1298C mutations in the *MTHFR* gene are no longer recommended in thrombophilia screening because several evidences have shown that, although they are associated with hyperhomocysteinemia, they do not increase the risk for thromboembolism, neither in the venous nor the arterial district. Similarly, genetic tests for 4G/5G PAI-1, I/D ACE, a/b HPA-1, V34L FXIII, and -455G>A FGB (beta-fibrinogen) mutations, previously considered as part of thrombophilia screening, should no longer be performed due to lack of evidence supporting their involvement in thrombosis risk. The activated protein C resistance test (APC resistance) can be performed as part of thrombophilia screening if the confirmatory genetic test (FV Leiden) is more expensive than the functional test or is not feasible. However, the diffusion of lower-cost genetic tests in many laboratories equipped for molecular diagnostics has made it possible to bypass the functional assay and resort directly to genetic testing.

The diagnosis of hereditary thrombophilia requires at least one of the following conditions: ATIII deficiency, protein C deficiency, protein S deficiency, FVLeiden, and FII G20210A mutations. However, a family history of venous thromboembolism without a corresponding biochemical or genetic finding should lead to the suspicion of hereditary thrombophilia in selected cases. The search for FV Leiden and FII G20210A mutations should be performed on first-degree relatives of subjects with these alterations and relatives of subjects with ATIII, protein C, and protein S deficiency. In any case, genetic tests should be preceded by genetic counseling.

Factors Interfering with Thrombophilia Screening

Functional tests included in thrombophilia screening should not be performed either during the acute phase of the thrombotic event or during anticoagulant therapy but at least 4 weeks after discontinuation of therapy with vitamin K antagonists since these tests may be falsified in these cases. Moreover, in the case of liver disease, many of the tests included in the thrombophilia screening may be altered (PT, ATIII, protein C, protein S, factor VIII). Even during estroprogestin treatment or pregnancy, thrombophilia screening should not be performed; it is necessary to wait at least 1 month after treatment discontinuation or 2 months after childbirth. However, these considerations do not apply to genetic tests. Moreover, tests for the diagnosis of antiphospholipid antibody syndrome can be performed during anticoagulant therapy.

Recommended Readings

Baglin T, Gray E, Greaves M et al (2010) Clinical guidelines for testing for heritable thrombophilia. Br J Haematol 149(2):209–220

Bates SM, Greer IA, Pabinger I et al (2008) Venous thromboembolism, thrombophilia, antithrombotic therapy, and pregnancy. Chest 133:844S–886S

Coppola A, Tufano A, Cerbone AM, Di Minno G (2009) Inherited thrombophilia: implications for preven tion and treatment of venous thromboembolism. Semin Thromb Hemost 35(7):683–694

Gohil R, Peck G, Sharma P (2009) The genetics of venous thromboembolism. A meta-analysis involving approximately 120,000 cases and 180,000 controls. Thromb Haemost 102(2):360–370

McCormack T, Harrisingh MC, Horner D, Bewley S, Guideline Committee (2020) Venous thromboembolism in adults: summary of updated NICE guidance on diagnosis, management, and thrombophilia testing. BMJ 369:m1565

Seligsohn U, Lubetsky A (2001) Genetic susceptibility to venous thrombosis. N Engl J Med 344(16):1222–1231

Simone B, De Stefano V, Leoncini E et al (2013) Risk of venous thromboembolism associated with single and combined effects of Factor V Leiden, Prothrombin 20210A and Methylenetethraydrofolate reductase C677T: a meta-analysis involving over 11,000 cases and 21,000 controls. Eur J Epidemiol 28(8):621–647

Testa S, Antonucci G, Intra E et al (2004) Gli screening per trombofilia. Riv Med Lab 5(2):118–120

(2020) Venous thromboembolic diseases: diagnosis, management and thrombophilia testing. National Institute for Health and Care Excellence (UK), London

Endocrine System

25

Marcello Ciaccio, Luisa Agnello, Giulia Bivona, Anna Maria Ciaccio, and Bruna Lo Sasso

Introduction

The endocrine system consists of a set of glands that, through the release of hormones and in synergy with the nervous and immune systems, regulate the various organs and apparatuses, contributing to maintaining homeostasis. The communication among the nervous, endocrine, and immune systems includes complex regulatory circuits operating at different functional levels. Laboratory medicine has an essential role in understanding pathophysiological mechanisms involving the endocrine glands and diagnosing diseases affecting the glands. Studying circulating hormone levels provides valuable information to the clinician.

Hypophysis

Marcello Ciaccio and Luisa Agnello

M. Ciaccio (✉) · B. Lo Sasso
Department of Biomedicine, Neurosciences and Advanced Diagnostics, Institute of Clinical Biochemistry, Clinical Molecular Medicine and Clinical Laboratory Medicine, University Hospital "P. Giaccone", Palermo, Italy

Department of Laboratory Medicine, University Hospital "P. Giaccone", Palermo, Italy
e-mail: marcello.ciaccio@unipa.it

L. Agnello · G. Bivona
Department of Biomedicine, Neurosciences and Advanced Diagnostics, Institute of Clinical Biochemistry, Clinical Molecular Medicine and Clinical Laboratory Medicine, University Hospital "P. Giaccone", Palermo, Italy

A. M. Ciaccio
Department of Health Promotion, Mother and Child Care, Internal Medicine and Medical Specialties (ProMISE) "G. D'Alessandro", University Hospital "P. Giaccone", Palermo, Italy

Anatomy

The pituitary gland is an endocrine gland located in the center of the skull base, within the sella turcica; it consists of two lobes, one anterior (adenohypophysis) and one posterior (neurohypophysis), which are anatomically and functionally distinct (Fig. 25.1).

The saddle is contiguous with several vascular and neurological structures and anchored to the lower region of the hypothalamus by the infundibulum, known as the pituitary peduncle. The hypothalamus synthesizes release and inhibitory factors that are secreted directly into the portal vascular system of the pituitary peduncle and, in turn, the pituitary gland, in response, releases its hormones (Fig. 25.2).

Pituitary hormones are pulsatile secreted and induce specific responses in various peripheral target tissues. In turn, hormones released from peripheral tissues regulate pituitary function through feedback at both hypothalamic and pituitary levels.

Diagnosis of pituitary disorders is mainly based on laboratory investigations to assess the function of the hypothalamic-pituitary-target gland axis.

Adenohypophysis

The anterior pituitary gland, also known as the adenohypophysis, accounts for about 80% of the entire gland and produces six main hormones: Growth Hormone (GH), Prolactin (PRL), Adrenocorticotropic Hormone or Corticotropin (ACTH), Luteinizing Hormone (LH), Follicle-Stimulating Hormone (FSH), and Thyroid-Stimulating Hormone (TSH). These hormones are also called pituitary tropins because they maintain the tropism of the target organs.

Growth Hormone

Growth hormone, or somatotropin, so defined for its ability to increase stature growth in the hypophysectomized animal,

M. Ciaccio (ed.), *Clinical and Laboratory Medicine Textbook*, https://doi.org/10.1007/978-3-031-24958-7_25

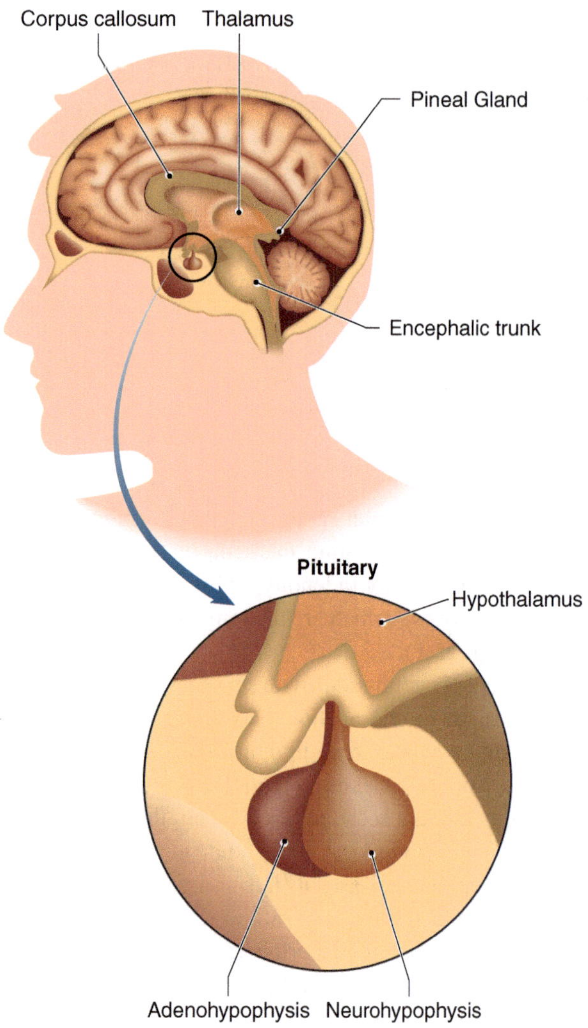

Fig. 25.1 Hypophysis. (Copyright EDISES 2021. Reproduced with permission)

is the most abundant hormone of the adenohypophysis. Its secretion occurs pulsatile every 3–4 hours, with more frequent and intense peaks in the early hours of sleep (REM phase [Rapid Eye Movement]); about 70% of daily GH secretion occurs during sleep.

GH secretion is inversely correlated with age, being elevated in children and during the pubertal period and gradually decreasing over the years. In about 50% of samples collected during the day in healthy and almost all elderly subjects, GH levels cannot be assayed by standard methods. GH levels are higher in women than in men and increase in both sexes following exercise, stress, trauma, or sepsis.

Physiologically, GH secretion is controlled by two hypothalamic hormones, Growth Hormone–Releasing Hormone (GHRH), with stimulatory action, and somatostatin (SRIH), with inhibitory action. Physiological, pharmacological, and pathological factors, regulating the secretion of GHRH and SRIH at the hypothalamic level, regulate the secretion of GH at the pituitary level. Dopamine, serotonin, acetylcholine, sex steroids, and endogenous opioids have stimulant actions. On the other hand, norepinephrine has a dual effect because it increases GH secretion through the α-adrenergic pathway and inhibits it through β-adrenergic receptors. The insulin-like growth factor (IGF-1), which is the target of GH, through a negative feedback mechanism, inhibits the production of the hormone.

GH has a half-life ranging from 20 to 25 minutes and is metabolized mainly in the liver and, to a lesser extent, in the kidney.

The main biological function of GH is to promote somatic and tissue growth by regulating many metabolic processes (Table 25.1). Many of the acute effects of GH are insulin-like.

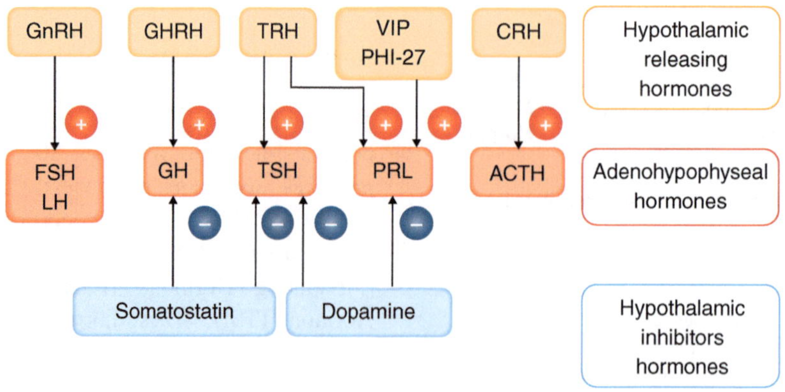

Fig. 25.2 Regulation of adenohypophysis hormone secretion. ACTH adrenocorticotropic hormone, CRH corticotropin-releasing hormone, FSH follicle-stimulating hormone, GH growth hormone, GHRH GH releasing hormone, GnRH gonadotropin-releasing hormone, PHI-27 histidine-isoleucine-27 peptide, PRL prolactin, TRH thyrotropin-releasing hormone, TSH thyrotropic hormone, VIP Vasoactive intestinal peptide. (Copyright EDISES 2021. Reproduced with permission)

Table 25.1 Metabolic action of growth hormone

Protein metabolism	Stimulates protein synthesis Reduces protein catabolism
Lipid metabolism	Reduces lipogenesis Stimulates lipolysis Stimulates the oxidation of fatty acids Stimulates hepatic expression of LDL receptors Promotes the production of HDL
Carbohydrate metabolism	Acute effects: stimulates insulin release and increases glucose oxidation, resulting in lower blood glycemia (insulin-like effect) Late effects: the prolonged stimulus causes a long- lasting anti-insulin effect characterized by reduced peripheral glucose uptake and reduced glycolysis, with consequent increase in blood glucose
Electrolytic and mineral balance	Promotes intestinal absorption of calcium and phosphorus Reduces the renal excretion of sodium and potassium causing water retention Stimulates the urinary excretion of calcium Stimulates the hydroxylation of 25(OH)D a 1.25$(OH)_2$D, promoting intestinal absorption of calcium
Immune system	Stimulates the trophism and function of the thymus

HDL high density lipoprotein, *LDL* low density lipoprotein

GH exerts its biological action directly through interaction with a receptor located on the surface of target tissues and indirectly through the IGF-1 factor, which stimulates its synthesis mainly in the liver. The GH receptor is highly expressed in the liver and cartilage. Once released into the circulation, IGF-1 acts as a hormone, interacting with receptors target cells' surface. IGF-1 receptors have an almost ubiquitous localization. In addition, IGF-1 secretion is also stimulated at target tissues, especially cartilage, where it stimulates growth by an autocrine/paracrine mechanism. Physiological factors influence serum concentrations of IGF-1. In particular, levels increase during puberty with a peak around 16 years of age and progressively decrease until they drop by 80% in elderly subjects. Women generally have higher levels than men. Since GH is the primary regulator of IGF-1 secretion, any alteration in its synthesis or action (hypopituitarism, GHRH or GH receptor deficiency) causes a reduction in IGF-1 levels. Furthermore, the activity of the GH/IGF-1 axis is strongly influenced by nutritional status and body composition. In particular, in obese subjects, GH secretion, both spontaneous and stimulated, is markedly reduced, with values overlapping those of patients with severe GH deficiency. At the same time, IGF-1 production remains preserved, thus suggesting an alteration of the negative feedback control threshold. In cases of prolonged fasting, anorexia nervosa, and, in general, in chronic malnutrition, there is an increase in GH and a reduction in IGF-1 levels.

Defects in the hypothalamic-pituitary axis can lead to low GH levels (GH deficiency) with significant consequences in children and adults or, vice versa, to hypersecretion of the hormone (acromegaly).

Growth Hormone Deficiency

Growth Hormone Deficiency (GHD) is a clinical condition caused by pathological GH deficiency and, therefore, characterized by reduced stature growth in children and increased body fat and reduced muscle mass in adults; adults with GH deficiency often complain of lack of energy and difficulty concentrating. GH deficiency can present in very variable forms with a clinical picture conditioned by the severity of the hormonal deficit (isolated or associated with other pituitary tropins) and by the age of onset. It is essential to identify GHD early, especially in children, as treatment with recombinant human GH is highly effective in inducing an increase in final height.

GH Deficiency in Children

The deficiency can be diagnosed at any age, but since infants with GHD have an average birth weight and length, the slowdown in growth rate is usually noticed relatively late. Characteristic signs of the deficit, in addition to short stature, are infantile facial features, delayed teething, small penis (micropenis), and increased fat in the subcutaneous tissue deposited on the hips and abdomen.

GHD can be congenital or acquired. Congenital causes include structural malformations of the brain, for example, abnormal development of the hypothalamus or pituitary gland, and genetic mutations resulting in alterations in the synthesis or function of GH or the receptor for GHRH or peripheral tissue insensitivity due to structural defects in the GH receptor or intracellular signal; in this case (peripheral insensitivity to GH), we speak of a syndrome known as Laron syndrome characterized by normal or elevated GH levels associated with reduced IGF1 values and, in some cases, reduced levels of GH-binding proteins (Growth Hormone–Binding Proteins, GHBP).

Acquired GH deficiency can be due to various causes. The most frequent are benign tumors of the pituitary gland (pituitary adenomas), which, as they grow, can damage the remaining gland; less frequent causes include cranial irradiation, brain trauma, and central nervous system infections and inflammatory conditions (e.g., sarcoidosis). GH deficiency can occur in isolation or combination. In the latter case, GH deficiency is associated with the deficiency of one or more pituitary tropins and is generally due to mutations of transcription factors such as Pit-1 and Prop-1, which regulate the differentiation of somatotropic cells of the adenohypophysis.

GH Deficiency in Adults

Growth hormone deficiency in adults is generally caused by hypothalamic or pituitary damage, usually secondary to pituitary surgery, tumors, radiation treatment, or a developmental age deficiency. Acquired pituitary insufficiency has a defined framework of functional deficit in which loss of adequate GH reserve precedes other tropin deficits (GH, LH/FSH, TSH, and ACTH).

GHD in adults is not defined by a characteristic clinical picture. However, it presents nonspecific signs and symptoms, including an increase in body fat, especially visceral, with an increase in the waist-to-hip ratio, a reduction in muscle mass, a reduction in bone density (osteopenia and osteoporosis) with an increased risk of fractures, reduced performance during physical activity, and finally, alterations in cardiovascular function. Thus, GHD in adults can manifest itself subtly and may not be easily recognizable on clinical observation alone. GHD in adults must be suspected in the presence of specific conditions.

Acromegaly

The term acromegaly identifies a chronic syndrome caused by GH hypersecretion during adulthood. An excess of GH during childhood and puberty is accompanied by excessive growth in height (gigantism). In adulthood, on the contrary, being impossible, any further increase in length of the diaphysis, the disease causes a volumetric increase of the distal parts of the body (bones and soft tissues); the term acromegaly derives from the Greek *akros*, extremities, and *megas*, large. Acromegaly is characterized by swollen feet, hands (the patient over the years is forced to change the size of the rings and shoes), increased volume of the tongue, prognathism with consequent dental malocclusion, enlargement of the facial bones, and organomegaly, i.e., the increase in the volume of internal organs (heart, liver, thyroid, intestine, kidney, etc.), with consequent alteration of their function. Patients with this pathology have a mortality rate three times higher than the general population due to the increased risk of cardiovascular, cerebrovascular, and respiratory diseases.

The process generally begins before age 40, but recognition of the disease can be very late, even after 10 years, due to the poor initial symptoms.

A GH-secreting pituitary adenoma sustains hormonal hypersecretion in 95% of cases. In contrast, in the remaining very rare cases, it is due to GHRH hypersecretion of intracranial (slow-growing hypothalamic adenomas, coristomas, ganglioneuromas) or extracranial origin, most frequently bronchial carcinoids, rarely neuroendocrine tumors of the gastrointestinal tract and thymic carcinoids. Cases of ectopic GH secretion by pancreatic, ovarian, pulmonary, or hematopoietic system tumors have also been described.

Laboratory Investigations

Basal Measurement

GH

The basal serum GH assay is indicated in diagnosing GH excess (acromegaly and gigantism) and controlling the pathologies mentioned above after pharmacological therapy; however, it has little diagnostic value in GH deficiency. Since GH is released in a pulsatile manner, performing the test at a single time of day is usually not clinically indicative. The most commonly used GH assay methods are based on monoclonal antibodies with enzymatic, fluorescent, and chemiluminescent tracers.

The whole blood sample should be taken in the morning on an empty stomach, in the patient in clinostat, and at rest for 30 minutes; exercise or stress conditions can significantly alter GH levels. Reference values in the healthy population are reported in Table 25.2, although it is not always possible to measure GH levels in healthy subjects.

In contrast to the healthy subject, in patients with acromegaly, GH is always measurable, although it may be at normal limits in some cases.

IGF-1

IGF-1 reflects GH deficiency or excess, but unlike GH, its levels are stable throughout the day. Therefore, measuring IGF-1 levels is a helpful estimate of average GH levels. Normal IGF-1 values vary with age and sex and depend on the assay method used. The most commonly used assay methods are radioimmunoassays and those based on chemiluminescent tracers.

Testing for the IGF-1 measurement may be required in conjunction with the measurement of other pituitary hormones (PRL, FSH, or LH) to diagnose pituitary disorders. Normal levels of IGF-1 indicate, with high probability, the absence of GH abnormalities. However, the patient's clinical context must be always considered; some people may have normal IGF-1 levels in the presence of GHD. Conversely, decreased IGF-1 levels are strongly indicative of GHD. Noteworthy, a decrease in IGF-1 can result from nutritional deficits (anorexia nervosa), chronic kidney or liver disease, production of inactive forms of GH, and intake of high doses of estrogen. Usually, elevated IGF-1 levels indicate increased GH production.

During puberty and pregnancy, it is normal to detect increased concentrations of GH and IGF-1; otherwise, they are very often due to pituitary tumors.

Table 25.2 Normal values of GH

Men	<5 ng/mL or <226 pmol/L
Women	<10 ng/mL or <452 pmol/L
Children	0–20 ng/mL or 0–904 pmol/L
Newborns	5–40 ng/mL or 226–1808 pmol/L

GH growth hormone

Finally, IGF-1 measurement helps monitor the treatment of GH-secreting tumors. If the tumor is surgically removed, measurement of IGF-1 and GH is an estimate of the actual complete tumor removal. If IGF-1 is still elevated after surgical removal of a pituitary tumor, then it may be that the surgery was not fully effective. Decreasing concentrations of IGF-1 during subsequent drug therapy and/or radiation therapy indicate that the treatment is lowering GH production. Increased IGF-1 levels may indicate pituitary tumor recurrence when a patient undergoes long-term monitoring.

Dynamic Measurement

The secretion of GH is pulsatile and episodic. Therefore, the finding of a basal value close to zero is not diagnostic, as it is also evident in normal subjects. In order to assess GH deficiency, it is preferable to use a stimulus test by different drugs administered alone or in combination.

The stimulation test is performed by collecting a blood sample after fasting for 10–12 hours. Then, under medical supervision, a solution that stimulates the release of GH by the pituitary gland is administered intravenously to the patient. Subsequently, following stimulation, multiple blood samples are collected at regular intervals to assess GH production by the pituitary gland.

The insulin-induced hypoglycemia test is the gold standard for diagnosing GH in adults because of its high sensitivity. When the hypoglycemic condition is reached, healthy subjects have a GH peak within 60 minutes, which persists for the next 2 hours; GHD in adults is diagnosed when GH peak response to hypoglycemia below 5 ng/mL. Although the induced hypoglycemia test is safe when performed under close medical supervision, it is contraindicated in elderly subjects and patients at high cardiovascular risk. In addition, obese normoglycemic or hyperglycemic patients with insulin resistance may not achieve an adequate hypoglycemic state. Finally, although it has good sensitivity, it has poor reproducibility. A viable alternative to the induced hypoglycemia test is the glucagon stimulus test. Other tests using arginine or clonidine as the stimulating agent are available. However, they present poor diagnostic accuracy. Glucagon is more potent than other agents in inducing GH secretion. Additionally, glucagon stimulates GH release more effectively when administered intramuscularly or subcutaneously than intravenously.

Finally, enhanced stimulus tests are based on administering GHRH with other agents, most commonly arginine. GHRH stimulates the synthesis and pituitary release of GH, whereas arginine potentiates the stimulatory effect of GHRH by inhibiting hypothalamic somatostatin release. The peak GH response depends not on age or sex but on body mass index (BMI) and central adiposity. Its main advantages are its reproducibility and high discriminatory power.

GH threshold values vary depending on the stimulus test used. Table 25.3 shows the main characteristics of the most common stimulus tests.

The GH suppression test is indicated for the diagnosis of GH excess and monitoring patients undergoing treatment. The oral Glucose Tolerance Test (OGTT) is performed by collecting a blood sample after fasting for 10–12 hours. The patient is then given a solution containing 75 g of glucose dissolved in 300 mL of water. Subsequent blood samples, collected at regular time intervals (30, 60, 90, and 120 minutes), make it possible to assess whether the release of GH by the pituitary gland is correctly inhibited (the inhibitory action is linked to the activation of endogenous secretion of somatostatin by glucose). In healthy subjects, after 1 hour, GH is less than 1 ng/mL, in patients with GH hypersecretion, levels never fall below this value. However, it should be kept in mind that the limit of normality for GH values after OGTT varies according to the sensitivity of the assay method used (1 ng/mL for conventional methods and 0.3–0.4 ng/mL for ultrasensitive methods). In addition, it should be considered that the following conditions may give false-positive results for lack of GH inhibition: malnutrition, depression, renal or hepatic insufficiency, diabetes mellitus, and obesity.

Diagnosis and Therapy

GH Deficiency in Children

The diagnosis of GH deficiency in children is based on auxological criteria (stature deficit associated with the reduced growth rate), clinical, laboratory (altered response of GH to pharmacological tests), and possibly neuroradiological findings (abnormalities of the hypothalamic-pituitary region). Based on these data, it will be possible to identify nonpathological conditions, defined as "normal variants" of growth, such as short stature and constitutional growth retardation, and diagnose the pathological forms.

Children with GHD may present a normal growth pattern in the first year of life, followed by a progressive slowing of growth speed in the subsequent period. Suspicion of GH deficiency should be raised in the case of a child with short stature and normal body proportions without dysmorphisms.

A child is defined as short in stature when his/her height is below the 3rd percentile or below 2 standard deviations (SD). Short stature is found quite frequently in clinical prac-

Table 25.3 Characteristics of the main GH stimulus tests

Test	Insulin-induced hypoglycemia	Glucagon stimulation test	Arginine	Enhanced test GHRH + arginine
Contraindications	History of seizures, coronary artery disease, pregnancy, or age >55 years	Malnourished patients or patients who did not eat for more than 48 hours Fasting hyperglycemia (glycemia >180 mg/dL)	Allergy to arginine	Allergy to GHRH analogue and arginine
Precautions	Some patients may develop neuroglycopenic symptoms during the test when glucose levels fall and should eventually be encouraged to report these symptoms	Patients could feel nauseous during test Possible late hypoglycemia (at the end of the test, patients should eat little and slowly)	Nausea, vomiting, headache and redness	Transient feeling of heat and/or redness due to GHRH Nausea, headache, transient hypotension
Procedure	Patient fasted from midnight for 8–10 hours (water is allowed) Discontinue all morning medications Weigh the patient Administer insulin (standard dose: 0.05–0.1 units/kg for non-diabetic subjects with BMI <30 kg/m^2 and high dose: 0.15–0.3 units/kg for subjects with BMI ≥30 kg/m^2)	Make sure the patient has been fasting for 8–10 hours Weigh the patient Place the patient in the supine position Administer 1 mg intramuscular glucagon (1.5 mg if the patient has a weight >90 kg)	Make sure the patient is fasting from midnight for 8–10 hours Discontinue all morning drugs Weigh the patient Administer arginine 0.5 g/kg (maximum 30 g) as an infusion for 30 minutes	Make sure the patient is fasting from midnight for 8–10 hours (water is allowed) Discontinue all morning drugs Weigh the patient Administer the GHRH analogue (1 μg/kg) as a bolus and arginine hydrochloride (30 g) simultaneously as an infusion for 0–30 minutes
Measurement	*Basal* Blood is drawn to measure glucose *During the test* Blood is drawn every 15 minutes to measure glucose levels When hypoglycemia (<40 mg/dL) is reached, glucose and GH are measured at 20, 25, 30, 35, 40, 60, and 90 minutes. When finished, the patient can drink fruit juice and eat to raise blood glucose *At the end of the test* Glucose levels should be >70 mg/dL	Measure serum GH and glucose levels at 0, 30, 60, 90, 120, 150, 180, 210, and 240 minutes	*Basal* Dosage of GH and IGF-1 *During the test* Serum GH levels are measured at 30, 60, 90, and 120 minutes	*Basal* Dosage of GH and IGF-1 levels *During the test* Serum GH levels are measured at 30, 60, 90, and 120 minutes
Interpretation	If adequate hypoglycemia (<40 mg/dL) is not achieved, GHD cannot be diagnosed A peak serum GH level <5 ng/mL at any time during the hypoglycemic phase of the test is diagnostic of GHD in the adult	Peak serum GH levels tend to appear between 120 and 180 minutes The GH peak <3 ng/mL and <1 ng/mL in patients with BMI <25 kg/m^2 and ≥25 kg/m^2, respectively, at any time during the test is diagnostic of GHD in the adult	The GH peak <0.4 ng/mL at any time during the test is diagnostic of adult GHD	GH values must be evaluated according to BMI; serum GH levels <11 ng/mL, 8 ng/mL, and 4 ng/mL at any time during testing in patients with BMI ≥30 kg/m^2, 25–30 kg/m^2 and <25 kg/m^2, respectively, are diagnostic of GHD in adults

GH growth hormone, *GHD* growth hormone deficiency, *GHRH* somatotropin releasing hormone, *IGF* insulin-like growth factor

tice but should be further investigated in the presence of the following conditions:

- Severe short stature (<−3 SD)
- Stature <−1.5 SD relative to the familial genetic target and growth velocity/year ≤−2 SD or ≤−1.5 SD after two consecutive years
- Stature ≤−2 SD with a growth velocity/year <−1 SD relative to normal for age and sex, assessed at a distance of at least 6 months, or a reduction in stature of 0.5 SD/year in children older than 2 years
- In the absence of short stature, growth velocity ≤−2 SD over one year or ≤−1,5 SD over two years

- Neuroradiologically demonstrated hypothalamic-pituitary malformations/injuries
- Neonatal signs or symptoms of GHD

In children with short stature and suspected GHD, the diagnosis is classically made when GH levels do not reach a cutoff value after two pharmacological stimuli.

Increased GH in response to a single drug test excludes the diagnosis of GH deficiency. A second stimulus test must confirm a positive response to a first stimulus test. If the second test is also positive, the hormone deficiency is confirmed, which, associated with the clinical and auxological features, allows the diagnosis of GH deficiency.

Extemporaneous GH measurements cannot distinguish healthy children from those with true GH deficiency. However, altered response to stimulus testing with normal auxologic and clinical findings cannot be considered probative of GH deficiency.

Determination of IGF-1 levels helps confirm possible GH deficiency. MRI of the pituitary gland can confirm the diagnostic hypothesis, highlighting any tumor lesions or structural defects.

An early diagnosis and an appropriate therapeutic approach allow a significant recovery of height growth and the achievement of a final height within the predictable target based on the height of the parents ("genetic target"), thus improving the patient's quality of life.

Therapy is based on administering recombinant human GH at a dosage of 0.2 mg/kg/week, subcutaneously, divided into daily evening administrations. A growth rate of up to 10 cm/year should be observed with replacement therapy.

However, the extent of the result depends on the age at which therapy begins. In some situations (onset of puberty, decline in growth rate), it may be helpful to modulate the dosage by increasing the dose to a doubling. In any case, the dosage should be modulated according to the clinical response regarding stature growth.

Replacement therapy also effectively corrects growth deficit in girls with Turner syndrome or chronic renal failure.

In the case of pituitary deficiency, any other tropin deficits should be corrected.

In children with GH insensitivity and GH receptor mutation, IGF-1 replacement therapy overcomes the receptor dysfunction.

GH Deficiency in Adulthood

Diagnosing GHD in adults is difficult because its symptoms are nonspecific.

Current guidelines recommend that the diagnosis of GHD in adults be based on clinical criteria, medical history, and biochemical confirmation of altered GH levels using appropriate stimulus testing.

Adult subjects must also have at least one of the following conditions:

- Previous pituitary surgery
- Tumors or granulomas of the hypothalamic-pituitary region
- Cranial irradiation
- Radiological evidence of a pituitary lesion
- Head trauma
- Diagnosis of GHD arising in childhood
- Reduced IGF-1 levels (rarer condition)

Stimulus testing is required to make the diagnosis. In particular, for diagnosing GHD in adults, the insulin-induced hypoglycemia test is the gold standard, with the highest sensitivity.

The GHRH + arginine test is now considered the alternative test of the first choice.

The IGF-1 assay is not sufficiently sensitive to be diagnostically valid; approximately 25% of adults with GHD have normal IGF1 levels. Instead, it is used as an index of therapeutic response.

Therapy is based on administering recombinant GH, initially at a dose of 0.15-0.3 mg per day; the dosage should be adjusted until IGF-1 levels reach normal levels according to sex and age. Generally, women require a higher dose than men. Long-term treatment with GH improves body composition, increasing lean body mass and reducing fat mass. However, adults are more susceptible to side effects than children, the most common of which are fluid retention, headaches, hypertension, and carpal tunnel syndrome.

Acromegaly

The diagnosis is made on average about 10 years after the onset of the first symptoms because the physical changes occur very slowly.

Once the clinical suspicion of acromegaly is raised, the diagnosis is confirmed by the finding of elevated basal GH levels, IGF-1, and failure to suppress GH below 1 ng/mL at 1 and 2 hours after oral glucose load (75 g).

In patients with acromegaly, GH secretion remains episodic but with increased peaks in number, duration, and amplitude. In fact, in these patients, basal GH is always measurable; to diagnose acromegaly, the repeated finding of basal GH levels >5 ng/mL in several samples is necessary.

The IGF-1 assay represents the most sensitive laboratory test for diagnosing acromegaly because it is always increased in these patients.

Subsequently, it is necessary to proceed to a localization diagnosis by radiological investigations (computed tomography [CT] and magnetic resonance imaging [MRI]), which generally allow for demonstrating the presence and extension of a pituitary adenoma. In such a case, it is necessary to investigate hypothalamic-pituitary function by assaying the other pituitary tropins (PRL, TSH, ACTH, LH, and FSH); for example, an increase in PRL is observed in 25% of patients. In rare cases, acromegaly is secondary to ectopic GH or GHRH secretion by an extracranial tumor that must be sought by appropriate instrumental investigations (chest X-ray, abdominal ultrasound, etc.).

Since acromegaly is a systemic disease, once the diagnosis is made, a series of investigations should be performed to define the disease complications, especially those cardiovascular.

Treatment of acromegaly can be surgical, radiotherapy, or medical and has the following goals:

- Normalize the secretion of GH and IGF-1 (regression and/or control of signs and symptoms of the disease);
- Remove the hypophyseal or extrahypophyseal tumor (correction of campymetric deficits);
- Maintain the integrity of normal function residual pituitary;
- Prevent recurrence;
- Improve and prevent complications (cardiovascular, respiratory, osteoarticular, and metabolic);
- Achieve a normal quality of life and life expectancy.

The disease's mortality is related to GH and IGF-1 levels.

The criteria for biochemical remission of acromegaly are:

- Basal GH <2–2.5 ng/mL
- GH after OGTT <1 ng/mL
- Normal IGF-1 according to age and sex

Therefore, the main therapeutic goal is to ensure optimal biochemical control.

Currently, surgical therapy represents the first choice, especially in pituitary microadenoma (i.e., with a diameter <1 cm). In this case, the results of neurosurgery are good, with a cure rate of about 80%. In the presence of macroadenoma (a much more frequent occurrence), the efficacy of neurosurgical therapy is more modest, with cure rates not exceeding 40% of cases.

Medical therapy is particularly indicated after insufficient neurosurgical intervention to maintain biochemical control of the disease or, in the first instance, in patients who refuse surgery. In addition, medical therapy is also often prescribed while awaiting surgery. Somatostatin analogs are the medical therapy of choice. They have a marked inhibitory effect on GH secretion, followed by a reduction in circulating IGF-1 levels.

Radiotherapy is currently the treatment of the third choice due to the high incidence of complications and slow biochemical response (5–15 years). It is indicated in cases where surgical therapy and/or medical therapy fail to control tumor growth and normalize hormonal secretion.

Prolactin

Prolactin is a peptide hormone synthesized in lactotroph cells constituting about 20% of the anterior pituitary gland. PRL secretion is pulsatile and exhibits a circadian rhythm, with the highest concentrations during sleep and the lowest in the morning, about 2–3 hours after waking up. Peak serum PRL levels are observed between 4 and 6 am. PRL has a half-life of about 50 minutes and is mainly metabolized in the liver and, to a lesser extent, in the kidney.

Serum PRL levels transiently increase following exercise, meals, sexual intercourse, minor surgery, general anesthesia, acute myocardial infarction, and other forms of acute stress. PRL represents a stress-related hormone. In addition, during the last two trimesters of pregnancy and the first months of lactation, marked hyperplasia of lactotroph cells occurs transiently, with hypersecretion of PRL; the sucking reflex induces an increase in PRL that lasts 30–45 minutes.

PRL is the only adenohypophysis hormone undergoing dopamine-mediated inhibitory neuroendocrine regulation. Thyrotropin-releasing hormone (TRH), vasoactive intestinal peptide (VIP), estradiol, oxytocin, and vasopressin, on the other hand, stimulate PRL release.

The PRL exerts its action mainly at the breast level, stimulating breast tissue development and promoting lactation. During pregnancy, in synergy with estrogen, progesterone, and placental lactogen, PRL stimulates further development of breast tissue aimed at milk production. After childbirth, the milk production process begins with the sudden fall of placental estrogen and progesterone, which antagonize the effects of PRL on lactation. PRL, in fact, stimulates the synthesis of milk proteins (lactalbumin), lipids, and carbohydrates that are poured into the alveoli and mammary ducts. Nipple sucking activates a nervous reflex, which, by stimulating the secretion of oxytocin from the neurohypophysis and PRL from the adenohypophysis, maintains the lactation process. Prolactin also plays an essential role in reproductive processes through direct and indirect mechanisms. In particular, it inhibits reproductive capacity indirectly by suppressing the synthesis and secretion of Gonadotropin-Releasing Hormone (GnRH) at the hypothalamic level, followed by a reduction in the secretion of gonadotropins and testosterone; directly, by reducing the sensitivity of LH and FSH receptors in the gonads. In this way, a state of anovulation (infertility) contributes to maintaining lactation, preventing it from being interrupted by a subsequent pregnancy. Prolactin also has a luteolytic effect, inducing a shortened or inadequate luteal phase of the menstrual cycle.

PRL exerts its biological effects through interaction with a receptor that belongs to the type 1 cytokine receptor family, to which GH and IL-6 receptors also belong.

Hyperprolactinemia

The term hyperprolactinemia refers to increased circulating PRL levels due to hormonal hypersecretion by the pituitary gland, generally associated with reproductive problems in both men and women. Hyperprolactinemia is found in physiological and pathological conditions (Table 25.4); according to the etiology, it can be classified into organic and functional. The most common cause of functional hyperprolactinemia is the intake of pharmacotherapeutic agents that reduce hypothalamic dopamine secretion or its inhibitory action on the pituitary gland (antipsychotics and antidepressants). Functional hyperprolactinemia is also common in pregnancy and chronic renal insufficiency due to reduced peripheral PRL clearance, polycystic ovary syndrome due to hyperestrogenism, or primary hypothyroidism due to increased TRH secretion. Organic hyperprolactinemia is, instead, due to lesions of the hypothalamic-pituitary region that compromise dopamine synthesis, its transport through portal vessels, or the response of lactotrophs elements; among these, the most common causes are pituitary adenomas (prolactinoma and GH/PRL and ACTH/PRL-secreting adenoma), hypothalamic tumors, and infiltrative conditions (sarcoidosis, craniopharyngioma, pituitary metastases, vascular malformations, empty saddle).

From a clinical point of view, the main manifestations of hyperprolactinemia are amenorrhea, galactorrhea (inappropriate secretion of milk or a lactescent liquid from the breast), and infertility in women and impotence, decreased libido, and infertility in men. The onset of symptoms in men is generally later than in women because very high PRL values are required. In the case of adenoma in both sexes, the symptoms due to the expansive lesion (headache and visual impairment) can be observed.

Table 25.4 Causes of hyperprolactinemia

Physiological	Pathological	Pharmacological
Breastfeeding	Pituitary adenomas	Histaminergic antagonists H_2: cimetidine
Exercise	Cirrhosis	Drugs that block the dopamine receptor: phenothiazines, butyrophenones, thiaprides
Pregnancy	Seizures	Inhibitors of catecholamines: reserpine
Sexual activity	Chronic renal failure	Serotonin reuptake inhibitors: fluoxetine
Sleep	Primary hypothyroidism	Inhibitors of dopamine synthesis: α-methyldopa;
Stress	Lesions of the pituitary stalk	Opioid
	Infiltrative diseases	Hormones: estrogen, antiandrogens and TRH
	Empty sella	
	Trauma	
	Hypothalamic tumors	

TRH thyrotropin releasing hormone

Prolactinoma

Prolactinoma, a prolactin-secreting pituitary adenoma, is the most frequent pituitary tumor, with a prevalence of 100 cases per million population. Depending on the diameter of the tumor, prolactinoma can be differentiate into microadenomas (<1 cm in diameter), which generally do not invade the parasellar regions, and macroadenomas (>1 cm in diameter), which can be locally invasive and compress adjacent structures. Generally, tumor size correlates with circulating PRL levels. Microprolactinomas are quite frequent in females and rare in males, with a male-to-female ratio of 1:20; this ratio becomes 1:1 for macroprolactinomas.

Laboratory Investigations

Blood sampling for PRL levels measuring should be performed in the morning in a fasting patient who has been awake for at least 2 hours. Since PRL secretion is pulsatile, a single determination is not sufficient to diagnose hyperprolactinemia; at least two blood samples should be taken 30–60 minutes apart, preferably using a needle cannula because puncture of the vein can also induce an increase in PRL secretion, or three samples taken on three different days. The normal range of serum PRL levels in adults is 10–25 ng/mL in women and 10–20 ng/mL in men. In patients with very high prolactinemia (>1000 ng/mL), values may be lower due to artifacts of the assay method; in these cases, the sample should be diluted.

Diagnosis and Therapy

Hyperprolactinemia

The finding of serum PRL levels >25 ng/mL indicates hyperprolactinemia.

The treatment of hyperprolactinemia depends on the etiology. However, regardless of the cause, it aims to normalize PRL levels. Dopamine-agonist drugs are used in many forms of hyperprolactinemia. In asymptomatic patients with hyperprolactinemia, treatment is not necessary.

Prolactinoma

Once physiologic and drug-induced hyperprolactinemia is excluded, the diagnostic suspicion of prolactinoma is raised for PRL values >100 ng/mL; PRL values >500 ng/mL are diagnostic of macroprolactinoma. Once a condition of hyperprolactinemia is identified, MRI or CT scan is necessary to define the presence of a lesion compatible with a pituitary tumor.

The first choice treatment in patients with a prolactinoma is dopamine agonist therapy aimed at reducing tumor size and PRL levels. In patients with asymptomatic microprolactinoma no treatment is necessary; these patients should have regular follow ups with serial measurements of PRL levels and MRI. Surgery is indicated in patients intolerant or unresponsive to long-term drug therapy.

Neurohypophysis

The neurohypophysis, or posterior pituitary gland, is an extension of the hypothalamus, consisting of axon extensions whose cell bodies are located in the supraoptic and paraventricular hypothalamic nuclei. The hypothalamic-neurohypophyseal system secretes: vasopressin, also known as antidiuretic hormone, and oxytocin.

Antidiuretic Hormone

Antidiuretic hormone (ADH), also known as vasopressin, is a nonapeptide synthesized mainly by magnocellular neurons of the supraoptic and paraventricular nuclei of the hypothalamus and in small amounts by some extra-hypothalamic tissues. Its primary function is maintaining hydroelectrolyte balance through regulating renal reabsorption of water. In particular, ADH has an antidiuretic effect by inducing an increase in the hydro-osmotic permeability of the cells lining the distal tube and medullary nephron collector ducts. In the absence of ADH, these cells are impermeable to water, but following the binding of the hormone to their receptors, they express on their apical membrane pores for water consisting of the protein aquaporin 2, which allows the reabsorption of water followed by the urine concentration and the increase in volemia, arterial pressure, and cardiac output. The magnitude of the antidiuretic effect is proportional to plasma ADH concentrations. ADH also maintains electrolyte balance by stimulating renal sodium reabsorption, mediated by ENaC (epithelial sodium channel) luminal channels, further promoting water reabsorption.

ADH at high concentrations also exerts other actions. It induces vascular wall smooth muscle contraction in the skin and gastrointestinal tract, reduces lipolysis in adipose tissue, stimulates glycogenolysis in the liver, and finally, acts synergistically with corticotropin-releasing hormone (CRH) for the secretion of ACTH. All of these effects are mediated by the interaction of ADH with its receptors.

The most critical factor that regulates the secretion of ADH is the "effective" osmotic pressure of liquids whose variations are detected by specialized hypothalamic cells known as osmoreceptors. There is a regulatory threshold below which ADH secretion is reduced, leading to the elimination of a large volume of very dilute urine (water diuresis) and above which, instead, ADH levels increase in direct proportion to plasma osmolality, up to a reduction in urine flow of 0.35 mL/min. The threshold for ADH release usually corresponds to a plasma osmolality of 280 mOsm/L and sodium levels of 135 mmol/L. The threshold value has interindividual variabilities, being genetically determined, and intraindividual variability, related to physiological factors such as posture, pregnancy, menstrual cycle, and aging.

Although plasma hyperosmolality represents the primary stimulus for ADH release, volemia and blood pressure reductions represent valuable inputs to its secretion. In particular, ADH levels increase in response to a >10–20% reduction in blood pressure and/or volemia. Finally, ADH secretion can be influenced by other variables such as nausea, vomiting, smoking, acute hypoglycemia, and various drugs (diuretics, morphine, carbamazepine, clonidine, etc.). Above all, nausea, even if transitory and not accompanied by vomiting, is a powerful stimulant that can induce a 100–1000-fold increase in circulating levels of ADH.

ADH exerts its effects through interaction with membrane receptors, which can be of three types:

- V1aR: predominantly expressed on vascular smooth muscle and responsible for a calcium-mediated vasoconstriction mechanism
- V2R: expressed mainly on collector duct cells and, to a lesser extent, at the endothelial level
- V1bR or V3R: mainly expressed on corticotropic cells of the adenohypophysis, where it modulates ACTH release via calcium signaling-mediated signal transduction

ADH has a half-life of about 10 minutes and is mainly eliminated renally and partly hepatically.

Oxytocin

Oxytocin is a nonapeptide whose primary biological function is stimulating milk secretion during lactation in response to sucking. The sucking of the nipple activates the neurohormonal reflex that stimulates the secretion of prolactin and oxytocin; PRL stimulates the synthesis and secretion of milk in the mammary ducts, while oxytocin promotes the contraction of the smooth muscles around the mammary glands, increasing the ejection of milk and, in addition, stimulates the secretion of PRL. Unlike PRL, oxytocin is sensitive to physical or psychological stimulation related to breastfeeding and independent of the stimulus on the nipple.

Oxytocin also stimulates contractions of the uterine smooth muscle, thus contributing to the onset of labor.

In males, no role of oxytocin is known.

Alterations in the neurohypophysis manifest as a hypo- or hyperfunction state. The laboratory plays a crucial role in the diagnosis of these disorders.

Diabetes Insipidus

Diabetes insipidus (DI) is a syndrome characterized by polyuria and polydipsia. DI reflects reduced action or secretion of

Table 25.5 Types of diabetes insipidus (DI)

Type of DI	Defect	Causes
Pituitary DI	Reduced synthesis and secretion of ADH	Acquired or genetic
Primitive polydipsia	Suppressed ADH secretion due to excessive fluid intake	Inappropriate thirst secondary to the reduction of the set point of the osmoregulatory mechanisms (dipsogen) Excessive thirst due to a psychiatric (psychogenic) disorder Intake of large amounts of fluids for purported health benefits (iatrogenic)
Gestational DI	Increased degradation of ADH during pregnancy	Increased production of placental N-terminal aminopeptidases
Nephrogenic DI	Partial or total loss of renal sensitivity to the antidiuretic action of ADH	Acquired or congenital

ADH antidiuretic hormone

ADH; four different types are distinguished (Table 25.5), described below.

Pituitary Diabetes Insipidus

The most common type is pituitary diabetes insipidus, also known as hypothalamic, neurogenic, or central diabetes insipidus, in which reduced ADH production and secretion are usually related to agenesis or irreversible destruction of the neurohypophysis that may be due to a variety of genetic or acquired alterations. However, it is idiopathic in about 50% of cases (Table 25.6). Reduced ADH secretion results in renal free water reabsorption failure with the consequent elimination of high amounts of poorly concentrated urine. As a result, there is an increase in plasma osmolality, which stimulates thirst for osmoreceptors, causing polydipsia.

Table 25.6 Causes of pituitary diabetes insipidus

Acquired
Head trauma
Neoplasms
- Primitive: craniopharyngioma, pituitary (suprasellar) adenoma, dysgerminoma, meningioma
- Metastatic: lung and breasts
- Hematological: lymphomas and leukemias
Granuloma
- Neurosarcoidosis
- Histiocytosis
- Disseminated xanthomas
Infections
- Meningitis
- Viral encephalitis
- Toxoplasmosis
Inflammatory diseases
- Lymphocytic neurohypophysitis
- Granulomatosis with polyangiitis
- Lupus erythematosus
- Scleroderma
Vascular causes
- Sheehan syndrome
- Internal carotid aneurysm
- Coronary artery bypass grafting
- Hypoxic encephalopathy
Toxins
- Tetrodotoxin
- Snake venom
Congenital malformations
- Septo-optic dysplasia
- Craniofacial defects
- Holoprosencephaly
Idiopathic
Genetics
Autosomal dominant (gene *ADH*)
Autosomal recessive
- Type A (gene *ADH*)
- Type B (gene *ADH*)
- Type C (gene *WFS-1*)
X- linked recessive (Xq28)

Primary Polydipsia

Primary polydipsia is a clinical condition characterized by an excessive and unmotivated introduction of fluids (ingestion of more than 3 liters of fluids/day), usually water, resulting in polyuria. The picture is observed mainly in patients with chronic psychiatric diseases, which is also defined as psychogenic polydipsia. The increased intake of liquids leads to a slight reduction in plasma osmolality, which determines a reduced secretion of ADH, leading to a reduction in urinary osmolality and an increase in urine volume; this triggers a compensatory increase in water excretion that prevents hyperhydration.

Gestational Diabetes Insipidus

Gestational DI is also a consequence of reduced ADH levels due to increased hormone degradation by an enzyme produced by the placenta (N-terminal aminopeptidase). Gestational DI typically develops in the third or fourth month of pregnancy and resolves 4–6 weeks after delivery. Generally, the changes are similar to those observed in pituitary DI. Plasma osmolality is usually in the normal range for a pregnant woman, which is generally about 3–4% lower than normal.

Nephrogenic Diabetes

It is caused by altered kidney sensitivity to the action of ADH that may be genetic or acquired (Table 25.7), resulting in the inability to concentrate urine, polyuria, increased plasma osmolality, and polydipsia. Plasma ADH concentration is normal or increased, according to plasma osmolality.

The severity of antidiuretic function defect is quite variable among patients with pituitary, gestational, and nephrogenic DI. In some individuals, the defect in ADH secretion or action is so severe as to induce high urinary flow, up to 10–15 liters per day.

Table 25.7 Causes of nephrogenic diabetes insipidus

Acquired
Drugs
- Lithium
- Demeclocycline
- Methoxyfluorane
- Amphotericin B
- Aminoglycosides
- Cisplatin
- Rifampicin
- Foscarnet
Metabolic alterations
- Hypercalcemia, hypercalciuria
- Hypokalemia
Vascular causes
- Sickle cell anemia
- Ischemia (acute tubular necrosis)
Granulomas
- Neurosarcoidosis
Neoplasms
- Sarcoma
Infiltrative diseases
- Amyloidosis
Pregnancy idiopathic
Genetics
Autosomal recessive (aquaporin-2)
Autosomal dominant (aquaporin-2)
X-linked recessive (ADH receptor 2 gene)

ADH antidiuretic hormone

Table 25.8 Main causes of inappropriate ADH secretion syndrome

Increased hypothalamic/pituitary secretion of ADH
Neurological/neuropsychic disorders
- Infections (meningitis, encephalitis, sarcoidosis, abscesses, herpetic infections, AIDS)
- Vascular causes (thrombosis, subarachnoid/subdural hemorrhage, temporal arteritis)
- Psychosis
- Hydrocephalus
- Post-surgical causes
- Guillain-Barrè syndrome
Drugs
- Antidepressants (carbamazepine, tricyclic antidepressants, phenothiazines, haloperidol, serotonin re-uptake inhibitors, etc.)
- MDMA/ecstasy
- Cyclophosphamide
- Non-steroidal anti-inflammatory drugs
Pulmonary disorders
- Tuberculosis, viral/bacterial pneumonia, asthma, atelectasis, pneumothorax, HIV
Ectopic production of ADH
Neoplasms
- Pulmonary microcytoma
- Nasopharyngeal tumors
- Gastrointestinal/pancreatic tumors
- Tumors of the genitourinary tract
- Mesothelioma
- Lymphomas, sarcomas
- Cyclophosphamide
- Chlorpropamide
Amplification of ADH effects at the receptor level
Drugs
- Cyclophosphamide
- Chlorpropamide

ADH antidiuretic hormone, *MDMA* methylenedioxymethamphetamine (ecstasy)

Inappropriate ADH Secretion Syndrome

Syndrome of Inappropriate Antidiuretic Hormone secretion (SIADH) is a rare disorder characterized by inappropriately elevated ADH levels associated with plasma hyperosmolarity and urinary hyperosmolarity due to an altered osmoregulatory mechanism. SIADH represents a frequent cause of hyponatremia (plasma sodium <135 mmol/L) in adults. Inappropriate ADH secretion can be of neurohypophyseal or ectopic origin. The latter represents the most frequent cause of SIADH due to neoplastic pathologies. On the other hand, the causes that can determine inappropriate secretion of ADH by the neurohypophysis are numerous and include infections, head traumas, vascular alterations, pulmonary disorders, or drug intake. The latter can determine SIADH by inducing an increase either in the release of ADH (morphine, carbamazepine, haloperidol, tricyclic antidepressants) or in the receptor sensitivity to the hormone (nonsteroidal anti-inflammatory drugs, cyclophosphamide, chlorpropamide) (Table 25.8).

Four types of altered osmoregulation have been described.

- Type a (40–70%): random secretion associated with markedly increased ADH levels, independent of plasma osmolarity. There is, therefore, a complete loss of the osmoregulatory mechanism.
- Type b (20–40%): it is the classic picture of osmostat reset (literally threshold reset) characterized by a reduction in the osmolar threshold for ADH secretion. There is, therefore, a defect affecting the inhibitory component of the osmoregulation mechanism.
- Type c (10%): low plasma osmolarity does not inhibit ADH secretion. Plasma ADH concentrations are, therefore, inappropriately high at low plasma osmolarity. There is a normal relationship between plasma osmolality and plasma ADH for physiological plasma osmolality values.
- Type d (<5%): it is characterized by normal osmoregulation (i.e., ADH secretion varies appropriately with plasma osmolality), but urine is concentrated even when ADH release is suppressed. The osmoregulatory mechanism of ADH secretion thus is conserved, suggesting that different alterations cause inappropriate antidiuresis. A constitutively activating mutation of the ADH receptor has been identified in some patients. There is no correlation between the etiology and the type of alteration; any of the four types described above can occur, regardless of the cause of SIADH.

The increased secretion of ADH causes increased reabsorption of free water in the distal nephron, which, in turn, causes extracellular hypervolemia, increased levels of the atrial natriuretic hormone, reduced plasma renin activity, and increased urinary sodium excretion. This natriuresis serves to balance hypervolemia but exacerbates hyponatremia. In addition, hyponatremia induces increased intracellular volume in all organs, including the brain. If hyponatremia develops slowly, patients do not have clear clinical manifestations; if, on the other hand, plasma sodium concentration drops abruptly, patients develop severe neurological manifestations.

Laboratory Investigations

The key laboratory tests for diagnosing neurohypophysis disease are listed below.

24-Hour Urine Test

The 24-hour urinalysis assesses urine volume, osmolarity, and specific gravity. Polyuria is the production of a volume of urine >2.5 L/day. Normal values of urine osmolarity are 300–800 mOsm/L. The normal specific gravity of urine is 1005–1020 g/L.

Plasma ADH Assay

The values of the hormone must always be evaluated in relation to plasma and urinary osmolality because, in the normal subject, they change according to the state of hydration. In the presence of good hydration conditions, ADH values are between 1.5 and 6 pg/mL; with free fluid intake, they suppress the assay method's sensitivity to the lower limit. The determination of ADH, however, is highly complex due to its low molecular weight (~1 kDa), its very short half-life (~20 minutes), and its binding to platelets (>90%). In addition, samples are affected by high centrifugation rates, and ADH levels are not very stable in vitro.

Plasma Osmolality Assessment

Normal plasma osmolality values are between 285 and 295 mOsm/L.

Water Deprivation Test

This test consists of water deprivation for an interval of 7–14 hours during which the patient's water balance is continuously monitored (hourly) through regular measurements of body weight, plasma osmolality and/or natremia, urinary volume, and osmolality. The test is discontinued in the presence of a weight loss greater than 5% or when a plasma osmolality greater than or equal to 300 mOsm/L is reached. If fluid deprivation does not result in urine concentration by the end of the test, the patient has severe pituitary or nephrogenic DI. In such a case, to make a differential diagnosis, desmopressin is administered at the end of the water deprivation test, and urine osmolality measurement is repeated every hour for two hours; an increase (>300 mOsm/L) is indicative of severe pituitary DI, whereas the absence of change is strongly suggestive of nephrogenic DI (Table 25.9).

Table 25.9 Water deprivation test

Step 1. Dehydration phase	
Objective	Differential diagnosis of polyuria states
Principle	Under physiological conditions, water deprivation causes an increase in plasma osmolality; the plasma hyperosmolality detected by hypothalamic osmoreceptors leads to the release of ADH and, consequently, the renal reabsorption of water
Procedure	It is preferable to perform it in the morning to control the patient - Bladder emptying by urine collection (urinary volume, osmolality and specific weight), blood sampling (hemoglobin, hematocrit, sodium, potassium, chlorine, protein and osmolality) and body weight detection - Absolute restriction of liquids between 8 and 16 - Hourly urine collection (urinary volume, osmolality and specific weight), blood sampling (osmolality) and body weight measurement - The test should be stopped if thirst becomes unbearable or if body weight has a reduction of >5%
Interpretation	Normal subject: water restriction causes diuresis contraction, urine excretion with increasing osmolality and specific gravity (osmolality >800 mOsm/L and urine specific gravity >1030) Pituitary and severe nephrogenic diabetes insipidus (DI): diuresis does not decrease, plasma osmolality increases (>290 mOsm/L) and urinary osmolality remains low (<300 mOsm/L) Partial pituitary and nephrogenic DI or primary polydipsia: normal plasma and urinary osmolality
Side effects	Excessive dehydration with possible hypovolemia, shock, and death. It is important to weigh the patient because a reduction in body weight >5% indicates severe dehydration and requires immediate termination of the test
Step 2. Desmopressin test	
Objective	Differential diagnosis of pituitary and nephrogenic DI
Principle	Desmopressin is a synthetic analogue of ADH; it binds to renal receptors and increases renal water reabsorption
Procedure	At the end of the water deprivation test, administer desmopressin intramuscularly (1–2 μg) and determine the urinary osmolarity after 1 and 2 hours
Interpretation	Pituitary DI: urinary osmolality >300 mOsm/L Nephrogenic DI: urinary osmolality <300 mOsm/L

ADH antidiuretic hormone

If the water deprivation test results in urine concentrations before plasma osmolality and/or natremia exceed the upper reference limits, in order to differentiate between primary polydipsia, partial pituitary DI, and partial nephrogenic DI, plasma ADH should be measured. Precisely, ADH concentration and plasma osmolality and/or natremia are measured, and fluid deprivation is continued while hypertonic saline (3% sodium chloride) is infused at a rate of 0.05 mL/kg/min. During the infusion, plasma osmolality and/or natremia are measured every 30 minutes; when the values rise above the normal limit, usually 60–90 minutes after the infusion, the ADH assay is performed, and the test is terminated. The results are interpreted using a nomogram depicting the relationship between plasma ADH values and plasma osmolality and/or natremia or urinary osmolality. This method differentiates between primary polydipsia, partial pituitary DI, and partial nephrogenic DI.

MRI of the Hypothalamic-Pituitary Region

MRI of the hypothalamic-pituitary region is an emerging approach that eliminates the need for water deprivation and provides a more straightforward but equally reliable way to differentiate the three major types of DI, regardless of the severity of the underlying defect. In most healthy adults and children, the posterior pituitary gland emits a hyperintense signal in T1-weighted mid-sagittal images. This bright spot is almost always present in primary polydipsia, but it is always absent or reduced in pituitary and nephrogenic DI. Thus, according to this method, the three types of DI can be differentiated in two steps. First, an assay of plasma ADH levels and determination of urinary osmolality without the restriction of fluid intake is performed; if plasma ADH levels are normal or high (>2 pg/mL) and urinary osmolality is low (<300 mOsm/L), the patient has nephrogenic DI. If ADH levels are low or undetectable (<1 pg/mL) and urinary osmolality is also low, an MRI of the brain may differentiate between pituitary DI (no or small T1 signal) or primary polydipsia (normal or enlarged T1 signal). In addition, MRI may also indicate the etiology of both forms of DI.

Copeptin

Another proposed method to the differential DI diagnosis is the measurement of copeptin. This glycoprotein originates from the pre-pro-hormone pre-provasopressin, consisting of ADH, neurophysin II, and copeptin. Pre-provasopressin is synthesized mainly in the paraventricular neurons of the hypothalamus and, during axonal transport, undergoes cleavage followed by the release of the three molecules. The latter are stored in the secretory granules of the neurohypophysis, waiting to be secreted in response to pathophysiological stimuli. The biological function of copeptin is still unclear, although its importance in mediating the correct conformation of ADH has been hypothesized. Copeptin and ADH are secreted in equimolar amounts. However, copeptin, compared to ADH, has a longer half-life, and greater pre-analytical stability, resulting in superior analytical accuracy, precision, and simplicity. The biological sample can be stored for at least 7 days at room temperature and up to 14 days at 4 °C. The assay is performed by immunometric technique. The reference values of copeptin in the general population are 1–14 pmol/L.

Finally, another way to distinguish between the three main types of DI is to closely monitor the effects of antidiuretic therapy on changes in water balance.

Once the diagnosis of the neurohypophysial disease has been made, instrumental investigations can be helpful to identify the underlying cause. MRI can detect tumors, infections, hemorrhages, infarcts, and hydrocephalus. In case of SIADH suspicion, a chest X-ray may reveal any lesions.

Diagnosis and Therapy

Diabetes Insipidus

In the presence of frequent urination, enuresis (involuntary emission of urine), nocturia (repeated need to urinate during the night), and/or persistent thirst, a 24-hour urine test should be performed; if the volume is >50 cc/kg/day, corresponding to a volume of 2.5 L/day, the diagnosis of polyuria is established. After excluding the most common causes of polyuria (antidiuretic drugs and metabolic causes such as diabetes mellitus) through appropriate laboratory investigations, a baseline copeptin measurement or the water deprivation test is performed (Fig. 25.3). In the case of the water deprivation test, if fluid deprivation does not lead to urine concentration (urinary osmolality <300 mOsm/L), the patient has severe pituitary or nephrogenic DI; to make a differential diagnosis, desmopressin is administered. If fluid deprivation results in the production of concentrated urine, the patient may have partial pituitary DI, partial nephrogenic DI, or primary polydipsia; to make the differential diagnosis, plasma ADH is measured before and after infusion of hypertonic saline, evaluating its values in relation to plasma osmolality/natremia and urinary osmolality (Fig. 25.4).

Figure 25.5 shows Timper's algorithm for differential diagnosis of DI by determination of baseline copeptin levels and/or after fluid deprivation. It has been shown that copeptin determination makes water deprivation testing unnecessary in cases of nephrogenic DI, correctly identifying all patients with reduced or absent peripheral sensitivity to ADH. In contrast, an 8-hour (overnight) fluid deprivation resulting in a copeptin value <2.6 pmol/L is suggestive of complete central DI. Finally, a baseline copeptin value <21.4 pmol/L in the absence of water deprivation requires evaluation of the peptide response to an osmotic stimulus that induces an increase in plasma sodium levels >147 mmol/L, such as infusion of 3% hypertonic saline; following the stimulus, copeptin values <4.9 pmol/L are suggestive of partial central DI; conversely, values ≥4.9 pmol/L indicate the presence of primary polydipsia. The algorithm proposed by Timper represents an evolution of the one proposed by Fenske (Fig. 25.6), which involves the calculation of the ratio of copeptin to serum sodium concentrations during an appropriate osmotic stimulus (fluid deprivation).

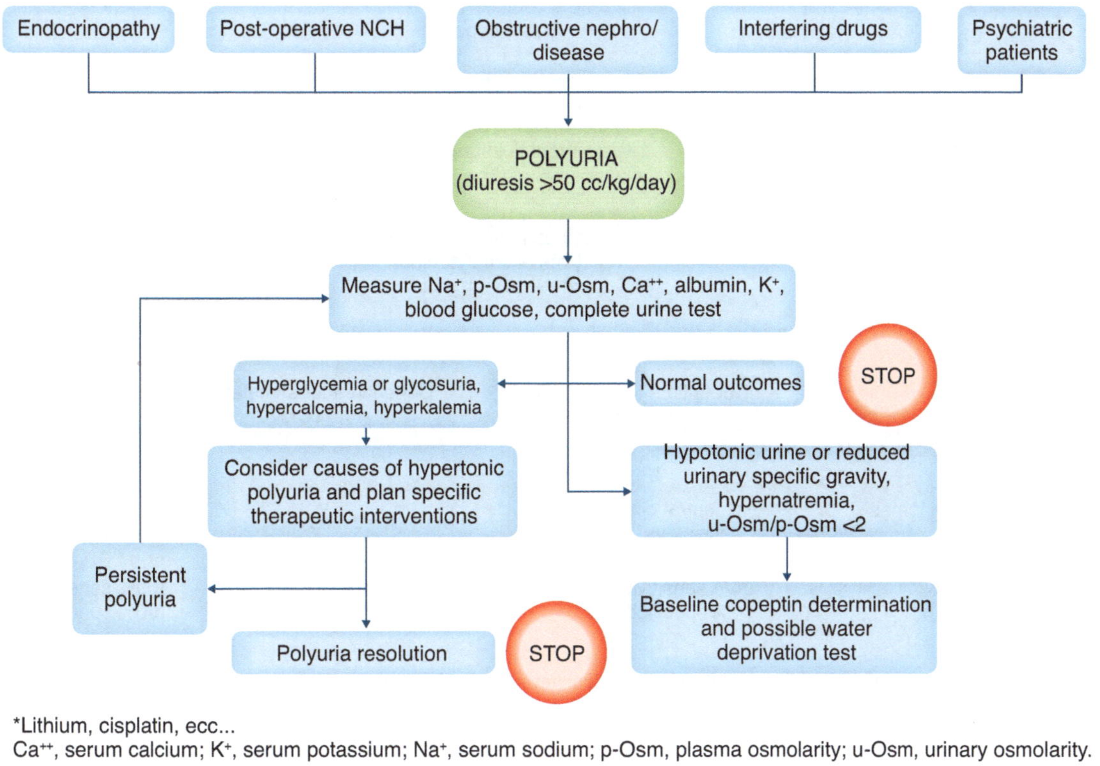

Fig. 25.3 Diagnostic algorithm for the evaluation of polyuria. (Copyright EDISES 2021. Reproduced with permission)

Fig. 25.4 Algorithm for the differential diagnosis of diabetes insipidus by water deprivation test. Na + natremia, PO plasma osmolality, UO urinary osmolality, DI diabetes insipidus. (Copyright EDISES 2021. Reproduced with permission)

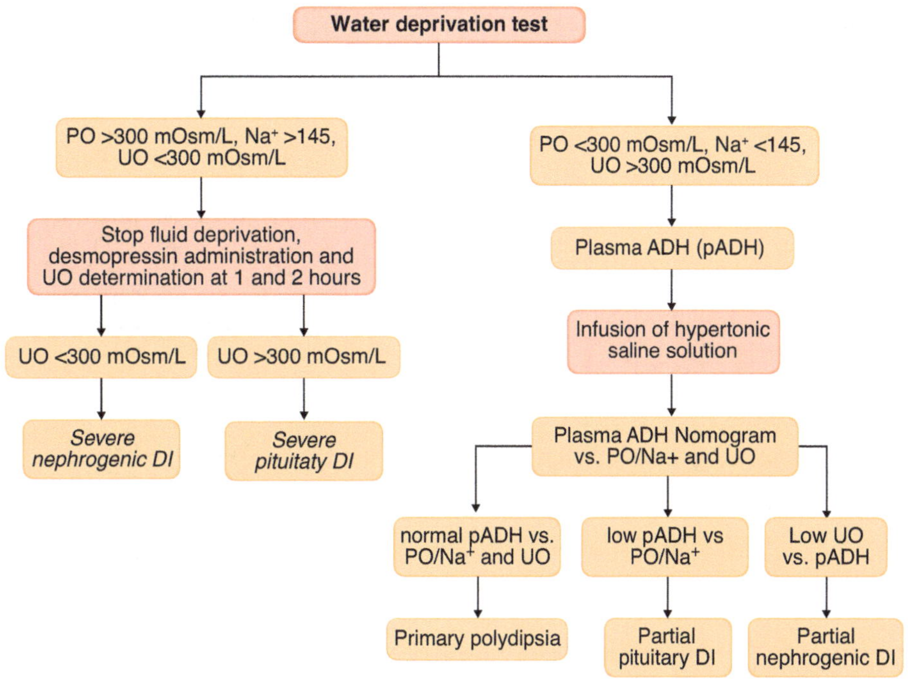

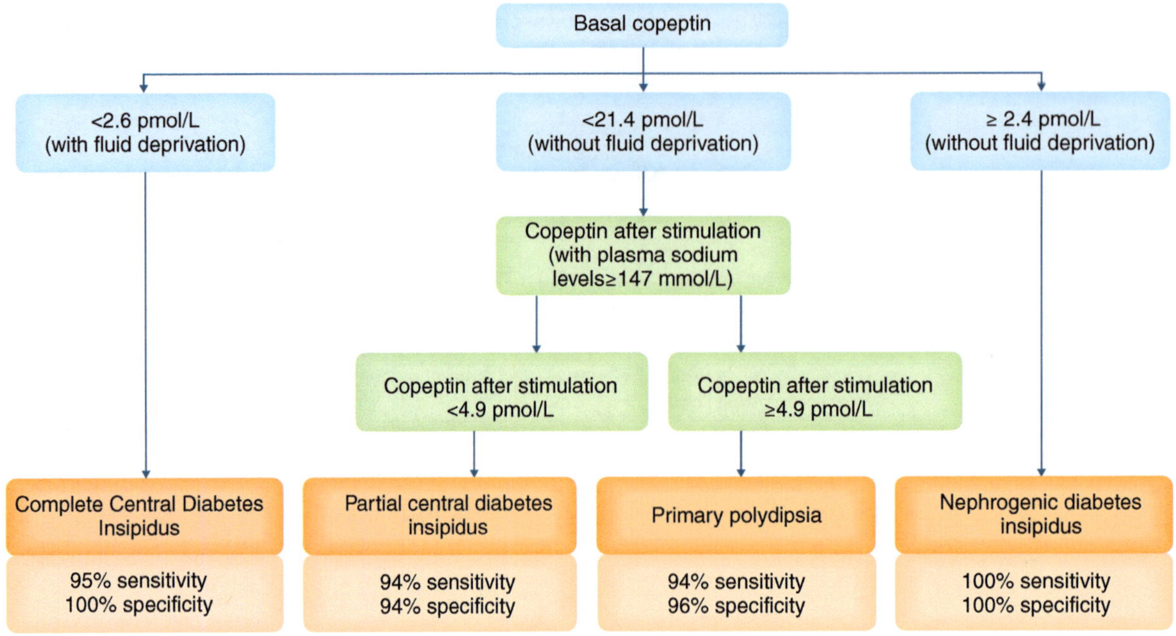

Fig. 25.5 Timper et al. algorithm for the differential diagnosis of diabetes insipidus by copeptin, in association or not with the liquids deprivation. (Copyright EDISES 2021. Reproduced with permission)

Fig. 25.6 Fenske et al. algorithm for the differential diagnosis of diabetes insipidus. (Copyright EDISES 2021. Reproduced with permission)

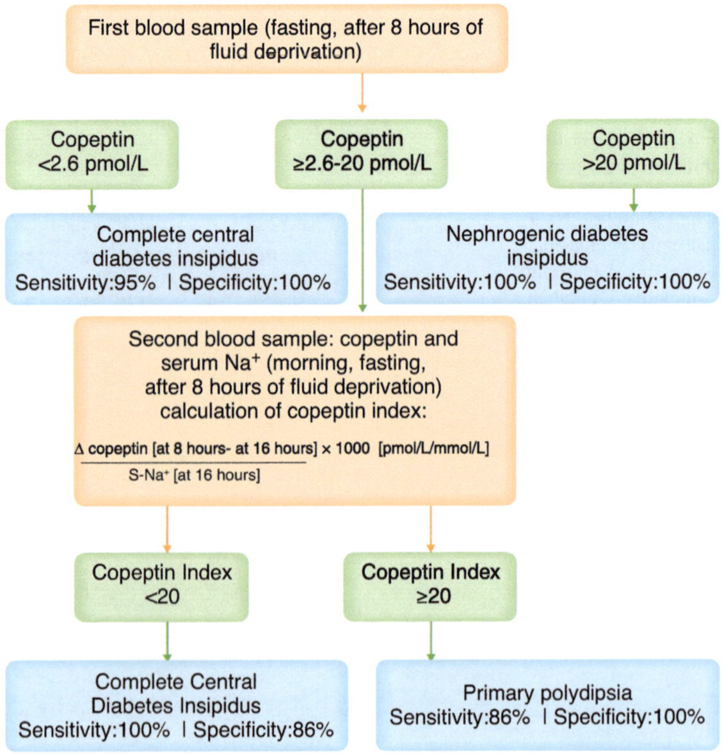

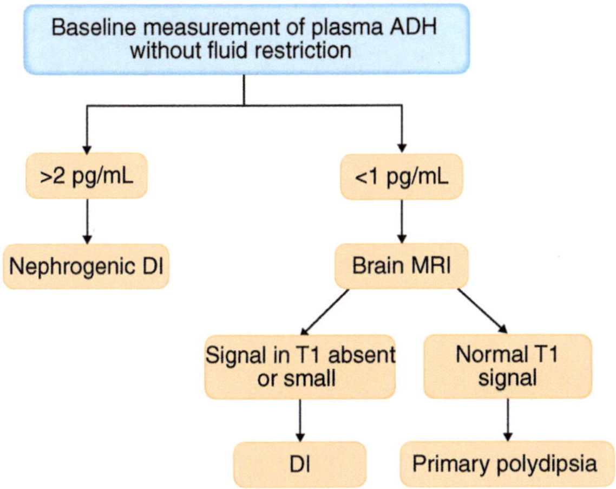

Fig. 25.7 Algorithm for the differential diagnosis of diabetes insipidus by ADH and magnetic resonance imaging. (Copyright EDISES 2021. Reproduced with permission)

Alternatively, a differential diagnosis of diabetes insipidus can be made based on plasma ADH and MRI values without water restriction (Fig. 25.7).

Signs and symptoms of uncomplicated pituitary DI can be treated with desmopressin, which selectively acts on ADH receptors, increasing the urine concentration and reducing urine flow dose-dependently. On the other hand, signs and symptoms of nephrogenic DI can be treated with a thiazide diuretic and/or amiloride in combination with a low-sodium diet.

Inappropriate ADH Secretion Syndrome

The diagnosis is made by exclusion, based on clinical history, laboratory data, and physical examination.

SIADH should be suspected in patients presenting with hypotonic hyponatremia associated with euvolemia, defined as the absence of clinical signs of hypovolemia (tachycardia, mucosal dryness) or hypervolemia (subcutaneous edema, ascites).

Table 25.10 Laboratory parameters for SIADH diagnosis

	Parameter	SIADH diagnosis
Serum	Sodium	<135 mmol/L
	Hypo-osmolality	<285 mmol/L
	Potassium	Normal
	Urea	Normal
	Uric acid	Normal
	Creatinine	Normal
	Thyroid hormones	Normal
	Cortisol	Normal
	Aldosterone	Normal
	Glucose	Normal
Urine	Osmolality	Urinary > plasma
	Sodium	>40 mmol/L
	Fractional uric acid excretion	12%

Table 25.10 shows the laboratory tests for the preliminary diagnosis of SIADH.

Other laboratory tests helpful for SIADH diagnosis are:

- Inability to eliminate a water load with urine (healthy subjects eliminate with urine more than 80% of the water load within 4 hours, with urinary osmolality <100 mOsm/L).
- Hypouricemia due to volume expansion and the action of ADH on renal ADH receptors that increases uric acid clearance. Recently, an algorithm based on assessing copeptin levels has been proposed to detect SIADH (Fig. 25.8).

Treatment of SIADH depends on the underlying cause. Hyponatremia is corrected by continuous infusion of hypertonic saline (NaCl 3%), which may be combined with loop diuretics. If the cause removal is not possible or has been followed by resolution of the SIADH, permanent treatment is indicated. The treatment of the first choice is fluid restriction.

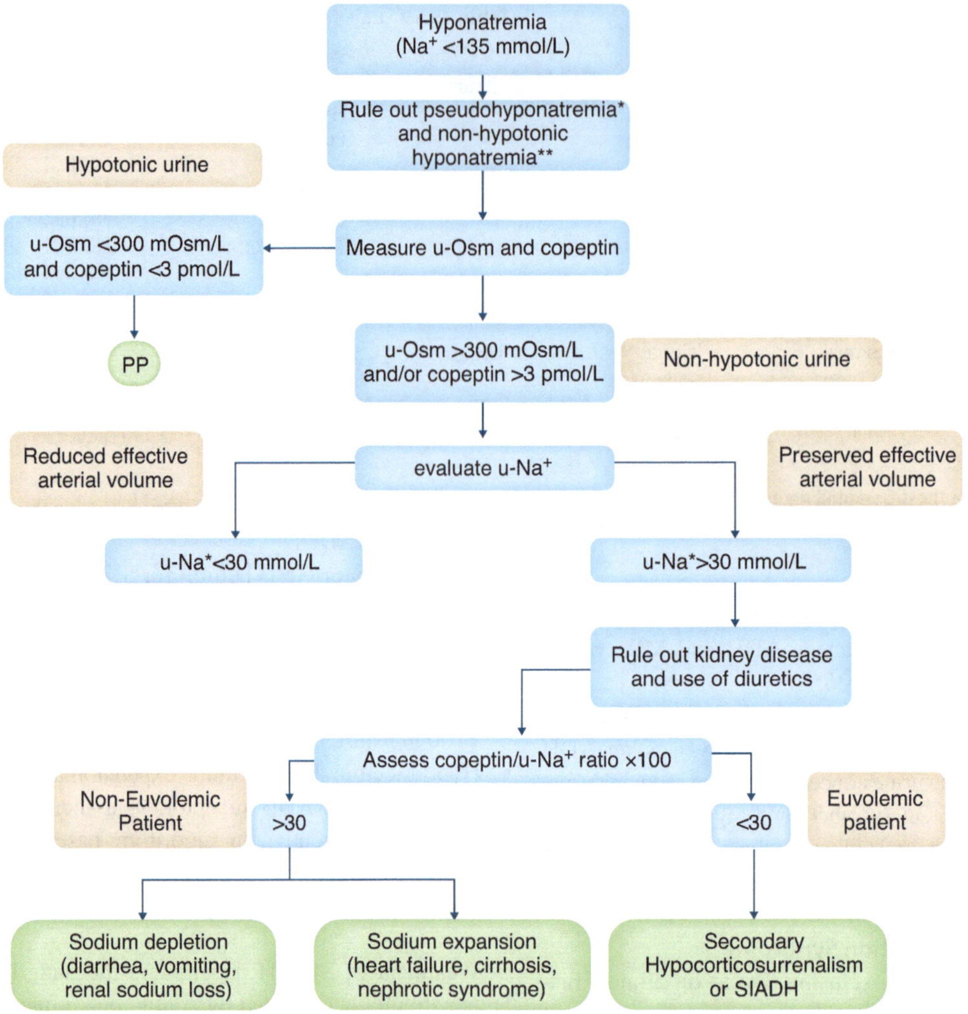

*Assess triglycerides, total cholesterol, total protein; **assess urea, blood glucose.
Effective p-Osm = 2 × (s-Na++ s-K+) + glycemia/18.
K+, serum potassium; Na+, serum sodium; PP, primary polydipsia; SIADH, inappropriate ADH secretion syndrome
u-Na+, urinary sodium; u-Osm, urinary osmolarity.

Fig. 25.8 Algorithm for the differential diagnosis of hyponatremia based on copeptin. (Copyright EDISES 2021. Reproduced with permission)

Thyroid

Marcello Ciaccio, Giulia Bivona, Luisa Agnello and Bruna Lo Sasso

Anatomy

The thyroid is a gland consisting of two lobes placed in the anterolateral part of the trachea, connected by an isthmus, immediately below the cricoid cartilage. Sometimes a pyramidal lobe originates from the isthmus, extending superiorly on the midline.

Two parathyroid glands are posterior to each upper and lower pole of each lobe.

The thyroid gland originates from the pharyngeal floor during the third week of gestation and is localized in the neck region, migrating from the foramen caecum along the thyroglossal duct. At the 11th week of gestation, the synthesis of thyroid hormones begins.

Vascularization of the gland is provided by the thyroid arteries, two superior and two inferior; the recurrent laryngeal nerve runs at the lateral margins of the thyroid gland.

Microscopically, the thyroid gland is made up of follicles composed of a layer of polarized cells (thyrocytes), which protrude into the lumen of the follicle; this is entirely filled with colloid, a liquid, homogeneous and viscous substance which contains high quantities of thyroglobulin, a protein indispensable for the synthesis of iodothyronines.

In the context of the entire glandular parenchyma, neural crest-derived cells, or C cells, are scattered throughout the glandular parenchyma; however, they are present in greater density at the junction of the superior and inferior thirds of the thyroid gland. C cells, also called parafollicular or medullary cells, produce a hormone, calcitonin, which, under certain paraphysiologic or pathologic circumstances, causes a reduction in blood calcium levels.

Thyroid Hormones

Biosynthesis and Secretion

Thyrocytes produce two hormones, triiodothyronine (T3) and tetraiodothyronine or thyroxine (T4). They are iodinated thyronines containing a phenolic ring and one to four iodine atoms. The biosynthesis of thyroid hormones requires tyrosine and iodine and consists of four steps:

- Active transport of iodide on the thyrocyte basement membrane
- Oxidation of iodide
- Covalent bonding with tyrosine residues (organification) leading to the production of tyrosines (monoiodotyrosine [MIT] and diiodotyrosine [DIT])
- Coupling of tyrosines and subsequent formation of T3 and T4

Thyroid-stimulating hormone (TSH) promotes these biosynthetic steps and secretion.

The daily iodine requirement to produce 100 μg/day of T4 is 100–150 μg. Most available supply comes from the deiodination processes. Although iodine is relatively scarce, small amounts can be taken up with water and food. Perchlorate and pertechnetate can act as substrates for the iodide transport system, acting as competitive inhibitors.

After entering the thyroid gland, iodine is transported to the apical membrane of the follicular cells, where the enzyme thyroperoxidase (TPO) mediates its oxidation. TPO is a heme-glycoprotein consisting of 933 amino acids, with a molecular weight of 107 kDa, located at the microsomal level and whose substrate is represented by hydrogen peroxide. In addition to catalyzing the iodine oxidation, which is followed by its organification in the tyrosine residues of thyroglobulin (Tg), TPO also mediates the formation of the ethereal bond between the iodotyrosine of Tg, leading to T4 and T3 synthesis. TPO is stimulated by TSH and inhibited by perchlorate, pertechnetate, and thiocyanate (Fig. 25.9). Thyroglobulin contains approximately 100 tyrosine residues, 2–3 T4 residues, and 0.2 T3 residues (one residue for every five molecules). Although several other proteins contain larger amounts of tyrosine, and although all of these can be iodinated, the synthesis of thyroid hormones is exclusive to thyroglobulin. At the origin of this mechanism lies the primary structure of the glycoprotein: the coupling of MIT and DIT takes place only at specific amino acid sequences present on thyroglobulin subunits.

In order to release thyroid hormones into the circulation, thyroglobulin must be reuptake from the thyrocyte to undergo enzymatic hydrolysis, following which T3 and T4 diffuse into the extracellular fluid and then into the vascular torrent. During this phase, T4 is deiodinated to T3, and iodide is largely recycled into the thyrocyte.

Approximately 80% of T4 is metabolized through deiodination. Of this, 40% is converted to T3, and the remaining 40% to reverse T3 (rT3). The bulk of T3 obtained by deiodination is formed at extratiroid sites (80%), mainly in the liver and kidney. Peripheral deiodase activity can be found in all tissues, where it locally modulates the bioavailability of T4; this peripheral regulation by deiodinases can determine systemic effects and represents an alternative control mechanism to the classic endocrine axis.

The process of deiodination will inactivate T3 with the formation of T2, which has no biological activity.

The deiodination process is catalyzed by 5′-deiodase, a microsomal enzymatic activity of which three isoforms are known. The isoform most commonly found in the liver, kidney, and thyroid is deiodinase I, which removes iodine at the outer ring of T4, and, then, converted to T3. In the brain and on the skin, deiodase II catalyzes the same reaction. Deiodinase III, distributed almost ubiquitously, has a fundamental role in the degradation of T3 into T2 and T4 into rT3, an inactive compound with two main functions: regulating the excess of thyroid hormones and glandular secretion by inhibiting deiodinases I and II (Fig. 25.10).

Specific deiodination pathways can be found in many organs and tissues; for example, deiodinase III is well represented in the placental and glial areas; deiodinase II is very active in the brain and functions even in conditions of reduced glandular secretion. For this reason, in hypothyroidism, a discrete production of T3 of encephalic origin is maintained, even in the presence of reduced serum T4.

Deiodinases regulate the activity of thyroid hormones locally in response to mainly environmental stimuli. The local control of the bioactivity and bioavailability of thyroid hormones, mediated by peripheral deiodinases, independently flanks the central control of the hypothalamic-pituitary

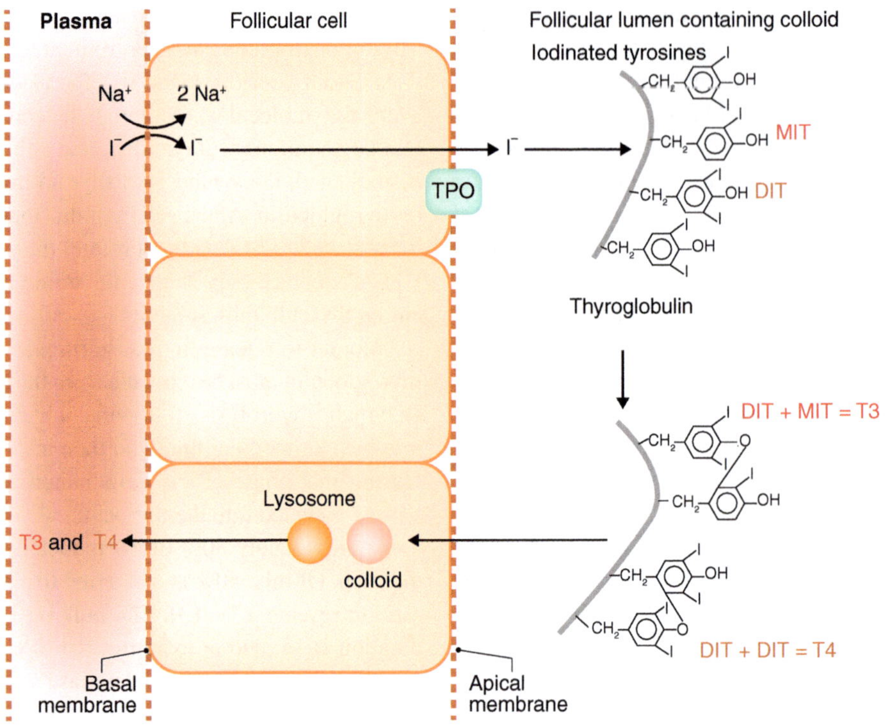

Fig. 25.9 Synthesis of thyroid hormones. A Na^+ symport protein on the apical membrane of follicular cells mediates the active transport within the follicular cell of two sodium ions and an iodine molecule. Pendrin, a protein expressed on the apical membrane of follicular cells, mediates the transport of iodide from the cytoplasm to the follicular lumen. Thyroid peroxidase (TPO), located on the apical membrane of follicular cells, mediates the oxidation of iodide and its organization in tyrosine molecules linked to thyroglobulin, leading to the formation of monoiodiothyroisin or diiodiotyrosin (MIT and DIT); the TPO also mediates the coupling of two iodized tyrosines leading to the formation of T4 (DIT + DIT) or T3 (MIT + DIT). Once iodized, TG accumulates in the follicular lumen as a colloid. Through the colloid pinocytosis at the level of the apical margin of the follicular cells, the colloid is reabsorbed and will appear as droplets within the cytoplasm. The colloid droplets move toward the basement membrane and fuse with the lysosomes; the lysosomal proteases will lead to the degradation of the colloid with the release of T3 and T4. The final phase is represented by the secretion of free iodothyronines T4 and T3 in the blood. (Copyright EDISES 2021. Reproduced with permission)

Fig. 25.10 Thyroid hormone metabolism. (Copyright EDISES 2021. Reproduced with permission)

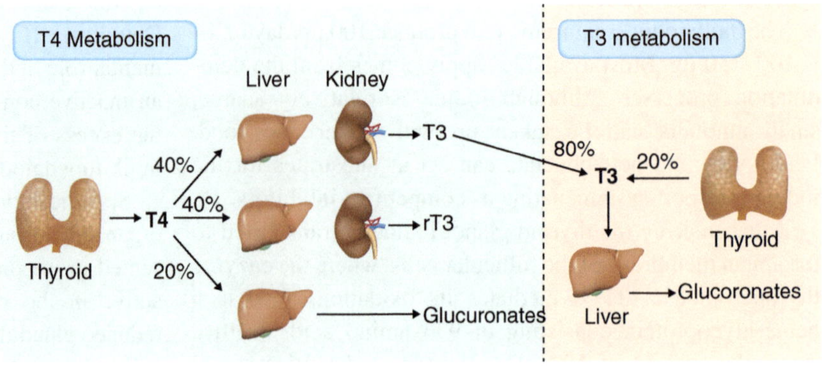

endocrine axis and can generate systemic effects, such as acclimatization and appetite. In other words, serum T4 concentrations are controlled by the central axis but functionally also by local deiodinases.

Catabolism of thyroid hormones occurs in the liver, where a process of glucuronidation allows their intestinal elimination as glycuronates (Fig. 25.10).

Regulation of Thyroid Hormone Secretion and TSH

Regulation of thyroid hormone secretion is an example of endocrine feedback of the hypothalamic-pituitary-target gland axis. The synthesis and release of T4 and T3 closely depend on the secretion of pituitary thyrotropin, TSH, which varies sensitively and linearly with changes in serum T4 and

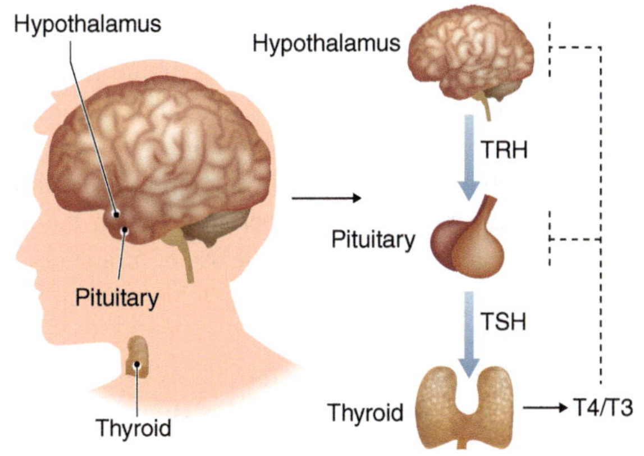

Fig. 25.11 Hypothalamic-pituitary-thyroid axis and retroregulation or negative feedback. (Copyright EDISES 2021. Reproduced with permission)

T3. This inversely proportional variation between TSH and T4 expresses the retroregulation, or negative feedback, upon which the functioning of the endocrine axis is based (Fig. 25.11). Secretion of TSH (and thus thyroid hormones) is stimulated by the thyrotropin-releasing factor, TRH, produced in the hypothalamus and inhibited by T4, T3, and TSH excess. Laboratory diagnosis of endocrine pathologies uses the negative feedback system: to detect alterations in thyroid secretion, serum TSH is initially determined.

TSH is a glycoprotein produced in the adenohypophysis that stimulates the synthesis and release of thyroid hormones. TSH consists of two distinct subunits, α and β; the former is common to other pituitary tropins (FSH, LH, human Chorionic Gonadotropin [hCG]), while the latter expresses the specific biological activity of the hormone. TSH secretion occurs by spontaneous fluctuations in a periodic pattern of 1–2 hours (pulsatile rhythm) and a morning peak before waking up (circadian rhythm). The normal reference values of TSH are between 0.4 and 4 mU/L; its half-life is 30 minutes.

Transport of Thyroid Hormones

More than 99% of T4 and T3 circulate reversibly bound to some serum proteins: Thyroxine-Binding Globulin (TBG), which carries the largest fraction of both T4 and T3 (75% and 80%, respectively); prealbumin, formerly known as Thyroxine-Binding PreAlbumin (TBPA) or transthyretin (TTR); albumin; and lipoproteins. TBG is a 415 amino acid polypeptide whose gene is located on the X chromosome. It is synthesized in the liver and belongs to a superfamily of serine protease inhibitors known as serpins (from serine protease inhibitors). The affinity of TBG for thyroid hormones is 50 times greater than that of prealbumin and 7000 times greater than that of albumin. TBG can undergo cleavage by leukocyte elastase at sites of inflammation, with a reduction in its affinity for T4 and T3. Transport proteins maintain an extratihyoid pool of T4 as stable as it is large to prevent sudden depletion of circulating quotas; they also provide protection in the presence of urinary iodine losses.

Alterations in TBG, albumin, and prealbumin affect total circulating T3 and T4 levels. Pregnancy and estrogen use cause an increase in serum levels, while nephrotic syndrome, androgens, and glucocorticoids cause a decrease. Many other drugs can interfere with the binding between thyroid hormones and transport proteins (barbiturates, heparin, furosemide, salicylates, 5-fluorouracil). Free circulating thyroid hormones (fT4 and fT3), which are not affected by changes in serum proteins, carry out biological function; fT3 and fT4 concentrations are, respectively, 2.5–4.5 pg/mL and 0.8–1.8 ng/dL.

Mechanism of Action of Thyroid Hormones

Thyroid hormones bind to specific receptors (TRs) in the nucleus of most cellular and tissue districts. They belong to the superfamily of receptors for steroid hormones and mediate most of the biological actions of T3 and T4; almost all of the hormone bound to the nuclear receptor is represented by T3, which constitutes the activated hormonal form, while T4 is a pro-hormone, although endowed with intrinsic biological activity. At least two types of receptors are known, α and β, each with two isoforms, α1 (TRα1) and α2 (c-erbAα2), β1 (TRβ1) and β2 (TRβ2). Some organs have numerous receptors (liver, pituitary gland), while others are sparsely represented (testis, spleen). Moreover, thyroid hormone's non-genomic (or early) actions are known, mediated by mitochondrial and membrane receptors, mainly related to the stimulation of second messengers; the mechanisms of action mediating genomic and non-genomic effects are not completely independent.

The actions of thyroid hormones are crucial for the proper performance of numerous physiological activities, such as fetal development, oxygen consumption, thermogenesis, positive inotropic and chronotropic effects, breath center control, intestinal motility, erythropoiesis, central nervous system (CNS) development, and function, skeletal turnover, energy substrate metabolism, cholesterol metabolism, hair system turnover; other effects of thyroid hormones include increased metabolic turnover of other hormones and expression of adrenergic receptors on skeletal muscle, heart, adipocytes, and lymphocytes.

Laboratory Investigations

TSH, fT4, and fT3

Serum TSH concentrations vary markedly in response to small changes in blood levels of T4 and T3. A rational approach to assessing thyroid function is to determine TSH, possibly accompanied by the measurement of thyroid hormones. The high sensitivity and specificity of serum TSH assays justify their use in clinical practice as a screening test for thyroid alterations. TSH measurement should be performed by high-sensitivity immunometric methods.

TSH assay may be required in the following cases:

- Screening of apparently healthy populations
- Diagnostic assessment of outpatients with clinical symptoms attributable to thyroid dysfunction
- Diagnostic investigation on the hospitalized patients with clinical symptoms ascribable to thyroid dysfunction
- Monitoring the course of the diagnosed thyroid disease and/or specific therapy (medical and/or surgical)

Adults with any of the following conditions are at high risk for thyroid dysfunction:

- Previous history of thyroid dysfunction
- Goiter
- Previous history of thyroid surgery
- Previous history of cervical radiation therapy
- Autoimmune diseases: diabetes mellitus type 1, celiac disease, etc.
- Use of drugs: lithium, cytokines, interferon, amiodarone, contrast media
- Family history of thyroid disease or other autoimmune diseases
- Chromosomal alterations: Down's syndrome, Turner's syndrome, Klinefelter's syndrome, etc.
- Alterations in laboratory tests suggestive of thyroid disease: hypercholesterolemia, hyponatremia, anemia, hyperprolactinemia, hyper-CK, hyper-LDH, hyper-AST/ALT, hyper-ALP, etc.
- Comorbidities: sleep apnea, depression, and dementia

For pediatric subjects, the following conditions are associated with a high risk of thyroid dysfunction:

- Children/adolescents with short stature or low growth
- Children with pubertal developmental disorders
- Hyperactive children and adolescents
- Children and adolescents with reduced school performance

With rare exceptions represented by TSH-secreting pituitary adenomas and pituitary resistance to the action of thyroid hormones, the finding of normal TSH values excludes the presence of alterations in glandular secretion.

Finally, when not mandatory, it is desirable to monitor glandular secretion in the elderly, infants, and nursing mothers.

The detection of an altered TSH value should be followed by the measurement of fT4 and possibly fT3 in the same sample (TSH-reflex) (Fig. 25.12). The use of profile or combination tests is reserved for the rare specific cases in which the TSH test alone has no informative value (TSH-secreting adenoma, thyrostatic drugs, suspected pituitary/hypothalamic diseases, resistance to the action of thyroid hormones).

The widespread availability and pronounced sensitivity of TSH immunoassays have rendered the TRH stimulus test obsolete; the diagnostic significance of non-elevation of TSH after bolus (200–400 μg) is thought to be the same as that of suppressed TSH. Although assays for total T3 and T4 are available, it is appropriate to determine free fractions (fT3 and fT4), unaffected by physiologic and pathologic conditions associated with changes in binding between total hormones and TBG1.

Physiological changes in fT3 and fT4 concentrations are frequently found with age, during pregnancy, and in prolonged fasting.

In some circumstances, TSH determination alone can be misleading, and its use as a screening test is contraindicated. In particular, TSH-secreting pituitary adenomas and pituitary resistance to thyroid hormones can lead to elevated TSH levels without hypothyroidism; on the contrary, TSH suppression can occur during the first trimester of pregnancy as a consequence of hCG secretion, or as a consequence of the pharmacological treatment of hyperthyroidism, or as a consequence of the administration of several other drugs, including dopamine and glucocorticoids. In any case, all serious extra-thyroidal diseases can lead to increased TSH levels.

It should be noted that the TSH value in thyroid function screening is null in the presence of known pituitary disease or only clinical suspicion of it.

Autoantibodies

The search for autoantibodies can complement the evaluation of thyroid function to make an etiologic diagnosis. Antibodies directed against thyroperoxidase (TPOAb), previously defined as antimicrosomal, have long been used to diagnose hyperthyroidism. Although their use in clinical practice is still widespread, the most updated literature recommends their determination only in subclinical hypothyroidism and in identifying the autoimmune nature of thyroiditis. The assay of TSH anti-receptor antibodies (TRAb) has reached high standards of analytical performance (third generation TRAb); in patients who do

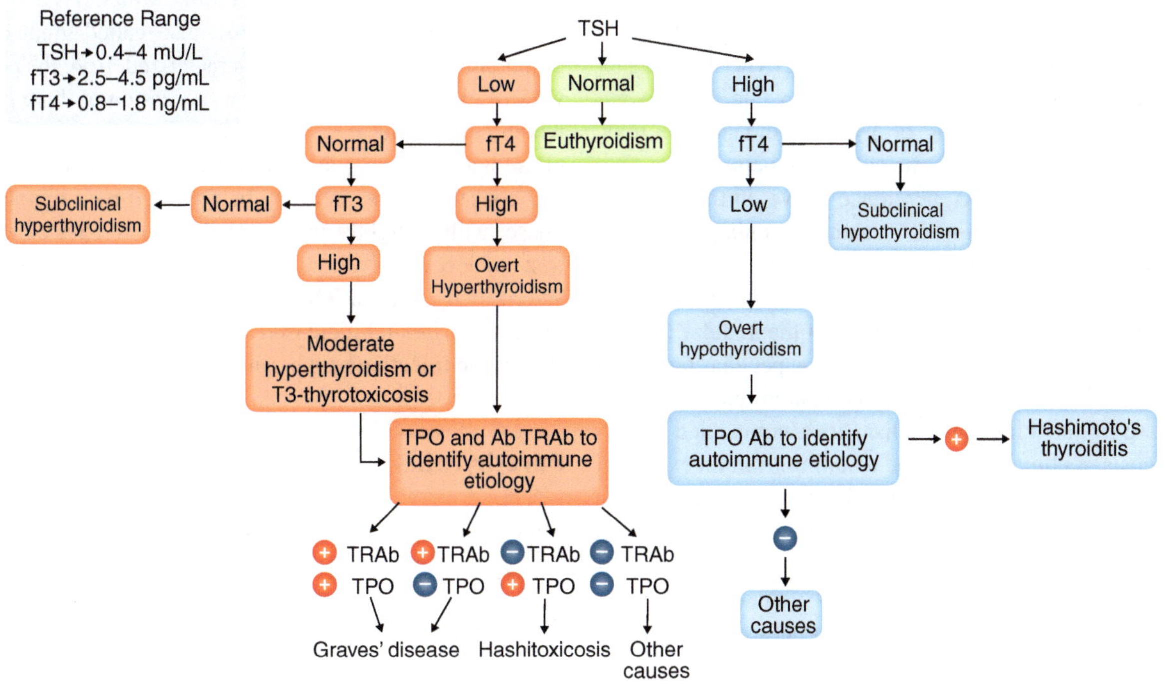

Fig. 25.12 TSH reflex: diagnostic algorithm of the main alterations of the glandular function. (Copyright EDISES 2021. Reproduced with permission)

not present an obvious clinic, they are used in the differential diagnosis between Basedow–Graves and other forms of hyperthyroidism. The measurement of TRAb is also indicated in therapy monitoring of hyperthyroidism to distinguish patients with a high probability of remission (80–100%) from those with a very low probability (20%).

It should be noted that TPOAb and TRAb are not infrequently detected in the serum of euthyroid subjects (2–5%).

Today, antibodies directed against thyroglobulin have no diagnostic value, and their determination is never appropriate.

Diagnosis and Therapy

The main laboratory pictures of the disease are represented by hyperthyroidism and hypothyroidism; both can present in different clinical forms, and the diagnosis is based on laboratory findings. Hypothyroidism has a prevalence of 4.6–8.9%, while hyperthyroidism has a prevalence of 0.9–2.0%. Hyper-/hypothyroidisms can be classified into primary, secondary, and tertiary based on the location of the lesion (thyroid, pituitary, or hypothalamic, respectively). The diagnosis of secondary hyperthyroidism is suggested by elevated fT3 and fT4 values together with increased TSH concentrations; tertiary forms of hyperthyroidism, as well as secondary and tertiary forms of hypothyroidism, are very rare.

The finding of suppressed serum TSH (<0.1 mU/L) raises the diagnostic suspicion of primary hyperthyroidism, and the finding of elevated fT4 levels is sufficient to confirm the diagnosis of frank hyperthyroidism. Evidence of normal fT4 levels associated with low TSH concentration are indicative of subclinical hyperthyroidism; the determination of fT3 must confirm it. In 2–5% of cases of primary hyperthyroidism, in addition to the decrease in serum TSH levels, there is only an increase in fT3 and not in fT4; this condition, termed T3-thyrotoxicosis, most likely represents an early phase of hyperthyroidism, i.e., moderate hyperthyroidism. On the contrary, a very high TSH value (>10 mU/L) suggests primary hypothyroidism, and its association with low fT4 levels indicates overt primary hypothyroidism. In the presence of elevated TSH concentrations and normal fT4 levels, the diagnosis of subclinical hypothyroidism can be made without resorting to the determination of fT3, which, in more than 25% of cases of hypothyroidism, does not show any alteration; therefore, the measurement of fT3 in hypothyroidism is not appropriate (Fig. 25.12). The anti-TPO antibody assay allows for the differential diagnosis between Hashimoto's thyroiditis (autoimmune-based hypothyroidism) and other forms of hypothyroidism.

Primary Hyperthyroidism

The most common clinical form is Basedow–Graves disease, a multisystem pathology on an autoimmune basis, characterized by pathognomonic clinical signs such as exophthalmos and pretibial myxedema. In most cases, Basedow's disease is accompanied by goiter, an enlargement of the gland which, in this case, is non-nodular or diffuse. The hyperthyroid subject is clinically thin, tachylalic, tachypsychic, and insomniac. Less common forms of hyperthyroidism are toxic multinodular goiter (TMG) and solitary hypersecreting nodule or toxic adenoma. As mentioned above, in the presence of a clinical blur or non-nodular glandular enlargement, the determination of TRAb allows the etiologic diagnosis. The conditions of TMG and toxic adenoma are distinguished by the character of functional autonomy of the nodular lesions; on a laboratory level, this gives rise to the peculiar picture of a disproportionate increase in fT3 compared to fT4.

These forms of hyperthyroidism are accompanied by thyrotoxicosis, a clinical syndrome resulting from an excess of thyroid hormones; hyperthyroid patients always present with thyrotoxicosis, but thyrotoxicosis may not result from hyperthyroidism.

Treatment of hyperthyroidism can be ablative, radiation, or pharmacological, depending on the nature of the lesion that caused it.

Primary Hypothyroidism

The most common cause of hypothyroidism is Hashimoto's thyroiditis, an autoimmune disorder common in countries with adequate iodine intake. Other causes of hypothyroidism are iatrogenic and iodine deficiency. Clinically, the hypothyroid patient is asthenic, overweight, bradylalic, and bradypsychic. Determination of serum TSH and fT4 is usually sufficient to diagnose. It is helpful to determine the presence of TPOAb to confirm the autoimmune etiology; the detection of these antibodies is frequent in hyperthyroid subjects, and their use in clinical practice also extends to Basedow–Graves disease. In the early stages, Hashimoto's thyroiditis may present with thyrotoxicosis, defined as hashitoxicosis, due to massive cell lesions with the conspicuous release of already synthesized thyroid hormones. From a biochemical-clinical point of view, it is characterized by hyperthyroidism and the presence of anti-TPO antibodies. It is short-lived and usually evolves toward spontaneous remission and, with time, toward hypothyroidism.

Laboratory alterations of the hypothyroidism may affect the lipid profile, up to frank dyslipidemia.

Treatment of hypothyroidism is always replacement therapy.

Thyroiditis

Thyroiditis is an autoimmune, inflammatory, or infectious disease, which can be divided into acute, subacute, and chronic diseases. Riedel's chronic fibrous thyroiditis is rare, as are acute suppurative forms due to pyogenes. Subacute forms, which are relatively frequent, include Hashimoto's disease and De Quervain's thyroiditis, which has a viral etiology. De Quervain's thyroiditis, also called giant cell or granulomatous thyroiditis, is characterized, from the point of view of glandular secretion, by a succession of three phases, which correspond to as many laboratory pictures. The first phase is thyrotoxic, with low TSH levels and high fT4 and fT3 concentrations, and may be followed by a hypothyroid phase, with a slightly increased or normal TSH concentration and low fT4 levels and, finally, the recovery phase, characterized by normal TSH, fT3, and fT4. The presence of fever and pain and increased indices of acute inflammation suggest an infection as the cause of the disease.

Parathyroids

Marcello Ciaccio, Giulia Bivona, and Luisa Agnello

Anatomy

The parathyroid glands, four in number and weighing approximately 40 mg each, are located in the anterior region of the neck, near the thyroid gland. They are usually arranged symmetrically and in pairs on each thyroid lobe's upper and lower poles (Fig. 25.13).

Microscopically, the gland consists of epithelial cells and abundant adipose and fibrous stroma. Parathyroid epithelial cells are divided into primary and oxyphilic, the former predominant. Each cytotype secretes a hormone that regulates calcemia and inorganic phosphate homeostasis, parathormone.

The parathyroids are supplied with blood by the parathyroid arteries, terminal branches of the inferior and middle thyroid arteries; waste blood from the parathyroids flows into the tributary veins of the thyroid gland.

Embryologically, the parathyroids originate from the dorsal endoderm at the end of the sixth week of gestation.

Parathormone (PTH)

Synthesis, Secretion, Metabolism, and Mechanism of Action

PTH is a polypeptide produced and secreted by parathyroid cells in response to certain secretagogue stimuli, the most potent of which is ionized calcium. PTH is formed by 84 amino acids and, like all secretion proteins, is synthesized from a larger precursor. This pre-pro-parathormone is cleaved in the wrinkled endoplasmic reticulum forming the pro-parathormone, a compound of reduced size that finally will be converted, in the Golgi apparatus, in the mature

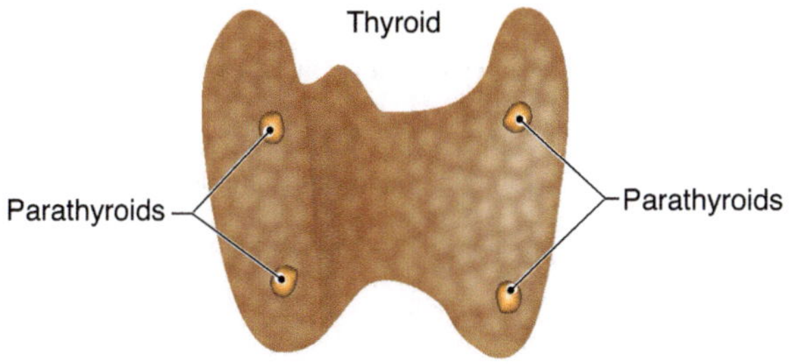

Fig. 25.13 Parathyroid glands. (Copyright EDISES 2021. Reproduced with permission)

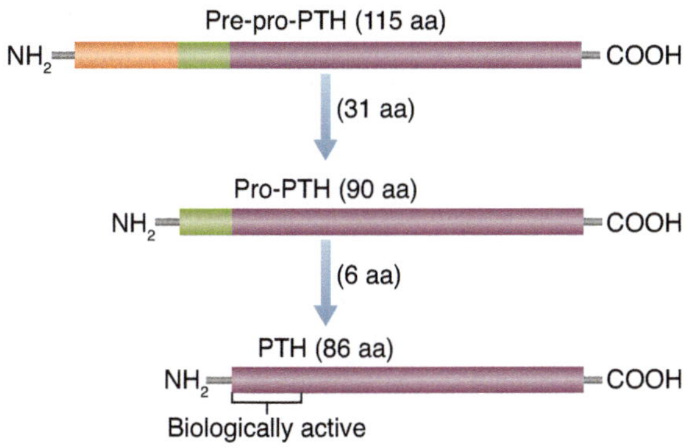

Fig. 25.14 Biosynthesis of parathyroid hormone. (Copyright EDISES 2021. Reproduced with permission)

metabolite, the parathormone. The hormone's biological activity lies in the amino acid sequence 1-34, which forms the amino-terminal fragment of PTH (Fig. 25.14).

PTH synthesis, not its secretion, strictly depends on intracellular calcium concentration.

Immediately after synthesis, PTH is stored in dense cytoplasmic granules from which it is secreted and in amino- and carboxy-terminal fragments. The half-life of PTH is 5–10 minutes; approximately 70% is metabolized in the liver and the remaining 30% in the kidney. A negative feedback mechanism regulates PTH secretion by plasma ionized calcium concentrations. Because plasma calcium concentrations must be maintained within very narrow limits, PTH is extremely sensitive to changes in its concentrations. Therefore, the relationship between the hormone and the cation is sigmoidal and represents a high-gain system since significant changes follow small changes in plasma ionized calcium in PTH secretion.

The sensitivity to changes in serum calcium concentration by PTH, and therefore its ability to promptly correct abrupt changes in calcemia, is due to the presence of a receptor on the membrane of the parathyroid cell (Ca2+ sensing Receptor [CaR]). The PTH/CaR interaction activates the adenylate cyclase and phospholipase C signal transduction systems. Via a CaR-coupled G protein, the hormone-receptor complex forms second messengers (cAMP, inositol triphosphate and diacylglycerol), responsible for intracellular biochemical signals transduction.

The pattern of PTH secretion has a circadian rhythm (with a peak around midnight) and an episodic rhythm (with fluctuations at intervals of about 60 minutes).

Although serum calcium is the primary regulator of PTH secretion, other secretagogue agents include magnesium, lithium, hyperphosphatemia, catecholamines, cortisol, histamine, dopamine, and alcohol.

In contrast, somatostatin, thyroid hormones excess, vitamin D_3, FGF23 (fibroblast growth factor 23), calcium channel blocker antihypertensive drugs, and aluminum intoxication, inhibit PTH secretion.

FGF23 is a phosphoglycoprotein secreted by osteoblasts and osteocytes in response to hyperphosphatemia and vitamin D_3. Noteworthy, parathyroid cell possesses the vitamin D3 receptor (VDR) and Klotho, which are sensitive to vitamin D_3 and FGF23, respectively.

The role of magnesium is not fully known. It is a less potent secretagogue than serum calcium, and under physiological conditions, its influence on PTH secretion is negligible.

Biological Actions

PTH's biological role is to rapidly buffer abrupt changes in extracellular fluid calcium levels. For example, the intermittent introduction of dietary calcium exposes one to fluctuations in calcemia that, in the absence of PTH, would be incompatible with life.

The biological activity of PTH is hypercalcemic and is realized through a direct action on the kidney and bone; indirectly, PTH also acts on the enterocyte, modulating calcium absorption. In particular, the hormone has a triple action at the renal level. In the proximal tubule, it induces the excretion of phosphates, inhibiting their reabsorption by the Na+/Pi co-transporter, and stimulates the activity of the enzyme 1α-hydroxylase, responsible for the conversion of 25(OH)-cholecalciferol (calcidiol) into 1,25$(OH)_2$-cholecalciferol (calcitriol), or active vitamin D_3, responsible for intestinal calcium absorption; in the distal tubule, determines the rapid reabsorption of calcium against the gradient and excretion of phosphate.

Note that much of the 24-hour calcium reabsorption occurs in the proximal tubule by a sodium-related and PTH-independent mechanism.

Slower, however, is the action of PTH on the bone, in which the final effect of the hormone signal is the stimulation of the reabsorption of the matrix, with consequent release of calcium salts into the circulation. This hormone action is mediated by a receptor on osteoblasts, which stimulates the anchorage of osteoclasts to the matrix through the secretion of several immune-endocrine mediators (some examples are IGF-1, IL-1, IL-6, TNF-α). As a result of the action of these cytokines on osteoclasts, an acidic microenvironment is created at the interface between the osteoclastic plasmalemma and the mineralized matrix, known as the rippled margin. This is functionally assimilated to a lysosome, in which the release of numerous lytic enzymes and, consequently, the extracellular matrix's erosion occurs. The calcium contained in the crystalline hydroxyapatite is thus released into the circulatory stream.

Finally, in the enterocyte, PTH induces the synthesis of active vitamin D3, which is responsible for intestinal calcium absorption.

Regulation of Calcemia and Calcium-Phosphorus Homeostasis

The total amount of calcium in the body is about 1–2 kg. Over 98% of this is found in the mineral phase, i.e., in the crystalline hydroxyapatite in the extracellular matrix of bone tissue. One percent of skeletal calcium is exchangeable with the organism in a cyclic alternation between the production and degradation of the newly formed matrix that constitutes bone turnover. In the formation phase, calcium is sequestered in the skeleton through incorporation into the matrix by osteoblasts; in the degradation phase, which immediately follows the previous one, calcium is released into the circulation by the degradation process of the same newly formed matrix by osteoclasts (bone resorption). Bone metabolism is based on the continuous alternation of the phases of formation and degradation described above; for this reason, it is considered an extremely dynamic metabolism.

Reference values for calcemia are generally between 8.5 and 10.2 mg/dL; however, depending on the determination method, they may vary up to ±0.5 mg/dL.

This circulating share is divided as follows: about 50% is free or ionized calcium and is biologically active; about 40% is bound to plasma proteins (mainly albumin) and is biologically inactive; finally, about 10% is complexed in salts and is also inactive. This distribution assumes importance in diagnosing calcemia alterations since variations in plasma proteins determine fictitious calcemia alterations.

Plasma calcium levels must be kept rigidly within the upper and lower limits of the reference range since even slight variations correspond to severe clinical effects.

The hormonal systems controlling calcemia are finely tuned to keep calcemia consistently within baseline limits.

Homeostasis is regulated, under physiological conditions, by PTH and vitamin D_3, both of which have a hypercalcemic effect. Calcitonin has a role only in paraphysiological and/or frankly pathological conditions and is hypocalcemic.

As mentioned above, PTH plays a crucial role in correcting changes in calcemia promptly.

Table 25.11 Actions of vitamin D

Calcemic actions	Maintenance of calcium and phosphorus homeostasis
No calcemic action	Control of cell proliferation (protein p21, p27) Inhibition of neo-angiogenesis Regulation of the immune system Induction of cell differentiation and apoptosis

Vitamin D is a steroid hormone involved in numerous processes, including maintaining calcium homeostasis (Table 25.11).

Vitamin D_3 can be synthesized in the skin, starting from 7-dehydrocholesterol, by the action of ultraviolet rays, or taken with the diet (Fig. 25.15). Once in circulation, vitamin D of endogenous and exogenous origin undergoes a first hepatic hydroxylation that converts it into 25(OH)-cholecalciferol, or calcidiol, a biologically inactive form, which undergoes a second renal hydroxylation, giving rise to the biologically active form of vitamin D, 1,25$(OH)_2$-cholecalciferol, or calcitriol. The latter increases serum calcium and phosphate concentrations by three mechanisms:

- At intestinal level, it determines the synthesis of calbindins, proteins involved in the absorption of calcium and phosphate
- At the level of osteoblasts, it stimulates the expression of the ligand RANK-L, which interacts with the RANK receptor, expressed on resting osteoclasts, thus accelerating the process of bone resorption
- At the level of the distal renal tubules, it increases the reabsorption of calcium and phosphate

PTH-Related Peptide

PTH-related protein (PTHrP) is a 141 amino acid peptide that exhibits a high degree of homology with PTH. It is involved in bone growth during skeletal development through regulating chondrocyte proliferation and differentiation. The same receptor and transduction system mediate the biological activity of PTHrP as PTH. Hypercalcemia during neoplasia is associated with abnormal hypersecretion of PTHrP.

Phosphorus

The total inorganic phosphorus in a healthy adult is about 1 kg, 85% in the skeleton. The circulating portion amounts to 2.8–4 mg/dL and is represented by free ions HPO42 and NaHPO4. Only 12% of circulating phosphorus is bound to plasma proteins, unlike calcium.

Although it is a constituent element of the skeleton, phosphorus is widely represented in all tissues, being involved in all metabolic processes.

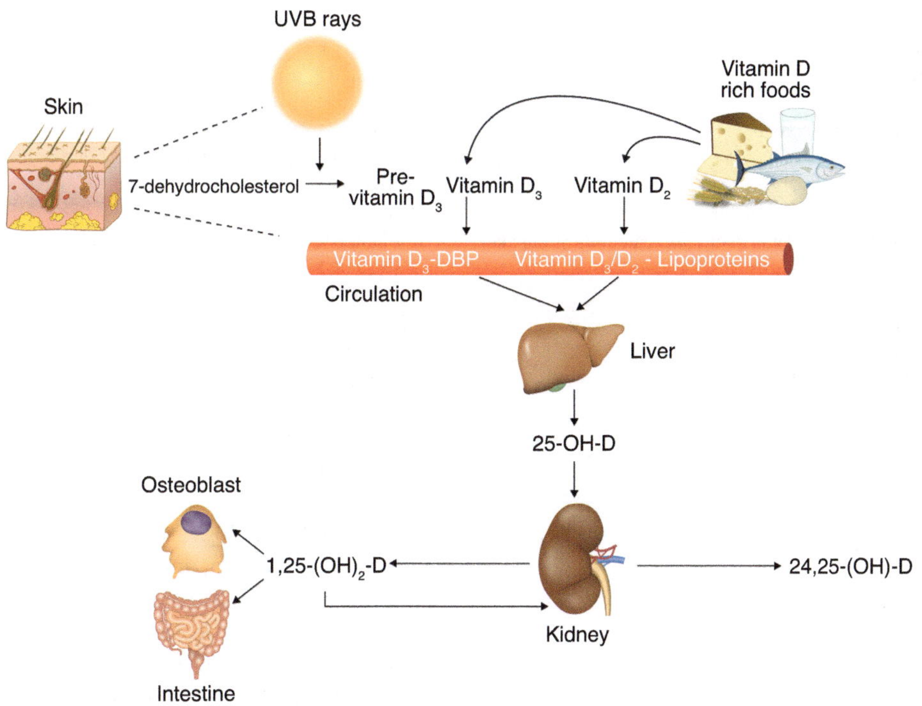

Fig. 25.15 Biosynthesis and calcemic actions of vitamin D. Vitamin D can be synthesized in the skin following exposure to ultraviolet rays (UVB) from the sun, which mediate the conversion of 7-dehydrocholesterol (pro-vitamin D) into pre-vitamin D_3 (pre-D3), which is converted into vitamin D_3 by heat. Vitamin D can also come from food as vitamin D_2 and D_3. After ingestion, they are incorporated in the chylomicrons and absorbed in the lymphatic system, through which they reach the circulation, where they are bound to the vitamin D binding protein (DBP) and lipoproteins. In the liver, vitamin D undergoes the first hydroxylation to 25-hydroxyvitamin D (25(OH)D), which, mainly in the kidney, undergoes a second hydroxylation which converts it into the active form, 1,25 dihydroxy vitamin D (1,25 $(OH)_2D$), by the enzyme 1α-hydroxylase. 1,25 $(OH)_2$-D performs numerous functions, including maintaining calcium homeostasis and acting on osteoblasts, enterocytes, and the renal tubule. At the renal level, there is another hydroxylase (24-hydroxylase) which converts 25(OH)D into the inactive form, 24.25-dihydroxyvitaminD [24.25(OH)D]. (Copyright EDISES 2021. Reproduced with permission)

Although intestinal phosphorus absorption is very efficient, the organ most involved in its metabolism is the kidney. 85–90% phosphorus filtered at the glomerulus is reabsorbed in the proximal and distal tubules. Proximal reabsorption of phosphorus is sodium-dependent and increases under decreased dietary intake; as mentioned above, PTH induces inhibition of this reabsorption with a final effect of urinary excretion of the element. Decreased dietary phosphorus intake results in increased phosphorus clearance.

Hypercalcemia

The clinical laboratory finding of increased serum calcium values, sometimes accompanied by hypophosphoremia, may be detected in asymptomatic individuals as an expression of chronic (hyperparathyroidism) or subtle (neoplasm) disease. Alternatively, it may be occasional. Clinically manifest hypercalcemias account for approximately 50% of cases of hypercalcemia and may present with a wide variety of signs and symptoms, including recurrent nephrolithiasis, peptic ulcers, and hypertension.

The etiologic classification of hypercalcemias distinguishes them macroscopically into parathyroid and extraparathyroid. Table 25.12 describes the main causes of hypercalcemia. Although these are numerous, most hypercalcemic syndromes depend on hyperparathyroidism and neoplasms. Forms caused by primary hyperparathyroidism (solitary adenomas of a gland) are often asymptomatic and have a benign course. In the forms resulting from neoplasia, this is rarely occult and the laboratory finding is included in the investigations.

In the presence of hypercalcemia (and altered calcemic values in general), it is crucial to ascertain that this is true and does not depend on an alteration of the plasma proteins (albumin). This is followed by the PTH measurement to assess the functionality of the parathyroids and the presence of hyperparathyroidism; the determination of 25-(OH)-vitamin D_3, and possibly of calcitriol, is recommended in the presence of PTH values within the reference values. Figure 25.16 describes the diagnostic algorithm for hypercalcemia.

Table 25.12 Causes of hypercalcemia

Parathyroid
Idiopathic hyperparathyroidism - Solitary adenoma - Associated with MEN1 and MEN2 Lithium Familial hypocalciuric hypercalcemia
Extra-parathyroid
Paraneoplastic hypercalcemia - Solid malignancies with metastases (breast cancer) - Solid neoplasms with humoral cause of hypercalcemia (lung and kidney cancer) - Hematological neoplasms Vitamin D-related hypercalcemia - Vitamin D intoxication - Sarcoidosis and other granulomatous diseases Endocrinopathies (hyperthyroidism) Iatrogenic Other - Milk-alkali syndrome - Aluminum intoxication

MEN multiple endocrine neoplasms

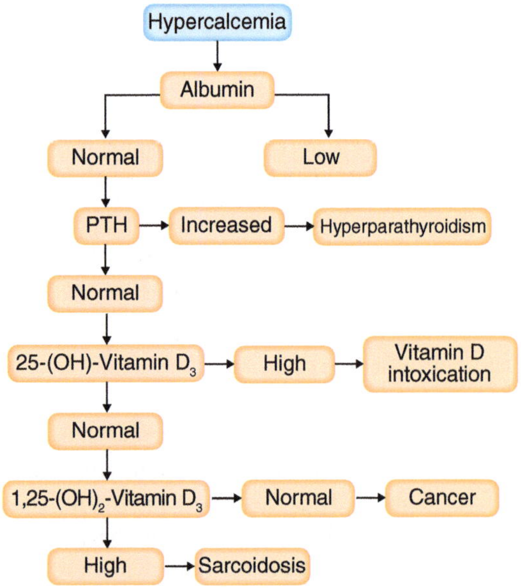

Fig. 25.16 Diagnostic algorithm of hypercalcemia. (Copyright EDISES 2021. Reproduced with permission)

Treatment of hypercalcemia is etiologic.

Hypocalcemia

Hypocalcemia is a less frequent clinical syndrome than hypercalcemia, characterized by decreased serum calcium levels, and often accompanied by hyperphosphatemia. The main causes of hypocalcemia are shown in Table 25.13.

A distinction is made between an acute and a chronic form of hypocalcemia. In the first one, the clinical relevance of the symptomatology varies according to the entity of hypocalcemia; indeed, mild hypocalcemia manifests with asthenia, paresthesias, cramps, and irritability, whereas severe hypocalcemia determines tetany and muscle spasms. Acute drug-induced forms are often transient and asymptomatic. On the contrary, chronic forms are characterized by striking symptoms, including neuromuscular signs, behavioral alterations, parkinsonian symptoms, and basal ganglia calcifications.

Table 25.13 Causes of hypocalcemia

Hypoparathyroidism - Idiopathic - Iatrogenic/surgical - Familiar - Functional (due to hypomagnesemia)
Resistance to PTH - Pseudohypoparathyroidism type 1 and 2 - Alcohol
Vitamin D_3 deficiency - Malabsorption - Chronic kidney disease
Vitamin D_3 resistance
Drugs
Other causes - Acute pancreatitis - Hyperphosphatemia - Blood transfusions

PTH parathyroid hormone

Again, albumin should be determined to ascertain hypocalcemia's nature (true or false). In true hypocalcemia, renal function assessment is recommended to exclude losses secondary to chronic renal failure. Magnesium also enters the diagnostic algorithm for hypocalcemia, hypomagnesemia being invariably associated with hypocalcemia. If magnesium levels are in the normal range, evaluation of PTH, phosphate, and vitamin D_3 levels, both the depot form and the biologically active form [25-(OH)-D_3 and 1,25-$(OH)_2$-D_3, respectively], is performed. In particular, the 1,25-(OH)-D_3 assay allows differential diagnosis between vitamin D_3-resistant and -dependent forms of rickets. The diagnostic algorithm for hypocalcemias is illustrated in Fig. 25.17.

In the presence of albuminemia <4 g/dL, it is necessary to correct calcemia, increasing its value of 0.8 mg/dL for each gram of albumin below the reference value, according to the formula:

$$\text{Calcemia}(\text{correct}) = \left[0{,}8 \times (4 \quad \text{albuminemia})\right] + \text{calcemia}(\text{measured})$$

Treatment of hypocalcemia is based on the administration of calcium and vitamin D_3 per os.

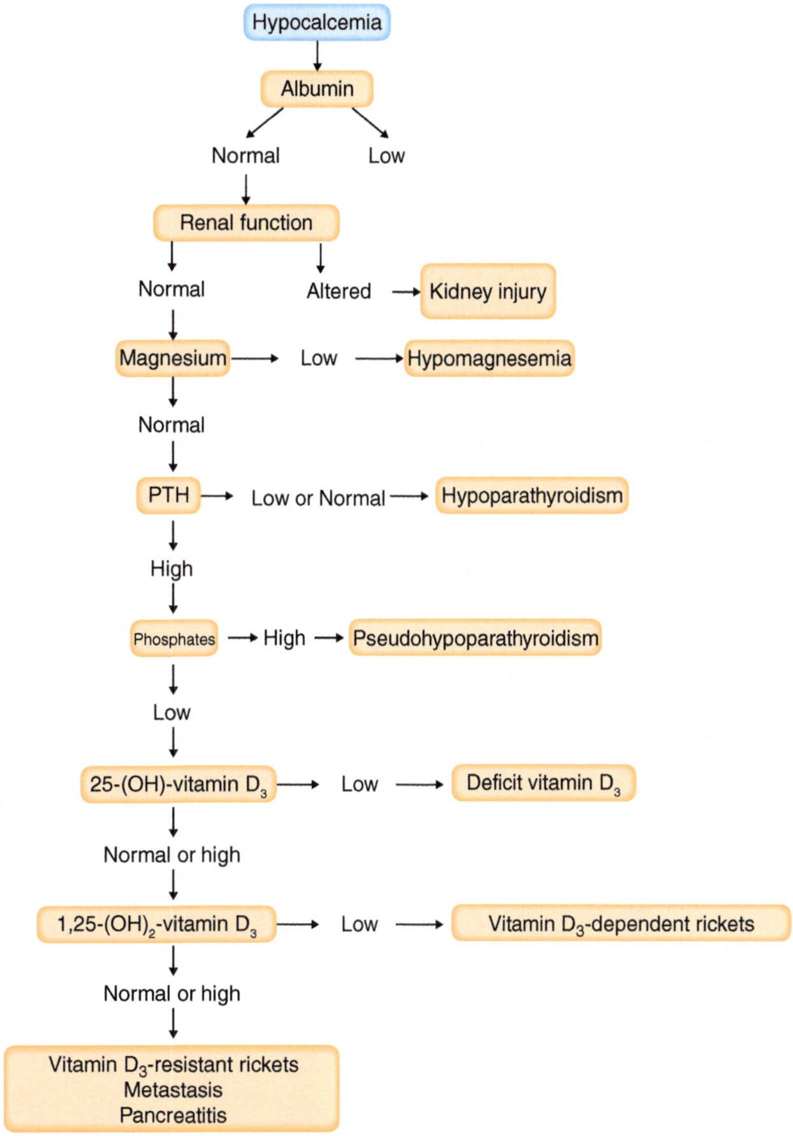

Fig. 25.17 Diagnostic algorithm of hypocalcemia. (Copyright EDISES 2021. Reproduced with permission)

Laboratory Investigations

Calcemia

Total calcemia is determined by colorimetric methods, which provide accurate and reproducible performance under well-controlled operating conditions; the assay does not require fasting or preparation for testing. However, if the patient is taking drugs that may interfere with calcemia values (vitamin D3, lithium), it is advisable to suspend their administration before performing the assay temporarily. In the presence of hypoalbuminemia, the above formula is required to obtain the corrected calcemia. Ionized calcium should be determined in selected cases (malabsorption, chronic disabling diseases).

Albumin

It is measured by separative methods (serum protein electrophoresis) or techniques that directly evaluate the concentrations, using specific reagents (methods using dyes; fluorimetric methods). Serum protein electrophoresis is the most reliable among the separative methods. The albumin assay does not require test preparation or fasting.

Renal Function

Creatinine measurement and clearance are used to assess kidney function.

Magnesium

It can be performed by atomic absorption spectrophotometry or colorimetric enzymatic methods, which are widely used in clinical practice. The reference values of plasma magnesium are 1.7–2.1 mg/dL.

Phosphate

It is performed by colorimetric enzymatic techniques and requires fasting from the previous 12 hours.

Table 25.14 Values of 25–(OH)-vitamin D_3 and their interpretation

Values of 25–(OH)-vitamin D_3 (ng/mL)	Interpretation
<20	Deficiency
20–29	Insufficiency
30–100	Sufficiency
>150	Toxicity

Serum or plasma should be separated as soon as possible from the corpusculate, which is very rich in phosphate esters. The reference values of plasma phosphates are 2.5–4.5 mg/dL. Phosphaturia in 24-hour urine is of little value if not related to dietary intake and tubular reabsorption; therefore, phosphate clearance, corrected with creatinine clearance (400–1000 mg/24 h), is more frequently calculated.

PTH

PTH determination is a diriment test in the etiologic diagnosis of hypercalcemia (for the differential diagnosis between PTH-dependent and non-PTH-related forms) and the differential diagnosis between hypoparathyroidism and pseudohypoparathyroidism. PTH is determined by immune chemiluminescence; the fragment tested is PTH 1-84, or intact PTH. Reference values are between 10–55 pg/mL.

Vitamin D_3

The most commonly assayed vitamin D_3 metabolite is 25-(OH)-vitamin D_3, which indicates available stores related to dietary intake and synthesized from the skin. It is appropriate to measure 1,25-$(OH)_2$-vitamin D_3 for the differential diagnosis of vitamin D_3-dependent and vitamin D_3-resistant rickets. Reference values of 25-(OH)-vitamin D_3 are shown in Table 25.14. The reference values of 1,25-$(OH)_2$-vitamin D_3 are: 20–60 pg/mL.

PTHrP

It is performed when a neoplastic origin of hypercalcemia is suspected. Values below 1 pmol/L are suggestive of tumor etiology. The PTHrP assay has limited uptake in clinical practice.

Adrenal Gland

Marcello Ciaccio, Luisa Agnello, and Giulia Bivona

Anatomy

The adrenal gland is an organ located retroperitoneally suprarenal. The cortical zone constitutes approximately 90% of the gland and surrounds the medullary (Fig. 25.18). In the cortical, we distinguish three histologically different zones: glomerular, fasciculate, and reticular, each deputed to synthesizing a different steroid hormone. Numerous small arteries serve the adrenal gland from the inferior phrenic and renal arteries. The venous outflow is ensured by a single vein that flows on the right into the inferior cava and, on the left, into the renal.

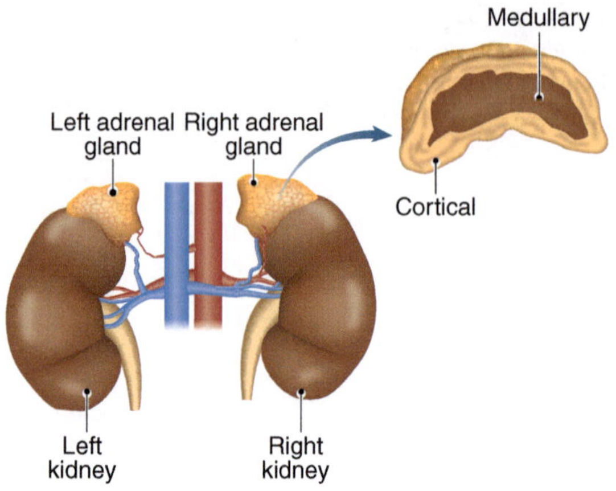

Fig. 25.18 Adrenal glands. (Copyright EDISES 2021. Reproduced with permission)

Embryologically, the cortical is mesodermal (gonadal crest sketch) derived, and the medullary ectodermal (neural crest) derived.

Adrenal Cortical Hormones

Biochemistry and Transport

Adrenal steroids share cyclopentanoperhydrophenanthrene in their chemical structure and may contain 19 or 21 carbon atoms. Hormones with 19 carbon atoms (C19) have androgenic activity and may have a ketone group at position 17 (17-ketosteroids). Steroids with 21 atoms (C21) regulate intermediate metabolism (glucocorticoids) or sodium homeostasis (mineralocorticoids).

Dehydroepiandrostenedione (DHEA) represents the primary adrenal androgen, cortisol among the glucocorticoids, and, finally, aldosterone among the mineralocorticoids.

The precursor of steroid hormones is cholesterol; it is taken up by adrenal cortical cells via the combined ApoB-100/ ApoE ligand on low-density lipoprotein (LDL). The synthesis of all adrenal steroids begins with transforming cholesterol into a common precursor, pregnenolone. Most of the biosynthetic steps of steroidogenesis are catalyzed by enzymes belonging to the supergene family of cytochrome P450 oxidases (Fig. 25.19).

The specificity of synthesis in the various adrenocortical zones depends on specific receptors for ACTH and on the distribution of specific enzyme systems. Indeed, cells in the glomerular zone, which lack 17α-hydroxylase, do not participate in the synthesis of cortisol and androgens, for which

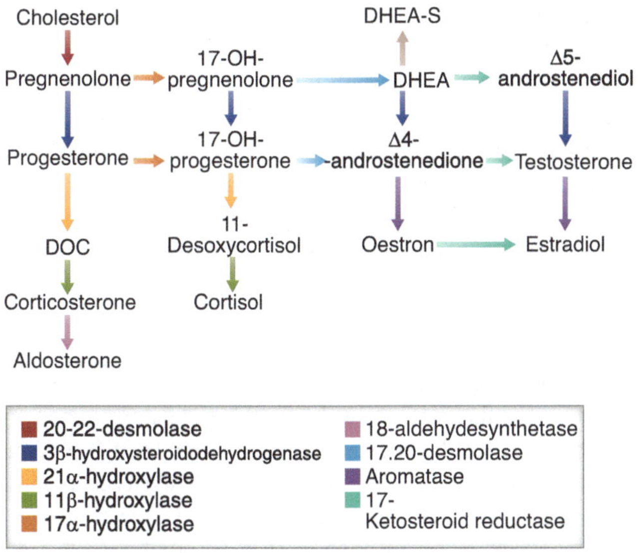

Fig. 25.19 Adrenal steroidogenesis. (Copyright EDISES 2021. Reproduced with permission)

this enzyme is essential. In contrast, the fasciculate and reticular zone cells, not possessing the 18-hydroxylase enzyme, cannot synthesize aldosterone. However, cells in the three zones can synthesize deoxycorticosterone, as the initial steps are common to both enzymatic pathways.

Almost all testosterone and cortisol (98%) circulate bound to plasma proteins, while only 60–65% of aldosterone circulates bound to nonspecific plasma proteins (15–20% corticosteroid-binding globulin [Corticosteroid-Binding Globulin, CBG] and 40–50% albumin). Testosterone binds with high affinity to sex hormone–binding globulin (SHBG) and low affinity to albumin. The transport protein with a higher binding affinity for cortisol is cortisol-binding globulin (CBG), an α-globulin that can bind up to 25 μg/dL of circulating hormone. When plasma cortisol concentrations exceed this concentration, the excess is distributed partly by binding to albumin and partly by increasing the free fraction.

Metabolism

The daily cortisol secretion is about 25–30 mg (8–10 mg/m^2) and follows a circadian rhythm, peaking in the first hours after waking up and decreasing during the day until reaching its lowest point at night. Cortisol metabolism strictly depends on the activity of the enzyme 11β-hydroxysteroidodehydrog enase (11β-HSD), of which two isoforms are known: 11β-HSD-1 and 11β-HSD-2. The first is liver-based and converts cortisone into cortisol (active corticosteroid); the second mediates the reverse transformation in the kidney and other tissues. At the hepatic level, cortisol and cortisone are conjugated with glucuronic acid (95%) or sulfuric acid and subsequently excreted renally.

A small portion of the cortisol produced (up to about 150 μg per day) is excreted and represents circulating cortisol in free form, filtered out by the kidney (urinary free cortisol).

The average daily aldosterone secretion is 100–200 μg/day, of which 30% is in free form, and 70% binds weakly to plasma transport proteins, CBG, and albumin.

Metabolism of aldosterone is characterized by an initial hepatic passage that considerably reduces its concentration (≈75%), followed by renal excretion; under adequate dietary salt intake, daily excretion of the hormone is approximately 50–250 μg.

Adrenal androgens are secreted as DHEA and its ester with phosphate (DHEA-S), 15–30 mg/day. The adrenal gland also secretes small amounts of androstenedione, 11β-hydroxyandrostenedione, and testosterone, from which the female urinary 17-ketosteroids originate. About one-third of them, in the male, is of testicular origin.

Glucocorticoid (GR) and steroid (MR) receptors are intracellular, and hormone binding activates or inhibits transcription factors. While GR binds only corticosteroids, MR binds both with equal affinity. Alterations in GR result in glucocorticoid resistance syndromes characterized by silent hypercortisolism.

Physiology

The synthesis and secretion of glucocorticoids and mineralocorticoids occur in very different homeostatic contexts. Cortisol secretion is an expression of the control exerted on the adrenal gland by the hypothalamic-pituitary endocrine axis; aldosterone, on the other hand, is secreted in response to the stimulus of angiotensin II, a potent vasoconstrictor regulated by the renin-angiotensin system.

Glucocorticoids

Plasma cortisol is secreted in response to pituitary corticotropin (ACTH) stimulation, which, in turn, responds to stimulation by the hypothalamic release factor (CRH). This mechanism of regulation of the endocrine axis is defined as positive feedback and is common to all the glands under hypothalamic-pituitary control. The inhibition by plasma cortisol on ACTH and CRH represents negative feedback. The resultant integration of these inhibition or release stimuli by each glandular product represents the basis for the foundation of the hypothalamic-pituitary-target gland endocrine axis. However, numerous other factors are involved in modulating these feedbacks, such as stress and hypoglycemia (Fig. 25.20).

Cortisol exerts negative feedback on ACTH in two ways: rapid, entrusted to GR, and delayed, due to suppressing synthesis of the precursor, pro-oppiomelanocortin (POMC). POMC is also a precursor of melanocyte-stimulating hormone (MSH) and endogenous opioids (met-enkephalins and β-endorphins).

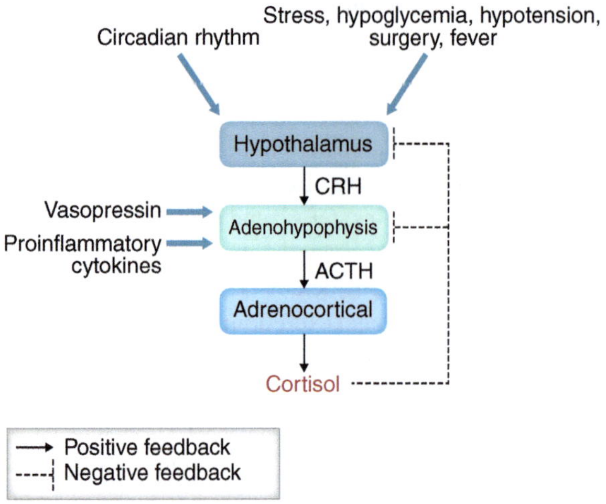

Fig. 25.20 Regulation of cortisol secretion. (Copyright EDISES 2021. Reproduced with permission)

It is helpful to remember that CRH also exerts a feedback control on the sympathetic nervous system, stimulating the locus coeruleus and inhibiting this activation by plasma cortisol.

Mineralocorticoids

Aldosterone secretion is regulated primarily by activation of the renin-angiotensin system. Renin, a hormone produced and secreted by the juxtaglomerular cells surrounding the glomerulus's afferent arteriole, mediates the circulation of angiotensinogen, produced in the liver, into angiotensin I, which becomes angiotensin II by the angiotensin-converting enzyme (ACE). Angiotensin II is a potent arteriolar vasoconstrictor that stimulates the biosynthesis and release of aldosterone from cells in the glomerular zone of the corticosurrene. Aldosterone secretion also follows a circadian rhythm, similar to that of cortisol.

Aldosterone regulates extracellular volume by inducing changes in renal hemodynamics and tubular sodium reabsorption. In addition to hypovolemia, other stimuli to renin secretion are activation of sodium load-sensitive macula densa chemoreceptors, orthostatism, and atrial natriuretic peptides. Intrarenal pressure and sodium load sensors increase renin secretion; natriuretic peptides reduce it.

Although to a lesser extent, potassium levels and ACTH are also involved in regulating aldosterone secretion.

Biological Actions

Glucocorticoids

The biological roles of cortisol are numerous and occur at various levels. Cortisol intervenes in the metabolism of glycides, lipids, and proteins by inducing hyperglycemia and stimulating protein and lipid catabolism. Besides exerting these counter-insular effects, cortisol can inhibit insulin synthesis. In addition, cortisol has numerous immunomodulatory effects, mainly through two mechanisms: reduction of the secretion of inflammatory cytokines and regulation of capillary permeability. Cortisol also controls the regulation of extracellular fluid by suppressing the secretion of vasopressin and inhibiting water entry into cells. Finally, cortisol is secreted in response to stressogenic stimuli of various kinds: surgery, trauma, exercise, mood deflections, anxiety, hypoglycemia, and fever.

Mineralocorticoids

A genomic mechanism mediates the regulatory action of mineralocorticoids on extracellular volume through binding to the MR receptor on the epithelial cells of the target organs. In particular, at the level of the renal cortical collecting ducts, sodium enters passively in favor of an electrochemical gradient through special channels; subsequently, it is actively excreted by the ATP-dependent Na^+-K^+ pump. Aldosterone acts directly on the transcription of genes encoding for sodium channels and the sodium pump, thereby promoting the reabsorption of sodium and water (water is passively reabsorbed along with sodium) and increasing the excretion of potassium and hydrogen ions. In addition, the hormone stimulates sodium reabsorption in the sweat, salivary and gastrointestinal glands. This increases blood volume and, consequently, blood pressure.

Androgens

Androgens promote the development of secondary sexual characteristics in men and are responsible for virilization in women. Adrenal forms have little effect in men, where gonadal androgens primarily control the reproductive function. Furthermore, the androgenic activity of DHEA, DHEA-S, and 11β-hydroxyandrostenedione is significantly reduced compared to testosterone. Adrenal androgens are under the control of ACTH.

Hypocorticosurrenalism

Hypocorticosurrenalism, or adrenal insufficiency, is characterized by the reduced or absent function of the adrenal cortex, either due to the inability of the adrenal to produce sufficient amounts of hormones (primary hypocorticosurrenalism) or secondary to hypothalamic-pituitary pathologies associated with impaired ACTH production or secretion (secondary hypocorticosurrenalism) (Table 25.15).

Primitive hypocorticosurrenalism is divided into an acute form, which represents a clinical emergency with the manifestations typical of hypovolemic shock, and a chronic form, known as Addison's disease, which is the result of progressive destruction of the adrenal gland characterized by asthenia, which is the most frequent symptom, weight loss, arterial

Table 25.15 Classification of adrenal insufficiency

Primitive
Organic forms
- Autoimmune
- Infectious (tuberculosis, mycosis, viral infections)
- Hemorrhagic
- Infiltrative
- Invasive
Iatrogenic forms
- Surgery (removal)
- Enzyme inhibitor therapy (metopyrone, ketoconazole, aminoglutethimide)
- Therapy with cytotoxic agents (mitotane)
Congenital forms
- Congenital deficiencies of adrenal steroidogenesis
- Familial glucocorticoid deficiency
- Congenital adrenal hypoplasia
- Adrenoleukodystrophy
Secondary
Organic forms
- Hypothalamic-pituitary disease
- Inhibition of the HPA axis after removal of glucocorticoid-secreting tumors
Iatrogenic forms
- Glucocorticoid therapy

HPA hypothalamic pituitary adrenal axis

hypotension, hypoglycemia, anemia, depression and hyperpigmentation of the skin. In this case, recognizing the disease in the early stages may not be easy because the signs and symptoms only manifest when more than 90% of the glands are destroyed. Primary adrenal insufficiency is relatively rare, with a higher incidence in subjects aged between 30 and 40 years. In the majority of cases (80–90%), it is due to autoimmune destruction of the glands, characterized by progressive degenerative atrophy of the cortical adrenal, with diffuse cytotoxic T lymphocyte infiltration, without changes in the medullary area; less frequently (10%), it is due to tuberculosis and, rarely (1%), other conditions are found (Table 25.15). Anti-adrenal antibodies are present in most affected patients, although their role is not yet fully understood. In addition, some patients have antibodies directed against other endocrine glands. In about half of the cases, Addison's disease is a component of polyglandular autoimmune syndromes (PAS) along with other autoimmune endocrinopathies such as hypogonadism, Hashimoto's thyroiditis, vitiligo, hypoparathyroidism, pernicious anemia, and celiac disease.

Hypercorticosurrenalism

The term hypercorticosurrenalism refers to a condition characterized by single or combined hypersecretion of hormones produced by the cortical adrenal gland. In particular, cortisol excess causes Cushing's syndrome, and aldosterone excess causes hyperaldosteronism.

Cushing's Syndrome

It is a rare disease caused by chronic exposure to cortisol excess due to endogenous (pituitary or extrahypophyseal tumors) and exogenous (chronic corticosteroid treatment) causes. All endogenous forms are characterized by increased cortisol production by the adrenals and can be distinguished into ACTH-dependent and ACTH-independent forms (Table 25.16). Cushing's disease is the most common among the ACTH-dependent forms. Cushing's disease is characterized by an increased secretion of ACTH, which in 90% of cases is due to a pituitary adenoma and only in 10% of cases to hyperplasia of ACTH-secreting cells; the latter may be the consequence of alterations in the central nervous system leading to excessive hypothalamic production of CRH or CRH-secreting intrasellar gangliocytomas.

ACTH-dependent forms are rarely due to ectopic secretion of ACTH or CRH from non-pituitary tumors; among these, the most common are small cell lung carcinoma or thymic, pancreatic, or ovarian carcinoid tumors, medullary thyroid carcinoma or bronchial adenomas. Endogenous ACTH-independent forms are, instead, due to adrenal neoplasms, usually unilateral and, in about half of the cases, malignant. Bilateral adrenal hyperplasias are characterized by a nodule with a diameter >1 cm (macronodular) or <1 cm (micronodular) and may be due to a hereditary disease (micronodular pigmented dysplasia) or to the stimulation of cortisol secretion by gastric inhibitory peptide (GIP) or luteinizing hormone, associated with the expression of specific receptors for these hormones in the adrenal cortex.

The most common cause of Cushing's syndrome is the administration of exogenous corticosteroids or ACTH for therapeutic purposes.

The clinical signs of Cushing's syndrome are variable due to the amount and duration of excess cortisol. In severe cases of hypercortisolism, the signs and symptoms are unmistakable. They include muscle weakness and easy fatigability due to hypotrophy of the skeletal musculature, striae rubrae,

Table 25.16 Endogenous causes of Cushing's syndrome

	Causes	Percentage
ACTH-dependent (70–80%)	Cushing's disease ACTH ectopic secretion syndrome CRH ectopic secretion syndrome	60–70 5–10 Very rare
ACTH-independent (20–30%)	Unilateral adrenal adenoma Unilateral adrenal carcinoma Bilateral macronodular adrenal hyperplasia Bilateral micronodular adrenal hyperplasia	10–22 5.7 <2 <2

ACTH adrenocorticotropic hormone, *CRH* corticotropin-releasing hormone especially in the abdomen, facial plethora, and centripetal obesity due to deposition of fatty tissue in the upper part of the face leading to the appearance of the typical full moon face, in the interscapular region with buffalo hump, in the supraclavicular area, and the abdomen. Other non-specific symptoms of Cushing's syndrome are osteoporosis, hirsutism, acne, hypertension, complicated in some cases by edema and heart failure, amenorrhea in women, and impotence in men. In addition, patients often present with emotional disturbances or psychosis (irritation, difficulty in concentrating, memory loss, and depression). It should be remembered that there are cyclical forms of the disease in which hormone overproduction occurs only periodically.

This is because many conditions such as obesity, chronic alcoholism, pregnancy, chronic exercise, anorexia nervosa, and depression can clinically mimic Cushing's syndrome with increased cortisolemia. This condition can be treated by correcting the underlying pathology.

Hyperaldosteronism

It is a syndrome characterized by an inappropriately high production of aldosterone that may be due to increased and autonomous production of the hormone in the adrenal cortex (primary hyperaldosteronism) or to non-pituitary and extra-surrenal stimuli with consequent hypersecretion of aldosterone (secondary hyperaldosteronism). Hypersecretion of aldosterone induces reabsorption of sodium ions leading to water retention, expansion of extracellular volume, and hypertension. It also causes increased excretion of potassium and hydrogen ions which may lead to hypokalemia (observed in 30–50% of cases) and metabolic alkalosis.

Primary hyperaldosteronism is characterized by hypersecretion of aldosterone, independent of activation of the renin-angiotensin system, which may be due to a unilateral aldosterone-secreting adrenal adenoma, known as Conn syndrome, (35%) or bilateral adrenal hyperplasia, also known as idiopathic hyperaldosteronism (65%). Other rarer causes are unilateral adrenal hyperplasia (2%), adrenal carcinoma (<1%), ectopic aldosterone-secreting tumors (ovarian and renal tumors), and familial hyperaldosteronism, of which three different forms are known (<2%). Familial type I hyperaldosteronism, or glucocorticoid-sensitive aldosteronism (GRA), is very rare (<1%) and has an autosomal dominant transmission; it is due to the fusion of the promoter region of the CYP11B1 gene, which codes for 11β-hydroxylase, and the coding sequences of the CYP11B2 gene, which codes for the aldosterone-synthetase enzyme, leading to the formation of the CYP11B1/CYP11B2 chimera gene. This form is characterized by ACTH-dependent aldosterone hypersecretion and elevated levels of the "hybrid" steroids 18-oxocortisol and 18-hydroxycortisol, which are under the control of ACTH. According to some authors, children or young adults with severe or resistant hypertension and a positive family history of early hypertension and/or premature hemorrhagic stroke are highly likely to be affected by GRA. In familial hyperaldosteronism type II, which also has autosomal dominant transmission, the underlying gene defect is not yet known; recent studies have identified an association with the chromosome 7:7p22 region. It represents the familial form of aldosterone-secreting adenoma and bilateral adrenal hyperplasia. Unlike GRA, type II hyperaldosteronism is not glucocorticoid-sensitive nor clinically distinguishable from the nonfamilial forms of primary hyperaldosteronism. Finally, familial type III hyperaldosteronism is due to mutations in the KCNJ5 gene, which encodes for the potassium channel and results in increased sodium conductance and cellular depolarization, with subsequent voltage-dependent calcium channel opening resulting in increased aldosterone production.

From a clinical point of view, hypertension is the main, and in most cases only, clinical sign.

Secondary hyperaldosteronism is characterized by increased aldosterone production in response to activation of the renin-angiotensin system. It may be associated with the accelerated phase of hypertension due to increased renin secretion, or as a result of a preexisting edematous syndrome or salt loss, as a compensatory mechanism of reduced volemia or cardiac output to maintain blood pressure within normal values (Table 25.17). In addition, during pregnancy, secondary hyperaldosteronism is physiologically observed due to estrogens inducing an increase in renin levels and plasma renin activity (PRA).

Laboratory Investigations

The diagnostic suspicion of corticosurrenal disorders is based on a careful personal and family history and a careful physical examination. Laboratory tests and imaging play a central role in confirming the diagnostic hypothesis.

In particular, laboratory investigations include plasma and urinary assays of hormones and their metabolites, both under basal conditions and following stimulation or inhibition.

Table 25.17 Secondary hyperaldosteronism causes

With arterial hypertension
- Renovascular hypertension
- Reninoma
- Malignant hypertension
- Essential hypertension with high renin
Without arterial hypertension
- Edema syndromes: liver cirrhosis, nephrotic syndrome, protein deficiency, heart failure, etc.
- Loss of salts: vomiting, diarrhea, abuse of diuretics, shock, nephropathies

Basal Measurements

Blood

Cortisol

Ninety percent of cortisol in the blood circulates bound to proteins, and the remaining 10% is free; therefore, the evaluation of cortisolemia is influenced by the concentrations of its transport proteins (CBG and albumin); alterations of the transport proteins, both physiological (pregnancy) and pathological (infections in the acute phase, hepatic cirrhosis, nephrotic syndrome) can determine false cortisolemia values. In these cases, the free cortisol index is calculated rather than evaluating the total cortisolemia. This involves the evaluation of the ratio between total serum cortisol and CBG. Other formulas are available that consider not only CBG but also albumin. However, they are not yet used in clinical practice. Moreover, since cortisol has a circadian secretion, it is necessary to define the sampling time; it is generally performed between 8 and 9 a.m., when hormone-concentration reaches its peak.

In normal subjects, serum cortisol concentrations are highest in the early morning (approximately 6:00 a.m.), ranging from 10 to 20 µg/dL. Serum cortisol concentrations range from 3 to 10 µg/dL at 4:00 p.m.; concentrations are lowest, less than 5 µg/dL, one hour after the usual sleep period.

Potassium

To increase the diagnostic accuracy of plasma potassium levels, increase the dietary sodium intake (2–3 g/day for at least 5 days) to identify latent hypokalemia. In addition, diuretic therapy or therapy with ACE inhibitors should be suspended for at least 3 weeks because it may give false positives. Normal potassium values are 3.5–4.5 mEq/L.

Aldosterone and Renin

Due to the circadian nature of aldosterone secretion, a single blood collection is not sufficient for measuring aldosterone and renin. In addition, aldosterone levels vary with body position (they increase with standing) and diet (with excessive potassium intake). Therefore, the patient should maintain the standing or supine position for at least 15–30 minutes before blood sampling. Normal plasma aldosterone values are 5–10 ng/dL in clinostat and <20 ng/dL in orthostatic.

Two techniques exist for determining renin: measurement of plasma renin activity (PRA) and direct plasma renin assay. PRA is based on the principle that renin mediates the activation of angiotensinogen to angiotensin I and is performed as follows:

1. Prepare two tubes with the same amount of plasma to be tested for PRA.
2. The two tubes are incubated for one hour but at different temperatures: the first at 37 °C and the second at 4 °C.
3. At the end of the incubation, measure the amount of angiotensin I present in the two tubes. The PRA is the difference between the amount of angiotensin I measured in the first tube and the amount of angiotensin I measured in the second tube.

Values are expressed as angiotensin I (in ng) produced per mL of solution in one hour. In patients with optimal sodium intake (2 g/day), normal PRA values are 0.5–2.5 ng/mL/h in clinostat and 2–4 ng/mL/h in orthostatic. This is the most commonly used method.

Alternatively, the concentration of active renin, which has less inter-laboratory variability than PRA, can be measured directly in plasma. However, renin is labile, so special care must be taken when preparing the sample; blood must be collected in a pre-cooled tube, the sample must be centrifuged at 4 °C, and then the plasma quickly separated and frozen at −20 °C until the assay is performed.

In either case, sampling should be performed in the orthostatic position (standing for at least two hours prior to sampling) or in the clinostat position (supine for at least two hours prior to sampling).

ACTH

Basal ACTH secretion shows a circadian rhythm with lower levels in the evening. Sampling for measurement of plasma levels should be performed in the morning, between 8 and 10 a.m. The ACTH assay is important in adrenal insufficiency because it allows differential diagnosis between primary and secondary forms.

DHEA Sulfate

It is secreted primarily by the adrenals and only in small amounts by the gonads. Thus, DHEA sulfate is a helpful indicator of adrenal androgen secretion.

Urinary Measurements

Free Cortisol

Measurement of free urinary cortisol is helpful in the assessment of adrenal function. Since free cortisol is filtered at the glomerular level and excreted by the kidney, the cortisol measurement in the 24-hour urine represents the free portion of serum cortisol and, at the same time, its "integrated" measurement, unaffected by variations due to the circadian rhythm. Immunometric methods perform the assay. One of the main limitations of the assay is the inadequacy of the 24-hour urine collection. It is, therefore, advisable to perform the urine creatinine measurement to verify the validity of the collection. Since cortisol secretion is subject to daily fluctuations, collecting 24-hour urine for 2–3 days is recommended.

Salivary Assays

Hormones are present in saliva in a free form and their concentrations correlate significantly with their free plasma fraction.

Saliva is usually collected by inserting a swab into the patient's mouth and waiting a few minutes until it is completely soaked in saliva; usually, two samples are collected.

Cortisol

Salivary cortisol represents a surrogate marker of free cortisol, in equilibrium with free serum cortisol. Furthermore, it is not affected by saliva production rate or transport proteins. The assay is performed by immunometric techniques or by mass spectrometry.

The determination of cortisol on a saliva sample collected around midnight (nocturnal salivary cortisol [NSC]) represents an important screening test for Cushing's syndrome, characterized by increased salivary and plasma cortisol levels at night. Salivary cortisol has the advantage that it can be collected at home by self-sampling and, being stable at room temperature, taken the following morning to a laboratory for analysis. However, it must be taken into account that the immunological assay could lead to false positives due to cross-reactivity with cortisone that salivary glands convert from cortisol by the enzyme 11-β-hydroxysteroid dehydrogenase type 2.

One of the main advantages of salivary cortisol is that levels tend to be altered when urinary-free cortisol is normal or only modestly increased in patients with Cushing's syndrome.

Salivary cortisol determination performed in the morning is a helpful screening test for adrenal insufficiency; values <1.5 µg/L is strongly indicative of the disease, while values >6.2 µg/L exclude adrenal insufficiency.

Dynamic Investigations

Dynamic laboratory investigations are aimed at assessing the integrity and function of the hypothalamic pituitary adrenal (HPA) axis; a distinction is made between stimulation tests, which are useful in diagnosing conditions of hormone deficiency, and suppression tests, which are useful in documenting hypersecretion of adrenal hormones (Tables 25.18, 25.19, 25.20, 25.21, 25.22, and 25.23).

Table 25.18 Rapid stimulation test with ACTH

Aim	Assess the functional reserve of the adrenal gland for cortisol production
Indications	Hypoadrenalism Differential diagnosis between primary and secondary hypoadrenalism
Principle	ACTH directly stimulates the adrenal secretion of cortisol
Method	Endovenous administration of 0.25 mg of a synthetic analogue of ACTH (cosynthropin) Dosage of basal plasma cortisol and aldosterone levels, 30 and 60 minutes after administration The test can be performed at any time of day
Interpretation	Healthy subject: cortisol peak >18–20 µg/dL Subject with hypoadrenalism: cortisol peak <18–20 µg/dL. Aldosterone levels allow for a differential diagnosis; in the secondary forms, but not in the primitive ones, there is an increase in aldosterone levels, at least up to 5 ng/dL

ACTH adrenocorticotropic hormone

Table 25.19 Stimulus test with CRH

Aim	Assess the function of pituitary corticotropic cells
Indications	Differential diagnosis between Cushing's disease (ACTH-dependent pituitary adenoma) and ectopic forms (hypercortisolism due to ectopic ACTH secretion) Differential diagnosis of hypoadrenalism
Principle	CRH stimulates the pituitary production of ACTH
Method	Endovenous administration of ovine CRH 1 µg/kg Basal and 15, 30, 45, 60, and 90 minutes ACTH and cortisolemia dosage after administration It is best to test in the afternoon or evening when baseline cortisol is at its lowest levels for the day
Interpretation	Normal response: an increase in ACTH of 20–50 pg/mL compared to baseline 15 minutes after administration, and an increase in cortisolemia >18 µg/dL after 30–60 minutes The ACTH peak confirms ACTH-dependence; in 90% of patients with Cushing's disease the response of ACTH and cortisol to the stimulus with CRH is normal or even higher than normal. In the ectopic forms, however, no variations are observed In secondary hypoadrenalism, the ACTH response to CRH is absent, while in the primary one the basal ACTH levels are elevated and further increase after CRH stimulation

ACTH adrenocorticotropic hormone, *CRH* corticotropin releasing hormone

Diagnosis and Therapy

Adrenal Insufficiency

In the presence of a patient with suspected adrenal insufficiency, with characteristic signs and symptoms of the disease (weakness, hypotension, weight loss, and hyperpigmentation), the morning baseline total serum cortisol or morning salivary cortisol or free cortisol index is determined, associated with the evaluation of ACTH levels; a serum total cortisol <5 µg/dL or free cortisol index <12 or salivary cortisol <1.5 µg/L, associated with an ACTH two times higher than the upper reference limit, is strongly suggestive of primary adrenal insufficiency. Elevated ACTH levels indicate primary adrenal insufficiency because the negative feedback of cortisol on hypothalamic-pituitary function is eliminated; in secondary forms, however, ACTH levels are low or "inappropriately" normal. Diagnostic confirmation should be per-

Table 25.20 Insulin tolerance test (ITT)

Aim	Evaluate the integrity of the function of the hypothalamus-pituitary-adrenal axis
Indications	Complementary test in the diagnosis of secondary hypoadrenalism
Principle	Insulin-induced hypoglycemia is a stressful condition that stimulates the hypothalamus to secrete CRH and the pituitary gland to secrete ACTH, thus leading to an increase in cortisolemia
Method	Endovenous administration, fasting, 0.05–0.1 IU/kg of rapid insulin Basal and 15, 30, 45, 60, 90, and 120 minutes glycemia and cortisol dosage after administration For the successful of the test, it is important that an adequate state of hypoglycemia is reached (glycemia <40 mg/dL)
Interpretation	Healthy subject: cortisol peak >18–20 μg/dL Subject with hypoadrenalism: cortisol peak <18–20 μg/dL

ACTH adrenocorticotropic hormone, *CRH* corticotropin releasing hormone

Table 25.21 Metyrapone test

Aim	Evaluate the function integrity of the hypothalamus-pituitary-adrenal axis
Indications	Diagnosis of Cushing's disease and secondary hypoadrenalism
Principle	Metyrapone induces hypocortisolemia by inhibiting the adrenal enzyme 11β-hydroxylase which mediates the conversion of 11-deoxycortisol to cortisol.
Method	Oral administration of 30 mg/kg of metyrapone at 24 hours Dosage of 11-deoxycortisol, cortisol and ACTH at 8 am the following day
Interpretation	Healthy subject: 11-deoxycortisol >7 μg/dL and ACTH >75 pg/mL Pathological subject: 11-deoxycortisol <7 μg/dL and cortisolemia <5 μg/dL

ACTH adrenocorticotropic hormone

formed by stimulation tests. The best screening test is rapid stimulation by ACTH. In subjects who respond to the test within normal limits, for example in subjects with chronic secondary insufficiency having reduced ACTH reserve, it is necessary to perform another stimulation test, such as CRH or metyrapone tests.

The etiological diagnosis is based on the determination of anti-adrenal antibodies and imaging. Therapy of adrenal insufficiency is gluco- and mineral-corticoid replacement.

Cushing's Syndrome

To confirm clinical suspicion of Cushing's syndrome, at least two of the following laboratory investigations must be performed:

- Measurement of free cortisol in 24 h urine
- Measurement of night-time salivary cortisol, performed by self-sampling at 24 h

Table 25.22 Low-dose dexamethasone test

Aim	Evaluate the function integrity of the hypothalamus-pituitary-adrenal axis
Indications	Endogenous hypercortisolism diagnosis
Principle	Dexamethasone is a potent glucocorticoid whose administration results in the suppression of the HPA axis in healthy individuals and a reduction in urinary and plasma cortisol. The test can be performed by one administration at night or serial administration over 2 days
Method	Night test - Oral administration of 1 mg of dexamethasone between 11 pm and midnight - Baseline cortisolemic dosage and at 8 am the next day Test in 2 days - Oral administration of 0.5 mg dexamethasone every 6 hours for 2 consecutive days - Dosage of creatinine and free cortisol in the urine of 24 hours collected before and during the test or determination of basal cortisolemia, at 24[a] and 48[a] time of the test
Interpretation	Night test - Healthy subject: cortisolemia <1.8 μg/dL at 8 a.m. - Hypercortisolism: cortisolemia> 1.8 μg/dL indicates hypercortisolism, regardless of the cause Test in 2 days - Healthy subject: cortisolemia <1.8 μg/dL and, on the second day of administration, a reduction in cortisoluria is observed >50% of baseline - Hypercortisolism: cortisolemia >1.8 μg/dL

- Dexamethasone suppression test (overnight or over two days)

In addition, they need to be repeated at different times because, to date, none of the available surveys have sufficiently high sensitivity and specificity.

If all the investigations performed are negative, Cushing's syndrome can be ruled out; if two or more tests are positive, they are indicative of the disease.

Once the diagnosis of Cushing's syndrome has been made, it is crucial to identify the cause to determine the specific treatment (Fig. 25.21). The first step in the differential diagnosis is the plasma measurement of ACTH levels. Plasma ACTH levels <10 pg/mL (or <2.2 pmol/L) in a patient with manifest endogenous hypercortisolism suggests an adrenal or otherwise ACTH-independent cause. CT or MRI of the adrenal glands can identify the etiology; typically, adenomas are <6 cm in size and homogeneous in appearance, whereas carcinomas are larger in diameter and irregular in appearance. The differential diagnosis between adenomas and carcinomas is also based on the plasma determination of DHEA-S, which is increased in carcinoma and reduced in adenoma.

Table 25.23 High-dose dexamethasone test

Aim	Evaluate the function integrity of the hypothalamus-pituitary-adrenal axis
Indications	Differential diagnosis of Cushing's disease (pituitary hypophyseal hypersecretion of ACTH) from ectopic ACTH syndrome and adrenal tumors. In Cushing's disease, the hypothalamic-pituitary axis is suppressible with supraphysiological doses of glucocorticoids, while in the case of adrenal tumors or ectopic ACTH syndrome, the secretion of cortisol is autonomous and, therefore, cannot be suppressed
Principle	In Cushing's disease, the hypothalamic-pituitary axis is suppressible with high doses of glucocorticoids (dexamethasone), while in the case of adrenal tumors or ectopic ACTH syndrome, the secretion of cortisol is autonomous and, therefore, cannot be suppressed
Method	Night test - Oral administration of 8 mg of dexamethasone between 11 pm and midnight - Baseline cortisolemic dosage and at 8 am the next day Test in 2 days - Oral administration of dexamethasone 2 mg every 6 hours for 2 consecutive days - Measurement of creatinine and free cortisol in the urine of 24 hours collected before and during the test or determination of basal cortisolemia, at 24ª and 48ª time of the test
Interpretation	Night test - Cushing's disease: suppression of cortisolemia> 50% of baseline - Subjects with adrenal tumors or ectopic ACTH syndrome show no change in cortisolemia following administration of dexamethasone Test in 2 days - Cushing's syndrome: suppression of cortisoluria> 50% of baseline - Subjects with adrenal tumors or ectopic ACTH syndrome show no change in cortisolemia following administration of dexamethasone

ACTH adrenocorticotropic hormone

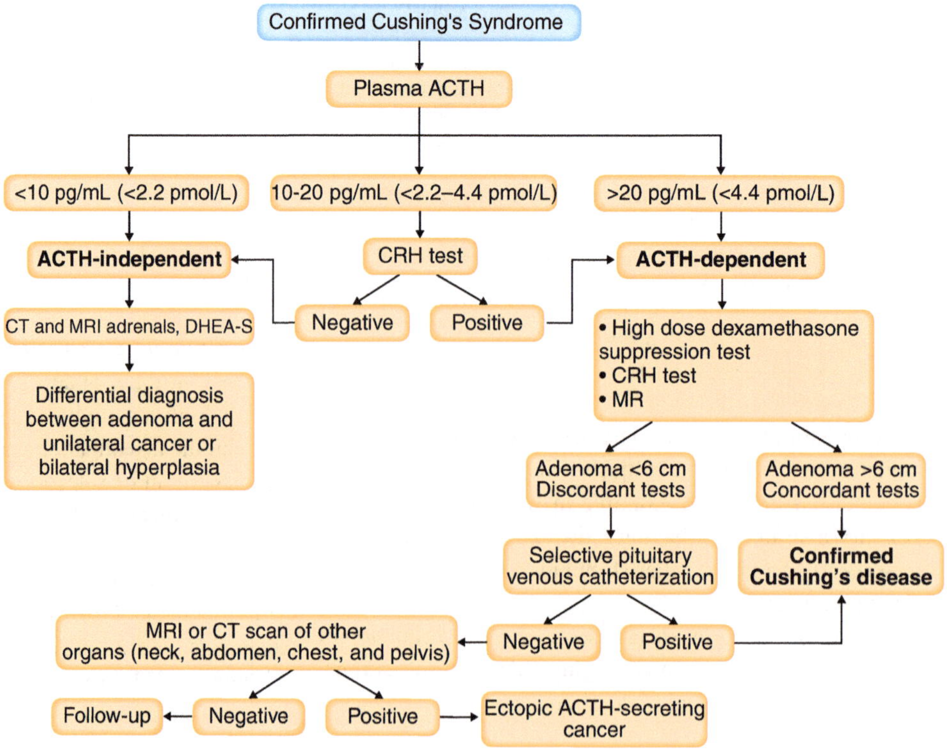

Fig. 25.21 Algorithm for the differential diagnosis of Cushing's syndrome. (Copyright EDISES 2021. Reproduced with permission)

ACTH values between 2.2 and 4.4 pmol/L (1020 pg/mL) are doubtful, and a CRH stimulation test may be helpful. Specifically, in cases of ACTH-secreting pituitary adenoma, testing with CRH will result in increased ACTH (positive test), whereas in the presence of an adrenal cause and ectopic ACTH, the test will be negative.

High ACTH values (>20 pg/mL or >4.4 pmol/L) indicate ACTH-dependent Cushing's syndrome. Although plasma ACTH levels are >40 pmol/L (200 pg/mL) in ectopic secretory forms and range between 6 and 30 pmol/L (30–150 pg/mL), in forms due to pituitary adenoma or hypothalamic-pituitary dysfunction, it is helpful to determine cortisol secretion in response to high-dose dexamethasone administration, perform CRH testing, and an MRI of the adrenals to differentiate between ectopic and pituitary ACTH production. Patients with pituitary adenoma (Cushing's disease) show a normal or increased response to CRH testing and cortisol suppression >50% of baseline values in response to testing with high doses of dexamethasone; positivity of both tests has high diagnostic accuracy (>95%). MRI with contrast medium can detect pituitary adenomas in only half of Cushing's disease patients. In addition, in some subjects (10–20%) without pituitary pathology, an occasional lesion ≤6 cm may be found on MRI. In subjects with questionable MRI and discordant laboratory tests, selective venous catheterization of the petrous sinuses (pituitary outflow venous pathways) can be performed by ACTH assay. If there is an ACTH-secreting pituitary tumor, ACTH levels measured in that area will be very high; if it is a tumor with ectopic ACTH production, the levels in this area will be the same as in the venous sampling. Selective venous catheterization of the petrous sinuses is the gold standard to discriminate between the ectopic or pituitary origin of increased ACTH secretion, with high sensitivity and specificity, although it is a very invasive technique.

Treatment of Cushing's syndrome can be surgical, medical, or radiation; in most cases, the surgical approach is preferred.

Primary Hyperaldosteronism

Diagnosis of primary hyperaldosteronism begins with a screening test, which includes the relationship between plasma aldosterone and PRA values, followed by a confirmatory test and, finally, investigations to assess the cause of hyperaldosteronism.

The screening test is recommended in patients with:

- Sustained blood pressure >150/100 mmHg, measured on three different days
- Hypertension (blood pressure >140/90 mmHg), resistant to three antihypertensive medications (including a diuretic)
- Blood pressure (<140/90 mmHg) controlled with four or more antihypertensive medications
- Hypertension and spontaneous or diuretic-induced hypokalemia
- Hypertension and adrenal incidentaloma
- Hypertension and sleep apnea
- Hypertension and family history of early-onset hypertension or cerebrovascular event at a young age (<40 years)
- Hypertension in subjects with at least one first-degree relative with primary hyperaldosteronism

Assessment of the aldosterone/PRA ratio (ARR) reduces the intra-individual variability in plasma aldosterone and renin levels due to changes in sodium intake, serum potassium levels, position (supine or standing), and age.

Determination of MRA must be preceded by discontinuation of antihypertensive medications for at least 6 weeks (varies by drug type). Currently, there is no uniformity in assay methods; therefore, there is wide variability in the cutoffs used for ARR. However, when sampling is performed in the morning in a seated patient, the most commonly used cutoff values are in the range 20–40 ng/dL aldosterone per ng/mL/h PRA; levels ≥40 ng/dL plasma aldosterone per ng/mL/h PRA, in addition to plasma aldosterone concentrations ≥15 ng/dL, are indicative of hyperaldosteronism. The MRA is a very sensitive but non-specific test; therefore, the diagnosis must be confirmed or excluded by a dynamic suppression test with a positive MRA. The most commonly used confirmatory tests are the oral sodium loading test, saline infusion test, fludrocortisone suppression test, captopril test, and losartan test.

Oral Sodium Load Test

It involves increasing dietary sodium intake (>200 mmol – about 6 g per day) for 3 days and the concomitant potassium chloride intake to keep potassium within normal limits. The 24-hour urinary aldosterone concentration (morning of day 3 to the morning of day 4) is then determined. In the absence of renal disease, aldosteronuria >12 μg/day (>33.3 mmol/day) supports the diagnosis of primary hyperaldosteronism.

Saline Infusion Test

It involves the infusion, in the patient supine for at least 1 hour, of 2 liters of 0.9% saline over 4 hours (500 mL/h). Blood samples for plasma aldosterone measurement are collected at time zero and, after 4 hours, at the end of the saline infusion. The test is positive if, after the infusion, plasma aldosterone is >10 ng/dL, and negative if plasma aldosterone is <5 ng/dL. Plasma aldosterone values between 5 and 10 ng/dL are questionable, and further testing is recommended.

Fludrocortisone Suppression Test

It is based on the administration of fludrocortisone acetate, 0.1 mg per os every 6 hours for 4 days, in combination with slow-release potassium chloride supplements (also every 6 hours and at doses sufficient to maintain serum potassium around 4,0 mEq/L), slow-release sodium chloride supplements with meals (30 mmol 3 times/day), and sufficient dietary sodium intake to maintain daily urinary sodium excretion of at least 3 mmol/kg body weight.

On the fourth day, at 8 a.m., a plasma cortisol sampling is performed; at 10 a.m., in the seated patient, plasma aldosterone and PRA sampling is performed, and the plasma cortisol sampling is repeated. The test is positive if plasma aldosterone is ≥6 ng/dL and PRA values <1 ng/mL/hour, serum potassium in the normal range, and plasma cortisol is lower than at eight o'clock.

Captopril Test

It involves administering captopril, 25–50 mg per os, to a patient seated for at least 1 hour. At time zero and after the patient has been sitting for 1–2 hours, blood samples are collected to determine PRA and plasma aldosterone. The test is positive if PRA remains suppressed while plasma aldosterone is not suppressed and, thus, remains elevated. In other forms of hypertension, however, there is a reduction in aldosterone levels and an increase in PRA.

Losartan Test

It is based on the administration of 50 mg of losartan. In cases of primary hyperaldosteronism, after 2 hours the ARR is >35 ng/dL of aldosterone/ng/mL/h PRA and plasma aldosterone concentration is >10 ng/dL.

Once the diagnosis of primary hyperaldosteronism has been confirmed, the cause must be identified. CT scanning is the most commonly used test for this purpose.

All patients with primary hyperaldosteronism who have to undergo surgical treatment (unilateral adrenalectomy) should perform adrenal venous sampling (AVS) to distinguish between unilateral and bilateral adrenal disease. This distinction is fundamental because adrenalectomy is indicated only in unilateral forms, while bilateral forms must be treated with medical therapy. AVS involves the determination of aldosterone and cortisol plasma concentrations in the two adrenal veins and in the inferior vena cava or in another peripheral vein; there are three different protocols:

- Unstimulated sequential or simultaneous sampling of the two adrenal veins
- Basal, unstimulated, sequential, or simultaneous sampling of the two adrenal veins, followed by another similar sampling stimulated by intravenous injection of a bolus of ACTH (or tetracosactide)
- Continuous ACTH infusion with sequential bilateral adrenal venous sampling

The ratio of plasma aldosterone concentration to plasma cortisol concentration in blood collected from the adrenal vein is termed the aldosterone corrected for cortisol ratio (A/C ratio).

When AVS is performed without ACTH stimulation, an A/C ratio of the adrenal veins of the two sides >2:1 indicates an unilateral form. Similarly, a diagnosis of lateralization can be made if the ratio of the A/C ratio of one side to the simultaneous A/C ratio of a peripheral vein is >2:1 and if, at the same time, the contralateral A/C ratio is less than or equal to the A/C ratio of a peripheral vein or inferior vena cava.

With the continuous infusion of ACTH, an A/C ratio of the two adrenal veins >4:1, indicates unilateral disease; a ratio <3:1 is suggestive of bilateral excess secretion of aldosterone and, therefore, the absence of lateralization; finally, a ratio between 3:1 and 4:1 may indicate either monolaterality or bilaterality of excess aldosterone production.

On the other hand, bolus infusion involves AVS being performed before and after intravenous injection of a 250 μg bolus of ACTH. However, this technique has shown poor diagnostic accuracy.

Finally, genetic testing for familial forms is recommended in subjects with primary hyperaldosteronism <20 years or with a family history of primary hyperaldosteronism or a cerebrovascular event occurring at <40 years. The diagnosis of familial hyperaldosteronism type I is based on detecting the chimeric gene CYP11B1/CYP11B2 by Southern blotting or PCR (Polymerase Chain Reaction). In the case of familial type II hyperaldosteronism, the diagnosis is based on confirmation of primary hyperaldosteronism in at least two family members and the exclusion of type I in hypertensive family members. Finally, tests for mutations in the KCNJ5 gene are not yet commercially available for familial type III.

Secondary Hyperaldosteronism

Elevated plasma aldosterone, PRA, and hypokalemia are suggestive of secondary hyperaldosteronism. In particular, elevated PRA levels allow for differentiating primary hyperaldosteronism from secondary hyperaldosteronism. However, since the determination of baseline values has low specificity and sensitivity, PRA should be determined after the captopril test (25 mg per os with samples taken after 60 and 90 minutes), which shows sensitivity and specificity >95%.

Renal scintigraphy associated with captopril administration allows the diagnosis of nephrovascular hypertension, while CT and MRI are essential for diagnosing renin-secreting tumors. Identifying the underlying pathology in other forms of secondary hyperaldosteronism allows for detecting the endocrine disturbance's cause.

Adrenal Medullary Hormones

The chromaffin cells of the adrenal medullary synthesize, accumulate, and release the catecholamine hormones: dopamine, norepinephrine, and adrenaline (epinephrine) (Fig. 25.22); of these, adrenaline represents the primary product of the adrenal medullary (80%) and the only one synthesized exclusively in the adrenal.

Adrenaline circulates bound to albumin or another low-affinity protein.

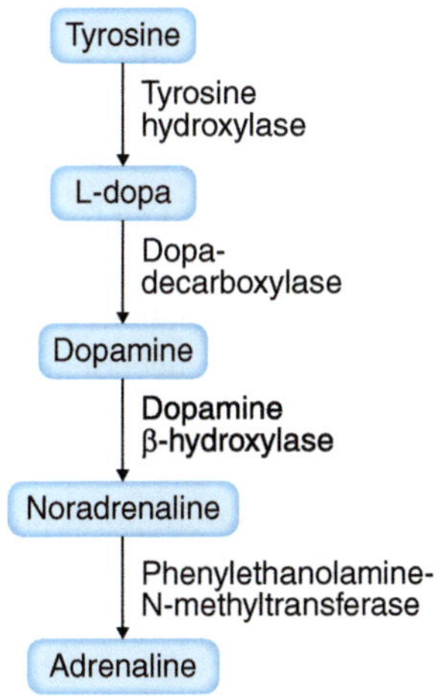

Fig. 25.22 Synthesis of catecholamines. (Copyright EDISES 2021. Reproduced with permission)

Catecholamines are released into the circulation in response to physical or emotional stress and mediate their biological action through interaction with α (α1, α2) and β (β1, β2, β3) adrenergic receptors; biological effects vary based on the receptor bound.

Endogenous catecholamines are involved in regulating cardiovascular function and circulatory hemodynamics. In particular, they mediate nervous control of vasomotor tone, influencing peripheral resistance; moreover, they have inotropic (increase in contractile force of cardiac muscle) and positive chronotropic (increase in heart rate) effects. Other effects of catecholamines unrelated to hemodynamics include stimulation of hepatic glycogenolysis and peripheral lipolysis, mydriasis, slowing of intestinal motility, uterine contractions and relaxation, and bronchodilation. The combined action on the cardiovascular function and the energetic substrates is related to the possibility of reacting in emergencies. Adrenaline is, by definition, the hormone of stress and fear (β-adrenergic effects).

Catabolism of catecholamines begins with degradation by the enzymes catechol-O-methyltransferase (COMT) and monoamine oxidase (MAO), which form inactive O-methylated and deaminated metabolites that are further modified by conjugation with glucuronic or sulfuric acid before being excreted renally. The primary catecholamine metabolites are metanephrine (derived from adrenaline), normetanephrine (derived from norepinephrine), and vanillyl mandelic acid (derived from norepinephrine and adrenaline). A minimal amount of adrenaline and noradrenaline (<5%) is excreted directly in the urine.

Pheochromocytoma and Paraganglioma

Pheochromocytoma and paraganglioma are rare neuroendocrine tumors arising from chromaffin cells. These cells originate from the neural crest and, upon migration, localize in the adrenal medulla and in the ganglia of the sympathetic chains of the thorax and abdomen, where they may give rise to tumors. In 85% of cases, the tumor originates in the adrenal medulla, where most of the chromaffin cells of our organism are located, and it is called pheochromocytoma; in the remaining 15% of cases, it can be found extra-adrenal, in the sympathetic paraganglia (mainly located in the retroperitoneum and the thorax), where it is called paraganglioma. Paraganglioma also refers to other parasympathetic tumors in the head, neck, and anterior thorax regions. However, unlike sympathetic paragangliomas and pheochromocytomas, they generally do not secrete catecholamines and are evident only by a compressive effect on surrounding vascular and nervous structures.

Pheochromocytomas and paragangliomas can occur as sporadic or as components of a genetically transmitted syndrome. Pheochromocytomas can occur at any age but are most common between the fourth and fifth decades of life and rare in infancy. 10% of cases are bilateral, 25% hereditary, and most are benign. Mainly catecholamine-secreting tumors produce noradrenaline, some noradrenaline and adrenaline, a few only adrenaline, and minimal part dopamine; the latter is more frequently associated with a malignant phenotype. These tumors are characterized by extreme variability in the clinical presentation, which includes signs and symptoms related to the production of catecholamines and their metabolites; among these, the most common are headache, sweating, and palpitations, which constitute a classical triad. The most relevant clinical feature is hypertension, which can be persistent (50%) or paroxysmal (45%). The classic triad and hypertension raises a strong diagnostic suspicion of pheochromocytoma. However, it may remain asymptomatic for years and, in some cases, grow to a considerable size before symptoms appear.

Laboratory Investigations

Laboratory investigations, both basal and dynamic, include assays of free catecholamines and their metabolites in plasma and 24-hour urine.

Basal Measurements

Plasma and urinary catecholamines, metanephrines, and vanillyl mandelic acid

Catecholamines have a short half-life; therefore, their determination, both plasma, and urinary, makes it difficult to discriminate between a pathologic overproduction and a peak secretion during sampling. Furthermore, their secretion is intermittent and independent of tumor size.

Metanephrines (metanephrine and normetanephrine) within the tumor are consistently produced through a process independent of catecholamine release. Furthermore, they have a longer half-life in circulation than catecholamines. Therefore, the metanephrine assay has a higher sensitivity and specificity than the catecholamine assay. Plasma and urinary concentrations of metanephrines correlate with tumor size and activity. Since some tumors produce only one type of catecholamine, it is preferable to measure metanephrines and normetanephrine separately; in this case, we speak of fractionated metanephrines. Finally, increased adrenaline or metanephrine indicates the adrenal gland origin, being the exclusive site of adrenaline production and contributing 90% to circulating metanephrine and only 35% to circulating normetanephrine.

Since some drugs, psychological stress, intense exercise, high protein meals, and some foods rich in catecholamines (banana, coffee, citrus fruits, cocoa, nuts, vanilla, pineapple) may alter the levels of catecholamines and their metabolites, blood sampling should be performed in the morning, after overnight fasting (8–12 hours) and in clinostat, to minimize the effect of the sympathetic nervous system. Similarly, for the determination of catecholamines and their metabolites in 24-hour urine, certain foods (see above) and drugs that might interfere should be avoided in the 48 hours before and during sample collection; in addition, emotional and physical stresses and strenuous exercise should be minimized before and during collection. Urinary catecholamines are unstable; therefore, collected urine should be stored in acid and away from light. On the other hand, urinary metanephrines are stable at room temperature for up to 3 days or longer if stored at 4 °C, without the need to add acid. Creatinine should always be assessed for adequacy in the 24-hour urine collection. Both plasma and urinary assays of catecholamines and metanephrines are performed by high-pressure chromatography (HPLC).

Vanil mandelic acid, the end-product of catecholamine (adrenaline and noradrenaline) metabolism, similarly to catecholamines, is measured on a 24-hour urine sample, after diet, abstention from strenuous exercise and stress in the 48 hours preceding collection; an acid, such as hydrochloric or muriatic acid, must be added to the container.

Chromogranin A

Chromogranin A is a glycoprotein secreted by medullary cells of the adrenal medulla and sympathetic nervous system. The measurement is performed by the immunoradiometric method (IRMA). In the case of pheochromocytoma, although its levels correlate with the tumor mass, plasma chromogranin A has a modest sensitivity, and, therefore, its use in clinical practice is limited. High levels of chromogranin A are also found in medullary thyroid carcinoma, small cell lung carcinoma, and epithelial carcinomas with neuroendocrine differentiation (prostate, breast, ovary, pancreas, colon).

Dynamic Investigations

Refer to Table 25.24 for the clonidine suppression test.

Diagnosis and Therapy

Patients with clinical signs and symptoms or at high risk of developing a pheochromocytoma/paraganglioma (Table 25.25) should undergo laboratory investigations to demonstrate excessive catecholamine production (Fig. 25.23).

Since catecholamines are metabolized within chromaffin cells to metanephrines (an intratumoral process that occurs independently of catecholamine release), the measurement of free plasma metanephrines is the test of choice to confirm the diagnosis of pheochromocytoma, with the highest sensitivity and specificity (99% and 89% respectively). Alternatively, fractionated urine metanephrines can be measured, but they have a slightly lower sensitivity and specificity (97% and 67%, respectively). The sensitivity and specificity of the plasma and urinary catecholamine assay are 84% and 81%, 86%, and 88%, respectively. The urinary vanyl mandelic acid assay has a very low sensitivity (68%), although the specificity is high (95%); the low sensitivity of this test makes it less useful in clinical practice.

Normal values of free plasma metanephrines exclude the presence of a pheochromocytoma. In contrast, a 4-fold increase in free plasma metanephrines above the reference limit indicates an almost 100% probability of tumor. In cases where plasma/urinary free metanephrines are only slightly increased (<3–4 times the reference limit), the assay should be repeated after discontinuing interfering drugs; persistently elevated values suggest the presence of tumor, whereas

Table 25.24 Clonidine suppression test

Aim	Inhibition of the sympathetic system
Indications	Urinary and plasma metanephrine concentration values increased, but not diagnostic for pheochromocytoma
Principle	Clonidine, agonist of α2-adrenergic receptors, in the absence of chromaffin tumor, inhibits the release of catecholamines by the sympathetic system causing a reduction of circulating catecholamines and metanephrines
Method	Determination of blood pressure and heart rate, 3 times at 1-minute intervals, and basal plasma levels of norepinephrine and adrenaline Administration of 0.3 mg of clonidine orally Evaluation of blood pressure, heart rate and plasma levels of norepinephrine or normetanephrine at 30-minute intervals for 3 hours
Interpretation	Normal response: decrease in plasma norepinephrine >50% from baseline and/or decrease in normetanephrine >40% from baseline Pheochromocytoma: reduction in plasma norepinephrine <50% from baseline and/or reduction in normetanephrine <40% from baseline

normal values make it less likely. The clonidine suppression test that does not suppress elevated plasma levels of normetanephrine after 3 hours of administration has high sensitivity and specificity (100 and 96%, respectively) for tumor diagnosis in doubtful cases.

Table 25.25 Patients to be screened for pheochromocytoma/paraganglioma

Condition/disease	Prevalence of FCC/PGL
MEN type 2A	≈50%
MEN type 2B	≈50%
VHL	10–20%
Neurofibromatosis type 1	≈2%
SDHB mutation	52%
SDHD mutation	29%
First degree relatives affected by the diseases listed above	–
Adrenal incidentaloma	4%
Resistant hypertension	0.5%
Unexplained heart failure	–
Headache, palpitations, sweating in association with hypertension	–
Patients who develop hypertensive crises during surgery or under general anesthesia	–
Patients with hypertension triggered by β-blockers, monoamine oxidase inhibitors	–
Orthostatic hypotension in a hypertensive patient	–
Diabetes onset in a young and thin subject with hypertension	–

FCC pheochromocytoma, *MEN* multiple endocrine neoplasm, *PGL* paraganglioma, *SDHB* succinate dehydrogenase subunit B, *SDHD* succinate dehydrogenase subunit D, *VHL* Von Hippel-Lindau disease

Tumors secreting only dopamine are rare, and, therefore, plasma dopamine and its metabolite 3-methoxytyramine are not routinely evaluated in most laboratories when pheochromocytoma/paraganglioma is suspected. However, these tests may be helpful in some cases, especially in metastatic disease where the metastatic tissue lacks the enzymes necessary for catecholamine synthesis. High levels of 3-methoxytyramine are a very sensitive marker of malignancy.

Once the diagnosis is made, radiologic evaluation allows localization of the tumor by anatomic imaging followed by functional imaging of the tumor tissue to distinguish between pheochromocytomas, paragangliomas, and other lesions. CT and MRI are recommended for initial anatomic tumor localization; both have high sensitivity (90–100%) but limited specificity (70–80%) due to the high prevalence of adrenal masses not associated with pheochromocytoma in the general population. MRI cannot expose the patient to radiation and, therefore, can be performed in children and pregnant women. Adrenal scintigraphy with iodine 123-labeled metaiodiobenzyl-guanidine (123I-MIBG) allows differential diagnosis between pheochromocytoma and paraganglioma because 123I-MIBG is concentrated in the chromaffin tumor tissue. Furthermore, scintigraphy can reveal the possible presence of multiple tumors and metastases. Scintigraphy has a high specificity (95–100%), especially in the case of malignant tumors and familial forms, but a limited sensitivity (85%), especially in the detection of metastases. Therefore, other investigations are recommended in cases where metastases are suspected, such as octreoscan with 111in-octreotide, since some tumors express somatostatin receptors, or posi-

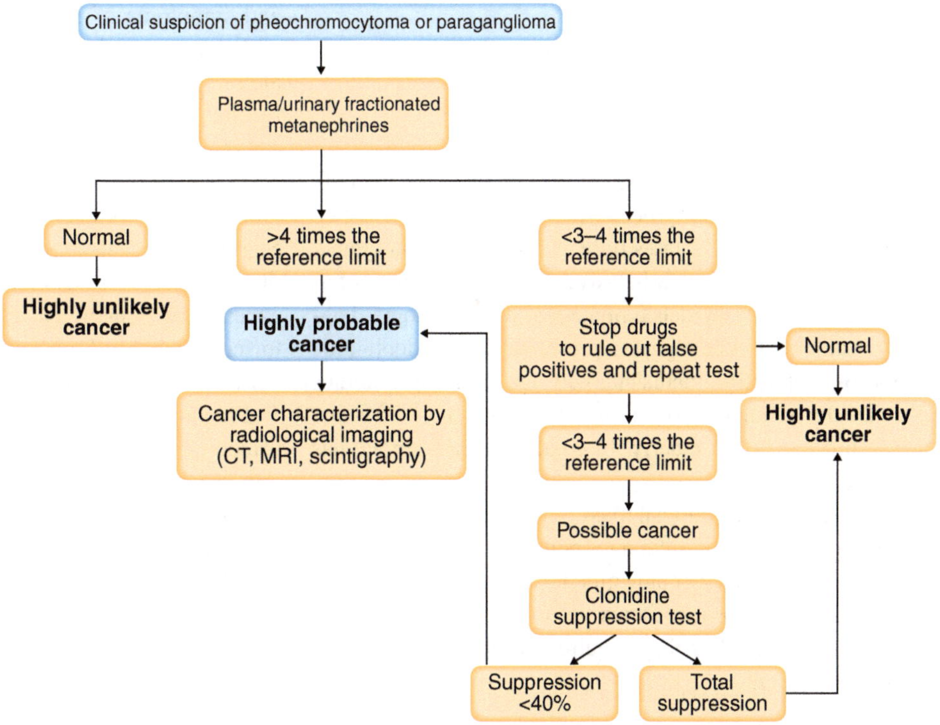

Fig. 25.23 Diagnostic algorithm for pheochromocytoma/paraganglioma. (Copyright EDISES 2021. Reproduced with permission)

tron emission tomography with 18F-dihydroxyphenylalanine (18F-DOPA), which appears to be superior to scintigraphy in detecting metastatic lesions.

Finally, genetic investigations aimed at identifying the mutated gene are indispensable for diagnosing hereditary forms of pheochromocytoma/paraganglioma and should be performed in all patients with the disease. At least one-third of patients have germline mutations responsible for the disease.

Since 1990, 14 different susceptibility genes have been identified. Among these, the primary are NF1 gene associated with neurofibromatosis type 1 (NF-1), RET gene associated with multiple endocrine neoplasia type 2 (MEN2), VHL gene responsible for von Hippel-Lindau syndrome (VHL), and the genes encoding the B, C, and D subunits of mitochondrial succinodehydrogenase (SDH), associated with pheochromocytoma/paraganglioma syndromes named PGL4, PGL3 and PGL1, respectively; SHDB mutations lead to the development of metastases in more than 40% of individuals with pheochromocytoma/paraganglioma.

In addition to family history, features suggestive of a hereditary form of the disease are young age, bilateral or recurrent pheochromocytomas, bilateral or multiple head/neck paragangliomas, or association between pheochromocytomas/paragangliomas and other malignancies. Since patients with pheochromocytoma or paraganglioma have a high prevalence of hereditary syndromes (MEN2, NF-1, and VHL), it is helpful to search for germline mutations even in patients without a known family history. Once an inherited syndrome is identified, genetic screening can be extended to family members.

The therapy of the first choice for pheochromocytomas is surgical excision.

Incidentaloma

Incidentaloma is a clinically silent adrenal mass incidentally found during imaging procedures (ultrasound, CT scan, MRI) performed for diseases unrelated to the adrenal gland. The increasing use of non-invasive imaging techniques has led to an increased incidence of adrenal incidentaloma diagnosis. Autopsy studies suggest a prevalence of incidentaloma of approximately 2%, increasing with age; radiologic studies estimate the frequency to be around 3% in patients <50 years, increasing to 10% in elderly. In childhood, incidentalomas are extremely rare.

In most cases, the adrenal mass is a non-secreting cortical adenoma, but sometimes it may represent a primary or metastatic malignant tumor or hide mild hormonal hypersecretion (Table 25.26). These rare forms need to be carefully evaluated and recognized as they require appropriate surgical or medical treatment.

Table 25.26 Classification and prevalence of adrenal incidentalomas

Incidentaloma	Prevalence (%)
Tumors of the adrenal cortex	70–94
- Non-functioning benign adenoma	71–84
- Subclinical Cushing's Syndrome	9
- Aldosterone-secreting adenoma	1.6–3.3
- Nodular hyperplasia	7–17
- Adrenal carcinoma	4
Tumors of the adrenal medulla	
- Pheochromocytoma	1.5–11
- Ganglioneuroma	0–6
- Gangioneuroblastoma, neuroblastoma, carcinoma	<1
Other adrenal tumors	
- Myelolipoma	7–15
- Lipoma	0–11
- Lymphoma, hemangioma, angiomyolipoma	<1
Cysts and pseudocysts	4–22
Hematoma and hemorrhage	0–4
Infections, granulomas	<1
Metastases (lung, liver, breast, kidney, melanoma)	2.5
Pseudo-adrenal masses (stomach, pancreas, kidney, liver, lymph nodes, vascular lesions, technical artifacts)	0–10

Diagnosis and Therapy

Following the incidental finding of an adrenal mass, the risk of mortality and morbidity due to hormonal hypersecretion or the presence of a malignant tumor must always be considered.

Although adrenal incidentalomas are clinically silent, they can sometimes present with hypersecretion of glucocorticoids, mineralocorticoids, or catecholamines associated with endocrine alterations. Therefore, patients should undergo a careful history and physical examination for signs and symptoms of hormonal hypersecretion, in association with laboratory investigations to document any hormonal alteration. The overnight dexamethasone suppression test is recommended to assess cortisol secretion. After stimulation, a cortisol value <1.8 μg/dL allows to exclude autonomous cortisol secretion; cortisol values between 1.9 and 5.0 μg/dL indicates autonomous cortisol secretion, while cortisol levels >5 μg/dL suggests autonomous cortisol secretion. The 5 μg/dL threshold dramatically reduces false positives at the expense of sensitivity. Therefore, guidelines recommend lowering the threshold to 1.8 μg/dL and possibly confirming cortisol hypersecretion with more specific tests.

Fractionated plasma or urinary-free metanephrines assay is recommended to exclude the presence of a pheochromocytoma. Evaluation of the aldosterone/PRA ratio to rule out primary hyperaldosteronism is recommended only in patients with concomitant unexplained hypertension or hypokalemia. Finally, sex hormone and steroid precursor assay is recommended only in patients with clinical (hirsutism or virilization) or imaging features suggestive of corticosurrenal carcinoma.

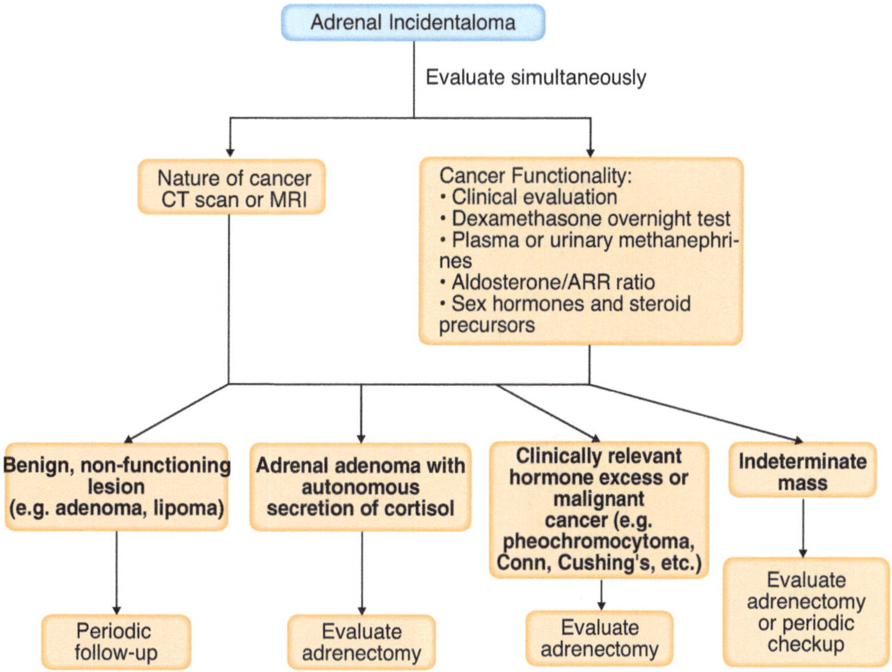

Fig. 25.24 Diagnostic algorithm of adrenal incidentaloma. (Copyright EDISES 2021. Reproduced with permission)

To evaluate the nature of the tumor (benign or malignant), the patient should undergo imaging procedures to assess the size and appearance of the lesion; the finding of a homogeneous mass, rich in lipids, with regular margins and <4 cm in diameter is indicative of benignity.

Therapy depends on the nature of the incidentaloma. No therapeutic measures are necessary for benign, non-functioning lesion, but periodic follow-up is recommended.

A diagnostic algorithm for adrenal incidentaloma is proposed in Fig. 25.24. In the case of malignant tumors, pheochromocytoma and hormone-secreting adenomas, adrenalectomy is indicated. Autonomous cortisol secretion should be considered a high-risk condition for the development of overt Cushing's syndrome. The guidelines recommend that patients with both possible and overt autonomic cortisol secretion should be screened for hypertension and type 2 diabetes mellitus, and the possible treatment of these conditions should be considered. In addition, for patients with (overt) autonomic cortisol secretion, a personalized approach should be considered to evaluate the possibility of surgical intervention, considering the patient's age, degree of excess cortisol, general well-being, presence of comorbidities, and patient preferences.

In patients with undeterminated adrenal masses (by imaging) who choose not to undergo adrenalectomy, repeat imaging (CT or MRI) is recommended after 6–12 months to rule out significant growth; if the lesion shows an increase >20%, surgical intervention is recommended.

Testis

Marcello Ciaccio, Bruna Lo Sasso, and Luisa Agnello

Anatomy

The gonads, male and female, are endocrine glands that, beyond the role in gametes, synthesize hormones called sex steroids, whose primary role is regulating reproductive function.

In particular, the male reproductive system is involved in several processes, such as sexual differentiation, hormonal changes, development of male sexual characteristics, spermatogenesis, and sexual reproduction. It consists of internal and external organs, the latter represented by the penis and testes. The internal organs include the vas deferens, seminal vesicles, ejaculatory ducts, prostate, and urethra. The testicle is an ovoid organ responsible for spermatogenesis and the production of male sex hormones. It develops during fetal life in the abdomen, from where it descends into the scrotum through the inguinal canal during the third trimester of life. This event is essential to ensure that the testicle is at a lower temperature than the body temperature, a necessary condition to perform its functions. Each testicle is located in the scrotal bursa, suspended at the extremity of the corresponding spermatic funiculus, surmounted by the epididymis, and lined almost entirely by the visceral leaflet of the tunica vaginalis propria in whose cavity it is contained; it is surrounded by the tunica albuginea consisting of smooth muscle and collagen. Fibrous septations from the tunica albuginea divide the testicular parenchyma into pyramidal lobules (Fig. 25.25).

The seminiferous tubules are within each lobule. The terminal part of the seminiferous tubules has a rectilinear course, forming the recti tubules, which converge in a complex canalicular system of ducts, the rete testis, from which originate a dozen of efferent canaliculi converging in one only canal, the epididymis. The epididymis distal portion

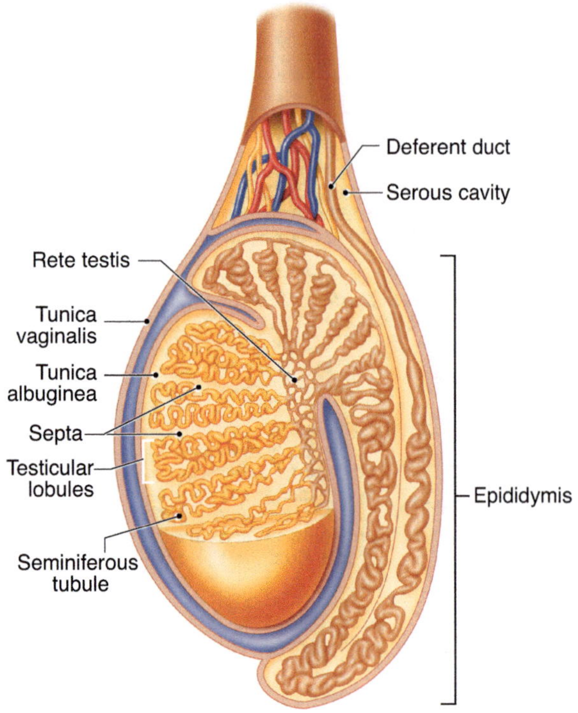

Fig. 25.25 Anatomical structure of the testis. (Copyright EDISES 2021. Reproduced with permission)

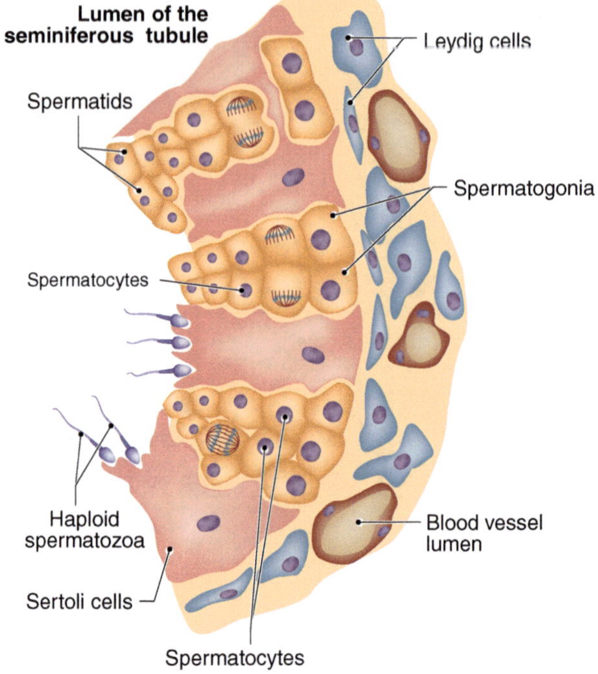

Fig. 25.26 Anatomical section of the seminiferous tubule. (Copyright EDISES 2021. Reproduced with permission)

functions as a spermatozoa storage. Additionally, it destroys "aged" spermatozoa or those showing structural anomalies. The epididymis is connected, through the ductus deferens, to the ejaculatory duct from which it receives the secretion from seminal vesicles, which are located behind the bladder and the prostate. The ejaculatory duct pours its contents into the prostatic urethra, where the prostatic secretion also flows.

The seminiferous tubules are formed mainly by two cell types: germ cells and Sertoli cells. Germ cells are responsible for spermatogenesis; these cells are arranged in layers in the wall of each tubule, according to a maturational order, from the periphery to the center (spermatogonia, spermatocytes, spermatids) (Fig. 25.26). The most immature cells, spermatogonia, go through two phases during their development; during the first, called spermatogenesis, DNA halving occurs, and through this phase, haploid cells, called spermatids, are obtained. Spermatogenesis is followed by spermiogenesis, characterized by a series of important morphological cellular changes, at the end of which mature spermatozoa are obtained. The wall of the tubules is covered by Sertoli cells, which support and nourish the germ cells. Between the seminiferous tubules are also the interstitial or Leydig cells, which are responsible for testosterone synthesis (Fig. 25.26).

Physiology

Male Hypothalamic-Pituitary-Gonadal Axis

Growth and sexual differentiation in the male result from a complex feedback circuit involving the hypothalamus, pituitary, and gonads. Neurons of the lateral ventricles of the hypothalamus synthesize and secrete the Gonadotropin Releasing Hormone (GnRH). It is a decapeptide that, through the pituitary peduncle, spills into the median eminence and reaches the anterior pituitary, where it binds to its specific receptors resulting in the release of Luteinizing Hormone (LH) and, to a lesser extent, Follicle-stimulating hormone (FSH) (Fig. 25.27). GnRH is released in a pulsatile manner every 60–90 minutes, and this pulsatility is critical for the normal functioning of the pituitary-gonadal axis.

Gonadotropins (FSH, LH) have two subunits: α and β. The α subunit is common to all glycoprotein hormones (FSH, LH, TSH, and hCG), while the β subunit is specific for the different hormones. LH and FSH levels vary according to the individual's age: high at birth, decrease during growth, low or undetectable levels throughout the prepubertal period, and increasing in later stages of development. FSH binds specific receptors on Sertoli cells, stimulating the synthesis and release of inhibin and other factors essential for regulating spermatogenesis (Fig. 25.27).

Inhibin is a heterodimeric protein hormone belonging to the Transforming Growth Factor-β (TGF-β) superfamily. It exists in two molecular forms: inhibin A and inhibin B, derived from the combination of an α subunit and two β subunits (βA and βB). Both isoforms are required in gonadal maturation processes in men and women. Inhibin B plays an essential role in spermatogenesis and controls, via negative feedback, the pituitary production of FSH. This feedback involves the inhibin B, which acts as an activin antagonist. Activin is structurally homologous to inhibin and stimulates FSH synthesis (Fig. 25.27).

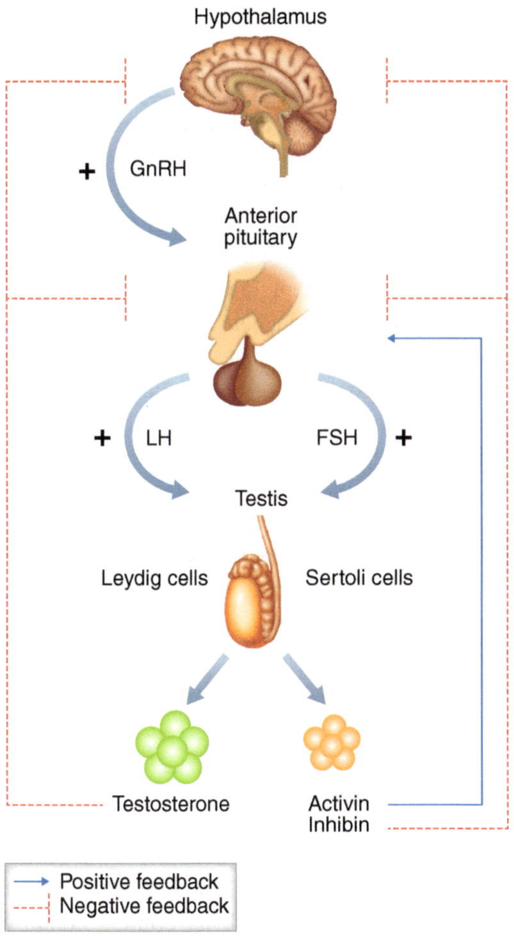

Fig. 25.27 Male hypothalamus-pituitary-gonad regulation. GnRH gonadotropin-releasing hormone, LH luteinizing hormone, FSH follicle-stimulating hormone. (Copyright EDISES 2021. Reproduced with permission)

Sertoli cells also produce Müllerian Inhibiting Substance (MIS), which, during male genital apparatus differentiation, will induce regression of Müllerian structures (fallopian tubes, uterus, upper portion of the vagina).

Leydig cells react to the LH-induced stimulus by converting cholesterol, either from LDL or synthesized within the cells, to pregnenolone and then to dehydroepiandrosterone (DHEA), androstenedione or androstenediol, which will eventually be converted to testosterone (Figs. 25.26 and 25.28). The testis produces 3–10 mg per day of testosterone. Once produced, testosterone can perform its functions by binding androgen receptors or be converted, at the target tissues, into a more potent metabolite, dihydrotestosterone (DHT), by the enzyme 5α-reductase, or into 17β-estradiol by the enzyme aromatase (Fig. 25.28).

Testosterone and estradiol exert a negative control mechanism on the hypothalamic-pituitary axis. Specifically, testosterone inhibits the production of GnRH, FSH, and LH, resulting in decreased production of testosterone by Leydig cells; estradiol blocks the production of LH and FSH (Fig. 25.26). Estradiol is essential in regulating bone development (inhibitory action), epiphyseal closure, and lipid metabolism; finally, it possesses a psychotropic action.

Testosterone is secreted in a pulsatile manner and circulates in the peripheral blood form bound to plasma proteins (i.e., SHBG and albumin) and, to a lesser extent, in free form. Testosterone is responsible for sexual differentiation, developing and maintaining secondary sexual characteristics, regulating erythropoiesis and libido control, and developing muscle mass and bone structure. DHT, produced by the conversion of testosterone, androstenedione, and DHEA at target tissues (mainly skin, liver, and urogenital tissues), is responsible for sexual differentiation (induces the development of external genitalia), sebum production, prostate development, and the appearance of hair. In addition to testosterone production, the testis synthesizes progesterone, 17-OH-progesterone, and androstenedione (D4).

Fig. 25.28 Synthesis of testosterone, estradiol and dihydrotestosterone in Leydig cells and peripheral tissues. (Copyright EDISES 2021. Reproduced with permission)

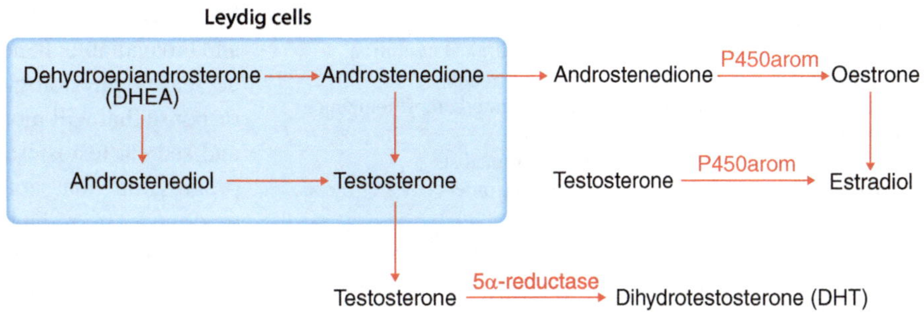

Small amounts of testosterone are present several years before puberty and produced in response to the adrenal androgens DHEA and dehydroepiandrosterone sulfate (DHEA-S), which are responsible for pubic and axillary hair (adrenarche) appearance.

Testis Pathologies

Hypogonadism

Male hypogonadism is characterized by a dysfunction of the hypothalamic-pituitary-gonadal axis marked by a testosterone deficit and/or an impairment in spermatogenesis. There are different clinical pictures due to congenital or acquired causes. In the primary forms, testes are the organs primarily affected and there is circulating testosterone decrease and gonadotropin increase with impaired spermatogenesis (hypergonadotropic hypogonadism). The alterations in testosterone synthesis are due to structural damage of the Leydig cells, which is sometimes associated with a deterioration of the seminiferous tubules leading to oligospermia or azoospermia. Testosterone deficiency, in turn, results in elevated levels of gonadotropins (FSH and LH) due to testosterone's failure to inhibit GnRH at the hypothalamic level.

Primary hypogonadism can result from congenital, acquired, and/or systemic disease-related causes (Table 25.27).

The most frequent congenital cause of hypergonadotropic hypogonadism is Klinefelter syndrome. This disorder causes dysgenesis of the seminiferous tubule and a 47,XXY karyotype. A rare form of hypogonadism results from mutations in the androgen receptors; in this case, the phenotype presents a partial or complete resistance to androgens.

Secondary hypogonadism is characterized by low testosterone levels, low or inappropriately normal gonadotropin concentrations, and reduced spermatogenesis (hypogonadotropic hypogonadism). This condition is characterized by inadequate production of FSH and LH by the hypothalamic-pituitary axis and may be of congenital or acquired origin.

Table 25.27 Causes of primary hypogonadism

Congenital causes	Acquired causes
Klinefelter syndrome (XXY)	Orchitis
Microdeletions of the Y chromosome	Bilateral surgical castration
	Testicular trauma
Mutations of the FSH receptor	Drugs, chemotherapy, ionizing radiation
Mutations of the LH receptor	
Myotonic dystrophy	Hemochromatosis
Cryptorchidism	General disorders (liver cirrhosis, CRI)
Noonan syndrome	
Bilateral anorchidia	Lymphoma or testicular cancer

FSH follicle-stimulating hormone, *CRI* chronic renal failure, *LH* luteinizing hormone

Table 25.28 Causes of secondary hypogonadism

Congenital causes	Acquired causes
Kallmann syndrome	Panhypopituitarism
	Hyperprolactinemia
Prader-Willi syndrome	Hemochromatosis
	Hypothalamic or pituitary tumors
Laurence-Moon syndrome	Drugs (opiates, glucocorticoids, GnRH agonists or antagonists)
Mutation of the β LH chain	Acute or chronic diseases (obesity, eating disorders, diabetes, stress, malnutrition, CRI)
Mutation of the β FSH chain	

FSH follicle-stimulating hormone, *GnRH* gonadotropin releasing hormone, *CRI* chronic renal failure, *LH* luteinizing hormone

Most acute or chronic systemic diseases, e.g., chronic renal failure, diabetes, anorexia nervosa, stress, malnutrition, obesity, hyperprolactinemia, hemochromatosis, and neoplastic or non-neoplastic masses in the hypothalamic-pituitary region, can induce gonadal hormone deficiency (Table 25.28).

Different clinical manifestations characterize androgenic deficiency, also depending on the degree and duration of the deficiency itself; in particular, a decrease in libido, erectile dysfunction, cognitive decline, sleep disorders, mood changes, reduction in lean body mass, increase in visceral fat, testicular atrophy, osteopenia, and sparse body hair can be observed. These individuals are predisposed to gynecomastia and present eunuchoid body proportions due to delayed epiphysis welding and the continued growth of long bones.

Late-Onset Hypogonadism

Late-Onset Hypogonadism (LOH) is a form of hypogonadism associated with aging, characterized by a reduction in blood levels of androgens, with or without a reduction in androgen receptor sensitivity. This clinical condition can lead to significant alterations in quality of life and dysfunction of various organs and systems, including sexual function. From about 30 years of age, the total serum testosterone level in men physiologically decreases by 1% per year. This occurs due to several age-related mechanisms, including an increase in the concentration of sex hormone–binding globulin (SHBG), which decreases circulating testosterone's free and bioavailable. In addition, the secretion of pulsatile GnRH decreases physiologically over time, attenuating the LH response that will more mildly stimulate the testis to produce and secrete testosterone. Generally, this process has a slow progression that may increase due to chronic disease or drugs. The reduction of testosterone leads to a clinical conditions, overlapping with primitive hypogonadism, characterized by loss of muscle mass, osteopenia, loss of libido, and age-related cognitive decline. The term andropause describes these age-dependent changes, although this is improperly used, as there is no rapid and abrupt drop in androgens but rather a slow and progressive change that has no correspondence with the female menopausal period. LH levels also undergo age-related changes, increasing over time, suggesting a primary testicular dysfunction underpinning the decrease in androgen levels.

Laboratory Investigations

Basal Hormone Measurements

Gonadotropin

Plasma and serum are the samples of choice for gonadotropin and androgen measurement.

The evaluation of gonadotropins, FSH, and LH, provides essential information on the origin of testicular dysfunction. They are currently measured by immunofluorometric or electrochemiluminescence assays, characterized by a high diagnostic sensitivity (Table 25.29). The values obtained by different methods vary considerably, and each technique has a different reference range.

Since gonadotropin levels change throughout the day, it is recommended that any pathologic or questionable results be evaluated on a second blood collection.

Testosterone

Circulating testosterone is carried 40% by sex hormone–binding globulin (SHBG) and 58% by albumin, while 2% is free and represents the biologically active fraction. Since the binding of testosterone to SHBG is very stable, only testosterone bound to albumin is bioavailable. "Bioavailable testosterone" refers to the proportion of free and albumin-bound testosterone that can be cleaved and made available in target tissues. Daily testosterone fluctuations require venous sampling between 8 and 10 a.m. However, several studies showed that the circadian rhythm fades with age and, therefore, in the elderly, this time window loses its significance.

The assay of total testosterone (including free and protein-bound portion) is performed by electrochemiluminescent immunometric method (Table 25.30) and liquid chromatography-tandem mass spectrometry (LC-MS/MS). Currently, LC-MS/MS is the best method, as it provides an accurate and precise estimate of total testosterone levels with high diagnostic sensitivity. However, alterations in SHBG levels may affect total testosterone levels. In such a case, the assay of free testosterone, i.e., the free portion representing approximately 2% of circulating testosterone, is recommended. Several methods have been proposed to measure the free fraction of testosterone. Equilibrium dialysis is the gold standard for quantifying the biologically active form of testosterone; however, it is a technique scarcely available in most clinical laboratories.

For this reason, it is recommended to use computational equations allowing free testosterone evaluation using SHBG, albumin, and total testosterone. The values obtained using the Vermeulen formula show a high correlation with those obtained by equilibrium dialysis, representing a simple and reliable indicator (Fig. 25.29). Alternatively, free testosterone can be measured by enzyme immunoassay (Table 25.30). Testosterone is also detectable in saliva at a concentration comparable to free testosterone in serum; however, its assay requires careful quality control because it may be subject to several artifacts.

Table 25.29 Reference values of the main hormones

Parameter	Reference values
LH[a] (IU/L)	1.7–8.6
FSH[a] (IU/L)	1.5–12.4
Total Testosterone[a] (μg/L)	
18–49 years	2.49–8.36
>50 years	1.93–7.40
free Testosterone[b] (ng/L)	3.84–34.17
Prolactin[a] (μg/L)	4.04–15.2

[a]Reference values using electrochemiluminescence method
[b]Reference values using the ELISA method

Table 25.30 Conditions associated with altered circulating SHBG levels

Elevated levels of SHBG
- Age
- Hyperthyroidism
- Hyperestrogenism
- Liver disorders
- HIV
- Antiepileptic drugs
Low levels of SHBG
- Obesity
- Insulin resistance and diabetes
- Hypothyroidism
- Nephrotic syndrome
- Glucocorticoids
- Androgens
- Progestogens
- Excess of GH

GH growth hormone, *SHBG* sex hormone–binding globulins

Inhibin B

The evaluation of inhibin B reflects the functionality of the seminiferous tubules. Reduced inhibin B concentrations are observed if spermatogenesis is impaired (e.g., after chemo-

$$TT = K_s \cdot [SHBG] \cdot FT/1 + (K_s \cdot FT) + $$
$$+ K_a \cdot [Alb] \cdot FT/1 + (K_a \cdot FT) + FT$$

TT = Total Testosterone (mol/L)
K_s = Association constant for T-SHBG binding (L/mol)
K_a = Association constant for T-Albumin bond (L/mol)
SBHG = SHBG concentration (mol/L)
FT = Free Testosterone (mol/L)
Alb = Albumin concentration (mol/L)

Fig. 25.29 Equation for calculating free and bioavailable testosterone (Vermeulen's formula). (Copyright EDISES 2021. Reproduced with permission)

therapy). The inhibin B assay is performed by ELISA (Enzyme-Linked ImmunoSorbent Assay).

Prolactin

High prolactin levels may be associated with hypogonadotropic hypogonadism. Increased prolactin can be detected in prolactin-secreting tumors (prolactinomas). Pituitary Magnetic Resonance Imaging (MRI) is essential in prolactinoma (macroprolactinomas or macroprolactinomas). The prolactin assay is performed by the electrochemiluminescent immunometric method.

Dynamic Hormonal Investigations

Stimulation Test with GnRH

In the past, the GnRH test was used in the differential diagnosis of hypogonadotropic hypogonadism, but it is currently no longer recommended because the assessment of testosterone and gonadotropins is sufficient to make the diagnosis. The functional test of stimulation with GnRH, measuring LH levels basal, 30 and 60 minutes after intravenous administration of 100 μg of GnRH, is used in the differential diagnosis of delayed puberty. An LH concentration >10 IU/L, 30 minutes after GnRH administration, allows the diagnosis of constitutional pubertal delay. Functional testing is helpful in patients preparing for pituitary surgery or with lesions in the hypothalamic-pituitary region.

Human Chorionic Gonadotropin Stimulation Test

Like LH, human chorionic gonadotropin (hCG) stimulates testosterone synthesis in Leydig cells. The hCG stimulation test is recommended to assess Leydig cell function, especially in young men with congenital anorchidism (an autosomal recessive syndrome with male karyotype and variable phenotype in the absence of testicular tissue) and to differentiate this clinical condition from cryptorchidism (failure of one or both testes to descend into the scrotal sac). The test involves the administration of 5000 IU of hCG intramuscularly and the dosage of testosterone at baseline and after 72 hours. Leydig cell function is considered normal if testosterone levels increase by 1.5–2 μg/L. This test is not recommended in adults because testosterone and gonadotropin assays are sufficient for diagnosis. In young men with congenital anorchidia, no increase in testosterone levels is observed after stimulation with hCG. Subjects with gonadotropin deficiency respond poorly to hCG administration due to a loss of sensitivity of Leydig cells.

Diagnosis and Therapy

Hypogonadism

The diagnosis of hypogonadism is essentially based on clinical signs and serum gonadotropin and testosterone levels measurement. The physical examination aim to identify morphological and structural alterations, with particular reference to skeletal proportions, height, weight, body fat distribution, and gynecomastia. The urogenital examination is of primary importance in evaluating the degree of development of secondary sexual characteristics.

Basal serum total testosterone is the recommended diagnostic test for hypogonadism. A second blood draw should confirm low total testosterone concentrations.

Numerous clinical and physiologic conditions may alter SHBG levels and, consequently, testosterone levels. Thus, evaluating this SHBG is recommended in certain patients (Table 25.30).

Free or bioavailable testosterone assay is recommended only when altered SHBG concentrations are suspected. Several variables such as intra-individual testosterone fluctuations, biological variability of the hormone's action on target organs, and limitations in currently available analytical techniques must be considered when assessing testosterone levels.

The combination of these factors has led to the lack of a standardized cut-off that can distinguish eugonadal and hypogonadal subjects; in this context, it is essential to integrate laboratory data with the clinical signs.

Gonadotropin evaluation provides helpful information about the origin of testicular dysfunction and helps in the differential diagnosis between primary and secondary hypogonadism.

The finding of low testosterone concentrations with elevated FSH and LH blood values suggests primary hypogonadism (Fig. 25.30). Karyotype analysis should be requested in the case of suspected Klinefelter syndrome, and further instrumental investigations help identify the cause. Elevated

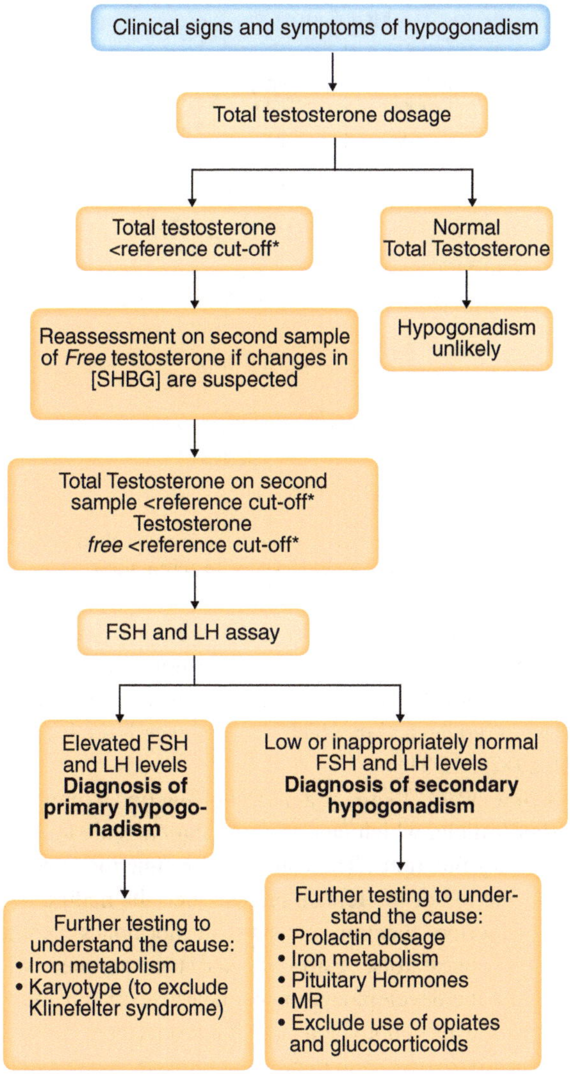

Fig. 25.30 Algorithm for the differential diagnosis of primary and secondary hypogonadism. *The values depend on the analytical methods used. (Copyright EDISES 2021. Reproduced with permission)

FSH values with normal LH and testosterone levels indicate impaired spermatogenesis with normal Leydig cell function. Low or inappropriately normal gonadotropin concentrations with below-normal testosterone values may suggest a secondary form (hypogonadotropic hypogonadism). In this case, further investigations, such as serum prolactin levels, transferrin saturation and pituitary hormones, MRI of the sella turcica; drug treatment with glucocorticoids are recommended. Excluding pathologies leading to secondary hypogonadism, idiopathic hypogonadotropic hypogonadism, or genetically determined hypogonadotropic hypogonadism should be considered, the latter recognizable through the study of the phenotype (Kallmann syndrome, X chromosome deletions, etc.).

The different clinical forms of hypogonadism require different therapeutic approaches; therefore, a correct and accurate differential diagnosis is essential. Moreover, it should be borne in mind that treatment is not always able to restore the functions compromised in hormonal decompensation. For subjects with primary hypogonadism, for example, hormone therapy cannot restore fertility because the seminiferous tubules are irreparably compromised. Furthermore, fertility in subjects with secondary hypogonadism is often restored by pulsatile GnRH therapy and gonadotropin therapy. When hypogonadism is diagnosed, testosterone replacement therapy can provide benefits in various psychophysical settings. The treatment can prevent or improve osteopenia, slowing down muscle mass loss, vasomotor instability, loss of libido, and erectile dysfunction. Testosterone replacement therapy is also contraindicated in individuals with prostate neoplasia or breast cancer. There are different pharmacological formulations of testosterone (intramuscular, oral, transdermal, gel formulation, subcutaneous implant).

Germ Cell Tumors of the Testis

Germ cell tumors represent a very heterogeneous group of diseases in terms of age of onset and clinical presentation, with a generally favorable prognosis. A distinction is made between gonadal (testicular or ovarian) and extra-gonadal (sacrococcygeal region, retroperitoneum, and mediastinum) tumors. In particular, testicular tumors, although very rare (1% of all male cancers), represent the most common malignancy in men aged 15–44 years in industrialized countries. Approximately 98% of testicular neoplasms are germ cell tumors; the remaining 2% includes Leydig cell tumors, Sertoli cell tumors, rhabdomyosarcoma, and lymphoma. The etiologic factors are not known. However, it is recognized that subjects with retained testes (cryptorchidism) and dysgenetic gonads (e.g., Turner syndrome, Klinefelter syndrome, hermaphroditism) are predisposed to develop germ cell tumors. Testicular germ cell tumors result from transforming germ cells, which arrests their maturation process. The pluripotency of germ cell tumors manifests itself in different histological types: undifferentiated germ cells (seminoma), embryonal carcinoma, embryonal-like differentiated cells (teratoma), extraembryonic differentiated phenotypes (choriocarcinoma). Most testicular tumors are clinically diagnosable at onset and present with a non-painful testicular mass, diffuse testicular pain, swelling, and induration. Abdominal ultrasound and CT scan of the chest and abdomen by the study of the retroperitoneum represent helpful diagnostic tests. Biomarkers, such as α-fetoprotein (αFP), β-chorionic gonadotropin (βhCG), and lactate dehydrogenase, are recommended for the diagnosis, staging of the disease, monitoring of therapy, and identification of recurrent forms of the tumor. αFP is a glycoprotein typically produced in the fetus and liver. Its circulating levels increase in different pathological conditions.

βhCG is a glycoprotein produced by syncytiotrophoblastic cells and consists of two subunits, α and β. The α subunit is common to other pituitary hormones (FSH, LH, TSH). The β subunit is specific and distinguishes it enzymatically and immunologically. βhCG is highly specific for testicular cancer and is produced specifically by choriocarcinoma cells. However, it can be detected in 5–10% of pure seminomas. Lactate dehydrogenase is a less specific marker whose increase reflects cell growth and proliferation rate. Increased plasma levels are observed in approximately 80% of advanced seminomas. These markers should be evaluated cautiously, and false positives due to physiological conditions, presence of other malignancies, or benign diseases should be considered. The persistence of elevated marker values at post-surgical follow-up, after they have been restored within the reference ranges, indicates occult metastases. The neoplasm, generally confined to the organ of origin, tends to disseminate via the lymphatic route, leading to lymph node metastases (especially in germinomatous forms), and via the blood, leading to mainly pulmonary metastases (especially in non-germinomatous forms).

Teratomas generally do not disseminate. Surgery represents the first therapeutic approach. In the case of testicular cancer, orchidectomy by inguinotomy is always indicated. The prognosis of germ cell tumors is generally favorable, even in situations with metastatic localization.

Ovary

Marcello Ciaccio and Luisa Agnello

Anatomy

The female gonads are two symmetrical oval-shaped organs. They have an intraperitoneal location between the external and internal iliac arteries. The size of the ovary of a woman of childbearing age is approximately 2.5–5 cm in length, 1.5–3 cm in width, and 0.6–1.5 cm in thickness, but may decrease during menopause. Histologically, a distinction can be made between a peripheral region, termed cortical, in which ovarian follicles at various stages of maturation and interstitial cells synthesing steroid hormones are found, and an inner region, termed medullary, which is richly innervated and vascularized and in which interstitial cells are less represented (Fig. 25.31).

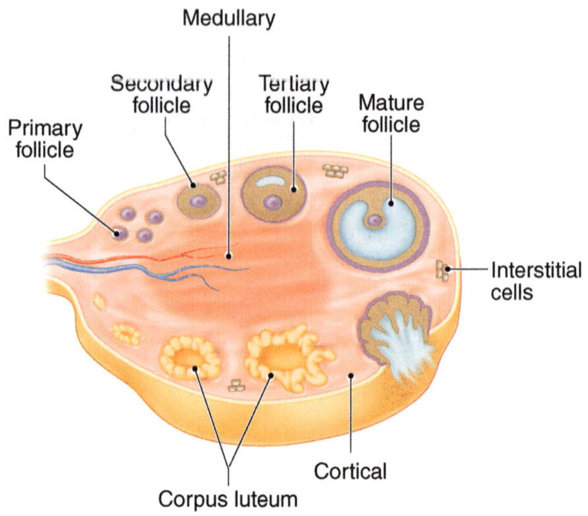

Fig. 25.31 Schematic representation of the ovary in childbearing age. (Copyright EDISES 2021. Reproduced with permission)

In the ovary, there is a morpho-functional unit called ovarian follicle, which can perform gametogenic and endocrine ovary functions. The gametogenic function consists in the production of the female gamete, i.e., the mature ovum. The endocrine function consists in hormones synthesis (mainly progesterone, estrogen, and testosterone), which determine the development of secondary sexual characteristics and the morphological and functional changes of the female reproductive system necessary for the ovum fertilization and implantation.

Physiology

Female Hypothalamic-Pituitary-Gonadal Axis

The ovary performs the gametogenic function during the ovarian cycle, which is temporally defined by the interval between one ovulation and the next and lasts on average 28 days. Given the difficulty in clinically identifying ovulation, we prefer to talk about the menstrual cycle, i.e., the period between one menstruation and the next. The menstrual cycle consists of a follicular phase, during which the follicle matures until ovulation, which occurs approximately halfway through the cycle, and a lutein phase, during which the follicle is transformed into the corpus luteum, which in turn is destined to degenerate in the absence of conception and implantation. In parallel with the follicle's transformations during the ovarian cycle, the uterus also undergoes

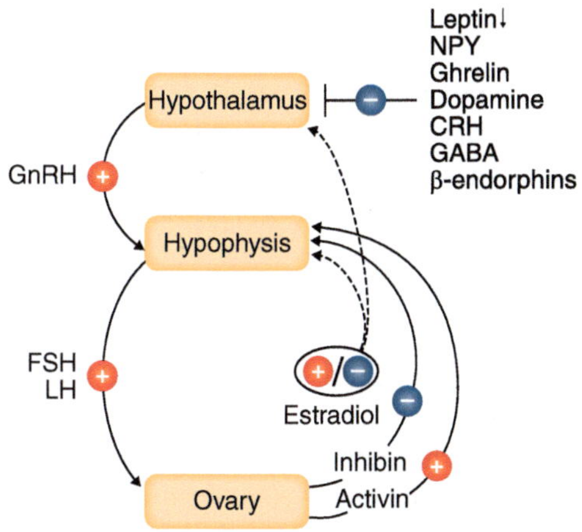

Fig. 25.32 Hypothalamic-pituitary regulation of ovarian hormone synthesis. (Copyright EDISES 2021. Reproduced with permission)

profound morphological changes that adapt to the fertilized ovum's implantation.

Overall, the ovary gametogenic function is regulated by a complex hormonal system consisting of the hypothalamic-pituitary axis and the related positive and negative feedback systems induced by ovarian hormones (Fig. 25.32).

Gonadotropins (FSH and LH) are secreted by pituitary gonadotropic cells in response to hypothalamic GnRH. During the follicular phase, there is a moderate FSH increase, which starts already in the late lutein phase of the previous cycle. FSH determines the selection of the dominant follicle, which manifests rapid growth capacity. In this phase, FSH, the predominant circulating gonadotropin, interacts with the receptors of the follicle's granulosa cells, inducing aromatase activity, followed by a massive production of estrogen. The progressive increase in circulating estradiol culminates with a peak around day 12 of the cycle, i.e., in the late follicular phase. Estradiol acts at the pituitary level by inducing the release of gonadotropins, especially LH, with a typical positive feedback mechanism. As soon as LH reaches discrete concentrations in the circulation, its inhibitory activity on the enzyme 17α-hydroxylase expressed in the interstitial cells of the theca of the follicle is manifested, thus, inhibiting the synthesis of androstenedione and consequently of estradiol, which in the ovulatory phase returns to low levels. The pre-ovulatory peak of circulating gonadotropins determines the maturation of the follicle leading to ovulation. At the beginning of the lutein phase, when the concentration of estradiol has decreased, the positive feedback exerted on the hypothalamus is also lost, and therefore the release of gonadotropins. Therefore, their concentrations return to pre-ovulatory levels. The lutein phase is characterized by a progressive increase in circulating levels of progesterone, whose synthesis is supported by theca lutein cells. This hormone acts in synergy with estradiol and inhibin A in maintaining low FSH and LH levels. The lack of gonadotropin stimulation will lead the corpus luteum to progressive degeneration and reduced estradiol synthesis in the late lutein phase. Consequently, as the negative feedback exerted by estradiol at the hypothalamic level on the release of GnRH is eliminated, the synthesis of FSH resumes, giving rise to a new cycle.

Synthesis and Functions of Steroid Hormones

The ovary produces estradiol, progesterone, and androstenedione, from which testosterone is derived.

The ovary's key feature of steroid biosynthesis is its compartmentalization into highly specialized cells. According to the bicellular steroidogenesis model, the interstitial cells of the theca of the follicle are sensitive to LH's action due to specific receptors. LH in these cells induces the synthesis of androgens, which, in the granulosa cells sensitive to the FSH action, will be converted into estrogens by aromatase activity. It should be noted that granulosa cells cannot synthesize androgens (Fig. 25.33).

The biosynthesis of steroid hormones occurs through the activity of enzymes belonging to the cytochrome P450 superfamily. The precursor common to all biosynthetic pathways is cholesterol, mainly derived from LDL lipoproteins and, to a negligible extent, from *ex novo* biosynthesis and mobilization of cytoplasmic lipid deposits.

The limiting step in steroid hormone synthesis is cholesterol transport within the mitochondrion. Once localized in the inner mitochondrial membrane, cholesterol is converted

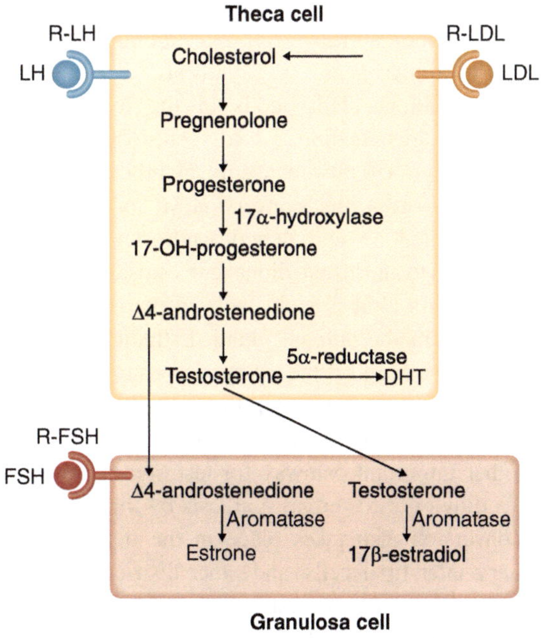

Fig. 25.33 Ovarian steroidogenesis. (Copyright EDISES 2021. Reproduced with permission)

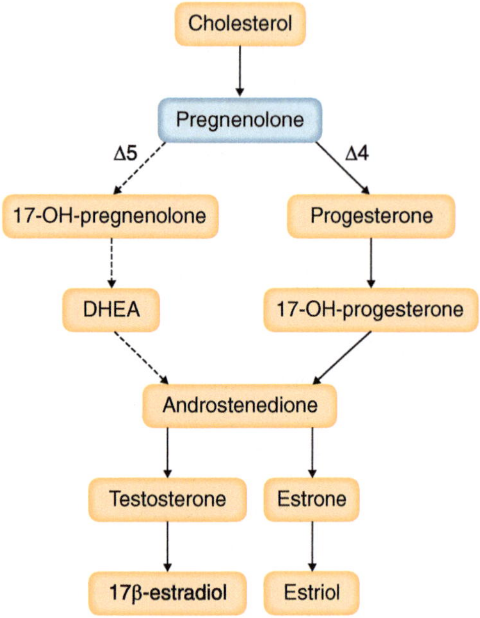

Fig. 25.34 Pathways of synthesis of androgens and estrogens. In the ovary, the Δ4 pathway is prevalent compared to Δ5, which occurs mainly in the adrenal gland. (Copyright EDISES 2021. Reproduced with permission)

to pregnenolone. This can follow two metabolic pathways, depending on cellular localization (Fig. 25.34). The Δ4 pathway represents the main pregnenolone metabolization pathway in the ovary and leads to the formation of progesterone. In interstitial cells, including those in the corpus luteum, pregnenolone is converted to progesterone and 17-OH-progesterone by a 17α-hydroxylase. The progressive removal of carbon atoms from progesterone (C21) leads to the C19-androgens synthesis, particularly Δ4-androstenedione. Hilar cells of the ovary can convert Δ4-androstenedione to testosterone. The Δ5-metabolization pathway occurs mainly in the adrenal gland and involves the conversion of pregnenolone to 17-OH-pregnenolone and dehydroandrosterone, which is then converted to androstenedione and testosterone. Estrogen synthesis occurs almost exclusively in the ovary and, to a lesser extent, in the adrenal gland. Estradiol, the primary estrogen synthesized by the ovary, and estrone are derived from Δ4-androstenedione and testosterone by aromatase expressed mainly in the granulosa cells.

Another important pathway for testosterone metabolism involves dihydrotestosterone synthesis by the 5α-reductase. This conversion also takes place in the ovary, especially medullary interstitial cells, and other districts, such as the skin in the pilosebaceous unit, which has the enzymatic kit for the synthesis and metabolism of androgens; in particular, it expresses the 5α-reductase, necessary for the conversion of testosterone into DHT (dihydrotestosterone, DHT).

The *ex novo* synthesis of ovarian hormones responds mainly to the control of LH. This hormone induces in the theca cells the expression of P450scc (human cholesterol side-chain cleavage enzyme), the enzyme converting cholesterol into pregnenolone, considered the limiting step of the entire biosynthetic pathway. Although not directly acting on the *ex novo* biosynthetic pathway, FSH has a critical regulatory role in converting androgens to estradiol by inducing the expression of aromatase in granulosa cells (Fig. 25.33).

Ovarian hormones are transported in the circulation by vector proteins such as SHBG (Sex Hormone Binding Globulin), which binds estrogens and androgens, corticosteroid-binding globulin, which mainly binds progesterone, and albumin, which is considered an aspecific transporter. The synthesis of these transport proteins occurs in the liver and is also modulated by insulin. Both estrogen and progesterone are catabolized in the liver.

The actions of estrogen and progesterone are mediated by interaction with nuclear receptors expressed in target cells, followed by modulation of gene expression. The biological effects of estrogens are ubiquitous. During puberty, they play a decisive role in developing the internal and external genitalia and in the induction of secondary sexual characteristics. In the fertile age, they are involved in fertilization, pregnancy, and childbirth. Estrogens also affect lipid metabolism by regulating the PCSK9 expression in the liver. However, many aspects of the action of estrogen (endogenous or exogenous) on lipoprotein metabolism remain unclear.

Synthesis and Function of Protein Hormones

In addition to steroid hormones, the ovary synthesizes peptides with regulatory action on ovarian function. Inhibins A and B, Anti-Müllerian Hormone (AMH), and activins belong to this class.

Inhibins were first isolated in the late 1980s from the follicular fluid as peptides capable of inhibiting pituitary FSH secretion. Two isoforms were identified, inhibin A and B, belonging to the TGF-β superfamily. Inhibins regulate ovarian function locally, through paracrine mechanisms, or by inhibiting pituitary FSH release. Inhibins act as antagonists of activins, constituting a structural homolog. The active form of the inhibins is a heterodimer consisting of an exclusive α subunit and a β subunit, shared with the activins, of which there are two isoforms, A and B. The combination of these subunits gives rise to inhibin A (α/βA) and inhibin B (α/βB). Given the structural homology with activins, inhibins bind to the pituitary receptor and, through the betaglycans, considere co-receptors, on the cell surface, inhibit the access of activins to the receptor, blocking the intracellular signal transduction pathway.

Inhibin B is mainly produced by the granulosa cells of mature follicles, while inhibin A is expressed by granulosa and theca cells. Circulating levels of the two molecules vary physiologically as a function of the ovarian cycle: inhibin B increases in response to FSH, increasing in the follicular and ovulatory phases, while inhibin A peaks in the lutein phase. Activins act oppositely to inhibins by stimulating the pituitary release of FSH and inducing the action of FSH-dependent aromatase.

AMH is a glycoprotein also belonging to the TGF-β superfamily. Its function has long been known in relation to its ability to direct sexual differentiation during the early stages of embryonic development. AMH is synthesized primarily by antral follicles. It has been proposed that AMH acts by inhibiting follicle sensitivity to FSH and may play a role in the process of dominant follicle selection. In the early stages of follicle development, AMH synthesis is high, inhibiting the recruitment of non-dominant follicles by paracrine mechanisms. As soon as the antral follicle increases in size, AMH synthesis rapidly decreases, making the follicle susceptible to the action of FSH and ensuring its subsequent development.

Given their biological actions, Inhibin B and AMH have now been considered ovarian reserve markers, together with FSH and estradiol assays.

Endocrine Disorders of the Ovary

Hirsutism and Virilization

Hirsutism is defined as excessive hair growth with male distribution in women. The leading causes of hirsutism are described in Table 25.31. Often, hirsutism is an expression of hyperandrogenism, of which it is the most frequent clinical sign. Sometimes, however, it can also be idiopathic and not associated with increased circulating androgens. On the other hand, it should also be mentioned that the absence of hirsutism does not exclude other conditions associated with androgen excesses, such as acne, alopecia, anovulatory cycles, or dysmenorrhea. The marked increase in circulating androgens may be accompanied by virilization, a rare condition defined by the progressive regression of female sexual characteristics and the appearance of male characteristics. Virilization manifests by clitoral hypertrophy, breast atrophy, increased muscle mass, changes in voice tone, and amenorrhea.

Testosterone and dihydrotestosterone bind to the androgen receptor. DHEA, DHEA-S, and androstenedione are prohormones. Given the peripheral synthesis of DHT, it should be noted that the circulating concentration of androgens may not reflect the actual concentration of active hormone in the pilosebaceous unit. Moreover, the local effect of androgens depends on specific receptors, whose expression is not yet defined either in terms of localization in the pilosebaceous unit or body distribution. Hirsutism, therefore, results from the complex interactions between the circulating levels of androgens, their local bioavailability, and the sensitivity of the target cells. This explains the poor correlation between circulating levels of androgens and hirsutism.

Table 25.31 Main causes of primary amenorrhea

Without sexual maturation
Ovarian causes (hypergonadotropic hypogonadism)
- Congenital ovarian dysgenesis
- Turner syndrome
- 17α-hydroxylase deficiency
- Mutations of ovarian receptors for gonadotropins
Hypothalamus/pituitary causes (hypogonadotropic hypogonadism)
- Functional hypothalamic amenorrhea
- Pituitary adenomas
- Head radiation therapy
- Head trauma
- Neurosurgery
- Inflammatory/infiltrative diseases
- Congenital GnRH deficiency (Kallman syndrome)
Constitutional delayed puberty
With sexual maturation
Congenital changes in the genital development
- Dysgenesis of the Müllerian ducts
- Androgen insensitivity syndrome
- Imperforate hymen
- Vaginal septum
- Stenosis of the cervix

Amenorrhea

Amenorrhea is defined as the absence of spontaneous menstrual cycles. It can be distinguished into primary amenorrhea when no menstrual cycles occur by age 16, and secondary or oligorrhea, when no menstrual cycles occur for at least 3 months in women with previous regular menstruation.

Primary Amenorrhea

Primary amenorrhea is a rare condition in about 1% of the female population and can occur in the absence or presence of sexual maturation. In the first case, the causes of amenorrhea should be investigated as early as 13 years. Primary amenorrhea can result from an alteration in the ovary, the hypothalamus, or the pituitary gland (Table 25.32).

Primary or hypergonadotropic hypogonadism is characterized by high circulating levels of gonadotropins. In ovarian failure, the negative feedback control on the hypothalamus is lost, resulting in an uncontrolled release of FSH and LH. Hypergonadotropic hypogonadism can be caused by congenital ovarian dysgenesis, Turner syndrome, deficiency of enzymatic activities involved in steroidogeneses, such as 17α-hydroxylase deficiency, or mutations in FSH or LH receptors causing ovarian resistance to gonadotropins.

Table 25.32 Main causes of secondary amenorrhea

Physiological causes
- Pregnancy - Breastfeeding - Menopause
Hypergonadotropic hypogonadism (premature ovarian failure)
Genetic disorders - Turner syndrome - Trisomy X - X Fragile syndrome - Mutations of estrogen receptors Autoimmune Iatrogenic Idiopathic
Hypogonadotropic hypogonadism
Organic causes - Pituitary adenomas - Head radiation therapy - Head trauma - Neurosurgery - Inflammatory/infiltrative diseases Functional hypothalamic amenorrhea Polycystic ovary syndrome (PCOS) Ovarian tumors
Anatomical abnormalities of the uterus and outflow tract
Asherman's syndrome Pathologies of the uterus

Secondary hypogonadism or hypogonadotropic hypogonadism is characterized by a deficit in the synthesis and release of gonadotropins in the pituitary gland. It can occur secondary to organic causes such as pituitary adenomas, radiation therapy, infiltrative diseases such as hemochromatosis, and sarcoidosis, or functional causes such as eating disorders and physical stress. After excluding organic causes by imaging, the differential diagnosis with constitutional growth retardation and delayed puberty should be considered. However, this distinction is complicated, and often, the most rational approach is based on a spontaneous clinical course. There are also congenital forms of hypogonadotropic hypogonadism, such as isolated GnRH deficiency. This condition can also be associated with other disorders characteristic of the underlying pituitary dysfunction, such as growth defects, diabetes insipidus, and galactorrhea. Usually, congenital hypogonadotropic hypogonadism is manifested by primary amenorrhea, which is the cause in about 20% of cases. Functional hypothalamic amenorrhea is due to a defect in GnRH synthesis and consequent release of FSH and LH.

Both forms of hypogonadism described above are associated with primary amenorrhea accompanied by the absence of sexual maturation. In the presence of sexual maturation, abnormalities in the development of the genital tract with alterations of the uterus and outflow tract should also be considered by gynecologic examination and pelvic ultrasonography. Congenital causes include agenesis of Müller's ducts (46,XX), Morris syndrome, and androgen insensitivity syndrome (46,XY). The treatment of primary amenorrhea is etiologic; therefore, the correct diagnostic framing is necessary before implementing any therapeutic intervention. The therapeutic approach may include surgical procedures, lifestyle changes, and hormone replacement therapy.

Secondary Amenorrhea

The most common cause of secondary amenorrhea is pregnancy. Other physiological causes of secondary amenorrhea are breastfeeding and menopause. Excluding the physiological causes, secondary amenorrhea can occur in hypergonadotropic hypogonadism, as in the case of premature ovarian failure, or chronic anovulation, which, by chronically altering the physiological pulsatility of ovarian hormone secretion, can lead to menstrual irregularity, up to amenorrhea. In the latter case, one can observe both low estrogen levels (hypogonadotropic hypogonadism), as occurs in functional hypothalamic amenorrhea or response to organic hypothalamic causes (brain tumors, head trauma, etc.), and normal/increased estrogen levels, as in the case of ovarian tumors or PolyCystic Ovary Syndrome (PCOS). Finally, secondary amenorrhea can occur due to anatomical defects of the efflux tract, as in Asherman's syndrome. The most frequent pathological causes of secondary amenorrhea are PCOS and organic and functional hypothalamic amenorrhea (Table 25.33).

In approximately 40% of cases, secondary amenorrhea is associated with PCOS. The condition's prevalence is not defined, mainly due to the lack of agreed diagnostic criteria.

Table 25.33 Reference values in women in relation to the menstrual cycle and post-menopause (measured by ECLIA, *electrochemiluminescence immunoassay*; *for the determination of free testosterone the reference values are measured by ELISA, *enzyme-linked immunosorbent assay*)

	Follicular phase	Ovulatory phase	Luteal phase	Post-menopause
LH (IU/L)	2.4–12.6	14–95.6	1–11.4	7.7–58.5
FSH (IU/L)	<12.5	4.7–21.5	1.7–7.7	25.8–134.8
Estradiol (ng/L)	12.5–166	85.8–498	43.8–211	5–54.7
Progesterone (μg/L)	0.2–1.5	0.8–3	1.7–27	0.1–0.8
Total Testosterone (μg/L)	0.084–0.0481			
Free Testosterone* (ng/L)	0.01–7.01			
DHEA-S (μg/dL)	65.1–368			
Prolactin (μg/L)	4.79–23.3			

FSH follicle-stimulating hormone, *LH* luteinizing hormone, *DHEA-S* dehydroepiandrosterone sulfate

However, it is the most frequent endocrinopathy in women of childbearing age. The difficulties in univocally defining PCOS stem from its etiology, which is still obscure in many respects, and the complex nature of this endocrinopathy. A shared opinion among experts in the field is that the term "polycystic ovary syndrome" is misleading for the complexity of the hormonal aspects that characterize this condition. There is a consensus that the diagnosis of PCOS can be made based on at least two conditions: chronic anovularity, hyperandrogenemia (clinical or biochemical), and polycystic ovary morphology.

From the combination of the different classification systems of PCOS, the European Society of Endocrinology has recently proposed a helpful approach to the disease, according to which it is possible to distinguish four different phenotypes:

- Hyperandrogenism (clinical or biochemical) and chronic anovularity
- Hyperandrogenemia and polycystic morphology of the ovary with ovulatory cycles
- Chronic anovularity and polycystic morphology of the ovary without hyperandrogenism
- Hyperandrogenism, chronic anovularity, and polycystic morphology of the ovary

Identifying specific phenotypes in women with PCOS is based on the detection of metabolic abnormalities, persisting even after menopause. Frequently, women with PCOS are overweight or obese and have different degrees of central and peripheral insulin resistance, dyslipidemia, reduced carbohydrate tolerance, and type 2 diabetes. In addition, a higher incidence of endothelial dysfunction, assessed by flow-mediated dilation of the brachial artery and increased intimal thickness, considered a morphological sign of atherosclerosis, has been documented in women with PCOS. However, it is still unclear whether these findings translate into increased cardiovascular morbidity and mortality since appropriate long-term prospective studies are not available. Both androgen excess and ovarian dysfunction are associated with the metabolic profile in patients with PCOS, especially if obese. PCOS acts synergistically together with obesity in altering insulin sensitivity. The metabolic profile of PCOS also includes atherogenic dyslipidemia, present in 70% of cases and characterized by hypertriglyceridemia and low HDL cholesterol. However, the impact of dyslipidemia on the different phenotypes of PCOS is still unclear. Several mechanisms underlying this association have been proposed involving both androgen receptors and lipoprotein lipase activity.

Less frequent causes of secondary amenorrhea are functional hypogonadotropic hypogonadism associated with eating disorders or pronounced physical stress, Premature Ovarian Failure (POF), hyperprolactinemia, and organic causes such as brain tumors.

Functional hypothalamic amenorrhea is frequent in young women and can be attributed to a defect in GnRH synthesis, resulting in a decrease in pituitary gonadotropins, although the pathogenetic mechanism remains obscure in many respects. Three primary forms of functional hypothalamic amenorrhea have been described, associated with stress, weight loss, and intense exercise, even in association. Maintaining ovarian function and reproductive capacity is related to preserving an adipose tissue share of about 20% of body mass. The influences of serum ghrelin and leptin levels on hypothalamic GnRH synthesis evidence this. Sudden weight loss, anorexia nervosa, or particularly intense sports training programs not accompanied by adequate nutritional intake may cause secondary amenorrhea. The estrogen deficit that characterizes this condition also compromises bone health. Indeed, several scientific societies have recommended the introduction of bone densitometry in the diagnostic workup of these patients.

POF is defined as a primary ovarian failure occurring before 40 years. It accounts for 10% of cases of secondary amenorrhea. The biochemical picture is hypergonadotropic hypogonadism, which is physiologically observed during menopause. The etiology has not yet been clarified. Some forms are genetic such as X chromosome rearrangements associated with ovarian dysgenesis, others autoimmune, or secondary to pelvic radiation therapy or chemotherapy. The natural history is unclear, although the recurrence of menstrual cycles and restoration of reproductive activity has been documented in some women.

Hyperprolactinemia is a frequent cause of female infertility and is often associated with galactorrhea. Hyperprolactinemia can be iatrogenic, associated with prolactinoma, PCOS, and primary hypothyroidism.

As in the case of primary amenorrhea, the therapeutic approach for secondary amenorrhea is etiologic and depends on the woman's age. Treatment can be either surgical or medical, depending on the etiologic agent.

Laboratory Investigations

Basal Measurement

Androgen

In decreasing order of serum concentration, the androgens present in a woman of childbearing age are DHEA-S (95% of adrenal origin), DHEA, androstenedione, testosterone, and DHT. The medical-scientific community has long debated which androgens should be measured to evaluate hyperandrogenism and the analytical methodologies employed. A reasoned laboratory approach is to assay total testosterone before initiating any medical therapy that may interfere with circulating levels. The total testosterone assay should be accompanied by free testosterone because the latter has greater diagnostic sensitivity for hyperandrogenism. However, there are critical analytical considerations that may limit the reliability of this test. Methodologically, free testosterone can commonly be assayed by ELISA as long as the inter-assay imprecision assessed internally within the laboratory remains below 10% and appropriate reference intervals are established in non-hyperandrogenic women. Alternatively, more accurate LC-MS/MS-based assays are now considered the gold standard for steroid hormones measurement, although their use is restricted almost exclusively to clinical research in specialized laboratories. Compared to the free testosterone assay, some authors have proposed an approach based on SHBG with total testosterone. The ratio between total testosterone and SHBG indicates the amount of free hormone indirectly and is, therefore, a valuable tool to identify possible states of hyperandrogenism, especially when accurate dosages of free testosterone are not available. Epidemiological studies have shown that high levels of total testosterone and low levels of SHBG have a high positive predictive value for polycystic ovary syndrome, a condition also characterized by hyperandrogenism. In addition, reducing circulating SHBG can be considered a marker of insulin resistance to predict the onset of metabolic syndrome and gestational diabetes in women with PCOS. The significance of Δ4-androstenedione and DHEA-S dosage in hirsutism patients is still poorly understood. Δ4-androstenedione may be considered a pro-androgen; therefore, its circulating concentration may not reflect the biological actions of the active hormone.

Moreover, plasma levels fluctuate with the ovarian cycle, higher in the lutein phase. For these reasons, its diagnostic value in the diagnosis of hyperandrogenemia may be limited. On the other hand, Δ4-androstenedione binds with lower affinity to SHBG than testosterone and, therefore, may have more significant androgenic potential. Also, for DHEA-S, there is no clear evidence of the usefulness of this assay in identifying hyperandrogenism, except for the suspicion of androgen-secreting adrenal tumors, since circulating DHEA-S is predominantly of adrenal origin. In general, the finding of elevated values of at least one androgen indicates a disorder associated with androgen excess. The reference values of the main androgens in women are described in Table 25.34.

Gonadotropin

Gonadotropin is essential to differentiate the various forms of hypogonadism; i.e., elevated FSH and LH levels (hypergonadotropic hypogonadism, primary ovarian insufficiency) or reduced levels of FSH, LH, and estradiol (hypogonadotropic hypogonadism, hypothalamic-pituitary alteration). Although not diagnostic, an LH/FSH ratio greater than 2 is frequently found in PCOS. It should be noted that the reference values vary considerably depending on the woman's age and, during productive life, and menstrual cycle phases. Physiologically, during menopause, there is a marked increase in FSH (Table 25.34). However, gonadotropin in amenorrhea should not be isolated due to their inherent variability but confirmed in subsequent measurements.

Table 25.34 Main diagnostic laboratory test for secondary amenorrhea

First-level investigations
- β-hCG
- 17β-estradiol
- Prolactin
- FSH, LH
- Pelvic ultrasound
- MAP test
Second-level investigations
- Test with estrogen progestin
- GnRH test
- Pituitary magnetic resonance imaging
Third-level investigations
- Karyotype
- Total testosterone, SHBG
- Clomiphene test

FSH follicle-stimulating hormone, *GnRH* gonadotropin releasing hormone, LH luteinizing hormone, *MAP* medroxyprogesterone acetate, *SHBG* Sex hormone–binding globulins

Estrogen

Estradiol represents the most common estrogen measured in clinical practice. It provides helpful information on hypogonadism, menstrual cycle alterations, and ovarian tumors. Also, in this case, the reference values vary according to the menstrual cycle and the woman's age. In women of childbearing age, estradiol levels show a pre-ovulatory peak around day 12 and anticipate the pre-ovulatory peak of LH. Throughout the rest of the cycle, estradiol levels remain low. The use of serum estrone assay has limited clinical significance.

Progesterone

Progesterone increases physiologically following ovulation. The serum dosage of progesterone in the lutein phase (21st day of the cycle) is used to confirm ovulation. In many cases, it is possible to have regular but anovulatory menstrual cycles, as in the case of PCOS, in which the finding of anovulation is one of the diagnostic criteria.

17-OH-progesterone, a progesterone derivative, does not have a precise biological role but is of particular diagnostic importance because its circulating concentration, both in basal conditions and after stimulation by synthetic ACTH, increases in cases of congenital adrenal hyperplasia and, in particular, 21α-hydroxylase deficiency. The 17-OH-progesterone assay is used in the etiological diagnosis of hirsutism to identify a 21α-hydroxylase deficiency.

Dynamic Investigations

GnRH Testing

The GnRH test is used in the differential diagnosis of amenorrhea to identify a possible pituitary defect. It is performed by intravenous administration of GnRH and serum LH and FSH measurement at basal and after 15, 30, 60, 90, and 120 minutes. There is no consensus on the interpretation of the test. Generally, a peak of at least three times baseline for LH and at least two times baseline for FSH is considered normal. Hypothalamic forms generally respond to the test, while in pituitary forms, the response is absent or reduced. In case of an absent or reduced response, it is necessary to repeat the test after a few days because pituitary cells not subjected to endogenous stimulation by GnRH may not be responsive to acute administration when the test is performed. This test has sometimes been used in differential diagnoses between constitutional growth retardation with delayed puberty and hypogonadotropic hypogonadism; however, its diagnostic performance in this context is poor. Overall, the clinical use of this test is limited because there is no agreement on the interpretation of the response to GnRH stimulation and its diagnostic cut-off.

Clomiphene Test

Although rarely used in clinical practice, this test allows assessing the degree of impairment of hypothalamic-pituitary axis function in patients with anovulatory cycles or oligorrhea. Clomiphene competes with estrogens to bind their receptor. When clomiphene binds the receptor, it blocks the negative feedback induced physiologically by estrogen on GnRH and gonadotropin synthesis. The test is performed by administering clomiphene orally for 5 days starting on day 5 of the menstrual cycle and assaying FSH and LH on days 5, 7, and 10 of the cycle; progesterone is assayed on day 21 to document ovulation. In patients with normal hypothalamic-pituitary function, an increase of FSH and LH greater than 50% of baseline and restoration of ovulation is observed, pointing toward ovarian pathologies. Conversely, an absent response to clomiphene (no increase in FSH and LH after stimulation) indicates hypothalamic-pituitary deficiency. A hypothalamic defect may be suspected if the response to clomiphene is absent, but the GnRH test is positive.

Diagnosis and Therapy

Hirsutism

The diagnosis of hirsutism is based on a clinical *scoring* system aimed at ascertaining the type and distribution of hair growth in body areas susceptible to the action of androgens. Several *scoring* systems have been developed, but today the *gold standard* remains the Ferriman-Gallwey scale (Fig. 25.35). This method involves the assignment of a score from 0 to 4 for 9 of the body areas susceptible to the action of androgens; a score above 8 depicts androgen-mediated hirsutism rather than hypertrichosis. However, this value depends significantly on the patient's ethnicity. It has been shown that the cutoff in Caucasian women of the Mediterranean area should be ≥10–11, while in the Asian race ≥6. A score up to 15 represents mild hirsutism, from 16 to 25 moderate, and above 25 severe. Although the Ferriman-Gallwey scale is an easy-to-use tool, it should be noted that the evaluation of hirsutism is subject to a high degree of operator-dependent variability. The history, essential in evaluating patients with hirsutism, should be aimed to investigate the age of onset of hair, symptoms or signs associated with hyperandrogenism (e.g., acne, dysmenorrhea, alopecia, infertility), and the speed of onset of hirsutism. Rapid onset may point toward an androgen-secreting neoplasm. In this case, signs of virilization are also often present.

Other endocrine disorders associated with hirsutism are polycystic ovary syndrome, Non-Classic Congenital Adrenal Hyperplasia (NCCAH), Cushing's syndrome, hypothyroidism, and hyperprolactinemia.

The diagnostic approach to pre-menopausal hirsutism is described in Fig. 25.36.

In postmenopausal women with hirsutism, the severity, the onset's rapidity, and the total testosterone dosage should be evaluated first. If the total testosterone level is elevated, endocrine conditions associated with hyperandrogenemia should be considered. If the hirsutism is mild and stable over time with normal testosterone levels, no further diagnostic investigations are required, and the most likely diagnosis is idiopathic hirsutism. In the case of normal total testosterone values but with moderate/severe hirsutism or clinical evidence of endocrine disorders associated with hyperandrogenism, it is necessary to proceed with free testosterone measurement.

Once the presence of hirsutism has been ascertained based on the clinical *score* and medical history, it is necessary to define its etiology. Hirsutism with ovarian polycystic morphology and hyperandrogenism is indicative of PCOS, which accounts for 70% of cases. If circulating androgen levels are normal, ovarian morphology and cycles are normal, hirsutism is idiopathic.

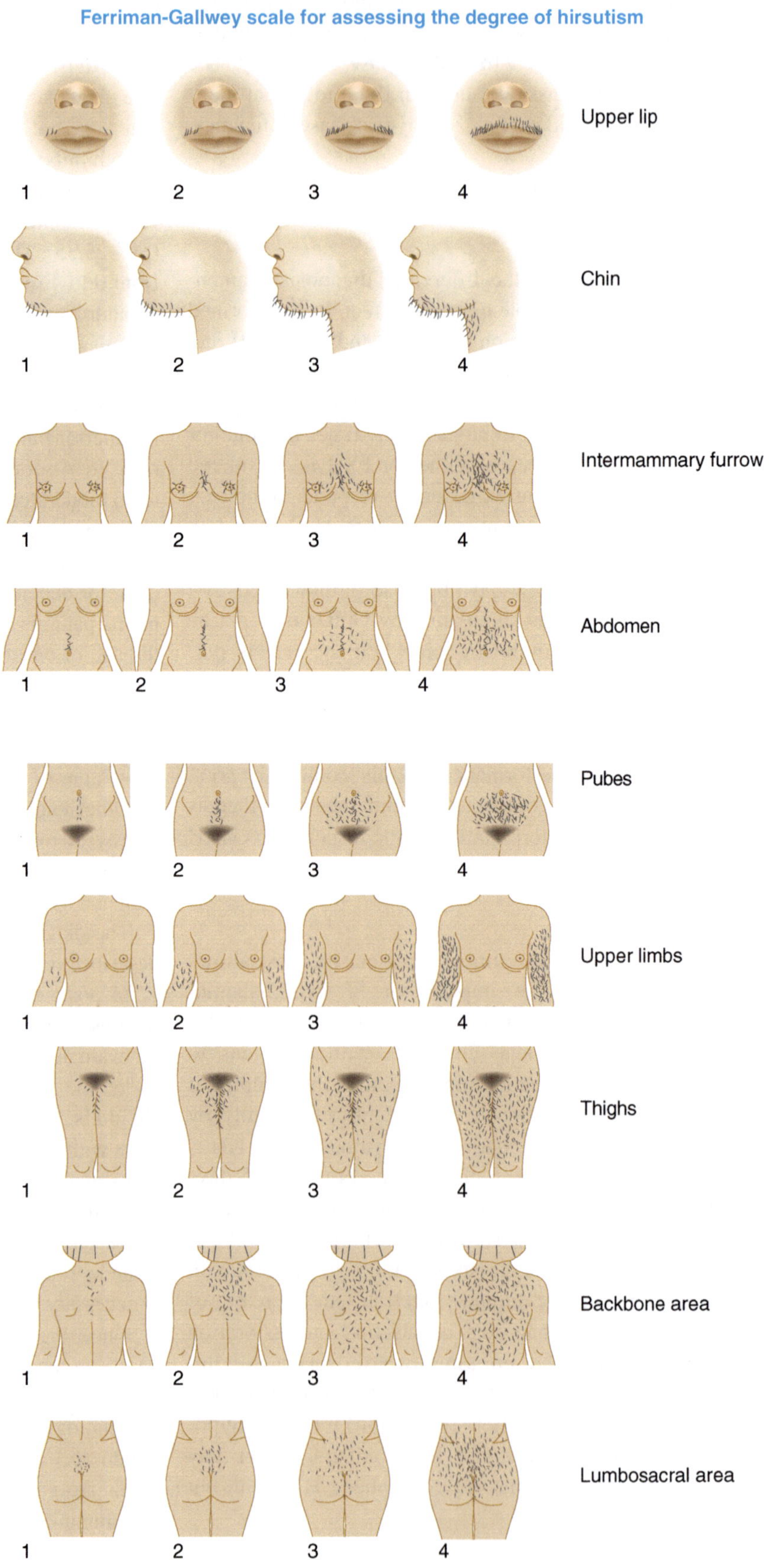

Fig. 25.35 Ferriman-Gallwey score for evaluating hirsutism. (Copyright EDISES 2021. Reproduced with permission)

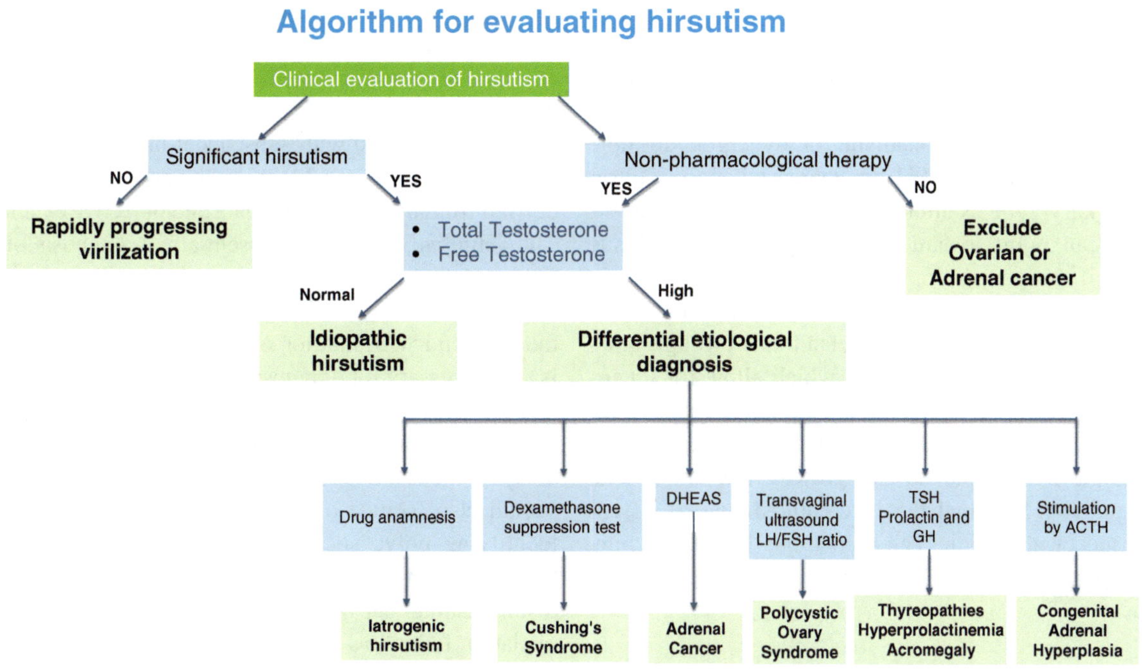

Fig. 25.36 Diagnostic algorithm of hirsutism. (Copyright EDISES 2021. Reproduced with permission)

In patients with hirsutism, FSH, LH, 17β-estradiol, and 17-OH-progesterone should be measured in addition to androgens. The prolactin and TSH assay excludes secondary forms of hyperprolactinemia and thyroid diseases. The most frequent cause of androgen excess is PCOS, which is associated with increased cardio-metabolic risk, especially if accompanied by obesity. In this regard, therefore, it is necessary to investigate the presence, even in the past, of oligorrhea or amenorrhea and to perform the dosage of progesterone in the lutein phase to document ovulation, since a significant proportion of women with hirsutism have anovulatory cycles. In addition, the diagnostic approach also includes transvaginal ultrasonography to assess any polycystic morphology of the ovary. An evaluation of the cardio-metabolic profile should further investigate PCOS. In particular, it is necessary to assess carbohydrate tolerance by Oral *Glucose Tolerance Test* (OGTT), BMI and abdominal circumference, complete lipid profile, and blood pressure.

As mentioned above, the diagnostic approach to hirsutism should also include the exclusion of less frequent causes. In this regard, it is necessary to exclude androgen-secreting tumor of ovarian or adrenal origin. A sudden increase in total testosterone can occur in these neoplasms; if a parallel increase in DHEA-S occurs, it is plausible that the neoplasm is of adrenal origin.

NCCHA due to 21-hydroxylase deficiency is frequent in some ethnic groups where screening is recommended. However, the introduction of the routine 17-OH-progesterone assay, basal or after stimulation with synthetic ACTH, a screening test for congenital adrenal hyperplasia, should be evaluated, taking into account the prevalence of the condition in the population and the availability and cost of the test. The abrupt onset and rapid progression of hirsutism indicate androgen-secreting ovarian tumors. Finally, rare causes of hirsutism such as acromegaly, hyperprolactinemia, pregnancy-related hyperandrogenism, and Cushing's syndrome should be excluded. Hirsutism can also be iatrogenic, for example, after administration of oral contraceptives containing androgenic progestins. Hirsutism is a clinical sign and does not necessarily require medical treatment, especially in mild or moderate forms, where cosmetic remedies can address it. More clinically relevant forms can be treated with estro-progestin therapy, which is considered the endocrinological approach of the first choice, or with antiandrogens in severe cases, or with insulin sensitizers. *Follow-up* of these patients involves clinical evaluation of treatment efficacy. It is not clear whether serum androgen dosage can help monitor therapy and, in particular, whether a possible decrease in circulating levels can predict a clinical improvement that, in any case, requires at least 4–6 months to be evident. It should be noted that during estro-progestin therapy, it is necessary to measure, in addition to total testosterone, SHBG, whose synthesis is induced by these drugs. The dosage of androgens is not indicated during treatment with antiandrogens.

Primary Amenorrhea

Hypogonadism diagnosis is based on a careful history and physical examination to evaluate sexual development. Once hypogonadism has been hypothesized, the differential diagnosis between the various forms is based on instrumental and laboratory investigations (Fig. 25.37). Without normal development of secondary sexual characteristics, the FSH and LH assay allows for distinguishing between hypo- and hypergonadotropic hypogonadism. In hypergonadotropic hypogonadism, FSH and LH are increased; pelvic ultrasound and karyotype examination are also indicated in this case. We can point toward a hypophyseal/hypothalamic pathology if the gonadotropins are normal or reduced. In these cases, it is necessary to deepen the overall function of the hypothalamic-pituitary-target gland axis through laboratory tests and instrumental tests, which allow for identifying any organic pathology. Once the organic causes have been excluded, it is necessary to consider functional hypogonadotropic hypogonadism and growth retardation with delayed puberty, for which the differential diagnosis, essentially based on the clinic, can sometimes be challenging. For this purpose, the GnRH stimulus test is of limited usefulness.

Secondary Amenorrhea

The most frequent cause of secondary amenorrhea is pregnancy. Therefore, the first test is the β-hCG assay (Fig. 25.38).

The diagnostic approach to secondary amenorrhea starts with an accurate personal history aimed at investigating eating habits, physical activity, any psychophysical stress, the age of menarche, the presence of galactorrhea, any signs of hypoestrogenism such as hot flashes and vaginal dryness, any trauma or surgical procedures or radiation therapies of the head and pelvis, signs of hyperandrogenemia. The decision to proceed with subsequent instrumental and laboratory investigations is based on the preliminary orientation derived from the anamnestic data and objective examination. The definitive diagnosis is essentially a diagnosis of exclusion. First-level biochemical investigations include the assay of FSH, LH, 17β-estradiol, prolactin, and TSH. In addition to the biochemical evaluation of primary amenorrhea, prolactin is also necessary for diagnosing secondary amenorrhea since it is a frequent cause (for a detailed discussion on the causes of hyperprolactinemia see section "Hypophysis"). The TSH assay excludes hypothyroidism. First-level investigations should also include a pelvic ultrasound, which allows for identifying polycystic morphology of the ovary, and the MAP test (test with progestin administration), which allows distinguishing between a progesterone deficiency due to ovulation failure, as frequently found in PCOS, and uterine abnormalities. The MAP test is performed by oral medroxyprogesterone to the patient for 5 consecutive days; the test result is evaluated 2–7 days after the end of administration by the presence/absence of bleeding. Second-level tests include the estroprogestin test, GnRH test, and pituitary MRI. The estroprogestin (oral contraceptive) test is per-

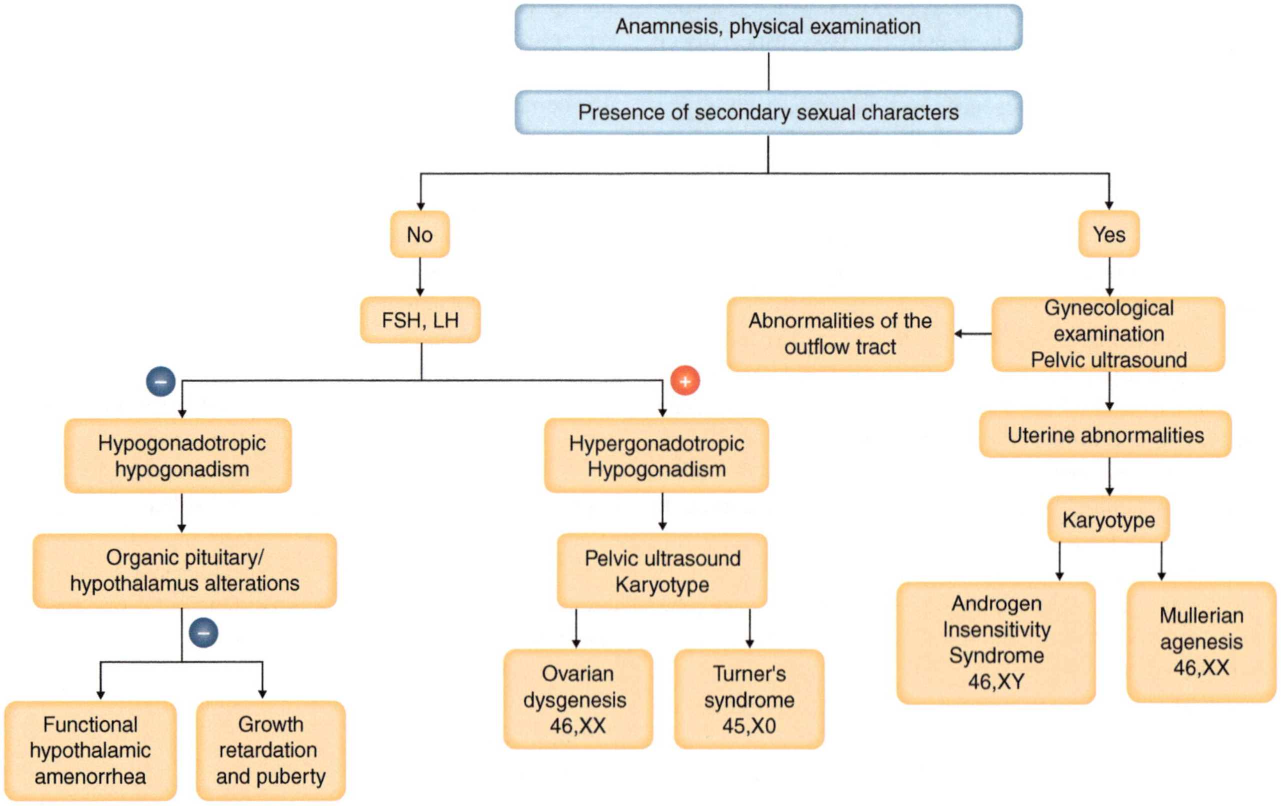

Fig. 25.37 Diagnostic algorithm of primary amenorrhea. (Copyright EDISES 2021. Reproduced with permission)

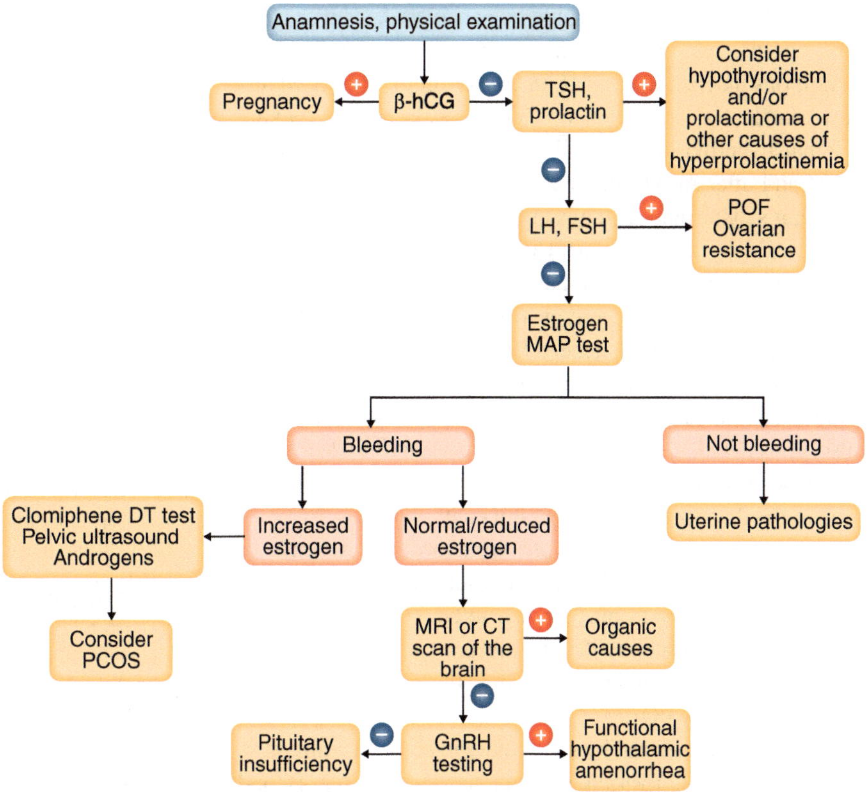

Fig. 25.38 Diagnostic algorithm of secondary amenorrhea. POF: premature ovarian failure. (Copyright EDISES 2021. Reproduced with permission)

formed by administering estroprogestin for 20 days; also, in this case, the test result is evaluated 7 days after stopping the treatment by the presence/absence of bleeding and allows to exclude uterine pathologies. The GnRH test shows pituitary insufficiency if the response is limited or absent. In the case of a normal response (increased FSH and LH), serious hypothalamic pathologies can be excluded. If the diagnostic picture remains uncertain, we should proceed with third-level investigations, i.e., androgen assay, karyotype, and clomiphene citrate test (Table 25.35).

In the case of suspected PCOS, first-line laboratory tests are essentially aimed at evaluating the presence of hyperandrogenemia, starting with the total testosterone assay. Free testosterone can be assayed directly or indirectly by the ratio of total testosterone to SHBG. There is no consensus on the validity of Δ4-androstenedione assay in diagnosing PCOS. However, recent studies have shown greater sensitivity and specificity of this steroid than testosterone, both assayed by LC-MS/MS, in diagnosing hyperandrogenism in women with PCOS. Similarly, DHEA-S is not routinely required in clinical practice, except when androgen-secreting adrenal cancer is suspected. In the case of elevated testosterone concentrations, causes other than PCOS should also be considered. If testosterone exceeds twice the upper reference limit, an androgen-secreting tumor should be considered, especially if accompanied by signs of virilization. Increased DHEA-S, as already mentioned, may point to an adrenal tumor; if DHEA-S

Table 25.35 Causes of hirsutism

Hyperandrogenism of gonadal origin - Ovarian hyperandrogenism - Polycystic Ovary Syndrome/Ovarian functional hyperandrogenism - Alterations of ovarian steroidogenesis - Insulin Resistance Syndrome (severe forms) - Ovarian Neoplasms
Adrenal hyperandrogenism - Premature adrenarchy - Adrenal functional hyperandrogenism - Congenital Adrenal Hyperplasia - Alterations in the action or metabolism of cortisol - Adrenal Neoplasms
Other endocrine disorders - Cushing's syndrome - Hyperprolactinemia - Acromegaly
Excessive peripheral production of androgens - Obesity - Idiopathic
Pregnancy-related hyperandrogenism - Hyperreactio luteinalis - Tecoma of pregnancy
Drugs - Androgens - Oral contraceptives containing androgenic progestins - Minoxidil - Phenytoin - Diazoxide - Closporine
True hermaphroditism

is normal, ovarian causes should be considered, such as ovarian hyperthecosis or androgen-secreting ovarian tumors, both conditions independent of LH. In case of mildly increased testosterone levels, in addition to PCOS, non-classical congenital adrenal hyperplasia should also be considered by assaying, basal and after ACTH stimulation, 17-OH-progesterone, and Cushing's syndrome. If testosterone is normal, free testosterone should be assessed. In PCOS, hypersecretion of LH is typically observed with normal or slightly reduced levels of FSH, thus resulting in the typical increase in LH/FSH ratio. However, it is not recommended LH/FSH ratio as a diagnostic criterion, given the influence of obesity on the pulsatility of their secretion. Beyond a marker of ovarian reserve, AMH is also considered a marker of the number of follicles in the ovary with polycystic morphology. However, its use in relation to the diagnosis of PCOS is still controversial.

In functional hypothalamic amenorrhea, normal or slightly reduced FSH, LH, and estradiol levels are observed, although the gonadotropin response to GnRH is preserved. It is considered, however, that in the patient with suspected functional hypothalamic amenorrhea, the FSH assay is sufficient to exclude POF.

POF is characterized by an increase in FSH. In addition to FSH, markers of ovarian reserve, namely inhibin B and AMH, which are reduced both in the transition to menopause and in POF, should be assessed. The diagnostic picture should be completed by the antral follicle count by ultrasound evaluation.

Recommended Readings

Hypophysis

Chanson P, Salenave S, Kamenicky P (2014) Acromegaly. Handb Clin Neurol 124:197–219. https://doi.org/10.1016/B978-0-444-59602-4.00014-9
Chinoy A, Murray PG (2016) Diagnosis of growth hormone deficiency in the paediatric and transitional age. Best Pract Res Clin Endocrinol Metab 30(6):737–747. https://doi.org/10.1016/j.beem.2016.11.002
Christ-Crain M, Fenske W (2016) Copeptin in the diagnosis of vasopressin-dependent disorders of fluid homeostasis. Nat Rev Endocrinol 12(3):168–176. https://doi.org/10.1038/nrendo.2015.224
Di Iorgi N, Morana G, Allegri AE et al (2016) Classical and non-classical causes of GH deficiency in the paediatric age. Best Pract Res Clin Endocrinol Metab 30(6):705–736. https://doi.org/10.1016/j.beem.2016.11.008
Dineen R, Stewart PM, Sherlock M (2016) Acromegaly. QJM:pii: hcw004
Glezer A, Bronstein MD (2015) Prolactinomas. Endocrinol Metab Clin North Am 44(1):71–78. https://doi.org/10.1016/j.ecl.2014.11.003
Peri A, Parenti G, Giuliani A et al (2012) Hyponatremia and syndrome of inappropriate ADH secretion (SIADH). G Ital Nefrol 29(6):661–673
Richmond E, Rogol AD (2016) Treatment of growth hormone deficiency in children, adolescents and at the transitional age. Best Pract Res Clin Endocrinol Metab 30(6):749–755. https://doi.org/10.1016/j.beem.2016.11.005
Richmond EJ, Rogol AD (2008) Growth hormone deficiency in children. Pituitary 11:115–120
Timper K, Fenske W, Kühn F et al (2015) Diagnostic accuracy of copeptin in the differential diagnosis of the polyuria-polydipsia syndrome: a prospective multicenter study. J Clin Endocrinol Metab 100(6):2268–2274. https://doi.org/10.1210/jc.2014-4507
Vilar L, Vilar CF, Lyra R et al (2016) Acromegaly: clinical features at diagnosis. Pituitary
Yuen KCJ Growth hormone stimulation tests in assessing adult growth hormone deficiency
Wong A, Eloy JA, Couldwell WT, Liu JK (2015a) Update on prolactinomas. Part 1: clinical manifestations and diagnostic challenges. J Clin Neurosci 22(10):1562–1567. https://doi.org/10.1016/j.jocn.2015.03.058
Wong A, Eloy JA, Couldwell WT, Liu JK (2015b) Update on prolactinomas. Part 2: treatment and management strategies. J Clin Neurosci 22(10):1568–1574. https://doi.org/10.1016/j.jocn.2015.03.059
TIROIDE
Felig P, Baxter JD, Frohman LA (1997a) La tiroide: fisiologia, tireotossicosi, ipotiroidismo e dolore tiroideo. In: Endocrinologia e Metabolismo, 3rd edn. Mc Graw Hill, Milano
Flamant F (2016) Future challenges in thyroid hormone signaling research. Front Endocrinol 7:58
Harrison TR (1999a) Malattie della Tiroide. In: Principi di Medicina Interna, 4th edn. Mc Graw Hill, Milano
Hulbert AJ (2000) Thyroid hormones and their effects: a new perspective. Biol Rev Camb Philos Soc 75:519–631
Janssen OE, Golcher HMB, Grasberger H et al (2002) Characterization of thyroxine-binding globulin cleaved by human leukocyte elastase. J Clin Endocrinol Metab 87:1217–1222
Kahaly GJ (2015) Bioassays for TSH receptor antibodies: quo vadis? Eur Thyroid J 4:3–5
Little AG (2016) A review of the peripheral levels of regulation by thyroid hormone. J Comp Physiol B 186:677–688
Mendel CM, Weisinger RA, Jones AL et al (1987) Thyroid hormone-binding proteins in plasma facilitate uniform distribution of thyroxine within tissues: a perfused rat liver study. Endocrinology 120:1742–1749
Monaco F (2011a) La Tiroide. In: Endocrinologia Clinica, 5th edn. Società Editrice Universo, Roma
Okosieme O, Gilbert J, Abraham P et al (2016) Management of primary hypothyroidism: statement by the British Thyroid Association Executive Committee. Clin Endocrinol (Oxf) 84(6):799–808
Pappa T, Ferrara AM, Refetoff S (2015) Inherited defects of thyroxine-binding proteins. Best Pract Res Clin Endocrinol Metab 29:735–747
Ross DS, Burch HB, Cooper DS et al (2016) 2016 American Thyroid Association guidelines for diagnosis and management of hyperthyroidism and other causes of thyrotoxicosis. Thyroid
Spandrio L (2006) Psico-neuro-endocrino-immunologia molecolare: ormoni e citochine. In: Biochimica Clinica Speciale/Trattato Italiano di Medicina di Laboratorio, vol II. Piccin, Padova
Yang X, Downes M, Yu RT et al (2006) Nuclear receptor expression links the circadian clock to metabolism. Cell 126:801–810

Parathyroid glands

Alpern RJ, Moe OW, Caplan M (eds) (2013) Seldin and Giebisch's the kidney, chap 66, 5th edn. Elsevier, Philadelphia

Ellegaard M, Jorgensen NR, Schwarz P (2010) Parathyroid hormone and bone healing. Calcif Tissue Int 87:1–13

Felig P, Baxter JD, Frohman LA (1997b) Metabolismo minerale. In: Endocrinologia e Metabolismo, 3a edn. Mc Graw Hill, Milano

Harrison TR (1999b) Metabolismo osseo, del calcio e del fosforo: ormoni calcio-regolatori. In: Principi di Medicina Interna, 4th edn. Mc Graw Hill, Milano

Monaco F (2011b) Paratiroidi, ormoni calciotropi e malattie metaboliche. In: Endocrinologia Clinica, 5th edn. Società Editrice Universo, Roma

Rodríguez-Ortiz ME, Canalejo A, Herencia C et al (2014) Magnesium modulates parathyroid hormone secretion and upregulates parathyroid receptor expression at moderately low calcium concentration. Nephrol Dial Transplant 29:282–289

Silver J, Naveh-Many T (2010) FGF23 and parathyroid glands. Pediatr Nephrol 25:2241–2245

Tenne M, McGuigan FE, Ahlborg H et al (2010) Variation in the PTH gene, hip fracture, and femoral neck geometry in elderly women. Calcif Tissue Int 86:359–366

Adrenal gland

Bornstein SR, Allolio B, Arlt W et al (2016) Diagnosis and treatment of primary adrenal insufficiency: an endocrine society clinical practice guideline. J Clin Endocrinol Metab 101(2):364–389. https://doi.org/10.1210/jc.2015-1710

Del Giudice A, Prencipe M, Aucella F (2013) L'iperaldosteronismo primitivo: algoritmo diagnostico. G Ital Nefrol 30(3):ISSN 1724-5590

Eisenhofer G, Goldstein DS, Walther MM et al (2003) Biochemical diagnosis of phaeochromocytoma: how to distinguish true- from false-positive results. J Clin Endocrinol Metab 88(6):2656–2666

Eisenhofer G, Timmers HJ, Lenders JW et al (2011) Age at diagnosis of phaeochromocytoma differs according to catecholamine phenotype and tumor location. J Clin Endocrinol Metab 96(2):375–384

Fassnacht M, Arlt W, Bancos I et al (2016) Management of adrenal incidentalomas: European Society of Endocrinology Clinical Practice Guideline in collaboration with the European Network for the Study of Adrenal Tumors. Eur J Endocrinol 175(2):G1–G34. https://doi.org/10.1530/EJE-16-0467

Funder JW, Carey RM, Mantero F et al (2016) The management of primary aldosteronism: case detection, diagnosis, and treatment: an endocrine society clinical practice guideline. J Clin Endocrinol Metab 101(5):1889–1916. https://doi.org/10.1210/jc.2015-4061

Garray A, Casey R, Wall D et al (2015) A review of the management of positive biochemical screening for phaeochromocytoma and paraganglioma: a salutary tale. Int J Clin Pract 69(7):802–809

Kola B, Grossman AB (2008) Dynamic testing in Cushing's syndrome. Pituitary 11(2):155–162

Lenders JW, Duh QY, Eisenhofer G et al (2014) Pheochromocytoma and paraganglioma: an endocrine society clinical practice guideline. J Clin Endocrinol Metab 99(6):1915–1942. https://doi.org/10.1210/jc.2014-1498

Nicolaides NC, Chrousos GP, Charmandari E (2000) Adrenal Insufficiency. [Updated 2013 Nov 18]. In: De Groot LJ, Chrousos G, Dungan K et al (eds) Endotext [Internet]. MDText.com, Inc., South Dartmouth. Available from: https://www.ncbi.nlm.nih.gov/books/NBK279083/

Pappachan JM, Raskauskiene D, Sriraman R et al (2014) Diagnosis and management of pheochromocytoma: a practical guide to clinicians. Curr Hypertens Rep 16(7):442. https://doi.org/10.1007/s11906-014-0442-z

Tsirlin A, Oo Y, Sharma R et al (2014) Phaeochromocytoma: a review. Maturitas 77:229–238

Testis

Basaria S (2014) Male hypogonadism. Lancet 383(9924):1250–1263

Corona G, Rastrelli G, Maggi M (2013) Diagnosis and treatment of late-onset hypogonadism: systematic review and meta-analysis of TRT outcomes. Best Pract Res Clin Endocrinol Metab 27(4):557–579

Ferlay J, Soerjomataram I, Ervik M et al (2012) GLOBOCAN 2012 v1.0, cancer incidence and mortality worldwide: IARC CancerBase. No. 11. International Agency for Research on Cancer, Lyon

Gion M, Trevisiol C, Rainato G et al (2016) Marcatori circolanti in oncologia. Guida all'uso clinico appropriato. I quaderni di Monitor. AGENAS

Lawrence KL, Stewart F, Larson BM (2017) Approaches to male hypogonadism in primary care. Nurse Pract 42(2):32–37

Paduch DA, Brannigan RE, Fuchs EF (2014) The laboratory diagnosis of testosterone deficiency. Urology 83(5):980–988

Peak TC, Haney NM, Wang W et al (2016) Stem cell therapy for the treatment of Leydig cell dysfunction in primary hypogonadism. World J Stem Cells 8(10):306–315

Trabert B, Chen J, Devesa SS et al (2015) International patterns and trends in testicular cancer incidence, overall and by histologic subtype, 1973-2007. Andrology 3(1):4–12

Ovary

Boehm U, Bouloux PM, Dattani MT et al (2015) Expert consensus document: European Consensus Statement on congenital hypogonadotropic hypogonadism-pathogenesis, diagnosis and treatment. Nat Rev Endocrinol 11(9):547–564

Broer SL, Broekmans FJ, Laven JS, Fauser BC (2014) Anti-Müllerian hormone: ovarian reserve testing and its potential clinical implications. Hum Reprod Update 20(5):688–701

Burger HG (2002) Androgen production in women. Fertil Steril 77:S3–S5

Conway G, Dewailly D, Diamanti-Kandarakis E et al (2014) The polycystic ovary syndrome: a position statement from the European Society of Endocrinology. Eur J Endocrinol 171(4):P1–29

Deligeoroglou E, Athanasopoulos N, Tsimaris P et al (2010) Evaluation and management of adolescent amenorrhea. Ann N Y Acad Sci 1205:23–32

Dwyer AA, Phan-Hug F, Hauschild M et al (2015) Transition in endocrinology: hypogonadism in adolescence. Eur J Endocrinol 173(1):R15–R24

Escobar-Morreale HF, Carmina E, Dewailly D et al (2012) Epidemiology, diagnosis and management of hirsutism: a consensus statement by the Androgen Excess and Polycystic Ovary Syndrome Society. Hum Reprod Update 18(2):146–170

Fanelli F, Gambineri A, Mezzullo M et al (2013) Revisiting hyper- and hypo-androgenism by tandem mass spectrometry. Rev Endocr Metab Disord 14(2):185–205

Ghosh M, Galman C, Rudling M, Bo A (2015) Influence of physiological changes in endogenous estrogen on circulating PCSK9 and LDL Cholesterol. J Lipid Res 56:463

Goodman NF, Cobin RH, Futterweit W et al (2015) American Association of Clinical Endocrinologists (AACE); American College of Endocrinology (ACE); Androgen Excess and PCOS Society. American Association of Clinical Endocrinologists, American College of Endocrinology, and Androgen Excess and PCOS Society Disease State Cclinical Review: guide to the best practices in the evaluation and treatment of polycystic ovary syndrome. Endocr Pract 21(12):1415–1426

Gordon C (2010) Functional hypothalamic amenorrea. New Engl J Med 363:365–371

Kushnir MM, Rockwood AL, Roberts WL et al (2011) Liquid chromatography tandem mass spectrometry for analysis of steroids in clinical laboratories. Clin Biochem 44(1):77–88

Luisi S, Orlandini C, Regini C et al (2015) Premature ovarian insufficiency: from pathogenesis to clinical management. J Endocrinol Invest 38(6):597–603

Martin KA, Anderson RR, Chang RJ et al (2018) Evaluation and treatment of hirsutism in premenopausal women: an Endocrine Society Clinical Practice Guidelines. J Clin Endocrinol Metabol 103:1233–1257

Master-Hunter T, Heiman D (2006) Amenorrhea: evaluation and treatment. Am Fam Physician 73:1376

Practice Committee of the American Society for Reproductive Medicine (2015) Testing and interpreting measures of ovarian reserve: a committee opinion. Fertil Steril 103(3):e9–17

Silva CA, Yamakami LY, Aikawa NE et al (2014) Autoimmune primary ovarian insufficiency. Autoimmun Rev 13(4-5):427–430

Vermeulen A, Verdonck L, Kaufman JM (1999) A critical evaluation of simple methods for the estimation of free testosterone in serum. J Clin Endocrinol Metab 84(10):3666–3672

Diabetes Mellitus: From Definition to Therapy

26

Marcello Ciaccio, Anna Maria Ciaccio, and Luisa Agnello

Introduction

This chapter provides an in-depth description of the various types of diabetes mellitus, from the definition to the classification, epidemiology, pathogenesis, diagnosis, monitoring, and therapy. In addition, a description of glucose metabolism will be provided to facilitate the understanding of the diabetic disease.

Definition

Diabetes mellitus is a chronic heterogeneous disease characterized by altered metabolism of glucose and other energy substrates. From a biochemical point of view, it is characterized by hyperglycemia due to a relative or absolute insulin deficit.

Historical Background

The first hints of this disease were found in Egyptian papyrus. However, the Greeks described it in detail. The term "diabetes" derives, indeed, from the Greek verb διαβήτης, which means "to pass through," alluding to the flow of water, since the most striking symptom is polyuria. The suffix mellitus derives, instead, from the Latin mel (honey, sweet) and alludes to the sweetish taste of the blood and urine of patients with diabetes, a characteristic already known by the ancient Egyptians, Greeks, and Indians.

Blood Glucose Regulation

Glucose is a monosaccharide (aldohexose) and represents the primary energy source for our organism's cells and the only energy source for central nervous system (CNS) cells and erythrocytes. The circulating glucose comes mainly from the diet, where it is present as monosaccharides and complex carbohydrates (polysaccharides, e.g., starch); a small amount comes from endogenous synthesis (gluconeogenesis). Glucose circulates in free form, and its levels are maintained within a relatively narrow range (70–100 mg/dL) by hormones, such as insulin and glucagon, which ensure that the balance between glucose production and utilization (Fig. 26.1).

In particular, in the postprandial period, most glucose derives from the diet and is metabolized by the body's cells. Glucose enters cells by facilitated diffusion mediated by glucose transporters (GLUTs), of which there are several isoforms with different characteristics (Table 26.1). The most important are GLUT-2, acting as a "blood glucose sensor" in pancreatic β-cells that synthesize and secrete insulin, and GLUT-4, which acts on insulin-dependent tissues (the adipose and muscle tissues), and its expression is induced by insulin. Following its entry into a cell, glucose is immediately phosphorylated at glucose-6-P (G-6-P), preventing its escape from the cell (Fig. 26.2).

M. Ciaccio (✉)
Department of Biomedicine, Neurosciences and Advanced Diagnostics, Institute of Clinical Biochemistry, Clinical Molecular Medicine and Clinical Laboratory Medicine, University Hospital "P. Giaccone", Palermo, Italy

Department of Laboratory Medicine, University Hospital "P. Giaccone", Palermo, Italy
e-mail: marcello.ciaccio@unipa.it

A. M. Ciaccio
Department of Health Promotion, Mother and Child Care, Internal Medicine and Medical Specialties (ProMISE) "G. D'Alessandro", University Hospital "P. Giaccone", Palermo, Italy

L. Agnello
Department of Biomedicine, Neurosciences and Advanced Diagnostics, Institute of Clinical Biochemistry, Clinical Molecular Medicine and Clinical Laboratory Medicine, University Hospital "P. Giaccone", Palermo, Italy

M. Ciaccio (ed.), *Clinical and Laboratory Medicine Textbook*, https://doi.org/10.1007/978-3-031-24958-7_26

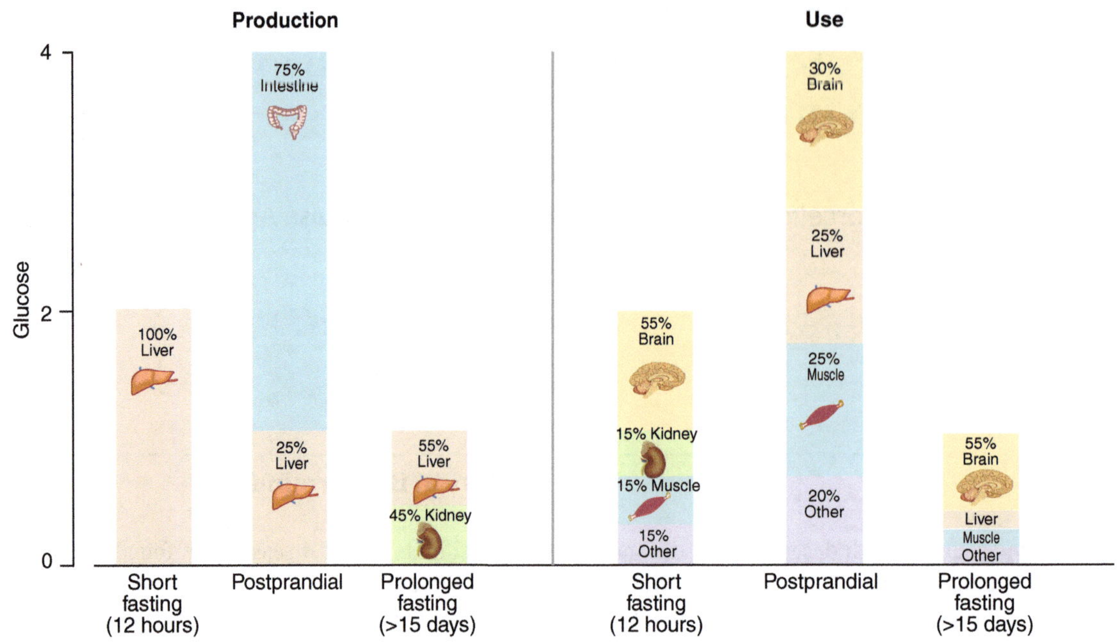

Fig. 26.1 Glucose production and utilization. (Copyright EDISES 2021. Reproduced with permission)

Table 26.1 Characteristics of glucose transporters (GLUTs)

	Location	Glucose affinity	Functions
GLUT-1	Erythrocytes, brain, all body tissues	High (Km = 1 mmol/L)	Glucose uptake to maintain basal intracellular glucose levels
GLUT-2 (glycemic sensor)	Liver, pancreatic β-cells, serous surfaces of the intestines and the kidneys	Low (Km = 15–20 mmol/L)	Glucose sensor for pancreatic β-cells
GLUT-3	Brain, kidneys	High (Km <1 mmol/L)	
GLUT-4 (insulin-ependent)	Insulin-sensitive tissues: adipose tissue and muscle	Average (Km = 2.5–5 mmol/L)	
GLUT-5	Intestine, liver, spermatozoa, kidneys, muscle, adipose tissue		Only fructose transport
GLUT-7	Endoplasmic reticulum of hepatocytes		Regulates the free glucose flow from the endoplasmic reticulum lumen following the action of glucose-6-phosphatase, an enzyme that dephosphorylates glucose-6-phosphate

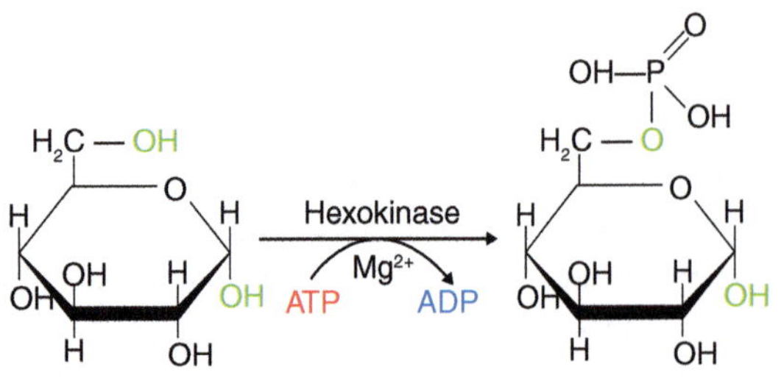

Fig. 26.2 Glucose phosphorylation reaction. (Copyright EDISES 2021. Reproduced with permission)

This phosphorylation reaction is mediated by hexokinases, of which four isoforms are known, with different characteristics (Table 26.2).

Table 26.2 Characteristics of hexokinases

	Location	Substrate	Affinity	Regulation
Hexokinase I	Brain	Glucose, galactose, Fructose, glucosamine	High	Inhibition by G-6-P
Hexokinase II	Muscle			
Hexokinase III	Ubiquitous			
Hexokinase IV or glucokinase	Liver	Glucose	Low	

In particular, hexokinase IV, also known as glucokinase, is expressed exclusively in the liver and, compared to other hexokinases, it is not inhibited by product (from G-6-P), and, therefore, even in the presence of high glucose concentrations, it mediates phosphorylation. This feature is important because the liver represents the main glucose storage organ.

The postabsorptive period (6–12 h after food ingestion, when the contents of the small intestine have been digested and absorbed) is characterized by a progressive accentuation of hepatic glycogenolysis. During short fasting (e.g., overnight fasting), circulating glucose derives primarily from the liver, by glycogenolysis (glycogen breakdown) and gluconeogenesis (synthesis of glucose from non-saccharide precursors).

During prolonged fasting (>15 days), circulating glucose derives from the liver and partly from the kidneys, where gluconeogenesis is activated (Fig. 26.1). After glycogen reserves depletion, free fatty acids from the adipose tissue are mobilized as the liver's and muscle's primary source of energy. The brain and anaerobic tissues receive glucose that derives primarily from gluconeogenesis. In addition, hepatic ketone bodies are synthesized from acetyl coenzyme A (acetyl-CoA), released into the circulation and used as an alternative energy source by tissues, including the CNS.

Several hormones are involved in blood glucose regulation (Table 26.3).

Specifically, insulin is a peptide hormone synthesized by pancreatic β-cells, consisting of an α-chain and a β-chain linked by a disulfide bridge. It is synthesized as a preprohormone (signal peptide + proinsulin, formed from insulin + C-peptide) and converted to proinsulin by the signal peptide removal in the endoplasmic reticulum; after S–S bridge formation, proinsulin translocates to the Golgi apparatus, where peptide is removed by proteolytic cutting (Fig. 26.3).

The mature insulin and C-peptide are stored in secretory granules within β-cells to be released into the circulatory stream by exocytosis in response to an appropriate stimulus. C-peptide is essential for the proper proinsulin folding. Hyperglicemia is the primary stimulus for insulin secretion. Insulin secretion is biphasic, with the first early peak due to secretion of preformed insulin and the second late peak due to *ex novo* synthesis. Insulin has a short plasma half-life of about 6 min, which determines its rapid circulating concentration variations; 40–60% of the hormone is catabolized in the liver and the remaining amount in the kidneys. The numerous cellular processes regulated by insulin depend on its binding to its receptor located on the membrane of the cells of the target organs, mainly the liver, muscles, and adipose tissue. The receptor is a heterodimer in which the subunits (α and β) are bound by S–S bridges (Fig. 26.4). Both subunits are extensively glycosylated. The α-subunit is extracellular and, therefore, it interacts with insulin; the β-subunit consists of a transmembrane and a cytoplasmic portion

Table 26.3 Characteristics of the hormones involved in blood glucose regulation

Synthesis	Hormone	Action	Effect on glycemia
Pancreas	Insulin	↑ Glucose input (muscle and adipose tissues) ↑ Glycolysis, hepatic glycogen synthesis, and fat synthesis ↓Lipolysis and gluconeogenesis	↓
	Glucagon	↑ Hepatic glycogenolysis and gluconeogenesis ↑ Lipolysis	↑
	Somatostatin	↓Release of insulin, glucagon, and pituitary hormones	
Adrenal gland	Adrenaline	↑ Glycogenolysis and lipolysis	↑
	Cortisol	↑ Gluconeogenesis and lipolysis Insulin antagonist	↑
Pituitary gland	Adrenocorticotropic hormone (ACTH)	↑ Release of cortisol and lipolysis	↑
	Growth hormone (GH)	Insulin antagonist	↑
Thyroid	Thyroxine	↑ Glycogenolysis and hepatic gluconeogenesis ↑ Intestinal absorption of carbohydrates	↑

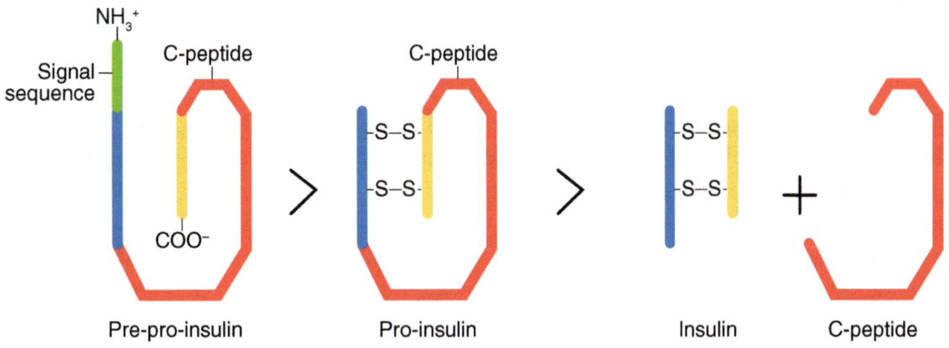

Fig. 26.3 Insulin biosynthesis. (Copyright EDISES 2021. Reproduced with permission)

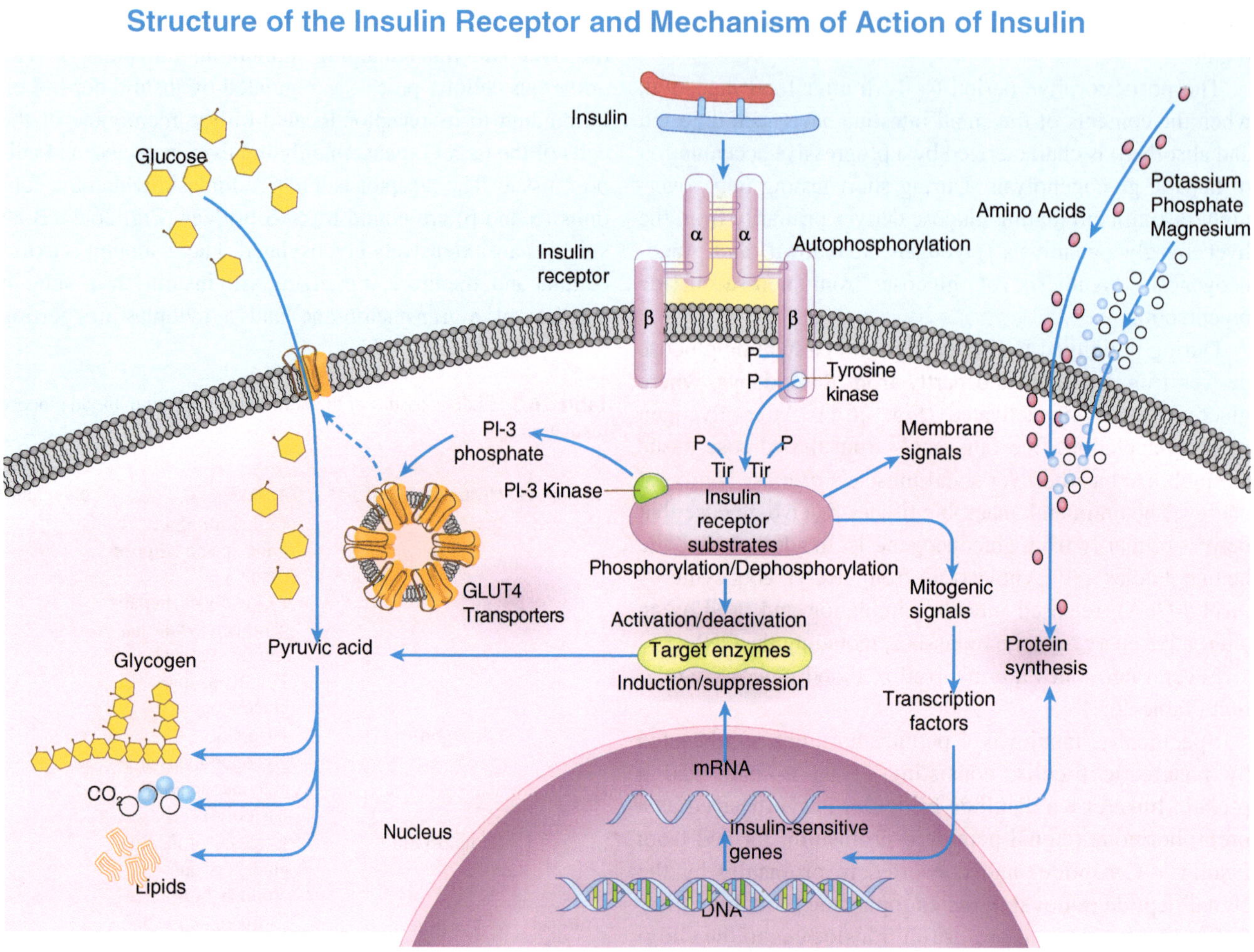

Fig. 26.4 Structure of the insulin receptor and insulin mechanism of action. The insulin receptor is a heterodimer, consisting of two extracellular α-chains and two transmembrane β-chains. The α-chains interact with insulin, whereas β-chains mediate intracellular signal transduction. In particular, insulin activates different cellular pathways leading to the GLUT-4 transporter expression on the plasma membrane, the regulation of the activity of different enzymes, and the transcription of insulin-sensitive genes. (Copyright EDISES 2021. Reproduced with permission)

Fig. 26.5 Anabolic and anticatabolic effects of insulin. aa, amino acids. (Copyright EDISES 2021. Reproduced with permission)

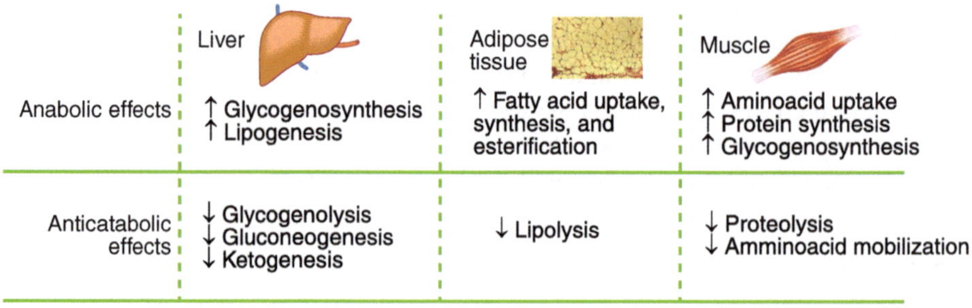

responsible for signal transduction. The receptor has a half-life of 7–12 h without and 2–3 h with insulin. Insulin is the primary hormone with anabolic and anticatabolic action, promoting the glucose and amino acids uptake by the cells of numerous tissues, stimulating the synthesis of glycogen, fatty acids, and proteins, and inhibiting catabolic processes, such as hormone-sensitive lipase in the adipose tissue and the process of β-oxidation of fatty acids (Fig. 26.5).

Glucagon is a protein hormone synthesized by the pancreatic α-cells in response to various stimuli, such as hypoglycemia. It is synthesized as a prehormone, accumulated in secretory vesicles, and is released by exocytosis in response to stimuli. It represents the hormone of energy emergency as it intervenes during substrate deficiency. Glucagon exerts its hyperglycemic action mainly by inducing glycogenolysis (demolition of glycogen) and gluconeogenesis in the liver.

Classification of Diabetes Mellitus

Since the term "diabetes mellitus" encompasses a group of metabolic disorders that share hyperglycemia, over the decades, these clinical conditions have been classified according to different criteria.

In 1979, a treatment-based classification of the various forms of diabetes was proposed. It distinguished between insulin-dependent type 1 diabetes mellitus (IDDM) and non-insulin-dependent type 2 diabetes mellitus (NIDDM). In 1997, following advances in the understanding of the etiology and pathogenesis of diabetes, the American Diabetes Association (ADA) revised this classification and decided to eliminate the terms insulin-dependent and non-insulin-dependent and the related acronyms (IDDM and NIDDM). On the other hand, the terms type 1 and type 2 diabetes mellitus are maintained, thus achieving a classification based on etiology. Currently, international guidelines classify diabetes mellitus into:

- Type 1 diabetes: It is an autoimmune disease caused by β-cell destruction. It is characterized by an absolute insulin deficiency (the latent autoimmune diabetes in adults (LADA) variant), has a slow course, and appears in adults.
- Type 2 diabetes: It is caused by insulin resistance and β-cell dysfunction.
- Gestational diabetes: It is diagnosed during pregnancy, and must be distinguished from overt diabetes. Gestational diabetes is caused by functional defects similar to type 2 diabetes; it is first diagnosed during pregnancy (commonly between the second and third trimesters) and usually regresses after delivery and then recurs, often at a distance, with the characteristics of type 2 diabetes.

Other types of diabetes may be due to genetic defects of the β-cells of the pancreas, such as maturity-onset diabetes of the young (MODY), a form of type 2 diabetes with juvenile onset; intake of drugs or toxic substances; genetic defects of insulin action; infections; diseases of the exocrine pancreas; alterations of the immune system (rare forms); endocrinopathies; and rare genetic syndromes.

Epidemiology

Diabetes mellitus represents one of the three health emergencies identified by the United Nations (UN) and World Health Organization (WHO), along with malaria and tuberculosis. The percentage of people with diabetes mellitus is estimated to be around 5% of the world's population, of which about 10% have type 1 diabetes mellitus and the remaining 90% have type 2 diabetes mellitus. The size of the problem, the spread to all age groups, and the severity of complications associated with the disease make diabetes one of the major health problems on a global scale. It has been estimated that every year there are 5–7 new cases of type 2 diabetes per 1000 people, without significant gender differences. Type 1 diabetes is a rarer condition, with a peak incidence between 5 and 15 years.

The prevalence of LADA, a slower-progressing form of autoimmune diabetes, is 4.5%; about 20% of patients initially defined as type 2 diabetes are affected by LADA.

Concerning gestational diabetes, it is estimated that about 7% of pregnancies are complicated by diabetes. The prevalence of gestational diabetes varies widely among different populations, being higher in women of Asian and Hispanic origin (9.7% and 8.3%, respectively) and lower in non-Hispanic white women (5.7%). Within each ethnic group, the prevalence of gestational diabetes mellitus has increased over time, in parallel with the increase in obesity in reproductive age women.

Etiopathogenesis

Type 1 Diabetes

Type 1 diabetes mellitus (DM1) is a multifactorial disease in which exposure to various environmental factors triggers, in a subject with genetic predisposition, an autoimmune response that destroys pancreatic β-cells, leading to an insulin deficit with a consequent chronic increase in glycemia (Fig. 26.6).

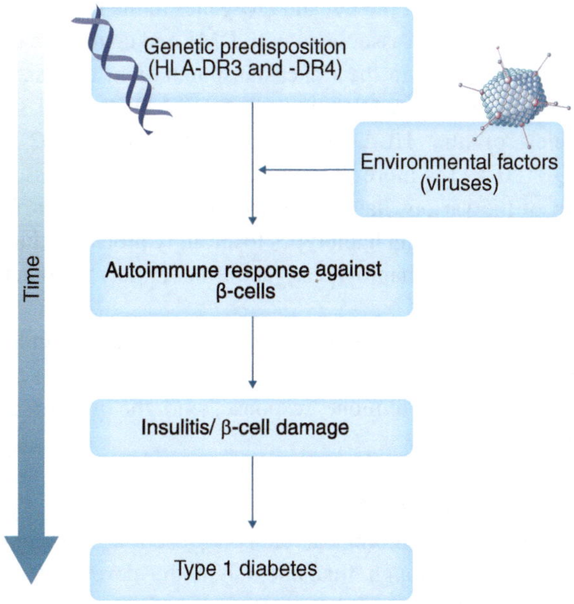

Fig. 26.6 Pathogenesis of type 1 diabetes. (Copyright EDISES 2021. Reproduced with permission)

In the initial phase, the autoimmune response, although not clinically detectable, is demonstrated by the presence of autoantibodies directed against specific β-cellular antigens leading to insulitis, an inflammation of the islets of Langerhans in which the β-cells reside, characterized by infiltration of lymphocytes. At the beginning of this phase, the β-cells still maintain their functionality. As the autoimmune response progresses, the number of β-cells decreases, and their capacity to produce insulin decreases leading to hyperglycemia. At stage, diabetes is diagnosed. The rate of β-cell destruction varies widely among individuals, as some cases progress rapidly to clinical diabetes, whereas others evolve more slowly. The rapidly progressing form is commonly seen in children, whereas the slow-onset form occurs in adults (LADA). Clinical manifestations do not become apparent until most β-cells (approximately 80%) are destroyed.

The DM1 etiopathogenesis, therefore, is the result of the interaction between genetic, environmental, and immunological factors.

Genetic Factors

Several genes or chromosomal loci associated with the disease have been identified. However, the human leukocyte antigen (HLA) system is the most important. The HLA system is a set of genes located on the short arm of chromosome 6 and forms a region known as the major histocompatibility complex (MHC). The HLA region includes more than 200 genes coding for 3 classes of proteins. Class I and II proteins are membrane glycoproteins that mediate the recognition of foreign peptides, whereas class III proteins play a crucial role in the inflammatory process.

Some alleles of the *DR* and *DQ* genes in the HLA class II locus are strongly associated with DM1 and contribute up to 50% of the risk. In the Caucasian population, the association with the HLA-DR3 and HLA-DR4 alleles and some alleles of the HLA-DQB1 locus (DQB1* 0302 and DQB1*0201) is more pronounced. In general, the association of the various haplotypes with the disease varies from highly predisposing haplotypes to strongly protective haplotypes, neutral haplotypes, and moderately protective haplotypes.

Other genes that individually confer a modestly increased risk of developing DM1 are the *PTPN22* gene, which regulates the innate immune response, and the insulin gene (*IDDM2*).

Environmental Factors

Numerous epidemiological data suggest that the genetic component, although fundamental in the development of type 1 diabetes, is not alone sufficient to determine the disease onset. It is, therefore, increasingly likely that the environment may play an essential role in the DM1 etiology. Environmental factors associated with the risk of developing DM1 are:

- Viruses, especially Enteroviruses, such as Coxsackie B4 virus
- *Mycobacteria*
- Feeding: Cow's milk, and diabetogenic substances in soy and wheat

Immunological Factors

In DM1, chronic hyperglycemia results from selective destruction of the islets of Langerhans β-cells mainly mediated by T-lymphocytes, both CD4 (T helper) and CD8 (T cytotoxic). In the lymphocyte infiltrates of the Langerhans islets (insulitis) of DM1 subjects, in addition to TCD8 (the most abundant) and TCD4, B lymphocytes, natural killer (NK) cells, dendritic cells, and macrophages have been identified. All these immune system cell types may contribute to the DM1 pathogenesis.

Proteins released from damaged or destroyed β-cells (e.g., during viral infection or exposure to toxins) are phagocytosed by antigen-presenting cells (APCs), such as macrophages or dendritic cells. APCs hydrolyze proteins into peptides to be presented by HLA class II molecules to proinflammatory T-helper 1 (Th1) lymphocytes. The latter triggers immune responses cascade, including the activation of:

- B lymphocytes, which produce autoantibodies against insular antigens
- Specific cytotoxic T lymphocytes against β-cell antigens

In addition, APCs may present antigenic peptides to regulatory T lymphocytes (T regs) that, under normal conditions, suppress the proinflammatory cascade and prevent β-cell destruction. Pancreatic tissue destruction is mainly due to cell-mediated immunity reactions, whereas autoantibody production is considered an epiphenomenon (i.e., they are not pathogenic), secondary to pancreatic β-cell destruction.

Both autoreactive T lymphocytes and autoantibodies can recognize different insular antigens, such as insulin, glutamic acid decarboxylase (GAD), tyrosine phosphatase-related islet antigen 2 (IA-2), and zinc transporter 8 autoantibody (ZnT8).

It has been hypothesized that the cell-mediated autoimmune response is initially directed toward a primary antigen, causing an initial tissue damage with the release of degradation products that induce secondary immune responses contributing to the extension and chronicization of the process.

The immune response trigger toward self-antigens is due to the loss of the physiological tolerance mechanism toward self-molecules. Several hypotheses have been proposed to explain this mechanism, among which the most accredited are the following:

- Defect in lymphocyte selection in the thymus
- Molecular mimicry
- Alteration of suppressor mechanisms

Defect in Lymphocyte Selection in Thymus

The physiological tolerance of the immune system toward self-antigens is mainly controlled by the thymus, where the selection of the lymphocyte repertoire takes place, preventing the maturation or activation of potentially self-reactive lymphocytes (negative selection); an alteration of this process could occur in patients with DM1. HLA molecules play an important role in negative selection because they present self-antigens to immature T lymphocytes that will undergo negative selection. HLA susceptibility alleles to DM1 bind peptides of insular antigens with low affinity, resulting in an inefficient presentation of self-antigens to self-reactive T lymphocytes that could escape negative selection and reach the periphery.

Molecular Mimicry

It consists of the immune response toward an exogenous antigen, such as a viral protein, which has an amino acid sequence common to a β-cell protein. Therefore, T lymphocytes also recognize the β-cell autoantigen, toward which they develop a reaction leading to its destruction. In this case, tolerance mechanisms are circumvented by the induction of an immune response against an exogenous antigen. For example, the Coxsackie B4 virus possesses sequence homology with GAD.

Alteration of Suppressor Mechanisms

Under physiological conditions, most of the self-reactive lymphocytes are eliminated by the thymus through the previously described mechanism of "clonal selection," or are actively suppressed by T-reg lymphocytes. Alterations in the latter can contribute to the development of the immune reaction against the self.

Type 2 Diabetes

Type 2 diabetes mellitus (DM2) is a multifactorial disease resulting from the interaction between genetic and environmental factors.

DM2 is characterized by variable degrees of insulin resistance, altered insulin secretion, and increased glucose production, leading to hyperglycemia (Fig. 26.7).

Insulin resistance is the reduced sensitivity of target tissues (muscles, liver, and adipose tissue) to insulin action, leading to:

- Reduced insulin-mediated glucose uptake in the adipose and muscle tissues
- Reduced insulin-mediated inhibition of hepatic gluconeogenesis
- Reduced inhibition of adipose tissue lipolysis due to the lacking insulin inhibition of hormone-sensitive lipase

In the early stages of the disease, insulin resistance leads to compensatory hyperplasia of pancreatic β-cells with

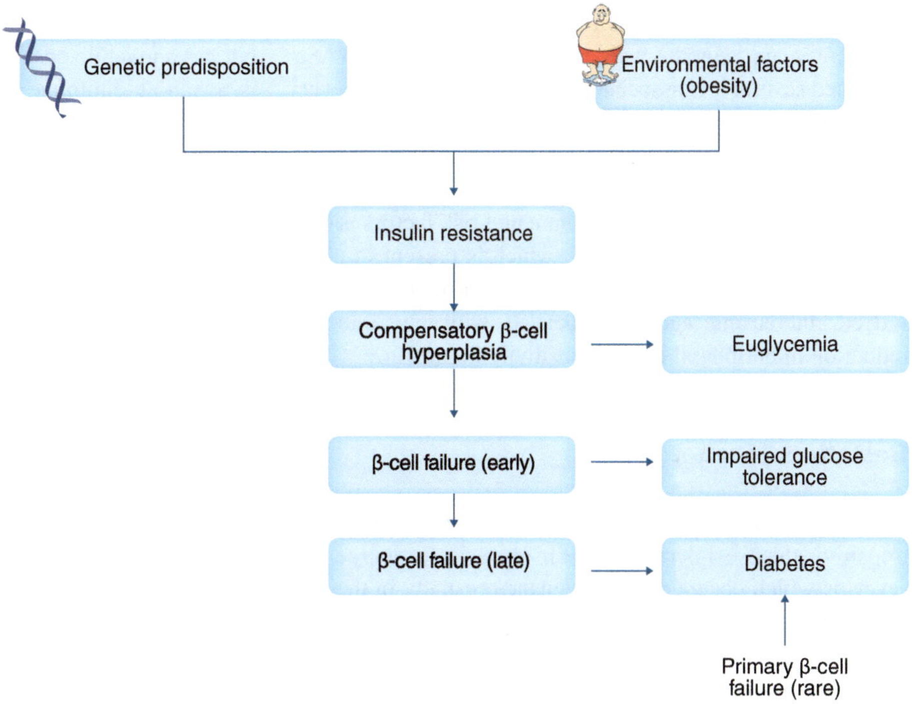

Fig. 26.7 Pathogenesis of type 2 diabetes. (Copyright EDISES 2021. Reproduced with permission)

hypersecretion of insulin, maintaining euglycemia (normal glucose levels). Thus, initially, hyperinsulinism compensates for peripheral insulin resistance. This condition can last up to several years. However, over time, β-cells will become insufficient, characterized by a progressive decline in cell mass and function, leading to hyperglycemia and overt DM2. In rare cases of primary β-cellular insufficiency, the onset of DM2 is not preceded by insulin resistance. The latter may be due to receptor alterations (reduced synthesis, increased degradation, reduced phosphorylation-dependent activation) or to alterations in post-receptor events. Among the various factors causing insulin resistance, obesity has a crucial role, with a dose–response relationship between visceral fat and insulin resistance degree.

DM2 results from the interaction between genetic and environmental factors.

Genetic Factors

The strong genetic component of DM2 is supported by some literature evidence:

- Studies on twins revealed that DM2 concordance is 70% in monozygotic twins and 20–30% in dizygotic twins.
- The risk of developing DM2 during lifetime is about 10% in the general population, 40% in subjects having an affected parent, and 70% in subjects having both parents affected.
- The risk of developing DM2 in a subject with a diabetic sibling is significantly increased compared to the risk of the general population.

However, unlike DM1, no genetic variants strongly predictive of the risk of developing DM2 have been identified. In recent years, genome-wide association studies (GWASs) identified several loci associated with DM2, each with a very modest effect on individual disease risk (10–40%).

Environmental Factors

One of the most important environmental risk factors is obesity, particularly visceral obesity, which is present in about 90% of DM2 patients. Age is another important risk factor. Indeed, increasing age is associated with physiological reduction in peripheral tissues sensitivity to insulin.

Gestational Diabetes

Gestational diabetes mellitus (GDM) is a condition of impaired glucose tolerance (IGT), of variable degree and severity, which occurs during pregnancy (usually in the second or third trimester) and generally regresses after delivery. However, it can recur at a distance, preferentially with the characteristics of type 2 diabetes.

GDM represents the most common metabolic alteration in pregnancy that, if not correctly recognized and adequately treated, is associated with high maternal–fetal morbidity, mainly related to excessive fetal growth (macrosomia).

During pregnancy, the organism undergoes a physiological adaptation, characterized by endocrine–metabolic changes necessary to ensure the supply of nutrients to the fetus and adequate preparation of the maternal organism for childbirth and lactation. Insulin resistance, which becomes more evident in the muscle and adipose tissues as pregnancy progresses, is a physiological condition aimed at fetal growth. The pathogenetic mechanisms of GDM are superimposable to those of DM2. An intolerance to carbohydrates develops when β-cellular secretion is no longer sufficient to compensate for peripheral insulin resistance, which is physiologically present during pregnancy. In GDM patients, the reduced action of insulin determines an excess of circulating nutrients, such as glucose, lipids, and amino acids. They can cross the placenta stimulating the fetal insulin secretion (hyperinsulinism), which in turn determines an increase in the adipose tissue with consequent organomegaly and macrosomia (Fig. 26.8). Furthermore, hyperinsulinism can determine the onset of respiratory distress syndrome in newborns due to the insulin inhibition on the phosphatidylcholine synthesis, which is the main constituent of lung surfactant.

Other Types of Diabetes Mellitus

aturity-onset diabetes of the young (MODY) deserves particular attention. MODY is a monogenic form of diabetes with autosomal dominant transmission, so defined because it phenotypically presents the characteristics of type 2 diabetes but has a juvenile onset (before the age of 25 years). MODY is a non-autoimmune form of diabetes caused by a point mutation or a deletion in genes encoding molecules involved in development or function of pancreatic β-cells, leading to altered insulin secretion. Currently, 14 different forms are known, but the most common are MODY 2, due to mutations in the gene encoding for glucokinase, and MODY 3, due to mutations in the gene encoding for hepatocyte nuclear factor-1α (HNF-1α) (Table 26.4).

Other less frequent monogenic forms are maternally transmitted diabetes with bilateral deafness, caused by mitochondrial DNA mutations or insulin gene mutations, which generally manifest as neonatal diabetes.

Diagnosis

The clinical laboratory plays a central role in diabetes mellitus diagnosis, which is based on two parameters: glycated hemoglobin (HbA1c) and blood glucose.

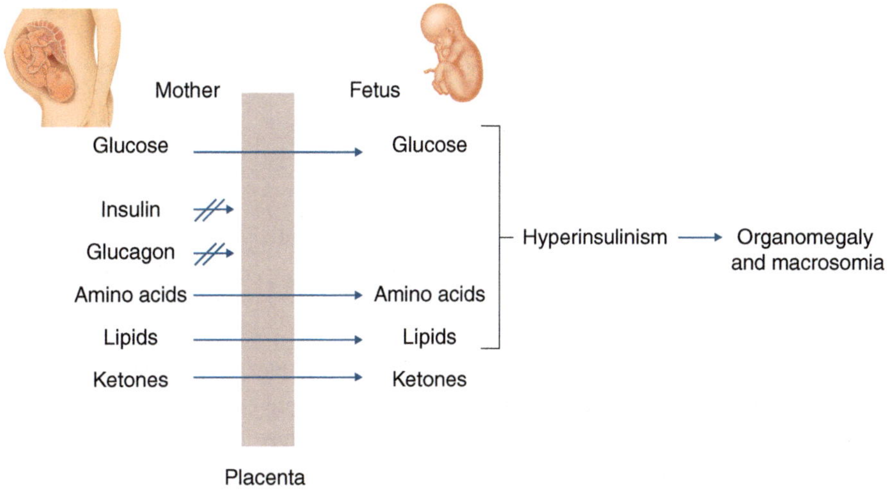

Fig. 26.8 Pathogenesis of gestational diabetes mellitus. (Copyright EDISES 2021. Reproduced with permission)

Table 26.4 Genetic–biochemical–clinical characteristics of MODY

	MODY 1	MODY 2	MODY 3	MODY 4	MODY 5	MODY X
Genetic locus	20q	7p	12q	13q	17q	Not known
Gene	HNF-4α	GCK	HNF-1α	IPF-1	HMF-1β	Heterogeneous?
Distribution (% of MODY households in Italy)	0%	35–40%	15–20%	0%	1–5%	10%
Age onset	>12 years	Pediatric	Post-pubertal	30 years	>12 years	25 years
Primitive defect	Pancreas	Pancreas Liver	Pancreas Kidneys	Pancreas	Pancreas	Insulin resistance
Severity of diabetes	Severe	Mild	Severe	?	Severe	Mild
Complications	Frequent	Rare	Frequent	Frequent	Frequent	Not known

In the presence of typical symptoms of the disease (polyuria, polydipsia, and weight loss), the diagnosis of diabetes mellitus relies on the finding, even on one occasion, of random blood glucose ≥200 mg/dL (regardless of food intake).

In the absence of typical symptoms of the disease, the diagnosis of diabetes mellitus relies on the finding, confirmed on at least two different occasions, of:

- Fasting blood glucose ≥126 mg/dL (fasting means at least 8 h without food)
- Blood glucose ≥200 mg/dL 2 h after oral glucose tolerance test (OGTT) (performed with 75 g)
- HbA1c ≥48 mmol/mol (6.5%)

For diagnostic and screening purposes, blood glucose should be measured in the plasma.

The use of a glucometer is not recommended because it generates non-standardized measurements.

The following measurements are not helpful for the diagnosis of diabetes:

- Postprandial blood glucose or glycemic profile
- Insulinemia measured at basal or during OGTT
- C-peptide
- Autoantibodies

In addition to diabetes, other states of dysglycemia are known. The following values of the main glycemic parameters are considered worthy of attention because they identify individuals at risk for diabetes and cardiovascular diseases:

- Fasting blood glucose of 100–125 mg/dL (impaired fasting glucose, IFG)
- Blood glucose of 140–199 mg/dL 2 hours after OGTT (impaired glucose tolerance, IGT)
- HbA1c of 42–48 mmol/mol (6.00–6.49%)

In subjects with IFG and/or IGT or HbA1c of 42–48 mmol/mol (6.00–6.49%), other diabetes risk factors (obesity, family history of diabetes, etc.) should be investigated to plan an intervention to reduce the risk of the disease. In these subjects, it is also appropriate to search for any other cardiovascular risk factors (dyslipidemia, hypertension, etc.) to define the overall cardiovascular risk and initiate appropriate therapeutic measures.

In subjects with IFG, especially with other diabetes risk factors, it is helpful to perform the OGTT. Moreover, metabolic syndrome is associated with high risk of diabetes.

OGTT is performed by administering to the patient 75 g of glucose dissolved in 300 mL of water; blood glucose sampling must be performed before glucose solution (basal glycemia) and 2 h after the solution administration (Table 26.5).

Table 26.5 Interpretation of the oral glucose tolerance test

	Test	
	Fasting glycemia	2-hours glycemia
Normal	<100 mg/dL (<5.6 mmol/L)	<140 mg/dL (<7.8 mmol/L)
IFG (impaired fasting glucose)	100–125 mg/dL (5.6–6.9 mmol/L)	
IGT (impaired glucose tolerance)		140–199 mg/dL (7.8–11.0 mmol/L)
Diabetes mellitus	≥126 mg/dL (≥7.0 mmol/L)	≥200 mg/dL (≥ 11.1 mmol/L)

Glycated hemoglobin is a more practical and reliable parameter than blood glucose; indeed, it has less pre-analytical instability than blood glucose, does not require preparation (e.g., fasting), and is not affected by acute events. The pre-analytical instability of glycemia is due to the in vitro glycolysis in red blood cells after blood sampling, causing a false reduction in blood glucose values. This phenomenon, defined as pseudo-hypoglycemia, can be avoided by centrifuging the samples immediately after the drawing or can be strongly limited by collecting blood in tubes pre-treated with an anti-glycolytic solution.

There is ` hemoglobin and fasting or 2 hours blood glucose levels. This may be partly due to laboratory variability but also, to some extent, could reflect different physiological processes. It is, therefore, plausible that a subject may have diagnostic blood glucose values for diabetes and normal glycated hemoglobin values, or vice versa. For this reason, an altered test must be confirmed by repeating the same test. If the patient has two different tests (i.e., fasting blood glucose and glycated hemoglobin) that agree, then a diagnosis can be made without further tests; if, instead, the patient has two different and discordant tests, then it is advisable to repeat the altered test and make the diagnosis based on this result.

Glycated hemoglobin has some limits because some clinical conditions, both physiological and pathological, can alter its levels, leading to false results (Table 26.6).

In these cases, it is possible to measure glycated albumin, which reflects the albumin glycation and is indicative of blood glucose in the 15–20 days before sample collection. Therefore, glycated albumin is a midterm indicator of blood glucose, earlier than glycated hemoglobin, which reflects blood glucose over the previous 35–45 days. Glycated albumin can also be used in poorly compensated diabetes, gestational diabetes, postprandial hyperglycemia, fluctuating diabetes, gastrectomy because provides information on the medium- to short-term glycometabolic compensation status.

The appropriate differential diagnosis of diabetes mellitus has important prognostic and therapeutic implications (Table 26.7).

Table 26.6 Conditions associated with altered glycated hemoglobin levels

Physiological conditions	Pathological conditions
Pregnancy, age, gender	Hemolytic anemia, iron deficiency anemia, transfusions, splenectomy, renal failure, antiretroviral therapy, alcohol addiction, thalassemia

Table 26.7 Differential clinical features of type 1 and type 2 diabetes mellitus

	Type 1 diabetes	Type 2 diabetes
Symptoms	Always present (polyuria, polydipsia, weight loss, muscle asthenia) Often striking and abrupt onset (ketoacidosis coma)	Often modest or absent
Tendency to ketosis	Present	Absent
Weight	Generally normal	Generally increased
Age at onset	Commonly <30 years; frequent in pediatric age	Commonly >40 years
Chronic complications	Not earlier than a few years after diagnosis	Often present at the time of diagnosis
Circulating insulin	Reduced or absent	Normal or increased
Autoimmunity	Present	Absent
Therapy	Insulin needed from onset	Diet, oral medications, insulin

The clinical picture is often sufficient for differentiating between type 1 and type 2 diabetes mellitus; however, in some cases, it may be necessary to assess autoimmunity biomarkers (insulin autoantibodies (IAAs), GAD autoantibody (GADA), tyrosine phosphatase-related islet antigen 2 (IA-2), ZnT8) and β-cell secretion. A modest percentage of patients initially diagnosed as type 2 diabetes mellitus is affected by LADA. According to the clinical criterion, such patients are classified as type 2 diabetes mellitus and begin treatment based on diet and oral hypoglycemic agents. However, they progressively manifest β-cell function deterioration up to insulin therapy is required. Clinically, LADA should be suspected if one or more of the following features are present:

- Age <50 years
- Body mass index (BMI) <25 kg/m^2
- History of autoimmune diseases, type 1 diabetes or autoimmune diseases family hystory
- The need for insulin therapy within 6–12 months of diagnosis

However, age of onset >50 years and obesity should not lead to a priori exclusion of the LADA diagnosis when the other criteria are met.

Practical diagnostic tests to confirm LADA clinical suspicion are:

- Biomarkers of autoimmunity (autoantibodies)
- β-cell function assessment by C-peptide

The antibodies evaluated in clinical practice are:

- Islet cell antibodies (ICAs): antibodies directed against pancreatic islet cells that bind various islet cell proteins. They have been the gold standard for more than 15 years, but due to the complexity of the analysis, their use is now limited.
- Glutamic acid decarboxylase autoantibodies (GADAs): autoantibodies directed against glutamic acid decarboxylases but not specific to β-cells because these enzymes are present in other organs, such as the brain. GADA is a very early marker.
- Tyrosine phosphatase-related islet antigen 2 (IA-2) antibodies: IA-2 is a transmembrane protein in the secretory granules of endocrine cells, where it is involved in insulin secretion.
- Insulin autoantibodies (IAAs): insulin is the only antigen highly specific for β-cells. These antibodies are detected in approximately 50% of pediatric patients with type 1 diabetes.
- Zinc transporter 8 autoantibodies (ZnT8): ZnT8 is a membrane protein of insulin-containing secretory granules. It is a very early and specific marker. These autoantibodies have been observed in 26% of DM1 patients previously classified as antibody-negative.

Table 26.8 shows when the measurement of antibodies is appropriate.

C-peptide should be measured after glucagon stimulation or a mixed meal. Testing after glucagon stimulation should be performed on fasting. It consists of blood collection at basal and after 6 min following endovenous injection of 1 mg glucagon. C-peptide values <0.2 nmol/L at baseline or < 0.6 nmol/L after stimulation indicates severe insulin secretion deficit and the need for insulin treatment. Blood glucose values >180 mg/dL contraindicate the execution of the test, as the resulting hyperstimulation of the cell would induce an insulin secretion overestimation. The test is helpful for the diagnostic and prognostic framing of cases of uncertain classification. However, it is not the only criterion to guide the therapeutic choice. Concerning MODY, the clinical criteria are:

- Age of onset <25 years
- Metabolic control maintained without insulin for more than 2 years
- Autosomal dominant inheritance (at least three generations of subjects affected by diabetes in the family pedigree)
- Absence of autoimmunity

A strong clinical suspicion of MODY requires the detection of the underlying genetic defect. The appropriate diagnosis of MODY is essential for the prognostic evaluation and family screening.

Concerning gestational diabetes, it is essential to distinguish between gestational diabetes and overt diabetes diagnosed in pregnancy. Pregnant women's normal fasting blood glucose reference value is <92 mg/dL. A fasting glycemia ≥126 mg/dL during the first trimester indicates overt diabetes; instead, a fasting glycemia between 92 and 125 mg/dL indicates gestational diabetes. If fasting glycemia is <92 mg/dL, based on the evaluation of the specific risk factors, the pregnant should undergo an OGTT between the 24th and 28th weeks of gestation, in the case of moderate risk, or between the 16th and 18th weeks in the case of high risk, and if OGTT is negative, it should be repeated between the 24th and 28th weeks (Fig. 26.9).

Moderate risk factors for GDM (OGTT at 24–28 weeks) are:

- Positive family history of diabetes in first-degree relatives
- Previous gestational diabetes (even if normal screening at 16–18 weeks)
- Fetal macrosomia in previous pregnancies
- Overweight or obesity (BMI ≥25 kg/m^2)
- Age ≥35 years
- High-risk ethnic groups (South Asia, Middle East, and Caribbean)

High risk factors for GDM (OGTT at 16–18 weeks) are:

- Obesity (BMI ≥ 30 kg/m^2)
- Previous gestational diabetes mellitus
- Fasting blood glucose of 100–125 mg/dL at the beginning of pregnancy or in the past

In pregnancy, an OGTT should be performed as in the general population; therefore, 75 g of glucose dissolved in 300 mL of water, but blood glucose sampling should be measured at basal (on an empty stomach), i.e., before glucose solution administration, after 1 hour, and 2 hours (Table 26.9).

Women with GDM should be screened for diabetes mellitus 2 by performing a classic OGTT with 75 g glucose after

Table 26.8 Conditions in which autoantibodies measurement is indicated

	Clinical utility
LADA	Confirm diagnostic suspicion
Patients with other organ-specific autoimmune diseases	Early identification of individuals at DM1 risk
First-degree relatives of patients with DM1	Early identification of individuals at DM1 risk

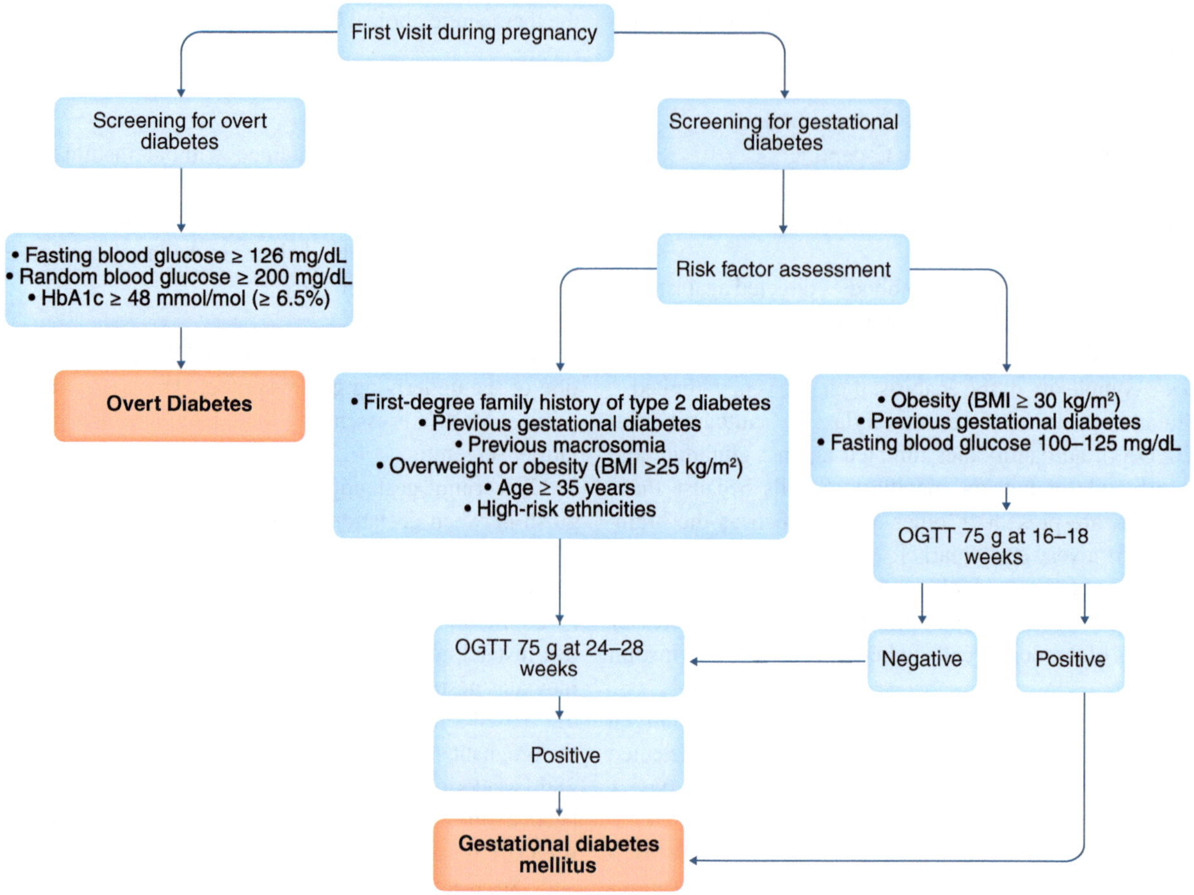

Fig. 26.9 Diagnostic algorithm for gestational diabetes mellitus. (Copyright EDISES 2021. Reproduced with permission)

Table 26.9 Oral glucose tolerance test in pregnancy

Times	Diagnostic glycemia (mg/dL)
0 min	≥92
1 hour	≥180
2 hours	≥153

6 weeks and within 6 months of childbirth. If the test is negative, then the OGTT should be repeated every 3 years; if impaired glucose tolerance (IFG or IGT) is found, then the test should be repeated every year.

Once the diagnosis of diabetes mellitus (any form) has been made, the initial assessment of the patient must include a complete medical examination aimed at defining the general clinical conditions, focusing on any possible chronic complications, through laboratory and instrumental tests. In particular, the initial assessment of a patient with diabetes mellitus is based on:

- Family history
- Physiological history (physical activity practiced, lifestyle, etc.)
- Pathological history
- Physical examination
- Laboratory tests: fasting lipid profile, including total cholesterol, high-density lipoprotein (HDL) cholesterol, triglycerides, and low-density lipoprotein (LDL) cholesterol; liver function tests and any further investigations, if steatosis or hepatitis is suspected; urinary albumin in all patients with type 2 diabetes and type 1 diabetes with disease duration >5 years; creatininemia (in children only in the presence of proteinuria) and estimated glomerular filtration; in patients with type 1 diabetes at diagnosis: screening for autoimmune thyroiditis and celiac disease; anti-insulin and/or anti-GAD and/or anti-IA-2 and/or anti-ZnT8 autoantibodies for the correct classification of the type of diabetes; and urine test to evaluate ketonuria, proteinuria, and sediment

Complications

Complications of diabetes mellitus can be classified into acute and chronic.

Diabetic ketoacidosis is the typical acute complication of DM1. It may represent the clinical onset manifestation, especially in children. Hyperosmolar hyperglycemic coma is the acute DM2 complication.

Diabetic ketoacidosis is due to insulin deficiency and excess counterregulatory hormones (catecholamines, glucagon, growth hormone, cortisol).

In particular, ketosis results from an increased hepatic synthesis of ketone bodies, which, in turn, results from the increased hepatic influx of fatty acids from the adipose tissue, in which the lack of inhibition by insulin of hormone-sensitive lipase determines the hydrolysis of triglycerides into glycerol and fatty acids, which are released into the circulation and reach the liver, where the fatty acids undergo β-oxidation leading to acetyl-CoA synthesis. The ketone bodies synthesis is favored when the acetyl-CoA concentration the Krebs cycle's oxidative capacity. At physiological pH, ketone bodies are present as ketoacids and are neutralized by bicarbonates; however, metabolic acidosis appears when the bicarbonate reserves are exhausted. The main symptoms are nausea, vomiting, abdominal pain, polyuria, polydipsia, asthenia, anorexia, mental status alterations, and dyspnea. The most frequent signs are tachycardia, dry skin, and mucous membranes, Kussmaul breathing, tachypnea, respiratory distress, dehydration, hypotension, fever, lethargy, sensory obnubilation, cerebral edema up to coma.

Hyperosmolar coma is a metabolic complication of type 2 diabetes mellitus that mainly occurs in elderly; the precipitating event is an inadequate water intake that, in subjects already at risk of dehydration due to polyuria, leads to hypovolemia, with a reduction in glomerular filtration that aggravates hyperglycemia. The main signs and symptoms are nausea, vomiting, dehydration, dry skin and mucous membranes, hypotonia of the eyeballs, hypotension, drowsiness, lethargy, and convulsions up to coma. It can be precipitated by infections, stroke, acute myocardial infarction, pancreatitis, uremia, parenteral nutrition, diuretics, peritoneal dialysis, or drugs (e.g., phenytoin, steroids). The diagnosis is based on:

- Severe hyperglycemia (>600 mg/dL)
- Hyperosmolarity (>320 mOsm/L)
- Marked dehydration
- Absent or trace ketones
- Absence of acidosis (pH >7.3; HCO_3 > 15 mEq/L)

Chronic complications can be divided into nonvascular and vascular. The latter, in turn, is classified into microangiopathies, affecting the small vessels, mainly in the retina and renal glomerulus, and macroangiopathies, affecting the peripheral and coronary circulation, with early atherosclerotic lesions formation (Table 26.10). Nonvascular complications include gastroparesis, infections, and skin lesions; in addition, long-term diabetes may be associated with hearing loss.

Table 26.10 Chronic complications of diabetes mellitus

Microvascular	Macrovascular	Nonvascular
Retinopathy, neuropathy, nephropathy	Coronopathy, arterial peripheral disease, cerebral vasculopathy	Gastrointestinal, genitourinary, dermatological infections, cataract, glaucoma, periodontal disease

The risk of complications increases according to the hyperglycemia duration; since type 2 diabetes is often preceded by a long period of asymptomatic hyperglycemia, many patients already have almost a complication at the time of diagnosis. Chronic hyperglycemia determines chronic complications through various mechanisms, including non-enzymatic glycation with the formation of advanced glycation end products (AGEs), activation of protein kinase C (PKC), and defects in the polyol pathway leading to increased oxidative stress.

Monitoring

The clinical laboratory also plays a fundamental role in monitoring patients with diabetes mellitus to evaluate the correct glycemic compensation and/or the possible onset of alterations allowing to promptly modify treatment.

The following laboratory tests must be performed at least once a year:

- Urinalysis to assess glycosuria (glycemia >180 mg/dL); ketonuria (frequent in decompensated DM1); and albuminuria (>30 mg/24 h or > 30 mg/g creatinine), which represents an important biomarker of diabetic nephropathy and is associated with a significant risk of cardiovascular disease
- Estimated GFR (eGFR) to assess renal function
- Ketonemia
- Complete lipid profile
- HbA1c to assess long-term glycometabolic control
- Glycated albumin to assess midterm glycometabolic control
- C-peptide to assess the residual secretory capacity of pancreatic β-cells
- Insulin, especially in patients with DM2, to assess the degree of insulin resistance. Insulin should be measured after stimulation (OGTT). In patients with DM2, the insulinemic response will be delayed and persistent.

Measurement of glycated hemoglobin is the gold standard for glycometabolic controlevaluation in diabetic patients (Table 26.11).

Table 26.11 Interpretation of glycated hemoglobin (HbA1c) values

HbA1c	Interpretation
20–41 mmol/mol (4–5.9%)	Normal values
42–48 mmol/mol (6–6.49%)	High risk of diabetes mellitus
≥48 mmol/mol (≥6.5%)	Diabetes mellitus diagnosis
<53 mmol/mol (<7%)	Therapeutic goal in patients with diabetes mellitus
<48 mmol/mol (<6.5%)	Low risk of complications in patients with diabetes mellitus
>64 mmol/mol (>8%)	Modify the diabetes mellitus therapy

The guidelines recommend to measure HbA1C:

- Two times per year in patients with stable metabolic control and who have reached the therapeutic target
- More than two times per year in patients with unstable metabolic control

It is inappropriate to measure HbA1C less than 2 months apart.

Therapy

Glycemic goals in patients with type 1 and type 2 diabetes are:

- Glycated hemoglobin <53 mmol/mol (<7%)
- Fasting and preprandial blood glucose of 70–130 mg/dL
- Postprandial blood glucose <160 mg/dL

Glycated hemoglobin represents the primary treatment target due to the close relationship between this index of glycemic control and micro- and macroangiopathic complications.

The diabetes treatment relies on a correct diet, physical activity, and, if necessary, glucose lowering drugs (insulin, oral hypoglycemic drugs).

DM1 requires therapy with exogenous insulin, due to the insulinopenia resulting from the pancreatic β-cells destruction; in DM2, on the other hand, treatment is based on lifestyle interventions (diet and exercise), oral hypoglycemic drugs, and, in some cases, insulin. In patients with DM2, metformin is the drug of the first choice. If monotherapy cannot achieve the glycemic goal, then combination therapy with metformin associated with other molecules can be evaluated. Insulin therapy is essential when glycemic control is unsatisfactory, even in polytherapy.

In a DM1 patient, the goal is to balance and combine caloric intake with the appropriate amount of insulin; therefore, diet and blood glucose monitoring must be integrated to define the optimal insulin regimen. In addition, weight gain often associated with intensive antidiabetic treatment should be minimized.

A DM1 patient can also undergo pancreas or islets of Langerhans transplantation, which aims to reduce the need for exogenous insulin while eliminating dangerous manifestations such as hypoglycemia.

Pregnant with gestational diabetes during should has the following glycemic goals:

- Fasting blood glucose <92 mg/dL
- Blood glucose 1 h after meals <140 mg/dL
- Blood glucose 2 h after meals <120 mg/dL
- HbA1C <42 mmol/mol (<6%)

The treatment of gestational diabetes is primarily nutritional, to provide adequate maternal and fetal nutrition, adequate caloric, vitamin, and mineral intake, and optimal glycemic control without ketonuria/ketonemia. If glycemic goals are not achieved after 2 weeks of a proper diet, then insulin therapy should be instituted. Oral hypoglycemic drugs are not currently recommended in pregnancy. Immediately after the diagnosis of gestational diabetes, pregnant should self-monitor glycemia by daily measurements.

Recommended Readings

Ahmed A (2002, April) History of diabetes mellitus. Saudi Med J 23(4):373–378

Ashcroft FM (2005) ATP-sensitive potassium channelopathies: focus on insulin secretion. J Clin Invest 115:2047–2058

Associazione Medici Diabetologi (AMD) – Società Italiana di Diabetologia (SID) Standard italiani per la cura del diabete mellito (2016)

Basta G, Del Turco S, De Caterina R (2004) Prodotti di glicosilazione avanzata: possibili implicazioni per l'aterosclerosi nel diabete. Febbraio 95(2)

Bellia C, Zaninotto M, Cosma C, Agnello L, Bivona G, Marinova M, Lo Sasso B, Plebani M, Ciaccio M (2018) Clinical usefulness of glycated albumin in the diagnosis of diabetes: results from an Italian study. Clin Biochem 54:68–72. https://doi.org/10.1016/j.clinbiochem.2018.02.017

Bellia C, Zaninotto M, Cosma C, Agnello L, Lo Sasso B, Bivona G, Plebani M, Ciaccio M (2017) Definition of the upper reference limit of glycated albumin in blood donors from Italy. Clin Chem Lab Med 56(1):120–125. https://doi.org/10.1515/cclm-2017-0179

Bernoville F, Beck-Peccoz P (1999) Diabetes Mellitus. Jain Publishers

Bonora E et al (2004) Population-based incidence rates and risk factors for type 2 diabetes in white individuals. The Bruneck Study. Diabetes 53:1782–1789

Brocco S et al (2007) Monitoring the occurrence of diabetes mellitus and its major complications: the combined use of different administrative databases. Cardiovasc Diabetol 6:5

Charbonnel B, Schernthaner G, Brunetti P et al (2005) Long-term efficacy and tolerability of add-on pioglitazone therapy to failing monotherapy compared with addition of gliclazide or metformin in patients with type 2 diabetes. Diabetologia 48:1093–1104

Chilelli NC, Piarulli F, Sartore G, Lapolla A (2016) Ruolo dei prodotti di glicazione avanzata nelle complicanze microvascolari del diabete di tipo 2: "attori protagonisti" nella memoria metabolica? G It Diabetol Metab 36:22–27

de Vegt F, Dekker JM, Stehouwer CD, Nijpels G, Bouter LM, Heine RJ (1998) The 1997 American Diabetes Association criteria versus the 1985 World Health Organization criteria for the diagnosis of abnormal glucose tolerance: poor agreement in the Hoorn Study. Diabetes Care 21(10):1686–1690

DeFronzo RA, Barzilai N, Simonson DC (1991) Mechanism of metformin action in obese and lean non insulin-dependent diabetic subjects. J Clin Endocrinol Metab 73:1294–1301

Drucker DJ (2006) The biology of incretin hormones. Cell Metab 3:153–165

Drucker DJ, Nauck MA (2006) The incretin system: glucagon-like peptide-1 receptor agonists and dipeptidylpeptidase4 inhibitors in type 2 diabetes. Lancet 368:1696–1705

Ferrera A, Kahn HS, Quesenberry CP, Riley C, Hedderson MM (2004) An increase in the incidence of gestational diabetes mellitus: northern California, 1991–2000. Obstet Gynecol 103(3):526–533

Fourlanos S, Perry M, Stein MS et al (2006) A clinical screening tool identifies autoimmune diabetes in adults. Diabetes Care 29:970–975

Groop L, Tuomi T, Rowley M et al (2006) Latent autoimmune diabetes in the adults (LADA) more than a name. Diabetologia 49:1996–1998

Group NDD (1979) Classification and Diagnosis of Diabetes Mellitus and Other Categories of Glucose Intolerance. Diabetes 28(12):1039–1057. https://doi.org/10.2337/diab.28.12.1039

Hasslacher C (2003) Multinational Repaglinide Renal Study Group. Safety and efficacy of repaglinide in type 2 diabetic patients with and without impaired renal function. Diabetes Care 26:886–891

Holst JJ (2006) Glucagon-like peptide-1: from extract to agent. The Claude Bernard lecture, 2005. Diabetologia 2:253–260

International Expert Committee (2009) International expert committee report on the role of HbA1c assay in the diagnosis of diabetes. Diabetes Care 32:1327–1334

ISTAT (2015) Annuario statistico italiano. http://www.istat.it/it/archivio/171864

Lapolla A, Dalfrà MG (2006) Screening e diagnosi del diabete gestazionale. Biochimica Clinica 30:5–6

Linee guida gravidanza fisiologica. Ministero della Salute (n.d.). http://www.salute.gov.it/imgs/C_17_pubblicazioni_1436_allegato.pdf

Rendell M (2004) The role of sulphonylureas in the management of type 2 diabetes mellitus. Drugs 64:1339–1358

Roberto T, Rita BA (2013) Le proteine glicosilate nel diabete. Riv Ital Med Lab 9:218–224

Thanabalasingham G, Pal A, Selwood MP et al (2012) Systematic assessment of etiology in adults with a clinical diagnosis of young-onset type 2 diabetes is a successful strategy for identifying maturity-onset diabetes of the young. Diabetes Care 35:1206–1212

Van de Laar FA, Lucassen PL, Akkermans RP et al (2005) Alpha-glucosidase inhibitors for type 2 diabetes mellitus. Cochrane Database Syst Rev:CD003639

Diabetes Mellitus: The Role of the Laboratory

27

Andrea Mosca and Martina Montagnana

Introduction

The laboratory plays a central role in diagnosing and monitoring diabetes mellitus. The list of the main tests that can be used in this setting is shown in Table 27.1. The information was taken from the American Diabetes Association (ADA) guidelines, the consensus document of the Italian standards of care for diabetes mellitus, and various documents prepared by the intersociety study group of the Italian Society of Clinical Biochemistry and Molecular Biology and the Italian Society of Clinical Pathology and Laboratory Medicine (SIBioC-SIPMeL) on diabetes mellitus.

Role of the Laboratory in Screening and Diagnosis

Concerning preclinical diagnosis (screening), it is useful to treat type 1 diabetes separately from type 2 and gestational diabetes.

For type 1 diabetes, various evidence from studies on animal models suggests that an immune therapeutic intervention before the onset of symptoms can delay or prevent their onset. However, the numerous clinical trials conducted on humans have produced disappointing results. Nevertheless, the ADA recommends that first-degree relatives of patients with type 1 diabetes be screened for autoantibodies, specific markers of immune aggression to β-cells, with the recommendation that those who screen positive should then be referred to specialist centers.

Table 27.1 Role of the laboratory in the management of patients with diabetes mellitus

	Type of exams
Diagnosis	
Preclinical (screening)	Immunological biomarkers (insula cytoplasmic antibodies, anti-insulin antibodies, glutamic acid carboxylase autoantibodies, anti-tyrosine phosphatase antibodies, zinc transporter 8 antibodies) Genetic biomarkers (e.g., human leukocyte antigens) Test for insulin secretion (fasting, in response to glucose load) Glycemia HbA1c
Clinic	Glycemia Oral glucose tolerance test HbA1c Ketone bodies (in urine and blood)
Surveillance	
Acute conditions	Glycemia Chetonic bodies Acid-base state (pH, bicarbonate) Lactate Other examinations related to pathophysiology or therapy (e.g. potassium, sodium, phosphate, osmolality)
Chronic conditions	Glycemia Glycated proteins (HbA1c, glycated albumin) Urinary proteins (urinary albumin excretion, albuminuria) Tests to evaluate complications (e.g., creatinine, cholesterol, triglycerides) Tests to evaluate pancreas transplantation (C-peptide, insulin) Eligibility for implantation of the insulin pump (C-peptide)

HbA1c glycated hemoglobin

Human leukocyte antigen genotype research, perhaps interesting from a research point of view, is not currently recommended for diabetes screening. Similarly, measurement of insulin secretion, the first functional abnormality found at the onset of both type 1 and type 2

A. Mosca
Department of Physiopathology and Transplantation and Center for Metrological Traceability in Laboratory Medicine (CIRME), University of Milan, Milan, Italy

M. Montagnana (✉)
Section of Clinical Biochemistry, University Hospital of Verona, Verona, Italy
e-mail: martina.montagnana@univr.it

M. Ciaccio (ed.), *Clinical and Laboratory Medicine Textbook*, https://doi.org/10.1007/978-3-031-24958-7_27

diabetes, is not currently recommended in routine laboratory testing.

Screening for type 2 diabetes in asymptomatic subjects has been a matter of much controversy. However, the rationale is that since it is estimated that, on average, about 30% of individuals with type 2 diabetes have the disease without having yet received a diagnosis, to prevent complications, the diagnosis should be made as early as possible. Currently, the ADA recommends screening for all asymptomatic individuals over 45 years of age. Screening can be done on younger subjects in the presence of risk factors (e.g., in the case of familial or ethnic group membership). In particular, based on the increasing prevalence of type 2 diabetes among adolescents, screening is recommended if there are conditions associated with increased insulin resistance or if there is a history of diabetes in the mother before pregnancy or gestational diabetes. Screening should be repeated every 3 years, starting at age 10.

In some countries, there may be significant differences. For example, in Israel, the prevalence of undiagnosed diabetes is negligible; the entire population is subjected continuously throughout their lives (at school age, during and after military service, etc.) to rigorous medical testing.

Finally, as regards the diagnosis, in the absence of symptoms, the measurement of fasting blood glucose, glycated hemoglobin (HbA1c), or the oral glucose tolerance test (OGTT) can be performed and should be repeated on a second occasion. See below for threshold values and interpretation.

Role of the Laboratory in Surveillance

The laboratory also plays an essential role in the surveillance of acute and chronic complications. For the former, especially in diabetic ketoacidosis, nonketotic hyperosmolar coma, and hypoglycemia, several molecules can be frequently measured to help clinicians implement appropriate therapy to return glycemic control to physiologic conditions and to correct other metabolic disorders. The National Academy of Clinical Biochemistry guidelines provide helpful information on the tests that can be performed in these acute conditions.

Regarding the surveillance of diabetes and the development of chronic complications, the Diabetes Control and Complications Trial (DCCT) and the United Kingdom Prospective Diabetes Study (UKPDS) have excellently documented the extent to which good glycemic control significantly reduces the development of micro- and macrovascular complications of diabetes. Therefore, the measurement of glucose and glycated proteins provides short- and long-term indices of the degree of such control. Detection and surveillance of complications are also achieved by measuring creatinine, urinary albumin secretion, and lipid balance. Finally, the success of new therapies, such as islet cell or pancreatic transplantation, can be followed by measuring C-peptide or insulin concentrations.

Antibodies

Autoantibodies are helpful markers in diagnosing type 1 diabetes mellitus, especially in the preclinical period (prediabetes). Indeed, it has been demonstrated that at least one autoantibody is present in the serum of 95–98% of these patients. The most important autoantibodies are anti-pancreatic insulin (ICA), anti-glutamic acid carboxylase (GAD), anti-insulin (IAA), anti-tyrosine phosphatase (IA-2A), and anti-transporter of zinc 8 (ZnT8).

In 1977, the presence of ICA was also identified in some patients diagnosed with type 2 diabetes. This type of autoimmune diabetes evolving towards insulin dependence was named latent autoimmune diabetes in adults. It is a form of diabetes that does not require insulin therapy for at least 6 months after clinical diagnosis and is characterized by the presence of autoantibodies against GAD, IA-2, and, more rarely, ZnT8 and insulin.

The following are the main autoantibodies used so far. From an analytical point of view, autoantibodies must be measured only in an accredited laboratory with a quality control program and an external quality assurance program.

Pancreatic Insula Cytoplasmic Antibodies (ICA)

They are a group of autoantibodies directed against pancreatic islet cells that bind to multiple islet cell proteins. They appear early and are present in 70–80% of new diagnoses of type 1 diabetes. They have long represented the gold standard for evaluating anti-islet antibodies. Currently, they are no longer evaluated, partly due to the difficulty of standardizing the immunofluorescence methodology used for their measurement but above all due to the difficult availability of the human pancreas necessary for their execution.

GAD Antibodies (GADA)

These are tests for autoantibodies directed against β-cell proteins (antigens) but not specific for β-cells. They represent the most frequent marker of autoimmunity found in new diagnoses of diabetes (about 70–80%), present in significant percentages in every age group.

Tyrosine-Phosphatase 2 Antibodies (IA-2A)

These are tests for autoantibodies directed against β-cell antigens, but they are not specific. They are positive in approximately 60% of patients with type 1 diabetes. Tyrosine phosphatase 2 (IA-2) is a transmembrane glycoprotein in the secretory granules of endocrine cells involved in insulin secretion. It has been demonstrated by radioimmunoassays that IA-2 autoantibodies are highly specific for type 1 diabetes and directed towards different epitopes depending on the age of onset.

Anti-insulin Antibodies (IAA)

These are autoantibodies directed against insulin; insulin is the only antigen considered highly specific for β-cells. These antibodies are found in approximately 50% of pediatric patients with type 1 diabetes, while they are not commonly measured in adults. A limitation of this test is that it does not distinguish between autoantibodies that recognize endogenous insulin and antibodies produced against exogenous insulin.

Zinc Transporter 8 Antibodies (ZnT8)

This autoantibody is directed toward a membrane protein, zinc transporter, ZnT8, which is found in the insulin secretory granules.

Glucose

Fasting Blood Glucose

The measurement of fasting blood glucose is a crucial test for the diagnosis of diabetes mellitus and for the screening of subjects at high risk of developing this disease. This measurement must be carried out in a laboratory, and great care must be taken in the appropriate collection and storage of the sample (preanalytical phase), in the choice of the method (analytical phase), and in the reporting and communication of results (post-analytical phase).

Preanalytical Aspects

Several preanalytical variables can affect blood glucose measurement.

As reported in the most recent recommendations on screening and diagnosis of diabetes mellitus, glycemia must be measured in venous plasma, respecting a fasting time of not less than 8 and not more than 12 h. The choice of the biological matrix to be used (whole blood, plasma, or serum) is essential because the results obtained on different matrices present a significant variation: glycemia measured in whole blood is lower than that measured in plasma since red blood cells contain less water than plasma and glucose dissolves in free water. The molality of glucose (i.e., the amount of glucose per unit mass of water) is identical in whole blood and plasma. Although red blood cells are freely permeable to glucose, the water content is about 93% in plasma and 73% in erythrocytes. Consequently, if the hematocrit is normal, plasma glucose is about 10–12% higher than whole blood glucose. Serum blood glucose is also about 5% higher than plasma.

Another critical aspect is the choice of glycolysis inhibitor, which must be contained in the tube if the plasma cannot be separated from the corpuscular elements within 30 min of collection.

To minimize glycolysis, guidelines recommend placing the tube in water with melting ice immediately after collection and separating the plasma from the cells by centrifugation.

The addition of sodium fluoride, an enolase inhibitor, produces an antiglycolytic action. However, a study has shown that this inhibition is lost within 4 h of collection. In this regard, SIBioC issued a formal recommendation in 2018 that the ternary mixture with citrate, sodium fluoride, and ethylenediaminetetraacetic acid (EDTA) should be used for blood glucose measurement for diagnostic purposes. Lithium heparin with sodium fluoride may still be used to monitor the ward's glycemic control.

The glucose consumption rate in vitro, at room temperature, averages 5–7% per hour, which can be quantified as approximately 0.6 mmol/L (10 mg/dL). However, some variables, such as glucose concentration, storage temperature, the number of white blood cells, and hematocrit, can increase or decrease the rate of glycolysis in whole blood. Proper storage of the sample allows blood glucose to remain stable for a more extended period. Glucose concentration stability in separated, unhemolyzed, sterile serum without glycolysis inhibitors has been demonstrated for 8 h at 25 °C and 72 h if the sample is stored at 4 °C.

Intraindividual (CVi) and interindividual (CVg) biological variabilities were 4.5% and 5.8% on plasma and 5.6% and 7.5% on serum, respectively.

There is, moreover, a circadian variability of glycemia, determined by the secretion of hormones that influence glucosidic metabolism (growth hormone, cortisol, and corticotropin); the average values of fasting glycemia are therefore higher in the morning than in the afternoon. For this reason, it is recommended to have blood glucose measurements between 7 and 10 a.m.

Analytical Aspects

Blood glucose measurement is performed in most laboratories using well-standardized enzymatic methods. Although enzymatic methods based on glucose oxidase and glucose dehydrogenase are also commercially available, guidelines recommend using methods based on hexokinase. The latter, using ATP, catalyzes the phosphorylation of glucose into glucose-6-phosphate. Glucose-6-phosphate dehydrogenase, in the presence of nicotinamide adenine dinucleotide phosphate (NADP), causes the oxidation of glucose-6-phosphate to gluconate-6-phosphate. The rate of NADPH formation during the reaction is directly proportional to glucose concentration and is measured photometrically.

The methods based instead on glucose oxidase are more subject to interference from various types of substances (e.g., uric acid, bilirubin), which reduce hydrogen peroxide (H_2O_2) with the risk of underestimating glycemia. The influence of hemolysis and lipemia is negligible.

Based on biological variability, the glucose determination should have an analytical imprecision of <2.9%, bias, <2.2%, and total error of <6.9%.

Postanalytical Aspects

Blood glucose values should be expressed in mmol/L, the unit of measure adopted by the international system and, as such, the international reference standard. However, blood glucose is still expressed in mg/dL in many laboratories.

Fasting plasma glucose concentration increases with increasing age between the third and sixth decades and then remains constant.

Extreme blood glucose values, that is, below 2.2 mmol/L (40 mg/dL) or above 27.7 mmol/L (500 mg/dL), should be reported as quickly as possible so that the clinician can implement measures to resolve the acute critical event (severe hypo or hyperglycemia).

Postprandial Blood Glucose

Postprandial hyperglycemia is a fairly common event in patients with type 1 and type 2 diabetes mellitus and can occur even in the presence of good metabolic control, as evidenced by acceptable HbA1c values.

Postprandial hyperglycemia is an independent risk factor for macrovascular disease, correlates with increased risk of retinopathy and carotid intima-media thickness, and causes oxidative stress, inflammation, and endothelial dysfunction.

Plasma glucose, which should be measured strictly 2 h after the meal, should not exceed a concentration of 7.8 mmol/L (140 mg/dL). Self-monitoring of blood glucose using point of care testing (POCT) is currently the most practical method of monitoring postprandial blood glucose.

Please refer to the section "Fasting blood glucose" for preanalytical, analytical, and postanalytical aspects.

Oral Glucose Tolerance Test

OGTT is one of the tests that can be used to diagnose diabetes mellitus or impaired glucose tolerance (IGT), and it is useful for a better diagnostic and prognostic assessment, especially in subjects with impaired fasting glucose (IFG), because a significant proportion of these subjects have a glucose response compatible with the diagnosis of diabetes.

The OGTT, an examination with low reproducibility, must be performed correctly and standardized. Two samples are collected, the first on an empty stomach and the second 2 h after oral intake of a glucose solution (300 mL) containing anhydrous glucose (75 g in adults and 1.75 g per kg of body weight in children, up to a maximum of 75 g).

Preanalytical Aspects

The test should not be performed in subjects with a documented history of diabetes mellitus or with investigations indicating the presence of diabetes mellitus; it should not be performed during acute pathologies and in the presence of factors that may influence the test itself, such as in the presence of recent infections, during convalescence from surgery, or during treatment with life-saving drugs; and finally, it should not be performed in gastro-resected subjects because of the risk of evoking the symptoms of the syndrome of impaired gastric emptying.

In the 3 days preceding the test, the subject should consume a standardized amount of carbohydrates (100–150 g per day) without changing the diet. The intake of nonlife-saving drugs (e.g., thiazides, contraceptives, and corticosteroids) should be postponed until after the OGTT has been performed.

The examination should be carried out on an empty stomach (at least 8 h and not more than 14 h).

The speed of intake of the glucose drink, administered at a temperature of 15–25 °C, should not exceed 5 min.

During the test, the subject must remain seated and is not allowed to drink, eat, or smoke. The test must be performed in the morning to avoid alterations in glycemia due to the previously mentioned circadian variability.

Throughout the procedure, the subject must remain in an area under the control of medical personnel, who can intervene in cases of illness and possibly decide to suspend the examination. The examination is not valid if vomiting occurs.

Analytical Aspects

Plasma blood glucose is measured using the methods described in section "Fasting blood glucose."

Postanalytical Aspects

As pointed out in section "Fasting blood glucose," basal and afterload blood glucose values during OGTT should be expressed in mmol/L. However, many laboratories still express blood glucose in mg/dL. Plasma glucose reading 2 h after OGTT of less than 7.8 mmol/L (140 mg/dL) is considered normal. If the value obtained is between 7.8 and 11.0 mmol/L (respectively 140 and 199 mg/dL), the diagnosis of IGT is made.

Diabetes mellitus is diagnosed (with a second diagnostic criterion met) if blood glucose 2 h after OGTT is greater than or equal to 11.1 mmol/L (200 mg/dL).

Blood Glucose Self-Monitoring Using POCT

According to the most recent position statement of the American Diabetes Association, self-monitoring of blood glucose using POCT is considered an essential component in the management of the diabetic patient since it helps monitor glycemic control, detects hypoglycemia, assists in adjusting therapy, helps in understanding glycemic fluctuations, and can reduce or delay the onset of vascular, neurological, renal, and retinal complications.

Self-monitoring of blood glucose is carried out by sophisticated electronic devices that are easy to use and take up little space. There are different types of instruments on the market based on different principles.

A small blood sample is taken from the diabetic patient by superficial skin puncture, usually on the outside of the fingers or, if preferred, on the fingertip, the sites with the best blood supply. Alternative sampling sites (forearm or ear lobe) may also be used, especially in children. The drop of capillary blood (a few microliters) is then deposited on a strip inserted into the machine's slot.

The principle on which the most common glucose meters are based is the quantification of an enzymatic reaction, the oxidation of glucose, which is proportional to the concentration in the blood. The reaction is carried out by placing a blood sample on a strip containing glucose oxidase (GOD) as the enzyme. During the oxidation reaction, gluconic acid and H_2O_2 are produced. Subsequent quantification of the reaction can be obtained either reflectometrically (measurement of light reflected from the colored test strip as a result of the chromogenic reaction catalyzed by glucose oxidase and peroxidase and subsequent conversion of the signal to glucose concentration) or impedance (electrical conductivity of blood induced by the electrical current generated by glucose oxidation). The electrons produced by the reaction form a current that is specifically calibrated to reflect the glucose concentration in whole blood.

In addition to devices based on the GOD reaction, there are others commercially available that use the reaction of glucose-dehydrogenase (GDH), an enzyme that catalyzes the formation of glucono-δ-lactone d NADH. Three different types of enzymes can be used in GDH-based glucose meters: glucose 1-dehydrogenase (1-GDH), glucose-6-phosphate 1-dehydrogenase (G6PDH), and quinoprotein glucose dehydrogenase (PQQ-GDH). As with all enzymes belonging to the class of dehydrogenases, they can have the natural cofactor (NAD or NADP) unbound or weakly bound to the enzyme (1-GDH and G6PDH) and therefore must necessarily be added to the glucose meter test strip or strongly bound to the enzyme as in the case of PQQ-GDH.

The heterogeneity of the devices, the modalities of use, and the related analytical techniques make necessary the training of the patient in the use of the glucometer (with particular attention to the preanalytical, analytical, and postanalytical aspects) and the periodic verification of the training. The choice of the device must be made based on a series of considerations, of which the economic aspect is only one of the points to be considered.

Preanalytical Aspects

Many preanalytical factors can affect the accuracy of blood glucose measurement, including patient preparation, storage, and use of strips. Preanalytical procedures are elementary but can be a source of important measurement errors, so, even if trivial, they must be explained very well to the patient. Before proceeding to the puncture site, a thorough cleansing of the skin is necessary, especially if foods with a high sugar content have been handled. The skin can be disinfected with alcohol, taking care that no residue remains that could hemolyze or dilute the drop of blood obtained. The choice of the type of lancing device (with a larger or smaller needle gauge) must be made by taking into account the skin thickness of the subject, who must undergo self-monitoring.

Other patient-related preanalytical factors, such as diseases restricting circulation, excessive hydration, or extreme hematocrit levels, may make the blood glucose measurement unreliable.

Test strips should be stored correctly in their original containers and carefully resealed after each use. They should not be exposed to high temperatures, humidity, altitude, or other environmental factors that can interfere with chemical reactions. In addition to the strips, the instrument should be maintained in good condition and serviced periodically.

Analytical Aspects

Although self-monitoring of blood glucose (SMBG) devices have traditionally been held to less stringent analytical standards than laboratory tests, there are guidelines that define analytical goals. To meet analytical accuracy requirements, 95% of capillary measurements must be within ±15 mg/dL (±0.83 mmol/L) of the mean of the values determined by the reference procedure for glucose concentrations below

100 mg/dL (<5.5 mmol/L) or ± 15% for values above 100 mg/dL. However, not all available instrumentation achieved the same analytical standard.

The patient undergoing self-monitoring must implement specific procedures to ensure proper meter operation and quality of results. It is essential that calibration be performed for those meters that do not have a self-calibration feature. The calibration is a procedure by which you put the signal provided by the instrument (instrumental reading) with the quantity to be measured and correct the differences between different batches of strips. It can be done by inserting a code, chip, or calibrating strip, which is different from instrument to instrument. It improves accuracy, so it is a process that must be repeated over time. Calibration errors are a persistent cause of error.

Control materials, consisting of control strips or solutions of known concentration simulating the analytical signal detected by the system reader, shall be used to verify the quality of the data produced. This must be carried out when the instrument is first used, and then a new lot of strips are opened if the device is damaged or if there is any doubt about the reliability of the data.

It is expected that a periodic check for correlation or interchangeability with the central laboratory results will also be implemented on the POCT analyzers, which serves both to check that the defined comparability requirements between the two systems are maintained and to transfer the documented quality of the laboratory analyzer to the POCT system.

There are many substances (e.g., drugs) that can interfere with the chemical reactions taking place on the test strip. The most important of these are ascorbic acid, paracetamol, and maltose.

Very high plasma triglyceride concentrations, typically above 2000 mg/dL, can lead to an underestimation of blood glucose by a glucometer because they decrease the amount of glucose present in the capillary volume. High bilirubin concentrations (usually greater than 500 mg/dL), as seen in jaundiced infants, patients with hepatopathy, hepatitis, or some forms of anemia, can create interference, especially when using glucose meters based on the GDH method.

Ascorbic acid, commonly known as vitamin C, has the potential to interfere with glucometer measurements, although levels of 1–2 mg/dL do not generally result in significant interference. However, overuse of vitamin C can cause this type of interference.

Finally, the type of interference may depend on the chemistry of each test strip, so it may also vary from one meter to another.

As new glucose meters continue to appear on the market, it is helpful to continually consult the literature to see comparative assessments.

Postanalytical Aspects

Two main postanalytical issues need to be known: the international unit of measurement (mmol/L) choice and the use of plasma-calibrated strips. In 2001, the International Federation of Clinical Chemistry and Laboratory Medicine (IFCC) recommended using plasma-calibrated strips, which transform the value obtained from whole blood into the value obtained from plasma. This allows the data obtained by the capillary blood glucose meter to be compared with the data obtained by measuring blood glucose in venous plasma.

Glycemic Variability

Various devices that continuously measure blood glucose in the subcutaneous tissue, known as continuous glucose monitoring (CGM), have recently become available. The availability of these devices now makes it possible to obtain a precise profile of a single patient's glycemic variations for some days (up to 14 days) and to precisely quantify this quantity. It has been known for a long time that, with the same glycemic control, subjects with higher peaks of glycemic fluctuations, more frequently, are at greater risk of complications than others.

The advantages of using these systems compared to traditional POCT for SMBG are unclear. However, it seems, based on the results obtained in a very recent clinical trial, that the use of CGM systems, compared to SMBG systems, resulted in an increase in the time spent in euglycemia, a reduction in the number of severe hypoglycemias in patients with type 1 diabetes, and a loss of sensitivity to hypoglycemia. This study, therefore, and the others that will undoubtedly follow, would seem to support using such tools, at least in a population of high-risk diabetics.

Ketone Bodies

Measurement of blood (ketonemia) or urine (ketonuria) concentrations of ketone bodies (acetone, acetoacetate, and D-β-hydroxybutyrate) is recommended to aid in the diagnosis of diabetic ketoacidosis in patients with type 1 diabetes and gestational diabetes. Indeed, the presence of ketones may indicate the onset of ketoacidosis, a condition that requires immediate medical action. In the patient with this condition, the increased concentration of ketone bodies leads to a lowering of the blood pH, resulting in acidosis (arterial pH <7.3 and/or HCO_3^- levels <15 mEq/L).

Under normal conditions and with a balanced diet, ketone bodies are produced in small amounts because acetyl-CoA is mainly used in the citric acid cycle. Conversely, under conditions (e.g., diabetes and prolonged fasting) that lead to accumulation of acetyl-CoA (as a result of increased gluconeogenesis in the liver and fatty acid oxidation in muscle and liver), the liver gets rid of excess fatty acids by producing ketone bodies, which it sends through the bloodstream to peripheral tissues where they are oxidized to produce energy (hydroxybutyrate).

The symptoms that characterize diabetic ketoacidosis are fatigue, general malaise, polyuria, polydipsia, cardiac arrhythmias, drowsiness, weight loss, and bradypnea. Other signs that appear are dehydration, hypotension, ECG abnormalities, brain dysfunction, loss of muscle mass, and Kussmaul's breath.

The concentration of ketone bodies can be measured both in plasma and in urine. However, we must emphasize that ketonuria does not express exactly the level of ketonemia in real time, but it provides indirect information about it, with a delay proportionate to the length of the interval between emptying the bladder and the other. Moreover, it does not detect the presence of all three ketone bodies but only of acetoacetate since β-hydroxybutyrate is not detectable by standard sticks, and acetone is eliminated by breath.

Preanalytical Aspects

Because acetone is volatile and acetic acid may be an energy source for bacteria, the analysis of fresh urine samples is recommended.

Increased levels of ketone bodies may be present in healthy subjects during fasting and in the morning urine of 30% of pregnant women.

Analytical Aspects

Urinary tests determine the presence of acetoacetate (which accounts for 10–25% of the total ketone bodies) using a colorimetric reaction with sodium nitroprusside. At alkaline pH, sodium nitroprusside reacts with acetone and acetoacetic acid (a product of acetone), forming a complex that results in color development. It is susceptible to acetoacetic acid and does not react with more complex ketone bodies.

False positives (in the case of the intake of drugs based on sulfhydryl compounds such as captopril, acetylcysteine, and penicillamine) and false negatives may occur, the latter related to an excess of vitamin C, excessive intake of liquids that can dilute the urine, and the degradation of the test strips over time.

In recent years, POCT methods have been developed for determining ketone bodies in capillary blood. These instruments measure β-hydroxybutyrate, which represents the predominant and earliest-appearing ketone.

Albuminuria

Urinary albumin is an essential biomarker for the development of diabetic nephropathy and is also a known marker of endothelial damage, which leads to a significant risk of cardiovascular disease. Therefore, its determination is indicated in diabetic patients at risk of renal (renal failure) and cardiovascular complications.

The progression of urinary albumin excretion increases the likelihood of developing chronic renal failure. In addition, the presence of subclinical amounts of albumin in the urine (previously identified by the term "microalbuminuria," to be dropped) is considered a cardiovascular risk factor for diabetic and nondiabetic patients.

Even after the institution of the treatment plan, measurement of albuminuria is recommended to evaluate the effectiveness of treatment. Guidelines indicate that urine albumin measurement should be performed annually in patients with type 1 diabetes with a disease duration greater than 5 years, all patients with type 2 diabetes regardless of disease duration, and pregnant patients. For the diagnosis of diabetic nephropathy, albuminuria must be positive in at least two out of three samples collected within 3–6 months.

In the DCCT and UKPDS trials, it was found that intensive therapy for diabetes reduces the risk of increased urinary albumin excretion (UAE) and overt nephropathy in diabetic patients. A new guideline on managing chronic kidney disease, which also covers the use of albuminuria measurement, has recently been published and is referred to for further discussion.

Preanalytical Aspects

The standardization of urinary albumin measurement is not optimal, so making clear and consistent recommendations is essential. Urinary albumin excretion can be increased by physiological factors such as exercise, posture, and diuresis, and samples should not be collected after strenuous exercise (up to 2–3 days before testing), in the presence of a urinary infection, during a period of convalescence, or after surgery. The urine sample from the 24-h collection and the first-morning specimen (on which to measure albumin and creatinine) are suitable. The collection of the first-morning sample is preferable because it is less affected by biological variability than the other methods listed.

Due to the high individual and daily variability, performing at least three different urine collections on three different days is advisable. The urine sample should then be stored at 4 °C (stable for up to 2 weeks) or at −80 °C if more extended storage is required (stable for up to 5 months). Freezing at −20 °C is not recommended. Freezing the sample may lead to a decrease in albumin. To avoid this problem, the sample should be thoroughly mixed before analysis. Visually turbid samples should be centrifuged to remove precipitates and cellular components. Possible bacterial contamination and glucose do not affect the investigation.

Analytical Aspects

Both semi-quantitative and quantitative tests are available for the determination of increased UAE. Semi-quantitative tests are recommended only in screening programs because they have low sensitivity since they cannot detect modest increases in urinary albumin excretion, which characterize the early stages of diabetic nephropathy. In contrast, quantitative assays are highly sensitive.

Several methods have been developed for measuring albumin in urine, including radioimmunoassay, enzyme-linked immunosorbent assay (ELISA), radial immunodiffusion, immunoturbidimetry, and nephelometry. Each method has advantages and disadvantages, so the choice depends on the laboratory. In general, however, these assays have similar inaccuracies, limitations, and reference ranges. In any case, the positivity for urinary proteins with the test strips must be confirmed by laboratory measurements. Regarding the measurement method for albumin, the recommendations for immunochemical techniques are a coefficient of variation (CV) <15% and a detection limit of approximately 2 mg/L. Based on observational clinical studies, other indications suggest a more stringent value for imprecision: CV <13% for albumin and CV <6% for the albumin/creatinine ratio.

Postanalytical Aspects

The reference intervals for making a diagnosis of diabetic nephropathy differ by type of sample collection and are shown in Table 27.2.

Units can be in mg/24 h for 24-h excretion or in mg/g creatinine, which gives the same result. The use of mg/mmol creatinine generates a different numerical value. The suggestion for deciding which unit to use is to establish uniformity of reporting among laboratories in the same geographic area. It is strongly discouraged to report albumin in mg/L.

There are no different decision levels for the sex of the diabetic patient, while for the albuminuria/cardiovascular risk association, the decision levels are still under study, but epidemiological studies suggest a lower concentration than that decided for diabetic nephropathy.

Table 27.2 Reference ranges of urinary albumin

	Urinary excretion of albumin		
Sample type and measurement units	Physiological or mildly increased (class A1)	Moderately increased (class A2)	Severely increased (class A3)
First sample of the morning			
mg/g creatinine	<30	30–300	>300
mg/mmol creatinine	<3	3–30	>30
24-h urine collection			
mg/24 h	<30	30–300	>300

Glycated Hemoglobin

The red blood cells of a healthy adult subject contain mainly hemoglobin A (HbA) (97%), hemoglobin A_2 (HbA_2) (2–3%), and traces of fetal hemoglobin (HbF) (0.5–1%). Chromatographic analysis of HbA led to the identification of several minor fractions: HbA_{1a}, HbA_{1b}, and HbA_{1c}, collectively referred to as HbA1. HbA_{1c} is a significant fraction and constitutes about 80% of HbA1. HbA_{1c} is the product of the nonenzymatic condensation reaction between the aldehyde group of glucose and the amino residues of some amino acids, mainly the terminal valines, of the hemoglobin β-chain.

Because red blood cells are permeable to glucose, the extent of HbA_{1c} formation is directly proportional to the concentration of glucose to which red blood cells are exposed in the circulation and the relative exposure time.

The measurement of HbA_{1c} in the blood is the gold standard for assessing glycometabolic control in diabetic subjects. From the results of the DCCT study, elevated HbA_{1c} levels increase the risk of micro- and macrovascular complications; consequently, this parameter is used as a target for possible therapeutic choices. Moreover, HbA_{1c} measurement is also used to confirm the diagnosis of diabetes: values ≥48 mmol/mol (6.5%), if confirmed on at least two subsequent occasions, would allow the diagnosis of diabetes to be made. In any case, it is not advisable to use HbA_{1c} for diagnosis in children and adolescents, in women within 2 months after childbirth, in subjects treated with glucocorticoids, immediately after surgery, or in the presence of acute pancreatitis.

The frequency of HbA_{1c} measurements is theoretically related to the average life of red blood cells (4 months) and, therefore, should be three times per year. The ADA recommends measuring HbA_{1c} twice a year in patients in stable metabolic control who have reached the therapeutic targets, with a greater number of determinations for patients in unstable control. Finally, in patients with gestational diabetes and in particular cases, measurements can also be made at shorter time intervals. In any case, the measurement carried out less than 2 months later is considered inappropriate for evaluating glycemic compensation.

Preanalytical Aspects

Table 27.3 shows the main interferences in the measurement of HbA_{1c}, specifically listed regarding the risk of false positives or negatives using this parameter for the diagnosis of diabetes. There is no doubt that there are differences between males and females in HbA_{1c} values in the absence of diabetes, with females showing slightly lower levels than males. Age appears controversial, but some authors reported an increase in HbA_{1c} of about 1 mmol/mol (0.1%) from age 30 to 70 for each decade of age starting at age 30.

Regarding ethnicity, it has now been ascertained that black subjects have, on average, higher HbA_{1c} values than Caucasians, with the same glycemic control, although the risk of developing microvascular complications in the long term does not seem to be affected. In the population, there are also subjects with a "fast" or "slow" glycation phenotype, depending on whether their HbA_{1c} values are higher or lower than average, respectively, with the same average daily glycemia. Other factors, such as lifestyle and diet between summer and winter, are related to possible seasonal variations in HbA_{1c} concentrations. Such variations can cause changes of 5–7% from the mean annual value.

Table 27.3 Limitations in the use of glycated hemoglobin for diagnosing diabetes

Factors	Risk in the diagnosis
Analytical interferences (potentially eliminable depending on the analytical method)	
Hyperbilirubinemia	Underestimate
Hypertriglyceridemia	Overestimate
Increase in white blood cells	Overestimate
Presence of hemoglobin variants	Over- or underestimate
Physiological factors (potentially known before analysis)	
Pregnancy	Underestimate
Age	Underestimate
Sex	Over- or underestimate
Seasonal variations	Over- or underestimate
Genetic determinants	Over- or underestimate
Presence of hemoglobin variants	Over- or underestimate
Pathological factors	
Type 1 diabetes of recent origin	Underestimate
Hemolytic anemias	Underestimate
Iron deficiency anemia	Overestimate
Recent blood loss, transfusions	Underestimate
Splenectomy	Overestimate
Kidney failure	Underestimate
Anti-retroviral therapy	Overestimate
Erythropoietin therapy	Underestimate
Alcohol abuse	Underestimate
Malaria	Underestimate

The intake of vitamin C or E seems to reduce HbA_{1c} due to an inhibition of the glycation process. Hypertriglyceridemia, hyperbilirubinemia, chronic salicylate intake, and opioid dependence may cause interference with the method, resulting in increased HbA1c values. Alcohol consumption also influences HbA_{1c} levels since it has been shown that abstinence from alcohol consumption in alcoholics in rehabilitation is associated with a decrease in HbA_{1c} of about 4 mmol/mol (0.4%). Finally, the presence of hemoglobin variants can cause positive and negative interference, depending on the method in use.

For sample collection, venous or capillary blood samples can be taken using a lancing device. The anticoagulant generally used is EDTA. The stability of the whole blood sample is 3 days at 20 °C, at least 5 days at 4 °C, and at least 6 months at −80 °C. If samples are stored at −80 °C, they should be tested within a short period of time after thawing. Some diagnostic kit manufacturers have introduced capillary blood collection systems that provide stability of approximately 1–2 weeks at room temperature. However, these systems are method dependent and cannot be adapted to other analytical systems.

Analytical Aspects

To date, more than 100 different methods have been used to measure HbA_{1c}. These methods can be grouped as follows: chromatographic methods based on the isoelectric point difference between HbA_{1c} and HbA (ion exchange, high-performance liquid chromatography [HPLC]) or the presence of glucose covalently linked to hemoglobin (affinity chromatography); immunochemical and enzymatic methods (detection of ketoamine). Usually, the results obtained by different methods are very well correlated, and there is no evidence that data obtained by one method are superior to data obtained by another.

It is important to remember that the HbA_{1c} result should always be contextualized; if it is in contrast with the patient's clinical picture, the test should be repeated, using a different method (if possible) or assessing glycemic control using other analytes (e.g., glycated albumin [GA]).

As with other tests, the measurements must be standardized for the HbA_{1c} figure to be usable. To this end, the IFCC established a working group in 1995 to standardize HbA_{1c} results globally. Subsequently, in 2007, another working group consisting of members of the ADA, the European Association for the Study of Diabetes, and the International Diabetes Federation (IDF), based on the IFCC's achievements, produced a consensus document on HbA_{1c} standardization, which dictated the strategy at the global level.

In the National Glycohemoglobin Standardization Program certified system, HbA_{1c} is reported as a percentage of total Hb. In contrast, the IFCC has recommended that HbA_{1c} be expressed in mmol/mol (HbA_{1c}/Hb). The comparison between the two systems has created an equation allowing conversion between the two units. In order to ensure a standardized result, it is considered appropriate that the inaccuracy be contained within 2% and that the bias for the target value (verifiable by participating in EQ programs with switchable materials and titer assigned by the IFCC reference method) be minimal, with a total error target within 6%.

Postanalytical Aspects

As previously reported, the measurement of HbA_{1c} is also used to confirm the diagnosis of diabetes: values ≥48 mmol/mol (6.5%), if confirmed on at least two subsequent occasions, would allow the diagnosis of diabetes to be made. HbA_{1c} values between 43 and 47 mmol/mol (between 6.1 and 6.4%) are associated with a high risk of diabetes.

The DCCT study found that every 10 mmol/mol (1%) increase in HbA_{1c} is associated with a worsening of mean blood glucose of approximately 35 mg/dL, and current ADA-recommended treatment targets indicate that the primary goal of therapy should result in an HbA_{1c} value of no more than 53 mmol/mol (7%). If HbA_{1c} is consistently above 64 mmol/mol (8%), therapeutic treatment should be promptly reassessed. Other organizations, such as the IDF, recommend achieving and maintaining HbA_{1c} values below 48 mmol/mol (6.5%) to minimize the risk of complications. Another typical use of HbA_{1c} is to measure the quality of care provided to patients with diabetes. The ADA criteria indicate that the percentage of patients with HbA_{1c} values greater than 75 mmol/mol (9%) should not exceed 20% of the total adult patients; the share of diabetics with values less than 53 mmol/mol (7%) should be at least 40% of the total; 84% of pediatric patients should have HbA_{1c} values less than 86 mmol/mol (10.0%) and 34% should have values less than 64 mmol/mol (8.0%).

Glycated Albumin and Fructosamine

While fructosamine represents an index of the concentration of stably glycated serum proteins (ketamine, mainly albumin, lipoproteins, and glycated globulins), GA represents a unique post-translational product of albumin alone with plasma glucose.

Although HbA_{1c} remains the gold standard for the diagnosis and monitoring of diabetes mellitus, several studies have shown that other biomarkers of nonenzymatic glycation, such as GA and fructosamine, can be used both in clinical conditions in which HbA_{1c} dosage does not reflect glycemic compensation (chronic renal failure, anemia, and hemoglobin variants) and in other clinical settings in which information on blood glucose in the previous 2–3 weeks is needed (poorly controlled diabetes, gestational diabetes, postprandial hyperglycemia, anemia, hemoglobin variants). Moreover, in other clinical settings, it is necessary to have information on blood glucose in the previous 2–3 weeks (poorly controlled diabetes, gestational diabetes, postprandial hyperglycemia, "fluctuating" diabetes, and gastrectomized patients). GA and fructosamine are particularly sensitive to recent changes in blood glucose and sudden fluctuations.

Preanalytical Aspects

Using these markers in clinical practice requires considering some preanalytical variables that may affect the result. The literature regarding this aspect is quite limited and sometimes conflicting. In summary, all those clinical conditions that alter the normal metabolism of albumin limit the use of fructosamine and GA: thyroid dysfunction, nephrotic syndrome, liver cirrhosis, and nonalcoholic steatohepatitis (NASH). Nephrotic syndrome, hyperthyroidism, NASH, Cushing's syndrome, and glucocorticoid intake decrease in AG and fructosamine values, while liver cirrhosis and hypothyroidism determine an increase in the values of these markers. Physiological or lifestyle variables such as age, gender, race, body mass index (BMI), hyperuricemia, and cigarette smoking also influence the values of these markers.

The intra- and interindividual biological variability of GA was 2.1% and 10.6%, while that of fructosamine was 2.3% and 6.3%.

GA values have an inverse correlation with BMI, fat mass levels, and cigarette smoking, while they correlate positively with age and are higher in females. Some authors also report an increase in GA values throughout pregnancy, while others report a decrease in values in the third trimester.

Fructosamine values show a modest decrease with gestational age and a modest increase with maternal age and correlate inversely with smoking. Higher values are found in males, and differences have been reported in different races.

Regarding the biological matrix used for assaying these markers, serum or plasma can be used. However, it has been reported that the values found in plasma are lower than those obtained in serum for fructosamine. It is therefore recommended that the same biological matrix is always used for monitoring.

As shown in two studies, GA is stable for 24 h on separated serum and plasma stored at room temperature, up to 1 week when stored at 4 °C, and up to 2 months when stored at −80 °C.

Stability for up to 2 weeks with serum stored at 4 °C and up to 5 weeks when stored at −20 °C has been demonstrated for fructosamine.

Analytical Aspects

Several methods have been developed to evaluate fructosamine in serum and plasma. The most widely used and best standardized are colorimetric assays that typically exploit the property of fructosamine to be a reducing agent under alkaline conditions. The first technique, developed in 1983, was based on reducing the nitroblue tetrazolium dye to formazan. The spectrophotometric method monitored the rate of formazan formation, which was directly proportional to the fructosamine concentration. Later, a nonionic detergent containing uricase was added to eliminate interference from uric acid and polylysine, thus allowing a more accurate and sensitive measurement. The assay thus modified is currently available and widely used in clinical laboratories. Although it is rapid, technically simple, inexpensive, and available for automation, the method is affected by ambient temperature variations and remains poorly standardized. In addition, many molecules with reducing activity, such as bilirubin and vitamins, may interfere with the measurement.

GA concentration can be measured by boronate affinity chromatography, ion exchange chromatography, HPLC, and immunoassays (ELISA or radioimmunoassays). An enzymatic method with good analytical performance and excellent specificity has recently been automated (Lucica GA-L kit, Asahi Kasei Pharma, Tokyo, Japan). The method is based on the initial elimination of endogenous glycated amino acids and peroxides by a ketamine oxidase and a peroxidase. The GA is then hydrolyzed by an albumin-specific proteinase, and the products of this reaction are oxidized by a ketamine oxidase. The H_2O_2 produced is then measured quantitatively by a colorimetric method. In parallel, the concentration of albumin is measured by the bromocresol violet method so that the final result can be expressed as the ratio of GA to total albumin. This method can be implemented on major automated analytical platforms and offers good analytical performance.

Insulin and C-Peptide

Routine testing of insulin and C-peptide is not recommended in most patients with diabetes. However, the assay of these analytes is indicated in the evaluation of fasting hypoglycemia to exclude the presence of an insulinoma. The diagnosis of insulinoma is based on the persistence of an elevated plasma insulin concentration associated with a decreased glucose concentration. Insulin determination is also helpful in evaluating and managing patients with polycystic ovary syndrome.

C-Peptide is a fragment of the original molecule from which insulin is formed. When endogenous insulin is produced, C-peptide is also released. In patients undergoing insulin therapy, to verify the secretory capacity of β cells, it is not possible to directly evaluate insulinemia because the insulin administered as a drug would also be measured. In these cases, it is indicated to evaluate the concentration of C-peptide (which is not contained in the pharmacological insulin) to verify the residual capacity of β-cells to produce endogenous insulin.

Measurement of the C-peptide response to glucagon may help in rare cases where it is difficult to differentiate the diagnosis between type 1 and type 2 diabetes.

The determination of C-peptide is essential in evaluating possible self-induced hypoglycemia due to nontherapeutic insulin intake.

Analytical Aspects

Immunometric methods can measure insulin and C-peptide, but despite the efforts of the scientific community, to date, there are no standardized methods.

One of the analytical problems affecting these analytes' measurements is cross-reactivity with insulin analogs.

Postanalytical Aspects

Insulin reference ranges depend on the method used and must therefore be defined by each laboratory. After an overnight fast, the insulin concentration in a healthy, nonobese subject is usually between 12 and 150 pmol/L. For C-peptide, reference ranges have not been defined with certainty, and different laboratories express C- peptide values with different units (nmol/L, pmol/L, or ng/mL). The concentration of C-peptide in the healthy population is usually between 0.78 and 1.89 ng/mL (0.25–0.6 nnmol/L).

Recommended Readings

American Diabetes Association (2016) Classification and diagnosis of diabetes. Sec. 2. In Standards of Medical Care in Diabetes – 2016. Diabetes Care 39(Suppl. 1):S13–S22

American Diabetes Association. 5 (2016) Glycemic targets. In Standards of Medical Care in Diabetes-2016. Diabetes Care 39(Suppl. 1):S39–S46

Bonetti G, Carta M, Lapolla A, et al (2018) Raccomandazioni per l'ottimizzazione della fase pre-analitica per una corretta determinazione della glicemia in ambito diabetologico. Biochimica Clinica (in press) https://doi.org/10.19186/BC_2018.037

Carta M, Mosca A, Lapolla A et al (2015) Raccomandazioni per l'esecuzione del test da carico orale di glucosio (OGTT). Biochimica clinica 39:135–140

Graziani MS, Caldini AL (2011) per il GdS intersocietario Diabete. Indicazioni per la misura dell'albumina nelle urine per l'accertamento e il monitoraggio della nefropatia diabetica. Riv Ital Med Lab 7:20–24 http://www.sibioc.it/gds/more/sezione-id/7/tag/Documenti%20 Ufficiali

International Diabetes Federation IDF guidelines for management of postmeal glucose in patients with diabetes. [Accessed 6 Dec 2016]. Available from: http://www.idf.org/sites/default/files/postmeal%20 glucose%20guidelines.pdf

Lapolla A, Mosca A (2012) Screening e diagnosi del diabete gestazionale: definite le raccomandazioni. Biochimica Clinica 36:12–14

Mosca A, Lapolla A (2008) Diagnostica del diabete mellito. In: Interpretazione degli esami di laboratorio, a cura di Mauro Panteghini. Trattato italiano di Medicina di Laboratorio, vol VII. Padova: Piccin Nuova Libraria, pp 85–133

Sacks DB, Arnold M, Bakris GL et al (2011) National Academy of Clinical Biochemistry; Evidence-Based Laboratory Medicine Committee of the American Association for Clinical Chemistry. Guidelines and recommendations for laboratory analysis in the diagnosis and management of diabetes mellitus. Diabetes Care 34:e61–e99

Sacks DB, Bruns DE, Goldstein DE et al (2002) Guidelines and recommendations for laboratory analysis in the diagnosis and management of diabetes mellitus. Clin Chem 48:436–472

Standard italiani per la cura del diabete mellito (2018). www.standarditaliani.it

Metabolic Syndrome

28

Marcello Ciaccio, Anna Maria Ciaccio, and Luisa Agnello

Introduction

Metabolic syndrome is a biochemical–clinical condition characterized by resistance to insulin action associated with a proinflammatory and prothrombotic state. It represents an important cardiovascular risk factor with a constantly increasing incidence in the general population; therefore, its early detection is fundamental to implement preventive interventions. This chapter describes the main characteristics of metabolic syndrome and the criteria used to diagnose it.

Definition

Metabolic syndrome is a condition in which the interaction among biochemical, clinical, and metabolic factors results in an increased risk of developing cardiovascular diseases, type 2 diabetes mellitus, and mortality. It is present in about 20–25% of the general population and is characterized by visceral obesity, hypertension, and lipid and glucose metabolism alterations.

M. Ciaccio (✉)
Department of Biomedicine, Neurosciences and Advanced Diagnostics, Institute of Clinical Biochemistry, Clinical Molecular Medicine and Clinical Laboratory Medicine, University Hospital "P. Giaccone", Palermo, Italy

Department of Laboratory Medicine, University Hospital "P. Giaccone", Palermo, Italy
e-mail: marcello.ciaccio@unipa.it

A. M. Ciaccio
Department of Health Promotion, Mother and Child Care, Internal Medicine and Medical Specialties (ProMISE) "G. D'Alessandro", University Hospital "P. Giaccone", Palermo, Italy

L. Agnello
Department of Biomedicine, Neurosciences and Advanced Diagnostics, Institute of Clinical Biochemistry, Clinical Molecular Medicine and Clinical Laboratory Medicine, University Hospital "P. Giaccone", Palermo, Italy

Beginning with its initial description as the "android obesity syndrome" by Vague in 1956, the syndrome has taken on several names over time, including "dyslipidemic hypertension syndrome," "syndrome X," "insulin resistance syndrome," "dysmetabolic syndrome," and "plurimetabolic syndrome."

Pathogenesis

The metabolic syndrome has a multifactorial etiology that includes genetic and environmental factors. The underlying pathophysiological mechanism is insulin resistance (IR), defined as an inadequate response of target organs (striated muscle, liver, and adipose tissue) to the physiological effects of insulin.

The consequences of reduced insulin sensitivity are:

1. Reduced insulin-mediated uptake of glucose in striated muscle and adipose tissue.
2. Reduced insulin-mediated inhibition of glucose production by the liver.
3. Significant reduction in lipolysis inhibition in adipose tissue.

Obesity, physical inactivity, and genetic factors predispose to the development of IR. In particular, the most important cause of IR is the increase of adipose tissue, especially at the abdominal-perivisceral level. Obesity alters mechanisms that regulate the activity of the insulin receptor. Specifically, mediators, such as the excess of circulating free fatty acids (FFAs) and tumor necrosis factor-alpha "shutdown" the receptor's activity.

Abdominal-visceral obesity results in an overflow of FFA into the circulation and a consequent increase in FFA uptake

M. Ciaccio (ed.), *Clinical and Laboratory Medicine Textbook*, https://doi.org/10.1007/978-3-031-24958-7_28

by muscle tissue, liver, and adipose tissue. FFAs are normally metabolized through two pathways:

- Oxidation
- Deposition (triglycerides)

However, when the overflow of FFAs exceeds the metabolic capacity of these two pathways, FFAs and their metabolites accumulate in the tissues, where they cause the activation of specific serine kinases, which, in turn, act by blocking the insulin signal.

Diagnosis

Currently, there are no single criteria for diagnosing metabolic syndrome, but scientific societies and organizations have proposed several. In general, the diagnosis of metabolic syndrome is based on the evaluation of biochemical parameters, such as glycemia and lipid alterations, and clinical parameters, such as arterial hypertension and obesity. The different diagnostic criteria proposed so far for metabolic syndrome are shown in Table 28.1.

Although IR represents the pathogenic mechanism underlying the metabolic syndrome, the assessment of this condition is not recommended for diagnosis. However, once the diagnosis of metabolic syndrome is made, several tests are available to assess the insulin resistance severity.

The fasting insulin measurement is not reliable because insulin may have normal values leading to false negative results. IR can be assessed by direct or indirect methods. The direct methods are the insulin measurement during the oral glucose tolerance test (OGTT) curve, the insulin suppression test, and the hyperinsulinemic-euglycemic clamp, which is the gold standard. Among the indirect methods, two indices are commonly used: the Homeostatic Model Assessment (HOMA) index and the Quantitative Insulin sensitivity Check Index (QUICKI). OGTT is performed after fasting for 8–10 h (overnight fasting) by giving the patient 75 g of glucose dissolved in 300 mL of water. Insulin measurement is performed before the administration of the glucose solution (basal) and then at precise time intervals (30′, 60′, 120′, 150′, and 180′). In healthy subjects, insulinemia reaches a concentration peak 6–10 times higher than the basal value after 30–60 min, then its levels begin to decrease, reaching a value approximately 2–3 times the basal value around 180 min and normal around 240 min (Fig. 28.1). In a subject with insulin resistance, the peak at 30–60 min will reach higher concentrations than in a subject and persist for a longer period because insulin is metabolized more slowly (Fig. 28.1).

The insulin suppression test is based on administering an octreotide and somatostatin solution to suppress insulin

Table 28.1 Diagnostic criteria of the metabolic syndrome.

	AHA-NHLBI	IDF	NCEP-ATPIII	WHO
Criterion	Any combination of three altered parameters	Central obesity and two other altered parameters	Any combination of three altered parameters	Glycemic alterations or insulin resistance and two further altered parameters
Glycemic changes	FPG >100 mg/dL or overt diabetes	FPG >100 mg/dL or overt diabetes	FPG >100 mg/dL or overt diabetes	FPG >100 mg/dL or IGT or overt diabetes
Insulin resistance				M-clamp of the lower quartile of the distribution
Central obesity	Waist circumference ≥102 cm in men and ≥88 cm in women	Waist circumference ≥94 cm in men and ≥80 cm in women	Waist circumference ≥102 cm in men and ≥88 cm in women	Waist/hip ratio >0.90 in men and >0.85 in women and/or BMI ≥30
Hypertension	Systolic ≥130 and/or diastolic ≥85 mmHg or antihypertensive treatment	Systolic ≥130 and/or diastolic ≥85 mmHg or antihypertensive treatment	Systolic ≥130 and/or diastolic ≥85 mmHg	Systolic ≥140 and/or diastolic ≥90 mmHg
Hypertriglyceridemia	≥150 mg/dL or lipid-lowering treatment	≥150 mg/dL or lipid-lowering treatment	≥150 mg/dL	
Low C-HDL	<40 mg/dL in men and < 50 mg/dL in women, or lipid-lowering treatment	<40 mg/dL in men and <50 mg/dL in women, or lipid-lowering treatment	<40 mg/dL in men and <50 mg/dL in women, or lipid-lowering treatment	
Dyslipidemia				Triglycerides ≥150 mg/dL and/or HDL-C <35 mg/dL in men and HDL-C <39 mg/dL in women
Microalbuminuria				Microalbuminuria (>20 μg/min) or albuminuria/creatinuria ratio >30 μg/mg

AHA, American Heart Association; BMI, body mass index; HDL-C, high-density lipoprotein cholesterol; FPG, fasting plasma glucose; IDF, International Diabetes Federation; IGT, impaired glucose tolerance; NCEP-ATP III, National Cholesterol Education Program Adult Treatment Panel III; NHLBI, National Heart, Lung, and Blood Institute; WHO, World Health Organization.

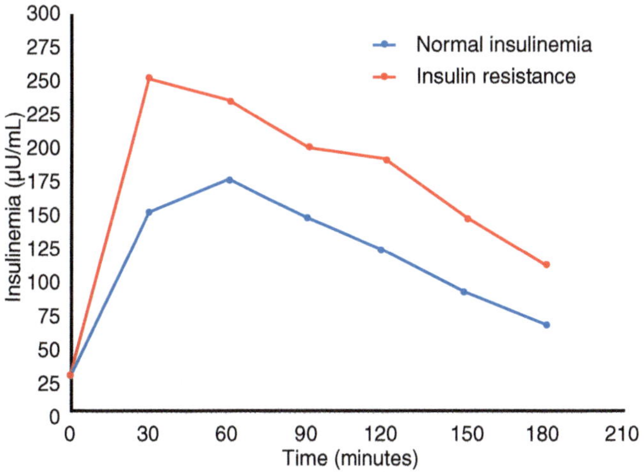

Fig. 28.1 Insulin curve after oral glucose tolerance test. (Copyright EDISES 2021. Reproduced with permission).

secretion; subsequently, blood glucose trends are assessed in response to a continuous infusion of glucose and insulin.

The hyperinsulinemic-euglycemic clamp is based on the infusion of an insulin quota to bring the hormone's blood values, for a time of 120 min, to a constant value of 100 µU/mL. At the same time, glycemia is constantly maintained at a basal level of about 90 mg/dL through a variable infusion of glucose, punctually adjusted by a feedback mechanism based on glycemic determinations repeated during the test. In conditions of constant insulinemia, in the last 40 min of the test, the amount of glucose infused in the unit of time to maintain euglycemia corresponds to the amount of glucose used by peripheral tissues, with transport dependent on the action of insulin. It is, therefore, an index ("*M* value" in mol/min/kg) of the tissues sensitivity to the insulin action. Although this test represents the gold standard for assessing insulin resistance, it is not commonly used in clinical practice because it is difficult to perform and requires time and expert personnel.

The HOMA index is based on a mathematical model that relates fasting serum glucose and insulin concentrations. The QUICKI (less used because it is more complex than the HOMA) is derived by calculating the inverse of the sum of the logarithms of fasting blood glucose and fasting insulin.

Therapy

In 2001, the Adult Treatment Panel III (ATP III) recommended two primary treatment goals for patients with metabolic syndrome. These goals have been reinforced by a report from the American Heart Association (AHA) and the National Institutes of Health (NIH) and clinical guidelines from The Endocrine Society:

- Treat the underlying causes (overweight, obesity, and physical inactivity) by modifying lifestyle and eating habits
- Treat cardiovascular risk factors if they persist despite lifestyle modifications

All components of the metabolic syndrome benefit from weight loss, maintained over time, achieved through caloric restriction and physical activity. There is no direct evidence that attempting to prevent type 2 diabetes and cardiovascular disease by treating the metabolic syndrome is effective. It is possible to treat IR with drugs that potentiate the action of insulin (e.g., thiazolidinediones and metformin).

Recommended Readings

Capurso C, Capurso A (2013) Dall'obesità alla resistenza insulinica: il ruolo degli FFA. Giornale Italiano dell'Arteriosclerosi 4(1):37–52

DeFronzo RA, Tobin JD, Andres R (1979) Glucose clamp technique: a method for quantifying insulin secretion and resistance. Am J Phys 237:E214–E223

Grundy SM, Hansen B, Smith SC Jr et al (2004) Clinical management of metabolic syndrome: report of the American Heart Association/National Heart, Lung, and Blood Institute/American Diabetes Association conference on scientific issues related to management. Circulation 109:551

Knowler WC, Barrett-Connor E, Fowler SE et al (2002) Reduction in the incidence of type 2 diabetes with lifestyle intervention or metformin. N Engl J Med 346:393

Meigs JB (2003) The metabolic syndrome. BMJ 327

Expert Panel on Detection, Evaluation, and Treatment of High Blood Cholesterol in Adults (2001) Executive summary of the third report of The National Cholesterol Education Program (NCEP) expert panel on detection, evaluation, and treatment of high blood cholesterol in adults (Adult Treatment Panel III). JAMA 285:2486

Genuth S, Alberti KG, Bennett P et al (2003) Follow-up report on the diagnosis of diabetes mellitus. Diabetes Care 26:3160

Grundy SM, Cleeman JI, Daniels SR et al (2005) Diagnosis and management of the metabolic syndrome: an American Heart Association/National Heart, Lung, and Blood Institute Scientific Statement. Circulation 112:2735

Rosenzweig JL, Ferrannini E, Grundy SM et al (2008) Primary prevention of cardiovascular disease and type 2 diabetes in patients at metabolic risk: an endocrine society clinical practice guideline. J Clin Endocrinol Metab 93:3671

29 Hypoglycemia

Marcello Ciaccio and Luisa Agnello

Introduction

Hypoglycemia is a metabolic condition characterized by reduced circulating glucose levels associated with clinical signs and/or symptoms such as alterations in mental status and/or stimulation of the sympathetic nervous system.

In healthy, fasting adults, the lower limit of plasma glucose concentration is 70 mg/dL (3.9 mmol/L). However,glucose levels may reach lower values away from meals. Hypoglycemia is clinically manifested when plasma glucose concentration is <55 mg/dL (3 mmol/L). Clinical manifestations of hypoglycemia are nonspecific. Additionally, low plasma concentration may be an artifact. Thus, hypoglycemia should always be confirmed by satisfying the criteria of Whipple's triad: (1) signs, symptoms, or both, consistent with hypoglycemia; (2) low plasma glucose concentration; and (3) resolution of signs and/or symptoms after the restoration of normal plasma glucose concentration (Fig. 29.1).

In the absence of the Whipple's triad criteria, hypoglycemia is unlikely.

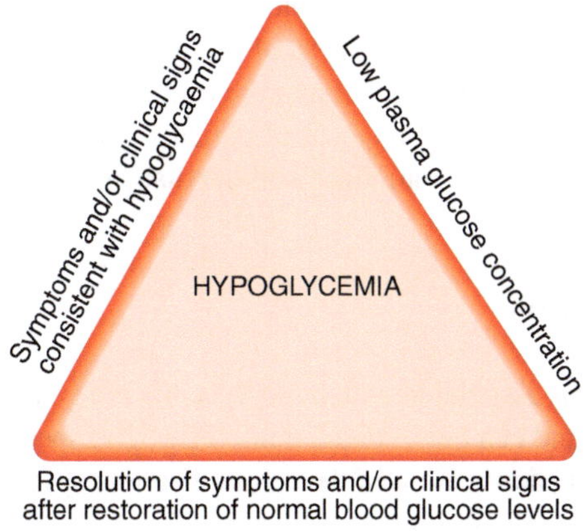

Fig. 29.1 Whipple's triad. (Copyright EDISES 2021. Reproduced with permission)

Epidemiology

Hypoglycemia is a relatively frequent condition in patients with type 1 diabetes mellitus (DM1), who develop, on average, two episodes of symptomatic hypoglycemia per week and at least one episode per year of severe hypoglycemia, temporarily disabling, often associated with epilepsy and coma. In DM1, hypoglycemia results from treatment with insulin or insulin secretagogues. A severe complication of hypoglycemia is sudden death secondary to cardiac arrhythmia.

In patients with type 2 diabetes mellitus (DM2), therapy is initially based on drugs, such as metformin or thiazolidinedione, that do not influence insulin levels or reduce plasma glucose concentrations, so they should not cause hypoglycemia. However, over time, patients with DM2 may develop endogenous insulin deficiency and, therefore, require treatment with exogenous insulin or insulin secretagogues. In this case, the frequency of development of hypoglycemia becomes superimposable to that of patients with DM1.

Hypoglycemia is a relatively rare condition in individuals who are not under treatment for diabetes. It is important to note that the plasma glucose concentration used to document Whipple's triad, in the absence of treatment of diabetes with insulin or insulin secretagogues, must be measured by a reliable laboratory method. Although the finding of a low plasma glucose concentration, measured by a reliable method, in the absence of signs or symptoms should not be ignored, this finding could be due to a condition of "pseudohypoglycemia," that is, an artifact of continued in vitro glucose metabo-

M. Ciaccio (✉) · L. Agnello
Department of Biomedicine, Neurosciences and Advanced Diagnostics, Institute of Clinical Biochemistry, Clinical Molecular Medicine and Clinical Laboratory Medicine, and Department of Laboratory Medicine, University Hospital "P. Giaccone,", Palermo, Italy
e-mail: marcello.ciaccio@unipa.it

M. Ciaccio (ed.), *Clinical and Laboratory Medicine Textbook*, https://doi.org/10.1007/978-3-031-24958-7_29

lism by the corpuscular elements of the blood after the sample has been drawn. This condition can occur when the blood sample is collected in a tube that does not contain a glycolysis inhibitor or the separation of plasma (or serum) from the corpuscular elements is delayed, especially in the presence of erythrocytosis, leukocytosis, or thrombocytosis.

Documentation of Whipple's triad confirms the presence of a hypoglycemic alteration, the etiology of which may be obvious (e.g., in a patient with diabetes treated with insulin or insulin secretagogues) or a diagnostic challenge (e.g., in a healthy individual with an insulinoma). On the other hand, in an individual without diabetes mellitus, an unequivocally normal plasma glucose concentration (>70 mg/dL (3.9 mmol/L)) during a symptomatic episode indicates that those symptoms are not the result of hypoglycemia.

The pathophysiology of hypoglycemia and the diagnostic approaches and management of patients with diabetes mellitus are different from those without diabetes.

Symptoms

Symptoms associated with hypoglycemia are generally nonspecific and can be subdivided into neuroglycopenic, related to glucose deprivation in the central nervous system (CNS), and neurogenic (or autonomic), related to the autonomic nervous system activation (Table 29.1 and Fig. 29.2). In a patient without diabetes, the onset of neuroglycopenic symptoms is strongly indicative of an underlying hypoglycemic disorder. Hypoglycemic symptoms are usually the same during each episode that may occur in the fasting or postprandial state. Patients often cannot describe their episodes in detail due to amnesia.

Hypoglycemia in Diabetes

In patients with diabetes, hypoglycemia is the result of the relative or absolute excess of exogenous insulin and the compromised physiological and behavioral defense against falling plasma glucose concentrations. The physiological defense against falling plasma glucose concentrations includes (1) reduced insulin secretion; (2) increased glucagon secretion; (3) and, in the absence of the latter, increased adrenaline secretion. On the other hand, behavioral defense mainly consists of carbohydrates assumption. This behavior is induced by the perception of symptoms, mainly neurogenic, mediated by the neural activation of the sympathetic system. All these defense mechanisms, not only insulin secretion, are impaired in DM1 and, in the long run, also in DM2. In overt DM1, circulating insulin levels does not reduce in response to decreased plasma glucose levels. In addition to the lack pancreatic β-cells response, α-cells also fail to produce glucagon in response to hypoglycemia. In the absence of the primary defense mechanisms (insulin and glucagon), DM1 patients are highly dependent on the third defense mechanism, i.e., the adrenaline secretion. The adrenaline response to hypoglycemia is often attenuated due to mechanisms that are not yet fully defined. In the context of a failure to decrease insulin and increase glucagon, when there is a fall in blood glucose levels in response to therapy-induced hyperinsulinemia, the attenuated adrenaline response causes the clinical syndrome of glycemic counterregulation deficiency.

Patients with this syndrome have a 25-fold increased risk of developing hypoglycemia. In addition, the attenuated sympathetic neural response causes the clinical syndrome of unconscious hypoglycemia, characterized by the impairment or complete loss of alarm symptoms that trigger behavioral defense. Unconscious hypoglycemia is associated with a sixfold increased risk of developing severe hypoglycemia.

The concept of hypoglycemia-associated autonomic failure in diabetes assumes that recent and antecedent

Table 29.1 Hypoglycemia symptoms

Neurogenic		Neuroglycopenic
Adrenergics	*Cholinergic*	Difficulty paying attention
Tachycardia	Sweating	Confusion
Tremors	Hunger	Diplopia
Pallor	Paresthesia	Blurred vision
Agitation		Convulsions
		Headache
		Coma

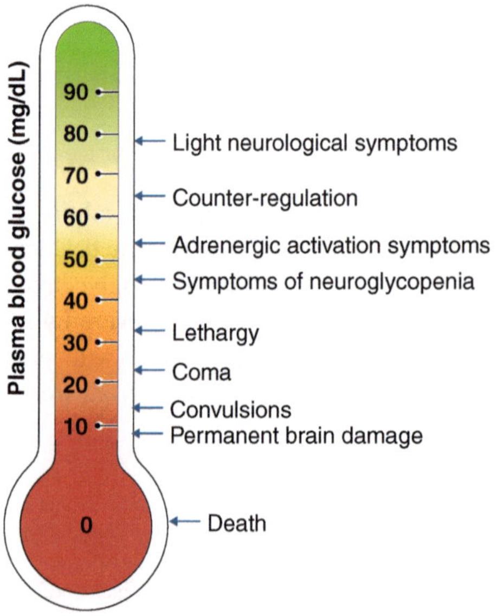

Fig. 29.2 Signs and symptoms of hypoglycemia. (Copyright EDISES 2021. Reproduced with permission)

hypoglycemia causes both deficits in glycemic counter-regulation (by reducing the response to adrenaline without insulin and glucagon response) and unconscious hypoglycemia (mainly by reducing the sympathetic neural response and the resulting neurogenic symptoms) and, therefore, triggers a vicious cycle of recurrent hypoglycemia. In subjects with DM1, unconscious hypoglycemia and adrenaline deficiency are reversible after 2–3 weeks of scrupulous attention to avoiding therapy-induced hypoglycemia. The glycemic counter-regulatory deficiency syndrome is primarily due to β-cell deficiency. Thus, DM2 patients may also develop it in the long run. In addition, the threshold for the sympathoadrenal response shifts toward lower plasma glucose concentrations due to antecedent hypoglycemia, similar to DM1. Table 29.2 shows the main risk factors for hypoglycemia and hypoglycemia-associated autonomic failure in patients with diabetes. There are many variables at stake. Thus,the insulin treatment must be carried out by adequately trained people who can recognize and manage the various and changing factors making the necessary corrections;for example, occasionally changing the insulin dose according to diet or after intense physical exercise. In this case, hypoglycemia can appear from 1–2 h up to 17 h after the end of the physical effort. Aerobic exercise causes:

1. The increase in the insulin-dependent and -independent muscular glucose uptake
2. The 40–60% reduction in the endogenous insulin secretion
3. The increase in insulin sensitivity in the next 2 h

Therefore, it is always necessary to adjust doses according to glucose consumption.

Table 29.2 Risk factors for hypoglycemia and hypoglycemia-associated autonomic failure associated with hypoglycemia in patients with diabetes mellitus

Conventional risk factors associated to a relative or absolute insulin excess	Risk factors for hypoglycemia-associated autonomic failure
Excessive, inappropriate, or incorrect measurement of exogenous insulin or insulin secretagogues Reduction of exogenous glucose release (e.g., after skipping a meal or during an overnight fast) Increased use of glucose (e.g., during exercise) Reduced endogenous production of glucose (e.g., after alcohol intake) Increased insulin sensitivity (e.g., following weight loss, an increase in regular exercise or better glycemic control, and in the middle of the night) Reduced insulin clearance (e.g., kidney failure)	Absolute deficiency of endogenous insulin History of severe hypoglycemia, unconscious hypoglycemia, or both, as well as previous exercise Aggressive glycemic therapy

Hypoglycemia without Diabetes

In subjects not affected by diabetes mellitus, hypoglycemia could be due to different causes. The Endocrine Society has proposed classifying the causes of hypoglycemia according to the subject's health status. (Table 29.3).

Drugs

Drugs represent the most common cause of hypoglycemia. Numerous drugs responsible for hypoglycemia have been described, with varying degrees of evidence (Table 29.4). In

Table 29.3 Classification of the hypoglycemic causes in non diabetic patients

Causes of hypoglycemia in patients with concomitant diseases	Causes of hypoglycemia in apparently healthy subjects
Drugs (Table 29.4): – Insulin and insulin secretagogues – Alcohol Chronic and acute diseases: – Hepatic, renal, and cardiac insufficiency – Sepsis and malaria – Inanition Hormonal deficiencies: – Cortisol – Glucagon – Adrenaline Non-insular tumors	Endogenous hyperinsulinism: – Insulinoma – Hypoglycemic syndrome of pancreatic origin not secondary to insulinoma – Hypoglycemia after gastric bypass – Autoimmune hypoglycemia Accidental, voluntary, or *factitia* hypoglycemia

Table 29.4 Drugs inducing hypoglycemia (excluding alcohol and oral hypoglycemic agents)

Moderate evidence
Cibenzolina Gatifloxacin Pentamidine Quinine Indomethacin Glucagon
Low evidence
Sulfonamides Artesunate/Artemisinin IGF-1 Lithium Propoxyphene/dextropropoxyphene
Very low evidence
Ace inhibitors Angiotensin receptor antagonists Beta-blockers Levofloxacin Mifepristone Disopyramide Trimethoprim sulfamethoxazole Heparin 6-Mercaptopurine

particular, drug-induced hypoglycemia is more common in elderly patients with underlying comorbidities. Moreover, the incidence of hypoglycemia in hospitalized patients is frequent when insulin or secretagogue drugs are administered to patients in whom parenteral nutrition is suspended or when insulin sensitivity increases (as when steroid therapy is interrupted). Another common cause of hypoglycemia is alcohol intake, as it inhibits gluconeogenesis. Consequently, alcohol-induced hypoglycemia usually occurs after a few days of alcohol abuse, especially when the subject does not adequately eat.

Chronic and Acute Illnesses

Among hospitalized patients, renal and hepatic failure and sepsis are common causes of hypoglycemia. In liver failure, hypoglycemia is due to the reduced endogenous glucose synthesis; in renal failure, it is caused by reduced insulin clearance and reduced mobilization of gluconeogenesis precursors. During sepsis, the recruitment of macrophages to the infected tissues leads to the release of cytokines that induce increased glucose utilization. In inanition, on the other hand, hypoglycemia may be due to loss of adipose tissue and subsequent reduction of gluconeogenesis precursors and increased glucose utilization.

Hormonal Deficiencies

Growth hormone (GH) and cortisol deficiency are associated with reduced availability of substrates for gluconeogenesis and increased peripheral glucose utilization.

Non-insular Tumors

Fasting hypoglycemia secondary to non-insular tumors is occasionally observed in patients with large mesenchymal or epithelial tumors. It is due to hyperproduction of insulin-like growth factor 2 (IGF-2) or immature precursors resulting in hypoglycemic episodes similar to those observed in patients with insulinoma.

Endogenous Hyperinsulinism

There are many causes of endogenous hyperinsulinism. but the underlying mechanism is the failure to reduce insulin levels during hypoglycemia. Insulinoma, a rare pancreatic β-cell tumor, benign in 90% of cases, is the leading cause of endogenous hyperinsulinism. From a clinical point of view, insulinoma is characterized by periods of neuroglycopenia due to endogenous hyperinsulinemic hypoglycemia occurring mainly in the fasting and only occasionally in the postprandial period.

Hypoglycemic syndrome of pancreatic origin not secondary to insulinoma, is clinically characterized by episodes of postprandial neuroglycopenia secondary to hyperinsulinism. From a histological point of view, it is characterized by β-cell hypertrophy and hyperplasia of the β-cell nuclei.

Autoimmune hypoglycemia is a rare condition characterized by anti-insulin antibodies without pancreatic β-cell alteration. In rare cases, anti-insulin receptor antibodies may also be found.

Finally, postprandial hypoglycemia can be observed in obese patients undergoing bariatric surgery. In these patients, increased levels of the intestinal incretin GLP-1 and subsequent glucagon suppression increase insulin secretion. In addition, the increased expression of IGF-1 and IGF-2 growth factor receptors results in increased β-cell activity, thus contributing to the insulin levels increase.

Hypoglycemia Factitia

Hypoglycemia can be caused intentionally, such as during the performance of a clinical test (insulin-tolerance test); accidentally, for example, as a result of a pharmacy error; or as a result of taking insulin as a potent anabolic agent, such as in athletes. Hypoglycemia *factitia* should always be suspected if hypoglycemia occurs suddenly in a healthy subject. In the case of voluntary insulin administration, the decisive diagnostic criterion is related to the discrepancy between the finding of elevated plasma insulin concentrations and normal C-peptide values.

Diagnosis

Once the suspicion of hypoglycemia has been confirmed by Whipple's triad criteria (Fig. 29.1), it is necessary to make a differential etiological diagnosis to start the most appropriate therapy.

Patients with diabetes should suspect the possibility of developing hypoglycemia when the self-monitored glucose concentration decreases rapidly or is not >70 mg/dL (3.9 mmol/L). This cutoff allows the patient to intervene preemptively to avoid symptomatic hypoglycemia. In addition, because self-monitoring is usually performed by analytically inaccurate devices, especially at low glucose levels, the 70 mg/dL glucose cutoff provides a margin for their analytical inaccuracy. Hypoglycemiain a diabetic patient must be treated by modifying therapy and then the improvement should be documented. If there is a history of unconscious hypoglycemia (recurrent hypoglycemia without symptoms),

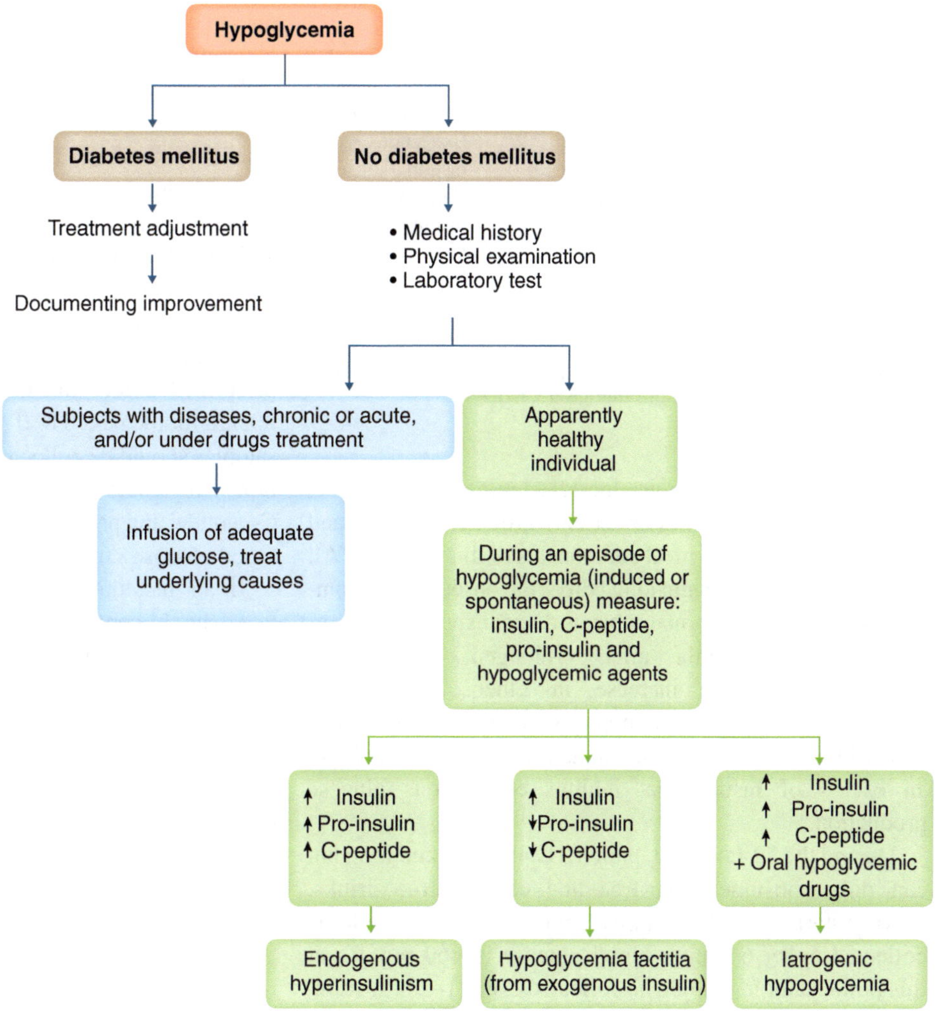

Fig. 29.3 Algorithm for the etiological diagnosis of hypoglycemia. (Copyright EDISES 2021. Reproduced with permission)

a 2–3-week period of scrupulous attention to avoid hypoglycemia is recommended.

Non-diabetes mellitus subjects with hypoglycemia often present with signs and symptoms, such as mental confusion, altered consciousness, epilepsy, or after the casual finding of low plasma glucose concentrations. In a symptomatic subject, the confirmation of hypoglycemia is based on the finding of low plasma glucose levels and the resolution of symptoms after glycemia correction.

The differential diagnosis is based on: (1) accurate anamnesis, which includes a detailed description of the events, the relationship with meal intake (hypoglycemia that occurs 2–3 h after meals is suggestive of hyperinsulinism, whereas if it occurs after prolonged fasting, it suggests an altered mechanism of gluconeogenesis), the presence of comorbidities, and the pharmacological therapy taken by the patient and family members; (2) physical examination; and (3) laboratory data to assess the role of insulin in hypoglycemia genesis (Fig. 29.3).

First, drug therapy, especially insulin, and alcohol, should be considered. The presence of diseases, chronic or acute, hormonal deficiencies, non-β-cell tumors, and previous gastric surgery should also be considered.

If all these conditions are absent, and thus the subject is healthy, further laboratory investigations should be performed. In particular, during an episode of hypoglycemia, blood glucose, insulin, C-peptide, and proinsulin should be measured simultaneously and a toxicological screening for oral hypoglycemic agents should be performed (Table 29.5). In some cases, patients experience symptoms of hypoglycemia after a short period of fasting; in such cases, continued observation, especially after an overnight fast, could result in an episode of symptomatic hypoglycemia. During the observation period, blood glucose should be measured continuously; if symptoms occur and hypoglycemia is documented (plasma glucose <55 mg/dL), the tests described above should be performed (Table 29.5). If this approach results in neither symptoms nor hypoglycemia and clinical suspicion remains high, the patient should undergo a stimulus test such as a 72-hour fast. The aim of this test is to induce homeostatic responses allowing the maintenance of plasma glucose concentration not associated with symptoms of hypoglyce-

Table 29.5 Differential diagnosis of hypoglycemia

Insulin (μU/mL)	C-peptide (nmol/L)	Proinsulin (pmol/L)	Hypoglicemic drugs	Interpretation
> > 3	<0.2	<5	No	Exogenous insulin
≥3	≥0.2	≥5	No	Endogenous insulin[a]
≥3	≥0.2	≥5	Yes	Oral hypoglycemic agents (drug-induced)

[a]Insulinoma, hypoglycemic syndrome of pancreatic origin not secondary to insulinoma; hypoglycemia after gastric bypass

mia under fasting conditions. Increased release of glucagon, adrenaline, and, to a lesser extent, growth hormone and cortisol are the most critical components of this response. All hormonal responses begin long before the onset of symptomatic hypoglycemia. Normal subjects do not experience symptomatic hypoglycemia after prolonged fasting due to hormone-induced increase in glucose production. Prolonged fasting results in hypoglycemia only if there is a defect in normoglycemia mainteinance due, for example, to an excess of insulin that inhibits endogenous glucose production.

During the 72-hour fast, patients cannot assume food but can only consume sugar-free and caffeine-free drinks. Fasting starts after the last meal. Blood samples should be collected every 6 h until blood glucose is <60 mg/dL; at this point, samples should be collected every 1–2 h. Insulin, C-peptide, and proinsulin should only be assayed in samples with blood glucose <60 mg/dL. Fasting ends when the glucose concentration is <45 mg/dL, the patient exhibits signs and/or symptoms of hypoglycemia, 72 hour have elapsed, or blood glucose is <55 mg/dL if Whipple's triad has been documented previously. The absence of signs and symptoms of hypoglycemia, as well as a low plasma glucose concentration during a 72-h fast, does not preclude a hypoglycemic disorder that may cause only postprandial symptoms. At the end of the test, 1 mg of intravenous glucagon is administered, and plasma glucose is measured after 10, 20, and 30 min; then, the patient can eat. Insulin is an anti-glycogenolytic, and hyperinsulinemia allows hepatic glycogen storage. Consequently, patients with hypoglycemia secondary to hyperinsulinism (as in the case of insulinoma) respond to glucagon administration (which is a potent glycogenolytic agent) by releasing glucose; in particular, a glycemia increase of at least 25 mg/dL will be observed compared to the glycemia at the end of the test. The healthy subjects, instead, will have released virtually all the hepatic glucose within the 72 hour of fasting and, therefore, will present a reasonable response to the stimulus of glucagon, with a much smaller increase of glycemia.

In case of non-insular tumors suspicion, autoimmune etiology, or hormonal deficiency, the measurement of anti-insulin antibodies, anti-insulin receptor, IGF-1/2, plasma cortisol, glucagon, and growth hormone can be performed.

If symptoms of hypoglycemia occur within 5 h of meals, patients should be assessed in the postprandial state by mixed-meal testing. Specifically, the patient take a nonliquid meal, resulting in the onset of symptoms, and be kept under observation for the next 5 h. Plasma glucose assay will be performed before the mixed meal and after intake every 30 min for the next 5 h. If severe symptoms occur before 5 h, samples should be collected before carbohydrate administration (to assess the correction of symptoms). Insulin, C-peptide, and proinsulin should be tested only in samples of blood glucose <60 mg/dL (Table 29.5).

If an insulinoma is suspected, radiological investigations should confirm the diagnosis. Computed axial tomography (CAT) and nuclear magnetic resonance (NMR) are the most commonly used noninvasive diagnostic procedures to detect the tumor site. However, more invasive techniques, such as endoscopic ultrasonography and venous sampling after arterial stimulation, are more accurate in preoperative localization of insulinoma and superior to standard noninvasive localization techniques. Finally, positron emission tomography (18F-DOPA-PET) is a technique successfully used for the localization of insulinomas, and, in children with hypoglycemia due to congenital hyperinsulinism, it represents the gold standard for the lesions localization before surgery.

Autoimmune hypoglycemia is a rare condition characterized by antibodies directed against insulin or the insulin receptor, resulting in hypoglycemia. Detection of anti-insulin and anti-insulin receptor antibodies is necessary to confirm the diagnosis of autoimmune hypoglycemia. Autoimmune hypoglycemia should be suspected when hypoglycemia is associated with high insulin levels (>100 μU/mL) and incompletely suppressed C-peptide levels. Finally, although elevated insulin levels may be observed following exogenous insulin administration, the associated C-peptide levels are usually extremely low.

Recommended Readings

Cryer PE, Axelrod L, Grossman AB, Heller SR, Montori VM, Seaquist ER, Service FJ, Endocrine Society (2009) Evaluation and management of adult hypoglycemic disorders: an endocrine society clinical practice guideline. J Clin Endocrinol Metab 94(3):709–728. https://doi.org/10.1210/jc.2008-1410

Desimone ME, Weinstock RS (2000) Non-diabetic hypoglycemia. [Updated 2017 Sep 23]. In: De Groot LJ, Chrousos G, Dungan K

et al (eds) Endotext [Internet]. MDText.com, Inc, South Dartmouth (MA). Available from: https://www.ncbi.nlm.nih.gov/books/NBK355894/
Dynkevich Y, Rother KI, Whitford I, Qureshi S, Galiveeti S, Szulc AL et al (2013) Tumors, IGF-2, and hypoglycemia: insights from the clinic, the laboratory, and the historical archive. Endocrine Rev 34:798–826
Galati SJ, Rayfield EJ (2014) Approach to the patient with postprandial hypoglycemia. Endocr Pract 20:331–340
Ito T, Igarashi H, Jensen RT (2012) Pancreatic neuroendocrine tumors: clinical features, diagnosis and medical treatment: advances. Best Pract Res Clin Gastroenterol 26:737–753
SID-AMD. Standard Italiani per la Cura del Diabete Mellito – 2016
Tesfaye N, Seaquist ER (2010) Neuroendocrine responses to hypoglycemia. Ann N Y Acad Sci 1212:12–28
Whipple AO (1944) Hyperinsulinism in relation to pancreatic tumour. Surgery 16:289–298

The Role of Laboratory in Pregnancy

30

Marcello Ciaccio and Luisa Agnello

Introduction

The clinical laboratory has a critical role in pregnancy, both in the phases preceding conception to verify the general health status, or to identify the presence of pathological conditions before conception, and during pregnancy to monitor the health status of the mother and fetus, and, in some cases, in the postpartum period to evaluate the evolution of pathologies related to the pregnancy (e.g., gestational diabetes).

Preconception Laboratory Test

In the period preceding conception and pregnancy, it is essential to be aware of any clinical conditions to intervene preventively and avoid maternal–fetal pathologies, thus allowing a normal pregnancy and correct fetal development.

Useful laboratory tests in this context are:

- Indirect Coombs test
- Toxoplasmosis, Others, Rubella, Cytomegalovirus, and Herpes Simplex (TORCH)
- Complete blood count (CBC)
- Hemoglobin levels
- Glycemia
- Cervical cancer screening (PAP test)

M. Ciaccio (✉)
Department of Biomedicine, Neurosciences and Advanced Diagnostics, Institute of Clinical Biochemistry, Clinical Molecular Medicine and Clinical Laboratory Medicine, University Hospital "P. Giaccone", Palermo, Italy

Department of Laboratory Medicine, University Hospital "P. Giaccone", Palermo, Italy
e-mail: marcello.ciaccio@unipa.it

L. Agnello
Department of Biomedicine, Neurosciences and Advanced Diagnostics, Institute of Clinical Biochemistry, Clinical Molecular Medicine and Clinical Laboratory Medicine, University Hospital "P. Giaccone", Palermo, Italy

Indirect Coombs Test

The Coombs test evaluates a possible incompatibility between maternal and fetal blood through the detection of maternal immunoglobulin (Ig)G antibodies that, crossing the placenta, can develop an immunological response against antigens of fetal erythrocytes, leading to severe consequences for the fetus, such as intrauterine death and hemolytic disease of the newborn (HDN), characterized by anemia, neonatal jaundice, edema, and hydrops. The antibodies involved may be different (anti-Kell, anti-C, anti-E, etc.), but in most cases, they are anti-D. In particular, the Coombs test is used to evaluate the possible presence of maternal antibodies directed mainly against the erythrocyte Rh antigen, also known as the D antigen.

The Rh factor is a protein present on the surface of red blood cells in about 15% of the population that is, therefore, defined as Rh+; the remaining 85% of the population does not present this antigen on red blood cells and is, therefore, defined as Rh−. Unlike the AB0 system, Rh subjects do not present anti-Rh antibodies in the serum unless exposed to Rh+ blood cells (e.g., transfusion). The Rh factor is inherited as an autosomal dominant trait, so if the mother is Rh− and the father is Rh+, the fetus will likely be Rh+ and, therefore, there may be an incompatibility between maternal and fetal blood. Under normal conditions, the placenta acts as a barrier between maternal and fetal blood; only small amounts of fetal blood can reach the maternal circulation. However, these are generally insufficient to induce an immune response producing antibodies to Rh+ fetal blood cells. However, special conditions, such as previous pregnancies, blood transfusions, or promiscuous use of syringes, can lead to maternal immunization to D antigen. In addition, invasive diagnostic procedures, such as chorionic villus sampling or amniocentesis, may result in the passage of fetal blood into the maternal circulation. To prevent an Rh− mother of an Rh+ fetus from developing antibodies to the D antigen, she should receive prophylaxis with anti-D immunoglobulin before undergoing invasive testing. At the time of delivery, mainly if

M. Ciaccio (ed.), *Clinical and Laboratory Medicine Textbook*, https://doi.org/10.1007/978-3-031-24958-7_30

traumatic obstetrical maneuvers are performed, or in the case of abortion, large amounts of fetal blood may come into contact with maternal blood, thus stimulating a process of immunization leading to the synthesis of anti-D antibodies. Therefore, subsequent pregnancies will be at high risk of developing HDN because exposure to the antigen (fetal Rh+ blood cells) triggers a secondary immune response with the production of IgG antibodies that can cross the placenta and damage fetal red blood cells. At birth, the Rh group of the newborn is determined. If the newborn is Rh−, then the mother does not need to undergo further immunoprophylaxis. On the other hand, if the newborn is Rh+ and the mother is Rh−, the mother will have to undergo immunoprophylaxis again.

Thus, Rh− women can get immunized:

- After delivery due to immunoprophylaxis with inadequate or absent anti-D Ig
- During pregnancy due to events associated with transplacental hemorrhages, such as miscarriage, amniocentesis, villocentesis, ectopic pregnancy, intrauterine fetal death, external version maneuvers, or abdominal trauma.

After sensitizing events, postpartum and prenatal anti-D prophylaxis reduces maternal immunization risk. Experimental studies have shown that routine prenatal prophylaxis in Rh− pregnant women results in an absolute reduction in the risk of immunization. Therefore, all nonsensitized Rh-pregnant women should receive immunoprophylaxis at 28 weeks of gestation.

There is certainly no risk to the fetus if both parents are Rh− or the mother is Rh+.

The Coombs test is performed by drawing maternal blood. Maternal serum/plasma is mixed with Rh+ blood at 37 °C for a sufficient period (45–60 min) for IgG class D antibodies, which may be present in the maternal serum/plasma, to recognize and bind to the D antigen. Coombs serum, an anti-human IgG antiserum, is then added and bridges between the attached antibodies to cause obvious agglutination of the red blood cells. In the presence of agglutination, the test is positive and, therefore, the mother will be Rh+; in the absence of agglutination, the test will be negative and, therefore, the mother will be Rh−.

TORCH

TORCH stands for Toxoplasmosis, Others (syphilis, hepatitis, HIV, HCV, etc.), Rubella, Cytomegalovirus, and Herpes Simplex. It refers to infectious agents (bacteria, viruses, and parasites) that can be dangerous to the fetus if contracted during pregnancy. The risk varies according to the infectious agent and the number of weeks of gestation, decreasing as the gestational period progresses. It is important to perform serological screening for antibodies against TORCH agents in the preconception phase to implement preventive measures to minimize the risk of infection during pregnancy (if the screening is negative), thus allowing proper fetal development.

Toxoplasmosis

Toxoplasmosis is a zoonosis caused by the intracellular parasitic protozoan *Toxoplasma gondii*. The cat is the definitive host of Toxoplasma, and humans, and other mammals and various birds, are intermediate hosts. The parasite multiplies in the intestine of felines and produces oocysts that are excreted with feces; accidentally, other animals (cattle or sheep) may ingest the oocysts and become infested. Therefore, the meat and viscera of these intermediate hosts, which contain Toxoplasma, can be ingested by other animals, where the cycle begins again. The disease can be contracted by coming into contact with oocysts contained in cat feces, raw or undercooked infected meat, vegetables, and fruit that have not been well washed. The acute infection is asymptomatic in 90% of cases, while the remaining 10% develop only mild lateral and retrocervical adenopathies. In Italy, 60% of the population is susceptible to infection. Maternal–fetal transmission occurs when the infection is contracted during pregnancy, that is, in a previously seronegative woman. The overall risk of transmission to the fetus and the clinical picture and severity of the congenital infection depend on the gestational age. In contrast, the risk of fetal infection increases with increasing gestational age because the placenta progressively thins out. Vice versa, the earlier the infection, the more serious the damage to the fetus. The major sequelae are chorioretinitis, blindness, hydrocephalus, neurological damage, and mental retardation.

Prenatal screening is based on a serological test aimed at establishing the immunological status of the woman through the detection of specific anti-toxoplasma IgG and IgM antibodies. In this way, it will be possible to determine if the woman is seronegative and, therefore, susceptible to infection, if she had a previous or ongoing infection (Table 30.1).

Table 30.1 Serological toxoplasmosis screening

IgG	IgM	Interpretation	Indications in pregnancy
–	–	Absent immunity	Follow food hygiene rules. Repeat the serological test every three months
±	+	Possible acute infection	Perform diagnostic investigations to accurately establish the stage of infection
+	–	Previous immunity	Do not repeat the serological test

Ig immunoglobulin

Table 30.2 Immunoglobulin G avidity test

Result	Avidity	Interpretation
0–19%	Weak	Probable current or recent infection (not more than 3 months)
20–30%	Intermediate	Probable recent infection
>30%	Heavy	Previous infection

IgM must be considered an alarm signal, but it is not a specific indicator of the acute phase of the infection. IgMs appear early, 10–12 days after infection, and reach a plateau after 2–4 weeks; in 72% of cases, they are still detectable after 9–10 months and can remain low titers for more than a year. Serum IgM does not necessarily indicate an ongoing or recent infection. However, it could be due to long-lasting IgM or nonspecific "natural" IgM, produced independently from the presence of Toxoplasma infection, recent or past, and often detectable in the first trimester of pregnancy. In order to confirm and date toxoplasmosis, level II investigations are performed, such as the IgA assay and the assessment of IgG avidity. IgA is generally produced after IgM and before IgG; initially, there is a rapid increase and subsequent decrease by 6–9 months, earlier than IgM; they may reappear after reactivation in the absence of IgM. The simultaneous presence of IgA and IgM allows a fairly reliable diagnosis of recent toxoplasmosis acquired in the last 6 months. The IgG avidity test is the most reliable method to date the infection. Avidity represents the strength of binding between an antibody and its antigen. The IgG avidity test, in general, is based on the principle that IgG undergoes an increase in antigen-binding affinity during the evolution of the immune response, which diagnostically translates into an antigen-antibody bond that is less sensitive to the action of a denaturing agent. Consequently, the closer the production of IgG in time, the lower the strength of antigen binding and the lower the avidity; this allows reliable dating of the infection. In other words, IgG with low avidity is usually present during the acute phase (Table 30.2).

Others

Syphilis

Syphilis is an infectious disease caused by the *Treponema pallidum* bacterium, transmitted mainly through sexual contact but sometimes by the transplacental route and during passage through the birth canal. Syphilis can therefore be defined as:

- Acquired, when contracted after birth
- Congenital or prenatal, when contracted transplacentally
- Connatal, when acquired at the time of passage through the birth canal

Syphilis develops in several stages, each characterized by different symptoms and courses.

In the primary stage, after an incubation period of about 3 weeks, a painless lump (a complicated ulcer or syphiloma) appears at the point of inoculation of the bacterium (generally the genitals). Usually, the syphiloma disappears even if it is not treated, but after 6 weeks, if the disease is not treated, it evolves into the secondary stage. The latter is characterized by the invasion of the bacterium throughout the body, resulting in unaccompanied rashes, itching, fever, enlarged lymph nodes, patchy alopecia, headache, weight loss, myalgia, and fatigue. These skin manifestations disappear after 1–2 weeks but recur over the years (latent stage). Without appropriate treatment, lesions in the internal organs, especially in the central nervous system, may occur after years; in the tertiary stage, the spinal cord will be affected (motor disorders), and in the last stage, the brain (dementia).

Maternal–fetal transmission can occur at any time during pregnancy, but the risk of transmission is most significant from the third month and increases as gestation progresses. The stage of the mother's syphilis is the most critical factor in determining the probability of vertical transmission. Primary or secondary syphilis, if untreated, carries a risk of fetal transmission of 70–100%, if treated, 2%; in case of early latent syphilis, 40–83%; and 2.5–10% in the late latent infection stage.

Early treatment of the mother with penicillin theoretically eliminates the risk of vertical transmission of the disease as long as the interval between initiation of therapy and delivery is >4 weeks. The severity of untreated fetal infection is greater the earlier the infection. Therefore, prevention and diagnosis of congenital syphilis depend on the infection's diagnosis and treatment in the pregnant woman.

The neonatal consequences of vertical transmission include not only congenital syphilis, which is characterized by developmental alterations in the child if left untreated, but also obstetrical complications such as late abortion, death in utero, fetal hydrops, and preterm delivery.

Two types of tests can be used for serological screening:

- Nonspecific treponema tests, based on the detection of IgG and IgM antibodies directed against substances released by the tissues due to the pathogenic action of treponema, defined as antilipoid antibodies. Among these, the most commonly used are the rapid plasma reagin (RPR) and the venereal disease research laboratories (VDRL).
- Treponema-specific tests, based on the detection of antibodies to *Treponema pallidum*-specific antigens.

VDRL is a simple test based on a flocculation reaction between the patient's serum and particular types of lipids, known as cardiolipins. RPR is a technical variant of VDRL in which the addition of micro carbon particles modifies the antigen. Both techniques are easy to perform, inexpensive, and characterized by high sensitivity and low specificity; a false-positive result can be observed in cases of infectious diseases such as malaria, tuberculosis, viral fevers, trypanosomiasis, and leprosy and in other conditions such as pregnancy, collagenopathies, advanced age, and drug addiction. Treponema-specific tests include the Treponema Pallidum Haemagglutination Assay (TPHA) and the Treponema Pallidum Particle Agglutination Assay (TPPA), which are positive 2–4 weeks after infection and have a sensitivity ranging from 60% to 99% (low sensitivity only at disease onset) and a specificity >99%, and immunoassays, such as the enzyme immunoassay (EIA) and the chemiluminescence assay (CLIA). EIA (with recombinant antigens) is positive 3 weeks after infection, with a sensitivity >98% and specificity >97%; CLIA is positive 3 weeks after infection, with a sensitivity and specificity >98% and >97%, respectively; fluorescent treponemal antibody absorption (an immunofluorescence test with adsorption, currently little used) has a sensitivity of 84% in primary syphilis and about 100% in the other stages, and a specificity of 96%. While for nonspecific tests, the titer of positivity correlates with disease activity and response to therapy, treponema-specific tests generally remain positive throughout life, regardless of therapy and disease activity.

Syphilis in pregnancy is a largely preventable cause of neonatal morbidity and mortality through screening and treatment of the infection in the mother. Diagnosis and staging of syphilis in pregnant women are based on clinical examination and serological testing. The serological screening test is a treponema-specific test (TPHA, TPPA, EIA, or CLIA). A second specific test must confirm positivity of one specific test. If the diagnosis of syphilis is confirmed, a nonspecific test (RPR or VDRL) should be performed to determine the stage of infection.

Hepatitis

Acute infectious hepatitis can be caused by various types of viruses (A-B-C-D-E) with hepatic tropism. Generally, from a clinical point of view, hepatitis is asymptomatic or associated with nonspecific symptoms such as nausea, vomiting, myalgia, and anorexia. While hepatitis A is not transmitted from mother to fetus (to date, only two cases have been reported), hepatitis B and C represent an important problem for Public Health and can be transmitted vertically. The hepatitis B virus (HBV) is a DNA virus of the Hepadnaviridae family that, after a long incubation period (1–6 months), causes an acute infection, often asymptomatic, which rarely evolves into fulminant hepatitis. The frequency of chronic infection depends on the age at which it is acquired, being low (1–5%) in adults and high in children (90%); all infants who become infected in the perinatal period will develop a chronic infection. In order to prevent vertical transmission of HBV, it is necessary to identify women who carry the virus by detecting the surface antigen (HbsAg) by serological tests; if the test is positive, the result must be confirmed. Then the search for "e" antigen (HBeAg), an index of active virus replication, is performed to assess whether the newborn should be subjected to immunoprophylaxis and postnatal vaccination.

Hepatitis C virus (HCV) is a RNA virus associated with a relatively low risk of vertical transmission (3–5%). The risk of vertical transmission increases significantly (up to 40–60%) if the mother is simultaneously infected with HIV. Screening is performed by the detection of HCV antibodies using serological tests (enzyme-linked immunosorbent assay [ELISA]); a positive test must be confirmed by another test. If positive, the diagnosis should be confirmed by the identification of HCV-RNA by polymerase chain reaction (PCR), which is the gold standard. Guidelines recommend prenatal screening for HCV only for women at risk.

Human Immunodeficiency Virus

The human immunodeficiency virus (HIV) is an RNA lentivirus with a reverse transcriptase that converts RNA to DNA. The virus can be transmitted vertically by the transplacental route, during childbirth, and during lactation. Most transmission episodes have been recorded during childbirth. Early identification of HIV-seropositive women aims to implement preventive measures to reduce the risk of maternal–fetal virus transmission. Preventive measures in HIV-positive women are associated with a 1% reduction in the risk of maternal–fetal transmission and include a planned cesarean section at 38 weeks of gestational age, prophylaxis with zidovudine (an antiviral drug), and artificial breastfeeding. Screening is based on detecting HIV 1 and 2 antibodies by ELISA. Since antibodies are detectable in the serum about 3 months after infection (the window period), a negative result does not exclude the presence of the infection. A positive result must be confirmed by a more specific test (Western blot). Infection in newborns is diagnosed through the search for nucleic acid by PCR or viral antigen because maternal antibodies can cross the placenta giving false-positive results.

Rubella

Rubella is an infectious disease caused by an RNA virus belonging to the Togaviridae family. The infection is generally contracted in childhood. In 1972, the vaccine was introduced in Italy, initially recommended only for girls in puberty but later extended to all children of both sexes aged <2 years. Acute primary infections and vaccination result in permanent immunity. If the infection is contracted during pregnancy, it can cause severe damage to the fetus, such as malformations or intrauterine death, because the virus can reach the fetal circulation and multiply rapidly in the embryonic tissues, causing an alteration of organogenesis. Vertical virus transmission can occur throughout gestation, with a greater risk of transplacental passage in the first trimester. The objective of screening is to assess the mother's immunity using the rubeotest (a serological test for detecting IgM and IgG anti-rubella antibodies) to identify women susceptible to infection. If a woman is negative for rubella before pregnancy, she can be vaccinated as a preventive measure. Unfortunately, there is no therapy to reduce vertical transmission, and the vaccine is not recommended during pregnancy because it could be teratogenic.

Cytomegalovirus

Cytomegalovirus (CMV) is a DNA virus from the Herpesviridae family. Humans are the only reservoir, and transmission can occur horizontally, from human to human by direct or indirect contact, or vertically, during pregnancy. The virus is present in all body fluids, including milk. Vertical transmission can occur by the transplacental route, either by primary infections (first acquired), secondary infections (reinfection or reactivation of latent virus), or by perinatal route, either during birth or lactation. The risk of vertical transmission is approximately 30–40% in primary infections and 0.5–2% in secondary infections. Infants with congenital infections are mostly (90%) asymptomatic, and a small percentage will present neurosensory alterations, such as mental retardation, psychomotor retardation, hypotonia, autism, hearing defects, and visual defects.

The diagnosis of CMV infection is based on serological investigations aimed at identifying the presence of IgG and IgM antibodies directed against CMV and the avidity of IgG antibodies. In particular, the detection of serum virus-specific IgG in a previously seronegative woman or virus-specific IgM associated with low-avidity IgG suggests a primary infection. The diagnosis of secondary infection, on the other hand, is based on the finding of an increased IgG antibody titer, with or without high avidity IgM and IgG. In the case of primary maternal infection, amniocentesis should be performed to assess whether the fetus is infected and to identify the presence of the virus by searching for its DNA. However, the diagnosis of fetal infection is not an indicator of disease. Unfortunately, no treatment is available to prevent vertical transmission of CMV infection.

Herpes Simplex

Herpes simplex is a DNA virus from the Herpesviridae family, and two different serotypes are known, characterized by common antigens and specific antigens (HSV1 and HSV2). Infection with both HSV1 and HSV2 can cause clinical manifestations at the genital or oral level. However, generally, HSV1 is contracted in childhood and causes orolabial manifestations, while HSV2 is typical in adults and causes anogenital lesions. Infection with HSV1 and HSV2 occurs through direct contact with someone who actively eliminates the virus. Following the primary infection, after the clinical phase is over, HSV1 and HSV2 remain latent in the nerve ganglia for several years of life. However, they can be temporarily reactivated with the appearance of clinical manifestations following various stimuli, such as physical or emotional stress, menstrual cycle, exposure to UV rays, etc.

Vertical transmission occurs mainly perinatally, during childbirth, by direct contact of the fetus with maternal secretions; it is more frequent in women with primary infections. The newborn may present with a localized infection of the skin, eyes, mouth, or central nervous system (encephalitis) or a systemic infection involving several organs.

Factors that influence the probability of vertical transmission are:

- Type of infection: the probability is high (up to 50%) in mothers with a primary infection contracted towards the end of gestation, while it is very low in cases where the primary infection is contracted in the first trimester of pregnancy or in cases of recurrent episodes towards the end of pregnancy.
- Presence of maternal antibodies can cross the placenta and reach the fetal circulation, thus preventing neonatal infection.
- Vaginal or cesarean section delivery; the former is associated with an increased risk of vertical transmission of infection. Cesarean section is recommended in cases of primary infection if lesions are present at the time of delivery or if they appear in the last 6 weeks of pregnancy.

Screening is based on the serological detection of IgG and IgM antibodies to HSV1 and 2. Seropositivity for HSV2 is almost exclusively due to an anogenital infection, while a positive result for HSV1 in an asymptomatic subject may be due to an orolabial or anogenital infection. In the case of active lesions, virus isolation by culture is indicated. However, it could be false positive in the case of recurrent or healing lesions. Viral DNA detection by PCR cannot always be performed on genital samples.

Complete Blood Count and Hemoglobin

It is crucial to perform a complete blood count as a screening test to assess the presence of any alterations and intervene preventively; for example, in the case of a reduction of the average cell volume and morphological alterations of the erythrocytes, the hemoglobin balance must be evaluated in order to identify a possible hemoglobinopathy.

Hemoglobinopathy is a blood disorder due to a genetic defect in one or more globins constituting the tetramer of hemoglobin (Hb), which can be qualitative (structural defects or Hb variants) or quantitative (synthesis defects or thalassemias). If hematological alterations due to hemoglobin defects are found in the woman, it is important to also evaluate the hemoglobin status of the partner to estimate the risk that the fetus could be affected by hemoglobinopathies, which in homozygosity can be associated with variable levels of anemia.

The hemoglobin balance, assessed by separative methods such as high-performance liquid chromatography or capillary electrophoresis, represents a first-level examination that provides qualitative and quantitative data regarding all hemoglobin fractions present in the circulation. In the case of suspected sickle cell anemia (the presence of the HbS variant), the sickle cell test represents a useful diagnostic confirmation tool. As far as the other hemoglobinopathies are concerned, the blood count results and the blood smear are valid diagnostic aids. However, the characterization of hemoglobin defects is only possible using second-level investigations based on molecular tests aimed to identify mutations in the genes encoding for globin chains.

Glycemia

It is important to evaluate glycemia in the preconceptional period, especially for a woman with diabetes, in order to ensure that she is in a state of good glycemic compensation and, therefore, reduce the risk of miscarriage or fetal malformations.

Pap Test

The Pap test is a diagnostic test for the prevention of cervical cancer. Its name comes from the surname of its inventor, Doctor Georgios Papanicolaou. The test consists of taking cells from the cervix and the cervical canal, smudged on a slide, stained according to the Papanicolau method, and examined under a microscope.

Ascertainment of Pregnancy

Following a delay in the menstrual cycle, when the possibility of pregnancy is suspected, investigations to confirm conception are based on detecting human chorionic gonadotropin (hCG) in blood or urine.

hCG is a glycoprotein hormone consisting of two subunits, α and β, noncovalently linked; the α subunit is identical to that of pituitary glycoprotein hormones, such as luteinizing hormone (LH), stimulating follicle hormone (FSH), and thyroid stimulating hormone (TSH), while the β subunit is specific for hCG. hCG is produced by trophoblastic tissue and therefore by the placenta during pregnancy, in trophoblastic pathology, and, to a lesser extent, by some poorly differentiated neoplasms. Indeed, following conception, hCG begins to be produced as early as the blastocyst, before implantation occurs.

In serum and urine, hCG may be present in different forms, such as the biologically active form and molecular forms that result from the dissociation or degradation of hCG. In particular, the modified heterodimer can be found as well as free α and β subunits or degradation products, such as the β-core fragment. In a physiological pregnancy, generally, the circulating concentrations of these molecular forms are low. However, some conditions, such asDown's syndrome, trophoblastic pathologies, and some neoplasms (testicle, ovary, breast, lung, and pancreas), are characterized by an increase in these forms.

The concentration of the biologically active form of hCG, both in serum and urine, increases exponentially during the first trimester of pregnancy, doubling every 48 h until it peaks around the 10th week of gestation. From the 10th week of gestation, the hCG concentration decreases progressively and will no longer be determinable a few weeks after delivery. The biologically active form of hCG has a half-life of about 40 h. During the first few weeks of pregnancy, hCG plays a vital role in maintaining the function of the corpus luteum. If an ectopic (extra-uterine) pregnancy occurs, hCG levels in the blood increase at a reduced rate. Therefore, hCG blood levels should be monitored over time if an ectopic pregnancy is suspected.

Furthermore, hCG levels can be altered if the fetus has chromosomal defects like Down's syndrome. The hCG assay is part of a screening protocol for detecting fetal chromosomal abnormalities. The minimum increase in hCG value in 48 h for an evolving pregnancy is 53%, while the minimum decrease in case of miscarriage is 21–35%, depending on the initial value. In 71% of cases of ectopic pregnancy, there is a lower increase or decrease in hCG for an evolving pregnancy or a miscarriage, respectively.

The quickest and easiest way to ascertain pregnancy is to perform a point-of-care test, which can be purchased in pharmacies or in large-scale retail trade and provides qualitative information with high sensitivity on the presence of hCG in the urine. To be reliable, the test should not be performed before 15–20 days from the date of presumed conception. Generally, point-of-care tests can identify an hCG concentration $\geq$25 mIU/mL. At low doses of hCG, the presence of FSH and LH may generate false positives; therefore, it is essential to perform two consecutive determinations. Positive tests should, however, be confirmed by a serum hCG assay.

The test is positive in the presence of physiological pregnancy, but also ectopic pregnancy, molar pregnancy (an unsuccessful pregnancy in which a nonviable fertilized egg implants in the uterus and results in the establishment of a pregnancy that will not be carried to term), and choriocarcinoma (a rare neoplasm of the trophoblast that develops inside the uterus).

The test is negative if the urine is diluted in the case of fetal death or miscarriage. Point-of-care tests and serum assays are performed by immunochromatographic methods using antibodies directed against different epitopes of hCG or its free subunits.

Pregnancy Laboratory Test

During pregnancy, it is important to carry out periodic laboratory tests, at least every 3 months, in order to evaluate the evolution of the pregnancy and to identify at an early stage any pathological alterations associated with the state of pregnancy and potentially dangerous to the health of the mother and the fetus, such as anemia or gestational diabetes; timely intervention through the administration of appropriate therapy can prevent complications. It is important to emphasize that pregnancy is a clinical condition characterized by physiological changes associated with altered levels of some laboratory parameters. It is therefore essential to know the reference values in pregnancy to correctly interpret the laboratory data.

Laboratory tests that should be performed during pregnancy are listed in Table 30.3.

Table 30.3 Tests to be performed during pregnancy

First trimester of pregnancy (within the 13th week of gestation)
Complete blood count with platelet count
TORCH agents (generally, the search for specific antibodies to toxoplasma, rubella, HIV, and syphilis are required)
Glycemia and glycated hemoglobin to identify the presence of overt diabetes
Chemical-physical and microscopic examination of the urine
Hepatic and renal function indices: transaminases and creatininemia
Indirect Coomb's test. If already performed in the preconception phase, repeat in case of Rh-negative women at risk of immunization
Second trimester of pregnancy (14th to 27th week of gestation)
Complete blood count
Complete urinalysis
Urine culture, in case of significant bacteriuria: at least 100,000 colonies per mL of urine
OGTT
Test for Rubella and toxoplasmosis, in case of seronegative women
Third trimester of pregnancy (since the 28th week of gestation to the pregnancy end)
Complete blood count
Complete urinalysis
Urine culture, in case of significant bacteriuria: at least 100,000 colonies per mL of urine
Test for Rubella and toxoplasmosis, in case of seronegative women
Ferritin, in case of reduction of the average globular volume
Hepatitis B virus antigen (HbsAg)
Antibodies to HCV and HIV, in case of risk assessed based on the history

OGTT oral glucose tolerance test, *HCV* hepatitis C virus, *HIV* human immunodeficiency virus

Pregnancy is associated with physiological anemia, characterized by hemoglobin levels lower than the reference values due to hemodilution. From the sixth week of pregnancy, it is generally observed a rapid and important increase in plasma volume and, to a lesser extent, an increase in the volume of red blood cells due to the erythroid hyperplasia that characterizes pregnancy (the bone marrow meets the oxygen demand that increases progressively with the production of red blood cells). These physiological changes lead to a dilution of the blood, resulting in a reduction in hematocrit levels, red blood cells, and hemoglobin. The fall in hemoglobin concentration reduces blood viscosity, thus improving placental perfusion and facilitating the maternal–fetal exchange of oxygen and nutrients.

A slightly increased plasma volume with an increased hemoglobin concentration increases blood viscosity, reducing blood flow to tissues (possible preeclampsia). In pregnancy, a hemoglobin value ≥11 g/dL is considered normal; hemoglobin values <11 g/dL should raise the suspicion of pathological anemia and, therefore, require further investigation. Ferritin is the most sensitive and specific biomarker to assess the extent of iron stores and establish the sideropenic nature of anemia. Hemoglobin values <8.5 g/dL are associated with an increased risk of poor neonatal outcomes. Plasma transferrin increases by 1.5–2 fold. In the last trimester of pregnancy, 90% of plasma iron is released via transferrin to receptors located on the membrane of syncytiotrophoblast cells, which are the same as those found on the membranes of reticulocytes. Therefore, ferritin decreases during pregnancy, partly due to hemodilution and partly due to the depletion of iron resources.

At the end of the third trimester of pregnancy, mild thrombocytopenia (with platelet counts not less than 100,000/mL) can be observed, which is generally a benign, asymptomatic condition that tends to resolve within 2 months of delivery. There may also be a slight increase in white blood cells, predominantly neutrophils, due to the action of estrogen. In general, pregnancy is characterized by a state of physiological hypercoagulability to ensure the deposition of fibrin between the uterine wall and the chorionic villi and maintain placental integrity; it is observed, therefore, an increase in fibrinogen, factors V, VII, VIII, IX, and X, while decreasing fibrinolytic activity, free protein S, and antithrombin III. This condition determines an increased thrombotic risk from 4 to 10 times throughout the pregnancy that remains until

2–3 weeks after delivery. In addition to the proteins involved in the hemostatic process, pregnancy is characterized by alterations in the serum concentration of other proteins produced in the liver. In particular, there is an increase in alkaline phosphatase, in part due to the placental production of one of its isoenzymes, a reduction in gammaglutamylpeptidase, and serum albumin. Additionally, total proteins and the albumin/globulin ratio are significantly reduced, while aspartate aminotransferase and alanine transaminase remain unchanged.

Pregnancy also induces adaptations in renal function to allow the gain of $Na+$ and water to ensure the expansion of extracellular volume and the growth of the fetus. During pregnancy, glomerular plasma flow increases progressively, reaching a maximum peak around the 20th week of gestation (up to 50% more) and remaining at this level throughout the pregnancy. In parallel, the glomerular filtration rate increases. Under these conditions, there is an increase in glucose excretion (sometimes glycosuria can be observed), creatinine, urea, proteins, and uric acid in the first trimester. Circulating glucose, creatinine, urea, and protein tend to decrease during pregnancy; uric acid is low in the first trimester, increases in the second, and is higher than normal in the third trimester. This occurs because tubular reabsorption changes normalize a few weeks after delivery.

The urine test allows to detect bacterial infections of the urinary tract, which are quite common in pregnancy, or other changes that may pose a risk to the fetus and the mother.

The presence of bacteria can be associated with the typical symptoms of cystitis (burning during urination and frequent urination) or be asymptomatic. Early detection of bacteriuria is important because it is associated with an increased risk of pyelonephritis due to the increased uterus volume, which compresses the bladder and urinary tract and causes stagnation of urine. In this condition, bacteria in the urine are more likely to travel up to the kidneys and cause the infection. In addition, untreated bacteriuria is associated with an increased risk of low birth weight and preterm birth due to premature rupture of membranes. Urinary tract infections in pregnancy could also trigger situations that lead to gestational hypertension or preeclampsia. The bacterium responsible for most infections is *Escherichia coli.*

The test is performed on a urine sample collected in the morning (intermediate micturition). The presence of nitrites, significant bacteriuria (>100,000 colonies per mL of urine), and increased leukocytes should raise the suspicion of urinary tract infection; to confirm the suspicion, a urinalysis with an antibiogram should be performed to isolate the specific pathogenic microorganism, determine antibiotic sensitivity, and direct the choice of therapy. The test is positive for a growth rate of ≥105 colony-forming units/mL. A urine culture is also performed in the presence of typical symptoms commonly referred to as cystitis.

Squamous epithelial cells derived from the vaginal epithelium can be found in urine. If present in low concentrations and in the absence of other altered parameters, the data is irrelevant, but if present in high concentrations and associated with bacteriuria and blood counts, it could indicate the presence of vaginitis. In this case, a vaginal swab should be performed to confirm the presence of an infection and proceed with antibiotic therapy.

Erythrocytes should not be detectable in urine. However, their presence in the urine sediment could indicate cystitis, a cervical polyp, a vaginal fold, or inflammation of the lower urinary tract; it could also be due to recent intense sexual activity.

Glycosuria should not be present. However, small amounts of glucose in the urine may be found in pregnancy due to physiological changes in the kidney; high concentrations may indicate gestational diabetes.

Protein should generally not be detectable in the urine, although a certain amount of protein is physiologically lost through the urine. Detecting proteinuria before the 20th week of gestation may indicate a renal alteration, while a later finding may indicate preeclampsia or gestosis.

Gestational Diabetes

Gestational diabetes mellitus (GDM) is classically defined as a condition of impaired glucose tolerance of variable degree and severity that occurs during pregnancy (usually in the second or third trimester) and generally regresses after delivery. However, it can recur at a distance, preferentially with the characteristics of type 2 diabetes.

GDM is the most common metabolic alteration in pregnancy that, if not correctly recognized and adequately treated, is associated with high maternal–fetal morbidity, mainly related to excessive fetal growth (macrosomia). Numerous studies have shown that early treatment of GDM reduces the incidence of adverse pregnancy outcomes.

GDM is, from a biochemical point of view, characterized by reduced insulin secretion accompanied by increased peripheral insulin resistance, two conditions typical of type 2 diabetes mellitus. In GDM, reduced insulin secretion cannot compensate for the insulin resistance characteristic of pregnancy. Indeed, as already mentioned, during pregnancy, the organism undergoes a physiological adaptation characterized by endocrine-metabolic alterations necessary to guarantee the supply of nutrients to the fetus and adequately prepare the maternal organism for childbirth and lactation (Table 30.4). These alterations are due to the action of hor-

Table 30.4 Metabolic changes in pregnancy

Glucose metabolism
Reduction of insulin sensitivity
Increased β-cell response
Modest increase in postprandial blood glucose
Lipid metabolism
Alteration of circulating levels of free fatty acids, triglycerides, cholesterol, phospholipids (initial reduction in the first 8 weeks followed by an increase)
Ketone bodies synthesis
Protein metabolism
Reduction of circulating amino acid levels in fasting and in the postprandial state

Table 30.5 Risk factors for gestational diabetes mellitus

Risk factors (OGTT at 24–28 weeks)
Family history of type 2 diabetes in first-degree relatives
Previous gestational diabetes
Fetal macrosomia in previous pregnancies
Overweight and obesity (BMI ≥25 kg/m^2)
Age ≥ 35 years
High-risk ethnic groups (South Asia, Middle East, Caribbean)
High risk factors (OGTT at 16–18 weeks)
Obesity (BMI ≥30 kg/m^2)
Previous gestational diabetes
Fasting blood glucose 100–125 mg/dL, in early pregnancy or in the past

BMI body mass index, *OGTT* oral glucose tolerance test

mones produced during pregnancy, such as estrogen, progesterone, cortisol, and placental lactogenic hormone.

The condition of insulin resistance that sets in with the progress of pregnancy, more evident in muscle and adipose tissue, is a physiological condition aimed at fetal growth. The reduced insulin-mediated utilization of glucose by the maternal organism determines, indeed, greater utilization of lipid substrates for energy purposes and, at the same time, guarantees the fetus a greater intake of carbohydrates. In addition, moderate but more prolonged postprandial hyperglycemia, which develops due to lower insulin sensitivity, favors the flow of nutrients from the mother to the fetus.

In GDM, the reduced action of insulin leads to an excess of nutrients in the circulation, such as glucose, lipids, and amino acids, which, crossing the placenta, cause hyperinsulinism in the fetus that, in turn, leads to organomegaly and macrosomia.

It is important to distinguish between overt diabetes in pregnancy and gestational diabetes. Manifest diabetes means the presence of diabetes that existed before pregnancy and was not recognized up to that moment; gestational diabetes means, instead, a form of diabetes that arose during pregnancy. The laboratory tests for diagnosing diabetes in pregnancy, both manifest diabetes and GDM, are fasting blood glucose, glycated hemoglobin, and the OGTT (oral glucose tolerance test). In particular, after pregnancy assessment, all women should be screened for overt diabetes by evaluating glycemia (fasting and random) and glycated hemoglobin within the first trimester of pregnancy. For the diagnosis of overt diabetes, the same criteria used in the general population are used:

- Fasting blood glucose ≥126 mg/dL (to be confirmed by a second blood test)
- Random blood glucose ≥200 mg/dL (to be confirmed by fasting blood glucose ≥126 mg/dL)
- HbA1c ≥ 48 mmol/mol (≥ 6.5%) (to be confirmed in a second sample)

Women diagnosed with overt diabetes should undergo intensive metabolic monitoring, as recommended for gestational diabetes.

Screening for GDM is based, first of all, on the assessment of specific risk factors (Table 30.5); women who present at least one risk factor for GDM should perform an OGTT with 75 g of glucose between the 24th and 28th weeks of pregnancy because GDM usually occurs in the second half of pregnancy. However, women with high risk factors should be screened early by performing a 75-g glucose OGTT between 16 and 18 weeks of pregnancy, to be repeated between 24 and 28 weeks if negative. Figure 30.1 shows the diagnostic procedure for screening and diagnosing diabetes in pregnancy.

Women with previous GDM should be screened for diabetes mellitus 2 by performing an OGTT with 75 g of glucose 6 weeks after delivery and within 6 months. If the test is negative, the OGTT should be repeated every 3 years; if impaired glucose tolerance (IFG or IGT) is found, the test should be repeated yearly.

The OGTT should be performed on an empty stomach in the morning by administering a solution consisting of 75 g of glucose dissolved in 300 mL of water. There are no particular indications to follow in the days preceding the test, but consuming at least 150 g of carbohydrates per day is recommended. During the test, assuming the sitting position and refrain from eating, drinking, and smoking is necessary. Blood sampling to obtain the plasma on which the glycemia will be performed by the enzymatic method will be done before the administration of the glucose solution (basal glycemia) and after 1 and 2 h from the administration of the solution. Table 30.6 shows the glycemia values that allow the diagnosis of gestational diabetes.

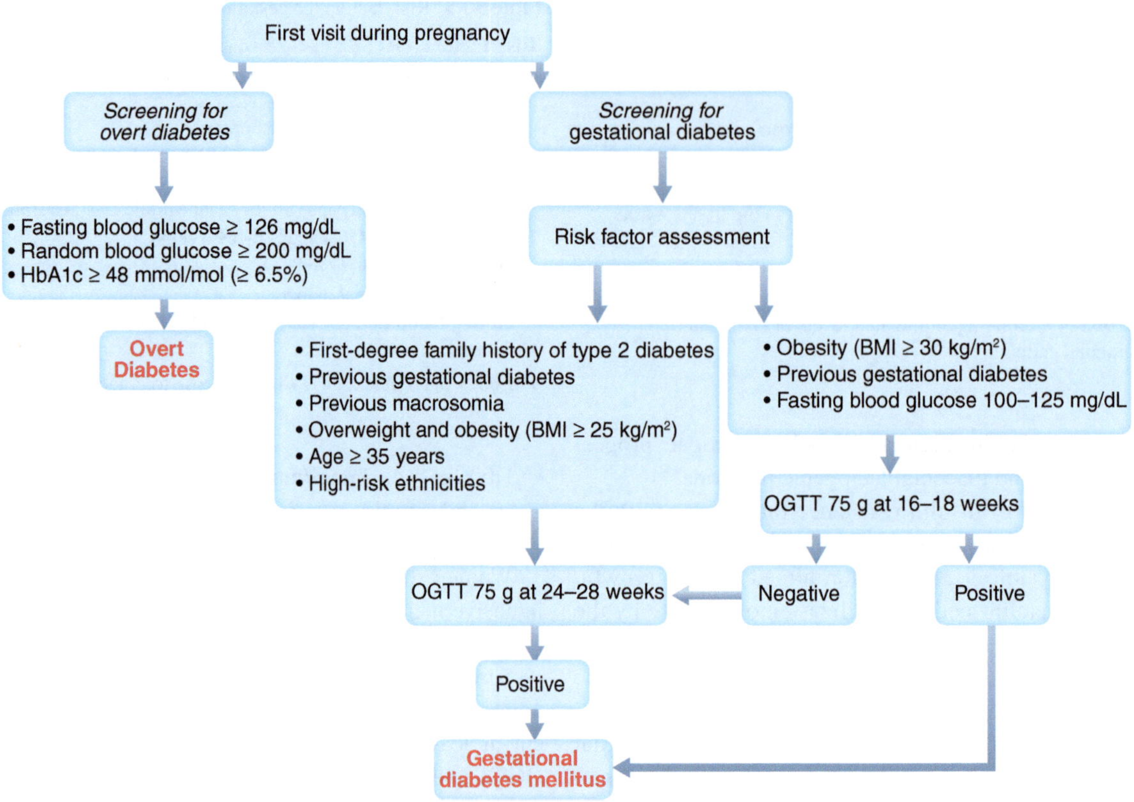

Fig. 30.1 Gestational diabetes diagnostic algorithm. (Copyright EDISES 2021. Reproduced with permission)

Table 30.6 Diagnostic criteria for gestational diabetes mellitus

Times	Glycemia (mg/dL)
0 min	≥92
1 h	≥180
2 h	≥153

Preeclampsia and Eclampsia

Preeclampsia is a clinical condition associated with hypertension and proteinuria, with or without edema, that occurs in the second half of pregnancy, no earlier than the 20th week of gestation, in previously normotensive, non-proteinuric women and usually disappears after delivery, within 6–12 weeks. Hypertension is a systolic blood pressure ≥ 140 mmHg or diastolic blood pressure ≥ 90 mmHg in two successive measurements at least 4 hours apart. Proteinuria is defined as an excretion ≥300 mg/24 h or a urinary protein/creatinine ratio ≥ 0.3 mg/mg (30 mg/mmol) on the extemporaneous sample, or, if quantitative measurement is not available, a dipstick finding of ≥1+ proteinuria. Preeclampsia can also be defined by the concomitant finding of hypertension without proteinuria if thrombocytopenia, increased serum creatinine, transaminases, pulmonary edema, and neurological symptoms such as headache and/or visual disturbances are present (Table 30.7).

Table 30.7 Diagnostic criteria of preeclampsia

Hypertension: Systolic blood pressure ≥140 mmHg or diastolic blood pressure ≥90 mmHg on two occasions at least 4 h apart after the 20th week of gestation in a previously normotensive woman
And
Proteinuria ≥300 mg/24 h or a urinary protein/creatinine ratio ≥0.3 mg/mg (30 mg/mmol) on the impromptu sample or, if quantitative measurement is not available, test strip detection (*dipstick*) of proteinuria ≥1+
Or
Hypertension: Systolic blood pressure ≥140 mmHg or diastolic blood pressure ≥90 mmHg on two occasions at least 4 h apart after the 20th week of gestation in a previously normotensive woman along with one or more of the following conditions (with or without proteinuria): - Platelet count <100,000 plt/microL - Serum creatinine >1.1 mg/dL or at least 50% increase in serum creatinine in the absence of known kidney disease - Transaminases ≥2 times the upper reference limit - Pulmonary edema - Neurological symptoms such as headache or visual disturbances

A recent meta-analysis estimated that approximately 4.6% of pregnancies worldwide are complicated by preeclampsia. Preeclampsia is defined as severe if the signs and/or symptoms listed in Table 30.8 are present.

Preeclampsia can have significant maternal–fetal consequences such as placental abruption, preterm delivery, low

Table 30.8 Conditions associated with severe preeclampsia

Neurological symptoms
Appearance of visual disturbances (scotoma, cortical blindness, and retinal vasospasm), severe headache or resistant to pain relief, altered mental status.
Liver disorders
Severe and persistent pain in the right upper abdominal or epigastric quadrant not justified by another diagnosis or finding of transaminases ≥2 times the upper limit of reference or both conditions.
Severe arterial hypertension
Systolic blood pressure ≥160 mmHg or diastolic blood pressure ≥110 mmHg on at least two occasions
Thrombocytopenia
Platelets <100.000 plt/microL
Kidney alterations
Serum creatinine >1.1 mg/dL or at least 50% increase in serum creatinine without known renal disease.
Pulmonary edema

birth weight, and fetal death; the most severe complication of preeclampsia is eclampsia, or gestosis, characterized by convulsive seizures and/or maternal coma. Preeclampsia is the result of an altered function of the placenta, the etiology of which is not yet fully known but could be linked to various factors such as poor development of the uterine placental spiral arterioles (which reduces uteroplacental blood flow in late pregnancy), immunological alterations, and placental ischemia or infarction. Preeclampsia may also be asymptomatic or cause edema associated with excessive weight gain. Identifying risk factors for preeclampsia is essential to initiating appropriate prophylaxis by week 20.

The main risk factors for preeclampsia are:

- Preeclampsia in previous pregnancies, especially those with adverse outcomes
- Pre-existing chronic hypertension
- Diabetes
- Multiple pregnancies
- Chronic kidney disease
- Autoimmune diseases (antiphospholipid antibody syndrome, systemic lupus erythematosus)

Moderate risk factors include nulliparity, obesity, maternal age <20 or >35 years, family history of preeclampsia, thrombophilia, or other thrombotic conditions. Right from the beginning of pregnancy, it is advisable to carefully assess the individual risk conferred by clinical and anamnestic factors and, if necessary, to start the pregnant woman with more stringent surveillance or pharmacological prophylaxis.

The monitoring of women at risk of preeclampsia includes the assessment of blood pressure, platelet count, serum creatinine, liver function indexes, and proteinuria within 24 h. These evaluations should be carried out from the beginning of pregnancy and help in more advanced gestational periods to identify new onset preeclampsia concerning preexisting pathological conditions.

Given the clinical relevance of preeclampsia for its complications on mother and fetus and the absence of adequate diagnostic screening tools, in recent years has been conducted an intense research activity aimed at identifying potential biomarkers predictive of preeclampsia.

Studies on the pathogenesis of preeclampsia have highlighted the importance of an aberrant synthesis of angiogenesis modulators in determining the endothelial damage and increased capillary permeability that underlie the pathogenesis. Angiogenetic factors of significant interest in this area include vascular endothelial growth factor (VEGF) and placental growth factor, along with some proteins with known anti-angiogenic activity such as Soluble Endoglin (sEng), the truncated form of the receptor for VEGF (Flt-1), known as soluble FMS-like Tyrosine Kinase-1 (sFlt-1). In the ischemic trophoblast, a characteristic feature of eclampsia, the production of anti-angiogenic factors (sEng and sFlt-1) increases, and that of angiogenic factors (VEGF, PIGF) is reduced. Alterations in the levels of these factors in the mother's plasma or urine precede the onset of preeclampsia by a few weeks, correlate with the severity of the disease, and normalize after delivery. These tests seem to be useful, especially in more advanced gestational periods, while they seem to be of little help before the 20th week. In particular, a recent meta-analysis has shown that the sFlt-1/PIGF ratio has the best diagnostic performance in predicting preeclampsia if used in the second-third trimester of pregnancy (after 20 weeks), with a sensitivity of 80% and a specificity of 92%. Growing scientific evidence has shown that increases in circulating proteins, such as pregnancy-associated plasma protein-A (PAPP-A), or changes in circulating free DNA in the first and second trimester predict adverse outcomes, including preeclampsia. However, this association is not strong enough to justify routine clinical use of these tests and, more importantly, to modify the clinical care pathway of women at risk for preeclampsia.

Women at high risk of developing preeclampsia (previous preeclampsia, prepregnancy diabetes, prepregnancy hypertension, and renal disease) should receive low-dose aspirin prophylaxis. Most risk factors are not modifiable, so interventions to reduce the likelihood of preeclampsia are limited to pharmacological prophylaxis and weight gain during pregnancy.

Prophylaxis aims to prevent severe and potentially fatal complications; uncomplicated preeclampsia is reversible and begins to resolve spontaneously after childbirth.

In the absence of maternal and/or fetal complications requiring immediate termination of pregnancy, the most important factor in the treatment decision is the gestational age. The safest treatment for the mother is delivery, but it may not be safe for the fetus, depending on the gestational age. In early gestational periods, conservative treatment aims to achieve fetal maturity (especially respiratory maturity) while keeping the maternal clinical condition under constant and careful control.

Pharmacological control of hypertension allows to avoid maternal complications directly related to blood pressure values, such as placental abruption, heart failure, and maternal cerebral hemorrhage. However, it cannot modify the disease's progression.

Eclampsia, or gestosis, consists of generalized tonic–clonic seizures in women with preeclampsia. It represents the convulsive manifestation of preeclampsia and is considered one of its most severe clinical manifestations.

Eclampsia most frequently occurs before delivery (almost always after the 28th week) but can also occur afterward (after 48 h and up to 4 weeks) or during labor. In the hours preceding the onset of convulsions, the woman may present signs and symptoms such as hypertension, headache, visual disturbances, and epigastric pain; in 25–40% of cases, there are no prodromal signs.

Eclampsia occurs in 2–3% of women with severe preeclampsia who do not receive prophylaxis and up to 0.6% of women with non-severe preeclampsia. In industrialized countries, it has an incidence ranging from 1.5 to 10 cases/10,000 deliveries. It is difficult to predict the eclampsia onset; the only way to reduce its incidence is to administer prophylaxis with magnesium sulfate in cases at risk, that is, in cases of severe preeclampsia, and especially when a headache, visual disturbances, and epigastric pain are present. Prophylaxis is not recommended in all other cases (moderate or mild preeclampsia). The diagnosis of eclampsia is essentially clinical and is based on the above-mentioned neurological symptoms. If the neurological deficit is not persistent, investigations are limited to those for screening for preeclampsia. Vice versa, it is appropriate to consider a differential diagnosis aimed at excluding other causes incidental to pregnancy (brain tumors, brain aneurysms) or conditions exacerbated by the gravid state (thrombotic thrombocytopenic purpura, hemolytic-uremic syndrome, cerebral venous thrombosis).

The therapeutic approach is based on maintaining maternal oxygenation and protection from accidental trauma during the acute crisis and treating hypertension and seizures with magnesium sulfate. Delivery is the only curative treatment.

Once eclampsia has been established, it is helpful to maintain intensive monitoring for 24–48 h after the convulsive crisis has ceased. Concerning the clinical condition, it is necessary to frequently monitor hemoglobinemia, hematocrit and platelets, liver function indexes (transaminases, lactate dehydrogenase, and bilirubin), and renal function indexes (creatininemia and serum electrolytes). Coagulation investigations are indicated in the case of hemolysis elevated liver enzymes, low platelet syndrome or disseminated intravascular coagulation.

Prenatal Screening for Down Syndrome

Down syndrome is, in about 95% of cases, the result of nondisjunction of chromosome 21 in oogenesis or spermatogenesis (trisomy 21). The gold standard laboratory test to diagnose a fetus with Down syndrome is karyotype analysis by chorionic villus sampling or amniocentesis. The latter are invasive techniques associated with a low risk of fetal loss. A non-invasive alternative, and therefore not associated with any risk to the fetus, is the application of tests that estimate risk by combining information from ultrasound examination and/or biochemical examination of maternal blood with maternal age and gestational age. These tests provide the result in a number; if this value exceeds a specific cut-off, the test will be positive, and the woman may undergo more invasive investigations, such as amniocentesis and villocentesis. In particular, ultrasonography evaluates nuchal translucency, which is the measurement of the thickness of the subcutaneous tissue at the nape of the neck of the fetus, that is, the space between the skin and the spine. The greater the thickness of nuchal translucency, the greater the probability that the fetus has Down's syndrome. Biochemical examinations, on the other hand, are based on the evaluation of serum markers such as human chorionic gonadotropin (total or free beta fraction, hCG), pregnancy-associated plasma protein A (PAPP-A), alpha-fetoprotein (AFP), unconjugated estriol (uE3), and inhibin A. Table 30.9 shows the main tests available for assessing the risk of Down syndrome. It is important to note that a positive result from these tests does not indicate that the fetus will have Down syndrome.

PAPP-A is a glycoprotein of placental origin, produced mainly in the syncytiotrophoblast.

AFP is a glycoprotein secreted first by the yolk sac and then by the fetus's liver It has the same functions as albumin; immediately after birth, its concentrations progressively decrease until it is absent. AFP can cross the placenta and reach the maternal circulation, where its levels will be directly proportional to those in the fetal circulation.

Estriol is a steroid hormone produced in unconjugated form exclusively during pregnancy by the placenta and fetal organs: the fetal adrenal glands produce dehydroepiandrosterone sulfate (DHEA-S), which is hydroxylated by the fetal liver to 16-hydroxy-DHEA-S and subsequently metabolized by the placenta to form unconjugated estriol. The uE3 secreted by the placenta then passes into the maternal circulation; after conjugation in the liver, it is subsequently eliminated in the urine. The levels of uE3 increase in the maternal

Table 30.9 Tests to assess Down syndrome risk

	NT 10th–13th week	hCG 10th–12th week	PAPP-A 10th–12th week	hCG 14th–20th week	AFP 14th–20th week	uE3 14th–20th week	Ininbin A 14th–20th week
Combined test	x	x	x				
Double test				x	x		
Triple test				x	x	x	
Quadruple test				x	x	x	x
Integrated test	x		x	x	x	x	x

NT nuchal translucency, *hCG* chorionic gonadotropin, *PAPP-A* plasma protein A associated with pregnancy, *uE3* unconjugated estriol

circulation as pregnancy progresses, especially in the third trimester; they are an index of the functionality of the feto-placental unit.

Inhibin A is a dimeric glycoprotein produced by the corpus luteum and placenta during pregnancy. Generally, in the circulation of a mother of a fetus with Down's syndrome, there should be, compared to a normal fetus, an increase in the levels of hCG and inhibin A and a decrease in the levels of PAPP-A, uE3, and AFP.

Recommended Readings

Abalos E, Cuesta C, Grosso AL et al (2013) Global and regional estimates of preeclampsia and eclampsia: a systematic review. Eur J Obstet Gynecol Reprod Biol 170:1

Agrawal S, Cerdeira AS, Redman C, Vatish M (2018) Meta-analysis and systematic review to assess the role of soluble FMS-like tyrosine kinase-1 and placenta growth factor ratio in prediction of preeclampsia: the SaPPPhirE study. Hypertension 71:306

American College of Obstetricians and Gynecologists; Task Force on Hypertension in Pregnancy Hypertension in pregnancy (2013) Report of the American College of Obstetricians and Gynecologists' Task Force on Hypertension in Pregnancy. Obstet Gynecol 122:1122–1131

Bartsch E, Medcalf KE, Park AL et al (2016) Clinical risk factors for pre-eclampsia determined in early pregnancy: systematic review and meta-analysis of large cohort studies. BMJ 353:i1753

Diagnosis of Perinatal TORCH Infections (1999) Newton ER. Clin Obstet Gynecol 42(1):59–70. quiz 174-5. Review

Duckitt K, Harrington D (2005) Risk factors for pre-eclampsia at antenatal booking: systematic review of controlled studies. BMJ 330:565

HAPO Study Cooperative Research Group (2008) Hyperglycemia and adverse pregnancy outcomes. N Engl J Med 358:1991

Linee guida gravidanza fisiologica. Ministero della Salute. http://www.salute.gov.it/imgs/C_17_pubblicazioni_1436_allegato.pdf

https://www.nice.org.uk/guidance/qs35/chapter/quality-statement-2-antenatal-assessment-of-pre-eclampsia-risk#what-the--quality-statement-means-for-service-providers-healthcare-practitioners-and-commissioners-2. Accessed on 30 Mar 2018

Lapolla A, Dalfrà MG (2006) Screening e diagnosi del diabete gestazionale. Biochimica Clinica 30:5–6

LeFevre ML (2014) U.S. Preventive Services Task Force. Low-dose aspirin use for the prevention of morbidity and mortality from preeclampsia: U.S. Preventive Services Task Force recommendation statement. Ann Intern Med 161:819

Levine RJ, Lam C, Qian C et al (2006) Soluble endoglin and other circulating antiangiogenic factors in preeclampsia. N Engl J Med 355:992

Levine RJ, Maynard SE, Qian C et al (2004) Circulating angiogenic factors and the risk of preeclampsia. N Engl J Med 350:672

Martin A, Krishna I, Badell M, Samuel A (2014) Can the quantity of cell-free fetal DNA predict preeclampsia: a systematic review. Prenat Diagn 34:685

Moore Simas TA, Crawford SL, Solitro MJ et al (2007) Angiogenic factors for the prediction of preeclampsia in high-risk women. Am J Obstet Gynecol 197:244.e1

Neu N, Duchon J, Zachariah P (2015) TORCH infections. Clin Perinatol 42(1):77–103. https://doi.org/10.1016/j.clp.2014.11.001. Epub 2014 Dec 20. Review

Poon LC, Nicolaides KH (2014) First-trimester maternal factors and biomarker screening for preeclampsia. Prenat Diagn 34:618

Scazzocchio E, Figueras F (2011) Contemporary prediction of preeclampsia. Curr Opin Obstet Gynecol 23:65–71

SID-AMD (2016) Standard Italiani per la Cura del Diabete Mellito

Zeisler H et al (2016) Predictive value of the sFlt-1:PlGF ratio in women with suspected preeclampsia. N Engl J Med 374:13–22

Hereditary Metabolic Diseases

31

Carlo Corbetta and Carlo Dionisi Vici

Introduction

Every year in the world, about eight million children (about 6% of those born) born with a severe congenital defect of genetic or partially genetic etiology, to which must be added a few hundred thousand more who present a congenital defect of postconception etiology due to maternal exposure to teratogenic agents (chemical, viral, bacterial, etc.). Many of these defects are incompatible with life or, in the case of survival, lead to serious neurological, motor, and sensory consequences; about three million children under 5 years of age die every year because of these congenital defects, and many have chronic and severe conditions of disability.

Among the defects with genetic etiology, congenital errors of metabolism, also defined as inherited metabolic diseases (IMD), constitute a group of diseases due to mutations of genes coding for proteins with an enzymatic or cellular transport function. IMDs represent a well-defined group among the so-called rare diseases (RDs) (pathologies with a prevalence in the population of less than 2.5 affects per 10,000 people), one of the chapters of greatest and current interest and attention in the medical and social fields. IMDs are associated with high morbidity and mortality and the possibility of recurrence in families at risk, with a high incidence of autosomal recessive inheritance forms (1:4 risk of recurrence at each conception).

Today, these hereditary diseases benefit from effective and targeted health interventions that have as their fundamental pivot the possibility of identifying patients through screening programs immediately after birth, in a clinical phase that is still asymptomatic, defined as the "free interval" that elapses between birth and the appearance of the first symptoms. Therefore, screening aims to activate a targeted therapeutic intervention before irreversible damage occurs in the patient, especially to the central nervous system, which can compromise the patient's future state of health.

In the field of public health management, it is now possible to define health pathways for these RDs based on the best strategies for early diagnosis (in particular through the action of neonatal screening systems) and then appropriate treatment. The latter is based on an integrated approach, which can vary from traditional diet therapy to the use of drugs and transplants (liver, kidney, and cellular therapies) to the correction of the genetic defect through gene therapy, not neglecting symptomatic treatments such as psychological support, neuromotor rehabilitation, etc.

Clinical Aspects

Definition and Etiology

IMDs, a well-defined category of recently recognized and defined rare diseases, are due to mutations in genes encoding for proteins with an enzymatic or cellular transport function. They are monogenic diseases, inherited in an autosomal recessive way, or linked to the X chromosome or mitochondrial DNA, or, more rarely, in an autosomal dominant way. Individually rare, IMDs collectively comprise more than 600 different diseases.

Classification

Altered functioning of a cellular biochemical pathway can cause:

- Accumulation of compounds upstream of the biochemical defect

C. Corbetta (✉)
Regional Newborn Screening Laboratory of Lombardy Region, Children's Hospital V. Buzzi, Milan, Italy
e-mail: carlo.corbetta@icp.mi.it

C. Dionisi Vici
Department of Pediatric Specialties and Liver-Kidney Transplantation, Division of Metabolic Diseases and Drug Biology, Bambino Gesù Children's Hospital, IRCCS, Rome, Italy

M. Ciaccio (ed.), *Clinical and Laboratory Medicine Textbook*, https://doi.org/10.1007/978-3-031-24958-7_31

- Absence of the end products of the metabolic pathway, which may result in the lack of production of essential substrates or alteration of energy homeostasis
- Activation of alternative metabolic pathways, with the production of secondary metabolites, usually not present in the human body, which exert a toxic action

The effect of these profound alterations in metabolic processes translates into the possibility of identifying and quantifying accumulated or poorly produced metabolites in the biological fluids of patients, which represent diagnostic biomarkers with high sensitivity and specificity. The analysis of biomarkers represents the instrument through which it is possible to decode the biological matrix of diseases; therefore, this type of investigation allows the diagnosis of IMD. Based on these assumptions, it is possible to classify IMDs according to clinical and/or pathophysiological criteria, as illustrated in Table 31.1, or systematically according to the biochemical pathway or subcellular organelle involved in the metabolic defect, as illustrated in Table 31.2.

Table 31.1 Clinicopathophysiological classification of metabolic diseases

Diseases that cause intoxication (acute and chronic)
Aminoacidopathies (phenylketonuria, leucinosis, tyrosinemia, homocystinuria)
Organic acidurias (propionic, methylmalonic, isovaleric, glutaric type I)
Defects of the urea cycle
Carbohydrates metabolism disorders (galactosemia, fructose intolerance)
Metabolic disorders of metals (copper, iron, manganese, and zinc)
Porphyrias
Diseases of energy metabolism
Mitochondrial diseases (mitochondrial respiratory chain defects, pyruvate metabolism defects, Krebs cycle defects,and fatty acid β-oxidation defects)
Cytosolic diseases (glycogenosis, defects in glycolysis/gluconeogenesis, defects in the metabolism of pentose phosphates)
Substrate deficiency diseases
Defects of creatine metabolism
Defects in neurotransmitter synthesis
Defects of amino acid synthesis (serine, glutamine)
Defects of cholesterol synthesis
Defects in the metabolism and transport of vitamins and cofactors (pyridoxine, vitamin B12, folate, and riboflavin)
Defects of metal metabolism (copper, zinc)
Disorders of the complex molecules metabolism
Lysosomal diseases
Peroxisomal diseases
Defects of glycosylation of proteins
Disorders of the phospholipids metabolism and complex lipids
Disorders of intracellular traffic

Table 31.2 Systematic classification of metabolic diseases

Disorders of amino acid and peptide metabolism
Disorders of carbohydrate metabolism
Disorders of the fatty acids and ketone bodies metabolism
Disorders of energy metabolism
Disorders of purine, pyrimidine and nucleotide metabolism
Disorders of sterol metabolism
Disorders of heme and porphyrin metabolism
Disorders of lipid and lipoprotein metabolism
Congenital defects of protein glycosylation and other disorders of protein modifications
Lysosomal diseases
Peroxisomal diseases
Disorders of neurotransmitter metabolism
Disorders of the vitamins and cofactors metabolism
Disorders of metals and trace elements metabolism
Disorders of xenobiotic metabolism

A detailed classification of IMDs is available at the following web address: www.ssiem.org/centralstore/ resources/ SSIEMClassificationIEM2011.pdf.

Clinical Features

The clinical picture that characterizes IMDs varies depending on the extent of the metabolic defect and is attributable to two main pathophysiological mechanisms:

- Accumulation of potentially harmful metabolites, leading to "intoxication"
- Lack of substrate synthesis, resulting in a "deficiency" condition

Given the extreme complexity of biological processes involved in IMD, the coexistence of the two mechanisms described above is possible, especially in diseases involving intermediate metabolism. From a clinical point of view, the involvement of the central nervous system is frequent, especially in clinical pictures of systemic type involving several organs or apparatuses; diseases with symptoms limited to a single organ or apparatus are rarer. Regarding the mechanism of intoxication, this can occur in an acute form or with a chronic, slowly progressive course. In the first case, the clinical manifestations are severe and require immediate and intensive therapeutic interventions.

IMD can generally occur at all stages of life, from the fetal period to adulthood. It is more common to observe the first symptoms in childhood, and, in most cases, the early onset is characteristic of the most severe forms.

Table 31.3 Clinical presentation of metabolic diseases

Newborn and early childhood
Metabolic diseases with dysmorphism, dysplasia, malformations, and intrauterine growth retardation
Acute neurological presentation (coma, lethargy, seizures, hypotonia, and movement disorders)
Hepatic (with and without hypoglycemia) and gastrointestinal presentation (acute liver failure, cholestasis, non-immune hydrops fetalis, hepatomegaly, and hepatosplenomegaly)
Cardiac presentation (cardiomyopathy, heart failure, and rhythm disturbances)
Later ages
Acute neurological presentation (coma, recurrent vomiting, lethargy attacks associated with recurrent vomiting, *stroke*, seizures, Leigh's syndrome, movement disorders, ataxia, psychosis, andpsychiatric disorders)
Progressive neurological presentation (mental retardation, dementia, psychiatric diseases, peripheral neuropathy, Leigh syndrome, and movement disorders)
Dehydration
Exercise intolerance and recurrent myoglobinuria
Gastrointestinal symptoms (abdominal pain, cyclic vomiting, diarrhea, intestinal pseudo-obstruction, and selective eating behavior)
Cardiac presentation (cardiomyopathy, heart failure, and rhythm disturbances)
Hepatic presentation (Reye-type syndromes, liver failure, ascites, hepatomegaly, hepatosplenomegaly, cholestasis, steatosis, cirrhosis, and hepatocellular carcinoma)
Manifestations affecting the skeletal system (dysplasia, *facies* characteristic/dysmorphism, pain, recurrent osteomyelitis, "inflammatory" symptoms, and osteoporosis)
Skin and adnexal manifestations (ichthyosis, *cutis laxa*, hypo/hyperpigmentation, hyperkeratosis, angiokeratosis, xanthomas, skin nodules, lipodystrophy, vesiculobullous lesions, rash, acrocyanosis, alopecia, hypertrichosis, and photosensitivity)
Ocular manifestations (cataracts, corneal opacities, chronic conjunctivitis, lens subluxation, cherry red spot, retinitis pigmentosa, optic atrophy, nystagmus, coloboma, vertical ophthalmoplegia, oculogyric crisis, and strabismus)
Renal manifestations (tubulopathy, renal failure, haemolytic-uremic syndrome, proteinuria, renal cysts, nephrolithiasis, nephrocalcinosis, and abnormal urine color/odor)
Hematological manifestations (anemia, leukopenia, thrombocytopenia, pancytopenia, myelodysplasia, dyserythropoiesis, acanthocytosis, vacuolated lymphocytes/granulocytes, macrophage/HLH activation, haemophagocytosis, haemolysis, coagulopathy, and selective immunodeficiency of coagulation factors)

Table 31.3 summarizes the most recent clinical classification of IMD.

Therapy

The first therapeutic approaches to IMD date back to the 1950s and 1960s of the last century; they were mainly based on dietary interventions. However, with the continuous development of medical knowledge and scientific research, therapeutic scenarios are now much broader, and it is possible to treat about 60% of patients affected by IMD (Table 31.4).

Table 31.4 Schematic representation of the therapeutic intervention in metabolic diseases

Reduction of precursors source of toxic substrates of exogenous origin: diet therapy
Reduction of toxic endogenous substrates: drugs, enzyme replacement therapy, molecules *chaperone*
Vitamins
Cofactors
Enzyme activators
Organ (liver, kidney) or cell (hematopoietic stem, cell therapy) transplants
Gene therapy

Early diagnosis is associated with a better prognosis in many cases, as it allows early specific treatments, avoiding the onset of severe disability and preventing death in the most severe cases.

Laboratory Strategies for Screening and Diagnostic Confirmation of IMD

Diagnostic Confirmation of Clinical Suspicion of IMD

In the case of clinical suspicion of IMD, the diagnosis is commonly based on first-level biochemical tests (available in all clinical biochemistry laboratories), followed by second-level tests (reserved for specialized laboratories) (Table 31.5), which must be followed, in a specialized clinical environment (clinical center of reference for IMD), by the definitive biochemical classification in genetic biochemistry laboratories. Diagnostic investigations are based on the qualitative or quantitative determination of specific biomarkers of the various diseases in biological fluids, usually plasma, urine, and cerebrospinal fluid (Table 31.6), followed by the demonstration of cellular enzymatic deficit and subsequent confirmation by genetic analysis.

From the point of view of laboratory medicine, the methodological and technical complexity of the differential diagnostic pathway for IMD has determined the progressive development, in many advanced health systems, of highly specialized laboratory facilities aimed at and integrated to support the clinical centers of reference deputies in the diagnosis and treatment of IMD. Therefore, within the laboratory medicine services, a new specialized branch has been created, which can be defined as genetic biochemistry: a technical and professional environment addressed to "rare diseases" that requires high and specific levels of efficiency, appropriateness, quality assurance, economies of scale, innovation and continuous research, expertise, and qualification of personnel. Below, in accordance with the recommendations

Table 31.5 First level and second level indicative laboratory investigations for the metabolic diseases diagnosis

Blood	
First level	Total cholesterol
Complete blood count, reticulocytes	HDL cholesterol
Peripheral smear	Creatinine, eGFR, urea,
Sideremia, ferritin	**Second level**
Glycemia	Insulin
Ammonemia	Ketonemia
Lactate	NEFA
Blood gas analysis	Cupremia, ceruloplasmin
Uricemia	Protein C, S, antithrombin III, factors VII, IX, XI
Coagulation, INR	Apoproteins
Transaminases	β2-microglobulin
CK	PTH
LDH	ACTH, cortisol
γ-glutamyltransferase	Vitamin B12
Total and direct bilirubin	Total homocysteine
ALP	
Triglycerides	
Urine	
First level	Reducing substances
Complete urinalysis	Sulfitest®
Ketonuria	Cupruria
Second level	Tubular reabsorption of P, uricuria
Myoglobinuria	
Liquor	
First level	Glucose
Microscopic examination	**Second level**
Cell count	Lactate
Proteins	

ACTH corticotropin, *ALP* alkaline phosphatase, *CK* creatine kinase, *LDH* lactate dehydrogenase, *NEFA* non-esterified fatty acids, *PTH* parathyroid hormone

Table 31.6 Investigations of biochemical diagnostic assessment for the diagnosis of hereditary metabolic diseases, which can be carried out in the different biological fluids

Exam	Method	Diseases group
Organic acids (urine)	GC/MS	Organic acidurias
Acylcarnitine (DBS, plasma)	MS/MS	Defects in the oxidation of fatty acids
Methylmalonic acid (plasma)	GC/MS, MS/MS	
Biotinidase (DBS)	Fluorimetric	
Pipecolic acid (plasma, liquor)	MS/MS	
Aminoacids (plasma, urine, CSF)	IEC, MS/MS	Aminoacidopathies
Total homocysteine (plasma)	Enzyme immunoassay, HPLC, MS/MS	Defects of the urea cycle
Orotic acid/orotidine (urine)	HPLC, MS/MS	
Succinylacetone (plasma, urine)	GC/MS, MS/MS	
Biopterins (serum, DBS, urine, liquor)	HPLC, MS/MS	
Mucopolysaccharides, total GAGs (urine)	Spectrophotometric, HPLC, MS/MS	Lysosomal diseases
Oligosaccharides (urine)	TLC, MS/MS	
Oxysterols (C-triol, 7-ketocholesterol) (plasma)	GC/MS, MS/MS	
Lysosphingolipids (plasma)	MS/MS	
Cystine (leukocytes)	HPLC, MS/MS	
Free and total sialic acid (urine)	Spectrophotometric, MS/MS	
Very long chain fatty acids (VLCFA) (plasma)	GC/MS, MS/MS	Peroxisomal diseases
Phytanic acid, pristanic acid (plasma)	GC/MS, MS/MS	
Plasmalogens (erythrocytes)	GC/MS, MS/MS	
Sugar chromatography (urine)	TLC	Disorders of carbohydrate metabolism
Galactose (DBS, urine), galactose-1-phosphate (erythrocytes)	Spectrophotometric	
Galactose −1-PUT (DBS, erythrocytes)	Enzymatic	
Polyols (urine)	MS/MS, MRI	

Table 31.6 (continued)

Exam	Method	Diseases group
Porphyrins (urine)	HPLC	Porphyrias
5-aminolevulinic acid (urine)	Spectrophotometric	
Neurotransmitters (liquor)	HPLC, HVE, MS/MS,	Defects in neurotransmitter synthesis
7-dehydrocholesterol (plasma)	GC/MS, MS/MS	Defects of cholesterol synthesis
Bile acids (plasma, urine)	GC/MS, MS/MS	Defects of metabolism of bile acids
Creatine and guanidinoacetic acid (plasma, urine)	MS/MS	Defects of creatine metabolism
Transferrin isoforms (serum, plasma)	Electrophoresis, MS/MS	Defects of proteins glycosylation
Purines and pyrimidines (urine, plasma)	HPLC, MS/MS	Defects of purine pyrimidines metabolism

DBS drop of dry blood, *GC/MS* gas chromatography-mass spectrometry, *HPLC* high performance liquid chromatography, *HVE* High Voltage Electrophoresis, *IEC* ion exchange chromatography, *MS/MS* tandem mass spectroscopy, *MRI* magnetic resonance, *TLC* thin layer chromatography

contained in the document "Good Laboratory Practices for Biochemical Genetic Testing and Newborn Screening for Inherited Metabolic Disorders", developed and published in 2012 by the Centers for Disease Control and Prevention, US Department of Health and Human Services, some key features are listed.

- Genetic biochemistry facilities represent a fundamental branch of laboratory medicine, aimed at the evaluation, diagnosis, therapeutic monitoring, clinical management, and, in some cases, the definition of the carrier status for congenital errors of metabolism. The tests performed by a genetic biochemistry laboratory require complex and highly specialized laboratory procedures aimed at the evaluation of enzymatic activities and biomarkers such as amino acids, organic acids, acylcarnitines, fatty acids, glycosaminoglycans, etc., by the use of a wide range of biological samples.
- The laboratory tests falling within the competence of this branch of laboratory medicine are classified as highly complex. They require more stringent criteria from a regulatory point of view (accreditation) regarding the characteristics of the analytical process, the quality management system, and the qualification and expertise of the personnel involved.

Neonatal Screening Process

The diagnosis of IMD is a model of integrated patient management resulting from an articulated and progressive pathway. Today, for many IMD, it is also possible to apply the strategy of mass neonatal screening to identify the subject at risk.

The term "neonatal screening" defines secondary preventive medicine programs, activated on a large scale in the first days of life, aiming at the early identification and timely treatment of infants at high risk for certain treatable diseases with a high risk of early mortality and/or severe morbidity in those not diagnosed early. Nowadays, implementing a neonatal screening program is an essential responsibility of the public health system and constitutes a crucial factor in protecting children's health status. Neonatal screening policies should be guided by an assessment of the overriding interests of the affected individual, with a secondary consideration for the interests of other stakeholders (healthy infants, families, professional areas, and health policy authorities). The screening program and recommendations on the appropriateness of including disease in a neonatal screening program should be based on scientific evidence and broad professional consensus.

In 1963, Robert Guthrie created the first laboratory test (which still today, although considered technically obsolete, bears his name: the Guthrie test) that allowed the semiquantitative measurement of the amino acid phenylalanine (Phe) in a drop of capillary blood, collected by heel prick, and allowed to absorb and dry on a special filter paper (the Guthrie card). This test, technically called the bacterial growth inhibition test, was based on the growth, in a particular agar medium deprived of phenylalanine, of bacterial spores in the deposition zone of a small disc of a few millimeters in diameter, obtained from the blood sample absorbed in the neonatal Guthrie card. The growth halo was proportional to the concentration of Phe in the sample and, through comparison with a series of samples with a known and progressive concentration of Phe (from 2 to 20 mg/100 mL), allowed to attribute, with sufficient accuracy for the use of the test, the value of Phe in the neonatal sample and select, according to a predetermined threshold or cutoff value (usually set at 2–4 mg/100 mL), the infants at risk (moderate or high) of phenylketonuria (PKU) to start the path of diagnostic confirmation and treatment. Because of its sensitivity, low cost, and easy application on a large scale, it allowed to carry out the first mass neonatal screening campaigns to identify newborns affected by PKU, the most common and frequent IMD that, if not treated early after birth with adequate diet therapy, is the cause of severe mental retardation, chronic and highly disabling.

In the following decades, many other diseases, mainly genetic, such as endocrinopathies, congenital errors of metabolism, hemoglobinopathies, and cystic fibrosis, resulted suitable to neonatal screening.

Traditionally, neonatal screening programs use analytes – mainly blood – as indicators or biomarkers of pathology, whose quantitative measurement or qualitative assessment allows, with sufficient efficiency, the selection of subjects at higher risk in the neonatal population. In some programs, laboratory analysis measures substrates accumulated in biological fluids by different mechanisms: (1) altered utilization or transformation, from enzymatic deficiency, of a substrate in a biochemical process (PKU: phenylalanine; galactosemia: galactose; congenital adrenal hyperplasia: 17-α-hydroxyprogesterone); (2) mechanical obstruction (cystic fibrosis: immunoreactive trypsin); (3) physiological activation of a feedback (hypothyroidism: TSH). In others, the deficiency or reduction of a substrate indicates a risk situation (hypothyroidism: T4) or, again, the presence of abnormal metabolites absent in normal conditions (hemoglobinopathies).

Finally, screening can be performed by measuring or qualitatively assessing a specific enzyme activity (galactosemia: galactose-1-P-uridyltransferase enzyme activity). In the first days of life, there is a "chronology" of the concentrations of the single biomarkers, which are strongly affected by the biological changes occurring in the delicate perinatal period of biochemical adaptation to autonomous life. The time of sample collection for neonatal screening must therefore be appropriately chosen in time windows that ensure, in the presence of pathology, optimal levels for the measurement or evaluation of the analyte in order to achieve the maximum efficiency (sensitivity and specificity) of the system.

Since the 1960s, technological evolution has offered increasing possibilities to expand the laboratory techniques applied to neonatal screening, providing the ability to use an increasing number of biomarkers for disease screening on a large scale. Table 31.7 highlights the main technologies applicable to neonatal screening. Since the 1990s, multiparametric screening allowing the simultaneous measurement of several analytes has been introduced. It is based on tandem mass spectrometry (MS/MS).

Many features make tandem mass spectrometry particularly suitable for the implementation of neonatal screening programs:

- Very high sensitivity: Extremely low blood volume is required
- High analytical speed: About 2–4 min/sample
- Possible automation of the analytical process
- High productivity
- Reduced cost per sample analyzed

Table 31.7 Main laboratory techniques applicable in DBS samples in neonatal screening programs

Microbiological
Radioimmunological (RIA)
Enzymatic
Colorimetric
Fluorimetric
Non-RIA immunometric: ELISA, TR-FIA
Electrophoretic: IEF
Chromatography: TLC, IEC, HPLC, GC/MS, MS/MS
Nucleic acid extraction and molecular analysis: PCR, *reverse dot blot*, NGS

DBS drop of dry blood, *GC/MS* gas chromatography-mass spectrometry, *HPLC* high performance liquid chromatography, *IEC* ion exchange chromatography, *IEF* isoelectric focusing, *MS/MS* tandem mass spectrometry, *NGS* Next Generation Sequencing, *PCR* Polymense Chain Reaction, *TLC* thin layer chromatography, *TR-FIA* Time-resolved fluorimmunoassay

Today, the most widespread, although not unique (especially in Europe), organizational model for IMD screening programs is the one developed in 2002 in the United States by the government offices of the Maternal and Child Health Bureau (MCHB), the Health Resources and Services Administration (HRSA), and the United States Department of Health and Human Services (DHHS), in collaboration with the American College of Medical Genetics (ACMG) and the American Academy of Pediatrics (AAP), which has redefined the set of diseases eligible for neonatal screening. They are divided into two panels, defined as:

- Core panel: severe diseases and
- Secondary target panel: conditions with minor clinical impact, which, in the screening process, are part of a differential diagnosis pathway for a condition included in the main panel

The main panel originally included 20 IMDs: nine organic acidemias (OA), five fatty acid oxidation defects (FAO), and six amino acidopathies (AA), all identifiable through the new technologies based on tandem mass spectrometry, which are associated, always in the field of hereditary metabolic diseases, with biotinidase deficiency and classical galactosemia, identifiable by different analytical technologies. The secondary panel includes six other OAs and eight FAOs (detectable by MS/MS technology) and two other nonclassical forms of galactosemia.

Table 31.8 shows the original description of the main and subpanels.

This panel is constantly updated according to new scientific evidence, which makes it possible to include new diseases according to the predefined criteria. Tandem mass spectrometry has introduced revolutionary progress in the screening and diagnosis of congenital errors of metabolism, allowing many screening laboratories in the world to extend the panels in use, by including ex novo, as new pathologies,

Table 31.8 ACMG Disease panels for extended neonatal screening

Tandem mass spectrometry				
Acylcarnitine		Aminoacids	Other technologies	
Organic acidemias (AO)	Defects of fatty acid oxidation (FAO)	Aminoacidopathy (AA)	Hemoglobinopathies (Hbpaties)	Other pathologies
Core panel				
IVA GAI HMG MCD MUT 3MCC Cbl A,B PROP BKT	MCAD VLCAD LCHAD TFP CUD	PKU MSUD HCY CIT ASA TYR I	Hb SS Hb S/βtH Hb S/C	CH BIOT CAH GALT HEAR CF
Secondary targets panel				
Cbl C,D MAL IBG 2M3HA 2MBG 3MGA	EXP GA2 M/SCHAD MCKAT CPT II CACT CPT IA DE RED	HPA not PKU TYR II BIOPT (BS) ARG TYR III BIOPT (BS) MET CIT II	Var Hb	GALK GALE

2M.3HBA 2-methyl-3-hydroxybutyric acidemia, *2MBG* 2-methylbutyryl-CoA dehydrogenase deficiency, *3MCC* 3-methylcrotonyl-CoA carboxylase deficiency, *3MGA* 3-methylglutaconic acidemia, *ARG* argininemia, *ASA* arginine-succinic aciduria, *BIOPT (BS)* biopterin biosynthesis cofactor deficiency, *BIOPT (REG)* deficiency of biopterin regeneration cofactors, *BIOT* biotinidase deficiency, *BKT* β-ketothiolase deficiency, *CACT* carnitine/acylcarnitine translocase deficiency, *CAH* congenital adrenal hyperplasia, *Cbl A, B* methylmalonic acidemia (Cbl A, B), *Cbl C, D* methylmalonic acidemia (Cbl C, D), *CF* cystic fibrosis, *CH* congenital hypothyroidism, *CIT* citrullinemia, *CIT II* citrullinemia II, *CPT IA* carnitine palmitoyltransferase IA deficiency, *CPT II* carnitine palmitoyltransferase II deficiency, *CUD* carnitine uptake deficiency, *DE RED* dienoyl-CoA reductase deficiency, *GA1* glutaric acidemia type 1, *GA2* glutaric acidemia type 2, *GALE* galactosemia due to galactoepimerase deficiency, *GALK* galactosemia due to galactokinase deficiency, *GALT* classical galactosemia, *HbS/BTh* HbS/β-thalassemia, *Hb SS* sickle cell anemia, *HCY* homocystinuria, *HEAR* deafness, *HMG* 3-hydroxy-3-methyl glutaric acidemia, *HYPER PHE* hyper-phenylalaninemia, *IBG* isobutyril-CoA dehydrogenase deficiency, *IVA* isovaleric acidemia, *LCCHAD* acyl-CoA dehydrogenase deficiency, long chain, *M/SCHAD* 3-hydroxy acyl-CoA deficiency, medium/short chain, *MAL* malonic acidemia, *MCAD* acyl-CoA dehydrogenase deficiency, medium chain, *MCD* multiple carboxylase deficiency, *MCKAT* ketoacyl-CoA thiolase deficiency, medium chain, *MET* hypermethioninemia, *MSUD* maple syrup urine disease (leucinosis), *MUT* methylmalonic acidemia (mutase), *PKU* phenylketonuria, *PROP* propionic acidemia, *SCAD* short-chain acyl-CoA dehydrogenase deficiency, *TFP* trifunctional protein deficiency, *TYR I* tyrosinemia I, *TYR II* tyrosinemia II, *TYR III* tyrosinemia III, *Var HB* other Hb-variants (Hb E), *VLCAD* acyl-CoA dehydrogenase deficiency, very long chain

fatty acid oxidation defects, and organic acidemias, and by significantly increasing the number of aminoacidopathies selectable with the program. The era of the so-called extended or expanded neonatal screening programs has begun.

Simultaneously, the development of the new technological model, starting from the 2000s, led to a new vision of health policy, which expands the organizational model of neonatal screening programs beyond the laboratory "border," transforming the single activities related to the selection of a newborn at risk for one of the pathologies included in the screening panel into an articulated systemic health model ("screening system"), sequential and multidisciplinary, which includes all the phases of selection, diagnostic confirmation (biochemical and genetic), taking charge and management at the clinical level (clinical diagnosis, therapy, genetic counseling), final evaluation (epidemiological, economic, efficacy) and all the functions and competences (neonatological, laboratory, clinical specialists) that interact with each other within the screening program and that respond, from a health point of view to two pivotal elements:

- To build an integrated and multidisciplinary service network
- To provide a timely and adequate response to a healthy demand that cannot be postponed

A neonatal screening program achieves its complete objective only when every newborn with a positive test has access to an efficient diagnostic evaluation and every newborn with a confirmed diagnosis has access to an appropriate care pathway, chronic, global, and centered on the social binomial "patient-family".

In more technical terms, neonatal screening can be equated, according to ISO 9000:2005, to a process, which can be defined as a sequence of related or interacting activities that transform an initial situation into a final one by adding value. In the field of public health, neonatal screening is a fundamental step in the evaluation of the health status of every newborn, allowing the newborn with a negative screening to exclude real-risk conditions for diseases included in the panel of pathologies and for the affected newborn to be promptly included in the most appropriate therapeutic care pathway. This is the added value of neonatal screening. The concept of the process allows a better understanding of neonatal screening as an action of definition in each newborn of a risk condition (usually classified into three levels: low, intermediate, and high) obtained through the quantitative measurement of biological markers (single or multiple) appropriate to the condition, which must follow the diagnostic confirmation phase. The classic operational flow model of the screening process is represented in Fig. 31.1, which illustrates the operational algorithm of an expanded neonatal screening program for IMD: the baseline test (amino acid and acylcarnitine analysis) is performed in all neonatal samples; if negative, it concludes the selection process; if positive (with respect to the reference interval, a statistically predefined threshold value in the reference population, or a risk score calculated by dedicated computer algorithms), it determines actions of deepening (control), articulated according to the risk score (low, intermediate, or high risk), and conducted chronologically with a timing defined by the characteristics of the disease and the level of risk. Once this second phase has been completed, all confirmed positives are generally reported to the clinical center of reference for diagnostic confirmation and for taking charge of the newborn at risk. This phase is carried out in concert among the screening laboratory, the laboratory for diagnostic confirmation (facilities that may coincide), and the clinical center. In any case, screening, diagnostic confirmation, and taking charge of the newborn at risk must be concluded, even for pathologies with less severe clinical onset, within the first month of life.

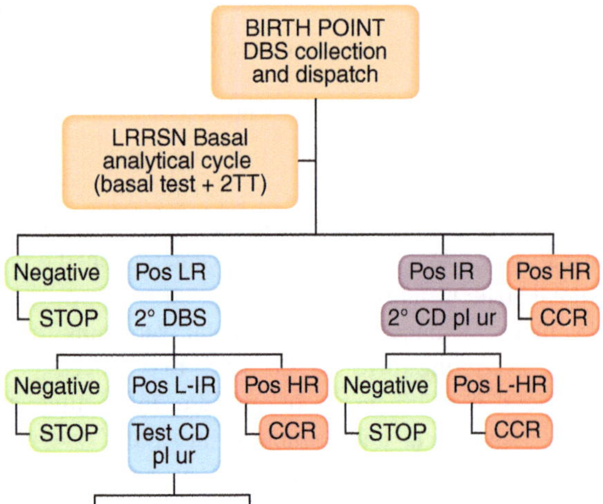

Fig. 31.1 Extended neonatal screening: operative-laboratory algorithm. (Copyright EDISES 2021. Reproduced with permission)

This classic design of operational flow is now enhanced by the possibility of performing in the basal sample collected in the first days of life, additional tests, called second-tier tests (2TT). The latter allows for improving the efficiency of the program, increasing especially the specificity and the positive predictivity; it results in a positive impact on both the health system (reduction in the number of required confirmatory tests, reduction of preanalytical and analytical costs) and the social system (reduction of parental anxiety for the outcome of the test).

Figure 31.2 highlights some examples of second-stage tests (2TT) and subsequent biochemical confirmatory tests in neonatal screening for congenital adrenal hyperplasia, cystic fibrosis, and expanded neonatal screening for IMD.

Concerning the current diffusion of extended neonatal screening programs, data from the international literature show a wide diffusion of these new programs in many geographical areas of the world with advanced healthcare systems (North America, Europe, Japan, Australasia, and, more recently, Russia and China).

In terms of health economics, the analysis of the overall results of extended neonatal screening programs shows a substantially favorable judgment. According to some authors, in terms of cost-benefit analysis, programs based on MS/MS technology for congenital errors of metabolism can determine an overall economic saving if compared to the high costs of care for subjects with long survival diagnosed in the

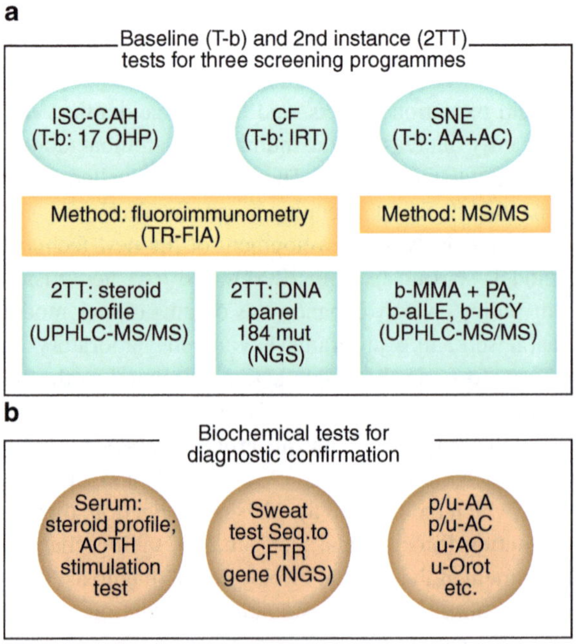

Fig. 31.2 (**a**) Baseline (T-b) and second instance (2TT) tests for three screening programs. (**b**) Biochemical tests for diagnostic confirmation. ISC-CAH, congenital adrenal hyperplasia; CF, cystic fibrosis; SNE, extended neonatal screening (Copyright EDISES 2021. Reproduced with permission)

absence of screening. Additionally, the use of MS/MS technology (compared to that of other technologies) can determine a higher level of economic savings because of its intrinsic characteristic to detect a panel of diseases in a single test. However, previous evaluations by other authors had instead shown lower economic benefits reserved for programs limited in the number of diseases screened.

Despite the now-prevailing evaluation of the effectiveness of extended neonatal screening programs, the persistence of critical issues or open problems related to the activation of these programs should not be underestimated. The main points of reflection can be synthetically listed:

- The selection of pathologies in which the therapeutic intervention, even if early and started in a presymptomatic period, does not modify the natural history – inauspicious for morbidity and mortality – of the disease.
- The identification of biological variants with low clinical "impact" or late-onset pathologies in which there is no evidence of the need or usefulness of therapeutic intervention.
- The inevitable, albeit limited, increase in false-positive newborns (especially in particular categories such as premature babies), with potentially negative repercussions (psychological, social, and economic) in the family.
- The risk that, in situations of false negativity (which, although rare, is still present in highly efficient programs), the false sense of security caused by the existence of a targeted screening program will further slow down the clinical diagnosis of disease.

As already mentioned, the paths of diagnosis and treatment of rare diseases (the pathologies with a prevalence of affected subjects present in a population of less than five cases in 10,000) have today, in the most advanced health systems, significant social and political attention. The technical and applicative potentialities offered by tandem mass spectrometry and the substantial international success of the extended neonatal screening programs (mainly addressed to "rare pathologies") determine a constant pressure, also in public opinion, for a further expansion of the number of pathologies considered in the screening panels. However, it must be considered that neonatal screening strategies are a tool for the presymptomatic selection of subjects at risk. However, they do not exhaust the entire pathway of diagnosis of hereditary metabolic diseases, as the panels, although large and further expandable, do not cover the entire set of diseases now identified and known.

In conclusion, it is possible to affirm that for IMDs, early postnatal diagnosis is the most effective tool for the prevention or reduction of both the risk of mortality and damage from severe morbidity, considering that for many rare congenital diseases, recent years have greatly improved the approach and the possibility of therapeutic intervention. The introduction of MS/MS technology and the application of molecular biology techniques have revolutionized the neonatal screening, allowing new scenarios of preventive medicine for IMDs.

In such a vast context of applicative potentialities, the temptation may arise in civil society and in the healthcare world (also intended as a response to legitimate social instances of lay groups supporting different pathologies) to adopt a global approach to neonatal screening for an ever-increasing number of congenital pathologies, even outside the criteria of choice based on strong scientific, epidemiological, and health economics evidence. Moreover, it must be considered the possibility that the interests related to the introduction of innovative therapies with high or very high economic value may determine an undue pressure towards operational choices that are effective from the point of view of selection and diagnosis but not supported by subsequent therapeutic interventions that are advantageous for the individual and economically and ethically sustainable for society.

Recommended Readings

American Academy of Pediatrics Newborn Screening Task Force (2000) Serving the family from birth to medical home. Newborn screening a blue print for the future. A call for a national agenda on state newborn screening programs. Pediatrics 106(Suppl 2):389–422

American College of Medical Genetics Newborn Screening Expert Group (2006) Newborn Screening toward a uniform screening panel and system. Executive summary. Pediatrics 117:S296–S307

Carroll AE, Downs SM (2006) Comprehensive cost-utility analysis of newborn screening strategies. Pediatrics 117:S287–S295

CDC (2012) Good laboratory practices for biochemical genetic testing and newborn screening for inherited metabolic disorders. MMWR Recomm Rep 61(2)

Chace DH, Kalas TA, Naylor EW (2003) Use of tandem mass spectrometry for multianalyte screening of dried blood specimens from newborns. Clin Chem 49:1797–1817

Feuchtbaum L, Cunningham G (2006) Economic evaluation of tandem mass spectrometry screening in California. Pediatrics 117:S280–S286

Gilbert-Barness E, Farrell PM (2016) Approach to diagnosis of metabolic diseases. Transl Sci Rare Dis 1:3–22

Guthrie R, Susi A (1963) A simple phenylalanine method for detecting phenylketonuria in large population of newborn infants. Pediatrics 32:338–343

Millington DS, Kodo N, Norwood DL et al (1990) Tandem mass spectrometry: a new method for acylcarnitine profiling with potential for neonatal screening for inborn errors of metabolism. J Inherit Metab Dis 13:321–324

Pandor A, Eastham J, Beverley C et al (2004) Clinical effectiveness and cost-effectiveness of neonatal screening for inborn errors of metabolism using tandem mass spectrometry: a systematic review. Health Technol Assess 8(12):1–121

Price CP, Christenson R (2008) Evaluating new diagnostic technologies: perspectives in the UK and US. Clin Chem 54:1421–1423

Rinaldo P, Tortorelli S, Matern D (2004) Recent developments and new applications of tandem mass spectrometry in newborn screening. Curr Opin Pediat 16:427–433

Saudubray JM, Garcia-Cazorla A (2016) Clinical approach to inborn errors of metabolism in pediatrics. In: Saudubray JM, Baumgartner M, Walter J (eds) Inborn metabolic diseases. Springer, Berlin/Heidelberg

Hyperphenylalaninemias, Tyrosinemias, Glycogenosis, Hyperammonemia

32

Marcello Ciaccio and Luisa Agnello

Hyperphenylalaninemias

It refers to all biochemical-clinical conditions characterized by elevated circulating levels of phenylalanine (>120 μmol/L or >2 mg/dL).

Hyperphenylalaninemias represent a heterogeneous group of autosomal recessive diseases characterized by an altered phenylalanine metabolism due to enzymatic deficiency of phenylalanine hydroxylase (PAH). This hepatic enzyme catalyzes the hydroxylation of phenylalanine (Phe) to tyrosine (Tyr), the precursor of several molecules, including dopamine, norepinephrine, and melanin. To perform its function correctly, PAH requires the presence of a cofactor, tetrahydrobiopterin (BH4). 2% of hyperphenylalaninemias are related to a defect in BH4 metabolism due to an alteration in its synthesis or regeneration.

Phenylalanine is an essential amino acid introduced into the body through the diet, which can be metabolized in five different ways:

1. Protein synthesis
2. Hydroxylation at the level of the benzene ring at the para position, leading to tyrosine synthesis
3. Hydroxylation of the benzene ring at the ortho position, leading to orthotyrosine synthesis
4. Transamination with α-ketoglutaric acid to form phenylpyruvic acid
5. Decarboxylation, leading to the phenylethylamine synthesis, a biologically active sympathomimetic amine

Under physiological conditions, about 50% of the amino acid introduced through the diet is transformed into Tyr by PAH, and the remaining 50% is used for protein synthesis. A small percentage of Phe undergoes transamination, leading to the phenylpyruvic acid synthesis, and decarboxylation, leading to phenylethylamine synthesis.

Phenylalanine is a toxic amino acid because it can cross the blood–brain barrier via a neutral amino acid transporter (LAT1), and its accumulation in the brain may impair psychomotor development. In addition, increased phenylalanine levels inhibit the intracellular transport of other amino acids (particularly tyrosine and tryptophan), leading to a maturational and myelination deficit in the central nervous system (CNS). Finally, phenylalanine at high levels causes an increase in the turnover of the myelin protein fraction that is not compensated by an increase in its synthesis. The accumulation of catabolic products has a direct toxic action on the intermediate metabolism of the CNS, as well as inhibits enzymatic activities related to the metabolism of tryptophan and tyrosine, causing a decreased synthesis of serotonin, γ-aminobutyric acid (GABA), dopamine, and sympathomimetic amines that are involved in the regulation of brain functions. Neuroimaging studies on patienta affected by hyperphenylalaninemia have shown white matter lesions associated with altered myelination. Untreated hyperphenylalaninemia manifests progressive retardation of psychomotor development associated with numerous other manifestations, such as rashes and convulsions.

The PAH gene was cloned in 1983, and since then, mutations in it have been studied. In 1996, McGill University created the "Phenylalanine Hydroxylase Locus Knowledgebase" (PAHdb) with the intent to collect more information on mutations in the PAH gene (http://www.pahdb.mcgill.ca/). To date, this database collects information on more than 500

M. Ciaccio (✉)
Department of Biomedicine, Neurosciences and Advanced Diagnostics, Institute of Clinical Biochemistry, Clinical Molecular Medicine and Clinical Laboratory Medicine, and Department of Laboratory Medicine, University Hospital "P. Giaccone", Palermo, Italy

Department of Laboratory Medicine, University Hospital "P. Giaccone", Palermo, Italy
e-mail: marcello.ciaccio@unipa.it

L. Agnello
Department of Biomedicine, Neurosciences and Advanced Diagnostics, Institute of Clinical Biochemistry, Clinical Molecular Medicine and Clinical Laboratory Medicine, and Department of Laboratory Medicine, University Hospital "P. Giaccone", Palermo, Italy

M. Ciaccio (ed.), *Clinical and Laboratory Medicine Textbook*, https://doi.org/10.1007/978-3-031-24958-7_32

mutations in the PAH gene, for 200 of which the residual enzymatic activity is also known. Of these 564 mutations, 70% are missense, 12% deletions, 11% splicing site mutations, 5% nonsense mutations, and, 2% insertions.

Most patients are compound heterozygotes, i.e., carrying several mutations in each allele, so the clinical phenotype depends on the effect and interaction of the mutations.

Hyperphenylalaninemias have an average incidence of about 1:10,000 infants and can occur in various forms that, based on phenylalaninemia levels, can be classified into:

- Moderate hyperphenylalaninemia: 120 and 600 μmol/L (2–10 mg/dL)
- Moderate phenylketonuria: 600 and 1200 μmol/L (10–20 mg/dL)
- Classical phenylketonuria: >1200 μmol/L (>20 mg/dL)

The classic phenylketonuria is the biochemical-clinical form that, for incidence and severity of the clinical picture, has greater importance.

Hyperphenylalaninemia is the only metabolic disease for which early diagnosis, neonatal screening, and effective therapy have been available throughout Europe for more than 30 years. The early detection and treatment of the disease allow for reaching adulthood, which is infrequent in the metabolic diseases field.

From a biochemical-clinical point of view, the main alterations of phenylketonuria are:

- Significant increase in circulating levels of Phe;
- Increased urinary excretion of phenylpyruvic acid, phenylacetic acid, and acid orthohydroxyphenylpyruvic;
- Reduction in circulating levels of Tyr,
- Decreased melanin biosynthesis due to competitive inhibition on tyrosinase;
- Decrease in circulating levels of 5-OH-tryptamine, dopamine, and GABA by phenylpyruvic acida inhibition of specific decarboxylases.

Diagnosis and Therapy

Three pivotal dates marked a turning point in the early diagnosis, treatment, and prognosis of hyperphenylalaninemia. In 1934, Dr. Asbjørn Følling identified hyperphenylalaninemia as a possible cause of neuropsychic deficits; in 1953, Dr. Horst Bickel demonstrated the effectiveness of a low phenylalanine diet in the treatment of phenylketonuria; and, in 1960, Dr. Robert Guthrie developed the first test that could be used as mass screening for early detection of hyperphenylalaninemia (Guthrie test).

Infants with hyperphenylalaninemia identified by neonatal screening should undergo in-depth diagnostics to confirm it.

The first diagnostic investigation involves plasma aminoacidogram to confirm an increase in phenylalanine levels, normal or low tyrosine values with a consequent Phe/Tyr ratio >2, and the absolute normality of the levels of other amino acids.

Once hyperphenylalaninemia is confirmed, it is necessary to verify its etiology, which may or not be due to a defect in BH4 metabolism. Thus, patients with hyperphenylalaninemia perform a BH4 load test in the neonatal period and before the start of diet therapy. This standardized test involves administering BH4 at 20 mg/kg. Phenylalanine is measured before and 8, 16, and 24 h after cofactor administration. Normalization of phenylalanine levels within 8 h of cofactor administration depicts a BH4 defect, whereas lack of or modest reduction in phenylalaninemia identifies an alteration in the PAH enzyme. Final diagnostic confirmation is based on molecular genetic investigations.

The definition of the clinical phenotype is essential to define the most appropriate therapeutic approach, establish the prognosis, and ensure proper counseling, especially for future pregnancies.

The treatment is based on the reduction, as early as possible and through a dietary approach, of the high circulating levels of Phe, without ever falling below the minimum requirement to ensure normal protein synthesis in the body. The intake of Phe must fluctuate between a minimum, below which normal growth is not possible, and a maximum that represents the limit of individual tolerance understood as the maximum amount of Phe compatible with maintaining a phenylalaninemia within the therapeutic target. The personalized diet plan will include:

- The assumption of amino acid mixtures with low or zero Phe content, enriched with vitamins, trace elements, and essential fatty acids;
- The restriction in the intake of natural proteins, in particular meat, eggs, and fish, preferring a vegetarian diet;
- The use of hypo/a-protein products.

The target Phe concentrations to be achieved by treatment are:

- <360 μmol/L (<6 mg/dL) for patients up to 12 years
- <600 μmol/L (<10 mg/dL) for patients older than 12 years.

Maternal Phenylketonuria

Maternal hyperphenylalaninemia exposes the pregnant to the risk of miscarriage and the fetus to an essential risk of embryofetopathy, the severity of which is directly related to maternal phenylalanine values. Phenylalanine can cross the placenta. In particular, the fetus could develop microcephaly, mental retardation, facial dysmorphia, intrauterine growth retardation, and structural heart defects.

Phenylketonuric embryofetopathy can be prevented by a strict phenylalanine-deficient diet before conception and throughout pregnancy. The woman with phenylketonuria must undergo strict metabolic monitoring during pregnancy, with phenylalaninemia measured twice weekly. Phenylalaninemia values should be maintained <360 μmol/L (<6 mg/dL).

Tyrosinemias

Tyrosinemia refers to pathological conditions characterized by increased levels of circulating tyrosine. Tyrosine, a nonessential amino acid, in mammals, is synthesized from phenylalanine by an enzymatic process of hydroxylation. Through complex metabolic pathways, thyroid hormones, catecholamines, and melanin are synthesized from tyrosine.

Tyrosinemias are distinguished into:

- Tyrosinemia type I
- Tyrosinemia type II
- Neonatal transient tyrosinemia
- Alkaptonuria

Tyrosinemia type I (or hepato-renal tyrosinemia) is due to the deficiency of the enzyme fumarylacetoacetate hydrolase and has an autosomal recessive inheritance. The deficiency of the enzyme fumarylacetoacetate hydrolase determines the accumulation of fumarylacetoacetate and maleilacetoacetate in the liver and kidney and succinylacetone systemically. These substances are responsible for hepato-renal damage and neoplastic degeneration of the liver. In addition, succinylacetone inhibits porphobilinogen synthase.

From a clinical point of view, it is characterized by acute manifestations in newborns and infants, such as severe liver failure, vomiting, bleeding, sepsis, hypoglycemia, renal tubulopathy, and chronic manifestations, such as hepatomegaly, growth retardation, rickets, hematomas, tubulopathy and neurological crisis.

The diagnosis is based on the following findings:

- Organic acids in urine: increase in succinylacetone (diagnostic), 4-hydroxyphenylpyruvate, 4-Hydroxyphenylacetate and 4-Hydroxyphenylacetate levels;
- Plasma amino acids: increase in tyrosine and methionine levels.

Tyrosinemia type II (or oculo-cutaneous tyrosinemia) is due to a deficiency of the enzyme tyrosine transaminase (TTA) and has an autosomal recessive inheritance. From a biochemical point of view, the increase in tyrosine levels, secondary to TTA deficiency, causes the accumulation of tyrosine crystals, which stimulate an inflammatory reaction and oculo-cutaneous signs; other symptoms include painful corneal lesions, hyperkeratosis (soles of the feet, palms of the hand), and mental retardation. Diagnosis is based on the clinical picture, increased plasma and urinary tyrosine levels, and increased levels of urinary tyrosine metabolites (such as 4-hydroxyphenylpyruvate, 4-hydroxyphenyllactate, 4-hydroxyphenylacetate, and N-acetyltyrosine), TTA activity assay, and liver biopsy.

Neonatal transient tyrosinemia is due to transient deficiency of the enzyme para-hydroxyphenylpyruvic oxidase. It has an incidence of 0.2–10% in newborns, with a higher frequency in males and preterm infants. From a clinical point of view, it is often asymptomatic and may be accompanied by prolonged jaundice and/or reduced motor activity; it resolves within 2–8 weeks after birth. It may be aggravated by high-protein diets.

The diagnosis is based on the following findings:

- Amino acids in plasma, and in particular, increased tyrosine (up to 40–50 mg/mL);
- Organicacidsinurine:increasein4-hydroxyphenylpyruvate and phenylacetate.

Alkaptonuria is due to a deficiency of the enzyme homogentisate oxygenase, which causes an increase in circulating homogentisic acid levels, reaching values above 3 mg/dL. Homogentisic acid is eliminated by urine and also excreted by sweat. It is precisely the homogentisic acid on the skin that, in the presence of oxygen and alkaline pH, is transformed into a brown pigment, called ochronotic pigment, which causes the typical dark coloration of the skin of the pubic and axillary region of the affected patients.

The same ocronotic pigment has a high affinity for cartilage and connective tissues and accumulates in costal cartilages, tendons, and ligaments, laryngeal and tracheal cartilages, but also in all connective tissues, including endocardium and vascular intima.

The diagnosis is based on finding organic acids in the urine, and in particular, increased homogentisic acid.

Glycogenosis

It indicates all the clinical pictures referring to diseases due to hereditary deficiency of enzymes involved in glycogen metabolism. In particular, glycogenosis can be due to:

- Alterations in glycogenolysis;
- Indirect glycogen synthesis increases, resulting in normal or atypical glycogen accumulation.

In vertebrates, glycogen is mainly found in the liver and skeletal muscle and plays different actions. In particular, muscle glycogen represents an exclusive source of energy for the muscle, rapidly usable, which is exhausted in less than 1 h during intense muscular effort. On the other hand, hepatic glycogen is fundamental for maintaining glucose homeostasis, acting as a glucose reserve for other tissues when exogenous glucose is unavailable (in the period between meals and in fasting); the reserve of hepatic glycogen is exhausted in 12–24 h.

The metabolism of glycogen consists of its degradation and synthesis, each divided into 3 stages catalyzed by a specific enzyme.

The synthesis takes place in the following 3 stages:

1. Glucose activation (UDP-glucose);
2. Addition of UDP-glucose units to the nonreducing ends on the glycogen by the formation of α-1,4-glycosidic bonds;
3. Formation of α-1,6-glycosidic bonds to create the branches.

Degradation takes place in the following 3 stages:

1. Breakage of α-1,4-glycosidic bonds with the release of glucose 1-phosphate;
2. Breakage of α-1,6-glycosidic bonds;
3. Conversion of glucose 1-phosphate to glucose 6-phosphate.

Depending on the mutated gene, glycogenoses can be classified into predominantly hepatic or muscular forms (Table 32.1).

All glycogenoses have autosomal recessive inheritance, except for glycogenosis IX characterized by X-chromosome-linked inheritance.

Glycogenoses have an overall incidence of 1 in every 20–25,000 births. Glycogenoses I and III are the most frequent hepatic glycogenoses, while glycogenosis IV is rare. In contrast, glycogenosis II, particularly the severe infantile form (Pompe disease), is the most common muscle glycogenoses.

Table 32.1 Classification of glycogenosis

Hepatic glycogenosis	
I	Defect of glucose-6-phosphatase
III	Defect of amyl-1,6-glucosidase (deramifying enzyme)
IV	Defect of 1,4–1,6-transglucosidase (branching enzyme)
VI	Defect of phosphorylase
IX	Defect of phosphorylase kinase
Muscle glycogenosis	
II	Defect of acid alpha-glucosidase (acid maltase)
V	Defect of myophosphorylase
VII	Defect of phosphofructokinase

Glycogenosis I, also known as von Gierke's disease, named after the Dutch physician who first described it in 1929, is due to mutations in glucose-6-phosphatase (G6Fase), an enzyme that is essential for glucose homeostasis regulation. Glucose-6-phosphate resulting from glycogen demolition and gluconeogenesis is transferred by a transporter (TG6F) from the cytoplasm into the endoplasmic reticulum (RE), where it is converted to glucose by G6Fase. In the enzyme deficiency, G6F cannot be converted to glucose and, therefore, cannot escape from the cell and enter circulation. In glycogenosis Ia, the gene encoding for the G6Fase enzyme is altered, while in glycogenosis Ib, it is the gene encoding for the TG6F transporter.

Glycogenosis III is due to a deficiency of the enzyme amylo-1,6-glucosidase or deramifying enzyme, leading to the hepatic accumulation of glycogen with an abnormal conformation (glycogen with very short branches) detrimental to the hepatocyte, progressively leading to periportal fibrosis and often to liver cirrhosis.

Glycogenosis II, also known as Pompe disease, is a lysosomal storage disease due to deficiency of acid α-glucosidase, also known as acid maltase. This enzyme hydrolyzes glycogen into glucose units. The deficiency results in an intralysosomal accumulation of glycogen. For this reason, this condition is considered a lysosomal storage disease.

Diagnosis

The diagnostic suspicion of glycogenosis is based on the biochemical-clinical picture; diagnostic confirmation is based on molecular genetic investigations and the dosage of enzymatic activity in muscle or liver tissue. From a biochemical-clinical point of view, hepatic and muscle forms present different characteristics. Patients with hepatic glycogenosis have:

- Hypoglycemia;
- Hepatomegaly (due to glycogen accumulation);
- Lactic acid: increased in type I, normal in type III.

Patients with muscle glycogenosis have:

- Intolerance to intense physical exertion with cramps, early pain, and myoglobinuria;
- No increase in lactic acid and excessive increase in amoniemia after exercise;
- Increased circulating levels of muscle enzymes, particularly the creatine kinase isoform MM (CK-MM).

Hyperammonemia

It refers to a clinical condition characterized by elevated blood levels of ammonium, frequent involvement of the central nervous system, and an uncertain prognosis. The severity of presentation varies significantly according to the patient's age, the level of ammonium elevation, and the timeliness of therapeutic intervention. Hyperammonemia represents a medical emergency. Ammonium (NH4+) is a low-molecular-weight metabolite produced by protein catabolism and eliminated hepatically through the urea cycle. In the urea cycle, a series of chain reactions in the mitochondrion and cytosol allows the catabolism of 2 nitrogen (N) atoms and the production of a urea molecule. Altered functioning of the urea cycle determines a reduced production of urea with an accumulation of the products upstream of the defect; therefore, hyperammonaemia and deficiency of those downstream.

Alteration of the urea cycle can be classified as :

- Primary, due to cycle enzyme deficiency;
- Secondary, due to enzymatic inhibition or substrate deficiency.

The severity of hyperammonaemia depends on:

- Enzyme involved;
- Residual activity of the enzyme involved;
- Rate of endogenous protein catabolism;
- Age of onset.

Ammonium is extremely toxic to the CNS due to:

- Glutamine-mediated induction of cerebral edema;
- Alteration of neurotransmitters;
- Energy deficit.

Causes of hyperammonaemia can be classified according to the presence of hypoketotic hypoglycemia (Table 32.2).

Among the causes of hyperammonemia with hypochetotic hypoglycemia, defects in beta-oxidation are the most common. Specifically, defects in beta-oxidation are associated with impaired acetyl-CoA synthesis, which in turn results in:

- Hyperammoniemia: acetyl-CoA is required for the synthesis of N-acetylglutamate, the cofactor of the enzyme carbamoyl phosphate synthetase that catalyzes step I of the urea cycle.
- Hypochetosis: ketogenesis begins from acetyl-CoA.
- Hypoglycemia: acetyl-CoA condenses with oxaloacetate to start the Krebs cycle, which in turn leads to the formation of oxaloacetate. Thus, if acetyl-CoA is missing, oxaloacetate will be missing for gluconeogenesis.

Table 32.2 Causes of hyperammonemia

Without hypoglycemia	With hypochetotic hypoglycemia (blood glucose <45 mg/dL and circulating 3-OH-butyrate <0.4 mM)
- Deficit of the urea cycle - N-acetylglutamate synthetase inhibitors (e.g., therapy with valproic acid and/or diet low in arginine; maple syrup urine disease; etc.) - HHH syndrome: Hyperammonemia, hyperornithinemia, homocitrullinuria - Defect in transport of dibasic aminoacids - Defects of the mitochondrial respiratory chain - Hepatic insufficiency - Defects in protein glycosylation	- Defects of ketogenesis: 3-OH-3-methyl glutaric aciduria - Defects of mitochondrial beta-oxidation of fatty acids - Multiple defects of acyl-CoA dehydrogenase or glutaric aciduria type II, complex deficiency ETF-Qo or ETF-DH inhibits mitochondrial oxidation of all fatty acids - Hyperinsulinism with persistent hyperammonemia

Particular care must be observed during blood collection, which can be venous or arterial, but it must be avoided prolonged stasis and, therefore, hypoxia (do not compress or block); the blood must be collected in a tube with EDTA, stored on ice and analyzed within 1 h.

If these rules are not followed, the test must be repeated. This is a highly URGENT test, which all laboratories must perform on an urgent/emergency basis. In the presence of encephalopathy, plasma ammonium levels should be measured, especially before a lumbar puncture.

Hyperammonaemia may occur in acute or subacute, chronic, and/or relapsing forms. Normal plasma ammonium values are <50 μmol/L (<90 μg/dL) and in the neonatal period <110 μmol/L (<190 μg/dL).

Hyperammoniemias are distinguished into neonatal, infantile, puberty, and adult forms.

Neonatal Form

It generally occurs in an acute form in a healthy newborn who, between 24 and 48 h of age, shows weak suction, hypotonia, progressive lethargy, and/or convulsions. Respiratory alkalosis is present in 50% of patients, allowing a differential diagnosis with neonatal sepsis, which is characterized by respiratory acidosis. If adequately treated, patients may survive, with variable risk of mental retardation and epilepsy.

Child Form

- Acute: episodes of irritability, refusal of food, drowsiness, acute encephalopathy, lethargy, altered state of consciousness, ataxia, and convulsions.
- Subacute: poor growth, recurrent vomiting, absence episodes, psychomotor retardation, and latent neurological symptoms.

Puberty and Adulthood Forms

Most patients do not have a cognitive delay, although it is possible to diagnose urea cycle deficiency in a patient with apparently isolated mental retardation or epilepsy.

- Acute: episodes of acute encephalopathy of varying degrees that may be confused with toxic or infectious encephalopathy.
- Subacute/chronic: learning disorders, behavioral alterations and/or symptoms of cerebellar lesions, psychic symptoms, episodes of disorientation, lethargy and/or acute encephalopathy following protein intake, fever and/or stress, and postpartum encephalopathy. Headache crises with repeated vomiting are frequent.

The possibility of an inherited metabolic disorder as a cause of hyperammonemia must always be considered at any age. A delay in diagnosis and treatment may result in brain death and/or irreversible neurological sequelae.

Table 32.3 shows the classification of hyperammonemia based on ammonium values.

In case of diagnostic suspicion of hyperammonaemia, II level tests include plasma aminoacidogram and the measurement of urinary amino acids, and plasma and urinary homocitrullinemia. Diagnostic confirmation can be obtained by molecular analysis.

After the diagnosis of hyperammonaemia has been made, treatment should be instituted immediately, even before a definitive diagnosis of the specific cause of the defect is established.

The aim of the treatment is to restore and maintain blood levels of ammonium within normal limits.

The principles of treatment are based on:

- Minimization of endogenous ammonium production;
- Administration of an adequate amount of nonprotein calories to arrest protein catabolism. Protein and amino acid intake should be discontinued until ammonium levels normalize.

Table 32.3 Classification of hyperammonemia based on NH_4^+ values

Age	Degree	Levels of NH_4^+ (μmol/L)	Clinical symptoms
Infants	Mild	110–180 (NV <110)	Food refusal, hypotonia, groans, drowsiness, and progressive lethargy
>30 days	Mild	90–150 (NV <50)	Drowsiness, irritability, vomiting, food refusal, medium-grade ataxia
Infants	Moderate	180–350	Lethargy, hypotonia, unresponsive pupils, no mydriasis. As the condition progresses, coma with pain response, middle pupils, hypotonia
>30 days	Moderate	150–350	
At any age	Serious	>350	Coma without pain response, hypotonia, lethargy, mydriasis, cardio-respiratory arrest
At any age	Very severe, with certain neurological sequelae in case of survival	>700	Severe coma, retinal hemorrhage. Breaks from apnea. Mydriasis. High risk of herniation after 12 months

NV normal values

The hypoproteic diet is also associated with arginine (in all defects except argininemia) and citrulline, and drugs that promote the elimination of nitrogen through alternative routes to the urea cycle, such as sodium benzoate and sodium phenylbutyrate.

Recommended Readings

Bembi B, Cerini E, Danesino C, Donati MA, Gasperini S, Morandi L et al (2008) Management and treatment of glycogenosis type II. Neurology 71(23 Suppl 2):S12–S36

Bickel H, Gerrard J, Hickmans EM (1953 Oct) Influence of phenylalanine intake on phenylketonuria. Lancet 265(6790):812–813

Blau N, Belanger Quintana A, Demirkol M et al (2010) Management of phenylketouria in Europe: survey results from 19 countries. Mol Genet Metab 99:109–115

Burlina A, Giordano L (2010) Diagnosi e trattamento dell'iperammoniemia neonatale. Edizione italiana del protocollo diagnostico-terapeutico del Gruppo di Studio Spagnolo-Portoghese per le Iperammoniemie neonatali. Depositato AIFA ottobre

Chace DH, Sherwin JE, Hillman SL, Lorey F, Cunningham GC (1998 Dec) Use of phenylalanine-to-tyrosine ratio determined by tandem mass spectrometry to improve newborn screening for phenylketonuria of early discharge specimens collected in the first 24 hours. Clin Chem 44(12):2405–2409

Chen YT (2001) Glycogen storage diseases. In: Scriver CR, Beaudet AL, Sly WS, Valle D (eds) The metabolic and molecular bases of inherited disease, 8th edn. McGraw-Hill, New York, p 1530

Corsello G, Bosco P, Calì F, Greco D, Cammarata M, Ciaccio M, Piccione M, Romano V (1999) Maternal phenylketonuria in two

Sicilian families identified by maternal blood phenylalanine level screening and identification of a new phenylalanine hydroxylase gene mutation (P407L). Eur J Pediatr 158(1):83–84

Corsello G, Cicero L, Giuffrè M, La Grutta S, Piccio- ne M, Pusateri F, Ciaccio M, Roccella M. (2006 Dec) Maternal hyperphenylalaninemia syndrome: neuropsychological evaluation of four subjects during childhood and adolescence. Minerva Pediatr 58(6):557–569

Folling I (1994) The discovery of phenylketonuria. Acta Paediatr Suppl 407:4–10

Guthrie R, Susi A (1963) A simple phenylalanine method for detecting phenylketonuria in large populations of newborn infants. Pediatrics 32:338–343

Hoang L, Byck S, Prevost L, Scriver CR (1996) PAH mutation analysis consortium database – A database for disease-producting and other allelic variation at the human PAH locus. Nucleic Acid Res 24:127–131

Hoeksma M, Reijngoud DJ, Pruim J, de Valk HW, Paans AM, Van Spronsen FJ (2009) Phenylketonuria: high plasma phenylalanine decreases cerebral protein synthesis. Mol Genet Metab 96:177–182

Kollberg G, Tulinius M, Gilljam T, Ostman-Smith I, Forsander G, Jotorp P et al (2007) Cardiomyopathy and exercise intolerance in muscle glycogen storage disease. N Engl J Med 357:1507–1514

Lindblad B, Lindstedt S, Steen G (1977) On the enzymic defects in hereditary tyrosinemia. Proc Natl Acad Sci USA 74:4641

Martynyuk AE, Glushakov AV, Sumners C, Laipis PJ, Dennis DM, Seubert CN (2005) Impaired glutamatergic synaptic transmission in the PKU brain. Mol Genet Metab 86(suppl.1):34–42

Oldfors A, DiMauro S (2013) New insights in the field of muscle glycogenoses. Curr Opin Neurol 26:544

Özen H (2007) Glycogen storage diseases: new perspectives. World J Gastroenterol 13(18):2541–2553

Pearsen KD, Gean-Marton AD, Levy HL, Davis KR (1990) Phenylketonuria: MR imaging of the brain with clinical correlation. Radiology 177:437–440

Pietz J, Kreis R, Rupp A et al (1999) Large neutral amino acids block phenylalanine transport into brain tissue in patients with phenylketonuria. J Clin Invest 103:1169–1178

Pisani A, Marchetiello C, Amicone M, Riccio E (2016) Il nefrologo nel trattamento dell'iperammoniemia da disordini del ciclo dell'urea. Giornale di Tecniche Nefrologiche e Dialitiche 28(4):233–238. https://doi.org/10.5301/GTND.2016.16390

Shefer S, Tint GS, Jean-Guillaime D et al (2000) Is there a relationship between 3-hydroxy-3methylglutaryl co-enzyme a reductase activity and forebrain pathology in the PKU mouse? J Neurosci Res 61:549–563

Surtees R, Blau N (2000) Neurochemistry of phenylketonuria. Eur J Pediatr 159:109–113

Tarnopolsky MA (2016) Metabolic myopathies. Continuum (Minneap Minn) 22:1829

van Wegberg MJ, MacDonald A, Ahring K et al (2017) The complete European guidelines on phenylketonuria: diagnosis and treatment. Orphanet J Rare Dis 12:162. https://doi.org/10.1186/s13023-017-0685-2

Weiner DL (2006) Metabolic emergencies. In: Fleisher GR, Ludwig S, Henretig FM (eds) Textbook of pediatric emergency medicine, 5th edn. Lippincott, Williams & Wilkins, Philadelphia, p 1193

Circulating Biomarkers in Oncology: Areas of Application, Critical Issues, and Perspectives

33

Massimo Gion, Chiara Trevisiol, and Aline S. C. Fabricio

Biomarkers in Oncology: An Evolving Concept

The concept of a biomarker as a biochemical signal associated with the presence of a malignancy dates back to 1845, when Henry Bence-Jones published the article "On a new substance occurring in the urine of a patient with mollities ossium." He accurately described the identification process of the "substance," but, more importantly, he hypothesized the role of the "new substance" as a disease biomarker, suggesting its clinical application and indicating its importance for understanding the pathophysiology of the associated disease.

However, the clinical importance of biomarkers in oncology emerged only in the early 1960s of the last century, thanks to two breakthroughs: the uncovering of immunoassays by Rosalyn S. Yalow and Solomon Berson, which led to the development of easy-to-use immunometric assays to analyze a large number of samples simultaneously and in a reasonably short working time, and the measurement of estrogen receptors by Elwood V. Jensen through the development of radioligand binding assay methods. The determination of estrogen receptors in the breast cancer tissue was first used by clinicians to predict the response to ablative hormonal therapies, such as ovariectomy, and then to additive therapies (progestins, antiestrogens, aromatase inhibitors, etc.) in women with breast cancer. Immunoassays were widely used to measure circulating levels of an increasing number of biomarkers, such as α-fetoprotein (AFP), carcinoembryonic antigen (CEA), carbohydrate antigen 125 (CA125), and many others progressively identified, initially classified as tumor-associated antigens and conventionally known as "tumor markers." The discovery of hybridoma technology by Georges Köhler and César Milstein, and the resulting availability of monoclonal antibodies in the 1980s, fueled the discovery of new biomarkers, the development of more sensitive and specific immunoassays, and the optimization of assay techniques. This was followed by a phase of great interest and enthusiasm, in which it was widely believed that the early diagnosis of several malignancies could be ultimately feasible through the determination of "tumor markers." The clinical role of tumor markers was progressively scaled down thereafter as their limitations began to be highlighted. The first one is the fact that tumor markers are not specific to cancer and can be elevated in many non-oncological conditions; the second one is related to the direct relationship between circulating levels of the marker and the extent of tumor tissue; therefore, it is unlikely that an early – and thus a small – malignancy can produce and release significant amounts of the marker. The awareness of these limitations has considerably constrained the potential role of circulating "tumor markers" for the early diagnosis of cancer. On the other hand, as concerns predictive biomarkers, the last two decades have witnessed extraordinary progress in the knowledge of the molecular mechanisms controlling cell growth, which has led to the development of many new anticancer drugs directed against specific molecular targets. The approach to the treatment of malignancies has therefore progressively changed with the introduction of an increasing number of molecular, targeted anticancer agents. This has led to an in-depth investigation of the association between the drugs and their biological targets, finally developing the concept of a mandatory drug–biomarker association, thus coining the term "companion diagnostics," to the extent that today several molecular targeted anticancer agents can be prescribed only if the specific target has been determined.

Thus, the concept of a biomarker has undergone a profound change over time, which has extended its biological meaning and broadened its areas of clinical application. The American Association for Cancer Research in partnership with the Food and Drug Administration and the National Cancer Institute (AACR-FDA-NCI) Cancer Biomarkers

M. Gion (✉)
Regional Center for Biomarkers, Department of Clinical Pathology, Azienda ULSS 3 Serenissima, Venice, Italy
e-mail: massimo.gion@aulss3.veneto.it

C. Trevisiol · A. S. C. Fabricio
Veneto Institute of Oncology IOV – IRCCS, Padua, Italy

M. Ciaccio (ed.), *Clinical and Laboratory Medicine Textbook*, https://doi.org/10.1007/978-3-031-24958-7_33

Collaborative consensus report has operationally defined a biomarker as "an objectively measurable characteristic and evaluated as an indicator of a normal biological process, of a pathological process, or the pharmacological response to a therapeutic intervention." From this definition, it follows that the term "biomarker" also includes, together with molecular indicators, clinical signs or metabolic information derived from imaging techniques. This more inclusive definition is intended to support the development and application of those new anticancer drugs directed at known biological mechanisms but still lacking a precise molecular target, as in the case of tumor neoangiogenesis. Thus, hypertension associated with bevacizumab administration has been considered a predictor of response in patients receiving this drug as the first-line therapy for metastatic colorectal cancer. Similarly, acneiform rash occurring at the first cycle of cetuximab therapy in patients with non-small cell lung cancer was found to be a clinical biomarker of response to the drug. Concerning imaging, changes of tumor metabolic activity evidenced by positron emission tomography–computed tomography (PET–CT) seems to be an early predictor of response to treatment. Humbert et al. reported that a maximum standardized uptake value (SUV) ≤2.0 after the first cycle of therapy is the most effective predictor of response to trastuzumab-containing regimen in women with breast cancer positive for human epidermal growth factor receptor 2 (HER2).

These biomarkers represent a pragmatic approach to clinical application, but they are only the epiphenomenon of a complex biological cascade. The need for their use testifies that more specific molecular biomarkers associated with the different steps of the signaling mechanism of interest suitable for clinical use have not yet been identified.

Critical Issues and Room for Improvement

Despite decades of research, thousands of published studies, and a significant amount of promising, preliminary results, only a limited number of biomarkers are recommended for clinical use. Several research groups have examined the pathway of biomarker translation from discovery to clinical practice, identifying numerous aspects that may cause delays or errors in the translation process. Individual research groups, scientific societies, and institutions have developed a variety of proposals to improve the quality of studies on biomarkers and accelerate the translational process to clinical practice.

A great deal of work has been carried out on classifying biomarkers to make the taxonomy more consistent with either the biological rationale or the needs for clinical application and to thus define the boundaries of the different research areas. Biomarkers are classified according to different criteria: (1) their biochemical nature; (2) the subcellular site of localization; (3) the biological material where they are mainly present and measured; and (4) the scope of their clinical application. The most commonly used classification is the scope of their clinical application, distinguishing between diagnostic, pharmacodynamic, prognostic, and predictive biomarkers.

Pepe et al. proposed study protocols organized into different phases, similar to the well-established approach used for the validation of a new drug. These protocols aim at standardizing the validation process of new biomarkers by reducing the variability of study design and thus improving the comparability of results from different studies. Other groups have suggested specific study designs for predictive biomarkers aimed at accelerating the early translation of results to the clinical practice. Furthermore, the different steps of the pathway from discovery to clinical application of a biomarker have been carefully examined to identify the points of weaknesses and confounding factors in the pre-analytical, analytical, and post-analytical phases.

The evidence provided by these methodological studies have led scientific societies and boards of experts to develop and disseminate specific guidelines to facilitate reporting of studies in a standardized, accurate, and reproducible form, such as the Standards for Reporting of Diagnostic Accuracy Studies (STARD) and REporting recommendations for tumour MARKer (REMARK) initiatives. Numerous scientific journals have adopted these guidelines as recommended requirements for article submission. An indirect and implicit objective of these guidelines is also to induce a progressive improvement of study design.

The briefly summarized abovementioned approaches, aimed at improving the translation of research results on biomarkers to clinical application, have been disseminated and implemented. They have contributed to the improvement of the quality of both study design and results reporting. However, they do not seem to have induced a meaningful improvement in the pathway of biomarker translation from discovery to clinical practice. In fact, even though basic research has discovered thousands of promising candidate biomarkers for clinical application, Anderson et al. have shown that the rate of introduction of new molecules into clinical practice has remained essentially stable with an average of 1.5 new biomarkers approved annually by the Food and Drug Administration (FDA), which is unsatisfactory for clinical needs.

Thus, the strategies adopted so far seem to have produced only a marginal improvement in the biomarker translation pipeline. The still limited availability of biomarkers for clinical practice suggests that crucial problems, probably not yet fully identified, persist in the multistage process that, starting from the discovery of a candidate biomarker, should transfer it into clinical practice through sequential steps of validation.

Biomarkers in the Blood and in Tissues: A Complementary Approach

This chapter deals with circulating biomarkers; however, it may be helpful to briefly examine some aspects related to the measurement of biomarkers in the tumor tissue, in order to identify the advantages and limits that the two matrices present and to better understand why the determination of biomarkers in tissues or in blood can answer different clinical questions.

Tissue biomarkers are receiving prevalent interest because of the need to identify molecular targets for new anticancer drugs. The determination of biomarkers in tissues has several points of strength. First, the biomarker is measured in the target tissue. Thus, no biomarker dilution occurs (as it does in the blood), and possible nonspecific or background signals should be mainly related to analytical aspects and eventually constrained. In the case of tissue biomarkers, diagnostic sensitivity and specificity are both associated with analytical specificity and sensitivity; in other words, the ability to correctly identify the molecule of interest is mainly dependent on the characteristics of the method of determination. The problem of tissue heterogeneity can also be addressed by several approaches, such as, for example, by selecting areas or cell types of interest by laser-assisted microdissection. Interpretation of biomarker expression at the tissue level thus appears relatively straightforward.

However, the determination of biomarkers in a tissue has some significant limitations associated with biological, conceptual, and technical aspects, which must be considered in both clinical application and research design.

The biological aspects concern: (1) the intrinsic complexity of the molecular mechanisms; (2) the complex interaction of tumor cells with the microenvironment; and (3) the redundancy of the regulatory mechanisms of the tumor cell. These variables are associated with the inherent biological complexity of development and progression of cancer and explain the difficulties in identifying appropriate biomarkers as drug targets for some signaling pathways. However, it can reasonably be expected that the increase of knowledge of cancer cell biology and the progressive development of technologies will lead to an increasingly wide variety of measurable biomarkers as well as to a progressive improvement of diagnostic sensitivity and specificity of the available methods.

The conceptual limitations of the determination of a biomarker in the tissue matrix are twofold: (1) the representativeness of the sample and (2) the timing of the sampling in relation to the possible variability of biomarker expression over time. Regarding representativeness, biomarker determination is performed with a minimal amount of tissue as only a few milligrams of the tumor tissue are usually examined by immunohistochemistry under routine conditions. Thus, it is difficult to consider the immunohistochemical determination of a biomarker as a truly quantitative measure in terms of representativeness of the examined tissue specimen. This issue might be of poor relevance in homogeneous tumors, but it should be considered in the case of intrinsically heterogeneous malignancies. The second conceptual aspect concerns the one-shot feature of the measure, which shows the expression of the biomarker in a single site and at a single time point. Therefore, clinical decisions during the disease course are based on the presence/absence of the biomarker measured in a sample examined even years earlier, rather than on the current phenotype of the malignancy. This entails a risk of error in the characterization of the disease and, consequently, in therapeutic decisions. For example, HER2 expression may differ in metastases and in primary breast cancer in approximately 25% of cases; nonetheless, the therapy of patients with metastatic breast cancer is based on HER2 determination obtained in the primary tumor; a comparable degree of variability between the primary tumor and metastases has been reported for estrogen receptors.

Finally, a technical issue must be considered. Biomarkers are measured in tissues mainly by molecular biology techniques or immunohistochemistry. The latter is extensively used for routine biomarker measurements for clinical decision-making. Immunohistochemical techniques are typically qualitative, and the assumption that biomarker determination (e.g., estrogen receptors) is quantitative stems from the fact that the result is often expressed as a quantitative score or a percentage. However, it has been shown that the routine immunohistochemical method performed on fixed tissue samples does not lead to accurate quantitative results, even when performed in highly qualified laboratories.

The measurement of biomarkers in a tumor tissue is crucial for prognostic assessment and the prediction of response to anticancer agents, but it is inherently unsuitable for risk assessment, diagnosis, early recognition of recurrence, and therapy monitoring. In these situations, biomarkers can only be measured in blood. Table 33.1 summarizes the scenarios in which biomarkers can provide clinical information if measured in a tumor tissue or in blood.

Biomarkers in the blood can be determined at all stages of the disease, can be easily repeated over time, and their measurement is typically quantitative and expressed on a continuous scale. The categorization into positive and negative results based on a threshold value is a conventional dichotomization operation. However, different from tissue biomarkers, they are measured in a biological fluid distant from the malignancy, and, therefore, they are a surrogate measure of the actual amount of biomarkers produced and released by the tumor tissue. In fact, the amount of biomarker in the

Table 33.1 Indications for biomarker use in different clinical scenarios according to the biological material: comparison between the tumor tissue and peripheral blood

	Indication for biomarker use	
Clinical scenarios	Tumor tissue	Blood
Assessment of occupational or environmental cancer risk	No	Yes
Screening of asymptomatic general population	No	Yes
Surveillance of subjects with increased cancer risk	No	Yes
Differential diagnosis in symptomatic patients	No	Yes
Initial diagnostic workup (prognosis)	Yes	No/Yes
Initial diagnostic workup (prediction of response to therapy)	Yes	No
Long-term monitoring after curative treatment of the primary tumor (early detection of relapse)	No	Yes
Therapy monitoring in patients with advanced or disseminated disease	No	Yes

blood results from several variables in addition to the production and release by the tumor tissue. Such variables include: (1) the possible production of the biomarker by normal tissues; (2) the possible presence in the blood of substances that cross-react with the measurement system; (3) the dilution in blood volume; and (4) metabolism and clearance. All of these factors may contribute to the blood concentration of a biomarker in a variable and unpredictable manner, both between different patients and in the same patient at different time points. Since the biomarker level in the blood results from different factors that add up algebraically (some factors cause an increase, others a decrease), it is necessary to use a threshold value to classify the biomarker result as positive or negative. The threshold value is usually determined from the distribution of the biomarker in a control group of healthy subjects or patients with non-oncological diseases. The classification of the biomarker results with reference to a positive/negative threshold value in subjects with the neoplasm of interest and in control subjects allows to classify patients and controls into four groups: (1) true positives (subjects with the disease and positive biomarker); (2) false negatives (subjects with the disease and negative biomarker); (3) true negatives (subjects without the disease and negative biomarker); and (4) false positives (subjects without the disease and positive biomarker).

The frequencies of patients and controls falling into the four groups enable the calculation of the diagnostic performance of the biomarker, such as sensitivity, specificity, predictive value, and likelihood ratio, as summarized in Table 33.2.

The non-specificity of a malignancy and the relatively low diagnostic sensitivity represent the commonly recognized limitations of the vast majority of, if not all, circulating biomarkers available for clinical use.

Table 33.2 Measures of diagnostic accuracy

Indicator	Formula	Definition
Sensitivity (SS)	TP/(TP+FN)	Proportion of subjects with the disease with a positive test result in the total of subjects with the disease
Specificity (SP)	TN/(TN+FP)	Proportion of subjects without the disease with a negative test result in the total of subjects without the disease
Positive predictive value (PPV)	TP/(TP+FP)	Proportion of subjects with the disease with a positive test result in the total of subjects with positive results
Negative predictive value (NPV)	TN/(TN+FN)	Proportion of subjects without the disease with a negative test result in the total of subjects with negative test results
Diagnostic accuracy	(TP+TN)/(TP+FN+TN+FP)	Proportion of subjects correctly classified by the test, in the total of examined subjects
Likelihood ratio of a positive result (LR+)	TP/FP	Proportion of subjects with the disease with a positive test result divided by the proportion of subjects without the disease with a positive test result
Likelihood ratio of a negative result (LR−)	FN/TN	Proportion of subjects with the disease with a negative test result divided by the proportion of subjects without the disease with a negative test result

FN false negatives, *FP* false positives, *TN* true negatives, *TP* true positives

Poor specificity is an intrinsic problem of known biomarkers. The so-called "tumor markers" are biochemical substances generally produced by several organs or cell types. They are therefore not specific to a given tumor but to the tissue or organ in which the tumor develops. The amount of biomarker produced and/or released by the tissue cells is low or very low in a physiological status but may increase under pathological conditions. These certainly include the development of a malignancy but also encompass several other conditions, such as, for example, the increase of the size of the organ (hyperplasia), inflammatory events, infections, autoimmune diseases, traumas, surgery, and administration of several drugs. In addition, the impairment of liver or kidney function can indirectly cause a spurious increase in circulating levels of some biomarkers by reducing their metabolism or clearance.

The poor diagnostic sensitivity is due to two major components: the frequently low concentration of the biomarker in the blood and the analytical sensitivity of the method of measurement. The low amount of biomarker present in the blood in patients bearing small tumors is an objective issue, which impairs the possible use of the biomarker in early diagnosis, and limits its value for the early detection of the relapse during the follow-up. When the amount of tumor tissue is still minimal, the circulating concentration of the biomarker may remain within the normal range. The analysis of biomarker increments over time seems a promising approach to improve the diagnostic efficacy for early cancer. The progressive development of the malignancy may lead to a gradual increase of the biomarker concentration, even when it is within the normal range. The variation of biomarker levels between serial samples has been investigated in several clinical scenarios, such as early detection of ovarian and prostate cancers or early recognition of relapse in different types of malignancies, including breast, ovarian, and colon cancers. The criteria to standardize the interpretation of serial sample-based biomarker measurements have been proposed to facilitate the conduct of clinical trials, but the available results are in general still inconclusive. This is due at least in part to the relatively low analytical sensitivity of many assay methods. Indeed, most commercially available methods for measuring "tumor markers" are designed to offer linearity of dose–response interpolation over a broad concentration range, ensuring reliable results in a wide variety of clinical situations. This choice, aimed at facilitating the routine use of biomarker determination in clinical laboratories, entails the assumption that the quantitative level of the biomarker below the threshold value – classified as negative – is clinically uninformative. The assay design often implies that analytical precision is optimal for biomarker concentrations close to, or above, the threshold value, with analytical variability tending to worsen as the biomarker concentration decreases. To be clinically meaningful, differences between serial samples must be more significant than the analytical and biological variability of the biomarker. The poor reproducibility and comparability between the results reported in different studies on biomarker variations over time may be due, at least in part, to unsatisfactory assay precision at low biomarker concentrations, which negatively affects both analytical sensitivity and the calculation of biological variability. The availability of differently designed assay methods, targeted toward low biomarker concentrations and therefore characterized by higher analytical sensitivity and high precision in the low value range, would allow to better explore the clinical use of decision criteria based on serial sample variations.

In conclusion, determining biomarkers in the tumor tissue and in blood is not an alternative but a complementary approach, presenting different advantages and limits. In both matrices, significant margins for improvement are recognizable, related to the development of knowledge, innovation of the study design, and improvement of technological aspects. Research policies in the field of biomarkers should consider the value of diversities between tissue and blood biomarkers, thus avoiding unnecessary redundancies and overlaps and emphasizing the advantage of complementarities.

Circulating Biomarkers in Clinical Practice

An increasing number of studies on candidate biomarkers for clinical application are continuously published. However, preliminary, often very promising results are rarely confirmed by further studies. Only for a few biomarkers evidence is sufficient to recommend their clinical use. These biomarkers, identified 30–40 years ago, are the so-called "tumor-associated antigens," whose circulating levels correlate with tumor bulk.

An essential aspect in the choice of biomarkers to be used in clinical practice concerns the types of literature to refer to. Published studies can be schematically classified into three categories with respect to the source of the information: (1) primary research, which concerns the original results of experimental studies; (2) secondary research, consisting of meta-analyses and systematic reviews; and (3) tertiary research, which leads to the production of guidelines. Primary studies and most meta-analyses and systematic reviews mainly deal with the efficacy of biomarkers; in other words, they assess diagnostic performance (e.g., sensitivity, specificity, predictive values, likelihood ratio) against a reference diagnostic criterion (usually, histological findings) or in comparison with other established biomarkers. The evaluation of efficacy is a necessary, but not sufficient, condition for recommending the clinical use of a biomarker, which requires an evaluation of the impact of the biomarker on clinical outcomes or organizational processes taking into account other context variables; for example, an effective biomarker should not be recommended if effective therapies to treat the diagnosed condition are not available. Clinical practice guidelines, as defined by the Institute of Medicine (IOM), are documents containing recommendations to optimize patient care based on the systematic review of evidence and the assessment of the benefits and harms of alternative care options. Therefore, it should be advisable to refer to guidelines when deciding whether to use a biomarker in clinical decisions.

In order to review the state of the art on the use of circulating biomarkers in clinical practice, guidelines on solid tumors produced worldwide from 2009 to 2015 have been searched and reviewed and recommendations on circulating biomarkers have been summarized. Information on biomarkers was extracted separately for 21 types of solid tumors and summarized for each malignancy into the following clinical scenarios: (1) screening of asymptomatic general population; (2) surveillance of individuals with increased cancer risk; (3) differential diagnosis in symptomatic patients; (4) initial diagnostic workup; (5) postoperative evaluation; (6)

long-term monitoring after curative treatment of the primary tumor; and (7) therapy monitoring in patients with advanced or disseminated disease.

The principal indications on the use of biomarkers in different clinical scenarios are summarized below. Please refer to the recommended lectures for detailed information.

Screening of Asymptomatic General Population

This scenario refers to the systematic search for the disease in asymptomatic subjects through organized, general population screening programs. The goal is to reduce cancer-specific mortality through early detection of the disease. It is offered by health-care organizations when evidence is sufficient to ensure that screening offers more advantages than disadvantages or risks to the population. No biomarker is currently recommended in this setting. In fact, the diagnostic specificity and sensitivity of available biomarkers is not sufficient in relation to the prevalence of the cancers being screened to ensure adequate predictive values. It should be noted that available guidelines recommend against the use of prostate-specific antigen (PSA) in organized prostate screening programs, based on the results of randomized trials of consistent size. Guidelines have a similar position in ovarian cancer, recommending against the use of CA125 to screen asymptomatic women. Similar considerations are reported by guidelines concerning the use of biomarkers to screen for cancer asymptomatic people referring to the physician for a checkup (i.e., spontaneous or opportunistic screening).

Surveillance of Individuals at Increased Risk

This scenario concerns the systematic search for cancer in people with a higher risk of developing a malignancy than the general population, due to familiarity, ethnicity, lifestyle, or environmental conditions associated with increased cancer risk. In these subjects, the predictive value of biomarkers may increase because the prevalence of malignancy is higher than that in the general population. Guidelines advocate using calcitonin (Ct) in the surveillance of relatives of patients with hereditary medullary thyroid cancer. The position of the guidelines regarding the surveillance of subjects with liver diseases associated with an increased risk of developing hepatocarcinoma is controversial; some guidelines recommend using AFP, while others are against it. Likewise, in the case of surveillance of patients with sclerosing cholangitis, carbohydrate antigen 19.9 (CA19.9) is recommended by some guidelines, while others recommend not to use this biomarker. Also, regarding the evaluation of PSA in subjects with an increased risk of prostate cancer due to familiarity or ethnicity, guidelines present controversial indications. Finally, guidelines do not recommend regular CA125 monitoring for women with a hereditary–familial risk for ovarian cancer.

Differential Diagnosis in Symptomatic Patients

This scenario concerns the differential diagnosis between a malignancy and other non-oncological conditions in subjects with suspicious symptoms for cancer. In the differential diagnosis, only a few biomarkers for a few malignancies are indicated. This is the case of Ct in the clinical suspicion of medullary thyroid cancer, PSA in the case of suspected prostate cancer, and CA125 in women with a suspected adnexal mass. Guideline recommendations are more controversial for other malignancies, in which biomarkers are recommended by some guidelines and discouraged by others. Examples are the use of AFP for space-occupying lesions of the liver, of CA19.9 in suspicious biliary tract or pancreatic cancers, and of nuclear matrix protein 22 (NMP22) in suspicious bladder cancer. Finally, determining lactate dehydrogenase (LDH) in kidney cancer is recommended only when metastatic dissemination of the disease is suspected.

Excluded from this scenario are patients presenting generic symptoms (asthenia, weight loss, inappetence, anemia) or clinical situations commonly considered "paraneoplastic" (recurrent thrombophlebitis, neurological symptoms); in these circumstances, biomarkers are not recommended.

To be noted, all guidelines explicitly recommend that biomarkers are not used as a standalone diagnostic procedure, but they must be considered in association with imaging techniques only.

Initial Diagnostic Workup

In this scenario, biomarkers can be measured to have a baseline value before any treatment. In addition, biomarkers can provide additional prognostic indications to help guide therapeutic decisions. Guidelines agree in recommending the determination of: CEA in colorectal cancer; AFP in hepatocarcinoma; CA19.9 in pancreatic cancer; CA125 in endometrial cancer; PSA in prostate cancer; AFP, human chorionic gonadotropin β-subunit (β-hCG), and LDH in testicular germ cell tumors and non-epithelial ovarian tumors; LDH in melanoma and lung cancer; Ct in medullary thyroid cancer; thyroglobulin (Tg) and anti-thyroglobulin antibodies (AbTg) in differentiated thyroid cancers; and LDH and C-reactive protein in mesothelioma. Biomarker results are included in algorithms to define risk classes in some malignancies, such as prostate cancer and testicular germ cell tumors.

Postoperative Evaluation

After the treatment with curative intent of the primary tumor, the patient may be reevaluated to confirm the radicality in the case of surgery or to estimate the probability of the treatment success in the case of radiotherapy. In this scenario, guidelines recommend: AFP in hepatocarcinoma; CA19.9 in pancreatic cancer; CA125 in epithelial ovarian cancer; AFP, β-hCG, and LDH in germ cell tumors of the testis and non-epithelial ovarian cancer; PSA in prostate cancer; CEA in non-small cell lung cancer; Epstein–Barr virus plasma DNA (EBV-DNA) in nasopharyngeal cancers; Tg, AbTg, and thyroid-stimulating hormone (TSH) in differentiated thyroid cancers; and Ct and CEA in medullary thyroid cancers.

Long-Term Monitoring after Curative Treatment of the Primary Tumor

Long-term surveillance (follow-up) is offered to patients treated with curative intent because early recognition of the relapse and a timely treatment should reduce mortality and/or improve quality of life. Guidelines agree in recommending: CEA in colorectal cancer; AFP in hepatocellular carcinoma; AFP, β-hCG, and LDH in germ cell tumors of the testis and non-epithelial tumors of the ovary; PSA in prostate cancer; LDH in kidney cancer; EBV-DNA in nasopharyngeal cancers; Tg, AbTg, and TSH in differentiated thyroid cancers; and Ct and CEA in medullary thyroid cancers. Guidelines also agree not to recommend any biomarkers in the follow-up of cervical cancer, bladder cancer, and melanoma. As concerns breast cancer and epithelial ovarian cancer, guidelines are against the use of biomarkers in the follow-up of asymptomatic women and recommend their use (carbohydrate antigen 15.3 (CA15.3) and CEA in breast cancer and CA125 in ovarian cancer) only in symptomatic cases or when clinically indicated. Finally, different guidelines provide inconsistent recommendations in endometrial cancer, with some documents recommending against using CA125, while others consider it as an optional test.

Therapy Monitoring in Patients with Advanced or Disseminated Disease

This scenario encompasses locally advanced malignances and distant metastases. In some cancers, such as differentiated thyroid cancers and germ cell tumors of the testis, therapies may impact the survival and probability of cure even in the metastatic spread. In other malignancies, the therapeutic intent is usually aimed mainly at symptom control and palliation of metastases. In both cases (probability of cure and palliation), monitoring the response to the therapy is crucial for the early detection of nonresponders or to recognize the onset of resistance, in order to promptly modify the therapeutic schemes. In this scenario, the use of biomarkers is relatively well established, and several guidelines advocate their use. Guidelines recommend: CEA in colorectal cancer; AFP in hepatocarcinoma; CA19.9 in pancreatic cancer; CA19.9 and CEA in biliary tract tumors; CA15.3 and CEA in breast cancer; LDH in melanoma; mesothelin (soluble mesothelin-related peptide (SMRP)) in mesothelioma; CA125 in epithelial tumors of the ovary; CEA, a fragment of cytokeratin 19 (Cyfra 21.1) and gastrin-releasing peptide (GRP) precursor (pro-GRP) in lung malignancies; PSA in prostate cancer; LDH in kidney cancer; AFP, β-hCG, and LDH in germ cell tumors of the testis and non-epithelial tumors of the ovary; TSH in differentiated thyroid cancers; and Ct and CEA in medullary thyroid cancers.

From a general point of view, biomarkers can be grouped into three categories with reference to the available recommendations:

1. Biomarkers indicated in only one malignancy: PSA in prostate cancer; Ct in medullary thyroid cancer; Tg, AbTg, and TSH in differentiated thyroid cancers; CA15.3 in breast cancer; SMRP in mesothelioma; EBV-DNA in nasopharyngeal cancer; and NMP22 in bladder cancer
2. Biomarkers indicated in a limited number of tumors characterized by a common or similar histological type: β-hCG in germ cell tumors of the testis and non-epithelial tumors of the ovary
3. Biomarkers expressed by more than one neoplasm; recommendations for the cynical use of biomarkers in this latter group are summarized in Table 33.3.

Neuroendocrine tumors constitute a separate subject as they encompass a very heterogeneous group of malignancies that can express and release various molecules, some of which are associated with specific symptoms. The number of possible histological types and anatomical localization, and the variety of biomarkers potentially measurable in relation to tumor site and histological variants, do not allow for an exhaustive summary here. Because of the relatively low incidence of these tumors, we prefer to direct readers to specific literature.

Measuring Methods

Biomarkers are routinely measured by immune assays, which are based on the high-affinity binding of a specific antibody to the target antigen. The majority of commercially available immunoassays are fully standardized and available on commercial platforms for routine use. Nevertheless, for

Table 33.3 Biomarkers expressed by different tumor types: recommendations of clinical practice guidelines

		Clinical scenario					
Marker	Malignancy	Screening	Diagnosis	Initial workup	Postoperative evaluation	Long-term follow-up	Therapy monitoring
AFP	Hepatocarcinoma	Yes[a]				Yes	
	Germinal tumors of the testis	No	Yes	Yes	Yes	Yes	Yes
	Non-epithelial tumors of the ovary		Yes[b]	Yes		Yes	
CA125	Endometrial cancer			Yes			
	Epithelial cancer of the ovary	No	Yes[b]			Yes[c]/No[d]	Yes
CA19.9	Colorectal cancer			No		No	
	Pancreatic cancer			Yes[e]			
CEA	Colorectal cancer			Yes		Yes	
	Breast cancer					Yes[c]/No[d]	Yes
	Lung cancer						Yes
	Medullary thyroid cancer		Yes	Yes	Yes	Yes	Yes
LDH	Melanoma			Yes		No	Yes
	Kidney tumors		Yes[f]			Yes	Yes
	Germinal tumors of the testis	No	Yes	Yes	Yes	Yes	Yes
	Non-epithelial tumors of the ovary		Yes[b]	Yes		Yes	

AFP α-fetoprotein, *CA19.9* carbohydrate antigen 19.9, *CA125* carbohydrate antigen 125, *CEA* carcinoembryonic antigen, *LDH* lactate dehydrogenase

[a]In patients with chronic liver disease and an increased risk of developing hepatocellular carcinoma
[b]Only in association with imaging techniques
[c]In symptomatic patients or with other clinical indications
[d]In the absence of symptoms or other clinical indications
[e]In the absence of jaundice
[f]In the suspicion of advanced disease

several biomarkers, international reference standards are not available. Moreover, evidence shows that results obtained with different immunoassays are not always interchangeable even if calibrated using the same reference standard. The advent of new immunoassays based on multiplexing technologies, capable of determining several biomarkers simultaneously, may further emphasize this issue. Therefore, from an operational point of view, patients who are to be followed over time should be advised to refer to the same laboratory for the determination of biomarkers.

Beside the indications on the type of biomarker recommended for clinical purposes, guidelines also provide general information on the proper use of biomarkers, which are summarized in Box 33.1.

Biomarkers Associated with Biological Mechanisms

Several innovative biomarkers associated with known molecular mechanisms, such as inflammation, angiogenesis, matrix degradation, microenvironment interactions, or deregulation of cellular signaling systems, have been intensively investigated.

The interest toward mechanism biomarkers is due to their association with the biological characteristics and functions of the malignancy. Several biomarkers in this category are related to either malignancy aggressiveness or cellular mechanisms, which may be a target of specific anticancer agents. Therefore, a solid biological rationale exists to assume that these biomarkers may have a prognostic (relationship with the aggressiveness of the neoplasm) or a predictive (relationship with the target of dedicated drugs) role. However, studies published so far on circulating levels of these biomarkers have failed to provide conclusive data to support their use for clinical purposes. This systematic failure of research on mechanism biomarkers at the circulating level is quite surprising, especially when dealing with biomarkers that are effective and useful in guiding therapeutic choices when measured in the tissue. A paradigmatic example is HER2/neu, a biomarker routinely measured in the breast cancer tissue to assess patients' suitability for treatment with monoclonal antibodies targeted against HER2/neu (trastuzumab). In contrast, the measurement of HER2/neu in blood has not led to consistent results for a possible clinical use. Studies on circulating levels of mechanism biomarkers deserve a general remark. In the majority of published studies, these biomarkers have been evaluated as classical "tumor markers." They have been examined with reference to the characteristics of the malignancy (e.g., histological type, disease stage) or clinical goals (e.g., diagnosis, monitoring, prediction of response to therapy). In fact, these biomarkers are linked to general pathophysiological mechanisms. Classical "tumor markers" are also produced in small amounts by normal cells, and their circulating levels may increase due to causes other than cancer. However, classical "tumor markers" are not the expression of well-defined pathophysiological mechanisms. Conversely, mechanism biomarkers are molecules that actively participate in known and well-defined biological functions. This is the case of chemokines, factors regulating neoangiogenesis, or proteases and their inhibitors. These biomarkers are expressed when a mechanism is activated, regardless of the cause. For this reason, these molecules cannot be specific to any given malignancy. Therefore, to evaluate these biomarkers, it is necessary to consider different approaches from those used so far for the classical "tumor markers." For instance, the pre-analytical phase should be differently considered. Many of these biomarkers circulate freely in blood, but they are also released from either blood vessel cells or cellular elements of blood, i.e., leukocytes and platelets. Blood represents a multi-compartmental matrix for these biomarkers. In the case of vascular endothelial growth factor (VEGF), it is well established that the concentration measured in the serum does not correspond to the amount of biomarker released by the tumor but also reflects the released biomarker by the cellular components of the blood. It has been observed that the VEGF concentrations in the serum may significantly increase up to hundreds of times by extending the time between collection and centrifugation from 1 to 3 h, a time interval compatible with ordinary specimen management and considered in many published studies as a "standard procedure" of serum preparation. This aspect represents a limitation if the biomarker is considered a "tumor marker," but it may be instead a source of information on the potential relationships of the host microenvironment with the malignancy. As a matter of fact, the host's production of inflammatory cytokines and neoangiogenic factors may be evaluated as a regulatory element of cancer development and progression. Therefore, mechanism biomarkers should be studied as both a possible signal of the tumor cell and an indicator of the host's interaction with the malignancy. In this perspective, research efforts should be focused on the pre-analytical phase, evaluating the biomarkers in the plasma obtained through different anticoagulants, serum collected by standardized and reproducible procedures, and cellular blood elements.

The second point of interest specific to mechanism biomarkers is their possible interactions in the blood with biological drugs targeted to the same biomarkers at the tissue level. In the case of the administration of a monoclonal antibody directed against a biomarker expressed in the tumor tissue, the concentration of the same biomarker in the blood should not be considered a "neutral" element in the complex drug/tissue biomarker/circulating biomarker interactions. High circulating biomarker levels could bind the drug, thus making it unavailable in adequate amounts to act on the tumor tissue. From another point of view, drug-induced inhibition of the signaling system in the tumor tissue could decrease the biomarker release in the blood in responder patients. Conversely, an increased biomarker production and release into the circulation could indicate the occurrence of resistance. Considering the complexity of the possible interactions, the mere determination of a mechanism biomarker as "tumor marker" is reductive, if not misleading. Some pilot studies examined the relationships between circulating marker levels and the response to biological drugs. An inverse relationship between blood HER2/neu concentration and response to trastuzumab has been described in women with advanced breast cancer. Also, the importance of determining free VEGF in the plasma to assess the pharmacodynamics of bevacizumab has been shown, suggesting the potential role of the biomarker in optimizing treatment and tailoring doses and/or the duration of administration. Recently, it has been found that bevacizumab induces a dose-dependent blockade of VEGF internalization, demonstrating that the increase in VEGF during bevacizumab administration does not represent a mechanism of tumor resistance but a response of the host. This mechanism, mediated by the biomarker–host interaction and not by the tumor, may significantly interfere with the response of the tumor itself to the biologically targeted drug.

Considering the growing number of biological drugs targeting known molecular mechanisms that research makes available to clinical oncology, the issue of measuring biological targets also in blood is becoming crucial. In fact, despite initial responses, the onset of resistance occurs at variable time intervals for most of these novel anticancer agents. The determination of target biomarkers in the primary tumor tissue has the limitations described above, the most critical of which is the possible occurrence of tumor phenotype variations, either incidental or induced by the therapy as a mechanism related to resistance. Frequently, accessing a metastatic lesion to confirm the actual biological characteristics of the malignancy is complex, invasive, and sometimes ethically unsustainable. In this scenario, the possibility of measuring the biomarkers of interest in the blood is a priority for research and clinical practice. Numerous studies address this issue by determining the target biomarkers in circulating tumor cells (CTCs) or by examining biomarkers associated with circulating nucleic acids. Collectively, this approach is referred to as "liquid biopsy," a topic covered in another chapter of this textbook. To be noted, the study of circulating protein biomarkers should not be seen as an alternative to the study of nucleic acid-derived components but as a complementary approach. Biological drugs do not act on nucleic acids but on protein targets. Moreover, the onset of resistance is not necessarily sustained by a variation in the mutation of interest but may be due to the activation of alternative signaling mechanisms, made available either by the biological redundancy of tumor cells or by molecules with a signaling function (cytokines, angiogenesis factors, etc.) expressed in the tissue microenvironment.

The microenvironment has attracted attention in recent studies attempting to identify predictive biomarkers for a new class of drugs targeting immune checkpoints such as cytotoxic T-lymphocyte-associated protein 4 (CTLA4), programmed cell death 1 receptor (PD-1), and programmed cell death ligand 1 (PD-L1), which has revolutionized the treatment of several solid tumors. These drugs act through an indirect stimulation of the immune system against neoplastic cells by inducing consistent and, generally, durable antitumor responses. These treatments exert profound effects on immune cells within the tumor microenvironment mainly by recruiting tumor-infiltrating leukocytes and modifying the production of inflammatory and immune-related factors, which may spread into the circulation, becoming possible disease markers. Given the cost of these drugs and the need to personalize therapies, more progresses with predictive markers are crucial. Therefore, the key challenge of immune checkpoint inhibitor therapies is the identification of reliable markers to guide disease management decisions and to advance in the understanding of the mechanisms of resistance to immunotherapy. Numerous biomarkers have already been tested for their ability to predict response to immunotherapeutic treatments. Although the cancer immunotherapy field is still evolving, tissue biomarkers such as PD-L1, mismatch repair, and microsatellite instability testing have already helped guide the use of US Food and Drug Administration-approved therapies.

A systematic analysis of the literature shows that several circulating inflammatory or immune-related proteins have also been implicated in the checkpoint immunomodulation and in the mechanisms underlying tumor pathogenesis or progression, including inflammatory cytokines, growth factors, C-reactive protein (CRP), and soluble PD-L1. Other circulating biomarkers that have been investigated in this context are circulating tumor DNA (ctDNA), circulating tumor cells (CTCs), circulating microRNAs and exosomes, and peripheral blood cells. However, a clear association between circulating biomarkers and clinical responses, resistance, and toxicities to immunotherapy remains to be shown.

As mentioned above, regarding the integration between the analysis of biomarkers in tissues and blood, the need for integrating the study of the different types of circulating biomarkers must be considered when designing research strategies. From the long history of biomarkers, we have learned that considering in a competitive way body districts, biological matrices, molecular classes, or technological approaches, significantly slows the translation of innovation to clinical practice. In the face of the impressive development of knowledge and technologies, reduction of mortality of advanced solid tumors is still unsatisfactory. To enhance the clinical usefulness of circulating biomarkers, it is crucial to move toward a comprehensive vision, to encourage and reinforce complementarity of knowledge and technologies, and to share research priorities.

Box 33.1 Circulating biomarkers: general directions

- Biomarkers recommended for diagnostic purposes should not be used as a standalone test but always in association with imaging techniques and followed by biopsy in case of confirmed suspicion.
- Biomarkers cannot be used as a standalone test for therapeutic decisions in any clinical scenario, but information provided by biomarkers must be confirmed by clinical criteria or imaging techniques.
- A single increased biomarker value in the absence of symptoms or instrumental evidence of disease should be confirmed by a subsequent determination before any further clinical decision.
- When monitoring a patient, serial biomarker determinations should be performed with the same method and preferably by the same laboratory.

Recommended Readings

Alidzanovic L, Starlinger P, Schauer D et al (2016) The VEGF rise in blood of bevacizumab patients is not based on tumor escape but a host-blockade of VEGF clearance. Oncotarget 7:57197–57212

Anderson NL, Ptolemy AS, Rifai N (2013) The riddle of protein diagnostics: future bleak or bright? Clin Chem 59:194–197

Arslan C, Sari E, Aksoy S, Altundag K (2011) Variation in hormone receptor and HER-2 status between primary and metastatic breast cancer: re- view of the literature. Expert Opin Ther Targets 15:21–30

Azzariti A, Porcelli L, Brunetti O et al (2016) Total and not bevacizumab-bound vascular endothelial growth factor as potential predictive factors to bevacizumab-based chemotherapy in colorectal cancer. World J Gastroenterol 22:6287–6295

Becker KF, Wipperfürth J, Herpel E (2018 Jul) Preanalytics and biobanking: influence of preanalytical factors on tissue sample quality. Pathologe 39(4):297–302

Bence JH (1848) On a new substance occurring in the urine of a patients with Mollities Ossium. Phil Trans R Soc Lond 138:55–62

Bossuyt PM, Reitsma JB, Bruns DE et al (2003) Towards complete and accurate reporting of studies of diagnostic accuracy: the STARD initiative. Standards for reporting of diagnostic accuracy. Clin Chem 49:1–6

Cooper L, Chen J (2022) Changes in companion diagnostic labelling: implementation of FDA's April 2020 guidance for industry for in vitro CDx Labeling for specific oncology therapeutic groups. Ther Innov Regul Sci 56(5):689–697

Drabovich AP, Martínez-Morillo E, Diamandis EP (2015) Toward an integrated pipeline for protein biomarker development. Biochim Biophys Acta 1854:677–686

Drucker E, Krapfenbauer K (2013) Pitfalls and limitations in translation from biomarker discovery to clinical utility in predictive and personalised medicine. EPMA J 4:1186–1178
Farkona S, Diamandis EP, Blasutig IM (2016) Cancer immunotherapy: the beginning of the end of cancer? BMC Med 14:73
Gion M, Trevisiol C, Rainato G et al (2016a) Marcatori circolanti in oncologia: guida all'uso clinico appropriato. I Quaderni di Monitor. AGENAS, Agenzia Nazionale per i Servizi Sanitari Regionali, Roma
Gion M, Trevisiol C, Rutjes AWS et al (2016b) Circulating tumor markers: a guide to their appropriate clinical use. Comparative summary of recommendations from clinical practice guidelines (PART 1). Int J Biol Markers 31:e332–e367
Gion M, Trevisiol C, Rutjes AWS et al (2017a) Circulating tumor markers: a guide to their appropriate clinical use. Comparative summary of recommendations from clinical practice guidelines (PART 2). Int J Biol Markers 32:e1–e146
Gion M, Trevisiol C, Rutjes AW et al (2017b) Circulating tumor markers: a guide to their appropriate clinical use. Comparative summary of recommendations from clinical practice guidelines (PART 3). Int J Biol Markers 32:e147–e259
Hayes DF (2021) Defining clinical utility of tumor biomarker tests: a Clinician's viewpoint. J Clin Oncol 39(3):238–248
Humbert O, Cochet A, Riedinger JM et al (2014) HER2-positive breast cancer: 18F-FDG PET for early predi- ction of response to trastuzumab plus taxane-based neoadjuvant chemotherapy. Eur J Nucl Med Mol Imaging 41:1525–1533
Indini A, Rijavec E, Grossi F (2021) Circulating biomarkers of response and toxicity of immunotherapy in advanced non-small cell lung cancer (NSCLC): a comprehensive review. Cancers (Basel) 13(8):1794
Ioannidis JPA, Bossuyt PMM (2017) Waste, leaks, and failures in the biomarker pipeline. Clin Chem 63(5):963–972
IOM (Institute of Medicine) (2011) Clinical practice guidelines we can trust. The National Academies Press, Washington
Jabbour MN, Massad CY, Boulos F (2012) Variability in hormone and growth factor receptor expression in primary versus recurrent, metastatic, and post-neo- adjuvant breast carcinoma. Breast Cancer Res Treat 135:29–37
Kaufman HL, Butterfield LH, Coulie PG et al (2016) Society for immunotherapy of cancer (SITC) statement on the proposed changes to the common rule. J Immunother Cancer 4:37
Khleif SN, Doroshow JH, Hait WN (2010) AACR-FDA-NCI cancer biomarkers collaborative. AACR-FDA-NCI cancer biomarkers collaborative consensus report: advancing the use of biomarkers in cancer drug development. Clin Cancer Res 16:3299–3318
Kulasingam V, Prassas I, Diamandis EP (2017) Towards personalized tumor markers. Npj Precis Oncol 1:17
Lord SJ, St John A, Bossuyt PM et al (2019) For the test evaluation working Group of the European Federation of clinical chemistry and laboratory medicine. Setting clinical performance specifications to develop and evaluate biomarkers for clinical use. Ann Clin Biochem 56(5):527–535
Lund F, Petersen PH, Pedersen MF et al (2014) Criteria to interpret cancer biomarker increments crossing the recommended cut-off compared in a simulation model focusing on false positive signals and tumour detection time. Clin Chim Acta 431:192–197
Mantovani A, Allavena P, Sica A et al (2008) Cancer-related inflammation. Nature 454:436–444
McShane LM, Altman DG, Sauerbrei W et al (2005) Statistics subcommittee of the NCI-EORTC working group on cancer diagnostics. Reporting recommendations for tumor marker prognostic studies (REMARK). J Natl Cancer Inst 97:1180–1184
Nadji M (2008) Quantitative immunohistochemistry of estrogen receptor in breast cancer: "much ado about nothing!". Appl Immunohistochem Mol Morphol 16:105–107
Niikura N, Liu J, Hayashi N et al (2012) Loss of human epidermal growth factor receptor 2 (HER2) expression in metastatic sites of HER2-overexpressing primary breast tumors. J Clin Oncol 30:593–599
Parker LA, Chilet-Rosell E, Hernández-Aguado I et al (2018) Diagnostic biomarkers: are we moving from discovery to clinical application? Clin Chem 64(11):1657–1667
Pepe MS, Etzioni R, Feng Z et al (2001) Phases of biomarker development for early detection of cancer. J Natl Cancer Inst 93:1054–1061
Ransohoff DF, Gourlay ML (2010) Sources of bias in specimens for research about molecular markers for cancer. J Clin Oncol 28(4):698–704
Seo MK, Cairns J (2021) How are we evaluating the cost-effectiveness of companion biomarkers for targeted cancer therapies? A systematic review. BMC Cancer 21(1):980
Vaught J (2015 Jun) Developments in biospecimen research. Br Med Bull 114(1):29–38
Walk EE, Yohe SL, Beckman A et al (2020) College of American Pathologists Personalized Health Care Committee. The cancer immunotherapy biomarker testing landscape. Arch Pathol Lab Med 144(6):706–724

Liquid Biopsy

34

Marcello Ciaccio

Introduction

Two types of tumor-derived biological material can be found in the blood: circulating tumor cells (CTCs) and circulating tumor DNA (ctDNA). This chapter describes the technologies used for their isolation and molecular characterization, their potentiality in clinical-diagnostic terms and the possible impact on patient outcomes. Developing technologies for isolationg and analysing CTCs and ctDNA has allowed us to deepen our knowledge of tumor biology and outline possible applications in the diagnostic field. Early cancer diagnosis and early identification of progression represent two very ambitious goals for liquid biopsy, even if the currently available technologies do not have the required sensitivity for such applications. To date, the major field of application for liquid biopsy is molecular analysis of CTC or ctDNA in the context of target therapy, especially when traditional biopsy is not feasible or informative.

Liquid Biopsy

Liquid biopsy refers to the possibility of obtaining tumor-derived material, such as cells, proteins, DNA, and RNA, from a body fluid, usually blood. In particular, two types of biological tumor material can be identified in blood: CTCs and ctDNA. CTCs are tumor cells that detach from tumor tissue and circulate through blood vessels. They may arise from the primary tumor and metastases; in some cases, they may represent metastatic precursors. CTCs are generally present in extremely low numbers in the blood of cancer patients. In addition, the isolation of these cells is complex and expensive. Therefore, the analysis of CTCs is not performed in clinical practice. Conversely, searching for circulating tumor DNA is widely used in the clinical field. However, it is essential to note that liquid biopsy does not require a cancer diagnosis, for which tissue biopsy is required. Instead, liquid biopsy can provide crucial information for prognostic or predictive purposes. CtDNA is released into the bloodstream from tumor cells and lysed CTCs. Once isolated from patients' blood, it can be used for molecular analysis. In particular, ctDNA analysis helps identify the possible presence of mutations in specific genes that may make cells sensitive to treatment with molecularly targeted drugs.

The diffusion of state-of-the-art sequencing technologies (Next-Generation Sequencing, NGS) and advanced computational methods have recently allowed a more detailed understanding of the tumor mutational profile by molecular analysis of CTCs and ctDNA.

There is no doubt that the technological advances will allow different clinical applications for CTCs and ctDNA, given their different biological significance in processes ranging from tumorigenesis to the most advanced stages of the disease.

Early studies on CTCs highlighted their negative prognostic significance in patients with metastatic breast, colon, and prostate cancer. Although of great therapeutic interest, the evidence deriving from these preliminary studies was indirect and did not sufficiently support the hypothesis that a more aggressive therapeutic regimen was more effective in the case of high levels of CTCs. More recent studies have evaluated the possibility that specific genetic mutations identified in CTCs or ctDNA could provide predictive indications on the efficacy of target therapy. For example, in the case of non-small-cell lung cancer, melanoma, and breast cancer, molecular analysis of EGFR, EML4-ALK, BRAF, and PIK3CA genes in patients with metastatic carcinoma is essential for targeted therapy and represents the most established diag-

M. Ciaccio (✉)
Department of Biomedicine, Neurosciences and Advanced Diagnostics, Institute of Clinical Biochemistry, Clinical Molecular Medicine and Clinical Laboratory Medicine, and Department of Laboratory Medicine, University Hospital "P. Giaccone", Palermo, Italy

Department of Laboratory Medicine, University Hospital "P. Giaccone", Palermo, Italy
e-mail: marcello.ciaccio@unipa.it

M. Ciaccio (ed.), *Clinical and Laboratory Medicine Textbook*, https://doi.org/10.1007/978-3-031-24958-7_34

nostic application for liquid biopsy. In all those cases where the traditional biopsy on a primary tumor is not feasible or informative, the molecular analysis conducted on CTC or ctDNA represents a valid and noninvasive alternative to characterize the tumor and select patients for targeted therapies.

Moreover, given the extreme dynamism with which the tumor evolves, it may develop resistance to targeted therapies by acquiring mutations that make the molecular target insensitive to the drug's action. In these cases, the possibility of longitudinal monitoring, i.e., repeating serial controls over time, represents an exciting opportunity to identify the appearance of resistance mutations early. Unlike traditional biopsy, liquid biopsy allows this approach due to the limited invasiveness of the diagnostic procedure, which involves a simple blood sample.

However, liquid biopsy may have the most significant impact in the early cancer detection and early identification of its progression. For example, detection of minimal residual disease after potentially curative therapy may lead to second-line therapies. In patients with localized tumors, evidence of CTC or ctDNA could help stratify the risk of recurrence to select patients who may benefit from adjuvant therapy effectively. This approach requires greater analytical sensitivity than is required by mutational analysis for targeted therapy, and most technology platforms available today cannot provide it. With the advent of technologies with increasingly higher analytical performance, it is conceivable that liquid biopsy could also be applied in screening high-risk subjects. Finally, the knowledge of CTCs and ctDNA can contribute to understanding the molecular mechanisms underlying tumor evolution and its molecular heterogeneity, as well as identifying molecular targets for new drugs.

Circulating Tumor Cells

CTCs represent a biological material of great relevance because they offer, on the one hand, the possibility to understand the biological mechanisms underlying the metastatic spread of the primary tumor and, on the other hand, to evaluate the response to primary treatment and to early recognize the progression.

CTCs are released into the circulation from both the primary tumor and metastases. How tumor cells are released into the circulation is not well understood. Active mechanisms, in which cells with more significant metastatic potential, such as those that have undergone epithelial–mesenchymal transition, and passive diffusion mechanisms of single cells or cell clusters may be involved. In the circulation, CTCs can be found either as isolated cells or in large clusters of more than 50 cells. Once in circulation, CTCs have a short half-life. In patients with localized cancer undergoing surgical therapy, CTCs disappear approximately 24 h after complete resection of the tumor. The morphology of CTCs is highly heterogeneous. They may be damaged, have an apoptotic phenotype, or a morphology similar to cells in the corresponding tissue biopsy. The proliferative power of CTCs is extremely variable, even within the same patient.

Technologies for the Isolation of CTCs

The main limitation in isolating CTCs is that they are extremely rare compared to the total number of blood cells in circulation.

In most patients with tumor, the concentration of CTCs is around less than 10 cells/mL, and only in a few selected cases is it possible to find several hundred of CTCs per mL of blood.

In recent years, several technologies have been developed to isolate CTCs. Among the most important technical aspects, in addition to the already mentioned need to select cells at very low concentrations, we must also emphasize the need to ensure the lowest possible degree of molecular or morphological alterations for subsequent characterization. Moreover, keeping the degree of contamination by blood cells low is desirable.

CTC isolation strategies can be classified generically based on cell selection mode. In particular, we distinguish:

- Methods based on the physical properties of CTCs (e.g., cell volume)
- Methods based on immunophenotyping (expression of specific immunological biomarkers)
- Methods based on removing leukocytes without specific molecular tags.

Methods Based on the CTCs Physical Properties

Tumor cells are often larger than leukocytes (mean diameter 15 μm vs. 10 μm). Based on this observation, filtration-based isolation methods have been developed. The platforms supporting these methods have the advantage of being easy to use, but with a reduced quality of the isolated cells, since the passage through a static filter generates hemodynamic stress that can compromise the integrity of CTCs. This approach is also limited by the extreme variability of tumor cell volume, sometimes even smaller than that of leukocytes, and by the fact that patients undergoing chemotherapy may present bone marrow megakaryocytes in the circulation. Other isolation methods based on the physical properties of CTCs exploit their different density, electrical charge, or photoacoustic resonance.

Immuno-phenotyping Methods

These CTC isolation methods are based on antibodies targeted to antigens expressed explicitly by tumor cells. This approach is among the most widespread and is exploited by some instrumental platforms, among which the best known is CellSearch®. The latter evaluates the contextual expression of several markers: a membrane marker, the epithelial cell adhesion molecule (EpCAM), and cytoplasmic markers, epithelial cytokeratins 8, 18, and 19 (CK). Based on this technology, CD45+ leukocytes are negatively selected and excluded from the analysis, while CTC nuclei are evaluated by DAPI staining. Briefly, after immunomagnetic capture by EpCAM recognition and enrichment, reagents are added to identify and count CTCs. Ultimately, the system recognizes as CTCs those cells whose characteristics correspond to those of a tumor cell, i.e., cells with EpCAM+, CK+, DAPI+, and CD45-phenotype.

Methods Based on the Leukocytes Removal

Among the new microfluidic CTC isolation strategies, the most promising is the removal of leukocytes from the whole blood sample, leaving unlabeled CTCs for subsequent molecular characterization. This technology's biological rationale is that leukocyte markers are well characterized while tumor cells can express different markers, even in the same patient. In addition, non-epithelial tumors, such as melanoma, do not express EpCAM, while others undergo epithelial–mesenchymal transition, which results in the loss of EpCAM expression and other epithelial markers. In the so-called CTC-iChip, an integrated microfluidic device, magnetophoresis first separates the nucleated cells from the other blood components (platelets, erythrocytes, plasma). In the second step, the leukocytes are discarded by immunological marking. With this technology, the yield, i.e., the recovery of CTCs, is 97%, starting from a sample of 8 mL of whole blood. Moreover, the methodology is compatible with the standard cytopathological and molecular analyses foreseen for CTCs.

Molecular Characterization of CTCs

Once isolated, CTCs should be counted and characterized. The classic characterization criterion is positivity for the cytokeratin marker and negativity for the leukocyte antigen CD45. Characterization of CTCs is performed by fluorescence microscopy, although not all platforms are equipped with a sufficient number of channels for staining with multiple markers.

Although early studies of CTCs have suggested that their absolute value is associated with poorer survival, the kinetics of CTCs assessed post-therapy are of great clinical significance. Indeed, it has been shown that CTCs decrease rapidly in patients with a good therapeutic response. Interestingly, while CTC count in individual patients with advanced tumors correlates with clinical response to treatment, it does not correlate with tumor mass as assessed both morphologically, by imaging techniques, and biochemically by circulating tumor markers.

Molecular characterization of CTCs can also be performed by evaluating the expression of specific RNAs. Preliminary studies have demonstrated the efficacy of the analysis of specific translocations (e.g., EML4-ALK in non-small-cell lung cancer and TMPRSS2-ERG in prostate cancer) on CTCs.

Finally, the recent introduction of NGS techniques has allowed an increasing characterization of CTCs, opening exciting scenarios on tumor biology and the diagnostic potential of these methods.

Circulating Tumor DNA

The presence of cell-free DNA in circulation has been widely documented. All living cells release discrete amounts of free DNA into circulation. Following normal cell turnover, cells in apoptosis release DNA fragments into the circulation. Under physiological conditions, apoptotic and necrotic cells are rapidly removed, and cfDNA levels are relatively low. More than 90% of healthy individuals have less than 25 ng of cfDNA per mL. Under certain pathological conditions, including inflammation, exercise, tissue injury, or surgery, cfDNA levels can increase significantly. Generally, cfDNA levels can increase up to magnitude even in patients with tumors. More specifically, ctDNA refers to the proportion of free DNA released from cancer cells. Thus, in patients with cancer, a fraction of cfDNA derives from the tumor and is referred to as ctDNA. The ctDNA can originate from primary tumors, metastatic lesions, or CTCs. The fraction of ctDNA relative to the total circulating free DNA can range from <0.1% to >10%. The mechanisms of ctDNA release have not yet been elucidated. Proposed hypotheses include passive mechanisms, such as cell necrosis, and controlled release mechanisms, such as apoptosis. The hypothesis that ctDNA may result from cell necrosis is supported by the observation that more advanced stage tumors, characterized by a higher degree of necrosis, are associated with higher levels of ctDNA. The hypothesis that ctDNA is released following apoptosis is supported by its high degree of fragmentation. Indeed, ctDNA is typically fragmented into segments of the same length, 160–180 bp, as the nucleosome-protected DNA observed in apoptotic cells. CtDNA can also be released within exosomes, although this mode appears more appropriate for low-molecular-weight nucleic acids such as miRNAs (Fig. 34.1).

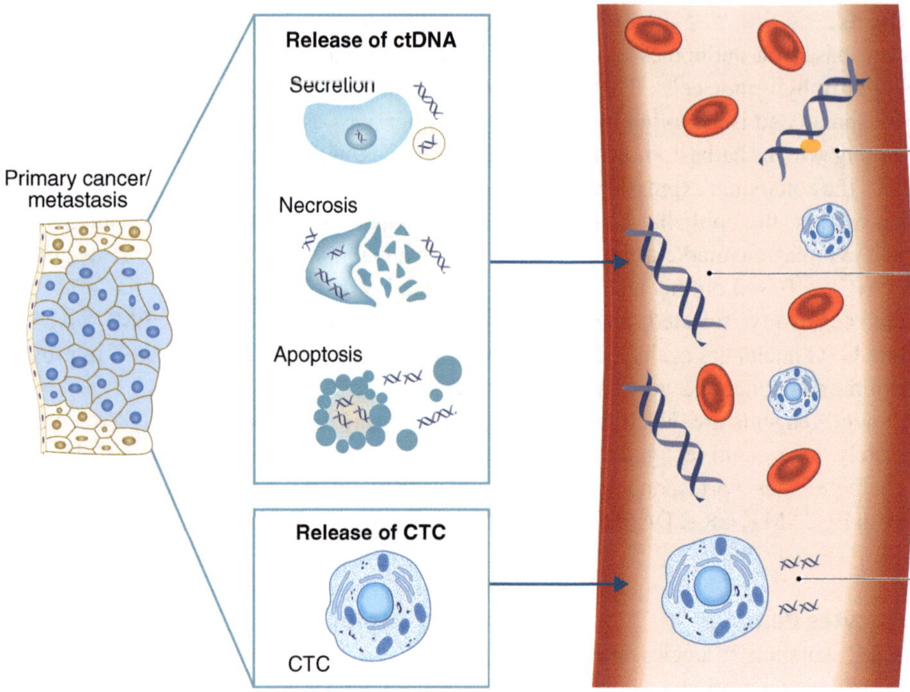

Fig. 34.1 Primary tumor and/or metastasis release ctDNA into the circulation by secretion, necrosis, and apoptosis. Circulating tumor cells (CTCs), another source of ctDNA, are also released. The ctDNA has the same alterations as the tumor cell of origin (point mutations, large rearrangements, insertions, deletions) and can be subjected to molecular analysis. (Copyright EDISES 2021. Reproduced with permission)

The clearance of ctDNA occurs in the spleen, kidney, and liver. Its half-life is very short, 16 min according to some Authors, about 1 h for the rapid phase, and 13 h for the late phase, according to others. In any case, the half-life is much shorter than many of the protein markers.

The ctDNA can provide quantitative information by measuring the number of copies in the circulation, and qualitative information, by searching for tumor-specific mutations. In the first case, given the extreme interindividual variability of ctDNA levels, it would seem to be the kinetics of ctDNA, and therefore its variations over time, rather than its absolute value, to provide clinically useful information about the effectiveness of therapy or the onset of relapse. In the second case, i.e., ctDNA mutational analysis, diagnostic applications are based on the fact that mutations in ctDNA correspond exactly to mutations in the primary tumor, including both point mutations and copy number variations and rearrangements.

Technologies for ctDNA Isolation

Theoretically, since the tumor-specific mutations sought in ctDNA are not present in normal cells, it could represent an excellent opportunity to identify the presence of a tumor mass noninvasively. However, ctDNA analysis to date is not helpful for diagnosis, detection of minimal residual disease, or prognostication because the sensitivity of currently available methods is often insufficient to identify ctDNA molecules, which in some cases represent little more than 0.01% of total cfDNA.

Isolation of ctDNA is far more feasible than that of CTCs; however, two factors make manipulation difficult:

- The extreme interindividual variability in ctDNA concentration;
- The "dilution" effect of circulating ctDNA because a large amount of cfDNA from healthy cells is present

The most variable stage is preanalytical, particularly sample collection, handling, and storage. It is preferable to use EDTA-plasma rather than serum since coagulation may result in the breakdown of leukocytes with subsequent release of wild-type DNA and further dilution of ctDNA. Commercially available methods typically require 1.5–2 mL of plasma. The stability of ctDNA is limited by the presence of circulating DNA activity, so sample processing should occur within hours of collection. Subsequent quantification is also not without criticality. In addition, other non-oncologic diseases may affect cfDNA levels.

ctDNA Mutational Analysis

The mutational analysis of ctDNA can be carried out basically by two approaches: a targeted one, i.e., aimed at finding known mutations that fall in mutational hot spots of specific genes known to be associated with cancer (e.g., KRAS, EGFR, and BRAF in lung cancer, colon, and melanoma, respectively), and a non-targeted one, based on NGS techniques. In the latter case, knowledge of specific tumor-associated mutations is not required However, any molecular alteration in the ctDNA is identified within the limits of the analytical sensitivity of the method. Among the disadvantages of massive sequencing methods are the limited analytical sensitivity, the still high costs, and the difficulties in interpreting the results for clinical purposes.

Initially, the analysis of known point mutations of ctDNA (targeted approach) was performed by end-point Polymerase Chain Reaction (PCR) and real-time PCR. More recently, digital-PCR, ARMS-PCR, or BEAMing technology have been developed to achieve higher analytical sensitivity. Most of these methods, however, allow the analysis of only a few loci; mutations in genes that do not present mutational hot spots, such as oncosuppressors, are not detected.

Application of Liquid Biopsy to Non-small-Cell Lung Cancer

Lung cancer is characterized by several driver mutations, the best known of which are EGFR point mutations, ALK-EML4 translocations, and RAS mutations. There are now targeted drugs, Tyrosine Kinase Inhibitors (TKIs), which act only if the tumor has some of these mutations (target therapy). These therapies have doubled patient survival. However, after treatment with these drugs, it is possible to observe a tumor progression caused by developing resistance mechanisms that make these therapies ineffective. In some cases, the molecular mechanisms of resistance have been identified and include specific mutations in the molecular targets of the drug itself. This is the case of EGFR in Non-small-Cell Lung Cancer (NSCLC). It has been shown that the molecular alterations of these tumors are not homogeneously distributed (tumor heterogeneity) and that metastases may present a mutational profile completely different from that of the primary tumor, making molecular analysis crucial when tumor progression is observed. Until now, tissue biopsy is considered the gold standard for molecular investigations but is limited by some factors, including its invasiveness, the impossibility of accessing some sites, tumor heterogeneity, and reduced patient compliance. In contrast, searching for specific mutations of the EGFR gene on plasma (ctDNA) is a noninvasive investigation that allows longitudinal monitoring by serial sampling over time. For these reasons, liquid biopsy is a valuable tool for selecting those patients with NSCLC who could benefit from treatment with TKIs.

The EGFR gene encodes for a transmembrane protein with an extracellular ligand-binding domain (EGF, TGFα), a single transmembrane hydrophobic domain, and an intracellular domain with tyrosine kinase activity. The activation of the receptor is followed by dimerization, autophosphorylation, and subsequent activation of several signal transduction pathways that regulate the processes of cell survival and/or apoptosis, angiogenesis (PIK3-AKT, JAK-STAT), cell differentiation and migration (RAS-RAF-MEK). The mechanism of action of TKI drugs involves their competition with ATP for binding to the receptor and blocking signal transduction. However, in unselected patients, the response to TKIs is limited to 10% of cases. These therapies' success depends on activating mutations, including exon 19 deletions (Ex19Del) and L858R, which account for 85% of EGFR mutations. These mutations are considered oncogenic. They constitutively activate the EGFR receptor without ligand, supporting the survival and antiapoptotic signal via PI3K-AKT and ERK-MAPK.

These mutations, in addition to being "activating" against EGFR, make the cell sensitive to TKIs because they increase the affinity of EGFR for the drug by 5–10 times compared to the wild-type receptor. This allows the inhibition of EGFR without toxicity at an epidermal and gastrointestinal level due to the blockade of EGFR in healthy cells. As already mentioned, using TKIs over time leads to the establishment of resistance in 60% of treated patients. From a molecular point of view, a possible cause is the occurrence of some mutations in EGFR, among which the best known is T790M. This mutation alters the ATP binding site to which TKIs also bind, thus canceling the sensitization induced by the other activating mutations. In addition to the mutations mentioned above, dozens of other mutations in the EGFR gene have been described whose effect on the activation status of the EGFR receptor and the action of TKIs has not yet been fully elucidated. These are uncommon and often associated with more frequent mutations, such as L858R or exon 19 deletions (Table 34.1).

In 2014, the European Medicine Agency (EMA) approved the use of plasma to assess EGFR mutational status in patient candidates for TKIs treatment. Patients with locally advanced or metastatic NSCLC and positive for L858R activating mutations or exon 19 deletions on plasma are eligible for treatment with TKIs. Patients with TKI-treated NSCLC who

Table 34.1 Mutations affecting *EGFR* and indications for the TKI use

Mutation	Exon	Indications for the TKIs use
L858R	21	Yes
Ex19Del	19	Yes
T790M	20	No
G719X	18	Yes?
Ex20Ins	20	No?
S768I	20	No?
L861Q	21	No?

TKI Tyrosine Kinase Inhibitors

have the T790M mutation on plasma are eligible for treatment with new third-generation TKIs, which have been shown to be effective in progression after treatment with first- and second-line TKIs caused by the occurrence of the EGFR T790M mutation.

Application of Liquid Biopsy to Other Neoplasms

The study of CTCs and ctDNA has broadened our knowledge of tumor biology, opening new perspectives on the potential diagnostic applications that can be derived from them. Among the most promising examples of the clinical-assistance implications are breast, ovarian, prostate, colorectal, hepatocarcinoma, and melanoma.

In breast cancer, it remains essential to identify those patients undergoing surgery who may benefit from adjuvant therapy. In these cases, identifying ctDNA or CTC after surgery could be decisive in planning appropriate follow-up programs. In particular, PIK3CA mutations in plasma before and after surgery have been associated with minimal residual disease. Moreover, the presence of CTC after the start of treatment correlates significantly with a lower PFS (Progression Free Survival) and OS (Overall Survival) and, therefore, with a worse prognosis.

Recommended Readings

Alix-Panabieres C, Pantel K (2013) Circulating tumor cells: liquid biopsy of cancer. Clin Chem 59:110–118

Ashworth TR (1869) A case of cancer in which cells similar to those in the tumors were seen in the blood after death. Aust Med J 14:146–149

Attard G, Swennenhuis JF, Olmos D, Reid AH, Vickers E, A'Hern R et al (2009) Characterization of ERG, AR and PTEN gene status in circulating tumor cells from patients with castration-resistant prostate cancer. Cancer Res 69:2912–2918

Bettegowda C, Sausen M, Leary RJ, Kinde I, Want Y, Agrawa N et al (2014) Detection of circulating tumor DNA in early- and late-stage human malignancies. Sci Transl Med 6:224ra24

Bidard FC, Peeters DJ, Fehm T et al (2014) Clinical validity of circulating tumour cells in patients with metastatic breast cancer: a pooled analysis of individual patient data. Lancet Oncol 15(4):406–414

Cho EH, Wendel M, Luttgen M, Yoshioka C, Marrinucci D, Lazar D et al (2012) Characterization of circulating tumor cell aggregates identified in patients with epithelial tumors. Phys Biol 9:016001

Cohen SJ, Alpaugh RK, Gross S, O'Hara SM, Smirnov DA, Terstappen LW et al (2006) Isolation and characterization of circulating tumor cells in patients with metastatic colorectal cancer. Clin Colorectal Cancer 6:125–129

Cristofanilli M, Budd GT, Ellis MJ, Stopeck A, Matera J, Miller MC et al (2004) Circulating tumor cells, disease progression, and survival in metastatic breast cancer. N Engl J Med 351:781–791

Danila DC, Heller G, Gignac GA, Gonzalez-Espinoza R, Anand A, Tanaka E et al (2007) Circulating tumor cell number and prognosis in progressive castration-resistant prostate cancer. Clin Cancer Res 13:7053–7058

Dawson SJ, Tsui DW, Murtaza M, Biggs H, Rueda OM, Chin SF et al (2013) Analysis of circulating tumor DNA to monitor metastatic breast cancer. N Engl J Med 368:1199–1209

Diehl F, Schmidt K, Choti MA, Romans K, Goodman S, Li M et al (2008) Circulating mutant DNA to assess tumor dynamics. Nat Med 14:985–990

Dressman D, Yan H, Traverso G, Kinzler KW, Vogelstein B (2003) Trans- forming single DNA molecules into fluorescent magnetic particles for detection and enumeration of genetic variations. Proc Natl Acad Sci U S A 100:8817–8822

Engelman JA, Settleman J (2008) Acquired resistance to tyrosine kinase inhibitors during cancer therapy. Curr Opin Genet Dev 18:73–79

Fan HC, Blumenfeld YJ, Chitkara U, Hudgins L, Quake SR (2010) Analysis of the size distributions of fetal and maternal cell-free DNA by paired-end sequencing. Clin Chem 56:1279–1286

Gascoyne PR, Noshari J, Anderson TJ, Becker FF (2009) Isolation of rare cells from cell mixtures by dielectrophoresis. Electrophoresis 30:1388–1398

Gertler R, Rosenberg R, Fuehrer K, Dahm M, Nekarda H, Siewert JR (2003) Detection of circulating tumor cells in blood using an optimized density gradient centrifugation. Recent Results Cancer Res 162:149–155

Karabacak NM, Spuhler PS, Fachin F et al (2014) Microfluidic, marker-free isolation of circulating tumor cells from blood samples. Nat Protoc 9(3):694–710

Leon SA, Shapiro B, Sklaroff DM, Yaros MJ (1977) Free DNA in the serum of cancer patients and the effect of therapy. Cancer Res 37:646–650

Liotta LA, Saidel MG, Kleinerman J (1976) The significance of hematogenous tumor cell clumps in the metastatic process. Cancer Res 36:889–894

Maheswaran S, Sequist LV, Nagrath S, Ulkus L, Brannigan B, Collura CV et al (2008) Detection of mutations in EGFR in circulating lung-cancer cells. N Engl J Med 359:366–377

Mandel P, Metais P (1948) Les acides nucleiques du plasma sanguin chez l'homme. C R Seances Soc Biol Fil 142:241–243

Ozkumur E, Shah AM, Ciciliano JC, Emmink BL, Miyamoto DT, Brachtel E et al (2013) Inertial focusing for tumor antigen-dependent and -independent sorting of rare circulating tumor cells. Sci Transl Med 5:179ra47

Perkins G, Yap TA, Pope L, Cassidy AM, Dukes JP, Riisnaes R et al (2012) Multi-purpose utility of circulating plasma DNA testing in patients with advanced cancers. PLoS One 7:e47020

Pilotto S, Rossi A, VavalaÃÄ T et al (2018) Outcomes of first-generation EGFR-TKIs against non-small-cell lung cancer Harboring uncommon EGFR mutations: a post hoc analysis of the BE-POSITIVE study. Clin Lung Cancer 19(1):93–104

Qian H, Gao F, Wang H, Ma F (2014) The efficacy and safety of crizotinib in the treatment of anaplastic lymphoma kinase-positive non-small cell lung cancer: a meta-analysis of clinical trials. BMC Cancer 19(14):683

Reck M, van Zandwijk N, Gridelli C et al (2010) Erlotinib in advanced non-small cell lung cancer: efficacy and safety findings of the global phase IV Tarceva lung cancer survival treatment study. J Thorac Oncol 5(10):1616–1622

Siravegna G, Bardelli A (2014) Genotyping cell-free tumor DNA in the blood to detect residual disease and drug resistance. Genome Biol 15(8):449

Stroun M, Anker P, Lyautey J, Lederrey C, Maurice PA (1987) Isolation and characterization of DNA from the plasma of cancer patients. Eur J Cancer Clin Oncol 23:707–712

Stroun M, Lyautey J, Lederrey C, Olson-Sand A, Anker P (2001) About the possible origin and mechanism of circulating DNA apoptosis and active DNA release. Clin Chim Acta 313:139–142

Stott SL, Lee RJ, Nagrath S, Yu M, Miyamoto DT, Ulkus L et al (2010) Isolation and characterization of circulating tumor cells from localized and metastatic prostate cancer patients. Sci Transl Med 2:25ra23

Tu HY, Ke EE, Yang JJ et al (2017 Dec) A comprehensive review of uncommon EGFR mutations in patients with non-small cell lung cancer. Lung Cancer 114:96–102

Vona G, Sabile A, Louha M, Sitruk V, Romana S, Schutze K et al (2000) Isolation by size of epithelial tumor cells: a new method for the immunomorphological and molecular characterization of circulating tumor cells. Am J Pathol 156:57–63

Yamada T, Nakamori S, Ohzato H, Oshima S, Aoki T, Higaki N et al (1998) Detection of K-ras gene mutations in plasma DNA of patients with pancreatic adeno- carcinoma: correlation with clinicopathological features. Clin Cancer Res 4:1527–1529

Yu M, Stott S, Toner M, Maheswaran S, Haber DA (2011) Circulating tumor cells: approaches to isolation and characterization. J Cell Biol 192:373–382

Wang JY, Hsieh JS, Chang MY, Huang TJ, Chen FM, Cheng TL et al (2004) Molecular detection of APC, K-ras, and p53 mutations in the serum of colorectal cancer patients as circulating biomarkers. World J Surg 28:721–726

Weight RM, Dale PS, Viator JA (2009) Detection of circulating melanoma cells in human blood using photoacoustic flowmetry. Conf Proc IEEE Eng Med Biol Soc 2009:106–109

Biological Fluids

35

Gaetato Bernardi, Cosimo Ottomano, and Sabrina Buoro

Introduction

Unlike blood, biological fluids are not always repeatable materials that provide clinical information otherwise unavailable to the caregiver. Despite their importance, their diagnostic power is not always known, nor are the pre-analytical precautions that should be part of the cultural baggage of anyone requesting a laboratory test. Biological fluids are divided into serous (peritoneal, pleural, pericardial), synovial, and cephalorachidian. For each of these, the biochemistry and cellular analysis tests of major interest in nonspecialist medicine are briefly described.

This chapter only touches on topics such as analytics, which are fundamental for specialists in laboratory medicine. However, it is worth mentioning here a fundamental concept in clinical biochemistry: the matrix. This is different between blood and biological fluids; for this reason, the determination of a measurand in the former creates different problems in the latter, forcing the specialist to verify the reliability of analytical methods in each area.

Reference values (RV) are critical to classify the individual health status. Unfortunately, the literature is not completely univocal in this regard, so it is highly advisable to always refer to the RV of the laboratory.

Pathophysiology

The serous (pericardial, peritoneal, and pleural) and articular cavities physiologically contain minimal quantities of fluid, similar to plasma in color and chemical-physical characteristics, whose function is essentially lubricating, favoring the endocavitary sliding of the viscera, covered by a serous membrane like the cavities that contain them.

The serous liquids are continuously formed by ultrafiltration of the plasma and, in part, are reabsorbed through the capillaries of the serosae, so that the quantity and composition remain constant. Liquids result from a dynamic process of formation and reabsorption in a game of pushes and counter-pushes of hydrostatic and colloid-osmotic pressure. In addition to this mechanism, there is also capillary permeability and lymphatic drainage.

Excess production or reduced reabsorption results in disease. These phenomena are pathognomonic of each cavity (pleural, pericardial, peritoneal, and articular).

In pathological conditions, for inflammatory, neoplastic, or traumatic lesions of the endocavitary organs, sometimes for an inflammatory or neoplastic pathology of the serosae themselves, the amount of intracavitary fluid increases, resulting in an effusion whose chemical-physical and cytological characteristics are modified according to the pathogenic noxa, with common etiopathogenic features.

It is good practice to obtain the concentrations of the measurands useful for diagnosis in the affected fluid and on the serum/plasma at the same time.

G. Bernardi (✉)
IRCSS Fondazione Carlo Besta Neurological Institute, Milan, Italy

C. Ottomano
Research Laboratory, Istituto di Ricerca e Cura a Carattere Scientifico (IRCCS) Synlab SDN Spa, Naples, Italy

S. Buoro
Regional Reference Center for the Quality of Laboratory Medicine Services, Milan, Italy

M. Ciaccio (ed.), *Clinical and Laboratory Medicine Textbook*, https://doi.org/10.1007/978-3-031-24958-7_35

Types of Analysis

The main laboratory tests useful for assessing any alterations in the endocavitary organs are macroscopic analysis and cytometric and biochemical examination.

Macroscopic Analysis

The color and appearance of fluids in pathologic effusions may vary based on their different etiologies. The appearance is usually clear and straw yellow in transudates.

Other colorations of pathological effusions include red, brown, green, white, and black. The presence of blood gives the fluid a red coloration, erythrocyte lysis, and oxidation of hemoglobin to methemoglobin result in a brownish (brown) coloration. If the fluid has a frankly hematic appearance, the determination of the hematocrit, or rather the count of the hematocrits, can help to distinguish between massive bleeding and serous effusion. A greenish coloration indicates the presence of bile. Fluids with a whitish coloration indicate pus or chyle; in subjects with melanoma, a blackish coloration may be observed. The ascitic fluids of jaundiced patients assume a particularly pronounced yellow coloration.

The appearance, in addition to being clear, maybe turbid or opalescent. The turbid appearance of the fluids is caused by increased cellularity or triglyceride concentration, while the chylous effusion may have an opalescent appearance.

Cytometric Examination

The cytometric examination consists in counting the total nucleated elements in the serous fluids and their differentiation according to morphological characteristics.

For cytometric examination of liquids, with the exception of CSF, the sample must be collected in tubes with anticoagulant (EDTA), mixed both after collection and before analysis. The liquid should be stored at room temperature, and the maximum time that can elapse between collection and analysis is about 2 hours.

In synovial fluid, the presence of hyaluronic acid, at a strongly acidic pH, causes the formation of a mucin clot that invalidates cell counts. The pre-treatment of the sample with hyaluronidase and the execution of the analysis at slightly acidic pH, or preferably neutral, improve the analysis.

The gold standard procedure for cell counting in biological fluids is light microscopy using a counting chamber (Burker or Thoma chambers for fluids of suspected high cellularity, or Fuchs-Rosenthal and Nageotte for low cellularity samples).

Differential cell counting is performed by light microscopy on cytocentrifugate, after panoptic staining (May-Grünwald Giemsa MGG or Wright). The cytocentrifugation allows the concentration of the cells, dispersed in the liquid, in a small circular area of the slide, where they appear in monolayer.

Each individual cell population is expressed as an absolute value and/or as a percentage, depending on the type of liquid examined and the decision thresholds.

Cytometric analysis can also be performed by automatic analyzers, reducing costs and reporting times. The dispersion of the replicates of the measurements (imprecision) also improves.

The cellular analysis of biological fluids represents one of the main diagnostic criteria in the evaluation of the effusion etiopathogenesis.

The cellular component of the fluids is generally made up of leukocytes (neutrophil granulocytes, lymphocytes, monocytes, more rarely eosinophilic and basophilic granulocytes), macrophages/histiocytes, mesothelial cells in cavity effusions, and synoviocytes in joint effusions (Figs. 35.1, 35.2, 35.3, and 35.4).

Mesothelial cells are large cells with basophilic cytoplasm on MGG staining (the intensity of basophilia varies according to the state of the cell), regular, slightly eccentric nucleus; sometimes it is possible to see the nucleolus. However, the morphology of mesothelial cells in effusions is varied and is closely influenced by the time of onset of

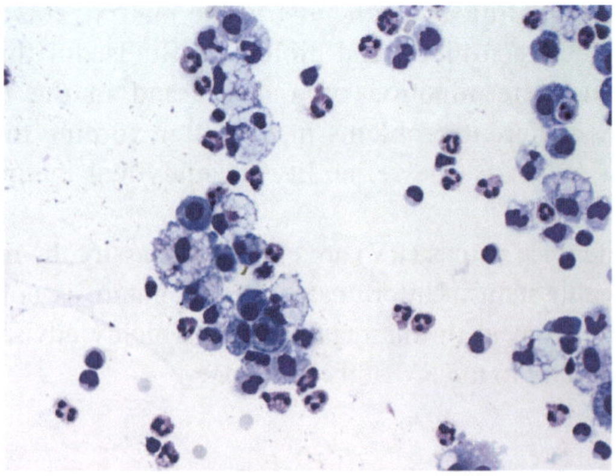

Fig. 35.1 Pleural effusion of an exudative nature in cytospin preparation with May-Grünwald Giemsa stain (400×), where it is possible to recognize neutrophilic granulocytes, lymphocytes, and monocytes. In addition, well-preserved mesothelial cells are present (large cells with abundant cytoplasm, strongly basophilic, and, in some cases, with small pericytoplasmic vacuoles; the nucleus is regular, slightly eccentric, and sometimes characterized by the presence of a nucleolus), malignant mesothelial cells conserved with numerous vacuoles, and finally macrophages (large cells, cytoplasm with many vacuoles, eccentric kidney nucleus and strongly thickened chromatin). Sometimes the morphological distinction among mesothelial cells, macrophages, and monocytes can be difficult and is influenced by the duration of the effusion and/or the time that has elapsed from the time of sampling to the execution of the cytometric analysis

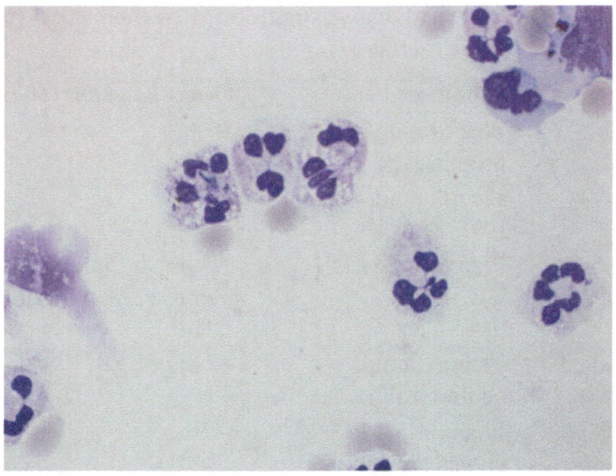

Fig. 35.2 Peritoneal effusion of a subject with cirrhosis and spontaneous bacterial peritonitis. Cytospin preparation with May-Grünwald Giemsa staining (1000×), where it is possible to recognize neutrophilic granulocytes, a lymphocyte in the upper right corner, while in the leftmost neutrophil granulocyte, in the central group of the image, the phagocytosis of some bacteria

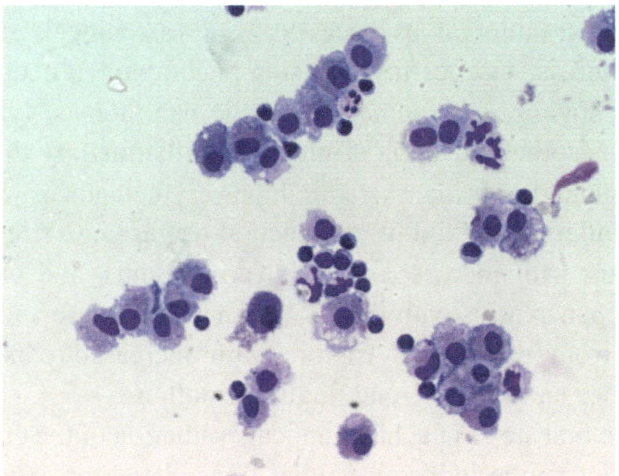

Fig. 35.3 Pericardial fluid from a healthy subject undergoing surgery. Cytospin preparation with May-Grünwald Giemsa staining (400×), where it is possible to recognize rare neutrophil granulocytes, lymphocytes, and monocytes. The predominant cell population is represented by mesothelial cells

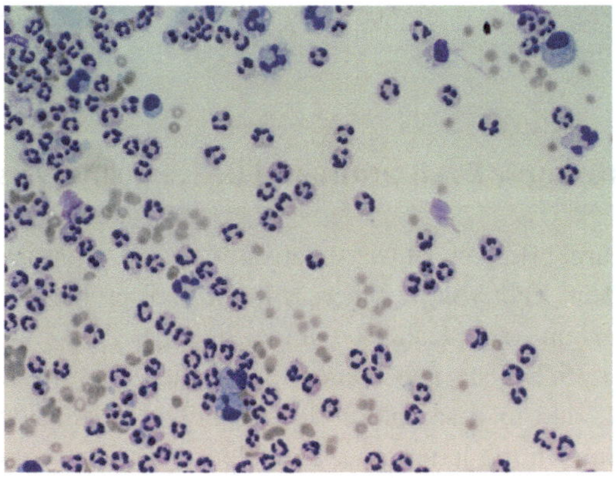

Fig. 35.4 Joint effusion of a subject with inflammation, characterized by a high number of well-preserved neutrophils and by some rare monocytes, synoviocytes with morphology similar to mesothelial cells. Cytospin preparation with May-Grünwald Giemsa staining (400×)

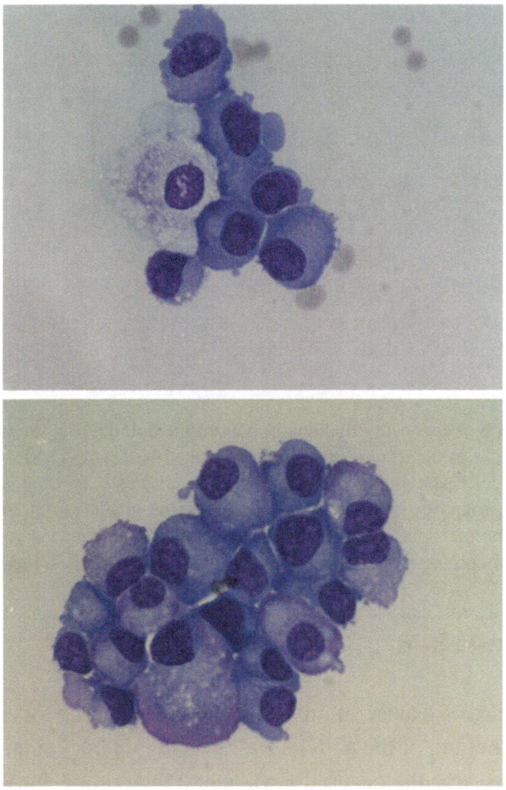

Fig. 35.5 Pleural effusion of a subject with multiple myeloma evolved into plasma cell leukemia. There are lymphoplasmacytic and plasma cells in the pleural effusion, as in the peripheral blood. The cells present in the pleural effusion are CD138, CD38 positive, and CD56 negative. Cytospin preparation with May-Grünwald Giemsa staining (1000×)

the effusion. It is possible to find mesothelial cells in mitosis or binucleated if the hyperplasia is benign. This heterogeneity sometimes makes it difficult to correctly identify mesothelial cells morphologically with respect to macrophages and neoplastic cells, so much so that cytochemical, immunocytochemical, or immunofluorescence techniques must be used. Finally, in the case of neoplastic effusions, malignant cells of various kinds are also present (Figs. 35.5 and 35.6).

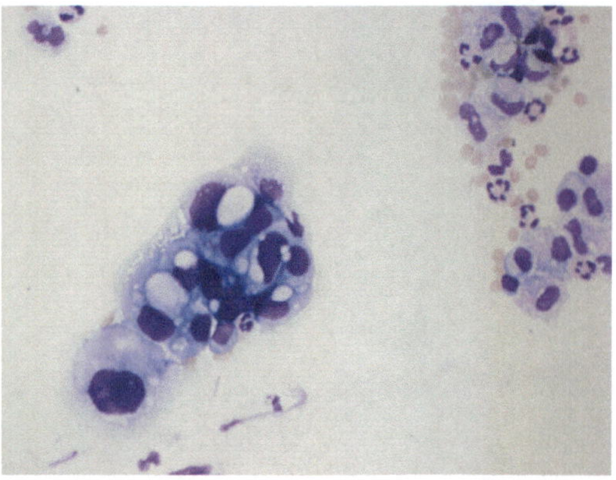

Fig. 35.6 Peritoneal effusion of undetermined diagnosis; the preparation contains neutrophil granulocytes, monocytes, lymphocytes, mesothelial cells, and atypical cells of probable neoplastic nature that make up the group of central cells. Subsequent cytological examination confirms the presence of undifferentiated ovarian cancer metastatic cells. Cytospin preparation with May-Grünwald Giemsa stain (400×)

Pleural Fluid

The pleural cavity is an almost virtual space between the mesothelium of the parietal and visceral pleura, containing a modest amount of fluid sufficient to facilitate movement between the two membranes.

Its increase, called pleural effusion, is due either to overproduction, mainly in congestive heart failure, or to reduced reabsorption, mostly from lymphatic obstruction during neoplasms. In summary, the most common conditions causing pleural effusion are heart failure, tumors, tuberculosis, and pulmonary embolism.

In the case of suspected pleural effusion, the patient's clinical picture can be correctly framed by radiological and/or ultrasound examinations to confirm its presence. In this case, thoracentesis is performed for both diagnostic and therapeutic purposes.

The distinction between exudate (increased capillary permeability or decreased lymphatic reabsorption), transudate (increased hydrostatic pressure or decreased oncotic pressure), and chylous effusion (caused by trauma or neoplasm, usually lymphoma involving the thoracic duct) is a crucial diagnostic step.

Transudates are generally bilateral because they have a systemic etiology.

Exudates are frequently unilateral, associated with disorders that increase capillary permeability or interfere with ipsilateral lymphatic reabsorption.

Effusions from malignant neoplasms behave ambiguously: unlike what we might expect, they are unlikely to be exudative due to the concomitant presence of congestive heart failure.

Table 35.1 Biochemical analysis performed on the pleural fluid and related interpretative criteria

	Parameter	Diagnostic *cutoff* for exudate
Light's criteria	Pleural fluid/serum protein ratio	≥0.50
	Pleural fluid/serum LDH ratio	≥0.60
	Pleural fluid LDH	>2/3 upper limit of the reference range of serum LDH
Additional criteria	Pleural liquid cholesterol	>45 mg/dL
	Pleural liquid cholesterol/serum ratio	>0.30
	Pleural fluid/serum bilirubin ratio	≥0.60

The main causes of a transudate are congestive heart failure, liver cirrhosis, hypoproteinemia (e.g., nephrotic syndrome), infections (e.g., bacterial pneumonia, tuberculosis, or other granulomatous diseases, such as sarcoidosis and histoplasmosis; viral or mycoplasma pneumonias), neoplasms (bronchogenic carcinoma, metastatic carcinomas, lymphomas, mesotheliomas), non-infectious inflammatory diseases (rheumatoid arthritis, systemic lupus erythematosus [SLE]), accumulation of fluid of extrapleural origin (e.g., from pancreatitis with increased lipid-lowering lipid), and exudate (from pancreatitis with increased pancreatic lipase and amylase, from esophageal rupture with increased pancreatic enzymes and acid pH, and finally, urinothorax).

The first step in the biochemical evaluation of the effusion is to determine its origin, either transudative or exudative. For this purpose, Light's criteria have been widely used in clinical practice over the past four decades (Table 35.1).

In their initial work, Light et al. reported a sensitivity of 99% and a specificity of 98% in identifying the type of effusion, although in subsequent studies the specificity was lower (65–86%). Despite this limitation, Light's criteria are considered superior to clinical judgment alone in characterizing effusions.

Macroscopic Evaluation and Cellular Analysis

In pleural effusions of transudative origin, the fluid is generally clear, light yellow, whereas in those of probable exudative origin (neoplastic or infected), the fluid is turbid (Table 35.2). The appearance of the fluid may be hemorrhagic, due to a sample collection not performed well or bleeding. In the latter case, the color of the fluid may vary from red to brown depending on the time elapsed between the start of the hemorrhage and the collection.

Table 35.2 Macroscopic and cytometric analysis of the pleural fluid and related interpretative criteria

Macroscopic examination	Cytometric examination
- *Transudate* = clear from pale to straw yellow - *Exudate* = from pale to straw yellow with a different degree of turbidity, sometimes milky; if the latescence persists after centrifugation, chylothorax occurs	The total cellularity is usually between 1395 and 3794 × 10^6/L, macrophages between 64% and 80%, lymphocytes between 18% and 36%, neutrophils between 0% and 1% and mesothelial cells between 0% and 2% In pleural effusions: - Total cellularity <1000 × 10^6/L is indicative of an effusion of probable transudative origin - Total cellularity ≥1000 × 10^6/L is indicative of an effusion of probable exudative origin. In this case a count of: - Neutrophils ≥50% is indicative of probable bacterial pneumonia, pulmonary infarction, pancreatitis, etc. - Lymphocytes ≥50% is indicative of probable tuberculous or viral infection, chylothorax, neoplasm, rheumatoid pleurisy, etc. - Eosinophils ≥10% is indicative of probable pneumothorax, trauma, pulmonary infarction, parasitic or fungal infections, drug reactions, rheumatic disorders, Hodgkin's lymphoma, and idiopathic eosinophilia *Comment*. About 80% of transudates have a cellularity <1000 × 10^6/L cells, while cellularity >10,000 × 10^6/L are usually associated with parapneumonic effusions. Finally, about 10% of transudates have a neutrophil count ≥50% and about 30% have a lymphocyte count ≥50%. Regardless of the number of cells found, it is the qualitative evaluation that allows their characterization in a neoplastic sense

In the case of neoplastic or chylous effusion, the fluid may be milky. Table 35.3 describes the differential characteristics of chylous and pseudochylous effusion.

Normally, the volume of fluid in the pleural cavity ranges between 4.1 and 12.7 mL, the total cellularity is between 1395 and 3794 × 10^6/L, macrophages between 64% and 80%, lymphocytes between 18% and 36%, neutrophils between 0% and 1%, and mesothelial cells between 0% and 2%.

In pleural effusions of exudative origin, on the other hand, the cell count is greater than 1000 × 10^6/L. In this condition, neutrophilia greater than 50% is indicative of an inflammatory process of bacterial etiology (e.g., parapneumonic effusion). A lymphocytosis higher than 50% is associated with tubercular infection, chylous effusions due to mechanical damage of the lymphatic system, and lymphoproliferative diseases. Lymphocyte characterization is especially useful in exudative effusions; in these cases, immuno-phenotyping can distinguish between benign and malignant lymphoproliferative diseases.

Eosinophilia (>10%) occurs in conditions as diverse as pneumothorax, pulmonary embolism, traumatic hemothorax, and Churg-Strauss syndrome.

The count of red blood cells distinguishes between serous effusion and hemothorax. The latter may be secondary to trauma, pulmonary embolism, and neoplasia.

Total cellularity is useful in the differential diagnosis between exudates and transudates: in the latter, approximately 80% have a cellularity of less than 1000 × 10^6/L cells; cellularities greater than 1000 × 10^6/L are usually associated with parapneumonic effusions.

Biochemistry

Light's criteria include assaying total protein and LDH on pleural fluid and serum to calculate the ratio. Table 35.1 summarizes the biochemical criteria for identifying an exudate with greater accuracy than using total protein alone.

The pH has a high positive prognostic value in non-pneumococcal pneumonia. The presence of creatinine or urea indicates the presence of urine in pleural fluid. C-reactive protein (CRP) of pleural fluid may be more sensitive than blood protein in diagnosing non-pneumococcal pneumonia.

The usefulness of assaying certain tumor-associated biological markers in pleural fluid is controversial. For example, the literature reports the case of a patient with a marked increase in prostate-specific antigen (PSA), which proved to be decisive in establishing the diagnosis of metastatic prostate cancer. The patient presented with severe anemia, peripheral edema, and pleural and pericardial effusions, but no neoplastic cells.

A milky pleural fluid suggests the presence of a chylothorax, which is always due to obstruction of the thoracic duct for mainly neoplastic or traumatic reasons. However, it is useful to distinguish this situation from a pseudochylothorax, which is usually due to rheumatoid pleurisy, tuberculosis, or myxedema. In these cases, triglyceride and chylomicron detection may be helpful; table 35.3 summarizes the criteria for the differential diagnosis between the two situations.

Table 35.3 Main characteristics of chylous and pseudochylous pleural effusions and related interpretative criteria

Characteristic	Chylous	Pseudochylous
Onset	Rapid	Gradual
Appearance	Milky or yellow tending to blood color	Milky or grayish or metallic
Microscopic exam	Lymphocytosis	Mixed cellularity and cholesterol crystals
Triglycerides	≥110 mg/dL	<50 mg/dL
Electrophoresis of lipoproteins	Chylomicrons present	Chylomicrons absent

Pericardial Fluid

In the healthy adult, pericardial fluid ranges from 10 to 50 mL, produced similarly as in the pleura.

Pericardial effusions are mainly due to phlogistic and/or neoplastic processes that compromise the integrity of the permeability of endothelial and mesothelial cells, and induce obstruction of the lymphatic pathways of the affected area.

Pericardial fluid is pale yellow and clear. Large effusions (>350 mL) are due to neoplastic infiltration (lymphoproliferative neoplasms or metastatic processes), uremia or microbial agents (bacterial, tubercular, or fungal infections), inflammatory processes, autoimmune diseases, cardiac infarction with mediastinal damage, and iatrogenic causes. The most frequent cause of idiopathic pericardial effusion is enterovirus infections. Patients with HIV infection often present with asymptomatic pericardial effusions that increase as the infectious disease progresses.

Pericardial fluid may be hematic due to incorrect sampling maneuver (with penetration of the heart chambers) or in the case of a hemorrhagic effusion, with correct sampling. In the latter case, the hematocrit or the number of red blood cells is lower than that of the blood. Moreover, in the first case the sample coagulates, in the second case this is very rare.

Post-pericardiotomy syndrome may occur in subjects undergoing cardiac surgery, who are still in hospital but sometimes already at home. It is characterized by fever, thoracic-pleural pain, and other signs of pleural, pericardial, and, less frequently, pulmonary inflammation. In more than 80% of cases, a pleural exudate develops, which is serous to frankly hematic, with pH > 7.4 and normal glucose levels. There are no specific laboratory tests except for detection of anti-myocardial antibodies and, serum C3, which may be decreased.

Table 35.4 shows the main laboratory tests performed in pericardial fluid.

Table 35.4 Brief description of the main laboratory tests on the pericardial fluid and related interpretative criteria

Macroscopic examination	Cytometric examination	Clinical biochemistry
- *Transudate* = clear from pale to straw yellow - *Exudate* = from pale to straw yellow with different degree of turbidity	Normally they are present from a minimum average value of 10×10^6/L cells up to a maximum of 1900–2210 × 10^6/L cells In effusions of an exudative nature, the average cellularity is generally ≥14.116 × 10^6/L In effusions of probable neoplastic origin, the average value is 3600 × 10^6/L *Comment.* In the above cases, the standard deviations detected are very large and sometimes overlap in the different categories of pericardial effusions Differential cell count has the following meaning: Neutrophils >70% are indicative of bacterial infection or rheumatic effusion Mononuclear cells/monocytes >75% are indicative of probable monocytosis secondary to hypothyroidism or malignant effusion *Comment.* Regardless of the number of cells found, the qualitative evaluation allows their neoplastic characterization	Diagnostic cutoff for exudate: - Total protein >3 g/dL and liquid/serum protein ratio >0.5 - LDH >200 U/L - Liquid/serum LDH ratio >0.6 Glucose <40 mg/dL is frequently found in bacterial, tuberculous, rheumatic, or neoplastic effusions. To date, there is insufficient evidence to indicate an appropriate use of other analytes in routine practice

Macroscopic Evaluation and Cellular Analysis

Pericardial fluid for diagnostic purposes is obtained by pericardiotomy or pericardiocentesis.

Normal pericardial fluid or fluid of exudative origin is generally clear yellow, whereas the fluid is turbid in exudative effusions, such as neoplastic or infected effusions. Finally, as described above, pericardial fluid may sometimes have a hemorrhagic appearance.

The mean value of cellularity can range from a minimum of 10 × 10^6/L to a maximum of 1900–2210 × 10^6/L cells. In contrast, the mean value ranges from 3600 × 10^6/L up to 14,116 × 10^6/L cells in pericardial effusions of probable neoplastic origin.

The effusions of bacterial or rheumatic etiology are generally characterized by the presence of about 70% of neutrophil granulocytes, while those secondary to hypothyroidism and/or neoplastic nature are characterized by 75% or more of monocytes or other mononuclear cells. Distinguishing activated mesothelial cells from neoplastic cells often requires further investigation, in addition to purely morphological ones.

Biochemistry

Light's criteria for distinguishing an exudate from a transudate also apply to pericardial fluid. However, the literature data do not show adequate specificity and sensitivity and, even if applied, have not been validated for this type of fluid; further studies, possibly multicentric and with large numbers, are necessary (Table 35.4). Glucose, total protein, pH, lipids, lactate dehydrogenase (LDH), and adenosine deaminase (ADA) should be assayed on this fluid.

A glucose concentration <40 mg/dL is frequent in bacterial, tubercular, rheumatic, or neoplastic effusions. Higher glucose values have less diagnostic value.

A protein concentration <3.0 g/dL has a sensitivity of 97% for an exudate, but a much lower specificity (around 22%), resulting in poor utility.

The pH of pericardial fluid can be markedly decreased (<7.1) in rheumatic and purulent pericarditis. Pathologies such as neoplasms, uremia, tuberculosis, and idiopathic forms show a moderate decrease in pH (7.2–7.3).

Triglyceride and cholesterol assays may be useful in distinguishing between chylous and pseudochylous effusions, along with lipoprotein electrophoresis, which is useful in checking for the presence of chylomicrons (Table 35.3). In pericardial fluid, LDH >200 U/L (i.e., pericardial fluid LDH/plasma LDH ratio >0.6) is characteristic of an exudate.

The ADA increases significantly in tubercular pericarditis, at a cutoff of 30 U/L the sensitivity exceeds 90%, while the specificity does not reach 70%. By raising the cutoff to 40 U/L, both sensitivity and specificity exceed 90%.

Ascitic Fluid or Peritoneal Effusion

The pathological accumulation of fluid in the peritoneal cavity is called ascites. The most common causes of ascites include cirrhosis of the liver, certain neoplasms, congestive heart failure, nephrotic syndrome, and pancreatic disease. The mechanism of ascitic fluid accumulation is controversial. Portal venous hypertension pushes fluid out of the vascular space, causing hepatic and interstitial edema. Both the liver and the intestine may, therefore, be the source of such accumulation. After the fluid enters the abdomen, it is reabsorbed by the lymphatic vessels. If the reabsorption capacity of the lymphatic system is exceeded, the fluid moves to the peritoneal space, causing ascites. History and physical examination provide clues to its possible etiology. Radiological and ultrasonographic evaluation can help detect small amounts of peritoneal fluid, and establish the possible etiology of ascites. Table 35.5 shows the main laboratory tests performed on this fluid.

Macroscopic Evaluation and Cellular Analysis

The ascitic fluid sample is usually obtained by paracentesis or drainage following paracentesis.

The appearance of the ascitic fluid is typical of cavitary fluid, while in case of jaundice it is generally clear and of an intense yellow color.

Table 35.5 Main laboratory test on ascitic fluid and related interpretative criteria

Macroscopic examination	Cytometric examination	Clinical biochemistry
- *Transudate* = clear yellow - *Exudate* = cloudy of variable color: - Yellow-green (bile contamination) - Iridescent green (acute pancreatitis) - Milky (chylous or pseudochylosis infections or effusions) In chylous effusions the milky appearance persists even after centrifugation	The diagnostic cutoffs are - Neutrophils >250 × 10^6/L, indicative of probable non perforated bacterial peritonitis - Total cellularity >1000 × 10^6/L with prevalence of lymphocytes, indicative of probable tuberculous peritonitis - Eosinophils >10%, indicative of probable chronic inflammation *Comment.* Regardless of the number of cells, the qualitative evaluation allows their neoplastic characterization	The albumin gradient values: - ≥11 g/L or 1.1 g/dL is indicative of portal hypertension ascites. This condition may include cirrhosis, fulminant hepatitis, vein-occlusive disease, hepatic vein obstruction (e.g., Budd-Chiari syndrome), congestive heart failure - <11 g/L or 1.1 g/dL is indicative of ascites from alterations of the peritoneum. This condition may include primary or secondary peritoneal carcinomatosis, tuberculous peritonitis, parasitic or fungal infection, pancreatitis

The reference range for cellularity of sterile ascitic fluid is 0 to 562 × 10^6/L. In sterile ascitic fluid, neutrophils have an average value of 27%. The absolute neutrophil count ranges from 0 to 2532 × 10^6/L with an average of 82 × 10^6/L cells.

To date, it is assumed that, in a peritoneal effusion, a neutrophil count <250 × 10^6/L cells is diagnostic of sterile ascites, whereas a value ≥250 × 10^6/L is diagnostic of spontaneous bacterial peritonitis. A total nucleated element count >1000 × 10^6/L cells, with lymphocytes predominating, is diagnostic of tuberculous peritonitis.

The count of nucleated elements, especially leukocytes, and their differentiation, has a substantial diagnostic value in the evaluation of various inflammatory conditions and is useful in the differential diagnosis and, therefore, in directing toward the appropriate therapy.

Biochemistry

The determination of the albumin gradient, i.e., the difference between the serum and ascitic fluid albumin allows to differentiate the etiology of the effusion (Table 35.6).

Fluids of biliary origin may be found in the peritoneal cavity during surgery on the biliary excretory tract or liver transplantation. A high bilirubin concentration in the abdominal/ascitic fluid is practically conclusive for damage to the biliary excretory tract, which almost always requires surgical correction.

Damage to the exocrine pancreas, from surgical maneuvers or medical causes, may result in pancreatic enzymes leaking into the peritoneum. The assay of pancreatic amylase, or lipase, in the fluid of peritoneal origin can confirm or exclude pancreatic damage, along with the presence or absence of an acute abdomen. If the same enzymes are also elevated in pleural fluid from the left lung, it can be concluded that pancreatic damage has also involved this organ.

Renal damage of various etiology (neoplastic, fibrosis, stone, surgical, traumatic) may result in urine leaking into the pleural and/or peritoneal cavity. Lithotripsy may involve uroperitoneum, as well as an unsuccessful external urinary diversion. Confirmation that the fluid in the two cavities is urine is obtained by measuring creatinine (2–10 mg/dL) and urea, which are significantly present. The creatinine to serum creatinine ratio will be greater than one. Finally, if fluid is inadvertently aspirated directly from the bladder, the urea/creatinine ratio should be higher than if the fluid were peritoneal due to the greater and faster reabsorption of urea itself.

The alkaline phosphatase (ALP) assay is useful in diagnosing damage to the abdominal hollow organs, making laparoscopy unnecessary. Values greater than 10 U/L have sensitivity and specificity greater than 95%. High ALP values are also diagnostic to differentiate a primary peritonitis from one secondary to intestinal perforation. In the latter,

Table 35.6 Clinical utility of some analytes in cavity liquids

	Pleural effusion			Ascites or peritoneal effusion			Pericardial effusion		
Analyte	Useful	Useless	Of limited utility	Useful	Useless	Of limited utility	Useful	Useless	Of limited utility
Light's criteria	X				X			X	
LDH	X					X		O	
Proteins	X					X		O	
Cholesterol	X					X		O	
Bilirubin		X				X		X	
Enzymes[a]		X			X			X	
Anti-trypsin		O				X		O	
ADA			X		O			O	
Albumin gradient			X	X				O	
Protein gradient			X						
Amylase			X			X		X	
PH			X		O			O	
Glucose			X			X			X
CEA			X			X		X	
CA19.9		X			X			O	
CA125		X			X			O	
CYFRA21-1		O			O			O	
CA15.3		X			X			O	
α-Fetoprotein		O				X		O	

X indicates that the diagnostic usefulness has been confirmed in several studies, *O* indicates that there is no robust evidence to really define usefulness in clinical practice, *ADA* adenosine deaminase, *CEA* carcinoembryonic antigen, *CA19.9* carbohydrate antigen 19.9, *CA125* carbohydrate antigen 125, *CA15.3* carbohydrate antigen 15.3

[a]Other enzymes besides LDH: amylase, ADA

ALP exceeds 240 U/L, a value that is almost never observed in a primary form of the disease.

LDH is often elevated in neoplastic effusions. A peritoneal LDH/serum LDH ratio >0.6 has a sensitivity greater than 80%.

Combined measurement of peritoneal LDH and cholesterol discriminates peritoneal carcinomatosis from hepatocarcinoma ascites.

In ovarian carcinomas, peritoneal fluid LDH is more useful in distinguishing malignant from benign forms than the serum value of the enzyme.

Finally, in spontaneous bacterial peritonitis, LDH allows a correct diagnosis in three out of four cases by using the 0.4 cutoff of the peritoneal LDH/serum LDH ratio.

The ADA distinguishes well, in endemic areas, a tuberculous peritonitis from those of different etiology, with sensitivity and specificity higher than 90%.

Synovial Fluid

Pathophysiology

Synovial fluid is in varying amounts in the joint space and consists of hyaluronic acid, glucose, proteins, lipids, electrolytes, enzymes, and cells. Its main functions are hydration and lubrication of the joint; it also provides a means for the passage of nutrients to the cartilage tissue. Synovial fluid is produced by ultrafiltration of peripheral blood through the synovium and the synthesis of hyaluronic acid by synovial cells.

Normal synovial fluid has the following characteristics: high viscosity, clarity and an opaque yellow color, low erythrocyte and leukocyte cellularity, with a predominance of lymphocytes and monocytes or macrophages; the neutrophil count is less than 10%. Normal synovial fluid is free of crystals and fibrinogen, and is sterile. Synovial fluid and its characteristics vary in different clinical conditions; its analysis can provide the clinician (orthopedist or internist) with helpful information to clarify the etiology of a joint pathology or its evolution.

Macroscopic Evaluation, Crystal Detection, and Cellular Analysis

Synovial fluid is collected through arthrocentesis in quantities of 3–5 mL.

Normal synovial fluid is generally light yellow, transparent, and viscous. In cases of rheumatic or infected effusions, it may be turbid; sometimes it may have a hemorrhagic appearance.

The search for crystals in synovial fluid is helpful in differential diagnostics; both the qualitative nature of the crystals and their semi-quantitative evaluation (rare, a few, or numerous) are of considerable clinical interest. The main constituents are monosodium urate (acute or chronic gout); calcium pyrophosphate (pseudogout or chondrocalcinosis); apatite (pre-articular calcification); and cholesterol crystals can also be found in joint effusions of subjects with chronic rheumatism.

The cell count in synovial fluid takes on different characteristics depending on the pathology involving the joint.

Bacterial Arthritis

Synovial fluid has a purulent appearance, and the cellular component consists predominantly of neutrophils, even in concentrations greater than 100,000 × 10^6/L cells. In Reiter's syndrome (often associated with Chlamydia trachomatis infection), synovial fluid has a cellularity of approximately 40,000 × 10^6 cells/L, with 80% neutrophils. Occasionally, Reiter cells (macrophages with basophilic cytoplasmic inclusions with panoptic staining) are also present.

Rheumatoid Arthritis

The average cell count in this disease is 30,000 × 10^6/L cells, which varies according to the disease phase. During the acute phase, lymphocytes and macrophages prevail, which are not pathognomonic of rheumatoid arthritis because they can also be found in other joint diseases.

Acute Joint Rheumatism

Synovial fluid is rich in fibrin, with lymphocytes, histiocytes, and sometimes synoviocytes predominating. The average cell count is 10,000 × 10^6/L cells.

Systemic Lupus Erythematosus

Synovial fluid is straw-colored, and the predominant cells are lymphocytes, in variable numbers. The characteristic cell is Hargraves' cell, or LE: a neutrophil twice or thrice the normal size with an eccentric nucleus surrounding an inclusion that appears red on panoptic staining. The average cell count is 12,000 × 10^6/L cells.

Gout

In the acute phase, the synovial fluid may appear cloudy to milky. Fresh urate crystals, with the appearance of needles or rods, can be observed either free or in the cytoplasm of neutrophils. The average cell count is 15,000 × 10^6/L cells, and neutrophils are the predominant cell population, followed by synoviocytes.

Pseudogout or Chondrocalcinosis

The synovial fluid is similar in cellular composition and appearance to that of gout, but the crystals are of calcium pyrophosphate, either in free form or phagocytosed. They are distinguishable by their rhomboidal shape and inhomogeneous size compared to urate crystals.

Traumatic Arthropathy

Synovial fluid is hemorrhagic or xanthochromic in color, with poor cellularity (average cell count <2000 × 10^6/L cells), predominantly neutrophils, sometimes also tissue cells such as chondrocytes or bone chips and connective tissue.

Villonodular Synovitis

It is a rare benign tumor of the synovial membrane, in which the synovial fluid is serohemorrhagic or brown, with few cellular elements (<3000 × 10^6/L cells), mainly histiocytes with, within them, hemosiderin, and sometimes plurinucleated giant cells.

Primary or Secondary Neoplastic Arthritis

The fluid is serohemorrhagic with poor cellularity (mean cell count <2000 × 10^6/L), possible presence of "foreign" cellular elements typical of the histotype of the primary tumor.

Prosthetic Infections

In orthopedics, cytometric analysis of synovial fluid is essential for the differential diagnosis between periprosthetic infections and other non-infectious diseases.

Many studies propose different decision thresholds, both for the leukocyte count and for the relative value of neutrophils and/or polymorphonucleates; unfortunately, the decision cutoffs are often not concordant, and this limitation has repercussions in clinical practice.

Pathological synovial fluid can be divided into four categories, as shown in Table 35.7.

Biochemistry

Although not routinely measured on synovial fluid, some measurands, such as glucose, uric acid, lactic acid, lipids (cholesterol and triglycerides), proteins, and some enzymes may be helpful in selected cases. Table 35.8 shows the reference values of the main measurands in synovial fluid.

Proteins are generally >3.0 g/dL in any type of inflammatory process, thus they have little diagnostic or prognostic usufulness.

Cholesterol and Triglycerides

The concentration of lipoproteins in synovial fluid is, on average, 40% lower than in plasma. In inflammatory processes and in crystal-induced arthritis (rheumatoid arthritis, SLE, gout), their levels increase significantly. Lipid effusions have been roughly divided into three groups: cholesterol-rich, lipid droplets, and chylous.

In synovial effusions, cholesterol can reach concentrations as high as 2600 mg/dL.

Chylous synovial effusions may be observed, rarely, in association with rheumatoid arthritis, SLE, trauma, filariasis, and pancreatitis (pancreatic-arthritic syndrome).

Glucose

The glucose level should be interpreted against the blood glucose level. Transient postprandial hyperglycemia is not reflected until several hours later in a physiologic increase in synovial fluid glucose. At equilibrium, normal synovial glucose is 10 mg/dL lower than plasma glucose, or even less. In general, in non-inflammatory or hemorrhagic joint diseases (osteoarthritis, villonodular synovitis pigmentosa, trauma, hemangioma, etc.), glucose may be, respectively, 10 and 20 mg/dL less than in plasma. In inflammatory diseases, instead, the distance between plasma glucose and joint glucose can vary from 0 to 40 mg/dL, while in infectious synovial forms, glucose can be 20–100 mg/dL less than in plasma; to a lesser extent (from 0 to 80 mg/dL), this is observed in crystal-induced forms.

Lactic Acid

The upper reference limit of the lactic acid is 25–30 mg/dL.

There is ample evidence that synovial fluid lactic acid is generally increased in monoarticular septic arthritis compared to non-septic forms.

In septic arthritis, lactic acid can reach concentrations of 1170 mg/dL, 40 times the value of the upper reference inter-

Table 35.7 Main laboratory tests on the synovial fluid and related interpretative criteria

		Normal	Group 1 – Not inflammatory	Group 2 – Inflammatory	Group 3 – Infectious	Group 4 – Hemorrhagic
Macroscopic exam	Appearance	Transparent	Transparent	Transparent/opaque	Opaque	Opaque
	Color	Light to pale yellow	Xanthochromic	Xanthochromic/white/blood	White	Red/brown or xanthromic
Cytometric exam	Leukocytes (×10^6/L)	0–150	<3000	3000–75.000	50,000–200,000	50–10,000
	Polymorphonuclear (%)	<25	<30	>50	>90	<50
	Red blood cells	Absent	Absent	Absent	Present	Present

Table 35.8 Reference values for the main biomarkers in the synovial fluid

	Reference range
Cholesterol	≤65% of the corresponding plasma
Triglycerides	≤40% of the corresponding plasma
Glucose (plasma-synovial)	≤10.0 mg/dL
Lactic acid	≤30.0 mg/dL
Proteins	≤3.0 g/dL
Uric acid (males)	≤8.0 mg/dL
Uric acid (females)	<6.0 mg/dL
Hyaluronic acid	3–4 g/L

val, contrary to what happens in degenerative forms. However, it has been observed that lactic acid does not always increase in bacterial forms; an example is *Neisseria gonorrhoeae* form. This bacterium does not induce an increase in synovial lactic acid. In general, a lactic acid >110 mg/dL indicates a septic form.

Uric Acid

The definitive diagnosis of gout is based on microscopic observation of monosodium urate crystals in synovial fluid.

Uric acid assay on synovial fluid proves to be a valuable aid in the definitive diagnosis of gout. The decision limits for the diagnosis are moderately controversial. For some authors, there is a substantial equivalence between the serum concentration of uric acid compared to the synovial fluid; for others, instead, its values in gout are higher than in the serum, thus constituting a situation without interpretative uncertainties. Probably both observations are true, due to a dynamic balance between the constitution and dissolution of crystals that would be reflected on the concentration of uric acid in this fluid.

Enzymes

Numerous enzymes have been studied in synovial fluid disorders: LDH, aspartate aminotransferase (AST), ALP, γ-glutamyltransferase, ADA, neuraminidase (lysozyme), cytidine deaminase, and several others. Their presumed clinical utility has not always been confirmed in practice. Among the enzymes of proven diagnostic help, LDH certainly stands out. Values between 400 and 700 U/L suggest a moderate-medium inflammatory state, while values >700 U/L strongly indicates severe inflammatory state.

C-Reactive Protein (CRP)

When the inflammatory phenomenon is very circumscribed, in particular clinical pictures, assaying the CRP in a biological fluid can be of considerable diagnostic and prognostic help as in the case of joint infections following implantation of prostheses, representing a more powerful tool than the same imaging.

Cerebrospinal Fluid

The cerebrospinal fluid (CSF), part of the central nervous system (CNS), is clear and colorless ("rock water"); it contains numerous solutes, a very low concentration of proteins, and is isosmotic with plasma, due to a higher concentration of chlorides. The concentration of proteins, under physiological conditions, is 20–45 mg/dL. Its total volume is 140 mL, although about 500 mL is produced in the course of a day.

CSF is produced in the ventricles by the choroidal plexuses, vascular formations of pial origin, projecting into the ventricles. The inner part of the ventricles, as well as the central canal of the spinal cord, is lined by a thin epithelial membrane called the ependyma. The choroid plexuses are equipped with unidirectional transport systems of ions from the periphery to the CSF (this involves by osmosis a movement of water) and, by dedicated mechanisms, glucose, and amino acids. Finally, the choroid plexuses synthesize the proteins of the CSF, such as transthyretin and asialotransferrin. The CSF passes from the ventricles to the subarachnoid spaces, where Pacchioni's granulations reabsorb it and ends up in the venous system and, according to recent studies, also in the lymphatic system.

The blood-liquor barrier (BLB) allows the passage of plasmatic molecules, especially proteins, into the CSF and should not be confused with the blood-brain barrier (BBB), which is, instead, impermeable, like a cell membrane, due to the presence of tight junctions that "weld" endothelial cells together.

CSF is enriched with solutes. Indeed, albumin, a protein used as a model for the passage from plasma to CSF because it is exclusively synthesized by the liver, has a concentration 2.5 times higher in the lumbar CSF than in the ventricular one.

CSF also has a cellular component (lymphocytes and monocytes) due to previously unknown connections between the meninges and the lymphatic vessels.

Circulating in a rigid bone structure, in case of increased production or reduced reabsorption, the amount of CSF cannot increase much, but its pressure increases, creating an endocranial hypertension that pathologically compresses the CNS. The CNS is reduced in volume and the CSF volume increases, a phenomenon called hypertensive hydrocephalus. Normotensive hydrocephalus occurs when CSF volume increases due to a primitive reduction of the cerebral mass, generally due to atrophic phenomena.

CSF is collected by lumbar puncture, usually performed by the neurologist through the intervertebral space between L4 and L5 or L5 and S1, where there is no spinal cord. This maneuver frequently results in blood contamination due to accidental rupture of capillaries in the needle path. For every

600 red blood cells present in the sample due to collection, there is one additional leukocyte in the CSF, and for every 1000 red blood cells, there is a 1 mg/dL increase in total proteins.

To discriminate between the presence of blood in the sample due to traumatic lumbar puncture and blood from subarachnoid hemorrhage, the CSF should be collected, whatever the quantity taken, in three tubes after having numbered them: in case of blood appearance from puncture, this will tend to disappear passing from the first to the third tube.

CSF is currently studied in the suspicion of the following diseases:

- Inflammatory meningitis on an infectious basis, or due to carcinomatosis by metastatic localization of tumors outside or within CNS (not to be confused with meningiomas, which are primary tumors of the meninges).
- Encephalitis, which can be infectious, inflammatory, autoimmune, often as a result of previous infection of the CNS or degenerative type, either on a genetic basis or acquired sporadic.
- Polyradiculoneuritis, inflammation of spinal and cranial nerve roots, acute or chronic.
- Myelitis, diseases confined to the spinal cord.
- Acute hemorrhagic phenomena, such as subarachnoid hemorrhages.

The molecules reach the CSF by simple passage through the blood-liquor barrier, or by secretion from the choroid plexuses, or by drainage from the central nervous tissue. Exit, not at all selective, occurs from the arachnoid granulations and most likely from the lymphatic structures of the meninges, as recently proposed.

CSF proteins fall into three categories according to their origin:

1. Plasma, by simple passage from the plasma to the CSF through the BLB, in an amount dependent on the state of the barrier.
2. Brain, only from the CNS; they increase in case of tissue damage (e.g., tau proteins and phosphorylated tau in Alzheimer's disease).
3. Mixed (plasma-cerebral).

In CSF diagnostics, the use of proteins has a three-pronged approach: assaying proteins as markers of barrier status, assaying proteins of brain origin, and assaying proteins of intrathecal synthesis.

The total protein assay is the most widely used test to diagnose barrier damage. As damage increases, its selectivity decreases. However, the test cannot detect simultaneous or exclusive increases in brain-derived proteins. The recommended method for assessing barrier damage is to assay albumin, either in CSF or serum, calculate the ratio, and multiply the result by 1000. This gives the albumin quotient (QAlb). The reference values of total protein and QAlb change with age: they are higher in infants, lower in children, and rise with age. It was determined that the reference value (RV) of QAlb, for subjects aged 11–80 years, should be calculated using the formula: QAlb = age/25 + 8.

The QAlb is a dimensionless figure, which does not require standardization of the albumin assay. The dosage of proteins of brain origin is expressed with absolute values, generally in mass concentration, with thresholds discriminating sick from healthy.

The dosage of intrathecal synthesis proteins is fundamental, because they are pathognomonic of CNS pathology. Electrophoretic fractionation is very useful, especially with isoelectrophoresis (IEF), as in the case of IgG of intrathecal synthesis, for which the patterns have long been encoded. The formulas quantifying the intrathecal fraction, the more complex ones, such as the Reiber equation, take into account the greater passage of specific proteins as the barrier damage increases and are to be preferred because they are more specific.

Analyses can be of the following types:

- Macroscopic analysis, usually performed by the neurologist at the time of the lumbar puncture
- Cytometric and cytological examination to count and identify the CSF cells
- Biochemical examination (metabolites, proteins, etc.)

Macroscopic Evaluation and Cellular Analysis

Under pathological conditions, the CSF loses its transparent and clear appearance, becoming increasingly turbid. The cornerstone of cellular analysis is not only cytometric analysis, by cell counting and differential counting of lymphocytes, monocytes, and granulocytes, but also qualitative evaluation by cytological examination for the search and reporting of particularly voluminous cells, probably of neoplastic origin, especially if in clusters. The cytometric examination is fundamental for diagnosing meningitis and allowing to define the type. Bacterial meningitis are characterized by a high number of neutrophil granulocytes (typically greater than 1000×10^6/L), while viral meningitis are characterized by a lower number of mononuclear, typically lymphocytes.

In the resolving stages of subarachnoid hemorrhages, siderophages may also be present: macrophages with iron inclusions.

Biochemistry

Glucose is used to estimate the extent of the presence of bacteria and cells; it should be reported as CSF to serum glucose ratio; reference value >0.4–0.5.

Lactates, a product of anaerobic glucose metabolism, increase under all conditions in which glucose is metabolized in the CSF. Since they are not affected by plasma concentration, they are reported in mmol/L; reference value <2.8.

Total protein is a crude indicator of blood-liquor barrier damage.

Immunoglobulin G, when intrathecally synthesized, has long been used, and still is, to make the diagnosis of multiple sclerosis (MS).

M-immunoglobulins have been defined as indicators of the type of CNS disease and not markers of recent infections: However, epidemiological studies and case reports have shown that they appear first and alone in acute infections.

Immunoglobulin A is a marker of purulent meningitis and neurotuberculosis.

Immunoglobulin free light chains are candidates to become the best indicator of immune activation in the CNS.

Specific antibodies should be investigated using criteria that allow their intrathecal synthesis fraction to be identified exclusively by calculating the Antibody Index (AI): the numerator is the quotient of the specific antibody and the denominator is the quotient of the total immunoglobulins of the same class. The search for specific liquid antibodies to infectious agents, especially viral, is also extremely useful in the era of molecular diagnostics. A particular application of the search for specific IgG antibody response is the so-called MRZ (measles, rubella, zoster) reaction search to discriminate MS patients from patients with viral infection, or affected by other CNS inflammatory diseases.

The search for tumor markers in the CSF is very useful for diagnosing meningeal carcinomatosis together with the search for tumor cells. CEA and mucin markers such as CA15.3, CA125, and CA19.9 are used. β-hCG is a specific marker of cerebral germinoma, but it may be present in highly undifferentiated tumors.

Autoantibodies against the CNS are responsible for well-defined neurological diseases. They are mainly to be sought in serum. Intrathecal synthesis can be demonstrated in the CSF, but the synthesis is not of particular clinical significance. Only anti-MNADr autoantibodies are to be sought in the CSF, because they are often higher in the CSF than in the serum, where sometimes they may be absent.

Biomarkers of Alzheimer's disease (AD), a neurodegenerative disease accounting for 50–60% of clinically diagnosed forms of dementia, are exclusively in the CSF. They include β-amyloid protein, resulting from the proteolysis of an amyloid precursor and constituting amyloid plaques; total tau protein (T-tau), associated with microtubules and mainly located in the axon, which has higher values in AD patients; phosphorylated tau protein (p-tau), a hyperphosphorylated form of tau protein and expression of neurofibrillary degeneration, which has higher values in AD patients.

Creutzfeldt-Jakob disease, the best known prion disease, does not have a specific biomarker. For its diagnosis, the protein 14-3-3, not related to the prion, but typical of extensive neuronal necrosis, is used.

Markers of subarachnoid hemorrhage are indicative of a rupture of a vessel in the subarachnoid space, representing an acute and dramatic event with low probability of survival for the patient. The reference examination is computed tomography of the brain, which demonstrates the presence of a hemorrhagic infarction. The CSF examination is performed only in doubtful cases, 2% of the total, especially in the post-acute phase, in search of hemoglobin and bilirubin, the main product of hemoglobin metabolism.

The markers of CSF, presence of CSF outside its physiological sites, indicate an abnormal communication of the cerebral spaces with the external environment, with a strong risk of infection. The most frequent form is rhinoliquorrhoea, a non-acute episode, often post-traumatic, of otorhinolaryngological scope. Post-neurosurgical CSF may represent an urgent complication of a recent neurosurgical procedure. The search for asialotransferrin, isoform of transferrin produced in the CNS, is the reference test, but the β-trace protein (β-TP), produced in the CNS, present in the CSF in concentrations much higher than in plasma, and rapidly measured on automatic analyzers is now the test of choice for these urgent requests.

CSF chlorides, which are higher in CSF than in plasma, decrease in chronic meningitis, such as tuberculosis.

CSF LDH is the only enzyme currently assayed, because it is produced by CSF cells. It is an absolutely non-specific marker of inflammation, if its isoenzymes are not assayed. Thus, it has a high negative predictive value. Physiological values are <21 U/L.

Recommended Readings

Adler Y, Charron P, Imazio M et al (2015) Guidelines for the diagnosis and management of pericardial diseases The Task Force for the Diagnosis and Management of Pericardial Diseases of the European Society of Cardiology (ESC) Endorsed by: The European Association for Cardio-Thoracic Surgery (EACTS). Eur Heart J 36(42):2921–2964

Ali Masoud A, Abdelkader NA, Bayiomy EM et al (2013) IgG-index as early predictor for neurological morbidity in Egyptian patients with acute meningitis. J Egypt Soc Parasitol 43(2):407–414

Andersson M, Alvarez-Cermeno J, Bernardi G et al (1994) Cerebrospinal fluid in the diagnosis of multiple sclerosis: a consensus report. J Neurol Neurosurg Psychiatry 57:897–902

Antoine L, Igor S, Timothy JK et al (2015) Structural and functional features of central nervous system lym- phatics. Nature 523(7560):337–341

Ben-Horin S, Bank I, Shinfeld A et al (2007) Diagnostic value of the biochemical composition of pericardial effusions in patients undergoing pericardiocentesis. Am J Cardiol 99:1294–1296

Bernardi G, Brunati P, Biagioli T et al (2014) L'analisi del liquido cefalorachidiano Biochim Clin 38:239–254

Bignardi GE (2015) Flow cytometry for the microscopy of body fluids in patients with suspected infection. Clin Pathol 68:870–878

Block DR, Algeciras-Schimnich A (2013) Body fluid analysis: clinical utility and applicability of published studies to guide interpretation of today's laboratory testing in serous fluids. Crit Rev Clin Lab Sci 50:107–124

Bourner G, De la Salle B, George T et al (2014) International Committee for Standardization in Hematology (ICSH). ICSH guidelines for the verification and performance of automated cell counters for body fluids. Int J Lab Hematol 6:598–612

Buoro S, Apassiti Esposito S, Vavassori M et al (2016a) Reflex testing rules for cell count and differentiation of nucleated elements in pleural and ascitic fluids on Sysmex XE-5000. Journal Laboratory Automation 21:297–304

Buoro S, Gustinetti R, Dominoni P et al (2012) Analytical evaluation of Sysmex UF-1000i for flow cytometric analysis of peritoneal fluid. Clin Biochem 45:1263–1265

Buoro S, Mecca T, Azzarà G et al (2016c) Mindray BC-6800 Body Fluid Mode, performance of nucleated cells and differential count in Ascitic and Pleural Fluid. Int J Lab Hematol 38:90–101

Buoro S, Mecca T, Azzarà G, Seghezzi M et al (2016b) Cell Population Data and reflex testing rules of cell analysis in pleural and ascitic fluids using body fluid mode on Sysmex XN-9000. Clin Chim Acta 452:92–98

Burgess LJ (2004) Biochemical analysis of pleural, peritoneal and pericardial effusions. Clin Chim Acta 343:61–84

Butch AW, Wises PK, Wah DT et al (2008) A multicenter evaluation of the Iris iQ200 automated urine microscopy analyzer body fluids module and comparison with hemacytometer cell counts. Am J Clin Pathol 129:445–450

Clinical and Laboratory Standard Institute (2007) Analysis of Body Fluids in Clinical Chemistry; Approved Guideline. CLSI document C49-A. Clinical and Laboratory Standard Institute, Wayne

Clinical and Laboratory Standard Institute (2006) Body fluid Analysis for Cellular Composition; Approved Guidelines. CLSI document H56-A. Clinical and Laboratory Standard Institute, Wayne

Corsini E, Bernardi G, Gaviani P et al (2009) Intrathecal synthesis of tumor markers is a highly sensitive test in the diagnosis of leptomeningeal metastasis from solid cancers. Clin Chem Lab Med 47:874–879

Corsini E, Gaviani P, Chiapparini L et al (2016) Intrathecal synthesis of onconeural antibodies in patients with paraneoplastic syndromes. J Neuroimmunol 290:119–122

Deisenhammer F, Bartos A, Egg R et al (2006) Guidelines on routine cerebrospinal fluid analysis. Report from an EFNS task force. Eur J Neurol 13:913–922

Deisenhammer F, Egg R, Giovannoni G et al (2009) EFNS guidelines on disease-specific CSF investigations. Eur J Neurol 16:760–770

Del Mese G, Nannini P, Zuin G et al (2015) Cerebro Spinal Fluid (CSF) Analysis and CSF Free Light Chains (FLC) indices in a patient with Varicella Zooster Encephalitis. Clin Chem Lab Med 53: 11:eA187- eA188

Fleming C, Russcher H, Lindemans J et al (2015) Clinical relevance and contemporary methods for counting blood cells in body fluids suspected of inflammatory disease. Clin Chem Lab Med 53:1689–1706

Galliera E, Drago L, Marazzi MG et al (2015) Soluble uroki- nase-type plasminogen activator receptor (suPAR) as new biomarker of the prosthetic joint infection: correlation with inflammatory cytokines. Clin Chim Acta 441:23–28

Graus F, Saiz A, Dalmau J (2010) Antibodies and neuronal autoimmune disorders of the CNS. J Neurol 257(4):509–517

Iwanaga T, Shikichi M, Kitamura H et al. Morphology and functional roles of synoviocytes in the joint. Arch Histol Cytol 2000; 63:17–31.

Jacobi C, Lange P, Reiber H. Quantitation of intrathecal antibodies in cerebrospinal fluid of subacute sclerosing panencephalitis, herpes simplex encephalitis and multiple sclerosis: Discrimination between microorganism-driven and polyspecific immune response J Neuroimmunol 2007; 187:139–46.

Karatolios K, Pankuweit S, Maisch B. Diagnostic value of biochemical biomarkers in malignant and non-malignant pericardial effusion. Heart Fail Rev 2013; 18:337–44.

Kitchongcharoenying P, Foocharoen C, Mahakkanukrauh A et al. Pericardial fluid profiles of pericardial effusion in systemic sclerosis patients. Asian Pac J Allergy Immunol 2013; 31:314–9.

Knight JA, Kjeldsberg CR. Cerebrospinal, synovial, and serous body fluids. In: McPherson RA, Pincus MR (eds). Henry's Clinical Diagnosis and Manage- ment by Laboratory Methods. Elsevier Saunders; 2006, pp. 437–41, 21st ed.

Kopcinovic LM, Culej J. Pleural, peritoneal and pericardial effusions a biochemical approach. Biochem Med 2014; 24:123–37.

Link H, Tibbling G. Principles of albumin and IgG analyses in neurological disorders. III. Evaluation of IgG synthesis within the central nervous system in multiple sclerosis. Scand J Clin Lab Invest 1977; 37: 397–401.

Lippi G, Danese E, Cervellin G et al. Laboratory diagnostics of spontaneous bacterial peritonitis. Clin Chim Acta 2014; 430:164–70.

MC Grath EE, Anderson PN. Diagnosis of pleural effusion: a systematic approac. Am J Crit Care 2011; 20:119–28.

Noppen M, De Waele M, Li R et al. Volume and cellular content of normal pleural fluid in humans examined by pleural lavage. Am J Respir Crit Care Med 2000; 162:1023–6.

Nussinovitch M, Finkelstein Y, Elishkevitz KP et al. Cerebrospinal fluid lactate dehydrogenase isoenzymes in children with bacterial and aseptic meningi- tis. Transl Res 2009; 154(4):214–8.

Paul HY, Cross MB, Moric M et al. Do serologic and synovial tests help diagnose infection in revision hip arthroplasty with metal-on-metal bearings or corrosion? Clin Orthop Relat Res 2015; 473:498–505.

Parviz J, Zmistowski B, Berbari EF et al. New definition for periprosthetic joint infection from the Workgroup of the Musculoskeletal Infection Society. Clin Orthop Relat Res 2011; 469:2992–4.

Redzic ZB, Segal MB. The structure of the choroid plexus and the physiology of the choroid plexus epithelium. Adv Drug Deliv Rev 2004; 56(12):1695–716.

Reiber H. Dynamics of brain-derived proteins in cerebrospinal fluid. Clin Chim Acta 2001; 310: 173-86. Reiber H, Felgenhauer K. Protein transfer at the blood cerebrospinal fluid barrier and the quantitation of the humoral immune response within the central nervous system. Clin Chim Acta 1987; 163:319–28.

Romero-Candeira S, Hernandez L, Romero-Brufao et al. Is it meaningful to use biochemical parameters to discriminate between transudative and exudative pleural effusions? Chest 2002; 122:1524–9.

Runyon BA; AASLD Practice Guidelines. Management of adult patients with ascites due to cirrhosis: an update. Hepatology 2009.

Sandhaus LM. Body fluid cell counts by automated methods. Clin Lab Med 2015; 35:93–103.

Seghezzi M, Buoro S, Manenti B et al. Optimization of Cellular analysis of Synovial Fluids by optical mi- croscopy and automated count using the Sysmex XN Body Fluid Mode. Clin Chim Acta 2016; 462:41–48.

Seguera RM. Useful clinical biological markers in diagnosis of pleural effusions in children. Ped Respir Rev 2004; 5:S205–12.

Tercic D, Bozic B. The basis of the synovial fluid analysis. Clin Chem Lab Med 2001; 39:1221–6.

Laboratory Diagnostics in Autoimmune Diseases

36

Tommaso Trenti, Alessandra Melegari, and Chiara Bonaguri

Introduction

The detection of circulating autoantibodies is crucial for autoimmune diseases, from diagnosis to prognosis and treatment.

Autoimmune phenomenology, one of the most studied topics in the field of immunology, has been the subject of many clinical, laboratory and experimental studies in the last decades, which have led to important clarifications about the immune physiopathology and have allowed to group in the same pathogenetic class many diseases whose cause was previously uncertain.

Nowadays, there are an increasing number of diseases for which a pathogenesis linked to abnormalities of the Immune Competent System has been ascertained or only hypothesized and which are often characterized by the presence of immunoglobulins in the circulation of autoantibodies, directed against "self" substances represented by constituents of the organism itself.

Autoimmune diseases develop when the immune system fails to recognize the "self" and activates cells or produces antibodies directed against cells, tissues and/or organs of the organism, causing an inflammatory process with tissue damage and consequent expression of an autoimmune disease. The pathogenetic role of immunological reactions resulting from this loss of competence in the recognition of the "self" together heterogeneous pathological conditions in terms of clinical features represents the common denominator of diseases with autoimmune etiology.

T. Trenti (✉)
Department of Laboratory Medicine and Pathology, Azienda USL/Azienda Ospedaliero-Universitaria di Modena, Modena, Italy
e-mail: t.trenti@ausl.mo.it

A. Melegari
Department of Laboratory Medicine and Pathology, Azienda USL of Modena, Modena, Italy

C. Bonaguri
Laboratory of Clinical Chemistry and Hematology, University Hospital of Parma, Parma, Italy

The natural history of all autoimmune diseases is divided into different phases or conditions, which can be basically ascribed to a potential phase, a latent subclinical phase and a clinical phase.

Until now, more than 80 pathologies have been described that derive from an autoimmune response with an estimated prevalence in Europe of 4000 affected persons per 100,000 inhabitants and an estimated 2,400,000 patients in Italy.

The availability of serological diagnostic tests, consisting of a panel of increasingly specific and sensitive biomarkers, has allowed the laboratory to play a role, not only of reference as central but irreplaceable in the diagnosis and prognosis of these diseases, resulting the laboratory medicine the only discipline that consider the autoimmunity in its globality, while, in the various clinical areas, appears divided by organ or specialty.

Indeed, the skills of the laboratory professional in relation to the entire clinical-diagnostic pathway are essential to offer an effective response to the request of the clinician, aimed at ensuring the health needs of the patient, and these skills must be performed with autonomy in the choice of diagnostic investigations useful to identify the evolution of autoimmune diseases in its complex and multiple expressions.

The remarkable progress in immunological diagnostics that has occurred in recent years, linked to the increased physiopathological knowledge of the role of autoantibodies and the development of new analytical technologies, together with the growing expectations of physicians and patients for investigations, has created the need for a more appropriate use of immunological tests for the detection of autoantibodies, both in terms of request and execution of the same.

Appropriateness is a complex objective that can only be achieved by integrating skills and agreeing on the best procedures to follow in each specific clinical situation. If, finally, appropriateness of requests and services also means less waste and better use of available resources, it is evident once again that this goal can be achieved more effectively, not so much through budgeting operations or restrictive economic

M. Ciaccio (ed.), *Clinical and Laboratory Medicine Textbook*, https://doi.org/10.1007/978-3-031-24958-7_36

measures but through collaboration between laboratory physician and clinician.

This chapter will discuss some autoimmune diseases and the related laboratory diagnostic procedures established according to the most recent recommendations and clinical evidence, which are continuously updated.

In this chapter, we present autoimmune rheumatic diseases, monitoring of biological drugs, celiac disease and autoimmune liver diseases, which have been selected in order to outline through the individual presentations the diagnostic complexity and the new promising perspectives that in the field of laboratory medicine affect the discipline of autoimmunity.

It was decided to present these themes because they are very emblematic of how much discussion, comparative activity and government there has been and still is and how the laboratory is the reflecting mirror that has gone through many changes and refined in recent years all its ability to diagnose in close collaboration with scientific research and technological instrumentation.

In fact, in autoimmunology, besides the indirect immunofluorescence technique, which keeps an irreplaceable role, new immunoassay tests, different in analytical technique (enzyme immunoassays, chemiluminescence immunoassays, immunoblots, etc.) but all potentially endowed with opportunities for results standardization, are now routinely used for single or multiple autoantibody assays and/or are currently object of numerous studies.

In this scientific context, which will probably see in the coming years the emergence of new diagnostic strategies and solutions, with consequent therapeutic effects and needs for clinical governance of processes, the role in terms of expertise of the autoimmunologist laboratory will be essential.

Autoimmune Rheumatic Diseases

The detection of pathology-specific antibody profiles is a key diagnostic criterion for the characterization of patients with connective tissue disease.

The most used serological tests are the detection of antinuclear antibodies (ANA), extractable nuclear antigens (ENA), and anti-dsDNA. The current nomenclature and acronyms for ANA and ENA are being re-evaluated; however, due to their widespread use, they will be retained as such in this chapter.

In recent years, the demand for these tests has increased significantly, due in part to the acquisition of new evidence on their clinical relevance but also to an inappropriate prescription, not well quantified, for:

- Insufficient communication between clinicians and laboratories, resulting in disagreement on the diagnostic significance of available tests and test results
- Different analytical techniques available for the same marker, availability of new target antigens
- Lack of both homogeneous terminologies and diagnostic algorithms, resulting in inhomogeneous behaviour

The in-depth evaluation of the clinical significance and diagnostic rationale of the laboratory's ANA, ENA and dsDNA tests is at the heart of the approach to connective tissue disease, identifying screening tests and in-depth tests that may allow an appropriate request to the clinical question.

The presence of ANA in serum is considered a marker of high diagnostic significance in collagen diseases.

The positive ANA test represents, in fact, a classificatory criterion in the diagnostic definition of non-organ-specific autoimmune diseases, such as systemic lupus erythematosus (SLE), scleroderma/progressive systemic sclerosis (SSc), mixed connective tissue disease (MCTD), Sjogren's syndrome (SS), dermatomyositis-polymyositis (DM/PM) and undetermined connective tissue disease (UCTD).

On the contrary, rheumatoid arthritis (RA) does not express this marker: This pathology frequently presents a serum negativity towards the autoantigens commonly evaluated in the diagnosis of connective tissue diseases. For this diagnosis, the combined evaluation of rheumatoid factors (RATest) and anti-cyclic citrullinated peptide antibodies (anti-CCP) is of particular interest.

Therefore, the detection of ANA in the serum of patients is an important diagnostic tool to allow the clinician a more precise classification of anamnestic and clinical pictures that are often blurred.

In recent years, commercial test offerings in this field have become numerous as the analytical techniques available for the same marker.

It is, therefore, necessary to make a reasoned choice of diagnostic tests and analytical methods in which the laboratory must be able to play a central role.

ANA are autoantibodies belonging to all classes of immunoglobulins, most often to the IgG class, directed against cellular constituents common to all cell types (non-organ-specific antibodies).

The study of the structural and functional characteristics of the target antigens and the characterization of the single antigen/antibody systems has made it possible to define close correlations between serological picture and clinical status, and in some conditions, ANA represent a real marker of disease, which, due to the precocity of appearance, assumes a high predictive and prognostic value.

Due to its high diagnostic sensitivity, the ANA test executed IFI method remains the gold standard and represents the first level of the diagnostic algorithm.

However, since these autoantibodies, especially at low titre, are also found in healthy subjects and/or those with non-autoimmune diseases, it is necessary to perform the

ANA test and the related in-depth examinations (ENA and dsDNA) according to a correct sequence to improve the diagnostic process.

Based on all the above considerations, the application of the ANA Reflex is justified. The ANA Reflex allows the autoimmunology laboratory to evaluate the fluoroscopic pattern, the positivity titre and the clinical information and to assess whether and which tests should be performed as in-depth examinations.

In order to be effective, the ANA Reflex must be an administrative tool valid for all prescribers both in hospitals and in the territory, but with the freedom to be applied in a "reflexive" way by the laboratory, which becomes responsible for the governance of the whole subsequent process.

In the presence of a positive ANA-IFI result, a greater diagnostic and prognostic power can be achieved with the characterization of antibody specificities directed against the different nuclear antigens through molecular tests.

The determination of specific antibodies directed towards the so-called ENA-, DNA- or RNA-associated proteins is useful for the diagnosis of SLE, neonatal lupus, SS, MCTD, PM/DM and SSc, with variable sensitivity and specificity, as reported in the literature. The variety of autoantigens recognized by ANA is extremely wide.

Anti-ENA antibodies in serum can be detected by several techniques currently used in the laboratory: enzyme immunoassay (ELISA), immunoblot (IB), and immunoassay (EIA).

For the choice of the method, it is necessary to keep in mind some factors: clinical location (reference laboratory or not), laboratory budget and, consequently, cost of reagents, organization and "experience" of the laboratory and level of dialogue between laboratory and clinician.

The ideal method should meet the criteria of clinical sensitivity, precision and accuracy, ease of execution, easy availability and low cost; at present, there is no method that can meet all these requirements alone.

Considering the characteristics of the available methods, a correct diagnostic procedure for the detection of anti-ENA can only make use of several methodological approaches that the autoimmunology laboratory must be able to put in place.

The diagnostic and prognostic usefulness of the anti-dsDNA antibody assay has led to the development of several techniques for their identification: radiobinding techniques (Farr technique), indirect immunofluorescence (IFI) on Crithidia Luciliae and enzyme immunoassay techniques (ELISA).

Anti-dsDNA antibodies are highly specific for SLE as they are almost exclusively found in SLE and are the tenth criterion for SLE according to the American College of Rheumatology.

If their concentration is high, they are prognostic of a relapse even if SLE is quiescent; moreover, their presence, in the absence of clinical symptoms, is indicative of subclinical SLE, being negative in drug-induced SLE and positive in less than 2% of cases with other autoimmune diseases. Numerous studies have documented that the concentration of anti-dsDNA antibodies correlates with the clinical course of SLE and lupus nephritis and that an increase in antibody titre may precede clinical flare by a few weeks.

The research of ANA in IFI is penalized by several critical issues, such as the subjectivity of the reader, the qualification and experience of the reader, the variability related to different substrates and the technology of the microscope. All this, together with the consolidation of the laboratories, has led to a workload of microscopic activity that is not always sustainable where there is a shortage of personnel dedicated to autoimmune diagnostics and there is no generational turnover and no insertion and training of new professionals. These problematic aspects have seen, in recent years, the emergence of new organizational and technological strategies to cope with the increasing numbers of tests and the contraction of healthcare personnel.

Recently, automated reading systems have been introduced in the laboratories, which mainly allow a screening between negative and positive samples and among the latter a "potential" attribution of the patterns of positivity and an estimation of the titre with the possibility of archiving the images as a partial solution to the problems mentioned above. The technological offer is of platforms with different characteristics that, however, must be dropped in their own organization, taking advantage of the benefits of standardization that, however, come only after an insertion and an evaluation that requires an alignment with their own microscopic practice. Certainly, the possibility of archiving images and, more generally, the integration of these systems in the management island of autoimmunity can make the daily work more efficient.

The scientific literature has produced several comparisons and considerations on the systems on the market, analyzing their strengths and weaknesses. Also these technologies, like the others, must be consciously included and governed by a structured and robust experience in the field of autoimmunology.

Considering the objective difficulties in numbers and organization, studies of comparative evaluation of ANA in IFI and solid phase ANA in immunoassay have also increased. There are different technologies and combinations of antigens on the market. It is a question of considering the comparison and possible combination of two different screening approaches. There are studies on the combination of the two tests in the clinical diagnostic pathway, and there are studies on the use of the solid phase as initial screening, followed by immunofluorescence. In addition to sensitivity and specificity reasoning, it is necessary for each laboratory to evaluate its own population, informatics and organizational strategy and clinical algorithm. Certainly, these choices, if shared across a regional territory, increase the strength of an algorithm.

Monitoring of Biological Drugs

Therapeutic Monitoring of Autoantibodies and TNFα Inhibitor Drugs in Patients Being Treated for Chronic Immune-Mediated Inflammatory Diseases

In the last decades, the therapeutic approach to chronic immune-mediated inflammatory diseases has been revolutionized by the advent of biological drugs, monoclonal antibodies capable of neutralizing the pro-inflammatory and immune-activating effect of tumour necrosis factor alpha (TNFα), the main cytokine involved in the pathogenesis of these diseases. This class of drugs includes chimeric (infliximab), humanized (adalimumab, golimumab), fusion protein (etanercept) and recombinant humanized antibodies (certolizumab). They are monoclonal antibodies with different structural and functional characteristics, different pharmacokinetic and pharmacodynamic properties and different efficacy and therapeutic safety profiles.

It is estimated that about 60–70% of patients treated with these drugs achieve favourable therapeutic responses in the phase of induction and maintenance of clinical remission of the disease, while the remainder experience therapeutic failure due to inefficacy of the therapy, loss of response over time or the appearance of adverse events, such as infections and reactions at the site of infusion. The main cause of therapeutic failure is the phenomenon of immunogenicity. Different epitopes of the anti-TNFα molecule are recognized as foreign elements by the immune system, which initiates the production of antibodies directed against the drug itself (antidrug antibodies, ADA).

ADAs belong to different classes of immunoglobulins (IgG, IgE, IgM) and are able to form a complex that prevents the binding of the drug to TNFα with a consequent increase in clearance and reduction in the functional concentration of the drug, which leads to a loss of response to therapy. ADAs can, therefore, interfere both at the pharmacokinetic level, reducing the active concentrations of the anti-TNFα drug, and at the pharmacodynamic level, preventing the drug from binding to TNFα to neutralize it. Biological drugs have different degrees of immunogenicity that vary depending on the route of administration, molecular structure, dose and any concomitant therapies. In general, the increase in ADA formation is associated with a reduction in the concentration of anti-TNFα drug measurable in the bloodstream.

In order to facilitate the most appropriate therapeutic choices and try to prevent immunogenicity, loss of response to therapy and the occurrence of adverse events, in recent years have been developed different schemes of therapeutic drug monitoring (TDM) specific for each disease, including the determination of the concentration of biological drug in use and the search for ADA.

From the point of view of the laboratory, the TDM presents some criticalities concerning: the methods of determination of the TDM concentration of the drug and ADAs, measurement of the concentration of the drug in the bloodstream and assessment of the presence of ADAs.

To date, the most widely used methods in the laboratory for the determination of TNFα inhibitory drugs and for the detection of ADAs are ELISA and the RIA method, but there are also other liquid phase enzymatic tests and tests that measure mobility, such as the Homogeneus Shift Assay (HSMA). There is no difference between liquid or solid phase assays in terms of measured concentration of the drug, but a difference has been observed in terms of specificity. The main limitation of these methods is the risk of having false positive results, due to the non-specific binding between the immunoglobulin and the drug. In the absence of standardization between the different tests, which contributes to limiting the reproducibility of the results, it is advisable to always monitor the patient with the same analytical method.

With regard to the determination of the concentration of the drug in circulation, the aspects that should always be taken into consideration are: (1) the sampling time, that is the number of days between the last infusion and the drug determination; (2) the number of previous infusions; (3) the characteristics of the patient that may influence the circulating levels of the drug (weight may modify the clearance of the drug, as well as gender, BMI value or concomitant use of immunosuppressants); (4) the characteristics of each drug (e.g. the kinetics and the route of administration, intravenous or subcutaneous). Moreover, drugs have specific trough levels and different concentration ranges depending on the type and duration of the disease and on the individual characteristics of the patients. For therapeutic monitoring purposes, it is useful to define therapeutic windows within which drug concentrations should fall.

The search for and determination of ADAs represent a critical aspect of TDM, especially since the factors associated with their development are not yet fully understood. ADAs can be neutralizing and non-neutralizing depending on the action they mediate on the drug itself. In fact, they can inhibit the anti-inflammatory action of the drug by binding directly to it, or they can form immune complexes, reducing the bioavailability of the drug by increasing its clearence. A distinction is then made between bound and free ADAs. The former are bound to the drug, while the free ones in circulation are detected when all the binding sites of the drug are saturated and, consequently, the concentration of free ADAs that could be determined is reduced.

Two other aspects must also be kept in mind in TDM:

- ADAs can be transient. In patients for whom minimal amounts of ADA are detected, these may disappear in a short time and, therefore, do not affect the activity of TNFα inhibitors; conversely, in the presence of high concentrations of ADA, these are usually suggestive of an ongoing immunogenic process;
- ADAs may give rise to cross-reactions with biosimilar drugs and may be interfered with by rheumatoid factor. Finally, the lack of a recognized and validated cut-off makes the interpretation of the laboratory data difficult.

Although the TDM of biologic drugs for autoimmune diseases presents some criticalities to date, the determination of the concentration of the drug in the bloodstream and ADAs is important at all stages of therapy, from induction and maintenance phase to remission, in order to ensure that patients receive adequate therapy and reduce the risk of loss of response.

Celiac Disease

The celiac condition is an autoimmune enteropathy (the most frequent autoimmune pathology since about 1% of the population presumably suffers from it) caused by the ingestion of cereals containing gluten and represents the final phase of a series of complex processes that involve both innate and adaptive immunity. It is believed that the prevalence of this disease is underestimated, since for every patient diagnosed, 3–10 subjects affected by the disease are not recognized, even though the diagnosis of the disease is significantly increasing both in Europe and in the United States, doubling every 15 years or so. The possibility of developing the disease is due to the presence of cofactors that cause the disease to appear at a particular time in the lives of predisposed individuals. Therefore, contrary to what was previously thought, this is not a prevalently paediatric disease; on the contrary, most subjects develop it in adulthood, and in 20% of the cases, the onset and diagnosis occur after the age of 60. From a clinical point of view, the manifestations of the disease are extremely variable and include pictures ranging from severe deterioration of general conditions (rare) to nuanced clinical symptoms (frequent), often with the absence of classic gastrointestinal symptoms. In children, the symptoms are usually more specific: gastroenterological symptoms (pain, abdominal bloating, etc.), irritability and mood swings and developmental deficits, etc., whereas in adults, the disease often presents in a nuanced manner or is detected incidentally.

From a pathogenetic point of view, what happens when gliadin penetrates the intestinal barrier is sufficiently well known. The damage occurs in the proximal tenue with complex alterations that occur at the moment of gluten introduction, affecting both innate immunity (with the production of IL-15) and adaptive immunity (with the activation of the immune system). The result is an alteration of biohumoral and cellular factors that cause both functional and anatomical deterioration of the intestinal villi with pictures up to complete atrophy. If the celiac condition is the only autoimmune disease for which an environmental trigger, gluten (a prolamin contained in wheat where, however, there are other toxic prolamins, present in spelt, kamut, rye and barley), is recognized, genetic predisposition is of extraordinary importance. Environment and genes, however, are necessary but not sufficient elements for the appearance of the coeliac condition: 30% of the general population has predisposing genes and comes into contact with gluten, but only 1% develops the celiac condition. There are groups at increased risk of developing the disease: first and second degree relatives of celiac persons, individuals with selective IgA deficiency, individuals with other autoimmune diseases or genetic abnormalities. For first-degree relatives, the risk is 10–15 times higher than for the general population. The conclusive diagnosis of the coelic disease may be proposed on the basis of clinical findings and the positive results of two intestinal biopsies, but now greater importance is being attributed to serological markers and genetic tests as they are of adequate diagnostic accuracy, cost efficacy and are widely used. Research and scientific evidence have questioned the value of the intestinal biopsy as the "gold standard," while the diagnostic, serological and genotyping tests are increasingly robust, although a joint evaluation of clinical, serological, genetic and histological data is necessary when present and deemed necessary.

Genetics in the Diagnosis of Celiac Disease: Role of HLA DQ2 and DQ8

The genetic background is a necessary condition for the development of the disease in both the classic clinical forms and the latent and silent forms. The main genetic factor predisposing to celiac disease is the HLA DQ2 haplotype (DQA1*0501-DQB1*0201), which is expressed in 90% of patients affected by the celiac condition, while the DQ8 haplotype (DQA1*0301-DQB1*0302) is expressed in 5% of patients and the remaining 5% are carriers of at least one of the two DQ2 allies (usually DQB1*0201). In a very concise manner, it can be stated that approximately 70–75% of the population is not a carrier of DQ2 or DQ8 and, therefore, has very little, if any, chance of suffering from the celiac condition. Of the remaining 25–30% of the population who are carriers of DQ2 or DQ8, only 4% will develop the disease. Therefore, the presence of the DQ2 or DQ8 haplotype identified by a positive genetic test is a necessary, but not sufficient, condition for diagnosing the celiac condition.

HLA DQ2/DQ8 molecules are expressed in the dendritic cells of the gut mucosa and have high affinity for deamidated gliadin peptides; the binding between DQ2/DQ8 and deamidated peptides is the basis of the immune response as gluten-derived antigens are exposed to T cells that trigger the pathological process of autoimmune response; this is the reason for the importance of the HLA DQ2/DQ8 genetic configuration.

HLA class II includes not only DQ2 and DQ8 but a wide variety of possible alleles, many of which are involved in susceptibility to autoimmune diseases or, in some cases, are protective. HLA enters the major histocompatibility complex (MHC) locus on the short arm of chromosome 6, where several genetic regions are located; the one of interest in celiac disease is the class II region, which contains a number of genes, including HLA DQB1 and HLA DQA1, which are in linkage with the DR locus. The A1 gene encodes for the α protein chain, the B1 for the β protein chain; in antigen-presenting cells, these proteins are expressed at the membrane level to form a heterodimer. About DQ2, the most important allele is *02 of the DQB1 locus, which encodes the β chain of the HLA DQ2 heterodimer. For the presence of DQ2 for the expression of celiac disease, it is necessary to have this allele. However, it is not sufficient: it is also necessary to have the *05 allele of the DQA1 locus. These alleles can be combined in a very variable and complex way. For DQ8, the combination of the *0302 allele from the DQB1 locus and the *03 allele from the DQA1 locus is required. Many techniques, now routine in the molecular biology laboratory, are available to perform HLA DQ2/DQ8 analysis. For example, an allele-specific PCR can be performed with analysis by agarose gel electrophoresis, or sequencing can be performed. In reality, however, the goal is to detect only the possible presence of the four alleles directly involved. One of the possible methods is a real-time PCR analysis, using specific sequences, primers and probes that recognize the alleles of interest. The DNA sample is analyzed, and if the main allele (*02 of the DQB1 locus) is detected, an amplification signal is detected. In order to confirm that the subject is positive for HLA DQ2, it is then necessary to detect also the *05 allele of the DQA1 locus, using the same method. It is then possible to verify if *02 of the DQB1 locus is in homozygosity or in heterozygosity. If the result is positive, the various allelic combinations translate into a different risk of disease. In fact, it is a different condition to have all the DQ2 and DQ8 alleles expressed compared to having only *02 of the DQB1 locus.

As far as family members are concerned, who are individuals at risk, there is debate in the scientific community about the advisability of carrying out the genetic test: If it is negative, the celiac condition can be excluded, and this is a reason for supporting the advisability of the test. However, the probability that family members of celiacs have this haplotype and that the test is positive is very high, and in this case, there is nothing that can be done to prevent the disease. The problem, therefore, is to translate the different risk associated with the results of the genetic analysis into clinical practice. There are many other genes other than the HLA genes mentioned above that play a role in the predisposition to celiac disease and other autoimmune diseases, but each non-HLA variant makes a minimal contribution to the genetic risk associated with HLA; therefore, it is difficult to think of analyzing all the variants, and, in any case, their clinical significance is not particularly relevant.

In conclusion, with regard to the diagnosis of celiac disease due to genetic predisposition, it can be concluded as:

- The HLA locus is the main genetic predisposition factor for celiac disease.
- The presence of the HLA DQ2 and DQ8 haplotypes is necessary, but not sufficient for the development of the pathology.
- The absence of HLA DQ2 and HLA DQ8 allows the exclusion of celiac disease, and these patients do not require further testing over time for suspected celiac disease.
- There is a risk gradient for celiac disease associated with the different HLA DQ haplotypes that are derived from various combinations of the HLA DQA1 and DQB1 alleles.
- The conclusive diagnosis of celiac disease in symptomatic children is made if there is the simultaneous presence of HLA DQ2 or DQ8 with IgA class antitransglutaminase levels greater than 10 times the threshold used by the laboratory.

Biomarkers and Algorithms for Diagnosis and Monitoring of Celiac Disease

It is essential to emphasize that all the diagnostic tests required for the diagnosis of the celiac condition must be carried out on an unrestricted diet: It is not necessary to suspend the intake of gluten before undergoing the diagnostic procedure in order to distort the results of the tests. For the patient, once the diagnosis has been made, an aglutinated diet followed regularly is fundamental; in fact, the elimination of gluten from the diet makes it possible to re-establish an intestine that is structurally almost normal and functionally intact. The tolerated amount of gluten is 10 mg/day. Early diagnosis of the pathology in a celiac subject, even if he or she is almost asymptomatic, is very important, since important complications may appear over time. The most significant of these are refractory celiac disease (a condition in which malabsorption symptoms and villi atrophy persist even when the strict aglutination diet has been in place for at least 6–12 months), ulcerative jejunalitis and collagenous sprue (which involve high mortality), lymphomas and epithelial tumours. The risk of developing a complication of the latter type has been reduced over time, thanks to the possibility of identifying (and then subjecting to the gluten-free diet) even subjects with milder forms than in the past, characterized by very nuanced symptoms.

The biomarkers of first choice for the diagnosis of celiac disease are IgA class anti-transglutaminase antibodies (anti-tTG IgA) and total IgA, where total IgA is performed mandatorily in the case of the first intake of the patient to assess any deficit of the same for potential false negatives. In the case of IgA deficiency, IgG class deamidated peptide antibodies (anti-DGP IgG) or IgG transglutaminase antibodies are considered. While determination of plasma concentration of anti-tTG IgA antibodies is generally considered more appropriate, detection of anti-DGP-IgA antibodies remains useful in children under 2 years of age or in cases of IgA deficiency.

Important national and international guidelines as well as national legislation now propose a definitive diagnosis of the celiac condition using only biohumoral data without performing an intestinal biopsy, in particular when values greater than 10 times the threshold value of the method's specific anti-tTG IgA antibodies are determined. In this case, the positive predictive value of the test is equal to that of biopsy in paediatric patients when associated with positivity for anti-endomysium EMA antibodies and HLA DQ2/DQ8 genetic testing. The European Society of Paediatric Gastroenterology, Hepatology and Nutrition (ESPGHAN) guidelines indicate that in symptomatic children with anti-tTG antibodies above tenfold threshold, confirmed by EMA and HLA DQ2–8 positivity, the diagnosis is conclusive and biopsy can be avoided (Fig. 36.1). Basically, the same diagnostic algorithm is applied in adults, that is anti-tTG IgA antibodies and total IgA determination are the first choice and initial tests, and anti-DGP-IgG (anti-tTG-IgG) antibodies are normally performed only in case of IgA deficiency; the confirmation of the diagnosis in the case of positivity of the initial tests is obtained by performing EMA antibodies and HLA DQ2/DQ8 genetic test. However, there is still the indication to perform intestinal biopsy in adult patients for a definitive confirmation of the diagnosis, or in any case, the primary studies present are not conclusive to exclude the usefulness of endoscopic examination. Genetic testing (presence of HLA DQ2/DQ8 class II histocompatibility antigen) is recommended as a surveillance test for at-risk subjects, prior to serology, because in the event of a negative result, the problem of diagnosing the celiac condition will no longer arise for the patient's life; in the event of a positive result, an investigation will be appropriate both in the event of symptoms and in asymptomatic patients every 3 or 4 years.

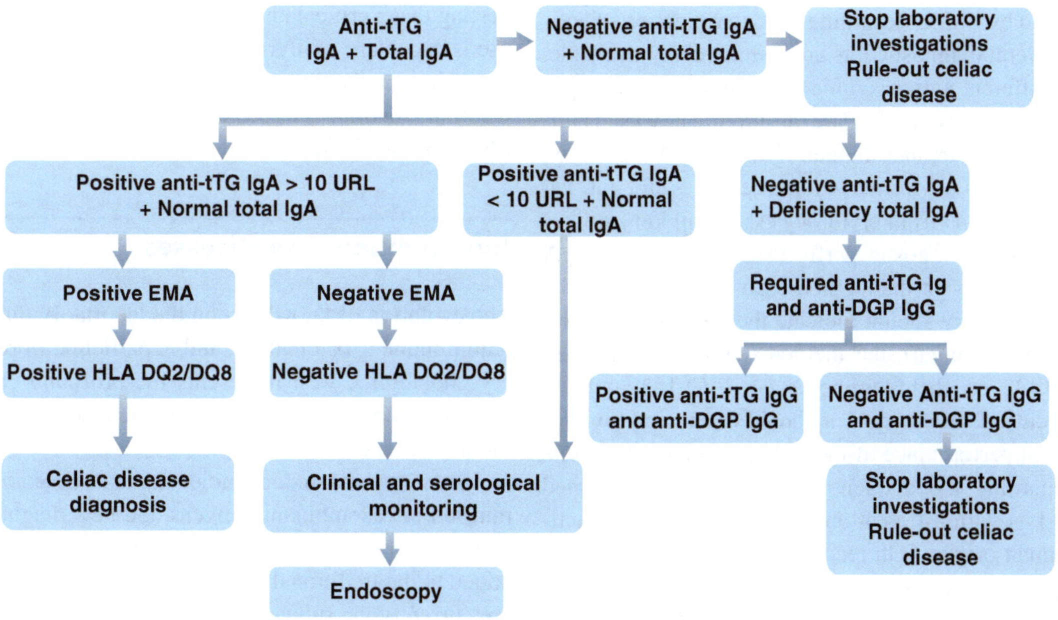

Fig. 36.1 Algorithm for the diagnosis of celiac disease in subjects according to clinical manifestations and age. (Copyright EDISES 2021. Reproduced with permission)

For the follow-up of patients with celiac disease, the current ministerial guidelines recommend a check-up within 6–12 months of diagnosis and, subsequently, every 1–2 years (except in the event of complications), with a blood count and anti-tTG IgA (or IgG in the case of IgA deficiency) at each check-up. It is also appropriate to perform blood tests for iron metabolism (sideremia, ferritin) and folatemia at the first check-up and repeat them at subsequent checks until normalization in case of abnormal values. Since autoimmune thyroiditis is frequently associated with celiac disease, TSH and TPOAb should be determined at diagnosis, and if both are negative, TSH should be repeated every 3 years; if both are positive, the patient should be taken to a specialist clinic for autoimmune thyroid disease. Bone densitometry is recommended only in adults, after 18 months of gluten-free diet. Other eventual haematological and instrumental tests are recommended on the basis of the clinical evaluation.

Quality Characteristics of Anti-transglutaminase Antibodies Assessment

The tests for diagnosing the celiac condition still have very differentiated and inhomogeneous analytical specifications where the process of harmonization has certainly not yet been concluded. The negative predictive value of all these diagnostic tests is rather high, but only anti-tTG class IgA and anti-endomysium have such high specificity and sensitivity as to be considered satisfactory. Some comparative studies have compared the EMA, anti-tTG and anti-DGP serological tests: the test with the best characteristics in terms of sensitivity and specificity appears to be the anti-tTG IgA, followed by EMA. Regarding the determination of anti-tTG IgA, several diagnostic kits are commercially available, which use different antigens, units and measurement ranges and suggest different cut-offs. The most frequently used antigens are of recombinant human origin on E. coli, on Baculovirus or derived from myeloma cell lines and purified human-type antigens from erythrocytes. In children, studies indicate that the diagnostic performance of anti-tTG IgA assays is dependent on the method used and the age of the child. The laboratory should indicate the antigen (recombinant or purified human), the method (enzyme immunoassay – ELISA, or chemiluminescence – CLIA) and the diagnostic cut-off. In addition, it should define and monitor the analytical performance of the tests performed within its scope. Regarding the antigen, a meta-analysis evaluated diagnostic kits with different antigens, indicating that antigens of human origin, both recombinant and purified, have superior specificity and sensitivity.

The heterogeneity of the results of the tests carried out in the various laboratories was highlighted in 2009 by a study whose results showed great variability in the sensitivity (69–93%) and specificity (96–100%) of the various methods and a rather important coefficient of intra-assay variation, suggesting that a critical knowledge of the analytical methods in use is necessary. With regard to the appropriateness of diagnostic tests in the diagnosis of celiac disease, a "decalogue of recommendations" has been proposed by SIBioC-Laboratory Medicine, see Table 36.1.

Table 36.1 Appropriateness of laboratory tests in diagnosising celiac disease: Decalogue of recommendations, from Sibioc-Laboratory Medicine

1.	The diagnosis of celiac disease is based on the integration of clinical, serological, and histological data. It is essential that all diagnostic tests are carried out on a free diet, avoiding the suspension of gluten intake, in order not to distort the results of the tests.
2.	The biomarkers of first choice in diagnosis are IgA anti-transglutaminase antibodies (anti-tTG IgA) and total IgA. If IgA deficiency is detected, antibodies to gliadin deamidated peptides (anti-DGP IgG) are measured.
3.	The genetic test for the detection of HLA DQ2/DQ8 is indicated for screening of subjects at risk before serology: Some experts indicate its use at birth for first degree relatives of celiac subjects.
4.	Diagnostic first level tests for celiac disease in subjects with clinical manifestations and over 2 years of age are anti-tTG IgA and total IgA.
5.	The test of choice for subjects with IgA deficiency are anti-tTG IgG and/or anti-DGP IgG.
6.	The most important test for asymptomatic individuals belonging to risk groups is HLA DQ2/DQ8.
7.	The EMA test in the diagnostic procedure remains controversial.
8.	In the diagnostic report of the biomarkers, the antigen used in the method, the method and the diagnostic threshold value must be clearly explained.
9.	In symptomatic children with concomitant presence of HLA DQ2/DQ8 and elevated levels of anti-tTG IgA (10 times the URL), the diagnosis can be made without biopsy.
10.	The absence of DQ2/DQ8 in subjects at risk allows for excluding celiac disease and to avoid its monitoring.

URL upper reference limit

Autoimmune Liver Diseases

The evolution of knowledge in the laboratory diagnostics of autoimmune hepatopathies, and in particular of primary biliary cholangitis, well represents the profound changes that have affected the field of autoimmune disease diagnostics in the last decade.

In recent years, in fact, molecular tests based on the use of purified or recombinant antigens and new diagnostic strategies and solutions, some of which are automated, have been added to the traditional methods in immunofluorescence and are currently the subject of different study protocols.

Autoimmune diseases of the liver are a heterogeneous group of immune-mediated pathologies characterized by an autoimmune aggression towards hepatocytes or cholangiocytes and can be classified into three major entities, primary biliary cholangitis, autoimmune hepatitis and primary sclerosing cholangitis, which contain within them some variants as well as overlapping syndromes, so-called Overlap, where both a cholestatic and an inflammatory component is present to varying degrees. In this field, the distinction between hepatocellular damage and cholestatic damage is fundamental for the purposes of differential diagnosis, and consequent therapeutic choices and antibody diagnostics plays an essential role in the correct classification of the different nosological entities.

Autoimmune hepatitis (AIH) is a chronic inflammation of the liver of unknown cause, in which tolerance to the hepatocyte is lost. The mechanisms underlying liver damage are cell-mediated and direct cytotoxicity.

The disease mainly affects females in both childhood and adulthood and may be associated with other autoimmune conditions (ulcerative colitis, autoimmune thyroiditis, primary biliary cholangitis and primary sclerosing cholangitis).

It is usual to classify AIH on the basis of specific autoantibodies, which for type 1 are mainly ANA and anti-smooth muscle antibodies (ASMA) (Fig. 36.2); type 2, on the other hand, usually presents LKM1 (liver-kidney microsomal antibodies). Other antibodies found in AIH are anti-SLA (soluble liver antigens), anti-LP (liver-pancreas), anti-ASGPR (asialo-glycoprotein) and anti-LC1 (liver cytosol type 1). Positivity for ANA, ASMA, LKM1 and SLA is included in the diagnostic criteria for AIH.

Primary biliary cholangitis (CBP) is a relatively rare chronic cholestatic hepatopathy with slow evolution and autoimmune aetiology characterized by an aggression of T lymphocytes against the cells of the intrahepatic bile ducts; the consequent damage makes the drainage of bile from the liver to the intestine difficult, thus causing stagnation in the liver and resulting in liver failure in a variable time.

Patients with CBP are primarily female (90%), usually diagnosed in the fifth or sixth decade of life. It has been suggested that genetic susceptibility is a predisposing factor for CBP, while environmental factors such as infections, chemicals and smoking may play a causal role.

The course of the disease is variable, and an accurate diagnosis in the early stages is crucial as early drug treatment can slow the progression to liver failure and improve survival.

The diagnosis is based on a combination of clinical/histological features, abnormalities of the liver biochemical profile in a setting of persistent cholestasis for more than 6 months and the presence of AMA and/or specific ANA antibodies in the serum.

AMA antibodies are detected in about 90% of CBP patients, and their high sensitivity and specificity makes their presence one of the three diagnostic criteria for CBP. The AMA are actually a heterogeneous group of antibodies and are directed with various percentages of frequency against different autoantigens that are part of the structural complex of the enzyme 2-oxyacid dehydrogenase, namely the subunits E2 of the enzyme pyruvate dehydrogenase (PDC-E2 – frequency 90%), the complex oxyglutaric dehydrogenase

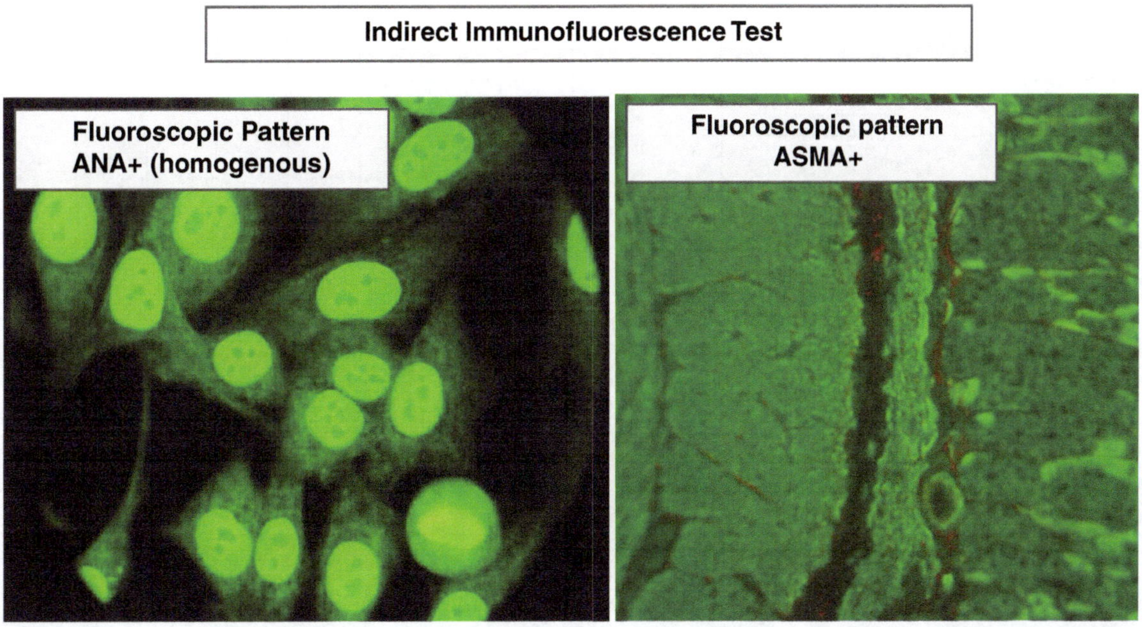

Fig. 36.2 Fluoroscopic pictures relating to positivity for anti-nucleus (ANA +, left) and anti-smooth muscle (ASMA +, right) autoantibodies

(OGDC-E2 – frequency 50%) and branched-chain ketoacid dehydrogenase (BCOAD-E2 – frequency 50%).

AMAs can be detected in the serum of asymptomatic patients with normal liver enzyme balance; follow-up studies have shown that over time most of these subjects develop CBP.

For a long time, the indirect immunofluorescence test (IIF) on cryostat sections of the liver, kidney and stomach of rats was considered the gold standard for routine screening of AMA. In recent years, on one hand, the limitations presented by this test (complex as execution, difficult to standardize, not fully automated, operator-dependent interpretation) and, on the other hand, the identification of molecular targets of AMA have led to the development, validation and introduction in the laboratory diagnostics of molecular antigen-specific tests using the technique of ELISA immunoassays or Immunoblotting.

The promising results obtained with the latest generation of ELISA tests using the MIT3 antigen (PDC-E2, OGDC-E2 and BCOAD-E2), capable of detecting positivity in about half of AMA IIF negative patients, suggest the use of these tests in place of or alongside IIF for AMA testing, particularly when laboratories are unfamiliar with IIF use and interpretation.

Immunoblotting assays also represent an interesting diagnostic alternative to the AMA in IIF because this multiplex assay allows the simultaneous detection of different CBP-specific autoantibodies using recombinant antigens.

In addition to AMA, ANA, usually associated with autoimmune rheumatic diseases, are frequently found in CBP patients. At present, the method considered the gold standard for the determination of ANA is IIF, using HEp-2 cells as substrate. CBP-specific ANA have been detected by several authors in 30–50% of patients, and two different fluorescence patterns have been described for these autoantibodies: rime-like/membranous (specificity for gp210 and Nucleoporin p62 proteins) and multiple nuclear dots (specificity for sp100, PML and small ubiquitin-like modifier proteins). The detection of these antibodies, characterized by a low sensitivity, allows to confirm the diagnosis of CBP in AMA-negative patients, frequently liable to misclassification. There is also some evidence that the presence of specific CBP ANA, particularly anti-gp210, is associated with a worse prognosis and a more aggressive disease. Finally, as for AMA, also for specific CBP ANA, the new solid phase tests (ELISA and Immunoblotting), especially those using recombinant antigens, such as MIT3, appear to be more sensitive and less subjective than IIF (Fig. 36.3).

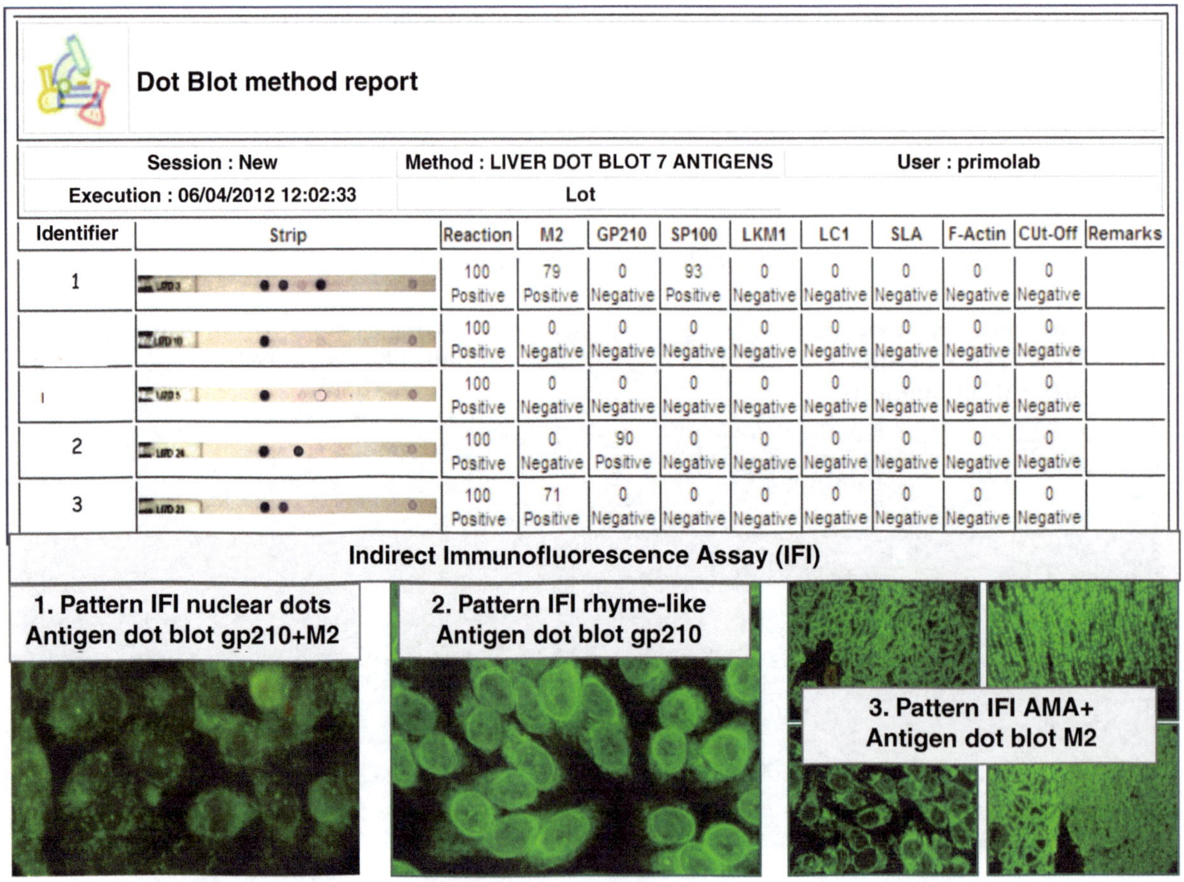

Fig. 36.3 Molecular antigens and analogous CBP-associated fluoroscopic pictures

Other non CBP-specific ANA, mainly anti-Centromere and anti-SSA/Ro – 52 kDa antibodies, are often detected (10–30%) in CBP patients, conferring, according to some authors, unfavourable prognostic features to the disease course. This finding correlates with the observation of the association of CBP with other autoimmune diseases, such as scleroderma, Sjogren's syndrome and CREST syndrome.

Finally, it is interesting to report new markers currently under study for the laboratory diagnosis of CBP, namely two new self-antigens (KLHL12 and HK1), genetic markers, particular metabolic profiles, miRNAs and epigenetic factors.

With regard to CBP, it is interesting to note, in summary, how the increased knowledge of the serological associations of the disease, together with the widespread use of non-invasive tests, such as laboratory tests, has significantly modified the initial clinical presentation in recent years, often allowing an early diagnosis to be made with respect to the detection of advanced liver disease.

Primary sclerosing cholangitis (PSC) is a chronic cholestatic syndrome characterized by inflammatory fibrosis of the intra- and extrahepatic bile ducts.

SPC typically affects young males and is commonly associated with inflammatory bowel disease, especially ulcerative colitis (CU).

From the point of view of laboratory diagnostics, most patients present an elevation of serum alkaline phosphatase and γGT that may be accompanied by a modest increase in transaminases; unlike CBP, in CSP, the search for AMA antibodies is negative. The determination of anti-neutrophil cytoplasmic antibodies (ANCA) with a pANCA panel in ethanol, an autoantibody marker also present in a high percentage of patients with CU (approximately 80%), was found to be useful for diagnosis.

In summary, laboratory diagnostics of hepatic autoimmune diseases, both in terms of diagnosis differential and early diagnosis, has made important progress in recent years and, without doubt, the prospect is open to new developments. In this field, an issue that could be of great interest in the near future is the improvement in the standardization of methods and the definition of new diagnostic algorithms.

At present, the detection of specific autoantibody profiles is essential, among autoimmune liver diseases, for the diagnosis of AIH and CBP, while the diagnosis of CSP is made by means of the picture provided by endoscopic retrograde cholangiopancreatography. The crucial value of autoantibodies for the differential diagnosis of EAI and CBP is expressed in the consequent therapeutic choices.

Failure to detect autoantibodies does not exclude the presence of an autoimmune hepatopathy, for example CBP AMA negative.

In this diagnostic field, in perspective, the percentage of patients negative for AMAs, which still represent the key immunological marker for CBP, will be significantly reduced by the use of new antigen-specific laboratory tests (ELISA and Immunoblotting tests) based on the use of recombinant antigens.

In addition, the increased use of automated systems that include a full panel of related CBP self-antigens, including the nuclear antigenic targets gp210 and sp100, will contribute to the optimization of new diagnostic algorithms. Conducting multicentre studies evaluating newly diagnosed CBP patients may effectively allow the evaluation of these new analytical tools in the screening and differential diagnostics of CBP. The results of these studies could contribute to redesign the entire diagnostic process, starting from the formulation of a reasonable clinical suspicion to the request for tests, up to the use of the most appropriate autoantibody tests and ensuring through the shared implementation of common algorithms a harmonization among different laboratories.

In other words, clinical governance and standardization in the field of immunological laboratory diagnostics of autoimmune liver diseases, as for other autoimmune diseases, are two essential elements to enhance the potential of autoantibody biomarkers both in diagnostic and prognostic fields.

Recommended Readings

Accordo, ai sensi dell'articolo 4, del decreto legislativo 28 agosto 1997, n. 281, tra il Governo, le regioni e le province autonome di Trento e di Bolzano sul "Protocollo per la diagnosi e il follow up della celiachia" (Rep. Atti n. 125/CSR) Gazzetta Ufficiale della Repubblica Italiana Serie generale n. 191 del 19 agosto 2015

Bizzaro N, Pasini P (1999) Correlazione tra titolo e pattern degli anticorpi anti-nucleo in immunofluorescenza indiretta e positività degli anticorpi anti-ENA. Med Lab 7:419

Bizzarro N, Wiik A (2004) Appropriateness in the anti-nuclear antibody testing: from clinical request to strategic laboratory practice. Clin Exp Rheumatol 22:349–355

Bonaguri C, Melegari A, Ballabio A et al (2011) Italian multicentre study for application of a diagnostic algorithm in autoantibody testing for autoimmune rheumatic disease: conclusive results. Autoimmun Rev 11:1–5

Bonaguri C, Melegari A, Dall'Aglio P et al (2009) An Italian multicentre study for application of a diagnostic algorithm in autoantibody testing. Ann N Y Acad Sci 1173:124–129

Catassi C, Gatti S, Fasano A (2014) The new epidemiology of celiac disease. J Pediatr Gastroenterol Nutr 59(Suppl 1):S7–S9

European Association for the Study of the Liver (2017) EASL Clinical Practice Guidelines: the diagnosis and management of patients with primary biliary cholangitis. J Hepatol 67(1):145–172

Floreani A, Caroli D, Variola A et al (2011) A 35 year follow-up of a large cohort of patients with primary biliary cirrhosis seen at a single centre. Liver Int 31:361–368

Ford AC, Sandborn WJ, Khan KJ et al (2011) Efficacy of biological therapies in inflammatory bowel disease: systematic review and meta-analysis. Am J Gastroenterol 106:644–659

Gabeta S, Norman GL, Liaskos C et al (2007) Diagnostic relevance and clinical significance of the new enhanced performance M2 (MIT3)

ELISA for the detection of IgA and IgG antimitochondrial antibodies in primary biliary cirrhosis. J Clin Immunol 27:378–387
Gatselis NK, Dalekos GN (2016) Molecular diagnostic testing for primary biliary cholangitis. Expert Rev Mol Diagn 9:1001–1010
Gershwin ME, Makay IR, Sturgess A, Coppel RL (1987) Identification and specificity of cDNA encoding the 70 kd mitochondrial antigen recognized in primary biliary cirrhosis. J Immunol 138(10):3525–3531
Granito A, Muratori P, Quarneti C et al (2012) Antinuclear antibodies as ancillary markers in primary biliary cirrhosis. Expert Rev Mol Diagn 12:65–74
Ha C, Mathur J, Kornbluth A (2015) Anti-TNF levels anti-drug antbosidues, immunosuppressants and clinical outcomes in inflammatory bowel disease. Expert Rev Gastroenterol Hepatol 9:497–505
Hirschfield GM, Dyson JK, Alexander GJM et al (2018) The British Society of Gastroenterology/UK-PBC primary biliary cholangitis treatment and management guidelines. Gut 67(9):65–74
Husby S, Koletzko S, Korponay-Szabo IR et al (2014) European Society for Paediatric Gastroenterology, Hepatology and Nutrition guidelines for the diagnosis of celiac disease. J Pediatr Gastroenterol Nutr 54:136–160
John M. Eisenberg Center for Clinical Decisions and Communications Science (2016) Diagnosis of celiac disease: current state of the evidence. Baylor College of Medicine, Houston, TX, Issued: July 26, 2016. Agency for Healthcare Research and Quality AHRQ USA. https://www.ncbi.nlm.nih.gov/books/NBK379842/pdf/Bookshelf_NBK379842.pdf. Accessed 12 December 2016
Kavanaugh A, Tomar R, Reveille J et al (2000) Guidelines for clinical use of the antinuclear antibody test and tests for specific autoantibodies to nuclear antigens. Arch Pathol Lab Med 124:71–81
Kupfer SS, Jabri B (2012) Celiac disease pathophysiology. Gastronintest Endosc Clin N Am 22(4):639
Leonard MM, Camhi S, Huedo-Medina TB et al (2015) Celiac disease genomic, environmental, microbiome, and metabolomics (CDGEMM) study design: approach to the future of personalized prevention of celiac disease. Nutrients 7:9325–9336
Lindor KD, Gershwin ME, Poupon R et al (2009) Primary biliary cirrhosis. Hepatology 50:291–308
Melegari A, Bonaguri C, Russo A et al (2012) A comparative study on the reliability of an automated system for the evaluation of cell-based indirect immunofluorescence. Autoimmun Rev 11:713–716
Mitrev N, Leong RW (2017) Therapeutic drug monitoring of anti tumor-necrosis factor α agents in inflammatory bowel diseases. Expert Opin Drug Saf 16(3):303–317
Nakamura M, Komori A, Ito M et al (2007) Predictive role of anti-gp210 and anticentromere antibodies in long-term outcome in primary biliary cirrhosis. Hepatol Res 37:5412–5419
Nielsen OH, Ainsworth MA (2013) Tumor necrosis factor inhibitors for inflammatory bowel disease. N Engl J Med 369:754–762
Ordás I, Mould DR, Feagan BG et al (2012) Anti-TNF monoclonal antibodies in inflammatory bowel disease: pharmacokinetics-based dosing paradigms. Clin Pharmacol Ther 91:634–626
Plebani M, Tozzoli R, Trenti T et al (2017) Appropriatezza della richiesta di esami ed esiti clinici: il caso delle malattie renali, tiroidee e della celiachia. Biochim Clin 41:266–285
Porcelli B, Alessio MG, Villalta D et al (2015) Linee guida per la diagnosi di laboratorio e istologica della malattia celiaca. Revisione 2015. Riv Ital Med Lab 11:76–95
Porcelli B, Cinquanta L, Barberio G et al (2016) Quanto e come vengono applicate le linee guida in Autoimmunologia nei Laboratori italiani? Riv Ital Med Lab 12:221–233
Rigby WF (2007) Drug insight: different mechanisms of action of tumor necrosis factor antagonists passive-aggressive behavior? Nat Clin Pract Rheumatol 3:227–233
Rossi E, Basso D, Zambon CF et al (2015) TNFA haplotype genetic testing improves HLA in estimating the risk of celiac disease in children. PLoS One 10:e0123244. https://doi.org/10.1371/journal.pone.0123244
Schaeverbeke T, Truchetet ME, Kostine M et al (2016) Immunogenicity of biologic agent in rheumatoid arthritis: lessons from clinical practice. Rheumatology 55:210–220
Strassburg LP (2010) Autoimmune hepatitis. Best Pract Res Clin Gastroenterol 5:667–682
Thalayasingam N, Isaacs JD (2011) Anti-TNF therapy. Best Pract Res Clin Rheumatol 25:549–567
Tighe D, McNamara D (2017) Clinical impact of immunomonitoring in the treatment of inflammatory bowel disease. World J Gastroenterol 23:414–425
Tonutti E, Bizzaro N (2014) Diagnosis and classification of celiac disease and gluten sensitivity. Autoimmun Rev 13:472–476
Tozzoli R, Bizzaro N, Tonutti E et al (2002) Guidelines for the laboratory use of autoantibody tests in the diagnosis and monitoring of autoimmune rheumatic diseases. Am J Pathol 117:316–324
Van der Windt DA, Jellema P, Mulder CJ et al (2010) Diagnostic testing for celiac disease among patients with abdominal symptoms: a systematic review. JAMA 303:1738–1746
Vande Casteele N, Gils A, Singh S et al (2013) Antibody response to infliximab and its impact on pharmacokinetics can be transient. Am J Gastroenterol 108:962–971
Villalta D, Sorrentino C, Girolami E et al (2015) Autoantibody profiling of patients with primary biliary cirrhosis using a multiplex line-blot assay. Clin Chim Acta 438:135–138
Vincent FB, Morand EF, Murphy K et al (2013) Antidrug antibodies (ADAb) to tumour necrosis factor (TNF)-specific neutralising agents in chronicinflammatory diseases: a real issue, a clinical perspective. Ann Rheum Dis 72:165–178
Wiik AS (2003) Appropriateness of autoantibodies testing in clinical medicine. Clin Chim Acta 333(2):177–180

Laboratory Diagnostics in Allergic Diseases

37

Alda Tiziana Scacchetti and Tommaso Trenti

Introduction

Currently, a considerable amount of evidence seems to indicate that the increased susceptibility of allergic individuals to produce allergen-specific IgE is under the control of multiple genetic and environmental factors. Although IgE is the immunoglobulin class quantitatively less represented in humans, the interaction between these antibodies, allergens, and effector cells gives rise to an impressive cellular immune reaction, which can trigger the well-defined clinical picture of allergic diseases.

In the past decades, immunoallergological diseases have registered a constant increase in their prevalence, sometimes with abrupt accelerations; they are characterized by heterogeneous and evolving clinical pictures.

Based on this general description, allergies could be considered an epidemic due to their inexorable progression, especially in the most industrialized areas of the world. Indeed, they are not contagious, yet they seem unstoppable; in most cases, they do not seriously jeopardize survival but weigh on the quality of life; they are treatable but sometimes not curable in their dynamic chronicity.

Discovery of Immunoglobulin E

The discovery of immunoglobulin E (IgE) in the late 1960s allowed the use of specific biomarkers of allergy in the diagnostic process. IgE was discovered in 1967 during the studies of Gunnar Johansson and Hans Bennich in IgE myeloma. It is structurally made up of two heavy chains and two light chains.

A. T. Scacchetti
Department of Laboratory Medicine and Pathological Anatomy, Baggiovara Hospital, Modena, Italy

T. Trenti (✉)
Department of Laboratory Medicine and Pathological Anatomy, Azienda USL/Azienda Ospedaliero-Universitaria di Modena, Modena, Italy
e-mail: t.trenti@ausl.mo.it

The possibility of having large quantities of purified IgE, obtained from the few cases of IgE myelomas, has allowed us to obtain, through animal immunization, purified anti-IgE antibodies, with consequent development of radioimmunological and enzyme immunoassay methods, which are now widely used in the diagnosis of allergic diseases.

Subsequent research has essentially focused on studying the molecular and cellular basis of the IgE response and subsequent IgE-mediated reactions, identifying the various chemical mediators, preformed or neoformed, and on the definition of their pathogenetic role in immediate and late reactions and the subsequent induced allergic inflammation.

Mast Cells and Basophils

IgE-binding cells were identified by incubating human leukocytes and tissue cells with iodine-125 (125I)-labeled myeloma IgE or with 125I-labeled IgE-specific antiserum. In both cases, autoradiography made it possible to demonstrate that the labeled probes used in the studies conjugated to blood basophils and tissue mast cells.

Basophilic granulocytes, in humans, represent 0.5–1% of circulating leukocytes. Their granule-rich cytoplasm is stained by basic dyes, hence the name of basophils. Under the electron microscope, they are characterized by a plurilobed nucleus, relatively few mitochondria, numerous glycogen granules, and membrane-bound electron granules distributed throughout the cytoplasm.

Paul Elrich first described mast cells in 1877. They were named so in reference to the numerous granules mistakenly believed to be exogenous and ingested by the cell. Mast cell precursors are formed in the bone marrow during hematopoiesis and then migrate to all peripheral vascularized tissues, where they differentiate into mature cells. They are present in the connective tissue contiguous to blood and lymph vessels and, in high numbers, in the skin and mucosal surfaces of the respiratory and gastrointestinal tracts.

M. Ciaccio (ed.), *Clinical and Laboratory Medicine Textbook*, https://doi.org/10.1007/978-3-031-24958-7_37

Allergens

First, the terms "allergenic source," "allergenic extract," and "allergen" should be appropriately defined.

The term allergen source refers to the material container of allergens; for example, peach, dog, egg, and milk are allergen sources.

Allergenic extracts, commonly used in in vivo and in vitro laboratory diagnostics, come from defined allergenic sources (peach, dog, mites, grass pollen, etc.) and are obtained by extraction and purification processes. Subsequently, the extracts are standardized according to different methods in the various diagnostic industries, with consequent results that are not comparable with each other due to the present lack of a harmonization process.

The quality of these extracts has improved substantially over the years; however, it still presents criticalities and limitations that are difficult to eliminate. The criticalities are intrinsically linked to the extraction processes, which cause the loss of some allergenic proteins, the acquisition of proteins from unknown sources, and the different concentrations and protein composition from one batch to another. Only recently, the concentrations (expressed in µg/mL) of major allergenic proteins have been measured in the extracts, but not of minor proteins, which might even be absent. The absence or low concentration of allergenic proteins in the extract may cause false negatives during diagnosis and consequent ineffectiveness of hyposensitizing therapies when the proteins contained in the extract are not present at the concentrations necessary to induce desensitization.

The results of in vitro specific IgE according to the extracts used vary according to the type of extract used, and the method used (CAP, Thermo Fisher Scientific; Immulite, Siemens etc.) and are therefore not comparable. To these variables, which can be defined as "dependent on the diagnostic technology," must be added those related to the clinical symptoms, the age of the patient, the time when the tests are performed (onset of the disease or follow-up), the prevalence of allergy in the population studied, the clinical complexity of the allergic pathology, and its globality.

The limitation, so far insuperable, consists in the impossibility of establishing, in a patient showing polysensitization to specific IgE in vitro, whether the polysensitization is due to cosensitization (sensitization to distinct and unique molecules from different allergen sources) or a mechanism of co-recognition (sensitization to different allergen sources containing homologous molecules).

The term allergen refers to a protein, glycoprotein, or carrier-conjugated haptene, with a molecular weight of 5–150 kDa and an isoelectric point between 2 and 10, capable of binding specific IgE and inducing an allergic reaction.

Thus, each allergen source contains different allergenic proteins, and each allergen may have a different number of antigenic determinants or epitopes.

An epitope is defined as an amino acid sequence recognized by a specific antibody; for example, milk-specific IgE recognizes specific epitopes contained in milk.

Generally, epitopes can be distinguished into linear, when IgE recognizes a contiguous amino acid sequence in the primary structure of the antigen, and conformational, when IgE recognizes a noncontiguous amino acid sequence characteristic of the three-dimensional structure of the protein (Fig. 37.1).

The primary sequence of an allergen can be easily found in online search sites such as Allergome (AllergomeAligner, www.allergome.org/ script/tools.php?tool = blaster) or BLAST in UniProt (www.uniprot.org).

The structural folds of a protein are of primary importance in provoking immunological sensitization and the relative antibody response. Many allergenic proteins, if subjected to heat or the action of proteolytic enzymes, as occurs during food preparation or the digestive process, undergo modifications that can determine the loss of conformational epitopes but also the possible unmasking of linear epitopes.

Food allergens can be divided into:

- Class 1 food allergens: these are made up of proteins resistant to digestion and heat and can act as sensitizers in the gastrointestinal tract. To this class belong, for example, the major allergenic proteins of milk, egg, fish, crustaceans, and some vegetables.

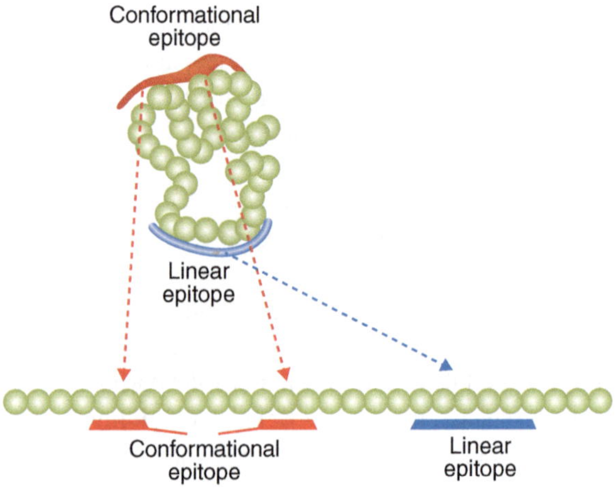

Fig. 37.1 Conformational epitopes consist of noncontiguous amino acids in the protein's primary structure, but they to be adjacent in the tertiary structure due to the folding of the protein chain. Their formation, therefore, depends on the three-dimensional structure of the allergen. Linear epitopes, on the other hand, are made up of amino acids that are adjacent in the primary structure of the protein. (Copyright EDISES 2021. Reproduced with permission)

- Class 2 food allergens consist of proteins that are not resistant to heat and digestion and are generally incapable of causing systemic symptoms. They are present in plants and foods of animal origin (thermolabile proteins of milk, meat, and eggs) and cause symptoms mainly localized to the oral cavity (oral allergy syndrome) as they lose their antigenic power following degradation in the stomach. The symptoms appear after sensitization to homologous allergens contained in pollens (nonsensitizing elicitors). This phenomenon, defined as cross-reactivity, explains why some patients can present even severe reactions when taking allergenic foods never before ingested.

The allergenicity of a single protein, therefore, depends on:

- its epitopes;
- its spatial conformation upon exposure to antigen-processing cells, such as macrophages, dendritic cells, or B lymphocytes;
- avidity (degree of reaction) between IgE and epitopes, which in turn depends on the number of allergenic epitopes on the molecule (valence), the size, and conformation of the molecule;
- degree of affinity between antibodies and epitopes, which increases in the course of the humoral immune response.

In recent years, 2503 molecular allergens have been characterized at the molecular level, and the latest update is January 6, 2017.

Identifying and characterizing allergenic sources have led to the subsequent industrial production and marketing of natural allergens purified or produced with recombinant DNA technologies.

The recombinant molecules thus obtained have a sensitivity of over 70% in mimicking the allergenic source.

Allergenic molecules are divided into genuine, true markers of a specific source (e.g., Ole e 1 is the marker protein of allergy to olive tree pollen and other Oleaceae), and panallergens, proteins shared by allergenic sources even taxonomically unrelated to each other, responsible for apparent polysensitization to tests performed with extracts. For example, profilin is a panallergen shared by pollens and plant foods. Its recognition by a patient allergic to pollens will cause positivity to all types of pollens and plant foods tested without the patient experiencing symptoms upon exposure to them.

The naming of the components observes an international convention:

- The first three letters (such as Phl, Bet, etc.) correspond to the first three letters of the Linnean name of the allergenic source (in the example, Phleum, Betula);
- The fourth letter (in lower case, as defined by the nomenclature of living organisms) indicates the first letter of the second name of the allergenic source. For this reason, a molecular component of Phleum pratense is defined as Phl p;
- A number is added to the letters to distinguish each component from all the others: Phl p1 indicates the first component identified (and usually cloned) in Phleum pratense;
- Other numbers can be used to define the component further: for example, Amb a 1, from Ambrosia artemisiifolia, has some isoallergens: Amb a 1.01, Amb a 1.02, Amb a 1.03, and Amb a 1.04. For Amb a 1.01, three different variants have been described (Amb a 1.0101, Amb a 1.0102, and Amb a 1.0103), characterized by a very high homology in the primary sequence;
- Finally, the letter "r" or "n" preceding the name of the component indicates its origin (r for recombinant or n for natural). Recombinant components are allergens cloned into eukaryotic or prokaryotic vectors using genetic engineering techniques; when a component is produced in prokaryotes (e.g., *Escherichia coli*), it does not have glycosylated chains.

Extractive natural molecules are highly purified (in this case, posttranslational modifications, such as glycosylations, are present).

Use of Recombinant Allergens in Allergology Diagnostics

The use of recombinant (or highly purified native) allergens to replace allergenic extracts represents a remarkable achievement in allergology as it improves the clinical-diagnostic process for several reasons. The first is that it allows overcoming one of the most critical hurdles related to the use of allergenic extracts, which is that of standardization, and this is because of their variability in composition and antigenic content, the diversity of supply sources, the presence of proteolytic enzymes and contaminated allergens, etc. Recently, the WHO/IUIS standardization committee (www.allergen.org) has embarked on a new program to introduce new standards from purified or recombinant native proteins, to be distributed to industries or academic bodies to prepare in vitro tests or to regulatory bodies for the comparison of allergen products. Using these standards allows for defining the allergenic content in mass units, which, if it can have a relative meaning in laboratory diagnostics, has an essential meaning in the preparations for specific immunotherapy. Moreover, the standard obtained using recombinant molecules has the advantage of being able to be reproduced unmodified and in practically unlimited quantities over time. Another essential advantage of molecular diagnostics is the ability to discriminate whether a state of polysensitization identified in vivo or in vitro, using extractive preparations, is due to cosensitizations, i.e., primary sensitizations toward specific major or minor allergenic proteins present in single allergenic sources, or to cross-sensitizations, i.e., cross-reactivities toward homologous molecules present in different allergenic sources, sometimes without clinical significance. The most important advantage, however, is at the level of diagnostic definition. A diagnosis based on traditional allergenic extracts can only lead to the identification of the allergenic source (e.g., allergy to birch, mites, etc.), but not of the molecular entity to which a patient is sensitized (e.g., Bet v1, Bet v2, Bet v4, Der p1, Der p10, etc.). Identifying the individual allergenic profile of a patient is not only a diagnostic refinement, but it has significant repercussions both prognostic and therapeutic, as described later.

In Vitro Allergy Diagnostics

The use of in vitro tests by various methods introduced in recent decades has undoubtedly led to significant progress in diagnosing IgE-mediated allergic syndromes.

Among the advantages of these tests, globally considered, we should point out above all the remarkably high sensitivity and specificity, the good reproducibility, the absolute harmlessness for the patient, and the possibility of being researched even while taking antihistamine therapy; the disadvantages are summarized, in practice, in the relatively high cost.

Detection of Total IgE

The determination of total IgE was introduced many years ago with the PRIST (Paper RadioImmunoSorbent Test) and is currently performed by various automated methods.

The serum concentration of total IgE varies, in non-atopic adults, from 10 to 200 IU/mL; in infants, this concentration is only a few IU/mL, and it progressively increases and reaches adult levels around the tenth year of life.

Total IgE is generally high in allergic syndromes due to IgE-mediated immunoreactions. Notably, the highest values of IgE are found in allergic diseases, but the finding of "normal" values of total serum IgE does not exclude the diagnosis of allergic disease. Indeed, many allergic patients have total IgE levels within the normal range.

It should also be mentioned that total IgE is found with altered values in various pathological conditions that are certainly not allergic and physiological or paraphysiological conditions (e.g., in smokers).

Therefore, it can be concluded that the determination of total IgE alone is of little clinical significance in allergic diseases.

Detection of Specific IgE

Since 1967, the year in which the first specific IgE assay was performed by Wide and collaborators, with the birth of the RAST (RadioAllergoSorbent Test), much progress has been made in the field of laboratory allergology diagnostics, through various stages consisting of the development of technologies based on solid phases with high binding capacity; use of enzymatic tracers linked to monoclonal IgE antibodies with calibration curves made with standards of known titre in quantitative units kUa/L and produced according to an international reference standard (now 3° IS WHO 11/234, formerly 2° IRP WHO 75/502 for total IgE); the introduction of diagnostic systems for the determination of specific IgE with high analytical sensitivity in total automation (third generation test).

Over the years, RAST has been followed by numerous other methods, which have replaced radioactive markers, and the techniques developed have been numerous (ELISA [Enzyme Linked ImmunoSorbent Assay], agglutination, precipitation). These in vitro methods differed significantly in terms of the type of detection system (colorimetric, fluorimetric, and chemiluminescence), the antiserum (mono or polyclonal, single or mixed), the support for carrying the allergen (cellulose polymers, polystyrene spheres, etc.), and the development in a solid or liquid phase, resulting in significant inhomogeneity among the different systems available in the laboratory setting. Since not all systems are the same, and not all are equally valid, the laboratory must guarantee the reproducibility of the data (system with a coefficient of variation <15% – NCCLS, 2004) and is certified by a national and/or international quality control (QC) program. Currently, the most used methods for detecting specific IgE are Cap FEIA and 3gAllergy.

Cap FEIA on Immunocap

It uses an immunofluorenzyme method and is based on a solid phase (cellulose polymer) covalently conjugated with the allergen, which reacts with the serum under examination so that the specific IgE, if present, binds to the allergen. After washing to remove unbound IgE, i.e., not specific for the allergen, labeled anti-IgE antibodies are added. Thus, a "sandwich" complex consists of the solid phase with allergen + patient-specific IgE + labeled anti-IgE. Using the fluorimetric detection method, the amount of marker present in the complex is measured, which will be directly proportional to the amount of specific IgE present in the sample under examination.

3gAllergy on Immulite 2000

It uses an enzyme-enhanced chemiluminescent method and is based on the use of a liquid phase. Patient serum-containing specific IgE and liquid allergen are incubated in the presence of antiligand coated beads. The liquid allergen, which acts as a ligand, binds to the specific IgE in the serum and the antiligand-coated beads. The ligand-specific IgE is detected by an anti-IgE antibody conjugated to an enzyme that catalyzes the chemiluminescent reaction in the presence of its specific substrate. Also, in this method, the concentrations obtained are proportional to the amount of specific IgE present in the patient's serum under examination.

Molecular Allergy Diagnostics

Starting from the mid-nineties of the last century, in parallel with the development of proteomic techniques, we have witnessed the birth of molecular allergology, which has provided various confirmations and numerous new findings in the field of allergology. In particular, the following must be mentioned:

- Most allergens possess high antigenic complexity.
- The individual response to an allergen depends on the genetic background of each subject. Several allergens can present structural homology, which can determine phenomena of cross-reactivity due to the recognition of different allergens by the same antibody.
- IgE generally tends to recognize conformational and non-linear epitopes. Indeed, in the case of food allergies, only molecules that maintain their three-dimensional structure even at high temperatures can induce allergy in sensitized patients.

The availability of recombinant allergens and/or highly purified extracts has allowed the introduction of molecular diagnostics (Component Resolved Diagnosis, CRD), which allows the analysis of IgE reactivity to individual allergens.

CRD allows the characterization of the specific allergological profile of an individual, allowing the distinction between primary or genuine sensitizations and cross-reactivities. The CRD has, therefore, revolutionized the clinical management of the patient allowing, for example, the selection of patients who can benefit from specific immunotherapy (ITS), or the assessment of the severity of the reaction to food allergens, allowing to distinguish between subjects sensitive to highly stable molecules and subjects sensitive to thermal and gastro-labile molecules, which will develop only oral allergy syndrome (OAS) after ingestion of raw food.

Analytical Methods for Molecular Diagnostics

CRD can be performed through the use of:

- Single molecular components (singleplex), which allow a targeted diagnosis
- Matrices, which consist of many allergens deposited on microarrays (multiplex)

Singleplex Molecular Diagnostics

Monoplex molecular diagnostics consists of a second-level test to confirm a diagnostic suspicion, allowing the identification of specific IgE toward a specific allergen. It has the advantage of being performed on the same analytical platform (quantitative, highly automated) where specific IgE toward extractive allergens are sought; this allows the use of algorithms by a reflex test approach so that the execution of few and targeted analytical assays determines an appropriate and efficient use of the available economic resources. In this approach, however, there is a risk of underestimating the presence of other unsuspected sensitizations by identifying only the components hypothesized a priori. The principle of the method is the same as the immunoassays described above for Cap FEIA and 3gAllergy consolidated on the same instruments used for the research of specific IgE, except allergens that in the determination of allergenic molecules will be recombinant or native allergens.

Multiplex Molecular Diagnostics (Microarrays)

Microarray diagnostics is a third-level test that allows the definition of the allergy profile of a patient. Currently, a microarray (ImmunoCAP ISAC, Thermo Fisher Scientific, Waltham, MA, USA) is available on the market that allows the simultaneous determination of IgE directed toward more than a hundred different allergens using a small amount of serum (30 μL). However, interpreting this test is somewhat complex and requires highly specialized personnel.

The test is also semi-quantitative, calibrated against an internal standard, and has lower diagnostic accuracy than the Cap FEIA test.

Table 37.1 describes the main characteristics of the two diagnostic systems. From the point of view of purely productive efficiency for the resources used, diagnostics using the microarray, based on the current costs envisaged for carrying out the tests, is advantageous if the patient's clinical situation requires the search for specific IgE toward several molecular components greater than 12–13 allergens.

Microarray Diagnostics (ImmunoCAP® ISAC).

Fluorescent antihuman IgE antibodies are used to detect antigen–antibody binding between specific IgE present in the patient's serum and antigens conjugated to a solid phase on a slide (chip). The fluorescence is subsequently measured by a scanner equipped with a laser excitation source. A densitometry software then analyzes the image and provides the test results as a function of the fluorescence intensity detected on each spot.

Table 37.1 Advantages and disadvantages of ISAC multiplexes and Cap FEIA singleplex

	Advantages	Disadvantages
ISAC	30 μL of serum or plasma from capillary or venous blood 112 allergens that can be measured simultaneously Natural and recombinant proteins Less amount of allergen required No interference due to high total IgE concentrations	Manual method Semi-quantitative results Less sensitivity For certain allergens, a high interassay variability has been reported Increased coefficient of variation Some allergenic sources are not included Inappropriate for monitoring sensitization Potential interference between IgE and other isotypes, mainly IgG
Cap FEIA	Automated method Quantitative results High sensitivity Low coefficient of variation Natural or recombinant proteins or crude extracts Appropriate for monitoring sensitization	40 μg of serum per allergen Each allergen is individually dosed Detection of low affinity antibodies which may have little clinical relevance

The test takes approximately 5 h to perform. The results are processed as ISAC classes (absent-low-medium-high) and international system units, providing a semi-quantitative IgE determination based on a specific reference curve. The system has high diagnostic reliability as each molecule is tested in triplicate.

Microarray-based diagnostics could be particularly useful in the diagnostic framing of complex clinical situations, such as:

- Patients with multiple sensitizations, not identifiable based on history and first-level tests
- Patients who do not respond to ITS therapy for whom it is appropriate to evaluate the presence of other possible allergens
- Patients with "idiopathic" anaphylaxis, to identify any sensitization not diagnosed by traditional tests
- Pediatric patients for whom it is not possible to obtain a sufficient sample to perform the analysis by monoplex diagnostics

However, there are some limitations to the use of multiplex technology:

1. The test is not automated and, therefore, is rather complex to perform and subject to a higher risk of errors.
2. Reduced analytical sensitivity and specificity characterize it compared to the single-plex method and a high rate of false positives.
3. The interpretation of the results is complex and must be performed by a specialist with a good knowledge of the various molecules and their diagnostic significance.

A new test system (ALEX2, MacroArrayDX, Wien, Austria) has recently been introduced for simultaneous detection of Total IgE and Specific IgE to 117 extracts and 183 molecules by solid-phase enzyme immunoassay. Extract and allergen molecules combined with nano-particles are sorbed on a solid-phase substrate, forming a macroscopic multiplex matrix - the immune allergy chip.

Conclusions

Molecular diagnostics in allergology has been a real revolution, having modified and transformed the concept of allergy, introducing a new language with a different semiological approach to diagnosis and therapy. The reference to allergenic families and their molecules, rather than to allergenic sources (plants, trees, fruit, etc.), can create quite a few difficulties in interpreting the diagnostic information produced, in particular, and understandably, for general practitioners, but sometimes also for clinical specialists in the field. For these reasons, the professionals working in the allergology laboratory must have the necessary skills to transfer the diagnostic information aimed at the efficacy of the results obtained from molecular tests in a clear language to the requesting physician and, possibly, to the patient himself. The interpretative comment of the in vitro molecular tests is a fundamental moment in the diagnostic process of the allergic patient.

The allergy diagnostic laboratory is fundamental in guaranteeing an effective diagnostic process of allergic pathologies only if it can receive adequate clinical information associated with the request for specific IgE. Without this information, the use of allergenic extracts and allergological molecules risks entering the great range of inappropriateness, generating not so much and not only an increase in costs but also wrong diagnoses or, even worse, delayed and confusing diagnoses.

The presence of an anamnestic card with the patient's clinical information substantially improves the research and selection of allergens to be prepared in the diagnostic session. The clinical pathologist must evaluate each request for specific IgE in light of the prescriber's request. This evaluation leads to a patient-specific choice of which panels or individual allergens to test. Based on the test result obtained with allergen extracts, the clinical pathologist with allergological expertise may decide to integrate the test with molecular allergens if this increases the informative value of the report. At the end of the diagnostic procedure, the clinical pathologist can produce an interpretative report that illustrates the results obtained and the diagnostic pathway used. This practice is undoubtedly burdensome in terms of time required, but it is certainly the most effective to optimize costs and resources and ensure patients' health and the overall quality of the clinical diagnostic-therapeutic process.

The evolution of allergy diagnostics has now arrived at a final and personalized mapping of the allergic patient. Although the molecular frontier is complex, it will become or already is a tool of fundamental importance in the armamentarium of the clinical allergist. An educational and training process of continuous updating with the interaction between laboratory medicine professionals and clinical allergology specialists is of paramount importance for the effective and appropriate use and interpretation of the diagnostic information produced by these innovative technologies and analytical methods.

Recommended Readings

Akkerdaas JH, Wensing M, Knulst AC et al (2003) How accurate and safe is the diagnosis of hazelnut allergy by means of commercial skin prick test reagents? Int Arch Allergy Immunol 132:132–140

Asero R (2010) Lo stato dell'arte della CRD: cosa sono e a cosa servono i componenti molecolari. Ligand Assay 15:12–13

De Knop KJ, Bridts CH, Verweij MM et al (2010) Component resolved allergy diagnosis by microarray: potential, pitfalls and prospects. Adv Clin Chem 50:87–101

Ferreira F, Hawranek T, Gruber P et al (2004) Allergic cross-reactivity: from gene to the clinic. Allergy 59:243–267

Gadisseur R, Chapelle JP, Cavalier E (2011) A new tool in the field of in-vitro diagnosis of allergy: preliminary results in the comparison of ImmunoCAP©250 with the ImmunoCAP©ISAC. Clin Chem Lab Med 49:277–280

Goikoetxea MJ, D'Amelio CM, Martínez-Aranguren R et al (2016) Is microarray analysis really useful and sufficient to diagnose nut allergy in the Mediterranean area? J Invest Allergol Clin Immunol 26:31–39

Hamilton RG, William PB (2010) Human IgE antibody serology: a primer for the practicing north American allergist/immunologist. J Allergy Clin Immunol 126:33–38

Harwanegg C, Hiller R (2006) Protein microarrays for the diagnosis of allergic diseases: state-of-the-art and future development. Eur Ann Allergy Clin Immunol 38:232–236

Hauser M, Egger M, Wallner M et al (2008) Molecular properties of plant food allergens: a current classification into protein families. Open Immunol 1:1–12

Iancovici-Kidon M, Tim CF (2007) Component-specific immunoglobulin E in the diagnosis of allergic disease in childhood: more of the same or something more? Isr Med Assoc J 9:476–478

Lack G (2004) New developments in food allergy: old questions remain. J Allergy Clin Immunol 114:127–130

Mari A (2008) When does a protein become an allergen? Searching for a dynamic definition based on most advanced technology tools. Clin Exp Allergy 38:1089–1094

Matson PNJ, Hamilton RG, Esch RE et al (2009) Analytical performance characteristics and clinical utility of immunological assay for human immunoglobulin E (IgE) antibodies and defined allergen specificities. Approved Guidelines – Second Edition. CLS Document I/LA20-A2 29:1–145

McCann WA, Ownby DR (2002) The reproducibility of the allergy skin test scoring and interpretation by board-certified/board-eligible allergists. Ann Allergy Asthma Immunol 89:368–371

Melioli G, Marcomini L, Agazzi A et al (2010) Diagnostica allergologica e CRD: il ruolo del laboratorio di patologia clinica. Ligand Assay 15:52–65

Paganelli R, Ansotegui IJ, Sastre J et al (1998) Specific IgE antibodies in the diagnosis of atopic disease. Clinical evaluation of a new in vitro test system, UniCAP, in six European allergy clinics. Allergy 53:763–768

Pastorello EA, Incorvaia C, Ortolani C et al (1995) Studies on the relationship between the level of specific IgE antibodies and the clinical expression of allergy: I. definition of levels distinguishing patients with symptomatic from patients asymptomatic allergy to common aeroallergens. J Allergy Clin Immunol 96:580–587

Roberts G, Lack G (2003) Relevance of inhalational exposure to food allergens. Curr Opin Allergy Clin Immunal 3:211–215

Sampson HA (2003) 9. Food allergy. J Allergy Clin Immunol 111(Suppl 2):S540–S547

Sastre J (2010) Molecular diagnosis in allergy. Clin Exp Allergy 40:1442–1460

Södeström L, Kober A, Ahlstedt S et al (2003) A further evaluation of the clinical use of specific IgE antibody testing in allergic diseases. Allergy 58:921–928

Thorpe SJ, Health A, Fox B (2014) The 3rd international standard for serum IgE: international collaborative study to evaluate a candidate preparation. Clin Chem Lab Med 52(9):1283–1289

Villalta D, Conte M, Asero R et al (2013) Isolated IgE reactivity to native walnut vicilin-like protein (nJug r 2) on ISACTM microarray is due to cross-reactive carbohydrate epitopes. Clin Chem Lab Med 51:1991–1995

Wang J, Godbald JH, Sampson HA (2008) Correlation of serum allergy (IgE) tests performed by different assay systems. J Allergy Clin Immunol 121:1219–1224

Biomarkers of Bone Remodeling

38

Franca Pagani and Martina Zaninotto

Introduction

This chapter describes the pathophysiological characteristics and clinical importance of determining circulating concentrations of skeletal tissue biomarkers. Biochemical markers play a critical role in the evaluation and differential diagnosis of these metabolic diseases since they are non-invasive, low cost, easy to perform, and represent dynamic indicators of the remodeling process, which occurs through a phase of resorption of old bone and a subsequent formation of new bone to have a condition of high-strength skeletal tissue.

In reality, the process is very complex, and thanks to recent biochemical and pathophysiological acquisitions, it has been possible to know in-depth and describe the molecular mechanisms that make the processes of neoformation and resorption closely coupled. The discovery of the RANK-RANKL-OPG complex, which will be described in detail in this chapter, represents one of the most critical advances in the knowledge of bone biology of the last decades.

The measurement of circulating molecules, products of the activity of osteoblasts and osteoclasts, has a considerable and recognized clinical importance providing an indirect measure of the possible decoupling of the processes of bone neoformation and resorption with important implications in monitoring the response of the patient to therapy or in the stratification of the risk of fracture. The biochemical characterization of these molecules, as well as the information that can be obtained based on their concentration, on the activity of osteoblasts and osteoclasts, will also be discussed in relation to the most common diseases associated with skeletal changes.

F. Pagani
Department of Experimental and Applied Medicine, University and Civil Hospital of Brescia, Brescia, Italy
e-mail: franca.pagani@poliambulanza.it

M. Zaninotto (✉)
Department of Laboratory Medicine, University-Hospital of Padova, Padova, Italy
e-mail: martina.zaninotto@aopd.veneto.it

It should be noted, however, that although the number of biomarkers suggested by scientific research is very high, biomarkers for which there is more scientific evidence in the literature and those recommended by national and international guidelines will be discussed in more detail.

Bone Tissue Biochemistry

Bone tissue consists of two main constituents: cells and the extracellular matrix. Compared to other tissues, the bone extracellular matrix is unique because it consists of an organic phase closely associated with a mineral phase. The organic phase consists of approximately 90% type I collagen and non-collagenous proteins. Although type I collagen is not specific to bone tissue as it is present in the skin, ligaments, and tendons, posttranslational modifications of the primary structure give it tissue-specific characteristics. Type I collagen is a protein formed by three filaments of about 1000 amino acids each, wrapped around each other to form an α-helix, which is synthesized as a precursor, the procollagen. The procollagen is excreted from the cell into the matrix, where specific enzymes remove the two globular ends, with the consequent formation of collagen that, without the globular ends, precipitates in the bone matrix and undergoes different maturation processes, including a process catalyzed by the enzyme lysine oxidase that determines the oxidation of lysine and hydroxylysine residues with the formation of cross-links of covalent type between the α chains of the same molecule and between adjacent molecules. Alongside this process, non-enzymatic mechanisms lead to the formation of glycation end products and racemization and isomerization phenomena at the terminal ends of the molecule. These modifications provide rigidity to the tissue's structure and increase the tissue's resistance when subjected to intense stresses. The collagen deposited in the bone matrix is closely linked to other non-collagenous proteins, and after a specific interval of time from its deposition, it undergoes complete mineralization.

M. Ciaccio (ed.), *Clinical and Laboratory Medicine Textbook*, https://doi.org/10.1007/978-3-031-24958-7_38

Osteoblasts and osteoclasts are the critical bone cells for producing biochemical markers of clinical interest. Osteoblasts, cells of mesenchymal derivation, are responsible for producing and depositing organic components of the matrix and activating the process of resorption of old bone tissue. Osteoclasts, large multinucleated cells derived from pluripotent cells of the monocyte/macrophage family, are responsible for the resorption of bone tissue. When activated, osteoclasts appear as polarized cells, with an apical zone containing mainly the cell's nuclei and a basal zone characteristically jagged (ruffled border) attached to the surface of the bone matrix. Osteoclasts are equipped with specialized adhesion structures, called podosomes, which allow them to delimit the area of bone tissue where resorption will occur, preventing damage to the surrounding newly formed tissue. In this compartment, the osteoclast pumps, through ion channels, hydrogen and chlorine ions (reducing the pH to 5) and proteolytic enzymes, such as acid phosphatase, carbonic anhydrase, and cathepsin K, leading to the formation of a gap that will be filled by new bone matrix produced by osteoblasts.

Bone tissue undergoes continuous remodeling, a process in well-defined units known as BMU (Bone Metabolic Unit). Bone remodeling has different characteristics at different stages of life. From birth and throughout the growth period, the activity of osteoblasts is separated from that of osteoclasts both temporally and spatially, and consequently, the formation process prevails over that of the resorption. The bones increase their size and mass and assume the form and structure of the adult skeleton. Around the age of 20, for both sexes, about 95% of bone mass has developed, with the remainder being completed between the ages of 28 and 36 years. The final level of bone mass reached this point in life determines peak bone mass, the magnitude of which correlates with the risk of developing osteoporosis later in life. After skeletal maturity is reached, the activity of osteoblasts and osteoclasts is, in physiological situations, "coupled." The formation/resorption balance is in equilibrium with the maintenance of skeletal mass. Following hormonal or other variations, starting from the peri-menopausal period in women and the fifth decade of life in men, the balance between the activity of osteoclasts and the activity of osteoblasts can be negative with the prevalence of the process of resorption over that of formation with progressive loss of bone mass. The whole process of bone remodeling is completed in 3–4 months and begins with the recruitment of osteoclast precursor cells to the site of resorption, under the action of cytokines and local growth factors. Osteoblasts play an essential role in osteoclastogenesis by producing stimulating factors, such as Receptor Activator of Nuclear factor-KappaB Ligand (RANKL) and osteoprotegerin, which, by preventing the action of RANKL on its osteoclastic receptor RANK, prevents overproduction of osteoclasts and, consequently, inhibits resorption. This is followed by the phase of resorption of the old bone tissue by osteoclasts and the phase of formation of new bone tissue by osteoblasts. The different phases of bone remodeling are controlled by numerous systemic and local factors. Systemic factors include calcium-tropic hormones (parathormone, calcitonin, vitamin D), sex hormones, and glucocorticoids. Local factors include growth factors, cytokines, prostaglandins, and leukotrienes. The role of these factors is very complex because none of them act independently, and many act in synergy to regulate the delicate balance on which bone remodeling is based. The knowledge of this process is, however, fundamental to understanding the rationale behind the measurement of circulating biochemical markers that constitute the evidence of the coupling or decoupling of metabolic processes that, as a whole, define the phenomenon of bone remodeling or "bone turnover".

These biochemical markers can be classified as biomarkers of bone formation, produced by the activity of osteoblasts, and biomarkers of resorption, produced by the activity of osteoclasts.

Bone Formation Markers

Bone Alkaline Phosphatase

Alkaline phosphatase (ALP) is a ubiquitous enzyme on the outer surface of all membranes, and this localization would make the hypothesis plausible that it is involved in the transport of specific substances across the cell membrane into the extracellular space. Its function is still unclear, but it has long been known that the enzyme plays an essential role in bone formation and mineralization.

The total circulating concentration is the sum of a group of different isoforms produced by various tissues (liver, bone, intestine, spleen, kidney, placenta), encoded by four genetic loci, three tissue-specific and one tissue-non-specific gene, located on chromosome 1. The hepatic/osseous/renal gene locus encodes for nonspecific ALP, which is present in liver cells, osteoblasts, and kidney cells. Since the phosphatases are encoded by the same gene and have the same protein nature, they should not be classified as isoenzymes. However, since they undergo specific, post-synthetic modifications of glycosylation and sialylation depending on the tissue involved, they give rise to three distinct isoforms, bone, liver, and intestinal, which have different chemical and physical characteristics.

In a healthy adult subject, about 50% of the enzyme activity in serum originates from the liver, while the remaining 50% is bone-derived, with a ratio of 1/1. In contrast, this ratio is modified in children and adolescents due to the significant bone growth activity, with a high serum concentration of bone ALP (BAP), which represents about 90% of the total activity. The bone isoenzyme is produced by osteoblasts and catalyzes the hydrolysis of phosphate esters by increasing the concentration of phosphorus, which is necessary for the mineralization process.

Currently, the determination of bone alkaline phosphatase is preferred to the measurement of total circulating ALP, which has been widely used in clinical practice as a marker of bone metabolism because it precisely reflects osteoblast activity. The relationships among ALP, sex hormones (e.g., estrogens), and bone growth are responsible for gender-related differences. Specifically, in adolescence, the peak of bone mass is different between males and females; moreover, in men compared to women, there is a higher activity between 20 and 50 years (mainly due to differences in the concentration of the bone isoenzyme), while in women, in menopausal age, there is an increase in total activity, maintaining an equal concentration ratio between the liver and the bone form.

Many techniques have been proposed for the separation of the two main circulating isoforms based on separative (electrophoresis), non-separative (heat inactivation), and immunochemical approaches. These methods exploit the principle of the different physicochemical properties of the two forms due to variations in the carbohydrate-binding site in the chain and the degree of sialylation. Immunochemical methods, which allow rapid quantification of enzyme activity or enzyme mass, show a certain degree of cross-reactivity between the bone and liver forms (15–20%). An immunochemical method that allows the determination of BAP with an automated system at high precision and sensitivity has been developed and is spreading in clinical practice.

Osteocalcin

Osteocalcin (OC) is a small protein of 5.8 kDa synthesized by osteoblasts, odontoblasts, and hypertrophic chondrocytes. The protein originates from a 75 amino acid propeptide which undergoes carboxylation of three glutamic acid residues through a vitamin K-dependent process. After carboxylation, a sequence of 25 amino acids is detached from the propeptide, and the mature protein is released into the cell. The high affinity of γ-glutamic residues for hydroxyapatite suggests an involvement of this protein in the process of bone mineralization. The newly synthesized OC is minimally released from the cell into the bloodstream while a greater amount is incorporated into the bone matrix, representing about 15% of the non-collagenous proteins of bone. Following bone resorption, osteocalcin is released into the bloodstream. By interacting with cell surface proteins and receptors, OC plays an active role in the organization of the extracellular matrix. However, although known for more than 20 years, its exact role is still to be defined. It has long been considered a marker of bone formation, but recent studies have re-evaluated its function, attributing it also an endocrine role and a regulator of energy metabolism. However, the results of clinical studies show a good correlation between serum OC levels and the rate of bone formation, verified by histomorphometry. Therefore, this biomarker is mainly used to evaluate the bone formation process. However, it is a protein subject to rapid degradation; therefore, the intact form and fragments of various sizes are found in the circulation. The intact molecule, consisting of 49 amino acids (aa), is present in serum for about 36% while circulating proteolytic enzymes can give rise to different fragments: an N-terminal fragment (1–19 aa), a fragment including the 20–43 aa tract, a C-terminal fragment (44–49 aa), a mid-N-terminal fragment (1–43 aa), and a mid-C-terminal fragment (20–49 aa).

Commercially available methods are immunometric using polyclonal and monoclonal antibodies directed against different epitopes of the marker. Consequently, numerous works in the literature that have compared the results of OC obtained by different methods demonstrate the impossibility of comparing the data obtained. The high heterogeneity of OC fragments in serum, associated with the instability of the molecule and the undefined role in the processes of bone turnover, determine significant limitations to using this marker in clinical practice. However, the methods that measure both the intact molecule and the fragment 1–43 seem to reflect the real synthesis of osteocalcin and, for this reason, would be preferred.

Procollagen Type I Terminal Peptides

Type I procollagen terminal peptides are the precursors of type I collagen, which in bone is synthesized by osteoblasts in the form of pre-procollagen. Although type I collagen propeptides may originate from other tissues, such as blood vessels, tendons, cornea, dentin, and skin, their turnover is slower than in bone and, therefore, they contribute very little to the circulating propeptide pool.

After the intracellular synthesis of the three strands of the procollagen molecule, their assembly, and excretion, two specific endopeptidases act on the NH2- and COOH-terminal portions of the molecule resulting in the formation of two peptides: an amino-terminal procollagen type I peptide (PINP) and a carboxy-terminal procollagen type I peptide (PICP). Both peptides are released into the circulation in equimolar concentrations during collagen synthesis and can be considered biochemical markers of bone formation. They are catabolized hepatically, but PICP is eliminated by mannose hepatic receptors, which are modulated in their action by hormonal and growth factors, whereas hepatic scavenger receptors deputed to PINP catabolism are unaffected by external factors. Therefore, PINP more accurately reflects bone formation.

The circulating concentration of amino and carboxy-terminal peptides of procollagen type I can be measured by immunometric methods. Some studies have shown a good correlation between serum PICP levels and the rate of bone formation, but the clinical relevance of PINP measurement is now absolutely validated and recommended by all clinical and biochemical guidelines. Indeed, the molecule is very well characterized, it is stable, and, as far as measurement methods are concerned, the structure and the calibrator have been defined, allowing accurate methods and comparable results.

Markers of Bone Resorption

Collagen Cross-Links

There are two types of cross-links in collagen: intramolecular cross-links between two chains of the same molecule and intermolecular cross-links between adjacent chains (Fig. 38.1). To this second group belong reducible links, which decrease in number as the collagen matures, and non-reducible links, which increase in number during the same process. These last bonds increase the fibers' mechanical resistance with age progression. Among the non-reducible bonds, two forms play an essential role in the study of bone remodeling: pyridinoline (hydroxylysylpyridinoline, PYD), formed by three hydroxylysine residues, and deoxypyridinoline (lysylpyridinoline, DPD), formed by two hydroxylysine residues and one lysine residue. The two forms have different tissue specificity: PYD is predominant in cartilage but is also found in bone, tendon, and blood vessel connective tissue; DPD is found only in bone tissue and dentin. Both forms are absent in the skin and are not affected by dietary intake of collagen-containing foods. Pyridinium cross-links are released due to the degradation of mature collagen fibrils that occurs during the phase of bone resorption by osteoclasts. They pass into the bloodstream and are subsequently eliminated by the kidney. In urine, about 60% of the cross-links are found in a form bound to peptides of various sizes and 40% in a free form. Measurement of circulating concentration, especially that of DPD, provides information on the osteoclastic activity. In 1978, Fujimoto et al. characterized the structure of pyridinium cross-links, and since then, many methods have been developed to determine these resorption markers, mainly in the urine. By high-performance liquid chromatographic procedure (HPLC), it is possible to determine the two cross-links pyridinoline and deoxypyridinoline, simultaneously. However, the demonstration that the direct measurement of the free portion of cross-links provides information on the state of bone resorption similar to that obtained by the measurement of the total cross-links has led to the development of direct immunological methods for measuring free PYD and DPD or only free DPD. RIA (RadioImmunoAssay), ELISA (Enzyme-Linked ImmunoSorbent Assay) on microplate or chemiluminescence immunoassays applied on automated instrumentation are now commercially available.

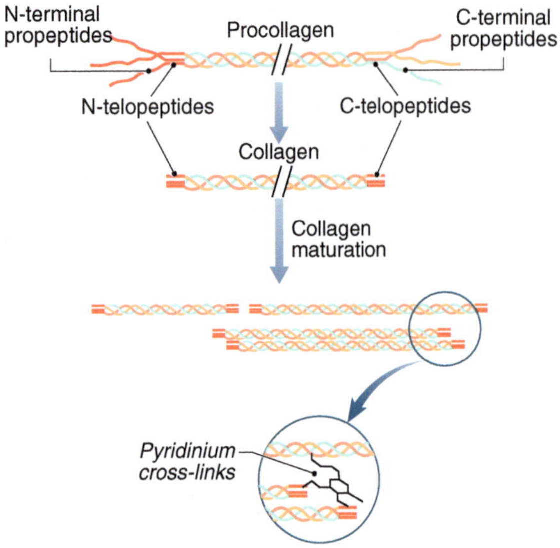

Fig. 38.1 Schematic representation of collagen and fragments. (Copyright EDISES 2021. Reproduced with permission)

Amino and Carboxy-Terminal Telopeptides of Collagen Type I

The development of antibodies able to recognize some sequences of the non-helical terminal portions of type I collagen has allowed the development of new analytical methods for evaluating biomarkers of bone resorption. In particular, a monoclonal antibody has been developed that recognizes a specific amino acid sequence (QYDGKGVG) of the amino-terminal portion of collagen I, which contains an essential region of intermolecular cross-links. This sequence, known as the amino-terminal telopeptide of type I collagen (NTX), is released due to the hydrolytic activity of cathepsin K produced by osteoclasts. In addition, a second peptide sequence consisting of an octapeptide (EKAHDGGR) located in the carboxy-terminal portion of the α1 chains of type I collagen, termed the carboxy-terminal telopeptide of type I collagen (CTX), which contains 40% of the deoxypyridinoline cross-links, has been identified. This octapeptide represents the result of non-enzymatic post-translational isomerization and racemization at the level of the aspartic acid residue contained in the sequence. Thus, in addition to the native αL-type form present in newly synthesized collagen, there are potentially three isomers present at more advanced stages of maturation: the isomerized form containing β-aspartate (βL), the racemized form containing the D-aspartate residue (αD), and the isomerized/racemized βD form. Theoretically, the ratio of the native αL form to each of the other three forms could be an index of the age of the bone tissue from which they are released.

Both NTX and CTX fragments, small, are filtered through the renal glomerulus. Their measurement can be performed on both serum and urine samples.

A larger fragment than the two previously described has been measured in serum. This is a hydrophobic fragment, relatively rich in phenylalanine, site of intermolecular linkages between α1 chains, located on the carboxy-terminal portion of type I collagen, known as ICTP and better defined by the abbreviation CTX-MMP. CTX-MMP and CTX, although derived from the carboxy-terminal fragment of collagen, are released, during the resorption phase, by different enzymatic processes. In fact, the action of cathepsin K, the major proteinase produced by osteoclasts, leads to the formation of CTX, while the action of several metalloproteinases produces CTX-MMP.

Some immunometric methods are available to determine NTX and CTX epitopes in serum and urine. For the measurement of CTX, these methods use antibodies specific for the octapeptide containing a β-isomerized aspartic acid residue. The presence of the β-isomer makes this marker more specific for bone tissue, correlating significantly with the age of the tissue itself. Serum CTX assay methods require the presence of a pair of β-D-isomerized octapeptides, making serum determination more specific than urine determination which requires the presence of only one octapeptide.

Tartrate-Resistant Acid Phosphatase

Acid phosphatase (ACP; EC 3.1.3.2) is a group of enzymes that hydrolyze phosphate esters in an acidic environment. ACP is a ubiquitous enzyme and, in cells, is localized predominantly in lysosomes. Using L-(+)-tartrate, the presence of a "tartrate-sensitive" and a "tartrate-resistant" fraction (TRACP) was detected in the serum of healthy subjects. The first comes from the prostate, lysosomes of various tissues and platelets, while the second comes from osteoclasts, Kupffer cells of the liver, macrophages of the pulmonary alveoli, spleen, and erythrocytes. Five isoenzyme bands can be identified by polyacrylamide gel electrophoresis. The band at position 5 is predominantly represented by the enzyme produced by osteoclasts. Fraction 5 is composed of an isoform whose optimal activity pH is 5, designated as TRACP 5a, and an isoform at optimal pH 5.8, designated as TRACP 5b. Studies of the biochemical composition of protein 5a have shown that it contains sialic acid residues absent in form 5b. The physicochemical characteristics of TRACP purified from osteoclasts correspond to the characteristics described for the 5b isoform, which, therefore, explicitly represents the enzyme produced during the bone resorption phase, whereas the origin of the 5a form remains unknown.

The demonstration of the existence in the serum of an isoform called TRACP 5b of specific bone derivation led to the development of a kinetic method based on the use of more specific inhibitors and an immunological method based on the use of a monoclonal antibody specific for this isoform.

New Biomarkers

The considerable number of studies carried out in recent years on bone biology led to the discovery of the physiopathological processes and biochemical mechanisms that underlie the complex process of remodeling that affects skeletal tissue. In particular, the description of the molecular mechanisms that couple the processes of formation and resorption has opened up new areas of knowledge and biochemical and therapeutic applications. In this context and from these discoveries, the proposal for new biochemical markers of bone remodeling may arise.

RANK-RANKL-OPG

It is a complex that results from the interaction between a 317 aa peptide, a member of the tumor necrosis factor (TNF) superfamily and expressed on osteoblast progenitor cells (RANK), and its receptor (RANKL) expressed on osteoclast precursors (Fig. 38.2). RANKL/RANK in-interaction results in the activation, differentiation, and fusion of osteoclast pre-

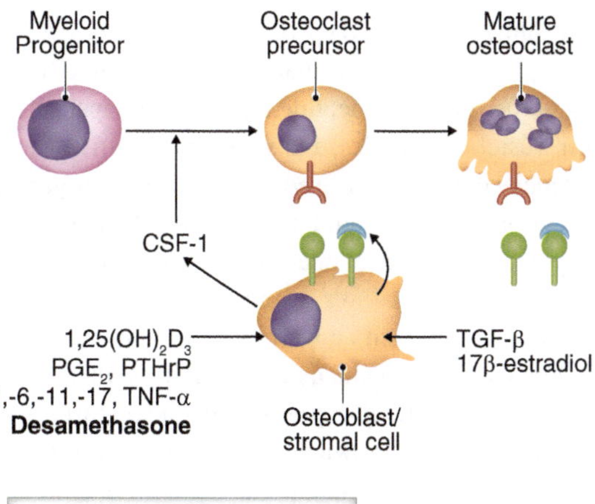

Fig. 38.2 The RANK/RANKL/OPG system and bone remodeling. (Copyright EDISES 2021. Reproduced with permission)

cursor hematopoietic cells and subsequent activation of the bone resorption process.

The progress of this process is inhibited by osteoprotegerin (OPG), a glycoprotein acting as a RANKL receptor antagonist, which inhibits the final differentiation and activation of osteoclasts, thus regulating the resorption process. This mechanism and, in particular, the demonstration that OPG is produced mainly by osteoblast progenitor cells represent further evidence of the close functional link between the mechanisms of neoformation and resorption. Interestingly, unlike all biochemical markers of bone remodeling characterized by a known circadian rhythm, OPG and RANKL present a constant circulating concentration. This characteristic could be attributed either to their long half-life as well as to a substantially stable release of these molecules from the bone tissue or to the fact that the measurement of the circulating pool could mask variations in tissue-specific isoforms. Based on these data, it can be hypothesized that the nocturnal increase in osteoclast activity is mediated by a mechanism independent of the RANKL/OPG ratio. The measurement of the concentration of these molecules has so far proved to be uninformative not only due to the presence in the circulation of both soluble and membrane-bound forms but also because of the analytical problems of the measurement methods and interpretation and expression of the results.

However, the critical role of the molecules described has led to the development of specific inhibitors of RANKL, some of which are used for therapeutic purposes, demonstrating real effectiveness in inhibiting the phase of skeletal resorption.

Cathepsin K

It is an enzyme belonging to the family of cysteine proteases, which, unlike other cathepsins, can act specifically on the regions of type I collagen characterized by triple helix structure and the linear telopeptide regions. Cathepsin K is localized in vesicles, granules, and vacuoles within the cytoplasm of osteoclasts and is released in the resorption gaps for the degradation of extracellular collagen. Its activity occurs mainly in the matrix of mature collagen, characterized by pyridinoline cross-links and telopeptide fragments C-terminal isomerized as well as numerous matrix proteins including osteocalcin and osteopontin. It is, therefore, a protein belonging to the group of biochemical markers of reabsorption. Although knowledge of its kinetics in circulation, its metabolism and elimination are still incomplete. The monitoring of the variation in circulating concentration could represent helpful information in the monitoring of therapies with specific cathepsin K inhibitors that seem to perform a mixed action: antiresorptive since they inhibit the activity of one of the essential osteoclastic enzymes, but without suppressing the activity of neoformation, with a consequent total anabolic effect on bone turnover. Recent data suggest that the measurement of some peculiar biochemical markers could allow a more detailed analysis of the variations that, in the resorption processes, are borne not only by the functionality of osteoclasts but also by their number. Among these, the measurement of cathepsin K, which may reflect the number of mature osteoclasts, and the measurement of fragments of degradation of collagen type I (NTX and CTX), generated by the activity of cathepsin K, which instead reflects the activity of osteoclasts. The different profiles of reduction of CTX and cathepsin K concentrations reported support the hypothesis that some therapies induce a decreased activity of osteoclasts followed by a decrease in their number. Finally, apoptosis occurs, as shown by the decrease in cathepsin K concentrations. The availability of methods for measuring biochemical markers whose circulating concentrations can provide separate information on the number and functionality of osteoclasts allows a more in-depth knowledge of the mechanism of action of anti-osteoporotic drugs.

Sclerostin

An essential source of information for preparing and testing new treatments is the study of the regulatory mechanisms of metabolic turnover.

In this area, to which also belongs the discovery of the RANK-RANKL-OPG system, biochemical studies are being extended on a glycoprotein, sclerostin, which belongs to a family of proteins that share the ability to express activity antagonistic to that of bone morphogenetic proteins, leading to a reduction in osteoblastic activity. It has been demonstrated that the expression of mRNA encoding for this protein is inhibited by parathormone (PTH) both in vivo and in vitro; moreover, recent studies have shown that the administration of monoclonal antibodies that block the activity of sclerostin results in an increase in bone mineral density, associated with an increase in circulating biochemical markers of neoformation and a significant decrease in those of resorption.

Since sclerostin is an excreted soluble factor, it has been possible to develop commercially available immunometric methods to measure its circulating concentration, but the results obtained so far seem somewhat conflicting. In particular, studies on healthy subjects show that the concentrations of sclerostin increase with age in both sexes but are modestly associated with bone mineral density and the concentration of biochemical markers of metabolic turnover. Addionally, they decrease in postmenopausal women treated with estrogen replacement therapy or PTH. Finally, they are not significantly affected by therapy with teriparatide. In studies evaluating the risk of fracture, which represents a critical clinical parameter in the evaluation of patients with osteoporosis, the associations between basal concentrations of sclerostin and incidence or prevalence of fracture risk do not seem to be statistically significant, highlighting the low effectiveness of the biochemical marker concerning this specific clinical information.

On the other hand, a significant positive correlation between protein concentration and bone mineral density has been demonstrated in all studies conducted in postmenopausal women, suggesting that circulating sclerostin mainly reflects the number of osteocytes. Thus, the concentration of sclerostin is proportional to the level of bone mass and not to the activity of individual cells or bone remodeling units. Also, a negative, statistically significant correlation with biochemical markers of bone remodeling (PINP and CTX) has been shown.

Before introducing sclerostin in clinical practice as a new and promising biochemical marker for the study of bone turnover, further studies are mandatory.

Fibroblast Growth Factor (Fibroblast-Growth Factor 23, FGF23)

Chronic renal failure determines significant complications on bone metabolism due to the decrease in vitamin D levels, in particular of 1,25-dihydroxyvitamin D (1,25(OH)2D), and the increase in PTH concentrations, leading to the onset of a bone mineral disease typically resulting from chronic renal failure (Chronic Kidney Disease-Mineral Bone Disease, CDK-MBD). In this population, it would be essential to have biochemical markers that allow early identification of the risk of fractures since the traditional parameters used in monitoring (Ca, P, PTH, and GFR) do not have sufficient diagnostic sensitivity. Recent studies have shown that in patients with chronic kidney disease (CKD), circulating FGF23 levels increase earlier than PTH and phosphorus levels, and their progression parallels that of the disease. Moreover, the evidence that FGF23 concentration is directly associated with fracture risk, correlated with decreased renal function (lower eGFR), increased PTH levels and decreased 1,25(OH)2D levels, makes this parameter a candidate biomarker. Indeed, it reflects the progression of bone disease following renal failure, can be easily monitored from the early stages of the disease, and can predict its outcome. Commercially available immunometric methods can easily measure the circulating levels of FGF23.

Genetic Determinants

Studies conducted on twins and families of subjects with bone diseases have shown that between 50 and 85% of the variability in bone mineral density and osteoporotic fractures is genetically determined and partially independent of bone mineral density. Moreover, recent studies to identify osteoporosis's genetic determinants have demonstrated the polygenic nature of inter-individual variations in bone mineral density and the fundamental role of numerous biological combinations in the onset of osteoporosis and susceptibility to fractures.

In an important and very recent study, 32 new loci have been identified and described in addition to 56 previously described, all strongly and significantly associated with changes in bone mineral density. Among these, several are located within the RANK-RANKL-OPG system, at the mesenchymal cell differentiation and endochondral ossification; 14 are significantly associated with fracturing risk. These data, which provide further evidence on the genetic architecture and pathophysiological mechanisms responsible for variations in bone mineral density and susceptibility to fractures, also allow understanding new metabolic mechanisms and potential targets for developing innovative drugs for the treatment of osteoporosis.

The knowledge of the biology of bone tissue, which is developing more and more, on the one hand, demonstrates the enormous complexity of the mechanisms that regulate metabolic processes and remodeling of a tissue considered inert until recently, and on the other hand, offers important areas of development for treatments aimed at substantially reducing the risk of fracture. The knowledge of the genetic determinants responsible for variations in bone mineral density and susceptibility to fractures potentially lead to the development of personalized therapies, whose effectiveness will still be monitored by the measurement of traditional biochemical markers of bone turnover, the only indicators that have the characteristics of specificity, accuracy, and easy application to routine clinical practice.

Sources of Variability in Bone Remodeling Biomarkers

Interpretation of laboratory results should always consider potential sources of variability in a data set. Even in the case of biochemical markers of bone remodeling, variability associated with the results of determinations can compromise their ability to characterize disorders of bone metabolism. Sources of variability may arise from a test's pre-analytical and analytical phases or the marker's features. Knowing and controlling the factors contributing to such variation is crucial for properly interpreting results.

Pre-analytical variability is influenced by patient preparation conditions (posture, exercise, fasting, etc.), biological sample collection and treatment methods (anticoagulants, acidifiers, etc.), storage methods (temperature, exposure to light), and the presence of hemolysis.

Pre-analytical conditions that cause important alterations in the results of bone remodeling markers are, for example, the use of hemolysed or lipemic sera for the determination of osteocalcin, leading to falsely high values or the determination of pyridinium cross-links on urine samples exposed to ultraviolet light, obtaining falsely low values.

Analytical variability results directly from the measurement procedure. In the field of biochemical markers of bone remodeling, analytical variability still represents an unsolved problem. Different methods for determining the same analyte are commercially available in many cases (bone alkaline phosphatase, osteocalcin, pyridinium cross-links). The different immunoreactivity of the ALP isoforms, the potential cross-reactivity with hepatic ALP in immunoassay methods, and the possibility of co-precipitation of the biliary and hepatic forms in the lectin precipitation method represent a problem for the bone alkaline phosphatase assay. For osteocalcin, it is necessary to consider the type of immunochemical method (competitive or non-competitive), the type of antibodies (polyclonal or monoclonal), the specificity of the antibodies (intact molecule or fragments), the source of the calibrators (bovine or human); for pyridinium cross-links, the specificity of the antibodies towards PYD or DPD, and the hydrolysis of the sample before measuring all molecules, free and bound.

The various methods use different calibration materials, and the antibodies used in immunochemical methods detect antigenically different molecules. This leads to extreme variability in the results obtained from different laboratories for the same marker, making it difficult to compare data. Hence the need to implement rigorous standardization strategies for the various methods and to make external quality assessment programs available to laboratories aimed explicitly at controlling accuracy and imprecision.

Last in descriptive terms, but no less important, is the biological variability of bone markers. In general terms, biological variability is defined as the variability of all biological parameters originating from homeostatic mechanisms that cause physiological fluctuations within individuals (CVi) and between individuals (CVg). Biological variability is quantifiable but cannot be reduced or eliminated. Knowing how much this type of variability can affect the result of an analyte is of great importance for interpreting the result itself. Essential information can be derived objectively only through its knowledge:

- The analytical objectives of the methods of determination
- The critical difference, i.e., the percentage difference necessary for two consecutive determinations in the same individual to be considered significantly different from each other
- The individuality index (II), which is calculated as the ratio between CVi and CVg, represents the estimate of the individuality of the biological variation, which provides indications on the usefulness or otherwise of the use of traditional reference limits
- The definition of the optimal biological sample

The data on biological variability allows answering many questions objectively. All bone markers considered, both serum and urine, have a high biological variability; this situation does not allow the use of a biochemical marker in the screening of any diseases affecting bone tissue, while it allows monitoring any changes in the individual, especially during drug treatment by applying the critical difference for the evaluation of the trend of the results. Many studies have demonstrated the existence of a marked circadian rhythm, with significant variations during the day in the concentration of both markers determined on serum and those determined on urine samples. This phenomenon makes it essential to carefully evaluate the time the biological sample is collected for determination, paying particular attention to ensure that samples are always collected at the same time of day in the patient being monitored.

Clinical Use of Biochemical Markers of Bone Remodeling

The determination, in blood and urine, of biochemical products derived from bone cell activity provides important information about the metabolic activity of bone tissue. The discovery of new bone-derived molecules and the development of more sensitive and specific analytical methods have significantly increased the ability of the laboratory to detect alterations in bone metabolism. To date, there is no ideal marker that possesses characteristics that can be used in the diagnostic phase due to a significant overlap in the values of these biochemical markers between the population of healthy subjects and subjects with bone disease and an intrinsic non-specificity towards the etiology of the alteration of bone metabolism. These molecules indicate an altered activity of the two bone cell lines, osteoblasts and osteoclasts. The determination of bone markers can be considered more as an estimate of a risk than a specific marker of disease. There are many primary or secondary pathologies of bone tissue in which increased concentrations of formation markers and resorption markers have been observed. Additionally, a large body of scientific evidence supports the use of these markers in the management of patients with bone disease.

Biochemical Markers of Bone Remodeling in Different Clinical Conditions Affecting Bone Tissue Metabolism

Bone Markers in Glucocorticoid-Induced Osteoporosis

Glucocorticoids (GCs) are commonly prescribed to reduce pain and the inflammatory component in inflammatory joints, bowel and chronic lung diseases, and autoimmune diseases. Chronic GC therapy involves, as a side effect, the occurrence of osteoporosis, a pathological condition defined as glucocorticoid-induced osteoporosis (GCIO). GCIO represents the primary cause of iatrogenic osteoporosis. Fractures, the most dreaded consequence of osteoporosis, may occur in 30–50% of patients on chronic GC therapy. The risk of experiencing fractures increases exponentially in the first 1–3 months after starting GC therapy. The bone loss rate depends on the dose of GC and the duration of therapy. However, some patients receiving low doses of GC lose more bone than those receiving higher doses, indicating a solid individual genetic component to the response, probably related to specific receptor polymorphism.

GCs in pharmacological doses interfere with bone cell metabolism responsible for remodeling the tissue. In GCIO, there is a decoupling of the two physiological processes of bone formation and resorption, with a consequent reduction in bone mass and deterioration of bone quality. It has been demonstrated that, during GC therapy, the bone formation phase is profoundly inhibited as the proliferation and differentiation of osteoblasts are reduced, and the phenomenon of cell apoptosis is increased. Furthermore, studies have shown a decreased production of collagen and osteocalcin.

Regarding the process of bone resorption, the data in the literature are less well known and sometimes discordant, probably due to the pathologies at which GC therapy is aimed, which may in themselves be the cause of increased resorption. In addition to the direct effect on bone tissue, GCs indirectly alter tissue metabolism through extra-skeletal effects. From a biochemical point of view, the marked decrease in serum osteocalcin is peculiar, accompanied or not by a decrease in other markers of osteoblastic activity such as bone alkaline phosphatase and PINP. Some studies have also shown increased resorption marker levels, such as urinary NTX. International guidelines on the treatment of osteoporosis recommend, especially if GC therapy is planned for more than 3 months and if the patient is at high risk of fracture, that antiresorptive therapy is instituted as soon as possible.

Bone Markers in Immobilization-Induced Osteoporosis

Studies in volunteers after immobilization have indicated that markers of bone resorption such as urinary CTX and NTX increase significantly due to a rapid increase in osteoclast activity. In a study conducted for 16 weeks, in addition to an increase in resorption markers, a reduction in formation markers was also observed, confirming that the mechanism underlying rapid bone loss after immobilization lies in the activation of remodeling cells. In a study conducted on a population of paraplegic patients with spinal cord injury, the biochemical picture of bone remodeling varied over time. During the first year after injury, urinary excretion of DPD markedly increased, while alkaline phosphatase and osteocalcin were normal or slightly increased. A recent study confirms in male subjects the change in biochemical markers following an experiment of bed immobilization for 14 days but hypothesizes a diverse effect on the risk of bone loss according to the age of the subjects studied.

Bone Markers in Inflammatory Bowel Disease

Patients with inflammatory bowel disease, especially Crohn's disease, have low bone mass and are at high risk of fractures due to several causes. Among these are steroid therapy, hypogonadism, vitamin D deficiency, low body weight, and high levels of circulating cytokines released by inflammatory bowel cells. In one study, high levels of NTX in urine correlated with bone mass loss at the lumbar vertebrae. High levels of inflammatory cytokines such as interleukin-1 (IL-1), interleukin-6 (IL-6), and TNF were reported in the same study. However, despite IL-1 and TNF being recognized inhibitors of the bone formation process, the formation markers were in the normal range. Bisphosphonate therapy led to an increase in bone density concomitant with a decrease in markers of resorption and formation. Inflammatory bowel disease therapy with infliximab, an anti-TNF-α drug, induces an increase in formation markers and a slight effect on resorption markers independent of the clinical response to therapy.

Bone Markers in Rheumatoid Arthritis

Patients with rheumatoid arthritis show an increased incidence of osteoporosis and fractures. Also, in this disease, several factors contribute to the loss of bone mass, from steroid therapy to vitamin D deficiency, decreased mobilization, and increased levels of cytokines active on bone tissue. In patients with rheumatoid arthritis, an increase in resorption markers was demonstrated, while no variation was shown in patients with osteoarthritis compared to control subjects. Regarding the markers of formation, the data in the literature are discordant, reporting in some cases an increase, in others a reduction. A comprehensive review of the literature reviewed the data on cartilage and bone damage biomarkers. Numerous studies have shown that bisphosphonate therapy prevents bone loss in these patients exposed to high doses of glucocorticoids to control disease symptoms and causes a decrease in circulating levels of bone markers. Infliximab therapy reduced the level of resorption markers with no adverse effect on formation, evidenced by the lack of reduction in PINP. A multicenter study on the effects of tocilizumab, an IL-6 monoclonal anti-receptor antibody, combined with methotrexate demonstrated a systemic reduction in bone resorption and cartilage turnover by measuring biochemical markers in patients diagnosed with moderate to severe rheumatoid arthritis.

Bone Markers in Primary Hyperparathyroidism

Excessive secretion of parathyroid hormone causes an increase in osteoclast activity and subsequent increase in osteoblast activity, which can lead to loss of bone mass and increased incidence of fractures, especially in women in the postmenopausal period. Patients with advanced hyperparathyroidism show high levels of markers of formation and resorption, which are not evident in mild cases. Increased bone remodeling returns to normal within 6 months to 1 year after parathyroidectomy. In patients on drug therapy with estrogen, raloxifene, and alendronate, levels of both bone formation and resorption markers were reduced.

Bone Markers in Cancer Pathologies

Bone metastases are a common complication in cancer diseases. These may be osteolytic, osteoadhesive, or mixed types. Biochemical markers of bone remodeling can be used in assessing the extent of metastases, response to therapy, and in some cases, in defining the prognosis in terms of survival. Osteolytic metastases are common in breast cancer, and their occurrence is associated with an increase in markers of bone formation, the level of which may decrease by about 70% after intravenous bisphosphonate treatment. Some authors showed that patients with the highest urinary NTX had the highest risk of skeletal complications and disease progression compared to patients with lower levels. Increased activity of serum alkaline phosphatase and other training markers was detected in prostate cancer patients who typically show osteoadhesive lesions. In a study conducted on the effects of zolendronic acid in patients with metastases from prostate cancer refractory to hormonal treatment, there was an increase in urinary NTX, which decreased significantly after intravenous infusion of the antiresorptive drug. According to some studies, the levels of bone markers correlated with an unfavorable prognostic index. In particular, in patients with prostate cancer, increased levels of PINP, bone alkaline phosphatase, and CTX were associated with a poor prognosis; this finding was also confirmed by a serial sampling of markers in the monitoring of hormone treatment. In another study, in patients with prostate cancer, conducting a multivariate analysis, only high levels of bone alkaline phosphatase maintained a significant association with shorter survival, while no significance was demonstrated for urinary NTX levels. In an animal model, in the presence of osteolytic and osteoadherent metastases, the correlation between tumor dissemination and biochemical markers of bone remodeling is represented by osteocalcin, PINP, CTX, and tartrate-resistant acid phosphatase isoenzyme 5b, was investigated. The best markers, in terms of diagnosis of osteoadhesive and osteolytic metastases, proved to be the bone formation markers PINP, especially osteocalcin. The results also demonstrated the usefulness of the markers in monitoring zolendronic acid therapy in both types of metastases.

Bone Markers in Paget's Disease

Paget's bone disease is a chronic skeletal pathology characterized by enlargement and deformation in one or more bone districts. In Paget's osteopathy, the two bone cell lines, osteoclasts and osteoblasts, do not interact in a balanced way. The altered cycle of bone resorption and neoformation produces a fragile bone, resulting in arthrosis, fractures, bone pain, and deformity. In the initial phases of the disease, an osteolytic type of phenomenon prevails, while in the second phase, the osteoblastic type is prevalent. In patients with extensive polyostotic disease, the formation and resorption markers can be very high. The biochemical feature is represented by the significant increase in alkaline phosphatase activity, up to 10 times and even more than the upper reference limit. The trend of the other markers of bone remodeling reflects the underlying bone pathological process. PINP and alkaline phosphatase correlate with the formation process, while serum osteocalcin does not correlate with the formation rate. The increase in resorption markers reflects the osteoclastic nature of the pathology. In untreated Paget's cases, αCTX, derived from newly formed bone, increased much more than βCTX (about 16-fold), a marker that, in contrast, reflects resorption of bone of normal composition. In one study, the response of biochemical markers to ibandronate therapy was investigated. Markers were measured on samples taken at baseline and after 6 months of therapy. Total alkaline phosphatase, bone phosphatase, and PINP were the most effective markers of formation in demonstrating the effect of bisphosphonate, while markers of resorption were less sensitive. The study concluded, however, that no marker could be shown to be superior to total alkaline phosphatase, which is already widely used in clinical practice.

Bone Markers in Multiple Myeloma

Skeletal involvement in multiple myeloma is characterized by extensive osteolytic lesions predominantly affecting the flat bones of the skull, vertebrae, pelvis, and long bones, including humerus and femur and is frequently associated with fractures and spinal cord compression. The development of lytic lesions is due in part to the interaction of myelomatous plasma cells with the bone matrix and, in particular, with stromal cells, which are stimulated to produce osteoclastogenic factors; in addition, myelomatous cells can directly activate osteoclasts, causing their differentiation through the release of various osteoclastogenic factors in the bone marrow environment. The progression of the osteolytic process is associated with an essential inhibition of the process of bone formation, a consequence of decoupling the two phases at the basis of remodeling. Biochemical markers of bone remodeling have been used to monitor, in these patients, the treatment with bisphosphonates, and it has been hypothesized that they can be used in the diagnosis of bone lesions earlier than conventional imaging or to identify that subgroup of patients who do not respond to therapy and therefore at higher risk of skeletal disease progression and death. These patients, identified by trends in biochemical bone markers, could benefit from alternative treatments.

In addition to the pathologic conditions discussed in the preceding paragraphs, many clinical studies have shown that biochemical markers of bone formation and resorption help test the skeleton's response to various endogenous or exogenous stimuli. Changes in bone metabolism may be induced by the administration of drugs such as warfarin, heparin, androgens, and growth hormone or resulting from chronic alcohol abuse.

Biochemical Markers of Bone Remodeling in Postmenopausal Osteoporosis

Most of the data in the literature concern the use of bone markers in postmenopausal osteoporosis, a condition with a high social impact both for its prevalence in the general population and for the importance of its most dreaded consequence: fracture.

Osteoporosis is a systemic skeletal disease characterized by reduced bone mass and macro and microarchitectural qualitative alterations accompanied by increased bone fragility.

According to the criteria of the World Health Organization, the diagnosis of osteoporosis is based on the assessment of bone mineral density through radiographic techniques, among which the most widely used is represented by the Dual-Energy X-ray Absorptiometry (DEXA), compared to the average of healthy adults of the same sex reached the peak of bone mass, i.e., around 20 years of age. The unit of measurement is the standard deviation from the mean peak bone mass (T-score). The risk of fracture begins to increase exponentially with densitometric values of T-score < -2.5 SD, the threshold value for diagnosing the presence of osteoporosis. However, skeletal and extra-skeletal factors must be considered in assessing individual fracture risk. Notably, osteoporotic fractures are demonstrated even in subjects with normal densitometry, and this supports the hypothesis that the deterioration of the architecture of the bone trabeculae also contributes to increasing the fragility of the skeleton. These pathogenetic mechanisms represent the rationale behind using biochemical markers to evaluate osteoporosis patients.

Prediction of Bone Loss

There is no doubt that in many clinical situations, it is very important to identify patients with a high rate of bone loss, a condition predisposing to increased bone fragility and an increased risk of fracture. The most common condition in this sense is certainly represented by the early postmenopausal period in women. The alterations typical of the involutional age, when the entity of osteoclastic resorption prevails over the new formation determining a progressive, albeit slow, loss of bone mass, are made more evident in women starting from menopause due to the loss of inhibition of osteoclastic activity exerted by estrogen. In the estrogenic deficiency, besides the prevalence of osteoclastic resorption over osteoblastic formation at the level of the single remodeling unit, an increased activation frequency and an increase in the number of the units themselves can occur. All this can result in two very different conditions in terms of prognosis: a condition of slow losers, in which the loss of bone mass is slow, and a condition of fast losers, in which the loss of bone mass can reach 2% per year in the first 3 years of menopause. Women categorized as fast losers might be candidates for antiresorptive therapy. Numerous studies have supported the hypothesis that bone marker values measured in women at the onset of menopause may predict the extent of bone loss in subsequent years, thus helping in deciding whether or not to undertake antiresorptive therapy and the extent of recovery after therapy. However, in order to make the use of markers in daily clinical practice effective, it is necessary to take into account the restrictive criteria used for the selection of patients that characterize such studies (age of the subjects studied, the magnitude of the risk calculated a priori, number of determinations made to obtain an average value) and apply them to individual patients.

Fracture Risk Prediction

Measurement of bone mineral density (BMD) by X-ray or ultrasound techniques demonstrates high specificity towards fracture risk assessment but relatively low sensitivity, meaning that many individuals who will fracture cannot be identified as high risk by BMD determination alone. This is because the risk of fracture is determined not only by the bone mass at the time of measurement but also by the bone loss over time. Emerging data from several prospective studies have found that biochemical markers of bone remodeling have a strong potential to predict fracture risk. The predictive value appears independent of bone mass values obtained by densitometry, age, and other clinical risk factors. In particular, an increase of the markers beyond the typical values of the pre-menopausal age, considered as reference, is associated with a doubled risk of encountering vertebral and non-vertebral fractures, particularly of the femur.

In order to identify all patients at risk of fracture and to undergo therapy, some algorithms have been developed in recent years that include, in addition to bone densitometry, the most important factors that contribute to an increase in the risk of fracture, including gender, age, previous fractures, family history, dietary habits, use of glucocorticoids. In 2008 a commission of the World Health Organization developed an algorithm known as FRAX® (http://www.shef.ac.uk/FRAX), which integrates multiple clinical risk factors. This tool calculates the probability of a patient sustaining a hip or other bone site fracture in the next 10 years. However, FRAX® presents some application issues. For this reason, national versions have been developed, such as the DeFRA by the Scientific Society for the Study of Skeletal Diseases (SIOMMMS; https://defra-osteoporosi.it). The inclusion in these algorithms for the determination of biochemical markers would increase their clinical sensitivity.

Therapy Monitoring

Osteoporosis is diagnosed with the determination of BMD by radiographic techniques, the most widely used of which is DEXA. Most of the antiresorptive drugs used for osteoporosis therapy produce a significant increase in bone mineral density shown by DEXA. However, for evaluating the effects of antiresorptive therapy, this procedure is of limited utility since at least 1–2 years must have elapsed since the start of treatment before the efficacy of any treatment can be defined with certainty. In addition, BMD does not provide information about the speed of bone remodeling, which may be necessary for choosing the most effective treatment for the individual patient. The measurement of BMD can therefore be complemented by determining biochemical markers of bone remodeling. The most validated indication for the use of bone markers concerns the assessment of the effectiveness of antiresorptive therapy with bisphosphonates or hormone replacement. The evaluation of bone markers can evidence a significant change in the efficacy of the treatment before and after just 3–6 months from the start of treatment; therefore, many months before any improvement in bone density is observed with DEXA, for which 2 years are required. Estrogen deficiency causes a rapid increase in bone remodeling, evidenced by increased circulating biochemical markers of bone formation and resorption. Hormone replacement therapy decreases biochemical markers that are dose-dependent and reaches a plateau maintained throughout treatment. Following discontinuation of treatment, biochemical markers return to baseline values. Numerous studies have documented these changes by demonstrating an initial significant decrease in reabsorption markers (3–6 months) followed by a decrease in formation markers that reach a plateau after 6–12 months of treatment.

Assessment of Adherence to Therapy

Oral bisphosphonates are first-line drugs for osteoporosis and, therefore, widely prescribed. Despite this, at least 1 in 2 patients who begin treatment do not show adequate compliance and even discontinue therapy altogether within 1 year of initiation.

This behavior, typical of many patients with chronic diseases, compromises the effectiveness of therapies and, in the case of osteoporosis, cancels the beneficial effect of therapy and exposes patients to high fracture risk. For this reason, the experts of the International Osteoporosis Foundation-European Calcified Tissue Society propose not to limit themselves to prescribing drugs but to carefully monitor adherence to therapy through the measurement of markers of bone remodeling in patients who have begun treatment with oral bisphosphonates for postmenopausal osteoporosis.

These recommendations stem from the results of the TRIO trial, a randomized, controlled trial that studied the effects of three oral bisphosphonates, alendronate, ibandronate, and risedronate, on biochemical markers and bone mineral density. Specifically, the authors recommend measuring serum PINP and CTX levels at baseline and 3 months after starting treatment. This time interval is optimal because the first few weeks after starting treatment are critical for nonadherence. When the reduction of these two markers does not reach significance, a reduction of 38% for PINP and 56% for CTX, respectively, it is necessary to evaluate patient compliance or investigate the presence of a secondary cause of osteoporosis.

Conclusions

Biochemical markers of bone remodeling offer many advantages in clinical practice: they do not require invasive procedures on the patient since they are determined on easily available biological samples; they can be repeated several times over time; their variations are detectable in a short time, thus allowing a timely intervention such as the discontinuation of a drug or encouraging the patient to continue therapy. However, all possible causes of variability in results must be carefully considered for their correct use. Implementing rigorous standardization strategies, external quality assessment programs for analytical methods, and defining gender- and age-specific reference ranges are also necessary.

Against this background, in 2009, the International Federation of Clinical Chemistry and Laboratory Medicine (IFCC) and the International Osteoporosis Foundation (IOF) formed a working group of experts to evaluate the clinical potential of biochemical markers in fracture risk prediction and treatment monitoring and to make recommendations for clinical use. The work was based on the analysis of results from prospective studies published in PubMed between 2000 and 2010. The results, published in 2011, support the role of bone remodeling markers in managing women with osteoporosis. However, more robust supporting scientific evidence is needed, especially for fracture risk prediction. The experts in the working group recommend using a formation marker, specifically serum PINP, and a resorption marker, serum CTX, as references in future observational studies and clinical trials aimed at studying the performance of biochemical markers and expanding international experience of their application in clinical practice. The criteria guiding the choice of PINP and CTX were the clear definition of the analyte, especially CTX, which allows the preparation of a reference standard; the specificity for bone tissue is not absolute for any marker, but in the case of PINP and CTX these are predominantly derived from bone metabolism, and their use in fracture risk assessment and therapy monitoring is widely supported by scientific evidence. There are commercially available methods for their determination within reach of all laboratories, from ELISA to RIA and automated immunometric methods; their biological and analytical variability and pre-analytical characteristics such as sampling conditions, type of biological sample, and stability are well known and therefore manageable. The standardization process would lead, as a direct consequence, to obtaining common and universally recognized decision levels and the possibility of defining therapeutic targets. The authors, recognizing the complexity of the standardization process, hope that a harmonization strategy can still be adopted as a short-term solution. In 2014, the IFCC-IOF group produced a meta-analysis on the two reference markers, serum PINP and CTX, as predictors of fracture. The articles included in the meta-analysis showed a significant association between PINP and CTX and fracture risk, with a hazard ratio per standard deviation increase in PINP of 1.23 and for CTX of 1.18 for all fractures and 1.23 for femur fractures. The National Bone Health Alliance (NBHA) in the United States has also stigmatized the need to promote a nationwide project involving all players in the field, from academics to government to laboratory technicians. Completing the project will help professionals make better use of biochemical markers of bone remodeling to aid in the treatment of osteoporosis, define standardized methods and common reference ranges, and, not least, have these markers recognized and therefore reimbursed by all health insurers in the United States. In Australia, the Australasian Association of Clinical Biochemists (AACB) has implemented a project to harmonize reference intervals for PINP and CTX. The working group members are clinical biochemists and endocrinologists with expertise in the field, and the initiative has been endorsed by the Royal College of Pathologists of Australasia (RCPA). The work has highlighted areas of uncertainty where additional data are needed to support reference intervals that are currently missing or insufficient. The authors conclude by highlighting that while biochemical markers of bone remodeling are widely used in bone diseases such as Paget's disease and show promise in the assessment of bone metastases, their utility in the management of postmenopausal osteoporosis is not as well defined. Nevertheless, many specialists are already using them mainly for treatment, and consensus on the adoption of reference intervals will help laboratory staff harmonize results.

Recommended Readings

Abildgaard N, Glerup H, Rungby J et al (2000) Biochemical markers of bone metabolism reflect osteoclastic and osteoblastic activity in multiple myeloma. Eur J Haematol 64:121–129

Adler RA, Fuleihan GE, Bauer DC et al (2016) Managing osteoporosis in patients on long-term bisphosphonate treatment: report of a task force of the American Society for Bone and Mineral Research. J Bone Miner Res 31:16–35

Anderson FH, Francis RM, Peaston RT, Wastell HJ (1997) Androgen supplementation in eugonadal men: effects of six months' treatment on markers of bone formation and resorption. J Bone Miner Res 12:472–478

Andreasen CM, Ding M, Overgaard S et al (2015) A reversal phase arrest uncoupling the bone formation and resorption contributes to the bone loss in glucocorticoid treated ovariectomised aged sheep. Bone 75:32–39

Apone S, Lee MY, Eyre DR (1997) Osteoclasts generate cross-linked collagen N-telopeptide (NTx) but no free pyridinolines when cultured on human bone. Bone 21:129–136

Atley LM, Mort JS, Lalumiere M, Eyre DR (2000) Proteolysis of human bone collagen by cathepsin K: characterization of the cleavage sites generated by cross-linked N-telopeptide neoepitope. Bone 26:241–247

Baron R, Neff L, Tran Van P et al (1986) Kinetic and cytochemical identification of osteoclast precursors and their differentiation into multinucleated osteoclasts. Am J Pathol 122:363–378

Bauer D, Krege J, Lane N et al (2012) National Bone Health Alliance Bone Turnover Marker Project: current practice and the need for US harmonization, standardization, and common reference ranges. Osteoporosis Int 23:2425–2433

Bauer DC, Sklarin PM, Stone KL et al (1999) Biochemical markers of bone turnover and prediction of hip bone loss in older women: the study of osteoporotic fracture. J Bone Miner Res 14:1404–1410

Bergmann P, Body JJ, Boonen S et al (2009) Evidence-based guidelines for the use of biochemical markers of bone turnover in the selection and monitoring of bisphosphonate treatment in osteoporosis: a consensus document of the Belgian Bone Club. Int J Clin Pract 63:19–26

Bernardi D, Zaninotto M, Plebani M (2004) Requirements for improving quality in the measurement of bone markers. Clin Chim Acta 346:79–86

Blair HC, Teitelbaum SL, Ghiselli R et al (1989) Osteoclastic bone resorption by a polarized vacuolar proton pump. Science 245:855–857

Blumsohn A, Colwell A, Naylor K, Eastell R (1995) Effects of light and g-irradiation on pyridinoline and telopeptides of type I collagen in urine. Clin Chem 41:1195–1197

Blumsohn A, Eastell R (1997) The performance and utility of biochemical markers of bone turnover: do we know enough to use them in clinical practice? Ann Clin Biochem 34:449–459

Body JJ, Dumon JC, Gineyts E, Delmas PD (1997) Comparative evaluation of markers of bone resorption in patients with breast cancer-induced osteolysis before and after bisphosphonate therapy. Br J Cancer 75:408–412

Bonde M, Garnero P, Fledelius C et al (1997) Measurement of bone degradation products in serum using antibodies reactive with an isomerized form of an 8 amino acid sequence of the C-telopeptide of type I collagen. J Bone Miner Res 12:1028–1034

Bonde M, Qvist P, Fledelius C et al (1994) Immunoassay for quantifying type I collagen degradation products in urine evaluated. Clin Chem 40:2022–2025

Bonnick SL, Shulman L (2006) Monitoring osteoporosis therapy: bone mineral density, bone turnover markers, or both? Am J Med 119:25S–31S

Borel O, Gineyts E, Bertholon C, Garnero P (2012) Cathepsin K preferentially solubilizes matured bone matrix. Calcif Tissue Intern 91:32–39

Boyce FB, Xing L (2008) Functions of RANKL/RANK/ OPG in bone modelling and remodelling. Arch Biochem Biophys 473:139–146

Brady JD, Ju J, Robins SP (1999) Isoaspartyl bond formation within N-terminal sequences of collagen type I: implications for their use as markers of collagen degradation. Clin Sci 96:209–215

Bravenboer N, Holzmann P, de Boer H et al (1997) The effect of growth hormone (GH) on histomorphometric indices of bone structure and bone turnover in GH-deficient men. J Clin Endocrinol Metab 82:1818–1822

Brown JP, Albert C, Nassar BA et al (2009) Bone turnover markers in the management of postmenopausal osteoporosis. Clin Biochem 42:929–942

Bruyere O, Collette J, Rizzoli R et al (2010) Relationship between 3-months changes in biochemical markers of bone remodelling and changes in bone mineral density and fracture incidence in patients treated with strontium ranelate for 3 years. Osteoporos Int 21:1031–1036

Buckwalter JA, Glimcher MJ, Cooper RR, Recker R, Bone biology. Part I. (1996a) Structure, blood supply, cells, matrix, and mineralization. Instr Course Lect 45:371–386

Buckwalter JA, Glimcher MJ, Cooper RR, Recker R, Bone biology. Part II. (1996b) Formation, form, modeling, remodeling, and regulation of cell function. Instr Course Lect 45:387–399

Buehlmeier J, Frings-Meuthen P, Mohorko N et al (2017) Markers of bone metabolism during 14 days of bed rest in young and older men. J Musculoskelet Neuronal Interact 17:399–408

Bull H, Murray PG, Thomas D et al (2002) Acid phosphatases. J Clin Pathol Mol Pathol 55:65–72

Canalis E, Mazziotti G, Giustina A, Bilezikian JP (2007) Glucocorticoid-induced osteoporosis: pathophysiology and therapy. Osteoporos Int 18:1319–1328

Cantini F, Niccoli L, Bellandi F, Di Munno O (1995) Effects of short-term, high dose, heparin therapy on biochemical markers of bone metabolism. Clin Rheumatol 14:663–666

Cavalier E, Rozet E, Carlisi A et al (2010) Analytical validation of serum bone alkaline phosphatase (BAP OSTASE) on liaison. Clin Chem Lab Med 48:67–72

Chailurkit L, Ongphiphadhanakul B, Piaseu N et al (2001) Biochemical markers of bone turnover and response of bone mineral density to intervention in early postmenopausal women: an experience in a clinical laboratory. Clin Chem 47:1083–1088

Chapurlat RD, Garnero P, Breart G et al (2000) Serum type I collagen breakdown products (serum CTX) predicts hip fracture risk in elderly women: the EPIDOS study. Bone 27:283–286

Chen P, Satterwhite JH, Licata AA et al (2005) Early changes in biochemical markers of bone formation predict BMD response to teriparatide in postmenopausal women with osteoporosis. J Bone Miner Res 20:962–970

Cheung CK, Panesar SR, Haines C et al (1995) Immunoassay of a tartrate-resistant acid phosphatase in serum. Clin Chem 41:679–686

Chopin F, Garnero P, Le Henanff A et al (2007) Long term effects of infliximab on bone and cartilage turnover markers in patients with rheumatoid arthritis. Ann Rheum Dis 67:353–357

Chow CC, Chan WB, Li JK et al (2003) Oral alendronate increases bone mineral density in postmenopausal women with primary hyperparathyroidism. J Clin Endocrinol Metab 88:581–587

Christenson RH (1997) Biochemical markers of bone metabolism: an overview. Clin Biochem 30:573–593

Christgau S, Rosenquist C, Alexandersen P et al (1998) Clinical evaluation of the serum CrossLaps one Step ELISA, a new assay measuring the serum concentration of bone-derived degradation products of type I collagen C-telopeptides. Clin Chem 44:2290–2300

Chubb SA (2012) Measurement of C-terminal telopeptide of type I collagen (CTX) in serum. Clin Biochem 45:928–935

Clemens JD, Herrick MV, Singer FR, Eyre DR (1997) Evidence that serum NTx (collagen-type I N-telopeptides) can act as an immunochemical marker of bone resorption. Clin Chem 43:2058–2063

Cloos PAC, Fledelius C (2000) Collagen fragments in urine derived from bone resorption are highly racemized and isomerized: a biological clock of protein aging with clinical potential. Biochem J 345:473–480

Coleman RE (2002) Efficacy of zoledronic acid and pamidronate in breast cancer patients: a comparative analysis of randomized phase III trials. Am J Clin Oncol 25(6 Suppl 1):S25–S31

Coleman RE, Major P, Lipton A et al (2005) Predictive value of bone resorption and formation markers in cancer patients with bone metastases receiving the bisphosphonate zoledronic acid. J Clin Oncol 23:4925–4935

Confavreux CB, Levine RL, Karsenty G (2009) A paradigm of integrative physiology, the crosstalk between bone and energy metabolism. Mol Cell Endocrinol 310:21–29

Cook RJ, Coleman R, Brown J et al (2006) Markers of bone metabolism and survival in men with hormone-refractory metastatic prostate cancer. Clin Cancer Res 12:3361–3367

Costa GA, Bilezikian JP (2012) Sclerostin: therapeutic horizons based upon its actions. Current Osteoporos Rep 10:64–72

Cremers SC, Lodder MC, Den Hartigh J et al (2004) Short term whole body retention in relation to rate of bone resorption and cartilage degradation after intravenous bisphosphonate (pamidronate) in rheumatoid arthritis. J Rheumatol 31:1732–1737

Demers LM (1997) Clinical usefulness of markers of bone degradation and formation. Scand J Clin Lab Invest 57:12–20

de Vernejoul MC (1996) Dynamics of bone remodelling: biochemical and pathophysiological basis. Eur J Clin Chem Clin Biochem 34:729–734

Devogelaer JP, Goemaere S, Boonen S et al (2006) Evidence-based guidelines for the prevention and treatment of glucocorticoid-induced osteoporosis: a consensus document of the Belgian Bone Club. Osteoporos Int 17:8–19

Diez-Perez A, Naylor KE, Abrahamsen B et al (2017) Adherence Working Group of the International Osteoporosis Foundation and the European Calcified Tissue Society. Recommendations for the screening of adherence to oral bisphosphonates. Osteoporos Int 28:767–774

Dovio A, Generali D, Tampellini M et al (2008) Variations along the 24-hour cycle of circulating osteoprotegerin and soluble RANKL: a rhythmometric analysis. Osteoporos Intern 19:113–117

McChung MR, Lewiecki EM, Geller ML et al (2013) Effect of denosumab on bone mineral density and biochemical markers of bone turnover: 8-year results of a phase 2 clinical trial. Osteoporos Int 24:227–235

Dresner-Pollak R, Karmeli F, Eliakim R et al (2000) Increased urinary N-telopeptide cross-linked type 1 collagen predicts bone loss in patients with inflammatory bowel disease. Am J Gastroenterol 95:699–704

Ekenstam E, Stalenheim G, Hallgren R (1988) The acute effect of high dose corticosteroid treatment on serum osteocalcin. Metabolism 37:141–144

Eriksen EF, Charles P, Melsen F et al (1993) Serum markers of type I collagen formation and degradation in metabolic bone disease: correlation with bone histomorphometry. J Bone Miner Res 8:127–132

Estrada K, Styrkarsdottir U, Evangelou E et al (2012) Genome-wide meta-analysis identifies 56 bone mineral density loci and reveals 14 loci associated with risk of fracture. Nat Genet 44:491–501

Fischer DG, Mischek A, Wolf S et al (2012) Paediatric reference values for the C-terminal fragment of fibroblast-growth factor-23, sclerostin, bone-specific alkaline phosphatase and isoform 5b of tartrate-resistant acid phosphatise. Ann Clin Biochem 49:546–553

Fledelius C, Johnsen AH, Cloos PAC et al (1997) Characterization of urinary degradation products derived from type I collagen. Identification of a beta-isomerized Asp-Gly sequence within the C-terminal telopeptide (alfa1) region. J Biol Chem 272:9755–9763

Franchimont N, Putzeys V, Collette J et al (2004) Rapid improvement of bone metabolism after infliximab treatment in Crohn's disease. Aliment Pharmacol Ther 20:607–614

Frenkel B, White W, Tuckermann J (2015) Glucocorticoid-induced osteoporosis. Adv Exp Med Biol 872:179–215

Fujimoto D, Moriguchi T, Ishida T et al (1978) The structure of pyridinoline, a collagen crosslink. Biochem Biophys Res Commun 84:52–57

Gage BF, Birman-Deych E, Radford MJ et al (2006) Risk of osteoporotic fracture in elderly patients taking warfarin: results from the National Registry of Atrial Fibrillation 2. Arch Intern Med 163:241–246

Garnero P (2009) Bone markers in osteoporosis. Curr Osteoporos Rep 7:84–90

Garnero P, Borel O, Delmas PD (2001) Evaluation of a fully automated serum assay for C-terminal cross-linking telopeptide of type I collagen in osteoporosis. Clin Chem 47:694–702

Garnero P, Cloos P, Sornay-Rendu E et al (2002) Type I collagen racemization and isomerization and the risk of fracture in postmenopausal women: the OFELY Prospective Study. J Bone Miner Res 17:826–833

Garnero P, Gineyst E, Riou JP, Delmas PD (1994a) Assessment of bone resorption with a new marker of collagen degradation in patients with metabolic bone disease. J Clin Endocrinol Metab 79:780–785

Garnero P, Hausher E, Chapuy MC et al (1996) Markers of bone resorption predicts fractures in elderly women: the EPIDOS prospective study. J Bone Miner Res 11:1531–1538

Garnero P, Shih WJ, Gineyts E et al (1994b) Comparison of new biochemical markers of bone turnover in late postmenopausal osteoporotic women in response to alendronate treatment. J Clin Endocrinol Metab 79:1693–1700

Garnero P, Sornay-Rendu E, Claustrat B, Delmas PD (2000) Biochemical markers of bone turnover, endogenous hormones and the risk of fractures in postmenopausal women: the OFELY Study. J Bone Miner Res 15:1526–1536

Garnero P, Sornay-Rendu E, Delmas PD (1995) Decreased bone turnover in oral contraceptive users. Bone 16:499–503

Garnero P, Sornay-Rendu E, Duboeuf F, Delmas PD (1999) Markers of bone turnover predict postmenopausal forearm bone loss over 4 years: the Ofely Study. J Bone Miner Res 14:1614–1621

Garnero P, Somay-Rendu E, Munoz F et al (2012) Association of serum sclerostin with bone mineral density, bone turnover, steroid and parathyroid hormones, and fracture risk in postmenopausal women: the OFELY study. Osteoporos Int 24:489

Garnero P, Thompson E, Woodworth T, Smolen JS (2010) Rapid and sustained improvement in bone and cartilage turnover markers with the anti-interleukin-6 receptor inhibitor tocilizumab plus methotrexate in rheumatoid arthritis patients with an inadequate response to methotrexate. Arthritis Rheum 62:33–43

Gennari C, Imbimbo B, Montagnani M et al (1984) Effects of prednisone and deflazacort on mineral metabolism and parathyroid hormone activity in humans. Calcif Tissue Int 36:245–252

Gerrits MI, Thijssen JHH, van Rijn HJM (1995) Determination of pyridinoline and deoxypiridinoline in urine, with special attention to retaining their stability. Clin Chem 41:571–574

Gineyts E, Cloos PAC, Borel O et al (2000) Racemization and isomerization of type I collagen C-telopeptides in human bone and soft tissues: assessment of tissue turnover. Biochem J 345:481–485

Glover SJ, Eastell R, McCloskey EV et al (2009a) Rapid and robust response of biochemical markers of bone formation to teriparatide therapy. Bone 49:1053–1058

Glover SJ, Gall M, Schoenborn-Kellenberger O et al (2009b) Establishing a reference interval for bone turnover markers in 637 healthy, young, premenopausal women from the United Kingdom, France, Belgium, and the United States. J Bone Miner Res 24:389–397

Gundberg CM, Markowitz ME, Mizruchi M, Rosen JF (1985) Osteocalcin in human serum: a circadian rhythm. J Clin Endocrinol Metab 60:736–739

Guo CY, Thomas WE, al-Dehaimi AW et al (1996) Longitudinal changes in bone mineral density and bone turnover in postmenopausal women with primary hyperparathyroidism. J Clin Endocrinol Metab 81:3487–3491

Hadjidakis DJ, Androulakis II (2006) Bone Remodeling. Ann N Y Acad Sci 1092:385–396

Halleen JM, Alatalo SL, Janckila AJ et al (2001) Tartrate-resistant acid phosphatase 5b is a specific and sensitive marker of bone resorption. Clin Chem 47:597–600

Halleen JM, Tiitinen SL, Ylipahkala H et al (2006) Tartrate-resistant acid phosphatase 5b (TRACP 5b) as a marker of bone resorption. Clin Lab 52:499–509

Hamwi A, Ganem AH, Grebe C et al (2001) Markers of bone turnover in postmenopausal women receiving hormone replacement therapy. Clin Chem Lab Med 39:414–417

Hannon R, Blumsohn A, Naylor K, Eastell R (1998) Response of biochemical markers of bone turnover to hormone replacement therapy: impact of biological variability. J Bone Miner Res 13:1124–1133

Hanson DA, Weis MA, Bollen AM et al (1992) A specific immunoassay for monitoring human bone resorption: quantitation of type I collagen cross-linked N-telopeptides in urine. J Bone Miner Res 7:1251–1258

Harris H (1990) The human alkaline phosphatases: what we know and what we don't know. Clin Chim Acta 186:133–150

Hassager C, Jensen LT, Johansen JS et al (1991) The carboxy-terminal propeptide of type I procollagen in serum as a marker of bone formation: the effect of nandrolone decanoate and female sex hormones. Metabolism 40:205–208

Heider U, Fleissner C, Zavrski I et al (2006) Bone markers in multiple myeloma. Eur J Cancer 42:1544–1553

Herrmann M, Pape G, Herrmann W (2002) Measurement of serum b-Crosslaps is influenced by proteolytic conditions. Clin Chem Lab Med 40:790–794

Herrmann M, Pape G, Herrmann W (2001) Stability of serum b-Crosslaps during storage: influence of pH and storage temperature. Clin Chem 47:939–940

Herrmann M, Seibel MJ (2008) The amino- and carboxyterminal cross-linked telopeptides of collagen type I, NTX-I and CTX-I: a comparative review. Clin Chim Acta 393:57–75

Hosada K, Eguchi H, Nakamoto T et al (1992) Sandwich immunoassay for intact human osteocalcin. Clin Chem 38:2233–2238

Huizenga NA, Koper JW, De Lange P et al (1998) A polymorphism in the glucocorticoid receptor gene may be associated with and increased sensitivity to glucocorticoids in vivo. J Clin Endocrinol Metab 83:144–151

Jakob C, Zavrski I, Heider U et al (2002) Bone resorption parameters [carboxy-terminal telopeptide of type-I collagen (ICTP), amino-terminal collagen type-I telopeptide (NTx), and deoxypyridinoline (Dpd)] in MGUS and multiple myeloma. Eur J Haematol 69:37–42

Janckila AJ, Takahashi K, Sun SZ, Yam LT (2001) Tartrate-resistant acid phosphatase isoform 5b as serum marker for osteoclastic activity. Clin Chem 47:74–80

Jilka RL, Takahashi K, Munshi M et al (1998) Loss of estrogen upregulates osteoblastogenesis in the murine bone marrow. J Clin Invest 101:1942–1950

Johansen JS, Brasso K, Iversen P et al (2007) Changes of biochemical markers of bone turnover and YKL-40 following hormonal treatment for metastatic prostate cancer are related to survival. Clin Cancer Res 13:3244–3249

Johansson H, Oden A, Kanis JA et al (2014) A meta-analysis of reference markers of bone turnover for prediction of fracture. Calcif Tissue Int 94:560–567

Johnell O, Scheele WH, Lu Y et al (2002) Additive effects of raloxifene and alendronate on bone density and biochemical markers of bone remodelling in postmenopausal women with osteoporosis. J Clin Endocrinol Metab 87:985–992

Kaddam IM, Iqbal SJ, Holland S et al (1994) Comparison of serum osteocalcin with total and bone specific alkaline phosphatase and urinary hydroxyproline:creatinine ratio in patients with Paget's disease of bone. Ann Clin Biochem 31:327–330

Kanda E, Yoshida M, Sasaki S (2012) Applicability of fibroblast growth factor 23 for evaluation of risk of vertebral fracture and chronic kidney disease-mineral bone disease in elderly chronic kidney disease patients. BMC Nephrol 13:122–140

Kanis J (2008) A on behalf of the World Health Organization scientific group. Assessment of osteoporosis at the primary health-care level. Technical Report. WHO Collaborating Centre, University of Sheffield, UK. http://www.shef.ac.uk/FRAX/index.htm

Karsdal MA, Woodworth T, Henriksen K et al (2011) Biochemical markers of ongoing joint damage in rheumatoid arthritis – current and future applications, limitations and opportunities. Arthritis Res Ther 13:215

Khan AA, Bilezikian JP, Kung AW et al (2004) Alendronate in primary hyperparathyroidism: a double-blind, randomized, placebo-controlled trial. J Clin Endocrinol Metab 89:3319–3315

Khosla S, Melton LJ, Atkinson EJ et al (1998) Relationship of serum sex steroid levels and bone turnover markers with bone mineral density in men and women: a key role for bioavailable estrogen. J Clin Endocrinol Metab 83:2266–2274

Khosla S, Melton LJ 3rd, Wermers RA et al (1999) Primary hyperparathyroidism and the risk of fracture: a population-based study. J Bone Miner Res 14:1700–1707

Kleerekoper M, Camacho P (2005) Monitoring osteoporosis therapy [editorial]. Clin Chem 51:2227–2228

Knott L, Bailey AJ (1998) Collagen cross-links in mineralizing tissues: a review of their chemistry, function, and clinical relevance. Bone 22:181–187

Lange U, Illgner U, Teichmann J, Schleenbecker H (2004) Skeletal benefit after one year of risedronate therapy in patients with rheumatoid arthritis and glucocorticoid-induced osteoporosis: a prospective study. Int J Clin Pharmacol Res 24:33–38

Langlois MR, Delonghe JR, Kaufman JM et al (1994) Posttranslational heterogeneity of bone alkaline phosphatase in metabolic bone disease. Eur J Clin Chem Clin Biochem 32:675–680

Lee AJ, Hodges S, Eastell R (2000) Measurement of osteocalcin. Ann Clin Biochem 37:432–446

Lee J, Vasikaran S (2012) Current recommendations for laboratory testing and use of bone turnover markers in management of osteoporosis. Ann Lab Med 32:105–112

Lee NK, Karsenty G (2008) Reciprocal regulation of bone and energy metabolism. Trends Endocrinol Metab 19:161–166

Lems WF, Gerrits MI, Jacobs JW et al (1996) Changes in (markers of) bone metabolism during high dose corticosteroid pulse treatment in patients with rheumatoid arthritis. Ann Rheum Dis 55:288–293

Lim V, Clarke BL (2012) New therapeutic targets for osteoporosis: beyond denosumab. Maturitas 73:269–272

Looker AC, Bauer DC, Chesnut CH III et al (2000) Clinical use of biochemical markers of bone remodelling: current status and future directions. Osteoporos Int 11:467–480

Lukert BP, Higgins JC, Stoskopf MM (1986) Serum osteocalcin is increased in patients with hyperthyroidism and decreased in patients receiving glucocorticoids. J Clin Endocrinol Metab 62:1056–1058

Magnusson P, Larsson L, Magnusson M et al (1999) Isoforms of bone alkaline phosphatise: characterisation and origin in human trabecular and cortical bone. J Bone Miner Res 14:1926–1933

Manolagas SC (1998) The role of IL-6 type cytokines and their receptors in bone. Ann N Y Acad Sci 840:194–204

Martín-Fernández M, Valencia K, Zandueta C et al (2017) The usefulness of bone biomarkers for monitoring treatment disease: a comparative study in osteolytic and osteosclerotic bone metastasis models. Transl Oncol 10:255–261

Martin TJ, Romas E, Gillespie MT (1998) Interleukins in the control of osteoclast differentiation. Crit Rev Eukaryot Gene Expr 8:107–123

Michael K, Harkonen PL, Vaananen HK, Hentunen TA (2005) Estrogen and testosterone use different cellular pathways to inhibit osteoclastogenesis and bone resorption. J Bone Mineral Res 20:2224–2232

Munoz-Torres M, de la Higuera Lopez-Frias M, Fernandez GD (2004) Advances in ostheoclast biology: the osteoprotegerin-RANK ligand system. Med Clin 122:75–77

Nakanishi M, Yoh K, Miura T et al (2000) Development of a kinetic assay for band 5b tartrate-resistant acid phosphatase activity in serum. Clin Chem 46:469–473

Naylor K, Eastell R (2012) Bone turnover markers: use in osteoporosis. Nat Rev Rheumatol 8:379–389

Nyquist F, Ljunghall S, Berglund M, Obrant K (1996) Biochemical markers of bone metabolism after short and long time ethanol withdrawal in alcoholics. Bone 19:51–54

Okabe R, Nakatsuka K, Inaba M et al (2001) Clinical evaluation of the Elecsys β-CrossLaps serum assay, a new assay for degradation products of type I collagen C-telopeptide. Clin Chem 47:1410–1414

Orr-Walker BJ, Evans MC, Clearwater JM et al (2000) Effects of hormone replacement therapy on bone mineral density in postmenopausal women with primary hyperparathyroidism: four-year follow-up and comparison with healthy postmenopausal women. Arch Intern Med 160:2161–2166

Pacifici R, Carano A, Santoro SA et al (1991) Bone matrix constituents stimulate interleukin-1 release from human blood mononuclear cells. J Clin Invest 87:221–228

Pagani F, Bonetti G, Stefini F, Panteghini M (2000) Evaluation of a fully automated assay to measure C-telopeptide of type I collagen in serum. Clin Chem Lab Med 38:1111–1113

Pagani F, Panteghini M (1994) Diurnal rhythm in urinary excretion of pyridinium crosslinks. Clin Chem 40:952–953

Palomba S, Orio F Jr, Manguso F et al (2005) Efficacy of risedronate administration in osteoporotic postmenopausal women affected by inflammatory bowel disease. Osteoporos Int 16:1141–1149

Panteghini M, Pagani F (1995) Biological variation in bone-derived biochemical markers in serum. Scand J Clin Lab Invest 55:609–616

Panteghini M, Pagani F (1996) Biological variation in urinary excretion of pyridinium crosslinks: recommendations for the optimum specimen. Ann Clin Biochem 33:36–42

Prummel MF, Wiersinga WM, Lips P et al (1991) The course of biochemical parameters of bone turnover during treatment with corticosteroids. J Clin Endocrinol Metab 72:382–386

Raisz LG (1999) Physiology and patophysiology of bone remodelling. Clin Chem 45:1353–1358

Rauchenzauner M, Schmid A, Heinz-Erian P et al (2007) Sex- and age-specific reference curves for serum markers of bone turnover in healthy children from 2 months to 18 years. J Clin Endocrinol Metab 92:443–449

Ravn P, Clemmensen B, Christiansen C (1999) Biochemical markers can predict the response in bone mass during alendronate treatment in early postmenopausal women. Alendronate Osteoporosis Prevention Study Group. Bone 24:237–244

Reid IR, Davidson JS, Wattie D et al (2004) Comparative responses of bone turnover markers to bisphosphonate therapy in Paget's disease of bone. Bone 35:224–230

Risteli J, Elomaa I, Niemi S et al (1993) Radioimmunoassay for the pyridinoline crosslinked carboxy-terminal telopeptide of type I collagen: a new serum marker of bone collagen degradation. Clin Chem 39:635–640

Rogers A, Hannon RA, Eastell R (2000) Biochemical markers as predictors of rates of bone loss after menopause. J Bone Miner Res 15:1398–1404

Rosen CJ, Chestnut CH III, Mallinak NJS (1997) The predictive value of biochemical markers of bone turnover for bone mineral density in early postmenopausal women treated with hormone replacement or calcium supplementation. J Clin Endocrinol Metab 82:1904–1910

Rosen HN, Moses AC, Garber J et al (2000) Serum CTX: a new marker of bone resorption that shows treatment effects more often than other markers because of low coefficient of variability and large changes with bisphosphonate therapy. Calcif Tissue Int 66:100–103

Rosen HN, Moses AC, Garber J et al (1998) Utility of biochemical markers of bone turnover in the follow-up of patients treated with bisphosphonates. Calcif Tissue Int 63:363–368

Rosenquist C, Fledelius C, Christgau S et al (1998) Serum CrossLaps One Step ELISA. First application of monoclonal antibodies for measurement in serum of bone-related degradation products from C-terminal telopeptides of type I collagen. Clin Chem 44:2281–2289

Ross PD, Knowlton W (1998) Rapid bone loss is associated with increased levels of biochemical markers. J Bone Miner Res 13:297–302

Ross PD, Kress BC, Parson RE et al (2000) Serum bone alkaline phosphatase and calcaneus bone density predict fractures: a prospective study. Osteoporos Int 11:76–82

Rossini M, Caimmi C, Giannini S et al (2012) The iniziative of hip fractures of the Veneto region. Clin Cases in Min and Bone Metab 9:45–49

Roux C, Garnero P, Thomas T et al (2005) Comite Scientifique du GRIO. Recommendations for monitoring antiresorptive therapies in postmenopausal women. Joint Bone Spine 72:26–31

Rubin MR, Lee KH, McMahon DJ, Silverberg SJ (2003) Raloxifene lowers serum calcium and markers of bone turnover in postmenopausal women with primary hyperparathyroidism. J Clin Endocrinol Metab 88:1174–1178

Saad F, Gleason DM, Murray R et al (2002) Zoledronic Acid Prostate Cancer Study Group. A randomized, placebo-controlled trial of zoledronic acid in patients with hormone-refractory metastatic prostate carcinoma. J Natl Cancer Inst 94:1458–1468

Scheld K, Zittermann A, Heer M et al (2001) Nitrogen metabolism and bone metabolism markers in healthy adults during 16 weeks bed rest. Clin Chem 47:1688–1695

Seibel MJ (2006) Biochemical markers of bone turnover. Part II: clinical applications in the management of osteoporosis. Clin Biochem Rev 27:123–138

Seibel MJ (2005) Biochemical markers of bone turnover. Part I: biochemistry and variability. Clin Biochem Rev 26:97–122

Seibel MJ, Lang M, Geilenkeuser WJ (2001) Interlaboratory variation of biochemical markers of bone turnover. Clin Chem 47:1443–1450

Seibel MJ, Robins SP, Bilezikian JP (1992) Urinary pyridinium crosslinks of collagen: specific markers of bone resorption in metabolic bone disease. Trends Endocrinol Metab 3:263–270

Shankar S, Hosking DJ (2006) Biochemical assessment of Paget's disease of bone. J Bone Miner Res 21:P22–P27

Siddhanti SR, Quarles LD (1994) Molecular and pharmacologic control of osteoblast proliferation and differentiation. J Cell Biochem 55:310–320

Silva-Fernandez L, Rosario MP, Martinez-Lopez JA et al (2013) Denosumab for the treatment of osteoporosis: a systematic literature review. Reumatol Clin 9:42–52

Smith MR (2006) Markers of bone metabolism in prostate cancer. Cancer Treat Rev 32:23–26

Stokes FJ, Ivanov P, Bailey LM, Fraser WD (2011) The effects of sampling procedures and storage conditions on short-term stability of blood-based biochemical markers of bone metabolism. Clin Chem 57:138–140

Tamura Y, Araki A, Chiba Y et al (2007) Remarkable increase in lumbar spine bone mineral density and amelioration in biochemical markers of bone turnover after parathyroidectomy in elderly patients with primary hyperparathyroidism: a 5-year follow-up study. J Bone Miner Metab 25:226–231

Tascioglu F, Colak O, Armagan O et al (2005) The treatment of osteoporosis in patients with rheumatoid arthritis receiving glucocorticoids: a comparison of alendronate and intranasal salmon calcitonin. Rheumatol Int 26:21–29

Ton FN, Gunawardene SC, Lee H, Neer RM (2005) Effects of low-dose prednisone on bone metabolism. J Bone Miner Res 20:464–470

Torres MM, Garcia RR, Raya PM et al (2009) Serum cathepsin K as a marker of bone metabolism in postmenopausal women treated with alendronate. Maturitas 64:188–192

Traba ML, Calero HA, Mendez-Davila C et al (1999) Different behaviors of serum and urinary CrossLaps ELISA in the assessment of bone resorption in healthy girls. Clin Chem 45:682–683

Valimaki MJ, Tahtela R (2005) Serum tartrate-resistant acid phosphatase 5B or amino-terminal propeptide of type I procollagen for monitoring bisphosphonate therapy in postmenopausal women? Clin Chem 51:2382–2385

van Staa TP, Cooper C, Brusse LS et al (2003) Inflammatory bowel disease and the risk of fracture. Gastroenterology 125:1591–1597

van Staa TP, Geusens P, Bijlsma JW et al (2006) Clinical assessment of the long-term risk of fracture in patients with rheumatoid arthritis. Arthritis Rheum 54:3104–3112

Vasikaran SD, Chubb SAP, Ebeling PR et al (2014) Harmonised Australian reference intervals for serum PINP and CTX in adults. Clin Biochem Rev 35:237–242

Vasikaran SD, Cooper C, Kanis JA (2011a) Recommendations for bone marker standards in osteoporosis: what, why and where to now ? Ann Clin Biochem 48:91–92

Vasikaran S, Eastell R, Bruyere O et al (2011b) Markers of bone turnover for the prediction of fracture risk and monitoring of osteoporosis treatment: a need for international reference standards. Osteoporos Int 22:391–420

Vasikaran SD, Morris HA, Cooper C, Kanis JA (2011c) Standardising biochemical assessment of bone turnover in osteoporosis. Clin Biochem 44:1033–1034

Vesper HW, Demers LM, Eastell R et al (2002) Assessment and recommendations on factors contributing to preanalytical variability of urinary pyridinoline and deoxypyridinoline. Clin Chem 48:220–235

Vesper HW, Smith SJ, Audain C, Myers GL (2001) Comparison study of urinary pyridinoline and deoxypyridinoline measurements in 13 US laboratories. Clin Chem 47:2029–2031

Viguet-Carrin S, Garnero P, Delmas PD (2006) The role of collagen in bone strength. Osteoporos Int 17:319–336

Watts N (1999) Clinical utility of biochemical markers of bone remodelling. Clin Chem 45:1359–1368

Weinstein RS, Parfitt AM, Marcus R et al (2003) Effects of raloxifene, hormone replacement therapy, and placebo on bone turnover in postmenopausal women. Osteoporos Int 14:814–822

Wilchers M, Schmidt E, Bidlingmaier F, Klingmuller D (1999) Diurnal rhythm of CrossLaps in human serum. Clin Chem 45:1858–1860

Woitge HW, Horn E, Keck AV et al (2001) Biochemical markers of bone formation in patients with plasma cell dyscrasias and benign osteoporosis. Clin Chem 47:686–693

Wong PK, Young L, Vaile JH et al (2004) Telopeptides as markers of bone turnover in rheumatoid arthritis and steoarthritis. Intern Med J 34:539–544

Worsfold M, Powell DE, Jones TJW, Davie MWJ (2004) Assessment of urinary bone markers for monitoring treatment of osteoporosis. Clin Chem 50:2263–2270

Yasui T, Uemura H, Tomita J et al (2006) Association of serum undercarboxylated osteocalcin with serum estradiol in pre-, peri- and early post-menopausal women. J Endocrinol Investig 29:913–918

Zwart SR, Wolf M, Rogers A et al (2009) Stability of analytes related to clinical chemistry and bone metabolism in blood specimens after delayed processing. Clin Biochem 42:907–910

Zehnder Y, Luthi M, Michel D et al (2004) Long-term changes in bone metabolism, bone mineral density, quantitative ultrasound parameters, and fracture incidence after spinal cord injury: a cross-sectional observational study in 100 paraplegic men. Osteoporos Int 15:180–189

Alzheimer's Disease and Other Neurodegenerative Dementias

39

Giulia Sancesario and Sergio Bernardini

Introduction

Diagnostic laboratory medicine tests guide up to 70% of clinical decisions and play a central role in modern clinical medicine. However, their use in diagnosing Alzheimer's disease (AD) and other neurodegenerative dementias is still limited, especially due to the inherent analytical issues in handling cerebrospinal fluid (CSF). Nevertheless, markers represent a valuable means to investigate neurodegenerative processes and guide clinical diagnosis in a field where histological confirmation is not feasible. Additionally, symptom overlap complicates the definition of an early diagnosis, relying on clinical and neuropsychological observation.

CSF analysis of four proteins (β-amyloid 1-42, β-amyloid 1-40, total tau [T-tau], and phosphorylated tau [p-tau]) offers significant added value to the diagnosis of dementia. These classical markers of neurodegeneration, together with other markers, such as 14-3-3, α-synuclein, and serum progranulin, represent a useful diagnostic panel for AD and several forms of dementia, such as vascular dementia, frontotemporal dementia, prion disease, and Parkinson's disease.

Cerebrospinal Fluid Biomarkers

The cerebrospinal fluid represents the best source to search for biomarkers of neurological and neurodegenerative diseases because, being in contact with the cells of the nervous system, it can reflect the changes that occur in the brain and that, in some cases, may precede the manifestation of cognitive impairment.

G. Sancesario (✉)
IRCCS Fondazione Santa Lucia, European Center for Brain Research, Rome, Italy
e-mail: g.sancesario@hsantalucia.it

S. Bernardini
Department of Experimental Medicine, University of Tor Vergata, Rome, Italy

In neurodegenerative diseases, the formation of intracellular inclusions of modified proteins (e.g., tau, α-synuclein, and TDP-43) and/or extracellular (e.g. β-amyloid and prion proteins) is associated with irreversible processes of neuronal death in specific areas of the brain, which identify different nosological entities characterized by cognitive impairment and dementia. Altered levels of these monomeric or aggregated proteins in biological fluids could be a helpful index that can be measured in CSF. This process implies that brain tissue may release different substances into the extracellular space and that these may diffuse into the CSF; conversely, sequestration of insoluble molecules and formation of fibrils in the brain would prevent their measurement in the CSF. However, CSF biomarkers represent only those soluble components released into the extracellular space, while the neurodegenerative processes that develop in the intracellular compartment remain hidden.

CSF collection is obtained by a lumbar puncture, which, if correctly performed, is relatively simple for an experienced operator and minimally invasive, with a low incidence of adverse events. However, patients' fear of lumbar puncture limits the spread of CSF-related diagnostic tests, although the diagnostic utility of CSF markers in dementia has recently renewed interest.

Alzheimer's Disease

Alzheimer's disease is a chronic neurodegenerative disease characterized by a long clinically asymptomatic phase, in which silent changes occur within the brain, and a symptomatic phase with the cognitive deterioration that inevitably evolves into dementia. The structural changes in the central nervous system are the formation of extracellular amyloid plaques, mainly composed of β-amyloid 1-42 protein, and intracellular neurofibrillary tangles of tau and p-tau. Amyloid plaques and tau tangles represent the distinctive histological markers of AD associated with the degeneration process leading to cognitive impairment. Identifying these mecha-

M. Ciaccio (ed.), *Clinical and Laboratory Medicine Textbook*, https://doi.org/10.1007/978-3-031-24958-7_39

nisms early, when cognitive function is not yet severely compromised, allows to improve the clinical course and, above all, the search for new personalized therapies. Clinically, the clinical onset of AD may occur by manifesting amnestic symptoms, such as the classic memory disorder, while atypical forms prevail in alterations of other cognitive domains, such as language, behavior, or movement.

Considerable evidence has shown that the use of biomarkers of neurodegeneration, β-amyloid 1-42 (Aβ42), β-amyloid 1-40 (Aβ40), T-tau, and p-tau protein, offers a significant added value to the diagnosis of AD. The CSF levels of these proteins allow investigating in vivo the pathological processes that characterize the disease and support the decision-making pathway by increasing the level of certainty in the clinical diagnosis.

According to the amyloid hypothesis, in AD the altered metabolism of Aβ42 causes an increase in the content of this protein due to the imbalance between its production and proper removal. Aβ42 forms soluble intermediates, the so-called "oligomers," which are highly toxic to neurons and synapses and progressively aggregate to form insoluble protofibrils and fibrils, which precipitate in plaques. This sequestering process is reflected in the CSF by a reduction in Aβ42 content, typical of AD pathology (Table 39.1). Aβ42 levels in AD patients are below 800–900 pg/mL, with wide variability in the values adopted among centers, due to different analytical issues.

However, there is considerable interindividual variability in physiological levels of total β-amyloid in subjects who may be hyper- or hypoproducers. These differences in global β-amyloid levels may affect the diagnostic accuracy of Aβ42, with falsely positive results (low Aβ42 levels) in non-AD individuals or falsely negative results (normal Aβ42 levels) in AD subjects. The assessment of Aβ40 content, whose levels do not vary during AD, thus allows investigating whether the reduction of Aβ42 is due to a real accumulation in the tissue or whether a physiological overproduction of total Aβ may hide the reduction of Aβ42 due to sequestration in plaques. Indeed, the use of the Aβ42/Aβ40 ratio improves the accuracy of Aβ42, especially in individuals with high total Aβ content. Aβ42/Aβ40 values below 0.055 distinguish patients with AD, whereas normal values may exclude the likelihood of cognitive impairment due to AD-type dementia.

Table 39.1 Cerebrospinal fluid (CSF) biomarkers in Alzheimer's disease

		Alzheimer's disease	
	Normal values	CSF	Tissue
Aβ42	>800–900 pg/mL	↓	Extracellular plaques of β-amyloid
T-tau	<350–450 pg/mL	↑	Intracellular neurofibrillary tangles
p-tau	<61 pg/mL	↑	Hyperphosphorylation
Aβ42/Aβ40	>0.069 pg/mL	↓	

Tau is a microtubule-associated protein and is involved in maintaining the stability of the cytoskeleton. The formation of neurofibrillary tangles composed of tau is one of the histological features of AD, along with hyperphosphorylation of tau at numerous phosphorylation sites. Following neuronal death, tau is released into the CSF with increased measurable levels. The increase in tau and p-tau is an index of neurodegeneration characteristic of AD pathology and is a helpful tool to assess the extent of neuronal damage and phosphorylation status. In patients with AD pathology, CSF T-tau levels are increased, with values above 350–400 pg/mL; in parallel, the content of phosphorylated tau exceeds the cutoff of 61–70 pg/mL (Table 39.1). Changes in tau levels can be nonspecific and are also observed in other forms of dementia or pathologies due to acute neuronal damage (e.g., ischemia). On the contrary, the process of hyperphosphorylation of tau represents a peculiar mechanism in Alzheimer's disease. Therefore, the evaluation of p-tau content represents a useful biochemical marker for the differential diagnosis, with high specificity, especially in cases of differential diagnosis between Alzheimer's disease and dementia with Lewy bodies.

Recently, several studies have demonstrated the diagnostic usefulness of CSF biomarkers even in the early stages of the disease, when silent pathological changes in the brain begin. In these stages, which are asymptomatic or characterized by mild symptoms, evaluating possible variations of measurable biomarkers in the CSF allows for identifying subjects at risk of developing Alzheimer's dementia. The redefinition of the concept of Alzheimer's disease and mild cognitive impairment (MCI) has considerably modified the diagnostic criteria of dementia, including the evaluation of biomarkers as an early in vivo index of AD. The reduction of Aβ42 is the first marker that is altered by the effect of aggregate formation, while the combination with t-tau and p-tau allows for predicting the progression of MCI toward AD dementia (prodromal AD). Moreover, even in asymptomatic individuals without cognitive impairment, biomarkers can identify preclinical stages of a neurodegenerative process that may evolve into cognitive impairment and dementia. Finally, biomarkers have a high negative predictive value and, in the elderly population, can be used to exclude a form of AD dementia.

Recently, the comprehensive panel of CSF markers, consisting of Aβ42, Aβ40, T-tau, and p-tau, has been included in the recommendations of the Alzheimer's Association and the National Institute on Aging (NIA), an agency of the US National Institutes of Health (NIH), for the diagnosis of AD and MCI, together with the assessment of topographic markers by neuroimaging (PET and MRI). The analysis of markers allows for assessing the probability of AD pathology through indicators associated with the pathogenetic mechanisms of the disease. In Italy, their use in clinical practice is limited to a few centers, mainly specialized in memory disorders, but a greater diffusion is desirable to promote early diagnosis.

Frontotemporal and Vascular Dementias

The clinical diagnosis of AD and the different forms of dementia is complicated by the manifestation of often overlapping clinical symptoms and by comorbidity, which is very frequent, especially in elderly individuals. The use of biomarkers helps the identification of pathological mechanisms responsible for cognitive deficits and guides the diagnosis, especially in doubtful cases. Indeed, in most forms of non-AD dementia, such as vascular dementia (VaD) and frontotemporal dementia (FTD), CSF biomarkers levels are normal, and their evaluation excludes the suspicion of Alzheimer's disease, directing the diagnosis toward a form of dementia with a different etiology.

Frontotemporal dementias are a heterogeneous group of syndromes characterized by cognitive impairment, which clinically manifests with behavioral or language disorders, symptoms also common in atypical AD forms. Normal levels of Aβ42 and tau characterize the CSF profile of patients with FTD even in those forms where the formation of tau protein inclusions is involved in degeneration processes. This profile excludes a suspicion of presenile AD and distinguishes mixed forms of dementia, where increased tau and decreased Aβ42 may indicate comorbidity between FTD and AD (Table 39.2).

Table 39.2 Cerebrospinal fluid (CSF) biomarkers in frontotemporal dementias

	Normal values	Frontotemporal Dementia	Dementias mixed
Aβ42	>800–900 pg/mL	Normal	↓
T-tau	<350–450 pg/mL	Normal	↑
Progranulin	100–250 ng/mL	↓Only shapes genetic with mutation *PGRN*	

Some genetic forms of FTD are due to a mutation in the gene for progranulin, a growth factor with neurotrophic properties, which causes a reduction in protein levels in the blood (normal values between 100 and 250 ng/mL). Therefore, the measurement of serum progranulin levels allows to identify carriers of the mutation, even asymptomatic ones, to be subjected to in-depth genetic investigation to highlight the presence of the mutated gene. However, normal progranulin values cannot exclude the presence of FTD of other origin and do not distinguish subjects with FTD from controls.

Often vascular events in different brain areas can cause neuronal damage and cognitive impairment (vascular cognitive impairment, VCI) that manifest themselves with different symptoms depending on the area or areas involved and do not necessarily evolve toward a form of dementia. Therefore, the clinical and pathological phenotype and the neuroimaging may be heterogeneous and coexist with other pathologies, particularly AD, leading to mixed dementia (MD). CSF analysis of biomarkers allows for investigating the probable cause of cognitive impairment and distinguishing VCI, VaD, AD, and MD (Table 39.3). Given the complexity of clinical and pathological pictures, which determines the difficulty in defining appropriate control groups, and the possible co-presence in VCI of neuropathological changes associated with age or AD, the results of studies on the usefulness of biomarkers have often been conflicting. Most studies agree that the CSF profile in patients with VCI is characterized by increased tau and normal levels of Aβ42 and p-tau (Table 39.3). Normal levels of Aβ42 allow us to distinguish VaD or VCI from AD or AD-like dementias with secondary vascular events. The presence of even clinically silent vascular damage in AD patients does not alter their CSF profile. While patients with AD and AD associated with vascular damage have a similar profile (reduced Aβ42, increased tau and p-tau), VCI patients without AD have a CSF profile similar to controls, sometimes associated with an increase only in T-tau protein.

Finally, the reduction of Aβ42 and the increase of tau allow the distinction between mixed dementias, permitting the correct classification of 85–89% of cases with VCI, AD, and mixed dementia.

Table 39.3 Cerebrospinal fluid (CSF) biomarkers in vascular dementias

	Normal values	Vascular dementia	Cognitive impairment vascular	Mixed dementia (VaD + AD)
Aβ42	>800–900 pg/mL	Normal	Normal	↓
T-tau	<350–400 pg/mL	Normal/increased	Normal/increased	↑
p-tau	<61 pg/mL	Normal	Normal	↑

AD Alzheimer's disease, *VaD* vascular dementia

Prion Diseases

Prion diseases are a group of neurodegenerative disorders caused by the misfolding of prion proteins, normal proteins found in the brain. The most common is the sporadic form, the so-called Creutzfeldt-Jakob disease (CJD), which differs from other dementias in its clinical course, rapidly leading to death. The broad spectrum of clinical manifestations and the overlap of symptoms with other acute-onset neurodegenerative diseases make in vivo diagnosis difficult, especially in the early stages of the disease. CSF analysis of tau protein and 14-3-3 has been included in the World Health Organization (WHO) diagnostic criteria and represents the gold standard for confirming the diagnosis of CJD.

CSF positivity for 14-3-3, identified by immunoblots or enzyme immunoassays, is highly specific for CJD. 14-3-3 has a very high negative predictive value for CJD since it is negative in patients with other neurological diseases and controls. It can also be found in CSF of patients with rapidly progressive dementia, vascular dementia, rapidly progressive AD, and frontotemporal dementia, reducing its diagnostic accuracy.

Although the role of this protein in CJD pathology has not been elucidated, its presence in the CSF suggests involvement in disease-induced molecular changes, and it may be released as an effect of neuronal death.

T-tau levels are significantly increased in CJD patients, as an effect of widespread neuronal damage, compared with both controls and AD patients, with values exceeding 1300 pg/mL (Table 39.4).

Table 39.4 Cerebrospinal fluid (CSF) biomarkers of prion diseases

	Normal values	CJD
Aβ42	>800–900 pg/mL	Normal
p-tau	<61 pg/mL	Normal
T-tau	<350 pg/mL	↑↑↑ > 1300 pg/mL
Protein 14–3-3	Negative	Positive

CJD Creutzfeldt–Jakob disease

Table 39.5 Cerebrospinal fluid (CSF) biomarkers of Parkinson's disease and synucleinopathies

	Normal values	PD	PDD	LBD
Aβ42	>800–900 pg/mL	Normal		Normal
T-tau	<350–450 pg/mL	Normal	Normal	Normal
p-tau	<70 pg/mL	Normal	Normal	Normal Differential diagnosis with AD
α-sin	Not standardized	↓	↓	↓

LBD Lewy body dementia, *PD* Parkinson's disease, *PDD* Parkinson disease with dementia

Parkinson's Disease and Synucleinopathies

Parkinson's disease (PD) diagnosis is based exclusively on clinical criteria, making it impossible to formulate an early and differential diagnosis between PD and other Parkinsonisms. PD, dementia with Lewy bodies (LBD), and multiple systemic atrophy (MSA) constitute the group of so-called synucleinopathies, diseases caused by the accumulation of the Lewy bodies, intracellular aggregates composed mainly of α-synuclein (α-sin). CSF levels of α-syn are reduced as an effect of tissue accumulation and allow to distinguish patients with PD, MSA, and LBD from those with AD and other neurological diseases, with a high predictive value (about 90%) in identifying a synucleinopathy, but with low accuracy for the diagnosis of PD (70.2% sensitivity and 52.8% specificity). The other markers of neurodegeneration, Aβ42, T-tau, and p-tau, are normal and, combined with α-sin, may improve diagnostic accuracy, especially in cases of doubtful diagnosis (Table 39.5). Very importantly, p-tau has the highest accuracy as a single marker in distinguishing LBD and AD, with a normality cutoff of less than 61 pg/mL. Finally, since cognitive impairment is a typical process in PD patients, both in the early stages and in those with more than 20 years of disease duration, the reduction of Aβ42 constitutes a predictive marker of dementia development in patients with PD (PDD).

However, the current panel of biomarkers is considered unsatisfactory due to its ineffectiveness in recognizing the different subtypes of PD, both in clinical and prognostic terms. The search for new markers is also necessary to aid the diagnosis, validation, and development of new drugs.

Variability and Standardization of CSF Examination

The analysis of nonstandard fluids, such as CSF, involves several critical issues to be resolved, including variability, instability, standardization of analytical and preanalytical procedures, and harmonization of normality values. Several studies have shown that the concentration of biomarkers has highly variable results, especially for Aβ, even using the same ELISA, leading to high variability even in cutoff values and complicating the interpretation of results among centers.

Several international initiatives aim to develop reference methods and procedures for analyzing CSF markers to achieve analytical uniformity and reproducibility among different laboratories.

The Alzheimer's Biomarkers Standardization Initiative (ABSI) has drawn up standard operating procedures (SOP) covering all preanalytical aspects and regulating the correct processing of the sample from collection to analysis. The standards to be adopted provide for collecting CSF in sterile tubes of polypropylene, material necessary to reduce the interaction between peptides and tube, without additives. Since the material of the tubes can alter the measurement of biomarkers, it is essential to standardize the assays, adopting a common type of tubes (brand, size, etc.) to avoid differences in results; the same attention should be paid to both the tubes used for collection and those for storage. In the case of traumatic sampling, since erythrocytes may interfere with the assay, it is necessary to discard the first 1–2 mL of CSF and proceed with a new tube.

Tubes should be transported on ice and centrifuged at 2000 rpm for 10 min at 4 °C within 2 h; the supernatant should be transferred to new polypropylene tubes and stored at −20 °C, if it will be used within 2 months or at −80 °C to preserve it for more extended periods. To avoid alteration of the marker content, it is important to avoid several freeze/thaw cycles that may interfere with the accuracy of the assay (Fig. 39.1).

The performance of CSF examination of biomarkers often takes place in centralized laboratories, with samples collected from different hospitals in neighboring areas being shipped. This requires careful observation of preanalytical procedures to avoid affecting the validity of the test. For this reason, it is necessary to use test tubes compatible with those used in the laboratory where the analysis is carried out. If it is impossible to transport the sample immediately after collection, the CSF can be centrifuged and frozen by dividing it into aliquots and transporting it on dry ice.

Observation of these rules prevents errors due to mishandling of samples, resulting in unreliable results, and promotes standardization among laboratories to obtain comparable biomarker results.

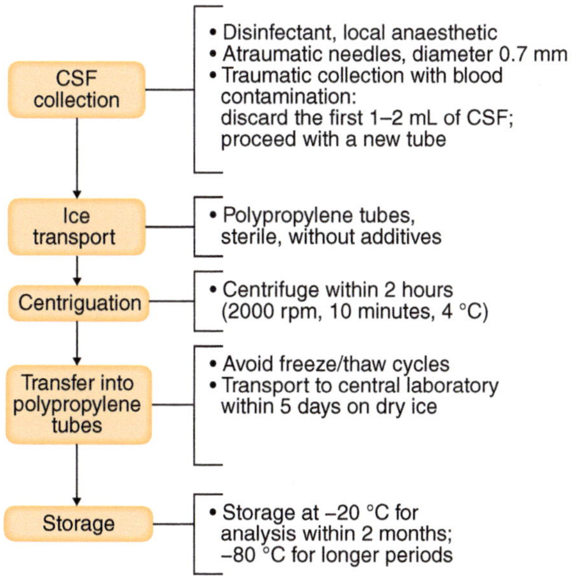

Fig. 39.1 Standard procedures for collecting cerebrospinal fluid for the analysis of AD biomarkers. (Copyright EDISES 2021. Reproduced with permission)

The final goal to be achieved is the standardization and harmonization in clinical laboratories not only of the operating procedures adopted but also of the units of measurement used, reference ranges, and decision criteria to reduce the differences among laboratories, which limit and compromise the value of guidelines, interpretation of results, and large-scale dissemination of CSF diagnostics in the field of neurodegenerative dementias.

Recommended Readings

Blennow K, Hampel H (2003) CSF markers for incipient Alzheimer's disease. Lancet Neurol 2:605–613

Brys M, Pirraglia E, Rich K et al (2009) Prediction and longitudinal study of CSF biomarkers in mild cognitive impairment. Neurobiol Aging 30:682–690

Clark CM, Xie S, Chittams J et al (2003) Cerebrospinal fluid tau and beta-amyloid: how well do these biomarkers reflect autopsy-confirmed dementia diagnoses? Arch Neurol 60:1696–1702

Davis DG, Schmitt FA, Wekstein DR et al (1999) Alzheimer neuropathologic alterations in aged cognitively normal subjects. J Neuropathol Exp Neurol 58:376–388

de Leon MJ, DeSanti S, Zinkowski R et al (2006) Longitudinal CSF and MRI biomarkers improve the diagnosis of mild cognitive impairment. Neurobiol Aging 27:394–401

Diniz BS, Pinto Junior JA, Forlenza OV (2008) Do CSF total tau, phosphorylated tau, and beta-amyloid 42 help to predict progression of mild cognitive impairment to Alzheimer's disease? A systematic review and meta-analysis of the literature. World J Biol Psychiatry 9:172–182

Fagan AM, Roe CM, Xiong C et al (2007) Cerebrospinal fluid tau/beta-amyloid(42) ratio as a prediction of cognitive decline in nondemented older adults. Arch Neurol 64:343–349

Fagan AM, Shaw LM, Xiong C et al (2011) Comparison of analytical platforms for cerebrospinal fluid measures of beta-amyloid 1-42, total tau, and p-tau181 for identifying Alzheimer disease amyloid plaque pathology. Arch Neurol 68:1137–1144

Galasko D, Chang L, Motter R et al (1998) High cerebrospinal fluid tau and low amyloid beta42 levels in the clinical diagnosis of Alzheimer disease and relation to apolipoprotein E genotype. Arch Neurol 55:937–945

Hansson O, Zetterberg H, Buchhave P et al (2006) Association between CSF biomarkers and incipient Alzheimer's disease in patients with mild cognitive impairment: a follow-up study. Lancet Neurol 5:228–234

Li G, Sokal I, Quinn JF et al (2007) CSF tau/Abeta42 ratio for increased risk of mild cognitive impairment: a follow-up study. Neurology 69:631–639

Mattsson N, Rosen E, Hansson O et al (2012) Age and diagnostic performance of Alzheimer disease CSF biomarkers. Neurology 78:468–476

Mattsson N, Zetterberg H, Hansson O et al (2009) CSF biomarkers and incipient Alzheimer disease in patients with mild cognitive impairment. JAMA 302:385–393

Mollenhauer B, Locascio JJ, Schulz-Schaeffer W et al (2011) Alpha-Synuclein and tau concentrations in cerebrospinal fluid of patients presenting with parkinsonism: a cohort study. Lancet Neurol 10:230–240

Paraskevas GP, Kapaki E, Papageorgiou SG et al (2009) CSF biomarker profile and diagnostic value in vascular dementia. Eur J Neurol 16:205–211

Shaw LM, Vanderstichele H, Knapik-Czajka M et al (2009) Cerebrospinal fluid biomarker signature in Alzheimer's disease neuroimaging initiative subjects. Ann Neurol 65:403–413

Snider BJ, Fagan AM, Roe C et al (2009) Cerebrospinal fluid biomarkers and rate of cognitive decline in very mild dementia of the Alzheimer type. Arch Neurol 66:638–645

Stefani A, Bernardini S, Panella M et al (2005) AD with subcortical white matter lesions and vascular dementia: CSF markers for differential diagnosis. J Neurol Sci 237:83–88

Stoeck K, Sanchez-Juan P, Gawinecka J et al (2012) Cerebrospinal fluid biomarker supported diagnosis of Creutzfeldt-Jakob disease and rapid dementias: a longitudinal multicentre study over 10 years. Brain 135:3051–3061

Vanderstichele H, Bibl M, Engelborghs S et al (2012) Standardization of preanalytical aspects of cerebrospinal fluid biomarker testing for Alzheimer's disease diagnosis: a consensus paper from the Alzheimer's biomarkers standardization initiative. Alzheimers Dement 8:65–73

In Vivo Clinical Biochemistry

40

Caterina Tonon, Raffaele Lodi, Claudia Testa, and Stefania Evangelisti

Introduction

Magnetic resonance spectroscopy is a non-invasive multinuclear technique that provides in vitro, ex vivo, and in vivo biochemical parameters. The main detectable nuclear species are hydrogen (^{1}H), phosphorus (^{31}P), carbon (^{13}C), sodium (^{23}Na), and fluorine (^{19}F). In vivo magnetic resonance spectroscopy is an advanced biochemical method complementary to conventional imaging that provides morphological information about the investigated tissue. The technological advancement of the hardware and software components of the instrumentation has led to an increasing diffusion of the magnetic resonance spectroscopy method, in particular of ^{1}H, in the clinical routine for the study of the brain. Thanks to the quantification of tissue-specific metabolites, it is possible to detect neurometabolic profiles related to multiple pathological substrates, such as, in the brain, cell proliferation, hypo and demyelination, neurodegeneration, necrosis, ischemia, inflammation, and infection; in the skeletal muscle, however, are assessable functional bioenergetic markers relevant to support the diagnosis of metabolic myopathies. In vivo magnetic resonance spectroscopy, although it is a technique with high diagnostic sensitivity, requires appropriate knowledge of the clinical context for a correct interpretation of the results. Overall, managing in vivo biochemical methods requires a multidisciplinary contribution of medical, biochemical, physical, and bioengineering expertise and technical and nursing personnel with specific training. Of crucial importance are the interactions with the specialists of the many fields potentially involved, from neurology to oncology, from neonatology to pediatrics, from genetics to pathological anatomy, as a further demonstration of the central role of laboratory medicine in clinical and applied research.

Magnetic Resonance Spectroscopy Technique: Basic Principles

The technique of choice for the study of clinical biochemistry in vivo is magnetic resonance spectroscopy (MRS or MR spectroscopy) because of its non-invasive characteristics and ability to reveal parameters of biochemical interest through direct biomarkers quantification in organs, tissues, or cells of biomedical interest.

MRS is based on the phenomenon of nuclear magnetic resonance (NMR), which is observed when the nuclei of atoms with a non-zero nuclear spin are immersed in a static magnetic field (B_0) and are excited by an oscillating electromagnetic field. Nuclei with zero spin, such as ^{12}C and ^{16}O, cannot generate magnetic resonance signals. Nuclei endowed with spin, such as ^{31}P, ^{1}H, ^{19}F, and less abundant isotopes in nature, such as ^{13}C and ^{17}O, have the property instead to orient themselves parallel or antiparallel to an externally applied static magnetic field, maintaining a precession motion around the axis of the magnetic field with a specific resonance frequency, called Larmor frequency ($\omega = \gamma B_0$). The slight excess of spin oriented in concordance with the magnetic field – a situation characterized by lower energy level and dependent on the nuclear species, the magnetic field strength, and the temperature of the system – determines the sensitivity of the magnetic resonance (MR) signal, which is lower than other diagnostic techniques based on radiation/

C. Tonon (✉)
Department of Biomedical and Neuromotor Sciences, University of Bologna, and Functional and Molecular Neuroimaging Unit, IRCCS Institute of Neurological Sciences of Bologna, Bellaria Hospital, Bologna, Italy
e-mail: caterina.tonon@unibo.it

R. Lodi
Department of Biomedical and Neuromotor Sciences, University of Bologna, IRCCS Institute of Neurological Sciences of Bologna, Bologna, Italy

C. Testa
Department of Physics and Astronomy, University of Bologna, Bologna, Italy

S. Evangelisti
Department of Biomedical and Neuromotor Sciences, University of Bologna, Bologna, Italy

M. Ciaccio (ed.), *Clinical and Laboratory Medicine Textbook*, https://doi.org/10.1007/978-3-031-24958-7_40

matter interaction and nuclear phenomena, such as spectrophotometry, fluorimetry, computed tomography (CT), and positron emission tomography (PET).

If the system is perturbed by an oscillating electromagnetic field, with a frequency equal to the Larmor frequency, of the order of MHz (radiofrequency), whose associated magnetic field (B_1) is perpendicular to the static one, it is possible to induce a variation of the spin orientation. The spin system reacts depending on each nuclear spin's chemical and physical surroundings due to the coexistence of different interactions among nuclear spins, magnetic fields, and electrons surrounding the nuclei. Consequently, the frequency of the signal obtained from the same nucleus, such as the proton (^1H) or the phosphorous (^{31}P), will differ depending on the molecule and/or the chemical group. Based on these principles, in the MR spectrum, it is possible to detect different chemical groups within the same molecule (Fig. 40.1) and/or different molecules containing the same nucleus, such as phosphocreatine, inorganic phosphate, and adenosine triphosphate (ATP) in the spectrum of ^{31}P (Fig. 40.2). This property, the basis of MR spectroscopy, is called chemical shift. The signal emitted by the sample, defined as FID (free induction decay), is made up of a superposition of radiofrequency, that is, sinusoidal signals which decay as a function of time, each at the resonance frequency of the nucleus which produces it. By applying a mathematical function, such as the Fourier transform, it is possible to convert the intensity of the signal as a function of time (FID) into a spectrum, in which the signal is expressed as a function of frequency and, therefore, the individual contributions of the spins belonging to the different chemical groups of the molecules of the sample are recognizable.

The width and, more generally, shape of the peaks depend on various factors and, above all, on the mobility and homogeneity of the MR system. The more mobile and homogeneous the system, the more the peaks of the spectrum are resolved, that is, well distinguishable and separated from each other.

The diagnostic technique based on the phenomenon of MR mostly applied in the clinical field is MR imaging (MRI). In magnetic resonance imaging, the described phenomenon is associated with magnetic field gradients to make the resonance frequency of the spin-dependent on their position in space. It follows that by selectively exciting certain intervals of resonance frequencies, it is possible to encode the signal of the spin belonging to a defined region of the sample. In this way, it is possible to generate images characterized by high resolution compared to other imaging techniques applied in the clinical field, such as CT and PET.

In the present paragraph, the physical principles essential for understanding the applications of the MR spectroscopy technique have been briefly reported; for a detailed explanation, please refer to specialized texts.

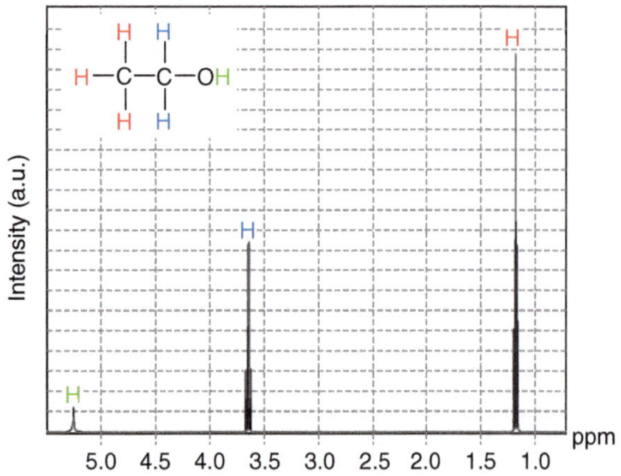

Fig. 40.1 ^1H-MRS spectrum of ethanol (CH_3–CH_2–OH) acquired in vitro in a magnetic field of 9.4 T. The number of peaks corresponds to the number of chemically non-equivalent protons; the amplitude of the peaks, whose intensity is expressed in arbitrary units (u.a.), on the ordinate axis, is proportional to the number of chemically equivalent atoms that resonate at the same resonant frequency (measured in ppm, or parts per million, on the abscissa axis). The different resonant frequency of the peaks, called chemical shift, depends on the electronic neighborhood surrounding the nuclear spin. Each peak's multiplicity (number of peaks in which each resonance frequency is divided) depends on the interaction between the nuclear spins, mediated by the electronic neighborhood, called scalar coupling or j-coupling

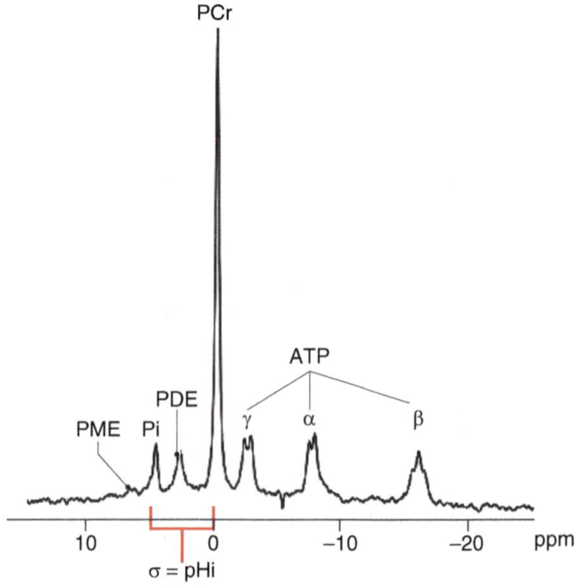

Fig. 40.2 ^{31}P-MRS spectrum of calf muscles acquired in vivo in a 1.5 T magnetic field. The peaks correspond to phosphorus nuclei within molecules present in at least 2 mM concentration in skeletal muscle tissue: β-, α-, γ-ATP phosphate groups of adenosine triphosphate; PCr, phosphocreatine; PDE, the mixture of phosphodiesters; Pi, inorganic phosphate; PME, the mixture of phosphomonoesters. From the chemical shift (σ) of Pi from PCr, it is possible to calculate the intracellular pH (pHi). ppm, resonant frequency expressed in parts per million

In Vitro, Ex Vivo, and In Vivo MR Spectroscopy

MR spectroscopy has numerous fields of application in both organic and inorganic areas. The organic applications most concern chemistry, biochemistry, and medicine.

Regarding organic chemistry, in vitro MRS is considered the primary tool for identifying molecules or compounds and determining their structure in samples of protein solutions, mixtures, drugs, or foods. In particular, MR spectroscopy allows to define the 3D structure of biomolecules and macromolecules, determining the following parameters:

- Distances between atoms in molecules
- Dihedral angles in chemical bonds
- Orientations of chemical groups in 3D space, both relatively and absolutely
- Kinetics of chemical exchange of the most labile atoms

The ability to obtain information related to the dynamics of chemical groups, regions, and even of the whole molecule is precisely the characteristic of MR spectroscopy for organic chemistry applications.

Regarding the application in the field of biochemistry and medicine, the biological samples used may be: (1) solutions and extracts of tissues and cells – in vitro MRS; (2) tissues from biopsy collection or operating pieces – ex vivo MRS; (3) perfused cells and organs or tissues studied directly in humans or animals – in vivo MRS.

In all biological samples listed above or in humans (and animals), the MRS technique allows the investigation of spectral profiles related to both physiological and pathological conditions, studying multiple metabolic pathways, such as glycolysis, oxidative metabolism, energy status, redox status, phospholipid metabolism, etc. Overall, the metabolic information that can be obtained in vitro, ex vivo, and in vivo are complementary, and the choice of the type of sample/system to investigate strictly depends on the clinical or scientific question.

Equipment

Magnets and Probes

The main component of an MRS instrument is the magnet, which generates the static magnetic field (B_0). The hardware component that generates the oscillating electromagnetic field with the B_1 component is constituted by probes (or coils), which can be defined as antennas able to emit and receive pulses in the radiofrequency range.

For in vitro clinical and preclinical biochemistry studies, magnets generally consist of coils of superconducting material, which are passed through by high currents without dissipating energy, allowing the generation of a magnetic field up to 100 times more intense than that of permanent magnets or electromagnets. The magnetic field strength is measured in Tesla (T).

In the case of in vitro or in vivo studies on cells, magnets with vertical conformation are used, with an intensity varying from 4 to more than 20 T, inside which the sample is inserted, contained in a calibrated glass or plastic tubes (5 or 10 mm in diameter) (Fig. 40.3).

For in vivo applications, magnets have a horizontal inner diameter (bore) which can vary significantly in width when animals are considered, depending on size. For human applications, the magnetic field strength varies from 0.2 to 7 T and the magnet's geometry is cylindrical, with a bore of 65–80 cm in diameter and a length of 120–150 cm (Fig. 40.4).

In addition to the substantial difference in the type of magnet used in preclinical and clinical settings, the coils that generate the B_1 field are also different, both in terms of size, depending on the type of sample to be investigated, and for the different resonance frequencies at which they can radiate the electromagnetic field. The frequency at which the coils must be tuned depends on the static magnetic field and the nuclear species to be excited and revealed.

Detectable Nuclear Species and Corresponding Metabolic Pathways

With the MRS technique, it is possible to detect different nuclear species whose sensitivity depends on the magnetic ratio (γ) value, characteristic of each atomic nucleus, and the abundance in nature of the isotope visible in MRS. In Table 40.1, the main nuclear species detectable by the MRS technique in the clinical and preclinical fields are described, with relative values of the gyromagnetic ratio, percentage of abundance in nature, and relative sensitivity, with the same number of nuclei. It can be observed that at the same natural abundance, the sensitivity of ^{31}P is about one-fifteenth of that of ^{1}H.

The nuclear species used in the clinical field is the hydrogen atom, commonly called proton (^{1}H), which allows the characterization of diseases of the brain, prostate, and breast. The phosphorus nucleus (^{31}P) is instead applied to support the diagnosis of muscular pathology linked to metabolic defects. In selected human research applications, the ^{13}C carbon nucleus, an isotope of the more widespread ^{12}C, is mainly used, while ^{23}Na and ^{19}F are the nuclear species that are widely used in preclinical studies.

In particular, with the ^{1}H-MRS, it is possible to study the metabolism of choline-containing phospholipids, glycolysis, and its products (lactic acid); energy metabolism (through the quantification of creatine and phosphocreatine);

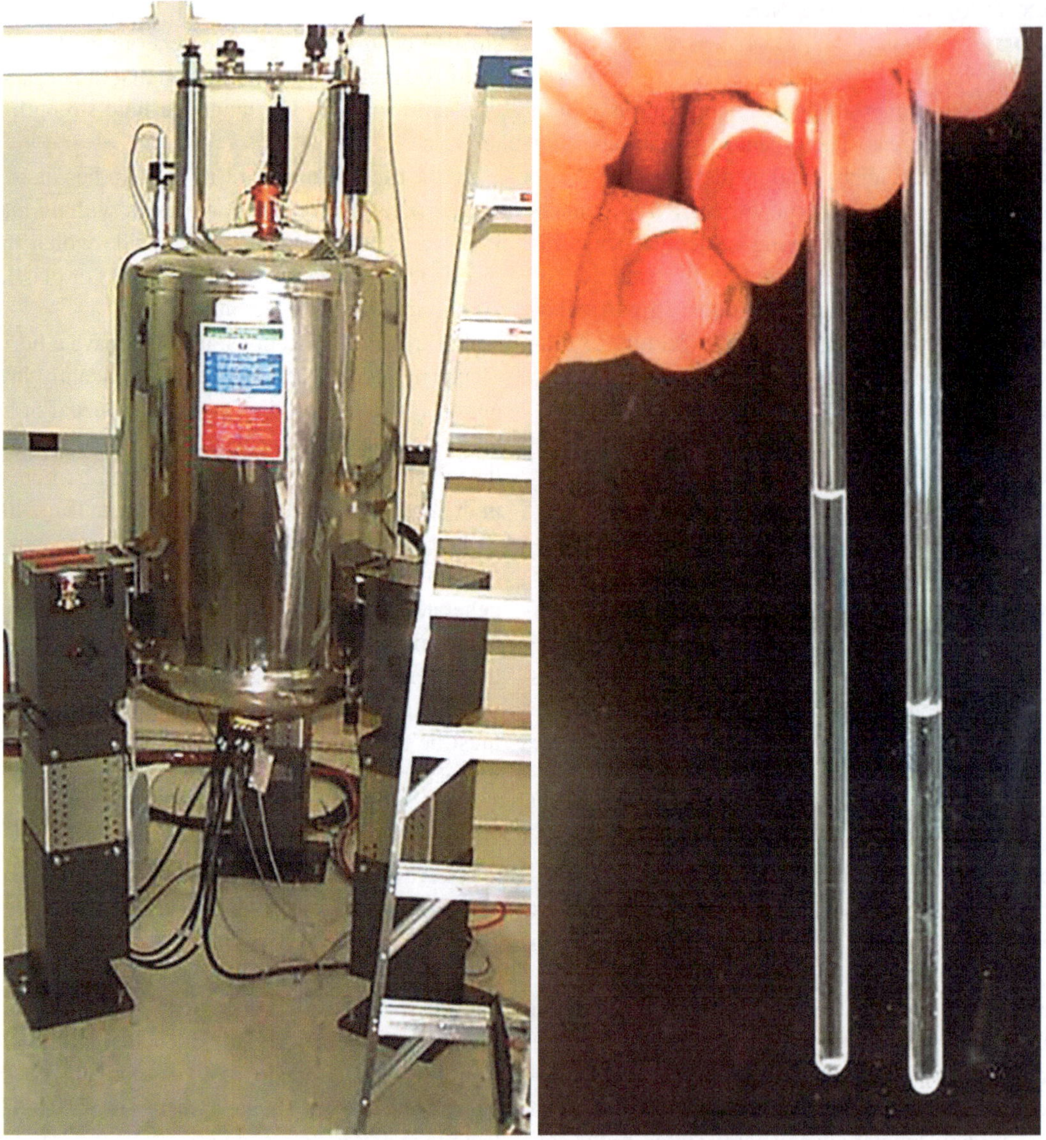

Fig. 40.3 Left, Bruker Advance System scanner with 9.4 T magnetic field for in vitro or in vivo cell studies. Right, relative sample holder tubes (Photo: Istituto Superiore di Sanità, Department of Cell Biology and Neuroscience, Rome, Italy)

neurotransmitters, such as gamma-aminobutyric acid (GABA) and its precursor; glutamic acid; the metabolism of neutral lipids (mobile lipids); amino acids; tissue-specific metabolites, such as N-acetyl aspartate, a neuro-axonal-associated metabolite, or mio-Inositol, glial marker. The ^{31}P-MRS can assess tissue bioenergetics (ATP, ADP, phosphocreatine, inorganic phosphate, phosphate sugars), phospholipid metabolism, and pH.

The ^{13}C-MRS allows for studying metabolite fluxes of biochemical pathways such as glycolysis, the tricarboxylic acid cycle, phospholipid metabolism, and the structure of carbohydrates.

The ^{19}F-MRS is used for evaluating the uptake and pharmacokinetics of fluorine-containing drugs and the metabolic pathways of ^{19}F-labeled substrates.

With ^{23}Na, altered cellular metabolism is studied by measuring sodium's intra- and extracellular concentrations.

Applications in Cells, Tissues (Biopsies), and Extracts

There is a growing interest not only in the study of single reactions or specific pathways but also in the knowledge of the metabolome of organisms, that is, the totality of metabolites consisting of low molecular weight molecules. Enzymes and proteins encoded by the genome have the ultimate effect of determining the concentration of an organism's metabolites, which is why knowledge of the metabolome is an important objective of the study in bio-

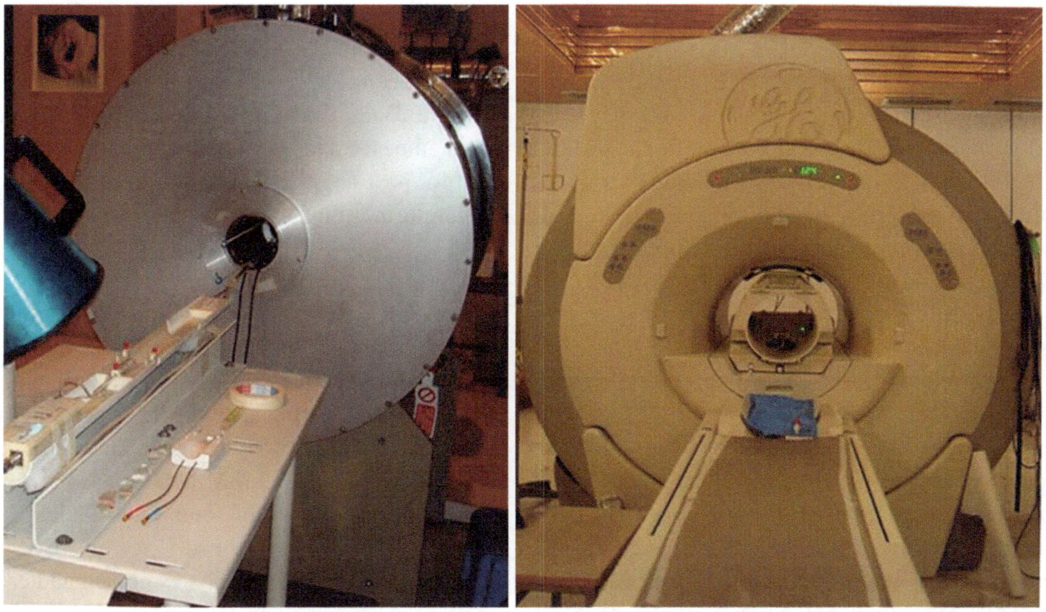

Fig. 40.4 Left, 4.7 T Varian/Agilent Anova MRI/MRS system scanner for in vivo animal studies. (Photo: Istituto Superiore di Sanità, Department of Cell Biology and Neuroscience, Rome). Right, 1.5 T GE Medical Systems Signa HDx 15, for human examinations (Photo: S. Orsola-Malpighi Hospital, Bologna, Italy)

Table 40.1 Nuclear species detectable by MRS and related gyromagnetic ratio values

Nucleus	$\gamma/2\pi$ (MHz/T)	Natural abundance (%)	Relative sensitivity (%)
^{1}H	42.58	99.98	100
^{31}P	17.25	100	6.64
^{13}C	10.71	1.11	1.59
^{23}Na	11.27	100	9.27
^{19}F	40.08	100	83

MRS magnetic resonance spectroscopy; γ gyromagnetic ratio expressed as MHz/T

Table 40.2 Range of the concentrations of main brain and muscle metabolites detectable by proton MRS (^{1}H) and phosphorus (^{31}P)

Metabolite	Concentration (mmol/L)	ppm
^{1}H-MRS (brain)		
NAA	7.9–16.6	2.02
Cr	5.1–10.6	3.02
CHO	0.9–2.5	3.22
mI	3.8–8.1	3.56
^{31}P-MRS (skeletal muscle)		
ATP	7.1–9.7	−2.52 (γ), −7.56
PCre	23–42	(α), −16.15 (β)
Pi	2.5–6.6	0
ADP	13–17 × 10^{-3}	4.91

ADP adenosine diphosphate, *ATP* adenosine triphosphate, *Cho* choline, *Cr* creatine, *me* myo-inositol, *MRS* magnetic resonance spectroscopy, *NAA* N-acetyl-aspartate, *PCr* phosphocreatine, *Pi* inorganic phosphate, *ppm* parts per million

chemistry and clinical biochemistry. The metabolome, similar to the proteome and unlike the genome, is state-specific so that one can refer to the metabolome of a single cell of an organism or the metabolome of a disease state. The metabolome is unstable. Indeed, it can change rapidly in response to stimuli. The goal of understanding the metabolome has given great incentive to the use of MRS in vivo and ex vivo.

In vivo MRS allows for the detection and quantification of metabolites present at higher concentrations (Table 40.2) due to limitations caused by the technique's sensitivity and the inherent inhomogeneity of tissues in vivo, which result in relatively broad peaks in the spectrum. Table 40.2 shows the range of concentrations of the main resting brain and muscle metabolites detectable in vivo in humans by proton (^{1}H) and phosphorus (^{31}P) MRS, respectively. The concentration of ADP is calculated by the creatine kinase reaction.

In order to detect the largest possible number of metabolites involved in the metabolic processes of an organ in physiological or pathological conditions, in vivo studies are also carried out on cell cultures, ex vivo on tissues and biopsies, and in vitro on cell extracts. In these cases, the obtained spectra are characterized by a better spectral resolution and sensitivity by using scanners with very high magnetic field and very long acquisition times compared to those of the clinical practice. However, even MRS of in vitro samples is less sensitive than the mass spectrometry (MS), which is the most widely used technique for metabolomics studies. However, the MRS remains very advantageous in sample preparation, which consists only of extraction or insertion of the sample into the sample holder. In contrast, for MS, sample preparation is generally based on a preliminary procedure, which often involves gas chromatography or liquid chromatography. In addition, the sample after MS is not reusable due to the ionization procedure it must undergo, unlike samples used for MRS studies. Given the high accuracy and reproducibility of ex vivo MRS in quantifying several metabolites (in the order of 20–30), this method is used to study variations between the physiological and a perturbed state of the metabolome. In the following sub-sections, we will mention the methods of in vivo MRS spectroscopy on cells and ex vivo MRS spectroscopy on biopsies or cell extracts to study the metabolome.

Cells

The characteristic broad lines of an MR spectrum are also typical of spectra obtained on cells in culture or bioptic samples. To obtain data with a better spectral resolution and, therefore, more informative from the point of view of the number of recognizable metabolites, it is possible to proceed either with the HR-MAS (high resolution-magic angle spinning) spectroscopy technique or with the extraction of metabolites in solution.

Biopsies

For the HR-MAS MRS technique, hardware allowing the rotation of the sample (spinning) at a precise angle relative to the magnetic field (54.74°) is required to eliminate spin-spin interactions that partly contribute to the width of the spectrum line. At the end of the MRS acquisition, which must be performed at a controlled temperature, the sample, whose integrity is preserved, can be used for other types of analysis, such as histological or gene expression studies.

Extracts

Collecting cells for MRS analysis should be based on a method that does not alter the concentration of metabolites or the resulting spectrum. First, the culture medium shall be removed from the cells, and if attached, an extraction medium shall be applied. Protein denaturation is then performed to inactivate enzymes that may alter concentrations. Denaturation can be performed in two ways: by an inorganic solution (5% perchloric acid, PCA) or an organic solution of chloroform-methanol-water to extract the lipid component from the tissue and leave only the polar component. In both cases, attention must be paid to possible contamination of the sample by the solvent. The extract obtained in both cases is liquid; therefore, the lines of the spectrum will be characterized by very low line widths.

In Vivo Proton MR Spectroscopy

Operational Context

The in vivo MR systems are installed in specially designed facilities, which include the following rooms: outside, the waiting room for patients and attendants; inside, the diagnostic room, where the magnet is located; the adjacent command or "control" room, with the acquisition console; the technical room; the nurse's station; the room for collecting the anamnesis and informed consent; and an outpatient room for post-anesthesia or sedation clinical and instrumental monitoring (Fig. 40.5).

Hardware

MR devices in clinical settings can be distinguished into "closed" or "open", depending on the geometric conformation of the magnets. Open devices are designed to improve patient comfort, particularly for those suffering from claustrophobia, but with a low magnetic field intensity (<0.5 T), which is insufficient for MR spectroscopy examinations. Closed devices consist of a tube into which the patient is inserted using a motorized bed. The coils used are different depending on the district to be examined and can be double (one for the emission and one for the reception of the signal) or single (both emitting and receiving). On the magnetic field gradients depend the signal's localization in circumscribed portions of space to allow both the localization of the volume of interest (or voxel) for spectroscopy and the generation of images. The electronic component of an MR instrument, including synthesizers, amplifiers, transmission and reception channels, and an electronic system for cooling, is very voluminous and occupies the technical compartment adjacent to the diagnostic room.

Software

The acquisition console of the MR scanner is a computer with high calculation and storage capacity because it must allow real-time analysis and storage of a large amount of data for each examined subject. Software is installed on this computer both for acquisition and real-time processing of the acquired data, so an immediate preliminary evaluation of the data is possible. The computer is usually connected to the internal hospital (or diagnostic center) network of the picture archiving and communication system (PACS), which is integrated with the radiology information system (RIS), which is the management software for booking, recalling the patient's identification at the time of the examination, and reporting. This equipment is housed in a room called the control room, accessed by physicians, technicians, and nurses, which is immediately adjacent to the diagnostic room.

Patient Preparation

No special preparation is required for an MR spectroscopy examination, and no contrast medium or radioisotopes are administered to the patient. Informed consent for performing the examination is unique and valid for conventional MR sequences and MR examinations performed in other districts (heart, abdomen, pelvis, etc.). In particular, the possible presence of absolute and relative contraindications, such as

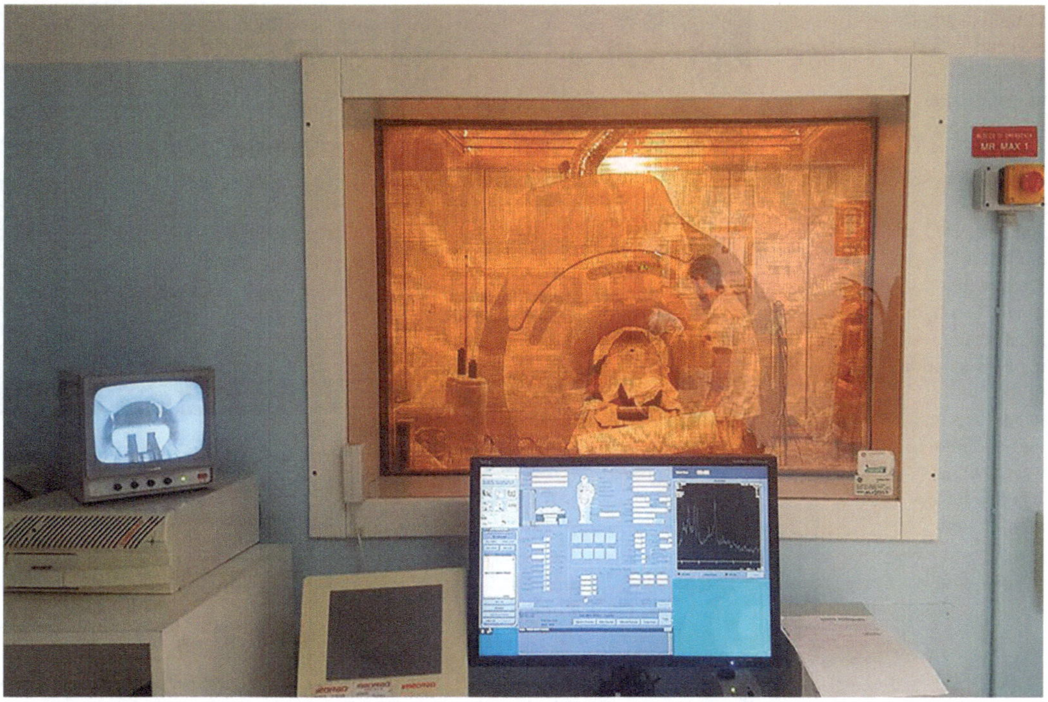

Fig. 40.5 Control room with console for the acquisition of imaging data and MR spectroscopy, with the observation window on the diagnostic room, where the technician has completed the positioning of the patient, supine on the movable bed, to perform a brain exam

endocorporeal ferromagnetic devices (pacemakers, metal clips, dental or orthopedic prostheses, etc.), must be carefully investigated. Therefore, the patient is asked to wear a lab coat and remove ferromagnetic objects such as analog watches, magnetic cards, keys, etc. Also, all operators entering the magnet room (medical staff, technicians, nurses, maintenance staff, cleaners, visitors) must not present contraindications to the MR examination or introduce ferromagnetic material (stethoscopes, scissors, etc.).

In the case of non-cooperative patients, sedation or general anesthesia is provided during the performance of the MR examination, with anesthesiological support to monitor vital functions. Causes of non-cooperation may be related to age (infants, children), cognitive impairment (dementia, encephalopathy), or claustrophobia.

For the success of the examination, the technical and nursing staff play a fundamental role since, among the various activities for which they are responsible, they explain to the patient that he must maintain the same position without moving during the examination and give him a bell, reassuring him that he will remain under constant visual and acoustic control.

Proton MR Spectroscopy

The most widely used MR spectroscopy technique in the clinical setting is proton spectroscopy to study the brain. Depending on the clinical question, during the examination, the voxel on which the biochemical investigation will be performed is manually selected (anatomical location and volume size). There are two types of voxel localization techniques: single voxel MRS (SV-MRS), in which a single spectrum is acquired from a single volume of tissue, and multi-voxel MRS (CSI-MRS) (Fig. 40.6), in which multiple spectra are acquired simultaneously. CSI-MRS is indicated for studying a large and/or heterogeneous composition, as in the case of a neoplastic lesion.

Qualitative and Quantitative Analysis of Proton Spectra

Because the concentrations of brain metabolites to be quantified are relatively low (even <1 mmol/L), the analysis step is critical. As illustrated in the previous paragraphs, the MR spectrum consists of a series of peaks corresponding to the different metabolites, representing the MR signal intensity as a function of resonance frequency, usually expressed in ppm (parts per million), a unit of measurement that allows frequencies to be made independent of magnetic field strength.

The height of each peak is proportional to the number of nuclei resonating at that particular frequency; therefore, the calculation of the subtended area allows the determination of the concentration of the corresponding metabolite. The quality of the spectrum is defined as spectral resolution, which is quantifiable by calculating the half-height width of each peak in the spectrum. The spectral resolution depends on the

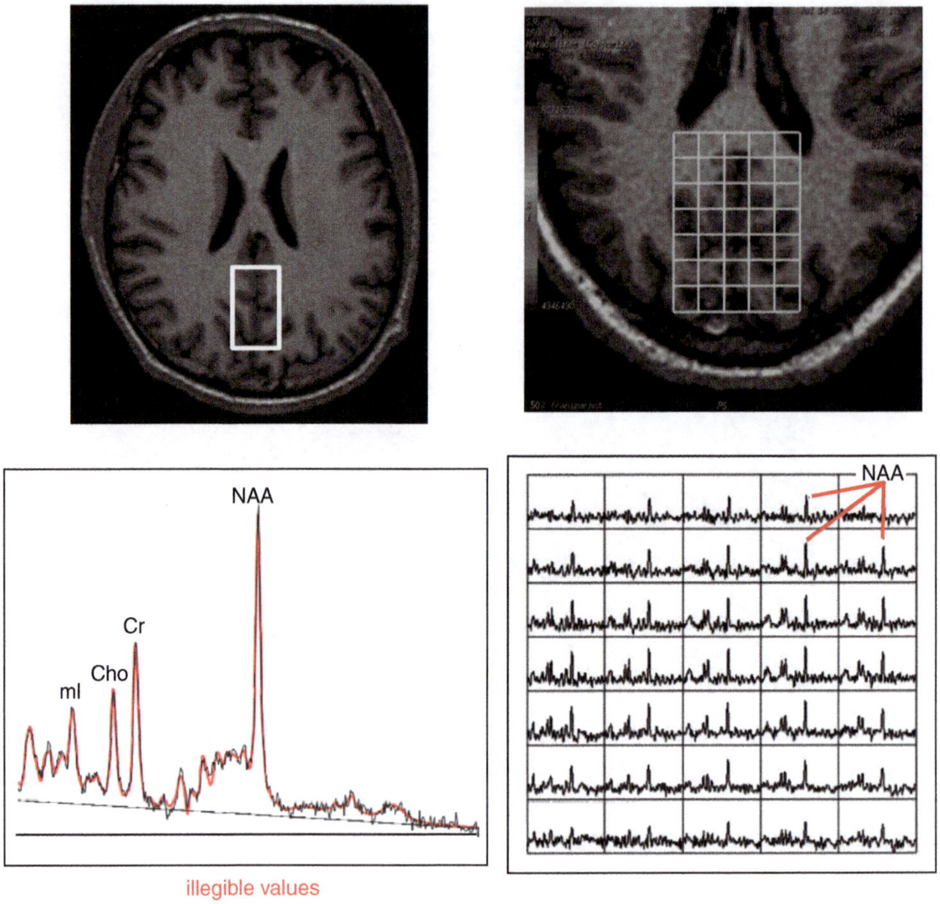

NAA= N-acetyl-a spartate, Cr= creatine, Cho= Choline, ml=myo-Inositol

Fig. 40.6 Example of single voxel (SV-MRS) and multi-voxel (CSI-MRS) ^{1}H-MRS technique. Above left, T_1-weighted axial MR image showing the localization of a single voxel including the median parieto-occipital cortex (volume: 18 cm^3) of a volunteer and below the spectrum obtained. Top right, T_1-weighted axial image showing the region's location examined with the multi-voxel technique at the same level, the parieto-occipital site, of a volunteer, whose spectra shown below have a spatial resolution of 1 cm^3

homogeneity of the tissue volume examined and the signal-to-noise ratio (SNR), which is mainly dependent on the size of the voxel volume, the intensity of the magnetic field, and any interference that may cause artifacts.

Similar to other diagnostic techniques, qualitative or quantitative analysis can be performed. The quantitative analysis requires fitting the peaks with appropriate models since the following criticalities can occur: overlapping of the resonances of the metabolites, distortions due to the baseline, and peak shapes not easily modeled. The most widespread method is obtained by a model based on prior knowledge in which the fitting is done through a model constituted by the sum of the spectra of the single metabolites simulated or measured through in vitro solutions (Fig. 40.7).

The quantitative analysis may be relative or absolute. In the case of relative quantitative analysis, the amounts of each metabolite are reported using that of Cr as a reference.

In absolute quantitative analysis, the concentration of metabolites is calculated using a reference (internal or external) whose concentration is known (or estimated), for example, intracellular water.

In addition to the supplementary instrumentation compared to conventional MR examinations in terms of hardware and software for spectra acquisition and processing, MR spectroscopy in the clinical setting requires the collection of reference parameters in a population of healthy control subjects. For an accurate comparison between the metabolite concentration values determined in the individual patient and the normative ones, several inclusion and exclusion criteria must be respected in the selection of control subjects, including the absence of neurological, psychiatric, and systemic pathologies, overlapping age and sex, same anatomical location and size of the selected voxel, same acquisition parameters, and spectra analysis protocol.

The value of the concentration of a metabolite is considered pathological, with a probability of 95.4%, if it is greater or lesser, +2 or −2 standard deviations (SD), than the mean value calculated in the control population.

Brain Metabolites

Biochemical Characterization and Biological Significance

At a magnetic field of 1.5 T, with short TE (30–40 ms), the following resonance peaks can be recognized in the healthy

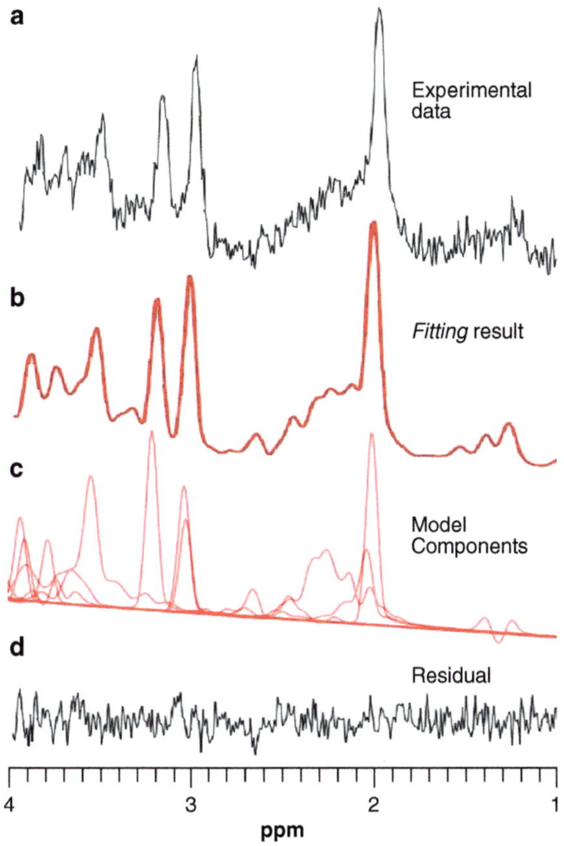

Fig. 40.7 Example of analysis based on prior knowledge of a ^{1}H spectrum localized in the cerebellar hemisphere of a patient with idiopathic Parkinson's disease. In (**a**) the acquired data, in (**b**) the result of the fitting obtained from the model shown in (**c**), consisting of the simulation of the signal of the cerebral metabolites. In (**d**) the residual signal after fitting the experimental data with the chosen model

adult from right to left (Fig. 40.8 and Table 40.2): the major one, located at 2.02 ppm, arising from the total N-acetyl-aspartate (NAA) content, including a proportion of N-acetyl-aspartyl-glutamate; at 3.02 ppm creatine (Cr), representing a mixture of Cr and phosphocreatine; at 3.22 ppm, the choline (Cho) peak, which is the sum of free Cho (3.21 ppm), phosphorylcholine (3.23 ppm), and glycerophosphorylcholine (3.24 ppm) signals; at 3.56 ppm, myo-Inositol (mI).

In addition, between 2.2 and 2.4 ppm, the peak indicated as Glx can be recognized, which originates from the signal contribution of glutamate (Glu), glutamine (Gln), and γ-aminobutyric acid (GABA), metabolites that at 1.5 T cannot be separated into individual peaks.

The spectral resolution improves with higher intensity clinical magnets (3 T and 7 T) compared to 1.5 T magnets. Due to the higher signal-to-noise ratio, the quantification of metabolites of interest significantly improves, making it possible to detect additional metabolites (Fig. 40.9). With a 3 T magnet, it is possible to separate the Glx peak into glutamate and glutamine and detect metabolites such as aminobutyric acid, glutathione, aspartate, and taurine. With ultra-high field clinical magnets (7 T), it is possible to quantify over 18 metabolites or mixtures of brain metabolites, which can be used in clinical practice.

The NAA metabolite is an amino acid expressed almost exclusively in the brain, in white and grey matter. Synthesized in the mitochondria of neurons, it is subsequently transported into the neuronal cytoplasm and along axons. Being contained in neurons and axons and, in much lower concentrations, in immature oligodendrocytes and astrocyte progenitor cells, it is considered a marker of neuronal-axonal integrity and density. Studies have shown that NAA levels reflect mitochondrial function and correlate with osmoregulatory function. NAA levels show a gradual and progressive increase in the neonatal period, reflecting the physiological maturation process of the brain. The reduction of NAA is an expression of neuronal-axonal damage, which can be secondary to neurodegenerative, ischemic, and inflammatory pathology or replacement of nerve tissue with neoplastic tissue.

Cr is involved in the creatine kinase reaction and, therefore, is a marker of cellular energy metabolism; due to the relative stability of its concentration under different metabolic conditions, it is used as an internal reference standard for the relative quantification of other brain metabolites. In some systemic conditions, such as chronic kidney disease with encephalopathy, its increase is detectable since Cr and phosphocreatine are metabolized into creatinine, which is excreted renally.

Cho is a mixture of water-soluble metabolites containing choline (characterized by the tetramethylammonium group). The content of Cho reflects the phospholipid metabolism of cell membranes and myelin and, typically, increases with increased membrane turnover and cell density (malignant neoplasms) or their destruction, as in demyelinating diseases (e.g., multiple sclerosis) in the acute phase. A reduction in the Cho peak may reflect hypomyelination or altered metabolism secondary to liver failure.

Myo-Inositol (mI) is a simple sugar, the precursor of phosphatidylinositol, the phospholipid with the highest inositol content, and phosphatidylinositol 4,5-biphosphate, a key molecule in signal transduction. mI is located in glial cells, almost exclusively in astrocytes, so it is considered an astrocytic marker and a regulator of osmotic (osmolyte) pressure. The increase of mI is detectable in different conditions: gliosis, astrocytic proliferation, conditions found in neurodegenerative processes (Alzheimer's disease), or in the case of an increased volume of astrocytic cells, such as during inflammation.

The mobile lipid complex comprises macromolecules detectable at 0.9 (CH_3) and 1.3 (CH_2) ppm in the form of fatty acids, triglycerides, and phospholipids. Their accumulation is associated with various pathological processes, such

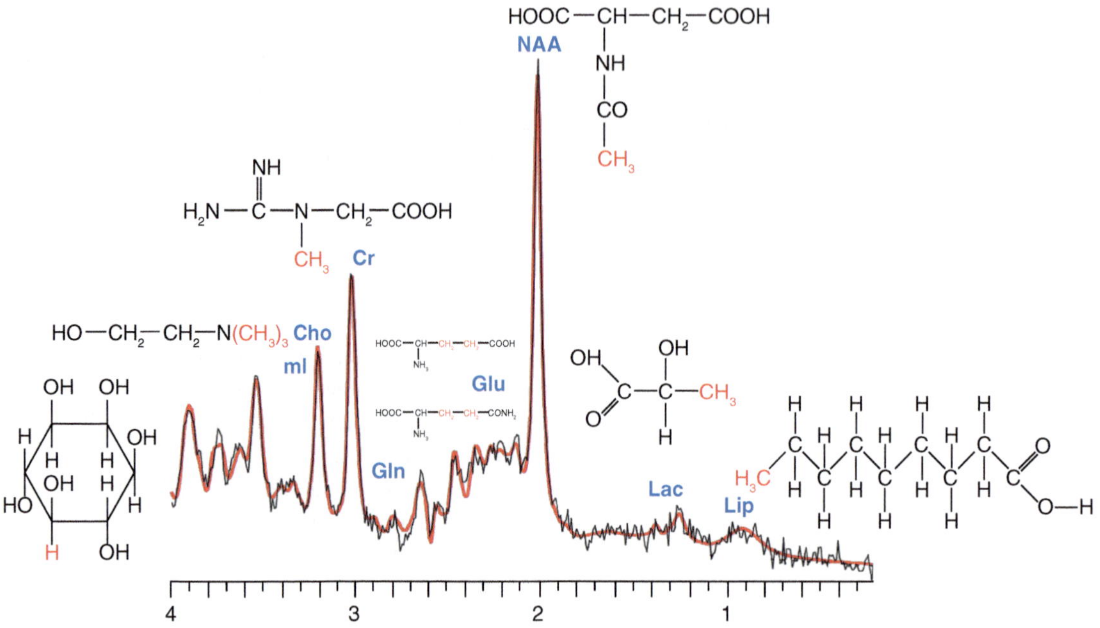

Fig. 40.8 ^{1}H-MRS spectrum acquired at the level of the parieto-occipital cortex of a control subject with a single voxel technique in a 1.5 T magnet. Each peak shows the structure of the corresponding molecule in which they have indicated the proton nuclei (^{1}H) from which the signal originates. The following metabolites are observed: N-acetyl-aspartate (NAA) at 2.02 ppm, creatine (Cr) at 3.02 ppm, choline (Cho) at 3.22 ppm, and myoinositol (mI) at 3.56 ppm. Furthermore, between 2.2 and 2.4 ppm, some peaks overlap the glutamate (Glu), glutamine (Gln), and GABA signals, metabolites that at 1.5 T are not detected separately. At 1.32 ppm in pathological conditions, the peak of lactic acid (Lac) can be observed and at 0.9 (CH_3) and 1.3 (CH_2) ppm, the complex of "mobile lipids", which includes macromolecules such as fatty acids, triglycerides, and phospholipids (Copyright EDISES 2021. Reproduced with permission)

as necrosis, inflammation, or infection. To avoid the misinterpretation of increased lipids as pathological, the voxel must be accurately localized to prevent contamination of the spectrum by the signal from lipids physiologically contained in the subcutaneous adipose tissue, scalp, and diploic space.

Since the physiological concentration of cerebral lactate (Lac) is below the sensitivity of MRS detection in vivo, its presence in the form of a double peak ("doublet") is always pathological in the adult subject (Fig. 40.10). Accumulation of Lac indicates a switch from an oxidative type of metabolism to anaerobic glycolysis, of which Lac is an end product. Therefore, an accumulation of Lac is attributable to the following conditions: hypoxia; ischemia; metabolic disorders, especially mitochondrial diseases; acute inflammation due to macrophage accumulation; arachnoid cysts; and normal pressure hydrocephalus.

Main Findings of Neurological Disorders Evaluated by the ^{1}H-MRS

The MRS in vivo is a useful clinical tool in many clinical conditions because, compared to other methods, it allows to investigate in a single patient both the white and the gray matter, detecting with accuracy and simultaneously time biochemical changes in the neuronal and glial population.

The clinical usefulness of MRS has been established for diagnosing cerebral neoplasms and other brain lesions, in particular demyelinating and infectious, acquired and genetically based metabolic pathologies, neonatal hypoxic-ischemic encephalopathy, and neurodegenerative diseases. In these conditions, it is possible to obtain biochemical markers with different clinical impacts: diagnostic, even at an early stage; prognostic; and surrogate of the effectiveness of a treatment (pharmacological, radiotherapy, or surgical).

However, because of the low specificity of metabolic patterns detectable by proton MRS in vivo, evaluating the laboratory test result in the clinical context of the individual patient is of primary importance for properly interpreting the result.

Congenital Errors of Metabolism

The proton MR spectroscopy technique allows the detection of specific diagnostic biomarkers of metabolic disorders, such as phenylalanine (phenylketonuria), glycine (non-ketotic hyperglycinemia), branched-chain amino acids, and keto acids (maple syrup urine disease). An increase in NAA peak is pathognomonic of Canavan disease, a condition due to aspartate-acylase enzyme deficiency, which manifests as macrocephaly and, radiologically, is characterized by diffuse leukoencephalopathy in T_2-weighted sequences (Fig. 40.11).

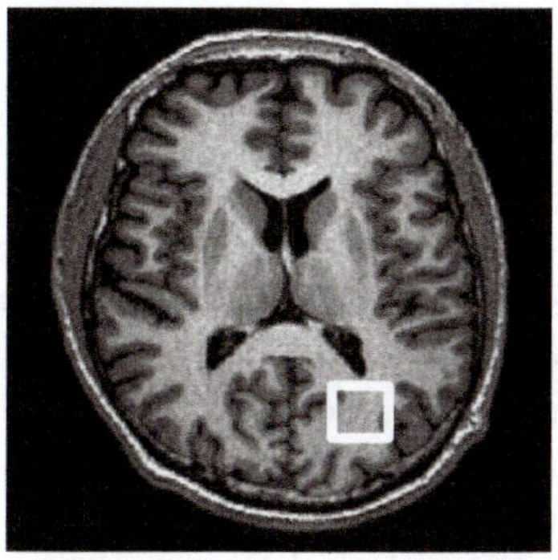

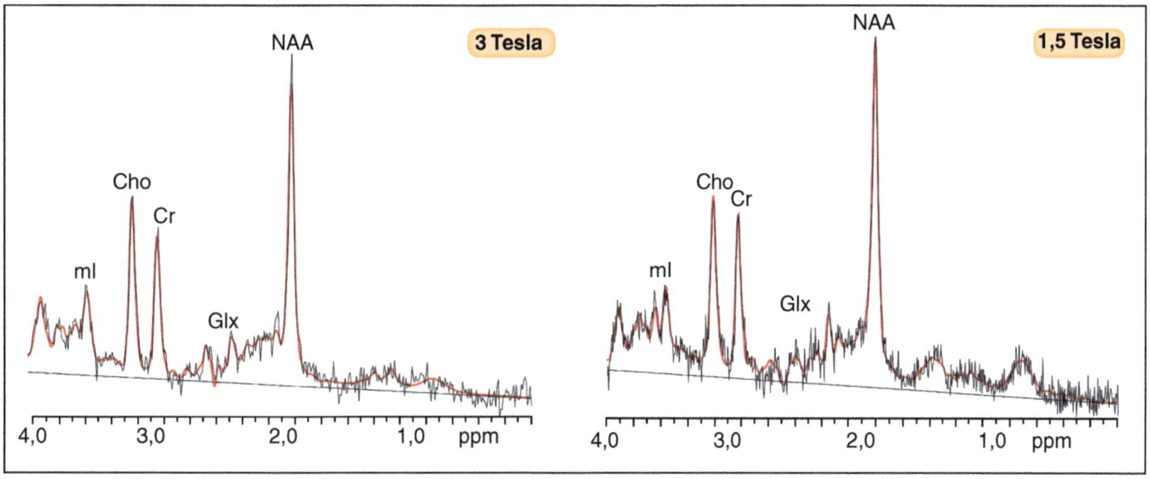

Fig. 40.9 ^{1}H-MRS spectra obtained from an 8 cm^3 voxel in the left parietal white matter of two volunteers (MR image on the left). The bottom left shows the spectrum acquired in a 3 T magnet, and on the right in one of 1.5 T. The spectrum acquired at 3 T shows a higher signal-to-noise ratio (SNR) and a better spectral resolution, allowing a fitting more accurate in the area between 2.0 and 3.0 ppm, where resonances of glutamate and glutamine, Glx, are detectable, with prominent peaks at 2.3 and 2.4 ppm, respectively

The peak of Cr is absent or markedly reduced in creatine deficiency syndromes, which are characterized by a relatively non-specific neurological manifestations (encephalopathy) and are caused by mutations that result in altered creatine synthesis, in most cases involving the enzyme guanidine-acetate methyltransferase (GAMT; OMIM 601240).

Most mitochondrial diseases from respiratory chain defects, disorders of pyruvate metabolism, or fatty acid oxidation present a pathological accumulation of lactate, which is quickly and directly detectable in the brain by using MRS technique (Fig. 40.10).

Perinatal Hypoxic-Ischemic Encephalopathy

Hypoxic-ischemic encephalopathy is the leading cause of pediatric mortality and morbidity, associated with a significant risk of developing long-term neurological sequelae, such as cerebral palsy, mental retardation, and epilepsy. The identification of accurate diagnostic and prognostic indicators, especially in cases with intermediate-grade clinical encephalopathy, is crucial, taking into account the evidence of the effectiveness of selective cerebral hypothermic treatment, which, if instituted in the first 6 hours of life, has shown a reduction in mortality and, in survivors, in significant disabilities. Several studies have shown that reduction in NAA and increase in Lac predict the prognosis at 2–5 years of life: The more severe these metabolic alterations, the prognosis (Fig. 40.12).

Neoplastic Pathology and Differential Diagnosis with Radionecrosis and Other Endocranial Masses

The typical pattern of a cerebral glioma (a primary tumor originating from cells of the glial series) is as follows: an increase in the concentration of Cho caused by cell proliferation and by an increase in membrane turnover; a reduction in NAA due to the replacement of parenchyma with neoplastic tissue; the increase in Lac and Lip depends instead on the presence of necrosis and/or hypoxia, determined by the imbalance between the high proliferation of neoplastic cells and the insufficient extent of neoangiogenesis (Fig. 40.13). The entity of the metabolic alterations described is corre-

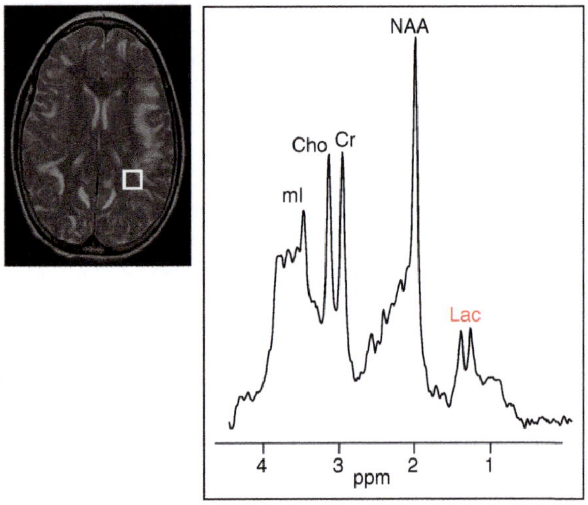

Fig. 40.10 On the right is the spectrum obtained in a 1.5 T magnet from a voxel located in the left parietal white matter of a patient with mitochondrial encephalopathy (Kearns-Sayre syndrome). The deficit of mitochondrial oxidative metabolism is demonstrated by the presence of the double peak of lactate (Lac) at 1.32 ppm, absent in normal brain tissue. On the left, the axial T_2-weighted MRI image shows a widespread alteration of the MRI signal in the subcortical white matter of the brain

lated to the degree of the neoplasm determined by hystopathological and molecular evaluation.

The absence of NAA is found not only in a high-grade primary tumor but also in a metastasis (secondary tumor), which, if the only lesion, may be included in the differential diagnosis. In the case of metastasis, there is no increase in Cho in the perilesional area, which is characteristic of high-grade primary glial tumors because of their infiltrative nature.

With 3 T scanners, it is possible to identify subtypes of grade II and III glial tumours that present an increased content of 2-hydroxyglutarate (2HG) determined by mutations that cause a loss of function of the isocitrate dehydrogenase enzyme in the cytoplasm (IDH1) and mitochondria (IDH2). Detection of this metabolic alteration in vivo is of great clinical relevance, as patients with IDH1/IDH2-mutated gliomas have longer overall survival than patients with gliomas without IDH mutation.

If clinical and morphological MR imaging do not provide elements for the differential diagnosis between radionecrosis and tumor recurrence, MR spectroscopy can differentiate between these two conditions: in the former, a tendency to

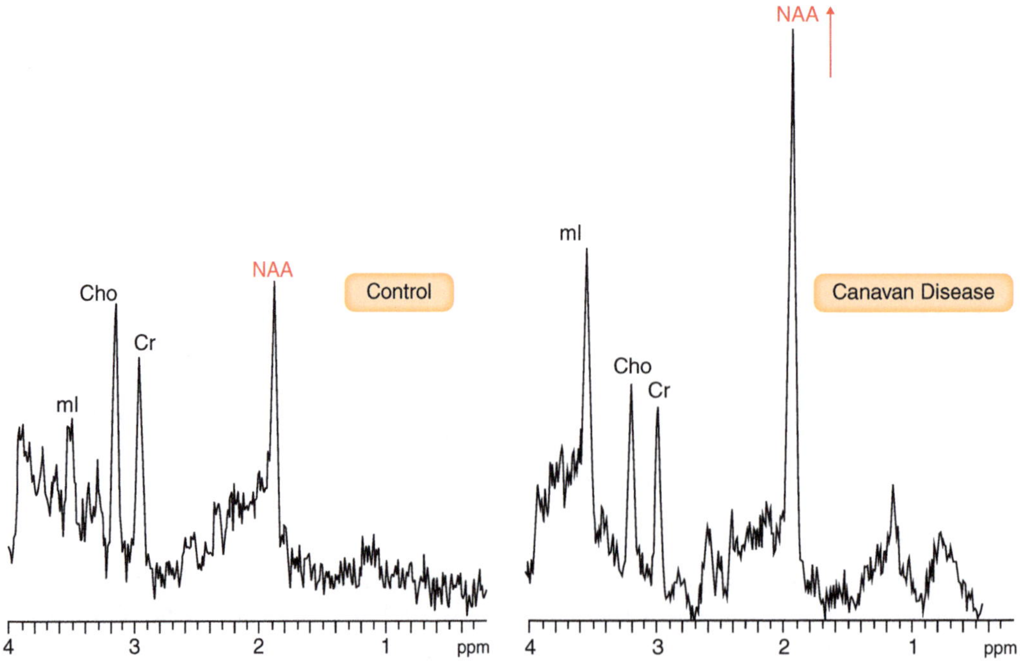

Fig. 40.11 On the right is shown the spectrum obtained in a 1.5 T magnet from a voxel located in the parietal white matter of a newborn with Canavan disease. Compared to the spectrum on the left obtained in a healthy newborn, the N-acetyl-aspartate (NAA) content increased more than twofold in the patient. In the patient, there is also an increase in the myo-Iinositol peak (mI) secondary to glial activation

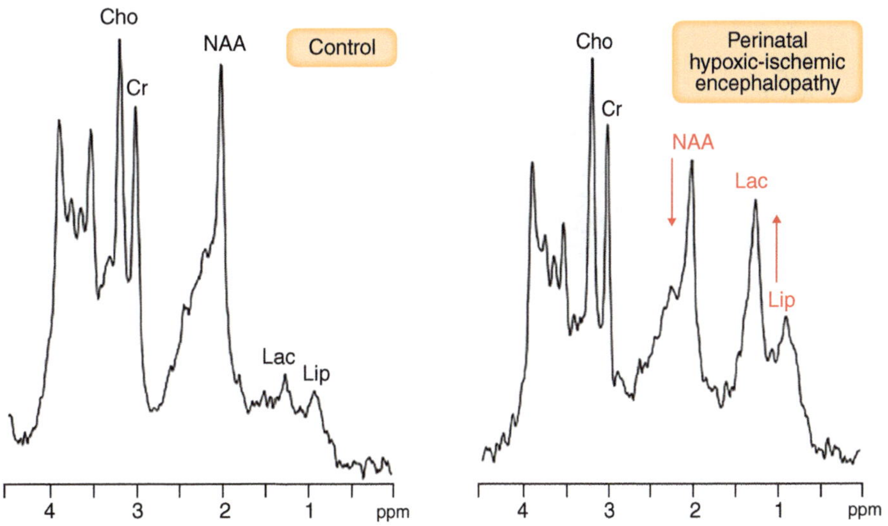

Fig. 40.12 On the right is the spectrum obtained in a 1.5 T magnet from a voxel located in the basal ganglia in a term infant with moderate perinatal hypoxic-ischemic suffering (Sarnat II). Compared to the left spectrum obtained in a healthy term newborn, the content of N-acetyl-aspartate (NAA) is reduced and that of lactate (Lac) and lipids (Lip) is increased in the patient

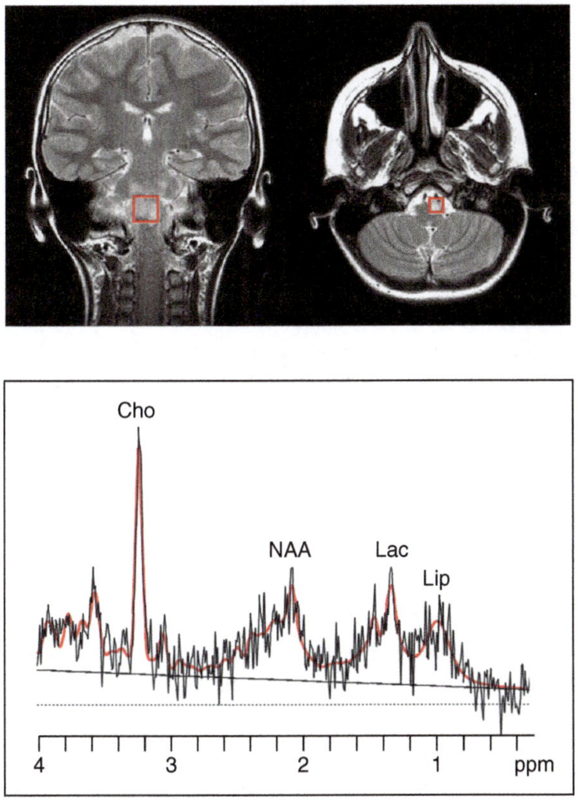

Fig. 40.13 11-year-old patient with the bulb-pontine lesion, shown in the T_2-weighted MRI sequence (coronal and axial planes). The ^{1}H-MRS spectrum (bottom) shows an increase in choline (Cho) content, a substantial reduction in N-acetyl-aspartate (NAA), and a pathological accumulation of lactate (Lac) and lipids (Lip). Overall, the picture is compatible with a high-grade glial neoplasm. The pathological examination of the sample obtained with stereotaxic biopsy concluded a form of multiform glioblastoma

disappear off all brain metabolites is observed, while in the recurrence, the metabolic profile of the neoplasm is observed, especially the increase of the choline peak.

In non-neoplastic lesions such as abscesses, metabolites absent in physiological conditions are detected. In bacterial abscesses (pyogenic), acetate, succinate, and cytosolic amino acids derived from the proteolytic activation of polynuclear elements, leucine, isoleucine, and valine (0.9 ppm), alanine (1.46 ppm) are found. In contrast, in parasitic abscesses, there is the presence of succinate and acetate, as well as an accumulation of lactate/lipids.

Due to the coexistence of different pathological substrates, such as cell proliferation, necrosis, and edema, gliomas may present a heterogeneous appearance in conventional MRI sequences. In the case of indication for brain biopsy, ^{1}H-MRS multi-voxel allows the identification of the most appropriate site for sample collection.

In advanced ultra-specialist centres, operating theatres with integrated MR equipment have been optimized, with which, in addition to conventional MRI sequences, it is possible to acquire MRS to obtain metabolic information, also in the intraoperative phase, to increase the accuracy in resecting a lesion and, therefore, improving the patient's prognosis and quality of life.

Neurodegenerative Disorders

In neurodegenerative disease, SV-MRS can provide information regarding the diagnostic framework, even in the early phase, obtained from the study of target structures in which, from the initial phases of the disease, metabolic alterations indicative of neurodegeneration (reduction of NAA) and glial activation (increase of mI) are observed. These alterations are particularly observed in the hippocampi and posterior cingulate gyrus in patients with mild cognitive impairment (MCI) and Alzheimer's disease (AD) (Fig. 40.14); in the thalami in patients with sporadic Creutzfeldt-Jakob disease (Fig. 40.15); and in the cerebellar hemispheres in patients with atypical parkinsonian syndromes (progressive supranuclear palsy, PSP; and

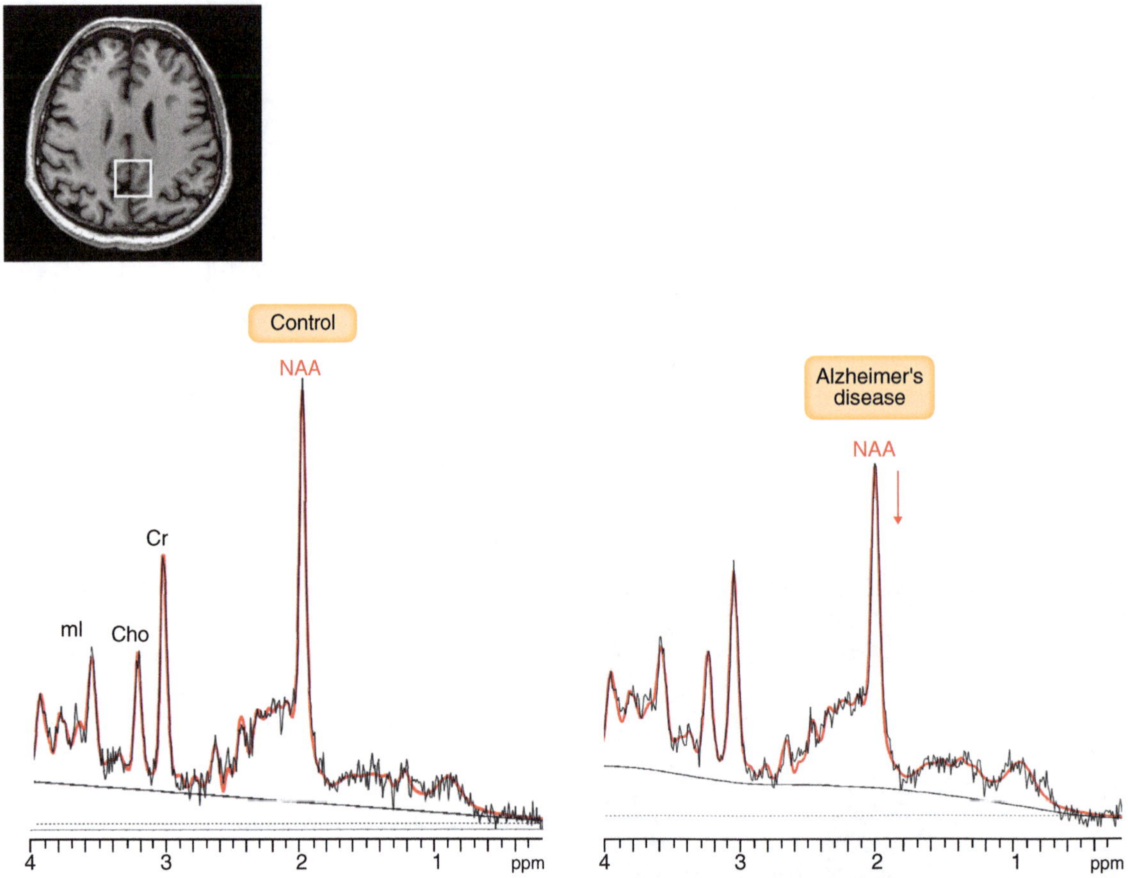

Fig. 40.14 Considering the potential efficacy of a targeted neurocognitive and pharmacological therapy, the differential diagnosis between the different forms of cognitive impairment on a degenerative basis is essential. Bottom right, ^{1}H spectrum acquired from a voxel located in the posterior cingulate cortex (axial MR image), showing a reduction in the neuronal marker N-acetyl-aspartate (NAA), compatible with Alzheimer's disease in a 69-year-old patient with multiple cognitive deficits. A spectrum obtained in healthy control of the same age is shown on the left

multi-system atrophy, MSA) and different forms of hereditary-degenerative ataxias. An ideal disease model in this field is the longitudinal study of subjects harboring mutations in genes encoding proteins such as presenilin 1 or 2 and, therefore, destined to manifest Alzheimer's disease, which is very rare in its genetic form. In the populations of subjects studied, although, in the asymptomatic phase, a reduction in NAA/Cr and an increase in mI/Cr inversely proportional to the number of years preceding the expected clinical onset was demonstrated.

Other Applications of Proton MR Spectroscopy

Prostate

In the Western world, prostate cancer is the most common neoplasm in males over 70 years of age, affecting about one in eight Europeans and one in seven Americans. Because of the epidemiology and the various potentially effective therapies available, in the last 15 years, metabolic diagnostic techniques have been developed to increase prostate cancer's diagnostic accuracy, particularly molecular imaging techniques such as proton MR spectroscopy and PET.

A prostate cancer diagnosis is based on clinical, biochemical, and instrumental criteria: prostate-specific antigen (PSA) serum concentration, rectal examination, and transrectal ultrasound. The definitive diagnosis is based on the histopathological study of the biopsy samples or the surgical specimen.

Recent European guidelines have defined that prostate MR is a second-level examination and that proton multivoxel MRS should be applied in selected cases, with the following recommendations: (1) localization of the primary tumor in patients clinically suspected but with previously negative prostate biopsies; (2) monitoring of treatment efficacy.

Since the aggressiveness of this neoplasm is highly variable, it is crucial to diagnose aggressive cases to set up an adequate therapy, reserving less aggressive forms to watchful waiting.

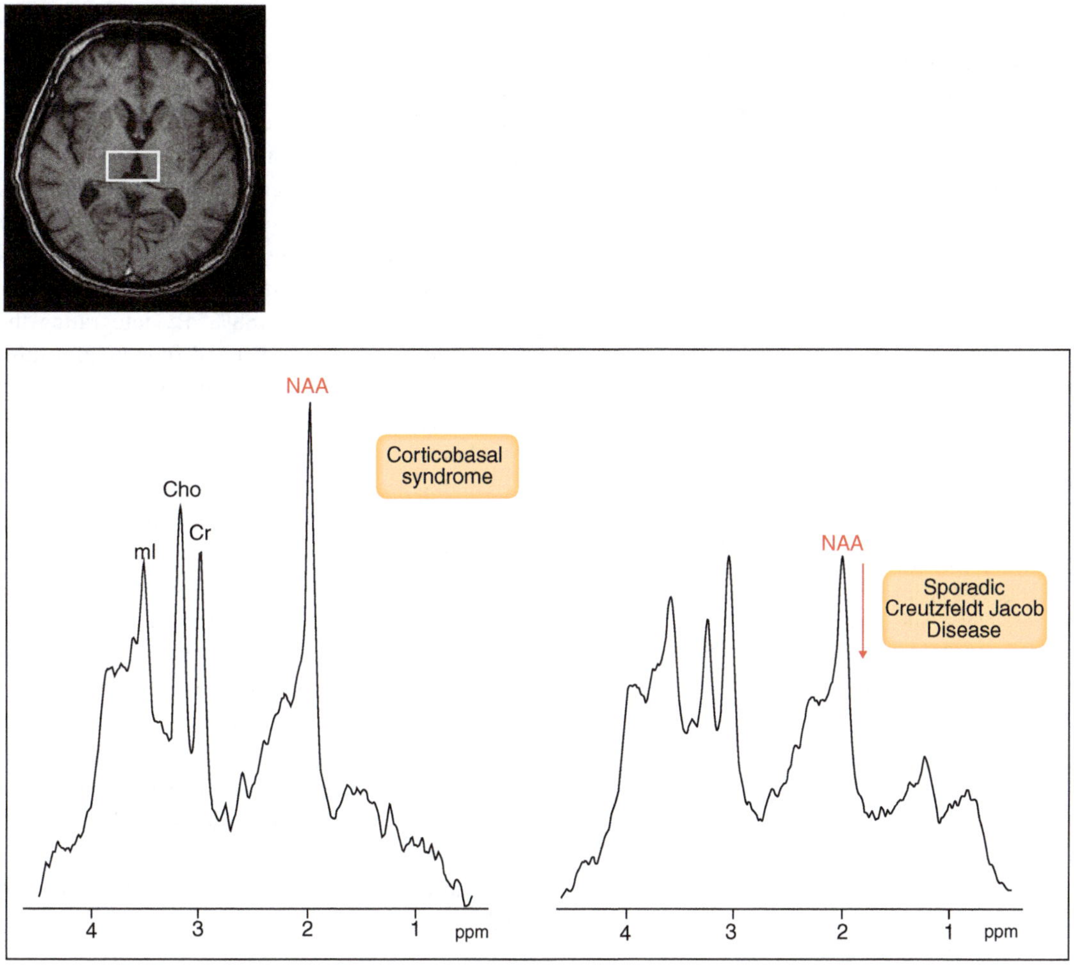

Fig. 40.15 Bottom right, ^{1}H spectrum acquired from a voxel located at the level of the medial thalami in a 65-year-old patient with a confirmed diagnosis of sporadic Creutzfeldt-Jakob disease on post-mortem examination, in which it is evident a reduction in the neuronal-axonal marker N-acetyl-aspartate (NAA), compared to a patient of the same age, with overlapping clinical manifestations, in which clinical follow-up allowed to confirm a form of atypical parkinsonism (cortico basal syndrome)

In the prostate gland, under physiological conditions in vivo, by proton MR spectroscopy, using a transrectal probe, it is possible to reveal the signal derived from the following metabolites: citric acid (Cit, at 2.6 ppm) and, as in other tissues, Cr at 3.0 ppm and Cho at 3.2 ppm (Fig. 40.16).

The metabolism of the prostate gland under physiological conditions is characterized by low oxygen consumption and minimal citrate oxidation due to an accumulation of zinc, which inhibits the activity of the aconitase enzyme and citrate metabolization, thus resulting in an accumulation of citrate in high concentrations in the prostate fluid (40–150 mM) (Fig. 40.17).

The transformation of the metabolic profile that occurs in cancer is characterized by an increase in citrate oxidation due to a reduction in zinc accumulation in the neoplastic tissue. The MR spectrum profile of prostate cancer is, therefore, characterized by a reduction, until disappearance, of citrate (Cit) and an increase, as in malignant tumours of other tissues, of Cho, related to the altered lipid metabolic pathway and increased cellularity of the neoplastic tissue. Ex vivo ^{1}H-MRS studies have shown that the reported peak in Cho is due to the contribution of signals derived from choline, phosphocholine, and glycerophosphocholine.

The ratio of Cho to Cit is used to discriminate cancerous tissue from healthy gland tissue and has been shown to correlate with tumour aggressiveness. A recent systematic review of the literature showed that the sensitivity and specificity of ^{1}H-MRS in combination with MR imaging are 68% and 85%, respectively, in diagnosing the primary tumor.

MR spectroscopy in the monitoring of radiation therapy has shown an increase in diagnostic accuracy of disease recurrence compared to the accuracy of MR imaging alone because it provides information on altered metabolism in addition to morphological data: accuracy measured as area under the curve (AUC) increases to about 80% compared to 50% for morphological imaging alone.

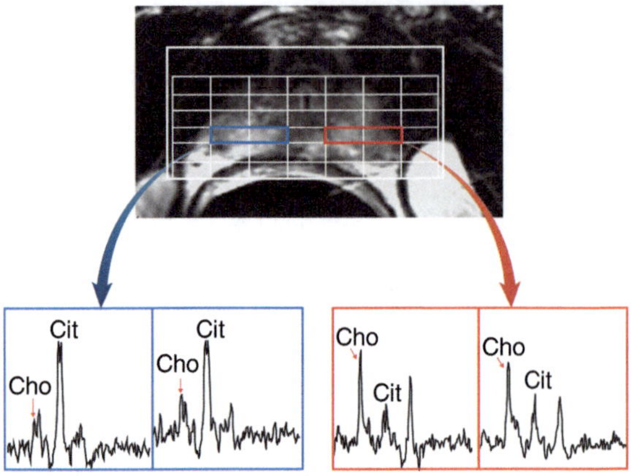

Fig. 40.16 Prostate cancer. Axial T_2-weighted image of the prostate gland of a patient with prostate cancer (Gleason 4 + 3 scale): in red, on the left side of the gland, two voxels are highlighted whose corresponding ^{1}H spectra show a metabolic alteration indicative of carcinoma (an increase of the choline peak [Cho] and reduction of that of citrate [Cit]), despite the absence of signal intensity alterations indicative of carcinoma on the conventional MRI sequence; in blue, on the right side of the prostate gland, two voxels are highlighted whose corresponding spectra not altered showing the citrate peak unchanged

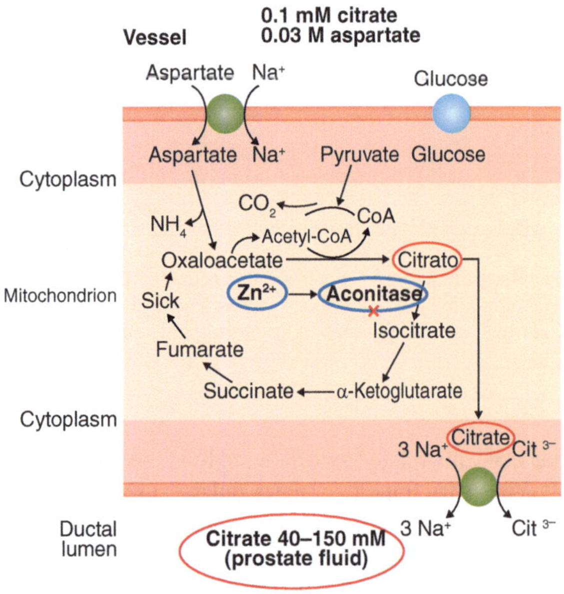

Fig. 40.17 Schematic representation of the citrate production pathway in prostate epithelial cells. The accumulation of zinc in healthy prostate tissue inhibits the activity of aconitase and therefore inhibits the metabolization of citrate in the Krebs cycle, thus causing a high accumulation of citrate in the prostate fluid

Breast

[1]H spectroscopy of the breast is a critical investigation that adds to the morphological and functional information of MR imaging, information on tissue metabolism that may be useful for the differential diagnosis between adenocarcinoma and benign lesions. The ^{1}H-MRS examination of the breast is performed in specialized centers. It is not yet integrated into clinical protocols due to several technical limitations, such as acquisition time; the difficulty of spectrum acquisition mainly due to the need to suppress the signal derived from the lipid component, which is particularly relevant in breast tissue; and the difficulty of spectrum interpretation. The ^{1}H spectroscopy of the breast is used for lesion characterization and in evaluating the effect of chemotherapy. Specifically, assessing the increase in choline (tCho) content, ^{1}H-MRS showed a sensitivity of 71–74% and a specificity of 78–88%. By integrating the ^{1}H-MRS results with morphological and functional results (dynamic imaging), the diagnostic accuracy of 96% of a malignant lesion is achieved.

Currently, the most widely used technique is SV-MRS, localizing the voxel on MRI images obtained post-contrast. In order to obtain reliable results, the voxel must include the suspected lesion as much as possible, avoiding perilesional adipose tissue. The most sensitive marker for breast cancer diagnosis is tCho. Discrimination between the presence/absence of a malignant lesion is performed based on a threshold value of tCho, identified by each laboratory in its case series. An acquisition by high spectral resolution and accurate post-processing allows to evaluate, with a specificity near 100%, if the tCho peak is centered at 3.22 ppm (corresponding to the phosphocholine signal), suggesting a malignant lesion, or shifted towards 3.28 ppm (corresponding to the taurine, myoinositol, and glycerophosphocholine signals), suggesting a benign lesion.

In Vivo Phosphorus MR Spectroscopy

Skeletal Muscle

Phosphorus MRS (^{31}P-MRS) evaluates skeletal muscle through quantifying metabolites with phosphate groups, metabolic pathways, oxidative phosphorylation, glycolysis, and lipid metabolism. In vivo, sufficiently resolved peaks with an adequate signal-to-noise ratio are detected for metabolites at least 1–2 mM concentrations. Consequently, while phosphates contained in mitochondria are not detectable, metabolites with cytosolic (sarcoplasm) localization can be detected, such as ATP, phosphocreatine (PCr), phosphodiesters (PDE), inorganic phosphate (Pi), and phosphomonoesters (PME). Furthermore, by measuring the chemical shift of Pi from PCr, it is possible to calculate the cytosolic pH, and by measuring the chemical shift of β-ATP from PCr, it is possible to calculate the concentration of free Mg^{2+}. For clinical or research pourposes, each laboratory standardizes the study protocol of one of the most typical muscle groups evaluated: calf muscles, quadriceps femoris, and finger flexors, using surface coils placed above or below the muscles to be tested.

ATP in the cytoplasm is in equilibrium with PCr through the reaction catalyzed by creatine kinase (CK): $PCr + ADP + H^+ \leftrightarrow ATP + Cr$. The concentration of ADP is not directly measurable, being of the micromolar order, but can be calculated based on the enzymatic creatine kinase reaction, which in vivo is close to equilibrium, knowing the K_{CK} (= 1.66 × 10^9 M^{-1}) and the total content of creatine plus phosphocreatine (= 42.5 mmol), as $[ADP] = ([ATP][Cr])/([PCr][H^+]K_{CK})$.

This parameter is of considerable importance since ADP is the main regulator of oxidative phosphorylation. Therefore, knowing its concentration, it is possible to calculate the phosphorylation potential PP = [ATP]/([ADP][Pi]), which in turn represents the energy reserve that can be readily used for the synthesis of ATP, a molecule with a high energy content that can be directly used in all cellular activities requiring energy.

Phosphorus MR spectroscopy technique provides information on energy metabolism in three phases: resting condition, aerobic muscle exercise, and post-exercise recovery (Fig. 40.18). In the first phases of muscular exercise, performed inside the MR scanner, the ATP consumed in the contraction is maintained at constant levels through the hydrolysis of PCr, whose consumption, associated with the increase in Pi, can be observed directly in real-time during the acquisition of the examination. In the initial phases of aerobic exercise, ATP is produced by glycolysis, which uses muscle glycogen and blood glucose to produce ATP and, at the same time, produces lactate, which leads to a progressive lowering of muscle pH. Then the activation of oxidative phosphorylation occurs and couples electron transfer to ADP phosphorylation in the mitochondria; it is the main pathway for ATP production. Pyruvate and fatty acids constitute the primary substrates of this oxidative process.

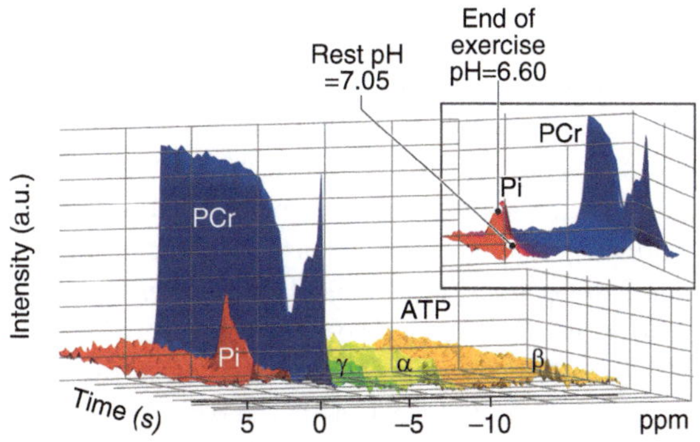

Fig. 40.18 Phosphorus spectra of the calf muscles acquired at rest, during an incremental aerobic exercise of plantar flexion and in the recovery phase following the interruption of exercise. During exercise, compared to rest, there is a progressive reduction in phosphocreatine (PCr, blue) and an increase in inorganic phosphate (Pi, red). During exercise, the resonanse frequency of the Pi approaches that of the PCr due to the progressive reduction of the sarcoplasmic pH (insert) caused by the production of lactic acid. In the post-exercise recovery phase, the PCr returns to resting values, while the Pi is progressively reduced as it is transported for the synthesis of ATP within the mitochondria, a compartment not visible by using MRI technique due to the high density of the matrix and poor mobility of molecules. During all phases of the examination, the content of β (orange), α (yellow) and γ-ATP (green) remains constant. ppm, resonant frequency expressed in parts per million

In the post-exercise muscular recovery phase, PCr is resynthesized through the creatine kinase reaction using ATP, which, in this phase, since glycolysis is interrupted at the same time as muscular contraction, is produced solely by mitochondrial oxidative phosphorylation. For this reason, the speed of post-exercise resynthesis of the PCr reflects in an exact way the speed at which the mitochondria produce energy. The existing hyperbolic relationship between the rate of resynthesis of PCr and [ADP] allows for calculating the maximum rate of oxidative synthesis of ATP (V_{max}), assuming a KM (Michaelis-Menten constant) of 30 μM:

$$V_{max} = V(1 + [KM / [ADP]end)$$

where V is the initial rate of resynthesis of PCr and [ADP] end is the concentration of ADP calculated at the end of the exercise.

Patient Preparation and Set-Up

In the week preceding the MRS examination, the patient should follow a similar preparation to that which precedes blood sampling for the dosage of biochemical markers of muscle damage (creatine kinase [CK], lactate dehydrogenase [LDH], aldolase): avoid physical exercise and traumatic muscle procedures (intramuscular injections, electromyography, biopsy).

In vivo assessment of muscle energy metabolism requires an ergometer placed inside the magnet. In most laboratories, the calf muscles are examined (medial and lateral gastrocnemi and soleus), which are exercised through plantar flexion obtained by pressing on a pedal, whose resistance is gradually increased through a compressed air system to which a piston is connected to the pedal. In the case of a protocol of aerobic exercise with the lower limb, the patient enters the bore of feet, supine, and with the arms along the sides in a comfortable position. Since the muscular exercise must be of increasing intensity and proportional to the muscle mass, the patient must be cooperative and understand the protocol before the examination (training phase).

The exercise duration varies with the exercise protocol and the condition of the subject but is typically between 2 and 5 minutes. The overall duration of a ^{31}P-MRS muscle examination, including exercise testing, is typically between 30 and 40 minutes.

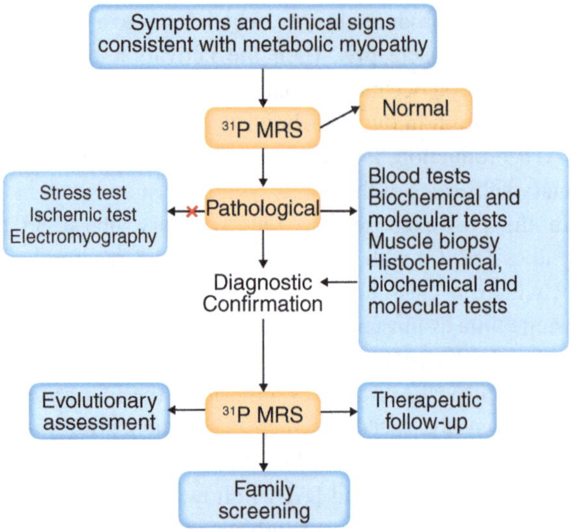

Fig. 40.19 Diagnostic algorithm showing the usefulness of muscle ^{31}P-MRS in the diagnostic process of patients with suspected metabolic myopathy

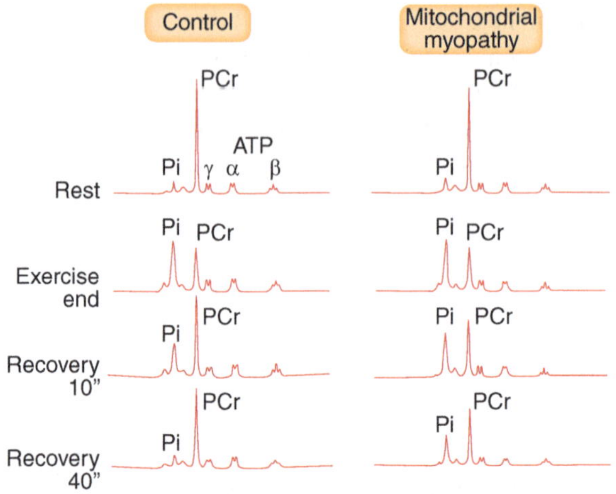

Fig. 40.20 Trend of changes in bioenergy markers (phosphocreatine [PCr] and inorganic phosphate [Pi]) in conditions of rest, of aerobic exercise, and at 10 and 40 seconds of the post-exercise recovery phase in healthy control and a patient with mitochondrial myopathy, carrier of the common deletion of mtDNA. At rest, a reduced PCr and increased Pi peak are observed in the patient compared to the control subject. The peaks of PCr and Pi are similar in the two subjects at the end of the exercise. In both spectra acquired in post-exercise recovery, it is evident that the peak of PCr is low and that of Pi high in the patient, indicating a deficit of mitochondrial oxidative phosphorylation

Main Pictures of Musculoskeletal Pathology Evaluated by ^{31}P-MRS Muscle

Muscle ^{31}P-MRS is a very sensitive technique used clinically to support the diagnosis for the main forms of metabolic myopathy, particularly mitochondrial myopathies and muscle glycogenosis (Fig. 40.19).

Mitochondrial Myopathies

Mitochondrial diseases are caused by a very high number of mutations that can affect both mitochondrial DNA (mtDNA) and nuclear DNA, resulting in a deficit of mitochondrial oxidative phosphorylation, which is the most important pathway for cellular ATP production. Mitochondrial diseases are characterized by a wide phenotypic variability due to their genetic and biochemical heterogeneity. Cerebral and muscle (skeletal and cardiac) involvement is typical, but diagnostic classification can be difficult due to the clinical variability observed in the same pedigree. Clinical manifestations depend not only on the gene involved and the type of mutation but also on the percentage of mutated mtDNA compared to wild-type mtDNA, which can vary in different tissues, and on the nuclear genetic makeup. Therefore, patients with the same mutation in the same gene may present different clinical manifestations, and mutations in different genes may result in the same clinical phenotype. In the context of laboratory diagnostics, the demonstration of a respiratory chain deficiency, the presence of specific morphological patterns on muscle biopsy, and molecular diagnostic confirmation are decisive steps for the classification. Muscle ^{31}P-MRS, in patients with mitochondrial myopathy, it is possible to detect, under resting conditions, an increase in inorganic phosphate and sometimes a reduction in PCr, compared to healthy subjects. Although this metabolic alteration is not specific, it can be used in screening, especially in cases where the subject cannot perform muscle exercise of sufficient intensity in the magnet. Approximately 90% of patients with mitochondrial myopathies have a slowdown in the rate of PCr resynthesis, measured in the post-exercise muscle recovery phase (Fig. 40.20), which is the most sensitive indicator of ATP synthesis deficit at the level of mitochondrial oxidative phosphorylation. The maximum rate of ATP synthesis (V_{max}) calculated from ^{31}P-MRS parameters in vivo is typically reduced in patients with mitochondrial disease, even without any muscle symptoms. Muscle enzymes are normal or only slightly increased in patients with mitochondrial myopathies. Typical biochemical alterations are represented by increased lactate and reduced oxygen consumption, which typically require the performance of an exercise test (e.g., exercise bike) to be detected. The typical biopsy alteration is represented by the presence of so-called Ragged-Red Fibers (RRF) to Gomori trichrome staining, consisting of abundant subsarcolemmal accumulations of mitochondria. Genetic analysis, necessary for a definitive diagnosis, is usually performed on a muscle tissue biopsy.

Muscle Glycogenosis

Diseases due to defects in carbohydrate metabolism frequently present with typically exercise-related disorders (cramps, contractures, rhabdomyolysis, that is massive necrosis of muscle cells) or muscle weakness. Most muscle glycogenoses are genetic diseases with autosomal recessive transmission due to deficits of cytoplasmic enzymes involved in glycogen breakdown or glycolysis.

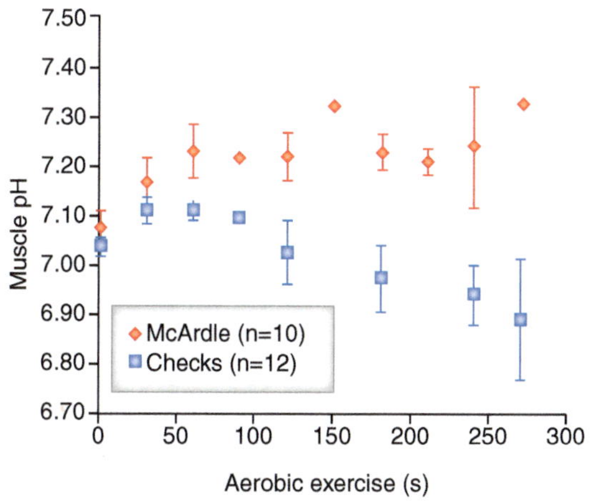

Fig. 40.21 Trend of the pH variation during aerobic muscular exercise in a group of healthy subjects (blue) and patients with McArdle's disease (red), in which a lack of acidification due to the absence of lactate production caused by myophosphorylase deficit is evident

McArdle's disease (glycogenosis type V), one of the most frequent forms of glycogenosis (1:100,000), is an autosomal recessive disorder of glycogenolysis caused by a deficiency of myophosphorylase, which results in a lack of glucose-6-phosphate synthesis. The 31P-MRS detects with high sensitivity a specific picture characterized by reduced ATP concentration in skeletal muscle at rest and failure to acidify muscle pH during muscle exercise due to lack of lactic acid production via the glycolytic pathway (Fig. 40.21).

Muscle enzyme levels (CK, LDH, and aldolase) are generally elevated even at rest and may increase markedly after muscle exercise. The ischemic test, performed on forearm muscles, does not detect lactic acid production, evaluated on venous blood after exercise. This test is not without risk for patients because it can cause an episode of rhabdomyolysis. On immunohistochemical investigation of muscle biopsy, myophosphorylase is absent.

Phosphofructokinase (PFK) deficiency disease (type VII glycogenosis) is a rarer form of muscle glycogenosis than McArdle's disease, which shares the same exercise-related clinical manifestations. The ^{31}P-MRS detects a specific picture, characterized by reduced ATP concentration of skeletal muscle at rest, the lack of acidification during muscle exercise of muscle pH due to the lack of lactic acid production, and the accumulation of intermediates of monophosphate glycolysis (PME), such as glucose-1-phosphate, glucose-6-phosphate, and fructose-6-phosphate, which accumulate upstream of the PFK enzymatic defect; this finding allows for a differential diagnosis with McArdle's disease. As in McArdle's disease, muscle enzyme levels (CK, LDH, and aldolase) are elevated at rest and increase after muscle exercise. On ischemic testing performed on forearm muscles that can trigger an episode of rhabdomyolysis, no lactic acid production is detected after exercise. Phosphofructokinase is absent in immunohistochemical investigation of muscle biopsy.

Recommended Readings

Choi C, Ganji SK, DeBerardinis RJ et al (2012) 2-Hydroxyglutarate detection by magnetic resonance spectroscopy in IDH-mutated patients with gliomas. Nat Med 18(4):624–629

Coakley FV, Teh HS, Qayyum A et al (2004) Endorectal MR imaging and MR spectroscopic imaging for locally recurrent prostate cancer after external beam radiation therapy: preliminary experience. Radiology 233:441–448

Costello LC, Franklin RB (2000) The intermediary metabolism of the prostate: a key to understanding the pathogenesis and progression of prostate malignancy. Oncology 59(4):269–282. Review

Danielsen ER, Ross B (1999) Magnetic resonance spectroscopy diagnosis of neurological diseases. Marcel Dekker, Inc, New York

De Graaf RA (2013) In vivo NMR spectroscopy: principles and techniques. Wiley

Heidenreich A, Bastian PJ, Bellmunt J, et al (2014) European Association of Urology. EAU guidelines on prostate cancer. part 1: screening, diagnosis, and local treatment with curative intent-update 2013. Eur Urol 65(1):124–137

Expert Panel on MR Safety, Kanal E, Barkovich AJ, Bell C et al (2013) ACR guidance document on MR safe practices: 2013. J Magn Reson Imaging 37(3):501–530

Godbolt AK, Waldman AD, MacManus DG et al (2006) MRS shows abnormalities before symptoms in familial Alzheimer disease. Neurology 66(5):718–722

Lodi R, Taivassalo T (2002) Phosphorus magnetic resonance spectroscopy in the clinical investigation of muscle disorders. In: Bertorini TE (ed) Clinical evaluation and diagnostic tests for neuromuscular disorders. Butterworth-Heinemann Publishers, pp 503–534

Near J, Harris AD, Juchem C et al (2021) Preprocessing, analysis and quantification in single-voxel magnetic resonance spectroscopy: experts' consensus recommendations. NMR Biomed 34(5):e4257

Opstad KS, Wright AJ, Bell BA et al (2010) Correlations between in vivo (1)H MRS and ex vivo (1)H HRMAS metabolite measurements in adult human gliomas. J Magn Reson Imaging 31(2):289–297

Oz G, Alger JR, Barker PB, MRS Consensus Group et al (2014) Clinical proton MR spectroscopy in central nervous system disorders. Radiology 270(3):658–679

Pinker K, Bogner W, Baltzer P et al (2014) Improved diagnostic accuracy with multiparametric magnetic resonance imaging of the breast using dynamic contrast-enhanced magnetic resonance imaging, diffusion-weighted imaging, and 3-dimensional proton magnetic resonance spectroscopic imaging. Investig Radiol 49:421–430

Provencher SW (2001) Automatic quantitation of localized in vivo 1H spectra with LCModel. NMR Biomed 14(4):260–264

Sardanelli F, Carbonaro LA, Montemezzi S et al (2016) Clinical breast MR using MRS or DWI: who is the winner? Front Oncol 6:217

Stanwell P, Gluch L, Clark D et al (2005) Specificity of choline metabolites for in vivo diagnosis of breast cancer using 1H MRS at 1.5 T. Eur Radiol 15(5):1037–1043

Testa C, Pultrone C, Manners DN et al (2016) Metabolic imaging in prostate cancer: where we are. Front Oncol 6:225

Tonon C, Gramegna LL, Lodi R (2012) Magnetic resonance imaging and spectroscopy in the evaluation of neuromuscular disorders and fatigue. Neuromuscul Disord 22(Suppl 3):S187–S191

Umbehr M, Bachmann LM, Held U et al (2009) Combined magnetic resonance imaging and magnetic resonance spectroscopy imaging in the diagnosis of prostate cancer: a systematic review and meta-analysis. J Eur Urol 5(3):575

Clinical Biochemistry of the Mind

41

Silvia Pellegrini and Pietro Pietrini

Introduction

The social dimension plays a fundamental role in human life, allowing interaction among individuals with shared life experiences and goals.

Parental care, friendships, and partnerships, as well as cooperation at work, are all examples of social life which have a positive impact on individual well-being. For example, the support of friends and family is a real shield against adverse life events and positively affects both physical and mental health. The creation of cooperative relationships with one's peers can lead to enriching experiences, thus stimulating an ever greater openness toward others.

Conversely, the tendency toward isolation, hostility, and aggression is often the result of reduced personal well-being, which is the common denominator of many mental illnesses.

Mental Disorders

About 450 million people worldwide suffer from mental disorders like major depression, bipolar disorder, psychosis, and pathological addictions. According to the World Health Organization, mental illness is a more critical health problem than cardiovascular disease and cancer: it is both a personal or social burden and a financial one.

Compared to the past, when mental disorders represented a real "stigma" to be kept hidden from society, with serious consequences for both patients and their families, in recent decades the concept that mental illnesses are to all intents and purposes pathologies whose biological causes can be investigated, in order to identify more and more targeted and effective treatments, in the same way as "physical" pathologies, has been increasingly affirmed.

Before the discovery of psychotropic drugs, around the middle of the last century, the possibilities of intervening in the natural course of psychopathological phenomena were limited, and patients were confined to dedicated structures, the asylums. In even earlier times, the patients were confined on the so-called "ships of the insane", boats that sailed from port to port, from which the patients were never disembarked. In this way, the action was taken only to protect society from the mentally ill, without any benefit to the patients, who were abandoned to their fate.

Despite the many advances in treating mental illnesses achieved since the 1950s, many problems remain. Some patients do not respond to therapies; others show not entirely satisfactory responses, and others still have a reduced tolerance to the drugs used. The reason is that knowledge of the origin and physiopathology of mental illnesses is still very limited today.

Despite the enormous efforts made to classify these disorders, there are still no absolute diagnostic criteria supported by solid neuroscientific evidence. The current diagnostic schemes of DSM-5 (Diagnostic and Statistical Manual of Mental Disorders) or ICD-11 (International Classification of the Diseases) are based on a purely descriptive classification of these diseases, which does not consider the causative mechanisms since they are still almost unknown. Moreover, this classification does not include the multiple comorbidities, both psychiatric and medical, nor the considerable prognostic instability of these conditions, both of which have been highlighted by longitudinal studies.

The central problem in understanding the etiopathogenetic mechanisms of psychiatric diseases lies in the fact that, except rare monogenic diseases such as, for example, Huntington's chorea or fragile X syndrome, they are, in most cases, complex multifactorial diseases. The molecular mechanisms involved are multiple and often shared by pathologies classified as different based on current diagnostic criteria. This explains why, in the case of mental illnesses, there are also

S. Pellegrini (✉)
Department of Clinical and Experimental Medicine, University of Pisa, Pisa, Italy
e-mail: silvia.pellegrini@unipi.it

P. Pietrini
IMT School for Advanced Studies Lucca, Lucca, Italy

M. Ciaccio (ed.), *Clinical and Laboratory Medicine Textbook*, https://doi.org/10.1007/978-3-031-24958-7_41

considerable difficulties in identifying objective biomarkers capable of recognizing affected individuals from those who are not, accurately distinguishing different diseases, or predicting the severity of a given pathological condition or the probability of response to therapy in each patient.

Biological Markers of Mental Illness

A biological marker, according to the definition of 1998 by the Biomarkers Definitions Working Group of the National Institutes of Health (NIH) in Bethesda (USA), is "something that can be measured quantitatively and that is an indicator either of normal biological functioning or of the presence of a disease or response to a drug treatment." It can be a gene, a protein, or any other molecule, as well as a morphological or functional characteristic associated with specific physiological or biological mechanisms. A biomarker, in any case, must be sensitive, i.e., able to detect the highest number of patients; specific, i.e., keep false positives to a minimum; accurate, i.e., its measurement must come as close as possible to the true value; precise, i.e., produce a value that can be repeated in subsequent measurements.

In the case of mental illnesses, the use of objective biomarkers to accompany the interview with the patient and clinical observation would make it possible to (1) increase diagnostic accuracy since the mere observation of symptoms sometimes does not allow, at least in some stages of the disease, to distinguish different pathologies (an example is given by bipolar disorder which, at onset as a depressive phase, can be misdiagnosed as unipolar depression and treated as such); (2) establish a correct diagnosis even at a very early stage of the illness, thus implementing the most suitable treatment as soon as possible; (3) identify individuals at risk and allow effective prevention; (4) immediately identify any relapses or changes in state (depression-mania in bipolar disorder) even before the reappearance of clinical symptoms; and (5) discover new potential therapeutic targets.

Currently, however, there are no biomarkers for mental illnesses that already have standardized clinical use, and the reason lies precisely in the complexity of these diseases.

However, there is much scientific evidence that can guide the selection of several factors that could become good biological markers.

Morphological and Functional Exploration Methodologies of the Central Nervous System

Since the mid-1980s of the last century, positron emission tomography (PET) and magnetic resonance imaging (MR), both structural and functional, have been used to measure phenotypic variations of the molecular or cellular targets of mental diseases or to highlight changes in brain circuits representative of specific behavioral alterations. These methods of investigation of brain structure and function can contribute to a nosological classification for diagnostic purposes, precisely as in other branches of medicine, such as oncology or cardiology.

It cannot be ruled out that, in the near future, psychiatric patients will be able to wear brain monitoring instruments to obtain clinically helpful measurements, in the same way, that heart patients wear cardiac monitoring instruments today.

However, the use of neuroimaging as a biomarker in clinical routine still presents difficulties.

A first obstacle is that many of the characteristics of mental illnesses are not evident from a purely structural point of view. The brains of these patients may appear morphologically normal and exhibit pathological phenotypes only during specific functional tasks. This makes the use of functional paradigms an essential requirement to detect distinct aspects of the disease, complicating the applicability of such measurements in clinical practice. Indeed, an ideal marker requires using a simple and easy-to-apply test.

The second problem is that, given the extreme phenotypic complexity of mental illnesses related to polygenic vulnerability modulated by multiple environmental factors, a single biomarker may not be sufficient to diagnose or monitor the response to treatment. It is necessary, therefore, to have multiple biomarkers to test, each indicative of a different phenotypic aspect of the disease. The best markers, therefore, will be those that require relatively simple instrumentation for their measurement.

The Concept of Endophenotype

Images obtained by MRI can be very useful as intermediate phenotypes or endophenotypes to identify other biomarkers.

An endophenotype is a quantitative, measurable trait intermediate between the "disease" phenotype and its underlying biological processes. Several endophenotypes combine to form the complexity of the disease, and each of them is more easily traceable, compared to the complex phenotype, to a precise molecular mechanism underlying the same disease.

For an endophenotype to become a biomarker, however, it must be as independent as possible from environmental influences. For example, an increase in the concentration of b-endorphin can be a biochemical marker of alcohol dependence, but b-endorphin levels can also be strongly altered by an isolated, albeit excessive, consumption of alcohol.

Therefore, the measurement of this biomarker must be contextualized to be clinically useful.

In the case of brain imaging, their use as endophenotypes may allow biochemical parameters or genetic variants to be related to complex disease phenotypes. For example, the ε4 allele of the ApoE gene is known to be a risk factor for Alzheimer's disease. An explanation for this association is provided by functional magnetic resonance imaging (fMRI) studies, in which more extensive brain activation during memory tasks was observed in healthy subjects carrying the ε4 allele than in subjects carrying the ε3 allele. This indicates a compensatory response in subjects with ε4, i.e., the need to use additional brain resources to cope with similar cognitive tasks. The endophenotype, consisting of fMRI images, shows, therefore, the existence of a link between the ApoE genotype and a deficit in memory functions, which increases the risk of Alzheimer's disease. This observation, therefore, confirms the validity of the ApoE genotype as a predictor of Alzheimer's risk.

Another example of the use of fMRI images as endophenotypes useful to identify biochemical markers of disease is given by a very recent work in which it was seen that, both in healthy subjects and in patients with mild cognitive impairment (MCI), the activation of the dorsal portion of the anterior cingulate cortex (dACC) in response to tasks of visuospatial attention correlates with the levels of vitamin B12. Again, an endophenotype, consisting of functional MRI findings, relates a biochemical parameter, vitamin B12 concentration, to cognitive abilities. The relationship between B12 and visuospatial abilities emerges only by including brain activity measured by functional magnetic resonance imaging in the correlation analysis. This finding explains the clinical utility of vitamin B12 dosage as a predictor of reduced cognitive ability.

Therefore, brain images represent excellent endophenotypes, which are useful for identifying the relationships between molecular mechanisms and complex phenotypes. This does not exclude, however, that they can be disease biomarkers in combination with other neurophysiological parameters.

Recent studies, for example, have found a reduction in hippocampal volume in depressed patients compared to healthy control subjects. This reduction has been observed in adult patients and adolescents, both at the first and subsequent episodes. Hippocampal atrophy may be due to neurodegenerative phenomena resulting from the increased glucocorticoid levels in depression. Studies in larger patient populations are needed to verify whether reduced hippocampal volume may be a valid biological marker of depression and risk of depressive episodes.

The Omics Sciences

Given the phenotypic complexity of psychiatric diseases, a promising approach for understanding the mechanisms underlying these diseases and identifying new biological markers is represented by omics sciences, i.e., genomics, proteomics, metabolomics, lipidomics, and epigenomics.

Genomics

Since the human genome sequence was completed in 2003, genomics studies have also been an essential resource for identifying new biomarkers in psychiatry. Mental illnesses are mainly heritable. In the last 20 years, genome-wide association studies (GWAS), in which whole genomes of thousands of patients and controls have been compared, have identified some genetic loci associated with schizophrenia, bipolar disorder, major depression, ADHD (attention deficit hyperactivity disorder), autism spectrum disorders, and, even, some loci associated with all five diseases, indicative of shared pathological mechanisms. Nevertheless, it has not yet been possible in these genomic regions to identify the genes responsible for the increased risk of disease, nor any other sequences involved. Researchers initially focused on coding regions and non-synonymous sequence variants. However, non-coding regions of the genome, such as those responsible for splicing mechanisms, may also play important roles in promoting behavioral disorders. About 93% of the variants identified by GWAS are located outside coding regions.

However, the genome study alone is insufficient for a complete characterization of the molecular mechanisms at the basis of mental processes. It is known that the same gene sequence can encode different proteins by post-transcriptional and post-translational modifications, so it has become necessary to flank the study of the genome with that of the proteome.

Proteomics

Studying the proteome means studying all the proteins a given cell expresses at a particular time or their changes in response to environmental factors or following the onset of disease or drug intake. Comparing the proteomes of psychiatric patients and healthy subjects or those of the same patient in different stages of disease aims to identify differentially expressed proteins, which could become good diagnostic or prognostic markers.

Proteomics uses different technologies, such as two-dimensional electrophoresis, image analysis, mass spectrometry, amino acid sequencing, and bioinformatics tools, to isolate, quantify, and characterize proteins of interest. In the case of mental diseases, the most informative biological material for proteomics studies is represented by postmortem brain tissue. However, in addition to the objective complexity of finding a large number of this type of sample, the postmortem brain tissue has the great difficulty of having to be processed immediately. After death, there is rapid dephosphorylation of proteins, which produces differences in results, the more evident, the greater the time elapsed between the time of death and the autopsy. This makes the feasibility of this type of study particularly complicated. This difficulty can be circumvented by studying the brains of animal models taken immediately after the sacrifice. In the case of mental illnesses, however, animal models have their limits since they cannot perfectly reproduce the pathological phenotype precisely because of the clinical complexity of mental illnesses. However, there are some animal models, such as the mouse with "social defeat," which has been described as a good model of depression even at the molecular level. Indeed, it shows differential protein expression patterns in the brain overlapping with those found in depressed patients studied postmortem.

Despite the difficulties described, the study of postmortem brains has allowed the identification of some proteins, which could represent good clinical markers. These are mainly growth factors, hormones, metabolic enzymes, and proteins involved in cellular architecture or synaptic transmission (Table 41.1). The same differential expression of some of these factors is observed in different pathologies, such as major depression, bipolar disorder, and schizophrenia, thus demonstrating a limited diagnostic specificity, whereas other proteins seem to be a disorder-specific phenomenon (Table 41.1). Finally, some of these proteins, such as NADPH oxidase 2 (NOX2) or others involved in (GABA) ergic neurotransmission, would seem to be good predictors of suicidal behavior.

The differential expression of many of these proteins can also be observed in cerebrospinal fluid (CSF) (Table 41.1), whose collection by lumbar puncture also allows the study in vivo in patients, contrary to what happens in studies on brain tissue taken postmortem. However, CSF collection is relatively invasive and not easy to apply to the diagnostic routine. It is unclear to what extent blood or saliva can represent the molecular processes occurring in the brain. Good indications in this sense derive from recent evidence showing the involvement of pro-inflammatory and immunoregulatory proteins in the pathophysiology of psychiatric disorders (Table 41.2). For example, increased levels of pro-inflammatory cytokines, such as IL-1b, IL-6, IL-12, IL-18, TNF-α, FN-g, and C-reactive protein, and reduced levels of anti-inflammatory cytokines, such as IL-10, IFN-α, and TGF-b, have been observed in both bipolar and schizophrenic patients. An increase in pro-inflammatory and oxidative stress response proteins, such as ceruloplasmin, ferritin, interleukin-1 receptor antagonist, macrophage migration inhibitory factor, and superoxide dismutase 1, has been asso-

Table 41.1 Major proteomic biomarkers detectable in the postmortem brain potentially useful in psychiatry

Biological processes	Biomarkers	Psychiatric phenotype
Metabolic enzymes	Aconitate hydratase	MD
	Enolase - ENO2; ENO3	MD
	Enzymes of glycolysis	SCZ
	Exokinase −2	MD
	L-lactic dehydrogenase	MD
	NOX2	Suicidal behavior
	Peroxiredoxin −6	MD with psychotic figures; SCZ
	Carboxypeptidase E prohormone converting enzyme	MD
	Prolyl-oligopeptidase convertase	MD
Mitochondrial metabolism	Mitochondrial dysfunctions	BD
Synaptic transmission and cellular architecture	Actin[a]	MD; suicidal behavior
	Actinin[a]	MD
	Calmodulin 1	MD with psychotic figures; SCZ
	Cathepsin D	Suicidal behavior
	Clathrin[a]	MD
	Dihydropyriminidase-type 2 or collapsin response mediator protein 2	MD; SCZ; Alzheimer's disease
	Dynamine[a]	MD
	Middle strand polypeptide	MD with psychotic figures; SCZ
	Spectrine[a]	MD
	Synapsin[a]	MD
	14–3-3-signal mediator	SCZ
Transport molecules	Triadnucleotide histidine binding protein	Mood and anxiety-like
Hormones and growth factors	BDNF	Initial stages MD
	Galanine	BD
	Pro-opiomelanocortin	BD
Other classes of molecules	Promoter region of the α1 subunit of the GABA-A receptor	DM; suicidal behavior

MD major depression, *SCZ* schizophrenia, *BD* bipolar disorder
[a]Also detectable in the cerebrospinal fluid

Table 41.2 Major proteomic biomarkers detectable in peripheral blood, potentially useful in psychiatry

Biological processes	Biomarkers	Psychiatric phenotype
Immunoregulation	α-2 macroglobulin	Early stages MD
	Ceruloplasmin	DM symptom severity; initial stages of MD
	ENRAGE	Severity of MD symptoms
	Ferritin	Severity of MD symptoms
	Haptoglobin binding protein	Severity of MD symptoms
	Vitamin D binding protein isoform 1	Initial stages MD
	Macrophage migration inhibiting factor (MIF)	Response to antidepressants in MD
	Pancreatic polypeptide	Severity of MD symptoms
	Prostaglandins	Initial stages MD
	Serotransferrin	Severity of MD symptoms
	Superoxide dismutase-1	Severity of MD symptoms
	Tenascina-C	Severity of MD symptoms
	Growth-regulated α protein	Severity of MD symptoms
Inflammation	Cytokines - IL-1; IL-6; IL-10, IL-13; IL-16; TNF-α	MD M, SCZ, BD; PTSD
	IL-1β; MIF mRNA	Response to antidepressants in MD
	C-reactive protein	MD; SCZ; BD
	Hemopexin	BD and SCZ
	Epsin	DB I
Hormones and growth factors	BDNF	Initial stages MD
	Growth hormones	Initial stages MD
	Growth differentiation factor 15	BD I
	Glucocorticoids[a]	MD; PTSD
	HPA axis	MD; PTSD
	Pigment epithelium-derived factor[a]	Sleep disorders
Lipid metabolism	Apolipoprotein-B100	Initial stages MD
	Apolipoprotein E precursor	Initial stages MD
Protein metabolism	Matrix metalloproteinases	BD I
	Von Willebrand factor	Severity of MD symptoms
	Complement cascade	Initial stages MD
Transport molecules	Vitamin D binding protein isoform 1	Initial stages MD
	Retinol binding protein 4	BD I
	Transthyretin	BD I

MD major depression, *SCZ* schizophrenia, *BD* bipolar disorder, *PTSD* posttraumatic stress disorder, *HPA* hypothalamus-pituitary adrenal
[a]Also detectable in the cerebrospinal fluid

ciated with depression. In contrast, alterations in the concentration of growth factors, hormones, transport molecules, and components of hemostasis would appear to be predictive of the severity of depressive symptoms (Table 41.2). Finally, peripheral cytokine levels have also been related to the treatment response with antidepressant drugs, in particular, tumor necrosis factor (TNF) or lithium in the case of bipolar disorder (Table 41.2). What is not yet known, however, is whether these changes in the expression of markers of inflammation are a consequence or cause of mental pathology. The possible co-occurrence of different inflammation causes represents a non-negligible confounding factor, which makes these markers too non-specific to have any actual diagnostic usefulness in psychiatry.

Another peripheral biomarker of altered brain function appears to be the growth factor BDNF (brain-derived neurotrophic factor). Its serum concentration is decreased in patients with bipolar disorder and seems to be inversely related to the patient's age and the disease's duration. BDNF has been reported to distinguish late stages from early stages of the disease with a sensitivity of 100%, specificity of 89%, and accuracy of 95%. BDNF also appears to be useful for the differential diagnosis between bipolar disorder and unipolar depression or as a predictor of treatment response. Patients treated with lithium who show the best response are those with the highest serum BDNF levels. Definitive evidence, however, regarding their usefulness as biomarkers does not exist for BDNF or other neurotrophic factors either.

Metabolomics

Even the combination of genomics and proteomics, although more informative than genomics studies alone, is still insufficient to represent the biochemical and molecular mechanisms underlying brain functioning comprehensively.

A further advance in this sense could derive, instead, from the study of the metabolome, i.e., the set of all metabolic processes essential for the organism's life, which reflect the complex interaction among genes, proteins, and environment. Analyses of the metabolome, conducted by NMR (nuclear magnetic resonance) spectroscopy or mass spectrometry on the serum of psychiatric patients, were found to be indicative of the presence of alterations in the citric acid cycle, the urea cycle, amino acid metabolism, or mitochondrial metabolism (Table 41.3), all essential processes for growth, proliferation, survival, and motility of cells and, therefore, likely indicators of altered neurogenesis and dendritic arborization. Metabolomics studies aim to define the so-called metabotypes, i.e., biochemical entities resulting from all metabolic interactions, predictive of a pathological state or the type of response to specific drug treatment.

Table 41.3 Main metabolomic biomarkers detectable in peripheral blood, potentially useful in psychiatry

Biological processes	Biomarkers	Psychiatric phenotype
Protein metabolism	Amino acid metabolism: β-alanine, serine, and arginine	BD
	Target of rapamycin in mammals (mTor)	Response to antidepressants in MD
	Branched chain amino acids (BCAAs): leucine, isoleucine, and valine	MD
Metabolic enzymes	Citric acid cycle, urea cycle: pyruvate, N-acetylglutamic acid	BD
	Cytochrome P450 2D6	BD
	Purinergic signal: xanthine	MD
	Pyroglutamic acid, sorbitol, and tocopherol-α	SCZ
	Quinolinic acid	Suicidal behavior
Mitochondrial metabolism	Glycolysis	SCZ
	Isocitrate dehydrogenase	BD
Hormones and growth factors	Testosterone and HPG axis; triiodothyronine and thyroxine	Aggressive behavior

MD major depression, *BD* bipolar disorder, *SCZ* schizophrenia, *HPG* hypothalamus-pituitary-gonads

Lipidomics

Lipids are also essential for brain function. Phospholipids are, in fact, the main components of cell membranes and constitute 60% of the brain's dry weight. Changes in lipid metabolism or lipid-mediated signal transmission processes alter cellular homeostasis, contributing to developing pathological states, including mental illness. In particular, polyunsaturated fatty acids such as omega-3 and omega-6 appear to play important roles in the pathogenesis of major affective and psychotic disorders (Table 41.4) and in response to treatment with antidepressants. Conditions such as metabolic syndrome, obesity, diabetes, and cardiovascular disease are frequently observed in psychiatric patients. However, it is unclear to what extent they are a consequence of treatment with psychotropic medications, which can cause weight gain, insulin resistance, and hypertriglyceridemia, or whether they are clinical conditions that preexist treatment.

Epigenomics

Epigenomics is a further discipline that can promote an understanding of molecular mechanisms underlying altered mental processes and the identification of new diagnostic or prognostic biomarkers. Epigenomics studies mechanisms that, without modifying the DNA sequence, affect the expression of genes. Among these, we find the addition of methyl groups (-CH3) at specific sequences in the promoters of genes, the CpG islands, capable of reducing gene expression, or post-translational modifications of histones – the basic proteins around which DNA is wrapped to form chromatin – such as acetylation which, by neutralizing the positive charges of histones, promotes the decondensation of chromatin and the consequent attachment of transcription factors to DNA.

Table 41.4 Main lipidomic biomarkers detectable in peripheral blood, potentially useful in psychiatry

Biological processes	Biomarkers	Psychiatric phenotype
Cellular architecture	Polyunsaturated fatty acids	MD; SCZ; BD; Alzheimer's disease
Lipid metabolism	Acetate, myoinositol	MD
	Docosahexaenoic acid (DHA), arachidonic acid (AA), docosapentaenoic acid, linoleic acid	Response to antipsychotics in SCZ
	Low and very low density lipoproteins	MD
Protein metabolism	Glycine, taurine, glutamine, alanine, valine, and leucine	MD
Energy metabolism	Creatine, creatinine	MD

Modified from Lozupone et al. (2017)
MD major depression, *SCZ* schizophrenia, *BD* bipolar disorder

Finally, microRNAs (miRNAs) are short non-coding RNA sequences, complementary to the 3′ untranslated ends (UTRs) of messenger RNAs (mRNAs), capable of temporarily or permanently blocking the synthesis of specific proteins by binding to their corresponding mRNAs and preventing them from being used as translation molds. miRNAs are tissue-specific and contribute to the development and plasticity of synapses, neurogenesis, and the arborization and formation of dendritic spines. Differential expression of some miRNAs has been observed in the prefrontal cortex of schizophrenic patients, while other miRNAs have been shown to be good predictors of response to treatment with antidepressants.

The Study of the Intestinal Microbiota

In recent years, the study of the intestinal microbiota has aroused growing interest, given the role it seems to play in mediating brain functions. There is, in fact, an axis of connection gut-brain, which closely relates these two organs. The intestinal microbiota, composed of over 100 trillion commensal bacteria living in symbiosis with the human organism, can produce most neurotransmitters and their precursors. Gut dysbiosis, an increase in pathogenic bacterial strains at the expense of symbiotic ones, has been linked with mood disorders, particularly depression (Table 41.5). A high-fat diet may

Table 41.5 Main microbic biomarkers detectable in the feces of psychiatric patients

Biomarkers	Psychiatric phenotype
Alistipes	MD
Bacteroidales	MD
Bacteroides	SR in MD
Bifidobatterium	MD
Clostridium	SR in MD
IgM and IgA against lipopolysaccharides (LPS)	MD
Lachnospiraceae	MD
Lactobacillus	SR in MD
Lactobacillus rhamnosus	Depression and anxiety
Lactococcus	MD
Oscillibatterium	MD
Parabacteroides	SR in MD
Porphyromonadaceae	SR in MD
Volatile fatty acids	MD

MD major depression, *SR* social rupture

promote anhedonia, pointing to the microbiota as a critical factor in the link between diet and depression.

The gut microbiota appears to be involved in the pathogenesis of both major depression and bipolar disorder. A different distribution of bacterial strains, with *Escherichia* and *Klebsiella* more abundant in bipolar patients and *Prevotella* prevalent in depressed patients, could allow a differential diagnosis between these two pathologies.

It has also been observed that depressed patients show high levels of IgM and IgA directed against lipopolysaccharides of gram-negative intestinal bacteria, suggesting that the passage of these bacteria into the bloodstream, due to increased intestinal permeability resulting from chronic inflammatory diseases, may promote the onset of psychiatric disorders.

Alterations in the gut microbiota have also been related to schizophrenia, as they cause a reduction in protective factors and increase neurotoxins and inflammatory mediators, which are responsible for the neuronal and synaptic damage observed in these patients.

Future Prospective

In summary, the studies reported so far in the scientific literature show that, although objective biomarkers of mental illness or response to treatment have not yet been identified and are clinically useful, many of the data produced help to underline the highly complex nature of psychiatric illnesses, reiterating the need to define their diagnostic framework better.

It is also becoming increasingly clear that genetics, in close relationship with the environment, plays a vital role in modulating human behavior.

Much of the current knowledge about the neurobiological basis of human behavior comes from studies of aggressive behavior. Aggression and antisocial behavior characterize, in fact, many psychiatric conditions, such as substance abuse, psychotic disorders, and borderline personality disorders.

The first concrete evidence of a relationship between genetics and violent behavior emerged in a study in the early 1990s. In a Dutch family, these authors identified a sequence variant in the gene encoding for the enzyme monoamine oxidase A (MAOA), which introduced a stop codon responsible for the production of a truncated and non-functional enzyme. MAOA is an enzyme responsible for the catabolism of neurotransmitters, so it plays a key role in regulating both serotonergic and dopaminergic transmission. Male subjects, carriers of this mutation, showed very violent behavior, guilty of serious crimes such as rape, murder, and arsonism. This behavior was present only in males because, being the MAOA gene on the X chromosome, the mutant allele was not counterbalanced by the wild-type copy on the other chromosome. Confirmation that the mutation in the MAOA gene caused the violent behavior came 2 years later, from the creation of knockout mice carrying the same stop mutation, which showed highly aggressive behavior toward their cage mates. This mutation has never been found again in the general population, being exclusive to the family studied by Brunner.

There is, however, a VNTR (variable number of tandem repeats) sequence variant in the promoter of the MAOA gene, which functional studies have correlated with reduced expression of the enzyme. This sequence variant is present in 33% of the Caucasian population. In a 2002 study, Caspi's group was able to demonstrate, in a sample of subjects studied longitudinally from childhood to adulthood, that the low alleles (with 2, 3, and 5 repeats) of this variant – so called because they are responsible for reduced expression of the enzyme – when associated with a hostile growth environment, i.e., characterized by abuse, mistreatment, or neglect during childhood, increase the risk of antisocial behavior (Fig. 41.1). Thus, it was shown that a genetic variant could predispose to violent behavior for the first time, but only when associated with a particularly negative growth environment. Subjects carrying the low alleles who grew up in a healthy environment without having suffered abuse and mistreatment were just as likely to behave correctly as those carrying the high alleles of MAOA.

Several subsequent studies, including some meta-analyses, have confirmed this relationship between low alleles of MAOA-VNTR and aggressive and antisocial behavior, again in association with a negative growth environment, so much so that MAOA has been called the "warrior gene" (Box 41.1).

Since the MAOA enzyme is responsible for the degradation of brain neurotransmitters, including serotonin, the association between aggression, a condition characterized by low serotonin levels, and reduced MAOA activity might

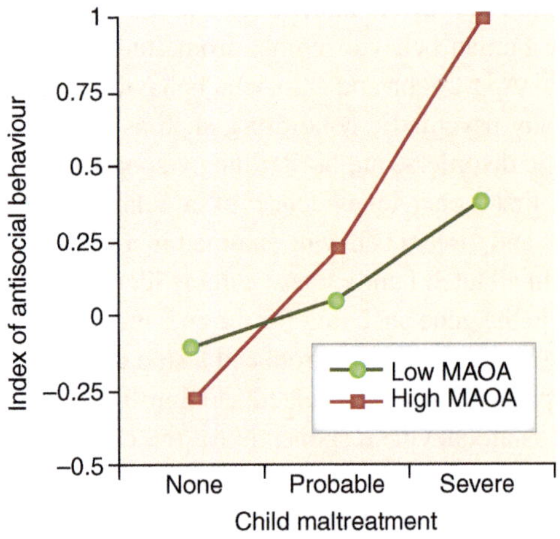

Fig. 41.1 First demonstration of a gene/environment interaction in antisocial behavior. (Copyright EDISES 2021. Reproduced with permission)

appear contradictory. This contrast, however, can be explained considering that increased serotonin availability during brain development may cause functional alterations in circuits critical for emotion processing.

Brain MRI studies have shown that males carrying the low alleles of MAOA, compared to carriers of high alleles, show a reduction of 8% of gray matter in limbic regions. These same authors observed in subjects with MAOA-low a decreased functional activation of the ventromedial prefrontal cortex and increased activation of the amygdala during emotional tasks of face recognition. Given the crucial role that the ventromedial prefrontal cortex plays in the control of aggressive behavior and that the amygdala plays in the modulation of emotions, an altered function of these brain structures may be responsible for a reduced ability to impulse control. This would explain why carriers of the low alleles of MAOA react to provocations more violently than carriers of the high alleles.

In addition to the MAOA gene, other polymorphic variants of genes coding for the main neurotransmitters, serotonin, and dopamine, for their receptors, transporters, and enzymes responsible for their degradation have been described in the literature as risk factors for violent behavior. In particular, strong evidence has emerged against a VNTR variant in the promoter of the serotonin transporter (5HTTLPR), an SNP variant (rs1800955) and a VNTR in dopamine receptor number 4 (DRD4), and a missense variant (Val/Met) in the enzyme catechol-methyl transferase (COMT).

These polymorphisms seem to modulate the effect that the environment exerts on personality traits, increasing the vulnerability of the individual to the development of impulsive and aggressive behaviors. Therefore, they exert a crucial role in regulating cognitive and emotional processes in response to stimuli and/or particularly negative events.

This evidence, which supports an increased genetic vulnerability to violent behavior, has been, in some cases, already used in forensics to reconstruct crime genesis and crime dynamics of violent crimes, both in Europe and the United States. Specifically, genetic profiles characterized by more than three or four of these risk alleles have been proposed as factors capable of influencing the capacity of the offenders at the time of the crime. The presence of an unfavorable genetic profile concerning the five previously mentioned sequence variants was recognized for the first time in the world, in 2009, by the Court of Appeal of Trieste, as a mitigating factor in a case of murder (https://www.nature.com/news/2009/091030/full/news.2009.1050.html). Two years later, in 2011, a judge of the Court of Como decided to include, among the reasons for the reduction of a sentence for murder, the genetic profile of the defendant who, again, possessed all the risk alleles mentioned. In the same year, the Tennessee Court of Appeals exempted a murderer from the death penalty in the United States because he carried a low allele of MAOA.

In the same period, other scientific evidence has led to the hypothesis that these polymorphisms, described as variants of "risk" for aggressive behavior, may be considered alleles of "plasticity," which make the brain more receptive to negative events, but also positive ones. The genetic variants associated with antisocial behavior in individuals with histories of abuse and maltreatment in childhood would be responsible for a greater propensity toward altruism and prosociality in people who grew up in particularly positive and protective environments (Fig. 41.3). This hypothesis reinforces the importance of the relationship between genes and environment in modulating human behavior and suggests using these genotypes as predictors of response to educational or rehabilitative treatment programs.

The new methodologies for studying the genetic and molecular factors underlying brain activity in physiological conditions and patients with mental disorders have led to the definition of a true "clinical biochemistry of the mind." Hopefully, this multidisciplinary and integrated approach will bridge the gap that separates psychiatry and clinical psychology from other medical disciplines, in which the specialist has the comfort of objective data from laboratory or instrumental tests during the diagnostic process and in any other moment of the clinical decision-making process. The chapters of this Treatise show the relevance and importance of laboratory medicine in the daily clinical activities of all branches of medicine; today, clinical biochemistry has developed a dialogue with border disciplines outside the scope of medicine in the strict sense. These studies' results outline new perspectives in psychiatry and forensic psychology and ethics, philosophy, and jurisprudence. Could we, therefore, one day arrive at measuring free will, just as we measure an individual's blood sugar today?

Box 41.1 MAOA – The Warrior Gene

The low 2-repeat allele of MAOA was also found in the DNA of Giovanni Dalle Bande Nere, a Renaissance leader who went down in history for his extremely reckless and violent behavior (Fig. 41.2). Giovanni Dalle Bande Nere was the son of Cosimo de' Medici, who died a few months after his birth, and Caterina Sforza, the "warrior lady" of the Renaissance, who spent her life defending her properties from Cesare Borgia. This suggests that the environment in which Giovanni Dalle Bande Nere grew up was anything but protective and, therefore, in combination with the low allele of MAOA, may have contributed decisively to the development of his particularly violent behavior. Giovanni Dalle Bande Nere carried out his first murder at the age of 12 and by the age of 28, when he died following the infection of a wound suffered in battle, he had already killed over ten people.

Fig. 41.2 Portrait of Giovanni Dalle Bande Nere by Francesco Salviati (Galleria Palatina, Florence)

Fig. 41.3 The concept of environmental plasticity genes. (Copyright EDISES 2021. Reproduced with permission)

Recommended Readings

Abi-Dargham A, Horga G (2016) The search for imaging biomarkers in psychiatric disorders. Nat Med 22(11):1248–1255

Anttila V, Bulik-Sullivan B, Finucane HK, Walters RK, Bras J, Duncan L et al (2018) Analysis of shared heritability in common disorders of the brain. Science 360(6395):eaap8757

Aronson JK (2005) Biomarkers and surrogate endpoints. Br J Clin Pharmacol 59(5):491–494

Bauer IE, Pascoe MC, Wollenhaupt-Aguiar B, Kapczinski F, Soares JC (2014) Inflammatory mediators of cognitive impairment in bipolar disorder. J Psychiatr Res 56:18–27

Belsky J, Jonassaint C, Pluess M, Stanton M, Brummett B, Williams R (2009) Vulnerability genes or plasticity genes? Mol Psychiatry 14(8):746–754

Bookheimer SY, Strojwas MH, Cohen MS, Saunders AM, Pericak-Vance MA, Mazziotta JC et al (2000) Patterns of brain activation in people at risk for Alzheimer's disease. N Engl J Med 343(7):450–456

Brunner HG, Nelen M, Breakefield XO, Ropers HH, van Oost BA (1993) Abnormal behavior associated with a point mutation in the structural gene for monoamine oxidase A. Science 262(5133):578–580

Buckholtz JW, Meyer-Lindenberg A (2008) MAOA and the neurogenetic architecture of human aggression. Trends Neurosci 31(3):120–129

Byrd AL, Manuck SB (2014) MAOA, childhood maltreatment, and antisocial behavior: meta-analysis of a gene-environment interaction. Biol Psychiatry 75(1):9–17

Caldwell HK (2012) Neurobiology of sociability. Adv Exp Med Biol 739:187–205

Cao X, Yeo G, Muotri AR, Kuwabara T, Gage FH (2006) Noncoding RNAs in the mammalian central nervous system. Annu Rev Neurosci 29:77–103

Cases O, Seif I, Grimsby J, Gaspar P, Chen K, Pournin S et al (1995) Aggressive behavior and altered amounts of brain serotonin and norepinephrine in mice lacking MAOA. Science 268(5218):1763–1766

Caspi A, McClay J, Moffitt TE, Mill J, Martin J, Craig IW et al (2002) Role of genotype in the cycle of violence in maltreated children. Science 297(5582):851–854

Cecchetti L, Lettieri G, Handjaras G, Leo A, Ricciardi E, Pietrini P et al (in press) Brain hemodynamic intermediate phenotype links vitamin B12 to cognitive profile of healthy and mild cognitive impaired subjects. Neural Plast 2019:6874805

Chan SW, Harmer CJ, Norbury R, O'Sullivan U, Goodwin GM, Portella MJ (2016) Hippocampal volume in vulnerability and resilience to depression. J Affect Disord 189:199–202

Collins SM, Surette M, Bercik P (2012) The interplay between the intestinal microbiota and the brain. Nat Rev Microbiol 10(11):735–742

Consortium C-DGotPG (2013) Identification of risk loci with shared effects on five major psychiatric disorders: a genome-wide analysis. Lancet 381(9875):1371–1379

Consortium SWGotPG (2014) Biological insights from 108 schizophrenia-associated genetic loci. Nature 511(7510):421–427

Cox DA, Gottschalk MG, Stelzhammer V, Wesseling H, Cooper JD, Bahn S (2018) Evaluation of molecular brain changes associated with environmental stress in rodent models compared to human major depressive disorder: a proteomic systems approach. World J Biol Psychiatry. 19(sup2):S63–S74

Cummings MA (2015) The neurobiology of psychopathy: recent developments and new directions in research and treatment. CNS Spectr 20(3):200–206

Dash S, Clarke G, Berk M, Jacka FN (2015) The gut microbiome and diet in psychiatry: focus on depression. Curr Opin Psychiatry 28(1):1–6

de Witte L, Tomasik J, Schwarz E, Guest PC, Rahmoune H, Kahn RS et al (2014) Cytokine alterations in first-episode schizophrenia patients before and after antipsychotic treatment. Schizophr Res 154(1–3):23–29

Denson TF, Dobson-Stone C, Ronay R, von Hippel W, Schira MM (2014) A functional polymorphism of the MAOA gene is associated with neural responses to induced anger control. J Cogn Neurosci 26(7):1418–1427

Fernandes BS, Gama CS, Kauer-Sant'Anna M, Lobato MI, Belmonte-de-Abreu P, Kapczinski F (2009) Serum brain-derived neurotrophic factor in bipolar and unipolar depression: a potential adjunctive tool for differential diagnosis. J Psychiatr Res 43(15):1200–1204

Ficks CA, Waldman ID (2014) Candidate genes for aggression and antisocial behavior: a meta-analysis of association studies of the 5HTTLPR and MAOA-uVNTR. Behav Genet 44(5):427–444

Gentili C, Valenza G, Nardelli M, Lanatà A, Bertschy G, Weiner L et al (2017) Longitudinal monitoring of heartbeat dynamics predicts mood changes in bipolar patients: a pilot study. J Affect Disord 209:30–38

Giusti L, Ciregia F, Mazzoni MR, Lucacchini A (2016) Proteomics insight into psychiatric disorders: an update on biological fluid biomarkers. Expert Rev Proteomics 13(10):941–950

Godar SC, Fite PJ, McFarlin KM, Bortolato M (2016) The role of monoamine oxidase A in aggression: Current translational developments and future challenges. Prog Neuro-Psychopharmacol Biol Psychiatry 69:90–100

Group. BDW (2001) Biomarkers and surrogate endpoints: preferred definitions and conceptual framework. Clin Pharmacol Ther 69(3):89–95

Hayashi-Takagi A, Vawter MP, Iwamoto K (2014) Peripheral biomarkers revisited: integrative profiling of peripheral samples for psychiatric research. Biol Psychiatry 75(12):920–928

Hodgins S, Piatosa MJ, Schiffer B (2014) Violence among people with schizophrenia: phenotypes and neurobiology. Curr Top Behav Neurosci 17:329–368

Horrobin DF (1998) The membrane phospholipid hypothesis as a biochemical basis for the neurodevelopmental concept of schizophrenia. Schizophr Res 30(3):193–208

Hou L, Bergen SE, Akula N, Song J, Hultman CM, Landén M et al (2016) Genome-wide association study of 40,000 individuals identifies two novel loci associated with bipolar disorder. Hum Mol Genet 25(15):3383–3394

Iofrida C, Palumbo S, Pellegrini S (2014) Molecular genetics and antisocial behavior: where do we stand? Exp Biol Med (Maywood) 239(11):1514–1523

Jensen AR (1967) Estimation of the limits of heritability of traits by comparison of monozygotic and dizygotic twins. Proc Natl Acad Sci U S A 58(1):149–156

Joëls M, Karst H, Sarabdjitsingh RA (2018) The stressed brain of humans and rodents. Acta Physiol (Oxf) 223(2):e13066

Kapczinski F, Fernandes B, Kauer Sant'Anna M, Gama C, Yatham L, Berk M (2009) The concept of staging in bipolar disorder: the role of BDNF and TNF-alpha as biomarkers. Acta Neuropsychiatrica 21(6):272–274

Kauer-Sant'Anna M, Kapczinski F, Andreazza AC, Bond DJ, Lam RW, Young LT et al (2009) Brain-derived neurotrophic factor and inflammatory markers in patients with early- vs. late-stage bipolar disorder. Int J Neuropsychopharmacol 12(4):447–458

Kékesi KA, Juhász G, Simor A, Gulyássy P, Szegő EM, Hunyadi-Gulyás E et al (2012) Altered functional protein networks in the prefrontal cortex and amygdala of victims of suicide. PLoS One 7(12):e50532

Kessler RC, Heeringa S, Lakoma MD, Petukhova M, Rupp AE, Schoenbaum M et al (2008) Individual and societal effects of mental disorders on earnings in the United States: results from the national comorbidity survey replication. Am J Psychiatry 165(6):703–711

Kim AH, Reimers M, Maher B, Williamson V, McMichael O, McClay JL et al (2010) MicroRNA expression profiling in the prefrontal cortex of individuals affected with schizophrenia and bipolar disorders. Schizophr Res 124(1–3):183–191

Kim-Cohen J, Caspi A, Taylor A, Williams B, Newcombe R, Craig IW et al (2006) MAOA, maltreatment, and gene-environment interaction predicting children's mental health: new evidence and a meta-analysis. Mol Psychiatry 11(10):903–913

Lee LH, Shui G, Farooqui AA, Wenk MR, Tan CH, Ong WY (2009) Lipidomic analyses of the mouse brain after antidepressant treatment: evidence for endogenous release of long-chain fatty acids? Int J Neuropsychopharmacol 12(7):953–964

Lee MY, Kim EY, Kim SH, Cho KC, Ha K, Kim KP et al (2016) Discovery of serum protein biomarkers in drug-free patients with major depressive disorder. Prog Neuro-Psychopharmacol Biol Psychiatry 69:60–68

Liu X, Shimada T, Otowa T, Wu YY, Kawamura Y, Tochigi M et al (2016) Genome-wide association study of autism spectrum disorder in the East Asian Populations. Autism Res 9(3):340–349

Lopez JP, Lim R, Cruceanu C, Crapper L, Fasano C, Labonte B et al (2014) miR-1202 is a primate-specific and brain-enriched microRNA involved in major depression and antidepressant treatment. Nat Med 20(7):764–768

Lozupone M, Seripa D, Stella E, La Montagna M, Solfrizzi V, Quaranta N et al (2017) Innovative biomarkers in psychiatric disorders: a major clinical challenge in psychiatry. Expert Rev Proteomics 14(9):809–824

Maes M, Kubera M, Leunis JC, Berk M (2012) Increased IgA and IgM responses against gut commensals in chronic depression: further evidence for increased bacterial translocation or leaky gut. J Affect Disord 141(1):55–62

Martin C, Tansey KE, Schalkwyk LC, Powell TR (2015) The inflammatory cytokines: molecular biomarkers for major depressive disorder? Biomark Med 9(2):169–180

Martins-de-Souza D (2013) Biomarkers for psychiatric disorders: where are we standing? Dis Markers 35(1):1–2

Maurano MT, Humbert R, Rynes E, Thurman RE, Haugen E, Wang H et al (2012) Systematic localization of common disease-associated variation in regulatory DNA. Science 337(6099):1190–1195

McDermott R, Tingley D, Cowden J, Frazzetto G, Johnson DD (2009) Monoamine oxidase A gene (MAOA) predicts behavioral aggression following provocation. Proc Natl Acad Sci U S A 106(7):2118–2123

Meyer-Lindenberg A, Buckholtz JW, Kolachana B, R Hariri A, Pezawas L, Blasi G et al (2006) Neural mechanisms of genetic risk for impulsivity and violence in humans. Proc Natl Acad Sci U S A 103(16):6269–6274

Naseribafrouei A, Hestad K, Avershina E, Sekelja M, Linløkken A, Wilson R et al (2014) Correlation between the human fecal microbiota and depression. Neurogastroenterol Motil 26(8):1155–1162

Ngounou Wetie AG, Sokolowska I, Wormwood K, Beglinger K, Michel TM, Thome J et al (2013) Mass spectrometry for the detection of potential psychiatric biomarkers. J Mol Psychiatry 1(1):8

Nilsson KW, Åslund C, Comasco E, Oreland L (2018) Gene-environment interaction of monoamine oxidase A in relation to antisocial behaviour: current and future directions. J Neural Transm (Vienna) 125(11):1601–1626

Palumbo S, Mariotti V, Iofrida C, Pellegrini S (2018) Genes and aggressive behavior: epigenetic mechanisms underlying individual susceptibility to aversive environments. Front Behav Neurosci 12:117

Paquet C, Magnin E, Wallon D, Troussière AC, Dumurgier J, Jager A et al (2016) Utility of CSF biomarkers in psychiatric disorders: a national multicentre prospective study. Alzheimers Res Ther 8:27

Parker G, Gibson NA, Brotchie H, Heruc G, Rees AM, Hadzi-Pavlovic D (2006) Omega-3 fatty acids and mood disorders. Am J Psychiatry 163(6):969–978

Pellegrini S, Pietrini P (2010) Siamo davvero liberi?Il comportamento tra geni e cervello. Sistemi Intelligenti

Pietrini P (2003) Toward a biochemistry of mind? Am J Psychiatry 160(11):1907–1908

Pietrini P, Guazzelli M, Basso G, Jaffe K, Grafman J (2000) Neural correlates of imaginal aggressive behavior assessed by positron emission tomography in healthy subjects. Am J Psychiatry 157(11):1772–1781

Polyakova M, Stuke K, Schuemberg K, Mueller K, Schoenknecht P, Schroeter ML (2015) BDNF as a biomarker for successful treatment of mood disorders: a systematic & quantitative meta-analysis. J Affect Disord 174:432–440

Reble E, Dineen A, Barr CL (2018) The contribution of alternative splicing to genetic risk for psychiatric disorders. Genes Brain Behav 17(3):e12430

Reus VI, Freimer NB (1997) Understanding the genetic basis of mood disorders: where do we stand? Am J Hum Genet 60(6):1283–1288

Reuter M, Frenzel C, Walter NT, Markett S, Montag C (2011) Investigating the genetic basis of altruism: the role of the COMT Val158Met polymorphism. Soc Cogn Affect Neurosci 6(5):662–668

Rigoni D, Pellegrini S, Mariotti V, Cozza A, Mechelli A, Ferrara SD et al (2010) How neuroscience and behavioral genetics improve psychiatric assessment: report on a violent murder case. Front Behav Neurosci 4:160

Rogers GB, Keating DJ, Young RL, Wong ML, Licinio J, Wesselingh S (2016) From gut dysbiosis to altered brain function and mental illness: mechanisms and pathways. Mol Psychiatry 21(6):738–748

Rong H, Xie XH, Zhao J, Lai WT, Wang MB, Xu D et al (2019) Similarly in depression, nuances of gut microbiota: evidences from a shotgun metagenomics sequencing study on major depressive disorder versus bipolar disorder with current major depressive episode patients. J Psychiatr Res 113:90–99

Sabol SZ, Hu S, Hamer D (1998) A functional polymorphism in the monoamine oxidase A gene promoter. Hum Genet 103(3):273–279

Sánchez-Mora C, Ramos-Quiroga JA, Bosch R, Corrales M, Garcia-Martínez I, Nogueira M et al (2015) Case-control genome-wide association study of persistent attention-deficit hyperactivity disorder identifies FBXO33 as a novel susceptibility gene for the disorder. Neuropsychopharmacology 40(4):915–926

Sartori G, Pellegrini S, Mechelli A (2011) Forensic neurosciences: from basic research to applications and pitfalls. Curr Opin Neurol 24(4):371–377

Sayana P, Colpo GD, Simões LR, Giridharan VV, Teixeira AL, Quevedo J et al (2017) A systematic review of evidence for the role of inflammatory biomarkers in bipolar patients. J Psychiatr Res 92:160–182

Scarpazza C, Pellegrini S, Pietrini P, Sartori G (2018) The role of neuroscience in the evaluation of mental insanity: on the controversies in Italy. Neuroethics 11(1):83–95

Scurich N, Appelbaum PS (2017) Behavioral genetics in criminal court. Nat Hum Behav 1:772–774

Sheline YI (2011) Depression and the hippocampus: cause or effect? Biol Psychiatry 70(4):308–309

Shyn SI, Shi J, Kraft JB, Potash JB, Knowles JA, Weissman MM et al (2011) Novel loci for major depression identified by genome-wide association study of Sequenced Treatment Alternatives to Relieve Depression and meta-analysis of three studies. Mol Psychiatry 16(2):202–215

Simons RL, Lei MK, Beach SR, Brody GH, Philibert RA, Gibbons FX (2011) Social environmental variation, plasticity genes, and aggression: evidence for the differential susceptibility hypothesis. Am Sociol Rev 76(6):833–912

Soni A (2009) in The Five Most Costly Conditions, 1996 and 2006: Estimates for the U.S. Civilian Non-institutionalized Population 1–3 (Agency for Healthcare Research and Quality)

Stelzhammer V, Haenisch F, Chan MK, Cooper JD, Steiner J, Steeb H et al (2014) Proteomic changes in serum of first onset, antidepressant

drug-naïve major depression patients. Int J Neuropsychopharmacol 17(10):1599–1608
Strimbu K, Tavel JA (2010) What are biomarkers? Curr Opin HIV AIDS 5(6):463–466
Sullivan PF, Geschwind DH (2019) Defining the genetic, genomic, cellular, and diagnostic architectures of psychiatric disorders. Cell 177(1):162–183
Sun X, Ma R, Jiang Y, Gao Y, Ming Q, Wu Q et al (2018) MAOA genotype influences neural response during an inhibitory task in adolescents with conduct disorder. Eur Child Adolesc Psychiatry 27:1159–1169
Sussulini A, Dihazi H, Banzato CE, Arruda MA, Stühmer W, Ehrenreich H et al (2011) Apolipoprotein A-I as a candidate serum marker for the response to lithium treatment in bipolar disorder. Proteomics 11(2):261–269
Taylor A, Kim-Cohen J (2007) Meta-analysis of gene-environment interactions in developmental psychopathology. Dev Psychopathol 19(4):1029–1037
Tuvblad C, Narusyte J, Comasco E, Andershed H, Andershed AK, Colins OF et al (2016) Physical and verbal aggressive behavior and COMT genotype: sensitivity to the environment. Am J Med Genet B Neuropsychiatr Genet 171(5):708–718
Uchino BN, Cacioppo JT, Kiecolt-Glaser JK (1996) The relationship between social support and physiological processes: a review with emphasis on underlying mechanisms and implications for health. Psychol Bull 119(3):488–531
van der Gronde T, Kempes M, van El C, Rinne T, Pieters T (2014) Neurobiological correlates in forensic assessment: a systematic review. PLoS One 9(10):e110672
Vescovi PP, Coiro V, Volpi R, Giannini A, Passeri M (1992) Plasma beta-endorphin, but not met-enkephalin levels are abnormal in chronic alcoholics. Alcohol Alcohol 27(5):471–475
Wang Y, Zhang Y, Hu W, Xie S, Gong CX, Iqbal K et al (2015) Rapid alteration of protein phosphorylation during postmortem: implication in the study of protein phosphorylation. Sci Rep 5:15709
World Health Organization (2008) The global burden of disease:2004 update. WHO, Geneva
Yuan X, Kang Y, Zhuo C, Huang XF, Song X (2019) The gut microbiota promotes the pathogenesis of schizophrenia via multiple pathways. Biochem Biophys Res Commun 512(2):373–380
Zhang-James Y, Fernàndez-Castillo N, Hess JL, Malki K, Glatt SJ, Cormand B et al (2019) An integrated analysis of genes and functional pathways for aggression in human and rodent models. Mol Psychiatry 24:1655–1667

Clinical Molecular Biology

42

Ettore Domenico Capoluongo and Giuseppe Castaldo

Introduction

Clinical molecular biology tests include a series of qualitative–quantitative analyses of biological molecules (DNA, RNA, proteins) to clinically evaluate patients with genetic diseases (congenital or acquired). They offer a diagnostic contribution at all stages of the natural history of the disease.

- Susceptibility diagnosis: looking for mutations to assess the risk of disease. Mutations do not cause the disease but carry a risk of developing it due to exposure to environmental or genetic triggers. This includes most multifactorial adult diseases and individual susceptibility to drugs.
- Early diagnosis: involves screening populations for early detection of disease. Symptoms may appear later, even at an advanced age. Early diagnosis is helpful for the early implementation of therapies that reduce the morbidity and mortality of the disease.
- Differential diagnosis: the tests contribute to the diagnosis, confirming or excluding, in a symptomatic subject, the presence of a disease.
- Carrier screening: involves searching for mutations in genes responsible for genetic diseases that are more frequent in specific ethnic groups or the relatives of subjects affected by hereditary genetic diseases.
- Prenatal diagnosis: helps to define the status of the embryo or fetus concerning specific diseases in the prenatal period.
- Staging of the disease: helps to define the extent of the disease and, therefore, contribute to formulating the prognosis. This group includes tests that assess the risk of distant metastases in a cancer patient by studying cells or circulating nucleic acids.
- Disease monitoring during therapy: biomarkers are used to monitor the response to therapy (e.g., to recognize the level of residual cancer cells after chemotherapy).
- Therapeutic choice: the tests are oriented to evaluate the response to specific drugs, usually "molecularly targeted."

Type of Test and Biological Sample

Most clinical molecular biology tests aim to look for germline mutations (i.e., present from birth) in specific genes responsible for an inherited genetic disease or susceptibility to disease. When looking for germline mutations, the analysis can be performed by extracting DNA from any cell type, such as peripheral blood, tissues (fresh or paraffin-embedded), cells from biological fluids, and, in the case of forensic analysis, biological traces containing cells, such as hair, seminal fluid, and bone fragments. Many adult diseases are multifactorial (or complex). They result from several genes' interactions with the environment (e.g., cardiovascular and psychiatric diseases, asthma and various respiratory diseases, diabetes, and some cancers).

In other pathologies, somatic mutations are sought, i.e., not inherited from the parents and present only in a cell line. In this case (very frequent in oncological molecular diagnostics), the analysis is carried out on the cells among which the mutant clone is presumably present and, therefore, on biopsy samples from different tissues, from peripheral blood or bone marrow (in the case of hemolymphoproliferative diseases).

Some clinical molecular biology tests are based on the analysis (including quantitative analysis) of mRNAs or proteins expressed by specific cells or tissues; in these cases, the analysis is performed on the cells or tissues expressing the mRNA or protein.

E. D. Capoluongo
Department of Molecular Medicine and Medical Biotechnology, Federico II University, Naples, Italy

Department of Clinical Pathology and Genomics, Cannizzaro Hospital, Catania, Italy

G. Castaldo (✉)
Department of Molecular Medicine and Medical Biotechnologies, University of Naples Federico II, Naples, Italy
e-mail: giuseppe.castaldo@unina.it

M. Ciaccio (ed.), *Clinical and Laboratory Medicine Textbook*, https://doi.org/10.1007/978-3-031-24958-7_42

Finally, some tests look for mutations of specific genes (or messenger RNAs) in circulating blood (e.g., in non-invasive prenatal diagnosis or in the search for circulating cancer cells); therefore, the analysis is performed on blood samples.

Diagnosis of Susceptibility

The diagnosis of susceptibility concerns diseases that are widespread in adults, and, in general, the molecular test can predict an increased or reduced risk of developing the disease. This susceptibility, but not the disease, can be transferred to descendants in a non-Mendelian and unpredictable way. The development of the various conditions associated with this risk is also affected by lifestyle and environmental factors and epigenetic modifications.

In recent years, the number of diseases for which susceptibility markers are known has increased significantly and continues to grow exponentially due to the availability of massive human genome sequencing technologies. Indeed, a complete genome sequencing analysis can be performed quickly and cost-effectively by selecting patients with a disease and healthy controls. Then, the myriad of data obtained is analyzed by appropriate bioinformatics systems, and the gene loci (usually single nucleotide polymorphisms [SNPs]) statistically associated with the disease are defined. The polymorphisms identified do not necessarily have to fall within a disease-gene and in themselves may also be devoid of a direct pathogenetic effect; in many cases, they are close to a disease-causing gene (or even present within the gene itself in non-coding intronic regions) that is not yet known and that contributes to the pathogenesis of the disease. The strict observance of the selection criteria of patients and controls and the stringency of the statistical criteria of data processing make the marker usable in routine, even without identifying the disease-gene. Indeed, for many multifactorial diseases (Table 42.1), panels for the routine analysis of susceptibility loci are already on the market.

Table 42.1 Multifactorial diseases for which panels for the routine analysis of predisposition loci are commercially available

Rheumatoid arthritis
Asthma
Type 1 diabetes mellitus
Type 2 diabetes mellitus
Hypertension
Bipolar disease
Celiac disease
Coronary heart disease
Crohn's disease
Multiple sclerosis
Breast cancer
Prostate cancer

Thrombophilia

In some cases, susceptibility genes are defined by a biochemical approach, i.e., by studying the proteins (and therefore the genes) involved in a biological process. For example, alterations in specific genes encoding proteins that inhibit the coagulation process may predispose to thrombotic episodes. Even these gene variants do not give rise to disease, but they predispose to developing thrombotic episodes that may be triggered in the presence of concomitant causes, or in patients immobilized for long periods, etc. The most known prothrombotic genetic factors are:

- Antithrombin III (ATIII) deficiency
- Protein C (PC) deficiency
- Protein S (PS) deficiency
- Factor V of Leiden (FV Leiden)
- G20210A prothrombin mutation (FII G20210A)

Hundreds of different mutations have been described for each of the three deficiency conditions (ATIII, PC, or PS deficiency) that can cause a more or less severe reduction in the functional activity of the protein. However, the search for mutations is carried out only for research purposes since, for diagnostic purposes, the functional activity of the three proteins can be assessed. The search for FV Leiden and FII G20210A mutations is, instead, part of the screening protocol for thrombophilia.

Pharmacogenomics

Clinical molecular biology has a key role in studying mutations of genes encoding for enzymes involved in drug metabolism. Along these lines, a new area of clinical molecular biology defined as pharmacogenomics has developed recently. It is well known that about 40% of the general population is resistant to one or more drugs and, at the same time, about 15–20% of hospital admissions are due to adverse drug reactions which, in the United States, represent the fourth cause of death (after cardiovascular disease, cancer, and stroke). All drugs (antibiotics, antineoplastics, etc.) are inactivated or, in some cases, activated through enzymatic reactions that occur mainly in the liver and constitute the metabolism of xenobiotics. The efficiency of the activation or inactivation reactions (and thus the more or less effective response to the drug) varies from individual to individual depending on the efficiency of the enzymes of xenobiotic metabolism, which in turn depends on mutations in the corresponding genes. One example, shown in Table 42.2, concerns cytochrome P450 2D6, which is involved in the metabolism of at least a

Table 42.2 Drugs metabolized by cytochrome P450 2D6

Antiarrhythmics (nimodipine)
Tricyclic antidepressants (imipramine, fluoxetine)
Antitussive (dextromethorphan)
β-blockers
Narcotics (codeine)
Neuroleptics (haloperidol)

quarter of all currently used drugs. Today, at least 70 variants of the gene encoding for cytochrome are known and, depending on the type of variant, individuals are divided into three groups:

1. Rapid metabolizers (about 90% of the population)
2. Slow metabolizers (about 5% of the population)
3. Ultrarapid metabolizers (who inactivate the drug early: 1% of the population)

The analysis of these variants can therefore allow to predict the individual response to the drug and modulate the drug's type (or dosage) to be administered. In other cases, the alterations mentioned above may concern the gene encoding for the protein targeted by the drug, as in the case of the β-adrenergic receptor 2, the target of anti-asthmatic drugs. For many last-generation chemotherapies and antibiotics, it is now possible to predict the patient's response by analyzing genes encoding for proteins involved in their metabolism, adopting more targeted therapeutic choices.

Guide Therapy in Oncology: The BRCA1/2 Test

The recent inter-societal recommendations of AIOM, SIGU, SIBIOC, and SIAPEC-IAP have indicated the diagnostic and therapeutic pathway for high-grade serous ovarian cancer (HGSOC) patients. For this neoplasm, the indications of the various international medical agencies (FDA, EMA, AIFA) provide for the execution of genetic testing BRCA1/2 to identify pathogenic variants (class 4 or 5) that give the patient access to treatment with PARP-1 inhibitors (Olaparib). The identification of a pathogenetic variant in one of the two BRCA genes has not only prognostic but also predictive importance. Also, it allows preventive planning strategies at a personal (e.g., assessing the risk of breast cancer) or family level (increased risk for breast, ovarian, prostate, and pancreatic cancer, mainly). Therefore, the BRCA test is essential for diagnosing and managing patients with ovarian cancer, mainly because about 22% (17–25% depending on the case series) of women with HGSOC are carriers of an alteration in one of the BRCA genes.

The test for BRCA1/2 is performed mainly using massively parallel sequencing technology (NGS), on the most common platforms, using CE-IVD certified kits coupled to bioinformatics software for data analysis. These methodologies allow the identification of most sequence variants, i.e., about 90% of those present in exonic regions and exon/intron junctions: from single nucleotide variants to deletions/insertions and truncating mutations.

NGS technology can fail in the detection of large rearrangements (partial or whole gene deletions), for which other analysis techniques, such as multiple ligation probe assay (MLPA) and multiplex amplicon quantification (MAQ), are required. The setting of these two methodologies requires using DNA with a high degree of integrity and a certain number of control samples; therefore, choosing reference samples is very important. Although apparently overlapping, these two technologies have peculiar characteristics making them complementary. Therefore, both should be used to thoroughly understand the CNV (copy number variation) status of BRCA1/2 genes.

The integration of NGS and MLPA/MAQ technologies allows an excellent coverage of all diagnostic needs related to the BRCA1/2 test. In some cases, however, it is necessary to resort to an analysis by CGH array (comparative genomic hybridization array) to characterize large rearrangements. The latter generally occurs in about 2% of patients analyzed and more frequently in BRCA1 due to the presence of the BRCA1 pseudogene, which has a homology of about 98%.

Finally, one of the significant problems for laboratory medicine professionals concerns the classification of variants of uncertain significance (VUS) that frequently occur in BRCA1/2. For a correct evaluation of such VUS, referring to the criteria of the Evidence-based Network for the Interpretation of Germline Mutant Alleles (ENIGMA; https://enigmaconsortium.org/) is necessary. It is important to follow these criteria because studies in mouse models have shown that variants in BRCA1/2 associated with increased risk of developing cancer are not necessarily indicators of better response to PARP-1 inhibitor therapy. It is, therefore, necessary for laboratories to develop very robust and sufficiently documented evaluation systems (in silico, in vitro, and/or in vivo) regarding the models of interpretation of BRCA1/2 variants, especially for a correct stratification of both the risk of disease and the possibility of access to targeted therapies. The laboratory should ensure a continuous update of the status of VUS to update the classification of these variants continuously.

Therefore, laboratories performing BRCA1/2 testing must have highly qualified personnel, complementary technologies (NGS, MLPA/MAQ, cGH array), and a bioinformatics infrastructure adequate to the test and workload. Finally, participation in recognized external quality assessment programs (EQA, EMQN) and rigorous internal testing controls guarantee the maintenance of the standards required by the significant accreditation systems (e.g., ISO15189, CPA).

Early Diagnosis

Molecular tests contribute to screening protocols aimed at the early diagnosis of diseases, especially in the neonatal period (neonatal screening) (see Chap. 31).

Diagnosis of Disease

To date, there are about 1800 clinical conditions for which the disease gene has been identified (mainly monogenic diseases), and it is possible to carry out molecular analysis, i.e., the search for mutations to confirm or exclude the presence of the disease. However, at present, only about a hundred diseases are included in the diagnostic pathways because of molecular diagnosis:

- In many genetic diseases, other laboratory tests with high diagnostic sensitivity and specificity allow rapid diagnosis (e.g., in α1-antitrypsin deficiency, the diagnosis is based on the nephelometric assay of the protein; Wilson's disease is currently diagnosed through the ceruloplasmin assay, etc.).
- Considering the complexity and costs of setting up molecular diagnostics for each disease (need to participate in international quality control programs, to analyze positive control samples at the same time as the patient sample, etc.), in many laboratories, molecular analysis is offered only for diseases that have a higher frequency, with the risk that there are no centers able to offer the conclusive diagnosis of many rare diseases (orphan diseases).
- The opportunities offered by the molecular diagnosis of genetic diseases are little known to many physicians.
- Some diseases are severe and there is no time to complete the diagnostic pathway.

Most hereditary genetic diseases are monogenic, i.e., there is a correspondence between a gene-disease and a clinical form (cystic fibrosis, hemophilia A and B, β-thalassemia, Duchenne dystrophy, etc.). In other cases, several genes-diseases (even some dozens) can cause the same syndrome (or very similar clinical forms), as in the case of congenital heart disease. Considering the high number of genetic diseases for which molecular analysis is currently performed, in the following, we will focus on some examples discussing the common aspects of all genetic diseases: cystic fibrosis, an example of autosomal recessive disease with a remarkable phenotypic heterogeneity; hemophilia A and B, examples of recessive X-linked diseases.

The Example of an Autosomal Recessive Disease: Cystic Fibrosis

Cystic fibrosis (CF) is a chronic, developmental, systemic autosomal recessive genetic disease. It is the most common severe genetic disease in the Caucasian population, with an incidence of approximately 1:2500. The disease is due to the alteration of a membrane protein called CFTR (cystic fibrosis transmembrane regulator) that regulates sodium and chlorine exchanges between the inside and outside of the cell. Dysfunction of the protein causes denser secretions (the disease was once called mucoviscidosis), leading to chronic lung damage sustained by a vicious cycle of bacterial infection by opportunists and immune response with the release of chemical mediators of inflammation and lytic enzymes. Ninety percent of patients has pancreatic insufficiency (due to gradual obstruction of the pancreatic ducts resulting in pancreatic fibrosis), for which it is necessary to take pancreatic enzymes to promote digestion; in addition, hepatopathy, diabetes, and congenital bilateral aplasia of the vas deferens (CBAVD) in males may be present. The mode of onset, severity, and course of the disease are variable. Some patients have early onset with severe disease symptoms; others have modest respiratory symptoms with a regular digestive picture. Discordance of clinical CF expression is also present in patients with the same CFTR genotype or pairs of siblings with the disease. This has prompted a search for genes inherited independently of CFTR that could act as modulators of clinical expression of the disease.

In addition to the classical form of CF, there are attenuated phenotype forms, characterized by modest clinical expression and normal pancreatic function, often identified in adult patients; in these cases, the disease affects only one organ, such as the vas deferens with CBAVD causing infertility in males or the pancreas with episodes of recurrent pancreatitis. These forms are now called CFTR-related diseases (CFTR-RD).

The diagnosis of CF is based on typical symptoms or familiality (sick siblings), which must be associated with increased chlorine levels in sweat (positive sweat test), or identification of two mutations in the CFTR gene. However, neonatal screening identifies almost all patients with a classical form of the disease at birth. In CFTR-RD, with a negative or borderline normal sweat test, molecular analysis becomes the only diagnostic tool.

Molecular testing of the CFTR gene should always be combined with appropriate multidisciplinary counseling, and the molecular diagnostic laboratory should work closely with the clinician to ensure that the appropriate tests are performed for each patient. Guidelines suggest performing sweat and molecular testing for CF in some cases (Table 42.3).

Table 42.3 Conditions in which molecular testing for cystic fibrosis is suggested

Classic clinical suspicion of cystic fibrosis (patient with typical symptoms, affected siblings, positive neonatal screening, or positive sweat test)
"Atypical" clinical symptoms and sweat test at the limit of the normality
Male infertility in the presence of congenital bilateral agenesis of the vas deferens
Adults with suspicion of other forms of CFTR-RD
Pregnancies in which the fetus has a hyperechoic intestine and/or dilation of the intestinal loop

CFTR-RD CFTR-related diseases

More than 2000 different mutations have been recognized in the CFTR, a medium-sized gene (27 exons). Some of these mutations are more widespread; others are specific to geographical areas or ethnic groups, and most are rare and specific to a few patients. The methods used for the molecular diagnosis of CF can be divided into level I and level II analyses. Level I tests evaluate a restricted panel of the most frequent mutations. In negative cases, methods called scanning (e.g., gene sequencing) are used, which analyze the whole gene (level II tests).

Level I molecular analysis are based on commercial kits or homemade methods that foresee the analysis of the most frequent mutations in the reference region of the laboratory; the detection rate of the first-level molecular analysis is linked to the panel of mutations sought and to the ethnic origin of the analyzed patient.

Level II molecular analysis uses scanning systems that allow the recognition of mutations in coding regions and the border regions between introns and exons involved in the splicing process. The most widespread technique is sequencing by automated instruments. Level II tests reach a higher detection rate (about 95%), but the clinical significance of the molecular result is not always easy to interpret because sometimes new mutations are identified for which it is difficult to establish the causative effect. In CFTR-RD patients, the mutations present are often different from those detectable by commercial kits, so a level I analysis has a lower diagnostic sensitivity than in classical CF patients; further investigation by gene sequencing is almost always necessary in these patients.

Once a new patient with CF or CFTR-RD has been identified, in addition to referral to a clinical referral center, it is important to carry out multidisciplinary family counseling and cascade mutation testing on blood relatives. In the case of couples in which both members are carriers of mutations, the opportunity of prenatal diagnosis can be offered.

The Example of Diseases Related to the X Chromosome: Hemophilia A and B

Hemophilia A (HA, due to FVIII deficiency) and hemophilia B (HB, due to FIX deficiency) are the most frequent congenital coagulation diseases (incidence 1:5000 and 1:30,000 in newborn males, respectively). They are transmitted as a recessive X-chromosome-related character and manifest clinically in males. In rare cases, hemophilia may occur in the female: (1) by selective inactivation of the non-mutated X chromosome; (2) in females born from a hemophilic father and a carrier mother; (3) in cases of Turner syndrome (X0) in which the hemophilic gene is mutated. Vice versa, the carrier female is asymptomatic. The characteristics of hemophilia have been known since ancient times. The Talmud, sacred text of Judaism, prescribed to exempt from circumcision boys who had lost a little brother to hemorrhage during the operation; in the eleventh century, the rule was extended to boys born from the same mother but different father, demonstrating the knowledge of the diagenetic transmission of the disease. Hemophilia was made famous by the descendants of Queen Victoria, some of whom were hemophiliacs—sadly known the story of the little Nikolas, son of the last Tsar of Russia, "cured" by Rasputin; famous the story of one of his sisters (escaped the massacre of the Romanovs) who toured the courts of Europe. Only DNA analysis of the remains of members of the Romanov family, found many years after her death, made it possible to establish that she was a braggart.

The diagnosis of hemophilia is suspected based on symptoms (or familiarity) and coagulation tests (normal prothrombin time [PT] and elongation of activated partial thromboplastin time [aPTT]) and confirmed by analysis of factor VIII and IX activity in plasma. Based on the activity of the deficient factor, hemophilia is classified as A or B and clinically as mild (activity between 5% and 25%), moderate (activity between 1% and 5%), and severe (activity <1%). The severe forms of hemophilia, which constitute more than half of the known cases, manifest themselves with hemorrhages that in 10% of the patients appear already in the perinatal period, putting the patient's survival at risk. The availability of prenatal diagnosis, in the cases in which there is familiarity and the pregnant woman is a carrier, helps to plan the delivery and to face the perinatal complications in the couples that choose to continue the pregnancy after a prenatal diagnosis of hemophilia.

The molecular analysis, in the case of hemophilia, is of little help to the patient (since the diagnosis is made based on traditional coagulation tests) but can identify the carriers (asymptomatic) in the family (mother, sisters, etc.). In fact, in front of a new hemophilic patient, there are two possibilities:

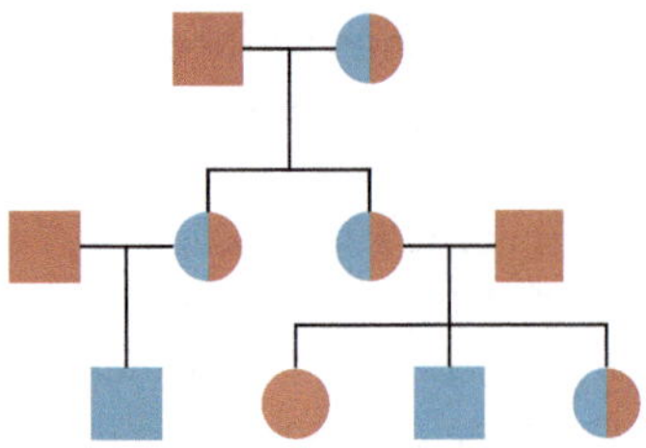

Fig. 42.1 Hemophilia: familiarity. (Copyright EDISES 2021. Reproduced with permission)

- There are other affected members in the family (Fig. 42.1), and, in this case, the mutation in the patient has been inherited from the mother (who is an obligate carrier), and the patient's sisters have a 50% risk of being carriers; in this case, it is helpful to identify the mutation in the patient and assess its presence in the sisters to establish whether they are carriers.
- The patient does not show familiarity (Fig. 42.2). In this case, the mutation may have been inherited from the mother who is a carrier (and therefore, the sisters are at risk), or the mutation may have arisen de novo in the patient, so that the mother is not a carrier and the sisters are not at risk of being carriers (and it is not necessary to test them for the mutation).

The hemophilia A gene includes 26 exons, while the F9 gene that encodes for factor IX includes 8 exons. Both forms of hemophilia are due to a myriad of different mutations, because the X chromosome, where the two genes are located, is much more susceptible to de novo mutations than the other chromosomes. This means that molecular analysis for HB in a new patient must be based on scanning all gene coding regions by sequencing (as shown in the two examples in Fig. 42.3 showing nonsense and a missense mutation, respectively). For HA, there is a difference between the different forms. One-half of severe HA cases are due to a complex rearrangement involving intron 22, called intron inversion (InvIVS-22), which can be analyzed by long-distance polymerase chain reaction (PCR). Approximately 5% of cases of severe HA have a similar mutation in intron 1 (intron 1 inversion), and the remaining 45–50% of cases, as well as all cases of mild to moderate HA, are due to different mutations involving the entire gene and, therefore, require sequencing analysis.

However, even using sequencing of the entire gene coding region, there is a small percentage (varying from case to case) of hemophilic patients in whom the mutation is not identified. A proportion of these patients may have extensive gene deletions, which may escape sequencing analysis and require quantitative PCR analysis. Even after rearrangement analysis, there is a residual number of hemophilic patients who do not show mutations: about 2% for HB and about 5% for HA. Presumably, these patients have mutations in noncoding areas, even far from the gene.

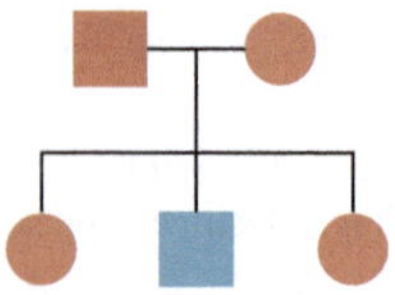

Fig. 42.2 Hemophilia: unfamiliarity. (Copyright EDISES 2021. Reproduced with permission)

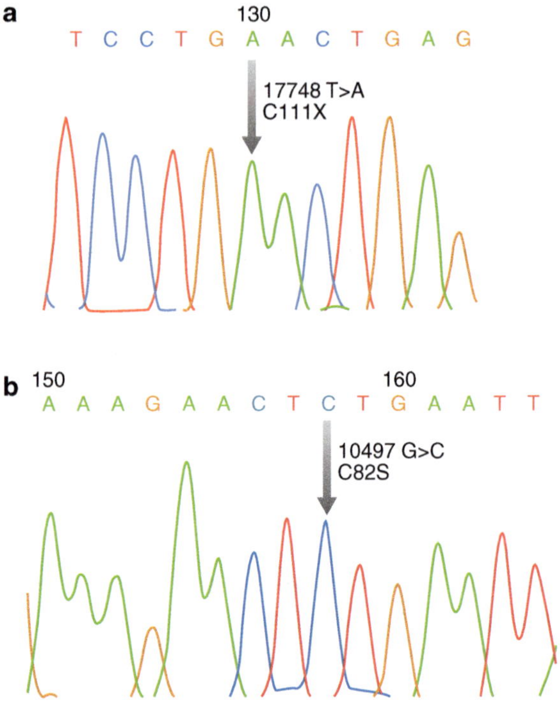

Fig. 42.3 Scanning of the coding regions of the gene by sequencing: (**a**) nonsense mutation and (**b**) missense mutation

A Critical Issue in Molecular Testing: Identification of New Mutations

The development of sequencing technologies for the analysis of disease genes has favored a substantial increase in the diagnostic sensitivity of the molecular analysis. However, it may happen to identify new mutations, never previously described, for which it is difficult to establish the pathogenetic role, and there are no univocal guidelines to follow. A first approach can be based on the type of mutation. If the mutation is nonsense or frameshift or involves the first nucleotides of the junction between exon and intron (involved in splicing) or finally a gene rearrangement (e.g., macrodeletions or macro-duplications), the probability that it is causative of disease is very high. Conversely, missense or intronic mutations further away from the junction may have no effect. A next step would be to search for the mutation in a fair number of unaffected subjects. If the mutation is present with a high frequency in the general population, the probability that it is pathogenic is reduced. Obviously, this study is expensive, and few laboratories have banks of biological samples. Among other things, healthy subjects should share the ethnicity of the patient in whom the new mutation was found, and this, in countries where globalization has provided a multi-ethnic population, is complicated. Alternatively, some databases have sequenced thousands of subjects, and then one can check the mutation frequency in these banks. Subsequently, one can refer to computerized prediction systems, which provide a score on the possible effect of the mutation based on the degree of phylogenetic conservation of the substituted amino acid or regarding the possible effect of the identified variant on splicing for intronic mutations. These prediction systems provide immediate but sometimes conflicting and inconclusive results; therefore, it is always necessary to use a minimum of three to evaluate each variant. These results cannot be used for reporting purposes but only to indicate whether or not to continue with subsequent in vitro studies. The last point is the experimental study of the effects of a mutation using in vitro or ex vivo systems. These studies are complex, expensive, and within reach of a few laboratories.

Bioethical Aspects: Late-Developing Diseases

In some cases, molecular diagnosis may concern diseases for which the development of symptoms occurs at an advanced age, and at present, there is no effective therapy. For example, the genetic analysis would allow – already at birth – to demonstrate alterations in the gene-disease of Huntington's chorea, a severe degenerative disease of the central nervous system that occurs around the fourth-fifth decade of life and for which no therapeutic approach has so far proved useful. In these cases, an early genetic diagnosis is of little therapeutic use (its only usefulness, as things stand, would be to inform the carrier of the risk of transmitting the mutation to possible offspring). On the other hand, early diagnosis would inevitably lead to a situation in which the patient knows for certain that will develop a serious and incurable disease, thus triggering devastating psychological consequences. In this and similar cases, significant ethical problems arise, as well as problems with genetic counseling and diagnostic communication to the patient and family. In countries where private insurance systems regulate healthcare, much more complex problems of an economic nature also begin to develop.

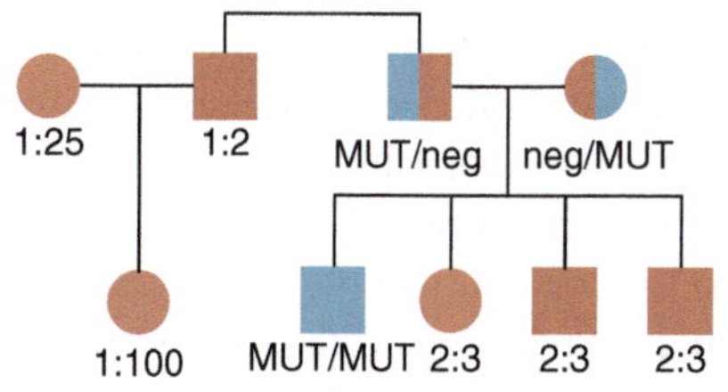

Fig. 42.4 Cystic fibrosis. (Copyright EDISES 2021. Reproduced with permission)

Carrier Diagnosis

For hereditary monogenic diseases, in addition to the analysis aimed at identifying affected patients (preferably at an early stage or asymptomatic), it is possible to carry out investigations aimed at finding healthy carriers, with two distinct modalities: cascade analysis and carrier screening.

Cascade Analysis

The first case is represented by the cascade analysis. Typically, an affected patient is identified, disease-causing mutations are identified, and then the search for mutations is extended to blood relatives based on the type of Mendelian transmission. For example (Fig. 42.4), in an autosomal recessive disease, such as cystic fibrosis, we study the patient's siblings (each of whom has an a priori 2:3 risk of being a carrier), then the parents' siblings (each of whom has an a priori 1:2 risk of being a carrier) and, in case of positivity, the children, and so on with a "cascade" mechanism. When a carrier is identified, it is essential to perform molecular analysis in the partner who has an a priori risk of 1:25 of being a carrier, as all subjects of the general population (in this case, not only the mutations of the family are analyzed, but an extended molecular analysis is performed). A targeted multidisciplinary genetic counseling must accompany each step of the molecular analysis.

Carrier Screening

For some diseases that are particularly frequent in specific geographical areas or ethnic groups with a strong tendency to intermarriage, carrier screening is possible even if there is no case in the family. For example, among Ashkenazi Jews, carrier screening (usually premarital) is performed for some more frequent diseases, such as Tay-Sachs disease and cystic fibrosis; in Italy, pregnancy guidelines suggest screening for thalassemia (in this case, screening is not based on mutation detection but hemoglobin analysis by chromatography or electrophoresis). For more than 10 years in some Italian regions (including Veneto), carrier screening (also in this case, in the premarital period) for cystic fibrosis has been started based on the analysis of panels of mutations most common in both partners and on the possibility of prenatal diagnosis if both partners are carriers. Preliminary data seem to indicate a reduction in disease incidence in screened areas. However, examples of carrier screening are few so far, both because the costs are still high and especially because screening should be accompanied by proper multidisciplinary counseling for couples, and, in large numbers, this is difficult to achieve.

Table 42.4 Incidence of some chromosomopathies

Chromosomopathy	Incidence
Trisomy 21 (Down syndrome)	1: 700
XXY (Klinefelter syndrome)	1: 1000
Triple X	1: 1000
Trisomy 18 (Patau syndrome)	1: 3000
Trisomy 13 (Edwards syndrome)	1: 5000
X0 (Turner syndrome)	1: 5000

The incidence is indicative since in some cases it increases with the age of the pregnant woman

Prenatal Diagnosis

Prenatal diagnosis of genetic diseases can be performed for two groups of diseases:

- Non-hereditary genetic diseases, such as Down's syndrome and other chromosomal diseases, in high-risk couples, such as pregnant women of advanced age or couples who have a family history of these diseases
- Inherited genetic diseases (usually monogenic) in couples at high risk because they already have an affected child or identified through carrier screening programs

Following are the differences between the two conditions:

- For non-hereditary genetic diseases, there is usually no evident familiarity, whereas for monogenic diseases, Mendelian-type inheritance is observed; therefore, it is often possible to identify at-risk couples before pregnancy.
- For non-hereditary genetic diseases, technologies aim to detect more or less extensive alterations of chromosomes, while for hereditary genetic diseases, technologies are used to search for mutations in single (or few) well-defined disease genes.

Prenatal Diagnosis of Non-hereditary Genetic Diseases

As shown in Table 42.4, some chromosomal diseases have a relatively high incidence, so they are the object of prenatal diagnosis, especially in conditions of high risk (advanced age of the pregnant woman, familiarity with the disease, suggestive fetal ultrasound signs). Until a few years ago, prenatal diagnosis of chromosomal diseases was performed exclusively in cells collected by amniocentesis (usually in the fifth month of pregnancy) through the classic analysis of the karyogram. The amniocytes are placed in culture; after a few days, cell growth in metaphase is stopped, and the chromosomes are studied under the microscope after staining, evaluating numerical or structural abnormalities. In addition to classic chromosome staining, more sophisticated techniques can be used, such as banding, which allows better resolution of the image of individual chromosomes and major defects (deletions, translocations), or gene probes associated with fluorescence detection systems that bind specific regions of individual chromosomes and allow their specific recognition (fluorescent in situ hybridization, FISH). Subsequently, chromosome analysis by karyogram was extended to chorionic villi, allowing prenatal diagnosis by chorionic villus sampling (in the first trimester of pregnancy). Amniocentesis and chorionic villus sampling, even in experienced hands, maintain a minimal risk of causing irreversible damage to the fetus or mother (around 1% overall), so they are currently defined as invasive prenatal diagnoses.

Alternatively, non-invasive diagnostic approaches are available. The first group of methods involves the analysis, at a specific time of pregnancy, of some markers in the maternal serum whose concentrations, assessed together with other parameters (age of the pregnant woman, family history of chromosomal diseases, etc.), allow an estimation of the risk of Down's syndrome for the fetus. When the risk exceeds a threshold, it is suggested to perform an invasive prenatal diagnosis.

More recently, non-invasive screening molecular tests have become available, defined by the acronym NIPT (non-invasive prenatal testing) based on the analysis of fetal DNA in maternal blood (which begins to appear from the fifth week of pregnancy). These tests are based on massive sequencing analysis of the entire fetal genome or comparative genomic hybridization techniques (see section "Preimplantation Diagnosis"), followed by analyzing the results using sophisticated bioinformatics programs. These technologies allow the identification of aneuploidies (numerical alterations of chromosomes) and structural alterations (unbalanced translocations, deletions, duplications, etc.) associated with the loss or gain of DNA. Some variants of these tests allow "molecular karyotyping" to be performed by massively parallel sequencing. Overall, these tests can identify more than 95% of all chromosomal alterations detectable by traditional karyogram.

Prenatal Diagnosis of Hereditary Genetic Diseases

In recent years, the demand for prenatal diagnosis of inherited genetic diseases has dramatically increased for several reasons:

1. Sampling techniques have become safer.
2. The number of diseases that can be diagnosed prenatally has increased due to the continuous identification of new disease-genes.
3. Analytical technologies have been developed allowing rapid and cost-effective diagnosis of the small amounts of DNA that can be obtained from prenatal sampling.
4. The improved knowledge among medical professionals of the opportunities offered by prenatal diagnosis has contributed to the increase in requests.

Moreover, alongside more widespread and well-known genetic diseases, such as cystic fibrosis or thalassemias, for which prenatal diagnosis has been routinely performed for several decades, there is now an increase in requests for prenatal diagnosis for rare diseases, such as those of amino acid metabolism, organic aciduria, and β-oxidation diseases, which are identified through the extensive screening.

Prenatal diagnosis, therefore, has become a valuable aid for the reproductive choices of many couples, provided that they have a range of complete, precise, and objective information that makes it possible to make informed choices. Therefore, the prenatal diagnosis must be accompanied by multidisciplinary pre-test and post-test counseling offered by professionals willing to give couples all the time they need. At the same time, it should be remembered that prenatal diagnosis is a medical act and, as such, should be used only when there are appropriate indications, without wanting to persist in "medicalizing" excessively the pregnancy that, in most cases, does not indicate the prenatal diagnosis. Also, because there are thousands of hereditary genetic diseases, only a small percentage can be diagnosed prenatally. Therefore, in many cases, couples who already have a child with severe genetic disease use prenatal diagnosis for hereditary genetic diseases, and usually the couples with the greatest desire to have a second child. In these couples, there is a high risk (already assessed in the couple through molecular analysis and multidisciplinary counseling) that a second affected child will be born.

From the methodological point of view, in almost all couples who perform prenatal diagnosis for hereditary genetic diseases, the mutation or mutations in the family are already known and are sought on DNA extracted from chorionic villi or amniocytes (possibly cultured). Before searching for pathogenic variants, however, it is necessary to analyze some polymorphisms located on different chromosomes to (1) confirm paternity from the fetal DNA sample since only parental mutations should be sought on the fetal DNA sample; (2) exclude that the fetal DNA sample is contaminated with maternal DNA; in cases of relevant contamination (Fig. 42.5), in which the fetal DNA sample shows small amounts of an allele of maternal origin, a new sample must be collected. However, this is very rare because before analysis, a macroscopic and microscopic cleaning of the fetal sample is carried out; (3) to define the sex of the unborn child. In the case of recessive X-linked diseases, such as hemophilia A or B, in the case of a female unborn child (who may be healthy or a healthy carrier), the search for mutations in the fetal DNA can be avoided.

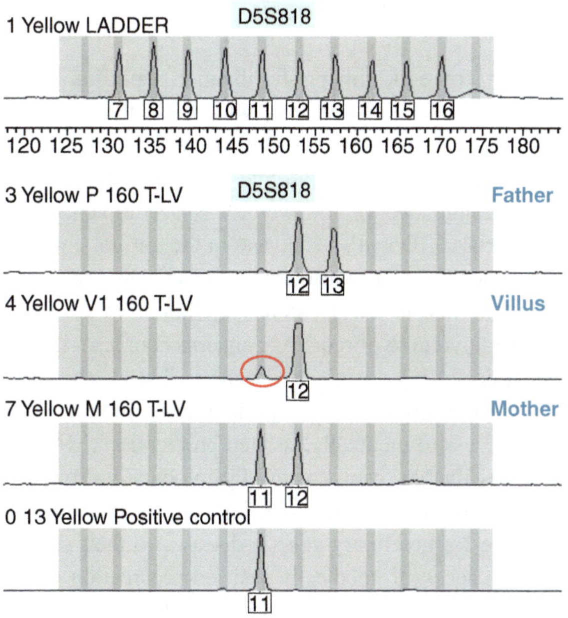

Fig. 42.5 Contamination analysis

Also, for hereditary genetic diseases, the analysis of fetal DNA in maternal blood offers new opportunities. It is impossible to replace fetal DNA analysis with invasive analysis because even if technologies are used to concentrate fetal DNA, a quantity of maternal DNA is still present in the maternal blood sample. Therefore, when a positive signal is obtained for the maternal mutation, it is impossible to ascribe it with certainty to the fetal sample. However, fetal DNA analysis may also offer some contributions to the prenatal diagnosis of inherited genetic diseases:

- In recessive X-linked diseases, Y-chromosome polymorphisms can be tested on fetal DNA; if they are absent, the fetus is female, and invasive prenatal diagnosis can be avoided.
- In cases of autosomal recessive diseases in which the maternal and paternal mutations are different, only the paternal mutation can be tested in fetal blood; if it is absent, invasive prenatal diagnosis can be avoided because the fetus has only a 50% risk of carrying the maternal mutation.

Pre-implantation Diagnosis

The reasons for requesting pre-fertilization or pre-implantation diagnosis coincide, in principle, with those that lead to the request for prenatal diagnosis. One part of the requests concerns the search for chromosomal anomalies by couples in which the high age of the woman is accompanied by an increased risk of Down's syndrome in the unborn child or by women in whom there have been repeated miscarriages suspected to be due to chromosomal alterations in the fetus. A second large volume of requests concerns couples with a high risk of transmitting an inherited genetic disease to their offspring. Until a few years ago, the legislation in force in Italy allowed access to medically assisted procreation (MAP) only for couples unable to procreate physiologically. This means that most couples at high risk of transmitting mutations responsible for hereditary genetic diseases to their offspring, not having access to medically assisted procreation (because they were able to procreate), had no possibility to carry out a pre-fertilization diagnosis or pre-implantation diagnosis. After a series of judicial sentences and in analogy with European legislation, today, all couples who request it have access to MAP and, consequently, to pre-implantation diagnosis. Therefore, couples at risk of genetic diseases can choose between the prenatal diagnosis pathway (invasive or non-invasive) and that of MAP with pre-implantation diagnosis. In these couples, the pre-test consultation is highly delicate because, in addition to the more strictly analytical aspects, it is necessary to make the couples understand that the path of MAP is not free from risks, including ovarian hyperstimulation syndrome. Women who resort to MAP receive an intense hormonal stimulation (which may not be tolerated due to side effects) aimed at inducing the production of numerous oocytes, among which will be selected those that morphologically appear more suitable for fertilization. There are ethical and religious reasons on the other side of the scale. Although prenatal diagnostic techniques are now commonly used for diagnosing genetic abnormalities at the fetal level (and non-invasive diagnosis is a further substantial contribution), if a sick fetus is detected, many couples will have the choice of continuing the pregnancy or therapeutic interruption. Only for very few diseases, in fact, in utero therapy is available (e.g., adrenal hyperplasia from 21-hydroxylase deficiency) or therapies that can be implemented from birth to counteract the effects of the disease (e.g., hemophilia can be treated from birth with recombinant deficient factor). Many couples at risk of genetic disease are forced to face repeated interruptions of pregnancy before they can have a healthy baby. Therefore, the possibility of choosing a prenatal diagnosis greatly benefits these couples, avoiding their recourse to therapeutic abortion. It must also be considered that, in some populations, prenatal diagnosis is not accepted because of the ethical/moral, or religious problems associated with the termination of pregnancy.

From a methodological point of view, the pre-implantation diagnosis does not offer particular difficulties for an advanced laboratory of clinical molecular biology. Once the in vitro fertilization is obtained, two cells (blastomeres) are taken from the embryo in the early stages of development, whose DNA will be analyzed. With the latest MAP protocols, it is also possible to take the sample later, obtaining more cells for analysis. After analysis, unaffected embryos can be implanted.

Depending on the type of diagnosis to be made, the most appropriate technology will be used. For example, to search for parental mutations in the gene-disease in the case of hereditary genetic diseases, the same methods used in prenatal diagnosis will be used (amplifying the gene of interest by PCR and sequencing to identify any mutation). On the other hand, the new technology of comparative genomic hybridization on microarrays (CGH-array) can be used to analyze chromosomal alterations. This technology makes it possible to identify all alterations linked to an increase or reduction in DNA content (trisomies and other aneuploidies, deletions, duplications, balanced translocations), which account for about 95% of all chromosomal alterations. In short, the entire DNA extracted from embryonic cells is analyzed compared to DNA from cells obtained from a healthy subject. The DNA is hybridized by a series of probes adsorbed on a tiny solid support that bind all the genome traits, labeled with different fluorochromes (for embryonic cells and DNA from healthy cells). Then, the image is acquired by a scanner and processed by a bioinformatics system. For each stretch of hybridized DNA, a light signal will be generated whose intensity will be proportional to the amount of DNA. By comparing the signal between the healthy sample and the sample from embryonic cells, it will be determined for each tract how many copies are present (higher signal in all probes that hybridize chromosome 21 in case of trisomy 21, the weaker signal in case of deletion of a tract, etc.).

Disease Staging: The Example of Circulating Tumor Cells

The contribution of laboratory medicine to cancer diagnostics is based on two groups of biomarkers:

- Traditional tumor markers analyzed in serum, such as AFP, carcinoembryonic antigen (CEA), and CA125. These markers do not contribute to the diagnosis of the disease but are used to monitor the patient during treatment.
- Molecular markers can provide diagnostic support at different stages of the disease. These include markers used to identify circulating tumor cells (CTCs). The search for circulating tumor cells is an approach initiated more than 20 years ago, based on the idea that such cells could sig-

nal the beginning of the metastatic process in patients with neoplasia.

Together with cellular markers, searching for mutations peculiar to the tumor (K-ra, BRAF, etc.) and giving access to target therapies directly on free circulating tumor DNA may further evaluate the tumor molecular phenotype.

The presence of distant metastases (perhaps not yet detectable by imaging methods) can radically change the therapeutic approach to the cancer patient. Today, for several neoplasms, the search for CTCs is included in the diagnostic protocols (also defined as liquid biopsy) and, obviously, their presence (independently from the metastatic process) is considered an unfavorable prognostic sign.

From the analytical point of view, the detection of CTCs can be performed by two approaches, both with very high analytical sensitivity (ability to identify one neoplastic cell out of about one million blood cells): (1) immunological, i.e., selecting CTCs by cytofluorimetry based on the presence of specific surface markers of the tumor cell and (2) molecular, i.e., analyzing by real-time PCR (RT-PCR) specific messenger RNAs produced by the neoplastic cell. Since the number of CTCs is small in both cases, it is helpful to perform an enrichment prior to the analysis to concentrate the CTCs. This enrichment is performed by filtering the blood sample on magnetic beads that expose an antibody targeted to an antigen specific to neoplastic cells (positive selection) or an antigen specific to normal cells (negative enrichment). Both analytical approaches also allow the quantification of CTCs, whose levels have a prognostic value.

Finally, the use of digital analysis technologies (digital PCR) will contribute in the future to quantifying markers associated with minimal residual disease in an increasingly precise manner. These technologies are today mainly used in translational research but promise to enter the routine practice of the diagnostic laboratory very soon.

Chronic Myeloid Leukemia, BCR-ABL Gene, and Disease Monitoring

An example of the use of clinical molecular biology in disease monitoring during therapy and therapeutic selection is described below.

Chronic myeloid leukemia (CML), typical in older adults, is a monoclonal expansion of bone marrow stem cells. The disease is due to an acquired genetic alteration, termed Philadelphia chromosome (described in 1961 in this city), which causes translocation of gene material between the long arm of chromosome 9 and the short arm of chromosome 22. The translocation causes the formation of a fusion gene (BCR-ABL) that encodes an oncogene with tyrosine kinase functions that promotes uncontrolled stem proliferation. The disease is characterized by a chronic course that can evolve toward a blastic phase. Until a few years ago, the diagnosis of CML was morphological (peripheral blood smear and marrow analysis), made more specific by the analysis (first cytological and then cytofluorimetric) of some specific surface markers of CML cells. Cytogenetics (in vitro culture of cells from peripheral blood or marrow and analysis of the karyogram), through the identification of the Philadelphia chromosome, has made the diagnosis more specific. Subsequently, clinical molecular biology technologies allowing BCR-ABL transcript analysis (i.e., messenger RNA analysis by RT-PCR) favored the strong increase of the analytical sensitivity, allowing recognition of a cell expressing the fusion gene among more than 100,000 normal cells. In the meantime, a class of molecular drugs capable of selectively inhibiting the BCR-ABL protein has been developed, drastically changing the disease's prognosis. At present, BCR-ABL transcript analysis has three main applications:

- Diagnosis of disease by searching for the transcript in circulating blood or bone marrow cells.
- Monitoring during therapy, which allows the identification of minimal residual disease. Currently, we speak of complete cytogenetic remission to indicate patients in whom the cytogenetic analysis for the Philadelphia chromosome is negative; significant molecular remission, when the molecular analysis (more sensitive than cytogenetics) shows a residue of positive cells below a threshold; complete molecular remission, when the molecular analysis is negative.
- Identification of therapy-resistant patients. During therapy with a tyrosine kinase inhibitor, some patients with CML develop resistance to therapy due to the development of somatic mutations in the BCR-ABL gene that make the protein resistant to the inhibitor. In these cases, there is a recurrence (also molecular) of the disease. In these patients, it is possible to analyze the gene in the leukemic cells (by PCR and sequencing) and identify the mutation and the percentage of mutated cells, planning to use a different inhibitor.

In addition to chronic myeloid leukemia, aberrant transcripts from fusion genes or other genetic abnormalities have been identified in several other hemolymphoproliferative diseases. In analogy to CML, analysis of these transcripts may be helpful in monitoring the patient during therapy.

Recommended Readings

Capoluongo E, Plebani M (2014) Circulating fetal cell-free DNA and prenatal molecular diagnostics: are we ready for consensus? Clin Chem Lab Med 52(5):609–611

Castaldo G, Scorza M, Elce A, Giordano S, Liguori R, Guerra G (2013) Omics in laboratory medicine. J Matern Fetal Neonatal Med 26 Suppl 2:13–16. https://doi.org/10.3109/14767058.2013.829694. Review

Drost R, Jonkers J (2014) Opportunities and hurdles in the treatment of BRCA1-related breast cancer. Oncogene 33:3753–3763

Hehir-Kwa JY, Claustres M, Hastings RJ, van Ravenswaaij-Arts C, Christenhusz G, Genuardi M, Melegh B, Cambon-Thomsen A, Patsalis P, Vermeesch J, Cornel MC, Searle B, Palotie A, Capoluongo E, Peterlin B, Estivill X, Robinson PN (2015) Towards a European consensus for reporting incidental findings during clinical NGS testing. Eur J Hum Genet 23(12):1601–1606

Maloy S, Hughes K (2013) Brenner's encyclopedia of genetics, Second edn. Elsevier

Paolillo C, Mu Z, Rossi G, Schiewer MJ, Nguyen T, Austin L, Capoluongo E, Knudsen KE, Cristofanilli M, Fortina P (2017) Detection of activating estrogen receptor gene (ESR1) mutations in single circulating tumor cells. Clin Cancer Res 23(20):6086–6093

Pinto C, Bella MA, Capoluongo E et al (2016) Recommendations for the implementation of BRCA testing in the care and treatment pathways of ovarian cancer patients. Future Oncol 12:2071–2075

Santonocito C, Scapaticci M, Guarino D, Bartolini A, Minucci A, Concolino P, Scambia G, Paris I, Capoluongo E (2017) Identification of twenty-nine novel germline unclassified variants of BRCA1 and BRCA2 genes in 1400 Italian individuals. Breast 36:74–78

Scaglione GL, Concolino P, De Bonis M, De Paolis E, Minucci A, Ferrandina G, Scambia G, Capoluongo E (2018) A whole germline BRCA2 gene deletion: how to learn from CNV in silico analysis. Int J Mol Sci 19(4)

Wallace AJ (2016) New challenges for BRCA testing: a view from the diagnostic laboratory. Eur J Hum Genet 24 Suppl 1:S10–S18

Zarrilli F, Sanna V, Ingino R, Santamaria R, Rocino A, Coppola A, Di Minno G, Castaldo G (2013) Prenatal diagnosis of haemophilia: our experience of 44 cases. Clin Chem Lab Med 51(12):2233–2238

Fever of Unknown Origin

43

Marcello Ciaccio, Luisa Agnello, and Anna Maria Ciaccio

Introduction

The term fever of unknown origin (FUO) indicates febrile illnesses with a temperature above 38.3 °C for which an etiological diagnosis has not yet been formulated after 3 days of hospitalization or after three specialist visits. Certainly, in this context, the clinical laboratory provides essential elements to guide the etiological diagnosis.

Definition of Fever

Fever is defined as an increase in body temperature associated with an increase in the hypothalamic set point (e.g., from 37 to 38 °C). The control of body temperature occurs by thermoreceptors, located on the skin and defined as peripheral thermoreceptors, which, together with the deep thermoreceptors in the hypothalamus and the large veins, send information to the temperature control centers, present at the level of the hypothalamus itself. The maintenance of body temperature occurs by the implementation of mechanisms aimed at balancing circadian variations; the excess of heat is balanced with dissipation through cutaneous vasodilation and sweating, while the lowering of temperature is balanced with the increase of heat production, mainly at a muscular level through shivering and the reduction of heat loss through vasoconstriction.

It is essential to distinguish between fever and hyperthermia, which are characterized by elevated body temperature.

Hyperthermia is a condition in which the uncontrolled increase in body temperature exceeds the body's ability to disperse heat and is potentially fatal. Furthermore, in hyperthermia, there is no alteration in the regulation of the thermoregulatory center at the level of the hypothalamus. Hyperthermia may be due to a primitive increase in endogenous heat production, as in the case of hyperthyroidism or heat stroke, or to an altered ability to disperse heat, as, for example, in the case of anhidrotic ectodermal dysplasia.

On the other hand, fever is defined as an elevation of body temperature beyond normal circadian variations due to alterations in the hypothalamic regulatory center. Fever is generally due to the action of pyrogens (any substance that causes fever) and, rarely, to cerebral alterations (brain tumors that compress the hypothalamus). Pyrogens can be of exogenous or endogenous origin. Most exogenous pyrogens are microorganisms, and their products, including toxins, bacterial endotoxins, and substances that form the cell membrane of bacteria, are the most important exogenous pyrogens. Most exogenous pyrogens induce fever by stimulating the synthesis of endogenous pyrogens by monocyte-macrophages. Specifically, endogenous pyrogens are specific cytokines, including interleukin-1 (IL-1), interleukin-6 (IL-6), tumor necrosis factor-alpha (TNF-alpha), ciliary neurotrophic factor (CNTF), and interferon alpha (IFN-alpha). Cytokines play a crucial role in the pathophysiology of fever by inducing the synthesis of prostaglandin E2 (PGE2) in specific brain areas and inducing metabolic, endocrinological, and immunological changes characteristic of fever. PGE2, through interaction with specific receptors present in the pre-

M. Ciaccio (✉)
Department of Biomedicine, Neurosciences and Advanced Diagnostics, Institute of Clinical Biochemistry, Clinical Molecular Medicine and Clinical Laboratory Medicine, University Hospital "P. Giaccone", Palermo, Italy

Department of Laboratory Medicine, University Hospital "P. Giaccone", Palermo, Italy
e-mail: marcello.ciaccio@unipa.it

L. Agnello
Department of Biomedicine, Neurosciences and Advanced Diagnostics, Institute of Clinical Biochemistry, Clinical Molecular Medicine and Clinical Laboratory Medicine, University Hospital "P. Giaccone", Palermo, Italy

A. M. Ciaccio
Department of Health Promotion, Mother and Child Care, Internal Medicine and Medical Specialties (ProMISE) "G. D'Alessandro", University Hospital "P. Giaccone", Palermo, Italy

M. Ciaccio (ed.), *Clinical and Laboratory Medicine Textbook*, https://doi.org/10.1007/978-3-031-24958-7_43

optic nuclei of the anterior hypothalamus, which are physiologically deputed to the control of thermoregulation, induces an increase in the equilibrium point (*set point*) of the hypothalamic thermostat; the production and loss of heat adapt to this new *set point*. The fever is, generally, characterized by three phases:

1. Prodromal or ascending phase: this coincides with the start of the production of prostaglandins which induce an alteration in the hypothalamic setpoint, which, in turn, triggers a series of reactions that lead to an increase in body temperature (chills, vasoconstriction, increased basal metabolism).
2. Acme or fastigium phase: lasts throughout prostaglandin production. Hypothalamic neurons maintain the new setpoint temperature, and the subject will have a warm sensation with hot skin, sweating, tachypnea, tachycardia, and headache.
3. Defervescence phase: it begins with a reduction in prostaglandins production. The hypothalamic setpoint returns to the normal temperature value, and a series of mechanisms are activated to lower body temperature (sweating, vasodilation). The defervescence phase can be gradual (by lysis) or immediate (by the crisis).

Based on the temperature value (measured by axillary method), fever can be classified into:

- Low-grade: 37–37.6 °C
- Moderate: 37.7–38.9 °C
- High: 39–39.9 °C
- Hyperpyrexia: > 40 °C

In addition, according to the course of the fever over time, we distinguish:

- Continuous fever: the fever remains high with slight variation.
- Intermittent fever: Characterized by wide fluctuations. In particular, there are one or two rises in temperature during the day, followed by a drop in temperature below 37 °C. This trend is typical of abscesses and sepsis.
- Remitting fever: characterized by daily fluctuations in temperature which, however, remains constantly >37 °C, typical of bacterial and viral infections.
- Recurrent fever: characterized by periods of continuous fever and intervals of days of normal temperature (pyrexia). This pattern is typical of quartan and tertian fever associated with the malaria parasite.
- Undulating fever: periods of 7–10 days alternate with periods without the fever of 7–10 days. This pattern is typical of brucellosis.

Definition of Fever of Unknown Origin (FUO)

Although bacterial and viral infections are the leading causes of fever, many factors can induce it, including neoplasms, rheumatologic diseases, drug treatment, etc. When the cause of fever cannot be identified in a patient who has had an elevated body temperature for several days, it is referred to as a fever of unknown origin.

In particular, the first definition of FUO dates back to 1961 by Petersdorf and Beeson and is based on the following criteria:

1. Fever with body temperature >38.3C° detected on several occasions
2. Duration of fever >3 weeks
3. Inability to make a diagnosis after a week of investigations in an inpatient

About 30 years later, due to the spread of the human immunodeficiency virus (HIV) and the use of immunomodulatory therapies, Durack and Street proposed a new classification of FUO into four categories:

1. Classic: fever with $T \geq$ 38.3C° that persists for >3 weeks without a recognized etiological cause after 3 days in a hospital or at least three outpatient visits.
2. Hospital or nosocomial: fever with $T \geq$ 38.3C° occurring on more than one occasion in a patient hospitalized in intensive care in which there was no infection, present or incubating, at the time of admission and which, after careful evaluation, remains undiagnosed after 3 days.
3. Neutropenic: fever with $T \geq$ 38.3C° found on repeated occasions in a patient with less than 500 neutrophils/µl for which the diagnosis remains uncertain after 3 days of appropriate investigations, including negative microbiological cultures after 48 h.
4. HIV-associated: Fever with $T \geq$ 38.3C° found on repeated occasions in an HIV-infected patient, for a period >4 weeks in an outpatient or > 3 days in an inpatient, if the diagnosis remains uncertain after case evaluation.

Causes of FUO

Table 43.1 summarizes the leading causes of FUO, which can be divided into four groups: infections, neoplasms, noninfectious inflammatory diseases, and mixed diseases. In approximately 10% of adults, no cause of FUO is identified. The most frequent cause of FUO is infection; in patients with HIV infection, the possibility of opportunistic infections (e.g., tubercu-

Table 43.1 Main causes of fever of unknown origin (FUO)

Infections (20–40%)	**Neoplasms (20–30%)**

Bacterial
  Intra-abdominal or pelvic abscesses
  Dental abscesses
  Appendicitis
  Endocarditis
  Sinusitis
  Tuberculosis (extrapulmonary)
  Urinary tract infections.
Viral
  Hepatitis A, B, C, D, E
  Cytomegalovirus
  HIV
  Epstein-Barr virus
Mycosis
  Candidiasis
  Histoplasmosis
  Aspergillosis
Parasitosis
  Leishmaniasis
  Malaria
  Toxoplasmosis
	Malignant
	  Colon cancer
	  Pancreatic cancer
	  Bladder cancer
	  Leukemia
	  Hodgkin's lymphoma
	  Non-Hodgkin's lymphoma
	  Sarcomas
	Benign
	  Renal angiomyolipoma
	  Atrial myxoma

Noninfectious inflammatory diseases (10–30%)	**Mixed (10–20%)**

Diseases of the connective tissue
  Still's disease in adults
  Rheumatoid arthritis
  Systemic lupus erythematosus; granulomatous diseases
  Crohn's disease
  Sarcoidosis
Vasculitis
  Giant cell arteritis
  Polymyalgia rheumatica
	From drugs
	  Antiepileptics (barbiturates; carbamazepine; phenytoin)
	  H2 antihistamines (cimetidine; ranitidine)
	  Antibiotics (carbapenems; cephalosporins; erythromycin; isoniazid; nitrofurantoin; penicillins; rifampicin; sulfonamides)
	Antihypertensives (captopril; hydralazine; hydrochlorothiazide; methyldopa; nifedipine)
	  Antiarrhythmics (procainamide; quinidine)
	  NSAIDs (ibuprofen; salicylates)
	  Heparin
	Factitic fever
	Thromboembolic diseases
	Gout
	Heart attack/tissue necrosis
	Changes in thermoregulation
	  Hypothalamic dysfunction
	  Encephalitis
	  Brain neoplasm
	  Pheochromocytoma
	  Hyperthyroidism
	Hereditary and metabolic diseases
	  Familial Mediterranean fever
	  Adrenocortical insufficiency
	  Fabry disease

losis, atypical mycobacterial infection, disseminated mycosis, and cytomegalovirus) should also be evaluated. However, FUO causes may differ geographically based on regional exposures, economic development, and available diagnostic tools. For example, infection may predominate in developing countries, whereas noninfectious inflammatory diseases and cancer are more common in developed countries.

Diagnosis

The initial evaluation is based on a careful history to identify focal symptoms and associated factors (such as travel, work, family history, exposure to animal vectors, and dietary habits) and on physical examination to identify potential diagnostic clues.

The history should investigate:

- Family history of infectious diseases (TB, HIV, viral hepatitis, etc.), autoimmune or inflammatory diseases, oncohematological diseases, hereditary causes of fever (familial Mediterranean fever), etc.
- Personal history of infections (consider any signs of immunodeficiency, especially for children <1 year old)
- Rule out a recurrence of the episode
- Evaluate associated signs and symptoms
- Duration and type of fever (periodic or continuous)
- Concomitant drug therapies (drug fever)
- Travel to an endemic area

The physical examination should include:

- Evaluation of auxological parameters
- Apparatus examination
- Inspection of the skin and mucous membranes (mucositis, aphthosis, condition of teeth, etc.) and all explorable lymph node stations

After a thorough history and physical examination, there are three different situations:

- Localizing symptoms or signs that were not present at previous visits, were not detected, or were not treated. Therefore, proceed with specific tests to confirm the diagnostic suspicion (Table 43.2).
- In most cases, the initial assessment identifies risk factors that may direct subsequent examinations, such as travel to

Table 43.2 Examples of diagnostic approach in case of suggestive findings

Cause	Potentially diagnostic indications	Diagnostic approach
CMV infection	History of blood transfusion from a CMV+ donor Mononucleosis-like symptoms	Anti-CMV IgM antibody titer
Abscesses	Abdominal or pelvic discomfort Previous surgery, trauma, peritonitis	CT or MRI
Systemic lupus erythematosus	Fatigue, arthralgia, nephritis, alopecia, peripheral edema, cutaneous vasculitis, erythematous lesions in areas exposed to the sun	Evaluation of clinical criteria, ANA and anti-dsDNA antibodies
Tuberculosis	History of exposure to mycobacterium Cough, weight loss, asthenia History of HIV infection	Chest x-ray, tuberculin test, IFN-γ release test Microscopic examination of the sputum for the detection of acid-fast bacilli, PCR for the detection of DNA, culture examination of body fluids (e.g., sputum)
Leukemia	Asthenia, weight loss, bleeding, paleness, bruising, anorexia, bone pain	CBC, bone marrow examination
Deep vein thrombosis	Pain, swelling, redness in the leg	Ultrasound D-dimer test
Drug fever	Fever coincides with medication taking	Drug discontinuation

CMV cytomegalovirus, *CT* computed tomography, *MRI* magnetic resonance imaging, *ANA* anti-nuclear antibodies, *anti-dsDNA* antibody directed against the DNA double helix, *Rx* radiography, *PCR* polymerase chain reaction, *CBC* complete blood count

an endemic area or exposure to possible disease vectors or weight loss not accompanied by anorexia, which could be indicative of infection (Table 43.2).

- If the initial assessment does not reveal clues but only nonspecific information, a series of laboratory investigations should be performed to identify potential diagnostic clues.

Following the initial assessment, nonspecific laboratory tests should be performed:

- Complete blood count
- Erythrocyte sedimentation rate (ESR)
- C-reactive protein (CRP)
- Liver and kidney function tests (ALT, creatinine)
- Blood cultures
- Tuberculin skin test or interferon-gamma release test
- Urine test and urine cultures

A chest X-ray may also provide vital information to identify a possible infection or neoplasm. The evaluation of preliminary laboratory results could help to formulate a differential diagnosis and direct toward more specific investigations to confirm the diagnostic suspicion, such as specific serological investigations (anti-HIV, HBV, HCV, Epstein-Barr, cytomegalovirus antibodies research, etc.), peripheral blood smear, electrophoresis of serum proteins, thyroid function tests, imaging studies (magnetic resonance imaging or computed tomography), echocardio, scintigraphy with marked leukocytes, and endoscopy or biopsy (liver, lymph nodes, pleura, pericardium, bone marrow). In general, computed tomography (CT) is helpful in delineating abnormalities localized to the abdomen or chest. Magnetic resonance imaging (MRI) is more sensitive than CT to detect most causes of fever of unknown origin involving the central nervous system and should be performed if a central nervous system etiology is suspected.

Figure 43.1 describes the diagnostic approach of FUO.

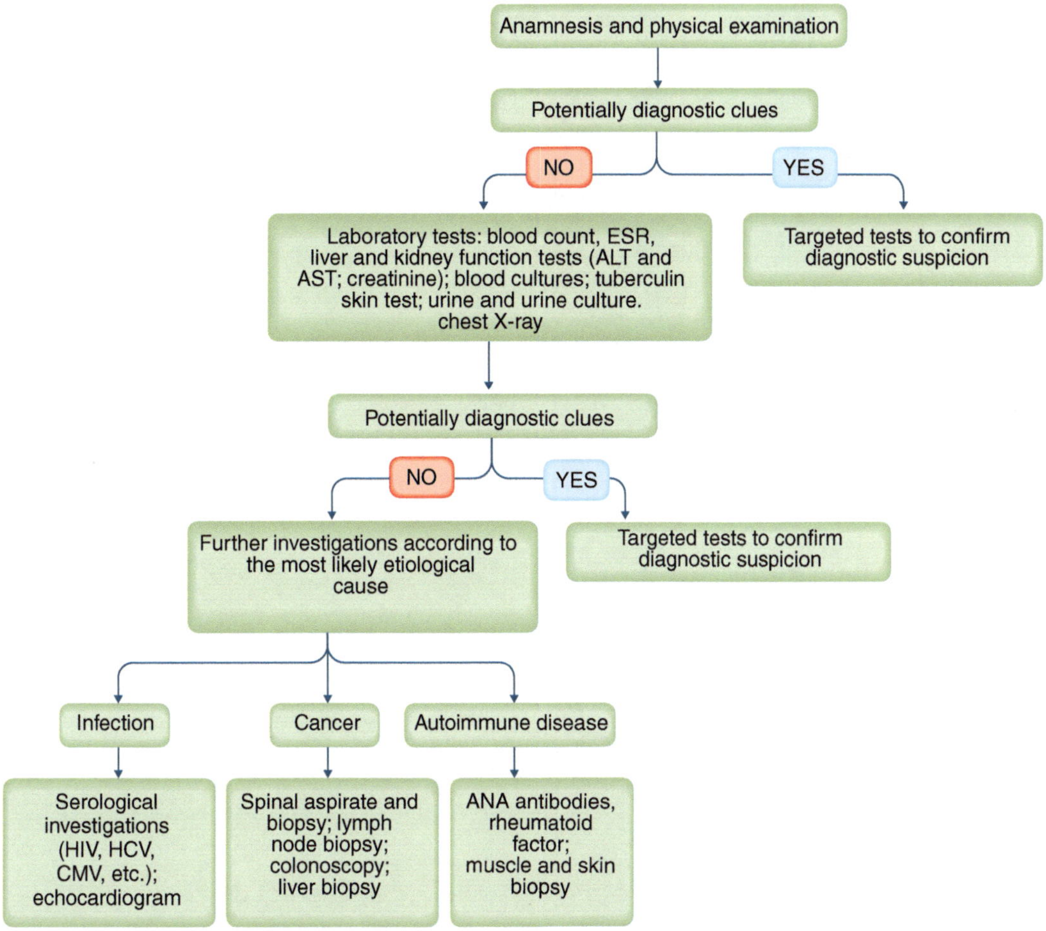

Fig. 43.1 FUO diagnostic pathway (Copyright EDISES 2021. Reproduced with permission)

Recommended Readings

Cunha BA (2007) Fever of unknown origin. Informa Healthcare, New York

Durack DT, Street AC (1991) Fever of unknown origin–re-examined and redefined. Curr Clin Top Infect Dis 11:35–51

Hersch EC, Oh RC (2014) Prolonged febrile illness and fever of unknown origin in adults. Am Fam Physician 90(2):91–96

Biomarkers of Inflammation

44

Marcello Ciaccio, Luisa Agnello, and Anna Maria Ciaccio

Introduction

Phlogosis, or inflammation, is an innate defense mechanism that our organism activates in response to various stimuli. Phlogosis is characterized by chemical mediators that play a fundamental role in various phases, from initiation to development and extinction or maintenance. Phlogosis markers are generally nonspecific because their levels increase during inflammation due to different causes. Therefore, their evaluation provides a general picture of the patient's clinical condition.

Definition of Inflammation

Inflammation, or phlogosis, is a nonspecific, innate defense mechanism representing an organism's response to a stimulus that causes cellular damage. Various exogenous and endogenous stimuli can cause cellular damage, triggering in the vascularized tissues a complex of reactions that together take the name of inflammation, whose objective is to eliminate or circumscribe the causes of the lesion and initiate reparative mechanisms.

M. Ciaccio (✉)
Department of Biomedicine, Neurosciences and Advanced Diagnostics, Institute of Clinical Biochemistry, Clinical Molecular Medicine and Clinical Laboratory Medicine, University Hospital "P. Giaccone", Palermo, Italy

Department of Laboratory Medicine, University Hospital "P. Giaccone", Palermo, Italy
e-mail: marcello.ciaccio@unipa.it

L. Agnello
Department of Biomedicine, Neurosciences and Advanced Diagnostics, Institute of Clinical Biochemistry, Clinical Molecular Medicine and Clinical Laboratory Medicine, University Hospital "P. Giaccone", Palermo, Italy

A. M. Ciaccio
Department of Health Promotion, Mother and Child Care, Internal Medicine and Medical Specialties (ProMISE) "G. D'Alessandro", University of Palermo, Palermo, Italy

Based on the duration over time, inflammation can be distinguished into:

- Acute: it has a short duration (from a few minutes to a few days), and it is characterized by the formation of an exudate consisting of liquids and proteins (edema), and by the migration and accumulation of leukocytes, mainly neutrophils, at the site of the lesion
- Chronic: it is long-lasting (days, months, years), and it is characterized by the presence of lymphocytes and macrophages, the proliferation of blood vessels, fibrosis, and tissue necrosis

The inflammation may remain localized at the initial site, or it may be accompanied by a moderate systemic reaction characterized by fever, malaise, hematological changes, and functional changes depending on the site of the process.

Inflammation can be divided into several phases, each characterized by mediators that regulate vascular and cellular responses:

1. Vasoconstriction: affects the arterioles of the affected areas and is stimulated by adrenaline and noradrenaline, chemical mediators of inflammation.
2. Vasodilation (or hyperemia): initially affects the arterioles and results in increased blood flow, which causes heat and redness. It is induced by histamines which secrete serotonin, the mediator of pain, together with platelets. Plasma proteases amplify the signal through cascade systems, and various cellular mediators are also called up in the inflamed tissue. The triggering and maintenance of vasodilation are, therefore, due to the release of rapid (histamine) and slow mediators (cytokines).
3. Exudation: the increase in vascular permeability of the microcirculation leads to the release of liquid, rich in proteins, called exudate, into the interstitial space of the tissues. This causes stasis, which is characterized by an increase in red blood cells, leading to an increase in blood viscosity and slowing circulation. Stasis triggers the next phase.

M. Ciaccio (ed.), *Clinical and Laboratory Medicine Textbook*, https://doi.org/10.1007/978-3-031-24958-7_44

4. Diapedesis: white blood cells escape from the vessels into the damaged tissue, according to the chemical concentration gradient (chemotaxis). Neutrophil granulocytes and monocytes (which then turn into macrophages) are recruited. They phagocytize the bacteria and degrade the tissue.
5. Passive hyperemia: it is induced by the slowing of the blood speed in the microcirculation. The blood, flowing less quickly, tends to clot (venous stasis, bluish veins).
6. Healing: after the exudate reabsorption, tissue regeneration can occur. It is characterized by the replacement of damaged cells and tissue (restituito ad integrum) or repair, which is characterized by the replacement of injured tissue with granulation tissue, which then matures into scar tissue.

If the damaged tissue cannot be regenerated or repaired, there is a transition from acute to chronic inflammation, characterized by a reduction in vascular phenomena and a qualitative change in the cellular component of the exudate.

Inflammation is characterized by chemical mediators that play a crucial role in the various stages, from initiation to development and extinction or maintenance (Table 44.1). Mediators can originate from circulation or cells and exert their biological activity through binding to specific receptors on target cells. One mediator can stimulate the release of other mediators by target cells, amplifying, modulating, or regulating inflammation.

The mediators of inflammation also have the function of activating the systemic response, characterized by the following alterations:

- Neuroendocrine (fever, drowsiness, secretion of catecholamines)
- Hematological (leukocytosis, thrombocytosis)
- Metabolic (decreased gluconeogenesis, cachexia, loss of muscle mass, increased adipocyte lipolysis, increased hepatic lipogenesis)
- Immunological (activation of the immune system)
- Biochemical (altered plasma concentration of specific proteins, defined as acute phase proteins)

Table 44.1 Mediators of inflammation

Vasoactive amines	Histamine Serotonin
Plasma proteases	Quinine system Complement system Coagulation system Fibrinolysis system
Acid arachidonic metabolites	Prostaglandins Leukotrienes Thromboxanes
Leukocyte products	Lysosomal enzymes Chemokines
Others	Free oxygen radicals Platelet activating factors Nitrogen monoxide

Acute Phase Proteins

A protein whose plasma concentration increases or decreases by at least 25% during inflammation is termed a positive or negative acute phase protein (PFA), respectively (Table 44.2).

Under physiological conditions, the liver synthesizes and maintains a constant concentration of several irculating proteins. Following an inflammatory stimulus, the released cytokines, especially IL-6, IL-1, and TNFα, induce an altered hepatic synthesis of specific proteins.

PFA alterations indicate the inflammatory process's presence, intensity, and duration.

The most commonly used indicators in clinical practice are C-reactive protein and erythrocyte sedimentation rate.

C-Reactive Protein

C-reactive protein (CRP) is a protein (alpha-globulin) that belongs to the pentraxin family, calcium-binding proteins consisting of five identical subunits that aggregate to form a ring-shaped pentameric molecule. It was discovered in 1930 due to its ability to bind and precipitate pneumococcal polysaccharide C. This phenomenon is due to the calcium-dependent reaction between the phosphatidylcholine residues of polysaccharide C in the bacterial wall and CRP. It can also bind other molecules that contain phosphatidylcholine residues, such as phospholipids, lipoproteins, or damaged cell membranes. Finally, CRP can activate the classical complement pathway and bind to chromatin, histones, and small riboprotein particles; CRP binding to cells or nuclear debris helps to eliminate material that could cause persistent inflammation and produce

Table 44.2 Acute phase proteins

Positive acute phase protein	Proteins whose concentration increases by 50% Ceruloplasmin Complement proteins C1q, C3, and C4 Fibronectin Serum P component of amyloid Protein whose concentration increases from 2 to 4 times Alpha 1 acid glycoprotein α1-antitrypsin Fibrinogen Haptoglobin Mannose-binding protein C4-binding protein β2-microglobulin Proteins whose concentration increases from hundreds to thousands of times C-reactive protein (CRP) Serum amyloid A (SAA) Procalcitonin (PCT)
Negative acute phase proteins	Albumin Transferrin Apo-AI Apo-AII

specific autoimmune reactions against nuclear antigens. Thus, CRP acts as an opsonin, capable of binding to many substances, both exogenous and endogenous, and facilitating their removal from the circulation by promoting their phagocytosis by leukocytes (opsonization).

CRP increases more markedly than other acute phase proteins during inflammation, bacterial infection, and trauma. Viral infections generally do not cause a significant CRP increase; therefore, in the absence of trauma, an increase in CRP indicates bacterial infection. However, if the viral infection is very "inflammatory," as in the case of adenovirus tonsillitis or mononucleosis, an increase in CRP may be observed.

CRP is synthesized by the liver mainly in response to IL-6; its levels increase within 4–8 hours after the onset of an inflammatory process, peaking within 48–72 hours and decreasing rapidly once the inflammatory state is resolved.

Increased CRP values are indicative of acute or chronic inflammation, autoimmune disease or disease caused by immune complexes, tissue necrosis, and neoplasms; however, normal values do not exclude the presence of small inflammatory foci, or systemic lupus erythematosus (SLE), progressive systemic sclerosis, dermatomyositis, and ulcerative colitis, in which the acute phase response is minimal. The increase in CRP, although indicating the presence of an inflammatory state, does not provide any information on the etiology and, therefore, it is not helpful for diagnostic purposes.

The CRP assay is, however, useful for:

- Assessing the course and severity of the inflammatory process (infections, acute myocardial infarction [AMI], deep vein thrombosis, rheumatic diseases, neoplasms)
- Determining the effectiveness of an anti-inflammatory therapy
- Checking the healing process of surgical wounds, burns, or organ transplantation
- Diagnosing and/or monitoring infection when it is necessary to have this information early, before the results of microbiological investigations (post-operative infectious complications, intensive care, neonatology, infections due to neutropenia from chemotherapy or post-transplantation, monitoring of response to antibiotic therapy)

Table 44.3 shows the normal and pathological CRP values.

Erythrocyte Sedimentation Rate

The erythrocyte sedimentation rate (ESR) represents the rate (space/time) at which red blood cells sediment in the blood sample after collection in a tube with anticoagulant. It is expressed in millimeters of sediment produced in 1 hour.

Table 44.3 Interpretation of circulating levels of C-reactive protein

[CRP] mg/L	Interpretation
<0.06	Normal values for newborns
<1.6	Normal values for children up to 1 month
0.06–10	Normal values for adults and children
10–50	Slight localized inflammation (cystitis, gingivitis), viral infections
50–100	More severe disease, Gram+ infections
>100	Bacterial Gram- infection

Under physiological conditions, red blood cells in suspension sediment little if they remain separated from each other. In contrast, if they aggregate and form cellular clusters, they sediment much more quickly. The aggregation of red blood cells is hindered by the electrical charges of their membranes and favored by inflammation proteins. The negative charge of the red blood cell surface causes erythrocytes to repel each other; however, this negative charge can be neutralized by positively charged plasma proteins, which therefore promote erythrocyte aggregation. Thus, ESR is regulated by the balance between pro-sedimentation factors (proteins, in particular, fibrinogen, CRP, IgM, and alpha 1-macroglobulin) and anti-sedimentation factors (negative membrane charge of erythrocytes). The aggregates that form are called rouleaux and consist of erythrocytes stacked on each other. The sedimentation process occurs in three stages:

1. Formation of rouleaux
2. Aggregation of rouleaux which tend to settle
3. Acceleration of sedimentation, with the accumulation of the rouleaux at the bottom of the tube

ESR indirectly indicates the degree of inflammation.

Generally, ESR is determined by quantitative capillary photometry and the value is obtained by dynamic measurement of the formation process of erythrocyte clusters.

ESR is a laboratory test commonly ordered in clinical practice but is highly nonspecific. ESR can increase in numerous pathological conditions. Moreover, the ESR value is not proportional to the extent and severity of the disease and its course. Finally, a negative ESR value does not rule out any disease. ESR is in great demand in clinical practice and represents a first-level test because its determination is easy to perform, is inexpensive, and provides a general indication of an inflammatory alteration. In clinical practice, high ESR values represents an "alarm signal" that should lead to more in-depth investigations.

Reference ESR values are reported in Table 44.4; since women and the elderly tend to have higher ESR values, these should be differentiated according to age and sex.

Table 44.5 shows the main pathological conditions associated with altered ESR values.

Table 44.4 Normal values of erythrocyte sedimentation rate (ESR)

	Age (years)	ESR (mm/h)	Range (mm/h)
Men	20–49	5	0–13
	50–69	7	0–19
Women	20–49	9	0–21
	50–69	12	0–28

Table 44.5 Conditions associated with altered erythrocyte sedimentation rate values

Infectious diseases
Bacterial
Mycotic
Systemic
Viral
Non-infectious inflammatory processes
Acute temporal arthritis
Rheumatoid arthritis
Fractures
Surgical intervention
Polymyalgia rheumatica
Anaphylactoid purpura
Trauma
Burns
Necrotic processes
Myocardial infarction
Acute pancreatitis
Neoplasms
Breast cancer
Leukemia
Lymphomas
Neuroblastomas
Tumors with metastases
Other conditions
Allergies
Severe anemias
Gastrointestinal bleeding
Pregnancy
Hypothyroidism
Hemolytic uremic syndromes

Procalcitonin

Procalcitonin (PCT) is a precursor of the hormone calcitonin. Under physiological conditions, calcitonin is produced and secreted by the C cells of the thyroid as a result of procalcitonin proteolysis; therefore, procalcitonin is almost absent or present at very low levels, on the order of picograms. However, under conditions associated with systemic inflammation, procalcitonin is produced and secreted almost ubiquitously. In summary, therefore, calcitonin is produced exclusively by C cells of the thyroid in response to hormonal stimuli. In contrast, several cell types and many organs can produce PCT in response to proinflammatory stimuli. Bacterial endotoxins and proinflammatory cytokines are the primary stimuli for PCT production. Viral infections and neoplastic and autoimmune diseases generally do not stimulate PCT production, except occasionally and at low concentrations. In addition, chronic non-bacterial inflammation and allergic reactions do not increase PCT levels.

Table 44.6 Interpretation of circulating levels of procalcitonin (PCT)

[PCT] ng/mL	Interpretation
<0.05	Normal
<0.5	Low, no significant systemic inflammatory reaction. Possible localized infection and inflammation (in the absence of systemic signs)
0.5–2	Moderate increase, indicative of moderate systemic inflammation. Infection is possible but several stimuli can induce PCT (severe trauma, cardiogenic shock, or major surgery)
>2–10	Severe systemic inflammation, possibly due to infection (sepsis)
>10	Marked systemic inflammation, almost exclusively due to septic shock and frequently associated with multiorgan dysfunction

The amount of PCT released and, therefore, the increase in circulation is related to the extent of the inflammatory reaction; very high values are generally found in severe acute bacterial infections and sepsis. Experimental studies have shown that PCT values increase with the severity of sepsis, reaching very high values in subjects with septic shock and being associated with increased mortality (see Chap. 45).

Localized bacterial infections not associated with symptoms of systemic inflammation (e.g., fever) or septic reactions result in low PCT values.

Table 44.6 shows the reference values of the PCT.

PCT is very sensitive and specific for bacterial infections. Experimental studies have shown that PCT levels increase rapidly (approximately 3 hours) following endotoxin administration, with a peak around 12 hours. At the end of the acute inflammatory reaction, PCT concentration decreases rapidly because it has a short half-life (20–24 hours). In clinical practice, PCT aids in guiding and monitoring antibiotic therapy; in this context, it is important not so much the absolute value but rather its variations over time.

PCT is a very stable protein *in vivo* and *in vitro*: in circulation, it is not degraded to the active hormone calcitonin.

α-1 Antitrypsin

It is a protein synthesized in the liver and is the primary inhibitor of serum proteases, leukocyte, and macrophage, together with antithrombin III and antichymotrypsin. Proteases are mainly produced during inflammation, so if their activity is not neutralized, they may cause excessive tissue damage. α1-antitrypsin increases nonspecifically in response to any inflammatory, infectious, or cell necrosis stimulus. In addition, increased values may be detected in pregnancy, therapy with oral contraceptives, inflammatory diseases (rheumatoid arthritis, bacterial infections, vasculitis), necrosis (myocardial infarction), and neoplasms (metastasized carcinomas, Hodgkin's disease).

The α1-antitrypsin assay is performed by the immunonephelometric method. Normal values are between 90 and 205 mg/dL.

β2-Microglobulin

It is a plasma protein belonging to the β-globulin class, present at a high concentration on the surface of nucleated cells, including lymphocytes and macrophages It represents one of the two chains that constitute the antigens of the HLA class I system. The serum concentration of this protein generally increases in all conditions of increased cell turnover, such as inflammation, autoimmune disorders, leukemias, lymphomas, and viral diseases. β2-microglobulin has a low molecular weight and easily crosses the glomerulus but is almost completely reabsorbed in the distal tubule. An increase in urinary β2-microglobulin with normal blood values indicates a reduction in tubular reabsorption due to a renal pathology with tubular damage on an inflammatory or degenerative basis (renal insufficiency, congenital or acquired tubulopathies, Wilson's disease). The β2-microglobulin assay is usually performed by chemiluminescence.

Fibrinogen

Fibrinogen, or coagulation factor I, is a glycoprotein synthesized mainly in the liver, with a dual role in the hemostasis: promoting the adhesion of platelets to the wall of blood vessels and forming the clot following its conversion into fibrin by thrombin. Fibrinogen levels increase markedly in response to an inflammatory stimulus; other possible causes include cancer, pregnancy, hepatitis, oral contraceptives (birth control pills), smoking, and obesity. High fibrinogen levels are associated with increased cardiovascular risk due to its procoagulant effect. Normal fibrinogen values are 200–375 mg/dL in men and 200–430 mg/dL in women.

Amyloid Serum Protein A

Serum amyloid A (SAA) protein is an apolipoprotein whose biosynthesis in the liver increases in response to the release of proinflammatory cytokines (TNFα, IL-6, and IL-1β) produced by activated monocytes during inflammation. SAA represents an essential acute phase protein because its levels increase significantly compared to physiological values, and this increase starts as early as 3–6 hours after the inflammatory stimulus, reaching a peak at 48–72 hours. Moreover, SAA is the precursor of amyloid A, the main fibrillar component of amyloid deposits in patients with amyloidosis.

Once synthesized and secreted, SAA binds to HDL isoform 3, displacing apolipoprotein I (Apo-AI), which plays a crucial role in reverse cholesterol transport. In response to inflammation, SAA-rich HDL have a reduced affinity for hepatocytes and an increased affinity for macrophages, neutrophils, and endothelial cells. SAA can modulate the inflammatory response through various actions, such as promoting chemotaxis, adhesion, and infiltration of monocytes, neutrophils, and lymphocytes; activating metalloproteases; inducing collagenase secretion; and, finally, opsonizing gram-negative bacteria by inducing phagocytosis. Thus, the increase of SAA during inflammation represents a defense mechanism of the organism. However, its chronic increase can have important pathological implications because it can lead to amyloid deposition and the onset of secondary or reactive amyloidosis, a chronic inflammation complication.

Under physiological conditions, SAA has a concentration <10 mg/L; in response to an inflammatory stimulus, its concentration can be >1000 mg/L. SAA is a sensitive parameter comparable to CRP for monitoring the acute phase of the inflammatory response, especially for identifying low-grade inflammation. SAA is helpful for acute viral infections.

Diagnostic Approach

The laboratory diagnosis for the evaluation of an inflammatory process is mainly based on the following:

- White blood cells (WBC) count and formula examination:
  - Bacterial inflammation: leukocytosis (>15,000) with neutrophils >80%
  - Viral inflammation: normal or slightly increased WBC, with reversal of neutrophil/lymphocyte ratio (increased lymphocytes)
  - Chronic inflammation: normal or slightly increased WBCs
  - Allergic reactions or parasitosis: normal or slightly increased WBC, eosinophilia
- Analysis of acute phase proteins and other biomarkers, especially ESR and CRP, as a screening test to assess the presence of an alteration and, subsequently, to monitor the progression of the disease or the effect of therapy

Recommended Readings

Agnello L, Bellia C, Di Gangi M, Lo Sasso B, Calvaruso L, Bivona G, Scazzone C, Dones P, Ciaccio M (2016) Utility of serum procalcitonin and C-reactive protein in severity assessment of community-acquired pneumonia in children. Clin Biochem 49(1–2):47–50. https://doi.org/10.1016/j.clinbiochem.2015.09.008

Bivona G, Agnello L, Scazzone C, Lo Sasso B, Bellia C, Ciaccio M (2015) Procalcitonin and community-acquired pneumonia (CAP) in children. Clin Chim Acta 451(Pt B):215–218. https://doi.org/10.1016/j.cca.2015.09.031

Ciaccio M, Fugardi G, Titone L, Romano A, Giordano S, Bivona G, Scarlata F, Vocca L, Di Gangi M (2004) Procalcitonin levels in plasma in oncohaematologic patients with and without bacterial infections. Clin Chim Acta 340(1–2):149–152

Jura J, Koj A (2011) Regulatory mechanisms controlling inflammation and synthesis of acute phase proteins. In: Veas F (ed) Acute phase proteins regulation and functions of acute phase proteins. InTech. https://doi.org/10.5772/18207. Available from: https://www.intechopen.com/books/acute-phase-proteins-re-gulation-and-functions-of-acute-phase-proteins/regulatory-mechanisms-controlling-inflamma-tion-and-synthesis-of-acute-phase-proteins

Kolaczkowska E, Kubes P (2013) Neutrophil recruitment and function in health and inflammation. Nat Rev Immunol 13:159–175

Laura O, Simona D, Riccardo A, Remigio M, Giampaolo M (2010) Siero amiloide A e flogosi. Biochimica. Clinica 34(5)

Biomarkers of Sepsis

45

Giuseppe Lippi, Gianfranco Cervellin, Marcello Ciaccio, and Luisa Agnello

Introduction

Sepsis is a significant cause of in-hospital and out-of-hospital mortality. Early diagnosis is a crucial aspect that influences its course, allowing promptly and targeted antibiotic therapy. Since clinical signs and symptoms of sepsis are often nonspecific, the role of the laboratory becomes crucial, both for determining biomarkers that can guide clinical decision-making and identifying the pathogen responsible. In the light of recent evidence, it seems, therefore, essential that the laboratory makes available (both in routine and especially in an emergency) one or more biomarkers of sepsis. The most significant consensus is now in favor of the determination of procalcitonin, possibly combined with a second biomarker (C-reactive protein and/or presepsin) to increase the diagnostic sensitivity and specificity. The procalcitonin value can also be used to guide antibiotic therapy, in this case, preferably using tests with high analytical sensitivity to identify early changes in concentration as a function of antibiotic therapy.

G. Lippi
Section of Clinical Biochemistry, University Hospital of Verona, Verona, Italy

G. Cervellin
Academy of Emergency Medicine and Care, Pavia, Italy

M. Ciaccio (✉)
Department of Biomedicine, Neurosciences and Advanced Diagnostics, Institute of Clinical Biochemistry, Clinical Molecular Medicine and Clinical Laboratory Medicine, University Hospital "P. Giaccone", Palermo, Italy

Department of Laboratory Medicine, University Hospital "P. Giaccone", Palermo, Italy
e-mail: marcello.ciaccio@unipa.it

L. Agnello
Department of Biomedicine, Neurosciences and Advanced Diagnostics, Institute of Clinical Biochemistry, Clinical Molecular Medicine and Clinical Laboratory Medicine, University Hospital "P. Giaccone", Palermo, Italy

Definition of Sepsis

About 1300 years ago, the Greeks adopted the term σήψις (sepsis) to define processes of decomposition or rotting, clinical conditions burdened by high mortality and often associated with infectious processes. However, during the twentieth century, there was a need to adopt a more accurate and functional definition, which has evolved substantially in three consecutive times (1992, 2001, and 2016) (Fig. 45.1). In the consensus conference held in 1992 between the American College of Chest Physicians and the Society of Critical Care Medicine, sepsis was defined as the result of the host's systemic inflammatory response syndrome (SIRS) to an infection, either established or suspected. Thus, in other words, sepsis was defined as the result of two conditions: suspected infection and SIRS. SIRS represents the first stage of a systemic inflammatory process of different etiologies (trauma, infection, burns, pancreatitis, etc.) and is diagnosed according to the presence of at least two of the following symptoms or signs:

- Body temperature >38 °C or <36°C
- Heart rate >90 beats/minute
- Respiratory rate >20 acts/minute or partial pressure of carbon dioxide ($PaCO_2$) <32 mmHg (4.3 kPa)
- White blood cell count >12,000/mm^3 or <4000/mm^3 or >10% immature neutrophil form

Sepsis complicated by organ dysfunction was defined as severe sepsis, which could progress to septic shock, defined as "sepsis-induced hypotension that persists despite adequate fluid infusion." In 2001, a task force of 29 experts, recognizing the limitations of these definitions, expanded the list of diagnostic criteria by adding further signs, symptoms, and biochemical tests but did not change them. The definitions of sepsis, septic shock, and organ dysfunction have remained unchanged for over two decades.

M. Ciaccio (ed.), *Clinical and Laboratory Medicine Textbook*, https://doi.org/10.1007/978-3-031-24958-7_45

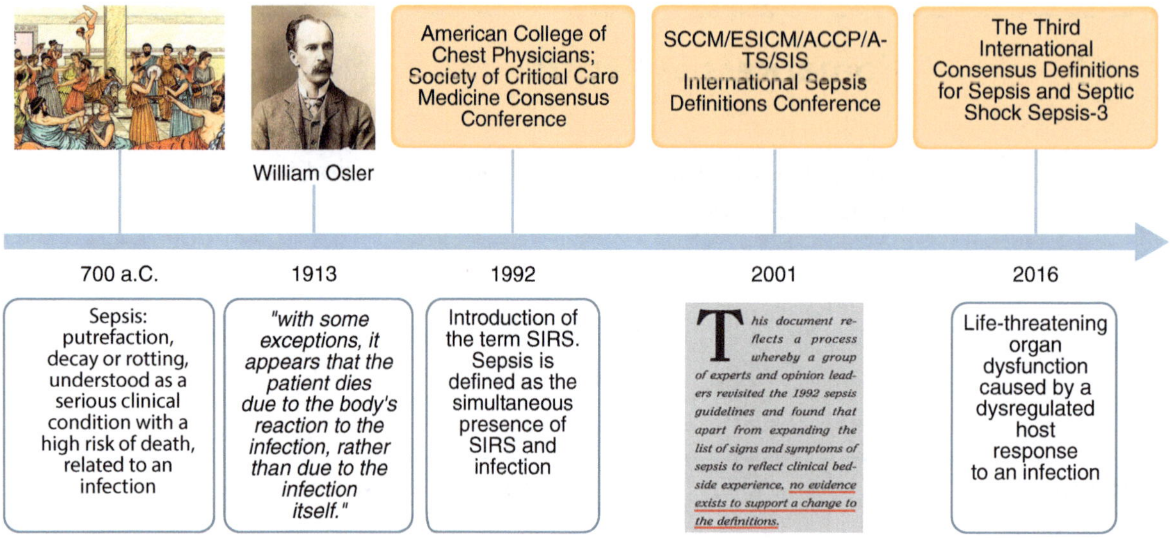

Fig. 45.1 Timeline of sepsis definition. Copyright EDISES 2021. (Reproduced with permission)

A better understanding of the pathobiology of sepsis as sepsis-induced alterations on organ function, morphology, cell biology, biochemistry, immunology, and circulation has led to the need to revise its definition. In 2014, a task force of 19 experts developed the consensus "Sepsis-3," which fundamentally revolutionized the old definition of sepsis, eliminating the concept of SIRS and deleting the term severe sepsis. Thus, the results of the Sepsis-3 consensus, published in 2016, defined the following conditions:

- Organ dysfunction: condition represented by an increase in Sequential Organ Failure Assessment (SOFA) score of at least two points.
- Sepsis: life-threatening organ dysfunction due to a host's dysregulated biological response during an infection.
- Septic shock: a specific type of sepsis in which severe metabolic, circulatory, and cellular alterations confer a higher risk of death than in the presence of sepsis alone.

The SOFA (Sequential Organ Failure Assessment) is a 0–4 score that explores the progressive deterioration of the respiratory, cardiovascular, hepatic, renal, neurological, and coagulation systems.

This new definition of sepsis ideally relates to a historic definition by William Osler, formulated early last century: "with some exceptions, the patient appears to die from the organic response to the infection rather than from the infection itself."

The original concept of sepsis, defined as an infection with at least two of the four SIRS criteria, focused exclusively on inflammatory excess. Today, however, it is recognized that sepsis is something more complex than an infection presenting along with inflammation. It is known that it involves the early activation of both pro- and anti-inflammatory responses associated with essential changes in several pathways, including cardiovascular, neuronal, hormonal, bioenergetic, metabolic, and coagulative. In other words, sepsis is now defined as an infection complicated by organ dysfunction, dangerous for survival, and caused by a dysregulated response of the organism.

Since the SOFA score appears rather complex, especially in urgent/emergencies, slowing down the early recognition of sepsis in a suspected patient, a simplified surrogate, called qSOFA (quick SOFA), has been developed based on three elementary parameters:

- Respiratory rate ≥22 acts/minute
- Mental alteration
- Blood pressure ≤100 mmHg

If the patient has at least two of the above parameters, there is a high probability of sepsis. Compared to the SOFA score, the qSOFA is based solely on clinical criteria and does not require laboratory testing, thus providing a more straightforward and faster assessment of the patient with suspected sepsis (Fig. 45.2).

However, it should be emphasized that whatever tool is used in the initial patient evaluation, none of the simple or complex scoring systems can make a definitive diagnosis of sepsis but rather serve as a screening tool to direct the clinician toward a targeted diagnostic workup. This is particularly important because various other diseases, such as polytrauma, acute pancreatitis, and venous thromboembolism, share the signs and symptoms of sepsis.

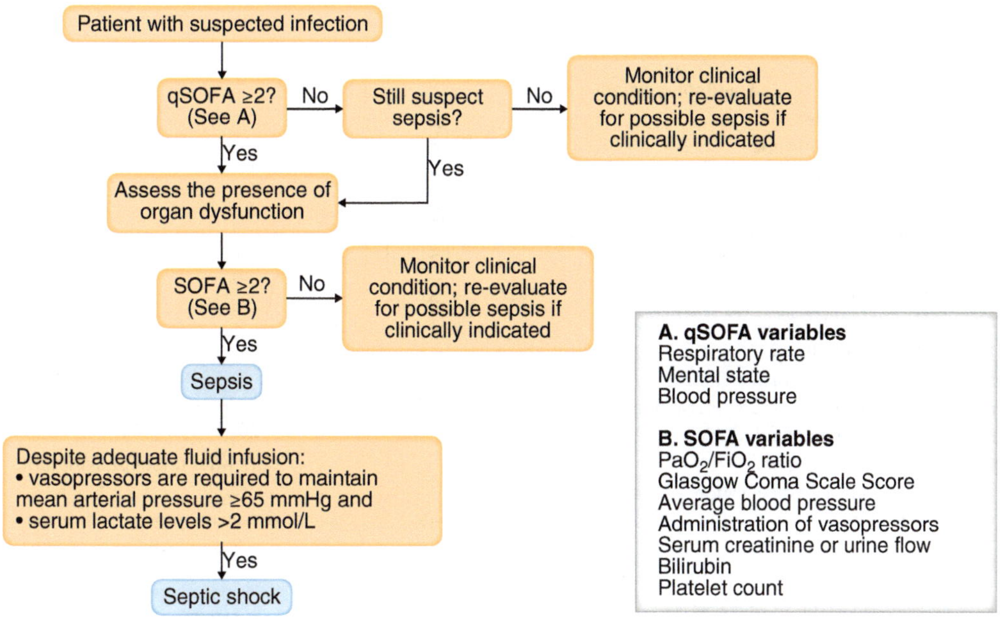

Fig. 45.2 Diagnostic algorithm for sepsis and septic shock. qSOFA quick Sequential Organ Failure Assessment, SOFA Sequential Organ Failure Assessment. (Copyright EDISES 2021. Reproduced with permission)

Epidemiology of Sepsis

Despite significant advances in diagnosis and treatment, sepsis remains a health problem with significant incidence and mortality.

It is believed that the frequency is currently around 2% in hospitals, although it increases considerably (up to 30–50%) in some categories of patients, such as those admitted to intensive care units. An epidemiological study in the United States predicts that the frequency of this disease in hospitals will increase by almost 10% per year. In the emergency department and emergency medicine, recent data show that almost all patients transferred to the ICU for severe sepsis have been through the emergency department before, and that in almost half of these cases, sepsis was not diagnosed during the stay in the emergency department, which justifies the need to improve the diagnostic tools available today. Early identification of the septic condition and prompt initiation of treatment (antibiotic and fluid therapy) are critical factors with a significant impact on the course of sepsis and septic shock, leading to a significant reduction in mortality of up to 18%.

These observations have led in recent years to amplify the search for new diagnostic tools that allow to:

1. Screen at the time of the appearance of signs and symptoms (or even earlier in high-risk patients)
2. Identify with high sensitivity the presence of sepsis
3. Recognize the pathogen responsible for sepsis
4. Obtain prognostic information based on the presumed severity of sepsis and the clinical characteristics of the patient
5. Promptly establish a targeted antibiotic therapy since the earliness and accuracy of the therapy heavily condition the prognosis

Role of the Laboratory in Sepsis

The laboratory plays a fundamental role in sepsis and septic shock. Clinical signs must always be integrated with laboratory tests to define the picture and immediate treatment. In addition to the laboratory investigations helpful in assessing the patient's overall clinical condition (blood count, creatinine, blood glucose, aminotransferase, bilirubin, cardiac troponins, coagulation tests D-dimer) (Table 45.1), lactic acid determination and blood culture play a critical role. Lactic acid levels represent the expression of peripheral tissue distress, and the concentration and kinetics correlate with the patient prognosis. Moreover, in preserved hemodynamic parameters, an important lactate increase (>4 mmol/L) could detect cryptic shock. Finally, lactate measurement is used to assess the clinical evolution of sepsis. The blood culture represents the fundamental test for correctly identifying the pathogenic species responsible for sepsis. However, the literature reports varying percentages of sepsis with negative blood cultures from 20% to 40%. Moreover, although blood

Table 45.1 Laboratory tests in sepsis

Lab test	Scope
Procalcitonin	Sepsis diagnosis and antibiotic therapy guide
Presepsin	Sepsis diagnosis
MR-proADM	Prognostic indicator
C-reactive protein	Sepsis diagnosis
Blood culture	Etiological diagnosis of sepsis
Complete blood count	Diagnosis of anemia and thrombocytopenia
Bilirubin	Hepatic impairment
Aminotransferase	Hepatic impairment
Creatinine	Renal impairment
Glycemia	Metabolic decompensation
Coagulation test	Early diagnosis of DIC (disseminated intravascular coagulation)
Cardiac troponins	Cardiac impairment
Lactic acid	Metabolic decompensation; prognostic indicator

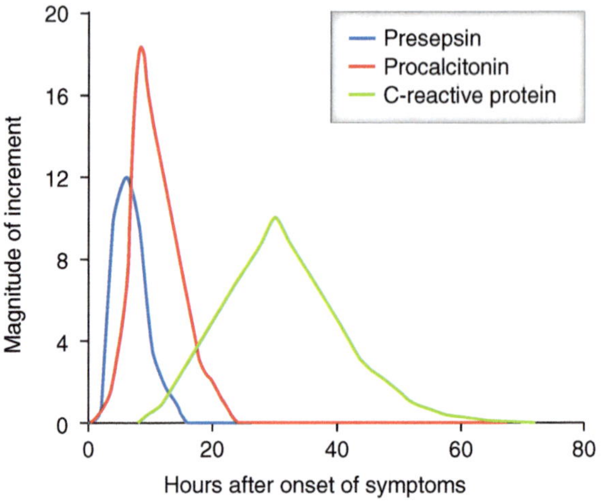

Fig. 45.3 Kinetics of presepsin, procalcitonin, and C reactive protein. (Copyright EDISES 2021. Reproduced with permission)

culture remains the reference method for the etiological diagnosis of sepsis, it should be remembered that its execution usually requires 24 to 72 hours, depending on the investigation's complexity. The recent introduction of molecular techniques that allow the identification of pathogens from culture directly on blood samples could shorten the etiological diagnosis by several hours. They are essentially based on the following: (1) DNA strip technology, a method characterized by high diagnostic efficiency, whereby DNA is isolated, amplified, and detected by strip hybridization and (2) polymerase chain reaction (PCR) and microarray-based microbiology, which allow DNA extraction, multiplex PCR, detection, and identification of multiple bacterial species simultaneously in less than 3 hours. Today, direct molecular tests would allow obtaining a precise etiological diagnosis (on average) in less than 6 hours, would not require dedicated personnel, and could be available at all times, even in emergencies. It is important to remember that blood cultures should be collected before antibiotic therapy in patients with suspected sepsis or septic shock because antibiotic therapy may alter the results.

In addition to common laboratory investigations, many biomarkers are used in clinical practice for managing patients with suspected or confirmed sepsis and septic shock to support clinical decision-making.

Biomarkers of Sepsis

As in other areas of medicine, the role of biomarkers in diagnosing sepsis can be crucial. The ideal sepsis biomarker should aid screening, facilitate early diagnosis, provide helpful information about prognosis, and, if possible, help select and monitor therapy. Three of these biomarkers have the best diagnostic performance: C-reactive protein (CRP), procalcitonin (PCT), and presepsin. In addition, another promising biomarker is adrenomedullin (ADM). On the other hand, concerning performance in guiding antibiotic therapy (especially in determining the optimal time to discontinue the same), the most solid evidence is in favor of PCT, albeit with some residual doubt. There is currently no solid evidence to suggest the use of other biomarkers, such as interleukin-6 (IL-6), lipopolysaccharide-binding protein (LBP), nCD64 (neutrophil CD64), sTREM-1 (soluble triggering receptor expressed on myeloid cells-1), and suPAR (soluble urokinase-type plasminogen activator receptor).

It is believed that PCT should be the biomarker of the first choice, possibly complemented by CRP to increase diagnostic sensitivity. Based on preliminary data, presepsin could replace PCT in the future, although data are currently insufficient. Whichever marker(s) is/are chosen, they should be measurable from the laboratory both routinely and on an emergency basis. Regarding the diagnostic threshold, it is preferable to use one that favors diagnostic sensitivity (e.g., 0.5 ng/mL for PCT) rather than a higher one that favors specificity. Regardless of its diagnostic performance, the biomarker value should always be interpreted according to the patient's clinical data to optimize its diagnostic efficiency. The timing of the biomarker request should also follow a rational approach based on in vivo kinetics and half-life, as illustrated in Fig. 45.3. Therefore, for PCT (increases 2–4 hours after symptom onset, peaks at 6–8 hours, and has a half-life of 20–24 hours) and PCR (increases 12–24 hours after symptom onset, peaks at 48–72 hours, and has a half-life of 20 hours), the time interval between one determination and the next should not be less than 24 hours, which is reduced to 5 hours for presepsin (which shows an increase 2 hours after symptom onset, peaks at 3–4 hours, and has a half-life of 4–6 hours).

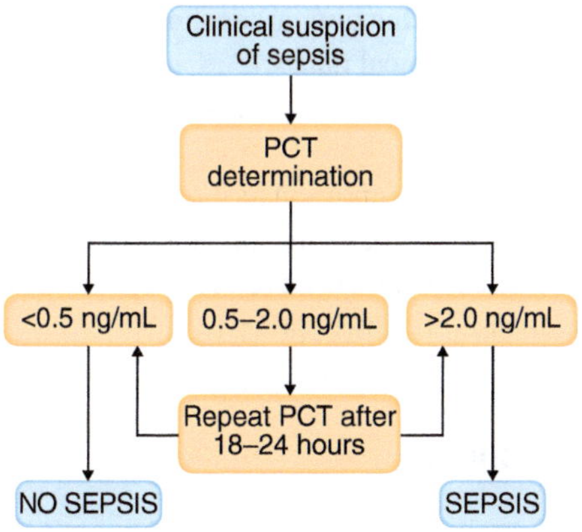

Fig. 45.4 Algorithm for the diagnosis of sepsis based on procalcitonin levels (PCT). (Copyright EDISES 2021. Reproduced with permission)

About antibiotic therapy, the use of PCT is helpful to identify early the possibility of suspending it, thus reducing both undesirable effects and the potential development of antibiotic resistance. Also, in this case, PCT monitoring should include an interval between two successive determinations of not less than 18–24 hours. A hypothetical PCT-based algorithm for the diagnosis of sepsis is described in Fig. 45.4.

C-Reactive Protein

CRP is commonly used in clinical practice to identify the presence of inflammatory and infectious processes. However, its specificity for diagnosing sepsis is relatively low. Indeed, CRP synthesis could be substantially enhanced by several stimuli other than sepsis (e.g., trauma). Moreover, CRP levels do not correlate sufficiently with the severity of the process; patients with severe sepsis might present only a moderate CRP increase. Finally, CRP is not considered an ideal sepsis biomarker because its values peak slowly and their decrease requires several days of treatment; therefore, CRP is not helpful for the early diagnosis of sepsis.

Procalcitonin

PCT is the precursor of the hormone calcitonin, which is involved in calcium homeostasis. Under physiological conditions, PCT has low blood concentration (usually <0.05 ng/mL). It is produced as pre-procalcitonin (141 amino acids) by the C cells of the thyroid, then converted to PCT (116 amino acids), and subsequently to the active hormone calcitonin (32 amino acids), N-terminal PCT (N-PCT; 57 amino acids), and catacalcin (21 amino acids) (Fig. 45.5). During severe bacterial infections (especially if generalized such as sepsis), PCT synthesis becomes predominantly extra-thyroid

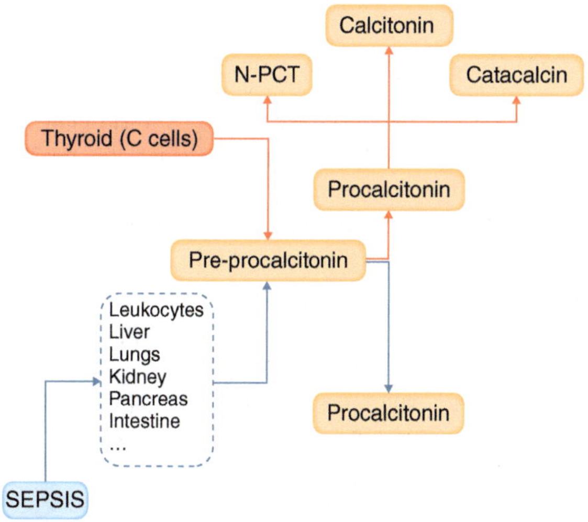

Fig. 45.5 Biosynthesis and biochemistry of procalcitonin (PCT). N-PCT N-terminal PCT. (Copyright EDISES 2021. Reproduced with permission)

(mainly by the liver, lungs, pancreas, kidney, intestine, and leukocytes), and its concentration increases from 100 to over 10,000 times. In subjects with sepsis, it is not uncommon to observe concentrations exceeding 10–100 ng/mL, with the extent of increase correlating to the severity of infection and prognosis. From a biological point of view, this production heterogeneity is justified by the restricted expression of the gene encoding for calcitonin (CALC-1) to thyroid C cells under physiological conditions. However, as a result of bacterial stimuli (either directly, due to endotoxin and other bacterial toxins, or indirectly, due to metabolic reactions of the body in response to infection), CALC-1 gene expression is considerably amplified in extra-thyroid tissues. In association with the increased extra-thyroidal synthesis, the degradation process of PCT into calcitonin, N-PCT, and catacalcin is considerably suppressed (Fig. 45.5).

Sepsis and systemic inflammation can be ruled out if the plasma concentration of PCT is ≤0.5 ng/mL. PCT secretion begins within 2–4 hours of sepsis onset, and its levels peak at 6–8 hours. Although normal PCT levels have a high negative predictive value, this biomarker cannot replace clinical evaluation. Therefore, if PCT levels are low in a patient with suspected sepsis, the patient should still be treated for sepsis. The kinetics of PCT provide more helpful information than the absolute value. Increasing PCT values over time is associated with a worse outcome, while decreasing PCT levels over time is associated with lower mortality, even if absolute PCT values are high.

Monitoring PCT levels in septic patients may also provide insights into the appropriateness of reducing antimicrobial therapy. In the literature, various algorithms based on PCT levels have been proposed to reduce the duration of antibiotic therapy in septic patients. Almost all algorithms support the discontinuation of therapy if PCT concentration is <0.5 ng/mL. As demonstrated by several randomized trials, the application of PCT protocols could reduce the length of antimicrobial treatment, thus leading to several benefits for patients (reduced antibiotic exposure) and the healthcare system (reduced resource consumption). However, conflicting evidence comes from a large retrospective cohort study that found no association between PCT-based strategies and improved antibiotic therapy use or reduced mortality. Therefore, further studies are needed to support the effectiveness of PCT-level-based algorithms.

Apart from its emerging role in the therapeutic monitoring of sepsis, an emerging aspect of PCT concerns its usefulness in the setting of severe localized infections, such as bacterial pneumonia. In this context, it has been suggested that the use of diversified cutoffs, always in combination with the clinic, can help in the diagnosis and initiation of antibiotic therapy (in the case of pneumonia, the cutoff should be set at 0.25 ng/mL, thus lower than the value of 0.5 ng/mL used in the diagnosis of sepsis).

Presepsin

The sCD14-ST (soluble cluster of differentiation 14-subtype), also known as presepsin, is a soluble fragment of CD14 (cluster of differentiation 14). CD14 is a receptor for complexes consisting of lipopolysaccharide (LPS) and lipopolysaccharide-binding protein on the membrane of phagocytes, whose primary function is the activation of the inflammatory cascade, as occurs in patients with sepsis. Presepsin levels in healthy subjects are <6 μg/L. However, during sepsis, they increase significantly, and this increase is related to the severity of the pathology.

Clinical studies have shown that presepsin has better sensitivity and specificity than other sepsis biomarkers. For example, compared to PCT, presepsin levels increase more rapidly, within 2 hours, peak at 3 hours, and decline at 4–8 hours. Thus, presepsin represents an early sepsis biomarker. However, it has a relatively low specificity because its levels increase significantly in sepsis and other pathological conditions, such as coronary artery disease, heart failure, liver cirrhosis, liver disease, and decompensated diabetes mellitus. Therefore, this biomarker should always be used in conjunction with clinical data.

Presepsin levels also correlate with sepsis severity and mortality risk.

Overall, presepsin represents a helpful biomarker for the management of septic patients allowing both early diagnosis and severity assessment. However, further studies are needed to validate its use in clinical practice.

Adrenomedullin

Adrenomedullin is a peptide hormone with a potent vasodilator effect, which was isolated and characterized in 1993 from pheochromocytoma tissue. Subsequent studies revealed the presence of ADM in other tissues, such as adrenal glands, heart, kidney, gastrointestinal system, and arterial vessels.

In addition to its vasodilator property, ADM exerts various physiological effects (autocrine and paracrine) on the cardiovascular, respiratory, renal, immune, and neuroendocrine systems; these are protective and supportive of the proper function of these organs. Therefore, the increased expression and activity of ADM could represent a defense response to organ damage and dysfunction.

High levels of ADM have been identified in patients with sepsis. However, its levels are often underestimated because they are not easily assayed due to various factors, such as rapid circulating binding to physiological receptors and proteins, a short half-life (approximately 22 minutes) due to rapid degradation by proteases, and the complexity of the analytical procedure.

Alternatively, a 48 amino acid fragment, MR-proADM (mid-regional pro-adrenomedullin), has been identified, produced from the same precursor and released in equimolar amounts to ADM. Specifically, ADM is synthesized as a 185 amino acid pre-pro-hormone, known as preproadrenomedullin, which is converted to a 164 amino acid peptide upon removal of the signal peptide known as proadrenomedullin. Proadrenomedullin consists of three vasoactive peptides: ADM, proadrenomedullin aminoterminal peptide (PAMP), and adrenotensin. In addition, there is a region whose biological activity is not known, called MR-proADM (Fig. 45.6). The MR-proADM reflects the lev-

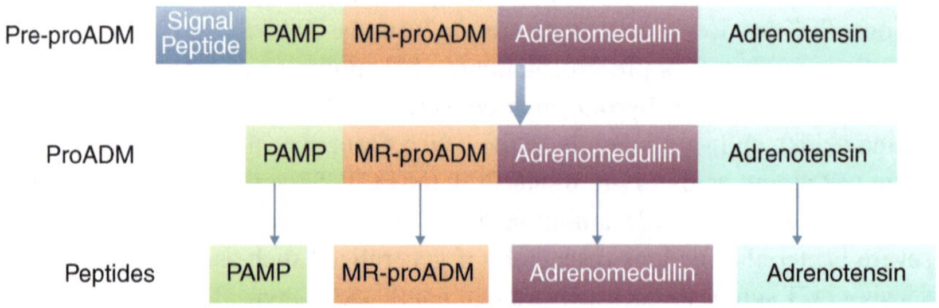

Fig. 45.6 Biosynthesis of adrenomedullin (ADM). PAMP aminoterminal peptide of proadrenomedullin, MR-proADM mid-regional pro-ADrenomeMedullin. (Copyright EDISES 2021. Reproduced with permission)

els and activity of ADM but has a longer half-life (several hours) than ADM, and its levels can be easily determined in clinical practice.

Compared to other biomarkers of sepsis, such as CRP and PCT, MR-proADM levels increase significantly in response to infections, indipendently of pathogen. In addition, several studies have shown that MR-proADM levels correlate significantly with sepsis severity and show high prognostic power in predicting mortality risk, superior to other biomarkers, including CRP and PCT.

Recommended Readings

Abraham E (2016) New definitions for sepsis and septic shock. Continuing evolution but with much still tbe done. JAMA 315:757–759

Chamberlain DJ, Willis E, Clark R et al (2015) Identification of the severe sepsis patient at triage: a prospective analysis of the Australasian Triage Scale. Emerg Med J 32:690–697

Chu DC, Mehta AB, Walkey AJ (2017) Practice patterns and outcomes associated with procalcitonin use in critically ill patients with sepsis. Clin Infect Dis 64:1509–1515

Churpek MM, Snyder A, Han X et al (2017) Quick sepsis-related organ failure assessment, systemic inflammatory response syndrome, and early warning scores for detecting clinical deterioration in infected patients outside the intensive care unit. Am J Respir Crit Care Med 195:906–911

Funk DJ, Parrill JE, Kumar A (2009) Sepsis and septic shock: a history. Crit Care Clin 25:83–101

Kitamura K, Kangawa K, Kawamot M et al (1993) Adrenomedullin: a novel hypotensive peptide isolated from human pheochromocytoma. Biochem Biophys Res Commun 192:553–560

Kopterides P, Siempos II, Tsangaris I et al (2010) Procalcitonin-guided algorithms of antibiotic therapy in the intensive care unit: a systematic review and meta-analysis of randomized controlled trials. Crit Care Med 38:2229–2241

Liu Y, Hou JH, Li Q et al (2016) Biomarkers for diagnosis of sepsis in patients with systemic inflammatory response syndrome: a systematic review and meta-analysis. Springerplus 5:2091

Perman SM, Goyal M, Gaieski DF (2012) Initial emergency department diagnosis and management of adult patients with severe sepsis and septic shock. Scand J Trauma Resusc Emerg Med 20:41

Prkno A, Wacker C, Brunkhorst FM et al (2013) Procalcitonin-guided therapy in intensive care unit patients with severe sepsis and septic shock a systematic review and meta-analysis. Crit Care 17:R291

Schuetz P, Chiappa V, Briel M et al (2011) Procalcitonin algorithms for antibiotic therapy decisions: a systematic review of randomized controlled trials and recommendations for clinical algorithms. Arch Intern Med 171:1322–1331

Schuetz P, Aujesky D, MuÃàller C et al (2015) Biomarker-guided personalised emergency medicine for all hope for another hype? Swiss Med Wkly 145:w14079

Scott MC (2017) Defining and diagnosing sepsis. Emerg Med Clin N Am 35:1–9

Seymour CW, Gesten F, Prescott HC et al (2017) Time treatment and mortality during mandated emergency care for sepsis. N Engl J Med 376:2235–2244

Shi J, Tang J, Chen D (2016) Meta-analysis of diagnostic accuracy of neutrophil CD64 for neonatal sepsis. Ital J Pediatr 42:57

Singer M, Deutschman CS, Seymour CW et al (2016) The third international consensus definitions for sepsis and septic shock (Sepsis-3). JAMA 315:801–810

Tamma PD, Avdic E, Li DX et al (2017) Association of adverse events with antibiotic use in hospitalized patients. JAMA Intern Med. Published online June 12, 2017. https://doi.org/10.1001/jamainternmed.2017.1938

Tang BM, Eslick GD, Craig JC et al (2007) Accuracy of procalcitonin for sepsis diagnosis in critically ill patients: systematic review and meta-analysis. Lancet Infect Dis 7:210–217

Tong X, Cao Y, Yu M et al (2015) Presepsin as a diagnostic marker for sepsis: evidence from a bivariate meta-analysis. Ther Clin Risk Manag 11:1027–1033

Valenzuela-Sánchez F, Valenzuela-Méndez B, Rodríguez-Gutiérrez JF et al (2016) New role of biomarkers: mid-regional proadrenomedullin, the biomarker of organ failure. Ann Transl Med 4(17):329

Wacker C, Prkno A, Brunkhorst FM et al (2013) Procalcitonin as a diagnostic marker for sepsis: a systematic review and meta-analysis. Lancet Infect Dis 13:426–435

Wang X, Li ZY, Zeng L et al (2015) Neutrophil CD64 expression as a diagnostic marker for sepsis in adult patients: a meta-analysis. Crit Care 19:245

Wu Y, Wang F, Fan X et al (2012) Accuracy of plasma sTREM-1 for sepsis diagnosis in systemic inflammatory patients: a systematic review and meta-analysis. Crit Care 16:R229

Wu J, Hu L, Zhang G et al (2015) Accuracy of presepsin in sepsis diagnosis: a systematic review and meta-analysis. PLoS One 10:e0133057

Xu L, Li Q, Mo A et al (2016) Diagnostic value of C-reactive protein in neonatal sepsis: a meta-analysis. Eur J Inflamm 14:100–108

Yuan H, Huang J, Lv B et al (2013) Diagnosis value of the serum amyloid a test in neonatal sepsis: a meta-analysis. Biomed Res Int 2013:520294

Zhang J, Hu ZD, Song J et al (2015a) Diagnostic value of presepsin for sepsis: a systematic review and meta-analysis. Medicine (Baltimore) 94:e2158

Zhang X, Liu D, Liu YN et al (2015b) The accuracy of presepsin (sCD14-ST) for the diagnosis of sepsis in adults: a meta-analysis. Crit Care 19:323

Laboratory Medicine and Exercise

46

Marcello Ciaccio and Luisa Agnello

Introduction

Exercise significantly influences laboratory parameters. This chapter describes the acute effects of physical performance on the main laboratory parameters that are observed during and immediately after sports performance, as well as the chronic effects, i.e., those persistent metabolic adaptations associated with the constant practice of physical exercise.

In particular, the workload, the degree of training, the recovery time, the degree of hydration, the diet, and the use of drugs and nutritional supplements with ergogenic effects strongly affect laboratory parameters. Therefore, it is important to correctly identify these variables to appropriately interpret the laboratory data.

In addition, a very ambitious goal is to identify biochemical factors that can predict athlete performance or possible overtraining.

It is necessary to distinguish between acute effects, i.e., those that occur during and immediately after the physical performance and are often transient, and chronic effects, i.e., those persistent biochemical adaptations associated with the constant practice of physical exercise.

M. Ciaccio (✉)
Department of Biomedicine, Neurosciences and Advanced Diagnostics, Institute of Clinical Biochemistry, Clinical Molecular Medicine and Clinical Laboratory Medicine, University Hospital "P. Giaccone", Palermo, Italy

Department of Laboratory Medicine, University Hospital "P. Giaccone", Palermo, Italy
e-mail: marcello.ciaccio@unipa.it

L. Agnello
Department of Biomedicine, Neurosciences and Advanced Diagnostics, Institute of Clinical Biochemistry, Clinical Molecular Medicine and Clinical Laboratory Medicine, University Hospital "P. Giaccone", Palermo, Italy

Acute Effects of Sports Performance on Laboratory Parameters

The physiological adaptations that take place during and after sports performance result in an increase in effective volemia (reduction in hematocrit and hemoglobin), increase in muscle mass (increase in creatine kinase [CK], aspartate aminotransferase [AST], γ-glutamyltransferase [GGT]), biochemical evidence of muscle trauma (increased CK and myoglobin, AST, alanine aminotransferase [ALT], lactate dehydrogenase [LDH], K^+), hypercatabolism, increased energy substrates demand (decreased glucose, triglycerides, and increased cortisol and testosterone), and, finally, cardiovascular overload (increased troponins).

It has been documented that after a marathon, athletes show a reduction in estimated glomerular filtration rate (eGFR) and an increase in creatinine and lipocalin-2, also known as neutrophil gelatinase-associated lipocalin (NGAL). The prolonged physical activity effect on renal function is transient, and values of these parameters tend to return to pre-marathon levels within 24–48 h.

Similarly, some traditional markers of hepatic damage also show transient increases following endurance sports competitions. In particular, the most marked increases are in total and direct bilirubin, LDH, and ALT. No significant variations are observed in AST, alkaline phosphatase (ALP), and GGT. These increases, however, are transitory, and the parameters involved tend to reduce again within 24 h of the competition end. The transitory nature of these variations raises questions about their pathophysiological nature. They may be the expression of physiological mechanisms of adaptation to physical stress.

Physical exercise, especially endurance practiced at an amateur and professional level (running, cycling, and swimming), involves increased cardiac output, heart rate, and systolic pressure for several hours. This increase in cardiac work sustained over time generates myocardial stress accompa-

M. Ciaccio (ed.), *Clinical and Laboratory Medicine Textbook*, https://doi.org/10.1007/978-3-031-24958-7_46

nied by the physiological release of oxygen-free radicals (OFRs) and alterations in intracellular pH, which, in turn, support and amplify the damage of myocardiocytes.

It has been extensively documented that an acute increase in cardiac troponins (cTn) is observed in subjects practicing endurance sports, immediately after a marathon. In most observations, circulating cTn levels return to baseline 24–48 h after the athletic performance end. The advent of high-sensitivity cTn assays made possible to confirm this phenomenon in several studies. However, it is not yet clear whether this increase is physiological or indicative of irreversible myocardial damage. Hypotheses have been put forward on the mechanisms leading to this transient increase in cTn, which involve the release of cTn from cell vesicles. It is known that in the myocardiocyte, there are two intracellular pools of cTn: cytosolic, consisting of soluble cTn (6–8% for cTnT and 3% for cTnI), and structural, consisting of cTn complexed with tropomyosin and actin. The cTn derived from the structural pool is responsible for the peak observed during myocardial necrosis, which is diagnostic of acute myocardial infarction. On the other hand, release from the cytoplasmic pool is responsible for the transient small increase observed following moderate- to high-intensity physical performance accompanied by transient myocardial stress. In this case, the release of cTn occurs following an increase in membrane permeability, accompanied by the detachment of membrane vesicles containing cytoplasmic proteins and not as a result of cell necrosis. When the appropriate tissue oxygenation is restored, this phenomenon ceases, with the consequent return of circulating cTn within the reference limits. Therefore, the myocardial stress observed during physical exercise is not so prolonged or severe as to cause myocardial damage irreversible but is sufficient to cause the release of vesicles and their contents, including cTn, into the circulation.

In addition, the reduced renal clearance observed during exercise could prolong the catabolic elimination of cTn and other proteins released from the myocardiocyte, increasing their stay in circulation.

Even the main hematological parameters obtained by blood count show altered values immediately after sports performance. In particular, an increase in white blood cells (WBC), neutrophils, and monocytes, together with a decrease in lymphocytes and eosinophils, have been documented in several studies. These changes are mainly attributable to the release of catecholamines and cortisol typically observed during exercise. In addition, a decrease in red blood cells (RBC) and hemoglobin (Hb) can be observed along with an increase in platelets (PLT) and mean platelet volume (MPV).

The main changes in biochemical-clinical parameters observed following a medium-high intensity sports performance are summarized in Table 46.1.

Table 46.1 Changes in the main biomarkers in response to endurance sports performance (assessed within 3 h of the competition end)

Renal function	
eGFR	↓↓
Creatinine	↑
NGAL	↑
Hepatocyte damage	
AST	↑
ALT	–
GGT	–
Direct bilirubin	↑↑
Total bilirubin	↑↑
Alkaline phosphatase	–
LDH	↑
Myocardial damage	
CK-MB	↑
cTnI	↑↑
hs-cTnT	↑
Myoglobin	↑
NT-proBNP	↑↑
Hematological parameters	
RBC	↓
Hb	↓
HTC	↓
MCH	n.d.
MCV	↑
RDW	n.d.
Reticulocytes	n.d.
PLT	↑
MPV	↑
WBC	↑↑
Neutrophils	↑↑
Lymphocytes	↓/n.d.
Monocytes	↑
Eosinophils	↓

n.d. = no difference

eGFR estimated glomerular filtration rate, *NGAL* neutrophil gelatinase-associated lipocalin, *AST* aspartate aminotransferase, *ALT* alanine aminotransferase, *GGT* gamma glutamyl transferase, *LDH* lactate dehydrogenase, *CK-MB* creatin kinase MB, *cTn* cardiac troponin, *NT-proBNP* n terminal brain natriuretic peptide, *RBC* red blood cells, *Hb* hemoglobin, *HTC* hematocrit, *MCH* mean corpuscular hemoglobin, *MCV* mean corpuscular volume, *RDW* red distribution width, *PLT* platelets, *MPV* mean platelet volume, *WBC* white blood cells

Medium- to Long-term Effects of Physical Activity on Laboratory Parameters

It has been widely demonstrated that there is a significant association between the frequency with which endurance or physical activity is performed and many laboratory parameters. In particular, performing physical activity several times a week, regardless of the type of activity, is associated with a reduction in glucose and HbA1c, LDL cholesterol, LDL/HDL ratio, triglycerides, eGFR, plasma proteins, C-reactive protein, as well as an increase in HDL, creatinine, sideremia, and TIBC (total iron-binding capacity).

The scientific community now accepts the beneficial effects of regular physical activity on circulating lipids and,

Table 46.2 Variations of the main blood chemistry parameters induced by regular physical activity

Lipid metabolism	
LDL cholesterol	↓
HDL cholesterol	↑
Triglycerides	↓
LDL/HDL ratio	↓
Glycemic homeostasis	
Glycemia	↓
HbA1c	↓
Renal function	
eGFR	↓
Creatinine	↑
Hematological and inflammatory parameters	
RBC	n.d.
Hb	n.d.
HTC	n.d.
Reticulocytes	n.d.
WBC	↑
Neutrophils	↑
Lymphocytes	↑/n.d.
Monocytes	↑
CRP	↓

n.d. = no difference

LDL low-density lipoprotein, *HDL* high-density lipoprotein, *eGFR* estimated glomerular filtration rate, *RBC* red blood cells, *HTC* hematocrit, *WBC* white blood cells, *CRP* C-reactive protein

more generally, on the cardiometabolic biochemical profile. Indeed, for several decades, physical activity has represented the primary tool for the prevention and control of diabetes and cardiovascular diseases. Moreover, these effects on biochemical parameters depend mass body composition changes due to physical activity than on the type of activity (aerobic, power, and combined). In particular, regular physical activity induces the reduction of total cholesterol, LDL, oxidized LDL, triglycerides, and apolipoprotein B, andthe increase of HDL and apolipoprotein A1 (Table 46.2).

Reductions in glycemia and HbA1c indicate improved glycemic homeostasis due to an improvement in beta-cellular secretory function and glucose uptake by skeletal muscle that is associated with increased muscle work. These effects have been documented mainly in overweight, obese, or diabetic subjects, in whom combined physical activity (aerobic and endurance) produces the most significant reduction in HbA1c levels.

The reduction in eGFR and increase in creatinine could be due to increased muscle creatine catabolism, especially in subjects with greater muscle mass, depending on sex and age.

In a large study, an association was documented between weekly frequency of physical exercise and iron metabolism, at least in men. This effect could express an adaptive mechanism in which iron requirements for erythropoiesis and myoglobin synthesis increase with to support increased muscle work. In addition, athletes lose iron much faster than nonathletes due to a moderate degree of hemolysis during the mechanical stress induced by muscle contraction. Literature evidence shows that sideropenic anemia is frequently found in female athletes, in which, a hormonal network involving hepcidin also takes part.

Recommended Readings

Baker P, Davies SL, Larkin J et al (2014 May) Changes to the cardiac biomarkers of non-elite athletes completing the 2009 London Marathon. Emerg Med J 31(5):374–379

Church TS, Blair SN, Cocreham S et al (2010 Nov 24) Effects of aerobic and resistance training on hemoglobin A1c levels in patients with type 2 diabetes: a randomio zed controlled trial. JAMA 304(20):2253–2262

Fragala MS, Bi C, Chaump M, Kaufman HW, Kroll MH (2017 Oct 23) Associations of aerobic and strength exercise o with clinical laboratory test values. PLoS One 12(10):e0180840

Katzmarzyk PT, Leon AS, Rankinen T et al (2001 Jul) Changes in blood lipids consequent to aerobic exercise training related to changes in body fatness and aerobic fitness. Metabolism 50(7):841–848

King AC, Haskell WL, Young DR et al (1995 May 15) Long-term effects of varying intensities and formats of physical activity on participation rates, fitness, and lipoproteins in men and women aged 50 to 65 years. Circuo lation 91(10):2596–2604

Lippi G, Salvagno GL, Danese E et al (2014a) Variation of Red Blood Cell Distribution Width and Mean Platelet Volume after Moderate Endurance Exercise. Adv Hematol. Article ID 192173, 4 pages

Lippi G, Salvagno GL, Danese E et al (2014b Nov 11) Mean platelet volume (MPV) predicts middle distance running performance. PLoS One 9(11):e112892

Lippi G, Banfi G (2010) Exercise-related increase of cardiac troponin release in sports: an apparent paradox finally elucidated? Clin Chim Acta 411:610–611

Lippi G, Schena F, Montagnana M et al (2011 Oct) Significant variation of traditional markers of liver injury after a half-marathon run. Eur J Intern Med 22(5):e36–e38

Nunes LA, Grotto HZ, Brenzikofer R, Macedo DV (2014) Hematological and biochemical markers of iron status in a male, young, physically active population. Biomed Res Int 2014:349182

Pitsavos C, Panagiotakos DB, Tambalis KD et al (2009 Sep) Resistance exercise plus to aerobic activities is associated with better lipids' profile among healthy individuals: the ATTICA study. QJM 102(9):609–616

Schwingshackl L, Dias S, Strasser B et al (2013 Dec 17) Impact of different training modalities on anthropometric and metabolic characteristics in overweight/obese subjects: a systematic review and network meta-analysis. PLoS One 8(12):e82853

Shave R, Baggish A, George K et al (2010 Jul 13) Exercise-induced cardiac troponin elevation: evidence, mechanisms, and implications. J Am Coll Cardiol 56(3):169–176

Sigal RJ, Kenny GP, BouleÃÅ NG et al (2007 Sep 18) Effects of aerobic training, resistance training, or both on glycemic control in type 2 diabetes: a randomized trial. Ann Intern Med 147(6):357–369

Zoppini G, Targher G, Zamboni C et al (2006 Dec) Effects of moderate-intensity exercise training on plasma biomarkers of inflammation and endothelial dysfunction in older patients with type 2 diabetes. Nutr Metab Cardiovasc Dis 16(8):543–549

Clinical Biochemistry of Exercise

47

Giuseppe Lippi and Nicola Maffulli

The Benefits of Movement

According to the Italian National Olympic Committee (CONI), apart from the many organic benefits, the main functions of movement concern the preservation and development of the organism as a whole, increasing information for the brain, and the expressive potential of the individual (in the form of language, culture, motor intelligence, and motor expressiveness). In a nutshell, movement involves growth, and acquisition of new information, which leads to increased knowledge of ourselves, the space around us, and the relationship between the body, space, and others. In other words, moving also means evolving.

Despite these undeniable advantages, progress has meant that we spend an average of 2 h a day traveling, 7–8 h in front of desks or behind public counters, 3–4 h sitting down to eat or watch television, and 6–7 h lying down to sleep. Therefore, we are likely to spend around 20 h out of every 24 h leading a sedentary lifestyle. Paradoxically, this triggers a vicious circle: sedentary life leads to a progressive deterioration of motor skills, which is then accompanied by a greater "effort" in overcoming sedentariness, in turn, followed by further deterioration of motor skills, the appearance of instability and fear of falling, worsening of motor disability, and loss of independence, finally culminating in social and psychological isolation, leading to the appearance or increase of frailty, understood as (age-dependent) organic and functional decay.

Therefore, hypokinesis deserves to be considered a serious "syndrome," which appears in sedentary individuals due to a lack of physical activity, bad life habits, and accumulation of psychophysical stress. It usually defines abstinent, hypotonic, and overweight individuals and manifests itself with negative consequences on all organic systems (weakening them and putting them at risk of pathologies typical of senile age), especially on the muscular ligamentous apparatus (loss of tone, strength, and muscular resistance), on the cardiovascular apparatus (poor blood circulation of the muscle districts, heart fatigue, tachycardia, and hypertension), on the respiratory system (reduced vital capacity, increase in residual volume), on the cytoskeletal system (reduction in density and appearance of degenerative phenomena, up to osteoporosis and greater susceptibility to fractures), on the endocrine/metabolic system (slowing of basal metabolism, imbalance of body mass, obesity, hypercholesterolemia, appearance of insulin resistance), and even on the neurovegetative system (alteration of mood, anxiety, and depression). All these have been brilliantly summarized by a very recent report of the World Health Organization (WHO), which shows that physical inactivity accounts for 5.5% of total mortality, preceded only by hypertension (12.8%), cigarette smoking (8.7%), and hyperglycemia (5.8%), therefore outweighing obesity (4.8%) and hypercholesterolemia (4.5%). On the contrary, physical exercise can contribute to reducing global mortality by up to 30%, mortality from 40% to 70% due to fracture of the femur, from 30% to 40% due to diabetes, by 30% due to cancer (especially colorectal), from 20% to 35% due to cardiovascular diseases, and from 20% to 30% due to degenerative diseases of the central nervous system (especially dementia). These enormous benefits are manifested through a myriad of positive effects on the organism, mainly represented by an improvement in the cardiovascular function and circulatory systems, an increase in muscle mass and efficiency, an increase in motor coordination, weight control, an improvement in metabolism (lower cholesterol, improved insulin tolerance, reduced constipation), and significant psychological benefits (reduced anxiety, increased energy and self-esteem, improved sleep patterns, a tendency to socialize, an improvement in the quality of life as a whole). Assuming that all the models demonstrate a close relationship (mostly dose-dependent) between the intensity and duration of physical exercise and subsequent changes, the

G. Lippi (✉)
Section of Clinical Biochemistry, University Hospital of Verona,,
Verona, Italy
e-mail: giuseppe.lippi@univr.it

N. Maffulli
Department of Musculoskeletal Disorders, Faculty of Medicine
and Surgery, University of Salerno, Salerno, Italy

M. Ciaccio (ed.), *Clinical and Laboratory Medicine Textbook*, https://doi.org/10.1007/978-3-031-24958-7_47

positive effects on the individual organs and apparatuses are briefly summarized in Table 47.1.

It is helpful to provide some definitions to understand the aspects dealt with in the following parts of this chapter. The term "physical activity" refers to muscular effort that results in energy consumption greater than that in resting conditions; the term "physical exercise" refers instead to a complex of repetitive, programmed, and structured actions aimed at carrying out a well-defined activity (e.g., running, cycling, lifting weights) that also leads to an improvement in physical fitness and health. Finally, the term "sporting activity" implies a form of physical exercise practiced systematically and/or continuously (both for competitive and noncompetitive purposes). In the collective imagination, physical activity and exercise are often considered synonymous, although this does not correspond to reality. Physical activity indicates any form of movement that is different from sedentariness and involves a greater expenditure of energy than when at rest (for example, fetching the newspaper, doing housework, washing). In contrast, physical exercise is structurally aimed at achieving an athletic performance and is constantly associated with a progressive improvement in functional capacities (athletic performance). In other words, although in both cases it is certainly possible to improve fitness and functional capacity compared to sedentariness, the gain due to physical exercise is substantially more significant.

Table 47.1 Positive effects of physical exercise

1. Cytoskeletal system (a) Stabilization or increase in bone density (b) Increased tendon and ligament strength (c) Reduction of osteoporosis risk (d) Reduction of fall and/or fracture risk
2. Muscular system (a) Increase in muscle mass and efficiency (b) Increased density of the capillaries, mitochondria, and myoglobin (c) Improved motor coordination
3. Cardiovascular system (a) Decreased blood pressure (5–10 mmHg) (b) Decreased heart rate (even at rest) (c) Increased recovery
4. Respiratory system (a) Increased strength and elasticity of the respiratory muscles (b) Increased circulating oxygen (c) Reduced respiratory rate (d) Increased ventilation efficiency (e) Improved functional recovery
5. Metabolism (a) Weight loss (b) Improved insulin tolerance and decreased risk of type II diabetes mellitus (c) Decreased low-density lipoprotein (LDL) and increased high-density lipoprotein (HDL) (d) Improvement of triglyceride metabolism
6. Psyche (a) Loss of anxiety (b) Increased energy and self-esteem (c) Improved sleep rhythm (d) Improved life quality
7. Malignant tumors (a) Global decline in the risk of developing tumors, especially: (i) Colorectal cancer (ii) Breast cancer

Quantification of Energy Expenditure

Before entering into the specific discussion of the clinical biochemistry of physical exercise, it is helpful to express some basic concepts regarding its main determinants, which are summarized in terms of intensity and duration. The first variable defines the volume of exercise per unit of time, whereas the second defines the overall exercise duration. Generally, the two variables are linked by an inverse relationship. Thus, the greater the intensity of the exercise, the lower the ability to postpone it; on the contrary, the lower the intensity of the exercise, the greater the ability to postpone it.

To summarize, the total volume of physical exercise (or activity) contributes to the unit of measurement "MET" (i.e., "metabolic equivalent of task"), which summarizes the consumption of calories in a unit of time (compared to the resting metabolism). It follows that 1 MET is equivalent to the consumption of 1 calorie per 1 kg of weight in 1 h or 0.21 liters of oxygen per 1 kilogram of weight in 1 h (also expressed in terms of 3.5 mL of O_2/kg/min). It is clear that the higher the MET value, the greater the intensity of physical activity. Thus, the definition of physical activity thresholds is based on the "cost" in terms of METs, as follows:

- <3 MET, light-intensity activities (walking the dog, doing housework, washing up)
- From 3 to 6 METs, aerobic physical activity of moderate intensity (usually consumes 3.5–7 calories/min; corresponds to medium-fast walking between 4 and 6 km/h or cycling at 15–20 km/h)
- >6 METs, high-intensity or "vigorous" aerobic physical activity (consumes >7 calories/min; corresponds to running more than 10–12 km/h or cycling more than 20–25 km/h)

In everyday sports practice, although there are detectors capable of defining energy expenditure in the real time (advanced heart rate monitors, potentiometers, especially for bicycles), it is also possible to empirically establish the actual intensity of exercise based on subjective physical characteristics. For example, during moderate exercise, the subject will feel a little warm and perceive an increase in the heart rate and breathing rate but will retain the ability to hold a conversation, although without being able – for example – to sing. On the contrary, during high-intensity exercise, the subject will start sweating (sometimes copiously), will perceive a substantial increase in the heart rate and breathing rate, and can hardly sustain a dialogue even of a few words.

According to the recent WHO guidance, the level of physical activity needed to maintain health in adults is equivalent to 150 min per week of moderate-intensity exercise (3–6 METs) or 75 min per week of high-intensity exercise (>6 METs), whereas in children (<17 years), the threshold increases to at least 60 min of moderate- or high-intensity exercise each day. However, the concept of dose response always remains valid; the greater the amount of physical activity (or exercise), the better the results in terms of disease prevention, psychophysical well-being, and improved individual performance.

Generalities on the Structure and Biochemistry of the Skeletal Muscle

For a better understanding of the clinical biochemistry of exercise, it is worth remembering that muscle fibers are not all the same but are divided into three types:

- Slow fibers (type I; also known as "slow-twitch fibers") are red, thin, resistant, and highly vascularized with predominantly aerobic metabolism.
- Fast fibers type A (type IIa; also known as "fast-twitch fibers – A") are white, coarse, and characterized by rapid/resistant contractions and glycolytic oxidative metabolism.
- Fast fibers type B (type IIb; also known as "fast-twitch fibers – B") are white, large, and characterized by explosive contractions and anaerobic metabolism.

The speed of contractions of the different types of fibers is different. Given a maximum value of 100%, the speed of white fibers of type IIB is just 100%, decreasing to ~50% for white fibers of type IIA and down to ~10% for red fibers of type I. The different fiber composition is partly defined by genetic inheritance and partly modifiable by exercise. It is therefore evident that marathon runners, road cyclists, and cross-country skiers are generally characterized by a high percentage of slow fibers (between 60% and 80%), whereas this percentage decreases substantially in sprinters (e.g., 100 meters) to 30–40%. Without entering into dangerous ethical considerations, the genetic and/or biochemical analysis of muscle fibers would (at least theoretically) allow us to predict the most appropriate sport for each subject. Nevertheless, as is well-known, the ability to excel in a sport is not only defined by the physical constitution of the athlete but also mainly depends on training, technique, and psychology (e.g., the ability to resist fatigue, the "will to win").

Regardless of muscle tissue composition, muscular work has a number of essential characteristics that explain how it works. They can be summarized as follows:

- Muscle strength, which represents the ability to develop tension and overcome an external resistance or oppose it. It is substantially dependent on mass, fiber composition, and speed of contraction.
- Muscle speed, which indicates the ability to contract in the shortest possible time. It is substantially dependent on fiber composition, enzymatic equipment, tension, and motor activation pattern.
- Muscular endurance, which indicates the ability to maintain a certain tension over time. It is substantially dependent on fiber composition, enzymatic equipment, and muscular circulation.

Clinical Biochemistry of Metabolic Substrates

Physical activity and the muscular contextual work involve an increased demand for energy substrates, depending on the intensity and duration. Muscular work is made possible only by consuming adenosine triphosphate (ATP), the main molecule of metabolism. Each muscular contraction consumes ATP, which must be resynthesized from scratch since the size of the molecule does not allow it to be accumulated in large quantities in muscles and cells. Therefore, the neosynthesis of ATP becomes possible starting from the so-called metabolic substrates, depending on the energy demand's rapidity and oxygen availability, as synthetically described in Table 47.2.

In general, and in line with the premises, it is therefore evident how the consumption of different substrates can vary substantially depending on the type of effort required. In this case, an exercise of high intensity and short duration (e.g., less than 10 s) will be based on purely anaerobic metabolism and will require almost exclusively the consumption of ATP and/or phosphocreatine, without affecting the stocks of other substrates. Nevertheless, as the duration of the effort increases and the intensity inevitably decreases, the stocks of ATP phosphocreatine will tend to be rapidly depleted and the metabolism will first become lactic anaerobic, with prevalent consumption of carbohydrates, and will then become progressively aerobic, with prevalent oxidation of carbohydrates (as long as available) and lipids. Should the availability of these stocks run out progressively, the organism would then be forced to use proteins as the last resort of metabolic substrates ("muscular cannibalism"), broken down into amino acids and then reconverted to glucose through gluconeogenesis. The dynamics of the evolution of muscle metabolism as

Table 47.2 Characteristics of metabolic substrates

Substrate	Power	Reserves	Metabolism	Cardiac frequency	Activation	Duration	Recovery
ATP	7.3 kcal/mol	1.2 kcal	Anaerobic alactic	100% or higher	Immediate	1–7 s	10–15 s
Phosphocreatine	10.3 kcal/mol	3.6 kcal	Anaerobic alactic	95–99%	Immediate	7–12 s	5–15 min
Carbohydrates	3.75 kcal/g	~2000 kcal (glycogen ~1900 kcal and glucose ~100 kcal)	Anaerobic alactic (sometimes aerobic)	80–95%	1–3 min	0.1–3 min	4–5 min
Lipids	9.0 kcal/g	~50,000 kcal	Essentially aerobic	60–80%	10–20 min	Variable	12–72 h
Proteins	4.0 kcal/g	~30,000 kcal	Aerobic	Variable	>30–120 min	Variable	Variable

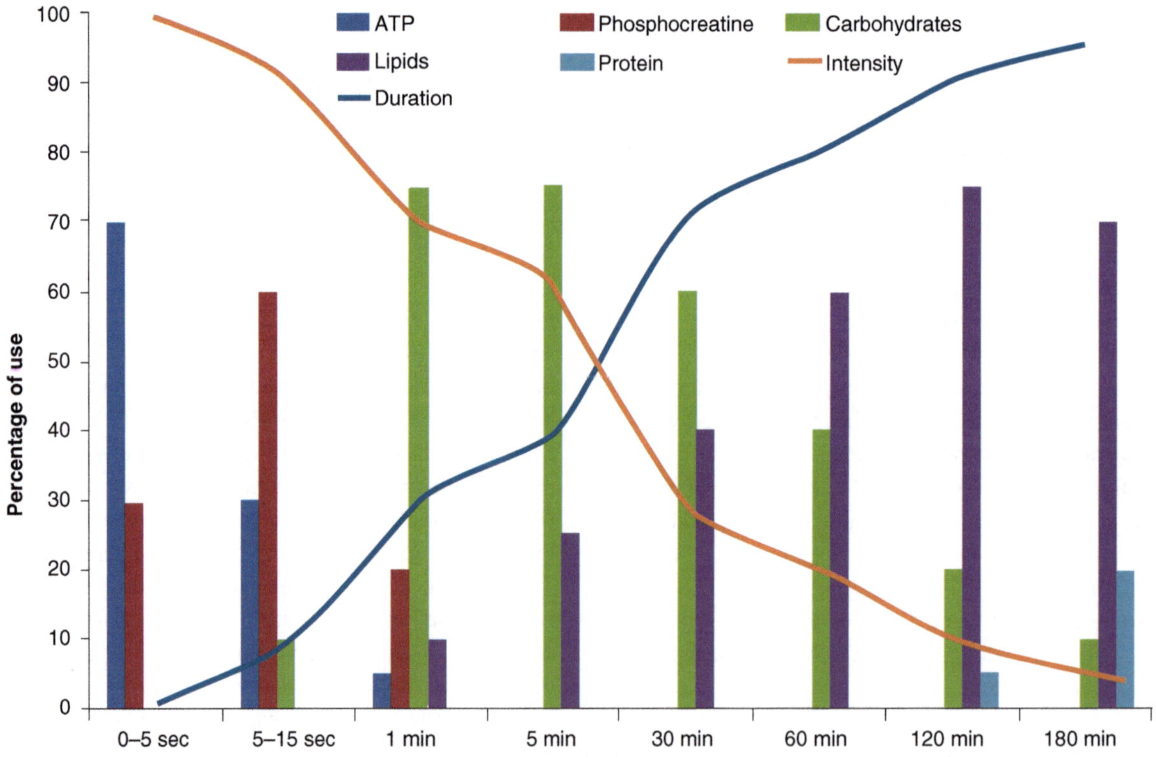

Fig. 47.1 The dynamics of the evolution of muscle metabolism as a function of intensity and duration of effort. (Copyright EDISES 2021. Reproduced with permission)

a function of intensity and duration of physical effort are well-summarized in Fig. 47.1.

It is essential to plan for the correct availability of energy substrates based on the programmed physical exercise. In this case, this means providing the body with the right food at the right time. If it, therefore, seems completely unreasonable to provide foods rich in lipids or complex carbohydrates (pasta, bread, rice, legumes) during intense but short efforts (e.g., weightlifting), then it is just as unreasonable to feed the athlete with large quantities of foods rich in simple and rapidly absorbed carbohydrates (sucrose, honey, fruit) in preparation for a marathon. Additionally, considering that the maximum degree of lipid oxidation occurs in correspondence with moderate-intensity exercise, the maximum effect on fat consumption in order to obtain a significant weight reduction will be obtained by dedicating oneself to moderate-intensity physical activity (60–80% of maximum oxygen consumption (VO_2max)) for a sufficiently long period (e.g., at least 30–60 min). A highly intense physical activity (90–95% of VO_2max) will produce a large consumption of carbohydrates (above all muscular glycogen), but it cannot be sustained for a long time and will therefore have little effect on lipid reserves.

As discussed above, physical exercise tends to progressively consume energy, even the proteins that make up the muscle itself, if other substrates (essentially carbohydrates or lipids) are unavailable. This is because physical exercise simultaneously stimulates the synthesis and degradation of proteins, the balance of which is defined by the intensity of the exercise (the more intense the effort, the greater the muscular catabolism) and by the availability of substrates for the neo-synthesis of amino acids and proteins (the lower the availability of substrates, the greater the muscular catabolism). In all subjects, especially those who are more physically active, it is therefore also necessary to supply the body with a constant quantity of proteins to maintain constant (or increase) muscle tissue. Protein intake can vary depending on body mass and needs. In particular, if a sedentary person generally needs 1 g of protein per kilogram of body mass, then this limit should be raised to 1.5–1.7 g of protein per kilogram of body mass for muscle maintenance and even to 2.0–3.0 g of protein per kilogram of body mass for muscle development ("strengthening"). It should also be remembered that the essential amino acids (phenylalanine, isoleucine, leucine, lysine, methionine, threonine, tryptophan, valine, histidine, and arginine) are pivotal for building muscle proteins and are mainly contained in foods of animal origin, such as meat, eggs, milk, and cheese.

Biochemical Parameters for Assessing Athletic Performance

In general terms, the biochemical parameters of exercise can be summarized in a discrete number of definitions, as follows:

- Maximum oxygen consumption (VO_2max): This defines the maximum sustainable exercise intensity by aerobic mechanisms of energy production, i.e., the maximum amount of oxygen usable in a unit of time per body weight (ml/kg/min). Used as a synonym for maximum aerobic power, it substantially depends on the availability of oxygen guaranteed by the integration of respiratory function (respiratory volume, lung ventilation, gas diffusion rate), cardiovascular function (heart rate, systolic volume, blood flow velocity), and blood transport capacity (amount of hemoglobin and hemoglobin saturation). The value of VO_2max can be plastically modified by training, usually increased up to 25%, but tends to physiologically decrease gradually with increasing age.
- Anaerobic threshold: This defines the maximum intensity of a physical exercise at which the ratio between generation and elimination of lactic acid in the blood remains constant (i.e., lactacidemia). It also expresses the intensity of work at which a blood concentration of lactic acid equal to 2–4 mmol/L is reached (i.e., in sedentary subjects or devoted to modest physical activity). It is therefore identified by the intensity of physical effort beyond which the metabolism remains aerobic in prevalence. Beyond this threshold, lactic acid begins to accumulate, and it is associated with fatigue and muscle pain. The volume of oxygen needed to bring the muscular system into aerobic conditions is therefore defined as an "oxygen debt."
- Aerobic capacity: Although there is no agreed-upon definition, as with anaerobic capacity, this term generally refers to the maximum amount of oxygen per minute required to perform muscular work. It depends on respiratory, cardiovascular, and blood functions as well as on the ability of the tissues to use oxygen.
- Anaerobic capacity: This essentially expresses the opposite of the previous definition and, therefore, the ability to perform muscular work under conditions of oxygen absence. It, therefore, depends to a large extent on the availability of the substrates of alactacidic and lactacidic anaerobic metabolism. In practice, it is defined as the ability to continue performing muscular work by simultaneously accumulating lactic acid. Some authors identify anaerobic capacity with the lactate concentration measured after a maximal middle-distance run (usually 800 m, with an average lactic acid concentration of 12.5 mmol/L).

The thresholds that distinguish the different areas can vary widely in the population, both as a result of genetic characteristics and following adaptations induced by training. For example, it is evident that a subject with a "congenital" high pulmonary capacity (e.g., >6 liters), a "naturally" low resting heart rate (e.g., 40 beats per minute), and "physiologically" high values of hemoglobin (e.g., >170 g/L) will have an extremely high VO_2max (e.g., >90 mL/min/kg), will be able to express a higher anaerobic threshold, and will have greater aerobic power. This will almost certainly translate into an increased ability to excel, especially in endurance sports. It is also not difficult to understand how all the parameters that express athletic capacity differ substantially between a hypothetical sedentary subject and a "champion" of endurance sports (resting heart rate: 60–80 vs. 35–55 beats per minute; threshold frequency: 150–170 vs. 170–190 beats per minute; lung capacity: 3–4 vs. 6–7 liters; VO_2max: 30–50 vs. 90–100 mL/min/kg; maximum aerobic power: 2–4 vs. 5–8 mL/min).

Biochemical Adaptation

Regardless of the "stigmata" of the champion (i.e., the genetic heritage), it is evident how a good training program can determine a substantial increase in athletic ability. It is a common experience that the more you train in a particular sport, the less effort you have to make to do the same work or, alternatively, the more work you would be able to do with the same effort. Trivializing, if a sedentary subject can cover an average of 5 km after 30 min of running (therefore, running at 10 km/h), the same subject, after a few weeks of constant training on the same distance, would be able to run much faster (e.g., at 15 km/h), to cover the same distance in less time (e.g., in 20 min) or covering a greater distance (e.g., 7.5 km) in the same time. This favorable evolution results from a well-defined biological principle named "adaptation" and involves almost all systems, organs, and apparatuses, as briefly summarized in Table 47.3.

Of particular interest in this context are undoubtedly the adaptations of the muscular tissue, which are essentially implemented by three mechanisms: (a) an increase in muscular mass (the transverse diameter of the muscle can increase by up to 30%); (b) an increase in vascularization (capillarization can increase by up to 25%); and (c) neurological mechanisms, which allow the muscle to contract more intensely (the coordination of impulses to the muscles improves, and the contraction becomes more "ordered" and practical). Concerning the first adaptation (an increase in mass), it has long been debated whether it is a consequence of hyperplasia (an increase in the size of the muscle fibers), hypertrophy (an increase in the number of cells), fibrosis (deposition of scar tissue), or a combination of all the three previous phenomena. The most accredited hypothesis is precisely the last one, involving a co-participation of all three mechanisms. If, on the one hand, it is indisputable how the continuous stimulus of a muscle determines a hypertrophic process, then it is equally clear that a muscle contraction

Table 47.3 Organic adaptations to physical activity

1. Musculoskeletal system Correct posture Muscle mass symmetry Joint mobility Osteoarticular hypertrophy
2. Endocrine/metabolic system Appetite control Basal metabolic reduction Decline in the weight-to-stature ratio Increase in lean mass Better diencephalic feedback Lipid structure improvement Neuroendocrine response improvement
3. Cardiovascular system Hypertrophy and hyperplasia Increase in diastolic and systolic range Decrease in blood pressure Increased peripheral perfusion Increased venous return
4. Respiratory system Bradycardia and bradypnea at rest Improved recovery after effort Lung volume increase
5. Emotional sphere Improved ability to withstand stress and fatigue Improved emotional control, adaptability Increased ability to socialize Improved self-esteem

determines an increased production of growth factors (some, such as insulin-like growth factor (IGF) and IGF-1, also produced as a result of a "physiological" muscle microtrauma, which determines the necrosis of some myocytes), which leads to the recruitment of muscle stem cells, thus promoting their subsequent differentiation into myocytes. Also not to be underestimated is the muscular enlargement linked to the progressive accumulation of the fibrotic tissue, consequent to repeated muscular microtrauma, especially in certain types of sports in which the muscular work becomes extreme (in weightlifting, for example). In these cases, although muscle regeneration is usually completed in 30–40 days, with restitutio ad integrum of the damaged muscle parenchyma, the deposition of the scar tissue is completed within 2–3 weeks. However, the reabsorption of the newly deposited fibrotic tissue is never complete, leaving more or less voluminous residues, which will amplify the muscle size as a whole. When following the decrease in the intensity of physical exercise (especially in old age), hyperplasia and hypertrophy will physiologically tend to regress and the fibrotic tissue will constitute part of the residual muscle volume.

Concerning circulatory changes within muscle tissue, blood flow at rest is approximately 3–4 mL/min/100 g. During intense exercise, the oxygen demand increases up to 20-fold and is associated with increased blood flow of up to 25-fold in exercising muscle (50–80 mL/min/100 g). This increase is not only unparalleled in other districts of the human body but also paradoxically involves the diversion of most of the blood flow in the muscle, increasing the irroration from 20% at rest up to 85–90% under conditions of maximal effort. Since most of the blood is then diverted to the muscle during exercise, the blood flow in other organs and systems decreases at the same time, contracting up to almost 90% (compared to the resting condition) in the kidneys, spleen, and digestive system. It is well documented that during a run at high intensity, the urination stimulus tends to stop progressively, only to resume at the end of the physical activity. This is a reflection of glomerular filtration, which tends to drastically decrease during physical activity (due to reduced vascularization of the renal parenchyma) and then gradually return to normal conditions at the end of training.

An increase in the oxygen-carrying capacity of the blood is a further adaptation to exercise. In individuals who regularly exercise in a moderate-to-intensive manner, the total blood volume (blood volume) and the amount of hemoglobin in the blood simultaneously increase with the training load. This has obvious logical assumptions (adaptive), when a greater blood mass (hypervolemia) allows the heart to reach a greater cardiac output, facilitating the transfer of oxygen to the tissues and also giving a thermoregulatory advantage (better cooling of the body), whereas a more significant amount of hemoglobin (due to the increased bone marrow synthesis of red blood cells) promotes a greater transport of oxygen to the muscle in exercise. However, in this context, it should be mentioned that hypervolemia and an increase in the amount of hemoglobin in the blood are rarely of equal magnitude since the former is greater than the latter. In summary, an aerobic training program can promote an increase in plasma volume by up to 25% (usually from about 5.2 to about 6.5 liters), at the same time as increasing the amount of hemoglobin in the blood by 20% (for example, from 800 to 950 g). Transforming these values into hemoglobin concentration in the blood, the balance appears paradoxically negative (from −1% to −2%). Despite the increase in total quantity, it is always physiological to record a progressive decrease in hemoglobin concentration in the blood in parallel with the increase in workload. Therefore, the rule is that when athletes start a 3-week cycling stage race (the Giro d'Italia, for example), the hemoglobin concentration always tends to be higher than the value recorded at the end. When this does not happen, it is reasonable to assume that the athlete has used some technique (more or less legitimate) to further increase the total amount of circulating hemoglobin, which will be discussed in a later section of this chapter.

The adaptation process now appears to result from a series of individual variables linked to the intensity of the exercise, its duration, and the character and duration of recovery intervals between one exercise and another. Programming adequate recovery allows the body to progressively raise the bar of performance (a phenomenon known as "supercompensation"), up to a hypothetical maximum defined by the theoretical limit of the athletic ability. Interspersing a series of exercises with an excessive recovery period (e.g., running only once a month) effectively nullifies the potential benefit of supercompensation. Conversely, continuing to perform highly intense exercise series without adequate recovery not only can negate the potential benefits of supercompensation but can also lead to negative consequences such as overreaching or overtraining. These three possibilities are summarized in Fig. 47.2. It is clear from the figure that, in the first case, adequate recovery allows a progressive increase in athletic performance. In the second case, where recovery is excessive compared to the load, athletic performance will

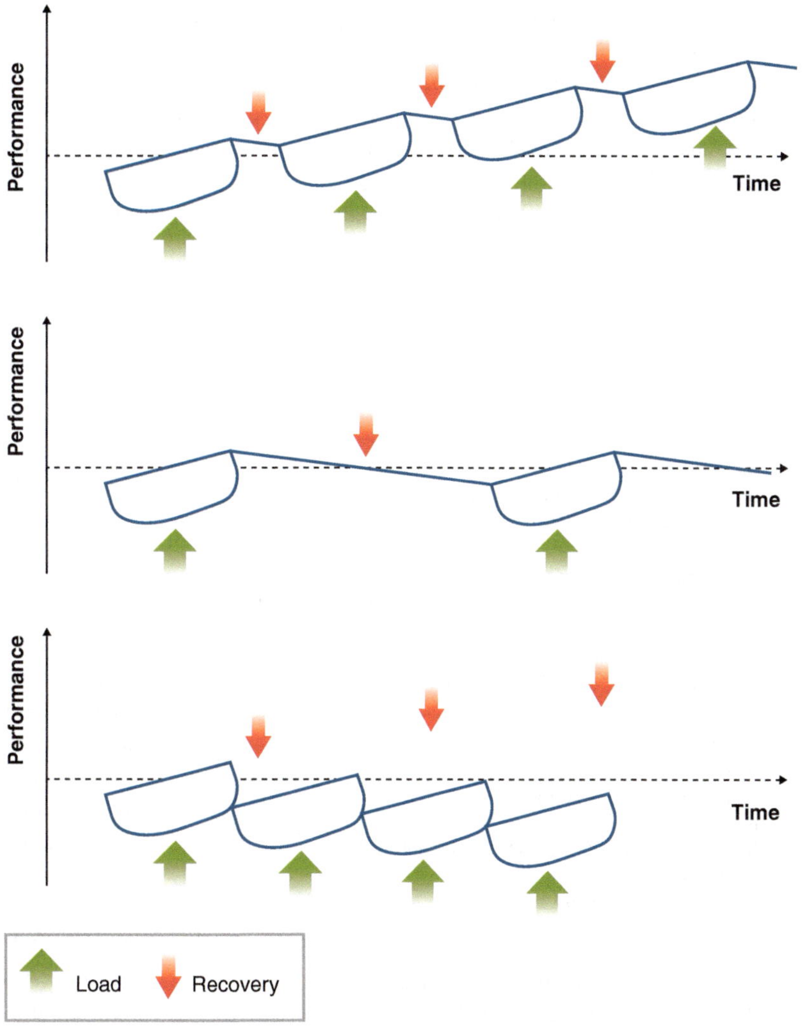

Fig. 47.2 Effects of the different cadences of workload and recovery on the progression of athletic performance (at the top, corrected program; in the middle, interval of excessive recovery, and at the bottom, interval of insufficient recovery). (Copyright EDISES 2021. Reproduced with permission)

remain unchanged. In the third case, on the other hand, characterized by too little or no recovery, fatigue will prevail over supercompensation, to the point of determining a paradoxical decline in athletic performance: the fateful, much-feared by professional athletes (and others), "overtraining syndrome."

The importance of adaptations, besides the practical effects on athletic performance, involves a real biochemical revolution. So much so that, after the Sydney Olympics, the debate arose as to whether medium/high-level athletes should be considered "normal" people because by doing a higher-than-normal physical activity, they may instead manifest many biochemical parameters that place them well outside the normal reference ranges, mostly established using sedentary or only moderately active populations. While not wishing to be all-inclusive, since the biochemical parameters that can be measured in the laboratory are almost infinite, the main biochemical alterations that occur as workloads (especially aerobic) progressively increase are summarized in Table 47.4.

All these biochemical alterations should be considered strictly "physiological," normal consequences of an adaptive metabolic process, which has progressively promoted a complex physical and biochemical change in the subject devoted to regular moderate/intense physical activity compared to a sedentary individual. The correct interpretation of these changes is essential for laboratory medicine professionals, sports physicians, and athletic trainers to be able to fully discern the concept of "pathological abnormality" from that of "physiological variation." For example, it should not be surprising to find significantly elevated creatine kinase (CK) and cardiac troponin (T or I) values after a marathon. This is in line with the physiological release during a relatively intense and prolonged physical activity. It becomes essential to know the behavior of the various laboratory parameters in athletes, since abnormalities of their kinetics instead represent highly sensitive spies of potential pathological processes (e.g., muscle damage). The variables that should therefore be considered include:

- Demographics (age, sex, ethnicity)
- Type of athletic performance (endurance, power or mixed sports)
- Workload (type of training and context, e.g., competitions or rest)
- Contingent situations (diet, hydration)
- Intake of drugs or nutritional supplements

A final consideration should be reserved for the clinical molecular biology of exercise. Reliable scientific evidence shows that specific genes' polymorphisms can significantly influence athletic performance. These genes mainly encode proteins and enzymes involved in determining the efficiency of muscle metabolism (e.g., oxidative and glycolytic enzymes), muscle performance (e.g., contractile proteins), tendon and ligament apparatus (tendon proteins), and psychological inclination. In the future, the role of laboratory medicine will be to demonstrate whether a panel integrating the most significant polymorphisms can help predict success in different sports disciplines.

Table 47.4 Biochemical adaptations to physical activity

Increased effective volume
Decrease in hematocrit (by 2–4%)
Decrease in hemoglobin (by approximately 10 g/L)
Increased muscle mass
Increase in creatine kinase enzyme (CK; total, MM, MB)
Increased myoglobin
Increased aspartate aminotransferase (AST)
Increased gamma-glutamyl transferase (GGT)
Muscle injury
Increased lactate dehydrogenase (LDH)
Increase in creatine kinase enzyme (especially MM)
Increased myoglobin
Increased potassium
Increased circulating free hemoglobin (intravascular hemolysis)
Hypercatabolism – high protein diet
Increased azotemia and azoturia
Increased uricemia and uricuria
Increased demand for substrates
Reduced resting blood glucose
Reduced resting triglyceridemia and cholesterol
Increased resting HDL cholesterol
Increased resting cortisol
Increased resting testosterone
Cardiocirculatory overload
Increased cardiac troponins after exertion
Increased natriuretic peptides after exercise

Doping

The World Anti-Doping Code ("The Code") is the principal document designed to harmonize anti-doping policies, rules, and regulations within sporting organizations and public authorities worldwide. Drawn up by the World Anti-Doping Agency (WADA), it functions as a "combination" of six international standards, which, in turn, aim to promote consistency among anti-doping organizations in various areas: types of tests, accredited laboratories, therapeutic use exemptions (TUEs), the list of prohibited substances and methods, protection of privacy and personal information, and enforcement of the Code by signatories. To date, more than 660 sporting organizations have accepted the World Anti-Doping Code, including the International Olympic Committee (IOC), the International Paralympic Committee (IPC), and all International Federations (IFs) recognized by the IOC, the National Olympic and Paralympic Committees, and the National Anti-Doping Organizations. It, therefore, represents the international standard to which all athletes (professional and amateur) must conform in the practice of competitive sport. The heart of the Code is the list ("The List") of prohibited substances and methods, published at the beginning of each year by WADA since 2004. This list summarizes all substances and methods prohibited during and outside competitions, primarily classified based on their effect or biochemical composition.

In this introduction to doping, a crucial element that has been omitted is its very definition. This is not the result of chance but rather the evolution that this concept has undergone over time, especially since WADA became the world control body. In the present case, the accepted definition of doping is now "the occurrence of one or more violations of the anti-doping rules laid down in Article 2.1 of the Code." In turn, Article 2.2 of the Code recognizes several paragraphs, which, at last, help clarify what is meant by doping today:

- Article 2.1: The presence of a prohibited substance or its metabolites or markers in a biological sample obtained from an athlete
- Article 2.2: Use or attempted use by an athlete of a prohibited substance or a prohibited method
- Article 2.3: Subtracting from collection, refusing to collect, or failing to transmit a biological sample for anti-doping examination
- Article 2.4: Athlete not present in the reported places of stay and/or absent during controls
- Article 2.5: Tampering or attempted tampering in any area of anti-doping control
- Article 2.6: Possession of substances or methods prohibited by WADA
- Article 2.7: Trafficking (or attempted trafficking) in substances or methods prohibited by WADA
- Article 2.8: Administration (or attempted administration) to any athlete of any prohibited substance or prohibited method, during or outside of competition
- Article 2.9: Complicity in activities that violate one or more of the preceding articles
- Article 2.10: Association of an athlete or other person subject to the control of WADA with persons who have been declared banned for a certain period of time by WADA

As an extension of the Anti-Doping Code, a list of prohibited substances and methods is published annually. This annual publication is functional to update the list according to new scientific evidence, with periodic inclusion of new substances and/or methods or, more rarely, exclusion of others for which a doping effect has not been unequivocally established (for example, alcohol was excluded from the list published in 2019). In the latest version of the list, prohibited substances and methods are subdivided, as shown in Table 47.5.

Table 47.5 List of prohibited substances and methods ("The List") by the World Anti-Doping Agency (World Anti-Doping Agency, WADA)

S0: Substances not yet included in the list; however, they are used to enhance athletic performance
S1: Anabolic agents (e.g., anabolic steroids)
S2: Peptide hormones, growth factors, and related and mimetic substances (e.g., erythropoietin, growth hormone)
S3: Beta-2 agonists (e.g., salbutamol, salmeterol)
S4: Hormones and metabolic modulators (e.g., tamoxifen, raloxifene, cyclofenil, insulin)
S5: Diuretics and masking agents (e.g., amiloride, spironolactone, furosemide)
S6: Stimulants (e.g., ephedrine, amphetamines)
S7: Narcotics (e.g., morphine, methadone, oxycodone)
S8: Cannabinoids (natural or synthetic)
S9: Glucocorticoids (e.g., cortisone, prednisolone)
M1: Manipulation of blood and its components (e.g., blood transfusions)
M2: Chemical and physical manipulation of the sample (e.g., adulteration of the urine)
M3: Genetic doping
P1: Beta-blockers (only in certain sports, such as motoring, archery, golf, billiards)

Without detracting from WADA, its Code, and its List, equally accepted and shared definitions of doping are "any practice designed to increase athletic performance in order to gain an unethical advantage over opponents" or "any practice designed to increase athletic performance that may damage the health of the athlete." Essentially, these further definitions have an ethical meaning, primarily devoid of practical consequences since the sanctionable practices are only those listed by WADA. Nevertheless, in the overall reasoning that considers the impact of doping practices not only on the outcome of competitions but also on the protection of athletes' health, the third definition should be an integral part of WADA's sanctioning practices. If it seems illegitimate to take anabolic agents to enhance muscle mass, then it should also be questionable to subject athletes to grueling training regimes, which could undermine their health in the long run.

Apart from the undisputed role of WADA in the management of anti-doping controls, it is clear that laboratory medicine plays a decisive role in the complex. The anti-doping laboratories accredited by WADA or by the various national and international federations use a wide range of analytical tools and methods borrowed from normal laboratory practice to analyze biological samples of athletes and identify the presence of illegal substances, sometimes even unknown. Detection in this regard mainly encompass high-performance separative techniques, such as liquid chromatography (LC) or gas chromatography (GC) combined with mass spectrometry (MS). An integral part of doping control is the counter-analysis, i.e., the indispensable and necessary confirmation of the anomaly found in the first and second samples (collected at the same time as the first) of the athlete. Each positive result is defined as an "Adverse Analytical Finding" (AAF), which will then possibly be converted into a clear violation of the Anti-Doping Code (Anti-Doping Rules Violation, ADRV), following an investigation during which the athlete will be allowed to provide explanations that justify the non-negativity previously found. If this is not possible, then the athlete will be disqualified commensurate with the violation and, where applicable (in Italy, for example), a report will be made to the judicial authorities to establish criminal proceedings against the athlete.

Recommended Readings

Lippi G, Mattiuzzi C (2018) Anti-doping testing: a moving target? J Lab Precis Med 3:60

Lippi G, Long UG, Maffulli N (2010) Genetics and sports. Br Med Bull 93:27–47

Lippi G, Schena F, Ceriotti F (2018) Diagnostic biomarkers of muscle injury and exertional rhabdomyolysis. Clin Chem Lab Med 57:175–182

Maughan RJ (2000) Nutrition in sport: Volume VII of the encyclopaedia of sports medicine. IOC Medical Commission Publication, Blackwell Science, Hoboken

Melzer K (2011) Carbohydrate and fat utilization during rest and physical activity. e-SPEN 6:e45–e52

Mul JD, Stanford KI, Hirshman MF, Goodyear LJ (2015) Exercise and regulation of carbohydrate metabolism. Prog Mol Biol Transl Sci 135:17–37

Wagner PD (1996) Determinants of maximal oxygen transport and utilization. Annu Rev Physiol 58:21–50

World Anti-Doping Agency (2018) The world anti-doping code. World Anti-Doping Agency, Montreal

World Anti-Doping Agency (2019) The prohibited list. World Anti-Doping Agency, Montreal

Preoperative Laboratory Test

48

Marcello Ciaccio

Careful patient evaluation before performing an elective, i.e., scheduled surgery, is essential to ensure the patient is adequately prepared for surgery, thereby reducing the morbidity associated with surgical procedures. In this context, the laboratory provides valuable information.

In recent years, we have witnessed an overuse of preoperative diagnostic investigations, resulting in increased health-care costs without significant benefits in terms of patient outcomes.

Although in many health-care settings, the choice of preoperative examinations is dictated by preestablished protocols, it would be more appropriate to modulate the choice of tests according to a careful assessment of surgical risk based on history, clinical data, and objective examination, as indicated by the American Society of Anesthesiologists (ASA) and the European Society of Anesthesiology. Recently, a Cochrane randomized controlled trial and meta-analysis has demonstrated the futility of predefined panels of preoperative examinations before cataract surgery. Thus, it is recommended that "routine investigations" should be replaced with targeted diagnostic tests tailored on the patient.

Furthermore, it should be noted that the objective of preoperative testing is not to define the risk of potential complications associated with surgical procedures.

Preoperative evaluation for elective procedures is usually performed 30 days to 12 weeks before surgery on an outpatient basis. It is always necessary when the surgical procedure requires anesthesia; it should not be performed for apparently healthy patients who undertake local anesthesia or sedation (dental, dermatological, and endoscopic procedures).

The National Institute for Health and Care Excellence (NICE) guidelines published in June 2003 have recently been updated. Once again, the tests to be performed must be evaluated according to the patient's disease status and the degree of complexity of the surgical procedure. However, the recommendation to abandon the approach for routine investigations of the whole preoperative population is confirmed.

A preoperative evaluation can also use instrumental investigations, such as chest X-ray, electrocardiogram, and spirometry, and laboratory tests.

The main laboratory tests before elective surgery are:

- Complete blood count (CBC)
- Kidney function tests
- Glycemia
- HbA1c
- Urine test
- Pregnancy test
- Immunohematological tests

In selected cases, blood gas analysis may also be included to complement the respiratory function assessment.

CBC should not be routinely requested but only when there is a high pretest probability of anemia, such as in chronic inflammatory diseases, chronic renal failure, or chronic liver disease, in the presence of clinical signs and symptoms of anemia. Thalassemia trait is particularly important in surgical procedures. The decision to perform further tests, such as hemoglobin electrophoresis, should be evaluated according to the clinical and family history. The platelet count must be monitored in relation to the coagulation.

M. Ciaccio (✉)
Department of Biomedicine, Neurosciences and Advanced Diagnostics, Institute of Clinical Biochemistry, Clinical Molecular Medicine and Clinical Laboratory Medicine, University Hospital "P. Giaccone", Palermo, Italy

Department of Laboratory Medicine, University Hospital "P. Giaccone", Palermo, Italy
e-mail: marcello.ciaccio@unipa.it

M. Ciaccio (ed.), *Clinical and Laboratory Medicine Textbook*, https://doi.org/10.1007/978-3-031-24958-7_48

The assessment of coagulation status includes prothrombin time (PT), activated thromboplastin time (aPTT), and fibrinogen. It is evaluated to identify possible disorders of hemostasis. In particular, these tests help identify acquired conditions associated with bleeding risk during surgical procedures. Similarly to CBC, there is consensus that these tests should be performed in patients with bleeding risk factors, i.e., chronic liver disease or oral anticoagulants.

Renal function tests include serum creatinine and electrolytes (Na^+, K^+, Cl^-, Ca^{2+}). These investigations should be performed in patients with a clinical and medical history indicative of renal impairment, such as, chronic renal failure, hypertension, heart failure, decompensated diabetes, and chronic liver disease. Moreover, they should be performed in case of treatment with diuretics, angiotensin-converting enzyme inhibitors, non-steroidal anti-inflammatory drugs, and digoxin. Age must be considered when evaluating renal function. However, in the absence of any of the above conditions, advanced age alone does not support the performance of these tests.

Glycometabolic decompensation should be investigated by measuring blood glucose and HbA1c in diabetic patients because perioperative glycemic control significantly influences the patient outcome. The prevalence of misrecognized diabetes in the preoperative population is low (0.5%), so routine performance of these tests is not recommended.

The main indication for urine testing in the preoperative patient is identifying a urinary tract infection, which may be followed by further investigations such as urine culture and antibiogram. These examinations should be requested if an infection may affect surgical or anesthesiological procedures.

The usefulness of pregnancy testing (serum β-hCG) in women of childbearing age before surgery has long been debated. The need to ascertain the presence of pregnancy is based on the perioperative risks for the fetus and the mother. In 5.8% of cases, surgical and anesthesiological procedures in pregnant women lead to spontaneous pregnancy termination; the percentage rises to 10.5% if referred to the first trimester of pregnancy. Pregnancy tests should be carried out when a pregnancy cannot be excluded based on anamnesis. This test should always be performed after the risks associated with the procedure have been adequately explained to the woman if she is pregnant and she has given her consent.

In patients undergoing thoracic surgery, pneumonectomy, esophagectomy, or head and neck surgery, it is recommended to perform blood gas analysis.

Immunohematological tests (blood group determination and transfusion compatibility) are prescribed in order not to delay a transfusion that may be necessary during surgery.

Infectious screening (HBV, HCV, HIV) should not be performed. Their relevance in this context is aimed exclusively at the safety of health-care workers because it is related to occupational exposure to potentially infected biological fluids.

Therefore, the choice of preoperative investigations must be evaluated according to the patient's health status and the degree of complexity of the surgical procedure.

According to the American Society of Anesthesiologists (ASA)'s classification, there are six classes of global operative risk:

- ASA 1: Normal healthy patient
- ASA 2: Patient with mild systemic disease
- ASA 3: Patient with a severe systemic disease that is not life-threatening
- ASA 4: Patient with a severe systemic disease that is a constant threat to life.
- ASA 5: Moribund subjects with life expectancy <24 hours regardless of surgery. Medical emergencies belong to this category.
- ASA 6: Brain-dead patient whose organs are being removed with the intention of transplanting them into another patient.

While the usefulness of preoperative screening is well-established for patients in categories 3–6, its use in healthy or asymptomatic patients in categories 1 and 2 is controversial.

The complexity of surgery is classified into three degrees:

- Grade 1: Minor surgery (e.g., excision of skin lesion, drainage of breast abscess)
- Grade 2: Intermediate surgery (e.g., reduction of inguinal hernia, saphenectomy, arthroscopy, tonsillectomy)
- Grade 3: Major or complex surgery (e.g., radical hysterectomy, endoscopic prostatectomy, total thyroidectomy, colon resection, lung surgery, cardiovascular surgery)

According to the most recent guidelines by NICE, preoperative laboratory assessment should be carried out according to the recommendations summarized in Table 48.1.

Table 48.1 Preoperative laboratory tests according to the complexity of the surgery and the ASA class according to the NICE 2016 guidelines

	ASA 1	ASA 2	ASA 3 or 4
Minor surgery			
Blood exam completed	Not routine	Not routine	Not routine
Hemocoagulation structure	Not routine	Not routine	Not routine
Renal function	Not routine	Not routine	Consider in individuals at risk of acute renal failure[a]
Respiratory function/blood gas analysis	Not routine	Not routine	Not routine
Intermediate surgery			
Blood cell count	Not routine	Not routine	Consider in subjects with cardiovascular or renal disease
Coagulation status	Not routine	Not routine	Consider in subjects with chronic liver disease and on oral anticoagulants therapy
Renal function	Not routine	Consider in individuals at risk of acute renal failure[a]	Yes
Respiratory function/blood gas analysis	Not routine	Not routine	Consider anesthetic consultation for patients in ASA 3 or ASA 4 class for suspected respiratory failure
Major surgery			
Complete blood count	Yes	Yes	Yes
Coagulation status	Not routine	Not routine	Consider in subjects with chronic liver disease and on oral anticoagulants therapy
Renal function	Consider in individuals at risk of acute renal failure[a]	Yes	Yes
Respiratory function/blood gas analysis	Not routine	Not routine	Consider anesthetic consultation for patients in ASA 3 or ASA 4 class for suspected respiratory failure

[a]Conditions associated with the risk of acute renal failure include intraperitoneal surgery, chronic renal failure (eGFR <60 mL/min), diabetes, heart failure, age >65 years, chronic liver disease, and nephrotoxic drugs in the perioperative period (NSAIDs after surgery)

Recommended Readings

Benarroch-Gampel J, Sheffield KM, Duncan CB et al (2012) Preoperative laboratory testing in patients undergoing elective, low-risk ambulatory surgery. Ann Surg 256(3):518–528

Bock M, Fritsch G, Hepner DL (2016) Preoperative laboratory testing. Anesthesiol Clin 34(1):43–58

Czoski-Murray C, Lloyd Jones M, McCabe C et al (2012) What is the value of routinely testing full blood count, electrolytes and urea, and pulmonary function tests before elective surgery in patients with napparent clinical indication and in subgroups of patients with common comorbidities: a systematic review of the clinical and cost-effective literature. Health Technol Assess 16(50):i–xvi, 1–159

Keay L, Lindsley K, Tielsch J et al (2012) Routine preoperative medical testing for cataract surgery. Cochrane Database Syst Rev 3(3):CD007293

Martin SK, Cifu AS (2017) Routine preoperative laboratory tests for elective surgery. JAMA 318(6):567–568

National Institute for Health and care Excellence. Routine preoperative tests for elective surgery. Pubblicatad Aprile 2016. https://www.nice.org.uk/guidance/ng45. Ultimaccess 2 Febbrai 2018

O'Neill F, Carter E, Pink N, Smith I (2016) Routine preoperative tests for elective surgery: summary of updated NICE guidance. BMJ 353:i3292

Rusci M (2004) Esami preoperatori: il documentNICE. Riv Med Lab JLM 5(2)

Rusci M, Cenci A, Milanesi B, Bracc G (2001) Linee guida per gli esami preoperatori. Riv Med Lab JLM 2(1)

Sui W, Theofanides MC, Matulay JT et al (2013) Preoperative testing before noncardiac surgery: guidelines and recommendations. Am Fam Physician 87(6):414–418

The Role of Laboratory in Urgency/ Emergency

49

Marcello Ciaccio, Luisa Agnello, and Paolo Carraro

Introduction

Laboratory tests play a crucial role in clinical decision-making, especially in the urgency/emergency setting, allowing timely and effective clinical interventions.

It is essential to make a distinction between "emergency" and "urgency." The difference, fundamentally, lies in the time required for intervention. Emergency refers to a condition in which immediate intervention is necessary because vital parameters are compromised and, therefore, the patient is life threatening; urgency, instead, refers to a condition in which rapid intervention is necessary but not immediate.

The introduction of automated analytical systems and their integration with computer systems allowed the clinical laboratory to manage and improve turnaround time (TAT). TAT is defined as the time interval between the specimens received in the laboratory to the time of reports dispatched with verification, and can be used for clinical purposes. All emergency laboratory tests must have a TAT <60 min.

Generally, laboratory tests in urgency/emergency are ordered from the emergency department, where a variety of clinical pictures may present, but those that commonly require laboratory support are as follows:

- Altered state of consciousness
- Asthenia
- Hemorrhagic diathesis
- Dyspnea
- Abdominal pain
- Chest pain
- Major trauma and polytrauma
- Burns

Altered State of Consciousness

An altered state of consciousness is a change in one's normal mental state. The person is not unconscious. One of the most common causes of presentation in the emergency department is a transient loss of consciousness, which may be complete (syncope) or incomplete (lipothymia)—permanent loss of consciousness ranges from a coma to psychiatric disorders. A careful history is essential to guide any laboratory tests request and diagnosis. The most common causes of altered consciousness are hypoglycemia, dehydration, hypovolemia, or anemia. Therefore, the following laboratory tests must be performed to exclude the abovementioned clinical conditions: glycemia, electrolytes (Na^+, K^+, Cl^-, Ca^{2+}), plasma creatinine, and complete blood count (CBC). In case of anamnestic suspicion, toxicological tests can be performed. In addition, a pregnancy test may be recommended in fertile women. In a diabetic patient, on the other hand, with hyperglycemia and altered consciousness, the laboratory parameters reported in Table 49.1 should be evaluated.

The Anglo-Saxon mnemonic anagram (TIPS-AEIOU-COMA) is a tool taught to first responders to deal with altered mental state patients. Each letter refers to one or more possible causes (Table 49.2).

M. Ciaccio (✉)
Department of Biomedicine, Neurosciences and Advanced Diagnostics, Institute of Clinical Biochemistry, Clinical Molecular Medicine and Clinical Laboratory Medicine, University Hospital "P. Giaccone", Palermo, Italy

Department of Laboratory Medicine, University Hospital "P. Giaccone", Palermo, Italy
e-mail: marcello.ciaccio@unipa.it

L. Agnello
Department of Biomedicine, Neurosciences and Advanced Diagnostics, Institute of Clinical Biochemistry, Clinical Molecular Medicine and Clinical Laboratory Medicine, University Hospital "P. Giaccone", Palermo, Italy

P. Carraro
Laboratory Medicine, Azienda ULSS3 Serenissima, Venice, Italy

M. Ciaccio (ed.), *Clinical and Laboratory Medicine Textbook*, https://doi.org/10.1007/978-3-031-24958-7_49

Table 49.1 Laboratory tests recommended in a diabetic patient with hyperglycemia and consciousness alterations

	Mild ketosis	Moderate ketosis	Severe ketosis	Hyperosmolar state
Glycemia (mg/dL)	>250	>250	>250	>600
pH	7.25–7.30	7.00–7.24	<7.00	>7.30
Bicarbonates (mmol/L)	15–18	10–15	<10	18
Urinary ketones	Positive	Positive	Positive	Weak positivity
Tonicity (mOsmol/kg)	Variable	Variable	Variable	>320
Sensory	Vigils	Clouded	Stupor or coma	Stupor or coma

Table 49.2 Mnemonic classification of the possible causes of consciousness alterations and the relative recommended investigations

	Cause of coma	Recommended exams
A	Alcohol and toxins	Blood gas analysis, toxicological tests, anion gap, plasma and urinary Osm, CK, myoglobinuria
E	Endocrinopathy Electrolytes Encephalopathy	Blood gas analysis, Na^+, K^+, TSH Blood gas analysis, Na^+, K^+, Cl^-, Ca^{2+}, P^+, Mg_2^+ Counts leukocytes, ammonium, PT, albumin
I	Insulin	Glucose, blood gas analysis, Hb, Na^+, K^+, Cl^-, P, Osm, lactate
O	Oxygen Opiates	Blood gas analysis, Hb, lactate, possible COHb Toxicological tests, blood gas analysis, anion gap
U	Uremia	Blood gas analysis, creatinine, urea, Na^+, K^+, Cl^-, chemical–physical examination of the urine
T	Trauma	Blood gas analysis, serial Hb
I	Infection	Blood gas analysis, leukocyte count, CRP, lactate, PT, aPTT, D-dimer, creatinine, urea, Na^+, K^+, Cl^-, glucose, blood/urine cultures, cerebrospinal fluid test
P	Psychogenic Porphyria	Glucose Complete blood count, glucose, Na^+, K^+, Ca^{2+}, TSH, any toxicological tests
S	"Seizure" (epilepsy) Syncope	Blood gas analysis, glucose, Na^+, K^+, Ca^{2+}, TSH, any toxicological tests Glucose

aPTT activated partial thromboplastin time, *CK* creatine kinase, *COHb* carboxyhemoglobin, *Hb* hemoglobin, *Osm* osmolality, *CRP* C-reactive protein, *PT* prothrombin time, *TSH* thyroid-stimulating hormone

Asthenia

Asthenia is an extremely nonspecific symptom indicative of various acute diseases, such as anemia, kidney failure, diabetes, or heart failure, as well as hyperparathyroidism, thyroid disease, rhabdomyolysis, and adrenal insufficiency. In addition, asthenia could be due to infectious (especially chronic infections) or pharmacological or physiological causes. In case of asthenia, the following laboratory tests are recommended:

- Hemoglobin (Hb)
- Glycemia
- Na^+, K^+, Cl^-, Ca^{2+}
- Creatinine
- Physical–chemical urinalysis

If electrolyte abnormalities are found, then a blood gas test is recommended. In the presence of fever, the evaluation of leukocyte count, transaminases (alanine transaminase (ALT) and aspartate transaminase (AST)), and C-reactive protein (CRP) are recommended. In case of muscle pain or statin use, creatine kinase (CK) should be tested. If thyroid disease is suspected, thyroid-stimulating hormone (TSH) should be tested. Finally, a pregnancy test should be performed if the woman is fertile.

Hemorrhagic Diathesis

Hemorrhagic diathesis, defined as the predisposition to develop bleeding, is generally the main symptom of a coagulation system alteration (coagulopathy), which can be congenital or acquired. If the coagulopathy is congenital, the diagnosis may rely on the anamnesis. In contrast, anticoagulant therapy is one of the leading causes of acquired coagulopathy and immune-based thrombocytopenia secondary to heparin treatment or myelotoxic therapy. Hepatic impairment, for example, due to fulminant hepatitis or paracetamol intoxication, may occasionally present with bleeding symptoms due to coagulation factors deficiency, such as factors V and VII, with consequent prolongation of the prothrombin time (PT). Another condition associated with bleeding diathesis, although extremely rare, is acquired hemophilia, characterized by the development of autoantibodies directed against clotting factors, called inhibitors. The most frequent condition is acquired hemophilia type A associated with a lengthening of the activated partial thromboplastin time (aPTT) and a deficiency of factor VIII (FVIII). Acquired hemophilia can have a rapid and fatal course. Thus, a mixing test can distinguish between FVIII deficiency or the presence of an inhibitor.

Useful tests for evaluating a hemorrhagic diathesis are:

- Platelet count
- Hb
- PT
- aPTT
- Fibrinogen

If PT or aPTT is altered, then a mixing test is recommended. A D-dimer assay is recommended if the condition is associated with disseminated intravascular coagulation (DIC). If laboratory tests are normal, but suspicion remains, then von Willebrand disease should be excluded by assaying Willebrand factor, factor VIII, and bleeding time.

Dyspnea

The term "dyspnea" refers to the subjective sensation of difficulty in breathing and is common to numerous pathologies, including non-respiratory ones. Blood gas analysis can estimate the severity of dyspnea and guide the etiological diagnosis (Table 49.3).

Other laboratory tests can support the etiological diagnosis. In particular, B-type natriuretic peptide (BNP) or NT-proBNP represents reliable markers for the differential diagnosis of the cause, since their levels are significantly elevated in patients with heart failure. D-dimer can be tested in case of suspected deep vein thrombosis (DVT) and pulmonary embolism (PE). It is characterized by a high negative predictive value, so if the levels are normal, then acute PE and DVT can be excluded; on the other hand, the positive predictive value of D-dimer is low, making it not reliable in confirming the suspicion of PE and DVT. In this case, a chest computed tomography (CT) scan is recommended to confirm the diagnostic suspicion. Cardiac troponin is recommended for high cardiovascular risk and ischemic heart disease. Polymerase chain reaction (PCR) and procalcitonin assays are recommended if community-acquired pneumonia (CAP) is suspected. Assessment of calcemia or ionized calcium is helpful if hyperventilation is associated with spasmophilia.

Table 49.3 Interpretation of blood gas analysis in dyspnea

Parameter	Pathophysiological orientation	Interpretation
PaO_2 ↓↓ or ↓	Hypoxia	Parenchymal involvement
PaO_2 ↓↓ $PaCO_2$ ↓	Hypoxemia	Cardiogenic pulmonary edema
PaO_2 = o ↓ $PaCO_2$ ↓	Hyperventilation	Possible pulmonary embolism with hyperventilation and shunt
PaO_2 ↓ $PaCO_2$ ↑↑	Hypoventilation	Exacerbated COPD leads to metabolic acidosis
PaO_2 ↓↓ $PaCO_2$ ↓↓	Hypoxemia with associated respiratory alkalosis	Attack of asthma or ARDS (if ventilation is efficient)

ARDS acute respiratory distress syndrome, *COPD* chronic obstructive pulmonary disease, *$PaCO_2$* partial pressure of carbon dioxide, *PaO_2* partial pressure of oxygen

Abdominal Pain

Abdominal pain could be a symptom of a clinical condition that does not require further examination or a severe condition that requires immediate surgery. Irritable bowel syndrome is the most common cause of abdominal pain. Generally, a careful history and physical examination of the abdomen allow discriminating between functional and organic disease. On the other hand, acute abdomen and massive digestive hemorrhage represent surgical emergencies that must be immediately recognized and treated. Table 49.4 shows the recommended laboratory tests according to the abdominal pain site.

C-reactive protein and blood count are the most useful parameters for identifying an urgent condition in a patient with abdominal pain and selecting patients for radiological investigation. The most useful biomarkers in abdominal pain due to acute intestinal ischemia are lactate, D-dimer, C-reactive protein, procalcitonin, and ischemia-modified albumin.

Table 49.4 Laboratory tests recommended in patients with abdominal pain

Pain site	Pathologies to consider	Level I tests	Level II tests
Right epigastrium/ hypochondrium	Hepatic colic Acute cholecystitis Hepatic distension Hydronephrosis Angina pectoris Acute myocardial infarction Pericarditis Pulmonary infarction Atypical or retrocecal appendicitis	CBC, creatinine, ALT, chemical–physical examination of urine, CRP, lipase, ALP	In case of fever: CRP, PT, aPTT In case of hepatomegaly: bilirubin, GGT In case of cardiovascular risk: troponin In case of cough: D-dimer, CRP In case of profuse vomiting: blood gas analysis, K^+
Epigastrium/ mesogastrium widespread	Appendicitis/ diverticulitis Hiatal hernia Gastritis, peptic ulcer Gastric perforation Gastric neoplasm Pancreatitis Angina, myocardial infarction Gastroenteritis Porphyria Aortic aneurysm Splanchnic thrombosis Metabolic cause	CBC, urea, creatinine, glucose, CRP, pancreatic lipase, ALP	In case of fertile woman: β-hCG In case of fever: PT and aPTT In case of cardiovascular risk: troponin In case of signs of ileus: Na^+, K^+, Cl^-
Flank	Kidney stones Hydronephrosis Cystopyelitis Renal neoplasm Colic flexure neoplasm Angina/myocardial infarction Aortic aneurysm Splenic infarction	CBC, creatinine, glucose, chemical–physical examination of urine, CRP, ALP	In case of fertile woman: β-hCG In case of fever: PT and aPTT In case of cardiovascular risk: troponin
Iliac fossa	Appendicitis/ diverticulitis Hernia	CBC, CRP, glucose	In case of fertile woman: β-hCG In case of signs of ileus: electrolytes
Lower abdomen	Appendicitis/ diverticulitis Enteritis/colitis Irritable colon Intestinal neoplasm Hernia Ureteral occlusion Ovarian cyst torsion Extrauterine pregnancy Urinary retention Cystitis Orchitis	CBC, creatinine, glucose, chemical–physical examination of urine, CRP	In case of fertile woman: β-hCG In case of rigid wall: PT and aPTT In case of signs of ileus: Na^+, K^+, Cl^-
Radicular irradiation	Shingles Root compression Diabetes Syphilis	CBC, blood sugar	In the case of a fertile woman: β-hCG
Post trauma	Organ disruption	CBC, PT, aPTT, chemical–physical examination of urine	In the case of a fertile woman: β-hCG

ALP alkaline phosphatase, *ALT* alanine aminotransferase, *aPTT* activated partial thromboplastin time, *β-hCG* β-subunit of human chorionic gonadotropin, *CBC* complete blood count, *GGT* gamma-glutamyl transferase, *CRP* C-reactive protein, *PT* prothrombin time

Table 49.5 Etiological classification and laboratory tests in chest pain

Cause	Pathologies to consider	Laboratory tests
Acute coronary syndrome	Acute myocardial infarction/angina	Troponin (kinetics)
Pericarditis	Infectious pericarditis Collagenopathy	Troponin, total leukocytes, CRP
Aortic dissection		Troponin, D-dimer
Pulmonary causes	Pneumothorax Pleurisy Pulmonary embolism	Troponin, total leukocytes, CRP, D-dimer
Gastrointestinal causes	Rupture of the esophagus Peptic reflux Gastric ulcer Cholecystitis or hepatic colic Pancreatitis	Troponin, pancreatic amylase
Other causes	Parietal pain Costochondritis Shingles	Troponin, total leukocytes

CRP C-reactive protein

Chest Pain

Chest pain is any pain between the lower dental arch and the epigastrium that does not have a traumatic or clearly identifiable cause. Acute myocardial infarction, aortic dissection, pulmonary embolism, and perforated peptic ulcers are among the most common causes. CBC and troponin assay may be performed if no other signs and/or symptoms are present. Na^+, K^+, and Cl^-, and creatinine measurements are recommended if the patient is on antidiuretic therapy.

BNP or NT-proBNP assay is recommended in a patient with heart failure signs. CRP should be measured for fever or ultrasound signs of pericardial effusion. D-dimer measurement is recommended if there are ultrasound signs of an aortic aneurysm or if the pretest probability for pulmonary embolism is low. A classification of the possible etiological causes of chest pain, with the relevant laboratory tests that should be performed, is shown in Table 49.5.

Major Trauma or Polytrauma

Trauma is defined as the effect of a force, usually mechanical, which is discharged on the part of the human body in a rapid and violent manner. Major trauma or polytrauma refers to injuries to two or more body districts with potential vital functions impairment. Laboratory tests are helpful to exclude organ damage, which was not initially evident. In addition, some tests must be repeated at 30–60 min intervals, such as hemoglobin and blood gas analysis, which are used to assess any loss of blood and fluid, hemodynamic stability, and peripheral perfusion. The initial assessment of a polytrauma patient, termed "ABCDE" (Table 49.6), allows for identifying life-threatening injuries. Specific laboratory tests are recommended depending on the injury site (Table 49.7).

Acute limb compartment is a consequence of trauma to one or more limbs. It consists of the collection of blood and/or edema in a muscle compartment, resulting in muscle ischemia. This condition is characterized by hyperkalemia, metabolic acidosis, and the release of myoglobin and CK.

Crush syndrome is characterized by a reperfusion injury resulting from traumatic rhabdomyolysis due to prolonged compression of the muscle tissue for at least 4–6 h. This compression causes compartment syndrome due to edema and swelling of the involved muscles compressing the vessels, causing ischemic changes, hypovolemic shock, and acute renal failure. Rhabdomyolysis lead to the release of metabolites and toxic substances from the damaged muscle tissue, such as amino acids contributing to acidosis, CK that is an excellent crush syndrome marker, free radicals that amplify tissue damage, histamine that causes vasodilation and bronchoconstriction, lactic acid that is mainly responsible for acidosis, and myoglobin that precipitates in the renal tubules and, especially with a low pH, contributes to renal failure. From a clinical laboratory point of view, it is possible to distinguish three phases:

1. Early phase, characterized only by hyperkalemia due to myolysis
2. Acute phase, characterized by metabolic acidosis, dark urine, creatininemia, increased levels of myoglobin and CK
3. Late phase, observed after 3–5 days, characterized by anuria and complications, such as DIC and sepsis

Creatinine values should be evaluated with caution because they are not a direct expression of renal function. Indeed, creatine and creatinine excess can result from injured muscles or dehydration, which may occur in the early stages. A feature of rhabdomyolysis-induced nephropathy is low urinary Na^+ excretion. The decrease of total calcium may be due to its necrotic muscle. Phosphate increases in parallel with K^+ by release from the injured muscle. In the urine, it is important to look for myoglobin and pigmented cylinders.

Table 49.6 ABCDE evaluation of the polytrauma patient

	Condition	Clinical evaluation	Laboratory tests
A	Airway	Obstruction control	No
B	Breathing	Breath control	Blood gas analysis, O_2 saturation
C	Circulation	Bleeding assessment	Hb, coagulation, blood group
D	Disability	Neurological evaluation	No
E	Exposure	Examine the whole body	No

Table 49.7 Laboratory tests recommended based on the trauma site

Site	Pathologies to consider	Laboratory tests
Craniofacial	Fractures Confusion of the brain Intracranial hemorrhages Pneumocephalus Spinal cord injuries	Serial Hb Blood gas analysis Creatinine Na^+, K^+ Pregnancy test, if the woman is fertile Toxicological tests, if the patient is a driver
Thoracic-cervical	Injuries of the esophagus–respiratory tract Lung lesions Pneumothorax Cardiac tamponade Myocardial contusion Vascular lesions DIC	Serial Hb and blood group Serial blood gas analysis Creatinine Na^+, K^+ Troponin PT, aPTT, fibrinogen Lactic acid, in case of hypotension Pregnancy test, if the woman is fertile Toxicological tests, if the patient is a driver COHb, if burns are present D-dimer, if coagulation is impaired
Abdominal	Vascular lesions Spleen–liver rupture Pancreatic injury Intestinal perforation Kidney injury Genitourinary lesions DIC	Serial Hb Blood group Serial blood gas analysis Creatinine Na^+, K^+ Amylase Urine chemical–physical examination PT, aPTT, fibrinogen Lactic acid, in case of hypotension Pregnancy test, if the woman is fertile Toxicological tests, if the patient is a driver D-dimer, if coagulation is impaired COHb, if burns are present
Limbs, with extensive muscular involvement	Acute limb compartment Crush syndrome DIC Sepsis	Serial Hb CK Blood group Blood gas analysis Creatinine Na^+, K^+, Ca^{2+}, P^+ Urine examination with sediment PT, aPTT, fibrinogen Pregnancy test, if the woman is fertile Toxicological tests, if the patient is a vehicle driver D-dimer, if coagulation is impaired COHb, if burns are present

aPTT activated partial thromboplastin time, *DIC* disseminated intravascular coagulation, *CK* creatine kinase, *COHb* carboxyhemoglobin, *Hb* hemoglobin, *Osm* osmolality, *PT* prothrombin time

Table 49.8 Laboratory tests in crush syndrome

Detection	Diagnosis	Monitoring
K^+	K^+	K^+
CK	CK	Urinary pH
Blood gas analysis	Chemical–physical examination of the urine (with cylinders detection) Plasma creatinine Blood gas analysis Plasma Ca^{2+} and P^+ Urinary Na^+ Serum and urinary myoglobin	Hourly urinary volume Osm Blood gas analysis Plasma creatinine

CK creatine kinase, *Osm* osmolality

Table 49.8 shows laboratory tests helpful in guiding the management of patients with crush syndrome.

Burns

The term "burn" indicates a skin lesion caused by irradiation or direct contact with thermal agents as well as all the pathophysiological alterations due to heat damage. The burns pathological insult is often the result of various effects, such as the damaging action on the external surfaces of the body, the inhalation of fumes (heat, toxic agents), and the consequences of the loss of liquids.

The treatment aim of a severe burn victim is fluid restoration. It is important to assess urinary volume, urinary osmolality, and plasma electrolytes every hour until stabilization. Hematocrit is a reliable marker of dehydration severity; burn dehydration is normotonic, and, therefore, other plasma parameters do not provide information on hemoconcentration. If an inhalation injury is suspected, then dyspnea, arterial desaturation, and increased carboxyhemoglobin (COHb) will be observed; the 15% COHb limit is used to confirm this condition requiring increased hydration.

In the case of burns caused by high-voltage electrocution or lightning, muscle damage should be assessed by CK and electrolytes (Na^+, K^+, Cl^-, and Ca^{2+}).

Recommended Readings

Carraro P (2011) Esami di laboratorio raccomandati in al- cune tipiche situazioni di Pronto Soccorso. Biochimica Clinica, 35(3)

Salapete C, Siniscalchi V, Ciampa A, Castelli G, Palermo E (2014) Due nuovi casi di emofilia acquisita A. Biochimica Clinica, 38(6)

Biological Drugs

50

Marcello Ciaccio, Bruna Lo Sasso, Salvatore Milano, and Caterina Maria Gambino

Introduction

The evolution of recombinant technologies and the current knowledge of molecular biology have shed light on the mechanisms underlying the functioning of our organism and the molecular pathways involved in the etiopathogenesis of many diseases. These advances have enabled identifying new "molecular targets" and developing targeted therapies. The latter are based on new-generation, so-called "intelligent", drugs targeting injured preserving healthy organs, tissues, and cells. Some are defined as "synthetic" when synthesized *ex novo* in laboratories, and others as "biological" when the active component is produced from a biological source. They act as endopharmaceuticals by mimicking the function of endogenous biological molecules (insulin and/or cytokines).

Biological drugs include hormones and enzymes, blood products, immunological drugs (sera and vaccines), immunoglobulins, allergens, and monoclonal antibodies.

Insulin, extracted from the pancreas of pigs and cattle in 1920, is the first example of a biological drug in the history of medicine, followed by the extraction of other proteins for therapeutic purposes (factors VIII and IX for the treatment of hemophilia and growth hormone obtained from human cadavers for the treatment of dwarfism). Initially, the use of "biological drugs" was limited due to the reduced availability of sources, high production costs, and considerable risk of spreading new diseases. During treatment, many hemophiliacs contracted hepatitis B and C virus infections and, in some cases, HIV.

Towards the end of the 1970s, advances in biotechnology made it possible to overcome these limitations, allowing the production of human proteins by transfection into bacterial cells; the synthesized molecules were called "biotech drugs" (Fig. 50.1).

In the same years, Geroge Kohler and Cesar Milstein identified a new technique to produce monoclonal antibodies (monospecific antibodies produced by a single cell type) using a "hybridoma." Hybridomas are engineered cells obtained by fusing murine B-lymphocytes isolated from the spleen or lymph nodes of the immunized animal with myelomatous cells that confer immortality and stability.

The hybridoma is propagated in culture as an individual clone and constitutes a stable and permanent source of monoclonal antibodies. It is used for diagnostic purposes or protein isolation and purification, both for research purposes and on an industrial scale (Fig. 50.2).

This discovery won them both the Nobel Prize in 1984. However, the antibodies produced had high immunogenicity (the ability to induce an immune reaction) due to the presence of murine fractions, which were therefore recognized as foreign by the human body. Subsequent research developed strategies to progressively "humanize" monoclonal antibodies, thereby reducing the immune response of the human organism (Fig. 50.3).

The identification of murine sequences in the structure of the molecule makes it possible to classify the antibodies into murine, identified by the suffix -omab; chimeric, consisting of a human and a murine component, identified by the suffix -ximab; humanized, in which the murine component is present in minimal fractions and identified by the suffix -zumab; and fully human, identified by the suffix -umab, which lack murine sequences and are, therefore, characterized by very low immunogenicity (Fig. 50.3).

M. Ciaccio (✉) · B. Lo Sasso · C. M. Gambino
Department of Biomedicine, Neurosciences and Advanced Diagnostics, Institute of Clinical Biochemistry, Clinical Molecular Medicine and Clinical Laboratory Medicine, University Hospital "P. Giaccone", Palermo, Italy

Department of Laboratory Medicine, University Hospital "P. Giaccone", Palermo, Italy
e-mail: marcello.ciaccio@unipa.it

S. Milano
Department of Laboratory Medicine, University Hospital "P. Giaccone", Palermo, Italy

M. Ciaccio (ed.), *Clinical and Laboratory Medicine Textbook*, https://doi.org/10.1007/978-3-031-24958-7_50

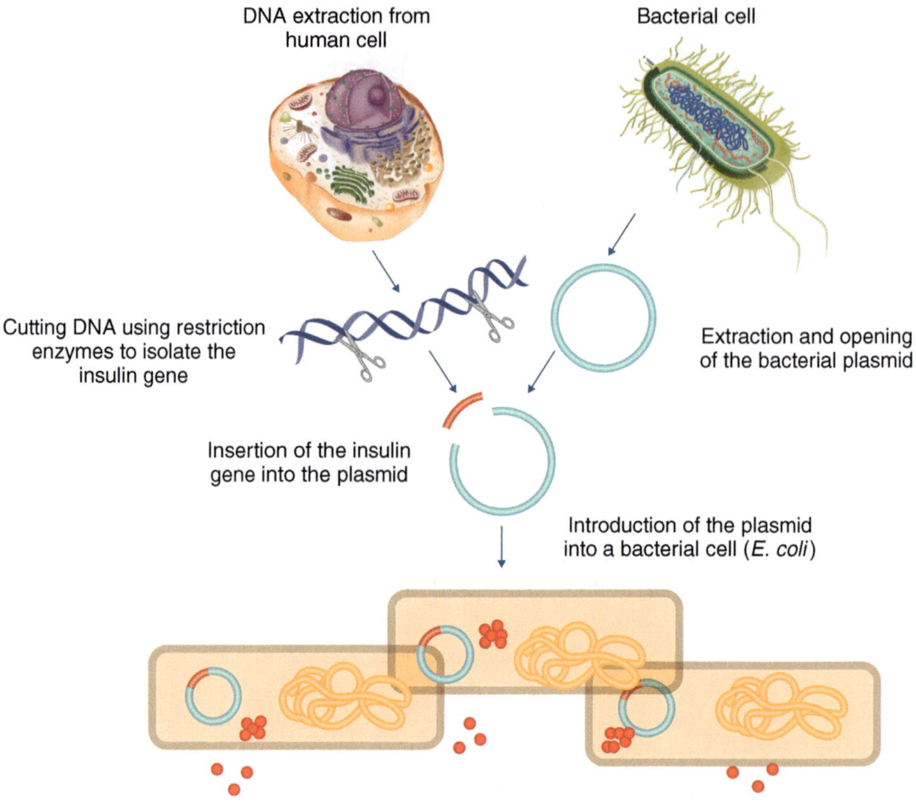

Fig. 50.1 Human insulin production via bacterial plasmid and transfection. (Copyright EDISES 2021. Reproduced with permission)

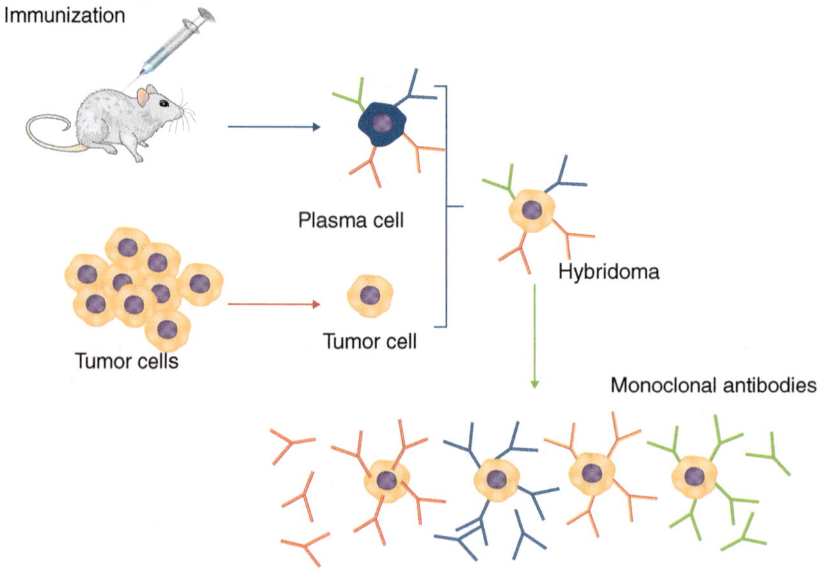

Fig. 50.2 Production of a monoclonal antibody from a hybridoma. (Copyright EDISES 2021. Reproduced with permission)

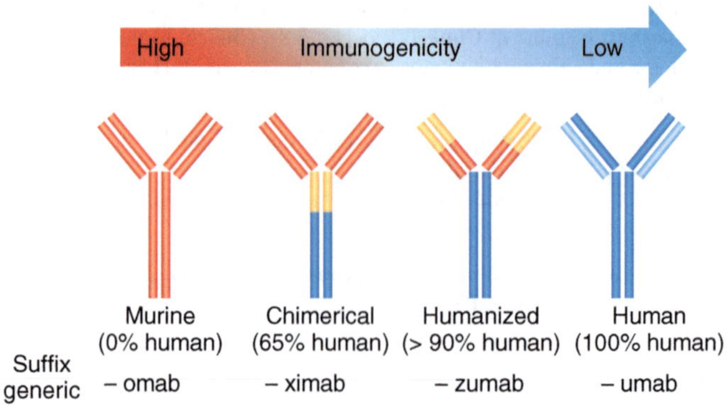

Fig. 50.3 Immunogenicity of monoclonal antibodies and identification suffix. (Copyright EDISES 2021. Reproduced with permission)

Biological and Traditional Drugs

Unlike drugs obtained by chemical synthesis, biological drugs require special attention concerning the manufacturing process on which their complex structure depends.

In particular, the production process determines biological medicine's uniqueness and specific characteristics. For this reason, the same biological drug produced by different companies, or even by the same company but using different methods, may have significant structural changes and, accordingly, different safety and efficacy.

Therefore, the national and international authorities granting the marketing of biological drugs are obliged to check every process in the production and distribution chain scrupulously. Drugs produced by pharmaceutical chemistry methods differ from biopharmaceuticals in several aspects, including the molecular dimensions that reflect the different structural complexity (tens of thousands of Daltons for biopharmaceuticals, hundreds of Daltons for traditional ones), the stability of the final product, and the presence of different co- and post-translational modifications (glycosylation) typical of biological drugs. In addition, most biopharmaceuticals present numerous aspects of heterogeneity related to the host cell used, the plasmids used to transfect the host cell, the growth conditions, fermentation processes, and different purification methods.

For these reasons, biological drugs are characterized by different levels of complexity that differentiate them into "first-generation" biotechs, such as erythropoietin, and "second-generation" biotechs, represented by more complex molecules, such as monoclonal antibodies.

Biotech drugs account for 20% of the drugs currently on the market and 50% of those in development. In many cases, they represent the only therapeutic option for significant and widespread pathologies such as anemia, cystic fibrosis, and some forms of neoplasia and represent the primary resources against rare genetic diseases.

Immunogenicity of Biological Drugs

Biopharmaceuticals, as previously mentioned, can be recognized by the body as "foreign," causing unwanted immune reactions due to their composition and molecular size; this intrinsic property of biological drugs is called "immunogenicity." Most immune responses are mild and usually do not cause adverse effects. However, sporadically, unwanted immune reactions can lead to severe and harmful outcomes for the patient's health. Immunogenicity modifies the efficacy and tolerability of the drug. It can be detected and quantified by anti-drug antibodies (ADA) that bind the drug in a stable complex, inhibiting the action of the biopharmaceutical on its molecular target. ADAs play twofold action: they reduce the circulating concentration of active drugs (pharmacokinetics) and decrease their ability to act by preventing drug-target molecule binding (pharmacodynamics). In particular, knowledge of the pharmacokinetics of biological drugs is of primary importance to obtain maximum efficacy with minimum adverse effects. The factors that influence the pharmacokinetics of biopharmaceuticals are multiple and include, in addition to ADAs, concomitant use of immunosuppressants, elevated tumor necrosis factor-alpha (TNF-α) concentration, body weight, and male gender.

ADAs vary in terms of i) synthesis that derives from the intrinsic immunogenicity of the molecule due to the presence of murine fractions and contaminants but also depends on the manufacturing process; ii) therapeutic protocol referring to drug dosage, duration, concomitant use of other immunomodulators, condition of the individual patient's immune system, and subjective pharmacokinetic variability.

Therefore, the concentration of circulating drugs and the determination of the ADAs, together with the evaluation of the patient's clinical status, represent the main parameters for biological therapy monitoring.

Fields of Application of Biological Drugs

Biopharmaceuticals represent a new therapeutic frontier for treating several immune-mediated inflammatory diseases, defined as "immune-mediated inflammatory diseases", which include different highly disabling pathological conditions, apparently unrelated to each other, but which share the same inflammatory pathways. In this clinical context, an unregulated immune response triggers inappropriate activation and production of proinflammatory cytokines (TNF-α, interleukin [IL]-12, and IL-6), whose actions can lead to pathological conditions (Table 50.1).

The primary aim of pharmacological therapies is to halt the course of the immune-mediated pathology and, in some cases, restore the initial conditions.

Traditional pharmacological treatments are based on symptomatic drugs, such as analgesics, non-steroidal anti-inflammatory drugs, and "disease-modifying anti-rheumatic drugs" (DMARDs). The latter include substances with mainly immunomodulatory action (chloroquine, hydroxychloroquine, and sulfasalazine) and immunosuppressive drugs (methotrexate [MTX] and cyclosporine A) that act on the pathogenetic mechanisms of inflammatory diseases to slow down the disease progression.

A new therapeutic approach is based on biological drugs to block the mediators of the immune-mediated inflammatory response.

Table 50.1 Major immune-mediated inflammatory diseases

Gastroenterological
Crohn's disease
Ulcerative colitis
Celiac disease
Whipple's disease
Chronic autoimmune hepatitis
Arthritis
Psoriatic arthritis
SAPHO
syndrome (Synovitis – Acne – Pustulosis – Hyperostosis – Osteitis)
Ankylosing spondylitis
Reactive arthritis
Juvenile chronic arthritis
Adult Still's disease
Vasculitis
Wegener's granulomatosis
Polymyalgia rheumatica/giant cell arteritis or temporal arteritis
Takayasu arteritis (aortic arch syndrome)
Kawasaki disease (mucocutaneous lymph node syndrome)
Polyarteritis nodosa
Cryoglobulinemic vasculitis
Behçet's disease (*silk road disease*)
Recurrent polychondritis
Urticarial vasculitis
Multicentric reticulum-histiocytosis
Erythema nodosum
Dermatological
Psoriasis
Gangrenous pyoderma
Pemphigus
Connective tissue
Systemic lupus erythematosus
Scleroderma
Sjögren's syndrome polymyositis/dermatomyositis
Mixed connectivity
Antiphospholipid antibody syndrome
Endocrinological/others
Chronic autoimmune thyroiditis
Graves' disease
Amyloidosis
Sarcoidosis
Myasthenia gravis
Multiple sclerosis

Biological Drugs in Gastroenterology: Crohn's Disease and Ulcerative Rectocolitis

Crohn's disease (CD) and ulcerative rectocolitis (UC) are frequently debilitating chronic inflammatory bowel diseases (IBDs), characterized by an inflammatory state of the intestinal mucosa with the release of proinflammatory cytokines by macrophages. The IBD etiology is unknown, and there are no specific curative therapies. The primary aim is to control the inflammatory state and maintain the disease in remission. The early disease onset, the influence on the social activity of patients, the costs of hospitalization, and the side effects of corticosteroid therapies have led to the development of new and potentially more effective therapies in recent years. Traditional therapies with corticosteroids and DMARDs are often flanked or replaced by biological drugs that inhibit the synthesis, release, and activity of proinflammatory mediators and, in particular, TNF-α inhibitors.

TNF-α is a proinflammatory and immunoregulatory cytokine, mainly produced by macrophages and other cell types (monocytes, lymphocytes, endothelial cells, keratinocytes, mast cells, and granulocytes) under different stimuli (trauma, radiation, bacteria, viruses, and tumor cells). The main biological functions of TNF-α are to promote the recruitment of neutrophils and monocytes in the foci of infection, to induce the expression of adhesion molecules (selectins and ligands of integrins) on endothelial cells, and to stimulate the production of IL-1, IL-6, and chemokines. High concentrations of TNF-α are found in some chronic inflammatory diseases due to an immune response.

The first biological drug used in the context of IBDs was Infliximab (trade name Remicade), which is also used in other clinical studies.

Infliximab is a chimeric G1 (IgG1) type monoclonal antibody whose ligand is TNF-α. It comprises a constant IgG1 region of human origin, which accounts for 75% of the antibody, and the antigen-binding part of murine origin, which makes up the remaining 25%. Infliximab can bind all forms of TNF-α (soluble, transmembrane, and soluble monomeric) and block the binding of TNF-α to its receptors, neutralizing its proinflammatory effects. The presence of a murine component in the molecular structure determines, however, a certain degree of immunogenicity and a decrease in the efficacy and tolerability of the drug.

Biotechnological advances have enabled the synthesis of a fully humanized antibody (adalimumab [ADL]), which significantly reduce the immune response. ADL, trade name Humira, is a monoclonal antibody that neutralizes the action of TNF-α and possesses excellent selectivity of action, not interfering with other body systems. The drug is administered subcutaneously, reducing the adverse events observed by intravenous administrations (burning, itching, and irritation of the skin at the site of infusion), promoting adherence to therapy, and improving the patient's overall quality of life. The drug is indicated not only for IBDs but also for treating psoriasis and psoriatic arthritis. It cannot be used if the patient is positive for *Mycobacterium tuberculosis*.

Etanercept, trade name Enbrel, is a biological drug to treat IBDs but with a different mechanism of action than monoclonal antibodies. It is a fusion protein obtained by recombinant DNA techniques of the human TNF-α factor receptor p75 with the Fc fraction of the human immunoglob-

Table 50.2 Biological therapies in use for IBD treatment

Active principle	Target molecular	Pathology	Route of administration	Dosage
Infliximab	TNF-α	Crohn's disease Ulcerative colitis	Intravenous	Initial phase: a first dose of 5 mg/kg followed by an additional infusion of 5 mg/kg 2 weeks after the first infusion; maintenance: infusion of 5 mg/kg at 6 week after the first dose, followed by repeated infusions every 8 weeks
Adalimumab	TNF-α	Crohn's disease Ulcerative colitis	Subcutaneous	Initial phase: a dose of 80 mg followed by doses of 40 mg at 2nd week; maintenance: 40 mg every other week
Golimumab	TNF-α	Ulcerative colitis	Subcutaneous	Initial phase: 200 mg dose followed by 100 mg at 2nd week; maintenance: 100 mg every 4 weeks
Certolizumab	TNF-α	Crohn's disease	Subcutaneous	Initial phase: 400 mg dose followed by 2nd and 4th week; maintenance phase: 400 mg every 4 weeks
Ustekinumab	IL-12 and IL-23	Crohn's disease	Intravenous Subcutaneous	Initial phase: infusion of 6 mg/kg followed by the same dosage after 8 weeks; maintenance phase: 90 mg after 8 weeks and one dose every 12 weeks thereafter

IBD inflammatory bowel disease, *IL* interleukin, *TNF-α* tumor necrosis factor alpha

ulin IgG1. The protein acts as a soluble receptor for TNF-α and has a high binding affinity, inhibiting the action of the pro-inflammatory molecule.

The drug is administered subcutaneously and is included in treatment protocols for IBDs, psoriasis, psoriatic arthritis, and rheumatoid arthritis (RA).

New biological drugs for IBDs treatment are Golimumab, trade name Simponi, a human IgG1κ monoclonal antibody produced from a murine hybridoma cell line using recombinant DNA technology; Certolizumab, trade name Cimzia, represented by a Fab' fragment of humanized recombinant antibody directed against TNF-α expressed in *Escherichia coli* and conjugated with polyethylene glycol (PEG); and Ustekinumab, trade name Stelara, a fully human IgG1κ monoclonal antibody that binds with high affinity and specificity the p40 protein subunit of interleukins IL-12 and IL-23, inhibiting the binding of these cytokines to the specific IL-12Rβ1 receptor expressed on the surface of immune cells. The latter drug, used in treating CD, is also indicated for psoriasis (Table 50.2).

Treatment of CD with biological drugs led to the healing of intestinal mucosal ulcers (so-called "mucosal healing") in 40% of cases within the first year of treatment and the maintenance of clinical remission in 60–70% of patients.

However, it is now known that biological therapies do not act in the same way on all subjects due to the occurrence of adverse events, the development of circulating ADA that neutralizes the effect of therapy, and individual variability in the pharmacokinetics of the molecule administered. In these cases, it is recommended to monitor the circulating drug concentration and simultaneously monitor the ADAs produced, modifying the therapeutic protocol in terms of dosage, frequency of administration, or, in some cases, replacing the anti-TNF-α drug with another of the same or different class.

Biological Drugs in Rheumatology: Rheumatoid Arthritis and Ankylosing Spondylitis

RA is a chronic inflammatory disease of the joints, characterized by a progressive course that can evolve towards ankylosis with disabling deformities. However, the involvement of other organs and apparatuses, such as the eye, skin, and lungs, is not uncommon. The onset and course are variable, with acute or slowly evolving forms resulting in numerous clinical variants. The etiopathogenesis of RA is not entirely known. It is hypothesized that an antigenic stimulus, a genetic predisposition (in particular, the presence of the human leukocyte antigen [HLA]-DR4), and the unregulated activation of the immune system with the consequent development of an inflammatory process are involved. The prevalence rate varies from 0.3% to 1.5%. The incidence is approximately six new cases per 10,000 people per year and occurs more frequently in women aged between 25 and 50 years, except in the elderly population, where the incidence is the same in both sexes.

Ankylosing spondylitis (AS) is a chronic inflammatory disease characterized by joint stiffness secondary to ankylosis of the joint surfaces and inflammation of the spine. Its prevalence is estimated to be around 0.1–0.2% and mainly affects males aged between 20 and 40 years, with a male/female ratio of 9:1. The pathogenesis of AS is not yet fully understood. Mechanisms of the immune-mediated response involving the HLA B-27, inflammatory cellular infiltrates, and cytokines with proinflammatory activity, such as TNF-α and IL-10, as well as genetic and environmental factors, are thought to play a critical role.

Traditional therapies for the described pathologies (RA and AS) involve the use of corticosteroids, non-steroidal antiphlogistic drugs, and DMARDs, among which we can mention slow-acting anti-rheumatic drugs (SAARDs), D-penicillamine, and chloroquine derivatives.

MTX is currently the drug of choice for the treatment of RA. Its mechanism of action involves the inhibition of dihydrofolate reductase and the reduction of adenosine levels, leading to a reduction in the number and activity of leukocytes. Although the causes of RA and AS are not yet known, it has been definitively demonstrated that chronic inflammation, which characterizes these clinical conditions, is determined by the loss of the physiological balance between proinflammatory and anti-inflammatory proteins with an excess of cytokines and the manifestation of inflammatory events in the joint sites. Research in recent years has evaluated the possibility of using molecules for mimicking the normal function of natural anti-inflammatory proteins, which are, for this reason, defined as "biological DMARDs" (bDMARDs). Biological therapies have proven to be able, in a short time compared to traditional drugs, to induce satisfactory control of arthritis and the evolution of joint erosive damage in many subjects not responding to traditional anti-rheumatic drugs. Currently, the administration of traditional anti-rheumatic drugs, especially MTX, is associated with biological drugs to enhance their action or prevent the loss of efficacy over time.

However, biopharmaceuticals are not exempt from inducing side effects (adverse drug reactions [ADRs]), mainly represented by an increased risk of infections due to the immune system response depression, particularly tubercular infections or non-severe infectious conditions that can be treated with antibiotics. Other adverse events specific to biopharmaceuticals are represented by reddening, itching, and swelling at the site of subcutaneous infusion and vasomotor crises (also anaphylactic); the synthesis and appearance in the circulation of autoantibodies, mainly antinuclear antibody and anti-DNA, have been observed (Table 50.3). In addition to the ADRs described above, biological drugs can present specific allergic reactions, different effects on the immune system, and an increased risk of the onset of some neoplasms.

The biological therapies currently available in the clinical context of rheumatic diseases are numerous (Tables 50.4 and 50.5). The biological drugs approved in the therapeutic protocols of RA are Abatacept (trade name Orencia), a fusion protein produced with recombinant DNA technology in Chinese hamster ovary (CHO) cells; Rituximab (trade name MabThera), a chimeric murine/human monoclonal antibody, consisting of a glycosylated immunoglobulin with the constant IgG1 regions of human origin and the variable region light chain and heavy chain sequences of murine origin (the antibody is produced using a culture of

Table 50.3 Main adverse events associated with biological disease modifying antirheumatic drugs

ADA synthesis[a]
Injection site reaction
Skin rash, urticaria, skin vasculitis
Headache, dizziness, behavioral disturbances
Gastrointestinal disorders
Leukopenia, thrombocytopenia
Increased liver enzymes
Alopecia
Upper airway infection
Skin infections
Chickenpox or Herpse Zoster
Pneumonia
Aseptic meningitis
Recurrence or onset of tuberculous infection

ADA antidrug antibodies

Table 50.4 Biologic disease-modifying antirheumatic drugs currently available in rheumatoid arthritis

Target molecular	Principle active	Route of administration	Dosage
TNF-α	Infliximab	Intravenous	Initial phase: infusion of 3 mg/kg followed by additional infusions of 3 mg/kg at 2nd and 6th weeks after the first infusion; maintenance: infusion of 3 mg/kg every 8 weeks
	Adalimumab	Subcutaneous	40 mg every other week
	Etanercept	Subcutaneous	25 mg to be administered twice a week or 50 mg once a week
	Certolizumab	Subcutaneous	Initial phase: 400 mg dose followed by 400 mg at 2nd and 4th week; maintenance phase: 200 mg every 2 weeks or 400 mg every 4 weeks
	Golimumab	Subcutaneous	50 mg once a month
CD80/86 CD28	Abatacept	Intravenous Subcutaneous	Initial phase: dose of 10 mg/kg followed by doses of 10 mg/kg at 2nd and 4th week; maintenance phase: 10 mg/kg every 4 weeks
B lymphocytes	Rituximab	Intravenous	Initial phase: infusion of 1000 mg followed by 2nd infusion 2 weeks later; maintenance phase: re-evaluation after 24 weeks for the next cycle
IL-6 receptor	Tocilizumab	Intravenous/ subcutaneous	8 mg/kg every 4 weeks
IL-1 receptor	Anakinra	Subcutaneous	100 mg/day

IL interleukin, *TNF-α* tumor necrosis factor alpha

Table 50.5 Biologic disease-modifying antirheumatic drugs for ankylosing spondylitis

Target molecular	Principle active	Route of administration	Dosage
Tumor necrosis factor-alpha	Infliximab	Intravenous	Initial phase: a first dose of 5 mg/kg followed by additional infusions of 5 mg/kg at 2nd and 6th weeks after the first infusion; maintenance: infusion of 3 mg/kg every 6–8 weeks
	Adalimumab	Subcutaneous	40 mg every other week
	Etanercept	Subcutaneous	25 mg to be administered twice a week or 50 mg once a week
	Golimumab	Subcutaneous	50 mg once a month

mammalian cells in suspension – CHO cells – and purified by affine chromatography and ion exchange); Tocilizumab (trade name RoActemra), a humanized IgG1 monoclonal antibody directed against the human IL-6 receptor, produced in CHO cells by recombinant DNA technology; and Anakinra (trade name Kineret), an IL-1 receptor antagonist produced in *Escherichia coli* cells by recombinant DNA technology.

The targeted therapies currently used in treating AS involve anti-TNF-α. In particular, treatment with the monoclonal antibody Etanercept showed a significant reduction in the AS disabling symptoms, revealing high efficacy (Table 50.5).

Biological Drugs in Dermatology: Psoriasis and Psoriatic Arthropathy

Psoriasis is a chronic inflammatory disease affecting mainly the skin and altering the immunological balance of the entire organism. Genetic, autoimmune, and environmental factors are involved in its pathogenesis, often making the course of the disease unpredictable. It affects both sexes equally and occurs at any age; however, two incidence peaks are observed in the adult population, at 20–30 and 50–60 years. Early cases identify more severe forms and are mainly associated with a positive family history. Psoriasis can occur in the same subject in different periods of life and in different forms, with erythematous patches, lesions on elbows, knees, scalp, buttocks, and genitals, alterations of fingernails and toenails, joint pain accompanied by redness, swelling, and limitation of movement, swelling of a toe or a hand (diphthylitis), and morning joint stiffness with pain in the dorsal and lumbar region that improves with movement. The treatment of psoriasis includes topical treatments, local corticosteroids, phototherapy, traditional systemic treatments, and systemic biological treatments (TNF-α inhibitors), according to its clinical manifestations (Table 50.6).

In addition to the biological drugs previously described, the pharmacological treatment of psoriasis involves using Alefacept (trade name Amevive), a human LFA-3/IgG1 fusion protein that acts by blocking the interaction between antigen-presenting cells and lymphocytes. Specifically, it binds the CD2 molecule by inhibiting the LFA-3/CD2 interaction and causes apoptosis of memory T cells. Although approved by the Food and Drug Administration (FDA), this drug has not received approval from the European Medicines Agency (EMA) due to the risk of infections, neoplasms, and allergies; therefore, further studies are required for final approval and marketing in Europe.

Psoriatic arthritis (PA) is a chronic inflammatory rheumatic disease of the joints associated with about 15–20% of cases of psoriasis.

It is classified as seronegative spondyloarthritis (absence of rheumatoid factor) and presents a wide heterogeneity in the type of clinical presentation, course, and joints affected. It is characterized by pain, swelling, warmth, and joint stiffness. In many cases, polyarticular erosive involvement is responsible for disability and reduced quality of life. In contrast to AS and parainfectious reactive forms, there is no close association with the HLAB antigen27.

Although the cause of PA is not yet well known, it is believed that in genetically predisposed individuals, under the influence of triggering factors (physical trauma, infection, and stress), the immune system can be activated against the joints and skin.

The peak incidence is between 20 and 40 years, with no sex-related differences.

An initial therapeutic approach involves cycles of therapy with non-steroidal anti-inflammatory drugs or cyclooxygenase 2 inhibitors; corticosteroids are limited to aggressive forms. In situations where symptomatic non-steroidal anti-inflammatory or steroidal drugs are ineffective, anti-rheumatic drugs, DMARDs, can be used. They have been proven to reduce the symptoms and progression of arthritis. The most commonly used DMARDs are sulfasalazine, cyclosporine (effective on both arthritis and cutaneous psoriasis), MTX, and leflunomide. In cases resistant to these therapies, where the arthritic clinical picture and/or the evolution of joint damage cannot be controlled, anti-TNF biological drugs are indicated. Infliximab, ADL, Etanercept, and Golimumab represent the most effective treatments for aggressive evolution and resis-

Table 50.6 Biological therapies for psoriasis and PA treatmet

Target molecular	Principle active	Pathology	Route of administration	Dosage
TNF-α	Infliximab	Psoriasis, PA	Intravenous	Initial phase: 5 mg/kg infusion followed by additional 5 mg/kg infusions at 2[a] and 6[a] weeks after the first infusion; maintenance: infusion of 5 mg/kg every 6–8 weeks
	Adalimumab		Subcutaneous	40 mg every other week
	Etanercept		Subcutaneous	25 mg to be administered twice a week or 50 mg once a week
	Golimumab	PA	Subcutaneous	50 mg once a month
IL-12 and IL-23	Ustekinumab	Psoriasis	Subcutaneous	Initial phase: dose of 45 mg; maintenance phase: 45 mg after 4 weeks and every 12 weeks thereafter
CD2	Alefacept	PA, psoriasis		In the testing phase; approved by the FDA, it has not received consensus in Europe from the EMA

EMA European Medicines Agency, *FDA* Food and Drug Administration, *IL* interleukin, *PA* psoriatic arthritis, *TNF-α* tumor necrosis factor alpha

tance to conventional therapies. These drugs also represent the most innovative and effective therapy for severe psoriasis (Table 50.6).

Biological Drugs in Oncology

Different biological drugs are currently available for treating colorectal, lung, renal, ovarian, and breast cancers, non-Hodgkin's lymphoma, and chronic lymphatic leukemia. Oncological biological drugs include monoclonal antibodies, inhibitors of growth factors or their receptors, vaccines, ILs, and interferons.

The first biological drug used in this field was Herceptin (IgG1 monoclonal antibody Trastuzumab), which reduced the risk of recurrence in operable breast cancer in more than 50% of cases. Among the most recent ones is Lapatinib, which is effective in contrasting metastases from breast cancer.

In oncology, research on therapeutic biologics is yielding exciting results, allowing the development of target-oriented biological drugs capable of inhibiting specific mechanisms that control tumor growth and spread.

However, applying these therapeutic protocols frequently requires precise and timely molecular characterization of the neoplasms and a careful selection of patients.

Molecularly targeted therapies can only be used if specific diagnostic "markers" indicative of one or more tumor molecular targets are detected in tumor cells or other biological samples collected from the patient. In their absence, targeted treatment cannot be applied.

Personalized treatment with monoclonal antibodies (cetuximab and panitumumab) in colorectal cancer acts through the control of the epidermoid growth factor receptor (EGFR), which, in turn, regulates the intracellular pathway of the RAS gene, which is responsible for numerous processes of tumor proliferation, angiogenesis, and metastasis.

In the presence of neoplasia, if the RAS gene is mutated, the encoded RAS protein (KRAS) is stably activated, regardless of the presence or absence of the stimulus from EGFR. In neoplasms with mutated RAS gene (KRAS), anti-EGFR drugs are ineffective. When the RAS protein is wild-type, it is activated only when it receives a stimulus, such as that induced by the EGFR receptor, and becomes inactive again when the stimulus ceases. In the latter case, the tumor will be more sensitive to the anti-EGFR biological drug, and the tumor cells survival will be compromised.

Patients with a mutation in the RAS gene may alternatively benefit from biotherapy with anti-angiogenic monoclonal antibodies, which interfere with the development of blood vessels that supply oxygen and nutrients to tumor cells, thereby hindering their growth.

The application of monoclonal antibodies to neoplasms is manifold. Table 50.7 shows the main molecu-

Table 50.7 Biological therapies in oncology

Biopharmaceutical	Target molecule	Neoplasm type
Antibodies		
Cetuximab-Panitumumab	EGFR	Colon, head
Trastuzumab	ERBB2	Breast
Bevacizumab	VEGF	Colon, lung
Rituximab	CD20	B-cell lymphoma
Gemtuzumab	CD33	Acute myeloid leukemia
Alemtuzumab	CD52	Chronic myeloid leukemia
Inhibitors		
Lapatimib	EGFR-HER2	Breast
Imatinib	BCR-ABL, KIT, PDGFR	Chronic myeloid leukemia, GIST
Gefitinib	EGFR	Lung
Erlotinib	EGFR	Lung
Sunitinib	VEGFR, PDGFR, FLT3	Renal, GIST
Sorafenib	B-Raf, VEGFR2, EGFR, PDGFR	Renal
Bortezomib	Proteasome	Multiple myeloma

[a]*GIST* gastrointestinal stromal tumors, *EGFR* epidermal growth factor receptor, *ERBB2* Erb-B2 Receptor Tyrosine Kinase 2, *VEGF* Vascular endothelial growth factor, *HER2* human epidermal growth factor receptor 2, *BCR-ABL* Breakpoint Cluster Region - Abelson murine Leukemia, *PDGFR* platelet derived growth factor receptor, *FLT3* FMS-like tyrosine kinase 3

larly targeted drugs currently used to treat some malignancies, either as single agents or combined with chemotherapy.

The use of these therapies is limited to neoplasms that present specific molecular alterations. In addition, the possibility to develop drug resistance must be considered.

The treatment with monoclonal antibodies in oncology can cause different side effects, including rashes, especially in the face and upper chest, nausea, vomiting, diarrhea, headaches, excessive growth of eyelashes and hair, dry skin, and redness of the eyes.

In some cases, these side effects may be momentary and reversible; in others, they are treatable.

Conclusions

In recent years, innovations in biological therapies have opened new horizons for treating patients with immune-mediated inflammatory and oncological diseases.

However, several aspects remain to be clarified. First of all, the indication for biological drugs remains selective, being currently limited to patients refractory to other types of treatment or, in the case of neoplasms, to selected patients. Moreover, the complexity of biotech drugs requires a tailored pharmacovigilance approach, which is necessary to monitor the dynamics of adverse events over time (increased risk of infections and increased incidence of neoplasms).

In light of current knowledge, therapeutic monitoring of anti-TNF-α inhibitors and ADAs is recommended to ensure that patients receive adequate therapy and address any loss of efficacy. In addition, to ensure optimal monitoring, it is essential to define the optimal drugs cut-offs to achieve the best performance. Currently, the cut-offs of Infliximab and ADL are known, which are 3.5 μg/mL and 6–7 μg/mL, respectively. This information is critical to correctly administering the biopharmaceutical, reducing immunogenicity and possible adverse events. Further studies are required to identify the cut-off values of the other biopharmaceuticals that are currently missing.

Biological drugs represent, today, a great opportunity in the treatment of multiple clinical conditions.

Recommended Readings

Atzeni F, Sebastiani M, Ricci C, Celano A, Gremese E, Iannone F, Meroni PL, Minghetti P, Sarzi-Puttini P, Ferraccioli G, Lapadula G (2015) Position paper of Italian rheumatologists on the use of biosimilar drugs. Clin Exp Rheumatol 33:1–4

Ford AC, Sandborn WJ, Khan KJ et al (2011) Efficacy of biological therapies in inflammatory bowel disease: systematic review and meta-analysis. Am J Gastroenterol 106:644–659

Ha C, Mathur J, Kornbluth A (2015) Anti-TNF levels anti-drug antibodies, immunosuppressants and clinical outcomes in inflammatory bowel disease. Expert Rev Gastroenterol Hepatol 9:497–505

Knight DM, Trinh H, Le J et al (1993) Construction and initial characterization of a mouse-human anti TNF antibody. Mol Immunol 30:1443–1453

Mav W, Murray DL, Snyder MR (2015) Tumor necrosis factor inhibitors: clinical utility in autoimmune diseases. Transl Res 165:270–282

McInnes IB, Schett G (2011) The pathogenesis of rheumatoid arthritis. N Engl J Med 365(23):2205–2219

Ritchlin CT, Colbert RA, Gladman DD (2017) Psoriatic arthritis. N Engl J Med 376(10):957–970

Sieper J, Braun J, Rudwailet M, Boonen A, Zink A (2002) Ankylosing spondylitis: an overview. Ann Rheum Dis 61(Suppl 3):1108–1118

Singh JA, Hossain A, Tanjong Ghogomu E et al (2016) Biologics or tofacitinib for rheumatoid arthritis in incomplete responders to methotrexate or other traditional disease-modifying anti-rheumatic drugs: a systematic review and network meta-analysis. Cochrane Database of Syst Rev 2016

Sirotti S, Generali E, Ceribellini A et al (2017) Personalized medicine in rheumatology: the paradigm of serum autoantibodies. Auto Immun Highlights 8:10

Strik AS, Bots SJA, D'Haens G et al (2016) Optimization of anti-TNF therapy in patients with inflammatory bowel disease. Expert Rev Clin Pharmacol 9:429–439

Vincent F, Morand E, Murphy K et al (2013) Antidrug antibodies (ADAb) to tumour necrosis factor (TN-F)-specific neutralising agents in chronic inflammatory diseases: a real issue, a clinical perspective. Ann Rheum Dis 72:165–178

Mass Spectrometry in Clinical Biochemistry and Laboratory Medicine

51

Giorgio Federici and Marcello Ciaccio

Introduction

Mass spectrometry measures the mass of a molecule after its ionization. Mass spectrometry originated at the dawn of the twentieth century from the observations of Joseph John Thompson, who showed the formation of ions from substances placed in a vacuum tube. He received the Nobel Prize in 1906 for this observation and related studies. In the following years, mass spectrometry was mainly applied in basic chemistry, physics, and industry studies, obtaining remarkable results in identifying the various isotopes of the elements. During this long period, ionization methods and various instruments were developed, which will be discussed later. Mass spectrometry made its debut in clinical biochemistry by coupling gas chromatography with the mass spectrometer. This technique was used around the eighties of the last century to identify congenital errors of organic acid metabolism by urine analysis. The laboratories that used this technique then were mainly university or polyclinic laboratories, as there were great difficulties in preparing samples and interpreting results. Technological development of more straightforward and more versatile mass analyzers made them more practical for clinical applications in later years. This was also facilitated by the availability of stable isotopes for use as internal standards for the exact quantification of the analyte under study. In the past two decades, new ionization methods have made it possible to analyze polar and water-soluble compounds such as peptides, proteins, oligonucleotides, and nucleic acids, as well as low-molecular-weight metabolites in bulk.

A mass spectrometer measures the mass-to-charge ratio for each molecule after the molecule has been ionized (m/z). Obviously, the value of the mass, i.e., the molecular weight, of the substance is defined using the unit of measure, the Dalton, abbreviated Da, which is 1/12 of the mass of the isotope 12C of carbon. For the quantitative determination of the charge z, it is opportune to express its value as multiples of the elementary charge possessed by an electron (negative) or a proton (positive). So, for each molecule, the ionic form is defined by the ratio with the number of elementary charges of its mass (m/z).

In order to analyze a sample in a mass spectrometer, the sample must form ions in the gas phase. For this reason, the first technology used for applications, including clinical applications, was gas chromatography coupled to the mass spectrometer. Recent technological developments have made it possible to use liquid, solid, or gaseous samples that, through special volatilization and ionization techniques, allow the introduction into the high vacuum system of the mass spectrometer. Figure 51.1 shows the basic schematic of a mass spectrometer in which the ions produced by the sample are separated by the analyzer (held under high vacuum conditions) according to their m/z ratio, sending a signal to the detection system that analyzes the electrical signal proportional to the number of ions present for that m/z signal. The connected processing system converts the various signals into a mass spectrum consisting of an ion abundance diagram as a function of m/z.

Ionization of the molecules to be analyzed by mass spectrometry can be performed as described below.

- Ejection of e^-: $M \rightarrow M^+ + e^-$
- Protonation: $M + H^+ \rightarrow MH^+$
- Cationization: $M + Cat^+ \rightarrow MCat^+$
- Deprotonation: $MH \rightarrow M^- + H^+$

Giorgio Federici is died.

G. Federici
Departments of Science and Chemical Technologies, University of Tor Vergata, Rome, Italy

M. Ciaccio (✉)
Department of Biomedicine, Neurosciences and Advanced Diagnostics, Institute of Clinical Biochemistry, Clinical Molecular Medicine and Clinical Laboratory Medicine, and Department of Laboratory Medicine, University Hospital "P. Giaccone", Palermo, Italy
e-mail: marcello.ciaccio@unipa.it

M. Ciaccio (ed.), *Clinical and Laboratory Medicine Textbook*, https://doi.org/10.1007/978-3-031-24958-7_51

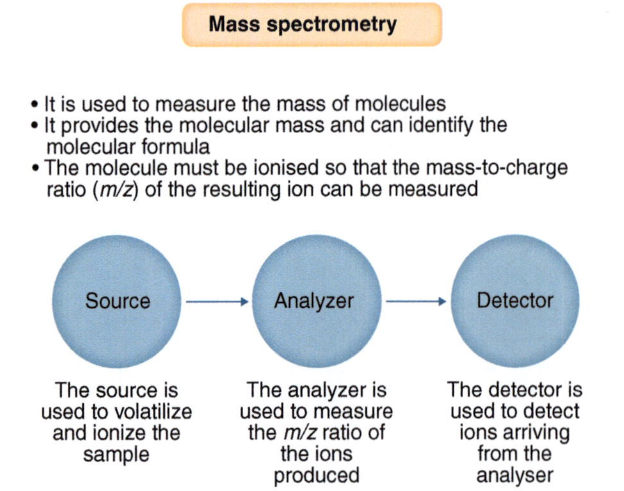

Fig. 51.1 Schematic representation of a mass spectrometer. In the instrumental part relating to the analyzer, a high vacuum is carried out to promote the mobility of the ions. (Copyright EDISES 2021. Reproduced with permission)

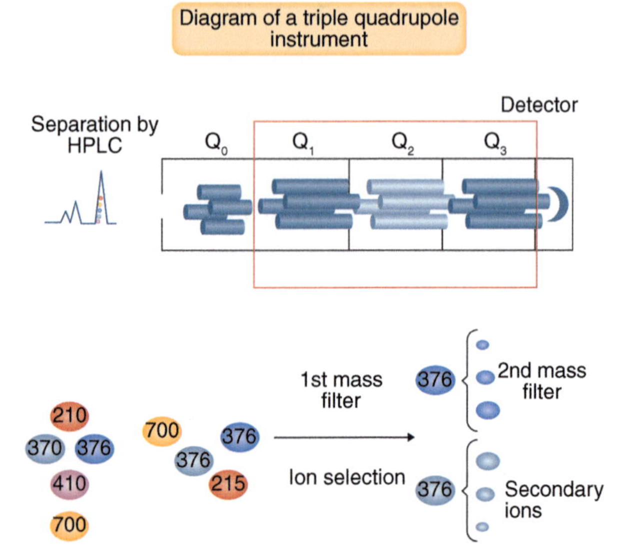

Fig. 51.2 Diagram of a triple quadrupole instrument. The small quadrupole Q0 focuses on the ionized molecules separated in the first quadrupole (Q1), fragmented generating child ions in the second quadrupole (Q2). The produced ions are then separated in the third (Q3). (Copyright EDISES 2021. Reproduced with permission)

The separation of mass ions takes place using a magnetic field and/or an electric field, and depending on how separation is achieved, mass spectrometers are distinguished into:

- Magnetic sector
- Time of flight (TOF)
- Quadrupole
- Ion trap
- Ion cyclotron resonance in transform of [Fourier transform-ion cyclotron resonance (FT-ICR)]

Depending on the required application, it is necessary to choose the type of mass spectrometer, considering the differences among those listed, mainly due to the maximum measurable m/z ratio; sensitivity; resolving power; analysis speed; and, finally, ease of use if the spectrometer is interfaced to a separation apparatus before the sample is introduced into the spectrometer (high-performance liquid chromatography, gas chromatography, EC).

The main analyzers used in laboratory medicine are currently quadrupole and time-of-flight (TOF) analyzers.

A quadrupole consists of four parallel bars of metal or ceramic. A direct voltage is applied to the bars; the diagonally opposite bars have the voltage of the same sign, and a radio frequency voltage is superimposed on this. The electric field forces the ions to follow a different oscillating trajectory for each m/z value. By adjusting the electric field, the value of m/z of the ions crossing the quadrupole can be selected; ions with m/z greater or less than the chosen value will be forced to travel trajectories that lead away from the electric field and, therefore, will discharge onto the quadrupole and will not reach the detector. This allows the selection of a particular ion or the scanning of m/z.

An instrument configuration consisting of three quadrupoles in series is used in so-called triple quadrupole instruments (Fig. 51.2). In this configuration, the first and third quadrupoles act as mass filters. IOn contrast, the central quadrupole acts as a collision cell where the ions from the first quadrupole are fragmented, thus generating the daughter ions that serve to characterize the molecule further.

It is possible to operate in a product ion scan, where the first quadrupole selects an ion of fixed m/z ratio; after fragmentation, the third quadrupole scans over a given mass range. In this way, it is possible from the fragment analysis to get information about the identity of a molecule. On the other hand, the parent ion scan mode is the third quadrupole that is permanently tuned to a given m/z value, while the first quadrupole scans over a mass range. This mode allows the identification of structurally similar classes of molecules.

In the neutral loss scan mode, on the other hand, both quadrupoles work in the scan, but a fixed difference in m/z is maintained between the first and third quadrupoles. In other words, while the first quadrupole is tuned to a given m/z, the

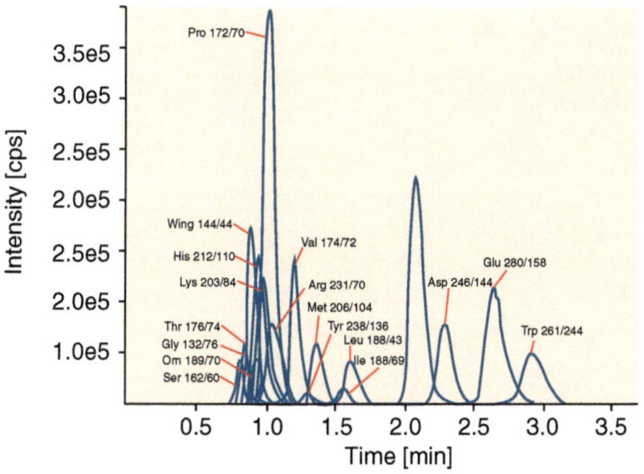

Fig. 51.3 Separating plasma amino acids in liquid chromatography-tandem mass spectrometry (LC–MS). The separation was performed on a 30 mm Supelcosil™ CN column. At 1.5 min, the chromatographic run allows the separation of the isobaric amino acids leucine and isoleucine

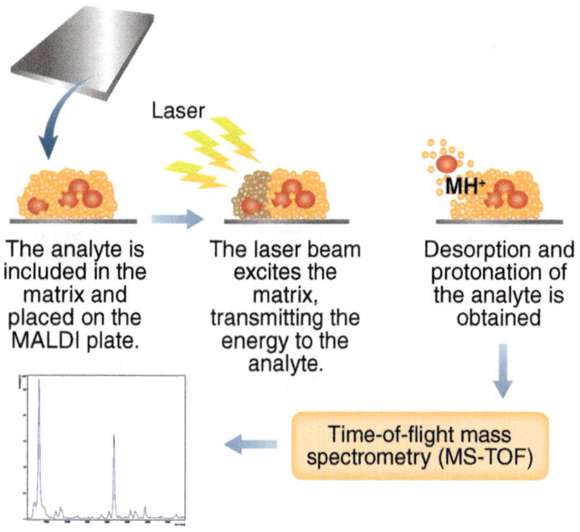

Fig. 51.4 Schematic of the desorption mechanism in a MALDI (matrix-assisted laser desorption/ionization) instrument. The laser beam excites the matrix, which passes energy to the analyte obtaining its protonation and release in the vacuum of the time of flight (TOF). (Copyright EDISES 2021. Reproduced with permission)

third is at (m/z – x), and x is kept constant during the scan. It is thus possible to easily identify classes of molecules that share some part of the structure.

A further mode, single reaction monitoring, is used for quantitative analysis in systems coupled with liquid chromatography or gas chromatography, using stable isotope-labeled standards. In this case, the molecule's identity is known a priori, and the first quadrupole is fixed on a determined value of m/z, while the third is on a fragment. This mode allows for increasing the specificity of the detection system. If more than one fragment is selected, the third quadrupole is referred to as multiple reaction monitoring. Figure 51.3 shows an experiment for separating amino acids in liquid chromatography-tandem mass spectrometry (LC–MS) on a triple quadrupole using deuterated standards to make the determination quantitative.

In mass spectrometry, matrix-assisted laser desorption/ionization, commonly referred to by the acronym MALDI (matrix-assisted laser desorption/ionization), is a soft ionization technique used in mass spectrometry since the 1990s. The MALDI technique is usually conducted under a vacuum. It is used in instruments with a ionization system consisting of a tube under a vacuum that measures the time of flight of the ions. This technique provides extremely accurate mass values and is referred to as MALDI-TOF. The technique consists of absorbing the sample on a matrix, which can be made of various materials, especially organic materials mainly with acidic characteristics (sinapic acid, anthranilic acid, nicotinic acid, etc.), and once brought into solution, is subsequently bombarded by a laser beam (often a nitrogen laser). The matrix is easily evaporated and, with the acidic behavior, provides a source of protons encouraging ionization of the analyte. The matrix-analyte system possesses strong absorption in the ultraviolet region to absorb the radiation typically generated by a laser. Under the pulse of the laser, the energy absorbed by the matrix is transferred to the analyte, which is ionized and vaporized by the excess energy secondarily released by the matrix itself (Fig. 51.4). This results in molecular ions that are generally singly charged, such as those created by the acquisition or loss of a proton. The MALDI-TOF technique is indicated for analyzing thermolabile and high-molecular-weight compounds, particularly for biological molecules such as proteins, peptides, complex carbohydrates, and polynucleotides nucleic acids that, by classical ionization techniques, would be partially destroyed or fragmented. It also provides extremely accurate mass values, by some instruments up to the fourth decimal place. Figure 51.5 shows a schematic of a MALDI-TOF instrument.

The next paragraphs describe some applications in laboratory medicine of mass spectrometry that contributed to a considerable development in the analytical capacity and the precision of laboratory analysis.

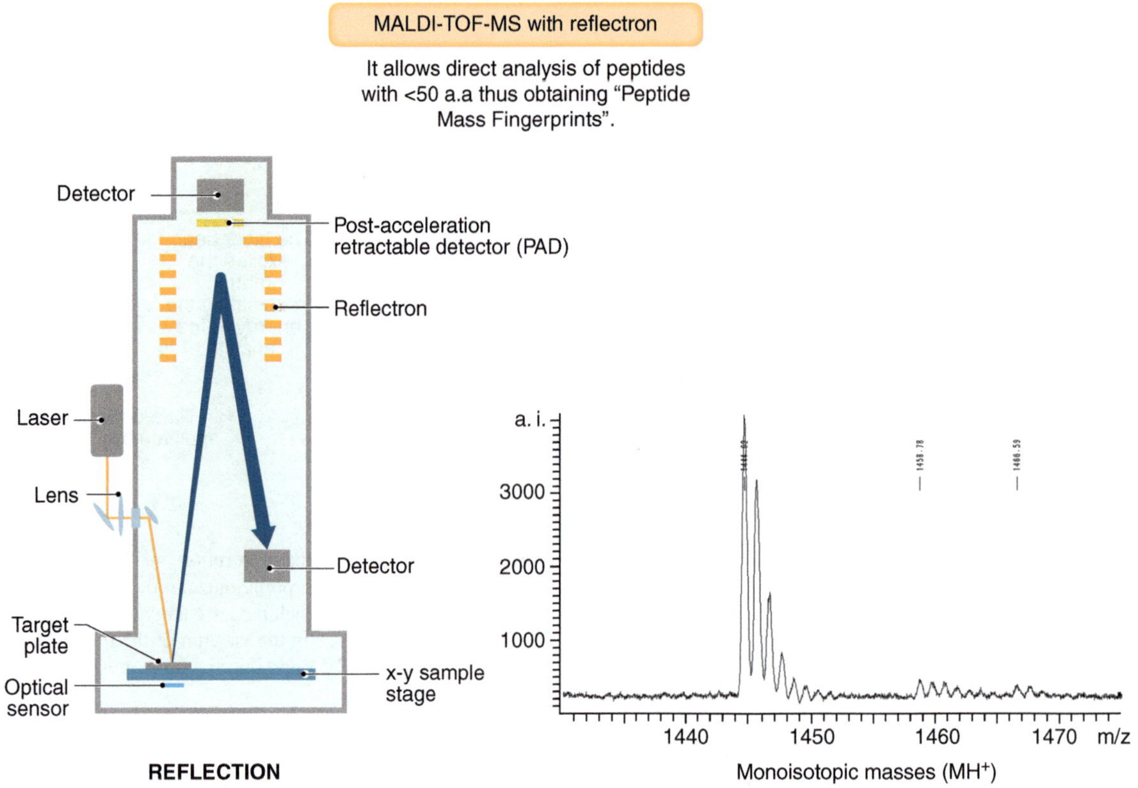

Fig. 51.5 Diagram of a MAL-DI-TOF (matrix-assisted laser desorption/ionization-time of flight) instrument. The laser beam hits the plate and the ions produced travel toward the detector. The presence of the reflectron system allows for lengthening the flight times and, therefore, the separation of the molecular ions. Exact mass values are obtained. (Copyright EDISES 2021. Reproduced with permission)

Therapeutic Dosage of Drugs

Therapeutic drug monitoring was one of the first fields of significant development for applying mass spectrometry coupled with liquid chromatography (LC–MS). Moreover, given this technology's specificity and sensitivity, it is the only to promote the use of precision medicine in clinical practice. This new approach to precision medicine, also known as personalized medicine, uses data from the patient's phenotype and genotype to define a personalized therapy.

In the past 15 years, techniques for determining drugs using triple quadrupole (LC–MS) have been developed with high sensitivity, absolute specificity, and high linearity ranges that allow, by appropriate standards, a correct quantitative determination of the circulating drug.

It is possible to consider therapeutic drug monitoring as a clinical specialty that achieves patient care by individual adjustment of drug therapy. This is necessary because most drugs have a very narrow therapeutic window of their effective doses, and, thus, a slight difference between the minimum effective doses and those at which they produce adverse toxic effects. For these reasons, for many drugs, an assay with absolute specificity and high precision is required, characteristics that the previously used immunometric methodologies cannot guarantee.

LC–MS methods are available for the following drugs classes:

- Antibiotics
- Antivirals
- Antifungals
- Antiepileptics
- Antipsychotics
- Antidepressants
- Cardioactive drugs
- Antineoplastic drugs
- Immunosuppressive drugs

Among these categories, immunosuppressive drugs, especially cyclosporine A, tacrolimus, sirolimus, and everolimus, are the first class of drugs for which LC–MS has entered the routine laboratory. At least 70% of laboratories worldwide currently use mass spectrometry. By a chromatographic run of about 2.5 min, including regeneration of the column, it is possible to obtain the separation of the four drugs and their quantitative determination using as internal standards the same molecules synthesized with some hydrogen atoms replaced by deuterium. This LC–MS technique allows the complete automation of the system through samplers with the capacity to analyze more than 20 samples per hour, making it possible to receive the patient in the clinic and have the analytical data quickly to accommodate the therapy.

Steroids

Steroid molecules were among the first to be analyzed by mass spectrometry. In the 1960s, gas chromatography coupled to the mass detector (GC–MS) allowed the analysis of steroid metabolism and urinary profiles, including both androgens and cortisol metabolites. In the late 1990s, the introduction of LC–MS and appropriate extraction techniques allowed the development of accurate and sensitive assay methods.

LC–MS is a more straightforward method to use in the routine laboratory, mainly because of the lower complexity of sample preparation. However, GC–MS is still a helpful tool for studying the steroid metabolome, identifying any new metabolites, for example, in conditions of 21-hydroxylase deficiency. In the same period, various manufacturers developed several immunometric assays. These methods for steroid analysis have been widely adopted in routine laboratories because of their simplicity, cost, and sensitivity, although these methods have several issues. First, cross-reactivity among similar analytes, then standardization problems between different laboratories, and, finally, issues related to the methods' sensitivity. These reasons have led some scientific societies to define immunometric methods as suboptimal, recommending using LC–MS by appropriate standards that make the determination quantitative. Much attention must be paid to the preanalytical phase of sample preparation because suboptimal results may depend on chromatographic conditions but never on the spectrometer's ability to measure the ions produced. Therefore, particular attention must be paid to the validation of the method to various parameters such as selectivity, sensitivity, stability, reproducibility, matrix effect, and conditions of the chromatographic run.

Notably, LC–MS allows multiparametric analysis, i.e., to obtain the quantitative detection of several molecules in the same run. Some manufacturers have put on the market kits to measure up to 12 hormones, both androgens and estrogens. The development of these ready-to-use methodologies will be greatly accelerated in the coming years, contributing to the standardization of analytical data. Research laboratories have developed multiparametric assays to simultaneously dose up to 1720 hormones, including metabolites and derivatives.

It is not the task of this chapter to deal with the various pathologies related to disorders of the metabolome of steroid compounds. However, it is necessary to point out that the availability of precise methods for the determination of androgens, estrogens, progestins, and corticosteroids, both in plasma and, possibly, also in cells, allows the medical specialist to optimally manage the various endocrine syndromes. In the metabolic pathway of steroidogenesis are expressed 17 steroid hormones that are coupled more than 50 possible metabolites, all determined by LC–MS.

Another compound that can be determined in LC–MS is vitamin D, which indicates a series of secosteroids characterized by the opening of a ring for the classical steroidal structure. This class's two most important representatives are vitamin D2 and D3, which have different biological origins and are nevertheless absorbed from the diet. Vitamin D3 in mammals is also photosynthesized in the skin due to the effect of radiation, starting from its precursor, 7-dehydrocholesterol. The measurement of circulating vitamin D metabolites is widely used for diagnosing and monitoring various diseases related to calcium metabolism: osteoporosis, renal osteodystrophy, parathyroid disorders, sarcoidosis, etc. Also, in this case, mass measurement is characterised by a remarkable sensitivity and specificity, especially considering the low levels of circulating concentration of these vitamins.

The bile acids also belong to this class. They are characterized by a carboxyl group in the side chain. These acid molecules are the primary catabolic product of cholesterol and are mainly involved in the absorption of dietary lipids and fat-soluble vitamins. The most abundant bile acids in the human body are cholic acid, chenodeoxycholic acid, lithocholic acid, and ursodeoxycholic acid. These compounds are present mainly as amidation derivatives with glycine and taurine, particularly taurocholic acid and glycocholic acid. The analysis of these compounds makes it possible to investigate hepatobiliary dysfunction. The group of bile acids is characterized by chemical–physical differences, such as the considerable difference in polarity among the free forms and those conjugated to taurine and glycine and the absence of volatility for most of them. In addition, the low values of circulating concentration and the minimal structural differences have made their measurement very difficult. Few studies in the scientific literature have grappled with this problem and now, the use of LC–MS allows their assay, although with some difficulties. Figure 51.6 shows an LC–MS separation of bile acids.

In addition, LC–MS allows the accurate study of hemoglobin (Fig. 51.7).

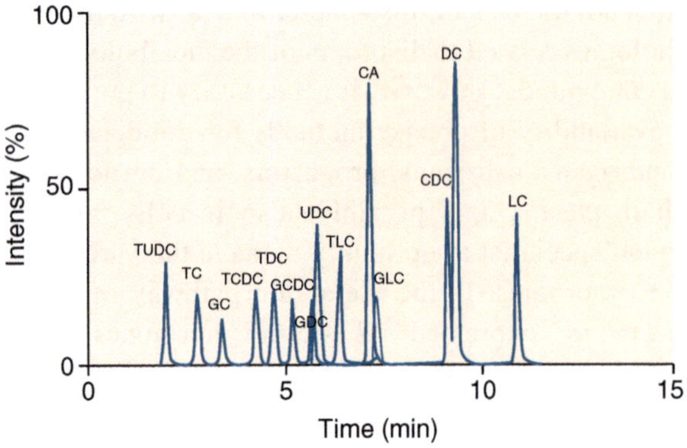

Fig. 51.6 Liquid chromatography-tandem mass spectrometry (LC–MS) separation of a mixture of 14 bile acids at a concentration of 0.5 μmol each. The run is carried out by an RPC18 column

Drugs of Abuse

One of the significant problems of modern society is the use of drugs of abuse. A recent study by the World Health Organization reports that around 300 million people worldwide use illicit substances, primarily cannabinoids, followed by cocaine, amphetamines, and opiates. Testing for drug abuse is typically required for clinical use, medico-legal purposes, antidoping certification of sports performance, or even for screening groups of workers. The drug of abuse assay is normally performed in two phases: the first screening followed by a second confirmation, if a result higher than the cutoff values of the analytical method is obtained. Urine is the biological matrix of the first choice for evaluating drugs

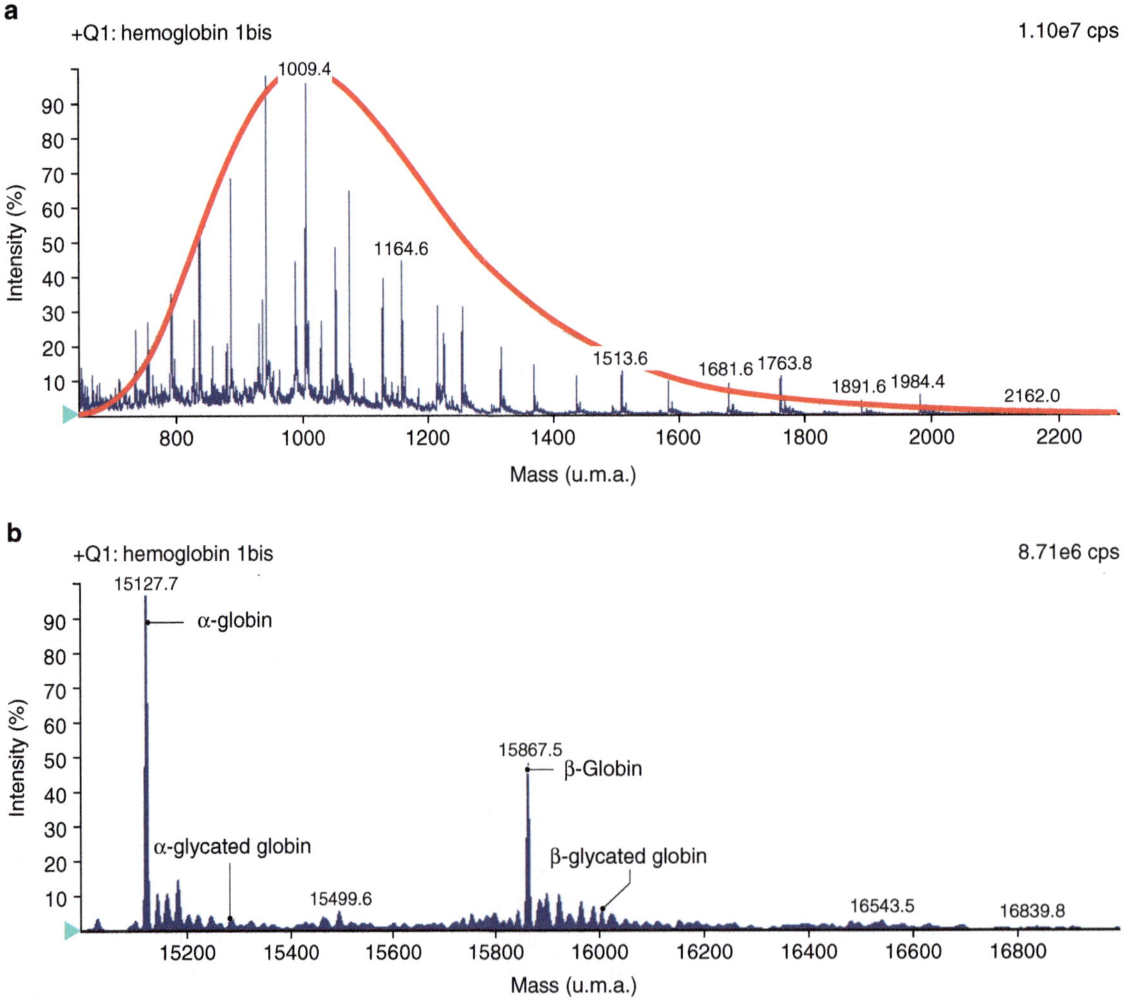

Fig. 51.7 Spectrum of hemoglobin run in liquid chromatography-mass spectrometry (LC–MS). In (**a**), the peaks due to the various ionization states of the molecule are visible, while in (**b**), after the deconvolution, operated by the software, the peaks of globins and glycates can be identified

of abuse. However, urine analysis has some disadvantages related to a possible lack of clinical relevance depending on the route of drug administration and the time elapsed between the assumption and the dosage. It can also affect the physiological state of the patient. The law makes special provisions for preserving the sample to be tested or, if already tested, for confirmatory testing. These provisions are called chain of custody and ensure the correctness of the analytical data from a medico-legal point of view. The analytical methods most commonly used in screening tests to identify the presence of drug abuse are immunochemical or immunochromatographic, based on the antigen–antibody reaction. These methods have some disadvantages based mainly on low specificity as they can give cross reactions with molecules similar to the analyte sought (e.g., drugs, substances taken with the diet, etc.). For this reason, in case of a positive result higher than the cutoff, a confirmation test to directly analyze the molecule in question must be performed.

For this purpose, the sample is analyzed by mass spectrometry, providing qualitative and quantitative information by comparison with internal standards. Internal standards consist of the same drugs obtained synthetically by modifying some H atoms in deuterium, thus increasing one molecular weight unit for each modified H. In this way, by exploiting the selectivity and sensitivity of the mass spectrometer, it is possible to identify the presence of the drugs even weeks after they have been taken, using, for example, the hair where the drug is believed to be a function of hair growth.

Neonatal Screening

Neonatal screening, one of the most critical public preventive medicine programs, aims to detect diseases for which a cure exists. Early diagnosis and initiation of treatment, whether pharmacological or dietary, can significantly improve prognosis in the first days of life, reducing disability and mortality.

For the analysis, a few drops of blood are collected from the newborn's heel (within 48 h of birth) and placed on a card made of special chromatographic paper on which some circles are imprinted to make the amount uniform of blood absorbed within each circle. The dried card, called the Guthrie card, will then be sent to the reference center for testing. If a test is positive, and, therefore, the presence of pathology is suspected, the newborn will undergo further investigations to confirm the suspect.

The last 20 years have witnessed a spectacular evolution in neonatal screening due to the development of mass spectrometry and its introduction into clinical laboratories. Mass spectrometry analysis of blood drops collected in Guthrie's cards makes it possible to identify diseases belonging to three main groups: aminoacidopathies, organic acidurias, and β-oxidation defects. There are more than 40 diseases in which aminoacidopathies are associated with an increase or decrease in amino acids. In contrast, organic acidurias and β-oxidation defects are associated with a variation in acylcarnitine values. Interpreting the results is complex and requires experienced personnel to analyze the metabolic profile. This type of analysis is referred to as extended or expanded neonatal screening and has so far been performed in different Italian regions in a very diverse manner.

Table 51.1 shows the diseases that can be analyzed by extended screening in Italy.

Table 51.1 List of diseases assessed by extended screening

Phenylketonuria
Benign hyperphenylalaninemia
Biosynthesis deficiency of the tetrahydrobiopterin cofactor
Regeneration deficiency of the tetrahydrobiopterin cofactor
Medium-chain acyl-CoA dehydrogenase deficiency
Glutaric acidemia type I
Isovaleric acidemia
Maple syrup urine disease
Type I tyrosinemia
Lack of carnitine transport
Long-chain hydroxyacyl-CoA dehydrogenase deficiency
Deficit of the trifunctional protein
Very-long-chain acyl-CoA dehydrogenase deficiency
3-Hydroxy-3-methyl glutaric aciduria
Deficiency of β-ketothiolase
Methylmalonic acidemia (CblA)
Methylmalonic acidemia (CblB)
Methylmalonic acidemia (Mut)
Propionic acidemia
Methylmalonic acidemia (CblC)
Argininosuccinic acidemia
Citrullinemia type I
Homocystinuria (CBS deficiency)
Tyrosinemia type II
Carnitine palmitoyl-transferase II deficiency
Multiple carboxylase deficiency
Methylmalonic acidemia (CblD)
Argininemia
Citrullinemia type II
Hypermethioninemia
Tyrosinemia type III
Carnitine palmitoyl-transferase (L) deficiency
Glutaric acidemia type II
Short-chain acyl-CoA dehydrogenase deficiency
Carnitine/acyl-carnitine translocase deficiency
Deficit of 3-methyl-crotonyl-CoA carboxyalsis
Deficiency of 2-methyl-butyryl-CoA dehydrogenase
3-Methyl-glutaconic aciduria (types 1, 2, 3, 4, and 5)
Deficiency of isobutyryl-CoA dehydrogenase
Malonic aciduria
Deenoyl reductase deficiency
Short/medium-chain 3-OH-acyl-CoA dehydrogenase deficiency
2-Methyl-3-hydroxy butyric aciduria
Ethylmalonic encephalopathy
Ornithine transcarbamylase deficiency
Deficiency of methylene tetrahydrofolate reductase
Ornithine transcarbamylase deficiency

Other diseases can be screened by mass spectrometry or other methods (lysosomal storage diseases, biotinidase deficiency, galactosemia, adrenocortical hyperplasia, congenital, etc.). However, there must be adequate therapy to introduce a pathology in the screening. It must be considered that the test is not "diagnostic" but only indicates the risk that the child is subject to the pathology. The preanalytical aspect of sample collection and processing is also critical to avoid false negatives as much as possible. For this reason, centers must have certified procedures and adequate personnel. The advantages of extended screening performed by mass spectrometry are due to the very high sensitivity of the methodology, the high analytical speed (less than 4 min for an analysis), therefore, also high productivity, the possibility of automation of the analysis, and last but not least, the low analytical cost.

Proteomics

The term proteome, coined by Mark Wilkins in 1994 during a conference organized in Siena by the University of Siena, describes the set of proteins of an organism or a biological system, i.e., the proteins produced by the genome.

The proteome differs from the genome because of its components' high dynamism and great molecular complexity. Different cells express different proteomes in a single individual, while the genome remains unaltered within the somatic line of the same organism. The remarkable molecular dynamism of the proteome is due to the adaptive capacity of the phenotype in response to variations in the individual pathophysiological state.

The study of the proteome is called proteomics and consists of the systematic identification of proteins and their characterization concerning structure, function, activity, quantity, and molecular interactions. It has long been practiced by separating proteins using two-dimensional gel electrophoresis. In the first dimension, proteins are separated according to their isoelectric point. In the second dimension, proteins are separated by molecular mass using SDS-PAGE. The gel is stained with Coomassie blue or silver salts to visualize proteins.

A cellular proteome is the set of proteins found in a cell type under particular environmental conditions, such as underexposure to stimulation of various kinds. It may also be helpful to consider the complete proteome of an organism, which can be imagined as the global set of proteins from all cellular proteomes. This is, roughly, the protein equivalent of the genome. The term "proteome" has also been used to refer to the set of proteins of a subcellular biological system, such as the mitochondrial proteome.

The proteome exhibits at least two levels of complexity that the genome lacks. While a sequence of nucleotides defines the genome, the proteome is not limited to defining the sequences of proteins present in the cell or tissue but also to the analysis of posttranslational modifications (glycosylation, phosphorylation, etc.) of individual proteins, which reflect on their biological function.

A proteomics experiment often generates the analysis of large numbers of proteins derived from complex biological samples, so separation methods must often be employed before performing mass spectrometry analysis. There are two types of approaches to the study of a protein system. The bottom-up approach in which the protein mixture is enzymatically digested (almost always by trypsin) to form a mixture of peptides which, after a chromatographic step, are analyzed by mass spectrometry providing information on the sequence of the peptides and, using various databases, allow identifying proteins. This is a highly informative approach, even if it does not provide information on posttranslational modifications. In the second approach, the top-down one, in which we analyze one or more target proteins, we act on the intact protein. Then, possibly through fragmentation in the spectrometer, we study the various ions formed. This approach, also called targeted, is the most helpful for identifying possible diseases biomarkers and comparing the profiles obtained in the tissue or the pathological cell to normal ones.

In 2007, clinical proteomics was born as an application of proteomic analysis to solve a specific clinical problem in the context of a clinical trial. This definition was born with the first scientific journal of clinical applications: Proteomics Clinical Applications. In the past 10 years, numerous scientific works have identified various proteins as possible diseases markers. However, many of them have not yet been fully validated clinically or by the authorities. Also the most used methods in clinical proteomics are LC–MS and MALDI-TOF associated with soft ionization methods. The use of hybrid instruments (Q-TOF) in which a triple quadrupole is associated with a TOF system is also emerging. An excellent combination of the triple quadrupole's high selectivity and the TOF's precision and mass accuracy is obtained in this case. Through these analyses, it is possible to identify biomarkers indicative or responsible for the pathological condition. However, a more important aspect of the proteomics evolution in recent years has been the development of quantitative techniques allowing the identification of a protein and its quantity in the biological sample. Various methodologies have been developed, but the most effective one for quantifying a protein is the addition to the mixture, after tryptic digestion, of a peptide, called proteotypic, equal to the natural one belonging to the protein's primary structure. The characteristics of the proteotypic peptide are (i) representative of the protein of interest; (ii) unique among the proteome under study; (iii) completely hydrolyzed in the proteolytic treatment; (iv) resistant to preparative artifacts;

Table 51.2 Workflow of targeted proteomics

1. Identification of interest proteins
2. Determination of quantotypic peptides
3. Development of selected reaction monitoring
4. Triple quadrupole analysis, or Q-TOF
5. Traces of the peptides of interest and the internal standard are obtained
6. The concentration of the protein is determined from the relationship between the signals
7. 50–100 proteins can be analyzed simultaneously

TOF Time Of Flight: flight time

and (v) and not susceptible to post-translational modifications. This proteotypic peptide is synthesized with stable isotopes, which increase its m/z value. In the mass spectrum, the precise value of the protein concentration can be obtained by making the ratio between the signal of the endogenous peptide and the proteotypic one. In recent work, more than 40 proteins were quantitatively assayed at the plasma level. Table 51.2 shows the experimental flow for quantitatively determining proteins in a biological sample.

In recent years, an innovative diagnostic approach has also been developed by directly analyzing histological samples and thus obtaining information about intact biological tissue. This technology, called mass spectrometry imaging (MALDI-TOF-imaging mass spectrometry, MALDI-TOFIMS), can also be applied to paraffin-embedded tissue samples, typical of pathological anatomy. Proteins, peptides, and metabolites can be identified by this method as hundreds of signals can be detected by a single spectrum, obtaining information on the localization of a protein and any interaction with colocalized proteins.

The development of MALDI technology has led to several clinical applications. However, the one that has forcefully entered the routine is the use of MALDI-TOF in microbiology to identify bacteria and fungi by exploiting the different peptide or protein profiles obtained from the sample subjected to the laser beam. Manufacturers have developed software for identification, and this technology dramatically reduces the time of identification of the bacterium with the consequent advantages of starting antibiotic therapy.

A further aspect that is developing with the use of mass spectrometry is metabolomics, i.e., the study of changes in the concentration of metabolites, in biological fluids or cells, in various physiological or pathological conditions of the individual. Through a single determination in MS, it is possible to quantify hundreds of metabolites.

In conclusion, evaluating the remarkable developments in proteomics and mass spectrometry over the past decade, it can be said that the use of clinical proteomics in medicine constitutes the future pathway to personalized precision medicine.

Recommended Readings

Abdel-Khalik J, Bjorklund E, Hansen M (2013) Simultaneous determination of endogenous steroid hormanes in human and animal olasna and serum by liquid or gas chromatography coupled to tandem mass soectrometry. J Chromatogr B 928:58–77

Amolatz B, Zohere E, Haas C et al (2017) Bile acid preparation and comprehensive analysis by high performance liquid chromatography hig resolution mass pectrometry. Clin Chim Acta 464:85–92

Annesley T, Cooks RJ, Herold DA, Hoofnagle AN (2016) Clinical mass spectrometry – achieving prominence in laboratory medicine. Clin Chem 62:1–3

Cifani P, Kentsis A (2017) Towards comprehensive and quantitative proteomics for diagnosis and therapy of human disease. Proteomics 17:1–2

Crutchfield CA, Thomas SN, Sokoll LJ, Chan DW (2016) Advances in mass spectrometry based clinical biomarker discovery. Clin Proteomics 13:1–12

Csosz E, Kallo G, Markus B et al (2017) Quantitative body fluids proteomics in medicine – a focus on minimal invasiveness. J Proteome 153:30–43

Duarte TT, Spenser CT (2016) Personalized proteomics: the future of precision medicine. Proteomes 4:29–49

Duncan MW, Nedelkov D, Walsh R, Hattan SJ (2016) Applications of MALDI mas spectrometry in clinical chemistry. Clin Chem 62:134–143

Focardi M, Pinchi V, Defraia B et al (2016) Newborn screening of inherited metabolic disorders: the italian situation. J Biol Regul Homeost Agents 30:909–914

Garg U, Zhang YV (2016) Mass spectrometry in clinical 23. Laboratory: applications in biomolecular analysis. Methods Mol Biol 1378:1–9

Jannetto PJ, Fitzgerald RL (2016) Effective use of mass spectrometry in the clinical laboratory. Clin Chem 62:92–98

Jenkinson C, Taylor AE, Hassan-Smith ZK et al (2016) High throughput LC-MS method for simultaneous analysis of multiple vitamin D analytes in serum. J Chromatogr B 1014:56–63

Jones G, Kaufmann M (2016) Vitamin D metabolite profile using liquid chromatography tandem mass spectrometry /LC-MS. J Steroid Biochem Mol Biol 164:110–114

Keevil BG (2016) LC-MS analysis of steroids in the clinical laboratory. Clin Biochem 49:989–997

Lehmann S, Brede C, Lescuyer P et al (2017) Clinical mass spectrometry proteomics (cMSP) for medical laboratory: what does future hold? Clin Chim Acta 467:51–58

Mak CM, Lee HC, Chan AY, Lam CW (2013) Inborn errors of metabolism and expanded newborn screening: review and update. Crit Rev Clin Lab Sci 50:142–162

Maurer HH (2010) Perspectives of liquid chromatography coupled to low and high resolution mass spectrometry for screening, identification and quantification of drugs in clinical and forensic toxicology. Ther Drug Monit 32:324–327

McShane AJ, Bunch DR, Wang S (2016) Therapeutic drug monitoring of immunosuppressant by liquid chromatography-mass spectrometry. Clin Chim Acta 454:1–5

Pandor A, Eastham J, Beverley C et al (2004) Clinical effectiveness and cost effectiveness of neonatal screening for inborn errors of metabolism using tandem mass spectrometry: a systematic review. Health Technol Assess 8:1–121

Sabbagh B, Mindt S, Neumaier M, FindeisenP (2016) Clinical applications of MS-based protein quantification. Proteomics Clin Appl 10:323–345

Sherl A (2015) Clinical protein mass spectrometry. Methods 81:3–14

Shipkova M, Svinarov D (2016) LC-MS as tool for TDM services: where are we? Clin Biochem 49:1009

Vestal LM (2016) Evolution of quantitative MALDI-TOF mass spectrometry for clinical applications. Clin Chem 62:20–23
Vincenti M, Salomone A, Gerace E, Pirro V (2013) Application of mass spectrometry to hair analysis for forensic toxicological investigation. Mass Spectrom Rev 32:312–332
Wallemacq P (2016) Mass spectrometry in laboratory medicine:high-tech now meeting the needs of routine laboratory testing. Clin Biochem 49:945–946
Wright I, Van Eyk JE (2017) A roadmap to successful clinical proteomic. Clin Chem 63:245–247

Toxicology and Analytical Pharmacology

52

Sergio Bernardini, Simona Martello, Antonello Nonnato, and Davide Farci Santarcangeli

Toxicology

Toxicology studies the undesirable effects of chemicals and their metabolites on the body and is a field of particular interest because of its implications in laboratory diagnostics.

The main areas of this discipline are molecular toxicology, toxicogenomics, descriptive toxicology, regulatory toxicology, forensic toxicology, environmental toxicology, developmental toxicology (with teratology), reproductive toxicology, pharmacotoxicology, and clinical toxicology; in particular, the latter deals with diseases caused by or associated with exposure to toxic substances.

Any chemical substance foreign to the body (food additives, dyes, emollients, drugs, pesticides, environmental pollutants, consumer products, etc.) is a xenobiotic. Xenobiotics are distributed to the different organs through the bloodstream, essentially bound to plasma proteins, where they can accumulate, thus causing toxic effects if the toxicity threshold (acute or chronic) is exceeded.

A poison can be defined as any agent capable of causing damage to a biological system. Virtually every chemical substance has the potential to cause damage or death if present in a sufficiently high concentration. A critical parameter in this respect is lethal dose, 50% (LD50), defined as the dose of a substance that causes death in 50% of exposed individuals. However, it should be noted that acute lethality measures such as LD50 do not always accurately reflect the full spectrum of toxicity or hazard associated with exposure to an agent because small doses that are not acutely toxic but accumulate over time can result in chronic toxicity.

Toxic agents are classified based on the interests and needs of those who classify them; they can be described according to their target organ, use, origin, physical and chemical characteristics, or effects.

The term "toxin" is generally used for toxic substances produced by biological systems such as plants, animals, fungi, or bacteria, whereas the term "toxic agent" is used in the case of toxic substances resulting from human activities.

The toxic effects of a substance (including those due to allergy or idiosyncrasy) may be immediate or delayed as well as reversible or irreversible; the reversibility or irreversibility of the effect primarily depends on the regenerative capacity of the affected tissue.

Another distinction among the various types of toxic effects concerns the site of action. Local effects occur at the site of first contact between the biological system and the toxic agent. Systemic effects require absorption and distribution of the toxic agent from its site of entry to a distant point where it can produce its toxic effects. Most substances produce systemic effects; for some substances, both local and systemic effects have been demonstrated. Generally, chemical agents that cause systemic effects produce their maximum toxicity in one or two organs, referred to as the target organs of toxicity; often, the target organ is not the site of maximum concentration of the chemical agent.

Interaction Among Chemicals

An interaction may occur by different mechanisms: alteration of absorption, protein binding, metabolism (or biotransformation), or excretion of one or more of the substances involved. In addition to the interaction phenomenon, the body's response to the combination of different toxic substances may be increased or decreased due to the toxic effects induced at the site of action.

S. Bernardini (✉)
Department of Experimental Medicine, University of Tor Vergata, Rome, Italy
e-mail: bernardini@med.uniroma2.it

S. Martello
Department of Biomedicine and Prevention, University of Tor Vergata, Rome, Italy

A. Nonnato
Clinical Biochemistry Unit, Department of Diagnostic Laboratory, A.O.U. Città della Salute e della Scienza Hospital, Turin, Italy

D. F. Santarcangeli
IRCCS Ca' Granda – Ospedale Maggiore Policlinico, Milan, Italy

M. Ciaccio (ed.), *Clinical and Laboratory Medicine Textbook*, https://doi.org/10.1007/978-3-031-24958-7_52

An additive effect occurs when two substances administered simultaneously produce an effect equal to the sum of the effects of each substance administered individually; this is the most common effect. A synergistic effect occurs when the combination of two substances produces effects greater than the sum of the effects induced by the substances administered individually. Potentiation occurs when a substance has no toxic effects, but, when administered simultaneously as another substance, it makes it more toxic.

Antagonism occurs when two or more substances interfere with each other. There are four basic categories of antagonism:

- Functional antagonism: When two substances have opposing effects on the same physiological function
- Chemical antagonism or inactivation: When a chemical reaction occurs between the substances involved, which generally results in a less toxic product
- Kinetic antagonism: When the absorption, metabolism, distribution, or excretion of one of the substances is altered, whose concentration and/or duration of the effect on the target organ will be diminished
- Receptor antagonism: When the combined effect of substances binding to the same receptor is less than the sum of the effects induced by each substance administered individually or when one substance antagonizes the effects of another substance

Tolerance

Tolerance is the condition of diminished response to the toxic effect of a substance due to previous exposure to the same or a structurally similar substance; the main mechanisms responsible for tolerance are a decreased amount of substance reaching the site where the toxic effect is produced (kinetic tolerance) and reduced tissue reactivity.

Exposure Characteristics

Toxic effects on a biological system cannot be produced without an agent or its metabolites reaching the appropriate sites in the body at a concentration and sufficient time to evoke the toxic effect. Thus, the toxic response depends on the physicochemical properties of the agent, the site of exposure, the metabolism of the agent, and the susceptibility of the biological system or individual.

The most important routes by which toxic agents gain access to the body are the gastrointestinal tract (ingestion), the lungs (inhalation), the skin (topical, percutaneous, or dermal route), and other parenteral routes. Generally, toxic agents produce the greatest effects and the most rapid responses when injected directly into the circulatory stream. The "vehicle" (i.e., the material in which the toxic agent is dissolved) and other factors in the formulation can greatly alter absorption; the route of administration can also influence the toxicity of agents.

In human exposure situations, the duration and frequency of exposure are often not clearly defined. However, in general, occupational and environmental exposures can be described as acute (a single incident or episode of exposure), sub-chronic (exposure for a few weeks or months), or chronic (repeatedly occurring over many months or years).

For many agents, the toxic effects obtained after a single dose differ greatly from those produced by repeated exposure. Acute exposure to agents that are rapidly absorbed may easily result in immediate toxic effects but may also cause delayed effects that may be similar to or different from the toxic effects resulting from chronic exposure; conversely, in the case of chronic exposure, in addition to the long-term effects observed with low doses, some immediate (acute) effects may also be observed after each administration. Another critical factor for the temporal characterization of exposures is their frequency: a single dose of a chemical agent that produces marked toxic effects may be without effect if the same dose is divided and administered at intervals. On the other hand, even if the substance does not accumulate, cellular or tissue damage may be present after each administration; therefore, it is important to assess whether the time interval between two successive doses is sufficient to allow complete recovery of the damage. Therefore, chronic toxic effects may occur if the substance accumulates in the biological system (because absorption exceeds metabolism and/or excretion), if it produces irreversible toxic effects, or if a sufficient time interval has not elapsed between exposure intervals to allow complete recovery of the damage.

Metabolism

Since each substance undergoes a process of biotransformation (called metabolism), which aims to transform foreign substances into more polar and water-soluble compounds to increase their excretion, the evaluation of the toxic potential of a xenobiotic must take into account not only the toxic effects of the toxic substance as such but also those of its metabolites. This is of particular importance in the evaluation of the action of a drug, where the metabolites may be the substances that exert the actual pharmacological action (since then, the administered substance is a pro-pharmaceutical) but may also have an essential toxic action.

The biotransformation of a substance can lead to the formation of inactive, active (with the same or a different spectrum of action than the original substance), and toxic metabolites.

The main site of metabolism is the liver, but the lungs, kidneys, and intestines also have a significant metabolic capacity, and every tissue has some metabolic capacity.

A feature of liver enzymes is that their synthesis and activity could be stimulated by repeated administration of substances such as drugs, pesticides, industrial chemicals, and foodstuffs (with ethanol being one example). Metabolic induction, which is particularly important in the case of drugs, results in an acceleration of metabolism and a reduction in the action not only of the inducing substance but also of other substances administered at the same time as the inducer.

Metabolism is carried out in two phases:

- Phase I or functionalization reactions have the purpose of inserting or highlighting functional groups in molecule such as –OH, $-NH_2$, or –COOH; reduction, oxidation, hydrolysis, and hydration are all phase I reactions, which, in most cases, are catalyzed by the cytochrome P450 system.
- Phase II or conjugation reactions are enzymatic reactions of biosynthesis by which an exogenous compound or a metabolite derived from a phase I reaction binds covalently with an endogenous molecule using the functional groups –OH, –COOH, $-NH_2$, or –SH. In general, conjugates are polar molecules that are easily eliminated; glucuronoconiugation, acetylation, and conjugation with glutathione are phase II reactions.

Through bioactivation process, xenobiotics are transformed into either xenobiotics with less or no activity than the parent (thus undergoing a detoxification process) or metabolites with more significant activity than the parent (thus undergoing a bioactivation process).

In this regard, it is essential to use appropriate biomarkers that are indicative of biochemical or molecular alterations due to xenobiotics and also that are measurable in samples of human body fluids:

- Biomarkers of exposure, defined as exogenous substances, or their metabolites or products of interaction between a xenobiotic and a target molecule or cell measured in a compartment of the body (e.g., urinary or plasma levels of exogenous compounds and/or their metabolites)
- Biologically effective dose biomarkers, which represent a measure of exposure closer to the target (e.g., DNA or plasma protein adducts of exogenous compounds)
- Biological effect biomarkers, which show a measurable biochemical, physiological, or other change in an organism that, as a result of exposure to a given risk factor and, depending on its magnitude, indicates actual or potential health damage or disease (e.g., chromosomal aberrations or sister chromatid exchange)
- Susceptibility biomarkers, which highlight the intrinsic or acquired decrease in the ability of an organism to respond to the possible effects of exposure to a particular xenobiotic (e.g., the study of genotypes that characterize the polymorphism of enzymes involved in the metabolism of toxicants)

The Metabolome and Metabolic Biomarkers

The use of metabolic biomarkers is a consolidated approach in medicine (and, therefore, also in clinical chemistry and analytical toxicology), in terms of not only clinical semeiotics but also early diagnosis and prevention.

In endogenous metabolism, the total or partial inactivation of an enzyme causes the reduction of the formation of the product and/or the accumulation of the substrate that can be metabolized by alternative metabolic pathways, causing an increase of upstream metabolites with inhibition of other metabolic pathways and, sometimes, impairment of cellular functions. This phenomenon can also be a component of the toxic effect of the xenobiotic of interest. Thus, the study of several related parameters, some increasing and others decreasing, offers a significant added value to the clinical informativeness of the data.

Analytical capacity is no longer a limiting factor for metabolomics. Technologies such as nuclear magnetic resonance spectroscopy or mass spectrometry (MS) have reached such levels of analytical functionality to allow the simultaneous assay of hundreds of metabolites on the same sample at acceptable costs. On the other hand, the knowledge of all the physiological and pathological metabolic networks and the relationships between the tissue and plasma metabolome are still incomplete.

Moreover, in toxicology, therefore, the study of the effects of toxic substances on the various metabolic pathways is of particular importance, as it is often necessary to search in biological matrices not only for the "parent" xenobiotic but also for its metabolites and/or biomarkers of damage related to the intake of the substance.

Dose–Response Relationship

The characteristics of exposure and the spectrum of toxic effects closely correlate with each other, and this is called a dose–response relationship. The response is gradual when the effect is continuous over a given dose range, whereas all-or-nothing effects are defined as quantal.

Each individual responds in a specific way to a substance, and the response of individuals to increasing doses of the same substance is distributed along a Gaussian; therefore, beyond the average response, we can distinguish hypersensitive from resistant individuals.

The development of a dose–response curve has important implications for assessing toxicity. For example, for those substances that are necessary for normal vital physiological functions (e.g., vitamins and essential trace elements such as selenium), the dose–response relationship in an individual has, over the entire dose range, a U-shaped pattern. This means that at extremely low doses (i.e., in a deficient situation), several adverse effects diminish as the dose increases. If the dose increases to a level at which no longer deficiency occurs, then the negative effects disappear and the body reaches homeostasis. However, increasing the dose to excessive levels will result in an adverse response (usually qualitatively different from that observed under deficiency conditions) that worsens with increasing dose.

Some non-nutritional toxicants may have beneficial or stimulatory effects at low doses, whereas at higher doses, they produce adverse effects. This phenomenon, termed "hormesis," can be described by a U-shaped dose–response curve.

Another critical aspect of the dose–response relationship is the concept of "threshold" defined as the dose below which the probability of an individual response is 0. In the case of individual dose–response relationships, there is certainly a threshold for most toxic effects. However, the variability of response between individuals and the qualitative variations in the type of response to a given dose make it challenging to establish a "no-effect" threshold for any substance. The presence or absence of a threshold is essential for identifying safe substance exposure levels.

In general, however, it is important to keep in mind that dose–response relationships often do not have a linear trend and that the increase in the toxic effect is almost always slower or faster than the increase in the dose.

When evaluating a dose–response relationship, some important parameters must be taken into account. For example, in the case of drugs, one of these is the therapeutic index (TI), defined as the ratio between the dose needed to produce a toxic effect and that needed to produce a desired therapeutic response. Similarly, the comparative toxicity index is obtained from the ratio between the doses of two or more different substances that produce the same response or between the doses of the same substance that are needed to have two or more different toxic effects.

The most used index of effect for both toxic and beneficial effects is the median dose, i.e., the dose required to achieve a response in 50% of a population (or to produce 50% of the maximum response). The therapeutic index (TI) of a drug represents an approximate measure of its safety and is expressed by the ratio of the toxic dose (historically the lethal dose) to the effective or therapeutic dose (TD): TI = TD50/TD50. However, median doses do not provide information about the slope of the respective dose–response curves for the therapeutic and toxic effects. One way to overcome this drawback is to use the effective dose in 99% of the population for the desired effect and to use the lethal dose (LD) for 1% of the population for a toxic effect; these parameters are used to calculate the margin of safety (MS): MS = LD1/TD99. For substances other than drugs, the term "margin of safety" is used to indicate the magnitude of the difference between the exposure dose for a given population and the highest dose that does not result in toxic effects in experimental animals (no-observed-adverse-effect level (NOAEL)).

To compare the toxic effects of two or more substances, it is necessary to establish the dose–response relationship for each substance. The potency of a substance refers to a range of doses within which it produces increasing effects, whereas the maximum potency of a substance reflects the limit of its dose–response relationship.

Variations in Toxic Responses

Selective toxicity refers to the ability of a substance to cause damage in one organism without affecting another, even if the two organisms live in close contact with each other. By using the advantages of biological diversity, it is possible to develop chemical agents that are lethal to an undesirable species and harmless to other species (e.g., pesticides). The selectivity of the toxic effect may be due to differences in the distribution or metabolism of the toxic substance in different organisms.

Although a basic tenet of toxicology is that "experimental results obtained in animals, when properly qualified, are applicable to humans," it is crucial to remember that quantitative and qualitative differences in response to toxic agents may exist in different species. Identifying the mechanisms underlying such species differences to humans in response to chemical agents helps gauge to what degree data obtained from laboratory animals to be tested are relevant to humans.

Even within the same species, individual variations in chemical response can occur, which depend on subtle genetic differences termed "genetic polymorphisms." These variations may be responsible for idiosyncratic reactions and individual differences in response to chemical agents.

Hazards and Toxicological Risks

Ideally, a toxicologist should be able to make quantitative judgments about the adverse or even toxic effects that substances or contaminants may cause in organisms, particularly in the human population. Realistically assessing the hazards to humans necessarily leads to a differentiation between absolute safety and acceptable risk. Achieving absolute safety would involve knowledge of measurable effects concerning the dose administered to each individual, the source of exposure, the type of toxic effect caused by the toxic agent taken singly or in a mixture, the existence or lack of a toxic effect threshold, and the variability of the toxic effect within the population. Very often, data are not qualitatively or quantitatively adequate to assess the toxicity of a substance, and, for this reason, it is necessary to have methodologies to objectively calculate the risk to humans following exposure to a toxicant. Although understandably, a specific risk can be accepted if it is linked to a real benefit, it is even inevitable when it derives from specific environmental contaminants. The issue is further complicated when examining the magnitude of risk concerning different population groups that show markedly different sensitivities; toxicological decisions generally tend to consider the biological response of a healthy adult individual.

Toxicological Evaluation

For compounds that have not demonstrated a carcinogenic effect, the first question to be answered is the recognition of the lowest dose (generally expressed in milligrams per kilogram of body weight) at which a toxic effect has been observed (lowest-observed-adverse-effect level (LOAEL)) or better still a dose that has shown no toxic effect in biological experiments performed (no-observed-effect level (NOEL)).

The NOEL is defined as the highest concentration of a substance that does not cause detectable morphological or functional changes and does not affect the growth, development, and life span of the biological organism under study. There is a range of toxicological experiments to be carried out in vitro on simple organisms such as bacteria and tests to be performed on laboratory animals or, when necessary, as in the case of pharmaceuticals, on humans.

Table 52.1 includes the main tests for formulating the toxicological evaluation of a compound with biological activity.

Generally, the most important study from which to derive the NOEL is the long-term and/or carcinogenicity study; the protocol for this test involves daily administration of the substance for a period equivalent to 75% of the lifetime of the animal under test and a route of administration by which humans will reasonably be exposed.

The focal point of the problem is the selection of doses to be used in toxicological experiments to detect a NOEL. The highest dose should produce an evident toxic effect, which should not be seen at the lowest dose. The higher dose should not cause high mortality that would invalidate the experiment nor should it cause an exaggerated pharmacological action that may cause toxic effects not directly attributable to the substance during the long treatment period. In order to obtain a dose–response curve, at least three dose levels plus, of course, an untreated control group should be planned. It would be desirable to conduct pilot pharmacokinetic and pharmacodynamic experiments for each substance to identify the animal species most similar to humans to obtain a more accurate NOEL from long-term experiments and avoid making the fundamental mistake of choosing the most sensitive species.

In the absence of a NOEL, the dose that produced no adverse effect (NOAEL) may be considered.

Figure 52.1 represents a hypothetical dose–response curve.

In order to establish the conditions of use that determine a potentially risk-free exposure to harmful substances that are not carcinogenic, since 1955, the World Health Organization has been using a method to calculate the acceptable daily intake (ADI) (also called admissible daily intake) of hundreds of substances present in the environment and in food.

The ADI estimates the amount of a substance (expressed in milligrams per kilogram of body weight) that can be ingested daily by an individual over a lifetime without an appreciable health risk. It is based on a safety factor (SF) applied to the experimentally determined NOEL. The value of the SF is set according to the nature of the toxic effect, the size and type of the population to be protected, and the quality of the available toxicological data. When long-term experimental results are available, a safety factor of 100 is traditionally used (a factor of 10 to address uncertainties

Table 52.1 Toxicological protocol

Physicochemical identification
Pharmacokinetics (absorption, distribution, metabolism and excretion (ADME))
Pharmacodynamics (interaction of the substance with its site of action)
Biochemical studies
Pharmacological studies
Acute toxicity studies
Short- and long-term studies
Genotoxicity studies
Studies of embryotoxicity and/or teratogenesis
Fertility studies
Carcinogenicity studies
Special toxicity studies (e.g., neurotoxicity)
Observations on humans
Ecotoxicology studies

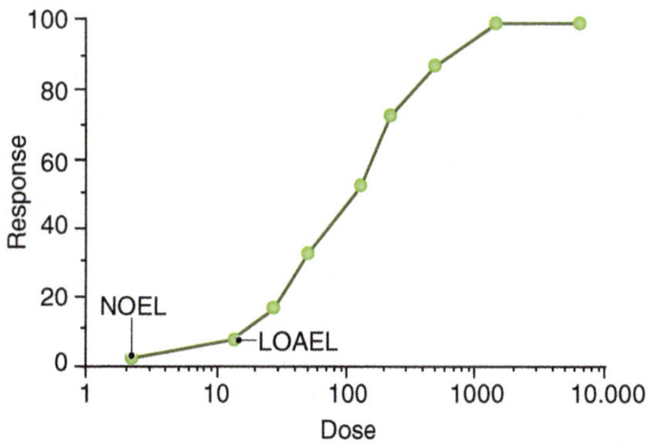

Fig. 52.1 Hypothetical dose–response curve of a given substance. LOAEL, lowest dose at which a toxic effect was observed; NOEL, dose that did not show any toxic effect in the biological experiments performed. (Copyright EDISES 2021. Reproduced with permission)

regarding the extrapolation of animal to human results and an additional factor of 10 to include in the ADI the population group that is also highly susceptible to the toxic effects of the compound).

An SF of 10 may only be used when adequate data are available in the human population. A factor of 1000 may be used when adequate data from short-term trials are available.

Using the ADI, it is possible to calculate the maximum concentration of a substance and the tolerable levels (TLs) to which an individual can be exposed daily through air, water, and food without incurring an appreciable health risk. This is done by multiplying the dose by the average body weight and dividing this product by the average daily consumption of specific food, water, or inhaled air. These quantities, set at the international level by the WHO, are in many cases deliberately "exaggerated" to be the most conservative to protect the exposed individual.

Finally, risk assessment includes calculating the margin of safety (MS), the ratio between the tolerable concentrations, and the measured concentrations of the substance in the relevant compartment. The margin of safety thus allows a quantitative assessment, using an absolute measure of the risk arising from exposure to a substance. Experience has shown that even a value close to unity does not represent a risk, as an examination of the entire calculation process shows that the margin of safety is used to reduce the dose that showed no toxic effect where the substance was administered repeatedly over long periods. In addition, it must be considered that the normal irregularity of the biological response, resulting from genetic differences even within the same species, and the analytical variability, based on the instrumental response that cannot be repeated constantly, must lead to considering NOEL, ADI, TL, and MS not in a strictly mathematical way. Indeed, exceeding the ADI does not necessarily determine an adverse effect.

In industrial toxicology, several exposure indices are recommended by the American Conference of Governmental Industrial Hygienists, such as the threshold limit values (TLVs), which represents the maximum concentration of a substance to which a worker can be exposed without the occurrence of adverse effects. The TLV can also be expressed in more specific ways: the time-weighted average (TWA) limit, which is the acceptable concentration for the period covering an 8-hour working day or a 40-hour working week; the short-term exposure limit (STEL), i.e., the maximum concentration to which a worker can be exposed over 15 minutes, provided that this does not occur more than four times in a day and at intervals of at least 1 hour and that the TWA is not exceeded; and, finally, the ceiling limit (CL), which is the value that cannot be exceeded even for a single instant during the working day.

The Philosophy of Risk Assessment

Highlighting the degree of toxicity of a substance is only the first essential step in defining the degree of risk from its exposure. The toxicity of formaldehyde, for example, varies radically depending on whether it is ingested, inhaled, or absorbed through the skin. In the first case, the molecule interacts directly with the genetic material or with the proteins of the cell wall with which it comes into contact; in the second case, up to a specific concentration, it is sequestered by the natural components of foodstuffs, which thus cancel or reduce its biological reactivity; lastly, at the skin level, this molecule mainly causes sensitization problems.

Therefore, it is clear that when judging the toxicological impact of a substance, different factors and skills come into play, which must be able to analyze the benefits derived from the use of such a substance and balance them against the possible risks, to propose valid alternatives based on solid knowledge of the toxic properties of the possible substitute, to take the responsibility of accepting the use of a certain molecule whose mechanisms of toxicity are fully understood rather than accepting as an alternative the use of a substance that seems risk-free only because the studies conducted are partial; and, finally, to regulate the quantities of a substance released into the environment in order to avoid ecological disasters, also bearing in mind the chemical transformations it may undergo.

Analytical Toxicology

Analytical toxicology is the application of methods of analytical chemistry for the qualitative and/or quantitative evaluation of substances that may exert toxic effects on living organisms.

In the clinical setting, analytical toxicology can assist in the diagnosis and treatment of intoxication as well as in monitoring the effectiveness of therapeutic protocols by clearly identifying the nature of the toxicant and determining the amount absorbed. This information, together with the patient's clinical picture, allows the clinician to correlate the signs and symptoms observed with the expected effects of the toxic agent.

A fundamental rule in treating poisoning cases is to remove the unabsorbed material, limit the absorption of different toxicants, and accelerate its elimination. The clinical toxicology laboratory plays its supporting role by monitoring the number of toxic agents in circulation or by measuring its excreted portion.

Significant is also the role of pharmacotoxicology and therapeutic drug monitoring (TDM), both for managing the overdose phenomenon (voluntary or not) and for identifying the optimal drug dose, to take into account individual variations in drug response and compliance.

Biological Monitoring

Biological monitoring performed directly on a worker (or in any case on a subject chronically exposed to a toxic agent) can be a better indication of exposure than simple environmental monitoring because it can indicate the percentage of the agent that has been absorbed. Environmental exposures often involve a mixture of chemical agents and/or compounds that are converted into major metabolites in the body. Therefore, analytical methods must be capable of detecting a broad family of chemical agents with their major metabolites and must also have sufficient specificity and sensitivity to detect small concentrations of analytes in complex biological matrices.

In addition to analyzing the agent or its metabolites in the subject's biological matrices, indirect methods can be used. Substances that interact with macromolecules may form adducts that persist for long periods; these adducts can be sampled periodically and potentially serve as a supplementary means of assessing long-term exposure to certain substances. Another helpful method of biological monitoring is to assess changes in normal metabolites induced by xenobiotics. However, monitoring changes in urinary excretion of these metabolites may not indicate exposure to a specific substance; this technique can only generically indicate a potentially toxic exposure. Early recognition of a toxicological problem may promote the protection of the subject before irreversible effects occur.

Clinical Toxicology

The following general points are essential elements in the initial clinical approach to a patient with intoxication:

- Stabilizing the patient
- Clinical evaluation (anamnesis, objective examination, laboratory tests, radiological investigations)
- Prevention of further absorption of the toxicant
- Increased elimination of the toxicant
- Administering antidotes
- Supportive symptomatic treatment and clinical follow-up

For laboratory tests, the number of substances for which analytical data can be rapidly obtained in an emergency is limited compared to the number of agents that may be the cause of intoxication; this further emphasizes the importance of recognizing clinical syndromes for intoxicated persons and the importance for the clinical toxicologist to set up general and supportive treatment in patients with intoxication from unknown causes. For substances that can be rapidly measured in emergency laboratories, quantitative data can often provide both prognostic and therapeutic indications.

For many drugs (salicylates, lithium, phenobarbital, etc.), there are predictive relationships between plasma concentrations and clinical evolution, just as for some agents, concentrations indicating the need for therapeutic interventions are known. Some authors have identified plasma concentrations of various drugs and substances that constitute "intervention levels" and threshold toxicity values; these values generally represent the average concentrations of the respective substances that, based on retrospective data, have been shown to have significant harmful effects.

Because of the limited in vitro diagnostic tests for poisons, clinical toxicologists use (where possible and indicated) specific clinical chemistry laboratory data in case of an emergency.

Role of the Laboratory

The laboratory is fundamental in toxicology. In general, the analytical process involves the following steps:

- General definition of the problem
- Analytical definition of the problem
- Choice of method, technique, procedure, and protocol
- Sampling
- Treatment of the sample
- Analysis
- Data evaluation and conclusions

It is crucial to keep in mind some definitions that help better evaluate the analytical data.

A qualitative analysis aims to verify the simple presence (or not) of one or more analytes; a quantitative analysis involves determining the concentration of one or more analytes in the matrix under analysis. The data quality is closely linked to a series of parameters (indicated below), which must be considered in the critical evaluation of the results. Accuracy expresses the closeness of agreement of a data item to the true or accepted reference value; it may include both systematic and random errors. In order to obtain a minor random error, it is advisable to calculate the result as the mean value of a large number of replications. The total systematic error, or bias, can be chosen as a measure of accuracy. Verification of accuracy is usually carried out using reference materials or, in their absence, by comparison with alternative methods.

Precision describes the agreement between two or more measurements obtained after replications and then comparing how close they are within a set of measurements.

The absolute or total error expresses the difference between the value obtained and the true value and is the sum of both random and systematic errors.

The objective of measurement is to estimate the "true value" of the quantity to be measured, termed "analyte" or "measurand," meaning by the true value that the ideal result that could be obtained through a perfect measurement. Every measurement is affected by errors; when all the known or suspected errors have been evaluated, and the result of the measurement has been subjected to all the possible corrections that derive from the said errors, there remains uncertainty about the correctness of the result, which, therefore, must be considered as only an estimate of the value of the measurand. The indication of a result on an analysis report, therefore, should be accompanied by the definition of the uncertainty of the measurement, defined as the value that characterizes the possible deviations of the result from the true value of the measurand.

Repeatability defines the accuracy of replicated measurements within the same laboratory, within a limited time interval, made by the same operator using the same equipment.

The term "reproducibility" refers to the precision of replicate measurements made using the same analytical method but under nonidentical conditions (different laboratories, operators, and equipment) and normally within unrestricted time intervals.

The sensitivity of an analytical method is defined as the smallest difference in analyte concentration that causes a significant difference in signal.

The detection limit is the limiting concentration below which the presence of the analyte cannot be detected, which must, therefore, produce a significantly different signal from that by a reagent blank.

Selectivity refers to the ability of an analytical technique to be unaffected by the presence of interferents or other components other than the analyte.

Traceability refers to the property that an instrument (or, in general, an analytical method) possesses when it is subjected to calibration using measurands whose measurements, regarding samples, have been categorized as "primary" in a given context. Once traceability is established, the instrument can produce measurements compatible with those produced by primary or reference samples and methods. Typically, traceability is established through validation procedures verifying the ability to produce analytical data by the methodology used.

When interpreting the results, some particular combinations of positivity or negativity to the test are possible in relation to the characteristics of the method used.

A false-negative result occurs when the analyte is present but in extremely low amounts compared to the detection limit of the method used or concerning pathophysiological variables of the subject (e.g., large amounts of liquids consumed shortly before sample collection may affect the detectability of the substance in urine samples).

On the other hand, a false-positive result is obtained when a substance is detected that is actually absent in the sample under examination and is due to the presence of interfering substances that cross-react with the reagents.

Analytical Techniques

Atomic Absorption Spectroscopy (AAS)

Atomic absorption spectroscopy (AAS) determines the concentration of inorganic elements in the air, water, soil, and biological matrices (blood, urine, tissues, etc.). The chemical–physical principle on which this technique is based is that atomic energy levels are discrete, so the electronic transitions allowed for radiative excitation ($h\nu$) are characteristic for each atom. Unlike molecules, however, atoms do not contain rotational or vibrational substructures. Therefore, the absorption of electromagnetic radiation by excitation at a higher energy level does not occur for a band of frequencies but at only one frequency and wavelength.

For monochromatic radiation, the consequence of the interaction between the photons and absorbing atoms is a decrease in the incident radius from $P0$ to P. The absorbance is related in logarithmic form to the incident ($P0$) and the emergent (P) radiation: $A = \log(P0/P)$.

All this implies that each atom will have its characteristic absorption spectrum and that for each wavelength corresponding to a sufficiently probable transition, it is possible to perform quantitative measurements by applying a formula analogous to the Beer-Lambert-Bouguer law and expressed as $A = \varepsilon bc$, where A is the absorbance (i.e., the fraction of electromagnetic radiation "absorbed" in the experiment), ε is the molar extinction coefficient (i.e., the specific absorbance of the element under examination at the wavelength used, per unit mass through a reaction cell of unit length; the wavelength at which the element under examination presents the maximum absorbance is chosen), b is the thickness of the reaction cell, and c is the concentration of the analyte.

Inductively Coupled Plasma (ICP)

Inductively coupled plasma (ICP) is an atomization source used in emission spectroscopy and ICP mass spectrometry. It is generated by the effect of a radio-frequency field on a gas current. The discharge is induced without contact between electrodes in an argon current rising through a quartz tube placed inside a spiral stimulated by a radio-frequency generator. The radio-frequency signal creates a magnetic field within the spiral and in the argon current. The magnetic field induces a circular current in the conductor (argon) to heat it. To initiate the ICP discharge, a spark or arc is applied to the argon. This discharge absorbs energy from the magnetic field and transforms the argon into a stable plasma whose temperature is 10,000 K in the base at the center of the plasma.

It should be noted that absorption and emission spectroscopy uses the same phenomenon in opposite ways; the former measures the electromagnetic radiation absorbed in an atomization process and the latter that reemitted after atomization and excitation.

Molecular Biology Techniques

Molecular biology techniques allow the detection, analysis, manipulation, copying, and cloning of nucleic acids; in toxicology, it is used, for example, to detect the presence of genetically modified organisms (GMOs) in food.

Electrophoresis

Electrophoresis uses an electric field to induce the migration of molecules in a buffered medium such as a gel. The migration rate depends on the molecular surface's mass and the number of electrical charges.

Classical Elementary Analysis

Classical elemental analysis is used to quickly evaluate the content of the elements (carbon, sulfur, trace metals, nitrogen, etc.) present in organic materials.

Fluorescence Spectroscopy

Fluorescence spectroscopy uses fluorescence to determine the chemical components of an organic sample. Fluorescence is obtained when a substance absorbs electromagnetic radiation, changes the electronic balance, and immediately reemits electromagnetic radiation. Each molecule absorbs and emits different amounts of electromagnetic radiation at various wavelengths.

Fourier Transform Infrared Spectroscopy

Fourier transform infrared spectroscopy (FTIR) is used to identify organic compounds in a sample. When electromagnetic radiation in the infrared spectrum passes through a sample, the chemical bonds in the molecule absorb energy at specific wavelengths. The data are collected and analyzed by a computer that, using the differential equation known as Fourier transform, mathematically separates the amounts of energy related to each wavelength and reports them in a spectrum. Since each organic compound has a particular spectrum, it is possible to compare the spectrum of the sample under examination with those contained in a library of known spectra, thus identifying the compound.

Mass Spectrometry

Mass spectrometry (MS) is ideal for identifying both organic and inorganic compounds. The sample is ionized in ionization chamber and serves to convert the components of the samples into charged particles. Often in this process, extensive fragmentation of analytes molecules occurs producing a spectrum of charged particles having different mass-to-charge ratios. Both positive and negative particles are produced in the ionization process, and one of these (usually the negative ion) is removed in the source. The mass analyzer is a dispersing device based upon the mass of the charged particles. A mass spectrometer contains an ion detector, a signal processor, and a read out. The finally result of separated mass charged particles is the mass spectrum that allows the identity and concentration of the analyte to be measured through its mass (each compound has a specific mass-to-charge ratio). Mass fragmentography is a particular application, with detection limits as low as nanograms/liter possible.

Nuclear Magnetic Resonance Spectroscopy

Nuclear magnetic resonance spectroscopy is based on the magnetic properties of certain atoms and isotopes. Since a nucleus bears a charge, its spin gives rise to a magnetic field that is analogous to the field produced when electricity flows through a coil of wire. The resulting magnetic dipole is oriented along the axis of spin and has a value that is characteristics for each type of nucleus.

Polarimetry

Polarimetry uses polarized light to analyze various organic compounds (sugars, proteins, steroids, etc.). When the sample is exposed to polarized light, its molecules "bend" the light in a characteristic way. Individual components can be identified and quantified by measuring the transmitted light's direction and angle. Polarimetry is mainly used to analyze substances that possess asymmetrical carbon atoms.

Ultraviolet/Visible Spectroscopy (UV–Vis)

Ultraviolet/visible spectroscopy (UV–vis) is a highly common technique for analyzing organic compounds. When UV radiation passes through the sample, it excites the molecules and causes them to absorb some of the UV radiation in a specific manner; the amount of UV radiation absorbed at a specific wavelength allows the determination of the substances present. In the medical field, UV–vis spectroscopy is mainly used to analyze proteins and nucleic acids.

Chromatographic Techniques

Chromatographic techniques are a set of techniques for separating the components of a homogeneous mixture based on the distribution of the mixture's components between two phases, one stationary and one (called the mobile phase) moving along a defined direction.

Gas Chromatography (GC)

Gas chromatography (GC) is a separative technique in which a mixture of compounds is evaporated and separated in a column consisting of inert support to which a stationary phase of different types and variable lengths is attached. The vaporized sample is pushed by an inert carrier gas (which thus acts as a mobile phase) through the column. During the passage, the individual compounds of the sample separate from each other, escape from the column individually, and arrive at a detector that measures their quantity. Gas chromatography can be used with mass spectrometry, allowing hundreds of compounds to be identified.

Ion Chromatography

Ion chromatography is an ion-exchange separation technique in which low concentrations of cations and anions are determined using low-capacity ion-exchange columns with weak buffers.

Liquid Chromatography (LC)

Liquid chromatography (LC), best known for its primary application, high-performance liquid chromatography (HPLC) and afterwards ultra-high performance liquid chromatography (UHPLC), separates and quantifies the components of a liquid mixture. Pumps in the system push a liquid's mobile phase containing the sample through a column, which is a hollow tube packed with extremely fine particles, to separate the components of the mixture, which are then measured by the detector. Liquid chromatography can also be coupled to mass spectrometry.

Table 52.2 presents a scheme of analytical technique choice as a function of analyte.

Table 52.2 Scheme for choosing the analytical technique according to the analyte's nature

Nature of the analyte	Analytical technique
Inorganic ions (cations and anions)	Gravimetry Volumetric precipitation Ox/redox volume Potentiometry Colorimetry Polarography Ion chromatography
Inorganic ions (mainly cations)	Overall volume Electrogravimetry Polarography Anodic stripping Atomic photometry
Acids and bases	Neutral volume
Organic substances	Spectrophotometry of absorption (UV/Vis/IR)
Heterogeneous materials (moisture, carbohydrates, lipids, proteins, fiber, etc.)	Spectrophotometry of absorption (NIR)
Molecular substances	High-performance liquid chromatography
Optically active electroactive volatile substances	Gas chromatography with dedicated detector Polarography Polarimetry
Radioactive substances	Radiometry

IR infrared, *NIR* near-infrared spectroscopy, *UV* ultraviolet, *Vis* visible

Forensic Toxicology

Clinical biochemistry laboratories are increasingly integrated with university laboratories of forensic toxicology. Forensic toxicology has expanded its field of interest. Its primary task is to study the relationship between humans and toxic agents in relation to the application of specific legal provisions. This peculiarity distinguishes forensic toxicology and leads it to investigate different issues: analysis of the corpse strictly related to forensic medicine, analysis of the voluntary use of drugs and alcohol, chemical risk in the workplace, doping, environmental pollution, and food toxicology.

Drugs and Psychoactive Substances

Individuals have always tried to escape from the difficulties of everyday life through the use of natural substances. In the last few years, new synthetic drugs, defined as new psychoactive substances (NPSs), have been introduced in the clandestine market, together with the drugs considered classic, as they have been present in the illicit market for a long time (cocaine, methadone, opiates, cannabinoids, amphetamines, ecstasy). Because of this, the list of drugs is continuously being updated. This has led to an important development of forensic toxicology applications through increasingly sophisticated instrumentation that has made it possible to discover new substances on the market and thus identify them.

Classic Substances

Cocaine

Cocaine is a methylbenzoylecgonine that comes from the leaves of *Erythroxylum coca* in which it is present along with other minor alkaloids. Clandestine extraction from the pulverized dry leaves is performed by specific treatments to form "coca paste" or raw cocaine. From raw cocaine, which can be taken by smoking, cocaine is obtained by treatment with different types of solvents.

The absorption of cocaine depends on the way that the substance is taken:

- Intravenous route, immediate
- Nasal route, takes a few seconds
- Inhalation, a few minutes (and in any case greater than 98% after 4 hours)
- Smoking, immediate with a plasma peak in a few minutes
- Oral route, slow, greater than 90% after about 2 hours

Based on the type of assumption, not only does the speed of absorption vary but also the times in which the plasma peak is reached (the maximum attainable concentration of cocaine after absorption) as well as the plasma half-life (the time required to halve the concentration in the blood). In the case of inhalation intake, the plasma concentration of cocaine increases rapidly, reaching a plasma peak in 15–60 minutes with a euphoric effect in 15–20 minutes; in the case of intravenous intake, the plasma peak is immediate with a euphoric effect in 4–8 minutes; and in the case of oral intake, the plasma peak is reached in about 50–90 minutes and a euphoric effect in 45–90 minutes.

Once taken, cocaine undergoes plasma and hepatic metabolism by esterases to form benzoylecgonine, the major metabolite, and other minor metabolites such as ecgonine methyl ester and ecgonine.

Benzoylecgonine is an inactive metabolite of cocaine and accounts for up to 50% of cocaine metabolism. Most of the cocaine consumed is metabolized and excreted in the urine as metabolites. In this biological matrix, only small amounts of cocaine are found (below 10%), whereas there is a significant presence of metabolites, in particular benzoylecgonine, which, together with the other metabolites, reaches peak concentrations in urine after about 4–5 hours of intake.

The effects of cocaine on the body depend on its action of activation of dopaminergic neurons, which leads to an initial euphoric action of the substance. In general, the psychic effects of cocaine are divided into three stages of different severity depending on the dose and frequency of use: an initial stage of euphoria with increased cognitive and motor performance and hypervigilance, followed by a phase of dysphoria, melancholy, apathy, difficulty in concentrating, and insomnia, until reaching a phase in which the subject is in a state of drowsiness, a state opposite to the initial one. It is precisely because of this type of effect that cocaine, like other stimulants, is defined as a drug with "on-off" effects on the body.

Cannabinoids

The pharmacokinetic profile of cannabis is poorly understood. The preferential route of administration as a recreational drug is in the form of smoking. When cannabis is smoked, the cannabinoids undergo a process of decarboxylation that makes them more pharmacologically active. This process is minor when the drug is administered orally. The main active ingredient of cannabis derivatives is delta-9-tetrahydrocannabinol (THC), but cannabidiol (CBD) and cannabinol (CBN) are also important.

In a cigarette containing cannabis, about 50% of THC is inhaled; the pharmacological effect appears rapidly, after 20–30 minutes, and disappears after 3–4 hours. Tetrahydrocannabinol passes into the bloodstream very quickly and , then, due to its high lipophilicity, is concentrated in tissues with a high lipid content, where it can be traced for up to 4 weeks after intake. It is then slowly released back into the bloodstream, which is the step that controls the speed of the catabolic process of the molecule (rate-limiting step).

After oral administration, absorption is slow and irregular: pharmacological effects appear within 30–60 minutes, reach their peak after 2–4 hours, and last for approximately 8 hours. Pharmacokinetic parameters show high interindividual variability.

The amounts of THC and CBD in the blood are lower when cannabis is administered as an oral spray (absorption through the mucosa of the mouth) than when it is inhaled (smoking) because oral absorption is slower and also because THC undergoes rapid distribution in high-lipid tissues and a hepatic first-pass effect with conversion to 11-hydroxy-THC (11-OH-THC), a metabolite with psychic effects.

After about 12–24 hours, plasma concentrations of THC, CBD, and 11-OH-THC are close to or below the diagnostic limit of detection due to hepatic and renal clearance and accumulation of cannabinoids in the lipid tissue. The lipid tissue's redistribution of cannabinoids and their slow release (after about 4 weeks) explains the high half-life of these molecules (>24 hours).

Cannabinoids undergo hepatic metabolism by cytochrome P450 enzymes. THC is converted to 11-OH-THC, pharmacologically active and equipotent to tetrahydrocannabinol, and then to inactive 11-nor-9-carboxy-delta-9-tetrah ydrocannabinol (THC-COOH), which is the major metabolite present in both blood and urine. The conversion of THC to 11-OH-THC is catalyzed by the cytochrome enzyme CYP2C9. Other metabolites have also been identified as minor hydroxylates obtained through the action of CYP3A4.

CBD is converted to 7-OH-CBD, the most important derivative; about 30 other derivatives have also been identified.

The slow excretion of cannabinoids influences the tolerability profile of the drug, especially regarding cognitive and psychomotor effects.

Amphetamines

Amphetamines are compounds whose chemical structure is similar to the body's natural stimulants: adrenaline, noradrenaline, dopamine, and serotonin. These psychostimulants are highly numerous (extremely easy to synthesize), mimic the action of the sympathetic nervous system, and therefore are also called sympathomimetics. Among them, we find, for example, amphetamine, methamphetamine, methylphenidate (Ritalin), modafinil, and ephedrine.

Amphetamines come in various forms and are taken in various ways, such as powders (snorted or injected), tablets (swallowed, snorted, or injected after crushing), or capsules (swallowed).

Methamphetamines most frequently occur as varying-sized crystals that are smoked, ingested, or injected as powders or tablets.

Amphetamines are used for various reasons: to stay awake for a long time, to improve performance in sports or work or to increase self-esteem and confidence, to reduce fatigue and endurance, and to reduce feelings of hunger. They are substances that provide physical and psychological tolerance and dependence.

When amphetamines are taken regularly, a high degree of tolerance to their effects develops and the user is forced to take progressively larger quantities to achieve the desired effects. Amphetamines (or amphetamines) and methamphetamines (or methamphetamines) activate the reward or pleasure system, which is also extremely high, leading to a continuous release of dopamine that induces addiction through a series of profound and persistent changes in brain function.

Opiates and Derivatives

The term "opiates" generally refers to natural opium products, synthetic derivatives, and compounds with narcotic analgesic action similar to that of morphine.

Opium is a latex made by incising the capsule of *Papaver somniferum* and drying it to form a brown paste.

Opium contains several alkaloids, with the main ones with narcotic action being morphine, codeine, and thebaine. Morphine is an addictive substance with physical and psychological dependence and tolerance. However, now, its abuse has been supplanted by heroin, a semisynthetic derivative obtained from the acetylation of morphine. The heroin found in the underground market is presented as a powder of different colors depending on the types of substances added. Adulterants (lidocaine, procaine, caffeine), sugars (lactose, mannitol, glucose), and impurities of origin and manufacture (papaverine, monoacetylmorphine, acetylcodeine, etc.) can be found.

Heroin induces tolerance and mental and physical dependence, which develop rapidly within a few weeks. It has also been found that the prolonged use of heroin leads to a simultaneous modification of organs and apparatuses. The scientific world has been trying to synthesize molecules with the same analgesic property as heroin and morphine without the adverse effects; this has led to the discovery of substances with cross-tolerance and, therefore, the ability to avoid a withdrawal crisis in drug addicts after their administration. An example of this type of substance is methadone.

Methadone

Methadone, an opiate agonist, once taken, is metabolized in the liver into a product that the body can use. Excess methadone is stored in the liver and the bloodstream. It acts for 24 hours and beyond. The higher the dosage, the more methadone is stored. This is also why patients taking blocking doses (over 70 mg per day) can go a whole day without their drug dose. Of course, the disadvantage is that a patient missing a dose is at risk of possible "destabilization." The patient slowly loses the blocking effect of the methadone and may begin to feel "hungry and craving for the drug."

Once the metabolized methadone is circulating in the plasma, it slowly passes to the brain where it occupies the opiate receptors.

Alcohol

Ethyl alcohol is the most widely used psychoactive substance; unfortunately, alcohol abuse has become a serious social problem. Ethyl alcohol ingested as an alcoholic beverage is rapidly absorbed by the stomach and intestines and is evenly distributed throughout the various tissues and body fluids. The elimination starts quickly and is carried out for 90–98% through metabolism whereas for the remaining 2–10% through the kidneys and lungs. Ethyl alcohol acts on the central nervous system with a substantially depressive effect. The ethanol concentration in the blood is in equilibrium with that in the encephalon, which is responsible for the toxic effects and is therefore a valid indicator of intoxication. The alcohol curve, which represents how the concentration of alcohol in the blood varies over time, presents at the beginning an increase corresponding to the absorption phase that usually lasts 30–60 minutes, then an equilibrium phase in which absorption and elimination are equivalent, and, finally, a phase of slow elimination that is a linear function of time. It has been studied that the blood concentration in this last phase constantly decreases at a value between 0.1 and 0.2 g/L per hour. This characteristic is of great medico-legal interest because it allows calculating, with good approximation, the concentrations of alcohol in the blood also in an antecedent moment to the drawing.

The determination of alcohol in the blood expresses a possible acute intoxication by this substance; however, we must not forget that in forensic toxicology, it is also helpful to know the possible chronic abuse of alcohol. In this case, different biomarkers are used, such as carbohydrate-deficient transferrin (CDT), which is currently the most reliable and specific.

New Psychoactive Substances (NPSs)

In recent years, the traditional substances (cannabis, cocaine, heroin, amphetamines, etc.) have been joined by new molecules, which are mainly synthetic, i.e., made in the laboratory. Some molecules derive from the design of potential drugs for therapeutic use, whereas others have been created as drugs for voluptuary use.

New psychoactive substances, therefore, comprise a broad spectrum of different molecules, many of which are not internationally controlled, all of which mimic the effects of traditional drugs. NPSs are chemical substances that come in solid or liquid form, such as tablets or capsules.

Substances found include synthetic cannabinoids, often contained in herbal mixtures, synthetic cathinones, phenethylamines, piperazines, tryptamines, ketamine, γ-hydroxybutyric acid (GHB) and γ-butyrolactone (GBL) acid, and others.

The Department of Anti-Drug Policies reports that from 2009 to 2013, more than 280 new molecules were reported by the European Observatory and Italian collaborating centers in the Early Warning System, mainly including synthetic cannabinoids, synthetic cathinones, and phenethylamines.

The appearance of new psychoactive molecules is a phenomenon that must be considered dynamic, responding to demands for new substances in a market that naturally reflects the legal status of the molecules that have been banned.

These new molecules were available until 2011 in smart shops in various Italian cities. However, the rapid updating of the tables in the Presidential Decree (DPR) No. 309 of 1990 as been amended and supplemented and law enforcement activity has greatly reduced their presence in these places, although NPSs are widely available on the Internet.

Recently (2013), the European Commission has proposed to make these substances illegal; in particular, the proposal distinguishes between moderate-risk substances, which are restricted to the consumer market, and high-risk substances, which are fully restricted and subject to criminal measures.

Choice of the Biological Matrix

The most commonly used biological matrices are urine, blood, and keratin. Of course, the choice of the biological matrix is related to the purpose of the investigation.

In particular, if it is necessary to demonstrate a current use, as in the case of traffic accidents, then the matrix of choice is blood, whereas if it is necessary to demonstrate an abuse of a specific substance for a period of time of a few days, then it is better to use urine. If it is necessary to demonstrate a chronic abuse of months or years, then the necessary matrix is keratin.

Blood

A blood sample indicates recent exposure to a substance. The use of this matrix is limited by various factors, in particular the invasiveness of the sampling and the difficulty in obtaining large volumes of samples. Blood analysis is critical in forensic investigations, where it is often necessary to know whether a subject is under the effect of a specific substance and whether there is an acute intoxication by the substance. Therefore, it is the first-choice matrix in traffic accidents and medico-legal cases with toxicological forensic interest.

Urine

Urine is the most used type of sample and has several advantages such as the noninvasiveness of the sample, the possibility of sampling large volumes, and analyzing the metabolites of substances even after days.

The residence time of substances of abuse in the urinary matrix depends on several factors: the substance's pharmacokinetics, its half-life, and the variability of the person's renal clearance as well as the time elapsed between intake and sample collection or the pH of the urine.

A limitation of the urine sample is its easy adulteration with various substances or its easy dilution with the aim, in both cases, of masking a possible intake and thus making the sample "negative." In this sense, as it is reported in the paragraph on the medico-legal value of toxicological analysis, the laboratory can adopt control procedures through the chain of custody and analytical procedures that highlight the authenticity and integrity of the sample collected.

Keratin

Keratin (hair and body hair) allows determination of the main drugs even after months and years, being able, in the case of hair, to analyze continuous segments of it and thus being able to make a history of a possible assumption, considering that hair under normal conditions grows at an average speed of about 0.5–1.5 cm per month. The possibility of reconstructing the history of the assumptions is an advantage in various fields of forensic toxicology, above all as an aid to various services and centers for treating drug addiction. On the other hand, the analysis of this matrix also has limitations, mainly due to cosmetic treatments that may interfere with the quality of the analytical result.

Hair is a type of tissue consisting of a hair follicle that contains the germinative center and a stem, where the keratinized and membrane cells together form three structures (from the outside to the inside): the cortex, cuticle, and pith. The surface of the hair is hydrophobic, whereas the inner part is hydrophilic. Hair has three different phases of growth that also have different durations: the anagenic phase, which is the active phase of growth, lasting about a few years; the catagenic phase, which is the quiescent phase, lasting a few weeks; and finally, the telogenic phase, which is the resting phase, lasting a few months.

The two hypotheses are accreditated by scientific literature to the incorporation of drugs into hair are hypothesized: one, the substances actively and/or passively diffuse from the blood to the germinal center and, two, the substances actively and/or passively diffuse from sweat and sebaceous glands.

A substance can be incorporated differently depending on the melanin content of the hair and the lipophilicity and basicity of the substance.

Saliva

For years, the salivary matrix has been studied in forensic toxicology. In particular, studies have focused on the timing of the appearance of drugs in this matrix. It is now undisputed that saliva has a significant advantage in its use. The concentration and the time of appearance/disappearance of drugs in the saliva is comparable with that of the same substances in the blood. Additionally, saliva can be sampled by a noninvasive method.

Problems that may arise in the use of saliva are the poor stability of the substances it contains, which, therefore, implies the need for special attention to the parameters of conservation, and the decrease in its production, with difficulties in collecting the sample in some individual psychophysical situations, such as anxiety and some pharmacological treatments.

Alternative Matrices: Sweat

Sweat is a type of matrix that can be sampled by rubbing the forehead or other body superfices on an adhesive plaster of absorbent material applied to the body surface for an extended period. Depending on the technique used for its collection, sweat can be considered as an alternative to single or serial urine collection or, in the case of a chronological correspondence between the duration of application of the absorbent patch and the period of growth, to hair. Its collection is a noninvasive technique and provides a cumulative measure of substance use covering a period ranging from a few days to several weeks. The disadvantages are due to the variability of sweat production by each subject and the real risk of accidental or nonaccidental removal of the collection device or its contamination. The analysis of this matrix is still in the experimental stage and, therefore, not yet applicable in a routine setting.

Medico-Legal Value of Toxicological Analyses

In forensic toxicology, the analytical data must have particular requirements of certainty and correctness.

Organizational quality is achieved by paying particular attention to the regulations concerning the handling of biological samples during their collection and also to the documentation needed to control the samples during all the handling phases, from collection to the issuing of the analytical report. In particular, laboratories must adopt a chain of custody for all those toxicological analyses with a medico-legal purpose. The chain of custody is a set of forms documenting all the steps that a sample undergoes, the place of its storage, and how it is stored, without forgetting that the sample is associated with a code that is reported at each stage and that at the time of acceptance it is associated with a person properly identified by an identification document.

Analytical quality is achieved by following the requirements of good laboratory practice (GLP).

A critical step in achieving analytical data quality is validating analytical methods. Before validation, certain factors must be respected when developing the analytical method. In particular, it is fundamental that the analyte for which the analytical method is being developed is detectable and quantifiable by the chosen analytical technique; the extraction method is the most suitable for extracting the analyte from the biological matrix to be analyzed; the limit of detection (LOD) and the limit of quantification (LOQ) are calculated; and a calibration curve of the analyte is constructed for each biological matrix by determining the linearity interval and, at the same time, the accuracy, precision, robustness, and selectivity of the method being developed are calculated.

Furthermore, it is necessary for each analytical procedure that reference cutoffs are specified depending on the diagnostic purpose and, therefore, on the matrix used. The cutoffs are previously established threshold values above which the sample is considered positive and below which it is considered negative.

Another critical element for the evaluation of data quality is the laboratory's adherence to external quality assessments (EQAs), whereby the laboratory analyzes unknown samples and sends the results to the EQA proponent company, which statistically processes the results of all participants and subsequently issues the certificate of participation in the programs.

Analytical Procedures

The type of strategy that a laboratory must adopt in terms of the choice of analytical technique essentially depends on the purpose of the analysis; it is understood that a positive result to a screening technique always requires a confirmatory analysis, as the screening techniques are not highly specific and, therefore, may produce false-positive results.

However, screening techniques are useful as they can conveniently and quickly screen a large number of samples.

Screening Analysis

Screening tests are essentially immunochemical and are based on a competitive reaction for a specific binding site of an antibody between a labeled antigen and an unlabeled one, i.e., the substance to be sought in the sample. The antibody binds both antigens, but the more antigen present in the sample, the less labeled antigen will be bound, so there will be a different amount of free labeled antigen, with the amount being inversely proportional to the amount of antigen in the sample. Therefore, at the time of saturation, it will be possible to calculate the amounts of both the bound labeled antigen and free labeled antigen. Measurement of the labeled antigen's signal will highlight the antibody's primary reaction and allow measurement of the extent of competition.

Immunochemical tests can be a homogeneous- or heterogeneous-phase. However, the most suitable for forensic toxicology are homogeneous tests, i.e., those in which, at the time of reaction measurement, it is not necessary to separate the free fraction of labeled antigen from that bound to the antibody. Homogeneous-phase immunochemical tests are enzyme immunoassays (enzyme-multiplied immunoassay technique (EMIT)), whereas heterogeneous-phase tests are radioimmunoassays (RIAs).

As mentioned, only enzyme immunoassays are now used in forensic toxicology because these methods are direct, automated, fast, easy to perform, require small sample volumes, and provide a semiquantitative response.

In EMIT tests, the labeled antigen is conjugated to an enzyme whose activity is measured when the reaction with the antibody reaches the equilibrium state. Lysozymes, glucose-6-phosphate dehydrogenase, and peroxidases are used as marker enzymes.

The analyte labeled with the enzyme binds to the antibody; the enzyme is in an inactive form because, in this condition, it cannot perform its catalytic activity. Any analyte present in the unknown sample competes with the antibody binding by displacing the enzyme-labeled analyte from binding to it. In this way, the enzyme, free from binding to the antibody, will be able to carry out its catalytic action. This activity is measured and directly proportional to the amount of analyte in the sample.

In forensic toxicology, immunochemical tests are helpful because, besides being simple and fast, they are highly sensitive. However, a limitation of these tests is their low specificity, which leads to false-positive results due to molecules with a similar chemical structure or interfering molecules that produce cross-reactivity. It is precisely because of these problems, and the medico-legal value of the results in forensic toxicology, that the results of immunochemical tests must be confirmed using techniques that use different analytical principles, such as chromatographic techniques.

Immunochemical tests (EMIT and RIA) use a threshold value, i.e., cutoff, above which the sample is considered positive and below which it is considered negative.

Confirmation Analysis

Confirmatory chromatographic tests are highly specific tests that distinguish individual substances within a class of substances, with a higher sensitivity than screening tests.

The most used chromatographic techniques in forensic toxicology laboratories are chromatography (gas or liquid) combined with mass spectrometry. In both cases, the compounds of a mixture are separated as they are retained by a stationary phase (a chromatographic column) and, at the same time, dragged by a mobile phase (gaseous or liquid) in different ways, depending on their chemical and physical characteristics. Depending on the type of analyzer or detector interfaced with the chromatograph, different analytical techniques will be used. However, the most suitable for forensic applications are those in which a mass spectrometer is interfaced with a chromatograph: gas chromatography combined with mass spectrometry (GC-MS) and liquid chromatography combined with mass spectrometry (LC-MS). These types of techniques provide a mass spectrum for each molecule, i.e., a specific graph showing the relative intensity of each ion for the mass–charge ratio (m/z, i.e., the mass of a given particle m with the number of z charges that the particle carries). These techniques can distinguish, for example, individual substances of the opiate class, such as morphine, codeine, and 6-monoacetylmorphine, and, by their different mass spectra and using LC-MS, they can also detect other metabolites, such as morphine-3-glucuronide and morphine-6-glucuronide. This is because LC-MS, unlike GC-MS, is a technique that allows the analysis of even thermolabile and hydrophilic molecules such as glucuronides.

The chromatographic methods adopted by each laboratory must be validated by the laboratory through the calculation of several parameters, including precision, accuracy, LOD, LOQ, robustness, and specificity.

Clinical Pharmacology

Therapeutic Drug Monitoring

The rationale of therapeutic drug monitoring (TDM) concerns drugs that, due to a restricted therapeutic index and to improve the clinical response, require periodic quantification in blood. TDM is effective in cases in which a correlation between a single dose or two doses near the maximum blood concentration reached by the drug and the exposure represented by the concentration–time curve is established.

Interindividual variability in metabolizing the drug leads to a wide distribution of plasma concentration and, consequently, to different efficacies at the site of action. The scheme of the dose–response process is exemplified in Fig. 52.2.

Requirements for Drugs to be monitored

Therapeutic drug monitoring is useful in the following cases:

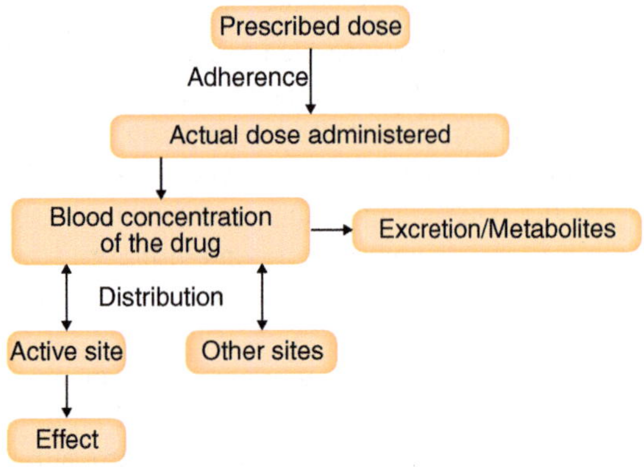

Fig. 52.2 Scheme of the dose–response process. (Copyright EDISES 2021. Reproduced with permission)

- Monitoring of drugs with poor correlation between the dose and clinical effect
- A narrow therapeutic index, which represents the margin between the therapeutic dose and the toxic dose; drugs such as oral anticoagulants, aminoglycoside antibiotics, antineoplastic drugs, immunosuppressants, and cardioactive glucosides are examples of drugs with a narrow range between the desired dose and the toxic dose and with susceptible individual variations
- If the blood concentration correlates with the clinical effect and the interindividual pharmacodynamic variability is minimal

Other aspects to be considered for the correct use of TDM include the evaluation of pharmacologically active metabolites and reversible action at the receptor site. The latter aspect ensures that the intensity and duration of the response is temporally correlated with the concentration of the drug at the receptor site.

In cases in which pharmacological efficacy is predicted by clinical biomarkers, TDM is not necessary. Examples are antihypertensive drugs monitored by blood pressure monitoring and hypoglycemic drugs monitored by glycemic monitoring.

Drugs That Meet the Requirements of TDM

Some drugs that meet the requirements of therapeutic monitoring are phenytoin, lithium, theophylline, and aminoglycoside antibiotics (gentamicin, tobramycin).

Other medications that are monitored but do not meet all TDM criteria are as follows:

- Carbamazepine (the effect–concentration relationship is not always linear due to the presence of active metabolites).
- Digoxin (fulfills most of the criteria, but there are some doubts about the effect–concentration relationship).
- Immunosuppressive drugs: For cyclosporine, being the first to be introduced into clinical practice, numerous studies have been performed to demonstrate the benefit of TDM-guided therapy management. This drug provides a model for subsequent immunosuppressive drugs, such as tacrolimus (FK506), sirolimus, everolimus, and mycophenolate.

Similar evidence has also been acquired to support the benefit of TDM for antiretroviral drugs.

Objectives of TDM

First, TDM must ensure sufficient drugs at the site of action to produce the desired effect. In addition, it must ensure that the concentration of the drug and its metabolites are not so high as to produce symptoms of toxicity and guide the clinician in adjusting the dosage, especially in clinical situations where pharmacokinetic changes are rapid (e.g., neonates, children, and patients with hepatic or renal dysfunction).

Appropriateness of Sampling

Sampling should be performed before a new drug administration, i.e., at the downstream point of the concentration vs. time curve.

Most drugs that meet TDM requirements need serum sampling (tubes without additives), except the immunosuppressive drugs cyclosporine, tacrolimus, everolimus, and sirolimus. In this case, international guidelines recommend whole blood dosing due to the 70–80% distribution of the drug in the red blood cell fraction. The only immunosuppressive drug with a low intra-globular prevalence is mycophenolate, and the guidelines recommend drawing it from a tube containing K2H2 ethylenediaminetetraacetic acid (EDTA) as an anticoagulant.

Recommended Readings

Ariens EJ, Simonis AM, Grana E (1978) Introduzione alla tossicologia generale. Piccin, Padova

Bertol E (2011) Analitica Tossicologica. Escupalio, Bologna

De Angelis L (1997) Elementi di tossicologia analitica. ISU UniversitaÃÄ degli Studi di Milano

Galli CL, Marinovich M, Restani P (1991) Tossicologia sperimentale. OEMF

Galli L, Marinovich M, Restani P (2008) Tossicologia. Piccin, Padova

Goodman J (2005) Goodman and Gilman's the pharmaco- logical basis of therapeutics. McGraw-Hill
Hallworth M, Royal Shrewsbury Hospital (UK), Watson I, University Hospital Aintree, Liverpoo UK (2010) Therapeutic drug monitoring – clinical guide, 3rd edn
Hayes AW, Kruger CL (2014) Principles and methods of toxicology. CRC Press
Klaassen CD, Watkins JB III (2013) Casarett&Doull's Elementi di tossicologia. Casa Editrice Ambrosiana, Milano
Lu FC, Kacew S (2015) Elementi di tossicologia. EMSI, Roma
Maines MD, Costa LG, Reed DJ et al (2000) Current protocols in toxicology. Wiley
Moffat AC, Osselton MD, Widdop B, Jo W (2011) Clarke's analysis of drugs and poisons, 4th edn. Pharmaceutical Press

Pharmacotoxicology of Substances of Abuse

53

Roberta Pacifici, Simona Pichini, Manuela Pellegrini, and Paolo Berretta

Introduction

The definition of "substance addiction" considers the presence in the consumer subject of at least three of the seven elements shown in Table 53.1.

Tolerance and physical addiction reflect the physiological adaptation to the effects of a psychotropic substance, while the other criteria define the uncontrolled desire to consume that substance. However, tolerance and physical addiction are not necessary and sufficient conditions for diagnosing dependence. Based on a study conducted in 2001, an organism tries to counteract the effects of a psychotropic substance through a vicious circle in which the hedonic point (the point at which pleasure is achieved) changes in response to the administration of the substance itself. Therefore, the state of drug addiction results from the individual's inability to regulate the mechanism of demand and subsequent allostasis, i.e., the ability to achieve stability through change.

Table 53.1 Elements characterizing substance addiction

Craving	(Formerly called psychological dependence). It is an intense desire to re-experience the effects of a psychoactive substance. Desire is the cause of relapse even after long periods of abstinence
Physical or psychological dependence	Outdated terminology that refers to a state of physical and/or psychological dependence and the consequent possible abstinence from the substance
Priming	It refers to a new use of a substance taken previously. This new exposure can lead to a rapid resumption of use of the substance even more intensively
Relapse	It is the resumption of the use of a psychotropic substance after a period of abstinence. Environmental cues (people, places, or things associated with past drug use) and stress can trigger an intense craving for intake and cause a relapse
Reward	It is a stimulus that the brain interprets as something inherently positive and as something to be achieved
Awareness raising	It is the increase in the expected effect after repeated administration of a psychotropic substance. Sensitization also refers to persistent hypersensitivity because of a substance in a person with a history of exposure (to the substance or stress). Sensitization is one of the neurobiological mechanisms involved in desire and relapse
Substance abuse	It is a phenomenon characterized by recurrent and clinically significant negative consequences associated with repeated use of the substance, such as the inability to lead a life with regular rhythms, use of substances in physically dangerous situations, legal problems related to the use of the substance and continued use despite continuing social and interpersonal problems
Substance dependence	It is a cluster of cognitive behavioral and physiological symptoms resulting in clinically relevant problems in a person who uses a substance continuously. To diagnose substance dependence, at least three of the following characteristics must be present: substance tolerance; withdrawal symptoms; use of the substance in increasing quantities and forever longer periods; persistent desire and unsuccessful attempts to reduce or control its consumption; continuous research of the substance; alteration of a normal lifestyle due to the consumption of the substance; continued use of the substance despite health, economic and social problems
Withdrawal syndrome	It is a set of signs and symptoms that follow the abrupt suspension or reduction of the use of a psychotropic substance or the consequence of the action of an antagonist substance used in a possible cessation treatment. This syndrome can be one of the causes of craving or short-term relapse

R. Pacifici (✉) · S. Pichini · M. Pellegrini · P. Berretta
National Centre on Addiction and Doping – Istituto Superiore di Sanità, Rome, Italy
e-mail: roberta.pacifici@iss.it

M. Ciaccio (ed.), *Clinical and Laboratory Medicine Textbook*, https://doi.org/10.1007/978-3-031-24958-7_53

Ionotropic and Metabotropic Mechanisms of Action of Substances of Abuse

Knowing the mechanisms that lead to addiction is very difficult because they are very complex, and one has to consider both cellular and molecular mechanisms and behavioral and social ones.

Substances of abuse generally act as receptor agonists of endogenous neurotransmitters in the central nervous system by binding to two types of membrane receptors, ionotropic (Fig. 53.1) and metabotropic (Fig. 53.2).

Ionotropic receptors (ligand-dependent ion channels) mediate fast synaptic transmission. The neurotransmitter or substance of abuse binds to the receptor, which undergoes a conformational change, opening the cell membrane and allowing ions to enter the cytoplasm.

This creates depolarization or polarization of the membrane and activation of different proteins.

Nicotine, for example, binds to nicotinic cholinergic receptors, which contain sodium channels, while benzodiazepines, barbiturates, and ethyl alcohol bind to the γ-aminobutyric acid (GABA) type A receptor, thereby facilitating chlorine entry. Ethyl alcohol and phencyclidine are inhibitors of N-methyl-D-aspartate sensitive glutamate receptors, which contain sodium and calcium channels.

Metabotropic receptors (G-protein-coupled), on the other hand, are receptors that mediate slow synaptic transmission. G proteins are trimeric structures composed of two functional units: an α subunit that catalyzes GTPase activity (i.e., the conversion of guanosyntriphosphate [GTP] to guanosyndiphosphate [GDP]) and a β/γ dimer that interacts with subunit α when it binds to the GDP (inactive state). A closer G protein is activated when the agonist binds to the receptor. The GTP-bound α-subunit subsequently dissociates from the β- and γ-subunits. Both can activate or inhibit enzymes (e.g., adenylate cyclase or phospholipase C), synthesizing second messengers such as cyclic adenosine monophosphate (cAMP), cyclic guanosine monophosphate, inositol triphosphate, and diacylglycerol. In addition, the β and γ subunits act directly by regulating calcium, sodium, and potassium channels. Second messengers are also responsible for activating the protein kinases responsible for channel phosphorylation. Protein kinases determine pharmacological effects and, at the same time, induce changes in transcription factors, such as the cAMP-sensitive binding protein, CREB (cAMP Response Element-Binding protein), and the δ-FosB protein.

For example, opioids bind to opioid receptors that reduce cAMP levels and cannabinoids to cannabinoid receptors. Classical hallucinogens are partial agonists of some specific serotonergic receptors. Amphetamines and cocaine indirectly affect dopaminergic, serotonergic, and adrenergic receptors by facilitating the release and inhibiting the reuptake of these neurotransmitters.

Psychotropic substances with the potential for abuse can increase the amount of dopamine in neurons of the ventral tegmental presynaptic area (VTA) and the nucleus accumbens located near the midline on the floor of the midbrain.

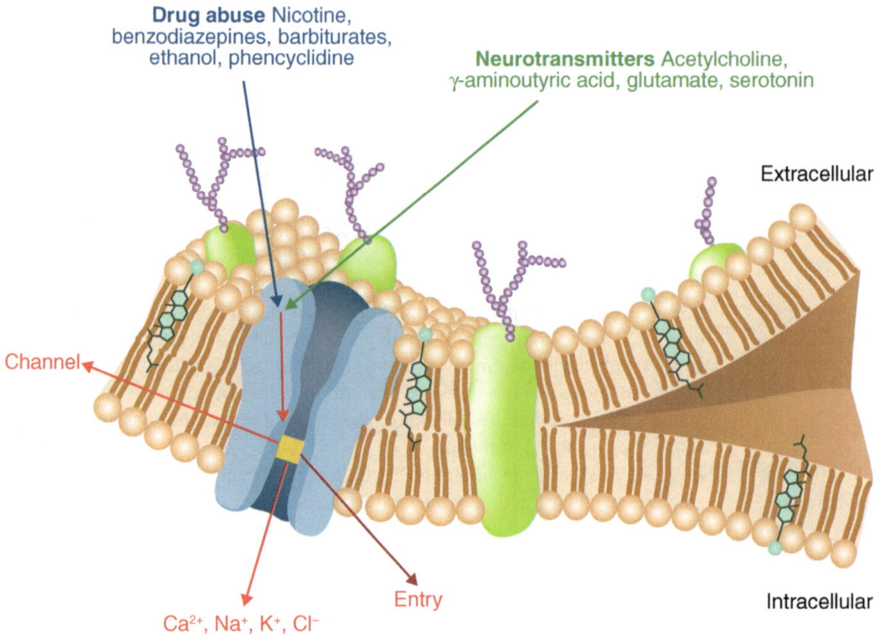

Fig. 53.1 Ionotropic mechanisms in the action of substances of abuse. (Copyright EDISES 2021. Reproduced with permission)

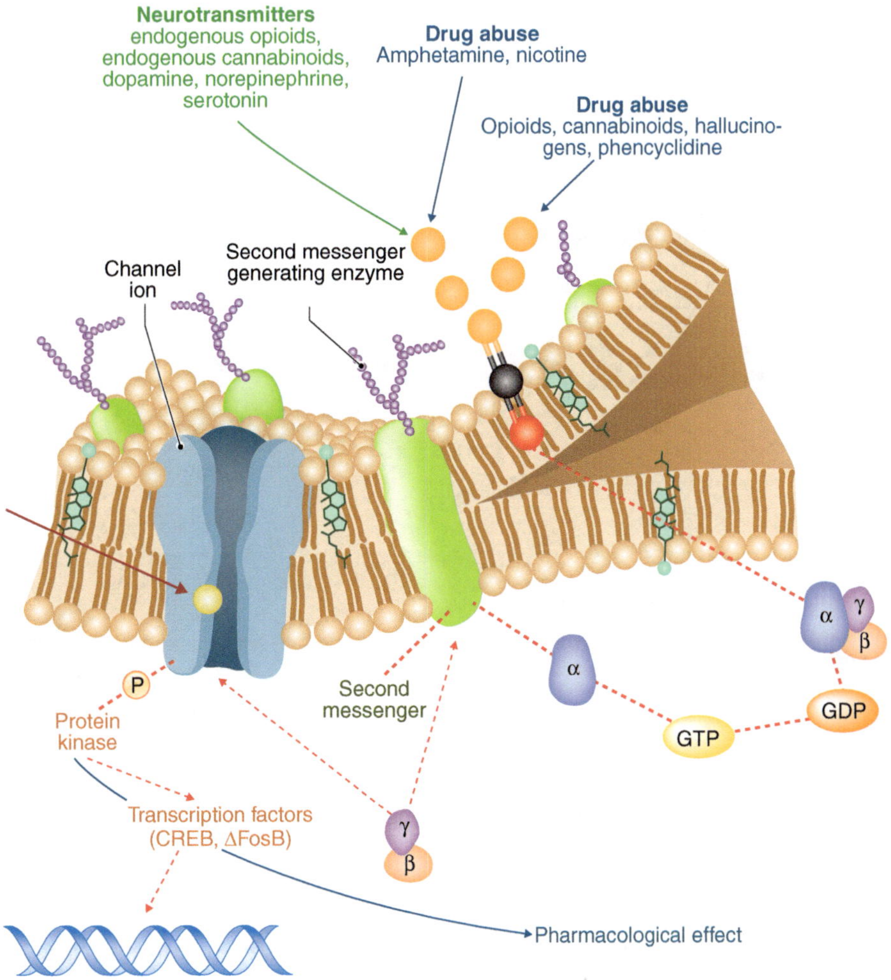

Fig. 53.2 Metabotropic mechanisms in the action of substances of abuse. (Copyright EDISES 2021. Reproduced with permission)

These areas involved in gratification, along with the amygdala and hippocampus, are the main components of the limbic system capable of controlling impulses, emotions, and memory (Fig. 53.3).

The mesocortical system is instead represented by the ventral tegmental area and by the prefrontal and orbitofrontal cortex; this system is involved in the conscious awareness of the effects of psychotropic substances, in the intense desire to experience again the effects of a psychoactive substance (craving) and in the compulsive phenomenon of the search for psychotropic substances. The two systems work in parallel, interacting with each other and thus increasing the extent of the different areas. Thus, while amphetamines, cocaine, and cannabinoids interact with the γ-aminobutyric acid neurons of the nucleus accumbens area of the prefrontal cortex, opioids, ethyl alcohol, barbiturates, and benzodiazepines interact in the ventral tegmental area, influencing the noradrenalin system in the locus coeruleus. Thus, the mesocortical-limbic dopaminergic systems originating in the ventral tegmental area include projections from the cell bodies of this area to the nucleus accumbens, the amygdala, and the prefrontal cortex; glutamatergic projections from the prefrontal cortex to the nucleus accumbens and the ventral tegmental area; projections from γ-aminobutyric acid neurons from the nucleus accumbens to the prefrontal area. Opioid interneurons modulate the inhibitory action of γ-aminobutyric acid on the ventral tegmental area by influencing the activation of noradrenergic neurons in the locus coeruleus. Serotonergic projections from the raphe nucleus extend to the ventral tegmental area and the nucleus accumbens.

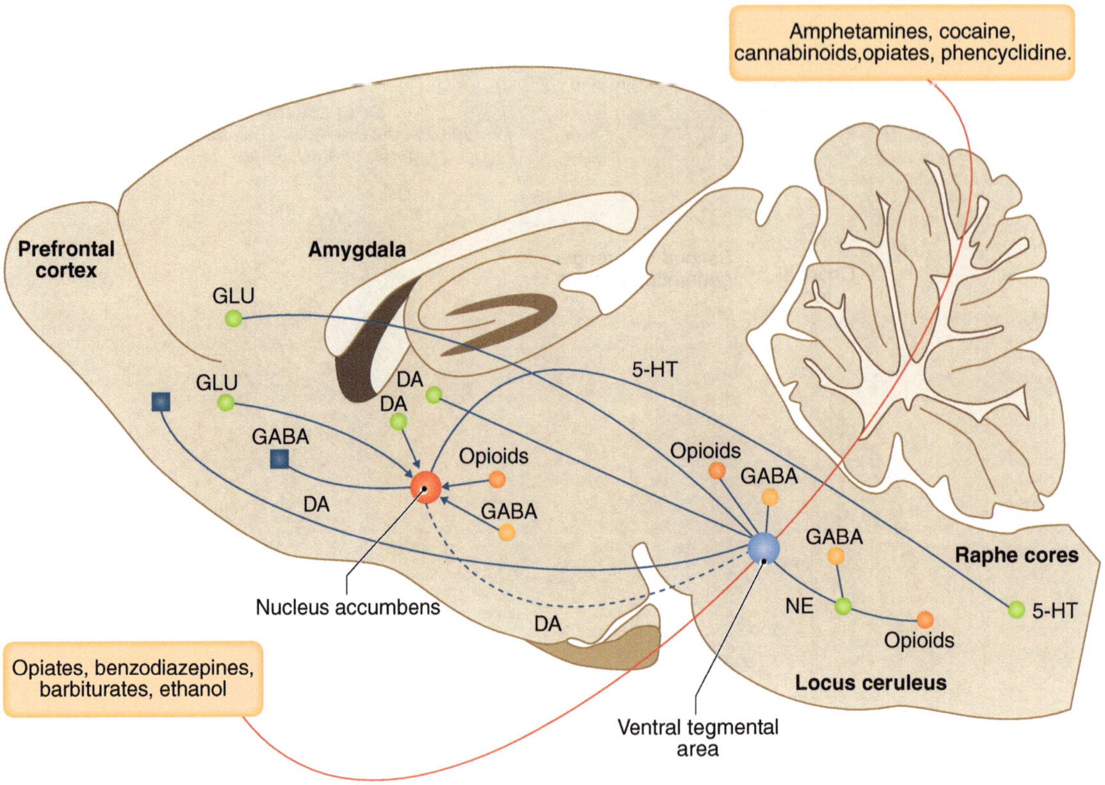

Fig. 53.3 Neuronal gratification circuits critical in the reinforcing effect of drugs of abuse. (Copyright EDISES 2021. Reproduced with permission)

Factors Influencing Substance Abuse

Pharmacological and Physicochemical Properties of Substances of Abuse

Pharmacological and physicochemical properties of substances of abuse are essential factors in the onset of addiction because substances can be taken in different ways based on specific characteristics such as:

- Fat solubility, which increases the passage of a substance across the blood-brain barrier
- The solubility in water, which facilitates the possibility of intravenous administration of substances
- Volatility, which favors the inhalation of substances in vapor form
- Resistance to heating, which favors the possibility of smoking the substance

These characteristics favor the rapid onset and intensity of the potential abuse-related effect, which is why substances that rapidly reach high concentrations in the brain are generally preferred. Thus, heroin is preferred to morphine, and smoking "crack" cocaine (cocaine hydrochloride in sodium bicarbonate, which allows the formation of "cocaine base") is preferred to intranasal administration of cocaine (in the form of hydrochloride salt).

Any short half-life of a substance (e.g., heroin) produces abrupt and intense withdrawal syndromes compared to a long half-life (such as methadone).

Personality and Psychiatric Disorders

Certain personality traits and mental disorders (present in drug users) are important conditioning factors in drug dependence, as are novelty seeking and risk-taking.

Polyaddiction is shared among users of substance abuse, and many of these individuals meet the criteria for abuse and/or dependence on more than one substance.

Many psychiatric disorders, particularly schizophrenia, bipolar disorder, depression, and attention deficit hyperactivity disorder, are associated with an increased risk of abuse. A dual diagnosis (substance abuse and mental disorders) has unfavorable implications for management and the possibility of detoxification.

Genetic Factors

Genetic factors that influence the metabolism and effects of psychotropic substances contribute to the risk of addiction. Although addictions are not heritable, strictly speaking, and it is not possible to predict the risk of inheriting susceptibility, it is possible at present to inherit a predisposition that may be linked to genetic variables of structures involved in neurotransmission. Most of the studies on these hypotheses have been carried out for alcoholism, but a genetic influence has also been demonstrated for other substances such as tobacco, opioids, cannabis, stimulants, and psychoactive drugs.

Studies conducted on adopted children have documented a higher incidence of alcoholism in subjects whose biological parents were alcoholics compared to subjects with alcoholic adoptive parents and non-alcoholic biological parents. Through new techniques of linkage and genetic association, it was possible to identify regions of the genome that seem more involved in these predispositions, underlining how addictions are complex behaviors influenced by both environment and genes. In other words, the genetics of an individual influence the activity of substances of abuse, including the risk of addiction.

Danger of Substances of Abuse

Substance abuse is a major social, legal, and public health problem in the modern world.

All psychotropic substances can be classified according to the risks, harms, and addiction they cause in humans. However, how these parameters are defined is not always clear. This lack of clarity is due to the great variety and complexity of factors that must be considered when estimating harm. There are three main factors whose association determines what is termed harm associated with a substance of abuse:

- Physical damage caused by the use of the drug
- Drug-induced addiction
- Effect that the use of such substances can have on the family and society

Physical Damage

Assessing the potential of a psychotropic substance to cause physical harm, i.e., damage to organs or systems involves a systematic consideration of the margin of safety of that substance in terms of acute toxicity and long-term health risks. The effect of a substance of abuse on physiological, respiratory, and cardiac functions is a major determinant of physical harm and the mode of administration. Substances taken intravenously (e.g., heroin) carry a high risk of sudden death due to respiratory depression. Tobacco and alcohol, on the other hand, have a high propensity to cause illness and death through chronic use. There is evidence, for example, that long-term cigarette smoking reduces life expectancy by an average of 10 years.

Generally, in physical harm that causes immediate effects (e.g., respiratory depression with opioids, acute cardiac crises with cocaine, and fatal poisonings), the toxicity of abuse substances is often measured by assessing the ratio of the lethal dose to the usual or therapeutic dose. Chronic physical damage is the health consequence of repeated use (e.g., psychosis with stimulants, a lung disease with cannabis). Intravenous administration of substances of abuse can cause the spread of viruses through the bloodstream with enormous implications for the health of the individual and for society.

Addiction

As already mentioned, addiction is a state characterized by the compulsive intake of the substance. This loss of control is chronic and is due to a significant alteration of the central nervous system, gradually caused by the substance taken with increasing frequency and dosage. The state of dependence can be both physical and psychological.

Physical dependence is due to the substance's lack of intake, which determines specific symptomatology in the consuming subject, depending on the type of substance used. Usually, physical dependence is present as a result of the abuse of opiates, alcohol, and sedatives and is always connected to a psychic dependency, but not necessarily to a state of tolerance.

Psychic dependence, on the other hand, is represented by the simple desire on the part of the user to continue taking the substance because of the sense of psychological well-being it causes. This dependence can occur in the user without a physical dependence, as with cocaine and other hallucinogens. However, it is now observed that poly-drug users have a psychological dependence associated with physical dependence. Furthermore, psychological dependence complicates any therapeutic and rehabilitative treatment of the user, producing a very high rate of relapse and failure in recovery programs.

Social Damage

Drugs also cause damage at a social level, significantly affecting the family and the subject's life (health care, school, etc.).

Substances that lead to acute intoxication are associated with considerable costs in terms of accidental harm to the user and others. Alcohol intoxication, for example, often leads to violent behavior and is a common cause of traffic accidents. Substance abuse causes serious harm to the fam-

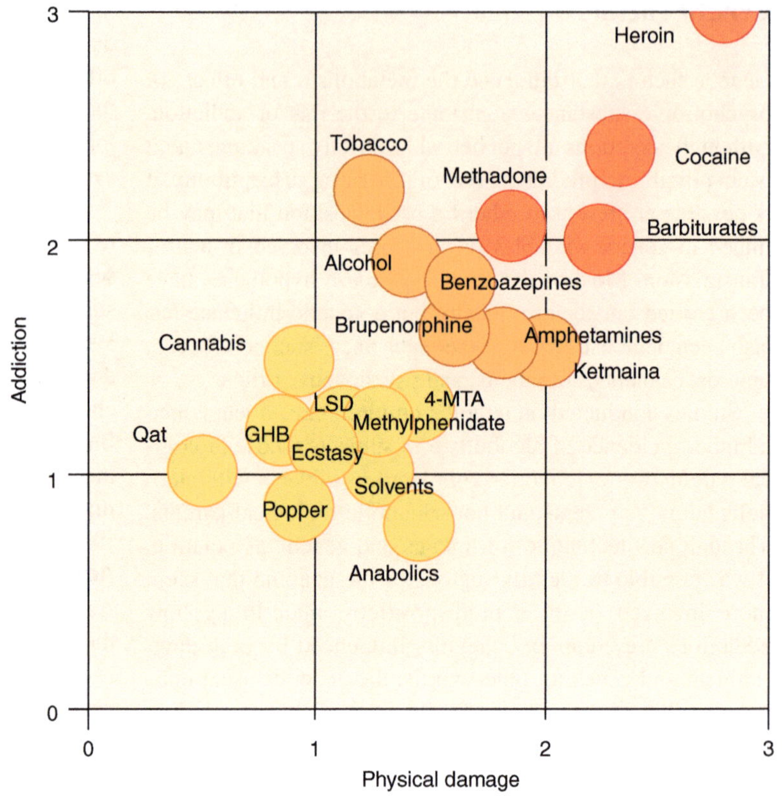

Fig. 53.4 Nutt diagram on the danger of the various substances of abuse. (Copyright EDISES 2021. Reproduced with permission)

ily, both because of the effects of intoxication and because they demotivate the user individuals with regard to work and affection, significantly increasing illicit money-seeking and crime rates.

The damage to society is due to a considerable increase in healthcare costs. For example, while tobacco is attributed to 40% of all hospital illnesses and 60% of deaths, alcohol is the cause of more than half of all injuries and admissions to emergency departments; and intravenous drug use can cause particular problems in terms of blood-borne virus infections, particularly HIV and hepatitis, and transmission, of infection in sexual partners.

Damage Assessment

Many scholars have attempted to define a scale of harm for different abuse substances. One of the best known and most widely used is the one formulated by Nutt et al., using nine different categories of harm (Fig. 53.4). For each of the harms, the researchers identified parameters with which to reclassify the individual substances (e.g., acute or chronic harm, psychological and/or physical dependence, social harm, and health costs), subsequently submitting their grid to both specialists in the field, essentially psychiatrists, and social scientists and medical researchers. Experts were asked not only to score dangerousness but also to discuss the method and the assessments in joint and repeated sessions. These cross-referenced and repeated assessments resulted in a pattern that is as follows.

Heroin and cocaine are the substances with the highest physical harm and risk of dependence according to this classification. This is followed by tobacco, ethyl alcohol, methadone, and cannabis.

Main Classes of Substances of Abuse and Mechanism of Action

Opioids

From the *Papaver somniferum*, an annual herb, 50/150 cm tall, with white, purple, or violet flowers, native to Turkey, but now present in all temperate climate zones, a whitish latex, the raw opium, is obtained by cutting the immature capsules. Usually, only one transversal cut is made with an incision knife, which extends almost entirely around the capsule. The initially white latex coagulates rapidly and turns brown. Once the incision has been made, the partially dried latex is collected for further processing. Opium contains about 25 alkaloids, the six most important being those with a phenanthrene core (morphine, thebaine, and codeine) and those with a benzylisoquinoline structure (noscapine, narceine, and papaverine).

Regarding the mechanism of action of opioids, three main receptor subtypes (μ, δ, and κ) have been identified, all belonging to the G protein receptor family. The endogenous ligands of these receptors are endorphins (α, β, γ, and δ), produced in the anterior lobe of the pituitary gland and endowed with analgesic activity due to their agonist action on these receptors. μ receptors mediate significant effects of opioids on the central nervous system, such as respiratory depression, supra-spinal analgesia, and euphoric properties, but are also responsible for numerous peripheral effects. In the cardiovascular system, morphine (the main natural opioid) and μ agonists do not have significant effects except at high doses, when they release histamine and depress the bulbar vasomotor center to the point of causing hypotension. Histamine may also be responsible for itching and skin redness. The activation of δ receptors (with inhibitory action), located on the nerve endings that carry the algogenic message to the spinal cord's dorsal horns, is responsible for the spinal analgesic action of morphine.

The supraspinal analgesic effects are instead produced by the action of the substance in the areas of origin of the descending anti-nociceptive fibers (locus coeruleus, periaqueductal grey, rostro-ventral bulb), in particular by the removal of GABAergic inhibitory control over these pathways. The increase in mood, decrease in anxiety, and fear followed by sedation, drowsiness, and confusional state are all direct effects on the central nervous system.

Respiratory depression depends on the dose taken and is caused by blockage of the bulbar respiratory center. In the presence of intense pain, the slowing of breathing is usually counteracted by the strong emotional alteration of the patient but may become heavy when the pain tends to disappear. Pupillary constriction is induced by the action of morphine on the Edinger–Westphal nucleus, which is attached to the parasympathetic autonomic system.

There are very strong structural similarities between some parts of opioid molecules and endorphins. For this reason, the functions of neurons that have receptors for endorphins can be modulated by opioids through the effects induced by their intake, causing functional alterations in brain structures in which the endorphins receptors are located. In general, opioids tend to inhibit the functions of the neurons of the structures that possess receptors for endorphins, which makes us understand how at the brain level, there are the biological premises for the major effects of opioids.

From morphine, the primary opium alkaloid, through a reaction with acetic anhydride (process of acetylation of the two hydroxyls in position 3 and 6) a semi-synthetic derivative is obtained: diacetylmorphine or heroin (Fig. 53.5).

Heroin is a lipophilic molecule that passes rapidly into the central nervous system with a plasma half-life of about 4–5 minutes. In the circulation, it is deacetylated in a short time to monoacetylmorphine and then again to morphine which carries out the psychotropic action.

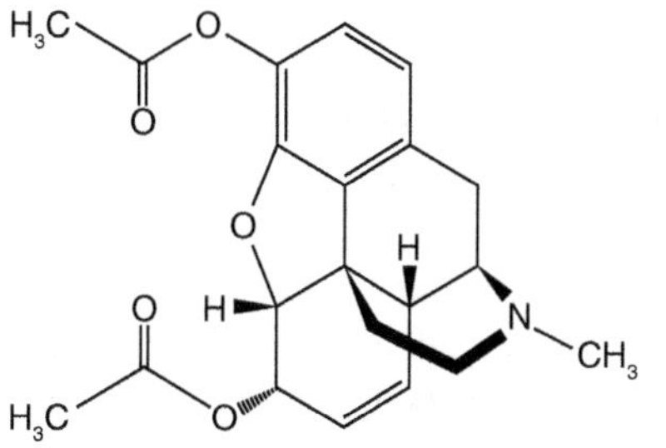

Fig. 53.5 Structure of the diacetylmorphine or heroin molecule. (Copyright EDISES 2021. Reproduced with permission)

Generally, the method of taking heroin is intravenously or intramuscularly. However, heroin is also "snorted" or smoked. These methods of taking heroin are chosen to avoid the risks of infection associated with the injection route, but also in the mistaken belief that they are less likely to lead to addiction. The intravenous injection produces the strongest intensity and the fastest onset of euphoria (7–8 seconds), while intramuscular injection produces a relatively slow onset of euphoria (5–8 minutes). When heroin is inhaled or smoked, the strongest effect is generally felt between 10 and 15 minutes.

The feeling of euphoria is accompanied by a lowering of the heart rate and blood pressure, itching, narrowing of the pupils, slowing of the ideomotor processes, and a state of drowsiness. The assumption of heroin or, more generally, opioids also inhibit the functions of the hypothalamus, through which the brain controls and directs the functions of the pituitary gland, the primary gland of the endocrine system, thus disrupting the hormonal balance that underlies the functioning of the reproductive system. The gastrointestinal action of opioids also produces a reduction in gastric secretion, bile, and pancreatic juice in the intestine, delaying all digestive processes, including intestinal motility, which is impaired over time, together with inhibition of the stimulus of defecation, with a general constipating effect of morphine and heroin. Opioid constipation is generated by decreased gastrointestinal secretory activity, blockage of peristalsis, and increased sphincter tone. Nausea and vomiting are effects of central origin, caused by stimulation of the CTZ (Chemoreceptor Trigger Zone) area, located on the floor of the fourth ventricle. The tone of the bladder and urethral sphincter muscles increases at therapeutic doses, and this causes urinary retention. Opioids have an antitussive effect and numerous neuroendocrine effects, such as the release of antidiuretic hormone (ADH), prolactin, somatostatin, and the decrease of luteinizing hormone (L.H.).

The risk that an individual will experience acute intoxication (overdose) is greatly increased if the person is concurrently taking alcohol, sedatives, and other drugs or if former users who have discontinued opioids for a long period resume opioid use. This phenomenon is due to the progressive attenuation of opioid tolerance, which makes the individual vulnerable to the risk of overdose. Finally, this can also be due to much more concentrated batches of heroin than those normally sold, introduced accidentally or deliberately into the illegal market. The acute intoxication provokes characteristic symptoms such as apathy, emotional detachment, drowsiness, nausea, and vomiting, followed by miosis, bradypnea, and, at high doses, coma, severe respiratory depression, and hypotension. Death occurs by breath arrest, and shortly before death, miosis disappears, and the pupil appears dilated due to anoxia. Treatment of heroin overdose requires assisted breathing and administration of intravenous naloxone, which must be carefully monitored to avoid withdrawal symptoms in the addict.

Heroin withdrawal syndrome is not life-threatening for the patient, contrary to ethyl alcohol deprivation. However, it has a substantial impact on the addicted subject, who is incentivized to seek and consume heroin, not only for the rewarding properties of the substance but also to avoid the unpleasant symptoms of withdrawal, which begins 6–12 hours after the last intake and lasts 7–10 days, with a peak of intensity around the third-fourth day. Such withdrawal is characterized by anxiety, irritability, dysphoria, tachypnea, tachycardia, mydriasis, sweating, insomnia, vomiting, diarrhea, and hyperalgesia. Heroin withdrawal is more intense than that induced by other opioids such as methadone and buprenorphine, which are used as heroin replacement drugs in the treatment of pathological addiction withdrawal.

Morphine, derived only from opium and taken mainly by oral administration, has a complex pharmacological physiognomy: its most important action is to exert powerful analgesic effects on the central nervous system. The abolition of pain sensitivity is highly specific since it does not imply any interference with other sensory mechanisms (visual, auditory, tactile, etc.). Another critical aspect of morphine action is abolishing pain "understood as suffering," that is, the set of reactive emotional responses that pain evokes, once perceived as a specific sensation: anxiety, dissatisfaction, depression, fear, etc.. The suffering subject treated with morphine often reports perceiving the painful sensation equally, but, at the same time, he has no more difficulty in tolerating it. Also, in the case of morphine, as for heroin, with repeated doses higher than the analgesic ones (10–15 mg), it is possible to reach stages of chronic intoxication whose gravity depends a lot on the assumed average dosage, the purity of the substance and the modality of assumption. In this case, the subject is apathetic, indifferent, and without initiative with intense dream activity. Also, in this case, the symptoms are represented by easy sweating, progressive weight loss, liver and lung problems, and alterations in the immune system. At doses of 25–30 mg (acute intoxication), there are depressing effects on the respiratory centers, which cause a decrease in the frequency and amplitude of breathing and the inhibition of respiratory automatism; moreover, there is miosis, and in most serious cases coma. As with heroin overdose, morphine requires immediate administration of naloxone or any opioid receptor antagonist and resuscitation to restore respiratory function. Lethal complications that may occur, especially if the subject is suffering from heart disease or is a drug addict, are pulmonary edema, right heart failure (acute pulmonary heart), infarction, and intestinal paralysis, which should be treated with resuscitative and symptomatic therapy.

Because of their central and peripheral effects, opioids generate a series of effects at an emotional level that vary from subject to subject and which depend above all on psychological factors and cultural conditioning rather than on the pharmacological action of the substance (motivations for drug abuse, expectations, character profile, the cultural value of abuse, etc.). Emotional traits that recur to some extent in every subject who uses opioids, and which are perhaps due to the inhibitory action of these substances on the anatomical structures involved in emotional and motivational behavior, are:

- The feeling of detachment from the world
- Disinterest in things
- Tendency to apathy and the absence of any concern

What probably makes opioids desirable is precisely this sort of emotional anesthesia typical of all people with adaptation problems, artificially attenuating the emotional connotation of the contingent existential circumstances.

Data analysis in the "European Drug Report 2016" reveals new concerns about increased overdose cases associated with heroin and other opioids. The average prevalence of high-risk opioid use among adults (15–64 years old) in 2014 is estimated at 0.4%, the equivalent of 1.3 million high-risk opioid users in Europe. Among the 185,000 patients undergoing specialist treatment in Europe who reported opioids as their primary drug, 34,000 were patients entering treatment for the first time (18%). Heroin is the most frequently reported substance by the European Hospital Emergency Network (Euro-DEN), accounting for 24% of admissions and 0.5% of deaths in all admissions, the majority of which are opioid-related. Opioids reported by patients in treatment include mainly methadone and buprenorphine.

Cocaine

Erythroxylum coca (Bolivian or Huanaco) and *Erythroxylum truxillense* (Peruvian or Truxillo) are plants belonging to the

Erythroxylaceae family cultivated in Peru, Bolivia, Colombia, Ceylon, and Indonesia. Their leaves contain between 0.7 and 1.5% total alkaloids, the most important of which are cocaine, cinnamyl cocaine (Fig. 53.6), and atruxillin.

The coca plant was introduced to Europe in the eighteenth century, and cocaine hydrochloride was first isolated in pure form in 1860 by the German chemist Niemann. Chewed coca leaves had been used in South America since ancient times for their psychostimulant properties and ability to reduce fatigue and hunger. However, the possibility of isolating the primary alkaloid, cocaine, in pure form radically changed how the substance was used and, therefore, its potential for abuse. In the second half of the nineteenth century, the intake mode was predominantly oral since the substance could be consumed by purchasing well-known beverages such as Mariani's wine. In the first decades of the twentieth century, endonasal intake became more popular, but in the 1970s, the production of the substance in the form of a free base, crack cocaine (cocaine hydrochloride and bicarbonate), which is easily vaporized and therefore, the alkaloid, which can be very easily absorbed through the lungs, began.

Cocaine has an indirect action on the receptors of several neurotransmitters by increasing synaptic levels mainly of dopamine but also noradrenaline and serotonin. In particular, the primary pharmacological effect on the central nervous system is to block the reuptake of neurotransmitters in presynaptic terminals. In particular, the enhancement of norepinephrine activity is the cause of the effects of this drug on the cardiovascular system, while the prolonged action of dopamine is the cause of most of the behavioral and psychological effects.

The main effects resulting from cocaine use are as follows:

- Central nervous system stimulant: perception of increased energy, both communicative and mental alertness, euphoria, inhibition of appetite and sleep
- Local anesthetic: anesthetic effect at the site of application
- Sympathomimetic: effect on the sympathetic nervous system with vascular and cardiac alterations

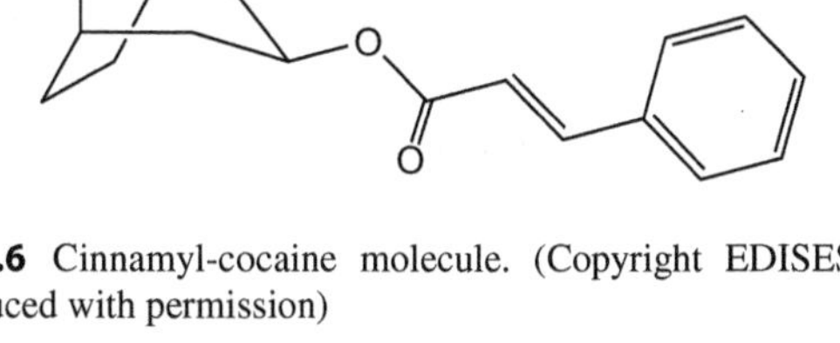

Fig. 53.6 Cinnamyl-cocaine molecule. (Copyright EDISES 2021. Reproduced with permission)

Specifically, the effects due to the use of cocaine vary depending on the dose and frequency of intake; therefore, it is possible to speak of short-term and long-term effects.

In the short-term case, the effects appear immediately after a single dose and disappear within a few hours. Taken at low concentrations, cocaine can cause euphoria, perception of increased energy, an increased state of mental alertness, sight, sound, touch, speech, and a temporary decrease in the need to sleep and eat. It should be noted that the duration of the euphoric effect depends on the route of administration: a rapid absorption of cocaine causes an intense but short-lived high. Thus, smoked cocaine produces an immediate effect that lasts between 5 and 10 minutes, while snorted cocaine acts more slowly but with a longer duration, about 15–30 minutes.

The physiological effects of taking cocaine in the short term are as follows: vasoconstriction of the large and small vessels, dilation of the pupils, increased body temperature, increased heart rate, and increased blood pressure.

High and frequent doses in the long term mark a decisive shift from euphoria to states of anxiety and great irritability, which dominate the clinical picture of the abuser. In this context, violent and aggressive behavior, panic attacks, and tactile hallucinations may appear, and the euphoric effects are overlaid by depression of mood, the intensity, and duration of which are related to the dose and the way the substance is taken. The increase in doses and frequency of use entails the risk of a significant increase in both psychological and physiological adverse effects. There is, in fact, an increased sensitization to the convulsant, anxiogenic and toxic effects caused by cocaine. For those who inhale cocaine, the most common problems are loss of sense of smell, epistaxis (nosebleed), swallowing problems, hoarseness and chronic irritation of the nasal mucous membranes. In addition, chronic cocaine use causes a significant loss of appetite, leading to frequent weight loss and generalized malnutrition.

The risk of overdose is high, as with heroin, due to variations in the purity of the street drug. This risk is highest when cocaine is taken with heroin (speedballing) or is taken by intravenous injection, or even when taken with alcohol or other drugs.

Severe cocaine intoxication can be life-threatening for the user; this condition requires careful medical intervention.

Medical emergency situations are characterized by dyspnea, hypertension, tachycardia, ventricular cardiac arrhythmias, and tonic-clonic convulsions. Acute cocaine intoxication should be treated intravenously with diazepam and a β-blocker such as propranolol to control seizures and ventricular arrhythmias that may lead to myocardial infarction.

Acute cocaine intoxication can cause severe and frequent cardiovascular disorders, such as arrhythmias, cardiac ischemia and myocardial infarction, dilated myocardiopathy, congestive heart failure, disseminated intravascular coagulation, and cerebral hemorrhage. In chronic endonasal cocaine use, vasoconstriction of the mucosa and perforation of the nasal septum may occur, while crack cocaine abuse causes severe irritation of the airways.

Abstinence from cocaine in the dependent subject is characterized by a first phase in which craving is not particularly intense and anxiety, need for sleep, and dysphoria prevail. In the second phase, which lasts a few weeks, depression of mood, apathy, anxiety, anhedonia, loss of motivation, suicidal ideation, and very intense craving appear, which can favor the resumption of the rhythms of assumption of the substance. The recovery by the subject of the normal hedonic and affective tone leads, in the third phase, to the normalization of sleep, but an occasional craving for cocaine (sometimes of considerable intensity) can be decisive in relapses.

The European Monitoring Centre for Drugs and Drug Addiction estimates of cocaine use report that cocaine use appears to be highest in western and southern European countries, with 11.9 million males and 5.3 million females. In addition, an estimated 2.4 million young people and young adults aged 15–34 years (1.9% of this age group) have used cocaine in the past year. Many users use cocaine in recreational settings, peak use at weekends and holidays. Spain, Italy, and the United Kingdom account for 74% of all treatment requests for cocaine dependence in Europe. Data on hospital emergencies offer particularly useful information for understanding severe health harms. A 2014 analysis by Euro-DEN, which monitors drug-related emergency admissions in 16 selected (sentinel) hospitals in 10 European countries, found that among the 5409 reported substance-related admissions, cocaine had a share of 17%.

Cannabinoids

Cannabis sativa, an angiosperm of the Cannabaceae family, is a shrubby, annual, and dioecious plant which presents specimens with male flowers and others with female flowers. It is widely spread and grows spontaneously in temperate and tropical areas or is also cultivated; it has a long taproot and a rough stem whose height varies from 80 centimeters to 3 meters.

The Cannabis sativa plant is a source of many active ingredients; it produces a complex mixture of chemicals, more than 400, 70 of which form the phytocannabinoid group. The main phytocannabinoid in cannabis is the psychoactive compound Δ9-tetrahydrocannabinol (THC, Δ9-THC) (Fig. 53.7). Other important compounds are cannabidiol (CBD), cannabinol (CBN), tetrahydrocannabivarin (THCV), and cannabichromene (CBC). The precursor of THC is tetrahydrocannabinolic acid (THCA), which, according to recent studies, is formed from cannabigerolic acid (CBGA) through oxidocyclization by the enzyme THCA-synthase. The corresponding cannabidiol (CBD) and cannabichromene (CBC) are also generated by decarboxylation of the corresponding acids. CBD and its acidic derivative, cannabidiolic acid, do not boast narcotic effects but possess various activities such as muscle relaxation, anxiolytic, antiemetic, pain-relieving, and many others. The conversion of cannabinoids from the acid form to the decarboxylated form occurs as the plant matures, through heating and/or in an alkaline environment. Since the only psychoactive phytocannabinoid is Δ9-THC, the potency of cannabis is defined in terms of the percentage concentration of this compound, which is also used as a comparative substance to evaluate the potency of cannabis preparations.

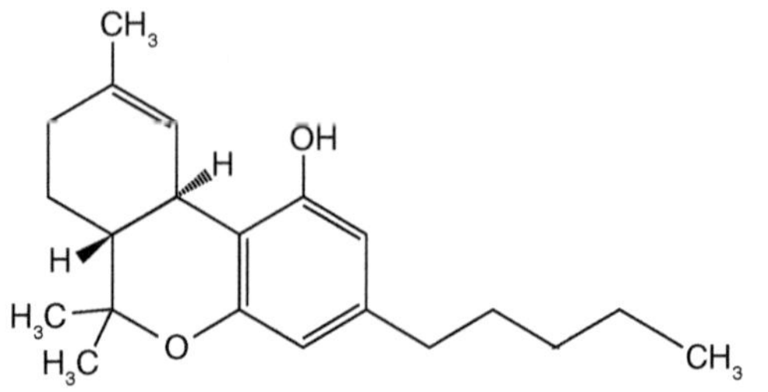

Fig. 53.7 Structure of the Δ9-tetrahydrocannabinol. (Copyright EDISES 2021. Reproduced with permission)

The highest concentration of cannabinoids is found in the flowering tips, less in the leaves, and only very small amounts are found in the stem and roots. The highest levels of THC are found in the resinous material produced by the glandular hairs located in the flowers. Fertilization and subsequent seed formation result in reduced THC levels; significantly lower amounts are present in the leaves and male plants and almost none in the seeds. While dried leaves and stems (Bhang) or young leaves and flowering ends (Ganja) have a low percentage of Δ9-THC (1–3% and 4–6%, respectively), marijuana consisting of dried female inflorescences has 5–10% (with up to 30% observed in some genetically modified marijuana species) and hashish (resin obtained from isolating the trichomes of the Cannabis sativa plant) has between 5% and 12% Δ9-THC. However, the highest percentage of Δ9-THC (about 30–40%) is found in hash oil, a resinous matrix of cannabinoids obtained by extraction with apolar (e.g., butane) or oily (e.g., olive oil, milk) solvents of the female inflorescences.

To date, two types of cannabinoid receptors have been identified in humans, the CB1 and CB2 receptors.

Endocannabinoids, a class of endogenous lipids, are present in humans and possess the ability to bind to cannabinoid receptors. The first endocannabinoid to be identified in 1992 was anandamide) followed by 2-arachidonoylglycerol, 2-arachidonyl-glyceryl ether, virodamine, N-arachidonoyldopamine, and palmitoylethanolamine.

The receptors have different signaling mechanisms and tissue distribution. Activation of cannabinoid receptors causes inhibition of adenylyl cyclase and thus inhibition of the conversion of adenosine triphosphate to cyclic adenosine monophosphate. Other mechanisms have also been observed, e.g., interaction with specific ion channels. Both receptors belong to the large family of G protein-coupled receptors.

The CB1 receptor is predominantly expressed in the central nervous system (cerebellum, hippocampus, and basal ganglia–corpus striatum, globus pallidum, and black substance), in the spinal cord but also peripherally. The presence of CB1 receptors in the hippocampus and cortex suggests their involvement in learning and memory processes, while localization in the basal ganglia and cerebellum seems to mediate effects on motor activity, coordination, and reaction. Also, in the central nervous system, CB1s are mostly present on axons and terminal nerves and to a lesser extent on cell bodies; their presence in glia is controversial. Several studies have shown that CB1 receptors, especially in the striatum, nuclei accumbens, and prefrontal cortex mediate all behavioral and neurochemical characteristics of THC and other cannabinoids. In particular, the effects of gratification, tolerance and physical dependence have been attributed to their interaction with opioids, glutamate, GABA, and, especially, dopaminergic systems. Cannabinoids act in the brain by triggering gratification processes very similar to those that result from the use of other types of substances of abuse. Like substances of abuse, cannabinoids activate dopamine production in the ventral tegmental area through presynaptic inhibition of GABAergic neurons in this area and its release in the accumbens nuclei and in the prefrontal cortex. Studies with CB1 antagonists have demonstrated the importance of these receptors in the phenomenon of craving. The presence of CB1 receptors in different peripheral organs—heart, eye, intestine, urinary bladder, and microcirculation—is related, for instance, to tachycardia, conjunctival reddening, urinary retention, and inhibition of intestinal motility following cannabinoid intake. CB1 receptors were also identified in the male and female reproductive system (ovaries, endometrium, testes, spermatozoa, vas deferens), where endocannabinoids may play a crucial role in embryo implantation.

The second type of cannabinoid receptor, CB2 (48% homologous to CB1), is predominantly expressed in peripheral tissues, mainly in the immune system, spleen, tonsils, thymus, bone marrow, and tissues involved in the production of immune cells.

In any case, the action of cannabinoids (generally inhibitory) in the modulation of neurotransmitter release occurs because CB1 receptors (which have a prevalent presynaptic presence) are coupled to inhibitory G-proteins. Thus the action of adenylyl cyclase is inhibited, followed by inhibition of calcium channels and activation of potassium channels. The final effect is precisely the presynaptic inhibition of neurotransmitter release, modulating neurotransmission and modifying neuronal excitability. In general, the prevalent modulatory and inhibitory activity of cannabinoids is expressed concerning immune function (through CB2), neurotransmission (through CB1), and, overall, respect to all associated functions.

In humans, the dose of Δ9-THC needed to produce psychotropic effects varies between 2 and 20 mg. When cannabis is smoked, Δ9-THC is rapidly absorbed, detectable in plasma within a few seconds, where it reaches a peak concentration after 3–10 minutes.

The compound reaches the central nervous system in about 5 minutes; it is also estimated that only 1% of smoked THC is found in the brain. The effects of smoking cannabis begin within a few minutes, peak around 20 minutes, and persist for 2–3 hours.

One of the characteristics of cannabinoids is the great variety of pharmacological effects they produce, which also depend on the subject, on expectations, the quality of the cannabis, and the way it is smoked. In small doses, cannabis produces euphoria, sedation, relaxation, and drowsiness, and in a sense, these effects are similar to those caused by ethyl alcohol. At high doses, Δ9-THC can produce hallucinogenic effects causing visual, auditory, time, and space distortions with the intensification of sensory experiences.

With the use of dosages higher than 20 mg (by inhalation), emotional detachment, difficulty in concentrating and loss of motivation, sedation, drowsiness, and a sense of loneliness and depression can occur, but if the user is living a happy moment and is in a particularly stimulating environment, a feeling of well-being, intoxication and loquacity can also be generated. In habitual users, withdrawal disorders are frequent and early; these include convulsions, palpitations, dyspnea, tachycardia, chest pain, craving, headache, intense anxiety, insomnia, nausea and vomiting, loss of appetite, and profuse sweating. Negative reinforcement promotes the continued use of the substance.

Prolonged cannabis use seems to lead to decreased psychophysical performance, attention, perception, reaction time, and motor coordination. Although it has not been shown to cause physical dependence, psychological dependence may be established in habitual users, perhaps conditioned by the environment they frequent, the group they belong to, and the "ritual" of collective use. Although there are conflicting opinions on whether long-term cannabis use impairs memory, amotivational syndrome seems to be proven in chronic users, accompanied by loss of energy and interest in work and normal daily routines (sleep–wake state, personal hygiene, appetite disturbances). Furthermore, chronic use of smoked cannabis leads to tobacco-like effects, i.e., respiratory tract irritation, bronchoconstriction, and cancer risk.

Cannabis is the most used drug in Europe (51.5 million men and 32.4 million women) and worldwide. In addition, an estimated 16.6 million young Europeans (15–34 years old), or 13.3% of this age group, have consumed cannabis in the last year, while about 1% of European adults are regular daily users of cannabis. A 2014 analysis by Euro-DEN, which monitors drug-related emergency admissions in 16 selected (sentinel) hospitals, found a 16% rate of admissions due to acute cannabis intoxication.

Amphetamines

Amphetamines are substances of synthetic origin with sympathomimetic action. The progenitor of this class of substances is amphetamine (Fig. 53.8), a substance synthesized in 1887 at the University of Berlin by a Romanian chemist, Lazar Edeleanu (also known as Edelean). However, the substance was not used for clinical purposes until 1920, when Gordon A. Alles synthesized the drug for use in the medical field (treatment of asthma, hay fever, and rhinitis). Amphetamines began to be widely used during World War II: they were administered to soldiers (mainly Japanese soldiers) to decrease their fear and increase their level of concentration. A few years later, in the 1960s and 1970s of the twentieth century, amphetamines became popular among the student population, who used them to increase concentration. In the same years, they were also widely used as anorectics.

The term amphetamines refer to a group of substances consisting mainly of amphetamine, methamphetamine, methylenedioxy-derivatives (of which the best known is 3,4-methylenedioxymethamphetamine).

Since the 1970s, methamphetamine (or methylamphetamine) has enjoyed great popularity, especially on the west coast (California) of the United States, completely replacing classical amphetamines in recreational use. Methamphetamine is a powerful, addictive stimulant with a high potential for abuse, resulting in severe damage to the central nervous system in the case of long-term high consumption. This drug is easily produced in clandestine laboratories from relatively cheap ingredients. On the illicit market, the substance is known as "ice" or "crystal" (the last two names refer specifically to the methamphetamine that is smoked). It is a white crystalline powder with a bitter, odorless taste that dissolves easily in water or alcohol and is used in recreational or "chemsex" contexts.

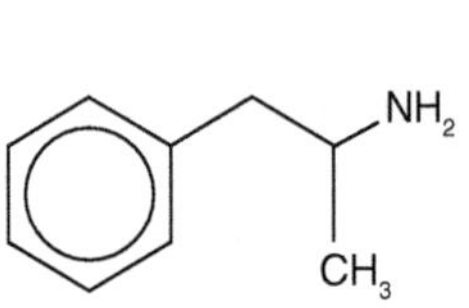

Fig. 53.8 Molecular structure of amphetamine. (Copyright EDISES 2021. Reproduced with permission)

Amphetamine and methamphetamine act at the level of the central nervous system, mainly by blocking the reabsorption of dopamine but also that of noradrenaline and serotonin. These molecules can block the protein involved in the presynaptic reuptake of these neurotransmitters. Indirect serotonergic stimulation and β-agonist action have been linked to the anorectic effect, while dopaminergic action to the psychostimulant and psychotoxic effect. The average doses of pure amphetamine abused are around 25 mg, those of methamphetamine from 5 to 40 mg of pure substance.

Amphetamine derivatives are generally taken orally, while methamphetamine, which possesses a relatively low evaporation point, is commonly smoked to produce an immediate and powerful stimulant effect.

Most of the effects of amphetamine derivatives are similar to those of cocaine. However, unlike cocaine, these substances have longer-lasting psychoactive effects.

Methamphetamine, a very powerful psychostimulant even in small doses, can increase insomnia and physical activity and decrease appetite. An immediate euphoric feeling has been reported by those who smoke or inject the substance. When ingested or snorted, methamphetamine (at an average dose of 25 mg) can produce a psychostimulant sensation lasting up to 12 hours. The use of the substance in high doses can cause rapid rises in body temperature, which can cause convulsions. Long-term substance abuse can be very harmful, with a high potential for abuse and morphological damage to dopaminergic neurons.

Chronic users may present with symptoms that include violent behavior, anxiety, confusion, and insomnia; they may also experience various psychotic symptoms, such as paranoia, auditory and tactile hallucinations, as well as mood disorders, and delirium. Although there are no physical manifestations related to withdrawal syndrome, when methamphetamine use is discontinued, many symptoms such as depression, anxiety, fatigue, paranoia, aggression, and intense craving for the drug may occur in the chronic user. In addition, methamphetamine abuse can also cause a variety of problems at the cardiovascular level, such as hypertension or even impairment of blood vessels in the brain. If not treated immediately, hyperthermia (elevated body temperature) and seizures caused by methamphetamine overdose can also be fatal. Currently, there are no specific drug treatments for addiction to amphetamine or similar drugs such as methamphetamine. The use of antidepressant medications may help reduce some of the previously described symptoms that may occur in former methamphetamine users. In addition, in cases of extreme agitation or panic, treatments with anti-anxiety drugs such as benzodiazepines are effective, while neuroleptics have been used in cases of methamphetamine-induced psychosis. The "European Drug Report 2016" shows amphetamines are the most prevalent substances in Northern and Eastern Europe. The report estimates that there are approximately 12 million amphetamine users in Europe (8.3 million males and 3.8 million females). In addition, while problems associated with long-term and chronic use and injecting methamphetamine have always been most apparent in northern European countries, problems associated with methamphetamine use have been most apparent in the Czech Republic and Slovakia. Methamphetamine was reported as the primary drug by approximately 32,000 patients undergoing specialist drug addiction treatment in Europe in 2014; approximately 13,000 were entering treatment for the first time, recording a 50% increase since 2006.

The 3,4-methylenedioxy-derivatives of amphetamine are psychostimulants that also have psychedelic-hallucinogenic properties. These substances act by inhibiting the reuptake of presynaptic serotonin, producing pronounced entactogenic effects (increased feelings of love and togetherness towards others, reduced judgment, and increased disinhibition).

In cases of acute intoxication, what is known as "serotonergic syndrome" may occur due to an exaggerated increase in stimulation at the central level, resulting from overstimulation of serotonergic 5-HT1A and 5-HT2 receptors.

According to data released in 2016 by the Lisbon Observatory on Drugs and Drug Addiction, among methylenedioxymethamphetamine derivatives, ecstasy is the "most popular" among both regular stimulant users and a new generation of young users. Synthesized for the first time in 1912 by Merck Laboratories, it was never produced as a drug and only became popular from the 1980s onwards, mainly in the United States.

The substance spread across Europe, mainly from the Spanish coast in the region of Catalonia, in the form of tablets containing average doses of 75–150 mg of the active ingredient. The substance was also inhaled in powder form. After a period of decline in consumption around the first decade of the twenty-first century, the consumption of the substance with its strong excitant and entactogenic effects, even if not precisely psychedelic, had a renewed spread, but at doses much higher than the previous ones (up to 800 mg of powder dissolved in soft drinks or water with a consequent increase in acute intoxications and sudden deaths). These last ones, which also occurred at the first assumption of the substance, happen for fulminant hepatitis, malignant hyperthermia, and cerebral edema.

MDMA (ecstasy), like other amphetamine derivatives, acts as a potent releaser in the synaptic loop of serotonin, responsible for its psychoactive effects.

The first subjective physical and behavioral effects (euphoria, hyperexcitability, nervousness, accelerated heartbeat, insomnia, anorexia, bruxism, and mydriasis) occur about 30 minutes after ingestion and last up to 6 hours.

A dose of 80–150 mg causes an increase in heart rate and body temperature, sweating, bruxism (teeth grinding), loss of appetite, and insomnia. The most frequently described behaviors are sensory excitement, lucidity and ability to concentrate, increased sensitivity to the environment's aesthetic aspects, disinhibition and ability to relate to others, increased introspective ability, decreased aggression, anxiety, and fear, and increased libido, visual, and space-time dispersions. Although less intense than those caused by high doses of psychostimulants, after MDMA use, "rebound" symptoms are characterized by fatigue, apathy, and mood depression. In the days following the use of the substance, difficulties in concentrating, memory disorders, loss of motivation, anxiety, irritability, deep sadness, and visual disturbances may occur.

Acute intoxication manifests itself with effects such as rhabdomyolysis with myoglobinuria, disseminated intravascular coagulation, hyperthermia, acute renal failure, and acute psychosis, especially if the substance is taken with alcohol. In predisposed subjects, severe hypertension and hepatic damage, even severe, may occur, with the consequent need for organ transplantation or death as reported above.

According to data reported in the "European Drug Report 2016," an estimated 2.1 million young people aged 15–34 (1.7% of this age group) have consumed MDMA/ecstasy in the last year, with national estimates ranging from 0.3% to 5.5%. However, MDMA use is rarely reported as a reason for undergoing specialist drug treatment. In 2014, the substance was reported by less than 1% (nearly 800 cases) of first-time treatment patients in Europe.

Hallucinogens

The term "hallucinogenic" refers to any substance of natural or synthetic origin that, acting as a serotonergic agonist on 5-HT1 a and 5-HT2A receptors and glutamatergic receptors of the neocortex, causes alterations in sensory perceptions and the visual sphere, the most significant of which is hallucination. These substances cause profound distortions of reality and sudden mood swings of varying duration and intensity depending on the substance's type, amount, and mode of intake, ranging from states of euphoria to feelings of blurred perception of self and the surrounding reality. The term "hallucinogen" includes true psychedelic substances, such as lysergic acid diethylamide (LSD) (Fig. 53.9), mescaline, dimethyltryptamine (DMT), and psilocybin; dissociative substances such as ketamine; and some plants with delirogenic active ingredients such as stramonium containing hyoscyamine, atropine and scopolamine or atropa belladonna containing atropine.

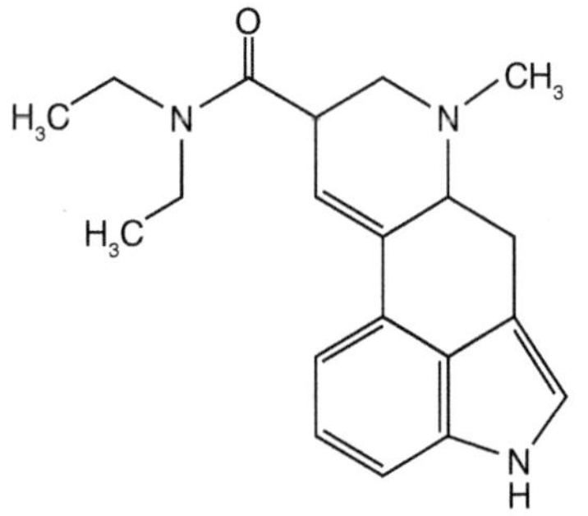

Fig. 53.9 Lysergic acid diethylamide molecule. (Copyright EDISES 2021. Reproduced with permission)

Although with some differences between the different substances, under the influence of hallucinogens, people see images, hear sounds, and feel sensations that are recognized as real but which do not exist in reality, up to the phenomenon of "synesthesia" in which a person seems to hear colors and see sounds. The effects of hallucinogenic substances are difficult to predict because the psychophysical state of the person strongly influences them at the time of intake, thus causing unexpected effects in the user, which vary from pleasant (good trip) to unbearable and terrifying (bad trip). The effects of hallucinogens occur from 30 to 90 minutes after intake and can last for more than 12 hours, without counting the presence of so-called flashbacks, i.e., the sudden reappearance of the effects even months after intake.

One of the first effects produced by hallucinogens is alterations in the visual field, with the appearance of colored spots, figures with undefined contours, and moving objects that leave a colored trail. The visual changes can be aggravated by the appearance of people or animals, up to real hallucinations, isolating the person from reality with the appearance of absurd landscapes and nonexistent figures.

The alteration of the perception system leads to the mutation of the sensations of the skin (which can become hypersensitive or anesthetized) and of the muscular proprioception, thus not being able to perceive the limbs and the rest of the body and arriving at the inability to determine in which position one is. This inevitably leads to a disturbance of balance and movement. A distorted perception of time may also occur, i.e., time passes very slowly.

Although indirectly, hallucinogens lead to changes in mood due to the type of visions caused by the ingested substances. Panic attacks and psychosis can occur, even lasting ones (such as fear of madness and death), states of trance and delirium, loss of self-control, and poor concentration. The altered sensitivity also leads to a high state of confusion. There is a risk of fatalities because the subject may perceive himself as another person and self-aggression to the point of death or think he can commit impossible actions such as flying into the void, the cause of many fatalities from hallucinogens.

Among the hallucinogenic substances, the one considered to be the most powerful is lysergic acid diethylamide (LSD), a semi-synthetic compound obtained from the lysergic acid contained in *Claviceps purpurea*, a parasitic fungus of grasses and synthesized by chance by Albert Hoffman in 1943 during research on the derivatives of ergot.

In humans, 1–2 μg/kg is sufficient to cause effects lasting 10–12 hours. Approximately 0.2–0.4 mg/kg of psilocybin, the active ingredient in mushrooms of the genus Psilocybe, and 5–8 mg/kg of mescaline contained in the peyote plant (*Lophophora williamsii*), a member of the cactaceae family, can cause effects similar to those of LSD. LSD is taken orally as an alcoholic LSD solution poured over sugar cubes, gelatine bars, paper towels, or other materials. The main symptoms of acute intoxication involve the emotional sphere and sensory perceptions, but sympathetic hyperactivity with tachycardia, moderate hypertension, mydriasis, and hypervigilance are present, accompanied by loss of appetite, asthenia, dizziness and nausea, tremors, and sweating.

There are no known cases of death as a direct consequence of taking LSD, but the real danger that derives from the use of this substance lies in the so-called "behavioral toxicity," i.e., the self-damaging behavior that the subject can put in place, driven by panic or delusions or even by the profound distortion of the perception of the world around him. Repeated use of LSD and other hallucinogens does not lead to dependence, and withdrawal symptoms are not observed after discontinuation, but there is a danger that repeated psychedelic experiences may lead to psychosis.

Psilocybin and psilocin are extracted from hallucinogenic mushrooms of the genera Psilocybe, Panaeolus, and Conocybe, which are the active ingredients of many species growing wild in the Americas, Southeast Asia, and Europe. Psilocybin, more potent than psilocin because it is more lipophilic, reaches the central nervous system more quickly and, when ingested, causes nausea and sometimes vomiting before the psychedelic effects occur, which appear after 30–40 minutes and last 5–6 hours and are similar to those produced by LSD. However, the appearance of behavioral toxicity is less than LSD.

From *Lophophora williamsii*, a small thornless cactus of the genus peyote that grows in the southern United States and Central America, mescaline is extracted from the top of the cactus, which contains an amount of 40–50 mg. The usual route of administration is orally, and the effects appear more slowly than other hallucinogens and last about 12 hours. Hallucinations caused by mescaline are described as more vivid and less dramatic than in the emotional sphere of the user and rarely cause depersonalization. The use of peyote cactus is usually accompanied by severe nausea and vomiting.

Dimethyltryptamine (DMT) is a hallucinogenic substance found in several plants, including the bark of *Virola calo-*

phylla; the seeds of *Anadenanthera peregrina* and *Mimosa hostilis*; the leaves of *Psychotria viridis*. DMT in low doses acts as a stimulant, while in high doses, it causes hallucinogenic effects that can last up to 48 hours.

In Europe, general prevalence levels of consumption of hallucinogenic mushrooms of the genus Psilocybe and LSD have been generally low and stable for some years now. Among young adults (15–34), estimates from national surveys of last year prevalence are below 1% for both substances, except Finland, with a prevalence of 1.3% for LSD; for hallucinogenic mushrooms, however, the prevalence is 1% for the United Kingdom, 1.3% for the Netherlands, 1.9% for Finland, and 2.3% for the Czech Republic. There are no data on dimethyltryptamine consumption.

Ethyl Alcohol

Ethyl alcohol (Fig. 53.10) is a linear alkyl chain alcohol of only two carbon atoms, whose condensed structural formula is CH3CH2OH. Produced in nature by the fermentation of sugars by a particular class of microorganisms, the *Saccharomyces*, it is the only alcohol suitable for food consumption. It is usually present in beers in percentages below 10% and on average between 2% and 5%; in wines in percentages between 5% and 20% (12% on average); in liqueurs and bitters in percentages from 20% to 40%; in distillates the percentage can reach 70%; and, finally, in alcoholic aperitifs with percentages that vary greatly depending on the alcoholic beverages used. Ethyl alcohol, when taken in low doses or during the initial phase of acute intoxication, can behave as a stimulant due to the suppression of central inhibitory systems.

However, when plasma ethanol levels increase, symptoms such as sedation, ataxia, impaired motor, and psychomotor activity, and even convulsions and delirium appear.

Ethyl alcohol acts on two types of neuronal receptors: γ-aminobutyric acid receptors and glutamate receptors. These two neurotransmitters are responsible for most of the inhibitory (GABA) and excitatory (glutamate) brain activity. Alcohol depresses brain activity primarily by increasing the inhibitory activity of GABA receptors and reducing the excitatory activity of glutamate. While the increase in GABA activity is probably responsible for most of the sedative

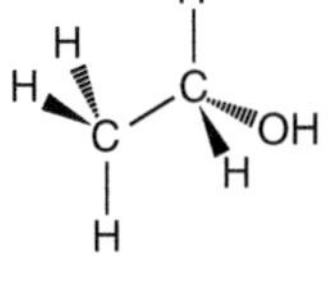

Fig. 53.10 Molecular structure of ethyl alcohol. (Copyright EDISES 2021. Reproduced with permission)

effects of alcohol, the suppression of glutamate NMDA receptor activity is probably responsible for the memory deficits people experience after drinking.

By acting on the GABAergic system, alcohol also causes an increase in dopamine in the brain centers of gratification.

Acute intoxication's characteristic signs and symptoms are well known: increased reaction time, decreased fine motor control, impulsivity, attention disorders, and memory deficits. As doses increase, deep sedation appears and, finally, coma, associated with respiratory depression, sometimes fatal. The treatment of acute intoxication depends on the severity of central nervous system depression and respiratory depression. Gastric lavage may be given, but care must be taken to prevent pulmonary aspiration of the return fluid. Since ethanol is readily miscible with water, it lends itself to removal by hemodialysis. In practice, acute alcohol intoxication is rarely associated with coma, and therapy is not required; it is sufficient to wait for the ingested ethanol to be metabolized by the tissues while providing supportive therapy to all organ functions.

The *Diagnostic and Statistical Manual of Mental Disorders* (DSM-5) distinguishes between alcohol abuse (or long-term use) and alcohol dependence (addiction or alcoholism). Alcohol dependence is a term used when there are classic signs of substance dependence, in particular, the development of tolerance associated or not with clinically evident withdrawal reactions. Tolerance can be defined as loss of response after repeated administration of a psychotropic substance for which a higher dose is required to produce the same effect as previously obtained at the lower dose. However, heavy users of alcohol not only acquire tolerance but also develop strong physical dependence. A person in this state of adaptation or physical dependence requires continuous administration of alcohol to maintain normal function. The alcohol withdrawal syndrome, or delirium tremens, generally depends on the daily dose size and is usually resolved by alcohol intake. It is manifested by symptoms such as an incoercible desire for alcohol, tremor, irritability, nausea, sleep disturbances, tachycardia, hypertension, sweating, perceptual distortions, seizures, severe agitation, confusion, visual hallucinations, fever, profuse sweating, diarrhea, and mydriasis.

Alcohol is an addictive psychoactive substance that causes illness and death in most countries. The data reported by the Organization for Economic Co-operation and Development (OECD), which was set up to promote policies to improve the economic and social well-being of people around the world and currently includes 34 member states, including Italy, show that among the 34 countries considered, the average per capita consumption of alcohol, based on available data, is around 8.9 liters for adults. Austria, Estonia, and the Czech Republic, as well as Lithuania, record the highest alcohol consumption with over 11.5 liters per adult in 2013, while the lowest alcohol consumption is recorded in Turkey and Israel, as well as Indonesia and India, where cultural and religious traditions probably limit alcohol consumption in part of the population.

As regards Italy, according to ISTAT (National Institute of Statistics) data, during 2015, 64.5% of Italians aged 11 years and over (equal to 35 million and 64 thousand people) consumed at least one alcoholic beverage, with a significantly higher prevalence among males (77.9%) compared to females (52.0%). Compared to 2014, there has been an increase in consumption: in 2014 63% of Italians aged 11 years and over (equal to 34 million 319 thousand people) consumed at least one alcoholic beverage.

Although alcohol consumption within the WHO European Region has decreased over the past 25 years, alcohol intake is a major mortality risk.

The report "Public health successes and missed opportunities. Trends in alcohol consumption and attributable mortality in the WHO European Region, 1990–2014 (2016)," published by WHO in October 2016, describes trends in alcohol consumption and deaths attributable to alcohol consumption in European countries and provides data from individual nations showing essential differences. Alcohol is the cause of more than 200 different illnesses and accidents that result in numerous deaths and injuries each year and is ranked as the third most significant risk factor for disease and premature death after smoking and hypertension. The WHO report highlights how the high level of alcohol consumption in Europe has led, over the years, to an increase in deaths attributable to the consumption of alcoholic beverages due to chronic diseases (such as cirrhosis of the liver, cardiovascular diseases, and cancer) and due to violent death (road accidents, homicides and suicides). As stated in the report, worldwide, 3.3 million deaths per year are caused by alcohol consumption, or 5.9% of all deaths, at a rate of 1 death every 10 seconds and, in addition, 5.1% of diseases are linked to alcohol consumption.

Benzodiazepines

Benzodiazepines are a class of psychotropic drugs (Fig. 53.11) with sedative, hypnotic, anxiolytic, anticonvulsant, and muscle relaxant properties with several pharmacological indications, including the main ones.

The main topics are anxiety and panic states, agitation, insomnia problems, acute intoxication by cocaine, amphetamines, hallucinogens, and treatment of alcohol dependence.

Although they are not legally considered illicit substances, they are nevertheless substances that, through chronic use, can lead to tolerance and dependence and have considerable potential for abuse. Tolerance manifests as a diminished pharmacological potency and develops relatively quickly regarding the sedative, hypnotic, anticonvulsant, and muscle relaxant effects. The subject, during treatment, no longer responds to the prescribed dose of the drug, initially sufficient to produce the desired effect. Therefore, a higher dose of benzodiazepine is required to achieve the same therapeutic effect. Moreover, the suspension of treatment or an abrupt reduction in the dosage of these drugs, even after a relatively short treatment cycle (3–4 weeks), can lead to two groups of symptoms: the rebound effect, which consists of the return of the symptoms for which the patient had been treated but in a more accentuated form, and withdrawal syndrome, the main sign of physical dependence on substances.

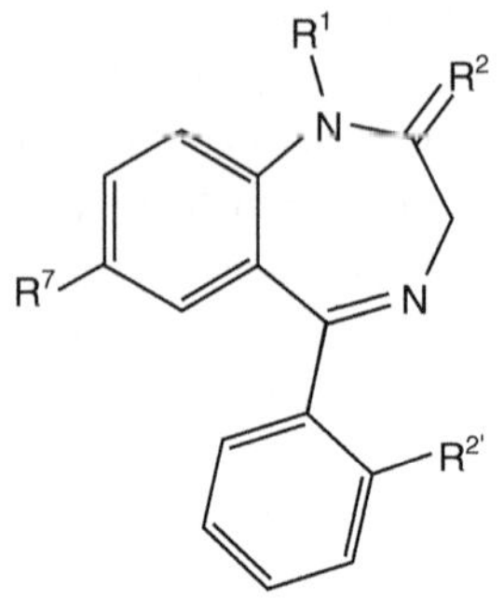

Fig. 53.11 Molecular structure of benzodiazepines. (Copyright EDISES 2021. Reproduced with permission)

Benzodiazepines act on GABA receptor complexes present in the central nervous system. These complexes consist of an α subunit (site to which benzodiazepines bind) bound to a γ subunit via the β subunit (active site to which GABA binds). Binding to the α subunit causes a conformational change in the receptor, which makes it more receptive to GABA with an inhibitory effect on neurons and consequent sedative action. The uncontrolled use of these psychotropic drugs can give rise to side effects related to their sedative effects and myorelaxant action. Effects may include drowsiness, dizziness, and decreased alertness and concentration. Lack of coordination can lead to falls and injuries, particularly in the elderly.

On the market, many molecules belong to this class of psychotropic drugs, so they are usually classified according to their plasma half-life, a parameter that provides information on the duration of action of each benzodiazepine:

- Short or very short half-life (2–6 hours), to this category belong triazolam and midazolam
- Intermediate half-life (6–24 hours), to this category belong oxazepam, lorazepam, lormetazepam, alprazolam, and temazepam
- Long half-life (1–4 days), to this category belong chlordiazepoxide, chlorazepate, diazepam, flurazepam, nitrazepam, flunitrazepam, clonazepam, prazepam, and bromazepam

Contrary to what one might think, there is no direct relationship between plasma half-life and rapidity of action since some drugs, although acting rapidly, are metabolized into other active compounds that significantly prolong their duration of action. Although cases of fatal overdose due to the use of benzodiazepines are rare, international drug agencies have recently published a safety communication on the risks of the combined use of opioid analgesics and benzodiazepines. They have requested that the warnings contained in the labeling of drugs belonging to these two classes be reinforced since opioids and benzodiazepines are both central nervous system depressants, and their combined use can lead to serious side effects such as respiratory depression, which, on rare occasions, can lead to coma and death. In the European Monitoring Centre for Drugs and Drug Addiction (EMCDDA) annual report on the state of the drug phenomenon in Europe in 2016, it is precisely the concomitant use of benzodiazepines and opiates is an important risk factor for drug-related deaths that is highlighted. Benzodiazepines have been identified in 40–80% of methadone-related deaths (France, the USA, Australia) and 50–80% of heroin-related deaths (Germany, Ireland, the UK). The current European Monitoring Centre for Drugs and Drug Addiction (EMCDDA) showed that benzodiazepines were implicated (i.e., had played a role in the death) in 28% of deaths in Scotland, 48% in France, 30–32.5% in Portugal, and 35% in Ireland. In 2016, the EMCDDA monitored 14 new and uncontrolled benzodiazepines, which counterfeiters sometimes use to produce fake medicines sold in Europe. Examples of this practice include counterfeit alprazolam tablets, intercepted in 2015, in which flubromazolam was found to be present, and fake diazepam tablets, which contained phenazepam; in some European countries, these counterfeit medicines have become an important part of the illicit drugs market.

New Psychoactive Substances

In recent years, we have witnessed a phenomenon assuming ever more significant proportions: the emergence of new psychoactive substances (NPSs), characterized by pharmacological and toxicological properties that are particularly dangerous to consumers' health. A new psychoactive substance is defined as a substance with narcotic or psychotropic effects, either in pure form or in a preparation, which is not covered by the 1961 United Nations Single Convention on Narcotic Drugs or the 1971 United Nations Convention on Psychotropic Substances, or by the laws of the various countries concerning the prohibition of the production, trade and sale of psychotropic substances, but which may pose a threat to public health comparable to that posed by the substances listed in these conventions.

These substances are sold freely in smart shops, and health shops, but above all on telematic sites where it is possible to distribute, buy, and receive the product entirely anonymously.

These substances, sold as party pills, bath salts, plant fertilizers, research chemicals (collectible and not for human use), or legal highs, have been called novel psychoactive substances worldwide.

Phenethylamines, tryptamines, piperazines, and piperidines are a part of these NSPs, which include cannabinoids and synthetic cathinone, ketamines and analogs, and new drug designers among the most widely used substances.

They change from year to year, and their diffusion, despite the adaptations of the tables of the European legislations, is very difficult to contrast, as drug designers can synthesize and distribute endless chemical classes of substances or chemical modifications on the market of existing psychotropic molecules.

NPSs are generally used in recreational situations, as euphoric drugs, hallucinogens, and even in combination with classic substances of abuse and alcohol. These new uses are different from classic addictions and are based on new models of risk behavior. New methods of consumption, which are very successful, especially among the very young, and are spreading at an impressive speed thanks to fast and illegal telematic sales systems, with fewer "management problems" than the classic substances. In the absence of controlled scientific support and studies in the international scientific literature on these substances, online forums (drugs forum), blogs, and social networks have developed where consumers exchange information, advice, directions, and "instructions for use" in a rapid and socializing way to the point of determining the success and spread or even failure of these new designer drugs. To date, over 500 new psychoactive substances have been intercepted by the National Early Warning System.

Synthetic Cannabinoids

Synthetic cannabinoids are molecules functionally and not structurally similar to natural cannabinoids, which are synthesized in chemical-pharmaceutical laboratories to identify molecules that preserve some properties of Δ9-tetrahydrocannabinol (THC), the psychoactive principle of cannabis. They represent one of the most significant examples of the phenomenon related to the consumption of so-called "emerging drugs." The first news related to them dates back to 2004, when several herbal mixtures called spices appeared in Europe and spread rapidly because they could be consumed as a legal alternative to cannabis. They were touted as legal herbal substances with cannabimimetic action.

Since the end of 2008, however, some European countries (Germany, Austria, Denmark, and the Netherlands), together with the U.S. Drug Enforcement Administration (DEA), have begun to report the surreptitious and undeclared presence in spice products of synthetic psychoactive compounds that act on cannabinoid receptors. Following reports and alerts from the various international agencies that monitor substances of abuse (European Monitoring Centre for Drugs and Drug Addiction [EMCDDA], Drug Enforcement Administration [DEA], National Institute on Drug Abuse [NIDA], etc.), several countries (European and non-European) have imposed legislative restrictions on the production and marketing of these herbal mixtures. Several countries (European and non-European) have imposed legislative restrictions on the production and marketing of these herbal mixtures.

As reported above, synthetic cannabinoids represent a large family of molecules structurally unrelated but functionally similar to Δ9-THC, the active ingredient in cannabis. Designated as cannabinoid receptor agonists, they have been developed over the past 40 years to study cannabinoid receptors in animal models. The Asian illicit market then thought to "use" the sale of these molecules for their psychotropic effects, for the legal holiday that allowed their free sale, and for the absence of being able to detect them in biological liquids due to the lack of analytical methodologies for their determination.

Although often referred to as "synthetic cannabinoids," most of these molecules constitute a heterogeneous group of substances, with several characteristics in common, including fat solubility, non-polarity, and the number of carbon atoms making up the molecule. A common feature of the various synthetic cannabinoids is the presence of a side chain that confers activity to the molecule: optimal activity requires the presence of more than four and up to nine saturated carbon atoms.

First-generation synthetic cannabinoids can be classified according to seven major structural groups:

1. Naphthoylindoles (e.g., JWH-018, JWH-073 and JWH-398, AM-2201)
2. Naphthylmethylindoles
3. Naphthoylpyrroles (e.g., JWH-307)
4. Naphthylmethylindanes
5. Phenylacetylindoles (benzoylindoles, e.g., JWH-250, AM- 694, AM2233)
6. Dichlohexylphenols (e.g., CP 47,497 and its analogs)
7. Classic cannabinoids such as HU-210

Recently, so-called second-generation synthetic cannabinoids have appeared on the market, as first-generation cannabinoids have already entered the international lists of prohibited substances. These new compounds are of indole or indazole chemical derivation: For example, 5F-APINACA (N-(1-adamantyl)-1-(5-fluoropentyl)-1H-indazole-3-carboxyamide also referred to as 5F-AKB-48), PB-22 (1-pentyl-8-quinolinyl ester-1H-indole-3-carboxylic acid) and 5F-PB-22 (1-(5-fluoropentyl)-8-quinolinyl ester-1H-indole-3-carboxylic acid) and are sold as pure standards or in herbal mixtures. The 5F-APINACA molecule has a structural modification that differentiates it from the well-known "JWH" series of synthetic cannabinoids by the presence of an indazole skeleton instead of an indole skeleton to which an adamantyl group is linked through a carboxamide function. PB-22 and 5F-PB-22 have a quinoline-type substituent group linked to the indole ring through an ester-type bond.

Furthermore, in Italy, the Early Warning System of the Department of Antidrug Policies in April 2014 reported numerous intoxications by a new series of cannabinoids with the name PINACA (e.g., AB PINACA, ADB-PINACA) analogous compound of AKB 48 also known as Apinaca where the adamantyl group is replaced by a 1-amino-3,3-dimethyl-1-oxobutan-2-yl group.

Synthetic cannabinoids bind in the brain and other peripheral organs to the same CB1 and CB2 receptors as Δ9-THC and the endogenous ligand anandamide, forming stronger bonds with the affected receptors than THC and therefore having a more potent effect. For this reason, the health effects are completely unpredictable. Since the exact composition of these substances is unknown, and since it varies greatly from one "supply" to another, it is possible that the effects are very different from what the consumer expects. Very often, the adverse effects due to the use of high dosages of these substances concern the gastrointestinal system (nausea, vomiting), the circulatory system (tachycardia, hypertension), and the central nervous system with visual changes and hallucinations.

Among the second-generation synthetic cannabinoids, the Lisbon European Monitoring Centre for Drugs and Drug Addiction in 2016 received reports of 10 deaths and 5 non-fatal intoxications related to the intake of the synthetic cannabinoid 1-(4-cyanobutyl)-N-(1-methyl-1-phenylethyl)-1H-indazol-3-carboxamide (CUMYL-4CN-BINACA) in Sweden and Hungary and, recently, 2 deaths related to 1-methoxy-3,3-dimethyl-1-oxobutan-2-yl 1-(cyclohexylmethyl)-1H-indazole-3-carboxylate (MO-CHMINACA) in Sweden.

As reported by the EMCDDA, in France, estimates indicate that 4% of people aged between 18 and 34 years old report having smoked synthetic cannabinoids once in their life. There are currently no estimates of consumption in other EU countries.

Synthetic Cathinones

In 2011, around two-thirds of the new designer drugs reported by the EMCDDA were synthetic cathinones, which are structural analogs of cathinone, an alkaloid of natural origin found in the leaves of the Catha edulis or Khat plant.

These are phenethylamine derivatives that have a ketone-like carbonyl group in the β-position of the alkylamine chain. The first to appear on the illicit and telematics markets was methcathinone synthesized in 1928 in the United States.

These substances are currently used as psychostimulants and hallucinogenic drugs among young people. Many synthetic cathinones are characterized by sympathomimetic activity, mainly related to the presynaptic release of catecholamines and inhibition of the reuptake of monoaminergic neurotransmitters. They can simulate the effects of cocaine.

Depending on the availability of different forms and formulations, cathinones can be ingested, snorted/smoked, or taken by injection or rectally. The most commonly reported clinical effects are: anxiety, reduced ability to concentrate and short-term memory, irritation of the nasal mucosa, headache, tachycardia, hypertension, hyperhidrosis, mydriasis, lockjaw, bruxism, hallucinations, severe psychomotor agitation and aggressiveness, convulsions. For many synthetic cathinones also identified in Italy, complete data are unavailable on their pharmacological and toxicological characteristics, precise mechanism of action and possible toxic effects. And very often, their effects from the clinical point of view are not easily distinguishable from the acute effects of MDMA or cocaine.

Literature data and information on specialized websites indicate that these compounds can produce severe acute intoxication with a high risk of fatal consequences related to potent stimulation of the catecholaminergic system. In addition, dopamine stimulation may explain the development of tolerance, abuse, dependence, and withdrawal syndrome after frequent consumption of these compounds.

The most popular synthetic cathinone at present, also because of the reports of numerous deaths recorded in some European countries and related to its intake, is mephedrone (4-methylmethcathinone, 4 [MMC]), but among the most sold and used, according to the news that appears on websites of users of these substances, there are butylone, methcathinone, methylone, methylenedioxypyrovalerone (MDPV). The substances generally appear as white powders, and when taken intranasally, orally, but also rectally, they cause euphoria, sociability, stimulation, sexual excitement, improved mood, hallucinations, decreased hostility and insecurity.

Although for mephedrone, similarly to other stimulants, it can be assumed that the risk of toxicity is related to the doses used, the necessary information to define a minimum toxicity dose is not available. Generally, for the most common synthetic cathinone, adverse effects occur about 15–45 minutes after intake and have an average duration of about 6 hours. Among the short-term adverse effects, users of these substances report increased body temperature and heart rate, dilation of the pupils, muscle tension and pain, dizziness, confusion, paranoia, and fear. The most common adverse effects reported after the intake of synthetic cathinones include cardiac hyperadrenergic symptoms (excitation, tachycardia, increased blood pressure, sweating), and psychiatric and neurological symptoms.

The popularity of these substances is linked to several reasons: the stimulant amphetamine-like effects with added hallucinogenic properties, the easy availability and affordable price (they can be easily purchased on the web with a credit card in complete anonymity), the free sale of new molecules put on the market when they are not yet included in the legislative tables that prohibit their marketing and use and, therefore, the absence of legal consequences. However, the impossibility to identify them in the biological matrices of consumers, since there are no rapid tests available to the Police or Emergency Rooms, makes them highly dangerous substances as they are difficult to treat clinically.

In Italy, in the recent past, some cases (aged between 18 and 36 years) of acute intoxication by synthetic cathinones have been reported, with symptoms such as mydriasis, anxiety, panic, visual and auditory hallucinations, psychomotor agitation and violent behavior. In one case, the intake (ingestion/sniffing) of the product, bought in a smart shop (as a plant fertilizer), caused, besides systemic effects, also hyperemia of the oral cavity, edema of the glottis and uvula. All patients were treated symptomatically and discharged after 24–48 hours of observation.

The European Monitoring Centre for Drugs and Drug Addiction in the "European Drug Report 2016" reports acute poisonings and deaths, by parenteral intake, particularly with stimulants such as mephedrone, α-PVP, MDPV, and pentedrone.

Fentanyl

Fentanyl and its derivatives (alfentanil, sufentanil, remifentanil, and carfentanil) are substances used as anesthetics and analgesics both in humans and in veterinary medicine (carfentanil) with a potency 40–80 times greater than morphine. Both fentanyl and its analogs can be taken for recreational purposes in ways other than the pharmaceutical formulation, by injection, by ingestion of sweets (transmucosal intake), or through the use of patches. Fentanyl in powder or patches is also smoked or taken intranasally.

The emergence of fentanyl on the illicit drugs market at the European level dates back to the first half of the 1990s. Illicit products restricted to the European market are fentanyl and 3-methylfentanyl and, recently, para-fluorophentanyl.

The primary mechanism of action of fentanyl is agonism on μ-opioid receptors, receptors distributed in the brain, spinal cord, and other tissues. The pharmacological effects of fentanyl are exerted primarily in the central nervous system. If fentanyl is taken in combination with heroin, cocaine, alcohol, or other substances capable of depressing the central nervous system, such as benzodiazepines, serious interactions may occur.

Since these are substances where the therapeutic and lethal indexes are very close, the risk of fatalities is very high. Many recent deaths of heroin addicts in the United States and England are suspicious deaths not from heroin but from fentanyl derivatives. Overdoses from fentanyl and fentanyl derivatives are characterized by reversible respiratory depression with naloxone. Sudden death may occur due to cardiac arrest or severe anaphylactic reactions. Although some fatalities have been reported following the therapeutic use of fentanyl and its derivatives, many deaths have occurred following non-medical use of the drug product.

In addition to the analgesic effect, fentanyl is capable of producing drowsiness and euphoria, although the euphoric effect is less pronounced than that caused by heroin and morphine. Major side effects include nausea, dizziness, emesis, fatigue, headache, constipation, anemia, and peripheral edema. After repeated use of fentanyl, tolerance and dependence quickly develop. When its use is discontinued, the characteristic symptoms of withdrawal occur, such as sweating, anxiety, diarrhea, bone pain, abdominal cramps, chills, and goosebumps.

New Designer Drugs

When Henderson first coined the term designer drugs in 1988, he was simply describing unofficial attempts to synthesize chemical analogs of psychotropic molecules subject to the control of use and therefore illegal under current legislation, which were also psychoactive analogs, but which, following chemical modification of the starting molecule, were legal and therefore freely available for sale.

New designer drugs are typically phenethylamines, tryptamines, and piperazines. The variety and chemical classes of substances that come onto the market as new designer drugs or, more generally, as new psychoactive substances changes from year to year, and it is therefore not possible with bans to counteract their spread because drug designers have the possibility to synthesize and distribute other chemical classes of substances on the market.

Phenethylamines

Phenethylamines are a relatively large class of molecules with psychoactive and stimulant action, structurally similar to amphetamine and its derivatives. Several subgroups are distinguished according to the different substitutions on the aromatic ring, alkylamine chain, and nitrogen itself. The "2C" phenethylamine series is characterized by the presence of two methoxyl groups on the aromatic ring with the possibility of a halogen in position 4 (e.g., 2C-B-2,5-dimethoxy-4-bromophenethylamine has a bromine atom in 4, 2CI-2,5-dimethoxy-4-iodophenethylamine has an iodine atom). This series is active on serotonergic 5-HT2C receptors at doses up to 100 mg with stimulant and entactogenic effects, but, at higher concentrations, these molecules cause hallucinations and psychotic reactions up to cases of death. The "D" series (2,5-dimethoxy-4-bromoamphetamine [DOB]; 2,5-dimethoxy-4-iodineamphetamine [DOI]; 2,5-dimethoxy-4-chloroamphetamine [DOC]), is analogous to the 2C series and presents two methoxyl groups on the aromatic ring with the possibility of a further substitution in position 4 and a methyl in α to the amine group. This type of substitution confers hallucinogenic properties through an agonist effect on all serotonergic receptors of the 5-HT2 series. Since the stimulating and psychedelic effects occur about 1 hour after intake, the most significant risk is the occurrence of repeated administration capable of causing, in addition to the psychedelic effects sought by the consumer also marked vasoconstriction.

The NBOMe series are analogs that possess a 2-methoxybenzyl group on the nitrogen atom and are agonists of the serotonin 5-HT2A receptor. They generally occur as drops or paper stamps (blotters) and, more rarely, as a powder. With few exceptions, compounds in this group are active at doses below 1 milligram and are taken orally. The onset of the effects (psychedelic, stimulant, empathogenic and entactogenic) occurs after a few minutes with several hours depending on the mode of consumption. NBOMe can cause hallucinations (especially visual, but less intense than LSD), alterations in the perception of time, and changes in the state of consciousness.

Benzofurans

Benzofurans are aromatic heterocyclic molecules with a benzene ring attached to one or more furan rings. They are phenethylamine derivatives structurally composed of two furanic or dehydrofuranic rings. Such molecules as 5-APB and 6-APB are marketed through web channels as powders, tablets or capsules. The latter are commonly ingested, while the powder can instead be inhaled or swallowed. Effects occur earlier when ingested (20–90 minutes) than when inhaled (30–60 minutes); the average duration of subjective effects reported is 6–12 hours, although prolonged hallucinogenic effects of up to 2–3 days are reported. Desired effects include increased empathic abilities, euphoria, alterations in visual perception, hypersensitivity to music and lights, a sense of well-being, and "love of the world." The adverse effects reported by users range from gastrointestinal symptoms (nausea, vomiting, diarrhea) to headache, high blood pressure, tachycardia, dizziness, muscle pain, alterations in sleep-wake rhythm, and anxiety. Their pharmacodynamic action is directed at serotonin 5-HT2A receptors and, to a lesser extent, at 5-HT1 receptors. It is thought that the interaction with both receptors may explain the important hallucinogenic effects. The most reported hallucinations are visual ones (distortions, lights, geometric patterns), but there are also alterations in auditory perceptions, connections with other realities, feelings of peace and well-being, and emotional stimulation. The psychedelic properties are LSD-like, but of longer duration. The toxic effects described in the literature are nausea, vomiting, headache, tachycardia, tremors, arrhythmias, panic attacks, amnesia, convulsions, pulmonary edema, and hepato-renal failure, which can also be fatal.

Triptamines

Tryptamines represent a group of molecules whose basic chemical structure is the natural alkaloid tryptamine. Some tryptamines are natural neurotransmitters such as serotonin. In contrast, others are present in plants and have hallucinogenic activity, such as dimethyltryptamine (DMT), a component of Ayahuasca decoctions, used by some populations of the Amazon for its hallucinogenic properties, or as psilocybin, present in some mushrooms genus Psilocybe. Finally, some tryptamines are instead of synthetic origin. They were designed and synthesized for research but were then placed on the drug market as new psychoactive substances. Among these, the most widespread are the AMT, the 5-MeO-DMT and its structural isomer 5-MeO-DPT, the 4-AcO-DMT, the 4-AcODiPT, the 5-MeO-DALT.

The most common intake routes include ingestion, smoke or vapor inhalation, and intravenous administration. This class of substances interacts predominantly with serotonin receptor subtypes, specifically 5-HT1A, 5-HT2A and 5-HT2C. Basically, it is believed that the hallucinogenic action of tryptamines is attributable mainly to the action that these molecules play on these serotonergic receptors. The effects on humans related to tryptamine abuse may vary from subject to subject. Those commonly reported by users include very intense hallucinatory visual experiences, euphoria, auditory and sensory hallucinations in general, change in perception of time, nausea, short-term memory loss, difficulty in concentrating, muscle spasms, dissociation from reality, fear, terror, panic attacks, insomnia and apparent death lasting up to 5 minutes. Their duration, as also reported by consumers, is relatively short, estimated at between 20 and 45 minutes.

Piperazines

Piperazines represent a huge class of chemical compounds in which the main part of the molecule is piperazine, to which different chemical groups are attached. Initially, they were described as failed pharmaceutical compounds since some of them were tested by industries as potential therapeutic agents without ever being subsequently marketed. In the last decade, piperazine derivatives have been introduced into the new substance of abuse market for their psychoactive properties.

In addition to the better known and more widely used 1-benzylpiperazine (BZP), other compounds with piperazine as the core molecule have been identified, of which 1-(3-chlorophenyl)piperazine (mCPP) and 1-(3-trifluoromethylphenyl)piperazine (TFMPP) are the most widely used.

BZP was initially developed as a potential antidepressant. However, the presence of features common to amphetamine limited its use due to the risk of developing conduct of abuse (in the 1980s, the antidepressant piberalin, tested in Hungary, was withdrawn from the market). In the late 1990s, BZP became popular, first in New Zealand and then in Europe, as a "legal alternative" to MDMA and methamphetamine. Heavy side effects, such as profuse vomiting for many hours, marked the substance's failure, as we read in forums of users of new designer drugs. MCPP is also used as an intermediate compound in the manufacture of various antidepressants such as trazodone and nefazodone. TFMPP is commonly used in combination with BZP to produce the entactogenic effect of MDMA.

The most common mode of administration is ingestion (tablets, capsules or soluble powder), but, although liquid formulations are rarer, they can also be taken by injection, smoked, or inhaled. They have a stimulant effect due to their properties as noradrenergic dopaminergic and, especially, serotonergic agonists in the central nervous system; in addition, they have toxic effects overlapping with those of amphetamines and other sympathomimetics.

They present severe side effects generally described by consumers, such as agitation, psychotic reactions, and hypertension but also disturbances such as gastrointestinal symptoms, including nausea and profuse vomiting. As already specified at the beginning, only for some substances of each class described, sufficient information and scientific literature have been collected to prohibit their sale, purchase, and use. Many others are freely available for sale, and many more will appear on the electronic market to be tested directly by the consumer.

Recommended Readings

American Psychiatric Association (2013) Diagnostic and statistical manual of mental disorders, 5th edn. American Psychiatric Publishing, Arlington. Edizione italiana: Manuale diagnostice statisticdei disturbi mentali. Milano: RaffaellCortina; 2014

Angoa-PeÃÅrez M, Anneken JH, Kuhn DM (2017) Neurotoxicology of synthetic cathinone analogs. Curr Top Behav Neurosci 32:209–230

Baik J-H (2013) Dopamine signaling in reward-related behaviors. Front Neural Circuits 7:1–16

Ballenger JC (2000) Benzodiazepine receptors agonists and antagonists. In: Sadock VA, Sadock BJ, Kaplan HI (eds) (a cura di). Kaplan & Sadock's comprehensive textbook of psychiatry, 7th edn. Lippincott Williams & Wilkins, pp 2317–2323

Banister SD, Moir M, Stuart J et al (2015) Pharmacology of Indole and Indazole synthetic Cannabinoid Designer drugs AB-FUBINACA, ADB-FUBINACA, AB-PINACA, ADBPINACA, 5F-AB-PINACA,

5F-ADB-PINACA, ADBICA, and 5F-ADBICA. ACS Chem Neurosci 6:1546–1559

Bird MK, Lawrence AJ (2009) Group I metabotropic glutamate receptors: involvement in drug-seeking and drug-induced plasticity. Curr Mol Pharmacol 2(1):83–94

CamiÃÄ J, FarreÃÄ M (2003) Drug addiction. N Engl J Med 349:975–986

Capriola M (2013) Synthetic cathinone abuse. Clin Pharmacol 5:109–115

Cottencin O, Rolland B, Karila L (2014) New designer drugs (synthetic cannabinoids and synthetic cathinones): review of literature. Curr Pharm Des 20(25):4106–4111

Cruickshank CC, Dyer KR (2009) A review of the clinical pharmacology of methamphetamine. Addiction 104(7):1085–1099

Das S, Shah N, Ghadiali M (2017) Intravenous use of intranasal naloxone: a case of overdose reversal. Subst Abus 38(1):18–21

de Boer D, Bosman IJ, Hidvegi E et al (2001) Piperazine-like compounds: a new group of designer drugs-of-abuse on the European market. Forensic Sci Int 121:47–56

de la Torre R, FarreÃÅ M, Navarr M et al (2004a) Clinical pharmacokinetics of amfetamine and related substances: monitoring in conventional and non-conventional matrices. Clin Pharmacokinet 43:157–185

de la Torre R, FarreÃÅ M, Roset PN et al (2004b) Human pharmacology of MDMA: pharmacokinetics, metabolism, and disposition. Ther Drug Monit 26(2):137–144

Ebersole BJ, Visiers I, Weinstein H, Sealfon SC (2003) Molecular basis of partial agonism: orientation of indoleamine ligands in the binding pocket of the human serotonin 5-HT2A receptor determines relative efficacy. Mol Pharmacol 63:36–43

Evans CJ, Cahill CM (2016) Neurobiology of opioid dependence in creating addiction vulnerability. F1000Res 5:1–11

file/0018/319122/Public-health-successes-and-mis-sed-opportunities-alcohol-mortality. 44

Frost JM, Dart MJ, Tietje KR et al (2008) Indol-3-yl-tetramethylcyclopropyl ketones: effects of indole ring substitution on CB2 cannabinoid receptor activity. J Med Chem 51:1904–1912

Gee P, Gilbert M, Richardson S et al (2008) Toxicity from the recreational use of 1-benzylpiperazine. Clin Toxicol 46:802–807

Giorgetti R, Tagliabracci A, Schifan F et al (2017) When "chems" meet sex: a rising phenomenon called "ChemSex". Curr Neuropharmacol 15(5):762–770

Greene SL, Kerr F, Braitberg G (2008) Review article: amphetamines and related drugs of abuse. Emerg Med Aust 20(5, 1ottobre):391–402

Heal DJ, Smith SL, Gosden J, Nutt DJ (2013) Amphetamine, past and present - a pharmacological and clinical perspective. J Psychopharmacol 27:479–496

Helander A, BaÃàckberg M, Beck O (2016) Intoxications involving the fentanyl analogs acetylfentanyl, 4-methoxybutyrfentanyl and furanylfentanyl: results from the Swedish STRIDA project. Clin Toxicol 54:324–332

Henderson GL (1988) Designer drugs: past history and future prospects. J Forensic Sci 33:569–575

http://www.emcdda.europa.eu

http://www.emcdda.europa.eu/edr2016

http://www.emcdda.europa.eu/system/files/publications/2637/TDAT16001ITN.pdf

http://www.emcdda.europa.eu/topics/pods/benzodiazepines

http://www.euro.who.int/__data/assets/pdf_

https://www.unodc.org/doc/wdr2016/WORLD_ DRUG_ REPORT_2016_web.pdf

https://www.unodc.org/documents/scientific/NPS_leaflet_2016_EN_LORES.pdf

Huang HH, Bai YM (2011) Persistent psychosis after ingestion of a single tablet of '2C-B'. Prog Neuropsychopharmacol Biological Psychiatry 35(1):293–294

Hyman SE, Malenka RC (2001) Addiction and the brain: the neurobiology of compulsion and its persistence. Nat Rev Neurosci 2:695–703

ISTAT (2015) Indagine Multiscop- Aspetti della vita quotidiana. Roma: IstitutNazionale di Statistica; Disponibile all'indirizzo: http://www.istat.it/it/archivio/91926; ultima consultazione 9/2/2016

Jebadurai J, Schifan F, Deluca P (2013) Recreational use of 1-(2-naphthyl)-2-(1-pyrrolidinyl)-1-pentanone hydrochloride (NRG-1), 6-(2-aminopropyl) benzofuran (benzofury/ 6-APB) and NRG-2 with review of available evidence-based literature. Hum Psychopharmacol 28(4):356–364

Jones JD, Mogali S, Comer SD (2012) Polydrug abuse: a review of opioid and benzodiazepine combination use. Drug Alcohol Depend 125(1–2):8–18

Karila L, Weinstein A, Aubin HJ et al (2010) Pharmacological approaches tmethamphetamine dependence: a focused review. Br J Clin Pharmacol 69:578–592

Karila L, Megarbane B, Cottencin O, Lejoyeux M (2015) Synthetic cathinones: a new public health problem. Curr Neuropharmacol 13(1):12–20

Kendler KS, Sundquist K, Ohlsson H et al (2012) Genetic and familial environmental influences on the risk for drug abuse: a national Swedish adoption study. Arch Gen Psychiatry 69(7):690–697

Kleber HD (2007) Pharmacologic treatments for opioid dependence: detoxification and maintenance options. Dialogues Clin Neurosci 9(4):455–470

Koob GF, Le Moal M (1997) Drug abuse: hedonic homeostatic dysregulation. Science 278:52–58

Koob GF, Le Moal M (2001) Drug addiction, dysregulation of reward, and allostasis. Neuropsychopharmacology 24:97–129

Le Boisselier R, Alexandre J, Lelong-Boulouard V, Debruyne D (2016) Focus on cannabinoids and synthetic cannabinoids. Clin Pharmacol Ther 101(2):220–229

Lkkola KT, Ahonen J (2008) Midazolam and other benzodiazepines, in Handb Exp Pharmacol. Handb 27 Exp Pharmacol 182(182):335–360

McKeen MJ, Quraishi SA (2013) Clinical review of intravenous opioids in acute care. J Anesthesiol Clin Sci 2:1–11

McLaren J, Swift W, Dillon P, Allsop S (2008) Cannabis potency and contamination: a review of the literature. Addiction 103(7):1100–1109

Meltzer PC, Butler D, Deschamps JR, Madras BK (2006) 1-(4-Methylphenyl)-2-pyrrolidin-1-yl-pentan-1- one (Pyrovalerone) analogues: a promising class of monoamine uptake inhibitors. J Med Chem 49:1420–1432

Meyer MR (2016) New psychoactive substances: an overview on recent publications on their toxicodynamics and toxicokinetics. Arch Toxicol 90(10):2421

Nugteren-van Lonkhuyzen JJ, van Riel AJ, Brunt TM, Hondebrink L (2015) Pharmacokinetics, pharmacodynamics and toxicology of new psychoactive substances (NPS): 2C-B, 4-fluoroamphetamine and benzofurans. Drug Alcohol Depend 157:18

Nutt D, King LA, Saulsbury W, Blakemore C (2007) Development of a rational scale tassess the harm of drugs of potential misuse. Lancet 369:1047–1053

Organisation for Economic Co-operation and Development (OECD) (2015) Health at a glance 2015. OECD Indicators. OECD Publishing; Disponibile all'indirizzo, Parigi. https://doi.org/10.1787/health_glance-2015-en. ultima consultazione 9/2/2016

Pacifici R, Marchei E, Salvatore F et al (2017) Evaluation of cannabinoids concentration and stability in standardized preparations of cannabis tea and cannabis oil by ultra-high performance liquid chromatography tandem mass spectrometry. Clin Chem Lab Med 55:1555

Papaseit E, MoltoÃÅ J, Muga R et al (2017) Clinical pharmacology of the synthetic Cathinone Mephedrone. Curr Top Behav Neurosci 32:313–331

Park TM, Haning WF (2016) Stimulant use disorders. Child Adolesc Psychiatr Clin N Am 25(3):461–471

Passie T, Halpern JH, Stichtenoth D et al (2008) The pharmacology of lysergic acid diethylamide: a review. CNS Neurosci Ther 14:295–314

Pertwee RG (2006) The pharmacology of cannabinoid receptors and their ligands: an overview. Int J Obes 30(Suppl 1):S13–S18

Prekupec MP, Mansky PA, Baumann MH (2017) Misuse of novel synthetic opioids: a deadly new trend. J Addict Med. in press

Public health successes and missed opportunities. Trends in alcohol consumption and attributable mortality in the WHEuropean Region, 1990-2014 (2016)

Quell SB, Brady KT, Sonne SC (2005) Mood disorders and substance use disorder: a complex comorbidity. Sci Pract Perspect 3(1):13–21

Rook EJ, Huitema AD, van den Brink W et al (2006) Pharmacokinetics and pharmacokinetic variability of heroin and its metabolites: review of the literature. Curr Clin Pharmacol 1:109–118

Schep LJ, Slaughter RJ, Vale JA et al (2011) The clinical toxicology of the designer "party pills" benzylpiperazine and trifluoromethylphenylpiperazine. Clin Toxicol 49:131–141

Schifan F (1994) Il profildell'Ecstasy. Salute e territori 91:37–40

Schiff PL Jr (2002) Opium and its alkaloids. Am J Pharm Educ 66:186–194

Schwartz MD, Trecki J, Edison LA et al (2015) A common source outbreak of severe delirium associated with exposure tthe novel synthetic cannabinoid ADB-PINACA. J Emerg Med 48:573–580

Scofield MD, Heinsbroek JA, Gipson CD et al (2016) The nucleus accumbens: mechanisms of addiction across drug classes reflect the importance of glutamate homeostasis. Pharmacol Rev 68(3):816–871

Scruggs JL, Patel S, Bubser M (2000) DOI-induced activation of the cortex: dependence on 5-HT2A heteroceptors on thalamocortical glutamatergic neurons. J Neurosci 20:8846–8852

Shukla RK, Crump JL, ChriscES. (2012) An evolving problem: methamphetamine production and trafficking in the United States. Int J Drug Policy 23(6):426–435

Staack RF (2007) Piperazine designer drugs of abuse. Lancet 369:1411–1413

Steele TD, McCann UD, Ricaurte GA (1994) 3,4methylenedioxymethamphetamine (MDMA, 'Ecstasy'): pharmacology and toxicology in animals and humans. Addiction 89:539–551

Suzuki J, Dekker MA, Valenti ES et al (2015) Toxicities associated with NBOMe ingestion-a novel class of potent hallucinogens: a review of the literature. Psychosomatics 56:129–139

Tasman A, Lieberman JA (2006) Handbook of psychiatric drugs. Wiley, p 151

Tittarelli R, Mannocchi G, Pantan F, Romol FS (2015) Recreational use, analysis and toxicity of tryptamines. Curr Neuropharmacol 13(1):26–46

Tournebize J, Gibaja V, Kahn JP (2016) Acute effects of synthetic cannabinoids: update 2015. Subst Abus 11:1–23

TOXNET, su toxnet.nlm.nih.gov, Meyer JS (2013) 3,4-methylenedio xymethamphetamine (MDMA): current perspectives. Subst Abuse Rehabil 4:83–99

Valente MJ, de PinhP G, de Lourdes BM et al (2014) Khat and synthetic cathinones: a review. Arch Toxicol 88(1):15–45

Vardanyan RS, Hruby VJ (2014) Fentanyl-related compounds and derivatives: current status and future prospects for pharmaceutical applications. Future Med Chem 6(4):385–412

Villalobos CA, Bull P, SaÃÅez P et al (2004) 4-Bromo-2,5-di- methoxyphenethylamine (2C-B) and structurally related phenylethylamines are potent 5-HT2A receptor antagonists in Xenopus laevis oocytes. Br J Pharmacol 141(7):1167–1174

Vuong C, Van Uum SH, O'Dell LE et al (2010) The effects of opioids and opioid analogs on animal and human endocrine systems. Endocr Rev 31(1):98–132

Weaver MF, Hopper JA, Gunderson EW (2015) Designer drugs 2015: assessment and management. Addict Sci Clin Pract 10:8

Wills TA, Vaccar D, Mc Namara G (1994) Novelty seeking, risk taking, and related constructs as predictors of adolescent substance use: an application of Cloninger's theory. J Subst Abus 6(1):1–20

Wood DM, Heyerdahl F, Yates CB et al (2014) The European drug emergencies network (Euro-DEN). Clin Toxicol 52:239–241

World Health Organization (1992) The ICD-10 classification of mental and behavioral disorders: clinical descriptions and diagnostic guidelines. World Health Organization, Geneva

Zameng L, Frison G, Bettin C, Sciarrone R (2015) Cannabis potency in the Venice area (Italy): update 2013. Drug Test Anal 7(3):255–258

Applications of "Omics" Sciences in the Laboratory

54

Valeria D'Argenio, Barbara Lombardo, Marcella Nunziato, and Lucio Pastore

Introduction

The availability of sensitive and accurate technologies able to combine high resolution and productivity has given rise to the spread of the so-called "omics" sciences in laboratory medicine. Since the completion of the Human Genome Project, the suffix "omics" has been used to indicate numerous new and different disciplines aimed at the study of a biological system as a whole. Genomics, transcriptomics, epigenomics, and the other "omics" sciences study different classes of molecules in different biological samples and in different pathophysiological situations. This less reductionist approach aims to understand higher-level operating principles and the interaction of different signals to answer more complicated biological questions.

The enormous amount of data generated by these approaches has revolutionized the approach to the study of living beings and poses new challenges regarding the methods of analysis and their correct interpretation. The integration of these disciplines allows us to identify molecular alterations related to different diseases and the development of new diagnostic and/or prognostic biomarkers, as well as the development of targeted therapies in the perspective of personalized medicine.

V. D'Argenio
CEINGE-Biotecnologie Avanzate Franco Salvatore, Naples, Italy

Department of Human Sciences and Quality of Life Promotion, San Raffaele Open University, Rome, Italy

B. Lombardo · L. Pastore (✉)
CEINGE-Biotecnologie Avanzate Franco Salvatore, Naples, Italy

Department of Molecular Medicine and Medical Biotechnology, University of Naples Federico II, Naples, Italy
e-mail: lucio.pastore@unina.it

M. Nunziato
Department of Molecular Medicine and Medical Biotechnology, University of Naples Federico II, Naples, Italy

"Omics" Sciences, Definitions, and Generalities

The revolution of "omics" sciences in laboratory medicine is a recent phenomenon, but the concepts on which it is based have a long history. The definition of genome, intended as the whole of the genetic heritage of an organism, has been present in the English language since 1926, while genomics, intended as the discipline that studies genomes, appeared in scientific literature in 1986. The suffix omics was then applied to different disciplines to mean the study of a biological system as a whole; for example, proteomics studies, the total protein content of a particular system, metabolomics studies, metabolites and small molecules, and so on.

The completion of the first draft of the human genome in 2001, with subsequent improvements, laid the foundation for whole-genome variant analysis. The HapMap project, completed in 2005, provided information on all polymorphisms in the genome, allowing the development of association analyses that set the stage for developing pharmacogenomics and other disciplines based on polymorphism analysis.

The primary impetus for using genomic analysis in laboratory medicine has come from the development of high throughput sequencing methodologies defined as next-generation sequencing (NGS). NGS methods are based on different principles characterized by different advantages and limitations, as will be discussed in the following sections. Another relevant method for the analysis of large variants at the genomic level is Comparative Genomic Hybridization (CGH), which allows the identification of deletions or amplifications of large genome tracts that may be the cause of, or associated with, different pathologies.

Proteome and metabolome analyses are instead based on mass spectrometry methodologies coupled with other separative methods; most of these technologies have been available for many years but only recently it has been possible to increase their processivity and, therefore, allow their application in laboratory medicine.

M. Ciaccio (ed.), *Clinical and Laboratory Medicine Textbook*, https://doi.org/10.1007/978-3-031-24958-7_54

From the Genome Project to the 1000 Genomes Project: Applications of Genomics to Laboratory Medicine

The sequencing of the human genome is, without a doubt, one of the greatest achievements of biomedical research. In 2001, with great fanfare, the completion of the first draft of the human genome was announced; the complete genome sequence was announced in 2003 and, although we are at draft number 19, some regions are still not fully characterized.

The completion of the genome project, like many significant achievements in biomedical research, was a starting point rather than an arrival point. Indeed, with sequence data it was possible to identify polymorphisms throughout the genome and characterize them; the international HapMap project has provided this information to the scientific community, significantly facilitating association studies and allowing the identification of DNA regions linked to specific phenotypes.

A significant step in the applications of genomics has been, as previously mentioned, the development of NGS sequencing techniques. Thanks to the possibility of obtaining millions or billions of bases for a single analytical session quickly and at a relatively low cost, this method has wholly overturned several aspects of biomedical research, giving rise to a new era of genomics. Compared to Sanger's method, or first-generation sequencing, NGS allows whole genomes to be sequenced in timescales and at costs previously unimaginable. For example, the Human Genome Project enabled the sequencing and annotation of the entire human genome sequence; its completion required an international consortium, about 10 years of work, and a cost of about $3 billion. Current NGS platforms allow an entire genome to be sequenced in less than a month and at a cost of less than $10,000; the goal of reducing costs to less than $1000 per genome is now achieved.

Since the market launch of NGS techniques in 2006, the U.S. National Institute of Health has seen a gradual reduction in sequencing costs, about 10,000 times less than the cost of the first human genome (https://www.genome.gov/sequencingcostsdata/). The availability of rapid, sensitive, accurate and cost-effective nucleic acid sequencing technologies paved the way for the 1000 Genomes Project. This project aimed to study human genetic variability; it lasted 7 years (from 2008 to 2014) and carried out the sequencing of 2504 genomes from 26 different populations. In this way, it was possible to identify over 88 million variants and follow their distribution in different geographical areas. Recently, the 100,000 Genomes Project has been launched with the ambitious goal of sequencing 100,000 genomes of patients and their families to correlate genomic data with clinical information that can provide new tools for the diagnosis and treatment of various diseases (www.genomicsengland.co.uk/the100000-genomes-project/).

The rapid and progressive development of increasingly sensitive and accurate technologies, combined with the drastic reduction of costs, has brought genomics into the medical realm. However, several aspects remain to be regulated. First of all, the high throughput of NGS systems requires the adaptation of IT platforms to store the vast amount generated data. Moreover, to be easily analyzed, catalogued, and compared, these data require computer systems capable of carrying out these operations and identifying any pathogenic alterations quickly and with a high level of reliability. A large number of software have been developed to analyze sequencing data, especially to identify the weight of variants or possible mutations in the presence of diseases with unknown genetic etiology. On the other hand, computer tools capable of defining the weight of variants and polymorphisms on less-defined phenotypes and on the risk of developing certain diseases are still lacking. Finally, issues related to privacy and the correct interpretation of variants with dubious meanings or the communication of incidental findings will animate the scientific community's debate in the near future.

Detection of Genomic Alterations: The CGH-Array

The methods of karyotype analysis have made it possible to identify alterations in the number and structure of the various human chromosomes and correlate them with the onset of various diseases. The advancement of knowledge in genetics has made it possible to identify many pathologies due to alterations in more or less extensive regions of the genome that cannot be identified by simple karyotype analysis.

For the identification of these alterations, a method called Fluorescent In Situ Hybridization (FISH) was first developed. It consists in hybridizing large DNA fragments of several thousand bases, marked with fluorescent pigments (probes), with the regions whose alteration is a possible cause of the pathology present in the patient. This method allows for analyzing with high sensitivity and specificity specific alterations, but it requires a diagnostic suspicion that directs the analysis to be carried out.

In the last 10 years, a method that allows the identification of alterations at the level of the entire genome, called comparative genomic hybridization (CGH), has been developed. CGH is based, like FISH, on DNA hybridization; in this case, the patient's DNA, marked with a fluorescent pigment, is hybridized to probes immobilized on a slide. In order to identify deletions or amplifications in the genome of the patient tested, a control DNA, labeled with a different pigment (usually a mixture of DNA from apparently healthy subjects of the same sex as the patient tested), is hybridized to the same slide. The fluorescence levels of the pigments corresponding to the patient and control DNA are compared at the level of each probe to identify possible alterations (Fig. 54.1). In early applications of CGH, large fragments of DNA obtained from Bacterial Artificial Chromosomes (BACs) were immobilized on slides as probes; more recently, the ability to synthesize high-density oligonucleotide probes directly on the media has allowed the identification of genomic alterations of even a few thousand bases.

The development of CGH has also made it possible to identify particular types of polymorphisms, called Copy Number Variants (CNVs), represented by deletions or the presence in multiple copies of regions ranging from a few to several thousand bases. CNVs are often polymorphisms with uncertain clinical significance; however, in some cases, they can predispose to pathological phenotypes such as mental disabilities and autism spectrum disorders. A particular characteristic of pathological CNVs is that they are more frequent in affected subjects than in the unaffected population. However, they cannot be considered a cause of disease given their presence also in normal subjects as in the example in Fig. 54.2.

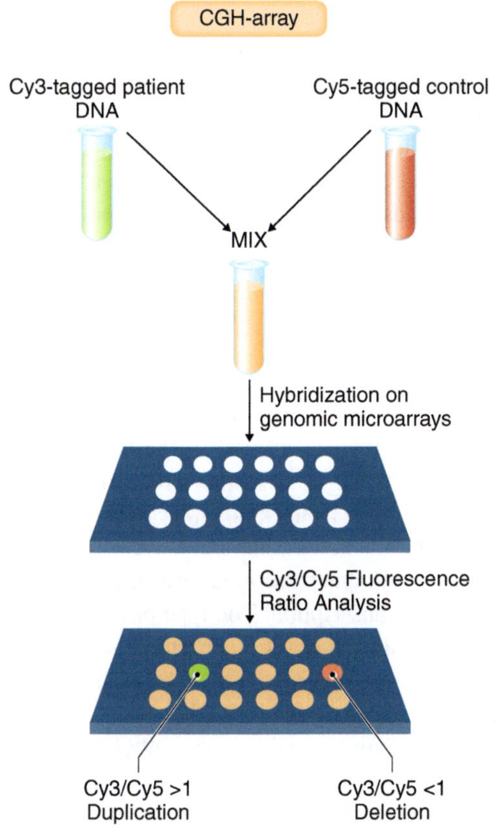

Fig. 54.1 Comparative genomic hybridization (CGH). Identical amounts of patient and control genomic DNA are labeled with different fluorescent pigments (Cy3 and Cy5, respectively) and hybridized with the probe oligonucleotides immobilized on the slide. The relationship between the fluorescence intensities of the patient and control DNA is the parameter that allows the detection of unbalanced genomic rearrangements. (Copyright EDISES 2021. Reproduced with permission)

CGH analysis can be considered an extremely useful genomic screening method for the identification of the molecular etiology of complex clinical pictures such as, for example, many neurological diseases. CGH can also be extremely relevant in prenatal diagnosis, together with or, as it is now commonly believed, in substitution of karyotype analysis. The possibility of identifying the molecular alterations causative of more than 150 pathologies makes CGH an extremely valuable diagnostic tool. Moreover, the possibility of analyzing even a limited number of cells after pre-amplification of the whole genome (Whole-Genome Amplification, WGA) has allowed its application to pre-implantation diagnosis with excellent results. Indeed, the identification of embryos with aneuploidy, and the consequent exclusion of the latter, allows to significantly increase the success of the implantation procedures.

CGH is now a consolidated technology with control and quality assurance procedures for the verification of genomic analysis procedures. However, genomic analysis using CGH has some limitations such as the inability to identify balanced translocations without loss of genetic material or small mutations and the low sensitivity in identifying low levels of mosaicism. Despite the great progress in improving the quality of this methodology, it is evident that the development of quantitative NGS methods will also allow the identification

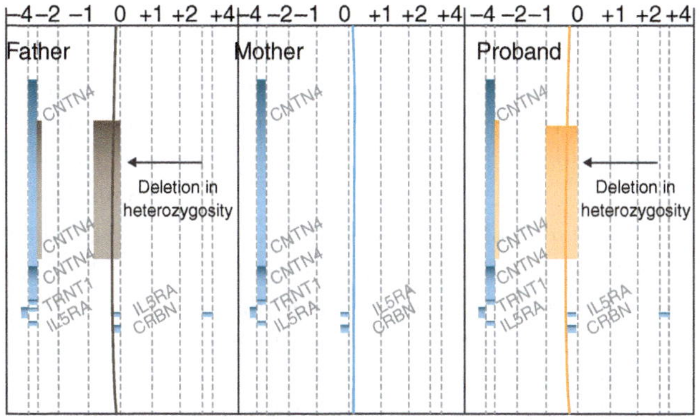

Fig. 54.2 Example of alteration associated with a pathology. The family analysis highlights the presence in the proband, suffering from psychomotor retardation, of a heterozygous deletion also present in the father who, on the other hand, does not show obvious clinical signs. The observed alteration involves the contactin 4 gene (CNTN4): the resulting protein mediates cellular interactions during the development of the nervous system and plays a role in synaptogenesis. Deletions of this region are more frequent in patients with neurological diseases than in healthy subjects. Therefore, they are considered to be associated with the disease but not causative. (Copyright EDISES 2021. Reproduced with permission)

of CNVs and genomic alterations in addition to the sequence analysis and therefore may significantly limit the applications of the CGH.

Transcriptomics and Gene Expression

The decoding of the entire sequence of the human genome, accompanied by the technological advancement of the last 10 years, has not only highlighted the presence of the tight link between gene expression and the molecular basis of human diseases, but also highlighted how the expression of specific genes can influence the phenotypic expression (e.g. the severity) of the same diseases.

Transcriptomics is the discipline that deals with the study of RNAs expressed in a cell or tissue. Unlike the genome, the transcriptome is not constant but, on the contrary, differs between cells and tissues with different characteristics and functions. In addition, the type and amount of RNA expressed in each condition is influenced by many factors, including the presence of specific disease conditions or particular treatments. RNA plays essential roles in many biological processes and, therefore, a qualitative or quantitative alteration of specific RNAs can lead to pathological processes. From a diagnostic point of view, the evaluation of one or more specific RNA molecules can be applied to several aspects related to human health, including the diagnostic phase, the prognostic stratification of patients, and the choice of targeted therapies. Therefore, genome-wide gene expression analyses are gaining a prominent role in biomedical research.

Traditionally, gene expression analysis is performed by quantitative polymerase chain reaction (PCR). Although this method is still widely used and demonstrates high accuracy and sensitivity, it does not allow the simultaneous analysis of a large number of targets. The use of microarrays, based on the retrotranscription of RNA from a system, and subsequent hybridization with probes corresponding to a panel of known RNAs immobilized on a slide, has allowed the simultaneous analysis of multiple target genes; however, low reproducibility and detection limitations have tempered the initial enthusiasm for this technology. With the advent of NGS technologies, RNA sequencing (RNA-Seq) is also becoming increasingly popular for molecular diagnostic applications. Molecular investigations based on RNA-Seq protocols allow to obtain the evaluation of gene expression with higher levels of sensitivity and accuracy compared to "traditional" technologies. In addition, it is possible to simultaneously evaluate different aspects, such as the presence of rare transcripts, splicing variants and non-coding RNAs. This type of approach is very useful in the field of molecular oncology as it allows to identify both possible fusion genes and splicing variants associated with tumor development and that, therefore, could have a diagnostic and/or prognostic significance. Numerous RNA-Seq protocols are currently available; the choice depends on the type of RNA to be investigated. As in the case of genome sequencing, the amount of data produced can lead to problems related to the management and correct interpretation of the results. The definition and the standardization of the methods concerning the reproducibility, the accuracy, and the precision of RNA-Seq in different clinical conditions and in different biological samples are certainly necessary for the routine use of these methodologies in laboratory diagnostics.

Epigenomics and Regulatory Mechanisms of Gene Expression

The genome of all cells present in a multicellular organism is fairly constant; however, hundreds of cell types have distinct morphological and functional characteristics. These differences also depend on specific mechanisms able to regulate gene expression and access to chromatin: the set of these DNA characteristics is called epigenome.

The epigenome, therefore, can be considered as a control system of the genome; it is partly inherited but can be modified by external factors, and is fundamental for the development of phenotypic characteristics of cells. Epigenetics is the scientific discipline that studies change in gene function that do not depend on DNA sequence variations but are fundamental to translate information (genotype) into function (phenotype). Essentially three main, closely interconnected mechanisms fall under this umbrella: post-translational modifications of histones, DNA methylation, and non-coding RNAs.

For years, histones were considered structural proteins with the sole function of assembling DNA into chromosomes. Only in the second half of the twentieth century it was discovered that histones could undergo post-translational modifications of various kinds (acetylation, methylation, phosphorylation, ubiquitination, ADP-ribosylation, and biotinylation) and that these modifications had effects on the control of the transcription. The type of consequence at the transcriptional level (activation or repression of gene expression) depends not only on the type of modification but also on the amino acid residue involved.

DNA methylation involves the methylation of the cytosine of the CpG dinucleotide by specific enzymes. The human genome contains approximately 3×10^7 CpGs and it has been shown that biologically relevant effects of DNA methylation occur when it involves clusters or "islands" of CpGs. CpG islands are, in general, located at the promoter level and their methylation is inversely correlated with transcription. Occasionally, CpG islands may also be intragenic or intergenic and, in these cases, are generally hypermethylated to prevent the formation of atypical transcripts by internal promoters.

Non-coding RNAs (ncRNAs) are a class of RNAs that are not translated into proteins and are responsible for regulating gene expression. Based on their length, ncRNAs are classified into small ncRNAs (between 20 and 40 bp), including miRNA, siRNA, piRNA, and XiRNA, and long ncRNAs (lncRNA, >200 bp). Generally, ncRNAs can regulate the function of other RNAs through complementarity with specific target sequences. miRNAs, in particular, are the most studied class of ncRNAs. miRNAs can inhibit the transcription of a target gene through binding to the 3′ UTR and, as a consequence, there will be premature degradation of the target messenger or the block of translation through the formation of a specific silencing complex. It is now known that miRNAs are involved in numerous biological processes. Moreover, not only can each miRNA have numerous targets, but each gene can be the target of different miRNAs and similar mechanisms are being identified for other ncRNAs. The lncRNAs are a relatively new class of ncRNAs identified thanks to international consortia (Fantom: fantom.gsc.riken.jp; ENCODE: www.encodeproject.org) whose aim is the functional characterization of cDNAs, the study of gene expression networks and the mechanisms involved in their regulation. These studies have shown that lncRNAs are involved in numerous biological processes, from the maintenance of telomere stability and X-chromosome inactivation to the regulation of the expression of specific target genes.

The various mechanisms of epigenetic regulation are interconnected: ncRNA expression, for example, can be regulated by DNA methylation, which in turn is influenced by histone modifications. Therefore, the epigenome regulates chromatin accessibility by influencing gene expression in a decisive manner. In this way, the different types of cells in an organism, despite having essentially the same genome, take different fates thanks to their epigenome, which regulates the gene expression of each cell. International consortia such as the International Human Epigenome Consortium (IHEC: www.ihec-epigenomes.org) have characterized the epigenome of many different cell types producing a reference database of different human cellular epigenomes to be used to identify specific alterations in the presence of different diseases. The realization of projects of this type is also made possible by the availability of high-resolution technologies, many based on NGS sequencing, which allow genome-wide epigenomic analysis with higher levels of sensitivity and resolution than other more traditional methodologies. For example, the identification of genomic regions that are the site of post-translational modifications of histones or other chromatin-associated proteins is traditionally performed by chromatin immunoprecipitation (Chromatin Immunoprecipitation, ChIP) with a specific antibody to isolate the genomic regions to which the protein of interest binds. After this step, the purified DNA fragments can be sequenced (ChIP-Seq) to identify all the binding sites of the protein of interest at the genomic level. The limitations of this method are represented by the availability of an antibody specific for the protein under study and the amount of cells needed for the experiment.

The study of DNA methylation is based on enrichment systems capable of isolating methylated DNA fragments or treatment of DNA with sodium bisulfite for the conversion of cytosines to uracil. The latter approach provides single-base resolution and is the most widely used method currently used in combination with both microarray and NGS sequencing. Similarly, the study of ncRNAs, traditionally performed by quantitative PCR or microarrays, can also benefit from the use of NGS techniques that allow the complete analysis (qualitative and quantitative) of ncRNAs present in a sample, including the identification of new molecules.

Epigenetic modifications have been associated with several diseases, including cancers and genetic disorders. DNA methylation, for example, was the first epigenetic modification associated with cancer diseases; generally, general hypomethylation (associated with the genomic instability typical of cancer cells) and hypermethylation of specific genes (usually oncosuppressors) are observed in cancer cells. Alterations of different classes of ncRNAs, including miRNAs and lncRNAs, have also been associated with numerous cancers but also with non-oncological diseases. The identification of epigenetic modifications improves the understanding of pathogenetic mechanisms underlying the development of numerous diseases; recently, some of these alterations have allowed the identification of new potential biomarkers both with diagnostic efficacy and relevant to the development of new therapeutic treatments.

Metagenomics, Microbiome Analysis, and Association with Human Diseases

Metagenomics is, in the field of "omics" sciences, the discipline that, by exploiting advanced biotechnologies, studies the complexity of microbial communities directly in their natural environment. In particular, the term microbiota defines the community of microorganisms, including bacteria, archaea, viruses, and some unicellular eukaryotes, living in a specific environment; the microbiome, instead, is the set of all the genomic elements of a specific microbiota.

In this context, if we consider the human body as an environment, the human microbiota can be defined as the set of all microorganisms living not only on the surface but also inside our body. These microbial communities perform fundamental functions in human physiology, immune system development, digestion, and detoxification reactions. The human microbiota produces proteins involved in functions critical to the health of the host, such as enzymes necessary for the hydrolysis of otherwise indigestible substances in food, as well as the synthesis of essential vitamins. This concept is at the basis of the definition of human beings as "superorganisms," as they are made up of two different genomes: the human genome, inherited from parents, and the acquired genome, represented by the microbiome.

The most important difference between these genomes consists in the dynamism of the microbiome, which can be influenced by a number of factors, such as age, diet, hormonal cycles, travel, therapies and diseases; the human genome, on the other hand, is fairly constant and significant variations are often associated with diseases. It follows that the maintenance of homeostasis of a "healthy" human microbiome (understood as qualitative and quantitative composition) appears essential for the maintenance of health; alterations, defined "dysbiosis," seem to be associated with the onset of many diseases. The study of the microbiome through metagenomics is therefore essential to define the qualitative and quantitative composition of the microbiota in a specific environment (organ, tissue, or biological fluid) under physiological conditions, but also to identify alterations associated with morbid conditions. The identification of a pathology-specific dysbiosis may allow to: (1) obtain information related to pathogenetic mechanisms; (2) identify possible diagnostic and prognostic markers; (3) identify possible targets for new therapeutic approaches. In this direction, the Human Microbiome Project, a multicenter project, aims to obtain the complete characterization of the human microbiome and to understand its role in the state of health and disease.

For a long time it was thought that the development of the human microbiota began in the postnatal period; recent evidence has shown that humans are not born sterile but microbial colonization occurs already in the antenatal period, although the type of delivery and the type of lactation strongly influence its composition. Most of the human microbiota resides in the gut (>1000 different species). It is estimated that in the human colon alone the density of microbial cells is greater than 1011 cells/g, equivalent to 1–2 kg of body weight, and that the human gut microbiome alone encodes for more than 5 million different genes, many more than the human genes themselves. All this explains the enormous interest in the study of the human microbiome and its functional role and accounts for the need for advanced technologies that allow the rapid and efficient analysis of microbial communities present in a specific environment and in a specific condition (such as the presence/absence of a disease of interest).

Early studies for the characterization of microbial communities were based on direct cultivation techniques to isolate microorganisms present in the sample. However, this strategy has had limited use because the growth conditions used favor the selection of one or more species over others; in fact, it is estimated that up to 99% of microbes are non-culturable. Other methods, such as quantitative PCR and polyacrylamide gel electrophoresis, are affected by the use of specific probes to detect certain bacteria. The rapid development of NGS has given a new impetus to metagenomics studies. These methodologies allow the qualitative/quantitative characterization of a specific microbiome, without the selection bias and limitations associated with culture methods. Currently, the most widely used strategy involves the analysis of bacterial 16S ribosomal RNA (rRNA) that allows to study the complexity of an entire bacterial microbiome in a relatively simple and efficient way. Indeed, the 16S rRNA gene is a molecule traditionally used for the phylogenetic classification of bacteria due to its peculiar structure characterized by hypervariable regions interspersed with ultra-conserved regions. Therefore, universal primers, complementary to the ultra-preserved regions, can be used to amplify, theoretically, in a single PCR reaction, the 16S rRNA gene sequence of all bacteria present in an environment. The identification of individual bacteria is then done, at the end of sequencing, by comparing the detected hypervariable sequences with those present in specific databases. Although this analysis strategy has the advantage of being easy to perform, rapid and relatively inexpensive, both the DNA extraction and amplification steps can introduce errors. Moreover, the taxonomic identification of bacteria depends on the reference databases used for the analysis and it is not possible to obtain data on other classes of microbes nor functional information. Other possible analysis strategies, again based on NGS techniques, are shotgun sequencing and metatracriptomics. Shotgun sequencing is the analysis of the entire microbiome. It is based on the extraction of DNA from the sample of interest, which is then used to prepare a library and sequenced. In this way, it has the advantage of excluding the biases deriving both from the cultivation of the microorganisms and from the amplification phase. In addition, bioinformatic analysis of the sequences produced makes it possible to identify not only the bacteria but also other microorganisms that may be present in the sample under examination. The high costs and the difficulties related to data analysis have so far limited the diffusion of this methodology only to research applications. Finally, metatracriptomics allows analysis of the entire transcriptome of a sample of interest to obtain a complete view of gene expression profiles and related functional data. Several factors, however, influence the large-scale application of metatracriptomics, such as the technical issues related to the extraction and storage of RNA samples, the quality and quantity of RNA and, most importantly, the enrichment procedures used to remove rRNAs. In addition, bioinformatics tools for the analysis of metatracriptomics data have not yet been validated on a large scale.

To date, alterations in the human microbiome have been associated with numerous diseases, both gastrointestinal and extraintestinal. Numerous evidences suggest that the microbiome may play an important role in numerous pathologies through interaction with the immune system as well as direct or indirect effects (mediated, for example, by epigenetic modifications) on the human genome and gene expression.

Nutrigenomics and Nutrition Medicine

The relationship among food, environment and health has been known since ancient times. The completion of the Human Genome Project and the simultaneous development of increasingly advanced technologies have made it possible to encode the relationship between certain genes and nutrition, as well as the effects of nutrition on the maintenance of health. Nutrigenomics is a discipline that, using molecular biology techniques, studies the effect of nutrients on the genome and on the production of specific proteins and/or metabolites. It is well known that mutations capable of altering specific enzymatic activities through the accumulation of toxic metabolites and/or the lack of production of essential metabolites, determine pathological states that can be treated through specific nutritional interventions. This is the case of numerous genetic diseases, classically defined as monogenic and including defects in the metabolism of amino acids, carbohydrates, and fatty acids, generally treated with targeted nutritional interventions. On the other hand, the effects of nutrients on the pathogenesis of specific diseases, as well as the beneficial effects of specific nutritional interventions, are also known in the case of multifactorial diseases such as celiac disease, diabetes and obesity. More recently, several studies have investigated the association between nutrition and cancer. In particular, it has been shown that micronutrient deficiencies, such as vitamin B12 and folic acid, can cause DNA damage similar to that induced by exposure to radiation and, therefore, lead to oxidative damage related to the development of cancer. On the other hand, toxic substances contained in food can induce the production of toxic metabolites which, by interacting with DNA, can cause mutations predisposing to the development of tumors. Other micronutrients, such as selenium, zinc, and prostacyclins, have been associated with an onco-protective effect.

Nutrigenomics, therefore, aims to promote the state of health by studying the mechanisms through which nutrients interact with DNA aiming to develop targeted therapeutic interventions. From this point of view, personalized nutrition is increasingly becoming an essential objective of nutritional sciences, not only in the presence of a specific pathology but also in healthy individuals to avoid food intolerances and/or negative effects related to the intake of certain nutrients and, thus, preserve the state of health.

Pharmacogenomics and Personalized Therapy

The term drug identifies any substance capable of exerting an effect on living systems. An ideal drug treatment leads to the improvement of the psychophysical state of the individual. In reality, drug treatments can have three types of effect: (1) healing or improvement of the patient; (2) no effect on the psychophysical state of the individual; (3) the occurrence of adverse drug reactions. The effect of adverse drug reactions on the incidence of new pathologies and on health expenditure is very high, causing 5% of hospital admissions.

In recent years, the field of clinical molecular biology has also been concerned with the analysis of DNA polymorphisms that may influence drug response. Single Nucleotide Polymorphisms (SNPs) often contribute to the variability of individual drug response. In addition, recently, several CNVs have also been implicated in the response to different drugs.

Starting from the 1990s of the last century, pharmacogenetics has been developed. It study the genetic predisposition to the response to certain drugs in terms of efficacy and side effects by investigating specific genes often involved in their metabolism. In addition, the completion of the Genome Project and the development of NGS methods has led to the development of "pharmacogenomics," which studies the drug response effect of variations at the level of the entire genome; in this way, many additional polymorphisms in genes involved in the cellular response to drugs have been identified, justifying the variety of drug-organism interactions.

The metabolism of many drugs passes through a common pathway consisting of liver enzymes belonging to the cytochrome P450 family. This superfamily of proteins has been so defined due to the presence of a heme group characterized by an absorbance peak at 450 nm. These enzymes increase the water solubility of drugs, allowing their excretion through bile and urine. Their mechanism of action consists in the conjugation of polar groups to the drug by means of oxidation reactions; subsequently, other enzymes add groups to the drugs that favor their excretion such as glutathione, glucuronic acid, glycine and sulfate, methyl, and acetyl groups.

Fifty-seven genes belonging to this superfamily are known, subdivided into various families and subfamilies. The nomenclature indicates all genes with the prefix CYP followed by a number indicating the family, a letter indicating the subfamily and a number specifying the enzyme.

Numerous allelic variants have been identified in the 57 CYP450 genes that alter the activity of various enzymes with effects in drug metabolization. For most drugs, the population can be divided into four groups: (1) normal metabolizers (MN); (2) slow metabolizers (ML); (3) rapid metabolizers

(MR); (4) ultrarapid metabolizers (MUR). Overall, ML subjects may experience adverse reactions due to accumulation, while in MR and MUR subjects, higher doses of drug are required to achieve therapeutic effects. Polymorphisms in these genes cause variations in protein conformation resulting in enzymatic hyper- or hypoactivity; gene deletions or duplications have also been observed for some of these enzymes.

An example of how the genomic make-up can influence drug response is given by adverse reactions to warfarin. Warfarin is an anticoagulant that prevents thrombotic episodes by inhibiting the action of vitamin K epoxide reductase, which is essential during coagulation. Warfarin is metabolized by the enzyme CYP2C9, which causes a hydroxylation of the drug of the S enantiomer. The latter has the greatest therapeutic value, while the R enantiomer is metabolized by the enzymes CYP1A1, CYP1A2 and CYP3A4. There are six known variants in the CYP2C9 gene with clinical significance, two of which are more widespread (CYP2C9*2 and CYP2C9*3) and give metabolic differences compared to the wild type form (CYP2C9*1). The CYP2C9*2 form is characterized by a C > T change at position 430 resulting in R > C substitution at position 144, whereas the CYP2C9*3 form is characterized by an A > C change at position 1075 resulting in I > L substitution at position 359. These variants cause a reduction in enzymatic activity compared to the wild-type form and, thus, a poor metabolization of the drug. The persistence of the drug causes a state of hypocoagulability characterized by hemorrhages of variable entity and severity. Therefore, the presence of the CYP2C9*2 and CYP2C9*3 forms involves the need to reduce the dosage of the drug. Also, other genetic factors (such as the subunit 1 of the vitamin K epoxide reductase complex [Vitamin K epoxide Reductase Complex 1, VKORC1]) and environmental factors influence the response to warfarin, so there is not a complete correlation between genotype and clinical manifestations.

The genomic analysis of drug response and the consequent identification of optimal treatment and dosage for each patient are fundamental for the development of the so-called "precision medicine" in which treatments, preventive or therapeutic, will be increasingly individualized based on genetic and clinical information.

Recommended Readings

1000 Genomes Project Consortium (2015) A global reference for human genetic variation. Nature 526:6874

Aleksandrova N, Shubina E, Ekimov A et al (2016) Comparison of the results of preimplantation genetic screening obtained by a-CGH and NGS methods from the same embryos. Gynecol Endocrinol 32:1–4

Allfrey VG, Mirsky AE (1964) Structural modifications of histones and their possible role in the regulation of RNA synthesis. Science 144:559

Althani AA, Marei HE, Hamdi WS et al (2016) Human microbiome and its association with health and diseases. J Cell Physiol 231:1688–1694

Berindan-Neagoe I, Monroig Pdel C, Pasculli B, Calin GA (2014) MicroRNAome genome: a treasure for cancer diagnosis and therapy. CA Cancer J Clin 64:311–336

Berná G, Oliveras-López MJ, Jurado-Ruíz E et al (2014) Nutrigenetics and nutrigenomics insights into diabetes etiopathogenesis. Nutrients 6:5338–5369

Berni Canani R, Paparo L, Nocerino R et al (2015) Differences in DNA methylation profile of Th1 and Th2 cytokine genes are associated with tolerance acquisition in children with IgE-mediated cow's milk allergy. Clin Epigenetics 7:38

Biamino E, Di Gregorio E, Belligni EF et al (2016) A novel 3q29 deletion associated with autism, intellectual disability, psychiatric disorders, and obesity. Am J Med Genet B Neuropsychiatr Genet 171B:290–299

Blasco-Baque V, Coupé B, Fabre A et al (2017) Associations between hepatic miRNA expression, liver triacylglycerols and gut microbiota during metabolic adaptation to high-fat diet in mice. Diabetologia. https://doi.org/10.1007/s00125-017-4209-3

Boyer SW, Barclay LJ, Burrage LC (2015) Inherited metabolic disorders: aspects of chronic nutrition management. Nutr Clin Pract 30:502–510. "Genome, n". Oxford English Dictionary (3rd ed)

Byron SA, Van Keuren-Jensen KR, Engelthaler DM et al (2016) Translating RNA sequencing into clinical diagnostics: opportunities and challenges. Nat Rev Genet 17:257–271

Calin GA, Dumitru CD, Shimizu M et al (2002) Frequent deletions and down-regulation of micro-RNA genes miR15 and miR16 at 13q14 in chronic lymphocytic leukemia. Proc Natl Acad Sci U S A 99:15524–15529

Cao W, Liu JN, Liu Z et al (2017) A three-lncRNA signature derived from the Atlas of ncRNA in cancer (TANRIC) database predicts the survival of patients with head and neck squamous cell carcinoma. Oral Oncol 65:94–101

Carreira IM, Ferreira SI, Matoso E et al (2015) Copy number variants prioritization after array-CGH analysis – a cohort of 1000 patients. Mol Cytogenet 8:103

Chan SL, Ang X, Sani LL et al (2016) Prevalence and characteristics of adverse drug reactions at admission to hospital: a prospective observational study. Br J Clin Pharmacol 82:1636–1646

Ciccia F, Ferrante A, Guggino G, Triolo G (2016) The role of the gastrointestinal tract in the pathogenesis of rheumatic diseases. Best Pract Res Clin Rheumatol 30:889–900

Cortese R, Lu L, Yu Y et al (2016) Epigenome-microbiome crosstalk: a potential new paradigm influencing neonatal susceptibility to disease. Epigenetics 11:205–215

D'Argenio V, Salvatore F (2015) The role of the gut microbiome in the healthy adult status. Clin Chim Acta 451:97–102

D'Argenio V, Precone V, Casaburi G et al (2013) An altered gut microbiome profile in a child affected by Crohn's disease normalized after nutritional therapy. Am J Gastroenterol 108:851–852

D'Argenio V, Casaburi G, Precone V et al (2016) Metagenomics reveals dysbiosis and a potentially pathogenic N. flavescens strain in duodenum of adult celiac patients. Am J Gastroenterol 111:879–890

David LA, Maurice CF, Carmody RN et al (2014) Diet rapidly and reproducibly alters the human gut microbiome. Nature 505:559–563

Dirks RA, Stunnenberg HG, Marks H (2016) Genome-wide epigenomic profiling for biomarker discovery. Clin Epigenetics 8:122

Doo M, Kim Y (2015) Obesity: interactions of genome and nutrients intake. Prev Nutr Food Sci 20:1–7

Flint HJ, Scott KP, Duncan SH et al (2012) Microbial degradation of complex carbohydrates in the gut. Gut Microbes 3:289–306
Gajer P, Brotman RM, Bai G et al (2012) Temporal dynamics of the human vaginal microbiota. Sci Transl Med 4:132–152
Goh LL, Lim CW, Sim WC et al (2017) Analysis of genetic variation in CYP450 genes for clinical implementation. PLoS One 12:e0169233
Human Microbiome Project Consortium (2012a) A framework for human microbiome research. Nature 486:215–221
Human Microbiome Project Consortium (2012b) Structure, function and diversity of the healthy human microbiome. Nature 486:207–214
International Human Genome Sequencing Consortium (2004) Finishing the euchromatic sequence of the human genome. Nature 431:931–945
Jamieson D, Sunter N, Muro S et al (2017) Pharmacogenetic association of MBL2 and CD95 polymorphisms with grade 3 infection following adjuvant therapy for breast cancer with doxorubicin and cyclophosphamide. Eur J Cancer 71:15–24
Karampetsou E, Morrogh D, Chitty L (2014) Microarray technology for the diagnosis of fetal chromosomal aberrations: which platform should we use? J Clin Med 3:663–678
Katara P (2014) Single nucleotide polymorphism and its dynamics for pharmacogenomics. Interdiscip Sci 6:85–92
Khan N, Vidyarthi A, Nadeem S et al (2016) Alteration in the gut microbiota provokes susceptibility to tuberculosis. Front Immunol 7:529
Khorkova O, Hsiao J, Wahlestedt C (2015) Basic biology and therapeutic implications of lncRNA. Adv Drug Deliv Rev 87:15–24
Koenig JE, Spor A, Scalfone N et al (2011) Succession of microbial consortia in the developing infant gut microbiome. Proc Natl Acad Sci U S A 108:4578–4585
Koren O, Goodrich JK, Cullender TC et al (2012) Host remodeling of the gut microbiome and metabolic changes during pregnancy. Cell 150:470–480
Kungulovski G, Jeltsch A (2016) Epigenome editing: state of the art, concepts, and perspectives. Trends Genet 32:101–113
Lee CK, Herbold CW, Polson SW et al (2012) Groundtruthing next-gen sequencing for microbial ecology-biases and errors in community structure estimates from PCR amplicon pyrosequencing. PLoS One 7:e44224
Lemon KP, Armitage GC, Relman DA et al (2012) Microbiota targeted therapies: an ecological perspective. Sci Transl Med 4:137–135
Li L, Mendis N, Trigui H et al (2014) The importance of the viable but non-culturable state in human bacterial pathogens. Front Microbiol 5:258
Lombardo B, Ceglia C, Tarsitano M et al (2014) Identification of a deletion in the NDUFS4 gene using array-comparative genomic hybridization in a patient with suspected mitochondrial respiratory disease. Gene 535:376–379
Marioni JC, Mason CE, Mane SM et al (2008) RNA-seq: an assessment of technical reproducibility and comparison with gene expression arrays. Genome Res 18:1509–1517
Morgan JL, Darling AE, Eisen JA (2010) Metagenomic sequencing of an in vitro-simulated microbial community. PLoS One 5:e10209
Nardelli C, Granata I, Iaffaldano L et al (2017a) miR-138/miR-222 overexpression characterizes the miRNome of amniotic mesenchymal stem cells in obesity. Stem Cells Dev 26:4–14
Nardelli C, Granata I, Iaffaldano L et al (2017b) Sex-comparative analysis of the miRNome of human amniotic mesenchymal stem cells during obesity. Stem Cells Dev 26:1–3
Palmer C, Bik EM, DiGiulio DB et al (2007) Development of the human infant intestinal microbiota. PLoS Biol 5:e177
Paparo L, Nocerino R, Cosenza L et al (2016) Epigenetic features of FoxP3 in children with cow's milk allergy. Clin Epigenetics 8:86
Parameswaran Nair N, Chalmers L, Peterson GM et al (2016) Hospitalization in older patients due to adverse drug reactions – the need for a prediction tool. Clin Interv Aging 11:497–505
Park ST, Kim J (2016) Trends in next-generation sequencing and a new era for whole genome sequencing. Int Neurourol J 20:S76–S83
Perez-Cobas AE, Gosalbes MJ, Friedrichs A et al (2013) Gut microbiota disturbance during antibiotic therapy: a multi-omic approach. Gut 62:1591–1601
Qin J, Li R, Raes J et al (2010) A human gut microbial gene catalogue established by metagenomic sequencing. Nature 464:59–65
Ramos L, del Rey J, Daina G et al (2014) Oligonucleotide arrays vs. metaphase-comparative genomic hybridisation and BAC arrays for single-cell analysis: first applications to preimplantation genetic diagnosis for Robertsonian translocation carriers. PLoS One 9:e113223
Romani M, Pistillo MP, Banelli B (2015) Environmental epigenetics: crossroad between public health, lifestyle, and cancer prevention. Biomed Res Int 2015:587983
Sales NMR, Pelegrini PB, Goersch MC (2014) Nutrigenomics: definitions and advances of this new science. J Nutr Metab 2014:202759
Sanger F, Nicklen S, Coulson AR (1977) DNA sequencing with chain-terminating inhibitors. Proc Natl Acad Sci U S A 74:5463–5467
Sharon I, Morowitz MJ, Thomas BC et al (2013) Time series community genomics analysis reveals rapid shifts in bacterial species, strains, and phage during infant gut colonization. Genome Res 23:111–120
Stewart CJ, Marrs EC, Nelson A et al (2013) Development of the preterm gut microbiome in twins at risk of necrotising enterocolitis and sepsis. PLoS One 8:e73465
Tarsitano M, Ceglia C, Novelli A et al (2014) Microduplications in 22q11.2 and 8q22.1 associated with mild mental retardation and generalized overgrowth. Gene 536:213–216
Tassano E, Biancheri R, Denegri L et al (2014) Heterozygous deletion of CHL1 gene: detailed array-CGH and clinical characterization of a new case and review of the literature. Eur J Med Genet 57:626–629
Venter JC, Adams MD, Myers EW et al (2004) The sequence of the human genome. Science 291:1304–1351
Walker AW, Duncan SH, Louis P et al (2014) Phylogeny, culturing, and metagenomics of the human gut microbiota. Trends Microbiol 22:267–274
Walsh CJ, Guinane CM, O'Toole PW et al (2014) Beneficial modulation of the gut microbiota. FEBS Lett 588:4120–4130
Wang L, Wang YX, Chen LP et al (2016) Upregulation of microRNA-181b inhibits CCL18-induced breast cancer cell metastasis and invasion via the NF-κB signaling pathway. Oncol Lett 12:4411–4418
Weber CM, Henikoff S (2014) Histone variants: dynamic punctuation in transcription. Genes Dev 28:672–682
Woese CR, Fox GE, Zablen L et al (1975) Conservation of primary structure in 16S ribosomal RNA. Nature 254:83–86
Wu GD, Chen J, Hoffmann C et al (2011) Linking longterm dietary patterns with gut microbial enterotypes. Science 334:105–108
Yadav SP (2007) The wholeness in suffix -omics, -omes, and the word om. J Biomol Tech 18:277
Yatsunenko T, Rey FE, Manary MJ et al (2012) Human gut microbiome viewed across age and geography. Nature 486:222–227
Zhang J, Chen Z, Chen C (2016) Impact of CYP2C9, VKORC1 and CYP4F2 genetic polymorphisms on maintenance warfarin dosage in Han-Chinese patients: a systematic review and meta-analysis. Meta Gene 9:197–209
Zhao S, Fung-Leung WP, Bittner A et al (2014) Comparison of RNA-Seq and microarray in transcriptome profiling of activated T cells. PLoS One 9:e78644

COVID-19

55

Marcello Ciaccio, Luisa Agnello, and Rosaria Vincenza Giglio

Introduction

In December 2019, several patients in Wuhan (Hubei Province), China, developed a form of pneumonia of unknown etiology. On January 7, 2020, the pathogen responsible for these forms of pneumonia was isolated from bronchoalveolar lavage fluid samples; it is a virus belonging to the Coronavirus family, which the World Health Organization (WHO) subsequently named SARS-CoV-2 (Severe Acute Respiratory Syndrome CoronaVirus-2) and COVID-19 the infection for which it is responsible (coronavirus disease). Since the first outbreak in Wuhan, the SARS-CoV-2 infection has rapidly spread throughout the country and worldwide. On January 30, 2020, WHO declared that the COVID-19 outbreak represented a "public health emergency of international concern," and on March 11, 2020, after assessing the severity and spread levels of the infection, it declared that the COVID-19 outbreak could be considered a pandemic.

The knowledge on the SARS-CoV-2 characteristics, infectivity, the response of our organism to the infection, as well as diagnostics, therapies, and vaccines are growing exponentially, so when this textbook is published, there will surely be further updates.

SARS-CoV-2

SARS-CoV-2 belongs to the Coronaviridae family, so called because of the crown-shaped tips on their surface. Coronaviruses are characterized by a single-stranded positive RNA genome, enveloped by capsid, and are classified into four groups: α-coronavirus (α-COV), β-coronavirus (β-COV), δ-coronavirus (δ-COV), and γ-coronavirus (γ-COV); SARS-CoV-2 is a β-COV. SARS-CoV and Middle East respiratory syndrome coronavirus (MERS-CoV), two highly pathogenic viruses whose infection in humans is associated with the onset of severe respiratory tract diseases, also belong to the β-COV group. In particular, SARS-CoV was identified in 2003 as the pathogen responsible for the SARS (severe acute respiratory syndrome) epidemic that began in China in late 2002, while MERS-CoV was identified in 2012 as the cause of MERS (Middle East respiratory syndrome).

Coronaviruses are common in many animal species (such as camels and bats), but they can rarely result in zoonotic infection by infecting humans. The SARS-CoV epidemic in China originated from bats, while the MERS-CoV epidemic originated from dromedaries. It has been speculated that bats, snakes, or pangolins can transmit SARS-CoV-2. In addition, the virus is highly transmissible from human to human via respiratory droplets and aerosols.

Coronaviruses consist primarily of four structural proteins: Spike (S, which contains the receptor binding domain), membrane (M), envelope (E), and nucleocapsid (N) (Fig. 55.1). Spike is a homotrimer transmembrane glycoprotein that protrudes from the viral surface and is responsible for coronavirus diversity and host tropism. The spike protein comprises two functional subunits: an N-terminal S1 subunit, responsible for binding to the host cell receptor, and a C-terminal S2 subunit, responsible for fusing the virus to the cell membrane. The S1 subunit is further subdivided into an N-terminal domain (NTD) and a receptor binding domain (RBD). The latter interacts directly with receptors in the host organism. The S2 subunit contains an FP (fusion peptide) domain and the HR1 (heptad repeat 1) and HR2 (heptad repeat 2) domains, which mediate virus-cell membrane fusion (Fig. 55.2). Notably, angiotensin-converting enzyme 2 (ACE2) has been identified as a functional receptor for SARS-CoV-2. ACE2 is expressed in various organs and tissues, such as the lungs, heart, kidneys, and bladder; in par-

M. Ciaccio (✉) · L. Agnello · R. V. Giglio
Department of Biomedicine, Neurosciences and Advanced Diagnostics, Institute of Clinical Biochemistry, Clinical Molecular Medicine and Clinical Laboratory Medicine, and Department of Laboratory Medicine, University Hospital "P. Giaccone", Palermo, Italy
e-mail: marcello.ciaccio@unipa.it

M. Ciaccio (ed.), *Clinical and Laboratory Medicine Textbook*, https://doi.org/10.1007/978-3-031-24958-7_55

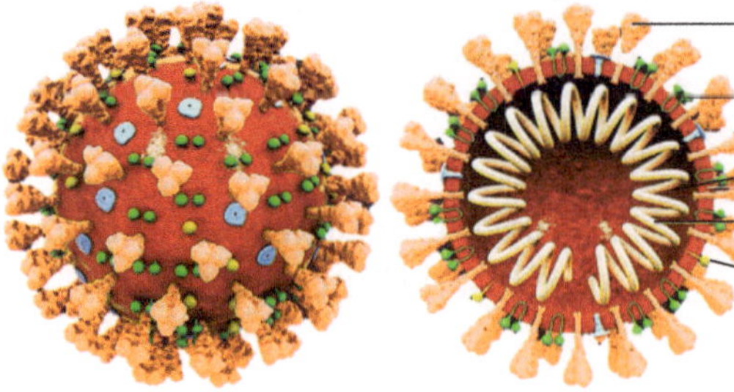

Fig. 55.1 SARS-CoV-2 structure. (Copyright EDISES 2021. Reproduced with permission)

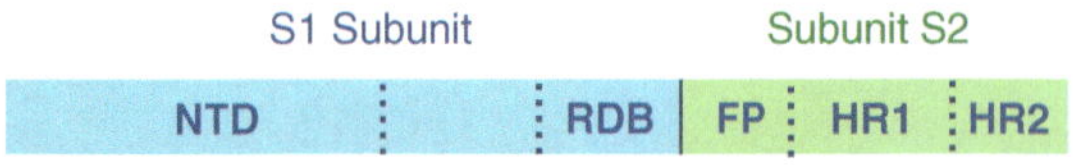

Fig. 55.2 SARS-CoV-2 spike protein sequence. *RBD* receptor binding domain, *NTD* N-terminal domain. (Copyright EDISES 2021. Reproduced with permission)

ticular, ACE2 is highly expressed on the apical side of pulmonary epithelial cells in the alveolar space.

SARS-CoV-2 Variants

Mutations in the SARS-CoV-2 genome can generate variants of concern (VOCs), which improve "viral fitness" through selective or survival advantages. Mutations can occur in any genomic region, but gene codifying for S protein has a high mutational rate. Most mutations are silent; thus, they do not modify the primary amino acid sequence, the function of the translated proteins, or viral infectivity. However, a single or a combination of mutations, can yield variants with selective and survival advantages and, thus, improved viral fitness. These mutational variants can present increased infectivity and/or transmissibility, human ACE-2 receptor binding affinity, viral replication, pathogenicity, and reinfection risk.

To date, the World Health Organization (WHO) has recognized five variants as VOCs: B.1.1.7 from the United Kingdom (501Y. V1, VOC 202012/01, alpha variant), B.1.351 from South Africa (501Y. V2, VOC 202012/02, beta variant), P.1 from Brazil (501Y. V3, VOC 202101/02, gamma variant), B.1.617.2 from India (VOC 202104/02, delta), and B.1.1.529 from multiple countries (omicron variant). Notably, the B.1.617.2 variant was associated with the fast spread of SARS-CoV-2 in several countries.

COVID-19

SARS-CoV-2 can infect humans, resulting in the disease known as COVID-19, which is characterized by an incubation period of 5–6 days, with an interval between 1 and 14 days after exposure.

COVID-19 is a respiratory tract infection characterized by a wide spectrum of clinical manifestations with varying degrees of severity, from asymptomatic infection to pneumonia, which can evolve into acute respiratory distress syndrome (ARDS) and multiorgan failure syndrome (MOF), up to death. The elderly (>65 years) and individuals with comorbidities, such as diabetes, hypertension, and chronic obstructive pulmonary disease, are more susceptible to developing a severe form.

The SARS-CoV-2 is transmitted through direct or indirect person-to-person contact. In most cases, transmission occurs through close contact (less than 1 m) with an infected person via droplets (respiratory secretions). Transmission may occur less frequently through indirect contact with a surface contaminated by the virus.

SARS-CoV-2 has also been detected in nonrespiratory samples, including feces, blood, eye secretions, and semen, but the role of these sites in transmission has not yet been clarified. According to a WHO-China report, transmission by the fecal-oral route does not significantly affect the spread of infection. Detection of SARS-CoV-2 RNA in blood has been reported in some studies. However, the likelihood of transmission by the bloodstream, i.e., through blood collected in test tubes, seems unlikely. Respiratory viruses are generally not transmitted by blood. Finally, there is no evidence that SARS-CoV-2 can be transmitted through contact with sites other than the mucous membranes, such as injured skin.

The precise interval during which an individual infected with SARS-CoV-2 may transmit the infection is uncertain. The potential for infection transmission begins before symptoms develop and peaks early in the course of the disease; after that, the risk of transmission progressively decreases.

The duration of viral RNA shedding varies and may increase with age and disease severity. In one meta-analysis, the median duration of viral RNA detection in respiratory specimens was approximately 18 days after the onset of symptoms; in some individuals, viral RNA was detected in the respiratory tract several months after initial infection. Detection of viral RNA, however, does not necessarily indicate the presence of an infectious virus, and there appears to

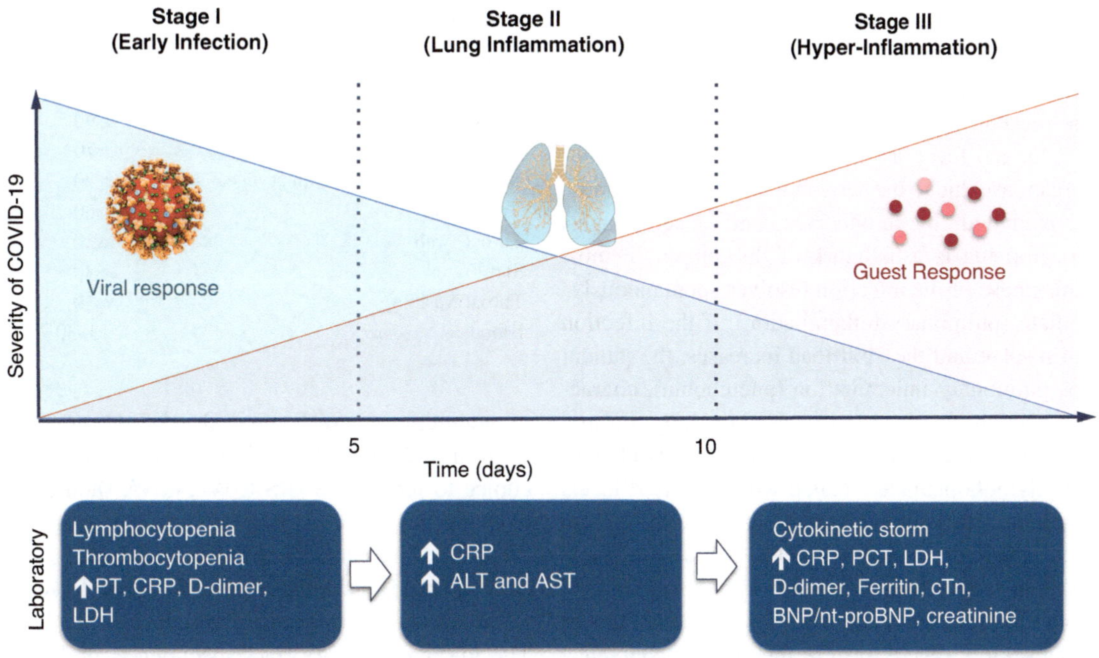

Fig. 55.3 Biochemical-clinical characteristics of COVID-19 infection. (Copyright EDISES 2021. Reproduced with permission)

be a threshold level of viral RNA below which infection is unlikely.

Virus transmission from infected but asymptomatic individuals, as well as from presymptomatic individuals, has been well documented. It has been shown that the levels and permanence of viral RNA in the upper respiratory tract of asymptomatic patients are similar to those of symptomatic patients.

The risk of transmission from an individual with SARS-CoV-2 infection varies depending on the type and duration of exposure, preventive measures, and individual factors such as the amount of virus in respiratory secretions.

The risk of transmission after contact with a COVID-19-infected individual increases according to the proximity and duration of contact and appears higher with prolonged contact indoors. Therefore, most secondary infections are most commonly contracted in the following settings:

- Family;
- Healthcare facilities, including hospitals, when personal protective equipment is not used;
- Environments where people reside or work nearby, such as cruise ships, homeless shelters, detention facilities, and university dormitories.

The virus on contaminated surfaces may be another source of infection if susceptible individuals touch these surfaces and then transfer the virus to the mucous membranes of the mouth, eyes, or nose. However, this type of transmission's frequency and relative importance remain unclear. This may be a potential source of infection, especially in environments with heavy viral contamination, such as in an infected individual's household or healthcare facilities.

To date, it is still unclear how long SARS-CoV-2 can persist on surfaces. However, SARS-CoV-2 appears to behave like other coronaviruses, which can persist for a few hours or up to several days, depending on the type of surface, temperature, and moisture content. Various disinfectants, including ethanol at 62% to 71%, can inactivate SARS-CoV-2 within 1 min. Therefore, disinfection environments in home and healthcare settings are essential. Sunlight has also been shown to inactivate SARSCoV-2 within 15–20 min.

Biochemical-Clinical Course

80–85% of infected patients present a favorable clinical course, being asymptomatic or developing only mild respiratory symptoms; 10–15% of patients, on the other hand, develop a severe or even critical form requiring mechanical ventilation. This probably depends on the presence of several factors, such as demographic characteristics (advanced age and male sex), clinical features (hypertension, diabetes, cardiovascular diseases, chronic respiratory disorders, cancer, and obesity), and genetic factors (polymorphisms in the sequence of the ACE2 gene, which can variably influence the virulence and pathogenicity of SARS-CoV-2 by affecting receptor binding).

Based on severity, the clinical course of infection can be ideally divided into three stages (Fig. 55.3):

1. Initial (early infection): the virus colonizes the lung parenchyma and infects the bronchial epithelium through interaction with the ACE2 receptor. At this stage, most patients present nonspecific symptoms, such as dry cough, cold, and fever, associated with an initial inflammatory response due to the activation of innate immunity, mainly mediated by monocytes and macrophages. Lymphocytopenia is a hallmark of this phase. In most cases, this phase of the infection resolves spontaneously.
2. Intermediate (pulmonary inflammation): if the infection does not resolve and the viral load increases, the patient develops pulmonary inflammation (pneumonia), characterized by cough, fever, and difficulty breathing. From a biochemical-clinical point of view, this phase is characterized by lymphopenia associated with increased levels of transaminases and biomarkers of systemic inflammation, such as C-reactive protein (CRP). At this stage, most patients require hospitalization.
3. Advanced (hyper-inflammation): a small percentage of patients develop a severe form of the disease, characterized by a state of systemic "hyper-inflammation," which can lead to the development of ARDS and MODS up to the death of the patient. At this stage, patients should be admitted to an intensive care unit. Biochemically and clinically, patients have a significant increase in circulating inflammatory biomarkers. In addition, since patients may develop complications in different organs, such as the heart and kidneys, there may be an increase in circulating biomarkers indicative of such complications, such as troponin, b-type natriuretic peptide (BNP), and creatinine.

Diagnosis

The diagnosis of COVID-19 is based on integrating clinical, laboratory, and radiological data. However, since clinical signs and/or symptoms and radiological findings are not specific to COVID-19, SARS-CoV-2 infection must always be confirmed by detecting viral RNA in biological specimens from potentially infected patients. Therefore, the clinical laboratory plays a crucial role in diagnosing COVID-19.

Tests used to diagnose infection are constantly evolving. To date, laboratory diagnosis uses techniques for direct or indirect virus detection. Molecular biology techniques for identifying viral RNA and rapid tests for identifying virus antigens belong to the first group. On the other hand, serological tests indirectly assess the infection by measuring antibody levels produced against SARS-CoV-2.

The success of the diagnostic procedure depends on the appropriate sample collection at the right time of infection.

Table 55.1 Biological matrices for detecting SARS-CoV-2

Biological matrix	Detection rate (%)
Bronchoalveolar lavage fluid	>90
Saliva	≈90
Sputum	≈70
Nasopharyngeal and oropharyngeal swab	≈70
Nasal swab	≈60
Throat swab	≈30
Stool	≈30
Throat washing	≈30
Blood	15–30

According to WHO and the US Centers for Disease Control and Prevention (CDC), the biological material of choice to look for SARS-CoV-2 RNA should be a sample from the respiratory tract. This may be from the upper respiratory tract (nasal swab, oropharyngeal swab, and nasopharyngeal swab) and more rarely, from the lower respiratory tract: endotracheal aspirate or bronchoalveolar lavage (BAL). The lower respiratory tract specimen is preferred due to a higher virus concentration in cases of pneumonia.

Other biological samples that can be tested including blood, stool, urine, saliva, and throat wash. However, the significance of virus identification in these matrices remains undetermined (Table 55.1).

Once correctly and accurately collected, biological specimens (particularly nasopharyngeal and oropharyngeal swabs) should be placed in separate sterile tubes containing 2–3 mL of viral transport medium and should be immediately analyzed or stored at 2–4 °C for less than 4 days or frozen at −80 °C until tested. Samples that do not meet these stringent preanalytical requirements may generate "false negative" results and should be discarded.

According to WHO and CDC, the definitive diagnosis of SARS-CoV-2 infection must be performed using molecular biology techniques, specifically, real-time reverse-transcription polymerase chain reaction (rRT-PCR) on respiratory material from the upper and lower tracts.

Detection by rRT-PCR is, to date, the reference method for diagnosing SARS-CoV-2 infection, but the test result is obtained between 1 and 6 h. Because rapid diagnosis is necessary for some settings, complementary methods have been developed for triage before obtaining a definitive diagnosis.

Rapid tests are used to assess the presence of the virus by detecting viral proteins (antigens) in respiratory samples collected similarly to molecular tests (nasopharyngeal swab). Although these tests significantly reduce turnaround times, the WHO suggests that they should only be used in cases where rRT-PCR is not readily available or shorter turnaround times are required. Since rapid tests have lower diagnostic accuracy than molecular tests, molecular testing must always confirm a positive result.

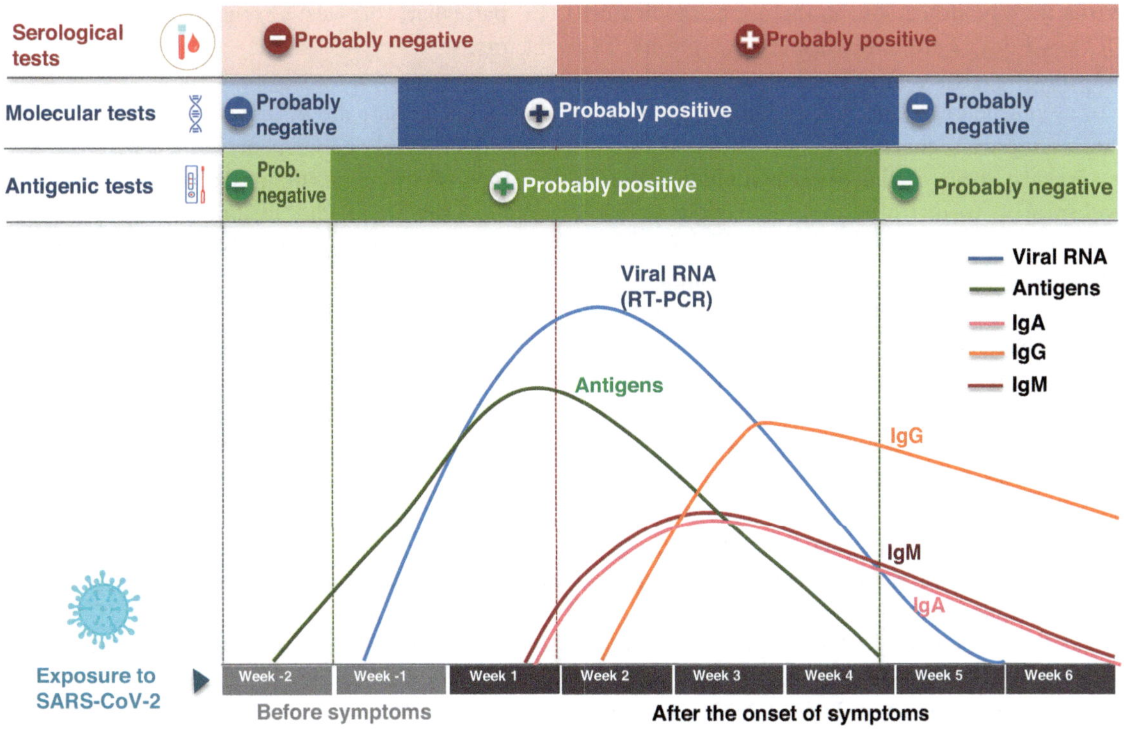

Fig. 55.4 Characteristics of serological and molecular tests for COVID-19 diagnosis. (Copyright EDISES 2021. Reproduced with permission)

Finally, the presence of circulating IgA, IgG, or IgM antibodies against SARS-CoV-2 can be detected to assess whether the infection has occurred.

The characteristics of currently available laboratory tests for the diagnosis of COVID-19 are summarized in Fig. 55.4.

Molecular Tests (NAAT, Nucleic Acid Amplification Test)

Molecular testing for viral RNA is the most reliable laboratory test for COVID-19 diagnosis.

The lower respiratory tract specimen is preferred due to a higher concentration of virus in cases of pneumonia. Lower respiratory tract specimens should be collected in a sterile container, wearing appropriate protective equipment and minimizing the possibility of generating aerosols. In particular, nasopharyngeal and oropharyngeal swabbing is a procedure based on the collection of mucus lining the superficial cells of the mucosa of the nasopharynx or oropharynx using a swab. The nasopharyngeal swab, in the current state of scientific knowledge acquired during the COVID-19 pandemic, represents the gold standard, with the most significant diagnostic sensitivity and specificity (Fig. 55.5). From the biological material obtained by the swab, RNA extraction and purification are performed for subsequent detection of viral RNA using a molecular biology method.

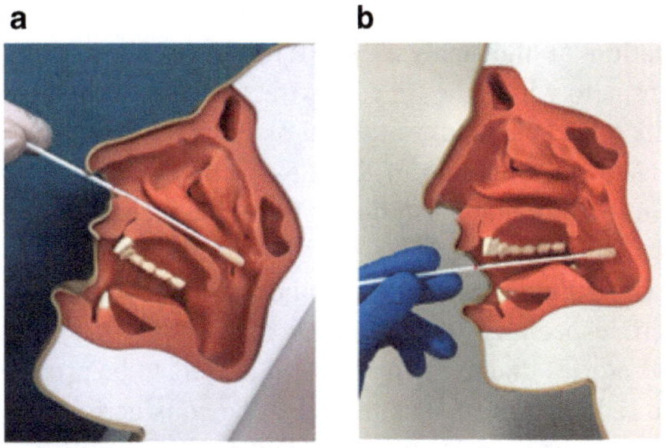

Fig. 55.5 Nasopharyngeal (**a**) and oropharyngeal swabs (**b**). (Copyright EDISES 2021. Reproduced with permission)

Correct nasal and oropharyngeal swabbing is essential to avoid false-negative results. rRT-PCR is the most commonly used method for detecting viral RNA from nasopharyngeal or oropharyngeal swabs. This method allows precise and sensitive identification of one or more virus target genes in the biological sample and real-time measurement of the initial concentration of the target sequence.

In light of the emergence of mutations in the gene encoding S protein, the use of tests based solely on the S gene for detecting SARS-CoV-2 infection by RT-PCR is not recommended. rRT-PCR can detect the pathogen even at low viral load in symptomatic, presymptomatic, or asymptomatic individuals.

The virus may be detectable in the upper respiratory tract (URT) 1–3 days before the onset of symptoms. The concentration of SARS-CoV-2 in the URT is highest at the time of symptom onset and then gradually decreases. Some studies report higher viral loads in patients who develop a severe form of the disease than in patients who develop mild forms; other studies report no such differences.

The presence of viral RNA in the lower respiratory tract (LRT) increases during the second week of illness. In some patients, viral RNA may only be detectable for several days, while in others, it may be detectable for several weeks and even months.

The prolonged presence of viral RNA is not necessarily associated with prolonged infectivity.

rRT-PCR is a technique that relies on recognizing and amplifying a region of the viral genome containing a gene of interest. Several diagnostic kits target different genomic regions, although most recognize genes encoding N, E, S proteins and RNA-dependent RNA polymerase (RdRp). In particular, rRT-PCR by the retrotranscription allows the synthesis of a double-stranded DNA (cDNA) molecule from an RNA template followed by PCR amplification. Technically, this procedure applies a "one-step RT-PCR," in which retrotranscription and PCR amplification are performed consecutively in the same reaction tube. The duration of the analysis in rRT-PCR is about 1 h and 30 min, to which the time of extraction and purification of RNA must be added, for a total of about 4 h. It is also possible to skip the step of RNA extraction and purification from the biological sample by using the technique of heat inactivation (95–98 °C) ("raw"). This allows simplifying, maximizing, and speeding up the first step of the analysis itself, especially in conditions of high pressure on the laboratories due to the increase of buffers to be processed. The risk of loss of sensitivity can be considered minimal and present only at high numbers of PCR amplification cycles.

Some less commonly used molecular assays rely on other techniques, such as loop-mediated isothermal AMPlification (LAMP), recombinase polymerase amplification (RPA), microarrays, clustered regularly interspaced short palindromic repeats (CRISPR) technology, or next-generation sequencing (NGS).

Molecular tests have good diagnostic accuracy. In particular, they are characterized by high specificity. Moreover, although they are characterized by high analytical sensitivity, and can identify even low levels of viral RNA, their clinical performance is variable. While false-positive results are rare, the rate of false-negative results varies from 5% to 40%. In particular, the test's sensitivity depends on several factors:

- Kit used: there are different kits available on the market with different analytical performances.
- Type of specimen: the nasopharyngeal specimen generally has a higher viral load than the oropharyngeal specimen. However, very recently, it has emerged that the viral load is higher in the oropharyngeal specimen for some SARS-CoV-2 variants. Thus, the latest recommendations suggest collecting the sample from both sites.
- Duration of the disease: the viral load peaks in the upper airways 3 days after the onset of symptoms.
- Sample collection: it is operator dependent.

Serological Tests

Serological tests are based on detecting serum antibodies against specific SARSCoV-2 proteins, such as glycoprotein S or protein N. They aim to identify whether the virus has infected an individual and whether an immune response against it has been developed. Therefore, serological tests detect exposure to the virus but are not able to confirm an ongoing infection. Thus, they should not be used for diagnostic but only for epidemiological and health surveillance purposes to assess the prevalence of infection in a given population.

Serological tests identify (qualitative) and/or measure (quantitative) different classes of immunoglobulins directed against SARS-CoV-2, typically IgA, IgM, and IgG.

The classic immune response to viral pathogens generally involves first the production of IgM, often accompanied by IgA, followed by IgG. In the serum of COVID-19 patients, it has been shown that IgA and IgM antibodies are detectable about 3–6 days after the onset of symptoms, while IgGs are detectable after about 10–18 days. In particular, IgA appears early in infected patients and peaks after 18–21 days. Several studies showed that IgG levels decay after 4 months after the peak.

Assay methods for antibodies can be classified into qualitative or quantitative.

Qualitative tests are generally rapid tests used as a point of care tests (POCT) that are based on the immunochromatographic technique and provide, in a short time (within 5–20 min), an indication of the possible presence of circulating anti-SARS-CoV-2 antibodies. According to current guidelines, a positive rapid serological test result should be confirmed by a quantitative serological test. If a positive result is confirmed, subjects should undergo clinical evaluation and viral RNA detection using molecular biology methods on nasopharyngeal swabs.

Quantitative tests give an accurate measure of the circulating levels of the various classes of immunoglobulins and are performed by ELISA (enzyme-linked immunosorbent assay) on semi-automated instruments or by chemiluminescence (CLIA) and electrochemiluminescence (EIA) on automated platforms.

Recently, methods to measure SARS-CoV-2 S-RBD class IgG neutralizing antibodies have been developed. In particular, among the 4 structural proteins of the virus, S and N proteins are the main immunogens. S is an important antigen that activates neutralizing antibodies and is essential in virus binding, fusion, entry, and transmission. The RBD within the S protein interacts directly with receptors in the host organism and is the target of neutralizing antibodies. The latter plays a crucial role because it can prevent the interaction between RBD and the ACE2 receptor. Therefore, IgG SARS-CoV-2 S-RBD antibodies correlate with the protective immune response of subjects infected with SARS-CoV-2 or who received the vaccine.

Antigen Detection Test

Rapid antigenic tests are qualitative, and within a short time (3–20 min), detect the presence of SARS-CoV-2 proteins (antigens) in the biological sample obtained by nasal or oropharyngeal swab.

Currently, different types of antigenic tests are available: first-generation tests, which are based on the solid phase immunochromatographic technique ("card") for the identification of specific viral proteins using polyclonal or monoclonal antibodies; second-generation tests, with immunofluorescent reading; and finally, third-generation tests, with immunofluorescence and microfluidic reading. The first- and second-generation methods are characterized by discrete sensitivity and specificity, which vary significantly depending on the time of sample collection. It is important to emphasize that the results of first- and second-generation rapid antigenic tests must be interpreted according to the epidemiological picture of the population studied. Therefore, rapid antigenic tests will have a high positive predictive value in a high-prevalence setting, such as asymptomatic health care workers with significant exposure. Conversely, in a low-prevalence setting, such as the asymptomatic general population, rapid antigenic tests will have a high negative but low positive predictive value.

On the other hand, the latest generation tests seem to show results superimposable to rRT-PCR assays, primarily when used within the first week of infection. Therefore, based on current knowledge, third-generation tests are a valid alternative to rRT-PCR. If the clinical condition of the patient shows discrepancies with the latest generation test, RT-PCR remains the gold standard for confirming SARS-CoV-2 infection. In patients with symptoms of SARS-CoV-2 infection, rapid antigen testing (preferably third generation) should be performed as soon as possible and, in any case, within 5 days of symptom onset. In case of a negative result, the test should be repeated by rRT-PCR or a second rapid antigenic test 2–4 days later. In subjects without symptoms, rapid antigenic tests should be used when a high positivity rate, e.g., approaching or exceeding 10%, is expected:

- In contact tracing activities, to test asymptomatic contacts with high-risk exposure.
- In community screening activities for public health reasons (e.g., school and workplace). In this situation, the risk of not detecting all cases or false-negative results is balanced by the timeliness of the results and the possibility of periodic testing.
- In health and social care/social health contexts such as closed or semi-closed communities (e.g., prisons, reception centers for migrants), in areas with high community transmission for periodic screening of residents/operators/visitors.

In general, rapid antigenic tests have many advantages over molecular tests, such as rapid response, low cost, and simplicity of execution. Moreover, the positivity of the antigenic test is closely related to infectivity. Unlike viral RNA, the antigenic test becomes negative at the end of the infectious phase, which remains positive for several weeks after clinical recovery. It has been shown that the spread of infection is entirely negligible in asymptomatic subjects with low viral load. In these subjects, molecular biology tests will give a positive result, while antigenic tests will give a negative result.

Antigenic tests, especially third-generation ones, are therefore useful laboratory tools for population screening because a high negative predictive value characterizes them and for diagnosing symptomatic subjects when there is the need to have a rapid response.

Figure 55.6 summarizes the characteristics of laboratory tests for diagnosing SARS-CoV-2 infection.

Choosing the most appropriate diagnostic test according to the clinical context is essential. Table 55.2 shows the indi-

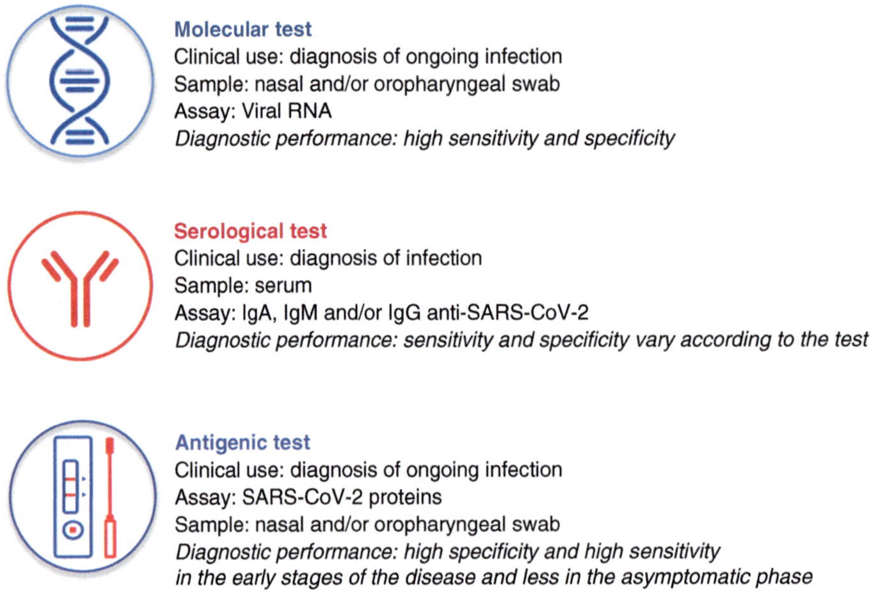

Fig. 55.6 Laboratory tests for diagnosing SARS-CoV-2 infection. (Copyright EDISES 2021. Reproduced with permission)

Table 55.2 Indications for the appropriate use of diagnostic tests based on the clinical context

	Test type	
Clinical context	*First choice*	*Alternative*
Symptomatic suspicion	Molecular test on nasal-oropharyngeal swab	Rapid antigen test on nasal-oropharyngeal or nasal swab + confirmation by molecular test on nasal-oropharyngeal swab for positive cases
Subject in quarantine if symptoms appear (e.g., close contact of confirmed case)		
Pauci-symptomatic suspicion	Rapid antigen test on nasal-oropharyngeal or nasal swab without molecular test confirmation for positive cases	Molecular test on nasal-oropharyngeal swab
Close contact of asymptomatic confirmed case		
Asymptomatic subject from a country at risk		
Close contact of confirmed case who regularly lives or is contact with fragile subjects at risk of complications	Molecular test on nasal-oropharyngeal swab	Rapid antigen test on nasal-oropharyngeal or nasal swab + confirmation by molecular test on nasal-oropharyngeal swab for positive cases
Asymptomatic due to planned hospitalization, entry into large closed communities (e.g., nursing homes and prisons)		
Screening of health care professionals/staff in high-risk settings (e.g., hospitals)		
Community screening (searching for people with ongoing infection in a large group of people) for public health reasons	Rapid antigen test on nasal-oropharyngeal or nasal swab + confirmation by molecular test on nasal-oropharyngeal swab for positive cases	Serological test + confirmation by molecular test on nasal-oropharyngeal swab for positive cases
Subject in isolation for confirmation of healing	Molecular test on nasal-oropharyngeal swab	
Asymptomatic contact in quarantine for the term of 10 days	Molecular test on nasal-oropharyngeal swab	Rapid antigenic on nasal-oropharyngeal or nasal swab
Asymptomatic performing the test on a voluntary basis, examination required for work or travel reasons or for requests not related to clinical or public health needs falling within the previous cases, etc.	Rapid antigen test on nasal-oropharyngeal or nasal swab + confirmation by molecular test on nasal-oropharyngeal for positive cases	Molecular test on nasal-oropharyngeal swab

cations for performing the various tests according to the clinical context.

Monitoring

The laboratory plays a fundamental role not only in the diagnosis but also in the monitoring of the disease. Biochemical, hematological, and immunochemical markers are essential to assess the severity of the disease, select the appropriate treatment, and monitor the response (Fig. 55.7).

Inflammatory Biomarkers

High pro-inflammatory cytokine levels, such as interleukin-6 (IL-6) and tumor necrosis factor-alpha (TNF-α), are associated with increased disease severity and mortality. In patients with severe forms of COVID-19, increased CRP, and erythrocyte sedimentation rate (ESR) are also observed. In patients who develop a bacterial coinfection, there is also an increase in procalcitonin (PCT). Lymphopenia, increased neutrophil/lymphocyte ratio, and platelets/lymphocytes have a negative prognostic value. In addition, increased D-dimer is associated with greater disease severity and a higher risk of developing a broad spectrum of thromboembolic events in COVID-19 patients, including pulmonary micro-thrombosis in situ, deep vein thrombosis, embolism, and even disseminated intravascular coagulation (DIC).

Globally, patients who develop a severe form of COVID-19 develop a significant pro-inflammatory response, termed "cytokine storm," which can evolve, leading to multiorgan damage. The biochemical monitoring of COVID-19 patients includes assessing the inflammatory profile and early recognition of cardiac, renal, and hepatic lesions through specific laboratory tests.

Cardiac Biomarkers

Evidence from the literature suggests significant cardiovascular dysfunction in patients who develop severe forms of COVID-19. Cardiovascular complications, such as acute myocardial infarction, may be due to various mechanisms, such as viral myocarditis, viral myocardial damage, cytokine-mediated myocardial damage, microangiopathy, and exacerbation of coronary artery disease. Cardiac troponin (cTn) and B-type atrial natriuretic peptide (BNP or NT-proBNP) represent important prognostic biomarkers to assess cardiac function in COVID-19 patients.

Liver Biomarkers

Patients with severe forms of COVID-19 may develop hepatic dysfunction, as evidenced by circulating increases in AST, ALT, and γ-GT. Although it is clear that hepatic damage is associated with an increased likelihood of developing a severe form of COVID-19, the pathogenetic mechanism underlying this alteration is not yet completely clear. Several hypotheses have been put forward, including immune-mediated damage due to the severe inflammatory response following infection, direct cytotoxicity due to active viral replication in biliary epithelial cells expressing ACE2, hypoxic hepatitis due to anoxia, and, finally, drug-induced liver damage. Liver function tests should be routinely performed in patients with COVID-19 to assess liver damage within the clinical setting, especially for those receiving antiviral treatment.

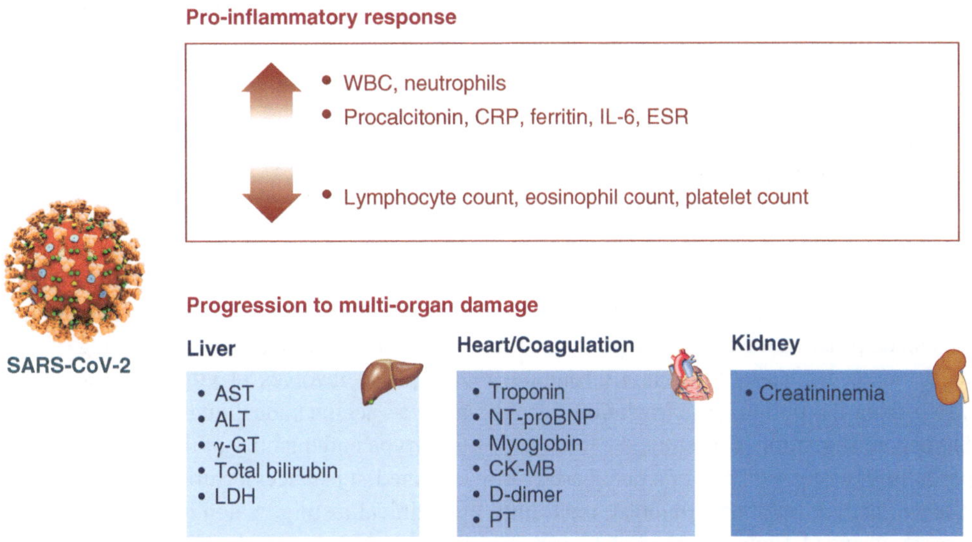

Fig. 55.7 Biochemical and hematological markers to assess the progression and severity of COVID-19. (Copyright EDISES 2021. Reproduced with permission)

Renal Biomarkers

Acute kidney injury (AKI) prevalence appears to be low among patients with COVID-19 (0.5–19%). However, literature data show that increased circulating creatinine levels are associated with an increased likelihood of ICU admission. Therefore, early identification of renal dysfunction is associated with a better prognosis. The pathophysiological mechanism underlying renal damage is not yet fully known but could be due to inflammation of the renal parenchyma due to cytokine storm, direct cytopathic effects on renal tissue cells, or cross-talk between organs, according to which, for example, cardiomyopathy and acute viral myocarditis may contribute to renal vein congestion, hypotension, and renal hypoperfusion.

Therapy

To date, there is no targeted therapy for SARS-CoV-2 infection.

According to the most recent guidelines, treating patients with COVID-19 varies depending on symptoms and the disease severity. In patients who develop severe forms of COVID-19 and require oxygen supplementation, administering corticosteroids, such as dexamethasone, is recommended, which, due to their potent anti-inflammatory effect, are associated with reduced mortality. In addition, clinical trials have demonstrated the efficacy of Remdesivir, the only antiviral drug currently approved by the Food and Drugs Administration (FDA) as a treatment for COVID-19 patients. Remdesivir was originally developed to treat diseases related to the Ebola and Marburg viruses. Remdesivir administration in hospitalized patients requiring supplemental oxygen has been shown to be associated with faster recovery and reduced mortality.

Paracetamol is recommended to lower the body temperature in patients who develop mild forms of the disease.

Low-molecular-weight heparin can be administered to prevent post-surgical venous thromboembolism and venous thromboembolism in the initial phase of the disease and in the advanced phase. In particular, in the initial phase, when pneumonia is present and there is reduced mobility of the patient with allurement, heparin can be used at a prophylactic dose to prevent venous thromboembolism. In hospitalized patients' most advanced phase of the disease, heparin should be administered at therapeutic doses to contain thrombotic phenomena starting from the pulmonary circulation due to disease-related hyper-inflammation. Currently, the therapeutic use of heparin is entering clinical practice based on incomplete evidence and critical uncertainties regarding safety. Therefore, randomized trials are needed to test its clinical efficacy and safety. Finally, in case of bacterial over-infection, it is possible to administer azithromycin alone or associated with other drugs. However, the lack of a solid rationale and the absence of evidence of efficacy in the treatment of COVID-19 do not allow recommending the use of azithromycin outside of possible bacterial over-infection. In addition, the use of hydroxychloroquine is strongly questioned.

Currently, several clinical trials are underway to validate the use of immunomodulatory therapies in patients who have developed moderate and severe forms. In particular, both products derived from the blood of patients who have been infected, such as convalescent plasma and immunoglobulins, and synthetic products, such as neutralizing monoclonal antibodies, are being tested.

Treatment of patients with convalescent plasma is based on the concept that an individual cured of infection will have developed antibodies against the pathogen. Therefore, infusion of plasma from an individual cured of infection for prophylactic purposes or in a patient who has become infected, especially in the early stages of the disease, may prevent the development of the disease or reduce its complications. Convalescent plasma treats some infectious diseases, such as Argentine hemorrhagic fever. Currently, studies are underway to evaluate the efficacy and safety of convalescent plasma administration in COVID-19 patients.

Neutralizing monoclonal antibodies is an attractive therapeutic option for treating infectious diseases. These are antibodies synthesized in the laboratory that recognize and neutralize SARS-CoV-2. The only monoclonal antibody currently cleared by the FDA is Bamlanivimab (LY-CoV555) manufactured by Eli Lilly. Bamlanivimab is a monoclonal antibody directed explicitly against the S protein of SARS-CoV-2, designed to block the virus from entering human cells. Bamlanivimab is authorized to treat mild to moderate cases of COVID-19 in adult and pediatric patients at least 12 years of age who test positive and are at high risk of progression to severe or hospitalization. Bamlanivimab should be administered immediately and within 10 days of symptom onset.

Vaccines

An intense research activity led to the rapid development of vaccines against SARS-CoV-2. Developing a vaccine is a long and elaborate process consisting of several phases. The first phase involves in vitro experimental studies to determine a vaccine's ideal qualitative and quantitative composition (type and quantity of the active component and all other expected substances). Subsequently, the vaccine undergoes preclinical testing, which includes in vitro studies and studies on animal models, through which the mechanism of action (i.e., the ability to induce an immune response), the toxicological profile, and the first evidence of efficacy and safety on a complex living organism are defined. Finally, there is the clinical trial phase, which involves the evaluation of the vaccine's effects on humans, especially in terms of efficacy and safety. This can be further subdivided into 4 phases: the first three precede the marketing authorization, and the fourth is conducted when the vaccine is already available on the market. In particular, in phases I, II, and III studies, the population treated with the vaccine progressively increases. Phase 4 studies, on the other hand, aim to verify the efficacy and safety of the vaccine in its actual conditions of use, assess its use in particular subgroups of populations and pathological conditions, and the cost–benefit ratio with respect to the disease and/or other vaccines.

There are several technologies for developing a vaccine:

- Whole viruses: the attenuated or inactivated virus is used directly. Examples of currently used vaccines developed using this technique are measles, yellow fever, and polio vaccines.
- Recombinant viral proteins: proteins found on the surface of the virus, or fragments thereof, or "virus-like particles" (VLP), i.e., the external envelope of the virus emptied of its genetic content, are used. An example is the vaccine against hepatitis B and the human papillomavirus.
- Recombinant viral vectors: it is used a harmless virus genetically engineered to carry the virus's proteins against which immunity is to be developed. An example is the Ebola vaccine.
- Nucleic acids (DNA or RNA): the genetic information of a virus protein, such as the S protein of SARS-CoV-2, is used.

All of the vaccines currently under study are designed to induce a response that blocks the S protein and, thus, prevents SARS-CoV-2 from infecting cells. In addition, vaccination also activates T cells that prepare the immune system to respond to further exposure to the virus. The European Medicines Agency and AIFA have licensed two mRNA vaccines, Pfizer mRNABNT162b2 (Comirnaty) and COVID-19 Vaccine Moderna mRNA-1273, and one viral vector vaccine, Vaccine AstraZeneca.

In the case of mRNA vaccines, genetic information is introduced into the body and then used by the cell to make copies of the protein S. If, at a later time, the vaccinated person comes into contact with SARS-CoV-2 again, the immune system will recognize the virus and be ready to "fight" it. The mRNA from the vaccine does not remain in the body but is degraded shortly after vaccination.

On the other hand, the viral vector vaccine is based on using a virus (generally an adenovirus incompetent to replicate) to carry the genetic sequence coding for the protein S inside the cell. The immune system is activated against the protein and produces antibodies that, should the subject come into contact with the virus, will protect from infection.

In addition, more than 200 candidates are currently in development.

It is possible to assess whether a vaccinated person has developed anti-SARS-CoV-2 antibodies by serological testing, which should be performed no earlier than 21 days after the last administered dose of vaccine.

Recommended Readings

Beigel JH, Tomashek K, Dodd L, et al. (2020) Remdesivir for the treatment of COVID-19—final report. N Engl J Med. Published online ahead of print. Available at: https://www.nejm.org/doi/full/10.1056/NEJMoa2007764

Ciaccio M, Agnello L (2020) Biochemical biomarkers alterations in Coronavirus Disease 2019 (COVID-19). Diagnosis (Berl). Jun 24:/j/dx.ahead-of-print/dx-2020-0057/dx-2020-0057.xml. https://doi.org/10.1515/dx-2020-0057

Curiale S (2020) Coronavirus: quello che c'è da sapere – 23 settembre 2020. Istituto Nazionale Malattie infettive "Lazzaro Spallanzani" IRCCS – Roma – © INMI 2020

Gambino CM, Lo Sasso B, Colomba C, Giglio RV, Agnello L, Bivona G, Ciaccio M (2020) Comparison of a rapid immunochromatographic test with a chemiluminescence immunoassay for detection of anti-SARS-CoV-2 IgM and IgG. Biochem Med (Zagreb). 30(3):030901. https://doi.org/10.11613/BM.2020.030901

Giri B, Pandey S, Shrestha R, Pokharel K, Ligler FS, Neupane BB (2020) Review of analytical performance of COVID-19 detection methods. Anal Bioanal Chem. 18:1–14. https://doi.org/10.1007/s00216-020-02889-x

Kampf G, Todt D, Pfaender S, Steinmann E (2020) Persistence of coronaviruses on inanimate surfaces and its inactivation with biocidal agents, Journal of Hospital Infection

Lippi G, Mattiuzzi C, Bovo C, Plebani M (2020) Current laboratory diagnostics of coronavirus disease 2019 (COVID-19). Acta Biomed. 91(2):137–145. https://doi.org/10.23750/abm.v91i2.9548

Nota tecnica ad interim. Test di laboratorio per SARS-CoV-2 e loro uso in sanità pubblica. Aggiornata al 23 ottobre 2020. Roma: Ministero della Salute, Istituto Superiore di Sanità; 2020.

Prabhakar PK, Lakhanpal J (2020) Recent advances in the nucleic acid-based diagnostic tool for coronavirus. Mol Biol Rep. 6:1–9. https://doi.org/10.1007/s11033-020-05889-3

Premraj A, Aleyas AG, Nautiyal B, Rasool TJ (2020) Nucleic Acid and Immunological Diagnostics for SARS-CoV-2: Processes, Platforms and Pitfalls. Diagnostics (Basel). 10(11):E866. https://doi.org/10.3390/diagnostics10110866.

RECOVERY Collaborative Group, Horby P, Lim WS, et al. (2020) Dexamethasone in hospitalized patients with COVID-19–preliminary report. N Engl J Med. Available at: https://www.ncbi.nlm.nih.gov/pubmed/32678530

Shi J, Han D, Zhang R, Li J, Zhang R (2020) Molecular and Serological Assays for SARS-CoV-2: Insights from Genome and Clinical Characteristics. Clin Chem. 66(8):1030–1046. https://doi.org/10.1093/clinchem/hvaa122

Vandenberg O, Martiny D, Rochas O, van Belkum A, Kozlakidis Z (2020) Considerations for diagnostic COVID-19 tests. Nat Rev Microbiol. 14:1–13. https://doi.org/10.1038/s41579-020-00461-z

www.CovidReference.com

From Aging to Healthy Longevity: Innovation in Medicine Including Predictive Medicine and Laboratory Medicine Approaches

56

Francesco Salvatore

Introduction

In the field of laboratory medicine, it is and will be increasingly important to find overall indicators of the individual health status. This short chapter tries to bring to the reader's attention the problem of aging and longevity of each individual, which are remarkably different from individual to individual, due to not only sex and ethnicity but also genetic (predictive medicine), as well as the environment and lifestyles to which each individual is subjected, including workplaces.

Therefore, it is necessary to distinguish personalized or individualized medicine, which concerns each subject, from other terminologies that can produce confusion and concern aspects of single diseases or clinical alterations or also relational attitudes to which obviously great attention must be also paid. Laboratory medicine is fully instrumental so that, through the data obtained by its methodologies and the best criteria to collect them, and together with the great contribution of in vivo imaging disciplines, it is possible to typify and monitor over time the health status of each individual, even creating an overall "score," also numerical, corresponding to it. Therefore, the final goal of laboratory medicine in due time will be to consistently contribute to the evaluation of the overall individual health status throughout the life. As a matter of fact, in the case of almost all the different chronic-degenerative diseases, the use of reference values and/or intraindividual signs and symptoms (i.e., belonging to the same individual) compared to those currently in use, which are interindividual, will allow to assess with greater precision the earliest moments of disease, so that they can be therapeutically addressed with the best probability of success. Indeed, suppose monitoring is started early, at the age of 18–20 years at the latest. In this case, it will be possible to detect the first signs or symptoms of the disease (early secondary prevention), allowing an early diagnosis, which represents the priority moment for any more effective therapy.

The Homeostatic Balance of the Organism: Genetic and Environmental Factors

The chronology of human life, as in all living beings, is marked by the calendar and the clock, but in reality, our metabolism, together with our genetic imprinting, defines the length and quality of life. These characteristics are continually modified during each individual's life by numerous mechanisms that have a favorable or unfavorable impact, thus modifying our health for the better or for the worse and, consequently, also the length and the best quality of our lives. The unfavorable mechanisms are, first of all, diseases, from the more strictly genetic ones to those acquired mainly from many factors present in the environment and that negatively affect our organism (from cosmic rays to other types of radioactivity and/or radiation that affect our cells, from all the other environmental noxae from harmful substances present in our food and/or in the air we breathe, as well as from mechanical, electrical, chemical accidents, and so on). The favorable mechanisms are, instead, those that can contrast all the above mentioned negativities, especially those continuously deriving from the progress of biology, especially that of man, as well as of medicine that, in its positive aspects both medical and surgical, continuously leads to the improvement of the state of health. Moreover, each one of us is capable of improving or worsening our health through our lifestyle (smoking, excessive alcohol, drug use, sedentary habits, exposure to environmental risks, excess food and harmful substances in the diet and in the air that we breath) (Fig. 56.1).

F. Salvatore (✉)
Department of Molecular Medicine and Medical Biotechnology, University of Naples "Federico II", and "Ceinge Biotecnologie Avanzate Franco Salvatore" scarl, Naples, Italy
e-mail: salvator@unina.it

M. Ciaccio (ed.), *Clinical and Laboratory Medicine Textbook*, https://doi.org/10.1007/978-3-031-24958-7_56

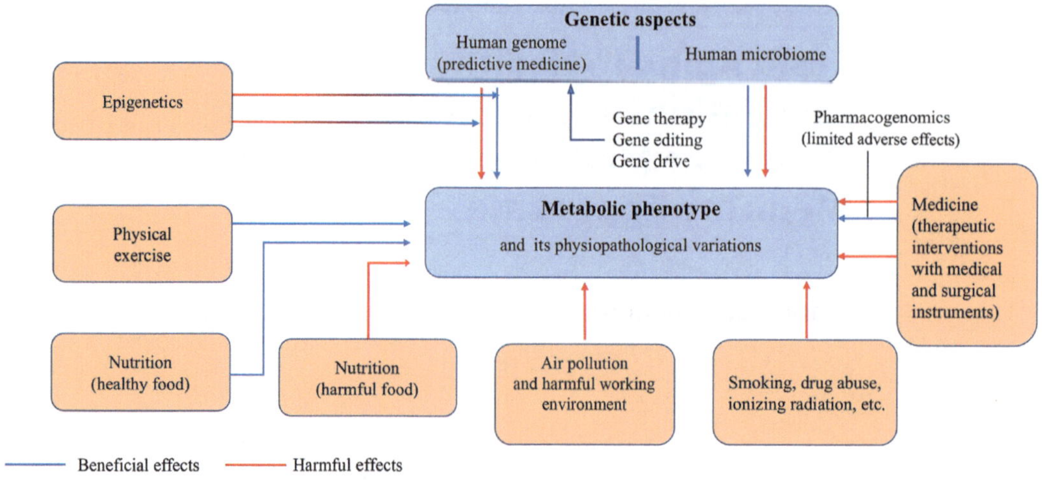

Fig. 56.1 Predisposing genetic factors (predictive medicine) and environmental factors (exposome) that modulate the phenomena of each individual in well-being status or/and in this diseased one along the course of life. Red arrows represent the harmful effects on human health by the exposome and the green arrows the beneficial effects. (Copyright EDISES 2021. Reproduced with permission)

Healthy Aging or Longevity

The balance between the processes described above leads inexorably toward aging, which is terminologically opposed to longevity, i.e., the length of the existence as longest it can be.

In reality, the efforts of medicine have succeeded in extending the life of the populations of the planet Earth, albeit with a remarkably variable population age, from over 80 years on average to only 45–50 years in the populations of some Asian islands. This obviously depends on the genetics of the population and above all on the environment and lifestyles, as well as on the presence or absence of medical sciences support devices (Fig. 56.2). Another deduction derives from the scientific knowledge of medicine and is the knowledge of the attitudes and the practice of the prevention of diseases and clinical alterations, which afflict or can afflict man, modifying the balance toward "healthy aging," today more appropriately called "healthy and active longevity." We have, thus, passed from the oxymoron healthy aging to the endiad, healthy and active longevity, replacing the pessimistic diction (aging) with the optimistic one (longevity). More and more, instead, it has been understood that prevention in medicine (both the primary one realized with the positive modifications of the noxious environmental noxae, and, above all, the secondary one aimed at the earliest possible diagnosis of a disease), even if unfortunately little practiced, is the most significant part of the medical work: it is here that the future also of laboratory medicine must turn to bring its most effective contribution to healthy and active longevity.

Biomedicine is more and more determining a series of benefits to the health of each individual and, therefore, the term aging should be abolished, not because it is not true that individual ages, in the ordinary sense of the word, but because in real life the diseases are producing aging and not vice versa. Indeed, unfortunately, all the causes that produce the deterioration of the health of the individual are actually diseases, and there is no deterioration of the organism linked to intrinsic causes independent from diseases, as it is still defined in other books, texts, and dictionaries as "intrinsic aging." Indeed, aging is a multifactorial disease in which many different morbid alterations converge toward the "fragility" of the individual, determining "aging." It is precisely this simple inversion of terminological conceptions that then modifies cause and effect (the diseases are the cause, the aging is the effect, and not the contrary) that is very important to underline, as it can lead us to the concept that aging is curable because the causes of aging, i.e., diseases, are curable. Certainly, this will not lead to immortality, but to the knowledge that the life is increasingly curable in its tendency toward positive health, that is to the positive balance in favor of good health as opposed to the negative balance tending toward aging. Prevention can be more effective and more personalized, addressing itself in its various aspects to individuals who will fall ill.

Therefore, a second point of inversion of tendency occurs, namely that of tending to couple the study of populations or groups of ethnic communities, etc., with that of personalized

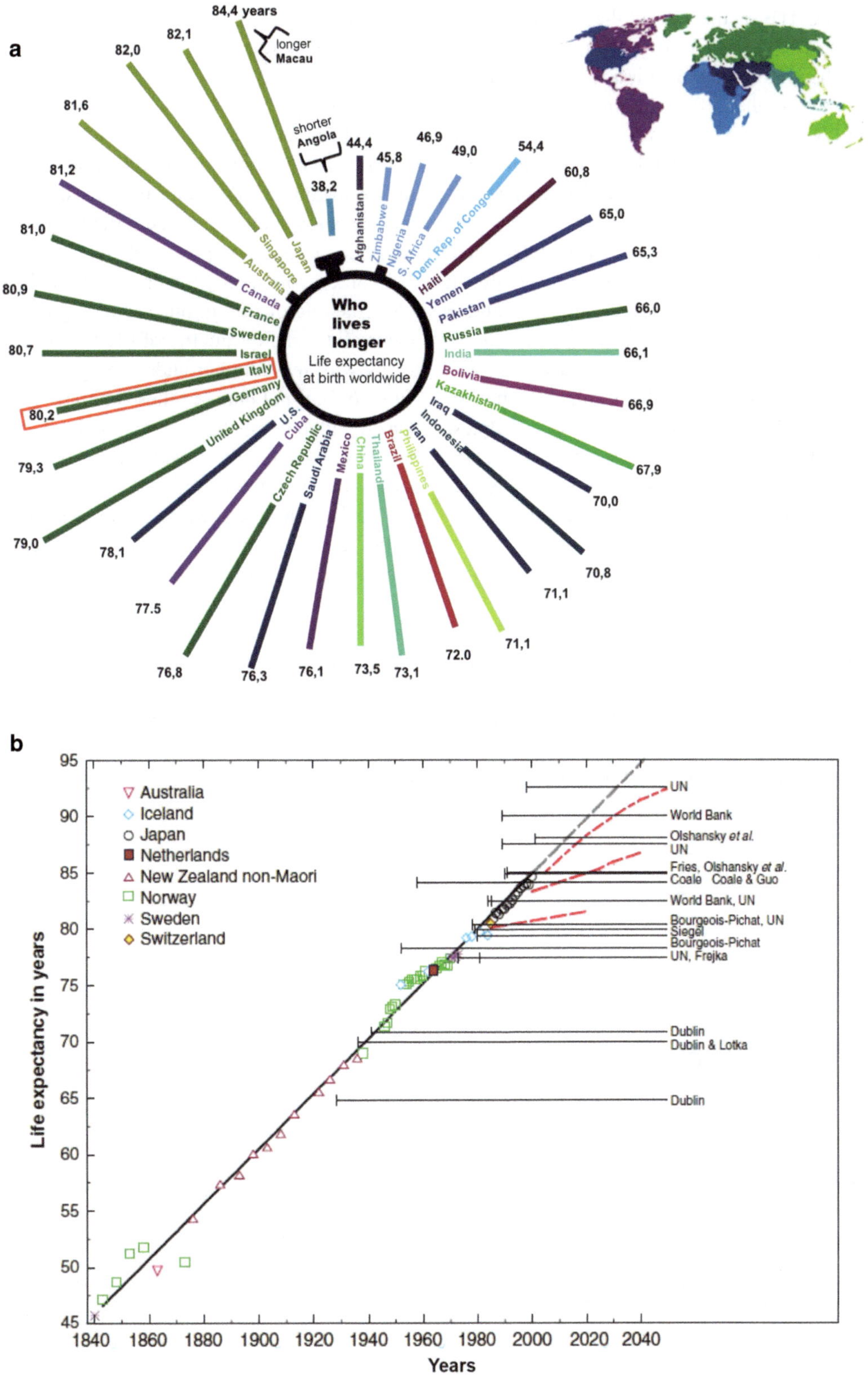

Fig. 56.2 Life expectancy of populations of various countries and areas of the Earth (Panel A): the Italian population is outlined in red; life expectancy over the past two centuries in various geographic areas. (Copyright EDISES 2021. Reproduced with permission)

aspects and measures (i.e., at the level of the individual), and therefore to the causes or first signs or symptoms of the various diseases that begin to afflict each individual in order to better mitigate or cancel their effects (aging), especially for chronic-degenerative diseases, to which this chapter essentially wants to refer. That is to say that personalized medicine should replace the population interindividual reference values, as we usually use in medicine.

Aging and so old age are ideological terms, today only measurable by the chronological time of the age of life. In other words, with the calendar, but this is certainly not valid to measure the individual state of health. Indeed, there are no precise chronological limits for separating the ages and seasons of life from the point of view of the health of the human being, but, when used, they represent approximate and almost fanciful limits, which do not correspond in most cases to states of physical and/or mental health, except in a broad approximation, which is scarcely useful in the medical practice.

Aging as a Multifactorial Disease

To date, neither a single marker nor a set of associated markers is known to constitute a "score" that can effectively indicate the state of health that overall determines how "old" a single person is. Today, the doctor immediately asks the patient about his/her age, for brevity, custom, and simplification.

The process of aging can be considered as a multifactorial disease and, therefore, it is necessary to look more and more at and analyze the individual in its entirety, which is not trivially the age (perhaps a measurer of the state of health in the pre-medicine era) but instead the sum of various functions, that is those of various organs and tissues, which are increasingly well known and which will increasingly be known. Their analysis allows not only to cure the pathological manifestations but, above all, to prevent them.

In other words, only the prevention will lead not only to a lengthening of life, in other words, greater longevity, but above all to a better quality of life for the individual, that is to say with less number and severity of morbid alterations (Fig. 56.3), if not many of these alterations have never been considered too much even by medical sciences because they have been fatally considered linked to aging, almost as aging, for some fatalistic ideas, would have been impossibile to be mitigated. Since almost no ideological belief seems to be in contradiction with medicine (vaccination, transplants, etc.), the finalists concept of aging (which certainly exists and is almost impossible to eliminate completely) should be abandoned in the sense of something completely inevitable, considering, instead, the measures to prevent the exposome (i.e., the deterioration of the organism due to environmental noxae) as the most important and topical, today, at the population level to slow down its progress.

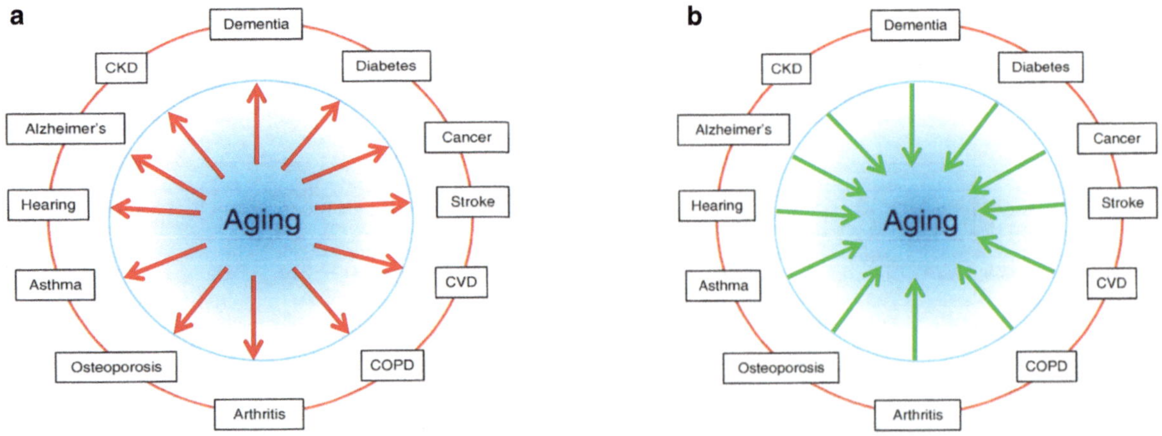

Fig. 56.3 (**a**) Aging as the main risk factor for chronic-degenerative diseases. (**b**) Series of numerous chronic-degenerative diseases as the main risk factors for aging. (Copyright EDISES 2021. Reproduced with permission)

Personalized Medicine and Laboratory Medicine

At the individual level, where so much the medical science is addressed today, it is necessary to change the register. That is, vaccination campaigns, cancer prevention campaigns, etc., are not enough, but it is also necessary to prevent all possible morbid alterations at the level of each individual, and this is the most important part of the personalized medicine of the near future. Prevention, which today is beginning to be known, must start from the first years of life, perhaps from birth, that is, from the knowledge of the genetic signs of real disease and/or predisposition derived from the knowledge of the human genome (predictive medicine). This will allow to look more accurately at the various functions and their alterations in the human organism during the various ages of life. Prevention will thus be able to limit the deterioration of many vital functions, such as skin withering, sarcopenia, bone fragility, diminished sight, hearing, smell, taste, tactile sensations, sexual activity, and many other morbid alterations that slowly and inexorably impair the organism. Therefore, it is on very early secondary prevention that one can and must seize the essential moments of medical intervention; one cannot, in other words, measure the global individual state of health only with the time but, identifying and reporting at a personal level, the areas and systems of the individual, with a series of multiparameter markers which may be signals of any diseases. It is from this aspect of personalized medicine that it will perhaps also be possible to implement a state of periodic personalized monitoring, which will make it possible to evaluate and measure a single multiparameter "score", which would indicate a global, but less approximate, state of health of each individual and, moreover, above all, through markers that can be attributed to organs or tissues or functions, to identify one or more diseases that increasingly require therapies and controls. In other words, through a statistically verifiable deviation from each intra individual value, it will be possible to precisely identify, measure, and treat single or multiple morbid alterations and, therefore, slow down their progression, thus improving the quality of life.

The Types of Medicine Referred to as Multiple P

Over the past few years, the identification of aspects of medicine that are more and more topical and have a catchy name has become more and more evident, essentially to use slogans concerning its objectives in a more specific way. Therefore, the specific adjective starting with the letter P has been used in addition to the already existing and long-standing one: preventive medicine (primary, secondary, tertiary). These adjectives, which I resume here, have become different and sometimes overlapping. Predictive medicine essentially refers to the one that everyone carries since birth in the DNA and that sometimes gives rise to well-defined diseases (the monogenic ones), but especially also only to predisposition with early and/or late expression mainly for chronic-degenerative diseases, including cancer. Precision medicine refers to the subgroup of disease types in which it is easier to recognize a common etiology and/or pathogenesis and, therefore, the possibility to have more precise and targeted therapies to try to treat, even early, any negative effects. Personalized medicine is the one most discussed in this chapter, concerns the single individual (therefore also called individualized, especially in the German literature), and is often confused and identified with precision medicine (which is instead addressed to the morbid alteration concerning more people, obviously having the same etiopathogenesis). Finally, participative medicine indicate, in short words, the part of medical behavior that is mainly addressed to the physician–patient relationship and that should be increasingly present and enlivened by the loyalty of the physician and by the attitude of participation (compliance) of the patient under treatment.

There is, therefore, an urgent need to modify the paradigm of the limits used up to now and to accelerate the development of diversified measures (medical preventive and then pharmacological and therapeutic) to achieve the longevity "dividend." In other and few words, it is very important to reverse the trend of treating diseases when they are already symptomatic and exploded, to when they are not yet present as such, but there are small typical signs that indicate the disease is starting to occur. This seems to be a feasible fact if the study and especially the prevention activity starts very early in the life of the individual, and there is more and more biological and medical research of markers, especially in laboratory diagnostics (including predictive medicine at the level of genes) and in imaging, which can, in systemic and mathematical models, even integrated, to unravel a warning of disease onset as early as possible.

Concluding Remarks and New Planning in Laboratory Medicine

Any method of primary or secondary prevention leading to a delay in aging is advantageous from the demographic-economic point of view. Many other discussions are necessary on the length and validity of good health up to the moment of the end of life, as well as in relation to the mechanisms to accompany any individual to its final outcome in the appropriate way, in tune with the will and belief of each individual person. It is precisely on this matter that the problems of this century and the next in the history of man will have to be debated, because the length of life, the longer well-being of the same, the problems of dualism between the population, today called old, and that of the young who have to replace the oldest, the demographic increase of the same world population (apart from the occupation of other planets) and the limitation of resources, together with the finding of new resources, are very difficult questions to study. It is also very difficult to think that the world population in all its social classes, ethnic groups, and even ideological differences, will stop thinking about wanting to live longer and in the best possible conditions. These problems are all of great importance and certainly more than topical, but it will be necessary to start thinking about them and find useful mediations to face them all together.

On the demographic aspect, apart from the life expectancy and the length of life of the various populations, even if subdivided between male and female, another important aspect, which has been evaluated and re-evaluated also very recently, is the incidence of mortality during the years spent by the individuals (Fig. 56.4). This parameter has always been and is on its increase along the life, and, on a semilogarithmic scale, it is represented as a straight line up to the age of about 80 years, after which a flattening trend is observed, but after 105 years, the recent research even indicates the reaching of a plateau (Fig. 56.4). The meaning of these data, though accurately obtained, lies at the intersection of different Schools of thinking and philosophies, but it certainly indicates that there is mortality that decreases its progress with increasing age.

Apart from the need to extend these difficult investigations to many other populations with very strict criteria and accurately recorded and validated data, also considering the difficulties of the divergence of opinions at international level among the various schools of thought, it is important to highlight that it will be very useful to try to understand at molecular and cellular level the profound reasons for the almost stabilized survival, allowing to understand many mechanisms and, therefore, provide a great contribution to the study of existence in good health.

Laboratory medicine has an important role in monitoring the state of health and, therefore, in the aspects of preventive medicine related to the individual. Indeed, an extensive project, in collaboration with a scholar expert in methodology and statistical culture, Matteo Vidali from Pavia, and other colleagues both in the field of laboratory as well as in the field of Imaging Diagnostics and Health Economics, will be promoted to participate in the verification of effectiveness from the diagnostic point of view. This will have to take place following a numerous group of biochemical, functional, and imaging parameters that can converge in deviations from the normal data related to the single subject recorded since the age of full maturity (18–20 years), also and above all in order to diagnose at the first beginning the incipient and progressively present diseases in every single individual, for the purpose of the earliest diagnosis.

The future is certainly in the hands of the physicians to find the initial signs of disease, through the most accurate and early diagnostic systems, but it is also in the culture of awareness of the population and of each individual and on the physician/patient relationship that should be as more participatory is possible.

In order to recapitulate in a synoptic way the concepts presented in this chapter, they are reported here below, in a summarizing as well as innovative form, also in order to push for reflection by colleagues and students who want to

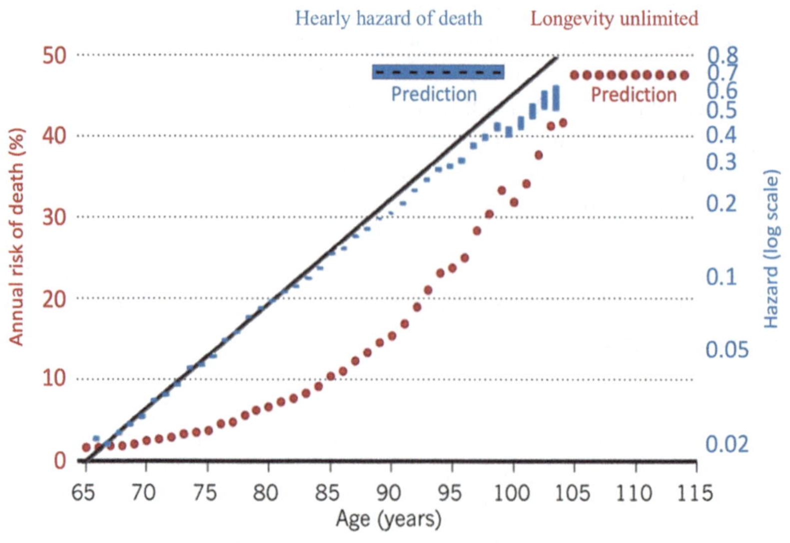

Fig. 56.4 The risk of death increases with age up to about 80 years, then the progression decreases up to 50% at around 105 years of age. (Copyright EDISES 2021. Reproduced with permission)

identify and transmit to the author of the chapter their reflections and criticisms.

1. The paradigm of aging as a cause of frailty, calculated solely on chronological age progression, must be abandoned. Instead, we must begin to use an index to indicate our state of health during the inexorable progression of chronological age in our lives.
2. The word "aging" is pessimistic. Indeed, it conjures up a sad and irreversible entity, while our physical and/or mental state (if the two can ever be divided in their terminology) can still be healthy and happy as our lives progress.
3. The passage of time per se is not the cause of increased frailty and/or status of a sick person (except, in part, for the mutations that occur with each cell division). The term "longevity" is much more optimistic, and we should constantly strive to maintain and enhance "active and healthy longevity."
4. It is therefore not correct that aging, taken simply by the individual age, is an important risk factor for chronic diseases. On the contrary, it is quite the opposite: chronic diseases are for the most part major risk factors for aging. We must not confuse or reverse "cause" with "effect" in the biomedical sciences.
5. Aging does exist, but it certainly depends, in many cases, on causes which are already known and which lead to illnesses. Therefore, in order to stop aging or delay its harmful consequences or counteract frailty, we must cure, or better still, strive to prevent the onset of disease.
6. Cellular and molecular alterations, hence diseases, adversely affect our phenotype:
   - We can counteract the negative effect of the exposome (as a set of environmental causes that negatively impact on our health) by biomedical sciences and also with a greater and more extensive use of secondary prevention through the analysis of early biomarkers and by modifying negative lifestyles (i.e., poor diet, exposure to air pollution and other toxic substances, insufficient physical exercise, etc.).
   - In the near future, it may also be possible to change the negative genetic predisposition (to the disease) using gene therapy, gene editing or 'gene drive' technologies.
7. Medical sciences and biomedicine, in particular, should focus much more on the triad: predictive medicine at the genomic level, primary prevention by eliminating external/environmental causes that strongly impact our organism (at the physical and mental level) and, finally, secondary prevention which means finding more and better signs or symptoms of diseases as early as possible. Certainly, we will have to die, but biomedical research should focus more and more on this triad to make life longer, but above all healthier and more active.
8. To enable the overall improvement of individuals' lives, we should seek to detect, at the level of each individual, early signs of disease using a range of markers in the fields of laboratory medicine, imaging, physical and instrumental signs, cognitive and neurosensory testing, etc. ("individualized or personalized medicine"), together with physician consultation.
9. We should, therefore, identify for each individual – starting at a young age (18–20 years) – parameters that should be periodically monitored according to medical advice, during an individual's lifetime to detect statistically significant or trending deviations that may be harbingers of diseases, especially chronic-degenerative ones.
10. More recently, the author of this chapter put forward a novel concept, that of the **biomedical person** which is represented by each human being from the birth to death. In any period or moment of his/her life each human being, in its entirety, is different, because his/her is changing, though slowly, also with respect to the alteration or diseases he/she bears on his/her body or mind. Thus, it is fundamental to switch since early life to intra-individual reference values to follow better initial signs and/or symptoms of diseases and, then, make better preventive measures to implement very precocious secondary preventions to each individual. This extensive procedure to the totality of human populations will produce finally a better well-being life and also a longer life span.

Recommended Readings

Barbi E, Lagona F, Marsili M et al (2018) The plateau of human mortality: demography of longevity pioneers. Science 360:1459–1461

Bodei R (2015) Limite. Il Mulino

Christensen K, Thinggaard M, Oksuzyan A, Steenstrup T, Andersen-RK, Jeune B, McGue M, Vaupel JW (2013) Physical and cognitive functioning of people older than 90 years: a comparison of two Danish cohorts born 10 years apart. Lancet 382:1507–1513

Dolgin E (2018) There's no limit to longevity, says study that revives human lifespan debate. Nature 559:14–15

Fontana L (2009) Modulating human aging and age-associated diseases. Biochim Biophys Acta (BBA) 1790:1133–1138

Fontana L, Partridge L (2015) Promoting health and longevity through diet: from model organisms to humans. Cell 161:106–118

Fontana L, Partridge L, Longo VD (2010) Extending healthy life-span from yeast to humans. Science 328:321–326

Fontana L, Kennedy BK, Longo VD, Seals D, Melov S (2014) Medical research: treat ageing. Nature 511(7510):405–407

Fried LP (2016) Interventions for human frailty: physical activity as a model. In: Olshansky SJ, Martin GM, Kirkland JL (eds) Aging: the longevity dividend. Cold Spring Harbor Perspectives in Medicine, pp 135–148

Luzzatto L, Pandolfi PP (2015) Causality and chance in the development of cancer. N Engl J Med 373:84–88
Olshansky SJ (2016a) Articulating the case for the longevity dividend. In: Olshansky SJ, Martin GM, Kirkland JL (eds) Aging: the longevity dividend. Cold Spring Harbor Perspectives in Medicine, pp 191–196
Olshansky SJ (2016b) Has the rate of human aging already been modified? In: Olshansky SJ, Martin GM, Kirkland JL (eds) Aging: the longevity dividend. Cold Spring Harbor Perspectives in Medicine, pp 221–237
Olshansky SJ, Martin GM, Kirkland JL (2016) Aging: the longevity dividend. Cold Spring Harbor Perspectives in Medicine
Omodei D, Fontana L (2011) Calorie restriction and prevention of age-associated chronic disease. FEBS Lett 585:1537–1542
Omodei D, Licastro D, Salvatore F, Crosby SD, Fontana L (2013) Serum from humans on long-term calorie restriction enhances stress resistance in cell culture. Aging 5(8):599–606
Omodei D, Pucino V, Labruna G, Procaccini C, Galgani M, Perna F, Pirozzi D, De Caprio C, Marone G, Fontana L, Contaldo F, Pasanisi F, Matarese G, Sacchetti L (2015) Immune-metabolic profiling of anorexic patients reveals an anti-oxidant and anti-inflammatory phenotype. Metabolism 64(3):396–405
Rattan SIS (2015) Biological basis of ageing, age-related diseases and interventions. Curr Pharmacogenomics Personal Med 13(2):90–92
Salvatore F (2009) Ultima lezione cattedratica – Healthy aging: ossimoro o endiade? Università di Napoli Federico II – Facoltà di Medicina e Chirurgia, Napoli
Salvatore F (2010) Healthy Aging: an Oxymoron or a Hendiadys. Rendiconti della Accademia Nazionale delle Scienze Detta dei XL – Memorie di Scienze Fisiche e Naturali. Serie V XXXIV(parte II):31–45
Salvatore F (2016) Longevità in buona e attiva salute. Rendiconti e Atti dell'Accademia di Scienze Mediche e Chirurgiche CLXIX: 59–74
Sierra F (2016) The emergence of geroscience as an interdisciplinary approach to the enhancement of health span and life span. In: Olshansky SJ, Martin GM, Kirkland JL (eds) Aging: the longevity dividend. Cold Spring Harbor Perspectives in Medicine, pp 1–13
Vaupel JW (2010) Biodemography of human ageing. Nature 464:536–542
Vaupel J, Lundstrom H (1994) Longer life expectancy? Evidence form Sweden of reductions in mortality rates at advanced ages. In: Wise DA (ed) Studies in the economics of aging. University of Chicago Press, Chicago, pp 79–102
Wild CP (2012) The exposome: from concept to utility. Int J Epidemiol 41(1):24–32

Biobanks of Biomedical Interest

57

Marialuisa Lavitrano, Barbara Parodi,
Maria Grazia Daidone, Mattia Barbareschi, Aldo Scarpa,
Matteo Macilotti, Sara Casati, and Rita Lawlor

Introduction

The progress of genomic and biomedical research and the availability of innovative technologies allow today to differentiate multifactorial diseases into precise diagnostic entities, an essential condition for the development of precision medicine, and to pursue the goal of effective treatments, reducing side effects, optimizing the design of clinical protocols, and improving prevention. To elucidate the etiology of complex diseases, it is necessary to have extensive homogeneous collections of biological samples and epidemiological, clinical, and biomolecular data on large numbers of patients and healthy subjects. The collection and analysis of biological samples is a necessary procedure for diagnosis. The close involvement of Laboratory Medicine and Pathological Anatomy services in material collection centers is essential for storing samples properly. In the not-too-distant future, the evaluation of various molecular parameters will become a prerequisite for diagnostics and patient follow-up, and the development of Biobanks and Biological Resource Centers (BRCs) will become an integral part of medical practice.

M. Lavitrano (✉) · M. Barbareschi · A. Scarpa · M. Macilotti · S. Casati · R. Lawlor
Department of Medicine and Surgery, University Milano Bicocca, Milan, Italy
e-mail: marialuisa.lavitrano@unimib.it

B. Parodi
Biological Resource Center, IRCCS S. Martino Hospital, Genoa, Italy

M. G. Daidone
Department of Experimental Oncology, IRCCS National Cancer Institute, Milan, Italy

Human Cells and Tissues for Research

Biological samples of human origin (cells: PBMC, stem cells; cell lines: primary, stabilized; frozen and fixed solid tissues; biological fluids: blood and derivatives, CSF, urine; DNA, RNA) and their use in in vitro systems can be a valid alternative to animal experimentation in many applications in biomedical and biotechnological research. The development of cell and tissue specimen collection centers is at the basis of three innovative fields of biomedical sciences:

- Molecular and genetic epidemiology (aimed at assessing the genetic and environmental basis of cancer in the general population and families)
- Molecular pathology (for the development of molecular procedures for classification and diagnosis in oncology)
- Pharmacogenetics (for predicting the individual response of patients to treatment)

A fundamental prerequisite for the reliability of in vitro systems is the quality of samples and data. Officially recognized biobanks and BRCs are the facilities that can guarantee the quality of samples and data, harmonization, and respect of participants' rights.

Definitions of Research Biobank, Research Specimen Collection, Diagnostic Documentation Archive

Research Biobank

A research biobank is a nonprofit service structure, organized according to criteria of quality, order, and purpose, aimed at collecting, processing, storing, and distributing human biological samples, and related data for research and diagnostic purposes. Biobanks are officially recognized by the competent health authorities, apply a quality system, and guarantee the rights of those involved. They must ensure the

M. Ciaccio (ed.), *Clinical and Laboratory Medicine Textbook*, https://doi.org/10.1007/978-3-031-24958-7_57

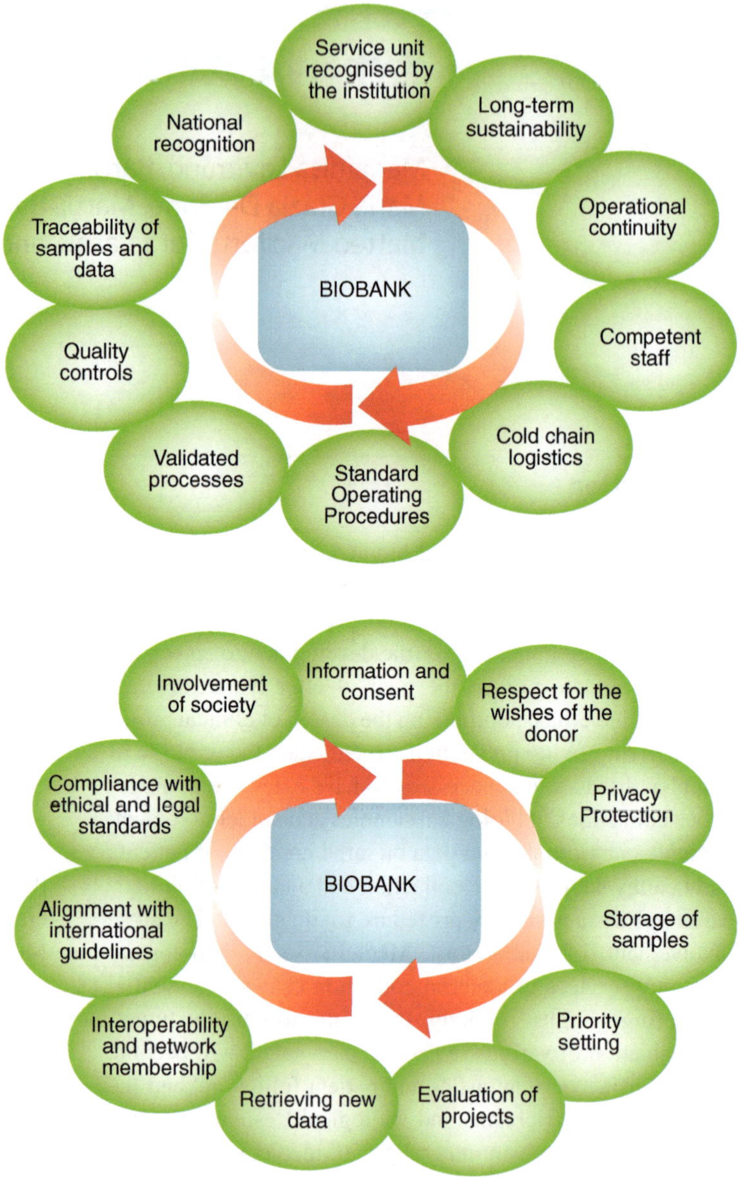

Fig. 57.1 Characteristics defining a biobank. (Copyright EDISES 2021. Reproduced with permission)

quality of the preservation of materials and information, guarantee the traceability of exchanges and the protection of the confidentiality of the subjects' sensitive data, and adhere to internationally recognized ethical principles, such as free and informed consent. The organization of biobanks, therefore, presupposes a consolidated quality system by international standards of reference, which are in addition to the criteria of Good Laboratory Practice and management quality.

Figure 57.1 shows the characteristics that define a biobank from the point of view of management and technical quality, professional skills, and attention to the ethical, legal, and social aspects underlying the collection and use of human biological samples for research.

Sample Collection

A biological research sample collection is a collection of biological materials generally related to specific projects and/or clinical protocols. These collections are generally limited to samples from a specific organ or disease.

Archives of Diagnostic Documentation

A diagnostic documentation archive is a collection of biological materials (e.g., the collection of tissues carried out at Pathological Anatomies or at Laboratory Medicine Centres) kept as "diagnostic documentation," in accordance with current legislation regarding materials collected for diagnostic purposes. In Italy, the Ministry of Health, Superior Council of Health, Session XXXIX Section III in the session of October 14, 1987, expressed the following opinion: "the histological diagnostic material (consisting of the preparation and its inclusion), regardless of the positivity or negativity of the report, is considered an integral part of the "remaining diagnostic documentation" referred to in the circular of 19.12.86 n.61." Thus, extending to tissues what is established in art. 9 of the DPCM of 20.2.84 concerning the "conservation of health documentation in public and private health institutions of hospitalization and care."

Biological Resource Centers

Biological Resource Centres (CRBs) are defined by the Organisation for Economic Co-operation and Development (OECD) as centers that "provide services for the preservation of living cells, genomes of organisms, and information relating to the heredity and function of biological systems. They conserve banks of cultivable organisms (microorganisms, plant, animal, and human cells), replicable parts of them (genomes, plasmids, viruses, cDNA), viable but no longer cultivable organisms, cells, and tissues, as well as databases containing molecular, physiological and structural information relevant to those collections."

Organizational Criteria

The biobank must belong to a public or private institution accredited at a regional or national level that can guarantee the structure's medium-/long-term support. The following figures must be identified among the personnel of the Institution where the biobank is located:

1. Biobank manager: the individual responsible for the operation of the biobank.
2. Quality manager of the samples and their associated data: a person in charge of controlling the quality of the stored samples and ensuring compliance with storage standards, who can control the accuracy and updating of the pathological and clinical data recorded in the biobank database.
3. Data Processor: a person in charge of checking that personal data is processed in compliance with current legislation, ensuring maximum confidentiality of the subjects involved.

The personnel working in the biobank, sized in relation to the volume of activity and whose continuity must always be guaranteed, must be qualified and possess specific training, with periodic updating, concerning the tasks they are assigned to perform.

Furthermore, a Scientific Committee must be foreseen to support the biobanking activity, that is, a committee in charge of evaluating the validity and relevance of the scientific projects requiring biological samples to the biobank. This committee has the task of assessing the priority of the various projects in relation to predefined and transparent criteria, taking into account any ongoing or competing research. Finally, in order for a biobank to be recognized by BBMRI, the infrastructure of biobanks and biomolecular resources will have to prepare a regulation in which the following must be specified:

- The organizational structure
- The function of the organs
- The standard procedures for preserving samples and data
- The policies for access to the biobank

Structural and Technological Criteria

The minimum structural and technological requirements of a biobank should be as follows:

1. Definition of a programmatic document with the aims of the structure, regarding the scope of the biobank, the specific functions to be performed, the type of stored material, the quantity of expected samples, the collection method, processing, and storage (also in terms of duration) of the samples, information management, transport and reception of the samples, management of the biological risk to which the personnel of the biobank are subjected and relative precautions, economic and scientific strategies for medium-/long-term sustainability, and the contingency plan in the event of closure.
2. Definition of dedicated logistics and rooms, with characteristics appropriate to the specific functions, adequate air conditioning and ventilation systems, and access control methods. Cryopreservation systems should be located in suitable premises, with controlled temperature, and, if liquid nitrogen is used, with appropriate safety systems to detect oxygen deficiency. The temperature in the cryopreservation systems must be monitored as well as that of the environment, recorded over time, and procedures must be defined for periodic checks, rooms and instruments ordinary maintenance, intervention in case of failures or malfunctions to ensure the continuity of freezing.

3. Preparation of a disaster recovery plan procedure, that is, a plan that defines the actions to be taken in the event of failure of the preventive measures. All equipment and systems for storing biological samples must be connected to systems that can guarantee their operation during a power failure. The availability and intervention of the biobank staff or other competent personnel must be guaranteed in case of necessity (floods, earthquakes, etc.) to secure the stored biological material.
4. Use of a certified quality system for all or the main phases of the various procedures, from the acquisition of informed consent of patient/citizen to the preservation of biological material, whether frozen or otherwise processed, including paraffin embedding, to the maintenance and calibration of the equipment used in the biobank.
5. Identification of an infrastructure dedicated to the biobank, possibly connected to the databases of the clinical departments, the pathological anatomy, the operative office for clinical studies, to manage the movement of biological samples and keep up-to-date the follow-up and associated data deriving from the research in which the samples have been used.
6. Availability of a procedure for a disaster recovery plan as a measure for the information infrastructure, in case of failure of the preventive measures, which includes periodic offsite backup of the recorded data to restore the information following a disaster.

Quality Criteria

Quality Criteria for Tissue Samples

The quality parameters of the samples are related to the three processes of sampling, treatment, and storage:

1. The collection of biological material for research purposes, subject to the informed consent of the patient, in the case of surgical specimens, refers to the collection of a portion of tissues, whether fresh, cryopreserved, or fixed, that are redundant for the diagnostic and/or therapeutic process ("advanced tissues"). The portion of tissue samples to be placed in the biobank is always subject to the collection of sufficient material to ensure the routine diagnostic process and to meet clinical and therapeutic needs. This element is essential to ensure a correct diagnostic process and the appropriateness of the sampling. For each case, representative samples of the pathology and possibly also healthy components (e.g., peritumoral samples) of the organ/tissue in question should be collected. Specimen collection should be performed as soon as possible after surgical excision (possibly within 20 min), and the time between excision and freezing or fixation should be recorded. In the context of specific research, it may be necessary to provide fresh, viable tissue for the isolation, propagation, and in vitro or in vivo treatment of cell populations. Again, sampling procedures must be carried out within a short period after collection, and traceability of such biological samples that do not fit into conventional fixation/freezing pathways must be provided in the biobank.
2. Freezing should be as rapid as possible, ideally using liquids with high thermal conductivity (e.g., isopentane precooled to −80° or liquid nitrogen). Samples should be stored in suitable containers with or without a cryopreserving medium (e.g., OCT®) or RNA-preserving liquids (e.g., RNAlater®).
3. Samples must be coded with a unique reference code for the biobank, possibly using two-dimensional barcodes, recorded in the computer database and associated with the patient master file. The allocation of samples must be identifiable in the management database and possibly recorded using barcode reading technology of the boxes where the samples are stored, whether they are room temperature cabinets for paraffin-embedded tissues or cryopreservation devices such as freezers or liquid nitrogen containers.

Quality Criteria for Clinicopathological and Patient Follow-Up Data

Quality parameters of clinical and follow-up data are related to the collection and updating processes:

(a) Data collection should be carried out through automated systems of data import from hospital operating systems of registry and management systems of Laboratory Medicine centers, pathological anatomy departments, medical oncology, surgery, or other departments involved. In case of impossibility, the collection must be carried out through a standardized procedure developed by the person in charge of the biobank. In order to standardize the information related to the processing of biological material and to facilitate intra- and intergroup communication, it is advisable to use the BRISQ system (Biospecimen Reporting for Improved Study Quality, Biopreservation, and Biobanking, 2011). For pathological data, it is appropriate to import the complete diagnosis and its coding (e.g., SNOMED or ICDO coding); for clinical data, a minimum dataset should be defined to allow a rapid evaluation of clinical and pathological parameters, survival, and response to therapy and the referral to specific treatment protocols.

(b) The updating of data, ensured by continuous input, is crucial. This process should be carried out automatically by computer or, if this is not possible, through periodic consultation of the medical records. Verification of data updating is the responsibility of the quality manager.

Collection and Sampling of Blood and Other Biological Fluids

In addition to the surgical/histological sample, the biobank can also store the patient's blood and its derivatives and/or other biological liquids, subject to informed consent. This collection can fall within the typology described above of the surgical sample; therefore, a part of the biological liquid redundant for the purposes of the diagnostic/therapeutic process is stored. Alternatively, when blood is involved, it is possible to proceed to a whole collection for research purposes, a collection "in addition" to the diagnostic/therapeutic needs. However, direct collection for scientific purposes is a delicate area of action, from the point of view of forensic medicine and bioethics, due to the risk of exposing the patient to inappropriate harm, for which the quality of informed consent and the authorization of an independent Ethics Committee are crucial.

Biobanking Processes

The biobank processes do not begin with specimen collection but much earlier and include how participants/patients are enrolled, how consent is obtained, and how clinical and surgical procedures, history, and treatments to which patients are subjected are recorded (Fig. 57.2). Beyond the distribution of the research samples, the biobank is involved in verifying that the elements contained in the consent are respected by the researchers and in retrieving the results of the studies, in order to enrich the scientific value of the stored samples.

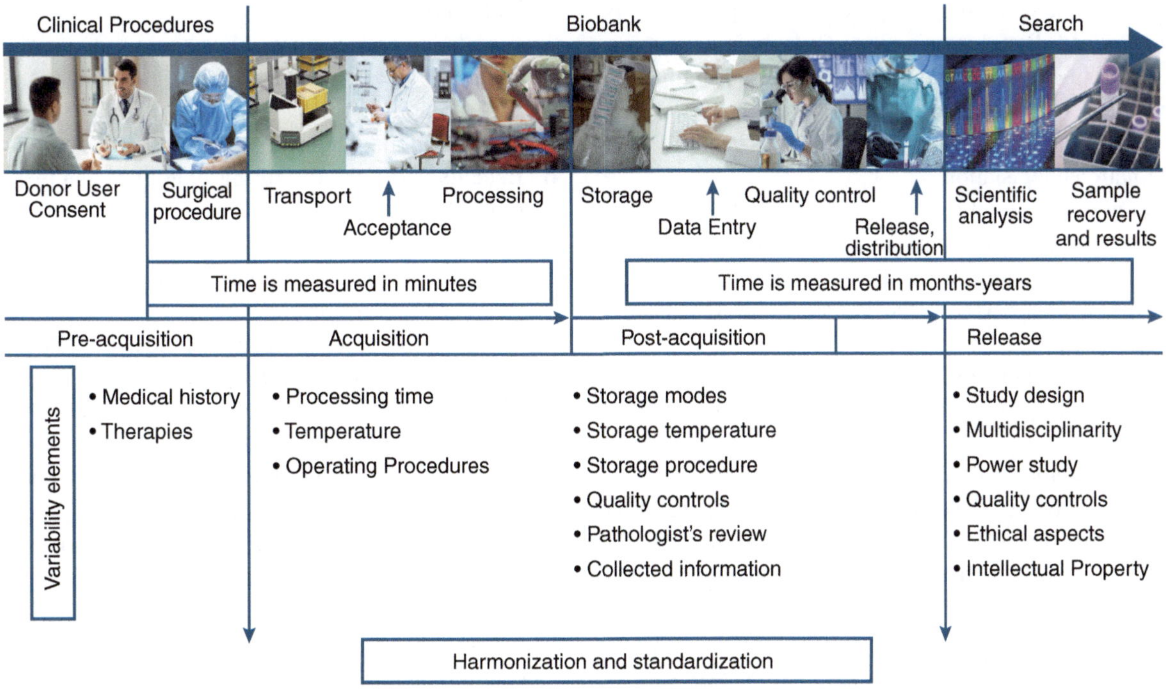

Fig. 57.2 Biobanking processes. (Copyright EDISES 2021. Reproduced with permission)

Privacy and Informed Consent

Privacy and Data Protection

The confidentiality of the individuals involved must be ensured throughout the biobanking process. To this end, all samples must be identified only by a code, which can be assigned automatically by computer algorithms. Only the person in charge of the biobank and the personnel expressly delegated shall be able to connect the personal data with the sample. The person in charge of the data processing shall prepare the necessary procedures to ensure the correctness of the data processing and update the data itself. Researchers should only use samples in a strictly coded form (anonymization or pseudonymization). The code should be designed so that researchers cannot indirectly recognize the identity of the person to whom the sample relates.

It is emphasized that anonymization is not an instantaneous event in which any link between the sample and the patient to whom it refers is abolished. However, it can be better described as a process during which the sample is identified, collected, and associated with a particular set of pathological and clinical data, at the end of which, before starting research activities, the link with the personal data is severed.

If additional follow-up information is to be acquired over time, as current regulations require, the samples must be pseudonymized. Pseudonymization is a technique that consists in storing data in a form that prevents the identification of the subject without the use of additional information (use of identification codes). All procedures must comply with current national legislation.

In Italy, the information systems of the biobank shall be equipped with all the technical standards necessary to ensure the confidentiality of the data, in accordance with the provisions of Annex (B) of the Privacy Code (Legislative Decree 196 of 2003):

1. Computer authentication
2. Adoption of procedures for managing authentication credentials
3. Use of an authorization system
4. Periodic updating of the identification of the scope of the processing allowed to the individual persons in charge of and responsible for the management or maintenance of the electronic instruments
5. Protection of electronic instruments and data against unlawful processing, unauthorized access, and specific computer programs
6. Adoption of procedures for safe keeping of backup copies and the restoration of data and systems availability
7. Maintenance of an updated security policy document
8. Adoption of encryption techniques or identification codes for certain processing of data liable to reveal health or sex life by health bodies

Informed Consent

In order to keep a biological sample in the biobank, the informed consent of the subject to whom the sample refers must be obtained. Informed consent is a central element in the relationship between the biobank and the person who gives the samples. Unlike the traditional consent given for a research project, the consent given to the biobank is not linked to a specific project, which is why the information preceding the consent must specify the field of study in which the sample is to be used in general terms, such as "research in the field of oncology." If the sample is used for a study in a different field of research after consent has been obtained, it will be necessary to contact the subject to obtain new consent.

In addition to specifying the scope of use of the sample, the disclosure should specify:

1. The organizational structure of the biobank and the purpose of the biobank
2. Rationale, purpose, and nature of the proposed research biobanking
3. Correlation between possible pathology and biobanking
4. Rights and responsibilities. Implications of participation
5. Nature and modalities of the collection/harvesting
6. Benefits and risks of storing biological samples in the biobank
7. Data management and data protection
8. Rules for access to and transfer of data and samples
9. Return and use of results
10. The technical and organizational measures to ensure the confidentiality of the persons involved
11. The technical and organizational measures to ensure the quality of sample preservation
12. The right to withdraw consent at any time, without prejudice to this for those involved
13. The right to access your sample at any time (if still in the biobank and not used up) and to know in which research projects it has been used
14. Names and contacts to interface with, with a clear indication of the person in charge of the biobank and how to contact the biobank
15. Mechanisms used by the biobank to assess the scientific quality and ethical acceptability of research projects using the samples

16. How long the biological sample can be kept in the biobank (unless it has been previously exhausted)

Access Modalities

The biobank will have to specify how researchers can access the biological samples in its rules. The criteria for access should be publicly available and in accordance with the principle of transparency.

Biological samples represent public resources, and, as such, they must be used to ensure their maximum exploitation. To this end, the biobanks must provide special procedures to control the quality of the projects through their evaluation by the Scientific Committee. In addition, for any organization to have access to the samples, the research project must be authorized by an independent ethics committee, which must approve the possibility of using human biological samples to realize the project itself.

The biobank may identify priority rules for access to the biomaterials, which, however, must not have the effect of excluding from the use of the samples third parties other than the institution where the biobank is established. The biobank may foresee and declare that they may constitute criteria for preferential assignment:

1. The active involvement in the research project of the institution where the biobank is established
2. The willingness of the researchers to share the data obtained, or part of the data obtained, from the project of research at the biobank
3. Having previously made available data acquired in previous research projects

The biobanks may not profit from the stored samples, but they may claim reimbursement of the costs incurred for storage and the work done by the biobank staff. If the project results in a scientific product (publication), the contribution of the biobank must be acknowledged in that product. The biobank must provide for the publication of an annual report indicating all the research projects undertaken as indicated in the following paragraph.

Biomaterials collected in biobanks are intended for research and are not collected for clinical purposes, as they represent redundant material for current diagnostic purposes. Therefore, access to these biomaterials is restricted to researchers as described above. However, the patients or treating physicians could require the sample cryopreserved in the biobank to carry out further clinical investigations.

Good Practices for the Use of Stabilized Cell Lines

Minimum Rules of Conduct for Cellular Handling

In order to perform cell manipulation correctly, the following rules should be followed:

- All human specimens and tissues should be considered potential carriers of pathogens and handled under a Class II biological safety hood.
- All tissues and resulting cell cultures must be processed aseptically using sterile materials and instruments.
- In the cell culture rooms, conditions must be guaranteed to prevent cross contamination among cells of different origins (physical and/or temporal separation of the different cultures).
- Keep the documentation provided with the cells purchased from certified sources (product sheet and material safety data sheet) in a known location and make it available to operators.
- Waste from the activities of cell culture laboratories that comes into contact with biological material (not necessarily infected) is considered medical waste at risk of infection and, as such, must be disposed of.
- Materials and cultures must be disinfected or autoclaved or decontaminated before being discarded.
- Liquid waste (including cell culture media) must be autoclaved before disposal or disinfected by disinfectant solutions (0.5% sodium hypochlorite) and placed in suitable leak-proof containers. Disposal in the sink and/or drains is forbidden.
- In the case of accidental spillage of cultures (e.g., spillage of flasks containing cell cultures), cover the material with a cloth or paper towel on which disinfectant should be poured and left for at least 30 min. The material, retrieved with forceps or shovel, should be autoclaved or soaked for 24 h in disinfectant.

Authentication of Human Cell Lines

"Evidence suggests that up to one-third of tumor cell lines used in scientific research are affected by interior intraspecies cross contamination or have been incorrectly identified, thereby rendering many of the conclusions doubtful if not completely invalid", Lancet Oncology, Contamination of cell lines—a conspiracy of silence vol. 2, July 2001, p. 393.

Misidentification of a cell line may result from cross contamination, mishandling, or technical error and implies a problem in good cell culture practices: the simultaneous presence of two cell lines under the biohazard hood, mislabeling of a vial, or thawing of the wrong line. Other sources of cross contamination may be the feeder-layer cells if they are still mitotically active, for example, due to insufficient irradiation.

Here are some simple precautions that minimize the possibility of misidentification:

- Carefully label all plastics used in the culture.
- Cultivate only one cell line at a time and clean the biohazard hood by a disinfectant liquid before introducing another cell line.
- Use dedicated bottles and aliquots of media and reagents for each cell line.
- Minimize aerosol formation.

STR profiling is the standard method for the authentication of human cell lines. The referenced standard is ANSI/ATCC ASN-0002-2011 Authentication of Human Cell Lines: Standardization of STR Profiling. This document develops a standardized procedure for the unique authentication and identification of human cell lines using STR profiling. The aim of the standard is the specification of the DNA extraction and STR profiling methodology, data analysis, data quality control, interpretation of results, and implementation of a searchable public database.

Recently, the main journals of oncological interest have introduced restrictive criteria for authors in terms of authentication of cell lines: for publication, each human cell line used must be accompanied by a certificate of authenticity provided by an authorized facility. In 2012, an International Committee of experts (International Cell Line Authentication Committee – ICLAC) (www.iclac.org) was set up to provide the scientific community with guidelines for the correct use of cell lines and the Database of Cross-Contaminated or Misidentified Cell Lines. The database lists cross-contaminated or misidentified cell lines and is updated regularly—https://iclac.org/databases/cross-contaminations/.

Recommended Readings

ANSI/ATCC ASN-0002-2011 Authentication of Human Cell Lines: Standardization of STR Profiling

Coecke S, Balls M, Bowe G, Davis J, Gstraunthaler G, Hartung T, Hay R, Merten OW, Price A, Schechtman L, Stacey G, Stokes W (2005) Guidance on good cell culture practice. A report of the second ECVAM task force on good cell culture practice. Altern Lab Anim 33(3):261–287

Comitato Nazionale Bioetica. Biobanche pediatriche. Aprile 2014

Comitato Nazionale Bioetica (2016) Gestione degli "incidental findings" nelle indagini genomiche con le nuove piattaforme tecnologiche. Marzo

Comitato Nazionale di bioetica. Comitato Nazionale per la Biosicurezza, le Biotecnologie e le Scienze della Vita. Raccolta di materiali biologici a fini di ricerca: consenso informato. Roma 2009

Comitato Nazionale per la Biosicurezza, le Biotecnologie e le Scienze della Vita. Linee guida per la certificazione delle biobanche. Aprile 2006

Consiglio d'Europa. Convenzione per la protezione dei diritti dell'uomo e la dignità dell'essere umano riguardo alle applicazioni della biologia e della medicina (Convenzione sui diritti dell'uomo e la biomedicina). Oviedo 1997

Council of Europe. Additional protocol to the Convention on human rights and biomedicine concerning biomedical research. Strasbourg 2005

Davis JM (2011) Animal cell culture: essential methods. Wiley-Blackwell/Wiley, Chichester. https://doi.org/10.1002/9780470669815

European Commission. Biobanks for Europe. A challenge for governance. 2012

European Council. Recommendation CM/Rec(2016)6 of the Committee of Ministers to member States on research on biological materials of human origin

Freshney RI (2010) Culture of animal cells: a manual of basic technique and specialized applications, 6th edn. New York, Wiley-Blackwell

Garante Privacy. Autorizzazione n. 2/2016 – Autorizzazione al trattamento dei dati idonei a rivelare lo stato di salute e la vita sessuale – 15 dicembre 2016

Garante Privacy. Autorizzazione n. 8/2016 – Autorizzazione generale al trattamento dei dati genetici – 15 dicembre 2016

Garante Privacy. Autorizzazione n. 9/2016 – Autorizzazione generale al trattamento dei dati personali effettuato per scopi di ricerca scientifica – 15 dicembre 2016

Geraghty RJ, Capes-Davis A, Davis JM, Downward J, Freshney RI, Knezevic I, Lovell-Badge R, Masters JR, Meredith J, Stacey GN, Thraves P, Vias M (2014) Guidelines for the use of cell lines in biomedical research. Br J Cancer, 2014

International Cell Line Authentication Committee (ICLAC) http://iclac.org/

Norma UNI ISO 20387–2018 "Biobanking – Requisiti generali per il biobanking"

OECD. Guidelines on Human Biobanks and Genetic Research Databases. 2009

OECI. From the biobank to the research biorepository. 2010

Per informazioni dettagliate su metodologie e Good Cell Culture Practice

REGOLAMENTO (UE) 2016/679 DEL PARLAMENTO EUROPEO E DEL CONSIGLIO del 27 aprile 2016 relativo alla protezione delle persone fisiche con riguardo al trattamento dei dati personali, nonché alla libera circolazione di tali dati e che abroga la direttiva 95/46/CE (Regolamento generale sulla protezione dei dati)

REGOLAMENTO (UE) N. 536/2014 DEL PARLAMENTO EUROPEO E DEL CONSIGLIO del 16 aprile 2014 sulla sperimentazione clinica di medicinali per uso umano e che abroga la direttiva 2001/20/CE

SIGU. Biobanche genetiche. Linee guida. 2003

Uniamo F.I.M.R. Vademecum per una buona pratica del biobanking. 2013

WMA. Declaration of Taipei on ethical considerations regarding health databases and biobanks. Ottobre 2016

Health Technology Assessment in Laboratory Medicine

58

Lucrezia Ferrario, Elisabetta Garagiola, Fabrizio Schettini, and Emanuela Foglia

Health Technology Assessment (HTA): From Theory to Practice

Within the healthcare sector, there are numerous examples of decision-making processes that are consistent with guidelines, primary evidence, or, more generally, addressed without a quantitative and evidence-based approach. A further critical element is represented by those decision-making processes that have been approached based on individual choice, and not in coherence with the achievement of objectives related to the optimization of resources and efficiency of healthcare processes.

However, the advent of the economic-financial crisis, which has affected all sectors worldwide, also impacted on the Healthcare Services of the most advanced economies, which were forced to revise their health expenditure. Hence, given the current healthcare settings characterized by a paucity of resources, the need to find evidence-based and sustainable criteria to meet the citizens' health need is strictly required, thus becoming a urgent priority.

Several factors contribute to an increase in healthcare expenditure, including the growth in life expectations, the increase in the population's average age, and the number of concomitant and chronic diseases affecting the population. While the abovementioned factors are difficult to manage, on the other hand, the healthcare expenditure devoted to investments in innovative healthcare technologies may be optimized.

For all these reasons, it becomes necessary to monitor, control, and manage accurately the financial resources allocated to healthcare, imposing that the decisions of health policy are at all levels "informed" and "conscious," especially those that can have a high economic and financial impact. It is also important to assess the validity of the technical-organizational aspects through approaches that guarantee the vision of the clinician and the proper management of the patient care pathway.

This can only be achieved through a vision of the healthcare process based on the principles of Clinical Governance: "*a system through which healthcare organizations are responsible for the continuous improvement of the quality of their services and the safeguarding of high standards of care through the creation of an environment in which excellence in healthcare can develop*". For the achievement of this challenging objective, a strong collaboration among physicians, hospital administrators, and managers is strictly required, thus supporting the decision-making processes, based on economic and clinical evidence.

Moving on from these promises, the most acknowledged and relevant tool related to Clinical Governance principles is Health Technology Assessment (HTA). HTA is a multidisciplinary process that summarizes information about the medical, social, economic, and ethical issues related to the adoption and diffusion of any innovative healthcare technologies, in a systematic, transparent, unbiased, and robust manner. It aims to inform the formulation of safe and effective health policies that are patient focused and seek to achieve the best value.

Technological innovations have become relevant for the protection and improvement of the health of the population, an objective that unites all European and international health systems.

Methodological and Applicative Aspects of HTA

By definition, a healthcare technology is represented by not only medical devices and drugs but also equipment, medical, surgical, and laboratory procedures, IT systems, artificial intelligence tools, that are available in the clinical practice to support healthcare organizations. In general terms, any ele-

L. Ferrario (✉) · E. Garagiola · F. Schettini · E. Foglia
LIUC Business School, LIUC - Università Cattaneo, Castellanza, Italy
e-mail: lferrario@liuc.it

M. Ciaccio (ed.), *Clinical and Laboratory Medicine Textbook*, https://doi.org/10.1007/978-3-031-24958-7_58

ments that could modify the organization of any healthcare process represent a healthcare technology and, as such, may deserve to be evaluated to understand their potential impacts, as well as advantages and criticalities.

This aspect can be assessed with an *ex ante* or an *ex post* approach. In other words, decision-makers may need an assessment before implementing a health policy choice and, therefore, before the health technology has been introduced in a health process, or they may need to understand after the technology's adoption if the choice implemented has performed according to both expectations and literature indications.

How could we limit the variability of the decision-making process and improve the reliability of the evaluation process itself? This could be possible by increasing the certainty that all the variables that can impact on a choice of health policy and modification of an organizational process have been considered. For this reason, focusing only on cost and effectiveness aspects is limiting. Based on this consideration, HTA was then created to provide a complete and evidence-based framework. As such, HTA represents "*a multidisciplinary and a multidimensional tool capable of studying the clinical, social, ethical, safety, efficacy, organizational and economic implications of the development, diffusion, and use of a health technology*". HTA, therefore, proposes a set of methodologies and tools through which Health Services, Regions, health facilities, and individuals can make choices concerning technologies that ensure appropriateness and higher quality of services with high levels of performance by means of specific methods based on scientific evidence but also understandable by the decision-makers.

Based on these premises, HTA is a very scalable tool that could support policy makers in identifying the correct health technology to be introduced on the market at different institutional decision-making levels: national and/or regional (macro level), healthcare organizations (meso level), and in clinical practice (micro level).

- The macro level is represented by the Institutional level (i.e., the level at which the NHS today deliberates). It is responsible for national and regional health policies planning, as well as for their implementation. The decision-making process, in this case, is oriented toward generating homogeneity within the national context of reference and ensuring that the main technological innovations entering the national context can have an adequate evaluation.
- The meso or hospital level has relevance because it represents the setting that, besides acquiring the technology, uses the technology itself, and therefore carries out the assessment from the preferred perspective. Evaluations at this level "allow to support the hospital management, thus favoring the diffusion of qualitative and quantitative standards".

Purpose		HTA Core Model® Domains
Complete	Quick	1. Health problem and current use of technology
		2. Description and technical characteristics
		3. Security
		4. Clinical Effectiveness
		5. Economic and financial evaluation
		6. Equity Analysis
		7. Organisational aspects
		8. Social aspects
		9. Legal aspects

Fig. 58.1 Dimensions of the EUnetHTA Core Model. (Copyright EDISES 2021. Reproduced with permission)

- The micro level is related to the real clinical practice, i.e., to the direct and daily decisions "on the field." Within this specific context, "HTA can be adopted based on the needs of professionals or professional associations, according to evidence-based medicine", for the definition of guidelines and/or protocols that can support the choice of the most appropriate diagnostic or laboratory procedures, or of the drug to be administered, as well as to verify the goodness of the services provided.

Based on the above considerations, HTA represents, at all levels, the tool to guide the process of evaluating healthcare technologies and the consequent decision-making phase.

Given the relevance of such methodology, the European Network of HTA (EUnetHTA) has proposed a standardized tool to assess any healthcare technologies. This framework is known as the EUnetHTA Corel Model, suggesting the examination of 9 dimensions or "domains," as indicated in Fig. 58.1. It should be noted here that EUnetHTA works to develop tools for the sharing of methodological frameworks and scientific evidence, reducing duplication and saving time and resources.

1. Health problem and current use of technology. This section includes a detailed description of the pathology, considering also the related prevalence and incidence data that, if available, should be as local as possible, referring to the national or local context where the decision is made. Furthermore, this dimension is aimed at defining the eligible target population that will benefit from the use of the innovative technology: this aspect is essential and should be addressed as quantitatively as possible to understand the need for technological use.
2. Description and technical characteristics. This "domain" focuses on the identification of the technical indications concerning the innovative technology and the comparators, i.e., those technologies that can be used as an alternative or as a replacement to the innovative technology.

Thus, it detects the main differences among technologies, thus detailing strengths and weaknesses, from a merely technical and scientific point of view.

3. Safety. This dimension includes information regarding the safety of the technologies investigated, such as adverse events or complications that may occur as a result of their use, and that may have a negative impact, not only from an effectiveness and outcomes point of view, but also from an economic perspective in terms of hospital activities required to solve and manage such adverse events. The information obtained from the analysis of this dimension is indispensable to building an overview of the use of the new technology in comparison with the standard of care, allowing the decision-maker to clarify which are the risks related to the use of all the technologies that can be used in the care pathway being assessed. In this sense, the sources of information can be represented by the available scientific evidence, the local registers of device vigilance, but also the same data on a national basis.
4. Clinical effectiveness. This domain allows defining the so-called clinical effectiveness and identifying the advantages related to the use of the innovative technology, from a clinical perspective. Clinical outcomes may be represented by mortality and morbidity rates, overall survival, specificity and sensitivity of a test, predictive ability of the test, as well as the technology capability to change the clinician's choices, etc.
5. Economic impact. This dimension allows the definition of the costs related to the implementation of the innovative technology and the use of the standard of care. After the definition of the economic resources' absorption of the patient's clinical pathway performed with and without the innovative technology, this domain also requires the conduction of a cost-effectiveness analysis and a budget impact analysis. On the one hand, the cost-effectiveness analysis is devoted to defining the technology presenting the best trade-off between costs sustained and outcomes achieved. On the other hand, the budget impact analysis is devoted to defining the potential emerging and ceasing costs, related to the introduction in the clinical practice of an innovative technology, over a specific time period, thus assuming the budget holder perspective.
6. Equity analysis. The examination of this aspect allows defining the availability and accessibility of the technologies under investigation within the local context of reference and identifying any groups of patients who might be disadvantaged by the introduction of new technologies. In general terms, this dimension would identify patients who may benefit from the innovative technology. In addition, the assessment of this domain would lead to the definition of a potential modification of the access to care at local level, generating an improvement in hospital waiting lists.
7. Organizational aspects. The assessment of this impact allows the decision-makers to understand the impact of the technology within the considered analysis unit (Operating Unit, Health Facility, etc.), thus defining the investments or disinvestments that the innovative technology may generate. In particular, an in-depth evaluation of personnel, training, and structural adjustments is required. It is necessary to assess how many resources are necessary to promote the organizational change in terms of emerging and ceasing costs. Furthermore, this domain should be evaluated by means of a qualitative approach, by collecting healthcare professionals' perceptions. Without the acceptability of the new technology perceived by the users, one cannot expect to achieve any significant result from its development, both in the first phase of its introduction, and in the continuous real practice adoption.
8. Ethical and social aspects. This dimension assumes the patients' point of view and is assessed by means of both a qualitative and a quantitative approach. At first, the patients' perspective should be analyzed to understand the patients' perceived satisfaction and quality of life. The social domain could thus also explore the patient's productivity loss, in terms of "out-of-pocket" expenditure.
9. Legal aspects. The analysis of the legal impact is devoted to the definition of regulations and legislative aspects related to the adoption of innovative healthcare technologies for supporting and defending the rights of the person and the interests of the society. In particular, the main aim is to understand whether there are legal factors that can encourage and/or discourage the setting of reference from using technology, but at the same time to understand the existence of any critical or positive factors.

In conclusion, through an accurate analysis of the abovementioned nine HTA dimensions, it is possible to provide decision-makers with evidence-based and scientific information, to adequately inform an investment process in health technologies.

This process could thus be applied for taking rational investments within the laboratory medicine setting, thus producing evidence-based processes and generating new knowledge.

Since HTA is a multidisciplinary activity, at first it is necessary to understand the role of the laboratory physician in defining a report and structuring valuable and valid information for decision-making process.

The Role of the Laboratory Physician in the Production of Useful Information for HTA

In dealing with this chapter, it is helpful to recall the above-mentioned concept of Clinical Governance as a tool that is widespread in a healthcare organization to create clarity of intent, sharing of objectives, and transparent decision-making process, in a multidimensional and multidisciplinary approach. Thus, clinical governance demands the re-examination of traditional roles and boundaries, between health professions, between doctor and patient, and between managers and clinicians, with an impact also on the Laboratory Medicine setting.

Since Clinical Governance tools, including HTA, are little implemented in the real-life practice, it would be relevant to increase the acceptance and the sharing of strategic hospital objectives through a profitable collaboration between the decision-makers and the healthcare professionals.

The coexistence of managers and practitioners may guarantee the usefulness of the evaluation process and, at the same time, an attempt to speak a common language that is halfway between the technical and organizational aspects and the strategic ones. The coexistence, therefore, of these two professional figures would establish the appropriate priorities for the use of technologies, and to allocate economic resources according to clinically acceptable criteria. Thus, the clinical and the managerial component should coparticipate to the decision-making processes.

At the organizational level, how the laboratory physician could be involved in this multidimensional and multidisciplinary process?

In the attempt to provide a practical role of the laboratory clinicians in the production of an HTA report, the following prevalent activities emerged.

- The laboratory physician should be the healthcare professional with a clear picture of the literature evidence available on the topic, regarding the efficacy and safety indicators of the technology beins assessed.
- In addition to efficacy and safety data, the laboratory physician should provide information about the volume of activity produced by a specific technology, within a specific time-horizon.
- The laboratory physician also assists in describing the processes of taking care of patients, thus providing quantitative information useful for both the economic and organizational assessment, thus defining the average economic resources absorption devoted to a specific clinical pathway involving laboratory activities.
- The laboratory physician could be also involved in the assessment of the perceptions related to any organizational, social, ethical, and equity changes in the patient's clinical pathway.
- Therefore, it is necessary to underline how an increase in the cost of laboratory technology could imply an optimization in the overall patients' clinical pathway, thus leading to a potential release of organizational resources as well as to a potential positive impact on the general patients' hospitalization.

Critical Aspects of HTA Development in Laboratory Medicine

In laboratory medicine, progress and technological innovation have a crucial role. Recent technological advances have revolutionized modern laboratory medicine and added significant value and visibility to its role in healthcare and clinical decision-making.

The growing application of innovative technologies within the laboratory setting, as well as their integration with microtechnology and point-of-care testing, has contributed to improved patient outcomes and narrowing of the clinical–laboratory interface to facilitate a patient-centered approach to healthcare.

However, the goal of continuing to ensure appropriateness and personalization of healthcare services can only be pursued through the availability of the most innovative healthcare technologies, which improve and optimize processes, both at the diagnostic and therapeutic level. Indeed, it is "*essential that innovation in laboratory medicine should be translated into clinical practice only if there is standardization and harmonization, to avoid confusion, risk of error and loss of credibility towards clinicians and patients*".

Despite the acknowledged importance of HTA in this context, few HTA evaluations have been conducted on laboratory diagnostics. The unavailability of practical information on this topic is primarily due to the following factors that hinder the diffusion of HTA in laboratory medicine.

1. Use of the same analytical platform for determinations concerning diversified diagnostic fields.
2. Limited literature evidence verifying the efficacy of laboratory tests, not only for innovative technologies, but also for those technologies that represent well-established procedures in clinical practice. In this regard, in most cases, there is a lack of high-quality evidence derived from meta-analyses or randomized control trials; however, it is important to verify how this aspect is strictly related to the nature of the technology investigated.
3. Low economic impact of this specific area if compared to the overall healthcare costs.

In general terms, HTA may help identifying obsolete or redundant testing (in terms of laboratory diagnostic procedures), but may also promote the introduction of innovative tests, which can ultimately help decrease the overall healthcare expenditure, thus optimizing the patients' clinical pathway and improving the overall quality of care. Predicting the development of certain pathologies within an appropriate timeframe and the therapeutic response will generate a benefit for the entire healthcare setting, since the incremental costs due to the introduction of new tests can be completely counterbalanced by decreasing the risk of disease development and by using the most appropriate therapeutic options in the single patient.

Based on this consideration, it is necessary to overcome the logic of thinking in budget "silos" to focus not exclusively on the cost of a single technological innovation but on the cost of the overall management of the patient clinical pathway.

Case Study: HTA Evaluation of New Technologies for Quality Control in Hemogasanalysis

The Research Context

Haemogas analyzers, mainly used in critical areas, have an essential role in emergency conditions providing information on the acid–base balance, the oxygen transport to the tissues, and the electrolytic and metabolic balance. They have a fundamental role, not only for the wide technological diffusion and the high number of analyses performed, but also for the clinical relevance and the urgency that haemogas analysis reports usually require.

Such analyzers are present in the critical area of the hospital, at the point of care, due to the immediate need for information regarding critical condition patients. In clinical laboratories, the analyses are performed by healthcare professionals, who are responsible not only for the quality of the samples. Indeed, the performance of point-of-care analyzers must be monitored through a system on board the instrumentation, able to automatically verify the analytical quality of the results (traditional quality controls). The equipment used in point-of-care should be supported by innovative technologies so that even operators with professional skills different from those of laboratory technicians can guarantee the same quality of the results. Moreover, the point-of-care setting is characterized by a high turnover, which makes it essential to have training and knowledge of these instruments to perform examinations in total operational safety. In this view, the quality of the results may be ensured, even by physicians without laboratory experience, in different contexts where the attention paid to the patient is essential and where there is a need to have an immediate result, one without the need of a successive quality test, in order to detect any possible error.

Moving on from these promises, with the aim to overcome the abovementioned drawbacks, an innovative technology – intelligent Quality Management (iQM) – is emerging in clinical practice. iQM has been conceived to provide continuous quality monitoring of the entire analytical process of blood gas analyzers, in real time, by automatically correcting nonconformities and producing documentation of all corrective actions, replacing traditional internal quality controls.

Based on the above, the present study aimed to quantify the potential advantages related to the application of iQM system at the point of care, compared with traditional quality controls that are performed in accordance with good laboratory practice, in order to verify that the introduction of this new technology would both improve patient care, and optimize the management of the instruments that generate analytical results.

The achievement of such objective, devoted to compare the systems used nowadays in the clinical practice, is relevant given the COVID-19 pandemic era. In fact, during the SARS-CoV-2 pandemic, these devices proved useful in emergency medical services for the early assessment of respiratory distress, allowing the appropriate care level to be determined for each patient. For example, in a study conducted in the Italian setting, blood point-of-care testing made it possible to evaluate and treat at home about half of the patients and to admit the others directly to the most appropriate ward, showing the generalizability of the results related to this technology also to different hospital settings.

Methods

For the achievement of the abovementioned objective, useful to manage the choices related to the introduction of iQM systems in clinical practice, an HTA evaluation was conducted.

As detailed in the previous chapters, the HTA grounded on the EUnetHTA Core Model framework, thus assessing the following aspects: (i) general and technical relevance; (ii) safety; (iii) efficacy; (iv) economic-financial impact, through an assessment of the impact on the budget (of the hospital providing the service); (v) social and ethical impact; (vi) equity impact; (vii) legal impact; and (viii) organizational impact (assessed both qualitatively and quantitatively).

A comparative approach was implemented, by estimating the advantages related to the introduction in the clinical practice of the innovative iQM technology (TO BE situation) with respect to the absence of such technology (AS IS situation).

For the assessment of the dimensions, data were gathered using the following data source. (i) evidence-based information, to retrieve effectiveness and efficiency data related to iQM implementation within the laboratory medicine setting; (ii) administration of a qualitative questionnaire filled in by 4 professionals (1 laboratory manager, 1 laboratory technician, 1 laboratory physician, and 1 clinical engineer), thus generating consensus on the social and ethical, equity, and legal impacts; and iii) quantitative approaches, useful for the economic evaluation as well as for the organizational assessment, by means of a real-life data collection performed with the involvement of a medium-size Italian hospital.

Results

General and Technical Relevance

Haemogas analysis is mainly implemented in critical area wards, with the evaluation of vital parameters of patients affected or potentially affected by different pathologies. The Emergency Department (ED) is one of the central hospital departments where emergencies are treated, and blood gas analyzers are used. In addition to ED admissions, blood gas tests are required in several surgeries to monitor O_2 and CO_2 concentrations in the blood.

Based on the surveillance source of ED accesses, the total number of annual accesses to ED services in Italy was about 21 million in 2019, and about 14 million in 2020, during the pandemic peaks. Of these accesses, in particular during COVID-19 pandemic, about 50% were due to severe respiratory infections, which has always represented the first cause of access to the Emergency Department. Thus, it is obvious that such technologies have been amply utilized in the Italian clinical practice.

From a technical perspective, iQM was introduced to provide a continuous monitoring of the quality of the entire analytical process of haemogas analysis, in real time, through the automatic correction of nonconformities and the automatic production of the documentation, with regard to all the corrective actions that were taken, thus replacing the use of traditional internal quality control.

Different from internal quality controls, iQM executes a monitoring of the complete system for all the use period of the cartridge on board the analyzer. The system verifications include fluidic controls, in order to check the integrity of samples, reagents, and peristaltic pumps. In addition, the system verifies the mechanical components of the cartridge, the position of the valves and of the sampling probe, and the module for the sample thermostat and electronic system. All the controls are automatic and are executed regularly during operation. If there is a malfunction in any components of the system, iQM verifies the nature of the malfunction and takes the appropriate corrective action, the main frequent causes of errors for sensors, such as micro-clots or any interference/malfunction, are tracked on the sensors.

The comparator (technology used for the comparison, representing the standard of care) requires quality controls according to traditional laboratory practice. Different from iQM, the control is not performed continuously, but at time intervals defined by the laboratory (e.g., every 8 h) and the obtained results (3 times in a working day) are then represented in control reports. The corrective actions may be partially automated or completely handled by users.

Safety

From a safety point of view, on the one hand, the technologies compared do not present any immediate risk for the patient since they are in vitro diagnostic devices. On the other hand, some specific aspects relating to the user's risk need to be examined in more detail, that is, risks of cutting, risks of biological contamination, and electrical safety.

From a qualitative perspective, the 4 experts reported, in an incremental logic, their perceptions with regard to the potential advantages and disadvantages of different aspects of safety using iQM (Fig. 58.2).

Effectiveness

The effectiveness parameter used is the average error detection time, derived from literature evidence, whose results favor the innovative iQM technology, as shown in Table 58.1.

Economic and Organizational Impact

From an economic perspective, the acquisition cost of iQM can be considered comparable to the acquisition value of traditional quality controls for haemogas analysis. However, in addition to this cost driver, the impact of efficiency generated by the different technologies, analyzing productivity, and costs–benefits generated by the different technologies should be taken into consideration.

For the achievement of the study objective, a haemogas analyzer with iQM, used at the point-of-care unit of a medium-sized Italian hospital was considered in relation with all the activities concerning the use of a haemogas analyzer without iQM as a comparator.

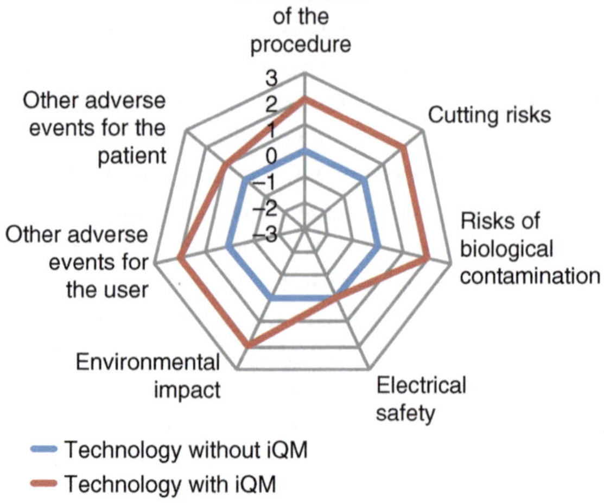

Fig. 58.2 Qualitative assessment of the safety dimension. (Copyright EDISES 2021. Reproduced with permission)

Table 58.1 Effectiveness

Average error detection	Time (range in hours) pH, PCO_2, PO_2, K, Calcium, Lactate, Hematocrit	Glucose	Sodium
Technology with iQM	0.05–0.5	0.12–1.2	0.17–1.7
Technology without iQM	8–24	8–24	8–24

The data collected at the Emergency Department produced the following parameters included in the evaluation.

- Annual workload: about 1,500 samples
- Healthcare professional's time spent to insert each cartridge: 5 min
- Average hourly cost of the healthcare professionals involved: € 22
- Recurrence of clot sample: 2/week
- 5% of CQ repetitions (4–5 QC repetitions per month) – and the requirement for manual documentation – 2 min per QC
- CQ manual resolutions out of range, once per month (20 min)

Using the described parameters, the time spent by the healthcare professionals in performing all the activities was estimated.

The examination of the economic and organizational dimension recorded a cost of the blood gas analyzer with iQM technology decidedly lower than the costs of the dedicated personnel, making the technological innovation more advantageous than the other technologies under examination without iQM (Table 58.2). The above advantages may be related to the fact that iQM requires fewer cartridges and less time not only for the replacement of consumables but also for maintenance activities, manual resolution, and manual documentation of nonconformities.

The evaluation of the economic impact then proceeded with an analysis of the impact on the budget, deriving from the use of the different technologies, excluding the purchase costs, since they are highly variable based on the workloads required and the accessory services included in the different supplies (Table 58.3). Specifically, the overhead costs were compared with the costs estimated in the 2016 budget for the hospital involved, which envisages a total cost equal to € 207,001,000 and a cost value related to the dedicated healthcare personnel equal to € 50,647,000. Thus, the innovative technology presents a lower impact on the costs of dedicated personnel, thus making it more advantageous and economically sustainable, than the other technologies under consideration.

Qualitative Dimensions: Impact on Equity, Social, Ethical, and Legal Impact

As mentioned in the methods section, a qualitative questionnaire was filled in by four experts, who gave their comparative perceptions, according to an evaluation scale ranging from −3 (less performant technology) to +3 (most performant technology), on equity, social and ethical, and legal aspects.

Table 58.2 Economic enhancement related to the use of a single blood gas analyzer in an emergency-urgency department, considering the execution of 1500 samples on an annual basis

Analyzer	Total annual commitment of operators (hours)	Annual cost (euro)
Innovative technology with iQM	1.4	31
Standard A technology, without iQM	75.39	1659
Standard B technology, without iQM	44.42	977
Standard C technology, without iQM	47.25	1040

Table 58.3 Impact analysis on budget, considering the point of view of the healthcare facility providing these services

	Cost per analyzer (EUR)	Average analyzer cost/ dedicated personal cost	Average total cost of analyzer/ costs structure
Innovative technology with iQM	31	0.00006%	0.00002%
Technology A standard, without iQM	1659	0.00328%	0.00080%
Technology B standard, without iQM	977	0.00193%	0.00047%
Technology C standard, without iQM	1040	0.00205%	0.00050%

In general terms, the investigated technologies have the same Essential Level of Care, thus guaranteeing equal access to care for the entire population that presents this health need (Fig. 58.3).

From an equity point of view, positive aspects of the adoption of iQM refer to timeliness of diagnosis, with a consequent shorter waiting time for the patient and the possible caregiver.

From a legal point of view (Fig. 58.4), all blood gas analyzers have CE marking, which guarantees compliance with all legal requirements. It is considered, however, that the improved accuracy of the data may have an impact on further legal aspects in terms of defensive medicine, which is now a widespread phenomenon among users.

Finally, from an ethical and social point of view (Fig. 58.5), thus considering the patients' perspective, no differences emerged, since the patient is not directly involved in this process.

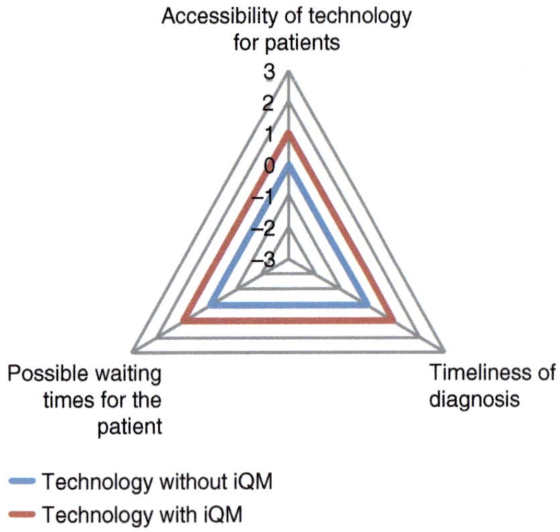

Fig. 58.3 Qualitative assessment of the equity dimension. (Copyright EDISES 2021. Reproduced with permission)

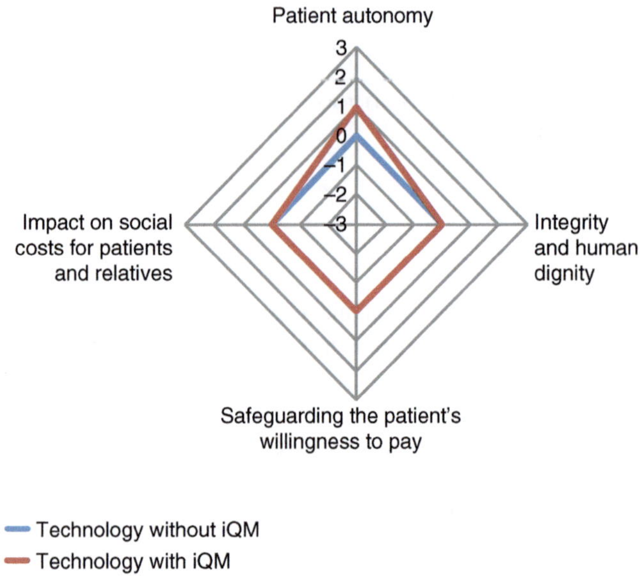

Fig. 58.5 Qualitative assessment of the ethical and social dimension. (Copyright EDISES 2021. Reproduced with permission)

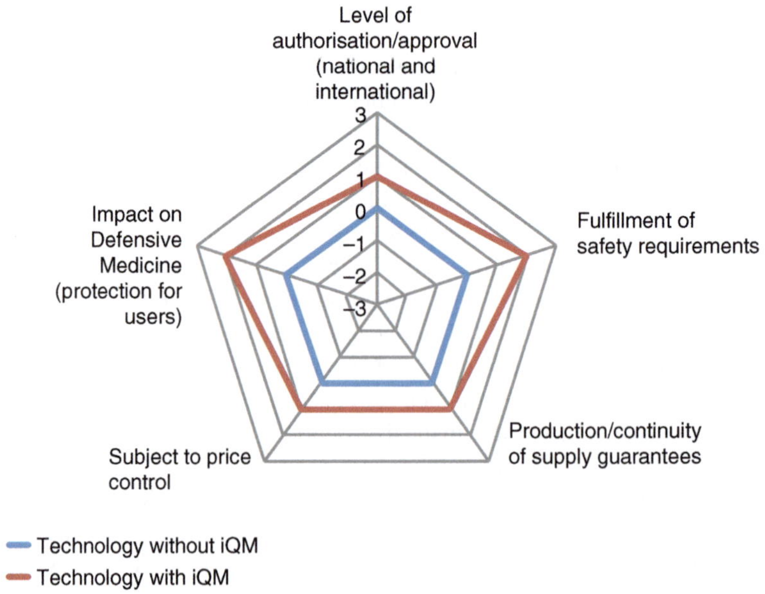

Fig. 58.4 Qualitative assessment of the legal dimension. (Copyright EDISES 2021. Reproduced with permission)

Conclusions

The HTA analysis showed how iQM technological innovation could be considered the preferable technological alternative to be adopted within the point-of-care setting.

In particular, the main advantages are related to a better safety profile, in terms of better-quality control of patients' results, as well as a significant improvement in the efficacy, in terms of capability of continuous control of sensors, with automatic restoration interventions, different from traditional quality controls that, under the same conditions, have more errors or limited performance.

In addition, although the acquisition value of the blood gas analyzer with iQM is comparable to that of the reference methods, the reduction of consumables, the higher number of tests that can be performed, and the lower number of working hours required to manage nonconformities have an indirect positive impact on the budget and, at the same time, is related to a saving in the time spent by the healthcare professionals in conducting the activities.

All these factors may lead to an optimization of the entire process, considering both the hospital's and the patients' perspectives.

Furthermore, the choice of the proper technology to be used within the point-of-care testing has become relevant, given the COVID-19 infection, where testing is essential for containment and mitigation purposes.

Recommended Readings

Abernethy MA, Vagnoni E (2004) Power, organization design and managerial behaviour. Acc Organ Soc 29(3–4):207–225

AGENAS. Portale Covid-19 AGENAS: Analisi temporale della pandemia 2022. 2020-2021. Available at: https://www.agenas.gov.it/comunicazione/primo-piano/1959-analisi-storica-pandemia-2020-2021

American Office of Technology Assessment (OTA) (1976) Development of MEDICAL TECHNOLOGIES: OPPORTUNITIES FOR ASSESSMENT. US Government Printing Office, Washington, DC

Cicchetti A et al (2010) Hospital based HTA worldwide: the results of an international survey, vol. 21. Health Technology Assessment International Web site. Available at: http://www.htai.org/fileadmin/HTAi_Files/Conferences/2008/Files/Sessions/Presentation_Mar-co_Marchetti.pdf

Cini S, Urbanelli A, Montemerani S, Ramacciani Isemann C, Righi L (2021) Point-of-care blood gas analysis as a tool for COVID-19 patients: a prehospital setting experience. Recenti Prog Med 112(12):821–823

Drummond MF, Schwartz JS, JoÃànsson B, Luce BR, Neumann PJ, Siebert U, Sullivan SD (2008) Key principles for the improved conduct of health technology assessments for resource allocation decisions. Int J Technol Assess Health Care 24(3):244–258

EUNetHTA, The HTA Core Model, Version 2.13.0. April 2016. Available online at: http://eunethta.eu/sites/5026.fedimbo.belgium.be/files/HTACoreMo-del3.0.pdf

Favaretti C, Torri E (2008) Valutazione della tecno- logia sanitaria e radiologia medica. Radiol Med 113:S37–S38

Garrid MV, Boerlum KF, Nielsen PC, Busse R, World Health Organization, Regional Office for Europe et al (2008) Health technology assessment and health policy-making in Europe: current status, challenges and potential/by Marcial Velasc Garrido, Finn Boerlum Kristensen, Camilla Palmhoej Nielsen and Reinhard Busse. Copenhagen: WH Regional Office for Europe. http://www.who.int/iris/handle/10665/107911

Gauld R (2014) Clinical governance development: learning from the New Zealand experience. Postgrad Med Journal 90:43–47

Liberati A (1997) La medicina delle prove di efficacia. Il PensierScientificeditore, Roma

Martelli F, La Torre G, Di Ghionn E, Staniscia T, Neroni M, Cicchetti A, Von Bremen K, Ricciardi W, NI-HTA Collaborative Group (2007) Health technology assessment agencies: an international overview of organizational aspects. Int J Technol Assess Health Care 23(4):414–424

Mauskopf JA, Sullivans SD, Annemans L et al (2007) Principles of good practice for budget impact analysis: report of the ISPOR task force on good research practices-budget impact analysis. Value Health 10:336–347

Mossialos E, McKee M (2002) Health care and the European Union – Profound but uncertain consequences for national health systems. BMJ. 324(7344):991–992

Pinnarelli L, Colais P, Mataloni F, Cascini S, Fusco D, Farchi S, Polo A, Lacalamita M, Spiga G, Ribaldi S, Magnanti M, Davoli M. Access to the Emergency Department in the time of COVID-19: an analysis of the first three months in the Lazio Region (Central Italy). Epidemiol Prev. 2020;44(5-6):359–366

Plebani M (2013) Harmonization in laboratory medicine: the complete picture. Clin Chem Lab Med 51(4):741–751

Plebani M, Lippi G (2016) Improving diagnosis and reducing diagnostic errors: the next frontier of laboratory medicine. Clin Chem Lab Med 54:1117–1118

Report of the Review of NHS Pathology Services in England. An Independent Review for the Department of Health (2006) London: Department of Health, p 6

Rizzotti P (2007) HTA e Little Ticket Technology. RIMeL/IJLaM 3(Suppl):44–47

Sackett DL, Rosenberg WM (1995) The need for evidence-based medicine. J R Soc Med 88(11):620–624

Scally G, Donaldson LJ (1998) The NHS's 50 anniversary. Clinical governance and the drive for quality improvement in the new NHS in England. BMJ 317:61–65. https://doi.org/10.1136/bmj.317.7150.61

Westgard JO, Fallon K, Mansouri S (2003) Validation of iQM active process control technology. Point Care J Near-Patient Test Technol 2(1):1–7. /iclac.org

Appendices

Appendix A: Recommendations for the Appropriate Venous Blood Collection

Introduction

These recommendations describe all steps in the venous blood collection procedure for both inpatients and outpatients. Blood collection from outpatients differs from inpatients primarily in preparation, patient position, and physical activity prior to collection. The information in the rest of the recommendations applies equally to inpatients and outpatients.

Using devices that incorporate disposable needles, holders, adapters, and primary vacuum tubes is recommended. Syringes are a possible alternative in case of (i) an emergency in which it is not possible to find the above-described devices; (ii) special anatomical and/or physical situations that make the use of the above-described devices impossible or inadvisable (e.g., veins that easily collapse when subjected to the negative pressure of the vacuum present in the primary tube).

These recommendations refer only to the use of a closed blood collection system (i.e., collection systems where the tube cap is not removed during the sampling process) and do not guide open-needle and syringe blood collection. Furthermore, they are limited to blood sampling using needles and do not consider catheter sampling. We discourage blood collection from an intravenous catheter because many studies show that this procedure increases the risk of hemolysis. When blood collection from the catheter is the only option, special precautions should be taken to minimize the risk of hemolysis and contamination of the sample due to intravenous mixing of fluids or washing solution.

Steps to Follow for Proper Blood Collection

1. Identify the patient.
2. Verify that the patient is fasting and adequately prepared.
3. Make sure to have all the materials for the blood collection.
4. Identify the tubes.
5. Wear gloves.
6. Apply the tourniquet.
7. Select the puncture site.
8. Disinfect the puncture site.
9. Vein puncture.
10. Collect blood in the first tube.
11. Remove the tourniquet.
12. Gently invert tubes immediately after collection.
13. Collect additional tubes in the order of collection.
14. Remove the needle from the vein and make sure the safety mechanism is activated.
15. Dispose of the needle.
16. Bandage the puncture site.
17. Instruct the patient to apply light pressure to the puncture site and not to bend the arm.
18. Reverse all tubes at least four times.
19. Remove gloves.
20. Advise the patient to rest for 5 minutes.

Before Collection

Patient Communication – Throughout the entire blood collection process, safe and empathetic communication with the patient is important.

Patient Position – It has been shown that changing position from supine to upright and vice versa can significantly affect the concentration of many laboratory parameters. Therefore, the patient should ideally not change the position for at least 15 minutes before blood sampling. Outpatients should ideally remain in a sitting position for 15 minutes prior to blood sampling. If the change in posture is unavoidable within this period, it should be documented to allow proper interpretation of laboratory results. If a patient has adequately rested for 15 minutes in the waiting area, a short walk from the waiting area to the collection area is considered acceptable and should not be documented.

Step 1 – Identify the Patient

1.1 The use of ID bracelets is recommended for all inpatients.

1.2 All patients should be actively identified through specific questions: "What is your name?"; "What is your date of birth?"

M. Ciaccio (ed.), *Clinical and Laboratory Medicine Textbook*, https://doi.org/10.1007/978-3-031-24958-7

1.3 At least two identifiers (patient's name and date of birth) and preferably one additional identifier should be used for proper identification. Additional identifiers can include addresses, health insurance numbers, patient identification numbers, ID card details, or any other personal item. The more data used to identify the patient, the less chance of misidentifying the patient.
1.4 The patient's identity must be matched with that of the blood test request.

Step 2 – Check That the Patient Is Fasting and Adequately Prepared

2.1 Blood sampling for all tests should be performed in the morning (between 7:00 and 9:00 am) in a fasting state (12 hours after the last meal). Water consumption is permitted during the fasting period. Patients should abstain from alcohol for 24 hours prior to the blood collection. Patients should not drink caffeinated beverages (coffee, energy drinks, and tea) in the morning before blood sampling. Cigarette smoking and chewing gum are also not allowed in the morning before the blood draw. Finally, morning medications should be avoided unless they are vital to the patient.
2.2 If blood sampling occurs during the day in a non-fasting state, only laboratory parameters may be assayed in an emergency or for which there is evidence that fasting is not required.
2.3 Strenuous physical activity should be avoided 24 hours prior to blood sampling.
2.4 Other potential factors, such as regular and/or recent physical activity, food and drug intake, over-the-counter medications, dietary supplements, and herbal preparations, are known to influence the concentration of specific analytes. If some of the above factors have been identified and the blood collection cannot be postponed, staff should, where appropriate, document all relevant information on pre-analytical conditions to interpret test results properly. In particular, physical activity is known to have acute and chronic effects on the body's metabolism.

Step 3 – Make Sure to Have All the Materials Needed for the Blood Collection

This section focuses primarily on drawing blood in an outpatient setting, not in a hospital ward where patients are bedridden.

3.1 Venous blood collection should be performed in a clean, quiet, and private environment. The area where blood collection takes place could have pictures of relaxing landscapes on the walls to make the space more comfortable.
3.2 There should be dedicated chairs and/or beds for blood sampling. The armrests of the chair should be adjustable to allow optimal positioning. If a dedicated chair is not available for blood sampling, blood sampling should be performed in a chair with armrests to prevent the patient from falling if they feel unwell.
3.3 Sanitizing or hand washing areas with soap and/or disinfectant and appropriate paper towels should be available and accessible to ensure proper hygiene.
3.4 Patient sample collection facilities should be separated from reception/waiting areas to ensure patient privacy. Patient privacy should be ensured throughout the entire blood sampling procedure. Conditions may differ in outpatient and inpatient settings and inpatients with different clinical conditions. However, care must be taken to ensure that blood sampling always takes place concerning patient privacy.
3.5 Equipment and supplies should be available in sufficient and appropriate quantity for the intended use during the collection. Equipment available may include the following:
- Utility cart
- Blood collection trays
- Gloves
- Blood collection system with safety features (needles and holders or needles with holders) integrated)
- Blood collection tubes (a full range of tubes with different volumes within the expiry date)
- Tourniquet (preferably disposable)
- Antiseptics to clean the puncture site
- Dressings
- Bandages
- Container for needles and sharps
- Sample mixer
- Watertight carrying bags

3.6 All required materials should be prepared before venous blood collection and for the required tests. The workplace should be arranged so that the operator performing the collection can reach all required materials without leaving his or her post.
3.7 Equipment should be stored appropriately and kept clean.
3.8 A stock management system must be in place to ensure supplies are used before expiration.
3.9 The needle, holder, and tube form an integrated blood collection system. Only individual components from the same manufacturer should be used as part of the same blood collection system because the safety of patients and caregivers may be compromised. Storing tubes under conditions inconsistent with the manufacturer's recommendations may affect extraction volume and stability of gels and additives. Environmental factors such as temperature, humidity, altitude, and light

exposure can significantly impact the quality of blood collection. Vacuum tubes past their expiration date may have a vacuum reduction that may result in the collection of a suboptimal volume of blood and, therefore, an incorrect blood/additive ratio. In addition, expired tubes may exhibit chemical deterioration of the additive. Blood collection tubes should be discarded after expiration to ensure sample quality.

Step 4 – Identify the Tubes

4.1 Tube labeling or tube identification (for pre-labeled tubes) must be performed in the presence of the patient. Otherwise, there is a risk that the tube will remain unlabeled and may be misidentified. Whether to label or identify tubes before or after blood collection should be based on a prospective risk analysis of the venous blood collection process in each institution.

4.2 Each institution should have a standard procedure to which all staff should adhere.

4.3 Essential specimen and patient information must be recorded within the laboratory so that the tube is traceable and unambiguously linked to the patient, the specimen collected, the test request, the requester, and the operator performing the collection. These records include, but are not limited to the following:
- Identification of an applicant, i.e., a person authorized (under national law) to apply for a blood test
- Full name of the patient
- Date of birth of the patient
- Patient's address (home address or hospital ward)
- Unique sample identification number
- Date and time of collection
- Identification of the operator performing the sampling

4.4 A minimum of two independent identifiers (patient's full name and date of birth) and preferably three (the two above plus one additional) should be used to identify the tube. Not all of the above data do not have to be recorded on the tube. If not on the tube, this information should be recorded on paper documents or linked to the laboratory information system and easily retrievable.

Blood Sampling

Step 5 – Wearing Gloves

5.1 A new pair of gloves should be worn to protect the patient and the staff performing the venous blood sampling.

5.2 Hands should be cleaned before donning gloves to minimize the risk of transmitting an infection when removing gloves and reassure the patient.

5.3 Assemble the needle and the holder (if not already preassembled) or the integrated holder with the tube.

Step 6 – Apply the Tourniquet

6.1 A tourniquet is conventionally defined as a constricting or compressive (elastic) device that can restrict venous circulation to one extremity (usually an arm) for a limited period. In the absence of other devices that can be used to make the veins visible, a tourniquet may be helpful, especially in those patients with minor or poorly visible veins.

6.2 It is recommended that blood collection be carried out without a tourniquet (especially in patients with prominent veins) and only used when necessary. If a tourniquet is used, the blood collector should ensure that the total tourniquet application time is 1 minute.

6.3 The tourniquet should be applied approximately one hand's width (7.5 cm) above the puncture site and should be tight enough to prevent venous but not arterial blood flow.

6.4 The use of disposable tourniquets is recommended to minimize the risk of infection and cross-contamination of the patient and healthcare personnel.

6.5 To minimize the risk of venous stasis, particularly if multiple tubes are to be filled, rather than tourniquets, vein illumination devices can be used to locate the veins. This is particularly useful in patients with problematic veins. It has been shown that vein illumination devices can be a helpful alternative to tourniquets to prevent venous stasis and subsequent alterations in blood concentrations of various biochemical, hematological, and coagulation parameters. Vein illumination devices may be a valuable prospect for the future, although further clinical trials are needed before widespread implementation can be recommended.

Step 7 – Select the Puncture Site

7.1 To select the venipuncture site, the patient's arm should be stretched downward.

7.2 If available, the most prominent veins in the cubital fossa (i.e., cephalic, basilic, median cubital, and median antebrachial veins) should be the first choice (Fig. A.1). The cubital vein is the preferred choice because it is usually the most prominent, does not roll under the skin, and can be found in the same place in most patients.

7.3 Dorsal veins may be used as an alternative if prominent veins are not available.

7.4 Collection of blood from wrist veins is discouraged.

7.5 Palpation of the vein may assist in assessing the appropriate venipuncture site. A cross-sectional graphic representation of the cubital fossa is shown in Fig. A.2. Understanding the anatomy of this region helps reduce the risk of injury during the blood collection procedure.

7.6 It is recommended that blood not be collected from previously inserted peripheral venous catheters, hardened veins, arteriovenous shunts, hematoma sites, and inflammation or swelling from an arm with a vascular graft, paretic arms, or arms with lymphatic drainage disorders.

7.7 Be sure to document when alternative venipuncture sites are used (e.g., hand and foot veins or any other site in addition to those mentioned above).

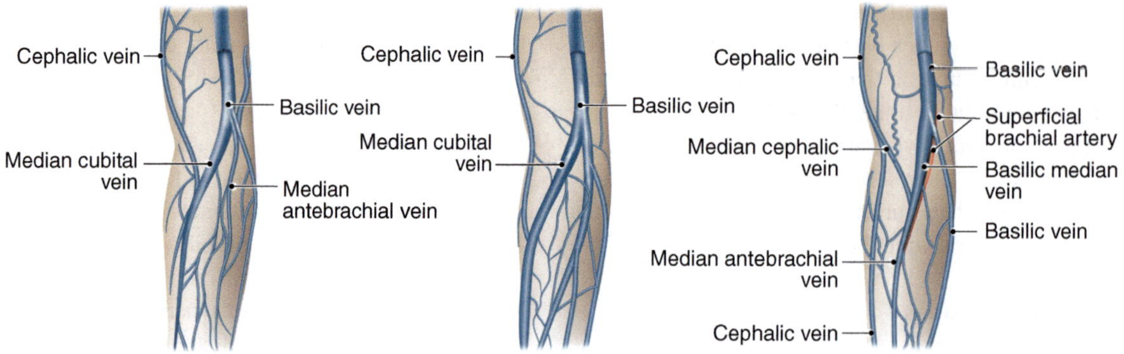

Fig. A.1 Forearm veins. (Copyright EDISES 2021. Reproduced with permission)

Fig. A.2 Topographical anatomy of the cubital fossa (cross-section at the elbow). (Copyright EDISES 2021. Reproduced with permission)

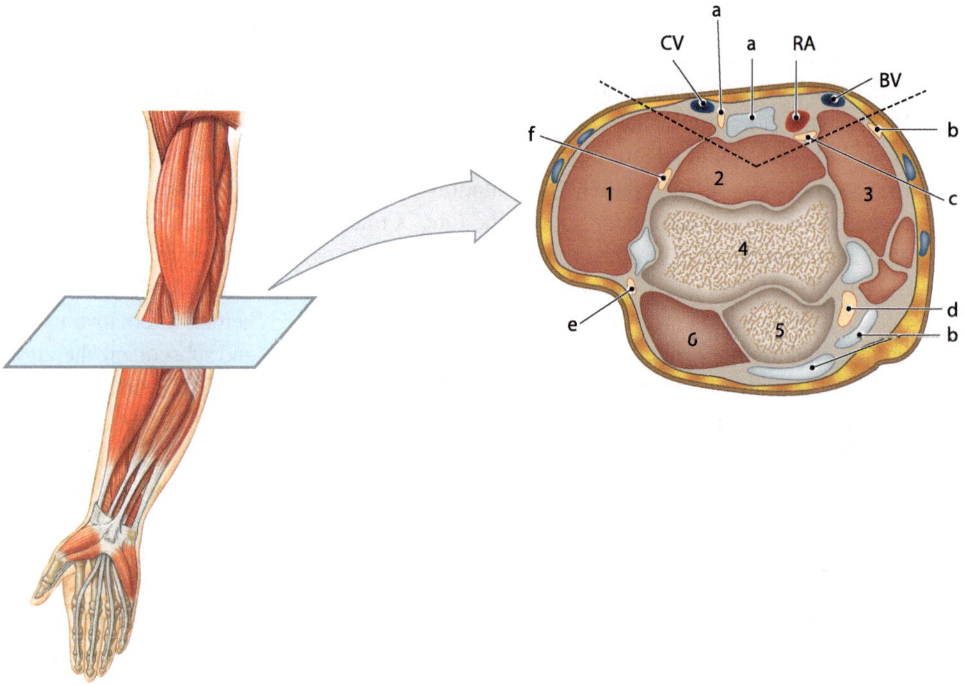

Step 8 – Disinfect the Puncture Site

8.1 The selected venipuncture site should be cleaned with 70% ethyl alcohol or any other appropriate disinfectant before blood sampling to prevent contamination with pathogens. Cleaning should be performed with a wipe, and the selected site should be allowed to dry. Do not wipe the sample site with the same gauze twice.

8.2 When collecting blood cultures, it is recommended to follow the instructions provided by the hospital microbiology department and/or the information provided by the disinfectant manufacturer. Cleaning the sample site by disinfecting twice with two separate gauze pads is recommended. Allow the disinfectant to dry for at least 60 s.

8.3 Do not touch the disinfected site after cleaning.

NB Alcohol evaporates quickly, and within 10 s, the amount of alcohol is reduced by half the initial amount. Although the inability to allow the alcohol to dry may indeed cause an itching sensation in some patients, it will not compromise the blood collection procedure or the quality of the sample. The presence of alcohol (should the venipuncture site not be allowed to dry) at the collection site is not a source of spurious hemolysis. Furthermore, under ideal blood collection conditions, the use of ethanol prior to venous blood collection does not interfere with the measurement of alcohol. However, to avoid the risk of false positive results, it is suggested that when collecting blood samples for forensic alcohol analysis, the alcohol should be allowed to dry before venipuncture is performed. Alternatively, a non-alcoholic antiseptic cleanser approved for use by the institution can be used to avoid the risk of contamination.

Step 9 – Puncture the Vein (Fig. A.3)

9.1 Puncture the vein by the bevel of the needle pointing upward, as this minimizes pain and the risk of perforation of the posterior wall of the vein.

9.2 Prevent vein rotation by extending the patient's skin.

9.3 Insert the needle longitudinally into the vessel, with determination and caution, at an angle of approximately 5–30° depending on the depth of the vein so that at least 0.5 cm of the needle is inserted into the vessel.

9.4 Hold the tube stand firmly by resting the hand against the patient's arm. Ensure the patient's fist is open and not clenched when blood arrives.

9.5 If a vein cannot be located, gently repositioning the needle (moving the needle back or forward) may help.

9.6 The use of rapid visualization devices can be helpful, especially with inexperienced staff or in children and patients with problematic veins. These devices allow seeing immediately when the needle has entered the vein (Fig. A.4).

Step 10 – Collect Blood in the First Tube

10.1 Collect blood either (a) by inserting the tube into the holder so that the cap is punctured and blood is drawn (vacuum technique) or (b) by slowly releasing the plunger (suction technique). Since blood collection techniques may differ from the manufacturer, the manufacturer's specific recommendations should always be followed, along with the recommendations in this document.

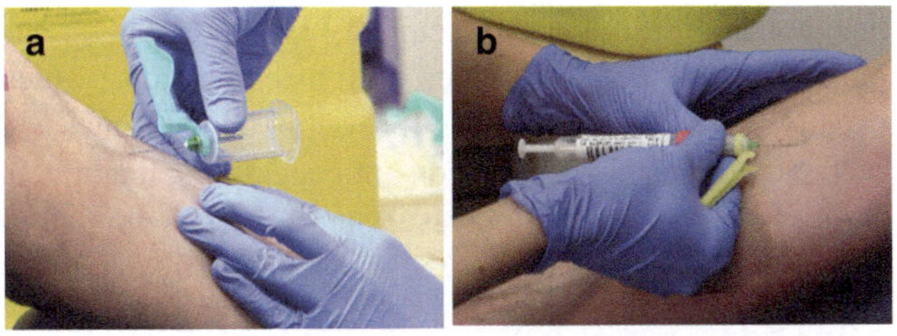

Fig. A.3 The needle must be inserted into the vessel at an angle of about 5°–30°, based on the depth of the vein. (Copyright EDISES 2021. Reproduced with permission)

It is recommended that blood be collected in the different tubes in the following order:

1. Blood culture tube
2. Citrate tube
3. Dry tube or tube with coagulation activator
4. Test tube with heparin
5. Test tube with EDTA
6. Test tube with glycolysis inhibitor
7. Other tubes

10.2 Ensure that the tubes are completely filled (e.g., to the level indicated on the tube). Underfilling tubes (tubes filled with less than 90% of the sample volume) should be avoided.

Step 11 – Remove the Tourniquet

11.2 The tourniquet should be removed as soon as blood flows into the tube.

11.3 If the blood draw is unsuccessful, the tourniquet must be released, and the blood draw must be carried out at an alternative site.

NB Tourniquets cause temporary occlusion of the veins and temporary venous stasis. If applied for an extended period (> 1 minute), a tourniquet induces a substantial change in blood composition due to the extravasation of water and small molecules such as ions into the subendothelial space. During this process, large molecules such as lipoproteins, proteins, protein-binding substances, cells, and coagulation factors remain within the vessel, so their concentration gradually increases. Most of these changes are negligible within 1 minute of tourniquet application but may become clinically significant later.

Step 12 – Gently Invert the Tubes Immediately After Collection

12.1 Immediately after sampling, gently invert the tubes. Any delay may affect the quality of the sample.

12.2 Gently mix each tube by inverting it once before collecting the next tube. An inversion turns the tube vertically 180° and returns it to its initial position (Fig. A.5).

Fig. A.4 Quick view blood collection device (left: butterfly; right: needle with a visible gap). (Copyright EDISES 2021. Reproduced with permission)

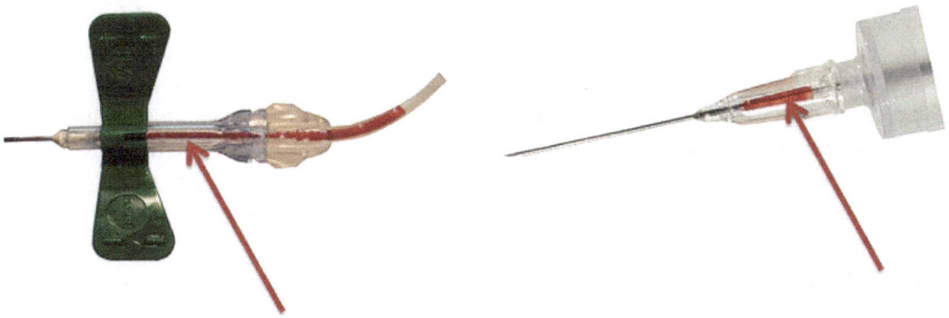

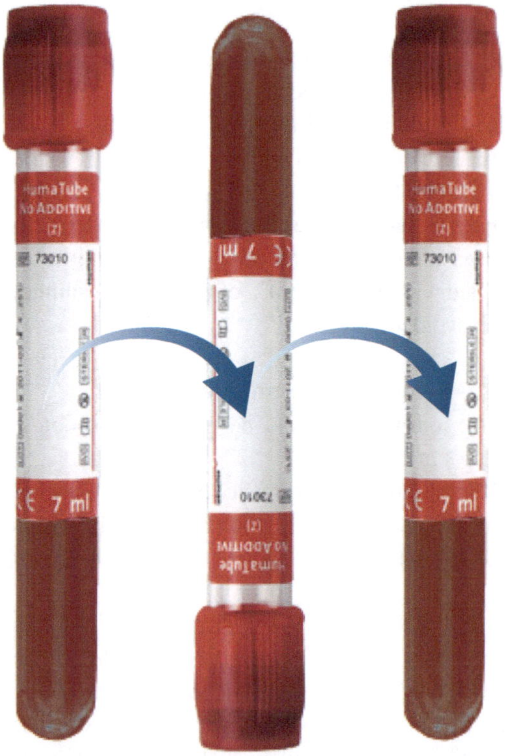

Fig. A.5 How to invert the tubes after collection. (Copyright EDISES 2021. Reproduced with permission)

12.3 The dominant hand should be used to hold the needle and a holder in place throughout the collection. In addition, the hand should not be changed during blood collection in the various tubes.

12.4 Avoid vigorously mixing samples (e.g., shaking) to prevent blood cell disruption, hemolysis, platelet activation, or blood clotting.

12.5 The use of automated mixing devices is recommended. Proper mixing of the tube after the blood has been drawn is an important step that ensures that the additive in the tubes (anticoagulant, clotting activator, etc.) is properly mixed, the blood samples are homogeneous, and the quality and integrity of the sample are maintained.

Step 13 – Collect Additional Tubes Following the Order of Collection

13.1 Collect all subsequent tubes and gently mix each tube once (a complete inversion), as explained in the previous step (see Step 12).

13.2 Collect the tubes in the recommended order (see Step 10).

Step 14 – Remove the Needle from the Vein and Make Sure the Safety Mechanism Is Activated

Gently remove the needle, trying not to cause injury, and press the puncture site with gauze to prevent bleeding without applying pressure.

Safety-engineered blood collection devices on the market may differ in how they are activated (e.g., while the needle is still inside the vein or after the needle has been removed from the vein). In accordance with EU Directive 2010/32 EU, it is recommended that only blood collection devices designed for safety should be used to prevent the exposure of healthcare workers and patients to a contaminated needle. The manufacturer's instructions for the device used should be followed.

Step 15 – Dispose of the Needle

15.1 Immediately after activation of the safety mechanism, the used blood collection device must be disposed of in a puncture-resistant sharps container.

15.2 Sharps containers must be nearby. Walking toward the sharps container is not an acceptable practice.

Step 16 – Bandage the Puncture Site

Check that the bleeding has stopped. Treat the wound by applying a plaster or bandage with tape to the dry area.

Step 17 – Instruct the Patient to Apply Light Pressure to the Puncture Site and Not to Bend the Arm

The patient is advised to apply gentle pressure to the puncture site and not to bend the arm to minimize the risk of hematoma or prolonged bleeding. Light pressure at the puncture site should be applied until bleeding stops, which usually takes about 2 minutes for routine blood draws and up to 10 minutes for patients on anticoagulation therapy. In addition, failure to apply pressure until the bleeding stops has been shown to increase the incidence and severity of bruising.

Step 18 – Reverse all Tubes at Least Four Times

18.1 After removing the needle from the vein and activating the safety mechanism, invert all tubes at least four more times so that the total number of inversions is five: the first inversion immediately after the tube is filled and four more times at the end of the collection when all tubes have been collected (after removing the needle from the vein). Ideally, the number of complete rotations should correspond to the manufacturer's instructions. Refer to Step 12 for information on the proper mixing procedure.

18.2 If only one tube is collected, invert it five times directly after collection.

18.3 After mixing, all tubes should be left upright before further processing.

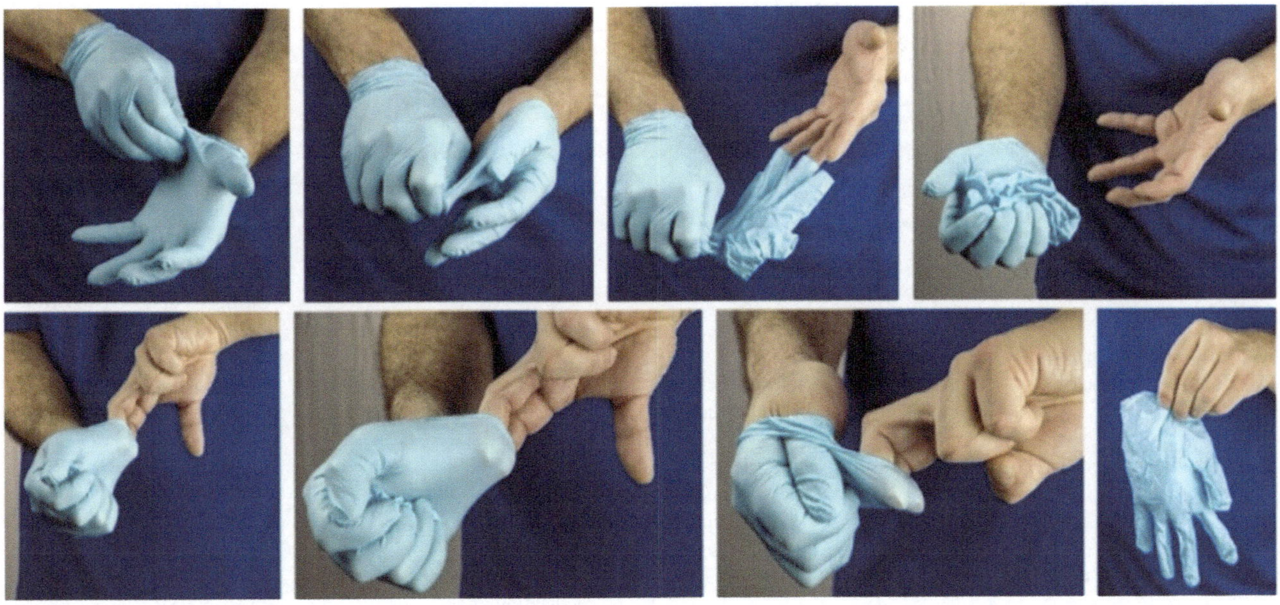

Fig. A.6 Procedure for the proper removal of gloves. (Copyright EDISES 2021. Reproduced with permission)

Step 19 – Remove the Gloves

19.1 As the gloves used may be contaminated with body fluids and/or microorganisms, changing the gloves after each venous blood sampling is recommended.

19.2 It is advisable to use the following procedure to remove the gloves: (1) take the outer part of the glove (at wrist level) with the opposite hand that is still wearing the glove and remove it by turning it inside out; (2) hold the removed glove in hand with the glove still worn; (3) slide the fingers of the hand without the glove to the wrist level of the glove still worn and remove it, constantly turning it inside out (Fig. A.6).

19.3 Discard gloves and clean hands.

After Sampling

Step 20 – Advise the Patient to Rest for 5 Minutes

20.1 Advise the patient to rest for 5 minutes or wait for the bleeding to stop (if more than 5 minutes have elapsed) before leaving the blood collection area.

20.2 You should ask the patient how he/she feels before leaving the blood collection center or in the waiting room. This can help identify patients who are at risk of fainting or syncope.

Appendix B: Table of Reference and Decisional Values of the Main Biomarkers

Reference values and decisional cut-off values

Biomarker	Reference values		Factors affecting levels	Decisional cut-off values
	Males	Females		
Serum creatine kinase (CK)	46–171 U/L	34–145 U/L	Gender, ethnicity, age, muscle mass, physical activity	
Alanine aminotransferase (ALT)	<59 U/L	<41 U/L		Acute hepatitis: >7 × URL Cirrhosis: up to 4–5 > URL AST/ALT>1
Aspartate aminotransferase (AST)	<34 U/L		Age	
Alkaline phosphatase (ALP)	43–115 U/L	33–98 U/L (childbearing age)	Age, sex, meals (sampling must be done on an empty stomach), pregnancy	Biliary obstruction: >4 × URL
Bone alkaline phosphatase (BAP)	7.5–26.1 μg/L	Fertile age: <22.7 μg/L Menopause: <31.6 μg/L	Age	
γ-glutamyltransferase (GGT)	<68 U/L	<40 U/L	Age, gender, body weight, medication intake anti-convulsants (e.g., phenytoin)	Biliary obstruction: >5–30 × LSR
Lipase	<45 U/L (method with diglycerides) <64 U/L (method with methylresorufine)		Renal pathology	Acute pancreatitis: >3 × URL
Pancreatic amylase	13–51 U/L			Acute pancreatitis: >3 × URL
Lactate dehydrogenase (LDH)	125–220 U/L		Age	
Total bilirubin	<1 mg/dL			Sub-jaundice: 2–3 mg/dL Frankish jaundice: >3 mg/dL
Unconjugated (indirect) bilirubin	<0.8 mg/dL			
Conjugated bilirubin (direct)	<0.3 mg/dL			
Serum albumin	35–52 g/L			
Transthyretin or serum prealbumin	20–40 mg/dL			
Total cholesterol	≤190 mg/dL			
LDL cholesterol	≤115 mg/dL in the absence of cardiovascular disease and cardiovascular risk factors ≤100 mg/dL in the presence of cardiovascular risk factors ≤70 mg/dL in the presence of cardiovascular disease			
HDL cholesterol	≥40 mg/dL	≥45 mg/dL		
Total cholesterol/HDL cholesterol	<5	<4.5		
Triglycerides	≤150 mg/dL		Smoking, exercise, alcohol intake	
Hemoglobin	≥120 g/L Pregnancy: ≥110 g/L	≥130 g/L		
Hematocrit	36–46%	40–54%		
Mean corpuscular volume (MCV)	80–100 fL		Age, pregnancy	Microcytic anemias: <80 fL Macrocytic anemias: >100 fL
Mean corpuscular hemoglobin (MCH)	27–31 pg		Age, pregnancy	Hypochromic anemias: <27 pg Hyperchromic anemias:>31 pg
Concentration corpuscular mean of hemoglobin (MCHC)	320–360 g/L (32–36 g/dL or %)			

Table of reference and decision values of the main biomarkers

Biomarker	Reference values: Males	Reference values: Females	Factors affecting levels	Decision values
Spread of distribution of the population erythrocyte (RDW)	RDW-CV: 12–15% RDW-SD: 40–55 fL		Age, pregnancy	
Mean platelet volume (MPV)	7–12 fL			
Platelet distribution width (PDW)	10–17.9%			
Fraction of platelets immature (IPF)	1–8%			
Plateletcrit (PCT)	0.22–0.24%			
Percentage of large platelets	15–35%			
Count reticulocytes	0.5–1.5%			
Immature reticulocytes fraction (IRF)	0.11–0.38%			
Reticulocyte hemoglobin content (CHr)	24–32 pg			Iron deficiency: <24 pg
Reticulocyte hemoglobin equivalent (RET-He)	28–35 pg			Iron deficiency: <28 pg
Sideremia (iron)	12–32 µmol/L	10–30 µmol/L		
Transferrin	1.9–3.7 g/L			
Saturation of the transferrin	18–40%			
Ferritinemia	18–250 µg/L	14–150 µg/L		
Total iron-binding capacity (TIBC)	45–66 µmol/L			
TIBC/transferrin	25			
Fibrinogen	1.5–4 g/L			
Urea	3.5–9.0 mmol/L (21–54 mg/dL)		Age	
Creatininemia	64–104 µmol/L (0.72–1.18 mg/dL)	49–90 µmol/L (0.5–1 mg/dL)	Age, gender, ethnicity, regular exercise, diet	
Proteinuria	100–150 mg/L			
Albuminuria	Spot sample: <30 mg/g creatinuria timed collection: <30 mg/24 hours		Gender, age, body mass, high protein diet	Microalbuminuria: 30–299 mg/24 hours Clinical albuminuria (frank nephropathy): ≥300 mg / 24 hours
Na^+ plasma	135–145 mmol/L			
K^+ plasma	3.5–5 mmol/L			
TSH (ECLIA technique)	0.4–4 mU/L			
fT3 (ECLIA technique)	2.5–4.5 ng/L			
fT4 (ECLIA technique)	0.8–1.8 ng/dL			
Calcemia	8.5–10.2 mg/dL			
Magnesemia	1.7–2.1 mg/dL			
Phosphatemia	2.4–4.5 m/dL			
Parathyroid hormone (PTH)	10–55 pg/mL			
25(OH)D3	30–100 ng/mL			Deficiency: <20 ng/mL Insufficiency: 20–29 ng/mL Toxicity: >100 ng/mL
Fasting plasma glucose	General population: 70–100 mg/dL Pregnancy: <92 mg/dL			Diabetes mellitus: ≥126 mg/dL Altered fasting blood glucose: 100–125 mg/dL Gestational diabetes: 92–126 mg/dL
Blood glucose 2 hours after OGTT	General population: <140 mg/dL Pregnancy: <153 mg dL			Diabetes mellitus: ≥200 mg/dL Impaired glucose tolerance: 140–199 mg/dL Gestational diabetes: ≥153 mg/dL
HbA1c	4–6% (20–42 mmol/mol)			Diabetes mellitus:>6.5% (>48 mmol/mol) Impaired glucose tolerance: 6–6.49% (42–48 mmol/mol) Therapeutic goal in diabetic patient: <7% (<53 mmol/mol)
CRP	Adult: <10 mg/L Newborn: <0.06 mg/L			
PCT	<0.05 ng/mL			

ECLIA electrochemiluminescence, *fL* femtoliters, *HbA1c* glycated hemoglobin, *URL* upper reference limit, *OGTT* oral glucose tolerance test, *CRP* C reactive protein, *PCT* procalcitonin, *pg* picograms, *LDL* low density lipoprotein, *HDL* high density lipoprotein, *RDW* red distribution width

GPSR Compliance

The European Union's (EU) General Product Safety Regulation (GPSR) is a set of rules that requires consumer products to be safe and our obligations to ensure this.

If you have any concerns about our products, you can contact us on ProductSafety@springernature.com

In case Publisher is established outside the EU, the EU authorized representative is:

Springer Nature Customer Service Center GmbH
Europaplatz 3
69115 Heidelberg, Germany

Batch number: 10732850

Printed by Printforce, the Netherlands